Introduction to
Environmental
Science

Introduction to Environmental Science

Arthur N. Strahler
Alan H. Strahler

Hamilton Publishing Company
Santa Barbara, California

Cover: Man's impact on the environment, Columbia River at Trail, British Columbia. Urbanization is seen against a mountainous backdrop of unspoiled forest. The downtown area of the city, seen at the left (front cover), was built on floodplain subject to seasonal inundations and has been protected by a massive concrete dike. River flow is now regulated by a succession of upstream dams. Beyond, the stack plumes of a smelter obscure the distant mountain slopes. On the right bank (back cover), development has utilized a river terrace, lying well above flood limits. (*British Columbia Government Photograph*)

Copyright © 1974, by John Wiley & Sons, Inc.

Published by **Hamilton Publishing Company,**
a Division of John Wiley & Sons, Inc.

All rights reserved. Published simultaneously in Canada.

No part of this book may be reproduced by any means nor transmitted, nor translated into a machine language without the written permission of the publisher.

Library of Congress Cataloging in Publication Data:

Strahler, Arthur Newell, 1918–
   Introduction to environmental science.

   Bibliography: p.
   1. Earth sciences. 2. Biosphere. 3. Ecology.
I. Strahler, Alan H., joint author. II. Title.
[DNLM: 1. Ecology. 2. Environment. QH541 S896i
1973]

QE26.2.S764   550   73-12024
ISBN 0-471-83161-1

Printed in the United States of America

10 9 8 7 6 5 4 3 2 1

Preface

The unprecedented explosion of public and academic interest in environmental problems within the last four years has stimulated the birth of a new discipline: environmental science. The components of this discipline are not new, for they are drawn from existing areas of science within biology, chemistry, physics, and the geosciences. What is really new about environmental science, however, is its viewpoint—its orientation to global problems, its conception of the earth as a set of interlocking, interacting systems, and its interest in Man as a part of these systems.

In the words of the National Science Board of the National Science Foundation, in its report to the president entitled *Environmental Science— Challenge for the Seventies,** environmental science is "the study of all systems of air, land, water, energy, and life that surround Man. It includes all science directed to the system-level of understanding of the environment, drawing especially upon such disciplines as meteorology, geophysics, oceanography, and ecology, and utilizing to the fullest knowledge and techniques developed in such fields as physics, chemistry, biology, mathematics and engineering. . . . Environmental systems contain the complex processes that must be mastered in the solution of such human problems as the maintenance of renewable resources (water, timber, fish), the conservation of nonrenewable recourses (fuels, metals, species), reducing the effects of natural disasters (earthquakes, tornadoes, floods), alleviating chronic damage (erosion, drought, subsidence), abating pollution by Man (smoke, pesticides, sewage), and coping with natural pollution (allergens, volcanic dust, electromagnetic noise)."

*Superintendent of Documents, U.S. Government Printing Office, Washington, D.C., 1971, pp. vii–viii.

Preface

Given such a broad definition of environmental science, no single textbook can offer even adequate introductory treatment of the whole field. As authors we must also be editors, and in this book we choose to stress the understanding of the *natural systems and processes* of the earth and their implications for and impact on Man. Thus, we omit such related subjects as noise pollution and nuclear warfare (as impacts of Man on Man); we also neglect the technologies of materials processing and pollution control (as man-made, rather than natural, processes); we do not treat consumerism, a subject sometimes confused with environmental science.

In this text we recognize two areas of interaction between natural systems and Man within environmental studies. The first area of interaction is in the realm of physical phenomena and is denoted *geoscience;* the second lies in the realm of biological phenomena, and is referred to as *ecoscience*. Whereas these two branches of environmental science are quite interdependent through system interactions, they form convenient divisions for organizing and structuring environmental principles and problems. This volume attempts to bring these two branches into a balanced treatment: first, geoscience, including the components and processes of the lithosphere, atmosphere, and hydrosphere; and second, ecoscience, including the structure and functions of the biosphere.

In addition, we can look at Man's interaction with natural systems from two viewpoints, corresponding to the two faces of a single coin. One face displays the impact of natural environmental forces upon Man (for example, floods, earthquakes, fires, landslides, and plagues). The other face displays the impact of Man upon the environment (for example, air and water pollution, species extinctions, and accelerated erosion). Our book would not be complete without treating both faces of this environmental coin, for the interaction of Man and environment is our central theme.

The consumption of natural resources—especially minerals and the fossil fuels—is an integral part of the subject matter of environmental science. Extraction, processing, and consumption of these resources bring about a large share of our problems of environmental pollution and degradation. Consequently, we give special attention to natural resources, their origin and occurrence, and the impact of their use.

We offer here an introductory text that interweaves the basic principles of science with environmental and resource problems. Topics in basic science are selected solely on criteria of relevance. We focus on the essential points of interactions between organisms and their environments in the atmosphere, hydrosphere, and lithosphere. Throughout, we use an *energy systems approach* examining and analyzing flows of energy and matter within and between open energy systems. The foundation for this approach is laid in our introduction and in our preliminary chapter on energy systems. Students with limited background in basic science will have some remedial work to do. To this end we include basic information on topics in elementary physics, chemistry, and biology, not only in the introductory chapter, but also at many points throughout the book. For example, elementary principles of electromagnetic radiation are explained in developing the radiation balance; principles of calorimetry are reviewed in developing the heat balance, and so forth. It is our hope that this treatment will accommodate a wide range of student preparation levels.

Environmental science expresses itself through the medium of the environmental problem. Each problem is real; it is actually experienced at a place and time in the biosphere. Therefore, our book includes numerous descriptions of real problems. Space has permitted only the briefest thumb-box sketch of each case, but we hope that each testifies to the reality of an environmental science.

When the student reaches the end of this book, he is introduced to new perspectives of broader scope in our epilogue, which concerns the relationships between society and environment. The future of mankind as embodied in the triad—*population, resources, technology*—is briefly discussed to show where the vital issues lie. Hopefully, our book will lead the student to inquire more deeply into problems of environmental planning and management. Hopefully, his curriculum will provide an organized program for this further study. Hopefully too, he will become involved in environmental issues and develop for himself the educational base needed to participate in the decision-making processes, helping to plan and manage our future environment wisely.

The bibliography placed at the end of the book has been the object of special attention. The rapid evolution in environmental science publications poses many problems in selection. We have, therefore, designed a working bibliography leading the student and instructor as quickly and directly as possible to literature in available works. Items are arranged as they occur in the text—chapter by chapter and topic by topic. Only recent works in print are included, and only a few widely distributed periodicals are used. In addi-

Preface

tion to supplementing the text, we intend this bibliography to serve as the starting point for a basic reference library of environmental science. The purpose of such a library is to allow students to broaden their knowledge on selected topics, to find the material needed to prepare term reports on selected topics, and to allow the instructor to round out his background in areas other than his specialty. The selections range in level from elementary, descriptive treatments to rather technical discussions, but we have stopped short of listing papers requiring advanced knowledge of mathematics, physics, and chemistry.

We wish to express our thanks and appreciation to several individuals who reviewed portions of the manuscript. We are particularly indebted to Professor F. Kenneth Hare,* presently serving as Science Policy Advisor, Environment Canada. Professor Hare critically read the manuscript and made many valuable suggestions for correcting and improving the text, not only in the atmospheric sciences—his field of special authority—but also over the whole range of subjects covered. Others who reviewed single chapters or groups of chapters are the following: Professor Helmut E. Landsberg, University of Maryland, Chapter 10 (Man's impact on the atmosphere); Dr. Martin Prinz, Institute of Meteoritics, University of New Mexico, Chapters 7, 8, and 10 (petrology and mineral resources); Professor Charles L. Drake, Dartmouth College, Chapter 9 (tectonic processes and continental evolution); Professor Donald F. Palmer, University of Southern California, Chapter 10 (Man's consumption of planetary resources); Professor S. Fred Singer, University of Virginia, Chapter 10, "Energy Systems," Appendix III (Darwin-Lotka law); Professor Joseph H. Butler, State University of New York at Binghamton, Chapters 12, 13 and 14 (hydrologic systems and water resources); Professor Nicolay P. Timofeeff, State University of New York at Binghamton, Chapter 12 (soil moisture balance); Professor Michael Barbour, University of California at Davis, Chapters 18 through 24 (energy systems of the biosphere); and Professor Preston Cloud, University of California at Santa Barbara, "Epilogue" (environmental perspectives).

ARTHUR N. STRAHLER
ALAN H. STRAHLER

*On leave, Department of Geography, University of Toronto.

About the Authors

Arthur N. Strahler (b. 1918) received his B.A. degree in 1938 from the College of Wooster, Ohio, and his Ph.D. degree in geology from Columbia University in 1944. He is a fellow of the Geological Society of America and the Association of American Geographers. He was appointed to the Columbia University faculty in 1941, serving as Professor of Geomorphology from 1958 to 1967 and as Chairman of the Department of Geology from 1959 to 1962. His published research has dealt largely with quantitative and regional geomorphology. Dr. Strahler pioneered in the development of environmental education by introducing an interdepartmental course in environmental science into the Columbia College curriculum in 1965. He is the author of several widely used textbooks of physical geography and the earth sciences.

Alan H. Strahler (b. 1943) received his B.A. degree in 1964 and his Ph.D. degree in 1969 from The Johns Hopkins University, Department of Geography and Environmental Engineering. His published research lies in the fields of plant geography and forest ecology. Dr. Strahler is presently Assistant Professor in the Department of Environmental Sciences of the University of Virginia, Charlottesville, where for the past four years he has directed development of the introductory college course in environmental science. In 1973-1974 he held appointment as Bullard Forest Research Fellow at Harvard University.

Contents

Introduction to Environmental Science   1

Energy Systems   7

PART I   Energy Systems of the Atmosphere and Hydrosphere

1. Atmosphere and Oceans   29
2. The Earth's Radiation Balance   38
3. Thermal Environments of the Earth's Surface   59
4. Circulation Systems in Atmosphere and Oceans   85
5. Atmospheric Energy Releases   113
6. Man's Impact upon the Atmosphere   143

PART II   Energy Systems of the Lithosphere

7. Igneous Processes and the Earth's Crust   167
8. Rock Alteration and Sediments   188
9. Tectonic Processes and Continental Evolution   209
10. Man's Consumption of Planetary Resources   235

PART III   Energy Systems at the Fluid-Solid Interface

11. Wasting of the Continental Surfaces   263
12. Subsurface Water of the Lands   288
13. Hydrology of Streams and Lakes   320
14. Fluvial Processes and Landforms   350
15. Waves, Currents, and Coastal Landforms   382
16. Wind Action and Dune Landscapes   411
17. Glacier Systems and the Pleistocene Epoch   427

PART IV Energy Systems of the Biosphere

18.	Life on Earth	461
19.	The Evolution of Life Forms	487
20.	Energy Flow in Organisms and Ecosystems	509
21.	Cycling of Materials in Organisms and Ecosystems	534
22.	Ecosystem Dynamics	559
23.	Aquatic Ecosystems	576
24.	Terrestrial Ecosystems	597

Epilogue: Environmental Perspectives	617
Appendix I: Dimensions, Definitions, and Equivalents in the Flow of Energy and Matter	A-1
Appendix II: Von Bertalanffy's Principles of Open Systems and Steady States	A-6
Appendix III: The Darwin–Lotka Principle of Energy Storage	A-9
Bibliography of Environmental Science	B-1
Index	I-1

Introduction to Environmental Science

For Londoners, December 8, 1952, was the fifth day in a row of heavy smog, better described as poison fog. Hospitals were crowded with seriously ill persons unable to breathe effectively. Many who had chronic respiratory diseases succumbed to an environmental stress from which they could not escape. This incident is not news; most of you have heard it before. The recurrent crisis of air pollution over urbanized areas is never very far from the minds of city dwellers. It is obvious that Man has played a leading role in degrading atmospheric quality over his cities. Understanding the smog problem requires a knowledge of basic principles of atmospheric science. These principles are explained in early chapters of this book and will help you to be more effective in formulating action programs that must be instituted to improve the quality of air over cities.

In May, 1971, the Food and Drug Administration announced that swordfish should not be eaten because an extensive sampling of canned swordfish in markets revealed the presence of mercury in amounts too great to be considered safe for human consumption. To what extent, if any, are Man's industrial processes responsible for the concentration of mercury in the tissues of swordfish? Is it possible that the mercury level in these fishes is no higher now than it has been for centuries in the past? To get the answers to these and other problems of mercury contamination requires that a number of sciences be tapped. First, of course, is biological science. The food chain must be traced back to its source. But one must also look to geology for the prime sources of mercury compounds and to hydrology (water science) for an understanding of the transport processes by which it enters the sea water.

As far back as 1947 the pumping of water from wells in Kings County, on Long Island, New York, (the Borough of Brooklyn) was finally discontinued because the fresh water beneath the surface had become contaminated by salt water drawn in from beneath the adjacent ocean. Since then, this large urban community has had to be supplied with fresh water brought from reservoirs in the Catskill Mountains, far upstate. Will the same fate overtake Long Island communities farther to the east? Planning for Long Island's future urban development depends very heavily

Introduction to Environmental Science

upon the question of availability of adequate supplies of fresh water. Already serious problems of contamination by salt water and detergents have arisen. To understand what is happening on Long Island requires a knowledge of the basic principles of water beneath the ground surface. We shall explain these principles, so that you can recommend effective action to head off the disaster of a failing water supply.

In November, 1970, a great surge of sea water, generated by a powerful tropical cyclone in the Bay of Bengal, swept over a densely populated low-lying coastal area in East Pakistan. The sudden, 20-foot rise of water level drowned many tens of thousands of persons and destroyed over 200,000 houses along with all rice crops in the area. Comparable disasters of the same origin are known to have occurred in this region in the past. Here is an area where the physical environment has dealt harshly with Man, and through little fault on his part. So we must recognize that Man, along with all other life forms, has always been subjected to environmental stresses over which he has no control. The list of severe environmental hazards and disasters is well known to all of us: volcanic eruptions, earthquakes, landslides, tornadoes, hurricanes and typhoons, "tidal waves" (misnomer for storm surges and seismic sea waves), river floods, blizzards, forest fires. But protective and evasive action is possible where advance knowledge permits. We shall be foolish if we don't take the opportunity to arm ourselves with information about these forms of environmental hazards. Where do such things occur? When can they be expected? Can warning systems be made effective? What should we do when disaster threatens? Broadly speaking, such information comes under the heading of *environmental protection*. Although we shall not give a full account of protection procedures, we will explain how and where these dangerous natural phenomena operate.

Less lethal in terms of human life toll, but nevertheless of severe economic importance, are environmental stresses whose effects are persistent in certain geographical regions—for example, drought, dust storms, and hail in the Great Plains region of the interior United States. To some degree, as yet not fully exploited, Man can act to alleviate these stresses. Rainmaking shows potentiality for success in favorable situations; hail formation may yet be reduced in intensity by artificial means; and treatment methods applied to the ground surface can to some extent reduce the blowing of dust and sand. Evidently, Man has at his disposal various means of exerting at least some limited controls over the environmental processes. We might wish to group these activities under the heading of *planned environmental modification*. Success in programs of this type requires knowledge of the fundamental processes of the atmosphere and of water on and beneath the lands.

It seems clear enough from the foregoing examples that there is a field of scientific investigation, actually an applied discipline, that can be labeled *environmental science*. Perhaps this label is just another term for applied science generally. In any case it is a field of such enormous scope that even a general overview would require many textbooks and many semesters of courses. The word "environment" is so common in the everyday vocabulary that we shall have to qualify its meaning very carefully to fit the sense of this book.

Environment, defined most broadly as "that which surrounds," requires a receiving object. What is surrounded? Surrounded by what? Primarily, our concern is with the environment of Man. But Man cannot exist or be understood in isolation from the other forms of animal life and from plant life. Therefore, we must deal with the environment of all life forms within the life-bearing layer, or *biosphere* of planet earth. This shallow

Introduction to Environmental Science

life layer lies at or close to vital *interfaces* between the basic earth realms: the *atmosphere* (gaseous realm), the *hydrosphere* (liquid water realm), and *lithosphere* (solid mineral realm). Interfaces are found between the atmosphere and the land surfaces, between the atmosphere and free water surfaces of oceans, streams, and lakes, and between water bodies and the solid earth surface below. Interfaces of the biosphere are vital because of the exchanges of matter and energy that take place between the superimposed realms.

Organisms, whether of one species or many, whether belonging to the plant kingdom or to the animal kingdom, interact not only with the physical environments which they occupy, but with each other as well. Study of these interactions—in the form of exchanges of matter, energy, and stimuli of various sorts—between life forms and the environment is the science of *ecology*, very broadly defined. The total assemblage of components entering into the interactions of a group of organisms is known as an *ecological system*, or more simply, an *ecosystem*. The root *eco* comes from a Greek word connoting a house in the sense of household, which implies that a family lives together and interacts within a functional physical structure. Ecosystems have inputs of matter and energy, used to build biological structures (the *biomass*), to reproduce, and to maintain necessary internal energy levels. Matter and energy are also exported from an ecosystem. An ecosystem tends to achieve a balance of the various processes and activities within it. For the most part these balances are quite sensitive and can easily be upset or destroyed.

While study of the environment of life forms is concentrated on the life layer itself, understanding of the fluxes of matter and energy that operate in that layer requires that we probe rather deeply into the overlying and underlying layers. To understand the exchanges of heat and water at the earth's surface requires understanding of processes operating in the entire lower atmospheric layer, and also of the action of upper atmospheric layers upon the sun's radiant energy as it travels earthward. To understand the properties of mineral matter exposed at the earth's surface requires study of the geologic processes of change that operate in a deep crustal layer of the solid earth.

For practical reasons, we must limit our study of Man's environment to a manageable area of concern, excluding other considerations that would make up a complete survey. For example, in its broadest sense the environment consists of all matter and energy capable of influencing life forms. A knowledge of the structure and activity of matter at the level of molecules, atoms, and subatomic particles and the behavior of that matter in the field of the earth's gravity (e.g., sciences of physics and chemistry) is the fundamental basis of all environmental study. While it's absolutely essential to apply principles of physics and chemistry to an understanding of the environmental processes, we can't undertake to supply that knowledge in this book. Hopefully, you have already acquired a good general science background.

Environmental influences include in the broadest sense forces and restraints that arise from Man's accumulated cultural resources contained in his elaborately developed social structures. These structures of society are industrial, political, religious, or esthetic in nature. No problem of the physical environment can be approached and solved without taking into account value judgments that are weighed against the consequences to our total culture. Cleaning up the air over cities and the water of lakes and streams requires an enormous output of human energy and tangible resources. Resources so expended must be drawn against other alternative resource uses. To what extent are we willing to change our life styles to restore quality of environment? Are we willing to relinquish the automobile in favor of mass rapid transit systems? Are we willing to put a larger share of our incomes into pollution abatement programs from which we will derive no pleasure or entertainment? Are we willing as a society to submit to rigid population control? The relationships between society and environment are beyond the scope of this book, since we plan to deal primarily with the physical and biological processes involved in interactions between Man and his environment of energy and matter.

To understand these processes you will need to take up a systematic study of each of the global realms: atmosphere, hydrosphere, lithosphere, and biosphere. This systematic study provides the understanding of natural processes, without which environmental problems can be neither understood nor solved. When you are planning programs of environmental action, there is no substitute for a real understanding of the workings of the earth's natural physical and biological systems. Perhaps it is accurate to say that most of the environmental problems facing the human race today have developed in the absence of the sustained application of principles of science and technology that have been at our disposal. Had society chosen to put this knowledge to use we might have headed off most of the troubles in which we now find ourselves. But now society is awakening and responding; it is calling upon research scientists to divert energy from pure science research and technological development to the alleviation of environmental problems. National priorities are changing in response to the demands of a troubled society. We see this change today in appeals to reconsider the necessity for indefinite rapid growth in the population and in the gross national product.

This last statement brings us to the subject of Man's resources of materials and energy and the prospects of their adequacy in the future. The raw materials of industry, in the form of mineral concentrations, have accumulated through exceedingly slow geologic processes acting over millions of years of time. Yet we are using these resources at an alarming rate and inevitably the world supplies of certain key minerals will run low or give out entirely. Use of industrial mineral resources, particularly the metals, has an important impact upon the environment. After use these materials are largely disposed of as wastes, with attendant problems of pollution of air, water, and soil. Extraction of the minerals in many cases leaves gaping pits and scars upon the land. The fossil fuels (petroleum, natural gas, and coal) now supply most of our energy for industrial processes, transportation, and heating. Here, again, we are rapidly consuming energy resources that required vast spans of geologic time to create. World supplies of fossil fuels will eventually run out and we must turn to other sources, such as solar energy and nuclear fuels. The combustion of fossil fuels has had an important impact upon the environment. One area of impact is upon the atmosphere; another is upon the face of the solid earth as huge strip mines scarify the lands and pipe lines cut across wilderness areas. Because of their environmental implications, we shall want to make a study of mineral and energy resources and their use.

Industrialization has added a new dimension to the environment, namely the introduction of new substances into the air, water, and ground—substances that were never present in the preindustrial era. One striking example is the radioactive substances (radioisotopes) disseminated into the atmosphere by nuclear explosions and eventually returning as fallout into soil and surface waters of the lands and into the oceans. Another example is the production and dissemination of synthetic compounds, such as insecticides and herbicides, which enter the natural cycles of water transport and the food chains of life forms. A third example is the dissemination of lead into the atmosphere through combustion of gasoline to which a lead compound has been added. So, we can recognize technological environmental science as a major field of study. Besides industrial pollutants, this science would deal with many phases of urban environmental problems that are in a large sense synthetic, such as disposal and purification of industrial wastes, the production of noise and heat, and the physical and psychological stresses of city life upon humans. Our concern with technological environmental problems will have to be limited to the effects of technology upon natural physical and biological systems, excluding the direct effects of technology upon Man.

Throughout this book you will find numerous case studies and examples of Man's impact upon environmental systems. These examples cover many facets of the environment and bring into play many areas of knowledge of processes of the atmosphere, hydrosphere, lithosphere, and biosphere. Typically, interactions occur in such a way that more than one field of science must be drawn upon for an understanding of the problem and its solution. Evaluating the problems presented by the case studies will test the strength of the systematic knowledge you acquire in mastering the principles of environmental science.

Energy Systems

Imagine yourself newly arrived on Moon Base Alpha. Sitting inside your life support compartment, you gaze through the spaceport at a majestic earthrise, your planet a patchwork of royal blues, verdant greens, and rich browns, superimposed with vast white cloud swirls, surrounded by the velvet blackness of space. From your lunar viewpoint the planet's environmental realms are easily identified by their colors. You note that the blue of the world oceans—the hydrosphere—dominates the area of the disc. Rich browns characterize the lithosphere—the solid portion of the planet, where it rises above the seas. Vast white cloud expanses dominate the gaseous portion of the planet—the atmosphere. The lush greens of plants are the visible portion of the biosphere, the realm of life on earth, which is confined to the life layer—the interface between land, sea, and air.

The dramatic contrast between the day and night portions of the globe strikes you immediately. The land and water masses of the illuminated side bask in the sunlight, absorbing its warmth and reflecting its light. You can anticipate the energy transformations occurring there: solar light energy is striking the earth and is being absorbed and converted to heat, increasing the temperature of the life layer. On the shadowed side of the globe, however, the reverse process is dominant: the heat energy stored by the oceans and continents is continuously being lost—radiated into outer space. Consequently, when sunlight is cut off the earth's surface cools.

As you watch the earth during the next few hours, its slow rotation becomes evident. The continent of South America glides from the center of the disc to its edge, then disappears. Australia emerges from behind the curtain of night and makes its way out onto the illuminated portion of the disc. Under your eye, each land and water mass takes its turn in the sun, absorbing and reflecting its quota of solar energy. Clearly, Africa and South America receive the lion's share, for they lie near the midline of the earth, where they meet the perpendicular rays of the sun. North America and Europe, on the other hand, are near the edges of the disc, and are short-changed because they intercept the sun's rays at a low angle. Not only are you witnessing great transformations of energy, but you are also seeing that while the sun's energy input to the whole earth is constant, there is a great variation, both through time and with position on the globe, in the quantity received by any given small area of surface.

Energy Systems

As the days go by (for you are still keeping earth time), you gaze often at your planet, noting the changes. Swirls of clouds form, dissolve, and reform in the atmosphere, making their way eastward across the disc, first obscuring one portion of a continent, then another. It seems reasonable that this atmospheric turmoil is related to unequal solar heating of the globe, and that it is part of a transport mechanism serving to carry excess heat of the equatorial belt to polar regions of heat deficit. Somehow these great swirls are tied in with the earth's rotation as it carries the turbulent atmosphere with it. Once again, you are witnessing a set of energy transformations: a portion of the radiant energy of the sun after being transformed into heat, is now being converted to kinetic energy of motion. Energy of matter in motion meets with resistance—friction of various sorts—again generating heat, and increasing the radiation of energy back into space.

In fact, from your point of view on Moon Base Alpha, you might visualize the earth as a single great *energy system* which receives solar energy as an input while it reflects light energy and radiates heat energy as an output. Within this system, many transformations of energy occur, and each transformation is associated with matter in one form or another. Thus, an energy system has (1) a body of matter, (2) an *energy input*, (3) an *energy output*, and (4) a set of *energy transports* and *transformations* which are produced where energy interacts with matter. In a system as complex as our earth, there are many energy transports and transformations associated with many different types of matter. Each of these is an *energy pathway* within the system. An example of an energy pathway is the absorption of light energy by matter at the earth's surface. This energy is converted to internal heat within the matter, and then radiated out to space. Thus, energy enters the pathway as light, interacts with matter, and then leaves the pathway as radiant heat.

Another basic principle of energy systems can also be observed from Moon Base Alpha, provided we use the proper instruments—the fact that *rate of energy input must equal rate of energy output,* unless energy is being stored somewhere in the system or is being removed from storage. The physical law behind this statement is that energy can neither be created nor destroyed: a fundamental law of the real world. (A seeming exception to this rule lies in the production of atomic energy, in which matter changes to energy—an environmentally significant process we will discuss later.) In other words, if we were to list in one column all the energy inputs to the system, including energy withdrawn from storage, and in another column all the energy outputs, including energy going into storage, we would find that both columns total alike. Since our list would resemble a household budget with its incomes, expenditures, and savings, we might call such a list an *energy budget* for our system. Analyzing the budget of an energy system will be an important tool in our future studies of planetary systems.

If you watched the earth for a longer time, you would also see matter entering and leaving the planetary surface. A meteoroid, no larger than a grain of sand, is deviated by our planet's gravitational field and plunges into the atmosphere, glowing as the heat of friction with the atmosphere vaporizes its components, providing new atoms for the atmosphere. Or, if the meteoroid is a very large one, it may impact the solid earth, increasing the planetary mass and providing a small additional chunk of rock for the action of erosive processes on the land or perhaps a contribution to the ocean's floor. Matter leaving the earth is not so conspicuous: a few atoms near the edge of the atmosphere diffuse into the emptiness of space; or perhaps a space probe blasts off for Mars, never to return.

Once again, since we know that matter, like energy, can neither be created nor destroyed, the matter input of an energy system must equal the matter output, unless matter is being stored or withdrawn from storage within the system.

Throughout this book we will use the energy system as a fundamental concept, for it provides the one framework common to all branches of environmental science, from the study of the earth's atmosphere to the study of its ecology. Man, through his great numbers and his ability to modify the face of the earth, is a significant part of the earth's energy system. All of Man's actions, from cultivating the earth to polluting it, can be viewed as changing inputs, outputs, or materials within natural energy systems which are oftentimes delicately balanced and all too sensitive. But before we can talk more about energy systems, some basic concepts and definitions have to be established. These are fundamentals of physics and chemistry—natural laws of the real world—perhaps familiar to you, perhaps not.

Matter and Energy—Some Fundamentals

By *matter* we mean atoms and molecules, both uncharged and charged as ions, occurring singly or as compounds, or as aggregations of compounds. All matter possesses *mass*, which is the property of being susceptible to gravitational attraction. Matter exists in three states: solid, liquid, and gas. A *gas* is a substance that expands easily to fill its container, is readily compressible, and usually much less dense than liquids and solids of the same chemical composition. The gas itself consists of atoms or molecules moving about freely in space as individuals, but continually colliding with one another. While the atmosphere is largely in the gaseous state, it also contains varying amounts of substances in the liquid and solid states.

A *liquid* is a substance that flows freely in response to unequal stresses, but characteristically maintains a free upper surface. Within a liquid, molecules or aggregations of molecules are in contact and move freely with respect to one another. Liquids are compressed only slightly under strong stresses. Liquids have densities closely comparable with solids of the same composition. Although the hydrosphere is largely composed of water in the liquid state, it also contains substances in the gaseous and solid states. Both gases and liquids belong to the class of *fluids*. Layers of fluids tend to assume positions of equilibrium at rest in which a less dense fluid overlies a more dense fluid. Such a layered arrangement represents *density stratification*.

Solids are substances that resist changes of shape and volume and are typically capable of withstanding large unequal stresses without yielding. When yielding occurs, it is usually by sudden breakage. We might say that strength is uniquely a property of solids. Although the earth's crust, the outermost portion of the lithosphere, is largely in the solid state, it also contains substances in both gaseous and liquid states.

One can't try to explain what matter is without including the phenomenon of *gravitation*, the attraction that every particle of matter in the universe exerts upon every other particle. We have, in fact, defined mass as the property of being susceptible to this attraction. Probably all of you have memorized at one time or another the *law of gravitation* formulated by Sir Isaac Newton. It goes as follows: any two bodies attract one another with a force that is directly proportional to the product of their masses and inversely proportional to the square of the distance separating them.

So far as environments of the earth's surface region are concerned, what counts is the gravitational attraction of the earth—a truly enormous mass—for very tiny masses, such as molecules of gas or liquid, and particles of soil, rock, or organic matter. (The infinitesimally small force with which the particle attracts the entire earth can be written off as inconsequential.) The earth's gravitational attraction for very small objects within its surface region is called *gravity*. Since the earth is very nearly spherical, gravity has an almost constant value over the entire surface. In other words, the earth's attraction for a given quantity of matter (such as a mass of 1 gram) will be the same over the entire globe, subject only to minor corrections that are not particularly important in understanding the planetary energy systems.

If a particle of matter is placed in a true vacuum (absence of matter) at the earth's surface, it will fall faster and faster toward the earth's center of mass. This uniform increase in the velocity with time constitutes an *acceleration*, and is a constant value. By careful measurement, we know that the velocity of fall will increase by 32 ft (980 cm) per second for each second of fall. The *acceleration of gravity is*, then, 32 ft (980 cm) per second per second. The acceleration of gravity multiplied by the quantity of mass upon which it acts constitutes the *force of gravity*. Since the acceleration of gravity is nearly constant everywhere at sea level, the force of gravity varies according to the quantity of mass. We can therefore measure the quantity of mass by the force registered upon a scale of the type that measures the amount of stretching of a coil spring.

Gravity is a fundamental and pervasive environmental factor for life at the earth's surface. All of you know from watching the *Apollo* astronauts on television, cavorting in the cabins of their spaceships in a condition of weightlessness, that the absence of a gravity field presents a new and disconcerting environment. Prolonged weightlessness is profoundly disturbing to physiological processes. Life forms on earth have evolved under a particular value of gravity that is unique for our planet. The maximum height of living things—an elephant or a tree—is dependent upon the structural strength of the organism by which it resists being crushed under the force of gravity. Because gravity is proportional to the mass of the attracting body divided by the square of its radius, lunar gravity is only about one-sixth of that on earth. On the moon much taller organic structures could stand and the upward vertical motion of masses against the force of gravity would require much less effort than on earth. The gravity force on Mars is only about two-fifths that of Earth and would be much less influential in life processes, should there be any.

Besides matter, the real physical world contains *energy*, which is often formally defined as the *ability to do work*. In mathematical terms of physics, energy is the product of force and distance. Energy, then, is the ability to move an object (exert a force) for a certain distance. Energy is stored and transported in a variety of ways. Some of the recognized forms of energy are: mechanical energy, heat energy, energy transmitted by radiation through space (electromagnetic energy), chemical energy, electrical energy, and nuclear energy.

Mechanical energy is energy associated with the motion of matter. There are two forms of mechanical energy: kinetic energy and potential energy. *Kinetic energy* is the ability of a mass in motion to do work. Thus, an automobile traveling down a highway possesses kinetic energy because it is a mass in motion. Should this mass strike a telephone pole, its ability to do work upon its own body and upon the telephone pole will become quite obvious. The energy it will release in collision will increase with the weight of the car, and it will also increase with the square of the auto's speed. Kinetic energy, then, is proportional to the

Matter and Energy—Some Fundamentals

quantity of mass in motion multiplied by the square of its velocity. In mechanical terms both work and energy are defined as force acting through distance (or force times distance).

Potential energy, or *energy of position,* is equal to the kinetic energy an object would attain if it were allowed to fall under the influence of gravity. Suppose a brick is balanced on the edge of a tabletop, then falls. The kinetic energy the brick possesses at the moment it hits the floor, as we have seen, is proportional to the mass of the brick multiplied by the square of its velocity. Since the brick is being accelerated by gravity at a constant 32 feet per second per second, we could find its velocity at impact by measuring the distance it will fall and using the acceleration formula from physics. If the brick is again lifted to the tabletop, the work done in lifting gives the block a quantity of potential energy. This energy will be released when the brick is again allowed to fall. It should be obvious at this point that the floor is merely a convenient stopping place for both the brick and our discussion; if we sawed a hole in the floor and allowed the brick to fall further, it would possess even more kinetic energy at its impact on the floor below. Therefore, with respect to the lower floor level, the brick would have a greater potential energy when we return it to the tabletop. Thus the three factors which determine the magnitude of potential energy are the mass of the object, the vertical distance it will be able to fall, and the acceleration of gravity.

A raindrop falling to the earth is an example of a potential-to-kinetic energy system in our environment. According to the principles we have just discussed, the potential energy of the raindrop within a storm cloud is proportional to its mass and the distance it can fall. As it falls, its velocity will at first increase as its altitude decreases. Because of air resistance, however, the raindrop will not continue to accelerate indefinitely. In other words, some of its kinetic energy is being drawn off through friction by air resistance. The greater the speed of fall, the larger is the resistance. A constant speed of fall (terminal velocity) is thus quickly reached. When the drop finally hits the earth, part of its kinetic energy will be passed on to the soil particles with which it collides, loosening them and setting the stage for erosion. Another part will be converted to heat, warming the drop and the soil particles. If we think of the raindrop and its fall as a simple energy system, the inputs, outputs and pathways of energy can be identified. (1) Energy input is supplied by the sun, which powers the rising air motions that lift the water into a high position over the land, providing it with potential energy. (2) Energy output occurs when kinetic energy and heat are passed on to surrounding air and to the soil particles. (3) Potential energy is transformed through one pathway to kinetic energy during the fall, and kinetic energy is transformed in a second pathway to heat energy through atmospheric friction. A third energy transformation pathway is the conversion of kinetic energy to heat in the collision of the drop with soil particles. A fourth pathway passes the kinetic energy of the drop into the soil particles, causing their movement. All these pathways have the drop as the common agent of energy transport or transformation.

Before leaving the subject of mechanical energy, we should point out that mechanical energy can be transmitted in *wave* form, in which kinetic energy is passed through impact from one particle of matter to the next. A sound wave in air is an example—a push on air molecules at one point will be transmitted outward to other molecules. Other examples are ocean waves and earthquake waves. In all of these wave forms, masses are displaced in a rhythmic manner and energy is propagated in ever-widening circles. However, frictional resistance within the medium withdraws energy and the waves gradually die out.

Heat is another form of energy of paramount importance within our environment. As noted already, kinetic energy can be converted into heat through the mechanism of friction. What is probably not apparent, however, is just how much heat is related to how much motion. Laws of physics tell us that "a little goes a long way." For example, the energy required to heat 1 cubic centimeter of water 1 degree Celsius (Centigrade) is about the same as that contained in a mass of nearly 100 tons moving at a speed of 1 centimeter per second!

Heat represents kinetic energy, but it is of an internal form, rather than the external form we see in moving masses. Thus a beaker of water resting completely motionless on the table has internal energy because of constant motion of the water molecules on a scale too small to be visible. This internal motion is the *sensible heat* of the substance and its level of intensity is measured by the thermometer. For gases, the internal motion is in the form of high-speed travel of free molecules in space but with frequent collisions with other molecules. The energy level in the gaseous state of water (water vapor) is thus higher than within the liquid water. Ice on the other hand, represents a lower energy level than liquid water, for here the molecules are locked into place in a fixed geometrical arrangement. This is the *crystalline* solid state. For these molecules the motion is one of vibration without relative motion.

When ice melts, work must be done to overcome the crystalline bonds between molecules. This work requires an input of energy, but it does not raise the temperature of the substance. Instead, it seems mysteriously to disappear. Since energy cannot actually be lost, it is actually placed in storage in a form known as *latent heat*. Should the water freeze and again become ice, the latent heat will be released as sensible heat. A similar transformation from sensible to latent heat takes place when a liquid evaporates into a gas since work must be done to overcome the bonds between molecules of the liquid. When water vapor returns to the liquid state, a process of *condensation*, latent heat is released as sensible heat. In Chapter 3 we shall analyze these changes in heat energy accompanying changes in state of water.

Transport of energy from one place to another in the form of latent heat is a very important process within our environment, particularly in the atmosphere. As an example, condensation of water in storm systems releases heat energy to the air, which, to a large extent, provides the storm with the energy necessary to sustain its strong winds. Thus the water molecule provides a pathway of energy transfer, absorbing solar energy in evaporation at the ocean's surface, and releasing it at a distant point in condensation to form clouds and rain.

All matter tends to lose heat. Heat may be lost directly to the surroundings through conduction, but even in a vacuum objects lose heat. A fundamental law of physics states that all matter at temperatures above absolute zero radiates *electromagnetic energy*. We can think of this radiation as taking the form of waves traveling in straight lines through space. The waves come in a very wide range of lengths, but all travel at the same speed—186,000 miles per second—regardless of their length. Together the total assemblage of waves of all lengths constitutes the *electromagnetic spectrum*. It includes *visible light* with all its rainbow colors, and also invisible shorter waves such as *ultraviolet rays*, *X rays*, and *gamma rays*. Besides these, the spectrum includes invisible long waves known as *infrared rays* (sometimes called *heat rays*), and still longer *radio waves*.

The temperature of the radiating objects determines which part of the spectrum carries the radiated energy. Thus, the sun, whose surface temperature is about three times greater than that of molten steel, radiates most strongly in the visible portion, although all other parts of the spectrum are represented. At normal earth temperatures, however, objects radiate mostly in the infrared part of the spectrum and not at all in the visible part. The total energy of radiation is also dependent on the temperature of the radiating object. Thus, one square inch or centimeter of the sun's surface emits about one and one-half million times as much energy each second as does one square centimeter on the earth. Details of electromagnetic radiation are given in Chapter 2. Since all energy to sustain life processes on earth comes by electromagnetic radiation from the sun, we shall go into the subject rather thoroughly.

Yet another form of energy transformation, of great importance in the atmosphere, is that taking place when a gas is forced to occupy a smaller volume or is allowed to expand into a larger volume. Those of you who have pumped up a bicycle tire with a hand pump may remember how hot the end of the pump cylinder becomes after a few strokes. Recall that a gas is composed of molecules or atoms moving freely in space. The higher the temperature of the gas, the faster its molecules (or atoms) move. By compressing the gas in the pump, we crowd them into smaller space, imparting additional kinetic energy to them, and thereby increasing the internal energy (and temperature) of the gas. In other words, the work we do in pushing the pump's piston is stored by the gas as heat energy. If the piston is released, the gas will push the piston back, returning the work done on the gas in the first place, and its temperature will be lowered correspondingly. As an example, the jet of air released by an auto tire valve feels cool; the expanding air pushes away the surrounding air, doing work on it—work which is removed from the internal energy of the jet by its cooling.

Chemical energy is absorbed or released by matter when chemical reactions take place. An example of importance in environmental science is the process of *photosynthesis*. Here radiant light energy from the sun is absorbed by molecules within a plant cell and used to combine carbon dioxide and water to form carbohydrate. Energy absorbed in this fashion is stored in carbohydrate compounds of the plant tissue. Should the plant matter be converted back to carbon dioxide and water, either through burning (chemical oxidation) or through the metabolism of the plant or of an animal feeding on the plant (biochemical oxidation), this energy will be released. In the case of burning, most of the energy will be radiated as heat to the surroundings. In the case of metabolism, some of the energy will be transferred to chemical energy in biochemical molecules needed for growth, development, and maintainance of the organism, while the remaining portion will be converted to heat and ultimately radiated away.

Electrical energy is familiar to all in the many forms of work done by electricity. Because positively and negatively charged objects have an attraction for each other, electrical energy can do work in much the same fashion as the mutual attraction of masses by gravitation. Except for lightning produced in thunderstorms, very few natural energy systems have pathways linked to electrical energy; rather, electrical energy is important for environmental science because of its wide use by Man. Production of electricity requires mechanical energy, energy which is usually derived from the heat of combustion of hydrocarbon fuels.

As environmentalists, we must be concerned with power generation, for it threatens to consume our finite supply of fossil fuels within the next few centuries. At the same time, byproducts of power generation can significantly lower environmental quality by air pollution. Fine, unburned particles of fossil fuels, if not trapped by special filtering systems, contribute to dust fallout from the air near generating plants. Sulfur dioxide, released by the combustion, is an extremely hazardous agent of air pollution. Because the conversion process from heat to electricity is so inefficient, large quantities of waste heat must be liberated to the environment, creating a potential for *thermal pollution*. Warm water discharged from power plants into lakes or estuaries can have harmful effects on the organisms inhabiting these waters.

Nuclear energy may well present the long-term alternative to the consumption of hydrocarbon fuels. Through the process of *nuclear fission* (splitting the atom's nucleus) or *nuclear fusion* (fusing two atoms into one), matter is converted to energy. Albert Einstein gave the exact relationship of matter to energy in his famed equation $E = MC^2$, where E is the quantity of energy, M is the quantity of mass, and C is the speed of light. The speed of light, 186,000 mi (300,000 km) per second, is a large quantity in itself, but when squared it becomes truly enormous. Consequently, the disappearance of a very tiny amount of mass produces a tremendously large amount of energy. For example, the mass which is converted to energy in the splitting of one pound of uranium is equivalent to the energy released by the burning of over 1500 tons of coal.

In the nuclear reactor, controlled fission or fusion produces an almost unlimited supply of heat which can be used to generate electric power. Under present designs, however, much more waste heat is generated by a nuclear plant than by a conventional plant of the same size, producing a higher potential for thermal pollution. In addition, plant effluents may contain radioactive wastes whose energy emissions are harmful to life. And, there is always the chance, however minute, that a reactor will go out of control, producing a mild explosion which could, if uncontained, release deadly fallout—a situation all that more disastrous because of the close proximity of many plants to population centers. The environmental hazards of controlling and exploiting nuclear energy rank close to the top of the list of problems that the human race must confront and solve in decades to come, assuming, of course, that life on this planet is not abruptly terminated by nuclear holocaust.

Energy Systems—Further Concepts

Thus far we have discussed some of the general characteristics of energy systems, and, in our discussion of forms of energy, presented a few examples. These systems are *open systems*, organized states of matter, which are open to the flow of energy and matter. Open energy systems have:

1. boundaries, either real or arbitrary;
2. inputs and outputs of energy and matter crossing those system boundaries;
3. pathways of energy transport and transformation associated with matter within the system.

We can also see that within an energy system,

4. matter may also be transported from place to place or have its physical properties transformed by chemical reaction or change of state.

Transport and transformation of matter requires that energy be absorbed or released, and thus always provides a pathway of transport or transformation of energy.

Energy Systems—Further Concepts

To demonstrate the working of an open system, get yourself a fish tank (home aquarium) and drill a small hole through the bottom. When the tank is placed under a running faucet, water accumulates in the tank, while water leaves the tank through the hole in the bottom. The rate of water flow out of the tank, however, will be related to the height of the water surface—the deeper the water, the greater the pressure at the hole, and the faster the outflow. Eventually the outflow rate will equal the inflow rate, and the water surface will be maintained at a constant level. In other words, the system has reached a *dynamic equilibrium*—an equilibrium because the amounts of energy and matter stored within the system as well as the rates at which energy and matter are leaving and entering it are constant with time—a dynamic equilibrium because the matter and energy themselves are constantly being transported and transformed. This device illustrates another important fact about open systems:

5. open systems tend to attain a dynamic equilibrium, or *steady state*, in which rate of input of energy and matter equals rate of output of energy and matter while storage of energy and matter remains constant.

In the example, storage of matter and energy is represented by the water within the tank.

The water tank example can be used to demonstrate another important fact about energy systems. Suppose the input is varied by shutting down the inflow from the faucet to some fraction of the original rate. What will happen? Outflow will remain the same at first, for it depends on the depth of water in the tank. As a result, matter and energy will be released from storage and exit from the system. Gradually, though, as the water level becomes lower, the outflow will again equal inflow, and a new steady state will be formed. In other words:

6. when the input or output rates of an energy system change, the system tends to achieve a new dynamic equilibrium, or steady state.

The period of change leading to establishment of the new steady state is a *transient state*. Thus, open systems are inherently stable, for if we change energy and matter inputs to a system it does not fall permanently out of balance and destroy itself, but instead quickly creates a new balance. If we compare the two steady states, another important fact emerges:

7. the amount of storage of energy and matter within an open system increases (decreases) when the rate of energy and material flow through the system increases (decreases).

Thus, when the flow of water was decreased, the amount of water in the tank also decreased. It should be obvious that if we were to increase the inflow, then storage would also increase.

To introduce yet another fact about open systems, our example must be modified slightly. Instead of a fish tank, we will place a large test tube under the faucet, after first boring a hole in the bottom of exactly the same size as in the tank. If the inflow rate is the same, the system will also come to an equilibrium in which the water depth will be the same as in our previous example. However, because the storage capacity of the test tube is much smaller, equilibrium will be reached much more quickly than in the fish tank. If the faucet is shut down by the same amount as previously, the system will come to a new equilibrium in which the water depth will be identical with that in the tank under reduced inflow. However, the new equilibrium will be reached much more quickly in the test tube because its storage capacity is much smaller. Thus, under the same change in input, the tank responds much more slowly because its capacity is much larger. Evidently, the tank is less sensitive to changes in input. We can, therefore, define *sensitivity* as the rapidity of the response of an open system to a given change in rate of input or output. Thus:

8. the greater the storage capacity within the system for a given input, the lesser is the sensitivity of the system.

Finally, if the faucet is turned off completely, the water in our fish tank will continue to flow out of the hole and the water level will drop without stopping until the tank is empty. We now have no storage and no input or output. In short, the system is destroyed. This leads to a final generalization concerning open systems:

9. when the inputs of energy or matter, or both, are cut off, the open system undergoes decay and is ultimately destroyed.

For example, in nature, a river system that no longer receives water input from its source region simply dries up and disappears. A plant cell that receives no input of light energy and essential nutrients simply dies and collapses. If the sun's radiant energy were to be abruptly cut off, the earth's processes of energy and matter transport and transformation would rapidly decline to zero intensity. Stored energy would leave the system by radiation into outer space and in a short time the planet would be totally inert and lifeless at a temperature close to absolute zero.

A special case of a system in decay is that in which there is no input of matter or energy but in which the rate of output of matter or energy itself depends upon the quantity of that matter or energy remaining in storage. An example of such a system is the radioactive decay of an isotope of uranium, U-238, to an isotope of lead, Pb-206. In the entire process, each atom of U-238 loses fourteen atomic particles: eight different helium nuclei (alpha particles) and six electrons (beta particles). With the loss of each particle, a new element is formed. The sequence of elements ends in lead-206, which is stable and emits no further bundles of matter and energy. If a gram of U-238 is set on a table, it will emit particles continuously until all the U-238 has been converted to Pb-206. Thus, it will spontaneously lose energy and "run down." Further, the table had best be made of granite set in concrete, or else it will be gone long before the conversion is complete, for it requires about 4.5 billion years to convert half of the U-238 to Pb-206. Note that the amount of energy the system loses will decrease through time, for it is dependent on the number of remaining atoms of U-238. This form of loss follows a schedule called *negative exponential decay*.

The U-238 received its total supply of energy at some initial point in time—presumably when the universe was created several billions of years ago. The stored energy continues to be the source for a larger open energy system surrounding the U-238, upon which the particles emitted can do work for, after all, they are masses in motion. Since no new U-238 is being formed, this energy input will slowly diminish, and so the open system which it drives must continually decay or run down, experiencing slow changes and adjustments of structure as its energy input wanes.

Thus far we have discussed only energy systems which are open to the flow of matter and energy. What about *closed* systems, systems where no energy or matter is allowed to enter or leave? Maintaining energy in storage after shutting off energy inputs and outputs is a difficult matter, for all objects at temperatures above absolute zero emit electromagnetic radiation. Thus, to form a completely isolated closed system at temperatures typical of the earth's surface, we would either need perfect insulation (impossible) or we would need to supply heat energy at a rate equal to the rate of loss, a process that violates the isolation of the system.

The universe, which by definition includes all matter and energy in existence, must be the only closed system. But lesser aggregations of matter within the universe—galaxies, stars, planets—cannot be closed systems, since energy will continue to flow from systems of higher energy levels to those of lower energy levels.

Types of Energy Systems

Variation of energy input with time is an important characteristic of natural energy systems; there are three types based on time-variations of energy inputs: (1) decay systems, (2) cyclic or rhythmic systems, and (3) random fluctuation systems. Although we have already discussed one example of a decay system, another is in order—that of the river system, consisting of a branching network of streams together with their water-contributing surfaces of sloping lands. Matter and energy enter the system in two ways: first, as rain or snow falling on the land, and second, as soil particles raised by geologic processes to a position above sea level, available for downstream transport. As water flows to lower levels, entraining solid particles with it, potential energy of position is transformed into kinetic energy of motion. Resistance is encountered in the flow of the fluid, and as resistance is overcome, kinetic energy of the masses in motion is transformed to heat energy through friction. This heat is lost by conduction or radiation, and is one of the system outputs. Other outputs occur at the river's mouth, where water and mineral matter exit from the system. As long as rain falls, the system will be supplied with matter and energy; in fact its supply may be quite constant as averaged over long periods of time. But what of the supply of soil and rock particles and their potential energy? Erosion and transportation processes diminish the input to the system itself. The land is lowered. Its store of potential energy is gradually reduced. Rain enters the system at progressively lower elevations. The river system is therefore self-consuming and must in the long run be brought to a state of virtual exhaustion. Thus, because its energy inputs slowly diminish, the river system provides another example of a decay system.

The second type of energy system is the *cyclic* or *rhythmic* system, in which energy input increases and decreases in a rhythmic manner. Most natural rhythmic systems are governed by astronomical controls. Thus, the rotation of the earth produces the day-night cycle of solar energy input at the surface. This cycle in turn produces the daily variations in the temperature, heat storage, and radiation output of the earth's surface and lower atmosphere.

Another rhythmic energy system is that of the earth-moon-sun tidal system. Neglecting for the moment the involvement of the sun in producing tides on earth, the moon and the earth are linked together by mutual gravitational attraction that deforms the shapes of the two bodies. As the earth turns on its axis these deforming forces continually change their centers of action, with the result that a continually changing set of forces act on the solid earth, the oceans, and the atmosphere. The waters of the earth yield easily to these forces and engage in a rhythmic flow which we see in the tidal rise and fall of ocean level and in the ebb and flood currents of tidal waters. In analyzing the tidal system, note that the energy input which causes water motion is part of the kinetic energy of rotation and revolution of the enormous masses of the moon and earth. This kinetic energy is inherited from the time of formation of the solar system, and performs work in moving water against resistance. The water movement creates friction, which warms the water; the heat is eventually radiated to space as an energy output. Although the tidal system is dominated by the rhythmic ebb and flow, we can also see a very long term decay aspect to the system, for the total amount of kinetic energy which can be input is limited by the initially stored energies of rotation and revolution of the earth and moon. Thus tidal friction is slowly sapping the rotational energy of the earth.

Random fluctuation systems are those in which input of energy and matter occur on an irregular time schedule. An earthquake and a volcanic eruption are good examples of randomly timed transformations of energy and matter in geologic energy systems. Although we may be able to define areas where earthquakes or volcanic eruptions are likely to appear, their exact location, size, and time of appearance are quite unpredictable. Irregular fluctuations of intensity are particularly characteristic of the flow of fluids—both liquid and gaseous. Fluids readily develop *turbulence,* a disturbed eddying in which vortices of flow continually form and dissolve. A cyclonic storm in the lower atmosphere is an example. Cold and warm air cannot flow smoothly past one another for any length of time without developing a vortex. This vortex grows and intensifies, constituting a storm. Ultimately the storm weakens and dissolves. Although the approximate time and place of the occurrence of such a storm can be guessed at, the precise schedule of events appears to be unpredictable. Thus the precipitation and winds of the storm present a randomly timed input of matter and energy to the surrounding atmosphere and land or water below.

Time schedules of inputs of all three types are commonly found to exist in a single natural energy system. The river erosion system is, over the span of geologic time, essentially a decay system, as we have seen. Yet the system is also subject to the rhythmic variation of annual floods and droughts—so if we are interested in how river systems transport their load from year to year, they may be approached better as rhythmic systems than as decay systems. There is an element of random fluctuation here, too. Late in a July afternoon, a series of thundershowers may dump large quantities of water into one stream basin in the system and not another. Thus, if the river system is studied over a period of days, it may behave more like a random-fluctuation system with large inputs of precipitation at different places and different times. To a large extent, then, the classification of energy input to a system will depend on the time scale through which the system is to be viewed.

Energy, Work, and Power

We can arrive at a deeper understanding of the planetary open systems, both inorganic and organic, and obtain an increased sense of unification of underlying basic concepts by a somewhat more sophisticated analysis of the flow of energy than we have thus far made. We draw this understanding from the field of *thermodynamics,* which can be broadly translated as "a study of heat as it does work."

The *first law of thermodynamics* has already been stated: Energy can neither be created nor destroyed. (The transformation of matter into energy by radioactivity is a special case.) We have stressed that all energy entering or leaving the boundaries of an open system must be accounted for, as well as all energy going in or out of storage within the system.

Energy, Work, and Power

Thermodynamics has its roots in the study of the *heat engine*, which is a device capable of transforming heat energy into mechanical energy. A perfect heat engine would be capable of transforming all of the input of heat energy into mechanical form; which is to say, the engine would operate with 100% efficiency. Suppose that we could construct an added device attached to the heat engine that would transform 100% of the mechanical energy back into heat. We could then recycle the energy without loss, and we would have created the *perpetual motion* machine, realizing the daydreams of countless searchers after a way to get something for nothing. The uselessness of our machine becomes obvious when we realize that to keep the heat supply from leaking out to the surroundings a perfect insulator must surround it. Moreover, the mechanical part of the device must operate without friction, either as drag on bearings or as air drag. There would be no way to make the machine do any useful work for us, unless it were capable of creating more energy than it recycled, and that possibility is ruled out by the first law of thermodynamics.

If fanciful musing over a perpetual motion machine has any value, perhaps it is to focus attention upon a very real question: What constitutes *useful work?** We have said that energy is the ability to do work. Both mechanical energy and work have the same basic definition, namely, they represent force acting through distance, thus: Energy or Work = Force × Distance. Suppose that our perpetual motion machine has as its mechanical section a wheel that turns, or a ball that rolls alternately down and up a sloping track. These actions qualify as work in the strict sense but are certainly not useful forms of work. Perhaps the reason that these activities are not useful work is that no energy flows out of the machine, i.e., there is no *energy drain* involved.

Take as a second example a pendulum swinging on a frictionless bearing in a perfect vacuum. Once set in motion, this pendulum will swing forever with no reduction in the length of its swing. The pendulum action represents mechanical work because it is a mass in motion. It exhibits acceleration, since it speeds up to maximum velocity at the bottom of its stroke and slows down to zero velocity at either end of its stroke. When moving at its fastest speed, the pendulum energy is all in kinetic form; at the instant of stopping to reverse direction, all of its energy is in potential form. Yet the pendulum performs no useful work. From this example, we derive the conclusion that processes involving only changes from potential to kinetic energy and back do not constitute useful work.

All useful work drains energy in the sense that energy flows out of the system as heat. Consequently, if not replaced from an external source, the total energy within the system declines as useful work is done. It is necessary to recognize two forms of useful work: First, is *processing work*, which rearranges matter to the accompaniment of energy drain through friction. For example, a person who spends an entire day removing books one by one from library shelves, dusting each book and returning it to the same place, is performing processing work. The energy to perform this work was derived from storage (chemical energy) and all of it was expended ultimately as heat, lost to the surroundings. Locomotion of organisms—fish swimming, birds flying, people walking—is largely processing work.

Second, we recognize *storing work*, in which part of the energy transformed in the activity is accumulated as potential energy. Pumping water

---

*Concepts of useful work, processing work, storing work, and the Darwin-Lotka law, discussed in this section, are adapted from Howard T. Odum, 1971, *Environment, Power, and Society,* Wiley-Interscience, New York, Chapter 2, pp. 26-37.

from a well into a storage tank on top of a building is a simple mechanical example of storing work. In the activities of plants and animals, accumulation of chemical energy within the tissues of the organism is the dominant form of storing work. It should also be evident that within organisms storing work must precede or accompany processing work. For natural physical systems the energy source may be either stored potential energy, in the case of a stream of water flowing from higher to lower levels, or a continuous input of flowing energy, as in the case of solar radiation maintaining the circulation of the atmosphere.

In all systems performing useful work some energy is drained off as heat. This lost heat does not, of course represent a disappearance or destruction of energy, but simply a dispersion of energy throughout a much larger region than the system itself occupies as the system assumes a lower temperature. The dispersed energy simply becomes unavailable to perform further work in the system.

Power is the rate at which work is done (or energy is transformed). When we calculate the rate at which a quantity changes, we divide the quantity by the elapsed time in selected time units. Consequently power is defined as follows:

$$Power = Work/Time$$

Suppose that two men are each supplied with a 100-lb load which is to be carried up a short flight of stairs. The first man requires 10 seconds to make the climb; the second man requires 5 seconds. Both men do the same amount of work, but the second man produces twice as much power as the first, since he does the work in half the time. (Mechanical units of force, work, energy, and power are given in Appendix I.) Throughout this book, we make use of the *calorie*, a unit of heat energy. It is the quantity of heat energy required to raise the temperature of one gram of pure water through a range of one Celsius degree, when the measurement is made for water close to its freezing point. Power can be stated in heat units as calories expended or stored per second, minute, hour, day, or year.

Power is of paramount importance in all environmental science, but most particularly in the study of life processes, since it measures the rate at which energy can be placed in storage, i.e., transformed into potential energy. Is there some optimum rate of energy storage with respect to total energy expended? Charles A. Coulomb, the French scientist (1736–1806), was interested in evaluating the mechanical power capability of the human male. This was a subject of general interest at a time when men provided a great deal of the power of industry, using devices such as treadmills that would turn wheels to grind grain, lift water, or operate a crane. Coulomb observed that a porter who brought firewood up to his apartment, a vertical ascent of about 40 ft (12 m), had a maximum work capacity of about six wagonloads per day. He made 66 trips per day up the stairs, carrying an average load of 150 lbs (68 kg). Suppose that the porter carried only one or two sticks of wood per trip. He could have made many more trips per day, but since much of his work consisted of lifting his own body weight, the ratio of stored potential energy (stored work) to energy expended in moving his body (processing work) would have been quite small. On the other hand, if he attempted to lift a much heavier load, his progress up the stairs would have been painfully slow, and the total energy stored would be small by the end of the day. Somewhere in between was an optimum ratio of load to body weight, such that the maximum stored work was done in ratio to processing work. Evidently, the porter had learned from experience the optimum load of firewood that would get the largest total weight upstairs in one day.

Without going into an explanation, we can simply state here that the maximum rate at which energy can be placed in storage is achieved when 50% of the energy is stored and 50% is expended in processing work needed to accomplish the storage function. This 50-50 division of power into the two functions for optimum storage is known as the *Darwin-Lotka law*. The law is named for Sir Charles Darwin, who stated the principle of natural selection in evolution, and A. J. Lotka, who in the early 1920s analyzed quantitatively the role of energy expenditure and storage in organic evolution. From an evolutionary viewpoint we simply comment that an organism capable of storing energy at the fastest possible rate would have advantages over other organisms of the same species with lesser abilities, since the food supply might be severely limited and it would be advantageous to be able to convert it into stored energy in the least possible time.

Order, Disorder, and Entropy

So far, we have been talking about useful work performed in open systems, which are the only existing systems of the real world. On the other hand, physicists have given a great deal of emphasis to idealized closed systems in which spontaneous changes take place in the way matter is structured and the direction in which energy is transformed. Even though closed systems cannot exist in nature, the concepts that are derived from their analysis are helpful, if only to show us why certain processes do not occur in open systems.

The classical case in thermodynamics assumes the existence of a container completely isolated from any exchange of matter or energy from the outside region. The space within this container is a perfect vacuum (matter is totally lacking), except that in one corner there is a very small capsule containing a gas under high pressure. All of the gas molecules are in random motion, traveling at high speeds, and having frequent collisions with other molecules. In a manner not explained, the capsule is imagined to be opened, releasing the gas. Molecules of gas quickly diffuse through the entire container and in a very short time are more or less uniformly distributed throughout the entire space. The gas pressure and temperature are also much lower than when in the capsule, for the same amount of molecular motion is now spread throughout a greater volume. The energy level of the system as a whole is represented by the average kinetic energy of motion of the molecules, measured by temperature. In dispersing from the concentrated state in the capsule to the dispersed state in the large container, there was a great decrease in energy level, and consequent decrease in temperature.

The physicist will tell you that because the molecule motions are randomized in speed and direction, there is a very, very minute possibility that at some instant of time all of the gas molecules will be found occupying a space equal to that of the original capsule. Although a real possibility, the odds against it happening are so very small that the event can be dismissed from consideration. The actual distribution of molecules, as represented by a constant density within the limits of measurement, is the most probable state of the system. From this imaginary case the conclusion is drawn that within any closed system spontaneous physical changes will always tend to go from higher to lower energy levels. This concept is expressed in the *second law of thermodynamics*, which says that a closed system always evolves spontaneously in the direction toward the most probable state of matter and energy within itself.

Energy Systems

To the physicist, the arrangement of all the gas molecules in the small capsule with none outside represents a condition of *order*; whereas the arrangement of molecules after dispersal in the large container represents *disorder*. The extent to which disorder is present in a closed system is known as the *entropy*. Thus the second law of thermodynamics can be restated as follows: Within any closed system the direction of spontaneous change is always from order to disorder, with a maximum of disorder and entropy as the ultimate equilibrium state. The direction of change may never be reversed and no decrease of entropy is possible.

The concept of order versus disorder also applies to the structural arrangement of units of matter. A simple physical model—perhaps a bit too simple—will illustrate. Suppose that an employee in a supermarket has painstakingly constructed on the floor a neat pyramid of a large number of oranges. After adding the final orange at the top he stands back to admire his work, which shows a high level of structural order. Suddenly an orange near the bottom slips out of place and the entire pyramid collapses, sending oranges rolling in every direction. After the rolling and rebounding motion of the fruit has ceased the oranges are all in contact with the floor at a common level. Their positions show no orderly geometrical pattern. In fact, we get the visual impression of complete disorder. Actually, if one were to repeat the experiment many times and make a study of the oranges as points distributed over a plane, we would find that they show a random distribution in space (allowing for dispersal from a central point source). The change from order to disorder involved a transformation of potential energy of position into kinetic energy of motion, and, since the model is not a closed system, kinetic energy in turn was dissipated by friction, leaving the system. The final arrangement has the maximum possible entropy, which is to say that the system has run down to attain the lowest possible energy level.

Returning to the real world of nature, consider the activities within a living organism, or better yet, within a single photosynthesizing cell. The cell is an open system—through the cell boundaries there are exchanges of energy and matter with the surrounding medium. Within the cell there operate biochemical processes whereby inputs of relatively simple molecules are converted with the aid of light energy into complex molecules capable of storing energy in concentrated form. It is this ability to concentrate chemical energy that sets apart living matter from nonliving matter. If the cell were isolated from that outside environment, becoming a closed system, the second law of thermodynamics would take over and the energy level in the cell would drop to the lowest possible level. The complex biological molecules on which life depends would slowly degrade to simpler, inactive chemical forms. The spatial organization of parts within the cell would deteriorate as molecules moved to a uniform distribution throughout it. The entropy of the cell would eventually reach a maximum level as life processes were destroyed. Thus, in a very real sense, the trend of organic evolution has been counter to the second law of thermodynamics—toward the maintenance of cellular order and toward the preservation of large, complex biological molecules. Because such a state is inherently unstable it can exist only in an open system where energy flows can constantly sustain it.

This introduction to energy, work, power, and energy systems may have proved to be a pretty heavy dose of theory, particularly for those of you who have not developed a strong background in the sciences. In the chapters to follow the many principles and concepts found in this introduction will be taken up one by one for detailed examination. Concepts that are not clear at this starting point will be clarified and applied to examples that will give them substance. Perhaps, then, after you have completed the last chapter you will want to reread this introduction as a summary of the role of energy in the environment.

Man's Impact on Energy Systems

All of Man's effects on natural energy systems can be viewed as (1) changing inputs or outputs of energy and matter, or (2) creating new pathways or altering existing pathways for the transport or transformation of energy and matter. Although many environmental problems created by Man are discussed in further chapters, two examples will provide a preview.

A river system interrupted by a dam is an example of an energy system in which Man alters existing pathways and varies outputs. Before the building of the dam, water and sediment move downstream. The dam, while filling, interrupts the streamflow and reduces water output. The filling increases energy and matter storage within the system. A greater quantity of water is now held between the upstream sources and the river mouth. Less kinetic energy is lost through friction in the quiet waters of the lake, and potential energy increases as the water level rises. Eventually the system forms a new equilibrium. Water reaches the top of the dam and spills over; downstream water output increases, and output once again equals input. But what of the incoming sediment? Because sediment is trapped in the lake behind the dam, sediment outflow is reduced. Once the lake fills with sediment, however, sediment will begin to spill over the dam, and sediment output will then equal sediment input. Once again, an equilibrium will be reached.

In the meantime, what benefits are gained from the dam? Since Man can raise or lower the flood gates on the dam, he can smooth the irregularities of the flood-drought cycle, accumulating water during high flow, releasing it during low flow. Thus municipalities and industries downstream can depend on a more constant supply for fresh water use or for making sure that water pollution is sufficiently diluted to minimize its impact on the organic life of the river. The fall of water can be used to turn turbines within the dam and create electric power. This action amounts to a harvesting of the potential energy of water within the flooded stretch, for this energy is not converted to kinetic energy which performs work in overcoming internal resistance, but to kinetic energy which performs work in generating electric power.

The energy system concept also helps to analyze further effects of the dam. Since water leaving the reservoir will be devoid of coarse sediment, it will expend less kinetic energy in friction, and hence flow faster in its old bed downstream. This faster flow will create more stress on the bed and banks, and therefore more sediment will be torn loose and added to the flow until the velocity is lowered and the stream comes to equilibrium again. This point, however, will be some distance downstream from the dam, and in the intervening stretch severe erosion can occur. In addition, the large surface area of the reservoir will produce greatly increased evaporation, representing a loss of matter from the river system to the lower atmosphere, while the river flow will be diminished. As an example, Lake Powell, on the Colorado River in Arizona and Utah, is backed up by Glen Canyon Dam; it loses 9 million cubic feet (270,000 cu m) of water to evaporation each year, which is nearly enough to serve the needs of the city of Phoenix. Immediately below the dam erosion has occurred, lowering the river channel by some tens of feet. And, as you might expect, the lake will have a limited lifetime; within 200 years it should be filled with sediment from the Colorado River.

The production of heat from power consumption gives rise to another environmental problem which is conveniently analyzed by the energy system concept. Since conversion of heat energy to kinetic energy or electricity can never be 100% efficient, heat energy is lost by every device which consumes power. As energy is passed on down the line from power plant to substation to transformer to TV set, electric range, or washing machine, heat energy is lost through electrical resistance or in conversion to mechanical energy. Thus, the entire output of an electric power plant must ultimately be released as a heat input to the life layer. This heat represents Man's rapid conversion of chemical energy held in storage by fossil fuels—storage of solar energy which has accumulated over millions of years of geologic time by preservation and conversion of plant and animal remains as coal, oil, and natural gas. We have already noted that all objects radiate heat energy, and that the amount of radiation increases with the temperature of the object. Taking the earth energy system as a whole, the amount of energy absorbed by the earth must equal that radiated away by the earth, except for energy in storage. Because the heat input from power generation warms the life layer, the amount of radiant heat emitted by the earth must increase slightly until a new equilibrium is reached. At this point, the life layer will be slightly warmer, and energy outputs will equal energy inputs.

At the present time, our input of additional heat to the life layer remains small—only about 25/1000 of 1% of the total which is radiated away. Thus, at the present time, Man's impact on the earth's heat budget by fuel consumption is negligible. But current projections show that power generating capacity is increasing by 7% each year, representing a doubling of Man-generated energy about every 10 years. If this increase continues, how long will it be until power generation has significant effects on the life layer? Experts predict that a 1 C° increase in the earth's mean temperature will be sufficient to cause real changes in the boundaries of deserts, plains, and forests. How soon might an increase occur? The answer, based on the above projections, is 91 years. Other experts predict that an increase of 3 C° could begin to cause melting of the ice sheets. The liquid water produced would raise sea level by some 100 ft (30 m), inundating most major cities and flooding substantial portions of continents. This water conversion would be complete in 108 years, if present trends continue. Of course, our reserves of fossil fuels will be exhausted before that time. But the prospect of developing breeder reactors, which produce more fuel from uranium than they consume, and even fusion reactors, which harness the power of the H-bomb, makes the prospect a very real one indeed, provided that Man's demands for power remain unchecked.

In actual fact, the situation is much more complicated. For example, carbon dioxide produced by fuel burning increases the *greenhouse effect* (see Chapter 2) which helps to make the life layer warmer. This effect could by itself account for an observed global air temperature increase of 0.6 F° (0.4 C°) in the interval between 1900 and 1940. Particulate matter, the fine dust particles which are released from smokestacks in the burning of fossil fuels, however, may have an opposite effect. Particles in the lower atmosphere help reflect incoming solar energy back out

Man's Impact on Energy Systems

to space before it can reach the surface, reducing total solar input. In short, many Man-influenced factors can produce significant effects on the global energy balance, in spite of its great magnitude. One of Man's greatest challenges in years to come will be in analyzing, predicting, and reducing his impact on the earth's energy system.

Dam-building and power generation are only two of many environmental problems of importance to Man, and throughout the rest of this book we will be discussing many others. Most, if not all, will be conveniently analyzed by keeping in mind the principles we have provided in this introduction to energy systems.

PART I

Energy Systems of the Atmosphere and Hydrosphere

1 Atmosphere and Oceans

Understanding the wide range of physical environments of the biosphere, or life layer, requires that we first take a very broad view of our planet. This is a view we might have if we could hold the spherical globe in one hand at arm's length and see the whole range of its surface from pole to pole. At this range, broad patterns of flow of air and water would be seen in full. We could see how the sun's rays fall upon the spherical surface with changing angles of attack as the globe is turned on its axis. The strangely irregular patterns of continents and ocean basins would stand in contrast to the orderly, belted patterns of atmospheric motion. We would be struck by the way in which continents divide up the world ocean into compartments, inhibiting the development of a uniform planetary system of water motions, in contrast to the atmosphere, which at upper levels enjoys an unrestrained freedom of motion over the entire globe in a continuous layer.

We should also be struck by the high speed of motion of the atmosphere, in contrast to the sluggish and often barely perceptible motion of the ocean waters. Evidently the atmosphere is capable of transporting heat and water vapor rapidly from any given point on the sphere to any other point. This thought might lead to the conclusion that atmospheric processes are the dominant environmental controls, with the oceans in a secondary role. The continental surfaces, which are motionless in comparison with atmosphere and oceans, must play the role of static receivers of heat and water from the atmosphere and of radiant energy from the sun. But things are not that simple, for interaction is a fundamental principle of environmental science. A great mountain chain strongly modifies processes of the lower atmosphere. Gross patterns of continents and oceans also are reflected in seasonal changes of atmospheric properties and motions.

Understanding environmental science taxes one's ability to keep tabs on many diverse happenings at the same time, but in the learning process, things must be taken up one at a time. So we will begin with the atmosphere and oceans

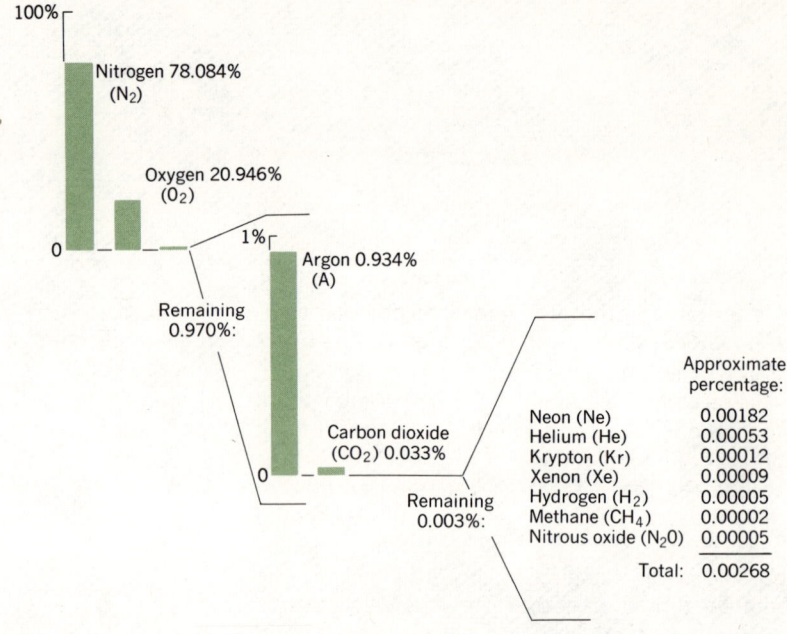

Figure 1.1 Component gases of the lower atmosphere (homosphere), as percent by volume. (Data from E. Gleuckauf, 1951, *Compendium of Meteorology*, Am. Meteorological Soc., Boston, p. 6, Table V. From A. N. Strahler, 1971, *The Earth Sciences*, 2nd ed., Harper & Row, New York.

and follow with the solid earth, building up a resource bank from which many categories of information can later be drawn upon to understand complex environmental problems affecting the biosphere.

Introducing the Atmosphere

The earth's atmosphere consists of a mixture of various gases surrounding the earth to a height of many miles. Held to the earth by gravitational attraction, this envelope of air is densest at sea level and thins rapidly upward. Although almost all of the atmosphere (99%) lies within 18 mi (29 km) of the earth's surface, the upper limit of the atmosphere can be drawn approximately at a height of 6000 mi (10,000 km), a distance approaching the diameter of the earth itself.

From the earth's surface upward to an altitude of about 50 mi (80 km) the chemical composition of the atmosphere is highly uniform throughout in terms of the proportions of its component gases. The name *homosphere* has been applied to this lower, uniform layer, in contrast to the overlying *heterosphere,* which is nonuniform in an arrangement of spherical shells.

Pure, dry air of the homosphere consists largely of *nitrogen* (78.084% by volume) and *oxygen* (20.946%) (Figure 1.1). Nitrogen does not easily enter into chemical union with other substances, but there are processes by which the gas is combined into nitrogen compounds vital to organic processes of the biosphere. In contrast to nitrogen, oxygen is highly active chemically and combines readily with other elements in the process of *oxidation*. Combustion of fuels represents a rapid form of oxidation, whereas certain forms of rock decay (weathering) represent very slow forms of oxidation.

The remaining 0.970% of the air is mostly *argon* (0.934%). *Carbon dioxide,* although constituting only about 0.033%, is a gas of great importance in atmospheric processes because of its ability to absorb heat and thus to allow the lower atmosphere to be warmed by heat radiation coming from the sun and from the earth's surface. Carbon dioxide is also an effective emitter of radiation and acts to cool the upper atmosphere.

Green plants, in the process of *photosynthesis,* utilize carbon dioxide from the atmosphere, converting it with water into carbohydrate. A pronounced rise in the carbon dioxide content of the atmosphere has been noted since 1900 and is a result of Man's combustion of vast quantities of hydrocarbon fuels. This example of Man's impact upon his environment is developed in Chapter 6. Cycles of replenishment and withdrawal of nitrogen, oxygen, and carbon (as carbon dioxide) from the atmosphere and oceans are explained in Chapter 19.

The remaining gases of the homosphere are *neon, helium, krypton, xenon, hydrogen, methane,* and *nitrous oxide,* listed in decreasing order of percentage by volume. Altogether, these constituents total slightly less than 0.003% by volume. All of the component gases of the homosphere are perfectly diffused among one another, so as to give the pure, dry air a definite set of physical properties, just as if it were a single gas.

The Heterosphere

The heterosphere, encountered at about 55 mi (90 km) above the earth's surface, consists of four gaseous layers, each of distinctive composition (Figure 1.2). Lowermost is the *molecular nitrogen layer* consisting dominantly of molecules of nitrogen (N_2) and extending upward to about 125 mi (200 km). Above this height lies the *atomic oxygen layer*, consisting dominantly of oxygen atoms (O). Between about 700 mi (1100 km) and 2200 mi (3500 km) lies the *helium layer*, composed dominantly of helium atoms (He). Above this region lies the *atomic hydrogen layer*, consisting of hydrogen atoms (H). No definite outer limit can be set to the hydrogen layer. A height of 6000 mi (10,000 km) may perhaps be taken as an arbitrary limit, for here the density of the hydrogen atoms is approximately the same as that found throughout interplanetary space. However, hydrogen atoms rotating with the earth, and hence belonging to the earth's atmosphere, may exist as far out as 22,000 mi (35,000 km).

The four heterosphere layers described above have transitional boundary zones, rather than sharply defined surfaces of separation. The arrangement of gases is in order of their weights: molecular nitrogen, the heaviest, is lowest; atomic hydrogen, the lightest, is outermost. Keep in mind that, at the extremely high altitudes of the heterosphere, the density of the gas molecules and atoms is extremely low. For example, at 60 mi (96 km), close to the base of the heterosphere, the atmosphere has a density of only about one-millionth that at sea level. Atoms and molecules of the heterosphere are neutral in charge and turn with the earth's rotation.

Subdivisions of the Homosphere

The atmosphere has been subdivided into layers according to temperatures and zones of temperature change. Three temperature zones lie within the homosphere; a fourth is assigned to the lower heterosphere. Figure 1.3 shows how temperature is related to altitude. Starting at the earth's surface, temperature falls steadily with increasing altitude at the fairly uniform average rate of $3\frac{1}{2}$ F° per 1000 ft (6.4 C° per km). This rate of temperature drop is known as the *normal environmental lapse rate*. Departures from this rate will be observed, depending upon geographical location and season of year. The layer in which the environmental lapse rate applies is known as the *troposphere*, the properties of which are discussed in detail below.

The normal environmental lapse rate gives way rather abruptly at a height of 8 to 9 mi (12.5 to

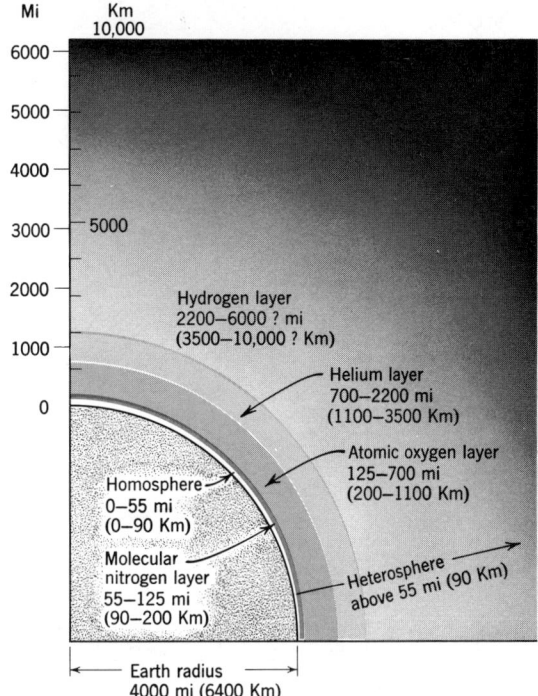

Figure 1.2 Homosphere and heterosphere. (Based on data of R. Jastrow, N.A.S.A., and M. Nicolet.)

15 km) to a layer known as the *stratosphere*, in which air temperature increases gradually with increasing height. The level at which the troposphere gives way to the stratosphere is termed the *tropopause* Figure 1.4 shows that the elevation of the tropopause is least at the poles, 5 to 6 mi (8 to 10 km), whereas at the equator, the tropopause is encountered at 10 mi (17 km). If the troposphere is thought of as a surface in three dimensions, it resembles an oblate ellipsoid with a polar flattening and an equatorial bulge.

Seasonal changes in the altitude of the tropopause are marked in middle and high latitudes. For example, at 45° latitude the average altitude in January is 8 mi (12.5 km), but rises to 9 mi (15 km) in July. Temperatures at the tropopause are markedly lower at the equator than at the poles. At first glance, this relationship may seem strange, accustomed as we are to considering the equatorial region to be hot and the poles cold. However, for a more-or-less constant temperature lapse rate, the higher the tropopause, the colder will be the air.

Upward through the stratosphere there is a slow rise in temperature until a value of about 32° F (0° C) is reached at about 30 mi (50 km). Here, at the *stratopause*, a reversal to falling temperature sets in. Temperature decreases through

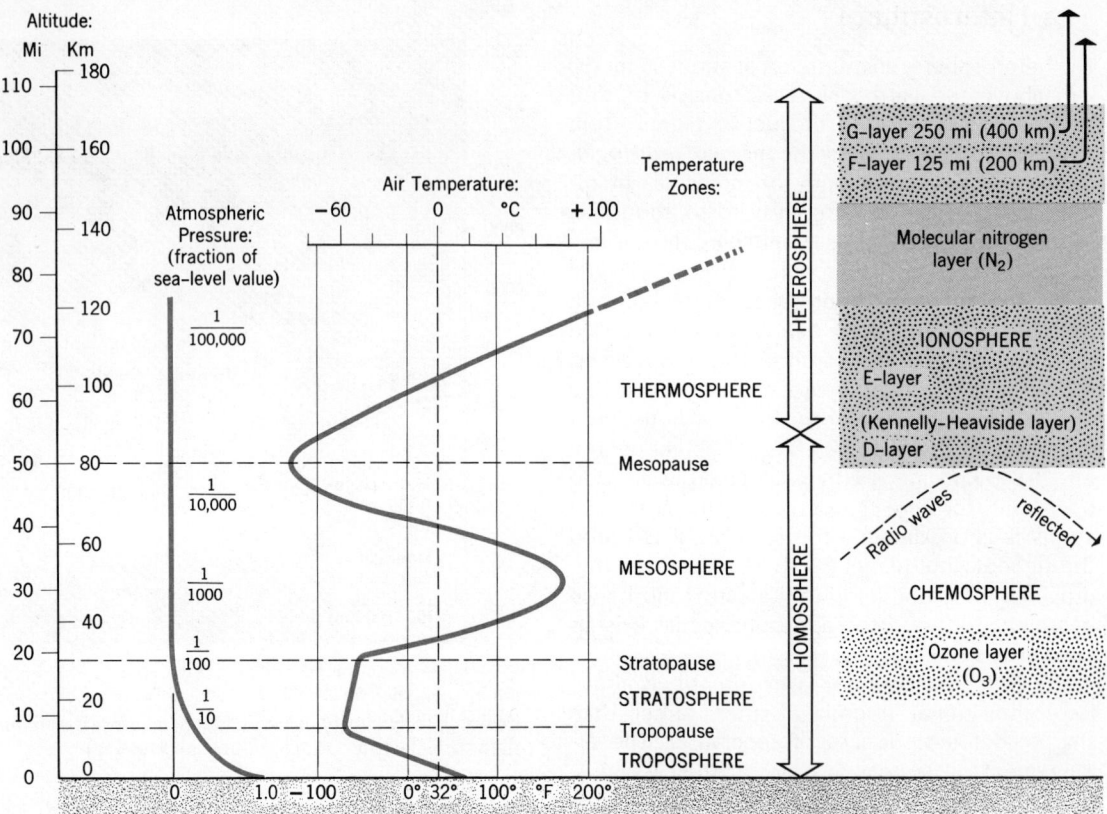

Figure 1.3 Structure of the atmosphere. (After A. N. Strahler, 1971, *The Earth Sciences*, 2nd ed., Harper & Row, New York.)

the overlying *mesosphere,* a layer extending upward to about 50 mi (80 km), where a low point of −120° F (−83° C) is reached. This level of temperature minimum and reversal is termed the *mesopause.* With further increasing altitude, a steep climb in temperature is observed within the *thermosphere.* As previously noted, the thermosphere lies within the heterosphere, so that the mesopause may be regarded as coincident with the upper limit of the homosphere. Within the thermosphere, temperatures reach 2000 to 3000° F (1100 to 1650° C), but such figures have little meaning when we consider that the density of the air is so slight as to approach a vacuum. Very little heat can be held or conducted by air of such low density.

The Troposphere

It is the lowermost atmospheric layer, the troposphere, that is of most direct importance to Man and other life forms in the life environment at the bottom of the atmosphere. Practically all phenomena of weather and climate that materially affect the biosphere take place within the troposphere.

In addition to pure dry air, the troposphere contains *water vapor,* a colorless, odorless gaseous form of water which mixes perfectly with the other gases of the air. The concentration of water vapor in the air is designated as the *humidity* and is of primary importance as an environmental

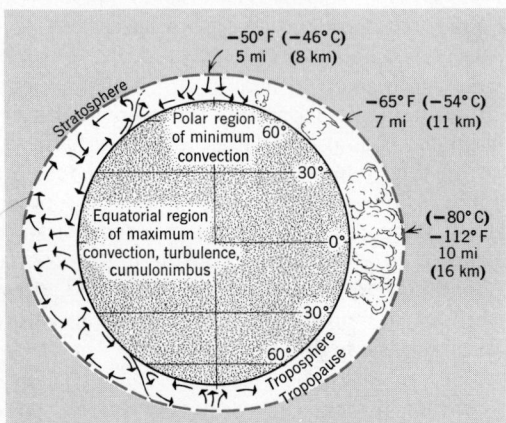

Figure 1.4 Schematic cross section of the troposphere. Figures give height and temperature of the tropopause.

factor. Water vapor can condense into clouds and fog. If condensation is excessive, rain, snow, hail, or sleet, collectively termed *precipitation,* may result. There is, in addition, a most important function performed by water vapor. Like carbon dioxide, it is capable of absorbing heat, which penetrates the atmosphere in the form of radiant energy from the sun and earth. Water vapor gives to the troposphere the qualities of an insulating blanket, which prevents the rapid escape of heat from the earth's surface.

The troposphere contains myriads of tiny dust particles, so small and light that the slightest movements of the air keep them aloft. They have been swept into the air from dry desert plains, lake beds and beaches, or explosive volcanoes. Strong winds blowing over the ocean lift droplets of spray into the air. These may dry out, leaving as residues extremely minute crystals of salt which are carried high into the air. Forest and brush fires are yet another important source of atmospheric dust particles. Countless meteors, vaporizing from the heat of friction as they enter the upper layers of air, have contributed dust particles.

Dust in the troposphere contributes to the occurrence of twilight and the red colors of sunrise and sunset, but the most important function of dust particles is not observable and is rarely appreciated. Certain types of dust particles serve as *nuclei,* or centers, around which water vapor condenses to produce cloud particles. As we shall see in Chapter 6, this process is intensified in the air over cities which discharge much chemically active dust into the air.

The stratosphere and higher layers are almost free of water vapor and dust. Clouds are rare and storms are absent in the stratosphere, although winds of high speed are observed and locally the air may be highly turbulent.

Atmospheric Pressure

Although quiet air seems to be intangible and to have no substance, it constantly subjects us to a considerable confining pressure. At sea level, the atmosphere exerts a pressure of about 15 lb per square inch (about 1 kg per square centimeter) on every solid or liquid surface exposed to it. Because this pressure is exactly counterbalanced by the pressure of air within liquids, hollow objects, or porous substances, its ever-present weight creates no special concern. The pressure on 1 square inch of surface can be thought of as the actual weight of a column of air one inch in cross section extending upward to the outer limits of the atmosphere. Air is readily compressible. That which lies lowest is most greatly compressed

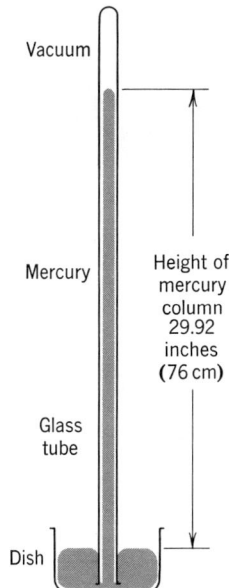

Figure 1.5 Principle of the mercurial barometer.

and is, therefore, densest. In an upward direction, both density and pressure of the air fall off rapidly, as shown in Figure 1.3.

The meteorologist uses another method of stating the pressure of the atmosphere, based on a classic experiment of physics first performed by Torricelli in the year 1643. A glass tube about 3 ft (1 m) long, sealed at one end, is completely filled with mercury. The open end is temporarily held closed. Then the tube is inverted and the end is immersed into a dish of mercury. When the opening is uncovered, the mercury in the tube falls a few inches, but then remains fixed at a level about 30 in (76 cm) above the surface of the mercury in the dish (Figure 1.5). Atmospheric pressure now balances the weight of the mercury column. Should the air pressure increase or decrease, the mercury level will rise or fall correspondingly.

Any instrument that measures atmospheric pressure is a *barometer.* The type devised by Torricelli is known as the *mercurial barometer.* With various refinements over the original simple device it has become the standard instrument. Pressure may be read in inches or centimeters of mercury, the true measure of the height of the mercury column. Standard sea-level pressure is 29.92 in. on this scale. In metric units this is 76 cm (760 mm).

Meteorologists use a pressure unit called the *millibar* (mb). One inch of mercury is equivalent to about 33.9 mb. Standard sea-level pressure is 1013.2 mb, and each one-tenth inch of mercury is equal to about 3 mb (0.1 in. = 3.39 mb).

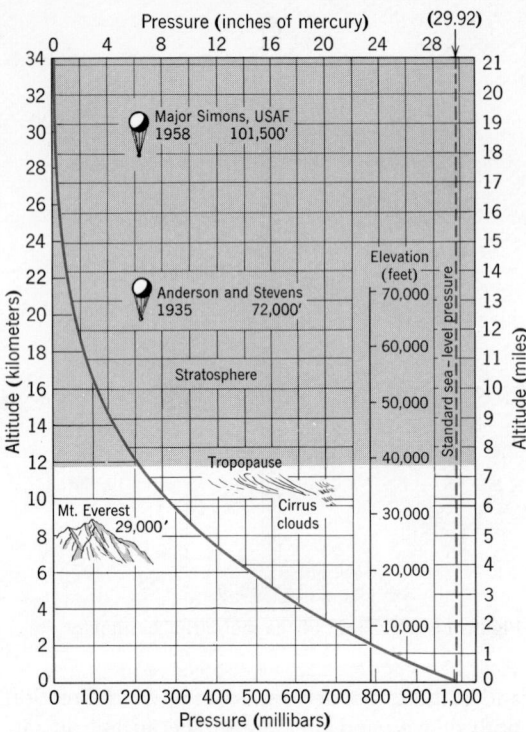

Figure 1.6 Decrease of barometric pressure with increase in altitude.

Figure 1.6 shows how pressure falls with increasing altitude. For every 900 ft (275 m) of rise in elevation, the pressure decreases by one-thirtieth of itself. As the graph shows (by a steepening of the curve), the rate of pressure drop becomes less and less with increasing altitude until, beyond a height of 30 mi (50 km), decrease is extremely slight.

Introducing the Oceans

Using the term *world ocean* to refer to the combined ocean bodies and seas of the globe, let us consider some statistics that emphasize the enormous extent and bulk of this great salt water layer. The world ocean covers about 71% of the globe; its average depth is about 12,500 ft (3800 m), when shallow seas are included with the deep main ocean basins. For major portions of the Atlantic, Pacific, and Indian Oceans the average depth is about 13,000 ft (4000 m). The total volume of the world ocean is about 317 million cu mi (1.4 billion cu km), which comprises just over 97% of the world's free water. Of the small remaining volume about 2% is locked up in the ice sheets of Antarctica and Greenland, and about 1% is represented by fresh water of the lands. These figures show that the hydrosphere, which is a generalized word for the total free water of the earth (whether as gas, liquid, or solid), is largely represented by the world ocean. To place the masses of the atmosphere and oceans in their proper planetary perspective compare the following figures (the unit of mass used here is 10^{21} kg):

Entire earth	6,000
World ocean	1.4
Atmosphere	0.005

What are the basic differences in properties and behavior between the world ocean and the atmosphere? How do atmosphere and ocean interact in the region of their interface? The answers to these questions are vital to understanding of environmental processes affecting the biosphere, because marine life depends upon the exchanges of matter and energy across the atmosphere-ocean interface. It's also significant that the earliest life forms originated and developed in the shallow layer of water immediately beneath the interface.

The atmosphere, being easily compressed, has no distinct upper boundary; it becomes progressively denser toward its base under the load of the overlying gas. The oceans, composed of liquid water that shows very little change of density under large compressional forces, has a sharply defined upper surface in contact with the densest layer of the overlying atmosphere. Whereas the most active region of the atmosphere is the lowermost layer—the troposphere—the most active region of the ocean is its uppermost layer. At great ocean depths water moves extremely slowly and maintains a uniformly low temperature. One reason for intense physical and biological activity in the uppermost ocean layer is that the input of energy and matter from the overlying atmosphere drives water motions in the form of waves and currents. The atmosphere is also the source layer of heat and of condensed fresh water entering the ocean. But the ocean surface also returns heat and water (in vapor form) to the lower atmosphere, a phenomenon of primary importance in driving atmospheric motions. Interaction between atmosphere and ocean surfaces is a topic we shall need to explore further in later chapters.

Already we mentioned that compartmentation of the oceans by intervening continental masses inhibits the free global interchange of ocean waters, whereas the atmosphere is free to move globally. Another difference in the two bodies is that the atmosphere has little ability to resist stresses and therefore moves easily and rapidly, changing its velocity very quickly from place to place. In contrast the ocean water can move only sluggishly and is very slow to respond to the changes in force applied by winds.

The two fluid layers tend to balance each other off in controlling the thermal environment of the earth's surface—the atmosphere imposing quick changes in temperature from day to night and from season to season, while the oceans tend to keep the thermal environment uniform and to suppress great swings of temperature from day to night and from summer season to winter season. The reason for this difference in environmental roles is that the capacity of air to hold heat is very small, that of water is very large. Consequently, the heat reserves of the atmosphere are small; those of the ocean are large. In this respect, the atmosphere is a fast spender, for whom life is "easy come, easy go"; the ocean is the banker who holds huge assets in reserve but is ready to loan funds to tide the spendthrift over a lean period.

Because for all practical purposes ocean water is incompressible, it undergoes only very slight changes in volume and density when it sinks to great depths or rises from great depths to the surface. In this respect ocean water is very different in behavior from air, which expands greatly as it rises and contracts greatly as it descends. Accompanying these changes in gas volume are changes in temperature that strongly affect atmospheric processes. The oceans experience only very slight temperature variations from the rising and sinking of water.

Composition of Sea Water

Sea water is a solution of salts—a brine—whose ingredients have maintained approximately fixed proportions over a considerable span of geologic time. Besides their importance in the chemical environment of marine life, these salts constitute a vast reservoir of mineral matter from which certain constituents may be extracted by Man for his use. One way to describe the composition of sea water is to state the principal ingredients that would be required to make an artificial brine approximately like sea water. These are listed in Table 1.1. Of the various elements combined in these salts, chlorine alone makes up 55% by weight of all the dissolved matter, and sodium 31%. Important, but less abundant than elements of the five salts in Table 1.1, are bromine, carbon, strontium, boron, silicon, and fluorine. At least some trace of half of the known elements can be found in sea water. Sea water also holds in solution small amounts of all of the gases of the atmosphere, principally nitrogen, oxygen, argon, carbon dioxide, and hydrogen.

TABLE 1.1 Principal Salts in Sea Water

Name of Salt	Chemical Formula	Grams of Salt per 1000 Grams of Water
Sodium chloride	NaCl	23
Magnesium chloride	$MgCl_2$	5
Sodium sulfate	Na_2SO_4	4
Calcium chloride	$CaCl_2$	1
Potassium chloride	KCl	0.7
With other minor ingredients, to total		34.5

The proportion of dissolved salts to pure water is the *salinity*, usually stated in units of parts per thousand by weight, and designated by the special symbol ‰. The figure 34.5‰, given as the total in Table 1.1, represents 3.45%. Salinity of sea water varies somewhat from place to place in the oceans. Where diluted by abundant rainfall over the equatorial oceans, the salinity may be between 34.5 and 35 ‰, whereas beneath desert belts evaporation raises the salinity of surface water to more than 35.5 ‰.

Salts of the ocean water have come from two sources through geologic time. One is from the chemical products of breakdown of minerals exposed on the lands to atmospheric weathering. The dissolved products are transported to the sea by streams. (The weathering process is explained in Chapter 8.) Contributions from rock weathering are principally oxygen (O), and the metallic elements, sodium (Na), magnesium (Mg), calcium (Ca), and potassium (K). A second source of elements is from the earth's interior by a process called *outgassing*, in which water and many dissolved gases, known collectively as *volatiles*, emerge from volcanoes, hot springs, and fumaroles (steam emissions). All of the water of the oceans and atmosphere is considered to have come from the earth's interior by outgassing. This process has also been the source of the element chlorine (Cl), which makes up 55% of sea water, and of sulfur, found in the sulfate radical (SO_4). Outgassing has also been the source of the atmospheric gases nitrogen (N_2), carbon (as carbon dioxide, CO_2), argon (A), and hydrogen (H).

You might be led to think that the salinity of the oceans would have risen steadily through geologic time as more and more of the constituents of the sea salts were received through processes of rock weathering and outgassing. But such is not the case, because the rate at which the various elements are added to the oceans is balanced by rate of the return of these elements to the solid state as mineral deposits on the sea floor, a process of *chemical precipitation*. As a result, the chemical composition and overall salinity of sea water may have been essentially constant over most of the time that life has existed—some 3 billion years. This chemical stability of the oceans is a remarkable phenomenon and is a basic environmental factor in the evolution of life forms.

Density of Sea Water

Density of any substance is the mass of a specified unit volume of the substance. For water, density is commonly stated in *pounds per cubic foot*, the value being 62.4 for water at near-freezing temperature. For scientific purposes, density is given in *grams per cubic centimeter*. Pure, fresh water at 39° F (4° C) is at its greatest density, 1 cubic centimeter of water weighing almost exactly 1 gram. Using the value of 1.000 as the density of pure fresh water, sea water has a density ranging from 1.027 to 1.028. Two factors determine sea water density: salinity and temperature. Greater salinity gives greater density. Colder temperatures give greater density down to the freezing point, which is about 28½° F (−2° C).

Density is a matter of prime importance in circulation of ocean waters because slight density differences cause water to move. Where denser water is produced by cooling or evaporation at the surface, it will tend to sink, displacing less dense water below. Such vertical currents may be described as *convectional*.

Layered Structure of the Oceans

As with the atmosphere, the ocean has a layered structure, the layers being recognized in terms of temperature or chemical composition. In the troposphere air temperatures are generally highest at ground level and diminish upward. In the oceans, temperatures are generally highest at the sea surface and decline with depth. This trend is to be expected, since the source of heat is from solar radiation and from heat supplied by the overlying atmosphere.

With respect to temperature, the ocean presents a three-layered structure in cross section, as shown in the left-hand diagram of Figure 1.7. At low latitudes throughout the year and in middle-latitudes in the summer there develops a warm surface layer. Here wave action mixes heated surface water with the water below it to give a warm layer that may be as thick as 1600 ft (500 m), with a temperature of 70° to 80° F (20° to 25° C) in oceans of the equatorial belt. Below the warm layer temperatures drop rapidly, constituting a second layer known as the *thermocline*. Below the thermocline is a third layer of very cold water extending to the deep ocean floor. Temperatures near the base of the deep layer are in the range of 32° to 40° F (0° to 5° C). In arctic and antarctic regions, the three-layer system is replaced by a single layer of cold water, as shown in the north-south profile of Figure 1.8. Temperature is a prime environmental factor controlling the abundance and variety of marine life, the bulk of which thrives in the shallow upper layer.

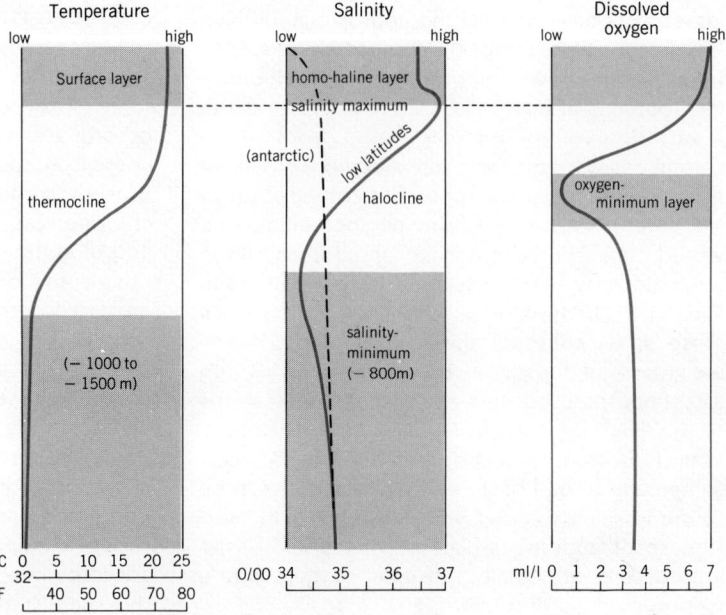

Figure 1.7 Changes with depth of temperature, salinity, and dissolved oxygen typical of oceans in low and middle latitudes. Depths are not to scale. Based on data of W. E. Yasso, 1965, *Oceanography*, Holt, Rhinehart and Winston, New York, Figure 2-4; and A. Defant, 1961, *Physical Oceanography*, Pergamon, New York, vol. 1, chap. 4. (From A. N. Strahler, 1971, *The Earth Sciences*, 2nd ed., Harper & Row, New York.)

Retrospect and Prospect

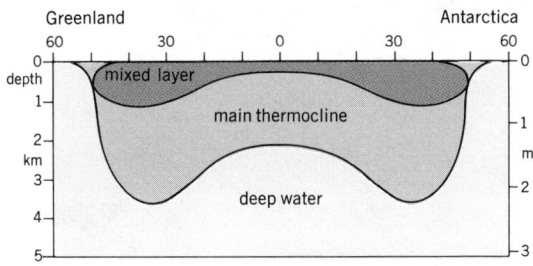

Figure 1.8 Schematic diagram of three-layered structure of oceans. (After J. Williams, *Oceanography*. Copyright © 1962, Little, Brown, Boston, p. 94, Figure 7-4.)

Salinity also has a three-layered structure in low latitudes, as shown in the middle diagram of Figure 1.7. Salinity is uniformly high in a shallow surface layer. Below this is a zone of rapid decrease, the *halocline,* and beneath it a deep layer of uniformly low salinity.

The content of free oxygen (O_2) dissolved in sea water shows an oxygen-rich surface layer, accounted for by the availability of atmospheric oxygen and the activity of oxygen-releasing plant life in the sea. As shown in the right-hand diagram of Figure 1.7, oxygen content falls rapidly with depth; over large ocean areas there is a distinct minimum zone of low oxygen content. Here oxygen has been consumed by biological activity. In deep water the oxygen content holds to a uniform and moderate value down to the ocean floor. The distribution of dissolved oxygen illustrates the interaction of organic and physical processes. Further details of the total oxygen cycle are given in Chapter 18 and will place this topic in its proper global perspective.

Retrospect and Prospect

The broad overview of atmosphere and oceans presented in this chapter has revealed the major elements of physical structure and chemical composition of those two great fluid layers. Most of the information has been about static conditions that one would encounter when probing upward into the atmosphere and downward into the oceans.

In the next four chapters we turn to the great systems of flow of matter and energy that continually involve the atmosphere and oceans, making them dynamic rather than static bodies. First, we shall trace the course of radiant energy from the sun as it passes through the atmosphere, reaches the earth, and is returned to outer space. The solar radiation system establishes the thermal environment of the biosphere, supplying it with the energy for biological processes. There follow accounts of the vast systems of transport of the atmosphere and oceans, whereby energy and matter are redistributed over the globe to provide more moderate and more favorable life conditions than would otherwise exist on our planet. Accompanying these circulation systems are intense disturbances of the air and sea—storms that constitute environmental stresses and hazards, often of phenomenal proportions.

Looking over the other planets of the solar system, the uniqueness of earth as a life environment is most striking. Only our earth has both a great world ocean and a comparatively dense, oxygen-rich atmosphere combined with favorable temperature range. Mars, our nearest planet, has practically no free water in any form and only a very rarified atmosphere with little oxygen. Venus, matching us closely in size, has a much denser atmosphere than earth. But while there is some free oxygen in the atmosphere of Venus, water in any form is apparently almost totally lacking, and surface temperatures are very much higher than on earth. Little Mercury with no atmosphere or water roasts in the sun's rays. The great outer planets—Jupiter, Saturn, Uranus, and Neptune—probably have huge quantities of water but it is frozen solid. Their atmospheres, composed largely of ammonia and methane, would be lethal to life such as ours even if the surface temperatures were not impossibly cold. The lunar environment has no free water or atmosphere to offer. So there is really no other place for Man to live but on planet Earth.

2 The Earth's Radiation Balance

All life processes as well as practically all exchanges of matter and energy at the interface between the earth's atmosphere and the surfaces of the oceans and lands are supported with radiant energy supplied by the sun. The planetary circulation systems of atmosphere and oceans are driven by solar energy. Exchanges of water vapor and liquid water from place to place over the globe depend upon this single energy source. It is true that some heat flows upward through the lithosphere to the earth's surface from internal radioactive and volcanic sources, but the amount is trivial in comparison with the energy which the earth intercepts from the sun's rays.

The flow of energy from sun to earth and then out into space is a complex system, since it involves not only electromagnetic radiation, but also energy storage and transport as heat in the gaseous, liquid, and solid matter of the atmosphere, hydrosphere, and lithosphere. However, we can simplify the study of this total system by first examining each of its parts. We will start with the radiation process itself and develop the concept of a *radiation balance*, which is perhaps the most important control of the earth's surface environment.

Organisms respond directly to the heating and cooling of the air, water, or soil that surrounds them. These temperature changes result from the gain or loss of energy by the absorption or emission of radiant energy. When a substance absorbs radiant energy the surface temperature of that substance is raised. This process represents a transformation of radiant energy into the energy of *sensible heat*, a physical property measured by the thermometer. Sensible heat, in the case of a gas, represents the kinetic energy of motion of the gas molecules, which are in constant high-speed flight and endlessly colliding with one another. A rise in temperature of the gas represents an increase in the average velocity of the molecules and in the frequency of their collisions.

Many of the biochemical processes taking place within organisms as well as many common inorganic chemical reactions are intensified by an increase in temperature of the solutions in which these reactions are occurring. Severe cold, which is simply the lack of kinetic energy within matter, may greatly reduce, or even completely stop bio-

The Earth's Radiation Balance

chemical and inorganic reactions. This is why the vital environmental ingredient of heat—heat in the air, water, and soil—needs to be thoroughly understood.

We are all familiar with the cyclic nature of temperature changes. There is a daily rhythm of rise and fall of temperature as well as a seasonal rhythm. There are also systematic average changes in air temperatures from equatorial to polar latitudes as well as from oceanic to continental surfaces. Correspondingly, the lower atmosphere and the surfaces of the lands and oceans must be receiving and giving up heat energy in daily and seasonal cycles. There must also be great differences in the quantities of heat received and given up in low latitudes as compared to high latitudes.

Despite the existence of thermal cycles and latitudinal contrasts in temperature, human history as well as the geologic record indicate an overall uniformity of the global thermal environment through time. It is apparent that the earth as a planet maintains within fairly narrow limits a certain average *planetary temperature* which has depended on maintenance of approximately the same distance from the sun and approximately the same planetary surface properties. If this were not the case, the gradual drift toward either increasing heat or increasing cold would ultimately render the earth's surface too hot or too cold to support life.

Solar energy is intercepted by our spherical planet and the level of heat energy tends to be raised. At the same time, our planet radiates heat into outer space, a process that tends to diminish the level of heat energy. Incoming and outgoing radiation processes are simultaneously in action. In one place and time more heat is being gained than lost; in another place and time more heat is being lost than gained.

There exists in combination with the global radiation balance a global *heat balance*, and the two together constitute the earth's total *energy balance*. Equatorial regions receive through solar radiation much more heat than is lost directly to space whereas polar regions lose by radiation into space much more heat than is received. So there must be included in the energy system mechanisms of heat transfer adequate to export heat from a region of excess and to carry that heat into a region of deficiency. On our planet, motions of the atmosphere and oceans act as heat-transfer mechanisms. A study of the earth's heat balance will not be complete until the global patterns of air and water circulation are described and explained in Chapter 4.

Storage of heat energy in a latent form is an important part of the earth's heat balance. We noted in Chapter 1 that changes of state between gaseous, liquid, and solid states are accompanied by the taking up of heat energy or release of heat energy. Water in its three states—as water vapor in the atmosphere and as liquid and solid water in the oceans and over the land surfaces—absorbs and liberates heat as it changes from one state to another. The processes of change of state of water in the atmosphere and hydrosphere are examined in Chapter 3.

To the inorganic, or physical energy cycle there must be added an organic phase, the *biochemical energy cycle*, in which a part of the incoming solar energy is used and given up by plants and animals. Briefly, this cycle consists of the absorption of solar energy by plants in the manufacture of carbohydrate compounds, which provide the food for animals. Eventually, after much recycling, most of these compounds are oxidized in a process known as *respiration*, and the energy is returned to the atmosphere. However, some fraction of the carbohydrate compounds may be stored in soil layers, for example, as peat in bogs. In the geologic past the total accumulation of hydrocarbon compounds in rock strata as coal and petroleum has been enormous, and represents a great storage bank of solar energy. While the biochemical energy cycle involves only a tiny fraction (about one-tenth of a percent) of the solar energy received by the earth, the process is of the first order of importance in the biosphere, since it represents the energy cycle of the whole organic world. This energy cycle of the biosphere is explained in detail in Chapter 18.

Upon further reflection, it becomes apparent that the movement of water through atmosphere, oceans, and upon the lands comprises a system of mass transport equal in importance to the flow of energy and that the activities of these two systems are closely intermeshed. The concept of a *water balance* can be developed and takes its place beside the heat balance (Chapter 5). Together, these two great systems of energy and matter form a single, grand planetary system and permit us to relate and explain many of the environmental phenomena of our earth within a single unified framework.

A systematic approach to the earth's energy balance begins with an examination of the input, or source, of energy from solar radiation. This radiation is traced as it penetrates the earth's atmosphere and is absorbed or transformed. We then turn to the mechanism of output of energy by the earth as a secondary radiator.

The Earth's Radiation Balance

Solar Radiation

Our sun, a star of about medium mass and temperature, as compared with the overall range of stars, has a surface temperature of about 11,000° F (6000° K). The highly heated, incandescent gas that comprises the sun's surface emits a form of energy known as *electromagnetic radiation*. This form of energy transfer can be thought of as a collection, or *spectrum*, of waves of a wide range of lengths traveling at the uniform velocity of 186,000 mi (300,000 km) per second. The energy travels in straight lines radially outward from the sun, and requires about $9\frac{1}{3}$ minutes to travel the 93 million miles (150 million kilometers) from sun to earth. Although the solar radiation travels through space without energy loss, the intensity of radiation within a beam of given cross section (such as 1 square inch) decreases inversely as the square of the distance from the sun. The earth thus intercepts only about one two-billionth of the sun's total energy output.

The solar radiation spectrum consists of (a) X rays, gamma rays, and ultraviolet rays, carrying about 9% of the total energy, (b) visible light rays, 41%, and (c) invisible infrared (heat) rays, 50%. Table 2.1 gives the wavelengths in microns, of the various parts of the spectrum. (A micron is equivalent to one ten-thousandth of a centimeter.) Hereafter, we shall apply the term *shortwave* radiation to the visible and ultraviolet portion of the spectrum (wavelengths less than 0.7 microns) as distinct from the *infrared* or *longwave* portion (longer than 0.7 microns). The total energy of the radiation spectrum is about equally divided between shortwave and longwave portions. Although our definitions of *shortwave* and *longwave* are in accordance with the leading American authority,*

*Glossary of Meteorology, R. E. Huschke, ed., 1959, American Meteorological Society, Boston, Mass. See definitions of solar radiation (p. 524), shortwave radiation (p. 508), longwave radiation (p. 349), and infrared radiation (p. 305).

it is common practice for meteorologists to use *shortwave radiation* as synonomous with *solar radiation*, including all wavelengths in the solar spectrum.

To understand and compare the radiation of energy from both the sun and earth, we refer to some principles of physics. For an ideal radiating surface, called by physicists a *black body*, the total quantity of energy emitted by electromagnetic radiation depends solely upon the surface temperature. To be more specific, the total energy radiated from this ideal surface increases as the fourth power of the absolute temperature, given in degrees Kelvin (°K). There is also a related law of radiation stating that as the temperature of the emitting surface increases, the wavelength of the peak level of energy emission is shifted toward the shorter wavelengths.

Let us apply these principles to the sun's emission of electromagnetic energy. Figure 2.1 is a graph on which the vertical scale shows the intensity of energy emission; the horizontal scale shows wavelength in microns. Both scales are logarithmic, meaning that the equal units of the scale represent powers of ten. Note that the ultraviolet, visible, and infrared regions of the spectrum are labeled at the top of the graph. The radiation of energy from the sun is represented by a solid line with minor irregularities (left side of the graph) arching to a peak at about 0.5 micron in the visible portion of the spectrum. A dashed line making a smooth curve above the solid line is the ideal curve of radiation of a black body whose temperature is 6000° K. The sun's emission curve fits the ideal curve nicely in the infrared region, but shows deficiencies in the visible and ultraviolet regions, because the sun's outer atmosphere absorbs part of the spectrum in these shorter wave lengths.

The source of solar energy is in the sun's interior where, under enormous confining pressure and high temperature, hydrogen is converted to helium. In this nuclear fusion process, a vast

TABLE 2.1 The Electromagnetic Radiation Spectrum.

	Wavelength (microns)	Total Energy (percent)
(Shortest)		
X rays and gamma rays	$\frac{1}{2000}$ to $\frac{1}{100}$	9
Ultraviolet rays	0.2 to 0.4	
Visible light rays	0.4 to 0.7	41
Total shortwave energy		50
(Longest)		
Infrared rays	0.7 to 3000	50

Insolation over the Globe

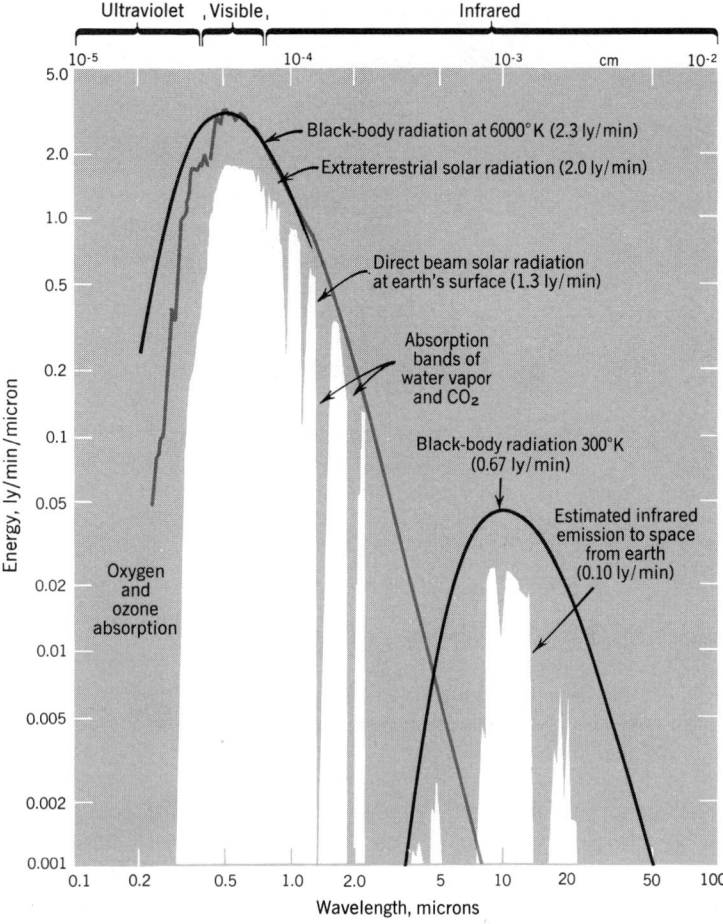

Figure 2.1 Spectra of solar and earth radiation. (From W. D. Sellers, 1965, *Physical Climatology*, Univ. of Chicago Press, Chicago, p. 20, Figure 6.)

quantity of heat is generated and finds its way by convection and conduction to the sun's surface. Because the rate of production of nuclear energy is constant, the output of solar radiation is also unvarying to any significant degree to the best of scientific knowledge. So, at the average distance of earth from the sun, the amount of solar energy received upon a unit area of surface held at right angles to the sun's rays is also unvarying. It is assumed, of course, that the radiation is measured beyond the limits of the earth's atmosphere so that none has been lost. Known as the *solar constant,* this radiation rate has a value of 2 gram calories per square centimeter per minute. One gram calorie per square centimeter constitutes a unit measure of heat energy termed the *langley*. Therefore, we can say that the solar constant is equal to 2 langleys per minute. (In English heat units, the solar constant is equivalent to 430 BTU per square foot per hour.) Orbiting space satellites equipped with suitable instruments for measuring electromagnetic radiation intensity have provided precise data on the solar constant.

Insolation over the Globe

Because the earth is a sphere (disregarding a very slight oblateness), only one point on earth—that upon which the sun's noon rays are perpendicular—presents a surface at right angles to the sun's rays. We can call this point of perpendicularity the *subsolar* point (Figure 2.2). In all directions away from the subsolar point, the earth's curved surface becomes turned at a decreasing angle with respect to the rays until the *circle of illumination* is reached. Along that circle the rays are parallel with the surface.

Let us now assume that the earth is a perfectly uniform sphere with no atmosphere. Only at the subsolar point will solar energy be intercepted at the maximum rate of 2 langleys per minute. (We will now use the term *insolation* to mean the interception of solar energy by an exposed surface.) At any particular place on the earth, the quantity of energy received in one day will then depend upon two factors: (1) the angle at which the sun's rays strike the earth, and (2) the length of time of exposure to the rays. These factors are

The Earth's Radiation Balance

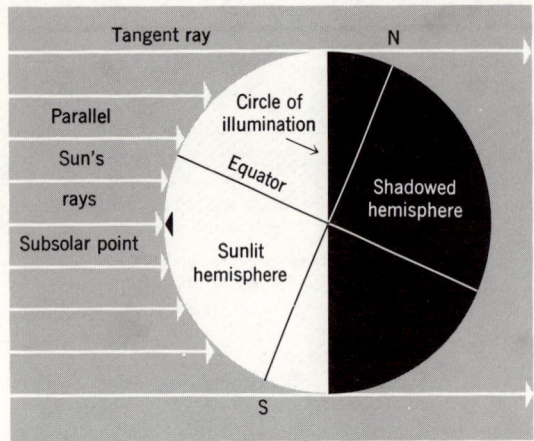

Figure 2.2 Relation of sun's rays to earth. Date is intermediate between equinox and solstice. (From A. N. Strahler, 1972, *Planet Earth; Its Physical Systems Through Geologic Time*, Harper & Row, New York.)

varied by latitude and by the seasonal changes in the path of the sun in the sky.

Figure 2.3 shows that intensity of insolation is greatest where the sun's rays strike vertically, as they do at noon in a belt near the equator. With diminishing angle, the same amount of solar energy spreads over a greater area of ground surface. So, on the average, the polar regions receive the least energy per unit area.

Consider first that, if the earth's axis were perpendicular to the plane of the earth's orbit as it revolves around the sun (*plane of the ecliptic*), the poles would not receive any insolation, regardless of time of year, whereas the equator would receive an unvarying maximum. But the earth's axis is not perpendicular to the plane of the orbit. As shown in Figure 2.4, the earth's axis

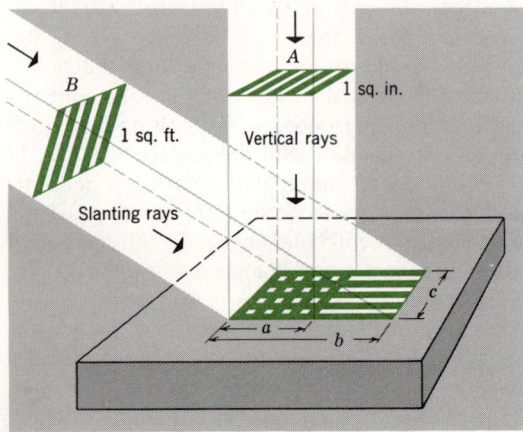

Figure 2.3 Relation of intensity of insolation to angle of sun's rays.

is tilted with respect to the orbital plane by an angle that measures 23½° from the perpendicular (or 66½° with respect to the plane of the orbit). Moreover, the axis at all times holds its orientation in space; the north polar axis always points to the same position among the stars. Consequently, as the earth travels in its orbit about the sun, the tilted globe assumes different positions with respect to the sun's rays.

On June 21 or 22 the earth is so located in its orbit that the north polar end of its axis leans at the full 23½° angle toward the sun. The northern hemisphere is tipped toward the sun, and the southern hemisphere is tipped away from the sun. This condition is named the *summer solstice*. Six months later, on December 21 or 22, the earth is in an equivalent position on the opposite point in its orbit. At this time, known as the *winter solstice,* the north polar axis again leans the full 23½° directly away from the sun; now it is the southern hemisphere that is tipped toward the sun.

Midway between the dates of the solstices occur the *equinoxes,* at which time the earth's axis makes a right angle with a line drawn to the sun, and neither the north nor south pole has any inclination toward the sun. The *vernal equinox* occurs on March 20 or 21; the *autumnal equinox* on September 22 or 23. Conditions are identical on the two equinoxes as far as insolation is concerned, whereas on the two solstices, the conditions of one hemisphere are the exact reverse of the other.

Referring to Figure 2.4 and examining first the globe at the left in its summer solstice position, you will see that a large polar region remains under the sun's rays during a full 24-hour day as the earth turns on its axis. This region lies poleward of the *Arctic Circle*, 66½° N latitude. Correspondingly, an equal south polar region, south of the *Antarctic Circle*, 66½° S latitude, lies in darkness (disregarding twilight) during the entire 24-hour day. At winter solstice, shown by the globe at the right-hand side of Figure 2.4, these conditions of insolation are exactly reversed; it is the north polar region that lies in darkness. At the two equinoxes, insolation is distributed as if the earth's axis had no inclination, so that the sun's rays graze both polar points simultaneously.

This analysis also shows that the subsolar point, representing the sun's noon rays, shifts through a total range 47° from one solstice to the next. This cycle does not make the yearly total of insolation for the entire globe different from an ideal situation in which the earth's axis would not be inclined, but it does cause a great difference in the quantities received at various latitudes.

Insolation over the Globe

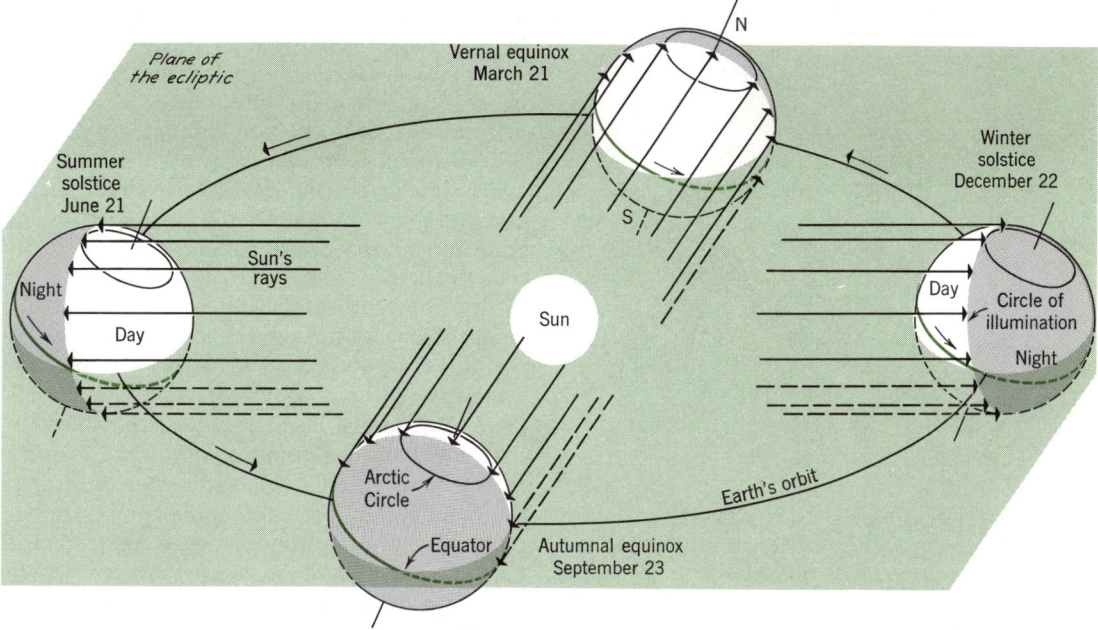

Figure 2.4 The seasons; equinoxes and solstices.

The total annual insolation from equator to poles in thousands of langleys (*kilolangleys*) per year is shown in Figure 2.5 by a solid line. A dashed line shows the insolation that would result if the earth's axis had no tilt. Notice how much insolation the polar regions actually receive—over 40% of the equatorial value.

A second effect of the axial tilt is to produce seasonal differences in insolation at any given latitude, and these differences increase toward the poles, where the ultimate in opposites (six months of day, six of night) is reached. Along with the variation in angle of the sun's rays there operates another factor, the duration of daylight. At the season when the sun's path is highest in the sky, the length of time it is above the horizon is correspondingly greater. The two factors thus work hand in hand to intensify the contrast between amounts of insolation at opposite solstices.

A three-dimensional diagram (Figure 2.6) shows how insolation varies with latitude and with season of year. Figure 2.7 shows graphs of insolation at various selected latitudes from equator to north pole. These diagrams show insolation at the outer limits of the atmosphere and would apply at the ground surface only for an earth imagined to have no atmosphere to absorb or reflect radiation. Notice that the equator receives two maximum periods (corresponding with the equinoxes, when the sun is overhead at the equator) and two minimum periods (corresponding to the solstices, when the subsolar point shifts farthest north and south from the equator). At the Arctic Circle, $66\frac{1}{2}°$ N, insolation is reduced to nothing on the day of the winter solstice, and with increasing latitude poleward this period of no insolation becomes longer. All latitudes between the Tropic of Cancer, $23\frac{1}{2}°$ N, and the Tropic of Capricorn, $23\frac{1}{2}°$ S, have two maxima and two minima, but one maximum becomes dominant as the tropic

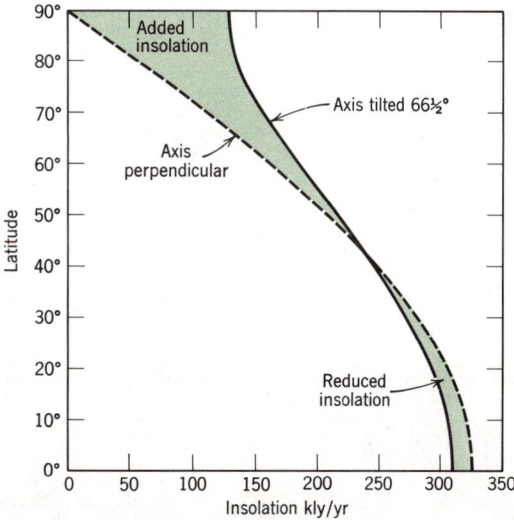

Figure 2.5 Total insolation from equator to pole (solid line) compared with insolation on a globe with axis perpendicular to the ecliptic plane.

The Earth's Radiation Balance

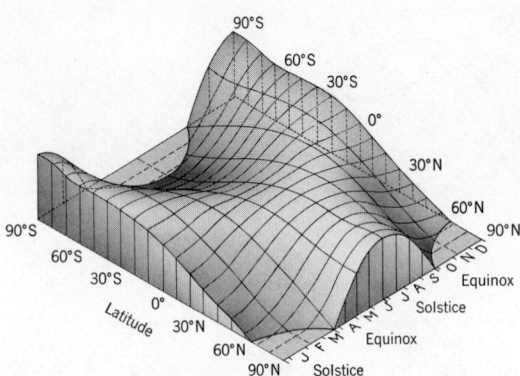

Figure 2.6 Insolation at various latitudes throughout the year. (After W. M. Davis.)

is approached. From $23\frac{1}{2}°$ to $66\frac{1}{2}°$ there is a single continuous insolation cycle with maximum at one solstice, minimum at the other.

World Latitude Zones

The angle of attack of the sun's rays, which determines the flow of solar energy reaching a given unit area of the earth's surface—and so governs the thermal environment of the biosphere—provides a basis for dividing the globe into latitude zones (Figure 2.8). We don't intend that the specified zone limits be taken as absolute and binding, but rather that the system be considered

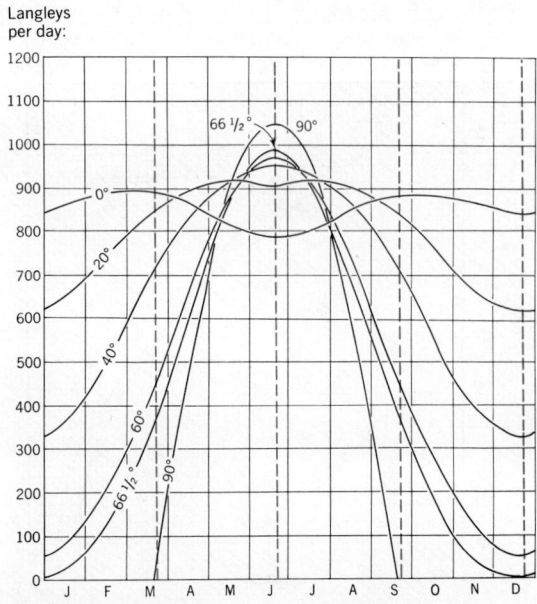

Figure 2.7 Insolation at selected latitudes in the Northern Hemisphere. (From A. N. Strahler, 1971, *The Earth Sciences*, 2nd ed., Harper & Row, New York.)

as a convenient terminology for referring to world belts throughout this text.

The *equatorial zone* lies astride the equator and extends roughly to 10° latitude north and south. Within this zone, the sun throughout the year provides intense insolation, while day and night are of roughly equal duration. Astride the tropics of Cancer and Capricorn are the *north tropical zone* and *south tropical zone* respectively, spanning the latitude belts 10° to 25° north and south. In this zone, the sun takes a path close to the zenith at one solstice and is appreciably lower at the opposite solstice. Thus a marked seasonal cycle exists, but is combined with a potentially large total annual insolation. Notice that literary usage and that of some scientific works differ from what is described here, for the word *tropics* has been widely used to denote the entire belt of 47 degrees of latitude between the tropics of Cancer and Capricorn. Such, indeed, is the definition of "tropics" you will find in most dictionaries. If classically correct, that definition is unsuited to environmental science, because it combines belts of very unlike properties.

Immediately poleward of the tropical zones are transitional regions which have become widely accepted in usage as the *subtropical zones*. For convenience, we have assigned these zones the latitude belts 25° to 35° north and south, but it is understood that the adjective "subtropical" may extend a few degrees farther poleward or equatorward of these parallels.

The *middle-latitude zones,* lying between 35° and 55° north and south latitude represent belts in which the sun's angle of attack shifts through a relatively large range so that seasonal contrasts in insolation are strong. Strong seasonal differences in lengths of day and night exist as compared with the tropical zones.

Bordering the middle-latitude zones on the poleward side are the *subarctic* (*subantarctic*) *zones,* 55° to 60° north and south latitudes, transitional between middle-latitude and arctic zones.

Astride the Arctic and Antarctic circles, $66\frac{1}{2}°$ north and south latitudes, lie the *arctic zones,* which may be further differentiated, if desired, into an *arctic zone* and an *antarctic zone*. We have specified as the latitudinal extent of the arctic zones 60° to 75° north and south, but these limits should not be imposed severely. The arctic zones have an extremely large yearly variation in lengths of day and night, yielding enormous contrasts in isolation from solstice to solstice. Notice that classical usage, as found in standard dictionaries, considers the "arctic" or "arctic region" as the entire area from the Arctic Circle to the North Pole, and "antarctic" in a corresponding sense for

Insolation Losses in the Atmosphere

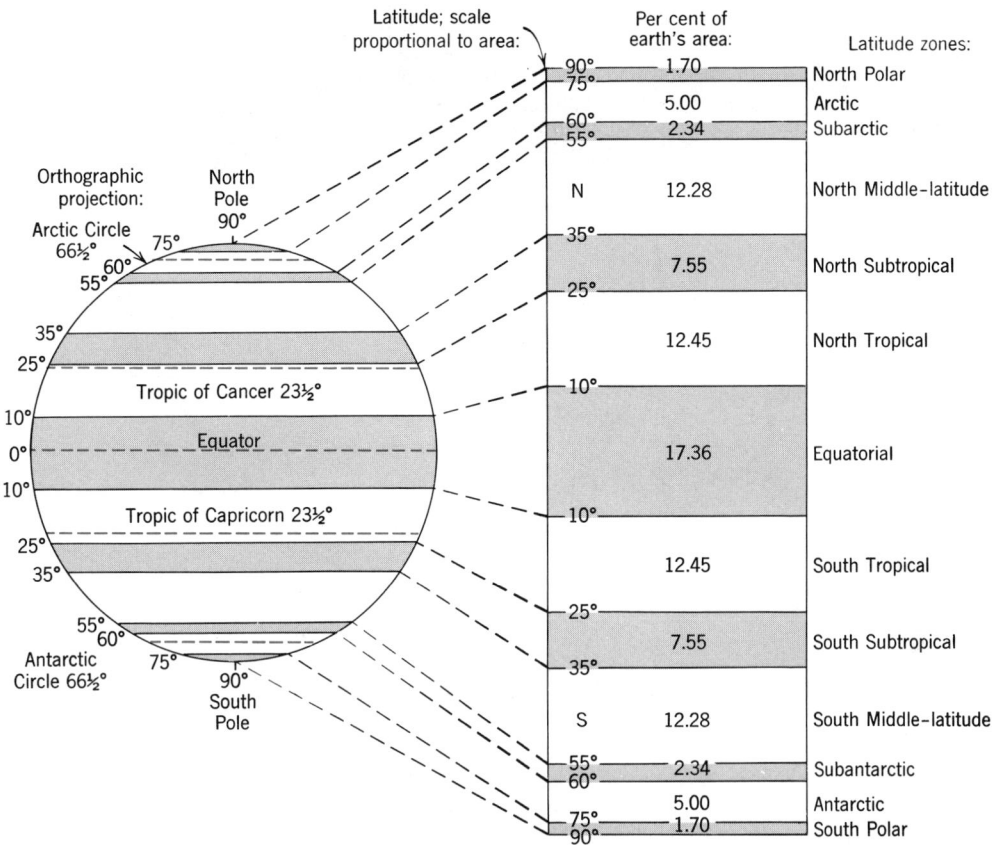

Figure 2.8   A system of latitude zones.

the Southern Hemisphere. As with "tropics," "arctic" and "antarctic" in the literary sense are not well suited to the needs of environmental science.

The *polar zones,* north and south, are circular areas between 75° latitude and the poles. Here the polar regime of a six-month day and six-month night is predominant and yields the ultimate in seasonal contrasts of insolation.

Insolation Losses in the Atmosphere

As the sun's radiation penetrates the earth's atmosphere, a series of selective depletions and diversions of energy take place. At an altitude of 95 mi (150 km), the radiation spectrum possesses almost 100% of its original energy, but in penetration to an altitude of 55 mi (88 km) absorption of X rays is almost complete and some of the ultraviolet radiation has been absorbed as well.

The absorption of shortwave radiation generates an environmental layer of great importance to life on earth and to the technological society of Man. This global layer is the *ionosphere*, located in the altitude range of 50 to 250 mi (80 to 400 km). Notice that this position coincides with the bottom of the heterosphere in which the molecular nitrogen layer and the atomic oxygen layer are found (see Figure 1.2). Furthermore, the ionosphere is essentially identical in position with the lower thermosphere (see Figure 1.3). The ionosphere consists of a number of layers in which the process of *ionization* takes place. Here highly energetic gamma rays and X rays from the solar radiation spectrum are absorbed by molecules and atoms of nitrogen and oxygen. In the absorption process, each molecule or atom gives up an electron, becoming a positively charged *ion*. The electrons thus released form an electric current that flows freely on a global scale within the ionosphere. Of particular interest in the field of radio communication is the ability of the layers of ions to reflect radio waves and thus to turn them back toward the earth. Most of the important reflection of longwave radio waves takes place in the lower part of the ionosphere, which bears the name of *Kennelly-Heaviside layer.* Without such reflection, long-distance radio communication would not be possible. Because the process of ionization requires direct solar

The Earth's Radiation Balance

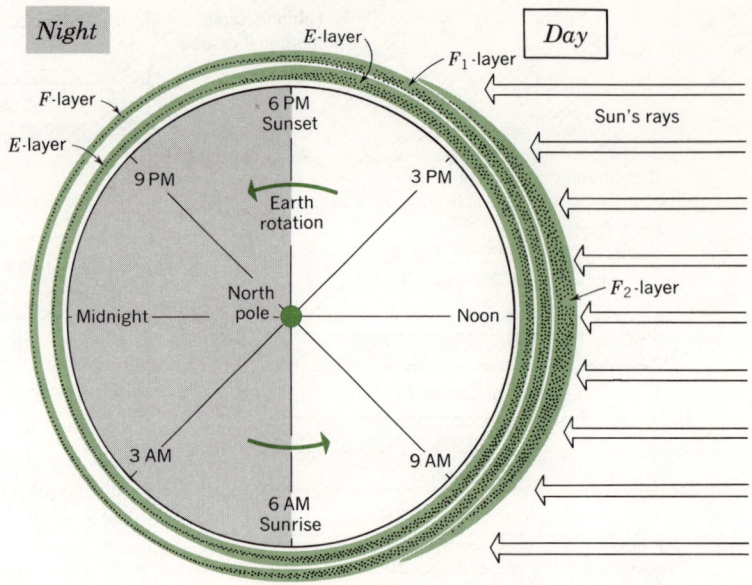

Figure 2.9 Ionospheric layers in a cross section through the earth's equator. Based on data of B. F. Howell, Jr., 1959, *Introduction to Geophysics*, McGraw-Hill, New York. (From A. N. Strahler, 1971, *The Earth Sciences*, 2nd ed., Harper & Row, New York.)

radiation, the ionospheric layers, of which there are five, are developed on the sunlight side of the earth (Figure 2.9). On the dark side, under nighttime conditions, the layers tend to weaken and disappear.

Yet another phenomenon, one of vital concern to Man and all other life forms on earth, is the presence of an *ozone layer*, largely concentrated in the region from 11 to 15 mi (18 to 25 km), but also extending down to an altitude as low as 9 mi (15 km) and up to 35 mi (55 km) (Figure 1.3). The ozone layer thus extends from the lower stratosphere into the lower mesosphere. The ozone layer is a region of concentration of the form of oxygen molecule known as *ozone*, (O_3), in which three oxygen atoms are combined instead of the usual two atoms (O_2). Ozone is produced by the action of ultraviolet rays upon ordinary oxygen atoms. The ozone layer thus serves as a shield, protecting the troposphere and earth's surface from most of the ultraviolet radiation found in the sun's radiation spectrum. If these ultraviolet rays were to reach the earth's surface in full intensity, all exposed bacteria would be destroyed and animal tissues severely burned. Thus the presence of the ozone layer is an essential element in the environment of the biosphere. It is also interesting to note that the high temperatures of the mesosphere are produced by the absorption of the ultraviolet rays in the upper part of the ozone layer. The level of greatest ozone concentration is at its highest altitude in low latitudes (28 mi; 48 km), but descends to the lowest altitude in arctic latitudes (22 mi; 35 km). There are also marked seasonal variations in altitude in middle latitudes.

As solar radiation penetrates into deeper and denser atmospheric layers, gas molecules cause the visible light rays to be turned aside in all possible directions, a process known as *Rayleigh scattering*. Where dust and cloud particles are encountered in the troposphere, further scattering occurs and may be described as *diffuse reflection* (Figure 2.10). That the clear sky is blue in color is explained by Rayleigh scattering of the shorter visible wavelengths. These predominantly blue light waves reach our eyes indirectly from all parts of the sky. The red wavelengths and infrared rays are less subject to scatter and largely continue in a straight-line path toward earth. The setting sun appears red because a part of the red rays escape deflection from the direct line of sight.

As a result of all forms of shortwave scattering, about 5% of the total insolation is returned to space and forever lost, while at the same time some scattered shortwave energy also is directed earthward. The latter is referred to as *diffuse sky radiation*, or *down scatter*.

Another form of energy loss, *absorption*, takes place as the sun's rays penetrate the atmosphere. Both carbon dioxide and water vapor are capable of directly absorbing infrared radiation. Absorption results in a rise of sensible temperature of the air. Thus some direct heating of the lower atmosphere takes place during incoming solar radiation. Although carbon dioxide is a constant quantity in the air (0.033% by volume) the water vapor content varies greatly from place to place, being as low as 0.02% under desert conditions to as high as 1.8% in humid equatorial regions. Absorption correspondingly varies from one global environment to another.

All forms of direct energy absorption—by molecules, including water vapor and carbon dioxide, and by dust—are estimated to total as little as 10% for conditions of clear, dry air, to as high as 30% when a cloud cover exists. A global average of 15% for absorption generally is shown in Figure 2.10. When skies are clear, reflection and absorption combined may total about 20%, leaving as much as 80% to reach the ground.

Yet another form of energy loss must be brought into the picture. The upper surfaces of clouds are extremely good reflectors of shortwave radiation. Air travelers are well aware of how painfully brilliant the sunlit upper surface of a cloud deck can be when seen from above. Cloud reflection can account for a direct turning back into space of from 30 to 60% of total incoming radiation (Figure 2.10). So we see that, under conditions of a heavy cloud layer, the combined reflection and absorption from clouds alone can account for a loss of from 35 to 80% of the incoming radiation and allow from 45 to 0% to reach the ground. A world average value for reflection from clouds to space is about 21% of the total insolation, whereas absorption by clouds is much less—about 3% (Table 2.2).

The surfaces of the land and ocean reflect some shortwave radiation directly back into the atmosphere. This small quantity, about 6% as a world average, may be combined with cloud reflection in evaluating total reflective losses. Table 2.2 lists the percentages given so far for the energy losses in insolation by reflection and absorption. Altogether the losses to space by reflection total 32% of the total insolation. Figure 2.11 shows the same data as in the upper part of Table 2.2 in graphic form.

The percentage of radiant energy reflected back by a surface is termed the *albedo*. This is an important property of the earth's surface because it determines the relative rate of heating of the surface when exposed to insolation. Albedo of a water surface is very low (2%) for nearly vertical rays, but high for low-angle rays. It is also extremely high for snow or ice (45 to 85%). For fields, forests, and bare ground the albedos are of intermediate value, ranging from as low as 3% to as high as 25%.

Orbiting satellites, suitably equipped with instruments to measure the energy levels of shortwave and infrared radiation, both incoming from the sun and outgoing from the atmosphere and earth's surface below, have provided data for

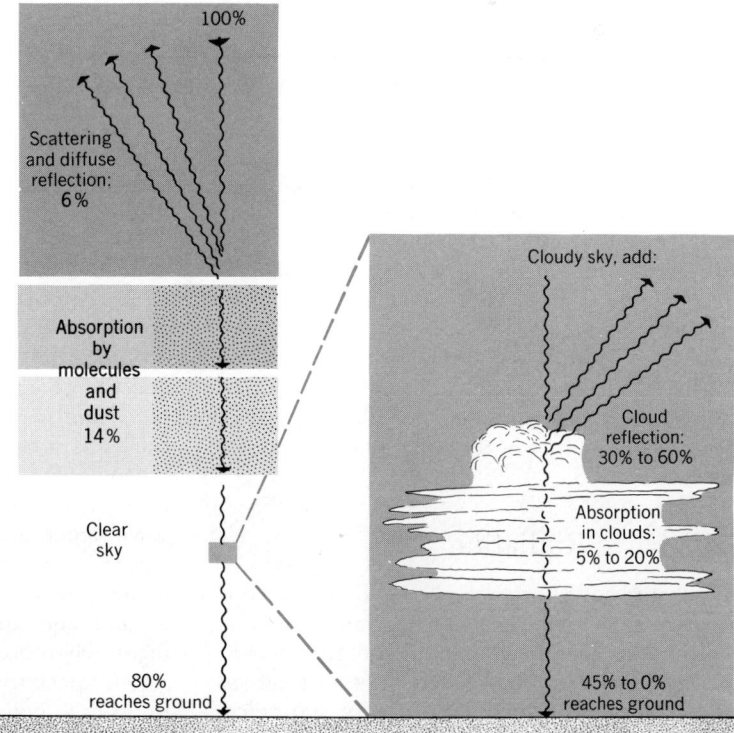

Figure 2.10 Losses of incoming solar energy. (From A. N. Strahler, 1971, *The Earth Sciences*, 2nd ed., Harper & Row, New York.)

The Earth's Radiation Balance

TABLE 2.2 The Planetary Radiation Balance

Incoming Solar Radiation	Percent Units	
Total at top of atmosphere (263 kilolangleys per year)	100	
Diffuse reflection to space by Rayleigh scatter and dust	5	
Reflection from clouds to space	21	
Direct reflection from earth's surface lost to space	6	
Total reflection loss to space from earth-atmosphere system (earth's albedo)		32
Absorption of energy		
By clouds	3	
By molecules, including water vapor and CO_2, and by dust	15	
By the earth's land-and-water surface	50	
Total absorbed by the earth-atmosphere system		68
Sum of reflection and absorption by entire earth		100
Outgoing Radiation (longwave)		
Radiation from earth's land-and-water surface	98	
Lost to space directly	8	
Absorbed by atmosphere	90	
Radiation emitted by atmosphere	137	
Lost to space	60	
Returned to earth's surface as counter-radiation	77	
Net outgoing radiation from earth's surface		21
Net outgoing radiation from atmosphere		47
Net outgoing radiation from earth-atmosphere system		68

SOURCE: W. D. Sellers (1965), *Physical Climatology,* University of Chicago Press, Tables 6 and 9.

estimating the earth's average albedo. Values between 29 and 34% have been obtained. The value of 32% given in Table 2.2 lies between these limits. The earth's albedo lies in a range intermediate between low values of the moon and inner planets (Mercury 6%, Mars 16%, Moon 7%) and high values of Venus (76%) and the great outer planets (73 to 94%). Here, again, we find another unique element of the environment of planet Earth as compared with that of the other planets and the moon.

Longwave Radiation

Recall that any substance possessing heat sends out energy of electromagnetic radiation from its surface, also, that the amount of energy thus radiated is directly proportional to the fourth power of the absolute temperature of the substance. Furthermore, the lower the temperature of the radiating substance, the longer are the wavelengths of the rays emitted.

The ground or ocean surface, possessing heat derived originally from absorption of the sun's rays, continually radiates this energy back into the atmosphere, a process known as *ground radiation* or *terrestrial radiation*. This infrared radiation consists of wavelengths longer than 3 or 4 microns and is referred to here as longwave radiation. The atmosphere also radiates energy both downward to the ground and outward into space, where it is lost. Be sure to understand that longwave radiation is quite different from reflection, in which the rays are turned back directly without being absorbed. Longwave radiation from both ground and atmosphere continues during the night, when no solar radiation is being received.

Refer back to Figure 2.1, which shows the relationship of radiant energy to wavelength. Toward the right of the diagram is a smoothly arched,

Longwave Radiation

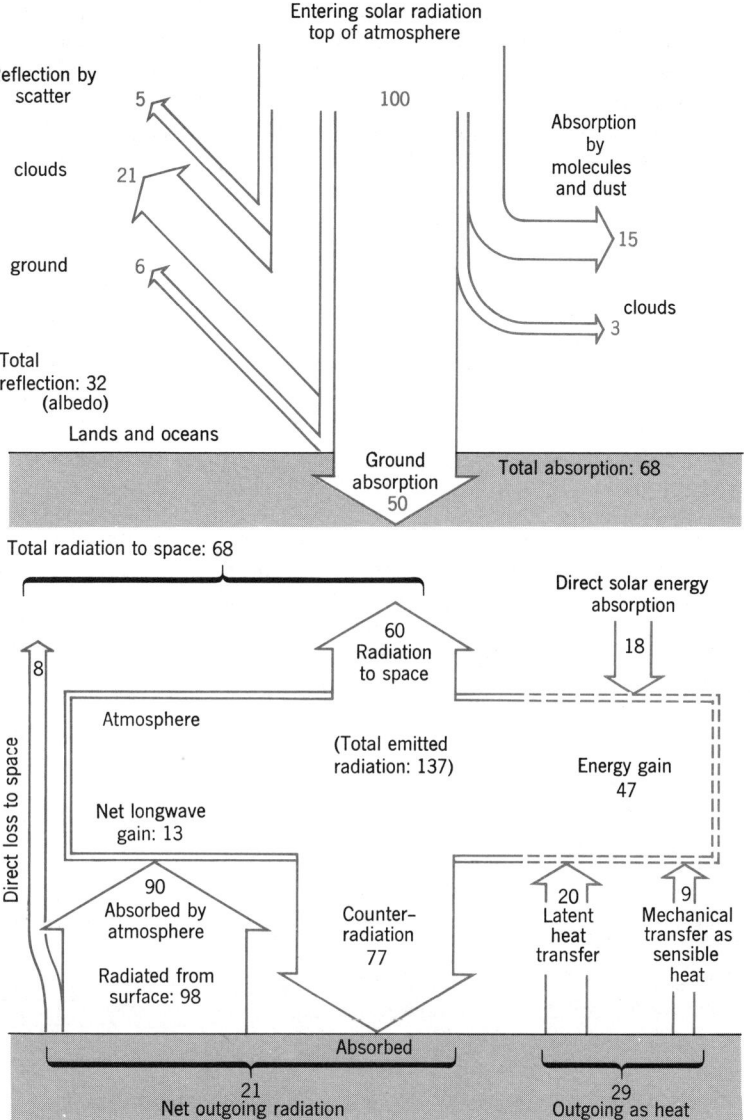

Figure 2.11 Energy-flow diagram of the global radiation balance. *Upper diagram*, solar radiation. *Lower diagram*, longwave radiation. (Refer to Table 2.2 for details and data source.)

dashed line representing the ideal black body curve of radiation for a temperature of 300° K. This theoretical planetary temperature, which is about 81° F (27° C), gives a peak of longwave energy at about 10 microns. The total energy emission for this curve would be 0.67 langleys per minute. The irregular solid line beneath the dashed curve shows that the actual longwave emission to space from the atmosphere takes place in a series of bands. Two of the gaps occur at from 5 to 8 microns and from 12 to 20 microns, since in these wavelength ranges outgoing longwave energy is absorbed by water vapor and carbon dioxide. Most of the longwave energy leaving the atmosphere passes through a *window* between 8 and 11 microns, with lesser windows at lower and higher wavelength ranges. Because of energy absorption over the entire longwave spectrum, the average outgoing longwave radiation of the planet has an estimated value of 0.10 langleys per minute. If the earth's radiation balance is to be maintained, the planet as a whole must lose to space each year a total amount of energy equal to that intercepted by the planet from the sun's rays. Yet we see from Figure 2.1 that the solar shortwave radiation curve contains a much greater value of energy than the longwave curve. You must realize that the sun shines on only one hemisphere at a time and that the angle of attack of the rays is highly oblique over much of that hemisphere. In contrast, longwave radiation constantly takes place directly outward from the entire spherical surface of the planet. Also, we have seen that 32% of the incoming shortwave energy is reflected directly to space and can be deducted at the outset. After such deduction is made, the absorbed shortwave energy averages 179,000 langleys per year and is balanced by an equal longwave emission to space from the planet as a whole.

Energy radiated from the ground is readily absorbed by atmospheric water vapor and carbon dioxide in wavelengths from 4 to 8 microns and 12 to 20 microns. Part of this terrestrial radiation is radiated back down to the earth surface, a process called *counter-radiation*. Thus the atmosphere receives heat largely by an indirect process in which the radiant energy in shortwave form is permitted to pass through, but that in longwave form is not all permitted to escape. For this reason, the lower atmosphere with its water vapor and carbon dioxide acts as a blanket which returns heat to the earth and helps to keep surface temperatures from dropping excessively during the night or in winter at middle and high latitudes. Somewhat the same principle is employed in greenhouses and in homes using the solar-heating method. Here the glass permits entry of shortwave energy. Accumulated heat cannot escape by mixing with cooler air outside. The expression *greenhouse effect* has been used by meteorologists to describe the atmospheric heating principle. Cloud layers are even more important in producing a blanketing effect to retain heat in the lower atmosphere, since they are highly effective absorbers and emitters of longwave radiation.

The lower half of Table 2.2 shows the components of outgoing, or longwave radiation for the planet as a whole, using the same percentage units as for the incoming radiation. Figure 2.11 shows the same data graphically. The total longwave radiation leaving the earth's land-and-ocean surface is equivalent in amount to 98 percentage units. Of this, 8 units are lost to space, while 90 units are absorbed by the atmosphere. In turn, the atmosphere emits longwave radiation. The total of this radiation is equivalent in amount equal to 137 percentage units of the insolation at the top of the atmosphere. This figure may at first seem absurd, but the next two lines show that this radiation is divided into two parts, one of which goes out into space (60 units) and the other of which is absorbed by the earth's surface as counter-radiation (77 units). The last three lines of the table show the net figures as follows: the earth's surface has a net outgoing longwave radiation of 21 units; while the atmosphere has a net outgoing longwave radiation of 47 units. Combining these two figures gives 68 units for the net outgoing radiation from the entire earth-atmosphere system, which equals the total energy absorbed by that same system (see upper total).

Further study of Table 2.2 will reveal what seems to be a discrepancy. In the upper part of the table the absorption of energy by the earth's land-and-water surface is given as 50%, whereas in the lower part of the table the outgoing radiation from the same surfaces is given as 21 percentage units, the discrepancy being 29 units. Actually, two other mechanisms transfer the missing energy back into the atmosphere. One is as latent heat, through the evaporation of surface water. A second is the mechanical transfer of heat from surface to air. The heat is first conducted as sensible heat from water or soil into the overlying air layer, then carried upward in turbulent eddies. (The reverse flow of heat can also occur in this way.) Evidently about 60% of the incoming energy is returned to the atmosphere by the combined processes of evaporation and sensible heat transfer. Of this 60%, about two-thirds returns as latent heat, about one-third as sensible heat.

Latitude and the Radiation Balance

Earlier in this chapter, in relating latitude to insolation, we showed that the tilt of the earth's axis causes a poleward redistribution of insolation, as compared with conditions that would apply if the axis were perpendicular to the orbital plane (Figure 2.5). Let us now look deeper into

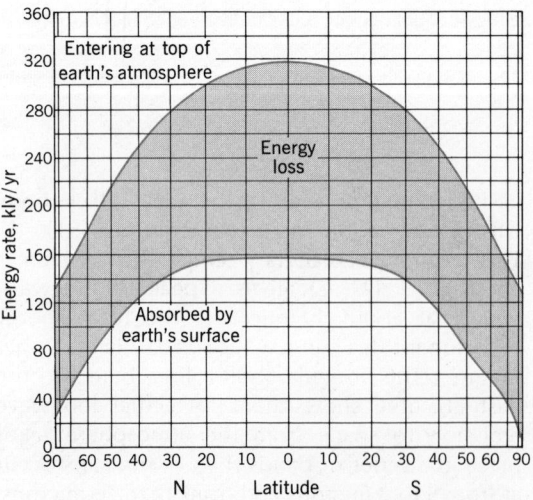

Figure 2.12 Mean annual entering solar radiation and radiation absorbed at surface of earth, shown in meridional profile. Based on data of W. D. Sellers, 1965, *Physical Climatology*, Univ. of Chicago Press, Chicago, p. 22, Figure 7. (From A. N. Strahler, 1971, *The Earth Sciences*, 2nd ed., Harper & Row, New York.)

the wide range in rates of incoming and outgoing energy in terms of a profile spanning the entire latitude range of 90° N to 90° S. In this analysis yearly averages are used, so that the effect of seasons is not seen. Figure 2.12 is a graph of the incoming solar radiation (insolation) plotted against latitude. The upper curve shows insolation entering the top of the atmosphere; it is essentially the same information as that shown in Figure 2.5. The lower curve shows incoming energy absorbed at the earth's surface. The intervening area between the two curves represents the loss of energy as insolation penetrates the atmosphere, combined with shortwave reflection from the surface. The horizontal scale is so adjusted that it is proportional to the area of earth's surface between 10-degree parallels of latitude. (It is the same type of latitude scale used in the right-hand part of Figure 2.8.) Notice that near the equator, about 50% of the entering insolation is absorbed by the ground, whereas near the poles less than 20% is absorbed. This difference is understandable in view of the low angle of attack of the sun's rays at high latitudes, causing those rays to pass through a much greater thickness of atmosphere than at the pole and thus to experience proportionately greater losses by reflection and absorption. Also, the surface albedo is much greater in high latitudes, so that a much greater proportion of shortwave energy is reflected from snow covered surfaces of arctic and polar regions than in low latitudes.

Figure 2.13 shows the average longwave radiation emitted by the earth as a planet. Because the data are recorded from an orbiting satellite, they represent the total radiation of the earth-atmosphere system. Obviously, the radiation is greatest from low latitudes, where the input of radiation is greatest and surface temperatures are highest. But notice that there are two maxima in the graph, one centered over each of the subtropical belts. There is a small, but pronounced dip close to the equator. Here we see the effect of two subtropical desert belts that girdle the globe, each representing a zone of reduced cloud cover and high ground surface temperatures. Over the equatorial belt cloud cover is denser on the average and reduces longwave radiation.

To make an assessment of the earth's radiation balance in terms of latitude, all forms of incoming radiation are combined and balanced against all forms of outgoing radiation. This difference is known as the *net all-wave radiation*, or simply the *net radiation*. As applied to the earth's surface incoming radiation includes both direct and indirect solar radiation as well as all downward

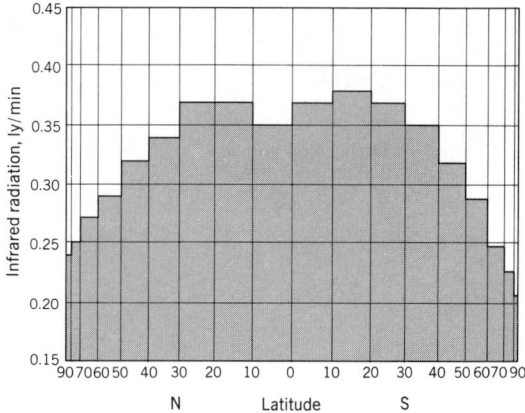

Figure 2.13 Average longwave radiation from the earth, as recorded by orbiting satellite. Based on data of T. H. Vonder Haar and V. E. Suomi, 1969, *Science*, vol. 163, p. 667. Figure 1. (From A. N. Strahler, 1971, *The Earth Sciences*, 2nd ed., Harper & Row, New York.)

longwave counter-radiation. Outgoing radiation combines both longwave radiation and reflected shortwave radiation. The net all-wave radiation is therefore the energy flux, whether downward (positive) or upward (negative).

Figure 2.14 shows net radiation for the earth's surface (upper graph), the atmosphere (lower graph), and for the combined earth-atmosphere system (middle graph). Notice that the net radiation is positive for nearly all of the earth's surface; whereas it has a negative value for all of the atmosphere. But when these two graphs are combined, there results a large region of surplus radiation from about 40° N to 30° S, and two high-latitude regions of deficit. On the diagram the areas labeled "deficit" are together equal to the area labeled "surplus", as the radiation balance requires.

It's obvious from Figure 2.14 that the earth's energy balance can be maintained only if heat is transported from the low-latitude belt of surplus to the two high-latitude regions of deficit. This poleward movement of heat is described as *meridional*, e.g., moving north (or south) along the meridians of longitude. We should expect the rate of meridional heat transport to be greatest in middle latitudes, and this fact is shown by the figures in Table 2.3. The unit used is the kilocalorie (kcal) multiplied by 10 raised to the 19th power (10^{19}).

The meridional flow of heat is carried out by circulation of the atmosphere and oceans. In the atmosphere, heat is transported both as sensible heat and as latent heat. This subject is developed in more detail in Chapter 4.

Figure 2.14 Average net radiation, shown in meridional profiles. Based on data of W. D. Sellers, 1965, *Physical Climatology*, Univ. of Chicago Press, Chicago, p. 66, Figure 19. (From A. N. Strahler, 1971, *The Earth Sciences*, 2nd ed., Harper & Row, New York.)

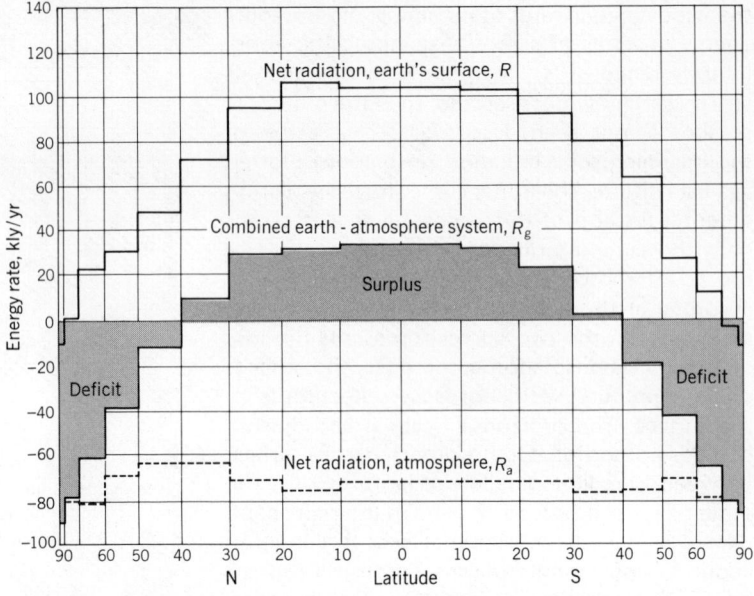

Annual and Daily Cycles of Radiation

The effects of changing angles of attack of the sun's rays upon the parallels of latitude from season to season are shown in a series of curves in Figure 2.15. The quantity scaled on the vertical axis is net radiation in units of langleys per day. Compare these curves with the insolation curves shown in Figure 2.7. There is a general resemblance in curve forms when similar latitudes are compared. Manaos, Brazil, located near the equator, has a large radiation surplus in every month, but with two minor maxima and two minor minima. Aswan, U.A.R., has a strong annual cycle but a large surplus of radiation in all months. For middle latitudes the situation is different. Both Toronto, Canada, and Yakutsk, U.S.S.R., have winter periods of radiation deficit, that of Yakutsk lasting through six months. But see how the summer peak at Yakutsk exceeds that of even the equatorial station! Later, in Chapter 3, the way in which air temperatures near the ground surface respond to the annual radiation cycle is brought into the picture to complete the annual cycle of the energy balance.

The daily cycle of total incoming solar radiation, as measured close to the ground surface, is shown in Figure 2.16 for Hamburg, Germany, 54° N latitude. Data are for June and December, the months of maximum and minimum values. Insolation begins close to sunrise and ceases close to sunset. Not only is insolation vastly greater in intensity in June, but it lasts for a nearly three times longer period of daylight. Near the equator, a similar insolation curve would differ very little from month to month and would always span approximately a 12-hour period.

Total incoming solar energy from both the direct solar beam and the indirect sky radiation are simultaneously measured by an instrument known as the *pyranometer* (Figure 2.17). A sensing cell enclosed in a glass bulb receives radiation from the entire hemisphere of the sky. This is the standard instrument for measuring solar radiation at observing stations.

Figure 2.16 also shows the daily curves of net radiation for June and December at Hamburg. During the hours that the sun is in the sky the curve rises and falls symetrically, peaking at noon. However, during the night hours, a negative value sets in and is held almost constantly while the

TABLE 2.3 Annual Meridional Heat Transport

Latitude (°N)	Heat Transport (kcal/yr × 10^{19})
90	0.00
80	0.35
70	1.25
60	2.40
50	3.40
40	3.91
30	3.56
20	2.54
10	1.21
0	−0.26

SOURCE: W. D. Sellers, 1965, *Physical Climatology*, University of Chicago Press, Table 12.

Man's Impact upon the Earth's Energy Balance

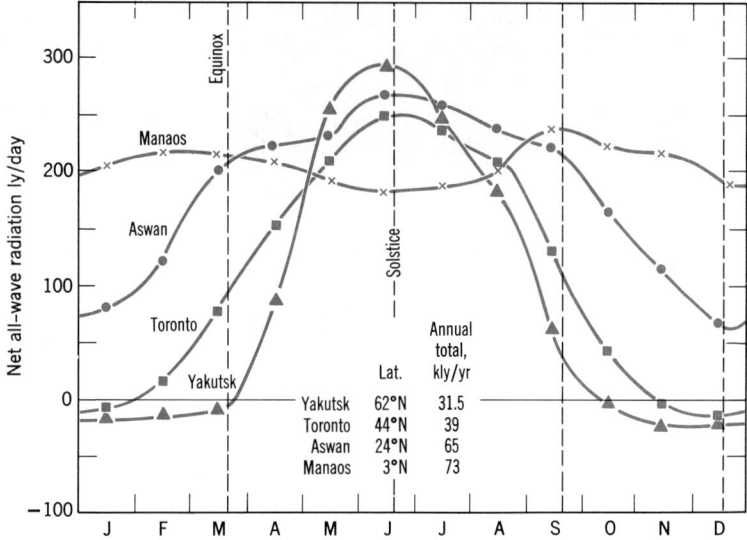

Figure 2.15 Net all-wave radiation throughout the year for four stations. (Data by David H. Miller and John E. Hay.)

sun is below the horizon. In Chapter 3 we shall see how the net radiation cycle is responsible for a daily cycle of air and soil temperatures.

Daily and seasonal cycles of radiation and heat at the earth's surface are a vital environmental factor of the biosphere and deserve careful attention. The responses of plants and animals to these rhythms of energy change explain in large part the worldwide ranges in life regions, from the rich life zones of the low latitudes to the sparse life zones of arctic and polar regions.

Man's Impact upon the Earth's Energy Balance

Although our analysis of the earth's energy balance is far from complete, it should be obvious by now that the balance is a sensitive one, involving as it does a number of variable factors that determine how energy is transmitted and absorbed. Has Man's industrial activity already altered the components of the planetary energy balance? An increase in carbon dioxide may be expected to increase the absorption of longwave radiation by the atmosphere. Will this change cause a rise in atmospheric temperature? Has such a change already occurred? If continued, where will it lead? An increase in stratospheric water vapor might cause that layer to become warmer. Increase in atmospheric dusts will increase scatter of incoming shortwave radiation and perhaps reduce the atmospheric temperature, but on the other hand increased dust content would act to absorb more longwave radiation and thus to raise the atmospheric temperature. Has either change occurred? Man has profoundly altered the earth's land surfaces by cultivation and urbanization. Have the accompanying changes in surface albedo and in the capacity of the ground to absorb and to emit longwave radiation resulted in

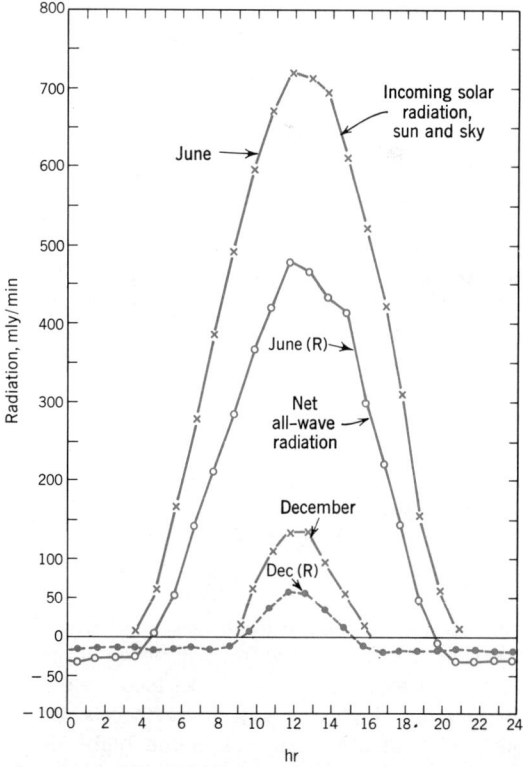

Figure 2.16 Mean hourly incoming and net radiation for June and December at Hamburg, Germany. Data by courtesy of Ernst Frankenberger. (From A. N. Strahler, *The Earth Sciences*, 2nd ed., Harper & Row, New York.)

The Earth's Radiation Balance

Figure 2.17 A pyranometer used to measure solar and sky radiation. (Photograph by courtesy of Weather-Measure Corporation, Sacramento, Calif.)

a changed energy balance? Answers to such questions require further study of the processes of heating and cooling of the earth's atmosphere, lands, lakes, and oceans; this will be the subject of the chapter to follow.

Cosmic Particles and Ionizing Radiation

Quite different in physical nature and source from solar electromagnetic radiation is that which comes to earth from interstellar space as *cosmic radiation* (commonly called *cosmic rays*). This form of radiation is important to the environment of life on earth, although its effects are to produce biological change, rather than to supply energy for life processes.

Cosmic radiation consists of *cosmic particles*, which are elementary particles traveling at speeds approaching that of light and carrying enormous energy. Cosmic particles are *protons*—portions of the nucleus of the atom. Most are nuclei of hydrogen atoms, others are of helium, and a very small number are of other heavier atomic nuclei. Their penetrating power is enormous and they can reach the earth's surface.

Cosmic particles arrive from all points in interstellar space and are believed to be accelerated to their high velocities and energy levels within our own and other galaxies. Some high-energy particles are derived from the sun. A single cosmic particle, entering the earth's atmosphere, impacts the nucleus of an atom of gas, giving rise to a complex chain of disintegrations and energy dissipations, collectively called a *cosmic shower*. Figure 2.18 shows some of the products of a cosmic shower. Products of the cosmic shower are neutrons, protons, and gamma rays.

The term *ionizing radiation* is used to cover radiation capable of tearing off electrons from atoms which intercept the radiation. In Figure 2.18 ionizing radiation is shown to occur by the action of gamma rays. In addition to cosmic rays extraterrestrial sources of ionizing radiation include solar flares, discussed in the next section of this chapter. X rays of the solar electromagnetic spectrum are also a form of ionizing radiation, since as we have already learned, they are responsible for development of the ionosphere. However, since cosmic particles penetrate the atmosphere with much greater energy than X rays, the effects of cosmic radiation are important at the earth's surface and make up a substantial part of the steady, or *background*, ionizing radiation to which life forms are exposed. Terrestrial sources of ionizing radiation include natural radioactivity of elements in crustal rocks (discussed in Chapter 9) and radioactivity from products of nuclear test explosions. Finally there are important sources from medical X rays and television tubes (cathode-ray tubes).

An absolute measure of the energy of ionizing radiation is the *rad*, representing the energy absorption of 100 ergs by 1 gram of the material exposed to the radiation. The *microrad* is one-thousandth of a rad. At sea level the level of cosmic radiation (also called *galactic radiation* by environmental scientists) is about 4 microrads.

The biological effect of ionizing radiation is to produce changes in genetic materials within the cells of organisms. Since Man is also a producer of ionizing radiation at intensity levels potentially much higher than natural levels, the total production of ionizing radiation to which life forms are exposed is a subject of great importance in environmental science. For purposes of evaluating these biological effects, the unit of ionizing radiation is the *rem*, which measures the biological effect of the radiation upon tissues of living matter. For very small amounts of natural radiation the *millirem* (*mrem*) is used; it is equal to one-thousandth of a rem (0.001 rem). Figure 2.19 shows the distribution of intensity of ionizing radiation of cosmic source (galactic radiation) from equator to pole, and from the earth's surface to an altitude of 120,000 ft (23 mi; 37 km). Radiation intensity, in millirems per hour (mrem/hr), is shown by curved lines on the field of the graph. The first point to note is that radiation increases

Cosmic Particles and Ionizing Radiation

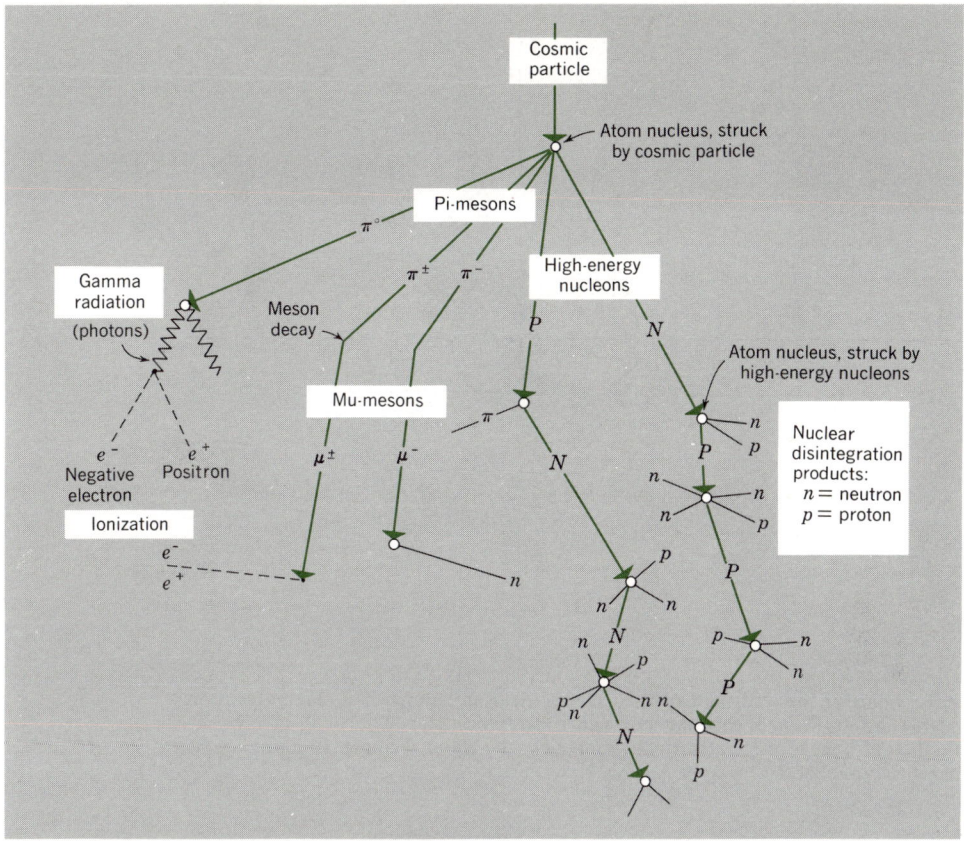

Figure 2.18 Diagram of a cosmic shower. Based on data of U.S. Air Force, 1960, *Handbook of Geophysics*, Macmillan, New York, Figure 18-1. (From A. N. Strahler, 1971, *The Earth Sciences*, 2nd ed., Harper & Row, New York.)

greatly with increase in altitude, due to the lessening of absorption of cosmic rays by collisions with atmospheric atoms. At sea level the background of cosmic radiation is extremely small, some 0.004 to 0.008 mrem/hr (this figure is not given on the graph), or about 32 to 73 mrem per year. Even at 20,000 to 30,000 ft (6 to 9 km), the intensity level is some 30 times greater than at ground level. The second point to note is that high latitudes have a much stronger level of cosmic radiation intensity than low latitudes, roughly by a factor of five. This latitudinal effect is explained by action of the earth's magnetic field and is discussed in the next part of this chapter.

From the standpoint of environmental science, the biological effect of cosmic radiation upon persons flying at high altitudes in jet aircraft is a matter of much concern, particularly since we face the decision as to development and operation of the supersonic transport (SST) aircraft at much higher altitudes than conventional subsonic jet aircraft. Figure 2.19 shows the levels of

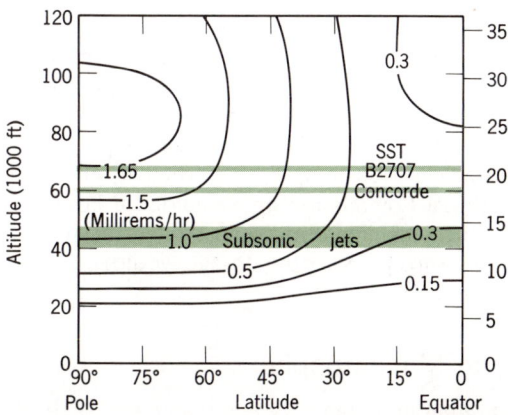

Figure 2.19 Intensity of galactic radiation from sea level to an altitude of 120,000 ft and from equator to poles. Data are for solar minimum, when galactic flux is the maximum. (After H. J. Schaefer, 1971, *Science*, vol. 173, p. 782, Figure 2.)

The Earth's Radiation Balance

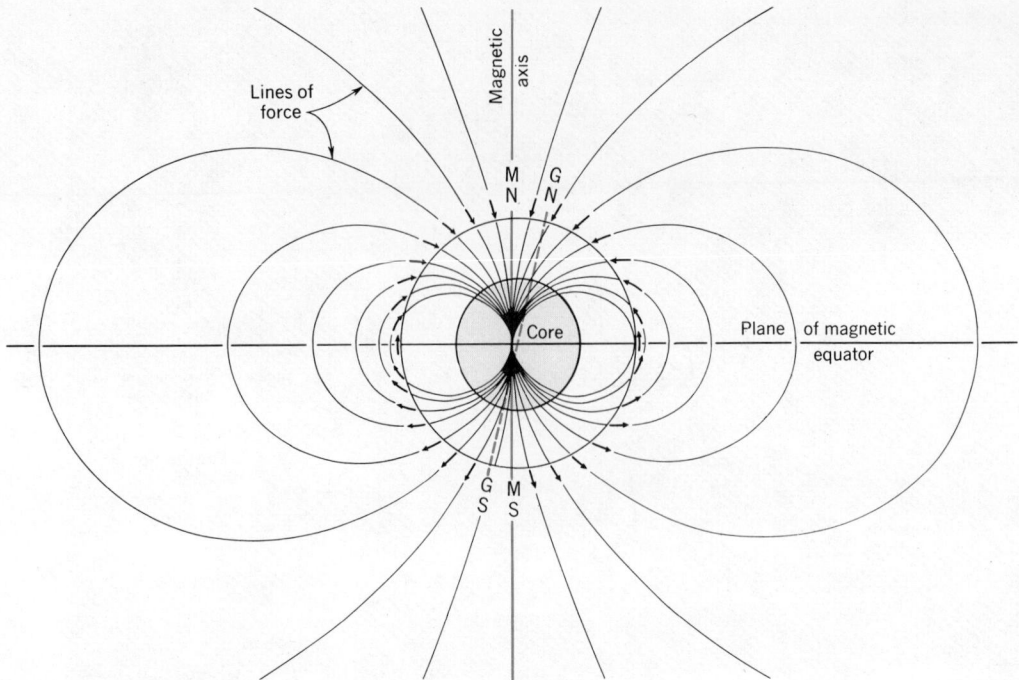

Figure 2.20 Lines of force of the earth's magnetic field are shown diagrammatically in a cross section drawn through the magnetic and geographic poles. The small arrows show the inclination of lines of force at surface points over the globe. (From A. N. Strahler, 1971, *The Earth Sciences*, 2nd ed., Harper & Row, New York.)

flight of the Concorde and Boeing 2707 aircraft compared with those of the Boeing 707 and 727 aircraft. In high latitudes the radiation at 60,000 to 70,000 ft (18 to 21 km) encountered by the SST aircraft is more than half again greater than for conventional jet aircraft at 40,000 to 50,000 ft (12 to 15 km). Great-circle routes between Northern Hemisphere cities pass largely over polar regions, where radiation levels are highest. Assuming that crew members of SST aircraft spend 480 hours per year at an altitude at which the radiation level is 0.1 mrem/hr, the yearly radiation dose will be 0.48 rem/yr, an amount approaching the maximum permissible dosage of 0.50 rem/yr established by the International Commission on Radiological Protection for the general public. For passengers making only a few trips per year the dosage would, of course, be much less.

The Magnetosphere

Of great importance in determining the behavior of ionized particles in the space region surrounding the earth is a vast field of magnetic lines of force. In combination with a flow of ionized particles from the sun, this magnetic field sets up a zone of flux of matter and energy that might be described as a magnetic atmosphere. In fact, this region has been named the *magnetosphere*.

The earth can be thought of as a simple bar magnet, the axis of which approximately coincides with the earth's geographic axis (Figure 2.20). Magnetism is generated within the earth's metallic core, a central spherical body about half of the earth's diameter. The earth's magnetic axis is inclined several degrees with respect to the geographic axis.

Lines of force of the earth's magnetic field, shown in Figure 2.20, pass outward through the earth's surface and into surrounding space. A magnetic compass needle, which is nothing more than a delicately balanced bar magnet, orients itself in a position of rest parallel with the lines of force. The lines of force, which extend out into space, comprise the earth's *external magnetic field*. If we assume, for purposes of comparison, that the earth's gaseous atmosphere extends outward to a distance equal to twice its own radius, or 8000 mi (13,000 km) it becomes evident that the magnetic field extends far beyond the outermost limits of the atmosphere. The effective limit of the external magnetic field lies perhaps 40,000 to 80,000 mi (64,000 to 130,000 km) from the earth.

Radiation Belts

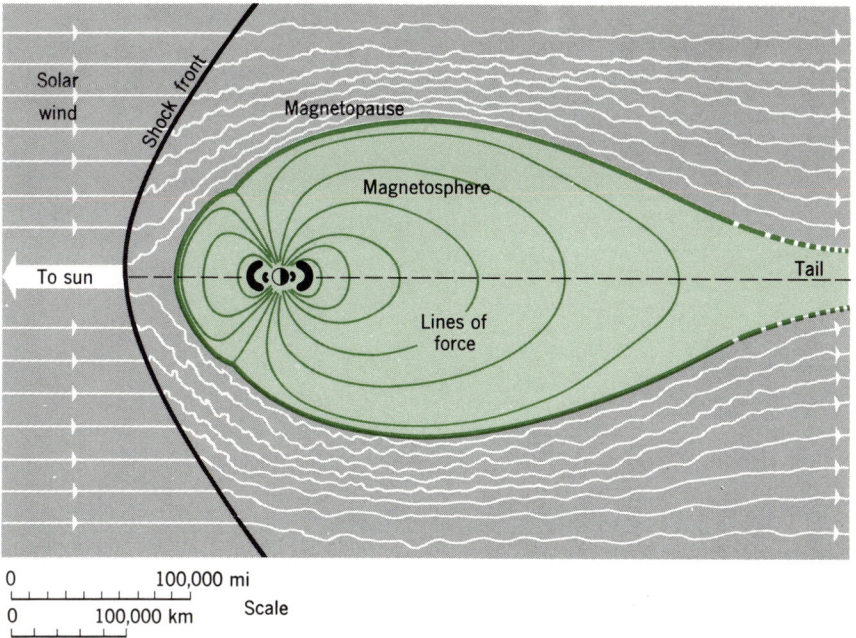

Figure 2.21 Magnetosphere and magnetopause. The Van Allen radiation belts are shown as black areas on either side of earth. (After C. O. Hines, *Science*, 1963, and B. J. O'Brien, *Science*, 1965.)

All of the region within this limit is magnetosphere; its outer boundary is the *magnetopause*.

The simplest geometrical model for the shape of the magnetosphere would be a doughnut-shaped ring surrounding the earth. The plane of the ring would lie in the plane of the magnetic equator, while the earth would occupy the opening in the center of the doughnut. Actually, this ideal shape does not exist because of the action of the *solar wind*, a more or less continual flow of electrons and protons emitted by the sun. Pressure of the solar wind acts to press the magnetopause close to the earth on the side nearest the sun (Figure 2.21). Here the distance to the magnetopause is on the order of 10 earthradii (about 40,000 mi, or 64,000 km). Lines of force in this region are crowded together and the magnetic field is intensified. On the opposite side of the earth, in a line pointing away from the sun, the magnetopause is drawn far out from the earth and the force lines are greatly attenuated. The extent of this *tail* is not known, but the entire shape of the magnetosphere has been described as resembling a comet. Length of the magnetic tail has been estimated to be at least 4 million miles (6,400,000 km) and is possibly vastly longer.

Radiation Belts

In 1958, space satellites carrying Geiger counters sent to earth information concerning the existence of a region of intense ionizing radiation within the magnetosphere. It was soon discovered that two ring-shaped belts of radiation exist, one lying within the other (Figure 2.20). These rings were named the *Van Allen radiation belts*, after the physicist who first described them. An inner belt was found to lie about 2300 mi (2600 km) from the earth's surface; an outer and much more intense belt at about 8000 to 12,000 mi (13,000 to 19,000 km) distance.

The Van Allen radiation belts represent concentrations of charged particles—protons and electrons—trapped within lines of force of the earth's external magnetic field. These highly energetic particles are derived from the solar wind and are trapped upon entering the magnetopause. Figure 2.22 shows how electrons and protons are held between force lines of the magnetic field, being turned back and forth so as to follow sinuous paths. The products of cosmic showers are also entrapped in this way. From the diagram, it

The Earth's Radiation Balance

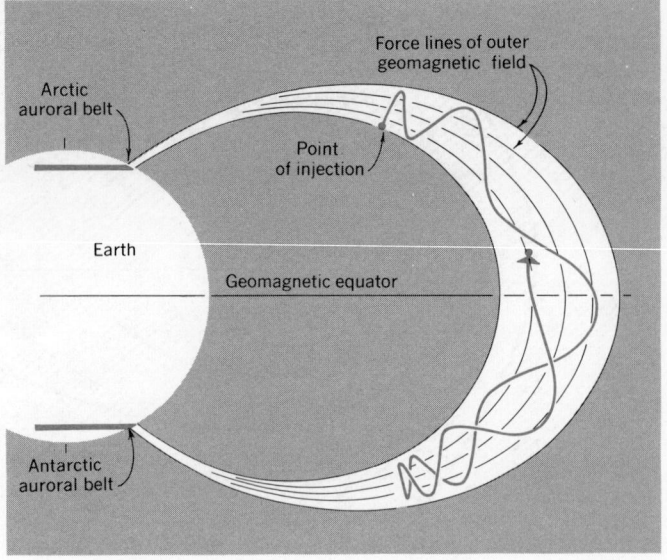

Figure 2.22 Entrapment of a particle in lines of force of the earth's magnetic field. Based on a diagram by R. Jastrow, 1959, *Jour. Geophys. Res.*, vol. 64, p. 1794. (From A. N. Strahler, 1971, *The Earth Sciences*, 2nd ed., Harper & Row, New York.)

is evident that the trapped particles are held far from earth over the magnetic equator, but can reach down close to earth's surface in the magnetic polar regions where the force lines enter the earth at a steep angle. This geometry explains the latitudinal variation in cosmic radiation background shown in Figure 2.19.

Intensity of trapped radiation fluctuates over a wide range. Solar flares from the sun's surface, occurring at irregular intervals, send bursts of ion clouds toward the earth. At such times, the intensity of trapped particle radiation is greatly increased. One manifestation of such events is the *aurora*, which is most intense over arctic and antarctic latitudes. On earth, severe disturbances to the magnetic field, known as *magnetic storms*, accompany the arrival of ion clouds from solar flares and seriously disrupt radio communication.

An important environmental role of the trapped energetic particles of the magnetosphere is to help screen the earth from penetration by cosmic particles. Observations have shown that when the concentration of trapped particles is high, following a solar flare, the incidence of cosmic particles measured at observatories drops appreciably. Should the earth's magnetic field weaken or disappear, as it may have for short periods in the geologic past, the intensity of cosmic radiation reaching the earth's surface would be increased, with possible important effects upon the genetic materials of organisms.

On the other hand, solar flares cause enormous increases in intensity of ionizing radiation at atmospheric levels in which jet aircraft now operate, and even higher levels are felt in the cruising range of SST aircraft. This radiation is from solar protons sent out from the flares and concentrated in the force lines of the magnetic field. One giant solar flare, occurring on 23 February, 1956, during the first hour of its arrival at the earth, is estimated to have produced a radiation intensity of more than 100 mrem/hr at an elevation of 35,000 ft (10.7 km). This figure is some 300 times the background level of cosmic radiation in high latitudes. Fortunately, these huge doses of radiation last generally for less than one hour and events of this magnitude average less than one per year. Detection of solar flares can be made hours in advance of their arrivals at the earth and forecasting service is available to warn of their coming. At such times, aircraft can be grounded to avoid subjecting passengers and crew to the radiation.

In this chapter we investigate the exchanges of energy across the interfaces between the lower atmosphere and the underlying surfaces of free water and solid land. In short, we are concentrating upon heat exchanges and heat storage in the biosphere, or life layer. Temperature of the air, the soil, or the surface waters of oceans, lakes, and streams is a measure of the heat energy level. Organisms require a favorable temperature range for growth and survival. Consequently the range of surface temperatures from equator to poles, and within daily and seasonal cycles is of vital concern to the environmental scientist.

Heat Flow Mechanisms

We have already found that the absorption of electromagnetic radiation by a gas, liquid, or solid tends to cause a rise of temperature of those substances, while outward longwave radiation tends to cause the temperature to fall as sensible heat is lost. Heat can also be transported by *conduction* through matter. Conducted heat travels from a region of higher temperature to a region of lower temperature. In such a case a *temperature gradient*, or *thermal gradient* exists, and the rate of heat flow increases as that gradient increases.

At this point you will want to clarify in your mind the distinction in meanings of the words *heat* and *temperature*. Heat is a measure of the total quantity of energy present in a given quantity of matter. Temperature, on the other hand, is a measure of the intensity level of internal energy. For example a bathtub full of cold water holds a much greater quantity of heat than a cupful of boiling water, yet the intensity level of molecular motion is much higher in the boiling water than in the cold water.

Rate of heat flow by conduction depends also upon an intrinsic property of the conducting substance, the *conductivity*, which is simply the relative ease of heat flow. Air, being a gas, has very low conductivity, in contrast to high conductivity of water and mineral solids. Compared with the

3
Thermal Environments of the Earth's Surface

pure metals copper and silver, thermal conductivities of some common substances are as follows:

Copper, silver	1.00*
Ice	0.005 – 0.007
Most rocks	0.007
Water	0.0014
Dry sand	0.001
Snow, compact	0.0005
Air	0.00006

Conductivity strongly affects the rate at which absorbed heat is passed downward into soil, rock, or standing water, or upward from those surfaces into the overlying air.

Heat is also transported by motion of air or water, which has gained this heat by absorption of radiation, or by conduction. Both air and water are fluids, which easily develop *turbulence*, an eddying motion within the fluid. This motion is also referred to as *convection*. Turbulence greatly aids in the carrying of heat through successive fluid layers, a process called *turbulent exchange*. The transport of heat by the combined mechanisms of conduction and turbulent exchange (convection) is referred to as the *sensible heat flux*.

Storage of sensible heat in air, water, soil, or rock depends upon another intrinsic property of the substance holding the heat. This property is the *specific heat*, defined as the number of heat units (calories) required to raise the temperature of one unit of mass (1 gram) of the absorbing substance by one temperature unit (1 Celsius degree). For some common substances, specific heats are as follows:

	Cal/g/C°
Water (15°C)	1.00
Ice (−2°C)	0.5
Air (100°C, pressure of 1 atmosphere at sea level)	0.24
Granite and basalt	0.2
Clay (dry)	0.2

From this table we see that considerably more heat must be added to a unit mass of water to raise its temperature by a given amount than to a unit mass of air, rock, or dry soil. In practical terms, this means that where a water surface and a surface of rock are both absorbing solar radiation at the same rate, the rock will experience a much more rapid temperature rise in the same period of time. However, after a given layer of

*The unit used is calories per second transmitted through a plate one centimeter thick over a cross sectional area of one square centimeter.

both substances has been heated to the same temperature, the water will be holding in storage a larger quantity of sensible heat energy than the rock.

These principles of heat transport and heat storage are needed if we are to understand the thermal environments of the biosphere. The same principles will help us to predict changes in the thermal environment of Man, as he transforms forests and cultivated lands into urban environments of asphalt pavements and masonry walls.

Heat Exchange during Changes of State of Water

A distinctly different form of heat transport takes place when water vapor is carried within air. This heat energy exists as *latent heat*, absorbed by the air when water vapor is taken up in the process of evaporation. Latent heat is given up by air in the reverse process of condensation.

To understand the role of latent heat, we must go back to some simple principles of physics. When water undergoes changes of state there are accompanying exchanges of heat energy. Figure 3.1 shows these changes of state in a schematic diagram and tells the accompanying heat exchanges.

Consider evaporation first. During evaporation from a free liquid surface, the more energetic water molecules fly off the liquid surface, taking with them kinetic energy of motion. As a result, the remaining liquid has lost energy and is cooled. This lost energy is now stored in the gas above in the form of internal kinetic energy, referred to as *latent energy*, or *latent heat*. For each gram of water evaporated, about 600 calories passes into the latent form and is held by the water vapor. When condensation occurs, the gas molecules of the water vapor come to rest upon the condensing surface and their kinetic energy of motion is transformed back into sensible heat. Thus the condensation of 1 gram of water vapor yields about 600 calories of sensible heat.

Melting of ice to form liquid water, like evaporation, absorbs sensible heat, which is then held in the water itself. This heat, known as the *latent heat of fusion*, amounts to about 80 calories per gram of water melted. Upon freezing, water liberates an equal amount of heat.

As Figure 3.1 shows, a third change of state is the direct transformation of water vapor into ice, a process called *sublimation*. We are familiar with this process through the build-up of ice on the coils of a refrigerator, and in the formation of hoarfrost on auto windshields. Sublimation releases an amount of heat approximately equal to

The Heat Balance Equation

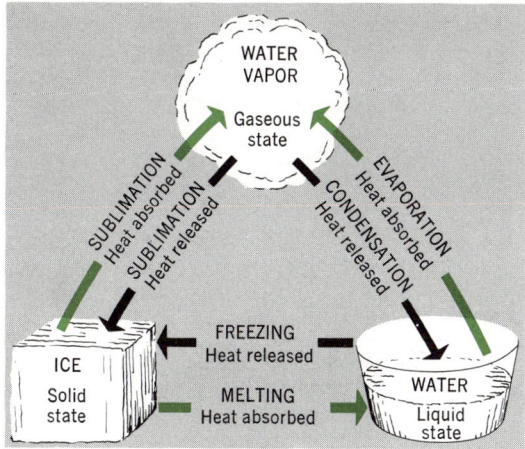

Figure 3.1 Three states (phases) of water.

the sum of the latent heat released by condensation and the heat released by freezing of water. Ice can also evaporate directly to yield water vapor. This process is also one of sublimation, but is usually referred to simply as evaporation.

Liberation and absorption of heat during changes of state of water are extremely important in the thermal environment of the biosphere. These forms of heat exchange are an essential part of the heat balance as it applies to soil and water surfaces and to the lower layer of the atmosphere.

Loss of heat energy by evaporation from exposed water surfaces, moist soil surfaces, and from plant foliage is known as the *latent heat flux*. Growing plants give up water to the air in a process known as *transpiration*, which is a form of evaporation taking place from specialized pores on the leaf surfaces. The combined loss of water from soil and plant surfaces by direct evaporation and by transpiration is known as *evapotranspiration*. This rather bulky term will be particularly valuable in analysis of the water balance, taken up in Chapter 12.

The Heat Balance Equation

Making use of the principles of heat transport and exchange, let us set down the various terms that describe the *heat balance* as it applies to a column of soil or water. This column has a cross-section of unit area, for example 1 cm² or 1 ft². Its upper limit is the surface of the soil or water in contact with the atmosphere. The column depth is indefinite, but sufficient to reach a region where exchanges of heat are close to zero. What we are looking for is a simple equation that will tell us the amount by which the heat content of the unit column changes per unit of time; let this term be designated by G. Units will be calories per minute. The equation is

$$G = R - H - LE - F$$

where R is net all-wave radiation at the upper surface (see Chapter 2),

H is flux of sensible heat upward through the surface into the atmosphere (e.g., the sensible heat flux),

LE is the latent heat flux by evaporation of water from the upper surface (L is latent heat of vaporization, about 600 cal/gm; E is quantity of water evaporated, gm/min.)

and F is horizontal flux of heat into or out of the column by conduction (for soil or rock) or in transport by water motions (for lakes or oceans).

Since the net radiation, R, represents the basic energy source for the whole system, it is best to rearrange the terms of the equation thus:

$$R = H + LE + G + F$$

Figure 3.2 will help you to visualize the four terms on the right side of the equation. Each of these terms has a positive value when heat is entering the column, a negative value when leaving. The horizontal heat flow, F, will be practically zero for soil or rock, since there is no motion of the solid matter and adjacent columns are presumed to be experiencing the same rates of gain or loss of heat. However, for a water column, currents moving horizontally in the water body can import or export heat in substantial quantities. The heat balance equation applies for any given time span, such as the second, minute, hour, day, or year.

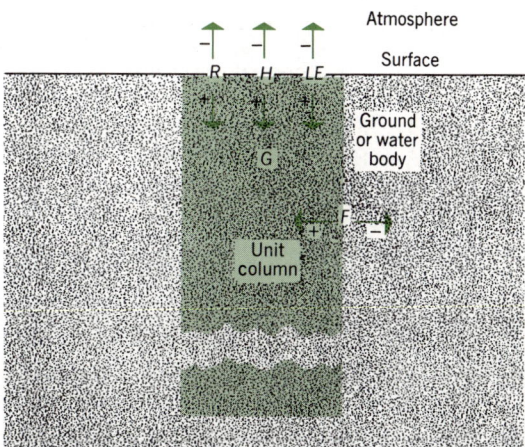

Figure 3.2 A diagram of the heat-balance equation for a unit column of soil or water (From A. N. Strahler, 1971, *The Earth Sciences*, 2nd ed., Harper & Row, New York.)

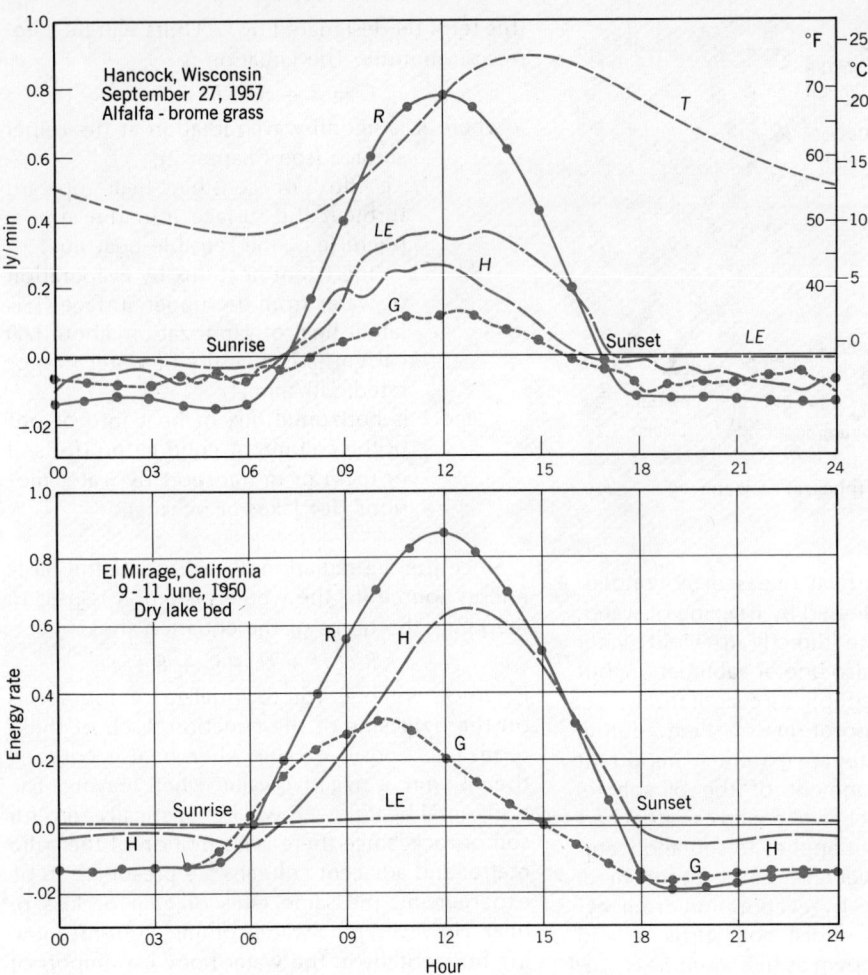

Figure 3.3 Daily cycle of the heat balance for representative days in a humid climate (Hancock, Wisconsin) and a desert climate (El Mirage, California). (Data from W. D. Sellers, 1965, *Physical Climatology*, Univ. of Chicago Press, Chicago, p. 112, Figure 33.)

The Daily and Annual Heat Balance Cycles

To understand the thermal environment of the biosphere we must consider both the daily and annual cycles of the heat balance at the surfaces of land and water bodies. We shall examine typical daily and annual cycles for locations on land. Here the horizontal transport term, F, drops out of the equation, leaving only

$$R = H + LE + G$$

Consider first the daily cycle, shown in Figure 3.3. Data are shown for two stations: one in a humid climate, the other in a desert climate. The vertical scale shows rate of heat flow in langleys per minute. (The unit column has a cross section of 1 cm².) The horizontal scale gives time through a 24-hour day. Readings of each of the components in the equation were taken at hourly intervals. A smooth curve has been drawn to connect hourly points. Consequently, there are four curves, each representing a term in the equation. Now, since the heat balance equation must be fulfilled at all times, the sum of H, LE, and G must always be equal to R. You can check this by using a pair of dividers on the graphs in Figure 3.3. For example, on the noon line, measure each quantity separately upward from the horizontal zero line, adding the lengths obtained for H, LE, and G. They should equal the length obtained for R.

Examine first the daily cycle for Hancock, Wisconsin. The station is located on a grass-covered surface. The time is late summer, close to the date of equinox. The net all-wave radiation curve, R, rises in a symmetrical, bell-shaped curve, peaking at noon. The most important heat change proves to be the latent heat flux, LE, since evaporation is rapid from the moist soil surface and through transpiration of moisture from the leaf surfaces of the grasses. Sensible heat flux is also important during the day. Consequently, the ground gains only a small amount of heat during the day, as shown by the gently rising and falling curve of G. During the night hours, all terms run evenly with values negative or at zero. Notice that the curve of LE runs close to zero most of the night, showing that there is no evaporation. However, just before sunrise there is some condensation as dew, making the value of LE negative as heat is liberated by condensation.

Look next at the graph for El Mirage, California, a desert station. The data are for a very hot summer day (lower part of Figure 3.3). The net all-wave radiation curve, R, is very similar to that for the Wisconsin station, but the other curves are very different. The sensible heat flux, H, increases rapidly during the morning hours and peaks with a very high value in the early afternoon. This effect is caused by intense heating of the bare, dry soil. In contrast, latent heat flux, LE, scarcely rises to a measurable level during the day, since the soil is bone-dry and there is no vegetation. The rapid gain in heat by the ground, represented by a steep rise in the curve for G during the early morning hours, is the cause of the steep rise in sensible heat flux. With almost no evaporation, the heating of the ground is intense. However, before noon, the rate of flow of heat into the ground begins to fall off and is reversed by mid-afternoon. Negative values of G during the hours of darkness show that the ground surface is giving up heat to the air above it.

Now it is an interesting observation that intense urbanization of an area in a humid climate, such as that of Wisconsin, will cause changes in the daily heat balance cycle in the direction of the desert conditions shown by the California station. Vegetation will be replaced by pavements, roofs, and masonry walls. In dry weather with evaporation and transpiration largely cut off, the latent heat flux falls to a low value, while the sensible heat flux climbs to high values. We shall say more about these effects of urbanization in Chapter 6.

Turning next to the annual cycle of the heat balance equation, look at the two graphs in Figure 3.4. Again, we are comparing a humid environment (Wisconsin) with a desert environment (Arizona). Units of energy flow are in langleys per day. Each point on a curve is the average value of one calendar month.

First consider the annual cycle for Madison, Wisconsin. All values of the heat balance terms

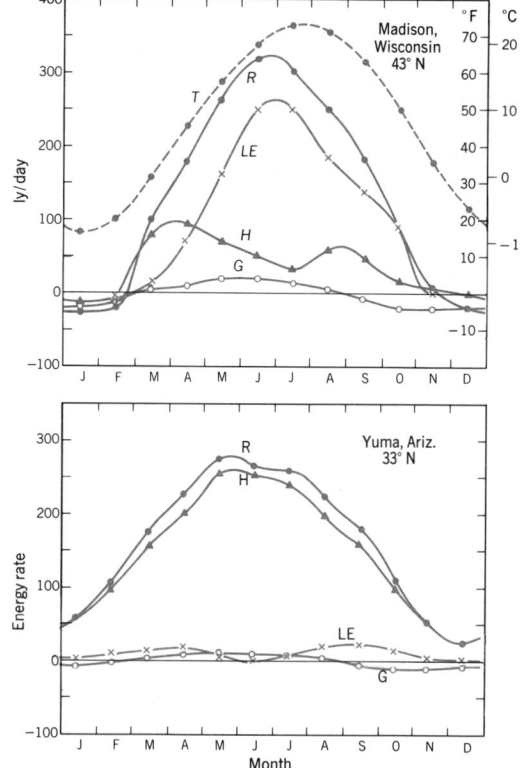

Figure 3.4 Annual cycle of the heat balance for a humid climate in middle latitudes (Madison), and for a desert in the subtropical zone (Yuma). (Data from W. D. Sellers, 1965, *Physical Climatology*, Univ. of Chicago Press, Chicago, p. 106, Figure 30.)

are negative during the winter months. In early spring net all-wave radiation rises rapidly, and with it the sensible heat flux, H. However, latent heat flux, LE, is delayed in its rise until foliage returns and transpiration is experiencing a rapid increase. This rise offsets the sensible heat flux and causes it to drop somewhat. The gain in heat by the soil, G, is quite small in consequence of the heavy rate of evapotranspiration.

Notice how strikingly different is the heat balance cycle for Yuma, Arizona. Here the net all-wave radiation is positive throughout the entire year, but with a strong annual cycle. Sensible heat flux, H rises and falls in a curve close to that of R. Latent heat flux, LE is very small at all times, since there is very little soil moisture and very sparse vegetation. Changes in stored soil heat, G, are small; the curve shows a weak annual cycle.

Let us now apply the information on daily and annual cycles of the heat balance equation to the heating and cooling of the ground, and then to water surfaces.

Thermal Environments of the Earth's Surface

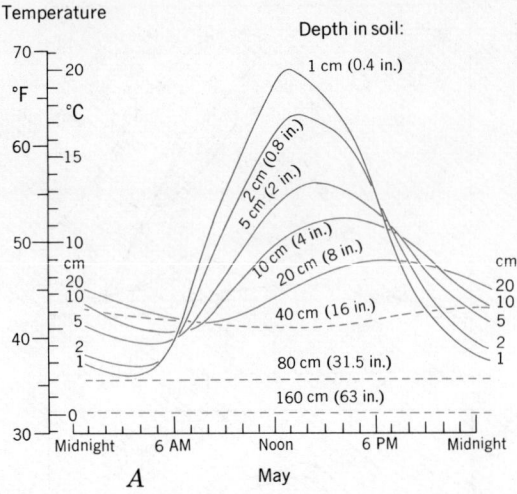

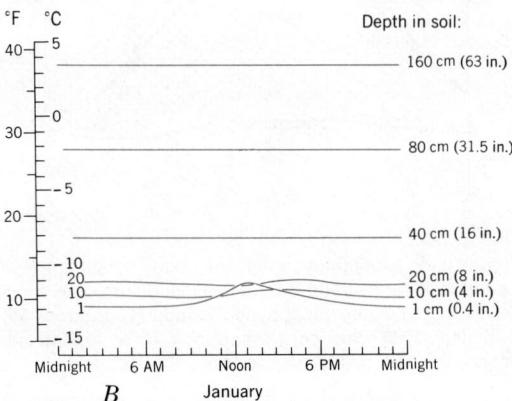

Figure 3.5 Daily cycle of soil temperature in May and January at Pavlovsk, U.S.S.R., 59½° N latitude. (Used by permission of the publishers from Rudolph Geiger, *The Climate Near the Ground*, rev. ed., Harvard Univ. Press, Cambridge, Mass. Copyright 1950, 1957, 1965 by the President and Fellows of Harvard College.)

Heating and Cooling of the Soil

Soil temperatures play a major role in the seasonal rhythms of plant physiology and in biological activity generally within the soil. We can learn much by placing a series of thermometers in the soil, spacing them at a series of depths, each about twice as deep as the one above it. Figure 3.5 shows temperature data for a locality in northern U.S.S.R. The observation depths run in a progression thus: 1, 2, 5, 10, 20, 40, 80, and 160 cm. Curves of temperature are averages for one month, the months being May and January. Looking first at the May graph, we notice immediately that the daily range of temperatures is greatest—about 18 C° (32 F°)—at the shallowest depth, and that this range falls off rapidly with increasing depth. Furthermore, the time of maximum temperature of the day is advanced with increasing depth. We can imagine a wave of heating advancing into the soil each day. As this wave passes down by conduction, heat is lost by absorption and the wave crest diminishes in height. There is also a time lag, so that the wave crest does not arrive at the 40-cm (16-in.) depth until after midnight. At a depth of 80-cm (31.5 in.) the wave has practically disappeared, so that no daily cyclic changes are preceptible below this depth.

Looking at the curve for January (lower graph in Figure 3.5) we find a very different situation. Close to the surface there is only a slight daily warming in the mid-day period. At this latitude—about 60° N—the sun is very low in the sky in the winter and daylight hours are few. The weak wave of warming does not penetrate very far and is imperceptible at a depth of 40 cm (16 in.).

Now, compare the May and January temperatures at Pavlosk for a depth of 160 cm (63 in.). Strangely enough, at this depth the soil is colder in May (near 0° C, 32° F), than in January (near 4° C, 38° F). What we are seeing is an annual cycle of heating and cooling that lags about half a year behind the annual surface cycle at this depth.

To look further into the annual cycle of heating and cooling of the soil, examine Figure 3.6, which shows temperatures at depths of 2.5, 5, 10, and 20 ft at an observing station in Brookhaven, Long Island, New York (41° N latitude). The soil here is sandy and porous. Again we find the annual wave of warming to decrease in amplitude with depth, while the peak of temperature comes progressively later with depth. At the 20-ft (6-m) depth the warmest time is in November. The many minor irregularities in temperature at the shallowest depth reflect the effects of rainstorms, which cool the soil, and the short periods of warmer or cooler air temperature that are normal at all seasons for this climate.

You might anticipate that for each climate region, there is a depth below which the soil or rock temperature is unchanging the year around, and that this constant temperature will be close to the average air temperature near the ground surface. The facts bear out this supposition. For example, the air temperature in Endless Caverns, Virginia, which is constant at 56° F (13° C) throughout the year, is only a little higher than the annual average air temperature of 52° F (11° C). The exceptional constancy of subterannean temperatures makes an unusual thermal environment for animals that live in the total darkness of limestone caverns.

Arctic Permafrost

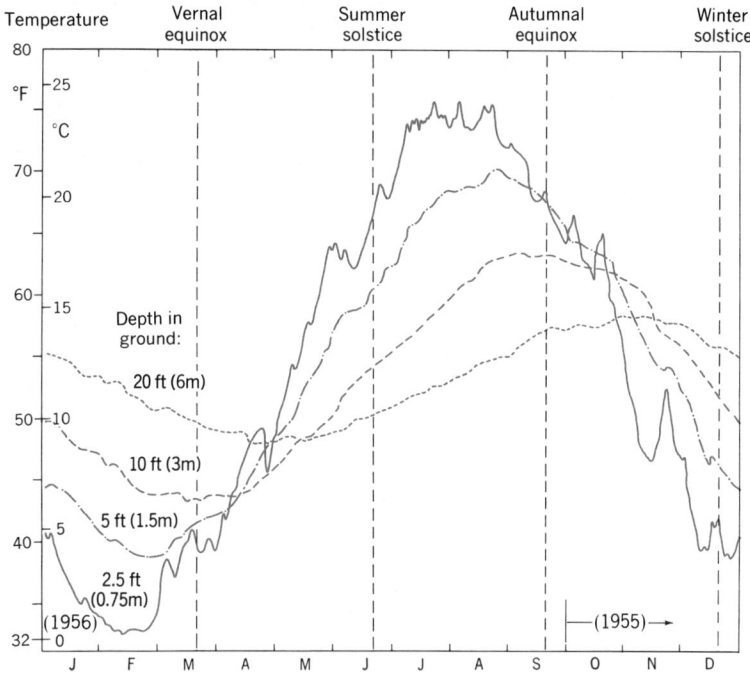

Figure 3.6 Soil temperatures recorded during one year at North Station, Brookhaven, Long Island, New York. (Data from I. A. Singer and R. M. Brown, 1956, *Trans. Amer. Geophysical Union*, vol. 37, p. 746, Figure 4.)

Arctic Permafrost

In arctic and subarctic lands of the Northern Hemisphere, winters are long and severely cold, because of a long-continued energy deficit in the radiation balance (see the net radiation curve for Yakutsk, U.S.S.R., 62° N latitude, Figure 2.15). As a result, subfreezing soil temperatures occur for six or seven consecutive months and all moisture in the soil and subsoil is solidly frozen to depths of many feet (Figure 3.7). Summer warmth is insufficient to thaw more than the upper few feet so that a condition of perennially frozen ground, or *permafrost*, prevails over large parts of lands lying poleward of the 60th parallel of latitude. Seasonal thaw penetrates from 2 to 14 ft (0.6–4m), depending on location and nature of the ground. This shallow zone of alternate freeze and thaw is termed the *active zone*.

The distribution of permafrost in the Northern Hemisphere is shown in Figure 3.8. Three zones are recognized. Continuous permafrost, which extends without gaps or interruptions under all topographic features, coincides largely with the arctic *tundra*, a treeless landscape, but also includes a large part of the forested subarctic region known as *taiga* in Siberia. Followed north beneath the Arctic Ocean, the continuous permafrost layer disappears under the protection of the overlying ocean waters. Discontinuous permafrost, which occurs in patches separated by frost-free zones under lakes and rivers, occupies much of the forested subarctic zone of North America and Eurasia. Sporadic occurrence of permafrost in small patches extends in places as far south as the 50th parallel.

Figure 3.7 This vertical river-bank exposure near Livengood, Alaska, reveals a V-shaped ice wedge surrounded by layered silt of alluvial origin. (Photograph by T. L. Péwé, U.S. Geological Survey.)

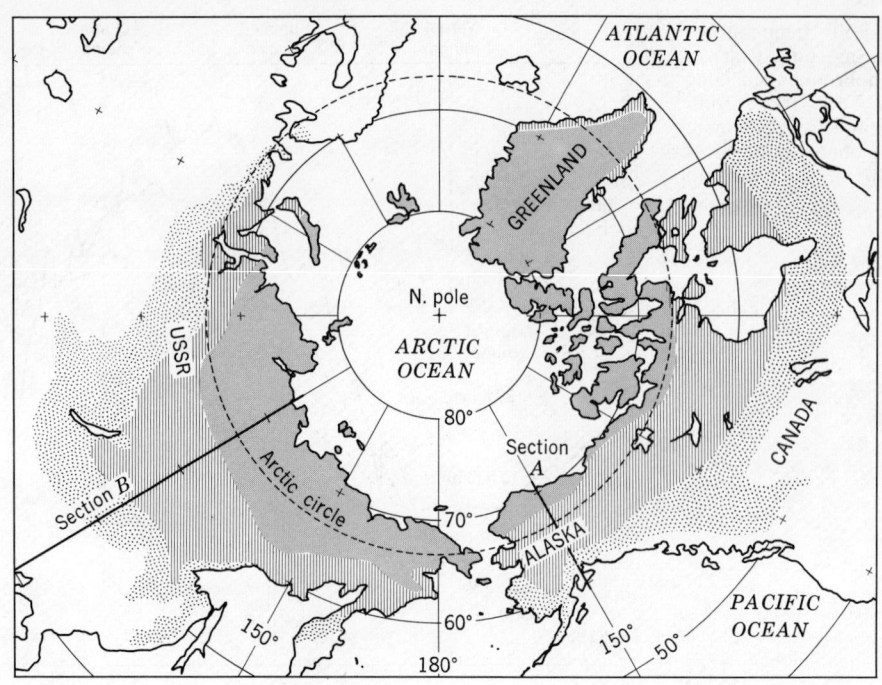

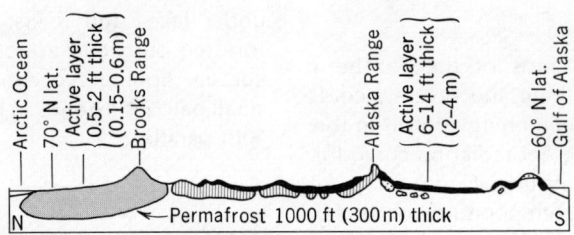

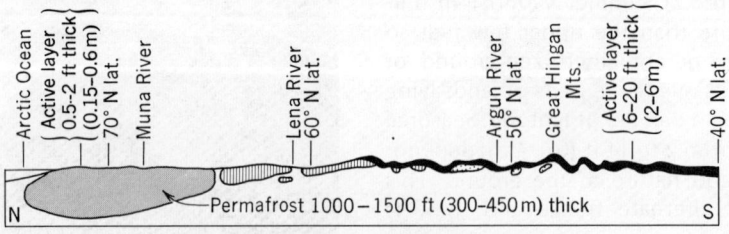

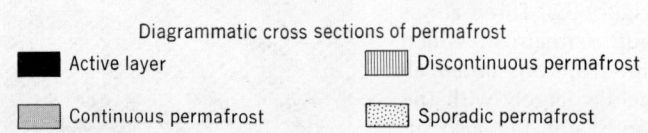

Figure 3.8 Distribution of permafrost in the Northern Hemisphere, and representative cross sections in Alaska and Asia. (From Robert F. Black, "Permafrost," Chapter 14 of P. D. Trask's *Applied Sedimentation*, John Wiley & Sons, New York, 1950.)

Depth of permafrost reaches 1000 to 1500 ft (300 to 450 m) in the continuous zone near latitude 70° (Figure 3.8). Perhaps much of this permanent frost is an inheritance from more severe conditions of the last ice age, but some permafrost bodies may be growing under existing climate conditions.

Permafrost presents problems of great concern in engineering and building construction in cold regions; these and other environmental factors of arctic regions are discussed in Chapter 11.

Energy Absorption by Water Layers

Insolation falling upon the surfaces of bodies of clear water—lakes or oceans—is disposed of in two ways. Direct reflection accounts for some loss. For vertical rays this amounts to only about 2%, but may be very high for low-angle rays. For the oceans as a whole (except where ice-covered), the average albedo ranges from 6 to 10%, a figure substantially less than for most land surfaces. Energy that is not reflected is transmitted through the upper zone of the water and is gradually absorbed, raising the temperature of the water. (The water also absorbs longwave counter-radiation from the atmosphere.)

Figure 3.9 shows how the solar energy spectrum is depleted as the rays pass down to greater depths. (The information is based on observations in distilled water, but applies fairly well to clear fresh-water lakes.) Notice that most of the infrared energy within the solar spectrum is absorbed within the top 4 in. (0.1 m); while at a depth of 40 in. (1 m) only the ultraviolet and visible wavelengths remain. Consequently, the warming process is concentrated in a very shallow surface layer, largely by absorption of infrared rays. Only about 10% of the energy reaches the 33-ft (10-m) level and this is largely in the violet, blue, and green range of the color spectrum. At 330 ft (100 m) all but 3% of the total energy has been absorbed; what remains is in the blue-green wavelengths.

The blue-green color of clear water bodies is explained by the fact that the blue and green wavelengths are easily scattered and reflected back to the eye, whereas the yellow and red wavelengths are absorbed without reflection. The explanation is similar to that for the blue sky color.

The lower limit of light penetration with sufficient intensity to permit photosynthesis by algae varies from about 200 ft (60 m) to 500 ft (150 m), depending upon water transparency. Above this limiting depth lies the *photic zone* of the marine environment. Below this depth lies the *aphotic zone*, in which light is insufficient for appreciable photosynthesis. With increasing depth total darkness sets in.

Heating and Cooling of Lakes and Oceans

The environmental properties of fresh water lakes in middle and high latitudes pass through an annual thermal cycle in response to the strong

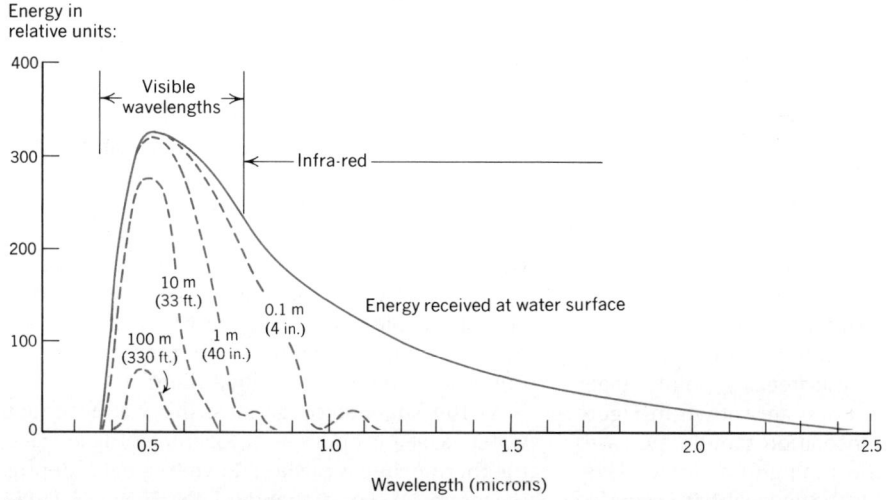

Figure 3.9 The solar radiation spectrum at various depths below the surface in pure, fresh water. (Data from H. U. Sverdrup, 1942, *Oceanography for Meteorologists*, Prentice-Hall, Englewood Cliffs, New Jersey, p. 54, Figure 8.)

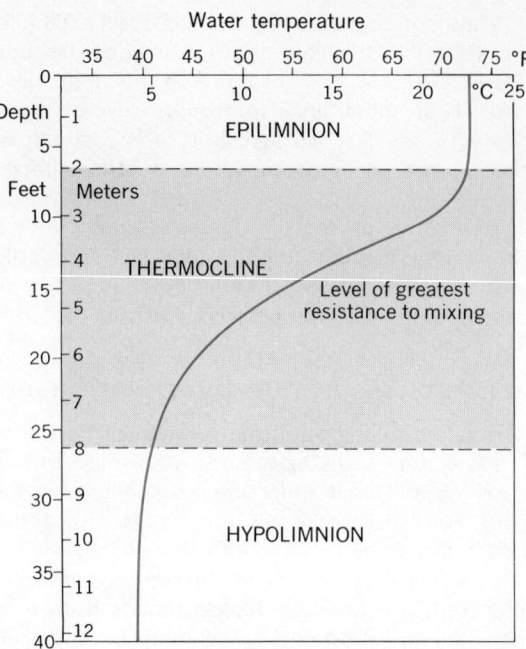

Figure 3.11 Summer temperature profile of a small lake in the middle-latitude zone. (From A. N. Strahler, 1971, *The Earth Sciences*, 2nd ed., Harper & Row, New York.)

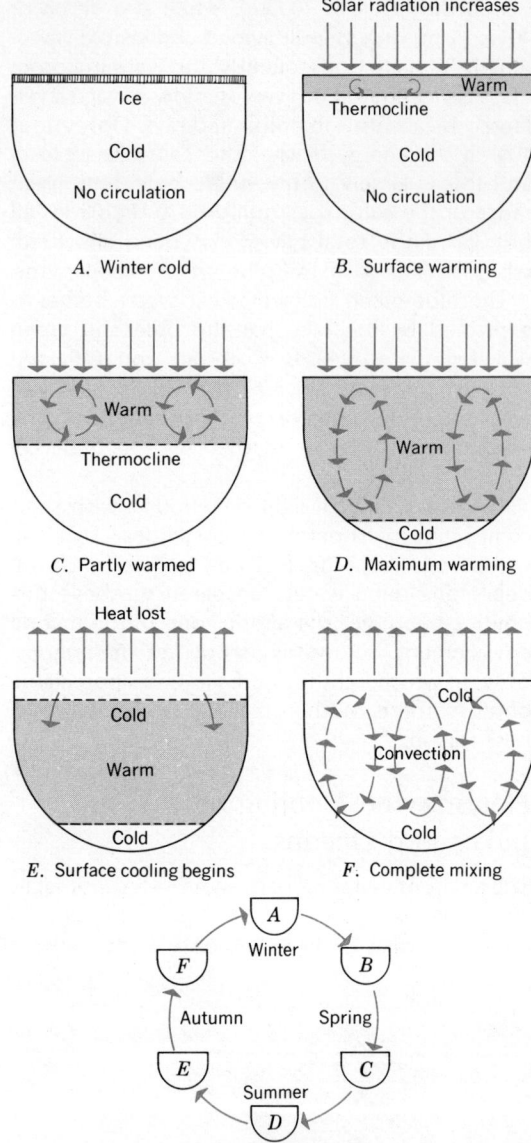

Figure 3.10 Schematic diagram of the annual cycle of heating, cooling, and mixing of lake water in the middle-latitude zone. (From A. N. Strahler, 1971, *The Earth Sciences*, 2nd ed., Harper & Row, New York.)

annual cycle of net radiation. This thermal cycle in turn strongly influences the growth of aquatic plant and animal life in lakes.

Let us start with late winter or very early spring, when the lake surface is still covered with ice and the water beneath lies stagnant at a temperature not far above the freezing point. Stages in the annual thermal cycle are shown in Figure 3.10 (Diagram A). As insolation rapidly increases the ice is melted and warming of a surface layer begins (Diagram B). Once the layer has exceeded 39° F (4.4° C), it is less dense than the colder water below and remains on top. Winds blowing over the lake surface create small waves, and these cause a mixing that results in thickening of the warm layer (Diagram C). If we were to take a series of temperature readings from the surface downward, the temperature profile would look like that in Figure 3.11. The upper warm layer, known as the *epilimnion*, has a uniform temperature throughout; it is said to be an *isothermal* layer. Below the epilimnion, temperatures drop rapidly. This zone is named the *thermocline*. Below the thermocline, temperatures again become uniform with depth, constituting the *hypolimnion*, a cold layer close to 39° F (4° C), the temperature at which fresh water is in its densest state. An important property of the thermocline is that water does not easily rise or sink through this layer, the reason being that water density changes with water temperature. A succession of horizontal fluid layers of rapidly changing density, whether they be of liquid or gas, tends to be stable and to resist mixing by vertical motions. (This principle is explained in Chapter 6 in connection with a related phenomenon known as an upper-level temperature inversion.)

As the summer progresses, the warm surface water layer, or epilimnion, becomes thicker and the thermocline is pushed down to greater depths (Diagram D). For a shallow lake, 50 to 75 ft (12 to 23 m) deep or less, the entire water body may be warmed, but for deeper lakes, a cold hypo-

Heating and Cooling of Lakes and Oceans

limnion persists. Inflowing water of streams or springs may add to the mixing of warm and cold water.

As winter approaches and the net radiation declines to become negative in value, e.g., a radiation deficit, heat is lost from the surface layer (Diagram *E*). Being denser than the warmer water below, this cold surface water sinks to the bottom, setting up a general overturning, or *convection*, within the entire lake. The thermocline is destroyed and gradually the water becomes uniformly cold throughout. Mixing ceases when the point of maximum density (39° F; 4° C) is reached. Further cooling produces a less dense surface water layer which remains on top, eventually freezing into a continuous ice layer.

In warm tropical and equatorial climates, where solar radiation is uniformly great throughout the year, lake water is comparatively warm down to the bottom. The coldest water that can sink to the lake floor can be no colder than the average surface water temperature at the coolest time of year. This seasonal cooling easily results in overturn and general mixing because the overall range of temperatures is small. For example, lakes at low elevation in Indonesia, situated near the equator, have a bottom temperature always close to 79° F (26° C), while the surface water temperature ranges annually from 79° to 84° F (26° to 29° C). When the surface water temperature is increasing, a weak thermocline develops, just as in lakes of middle latitudes, but this is destroyed when the seasonal cooling takes place. Because insolation at the top of the atmosphere is uniformly great throughout the year at low latitudes, the cooling of the surface water seasonally is brought about by the occurrence of a rainy season (a rainy monsoon) in which cloud cover reduces incoming radiation and copious rains have a cooling effect. In the dry season, insolation again becomes intense and the surface water is warmed.

The three-layer thermal structure of the oceans was described in Chapter 1 and pictured in Figures 1.7 and 1.8. Recall that a warm surface layer, a thermocline, and a cold deep layer are typical of middle-latitude oceans. The annual cycle of change in these latitudes is illustrated in Figure 3.12. During the winter (Diagram *A*) the upper layer is mixed by wave action under the force of strong winds. Temperatures are nearly isothermal in the upper 500 ft (150 m) although a gradual decrease with depth takes place until bottom temperatures close to the freezing point of fresh water (32° F, 0° C) are reached. With the arrival

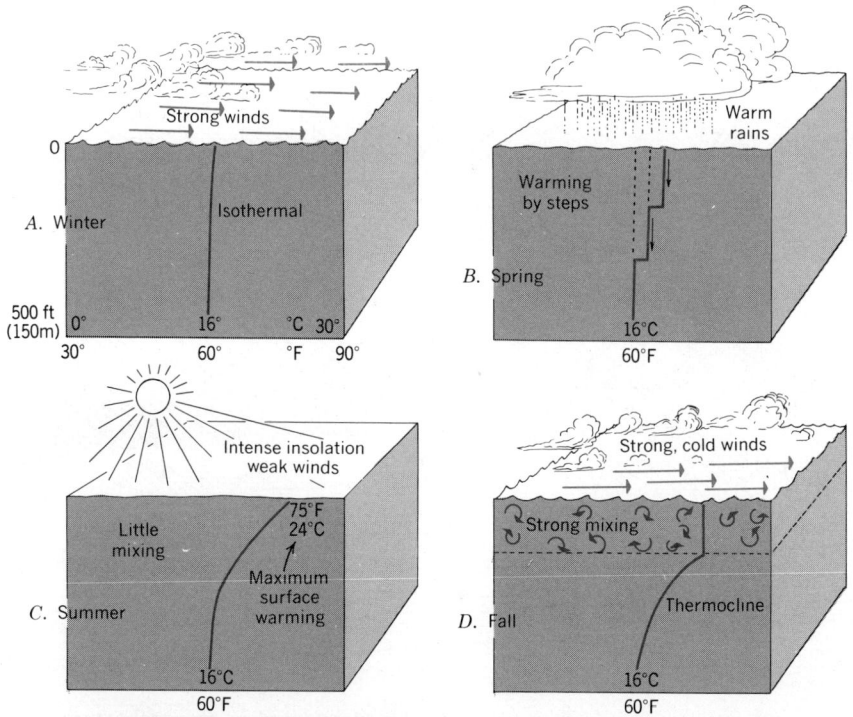

Figure 3.12 Seasonal water temperature changes and mixing typical of the ocean in the middle-latitude zone. Based on data of E. C. LaFond, *Scientific Monthly*, August, 1954. (From A. N. Strahler, 1971, *The Earth Sciences*, 2nd ed., Harper & Row, New York.)

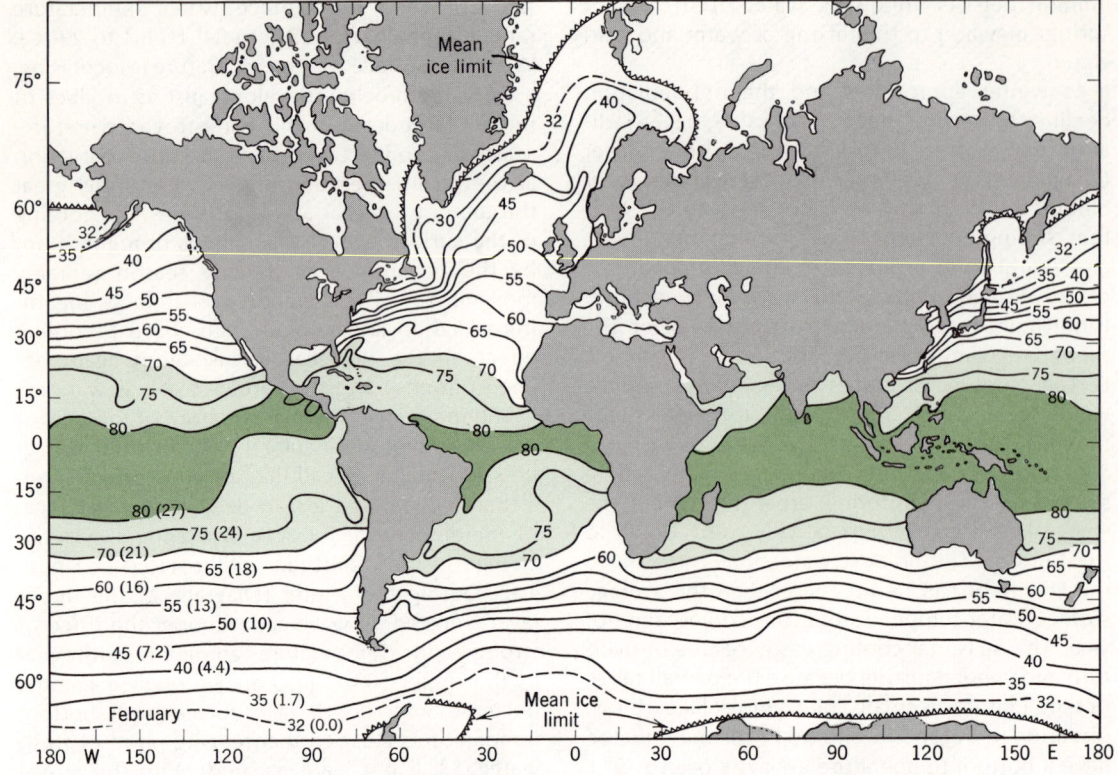
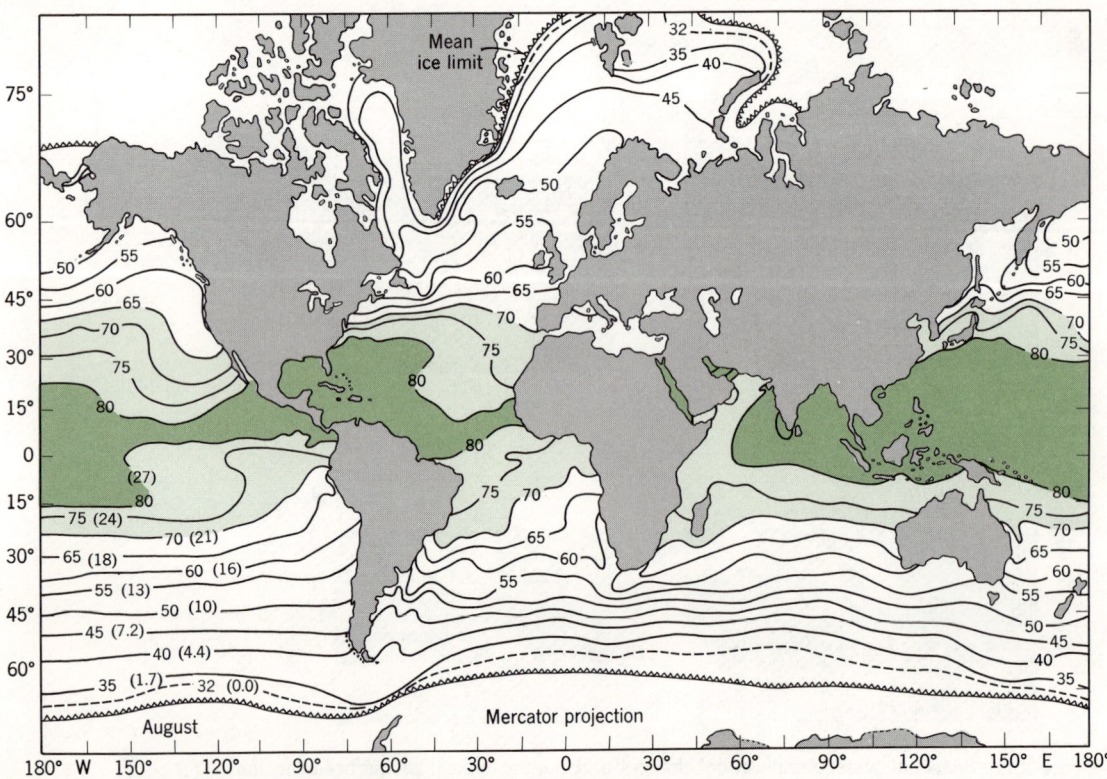

Figure 3.13 August and February mean sea-surface temperatures, °F. Celsius degrees in parentheses. (Based on data of U.S. Navy Oceanographic Office.)

Sea Surface Temperatures

of spring (Diagram B) increased insolation and the effects of warm rains causes warming by steps. Throughout the summer (Diagram C) intense insolation and lack of vigorous mixing develops a warm surface layer. In the fall (Diagram D) strong winds result in a mixing of the surface layer and the development of a pronounced thermocline. Radiation and conduction of heat from sea to atmosphere then cools the surface layer and the isothermal conditions of winter again set in.

Sea Surface Temperatures

Thermal environments of the ocean surface are summarized on a global basis by two maps, one showing surface-water temperatures for August, the other for February (Figure 3.13). Because a positive radiation balance at the sea surface continues to add heat to the sea surface well past the summer solstice, sea surfaces reach their maximum temperatures in middle and high latitudes during the month of August. Correspondingly, there is a continuing deficit, or heat loss, well past the winter solstice, so that February sees the lowest surface-water temperatures generally.

Temperature maps make use of lines of equal temperature, or *isotherms*. As you would expect, the isotherms trend east-to-west around the globe generally, with maximum values in a low-latitude belt and strong poleward decline both north and south. Large areas of equatorial oceans maintain a surface temperature higher than 80° F (26° C). In arctic and antarctic waters temperatures remain close to the freezing mark (32° F, 0° C). But in middle latitudes there is a large annual range. These relationships are shown in Figure 3.14 by two generalized profiles giving the annual range in sea-surface temperatures for the Atlantic and Pacific oceans. Notice the peak ranges between 40° and 50° N latitude, and between 30° and 40° S latitude. That the range is greater and peaks at a higher latitude in the Northern Hemisphere is a reflection of the land-locked nature of the North Atlantic and North Pacific oceans, inhibiting mixing by ocean currents. Another factor, which we shall investigate in later pages of this chapter, is that the adjacent landmasses of North America and Eurasia accentuate the temperature contrasts between summer and winter. The South Atlantic and South Pacific, poleward of the 45th parallel of latitude, are joined with the Indian Ocean in a continuous ocean belt—the Southern Ocean—in which heat exchange by currents is free and no large land areas intervene. As a result, the sea-surface temperatures have a much smaller range there than in the Northern Hemisphere.

Life forms are strongly responsive to the world distribution of ocean water temperatures. For example, the reef-building corals require temperatures at least as high as 68° F (20° C), and usually between 77° and 86° F (25° and 30° C). As a result coral reefs are limited to a latitude zone between 25° S and 30° N. In general, invertebrate marine animals show a much greater diversity of species in warm low-latitude waters than in cold water

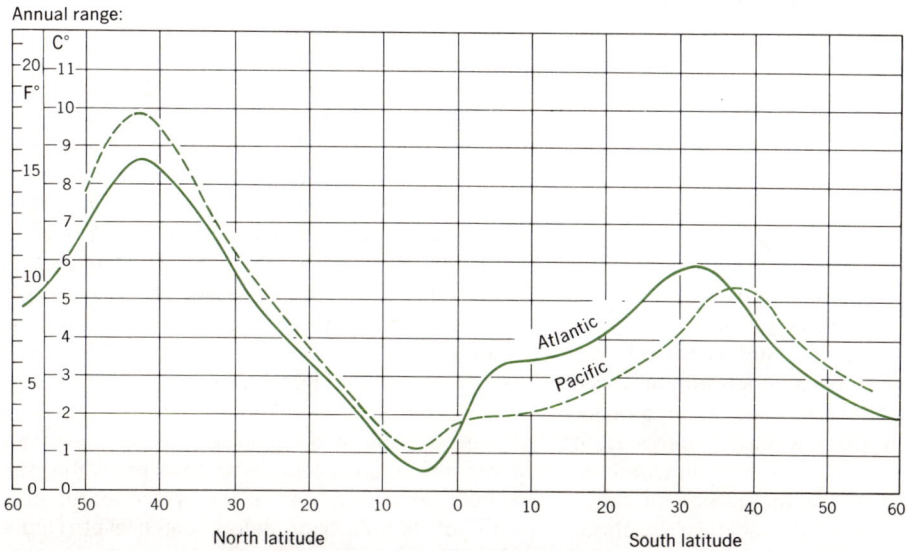

Figure 3.14 Mean annual range of sea-surface temperature shown in meridional profiles of the Atlantic and Pacific oceans. (After H. U. Sverdrup, M. W. Johnson, and R. H. Fleming, 1942, *The Oceans*, Prentice-Hall, Englewood Cliffs, N.J., p. 130, Figure 31.)

Figure 3.15 The U.S. Coast Guard icebreaker *Northwind* forces a passage through McClure Strait, Banks Island, Canadian Northwest Territory, in mid-August. To the left is an open lead. Cutting across from left to right is a rugged zone of pressure ridges. (Official U.S. Coast Guard photo.)

of arctic coasts. Production rates of minute floating organisms, known collectively as *plankton*, are also strongly influenced by water temperatures.

Sea Ice and Its Distribution

In arctic and polar regions an environmental phenomenon of importance to Man and other forms of animal life is *sea ice*, a solid floating layer that in winter transforms the marine environment into a quasi-terrestrial environment. Predatory terrestrial animals (Man, polar bear) can move over the ice layer and through openings in it bring up animals whose meat is the principal winter food (seal, walrus, fish). The ice cover also has a major climatic role, for evaporation from the free ocean water surface is cut off by an ice cover. There is only minimal evaporation from the cold, solid upper surface of the ice. Supply of arctic and antarctic outposts by ship, maintenance of observing stations on floating ice masses, and submarine operation in the Arctic Ocean are forms of Man's activity influenced by sea ice.

The oceanographer distinguishes *sea ice*, formed by direct freezing of ocean water, from *icebergs*, which are masses of land ice broken free from tide-level glaciers and continental ice shelves. Aside from differences in origin, a major difference between sea ice and icebergs is in thickness. Sea ice, which begins to form when the surface water is cooled to temperatures of about $28\frac{1}{2}°$ F ($-2°$ C) is limited in thickness to about 15 ft (5 m), because heat is supplied from the underlying water as rapidly as it is lost upward, once an insulating layer of floating ice has been formed.

Pack ice is the name given to ice that completely covers the sea surface (Figure 3.15). Under the forces of wind and currents, pack ice breaks up into individual patches, termed *ice floes*. The narrow strips of open water between such floes are *leads*. Where ice floes are forcibly brought together by winds, the ice margins buckle and turn upward into pressure ridges resembling walls of irregular hummocks (Figure 3.15). The difficulties of travel on foot across the polar sea ice are made extreme by the presence of such obstacles. The surface zone of sea ice is composed of fresh water.

The Arctic Ocean, which is surrounded by landmasses, is normally covered by pack ice throughout the year, although open leads are numerous in the summer (Figure 3.16). The situation is quite different in the antarctic, where a vast open ocean bounds the sea ice zone on the equatorward margin (Figure 3.17). Because the ice floes can drift freely north into warmer waters, the antarctic ice pack does not spread far beyond 60° S latitude in the cold season. In March, close to the end of the warm season, the ice margin shrinks to a narrow zone bordering the Antarctic continent.

Icebergs, formed by the breaking off, or *calving*, of blocks from a valley glacier or tongue of an icecap, may be as thick as several hundred feet. Being only slightly less dense than sea water, the iceberg floats very low in the water, about five-sixths of its bulk lying below water level (Figure 3.18). The ice is fresh, of course, since it is formed of compacted and recrystallized snow.

In the Northern Hemisphere, icebergs are derived largely from glacier tongues of the Green-

Sea Ice and Its Distribution

Figure 3.16 Sea ice of the Arctic Ocean. Common tracks of icebergs are shown by arrows. (From A. N. Strahler, 1971, *The Earth Sciences*, 2nd ed., Harper & Row, New York.)

land Ice Sheet (Figure 3.16). They drift slowly south with the Labrador and Greenland currents and may find their way into the North Atlantic near the Grand Banks of Newfoundland.

Icebergs of the antarctic are distinctly different. Whereas those of the North Atlantic are irregular in shape and therefore present rather peaked outlines above water, the antarctic icebergs are commonly *tabular* in form, with flat tops and steep clifflike sides (Figure 3.19). This is because tabular bergs are parts of ice shelves, the great, floating platelike extensions of the continental ice sheet (Chapter 17). In dimensions, a large tabular berg of the antarctic may be tens of miles broad and over 2000 ft (600 m) thick, with an ice wall rising 200 to 300 ft (60 to 90 m) above sea level.

While icebergs play an insignificant role in organic processes of the life layer, their presence in lanes of ocean traffic constitutes an environmental hazard to Man. You have doubtless read about the sinking of the steamship *Titanic* in 1912, after collision with an iceberg in the North Atlantic. In all, 1500 lives were lost in this marine disaster.

Thermal Environments of the Earth's Surface

Figure 3.17 Sea ice of the antarctic region. Based on data of National Academy of Sciences and the American Geographical Society. (From A. N. Strahler, 1971, *The Earth Sciences,* 2nd ed., Harper & Row, New York.)

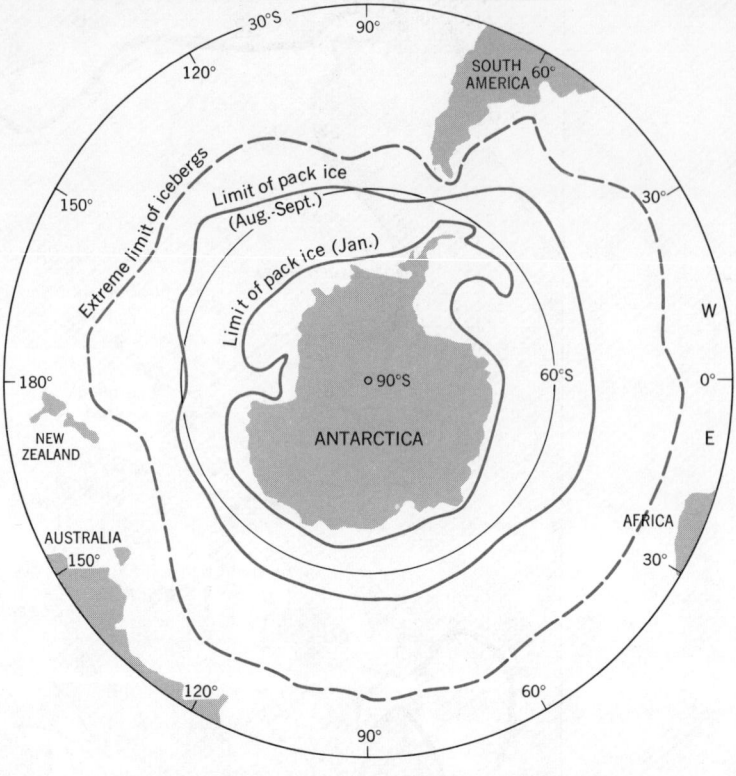

Daily Cycle of Air Temperature near the Ground

At every recording weather station, the temperature of the air is read at regular intervals from thermometers mounted inside a boxlike shelter built several feet off the ground (Figure 3.20). Instrument height is usually 4 to 6 ft (1.2 to 1.8 m), or approximately at the observer's eye level. The instruments are protected from direct sunlight, but air is allowed to circulate freely through the shelter. Standard equipment consists of a pair of *maximum-minimum* thermometers, one of which shows the maximum temperature, the other the minimum, that have occurred in the period since last reset. In addition, an automatic recording thermometer, called a *thermograph* may be used to draw a continuous temperature record on a piece of graph paper (Figure 3.21).

When hourly temperature readings are plotted on a graph, the curve for a clear day typically shows one low point near sunrise and one high point in midafternoon with a fairly smooth curve throughout. This rhythmic rise and fall of air temperature is termed the *daily,* or *diurnal, air temperature cycle.*

Figure 3.22 relates the typical diurnal air temperature curve (*bottom*) with cycles of insolation and net all-wave radiation. (Refer to Figure 2.16 for similar graphs.) The diagrams are schematic

Figure 3.18 This great iceberg in the east Arctic Ocean dwarfs the U.S. Coast Guard icebreaker *Eastwind.* (Official U.S. Coast Guard photo.)

Daily Cycle of Air Temperature near the Ground

Figure 3.19 This tabular iceberg was observed near the Bay of Whales, Little America, Antarctica, in January 1947. (Official U.S. Coast Guard photo.)

and apply to a typical middle-latitude location (40° to 45° latitude). We assume equinox conditions (March 21 or September 23), with the local time of sunrise and sunset as 6:00 A.M., and 6:00 P.M., respectively. The uppermost graph shows insolation throughout a clear day. Disregarding small amounts of indirect incoming energy before sunrise and after sunset, the curve for equinox begins at 6:00 A.M., reaches a maximum at noon, and ends at 6:00 P.M., local time.

The middle graph shows the net all-wave radiation. Where the curve is above the zero line, excess energy is passing upward from ground to atmosphere; where it is below the zero line, excess energy is passing from atmosphere to ground. These positive and negative values are labeled *surplus* and *deficit,* respectively. When a surplus is present, the temperature of the air layer above the ground tends to rise; when a deficit exists, the air tends to be cooled. The net all-wave radiation curve tends to be symmetrical with respect to noon (the maximum point) and to be nearly flat during hours of darkness. As the graph shows, a surplus normally sets in about one hour after sunrise and thereafter increases rapidly. A deficit sets in about an hour or so before sunset.

The typical diurnal air temperature curve, shown in the lowermost graph of Figure 3.22, is not symmetrical. The minimum point occurs at about sunrise. Temperature rises steeply as the radiation surplus increases rapidly. Air temperature continues to rise past noon, for the radiation surplus, although beginning to diminish, is still large. The air temperature maximum typically occurs between 2:00 and 4:00 P.M. Thereafter, the

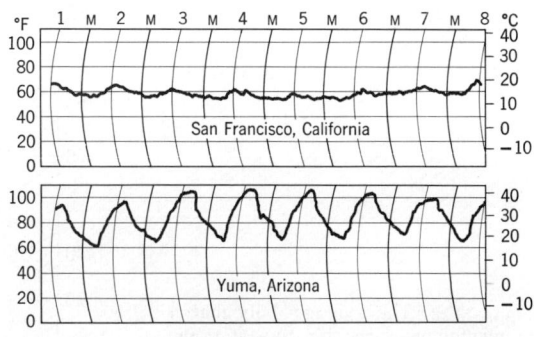

Figure 3.20 Standard National Weather Service instrument shelter. Rain gauge at left. (National Weather Service.)

Figure 3.21 Thermograph trace sheets show temperatures for one week. (After Kincer, U.S. Dept. of Agriculture.)

Thermal Environments of the Earth's Surface

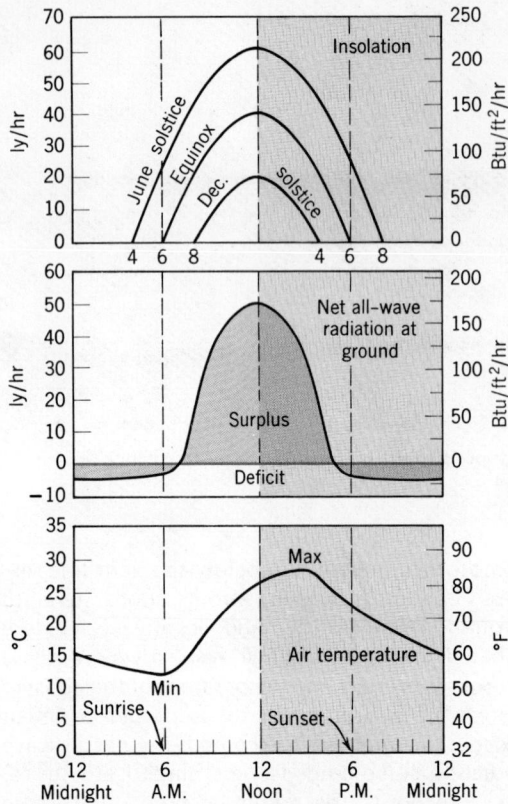

Figure 3.22 Relation of diurnal air temperature curve to insolation and net ground radiation, middle latitudes at equinox.

lent motions increases in intensity, carrying the heated air upward and replacing it with cooler air. The effect of this mixing is to cause the air temperature curve to begin to drop long before the radiation surplus has ended. (Referring back to Figure 3.3, lower graph, notice that the sensible heat flux, H, increases to a strong peak shortly after noon, with the result that the rate of heat gain, G, falls steadily during the afternoon.

Under conditions of June solstice, insolation is greatly increased (upper graph in Figure 3.22); the insolation commences much earlier and ends much later. The surplus part of the net ground radiation curve (not shown) is similarly broadened and raised in height. The hour of minimum temperature is set back correspondingly to perhaps 4:00 A.M. However, the hour of maximum temperature remains essentially the same. At December solstice, corresponding reductions of incoming energy and a narrowing and lowering of the surplus part of the net radiation curve occur. The time of minimum temperature is advanced correspondingly.

Figure 3.23 shows the average diurnal cycle of air temperature at two places, one of interior continental location in a dry climate, the other very close to the ocean water on a windward coast. Notice that the hour of minimum temperature changes with solstice and equinox, but that the hour of maximum temperature remains fairly constant.

air temperature begins to fall, even though a radiation surplus exists, as the middle curve shows. If air temperature depended only upon ground radiation, the maximum temperature would occur later in the day, perhaps at about 5:00 P.M. under the equinox conditions illustrated. However, another factor has entered the picture. During the early afternoon, mixing of the lower air in turbu-

Thermal Extremes near the Ground

We are now ready to combine the daily cycle of heating of the soil with that of the air layer immediately above it. Figure 3.24 shows temperature profiles observed through a range of heights above the ground surface and depths into the soil beneath. These observations were made at Death

Figure 3.23 The average daily march of air temperatures for two stations. (After J. B. Kincer, U.S. Dept. of Agriculture.)

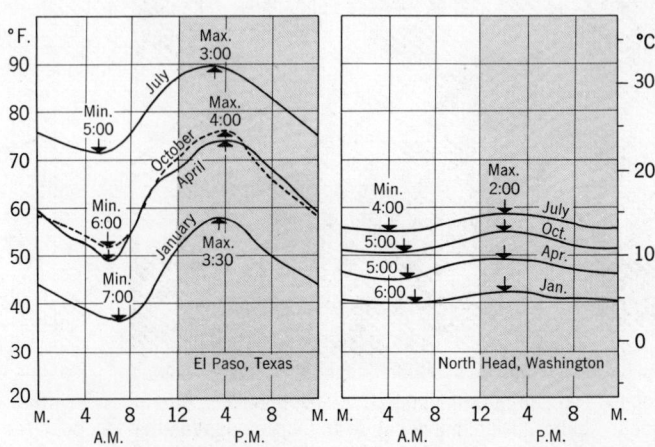

The Annual Cycle of Air Temperature

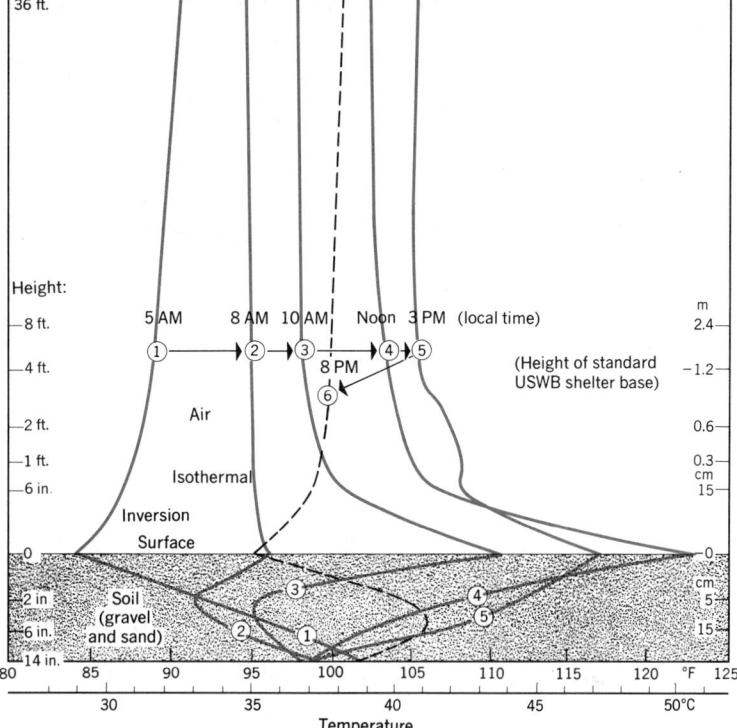

Figure 3.24 Air and soil temperatures at various levels above and below the ground surface throughout the day and night. Data are averages for July and August 1950. Height follows a square-root scale. Based on data of Quartermaster Research and Development Branch, U.S. Army. (From A. N. Strahler, 1971, *The Earth Sciences*, 2nd ed., Harper & Row, New York.)

Valley, California; they are averages for the months of July and August. Although they represent the extremes of the thermal environment of a hot, dry desert, the same principles apply generally, although to a less pronounced degree, for all land surfaces when soil moisture is low and insolation is strong.

Starting with Profile 1, taken at dawn, we find that the air layer close to the ground has reached its lowest temperature of the day. Air temperature is higher as we move up from the surface, a condition known as a *ground-level temperature inversion*. Temperatures also increase strongly with increasing depth into the soil. By 8:00 A.M. (Profile 2) the lower air layer is considerably warmed, along with a shallow soil layer. At 10:00 A.M. (Profile 3) heating of the ground surface is becoming intense, with rapidly declining temperature both upward and downward. Surface heating reaches its most intense level in Profile 4, taken at noon. By 3:00 P.M. (Profile 5) the ground has cooled somewhat but the air layer above 1 ft (0.3 m) is now at its warmest point of the day. By 8:00 P.M. (Profile 6) cooling of the soil surface has been rapid and an inversion has already developed in the overlying air layer. The wave of warming in the soil can be seen in its downward progress, peaking at about the 6-in. (15-cm) level.

The important lesson to be learned from this rather complex series of profiles is that the ground surface and the shallow layers of air and soil immediately adjacent to it constitute a severe thermal environment for plants and animals subjected to its stresses. Conditions at tree-top height in a forest, and in the lower root zone of the soil are much more moderate in terms of temperature extremes. Unfortunately, the data read from a thermometer at the standard height of 4 ft (1.2 m) may be quite misleading in terms of temperatures near the ground.

Microclimatology, the science of climate close to the ground surface, is a branch of the environmental sciences that attempts to assess the differences of climate from place to place over the ground within relatively short horizontal and vertical distances. In short, microclimatology seeks to describe the real climate of the life layer, as it departs from the generalized overall average of climatic conditions of a large area.

The Annual Cycle of Air Temperature

In order to build statistical information about temperatures for longer periods of time than a single day, a unit known as the *mean daily temperature* is used. The National Weather Service follows a very simple method of obtaining the mean daily temperature, using readings made

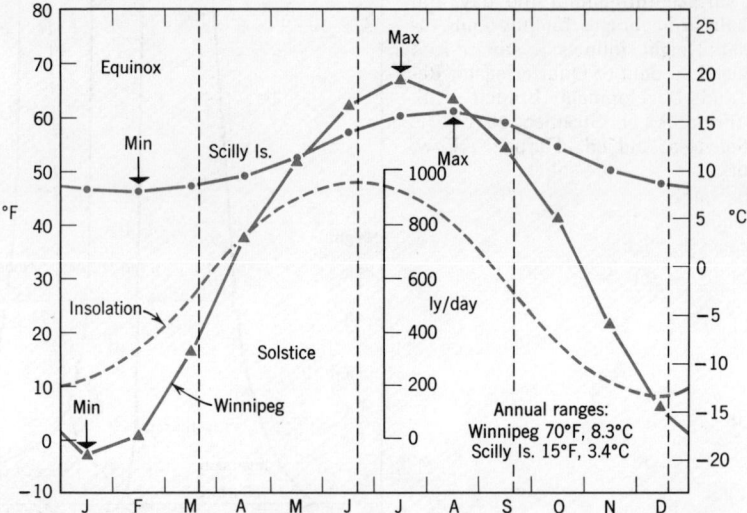

Figure 3.25 Annual cycles of monthly mean air temperatures for two stations at 50° N latitude: Winnipeg, Manitoba, Canada, and Scilly Islands, England. (Data of Meteorological Office, Air Ministry, Great Britain.)

once a day from the maximum-minimum thermometers. The maximum and minimum temperatures for one day are added together and divided by 2. If the mean daily temperatures are collected for many years and averaged for each month, then plotted on a graph, a temperature curve for the year is obtained. Figure 3.25 shows such curves for two places at 50° latitude. Winnipeg, Manitoba, has a midcontinent location; the Scilly Islands, England, lie to the lee of the North Atlantic Ocean.

Although insolation reaches a maximum at summer solstice, the hottest part of the year for inland regions is about a month later, since heat energy continues to flow into the ground well into August (as shown by curve G in Figure 3.4, and by the uppermost line in Figure 3.6). Air temperature maximum, closely coinciding with maximum ground output of longwave radiation, is correspondingly delayed. (Bear in mind that this cycle applies to middle and high latitudes, but not to the region between the tropics of Cancer and Capricorn.) Similarly, the coldest time of year for large land areas is January, about a month after winter solstice, because the ground continues to lose heat even after insolation begins to rise.

Over the oceans and at island or coastal locations there are two differences in the annual temperature cycle. (1) Maximum and minimum temperatures are reached about a month later than on land—in August and February, respectively. (2) The yearly range is much smaller over oceans than at inland continental stations. Coastal regions and offshore islands are usually influenced by the oceans to the extent that maximum and minimum temperatures occur later than in the interior. This principle shows nicely for the monthly temperatures of the Scilly Islands, off the Cornwall coast of southeast England. February is slightly colder than January.

Why does the annual air temperature cycle over the oceans show a much lower range and later occurrences of maximum and minimum values than at inland continental locations at the same latitude? The cycle of insolation is about the same for both, yet the lower air layer shows a different thermal response. The reasons lie in basic differences in physical properties of the ocean surface as compared with the land surface. After all, it is the annual cycle of the heat balance of the ocean or land surface that largely determines temperature of the overlying air.

A general law may be stated as follows: Land surfaces are rapidly and intensely heated by a given quantity of solar radiation, whereas under equal insolation water surfaces are heated more slowly and to a lesser degree. On the other hand, land surfaces cool off more rapidly and reach much lower temperatures than water surfaces, when insolation is reduced or cut off. Temperature contrasts are therefore great over land areas,

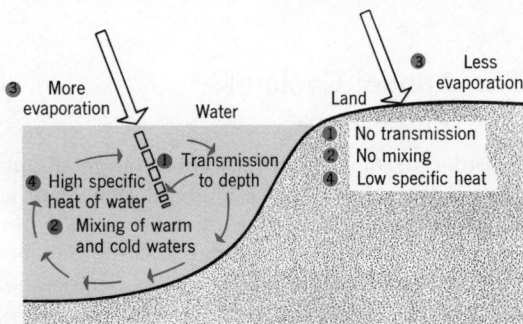

Figure 3.26 Constrasts in heating of large land and water bodies.

but only moderate over water areas. It is further true that the larger the mass of land, the greater are seasonal temperature contrasts in both ground and air.

An explanation of the law of land and water contrasts may be found in the application of certain principles of physics (Figure 3.26). As we have seen, water is transparent and permits solar radiation to penetrate many feet, thus distributing the heat through a thickness of several feet of water. The ground surface, being opaque, absorbs heat only at the surface, which can thus attain a higher temperature than the water surface. Ocean waters are mixed by vertical rising and sinking motions in the surface layer, allowing heat to be distributed and stored through a great mass of water, but no such mixing can occur in the solid ground. Water surfaces permit continual evaporation, the latent heat flux, which is a cooling process and serves to reduce surface heating. Ground surfaces, which are moist and covered by vegetation also permit cooling by evaporation and transpiration but to a lesser degree on the average than ocean surfaces. As a further cause of contrast, the specific heat of water is about five times as great as dry soil or rock. If heat is being applied equally to both substances, the dry ground will attain a high temperature long before the water will. For soils holding a large moisture content, this effect is minimized, but it is important in deserts and in periods of summer drought in otherwise moist climates.

Basic differences between the annual temperature cycle of an inland location, as compared with a marine location are nicely summarized by the graphs in Figure 3.23, previously used to show the daily temperature cycle. El Paso, an inland desert station, has an annual range of about 30 F° (17 C°), seen in the large spread between the January and July curves. North Head, Washington, in contrast, has a range of only about 16 F° (9 C°). This coastal station is strongly influenced by the presence of the adjacent Pacific Ocean, for prevailingly westerly winds bring air from over the ocean surface to the coastal zone. Correspondingly, these two stations show how the daily temperature range is accentuated by an inland desert location, but greatly reduced by the marine location. Daily air temperature range at El Paso averages about 20 F° (11 C°); that of North Head, only about 5 F° (3 C°).

The biological significance of the thermal contrasts between marine and inland continental locations in middle and high latitudes is that plants and animals are typically subjected to strong daily and annual thermal stresses in the inland locations, but enjoy a rather more benign, or equable, environment both in the shallow ocean waters and on lands bordering the ocean. In combination with a second prime environmental factor—that of availability of water—the thermal environment largely determines what forms of vegetation (such as forest, grasslands, or desert shrubs) can exist.

Global Distribution of Air Temperatures

As in the case of sea surface water temperatures, the global distribution of air temperatures, measured close to the surfaces of the lands and oceans, shows up well on world maps using isotherms (Figure 3.27). These are drawn for the two months of maximum and minimum values: January and July (as compared with February and August for the sea-surface temperature maps). Figure 3.28 shows annual temperature range.

Isotherms have a general east-west trend around the earth because of the general decrease of insolation from equator to polar regions. The east-west trend and parallelism of isotherms are best developed in the Southern Hemisphere, south of the 25th parallel, where land areas are small. In the Northern Hemisphere, isotherms show wide northward and southward deflections where they pass from a land area to an ocean area, particularly in January, when land and ocean surface temperatures are brought most strongly into contrast.

The land-water effect is represented diagrammatically in Figure 3.29 for the Northern Hemisphere. The January isotherm is deflected southward over the land, northward over the water. Temperatures along a single parallel are low on land but high on water. In July the reverse is true, for in summer the isotherm is pushed far north over the continent.

Throughout the year, isotherms shift through several degrees of latitude, following the changing path of the sun but lagging behind a month or so in time. Over large water areas, such as the South Pacific, the annual shift amounts to only about 5° latitude, whereas over landmasses, such as Africa, this shift is as much as 20° latitude. (Examine the change in position of the 70° F, 21° C, isotherms over southern Africa.) This difference in amount of latitude shift is also explained by the rapidity and intensity with which lands are heated and cooled as compared with ocean areas.

Certain definite centers of high and low temperature occur and are shown by isotherms which are completely closed to form oval or irregular-shaped enclosures. Notice that all of them are

Thermal Environments of the Earth's Surface

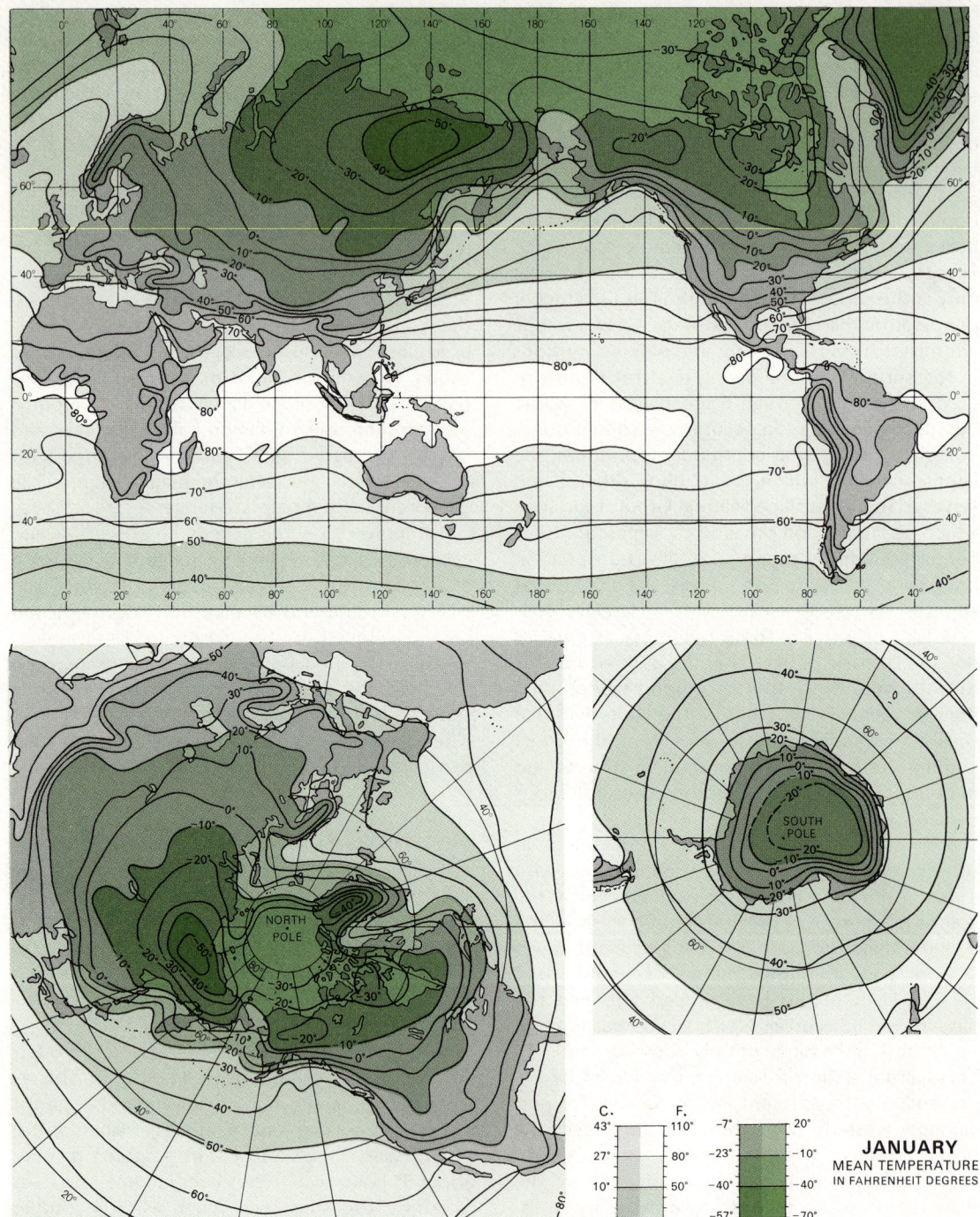

Figure 3.27 Above and right: January and July temperatures in degrees Fahrenheit. (Compiled by John E. Oliver from data by World Climatology Branch, Meteorological Office, *Tables of Temperature*, 1958, Her Majesty's Stationery Office, London; U.S. Navy, 1955, *Marine Climatic Atlas*, Washington, D.C.; and P. C. Dalrymple, 1966, American Geophysical Union. Cartography by John Tremblay.

over landmasses. In July, high-temperature centers occur over the southwestern United States, North Africa, and southwestern Asia. In January, a continental center of low temperature occurs over Siberia and is strongly developed with the average January temperatures lower than −50° F (−46° C). A corresponding region of low temperature, marked off by the closed isotherm of −30° F (−34° C), occurs in northernmost North America. It is not so well developed as that of

Global Distribution of Air Temperatures

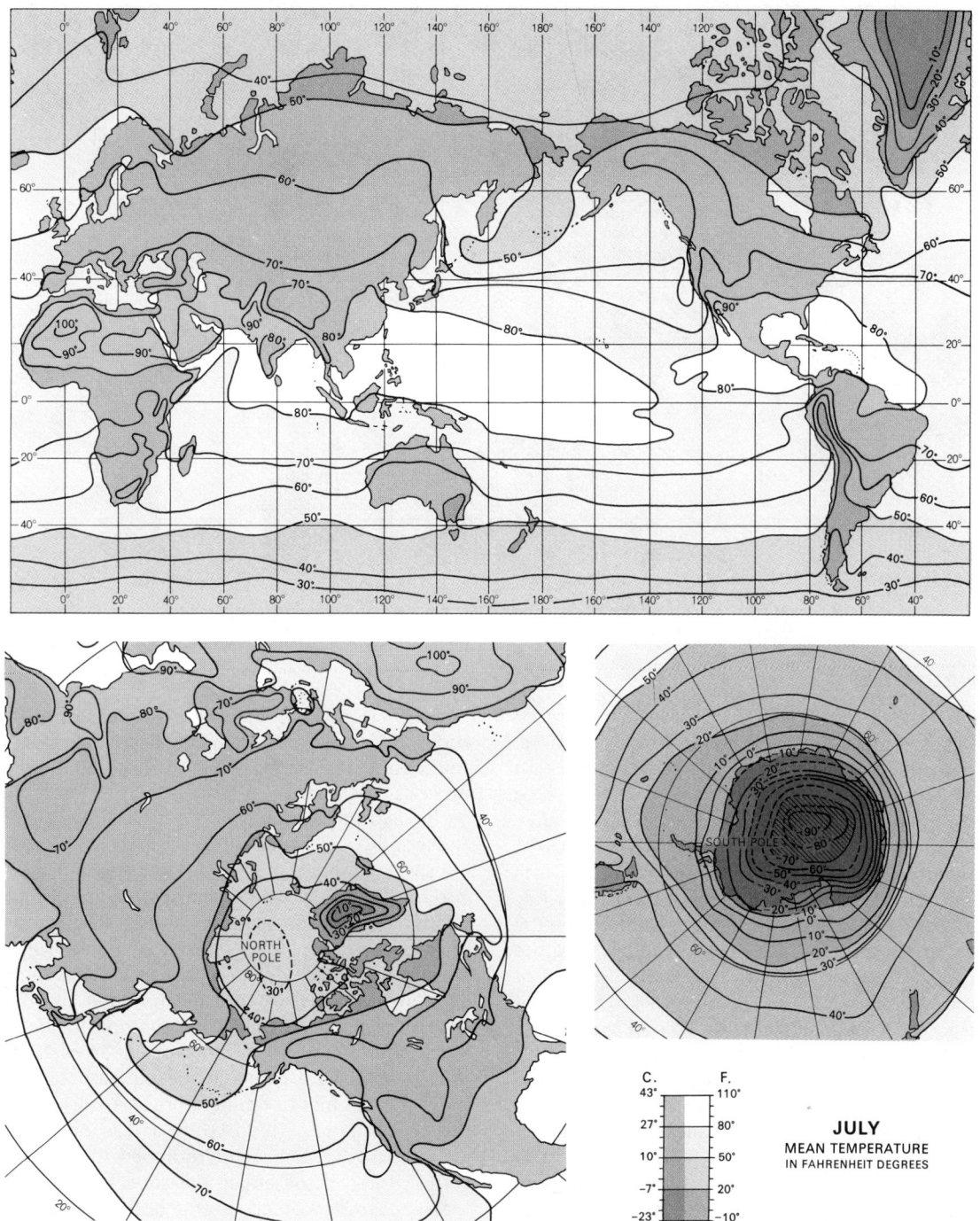

JULY
MEAN TEMPERATURE
IN FAHRENHEIT DEGREES

Asia because of the presence of considerable areas of Arctic Ocean among the islands of the northern fringe of the landmass and the smaller size of the North American landmass.

Permanent centers of low temperature exist over Greenland and Antarctica, the two regions of massive ice sheets. Temperatures over Greenland do not, however, reach the extreme low of northern Siberia in January, although the annual average temperature of the ice sheet is much lower. Analyzed from the polar projection in Figure 3.27, the January region of extreme cold is a distorted ellipse with its long axis extending across the Arctic Ocean from eastern Siberia to northern Canada, but with a sharp, deep reentrant over Greenland. By contrast, the antarctic region yields concentric isotherms neatly centered on the South Pole.

Thermal Environments of the Earth's Surface

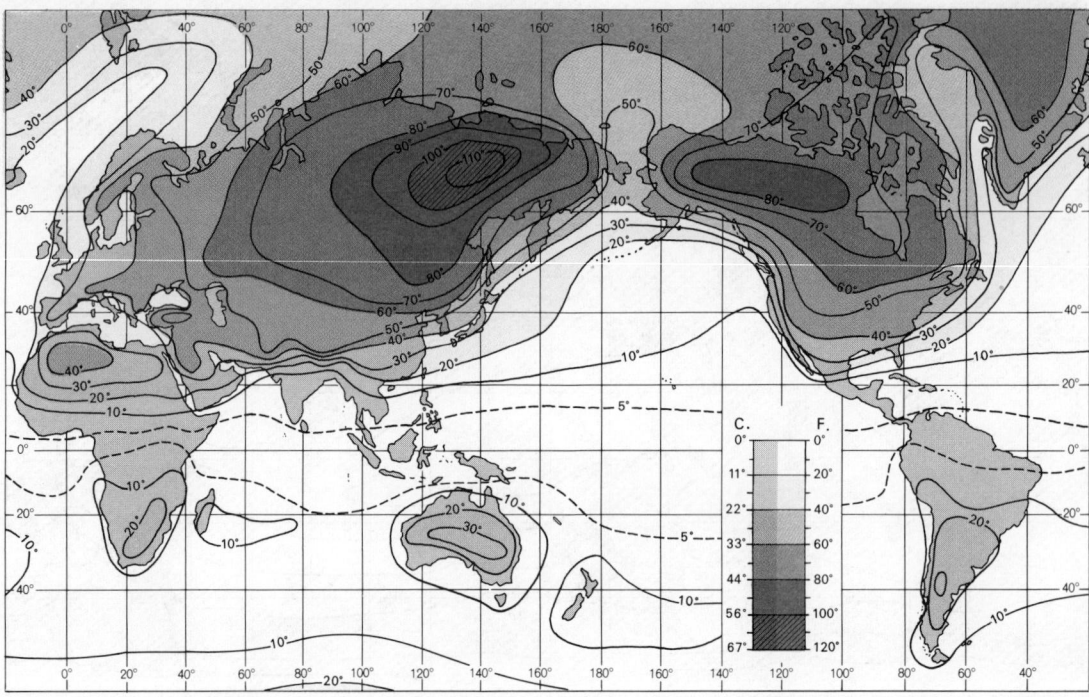

Figure 3.28 Annual range of air temperature. (Same data source as in Figure 3.27.)

A comparison of winter temperatures at the two poles is instructive, since one lies in a region of deep ocean and the other in the heart of a continent at high elevation. North-polar January mean temperature is probably about −30° F (−35° C), or about the same as for a sea-level land surface at that latitude. By contrast, the July average for the South Pole is below −80° F (−62° C), because heat radiates rapidly from the elevated plateau surface. The true cold pole of the antarctic continent is centered 7° to 8° distant from the geographic South Pole (Figure 3.27). Here the July mean is below −90° F (−68° C).

The annual range of monthly mean temperatures at any desired location may be roughly computed from the January and July maps but is more conveniently analyzed on a world map of annual range of monthly mean temperatures. (Figure 3.28). The lines on this map are drawn through points of equal range and may be called *corange lines*. In northern Siberia, the range between January and July means is about 110 F° (61 C°), greatest of any place on earth. Next are north-central Canada, just west of Hudson Bay, and Greenland with ranges of 70 F° (40 C°). The Arctic Ocean and surrounding continental fringes have a comparable range. Then follow Africa, South America, and Australia, with maximum ranges of about 30 F° (17 C°). An equatorial belt, about 35° of latitude in width over the oceans and about 10° wide across Africa and South America, has a range of 5 F° (2.8 C°) or less.

The monotonous uniformity of average daily and monthly air temperatures throughout the year is a striking feature of the equatorial zone. Figure 3.30 shows the daily high and low readings of air temperature at Panama (9° N latitude) throughout the months of July and February of a given year. These are the months of highest and lowest temperature averages, yet the means differ by less than one Fahrenheit degree. In contrast, the daily temperature range is on the order of 15 to 20 F° (8 to 11 C°). As a result in the equatorial zone the daily extremes of temperature far overshadow the annual range. This is truly a seasonless climate, but it has been aptly quipped that "night is the winter of the equatorial zone."

Ocean currents locally exert a noticeable modification upon the isotherms. The North Atlantic current, which runs northeastward close to the British Isles and the Norwegian coast, causes a sharp northeastward bend of isotherms in winter, when temperature contrasts are generally most marked. An opposite, or equatorward, deflection of isotherms occurs along the western coasts of South America, North America (July), and Africa, where cold currents cause the air temperatures to be lower than usual for tropical latitudes.

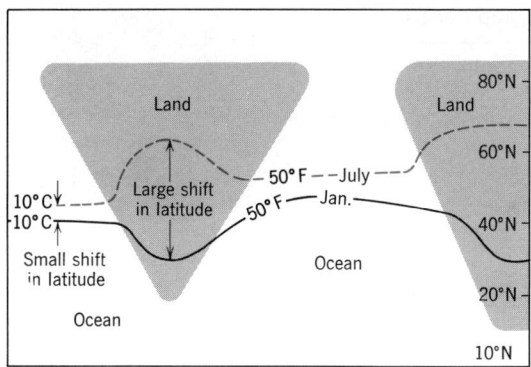

Figure 3.29 Seasonal shift of an isotherm.

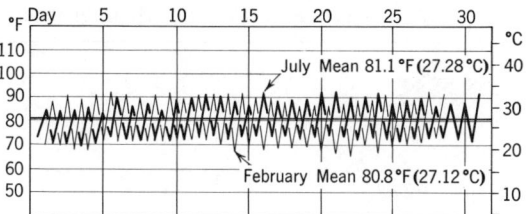

Figure 3.30 July and February temperatures at Panama, 9° N latitude. (After Mark Jefferson, *Geographical Review*.)

Radiation and Heat Environments of High Altitudes

The rapid decrease in density of the atmosphere with increasing altitude brings with it major changes in the environments of radiation and heat at the ground surface. Recall from Chapter 1 and Figure 1.7 that barometric pressure decreases by about one-thirtieth of itself for every 950 ft of altitude increase. Thus at 15,000 ft (4.6 km), an elevation representative of mountain summits in the higher ranges of the western United States, pressure is only about 570 mb, or a little over half the sea-level value. Air density is correspondingly reduced, with the result that the overlying atmosphere reflects and absorbs a much smaller portion of the incoming solar radiation than at sea level.

Measurements of insolation at elevation 12,000 to 14,000 ft (3.6 to 4.3 km), taken near the summit of a mountain range in the desert region along the border between California and Nevada, showed noon peak values approaching 2.0 langleys per minute under clear-sky conditions. Compare this value with the June maximum of about 0.7 for Hamburg, Germany, shown in Figure 2.16. At the mountain location, the peak of net all-wave radiation reached almost 1.6 langleys per minute, a value about triple that for Hamburg. Of course, the Hamburg data include days with cloud cover as well as clear days, while the mountain measurements are for clear skies only. Nevertheless, the great daytime intensity of incoming and outgoing radiation at the high altitude position is truly remarkable.

Increasing intensity of insolation with higher altitude has a profound influence upon air and ground temperatures. Surfaces exposed to sunlight heat rapidly and intensely, shaded surfaces are quickly and severely cooled. This results in rapid air heating during the day and rapid cooling at night at high-mountain locations as shown by the increasing spread between high and low readings from left to right in Figure 3.31.

The contrast between exposed and shaded surfaces is particularly noteworthy at high altitudes. It has been found that temperatures of objects in the sun and in the shade differ by as much as 40 to 50 F° (22 to 28 C°).

Increased intensity of insolation is accompanied by an increase in intensity of violet and ultraviolet rays; thus sunburn is more severe.

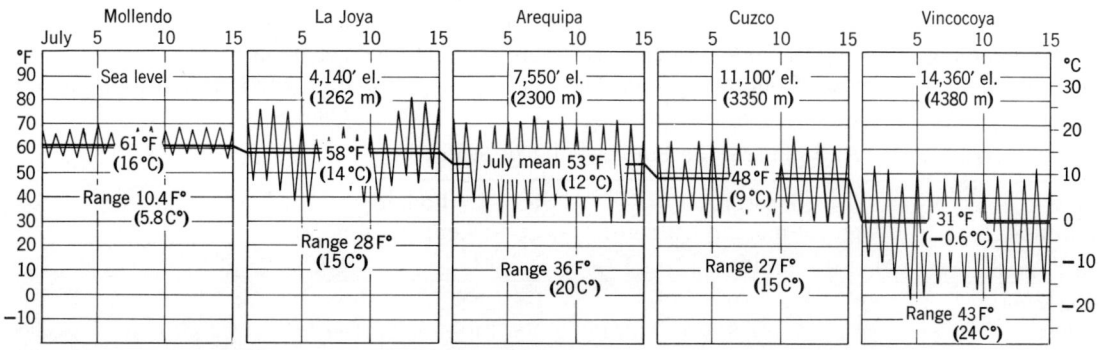

Figure 3.31 In Peru, on the west side of South America in the region of latitude 15° S, one can find stations of great differences in elevation. Increasing altitude brings not only a lower mean temperature, but also greatly increased daily temperature range. (After Mark Jefferson, *Geographical Review*.)

Thermal Environments of the Earth's Surface

TABLE 3.1 Altitude and Mean Annual Air Temperature

	Approximate Altitude		Approximate Mean Annual Air Temperature	
	Ft	Km	°F	°C
Belém, Brazil	0	0	80	27
Takengön, Indonesia	4000	1.2	70 (66)	21 (19)
Quito, Ecuador	9500	2.9	59 (47)	15 (9)
Jauja, Peru	11,000	3.4	54 (41)	13 (5)

The general decrease in air temperature with altitude has been discussed in Chapter 1, the average environmental lapse rate being about 3½ F° per 1000 ft (2 C° per 300 m). Using that rate, we might expect a station at an altitude of 10,000 ft (3000 m) to have a temperature about 35 F° (20 C°) below that of a nearby sea-level station. Actually the difference is somewhat less than this amount. In equatorial regions, where the average air temperature of all months of the year is almost a constant figure, the effect of increasing altitude can be seen in the sample figures of Table 3.1. Figures in parentheses are calculated by means of the normal environmental temperature lapse rate, assuming 80° F (27° C) to be the sea-level value. That the higher places are considerably warmer than the calculated value can be explained through the absorption and reradiation of solar energy by the ground surface.

Besides the thermal effects we have found, the high-altitude environment has other physical effects important to Man and other life forms. At high altitudes, intensity of surface bombardment by cosmic particles is increased, with the result that organisms are subjected to larger doses of ionizing radiation than at sea level. For a mile-high city such as Denver, Colorado, the intensity of ionizing radiation from cosmic sources is about double that at sea level, while at an altitude of 20,000 ft (6 km) the intensity is about seven times greater.

The physiological effects of a pressure decrease on humans are well known from the experiences of flying and mountain climbing. The principal influence is through an insufficient amount of oxygen to supply the blood through the lungs, a condition known as *environmental hypoxia*. At altitudes of 10,000 to 15,000 ft (3000 to 4500 m) mountain sickness (altitude sickness) occurs, characterized by weakness, headache, nosebleed, or nausea. Persons who remain at these altitudes for a day or two normally adjust to the conditions, but physical exertion is always accompanied by shortness of breath.

At reduced pressures the boiling point of water or other liquids is reduced so that cooking time of various foods is greatly lengthened. Table 3.2 gives some data on pressure and boiling point relationships. From these figures it is obvious that the use of pressure cookers will be of great value above 5000 ft (1500 m) wherever the cooking involves boiling of water. Other environmental effects of increasing altitude are related to availability of water and have profound influence on plant life.

In summarizing the relationship of altitude to radiation and the thermal environment, the important points are that while the average air temperature falls with increasing altitude, the intensities of both incoming and outgoing radiation of energy show a strong increase. You might be tempted to say that climbing to a higher altitude near the equator is like traveling from the equator to the poles. This may be a valid analogy when restricted to average temperatures, but it does not apply at all for the daily ranges in temperature.

TABLE 3.2 Altitude and the Boiling Point of Water

Altitude Ft	Pressure (height of mercury column)			Boiling Temperature	
	M	In	Cm	°F	°C
Sea level	0	29.9	76	212	100
1000	300	28.8	73	210	99
3000	900	26.8	68	206	97
5000	1500	24.9	63	203	95
10,000	3000	20.7	53	194	90

Two great fluid systems of energy and mass transport operate on a global scale on planet Earth. These circuits operate within the atmosphere and oceans and are powered by solar energy; they serve to transport heat as well as matter across the parallels of latitude (Figure 4.1). In terms of the energy balance of the earth, the circulation systems transfer heat from the low-latitude region of net radiation surplus to both polar regions of net radiation deficit. Rise of heated air at low latitudes also aids in heat removal by lifting that air to upper troposphere levels where it radiates energy freely into outer space. Without these forms of transport surface temperatures would be much higher than they presently are over the equatorial zone, and much lower than they are over the polar zones. By reducing thermal extremes from equator to poles, the fluid circulation systems make a much more favorable terrestrial environment for life than might otherwise exist.

With respect to their general circulation patterns, both atmosphere and oceans exhibit the properties of *heat engines*, which is to say that they are driven fundamentally by differences in heat intensity set up by differences in the net radiation from one region to another. When the

4
Circulation Systems in Atmosphere and Oceans

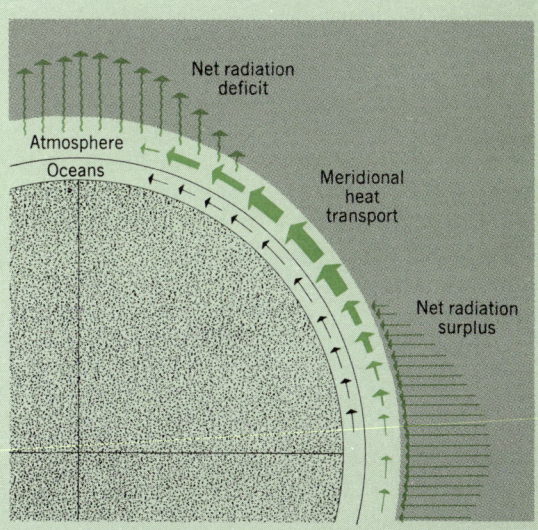

Figure 4.1 Heat transport and the earth's heat balance. (From A. N. Strahler, 1971, *The Earth Sciences*, 2nd ed., Harper & Row, New York.)

atmospheric layer is warmed it tends to expand, as do all gases when heated. Expansion leads to changes in air density and therefore in barometric pressure. When barometric pressure differences exist at a given level in the atmosphere, a force is set up causing the air to begin to move in the direction from higher to lower pressure. In responding to this force the moving air takes on kinetic energy, and this in turn is dissipated as heat through friction, both within the mass of moving air and in contact with the earth's surface. Eventually, the heat of friction is disposed of as longwave radiation in the total planetary radiation loss to space. Some of the kinetic energy of air motion is transformed to kinetic energy of the ocean surface in the form of waves and forward water motion. In this way, the broad patterns of oceanic circulation are set up and maintained. Kinetic energy of water motions in turn is transformed by friction into heat and is returned to the atmosphere for eventual dissipation into space.

Although the speed and direction of air and water currents that make up the total circulation of the fluid layers vary greatly from place to place and season to season, the average activity of the entire system over long periods of time operates uniformly in the transformation of energy and the transport of energy and mass across the parallels of latitude. This uniformity constitutes a *steady state* in an open system that receives energy and gives up energy at an average constant rate.

Barometric Pressure and Winds

Broadly defined, *wind* is any air motion relative to the earth's surface, but in practice the word refers to dominantly horizontal air motion. Localized vertical motions, such as occur in certain storms, are designated by such terms as *updraft* or *downdraft*. Description of wind consists of two quantities: speed and direction. Speed, relative to the earth's surface is given in miles per hour, knots, or meters per second; direction as the compass bearing from which the wind blows. Loosely speaking, direction is given as a cardinal point of the compass, such as N, NE, or E, but more precisely the direction is stated as the number of degrees between 0° and 360° read clockwise from geographic north.

If the rate of barometric pressure decrease with altitude were precisely the same over the entire earth, surfaces of equal pressure, termed *isobaric surfaces*, would be found to lie at constant altitudes above the surface. Such uniformity is rarely found to exist over any substantial area for very long periods at a time. Instead, the rate of pres-

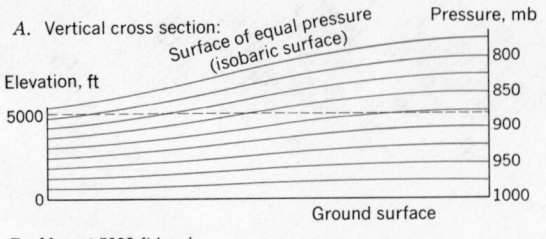

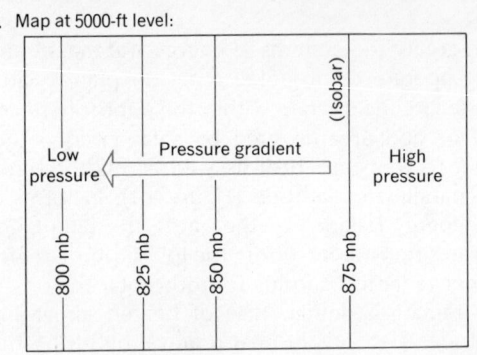

Figure 4.2 Isobaric surfaces in vertical cross section; an isobaric map drawn at an altitude of 5000 ft (1500 m). (From A. N. Strahler, 1971, *The Earth Sciences*, 2nd ed., Harper & Row, New York.)

sure decrease with altitude changes from place to place, with the result that the isobaric surfaces change in altitude, sloping either downward or upward. Sloping isobaric surfaces are shown in cross-section in Figure 4.2. Note that the rate of upward decrease in pressure is more rapid at the left side than at the right side, so that the surfaces of equal pressure slope down from right to left. Consequently, if one were to follow the 5000-foot level from right to left, he would cross isobaric surfaces of progressively lower pressure. A map drawn at the 5000-foot level is shown in the lower part of the figure. Isobaric surfaces now appear as lines, known as *isobars*: they are arranged in parallel positions. Pressure designated as *high* exists at the right side of the map; *low* pressure at the left. A *pressure gradient* exists in a direction always at right angles to the isobars and always from higher to lower pressure.

Where a pressure gradient exists, there also exists a force tending to cause the air to accelerate in the direction of the pressure gradient; this is the *pressure-gradient force*. Magnitude of this force is directly proportional to the steepness of the pressure gradient. With this information in hand, let us analyze the mechanism of a simple wind system set up by differences in heat.

Figure 4.3 breaks down the system into developmental stages for the sake of simplicity. Diagram *A* shows initial conditions of a uniform pressure decrease with altitude and horizontal

Barometric Pressure and Winds

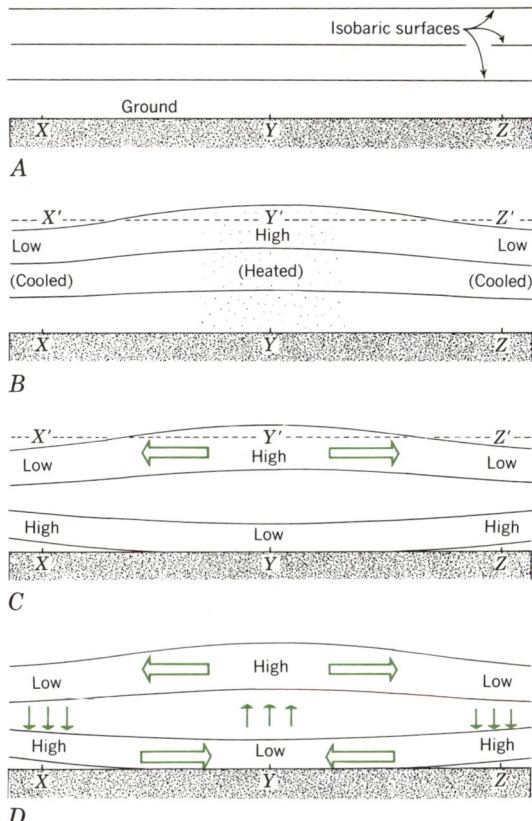

Figure 4.3 Hypothetical model of a simple convective circulation system. (From A. N. Strahler, 1971, *The Earth Sciences*, 2nd ed., Harper & Row, New York.)

isobaric surfaces. There is no pressure gradient and no force tending to produce air movement. Diagram B shows that the air layer at Y, near the center of the region, has been uniformly warmed and has expanded, lifting the isobaric surfaces; whereas air at X and Y on either side has been cooled, lowering the isobaric surfaces. At a given upper level shown by the horizontal dashed line, barometric pressure is now relatively high near the center and relatively low at the sides, giving a pressure gradient outward from the central region to either side. Air now begins to move in the direction of this gradient, as shown by the broad arrows in Diagram C. However, the accumulation of incoming air at high level in the low-pressure regions causes high barometric pressure to develop at low level, while removal of air from the central region causes lower pressure to develop there at the surface. These two surface lows and one high are labeled in Diagram C. Again, a horizontal pressure gradient is created, but this time in reverse direction from that at upper levels. In consequence, air begins to move from the sides (positions X and Z) toward the center (position Y), as shown by the broad arrows near the surface in Diagram D. Once these air motions are initiated, the circuit is completed by rising of air in the central region and sinking of air in the two regions at the sides. The complete circulation is now formed into two *cells*, in which circulation will continue as long as heat is applied at the center and lost at both sides. Air in motion encounters resistance by internal friction, as for all fluids in motion. Work must be done to overcome fluid resistance. Consequently the wind speed reaches a limiting value, or terminal speed. Energy thus converted into heat is conducted and radiated from the flow system. It is important to know that the winds in this system are quite strong (fast) because air motion is in a thin layer, in comparison to the rising and sinking motions, which are very slow because they take place over a zone of broad cross section.

The simple thermal circulation described above fits rather well into a real system of local winds known as land and sea breezes (Figure 4.4). Conditions of early morning are calm, with no pressure gradients (Diagram A). By afternoon air over the coast has been warmed more than over the adjacent ocean, lifting the isobaric surfaces and causing an upper-air movement from land to ocean, and a low-level movement from ocean to land. The latter is the *sea breeze*, a welcome relief on hot summer days. At night, more rapid cooling of the air layer over the land causes the isobaric surfaces to subside, setting up a reversal of pressure gradient and winds. Now the *land breeze* sets in, transporting cooler air from the coast over the adjoining water. Sea breezes are a distinctive

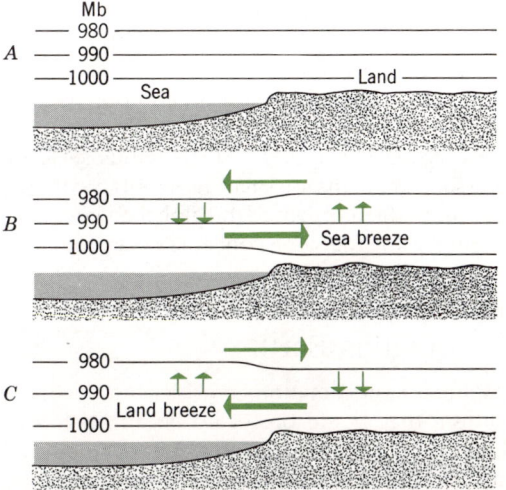

Figure 4.4 Sea breeze and land breeze. (After S. Petterssen, 1969, *Introduction to Meteorology*, 3rd ed., McGraw-Hill, New York, p. 171, Figure 10.1.)

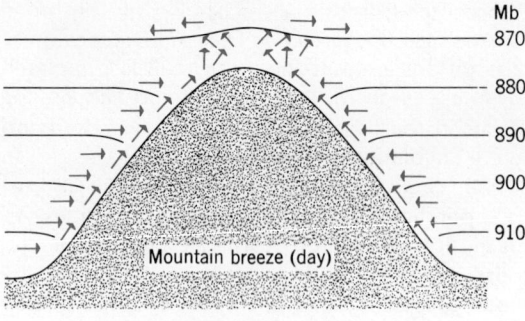
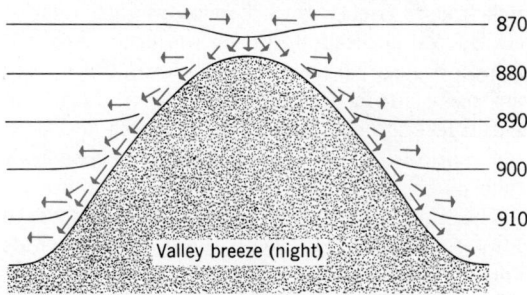

Figure 4.5 Mountain breeze and valley breeze. (After S. Petterssen, 1969, *Introduction to Meteorology*, 3rd ed., McGraw-Hill, New York, p. 172, Figure 10.2.)

environmental feature of coasts during the summer season. Close to the shoreline the breeze is often persistent and strong, but only in a very narrow zone.

A second and related example of the simple wind system is seen in mountain and valley breezes, illustrated in Figure 4.5. Daytime heating of the air column over mountain slopes causes a local pressure gradient toward the mountainside (Diagram A). Air moving in this direction is forced to flow up the mountain slope as a *mountain breeze*. The air can then rise to higher levels over the mountain summit. At night the gradient is reversed and air flows outward and downward in a *valley breeze*.

From these examples of local winds set up by pressure gradients of small horizontal extent let us turn to the problem of the global air circulation system, using the same concepts on a grand scale.

Idealized Circulation on a Nonrotating Earth

As a first approximation, consider the circulation pattern that would be set up on an earth imagined not to rotate, or perhaps to rotate very slowly. The reasons for specifying nonrotation will become evident in later paragraphs. Figure 4.6, Diagram A, shows the setup as a spherical earth surrounded by a uniform atmospheric layer

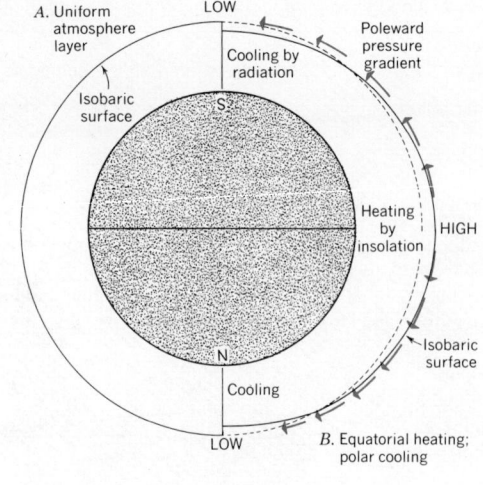
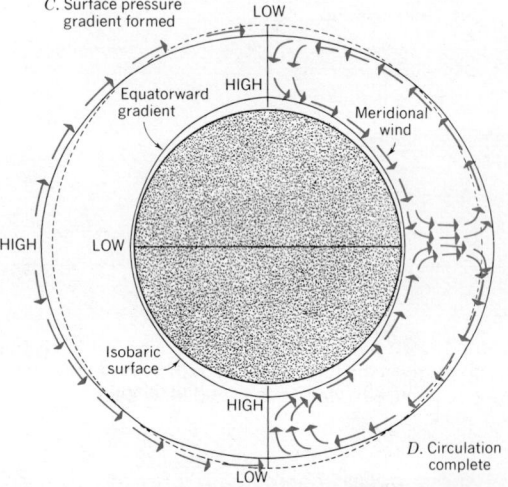

Figure 4.6 Convective wind system on an imagined nonrotating earth. (From A. N. Strahler, 1971, *The Earth Sciences*, 2nd ed., Harper & Row, New York.)

with no horizontal pressure gradient in existence. Imagine now that heat is applied to the equatorial belt, creating a net radiation surplus, while longwave radiation to space over the polar zones yields a deficit and results in polar cooling. If the earth could not rotate, the heat source would need to be in the form of an encircling ring in the plane of the equator. However, we can postulate a very slow earth rotation (as with planet Venus) and accomplish much the same result. As shown in Diagram B, a pressure gradient from equator to poles has been set up at high levels, causing poleward air movement. The redistribution of mass then causes formation of an equatorial belt of low pressure and two polar highs at low levels (Diagram C). Here the pressure gradient is equatorward and causes surface winds to blow from polar zones toward the equator. We can describe as *meridional* both the high-level and surface winds, since they follow the meridians of longitude and have no east-west component. Circulation is completed by a rise of warmer air over the equatorial zone and a sinking of cooler air over the two polar zones. As noted earlier, air flow meets with resistance, which increases to the point that wind speeds reach constant values. A steady state would then be reached in which energy transfer as sensible heat exactly matched the net input and net output of radiant energy. Isobars would run in true east-west circles, as do the parallels of latitude. At low levels there would be an *equatorial trough* of low pressure and two *polar highs*; these do, in fact, exist on the earth. Our simple model is, of course, unworkable, but it does provide a first step in explaining the global circulation.

Coriolis Effect and the Geostrophic Wind

The earth's rotation introduces a profound modification in flow of air in response to the pressure gradient force. This modification is the *Coriolis effect*, named for G. G. Coriolis, a French mathematician who developed the concept fully in 1835. In application, the Coriolis effect causes all matter in motion to be deflected toward the right in its path, if in the Northern Hemisphere, and toward the left, if in the Southern Hemisphere. The effect applies to gases, liquids, and to solid masses, although the response is particularly great in the fluids, since they offer little resistance to unequal stresses.

To explain the Coriolis effect is not easy, for a great deal of rather complex mathematics is required. Instead, you may need to settle for an appreciation of the basic cause of the phenomenon, which is simply that any particle of matter in motion in a straight line tends to maintain that path with respect to space coordinates. However, a particle in motion at the earth's surface is constrained by the earth's gravity to follow a curved path parallel with the earth's surface. Because the earth rotates on an axis, carrying with it the atmosphere and oceans, a particle in motion is subjected to the equivalent of a force acting always at right angles to the path of motion. Physicists refer to this force as being "fictitious," but for practical purposes the Coriolis effect can be treated as a force, meaning that it can be assigned a value of intensity and a value of direction, just as for the force of gravity and the pressure-gradient force.

Figure 4.7 is a schematic diagram of the Coriolis force, showing that the force always pulls at right angles to the path of motion (small arrows sticking out from broad arrows), and that the force for a given mass at a given location is always present in constant intensity for a given speed, whether the compass direction of motion is east, west, north, south, or any intermediate direction. However, there is a strong effect of latitude, such that the force has zero value at the equator and increases in intensity to a maximum at either pole, as shown by the increasing lengths of the small arrows in the poleward direction. Force intensity increases as the sine of the latitude. There is one more variable to consider; that of speed of motion. The Coriolis force increases in intensity in direct proportion to the linear speed of motion of the particle. The faster the motion, the stronger the force tending to deflect the particle from its straight-line path of motion.*

Now, let's put these rules to work in following the path of a small mass (a parcel) of air as it begins to move horizontally across the isobars in response to a pressure gradient. Figure 4.8 shows what happens. Starting at A the parcel begins to move at right angles across the isobars, gaining

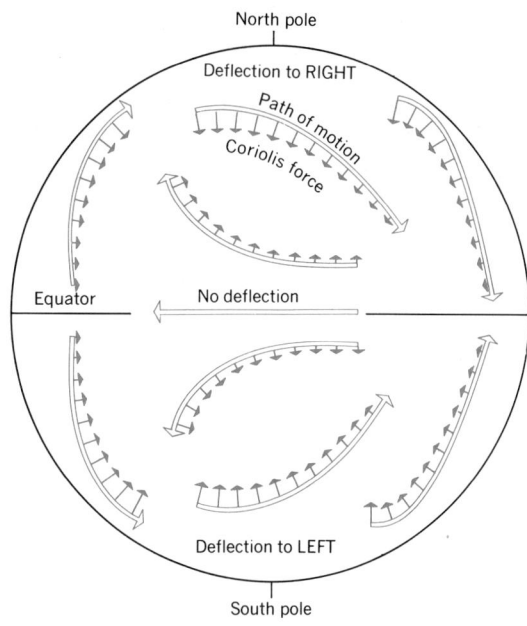

Figure 4.7 Direction of application of the Coriolis force is shown by small arrows, always at right angles to the path of air motion. (From A. N. Strahler, 1971, *The Earth Sciences*, 2nd ed., Harper & Row, New York.)

---

*The Coriolis effect can be stated formally in the following equation: Coriolis force per unit of mass = $V2\Omega \sin \phi$, where V is horizontal linear velocity (cm/sec); Ω is angular velocity of the earth's rotation (a constant equal to 0.00007292 radian per mean solar second); and ϕ is latitude in degrees. Force per unit of mass represents an acceleration.

Circulation Systems in Atmosphere and Oceans

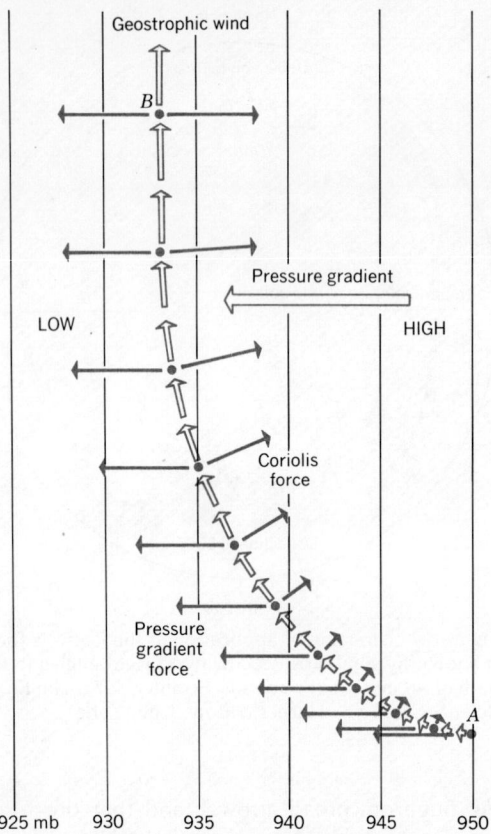

Figure 4.8 Deflection of a parcel of air by the Coriolis force, leading to development of the geostrophic wind. (From A. N. Strahler, 1971, *The Earth Sciences*, 2nd ed., Harper & Row, New York.)

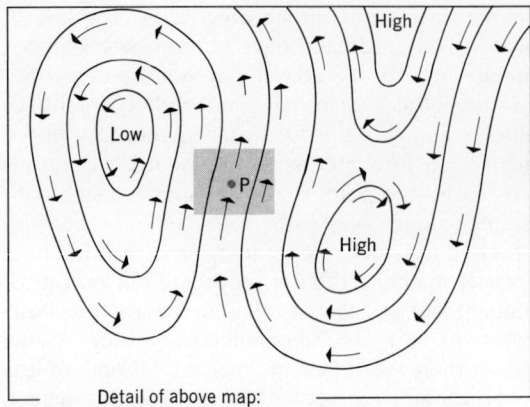

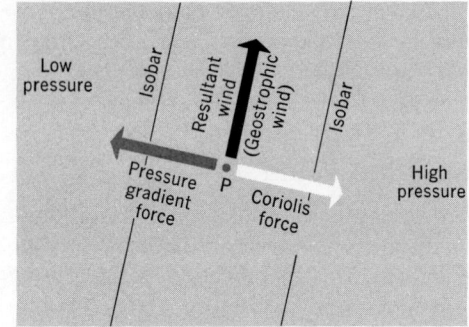

Figure 4.9 Wind follows the isobars at high levels.

speed as it goes. At once, the Coriolis force begins to act, pulling to the right and causing the path to turn right. As the parcel moves faster, the Coriolis force increases, as shown by the increasing length of arrows, intensifying the pull to the right. As the path is turned right, the motion becomes rotated to the point that ultimately it parallels the isobars and is at right angles to the barometric gradient. At this point (labeled R) the Coriolis force exactly balances the pressure-gradient force. Beyond this point the speed of the parcel is constant and a steady state of motion exists. When there is a perfect balance between Coriolis force and pressure gradient force, air traveling parallel with a set of straight, parallel isobars at constant speed constitutes the *geostrophic wind*.

A third force acting upon moving air to influence its direction and speed is *centrifugal force*. We shall not go into details of the effect of this force, except to note that it influences the speed of the wind when the motion follows a curved path. But the direction of movement in steady state is nevertheless parallel with the curved isobars as long as they maintain the same separating distance.

Figure 4.9 shows how wind direction follows the isobars at high levels in the atmosphere. In the Northern Hemisphere the lower barometric pressure always lies to the left of the flow direction; the high pressure to the right. These relationships are shown schematically in Figure 4.10.

Cyclones and Anticyclones

Near the earth's surface a fourth force comes into play to modify the direction and speed of wind: the force of friction between the moving air layer and surface of the ocean or land. The action of this *frictional force* is illustrated in Figure 4.11 by a group of arrows. The frictional force, always opposite in direction to the direction of air movement, reduces the wind speed and consequently brings on a reduction in the Coriolis force. As the broad arrows in Figure 4.11 show, the frictional and Coriolis forces, at right angles to one another, form a parallelogram of forces in which the resultant force is equal and opposed to the pressure-gradient force. Consequently, the direction of the wind is at an angle to the isobars. For fairly smooth surfaces, such as water or a flat prairie, the angle is small—some 20° to 25° as

The Planetary Circulation

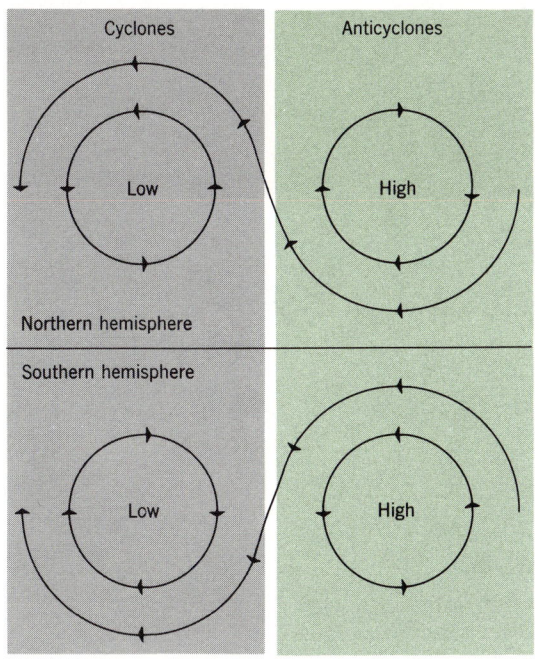

Figure 4.10 Winds at high levels around cyclones and anticyclones.

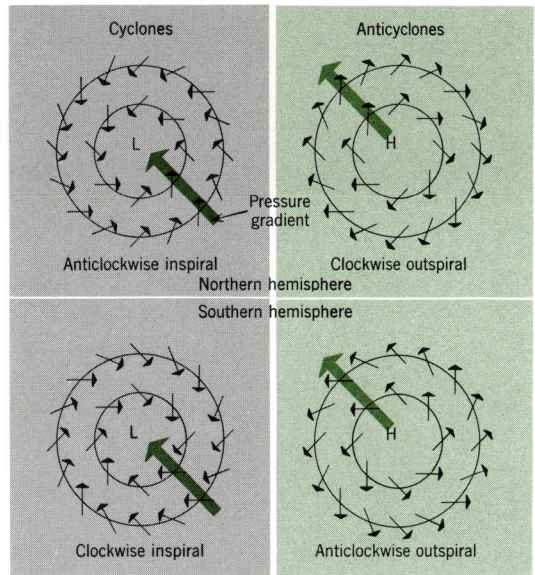

Figure 4.12 Surface winds within cyclones and anticyclones.

measured from the isobars—and the wind speed is reduced to some 60 to 70% of the geostrophic value. For rough terrain the angle may be as much as 45° and the wind reduced in speed to as little as one-third of the geostrophic value. The influence of surface friction dies out rapidly with increased elevation and is negligible above a height of about 2000 to 3000 ft (600 to 900 m). You can observe this effect by noting that the wind direction as shown by your wind vane usually differs from that of low clouds moving above you.

From the standpoint of an energy system, the reduction of wind speed near the surface represents the flow of momentum of the moving air into the ground beneath it. Kinetic energy is transferred into surface motion of water or is spent in warming the air layer at the boundary. (The warming effect is unimportant.)

Movement of air with respect to isobars at low levels is shown schematically in Figure 4.12. The pattern of wind arrows spirals inward toward the center of low pressure, constituting a *cyclone*, and spirals outward from the center of high pressure, or *anticyclone*. Spiral directions are reversed in the opposite hemisphere.

Air spiralling inward within a cyclone represents a *convergence* and requires that the excess air escape by having an upward component of motion. Air spiralling outward from an anticyclone constitutes a *divergence*, and must be supplied by a downward component of motion, drawing in air from higher levels. These motions will be tied together in further exploration of cyclones and anticyclones in Chapter 5.

Now we can get back to the investigation of the earth's large-scale atmospheric circulation, applying principles of forces affecting winds.

The Planetary Circulation

Introducing the Coriolis force into the simple model of meridional circulation already developed, consider what happens to air that begins to move poleward at high levels from the equatorial belt. As Figure 4.13 shows, a parcel of air starting at point A and beginning to move north begins to be affected by the Coriolis force a short distance north of the equator. Its path is then turned eastward in traveling from A to B. Upon

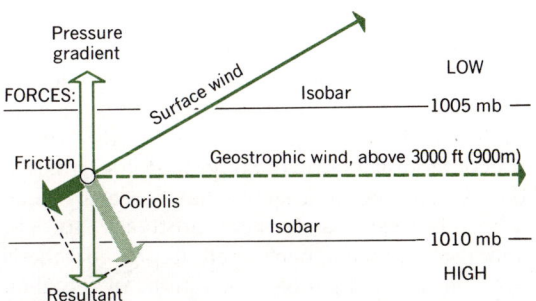

Figure 4.11 The surface wind direction in relation to three controlling forces. (From A. N. Strahler, 1971, *The Earth Sciences*, 2nd ed., Harper & Row, New York.)

Circulation Systems in Atmosphere and Oceans

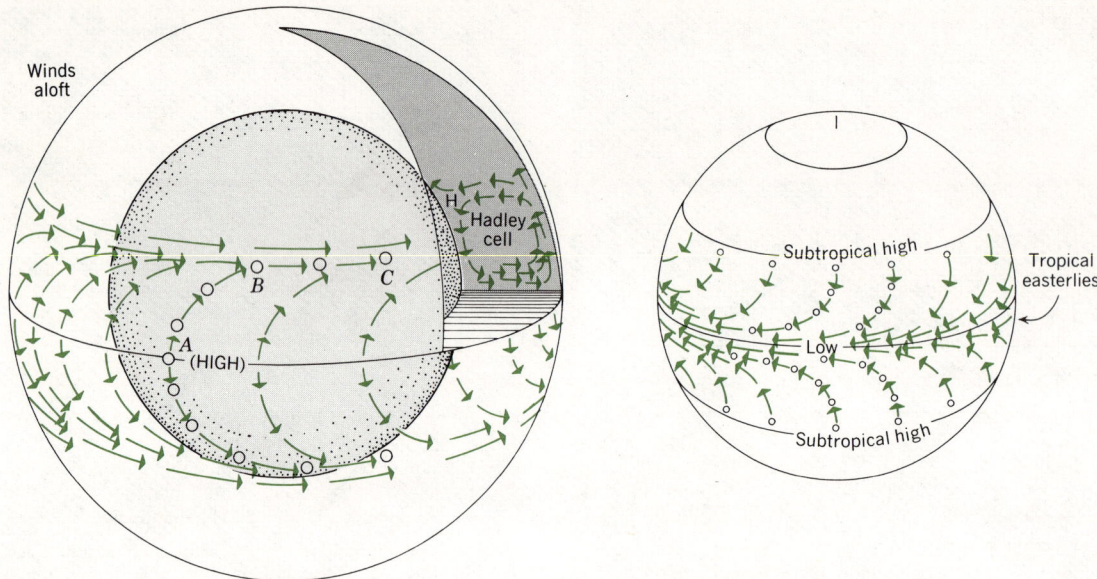

Figure 4.13 Idealized diagram of formation of the Hadley cell circulation and the tropical easterlies. (From A. N. Strahler, 1971, *The Earth Sciences*, 2nd ed., Harper & Row, New York.)

reaching point C, the flow is geostrophic and runs due east, parallel with the isobars. (In the Southern Hemisphere deflection toward the left also results in eastward flow.) With no further poleward progress possible, the air tends to accumulate in a subtropical zone at about latitude 20° to 30°, constituting a convergence. As a result a sinking motion, or *subsidence,* develops in this zone, creating a belt of increased pressure, the *subtropical high-pressure belt.* Air reaching low levels is now required to move away from the convergence zone. Part of this air moves poleward at low level, but much of it moves equatorward, as shown in the lower diagram of Figure 4.13. Equatorward motion results in deflection to the west, setting up an easterly wind system. These winds are known as the *tropical easterlies.* Air converges over the equator, in the *intertropical convergence zone,* and is accompanied by a general rise to complete the entire circuit.

Taking into account only the north-to-south, south-to-north, and vertical components of air movement—in other words only the meridional flow—we find a cell of atmospheric circulation dominating the tropical and equatorial zones. This system has been named the *Hadley cell* in honor of George Hadley, who postulated its existence in 1735. Ideally there should be two matching Hadley cells, one for each hemisphere. Actually, the cell develops strongly in the winter season of the respective hemisphere, but weakens during the summer season. Consequently, the Hadley cell is developed strongly only on one side of the equator at a time (November through April in the Northern Hemisphere; April through October in the Southern Hemisphere). Following the annual cycle of the sun, the cell of the Northern Hemisphere grows and transgresses southward over the geographic equator during the winter months, because then the zone of most intense insolation has shifted into the tropical zone near 20° S latitude.

The Hadley cell involves some exchanges of air with regions lying poleward of it, for otherwise surplus heat could not be carried into high latitudes. Air subsiding in the subtropical high-pressure belt divides at low level, part of it escaping poleward and carrying with it sensible and latent heat. In compensation, some air drawn from higher latitudes must enter the cell at high levels. You might be led to suppose that another cell, similar to the Hadley cell, but reversed in direction of motion, lies on the poleward side, and that the two are meshed together like two gears. Such a neighbor cell was formerly postulated to exist and even given a name, but it proved largely fictitious in the light of observations of actual air movements, being weakly developed at best. What, then, is the remainder of the global circulation like?

Angular Momentum Transport by Air Motions

We now change the frame of reference from that of four forces influencing winds to one of *momentum* of air masses in circular motion following the earth's rotation. Momentum can be thought of as *quantity of motion*. For a body moving in a straight line momentum, L, is equal to the product of the mass, M, and its linear velocity, V. Thus, *linear momentum* is defined as

$$L = M \cdot V$$

(Energy would be represented by $M \cdot V^2$, if stated in equivalent terms.)

For motion in a circular path, momentum becomes *angular momentum,* and is equal to the product of the linear momentum and the radius of the arc, R, in the path of motion, thus:

$$L = M \cdot V \cdot R$$

One of the fundamental laws of mechanics states that within a rotating system to which no energy is added or from which none is taken away the total angular momentum remains constant. This *law of conservation of angular momentum* is illustrated by a simple device. A small weight is attached by a length of string to a stick which serves as a handle. We now swing the weight in a circle. If we allow the weight to wind up the string around the stick, we will see that as the string gets shorter, the speed of the whirling weight undergoes a marked increase. What is happening is that as the radius, R, decreases, the linear velocity, V, must increase proportionately in order to keep the angular momentum constant.

Turning now to the earth as a spherical globe rotating on its polar axis, every particle at the earth's surface possesses angular momentum because of that rotation. As shown in Figure 4.14, a small mass situated at point A on the earth's equator has a greater linear velocity V_1, than the same mass at point B, located at 30° N, because the circumference of the circular travel path at that parallel is shorter than at the equator.

To conform with the law of conservation of angular momentum, the sum of angular momentum of all particles of matter comprising the earth must remain a constant value. Consider that the earth's angular momentum is divided into three major parts: that of the oceans; that of the atmosphere; and that of the lithosphere. Now, when dealing with day-to-day and seasonal changes in mass, we can consider that the angular momentum of the lithosphere does not change appreciably through motions of solid matter from one place to another. However, the atmosphere and oceans, moving freely, can easily gain or lose

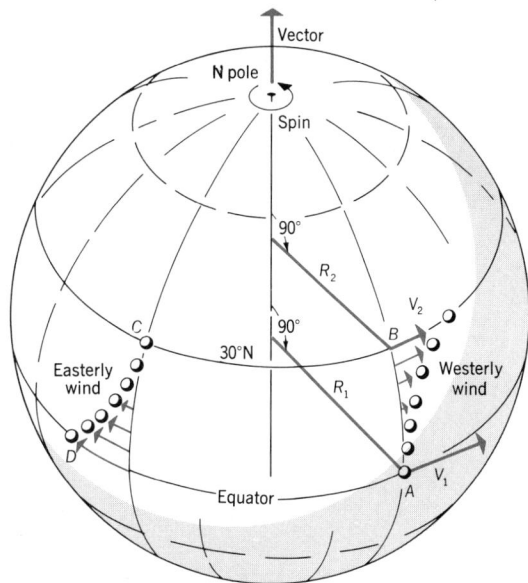

Figure 4.14 Meridional transport of angular momentum to produce westerly winds. (From A. N. Strahler, 1971, *The Earth Sciences*, 2nd ed., Harper & Row, New York.)

angular momentum, but in that case there must be a compensating change in momentum of one or both of the other parts to maintain the total constant.

Let us turn again to Figure 4.14 and find out what happens when a parcel of air located at point A travels north toward point B. Since the air parcel represents a constant unit of mass, changes can occur only in radius and in linear velocity. As the parcel travels north it passes across parallels of latitude of decreasing radius, which means that fixed points on the ground beneath have slower speeds of eastward motion as successive parallels are encountered. The air parcel carries with it the initial quantity of momentum which it had at the equator (point A), but since it is forced to move into a smaller radius as it moves north, its eastward velocity must increase in order to keep the momentum constant. Consequently, the air parcel moves eastward more rapidly than the ground beneath it. This relative motion constitutes a westerly wind. Momentum is thus transported poleward into higher latitudes.

The reverse direction of travel, equatorward, is illustrated in Figure 4.14 by a parcel of air originating at point C and moving to point D. In moving across parallels of increasing radius, the parcel experiences a decrease in eastward speed, falling behind the eastward-moving ground sur-

Circulation Systems in Atmosphere and Oceans

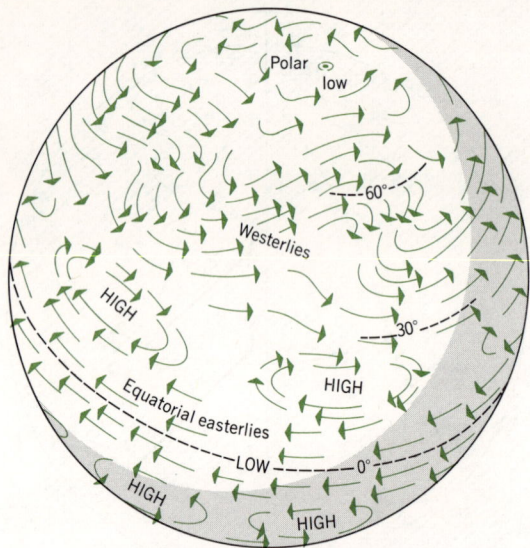

Figure 4.15 Idealized patterns of global circulation at high levels in the troposphere. (From A. N. Strahler, 1971, *The Earth Sciences*, 2nd ed., Harper & Row, New York.)

face beneath. Consequently, the motion of the parcel relative to the ground is westward, constituting an easterly wind.

Going back to the Hadley cell circulation, the high-level air flow from equatorial zone to subtropical zone results in poleward transport of angular momentum and produces a strong westerly wind at high levels in the region of the 30th parallel. Within the same cell, air moving equatorward near the surface is forced into an easterly flow, which we have already identified as the tropical easterlies. These latter winds, because of surface friction, cannot develop high speeds.

By means of the momentum concept, we have explained the existence of westerly and easterly winds associated with the Hadley cell circulation. But what about circulation in middle and high latitudes?

Atmospheric Circulation in Middle and High Latitudes

Observations developed largely during and following World War II have shown that the high-level air flow in the troposphere of middle and high latitudes of both hemispheres is dominated by the *upper-air westerly winds*. These winds move ceaselessly around the globe, forming a huge vortex, as shown in Figure 4.15. The vortex is centered around a *polar low*. (Recall that in the Northern Hemisphere, the low pressure lies to the left of the direction of air motion.)

Although the upper-air westerlies represent the persistent motion, there are undulations and sec-

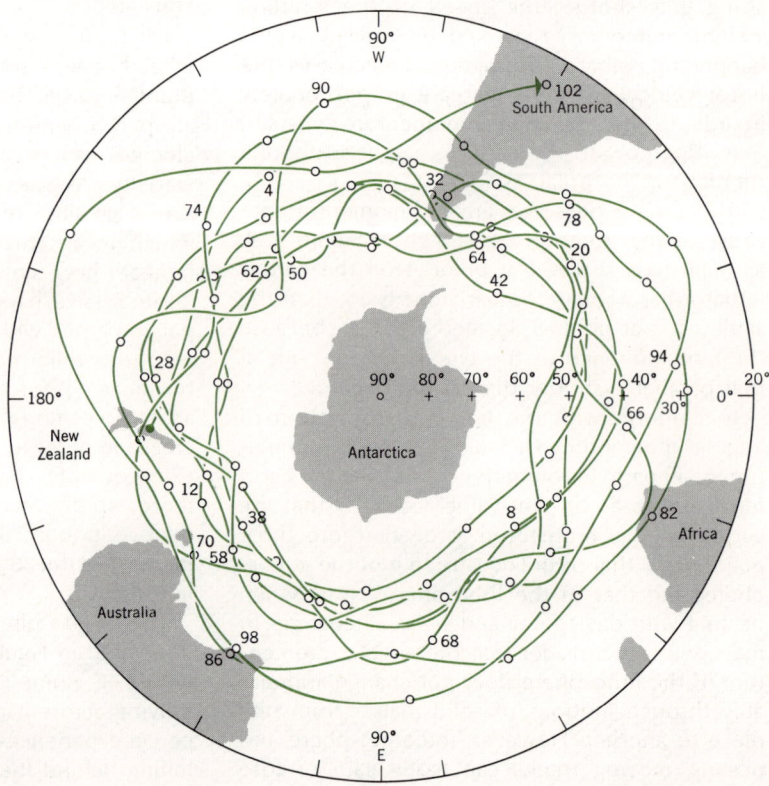

Figure 4.16 Path of a GHOST balloon traveling at the 200-mb (40,000-ft; 12-km) level. Numbers show its position on 102 consecutive days following launching in New Zealand. (Data of V. E. Lally and E. W. Lichfield, NCAR, as reported in *Amer. Meteorological Soc. Bull.*, 1969, vol. 50, p. 868, Figure 1.)

Atmospheric Circulation in Middle and High Latitudes

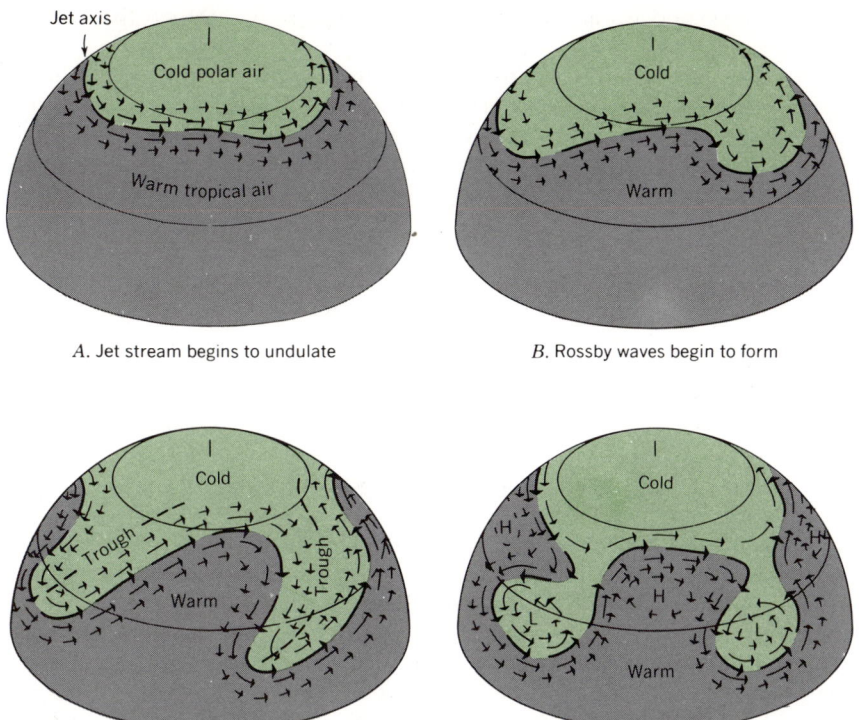

Figure 4.17 Development of upper-air waves in the westerlies. Modified from diagrams by J. Namias, NOAA, National Weather Service. (From A. N. Strahler, 1971, *The Earth Sciences*, 2nd ed., Harper & Row, New York.) *A.* Jet stream develops undulations. *B.* Rossby waves begin to evolve. *C.* Strongly developed waves. *D.* Occlusion of cells of cold and warm air.

ondary vortices (cyclones and anticyclones) which form from time to time, in which the direction of air motion may be temporarily reversed. Undulations in the westerly flow take the form of *upper-air waves*, shown in Figure 4.15. These waves have been named *Rossby waves* in honor of a meteorologist of that name who developed the mathematical analysis of the wave motion. The reality of the upper-air westerlies, with undulations in the flow pattern, is seen in the trajectory of a high-altitude balloon, known to researchers as GHOST balloon, followed throughout 102 days of travel in the Southern Hemisphere (Figure 4.16). The path was at an altitude of about 40,000 ft (12 km), where barometric pressure is about 200 mb, or about one-tenth the sea-level pressure. This map has real environmental significance, since it represents the travel path of foreign gas and suspended dust particles in the atmosphere. Any contaminants, such as radioactive particles, introduced into high levels of the atmosphere in these latitudes, will be carried around the globe from west to east, gradually becoming distributed in a broad zone, and capable of settling out over a huge expanse of the earth's land and ocean surface.

Figure 4.17 is a set of four schematic diagrams showing how the smooth westerly flow develops into a set of Rossby waves. The waves deepen and are finally pinched off to form isolated centers of low pressure known as *cut-off lows*, alternating with centers of high pressure, known as *cut-off highs*. These vortices eventually weaken and dissolve, so that the smooth west-to-east flow is resumed. The cycle of growth and decay of upper-air waves and their cutoff may require many days to complete, but seems to follow no regular schedule.

Looking further into the Rossby waves, let us find out more about the temperature relationships involved. Meteorologists speak of enormous bodies of atmosphere, involving substantial thicknesses of the troposphere and having subcontinental extent, as *air masses*. An air mass is characterized by its temperature, whether cold or warm, and by its moisture content. For purposes of generalization we will designate on a hemispherical basis two kinds of air masses: *polar* and *tropical,* the first being cold, the second warm. As you would expect, the cold polar air masses constitute the troposphere of the high latitudes and polar regions, while the warm tropical air masses constitute the troposphere of the equatorial and tropical latitudes. These air mass regions are labeled in Figure 4.17. The zone of contact of the polar and tropical air, called the *polar front,* is one of strong barometric pressure gradient. The steep decline in both pressure and temperature from south to north across the boundary zone is shown in cross section in Figure 4.18. As you already know, the pressure gradient force increases as the barometric gradient itself increases. As the force increases, so does the wind speed. Consequently, the zone of contact between cold and warm air is one of very high westerly wind speeds developed in a narrow zone. This intensive wind phenomenon is called the *jet stream.* (Its position is shown in Figure 4.18 by a circle labeled "Jet" and in the diagrams of Figure 4.17 by a line of heavy arrows.)

The jet stream is strongly developed in the Rossby waves, as shown in Diagrams *B* and *C*. Air speeds of 200 mi (320 km) per hour or higher are at times developed in the jet stream. In some respects, the jet stream is like a water jet that would result from a nozzle thrusting a narrow stream into a body of standing water. There is a central core of highest speed, surrounded by zones of decreasing speed. The jet stream that forms between polar and tropical air in the latitude range 40° to 60° is known as the *polar front jet stream.* Its core lies about on the tropopause (contact zone between troposphere and stratosphere), as shown in Figure 4.18.

In earlier paragraphs, we learned that there is a strong westerly wind developed at high levels about at 30° latitude, where poleward-moving air banks up to form a convergence zone at the poleward limit of the Hadley cell. Here, too, a jet stream is formed. It is known as the *subtropical jet stream* and is located in the altitude range of 43,000 to 46,000 ft (13 to 14 km), which is at tropopause level in these tropical latitudes. Figure 4.19 is a Northern Hemisphere map showing the approximate location of both jet streams. Other jet streams exist, but can only be mentioned here. One, the *tropical easterly jet stream,* occurs seasonally in very low latitudes and is an easterly wind; another, the *polar-night jet stream,* occurs high in the stratosphere over high latitudes. (The locations of these jet streams are shown in Figure 4.21.)

Getting back to the development of Rossby waves, we look for an explanation of the prevailing westerly winds. What drives them? What is the energy source? Is there another thermally driven cell, such as the Hadley cell, operating in middle latitudes? The answer does not lie in a thermally driven cell because the persistent west-to-east air movement occupies the entire troposphere in middle and high latitudes. In other words, there is no prevailing easterly flow to complete a cell of circulation, as there is in the Hadley cell.

The westerlies represent air moving over the earth's surface faster than the earth itself rotates on its axis. Consequently, friction between the westerlies and the surface withdraws angular momentum from the westerly winds and should bring them to a halt. Evidently, there is some

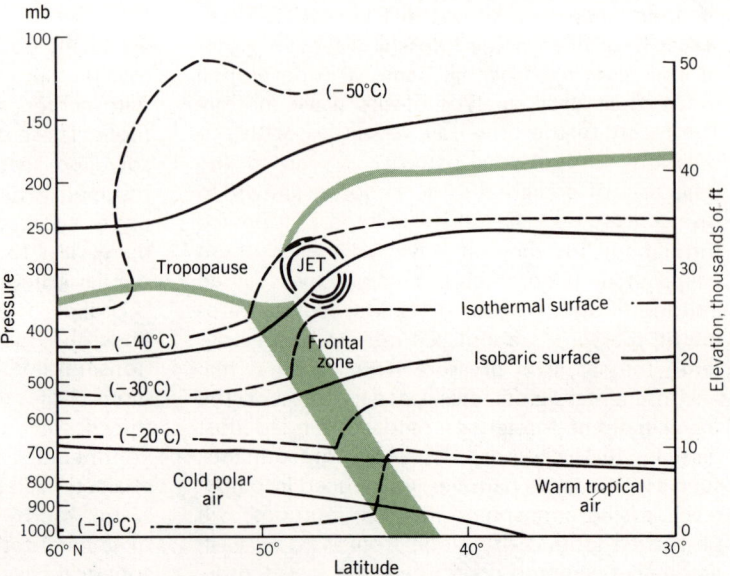

Figure 4.18 Idealized cross section through a frontal zone with jet stream core at the tropopause. (After E. R. Reiter, 1967, *Jet Streams,* Doubleday, Garden City, N.Y., p. 122, Figure 57.)

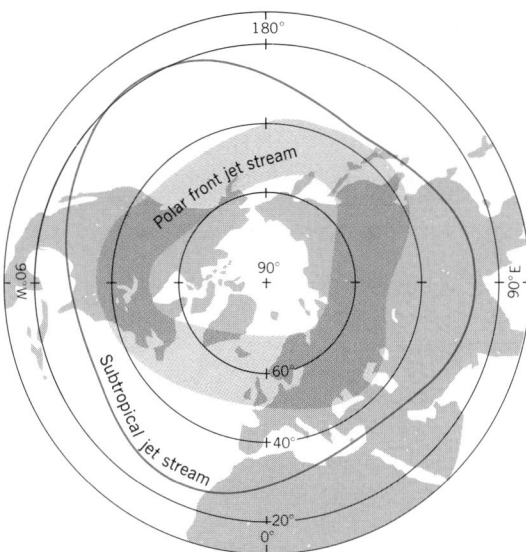

Figure 4.19 Average position of the subtropical jet stream axis in winter and belt of principal winter activity of the polar front jet stream. (After H. Riehl, 1962, *Jet Streams of the Atmosphere*, Colorado State University, Fort Collins, p. 2, Figure 1.2.)

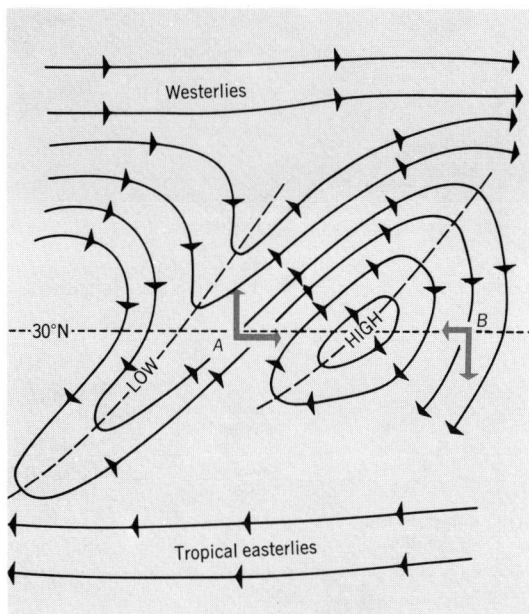

Figure 4.20 Angular momentum transfer by upper-air waves in the zone of westerly air flow. (From A. N. Strahler, 1971, *The Earth Sciences*, 2nd ed., Harper & Row, New York.)

mechanism that continually transports angular momentum poleward within the westerly wind belt in order to sustain these winds. The answer lies in the Rossby waves which develop in the polar front zone. Figure 4.20 is a schematic map of a Northern Hemisphere region centered about on the 30th parallel. An upper-air wave takes the form of a trough of low pressure. Note that the axis of the trough is slanted from southwest to northeast, as the dashed line shows. Adjoining the trough to the east is a cut-off high; its axis of elongation also slants southwest-to-northeast. At point A, air moving toward the northeast (streamline arrows) has two components of motion, one due north and one due east (broad arrows). At point B, the air moving toward the southwest also has two components of motion, one south and one west. However, because of the earth's eastward rotation, the westward motion at B is less than the eastward motion at A, so that the net motion of the two systems is eastward and represents a northward transport of angular momentum across the parallels of latitude. This *eddy transport mechanism*, as it is known, is the basis for the *wave theory* of cause of prevailing westerly winds. Most of the momentum transport occurs in the region of the 30th parallel, but the mechanism acts with decreasing effectiveness into high latitudes.

In the two polar zones, observations show a large frequency of easterly winds at the surface and in lower levels of the troposphere, and these are sometimes classed as a system of prevailing winds, the *polar easterlies*. In Antarctica particularly, where a strong center of high barometric pressure exists at low levels over the cold ice surface, prevailing easterly winds are perhaps a reality. If so, we are dealing with another thermally driven circulation system, in which denser cold air at low levels spreads equatorward and is replaced by air brought into the upper-air low (polar low).

Figure 4.21 is a cross-sectional diagram from pole to pole in which an attempt has been made to show in a schematic way the east-to-west and west-to-east components of the general atmospheric circulation. Obviously the meridional components of motion are omitted, and consequently the Hadley cell is not represented.

Heat Transport across Parallels of Latitude

A central theme of this and earlier chapters has been the earth's heat balance. We now have at our disposal some facts about the general atmospheric circulation; these can tell us how heat in sensible and latent forms is transported from the equatorial zone to polar zones. By means of the Hadley cell circulation heat can be transported to the subtropical zone, and from there it travels poleward in the Rossby waves and the cut-off

Circulation Systems in Atmosphere and Oceans

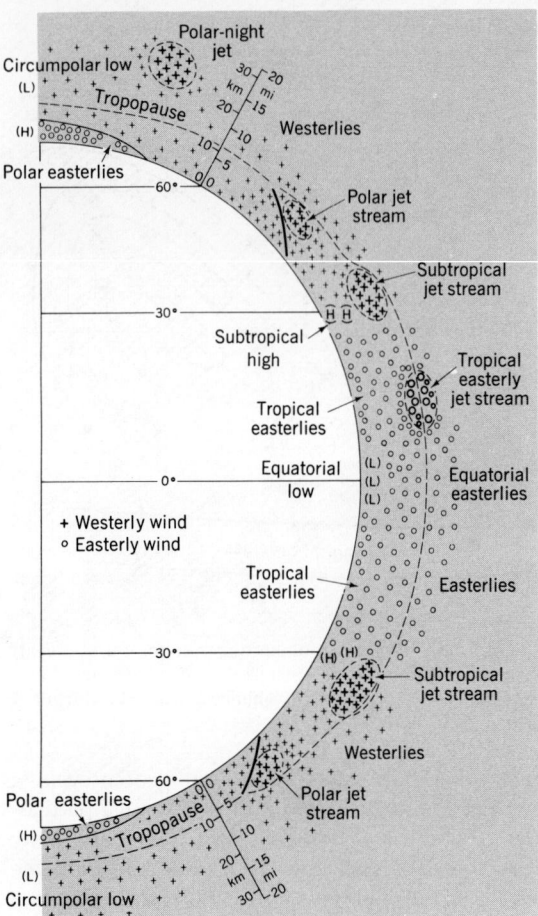

Figure 4.21 Idealized diagram of zonal wind directions and jet streams in a meridional cross section of the atmosphere. (From A. N. Strahler, 1971, *The Earth Sciences*, 2nd ed., Harper & Row, New York.)

lows and highs pictured in Figure 4.17. Warm tropical air moves poleward on the eastern sides of low-pressure systems; cold polar air moves equatorward along their western sides. This horizontal exchange of air masses and heat is described as *advection*. While gradual rising and sinking of air masses accompanies advection, the dominant motion is in the horizontal direction. (Dominantly vertical atmospheric motions are described as *convection*.)

It is a bit premature at this point to bring in an equation for the earth's total heat balance, because yet another mechanism of heat exchange exists; that of the ocean currents. Since most of the energy that goes to produce ocean currents comes from the frictional drag of winds blowing over the ocean surfaces, we will need to examine the oceanic circulation system. First, however, we shall want to get a general picture of the system of prevailing surface winds of the globe, since these are the direct driving agents that produce water movement.

Global Patterns of Barometric Pressure and Surface Winds

Prevailing surface winds are closely tied to prevailing barometric pressure patterns, which can be shown by isobaric maps (Figures 4.22 and 4.23). Isobars are drawn for the months of January and July because these are the months exhibiting the maximum and minimum air temperatures over large land and ocean areas of middle and high latitudes. Mean monthly pressures are compiled from long-term averages of daily readings. However, these readings must first be reduced to sea-level values for stations having appreciable elevations above sea level. If this adjustment were not made, the strong effect of altitude on pressure would completely mask the small horizontal pressure differences and gradients that control wind direction and strength. Isobars are shown on the world maps for intervals of 3 millibars (mb). The 1014-mb isobar represents approximately the standard sea-level pressure of 1013 mb. Pressures higher than this value are described as *high*, lesser pressures as *low*. (Figure 4.24 is a conversion scale for obtaining equivalent values in inches of mercury.) Isobars thus define the belts and centers of high and low pressure, but you should keep in mind that these data represent long-term averages only. In middle and high latitudes particularly, moving cyclones and anticyclones bring large day-to-day fluctuations in pressure, but these are not seen in the averages.

Occupying the equatorial zone is a belt of somewhat lower than normal pressure, between 1011 and 1008 mb, which is known as the *equatorial trough*. Lower pressure is made conspicuous by contrast with belts of higher pressure lying to the north and south and centered on about latitudes 30° N and S. These are the *subtropical belts of high pressure*. In the Southern Hemisphere this belt is clearly defined but contains centers of high pressure, termed *pressure cells*. In the Northern Hemisphere in summer the high-pressure belt is dominated by two oceanic cells, one over the eastern Pacific, the other over the eastern North Atlantic. Average pressures exceed 1026 mb in the centers of the cells. High pressure is explained by the upper-air zone of convergence on the poleward margin of the Hadley cell, while the equatorial trough is explained by upper-air divergence above the rising-air zone of the Hadley cell. These belts of high and low pressure persist far up into the atmosphere.

Global Patterns of Barometric Pressure and Surface Winds

Poleward of the subtropical high-pressure belts are broad belts of low pressure, extending roughly from the middle-latitude zone to the arctic zone but centered and intensified in the subarctic zone at about the 60th parallels of latitude. In the Southern Hemisphere, over the continuous expanse of southern ocean the *subantarctic low-pressure belt* is especially well defined with average pressures as low as 984 mb. The polar zones have permanent centers of high pressure, the *polar highs*, better illustrated by the South Polar zone where the high contrasts strongly with the encircling subantarctic low. Keep in mind that at upper levels there is a polar low above the surface high.

The pressure belts shift seasonally through several degrees of latitude, just as do the isotherm belts that accompany them.

The vast landmasses of North America and Asia, separated by the North Atlantic and North Pacific oceans, exert a powerful control over pressure conditions in the Northern Hemisphere. As a result, pressure centers replace the belted arrangement typical of the Southern Hemisphere.

Continents develop high-pressure centers when winter temperatures fall far below those of adjacent oceans. In summer, continents develop low-pressure centers, since this is the season when land-surface temperatures rise sharply above temperatures of the adjoining oceans. Ocean areas show centers of pressure opposite to those on the lands, as seen in the January and July isobaric maps. In winter, pressure contrasts are greater, just as temperature contrasts are greater. Over north central Asia there develops the *Siberian high,* with pressure average exceeding 1035 mb. Over central North America is a clearly defined, but much less intense, ridge of high pressure, called the *Canadian high.* Over the North Pacific and North Atlantic oceans are the *Aleutian low* and the *Icelandic low*, named after the localities over which they are centered. These two low-pressure areas are characterized by a high incidence of deep cyclonic storms, accounting for the low average pressure.

Figure 4.25 shows diagrammatically the pressure centers as they appear grouped around the North Pole. Highs and lows occupy opposite quadrants.

In summer, pressure conditions are exactly the opposite of winter conditions. Asia and North America develop lows, but the low in Asia is more intense. It is centered in southern Asia where it is fused with the equatorial low-pressure belt. Over the Atlantic and Pacific oceans are two well-developed cells of the subtropical belt of high pressure, shifted northward of their winter position and considerably expanded. These are termed the *Azores* (or *Bermuda*) *high* and the *Hawaiian high* respectively.

Prevailing surface winds during the months of January and July are shown by arrows on the pressure maps, Figures 4.22 and 4.23. A highly diagrammatic representation of the pressure and wind systems in Figure 4.26 shows the earth as if no land areas existed to modify the belted arrangement of pressure zones.

The equatorial trough of low pressure, lying roughly between 5° S and 5° N latitude, was long called the *equatorial belt of variable winds and calms,* or the *doldrums.* Centrally located on a belt of low pressure, this zone has no strong pressure gradients to induce a persistent flow of wind. In many parts of the equatorial zone there are no prevailing surface winds, but there is a fair distribution of wind directions around the compass. Calms prevail as much as a third of the time. Violent thunderstorms with strong squall winds are common.

North and south of the doldrums are the *trade wind* belts, covering roughly the zones lying between 5° and 30° N and S. The cause of these winds, which are a surface expression of the tropical easterlies, has already been covered in earlier paragraphs. In the Northern Hemisphere the prevailing wind is from the northeast and the winds are termed the *northeast trades;* in the Southern Hemisphere, they are the *southeast trades.* Trade winds are noted for their steadiness and directional persistence; most of the time the wind comes from one quarter of the compass.

The system of doldrums and trades shifts seasonally north and south, through several degrees of latitude, as do the pressure belts that are associated with them. Because of the large land areas of the Northern Hemisphere, there is a tendency for these belts to be shifted farther north in the Northern Hemisphere in summer (July) than they are shifted south in winter (January). The trades are best developed over the Pacific and Atlantic oceans, but are upset in the Indian Ocean region by the proximity of the great Asiatic landmass.

The trade winds provided a splendid avenue for westward travel in the days of sailing vessels. Steadiness of wind and generally clear weather made this a favorite zone of mariners. Crossing of the doldrums was hazardous because of the possibility of being becalmed for long periods

Figure 4.22 Following pages: Average barometric pressures (millibars, reduced to sea level) and surface winds. Wind arrows in polar regions largely inferred from isobars. (Compiled by John E. Oliver from data by Y. Mintz, G. Dean, R. Geiger, and J. Blüthagen. Cartography by John Tremblay.)

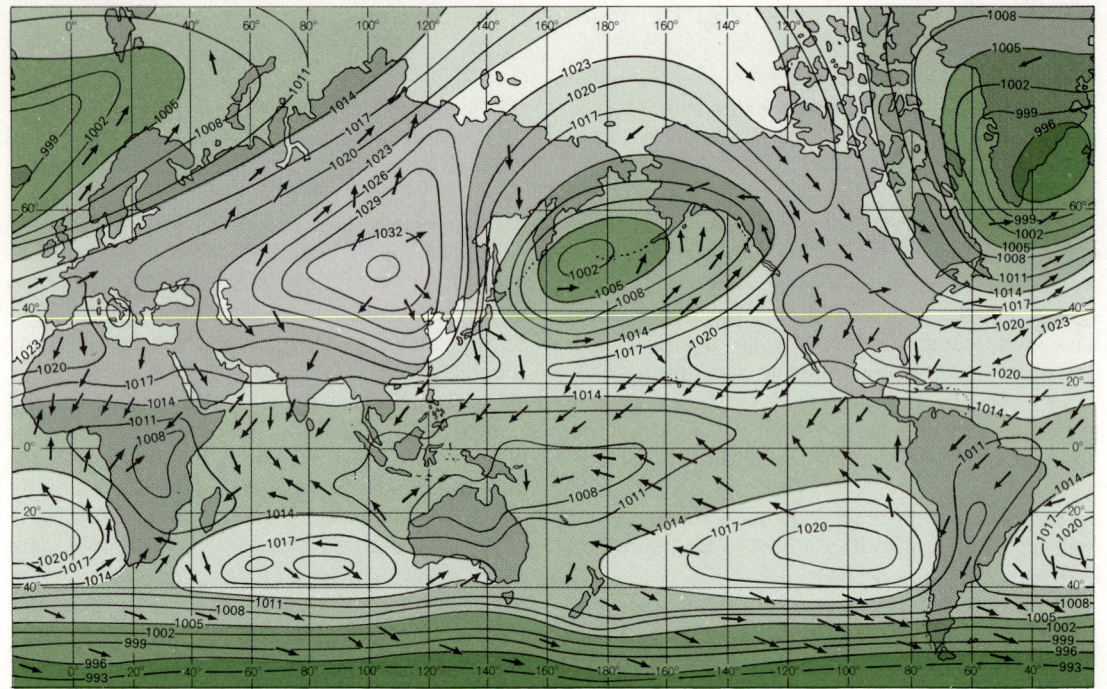

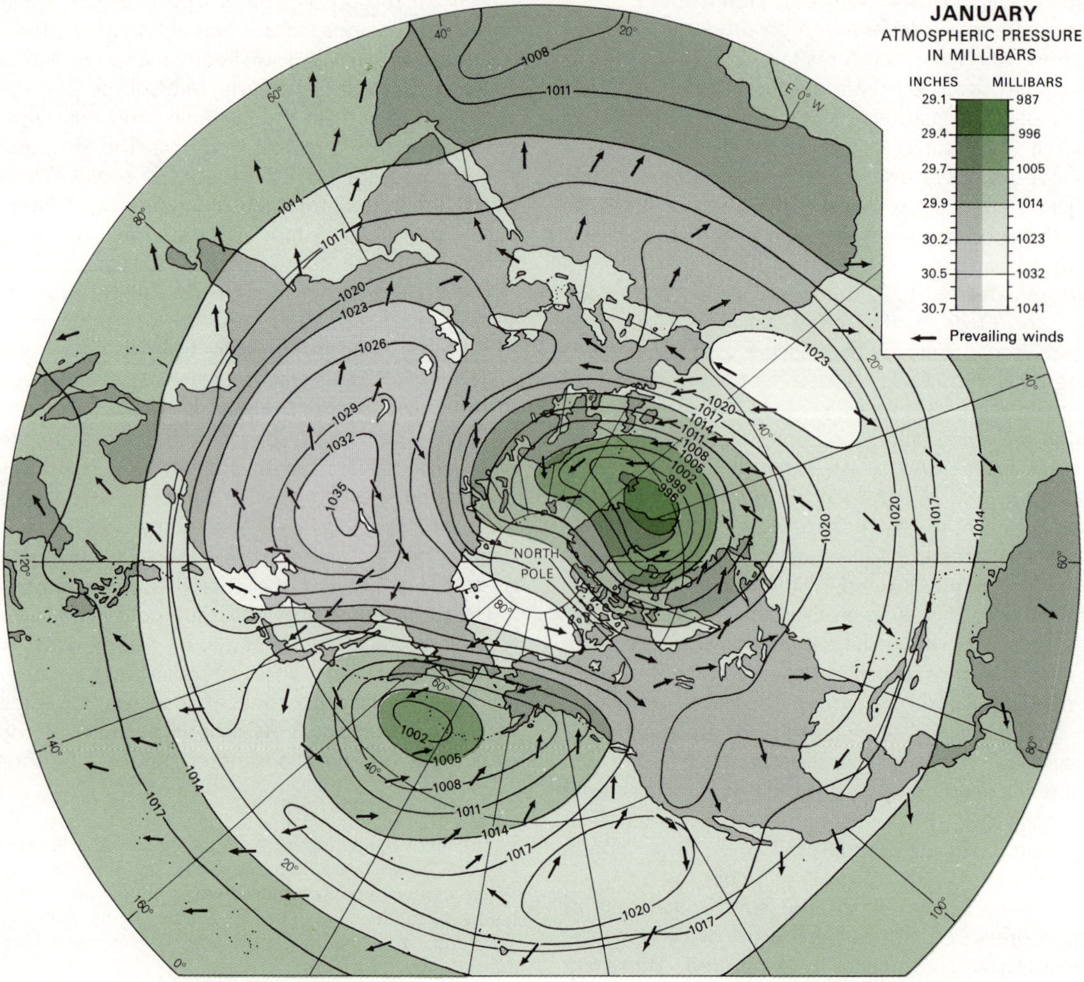

JANUARY
ATMOSPHERIC PRESSURE
IN MILLIBARS

INCHES	MILLIBARS
29.1	987
29.4	996
29.7	1005
29.9	1014
30.2	1023
30.5	1032
30.7	1041

← Prevailing winds

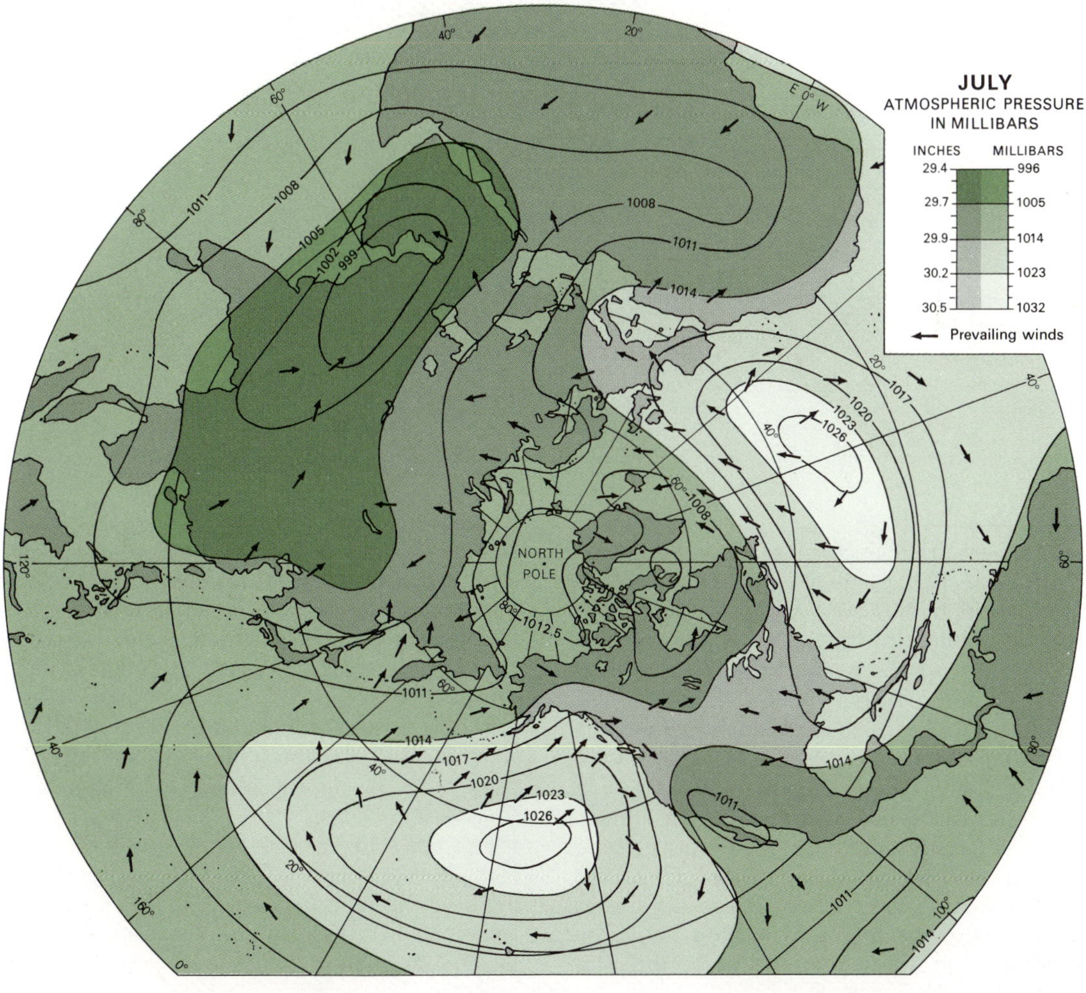

JULY
ATMOSPHERIC PRESSURE
IN MILLIBARS

INCHES	MILLIBARS
29.4	996
29.7	1005
29.9	1014
30.2	1023
30.5	1032

← Prevailing winds

Circulation Systems in Atmosphere and Oceans

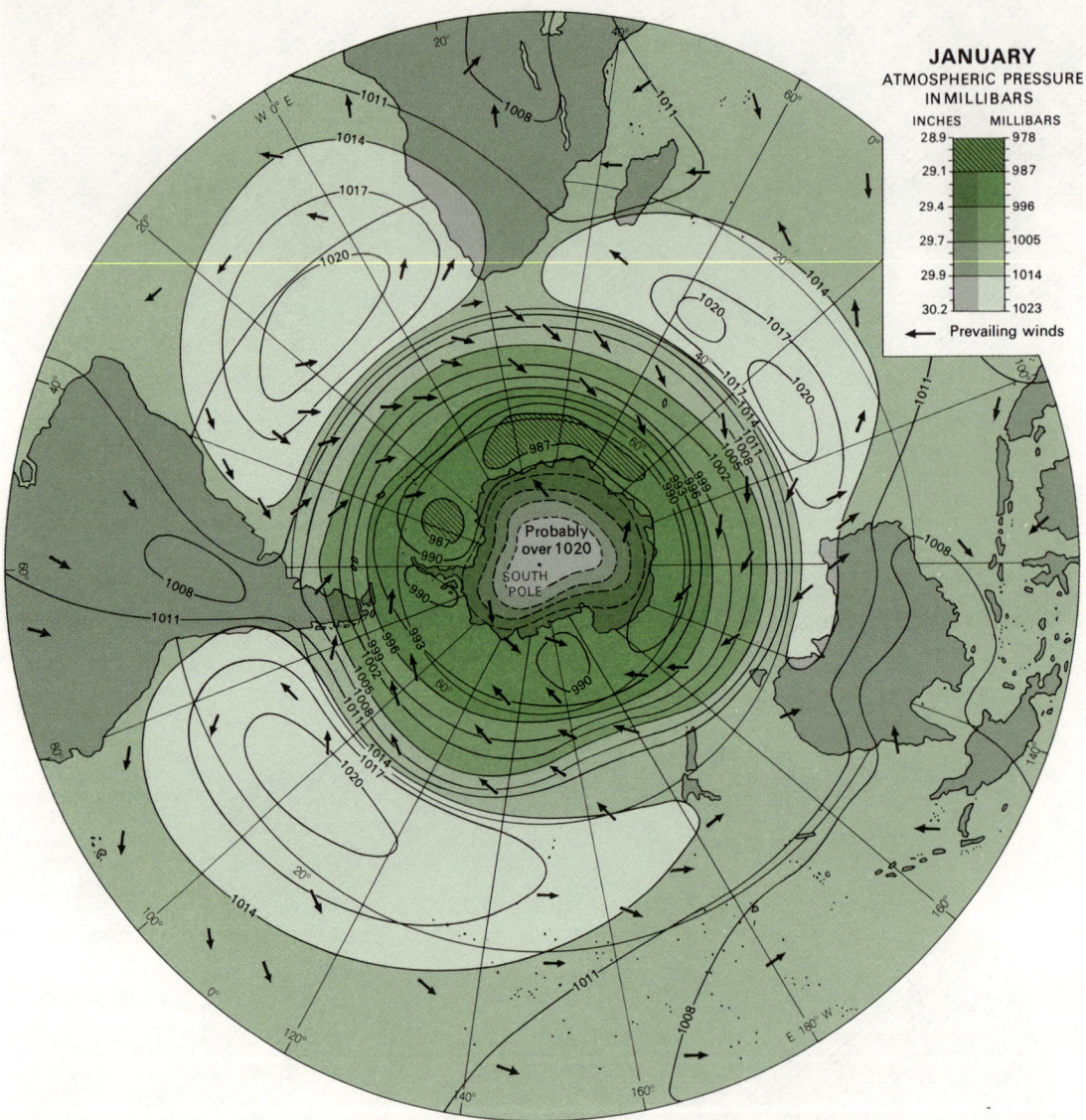

Figure 4.23 Average barometric pressures and surface winds, Southern Hemisphere. (Same data sources as Figure 4.22.)

and because of the uncertainty of wind direction. The trade wind belts are not altogether favorable for navigation, however, because over certain oceanic portions, at certain seasons of the year, terrible tropical storms known as hurricanes or typhoons occur (Chapter 5).

Between latitudes 30° and 40° are zones that have long been called the *subtropical belts of variable winds and calms*, or *horse latitudes*, coinciding with the subtropical high-pressure belts. As already noted, the high-pressure areas are concentrated into distinct anticyclonic cells, located over the oceans. There is also a latitudinal shifting following the sun's changing path in the sky from one solstice to the next. This shift amounts to less than 5° in the Southern Hemisphere, but it is about 8° for the strong Hawaiian high located in the northeastern Pacific.

Winds in the horse latitudes are distributed around a considerable range of compass directions. Calms prevail as much as a quarter of the time. The cells of high pressure have generally fair weather, with a strong tendency to dryness because the air is subsiding, as explained in Chapter 5.

Between latitudes 35° and 60°, both N and S, is the belt of the *prevailing westerly winds*. Moving from the subtropical anticyclones toward the subarctic lows, these surface winds are shown on Figure 4.26 to blow from a southwesterly quarter

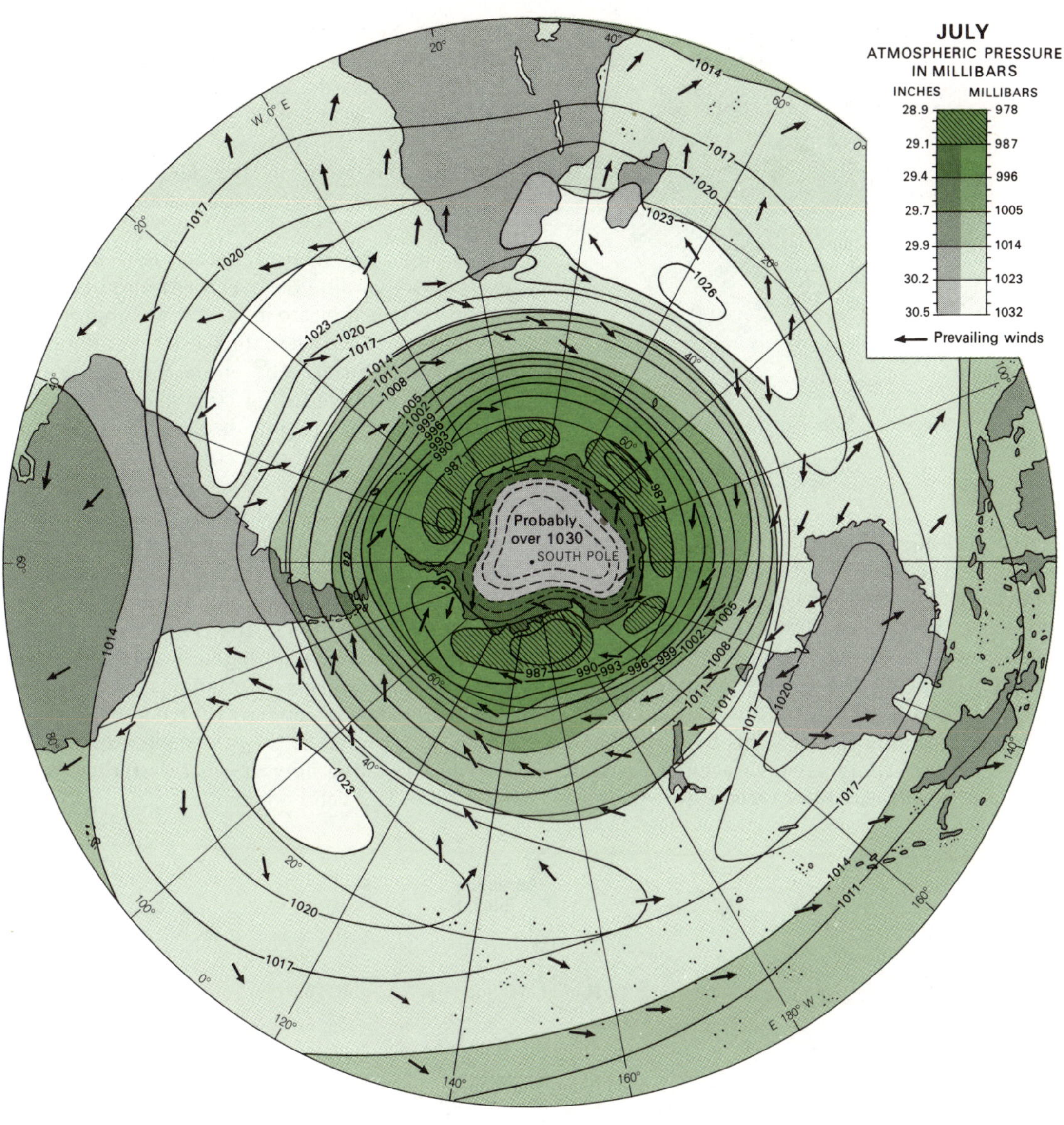

Global Patterns of Barometric Pressure and Surface Winds

in the Northern Hemisphere, from a northwesterly quarter in the Southern Hemisphere. This generalization is somewhat misleading, however, because winds from polar directions are frequent and strong especially in the winter, in moving cyclones and anticyclones. It is more accurate to say that within the westerly wind belt, winds can blow from any direction of the compass, but that the westerly components are definitely predominant.

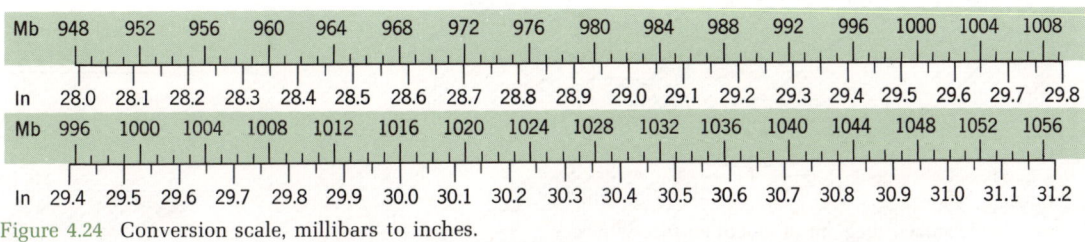
Figure 4.24 Conversion scale, millibars to inches.

Circulation Systems in Atmosphere and Oceans

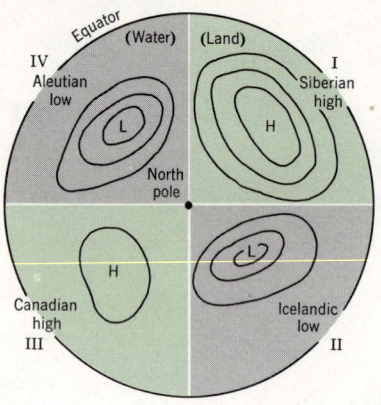

Figure 4.25 Northern Hemisphere pressure centers in winter.

In the Northern Hemisphere, landmasses cause a disruption of the westerly wind belt, but in the Southern Hemisphere, between the latitudes 40° and 60° S, there is an almost unbroken belt of ocean. Here the westerlies gain great strength and persistence, giving rise to the mariner's expressions, "the roaring forties," "the furious fifties," and "the screaming sixties." This belt was extensively used for sailing vessels traveling eastward from the South Atlantic Ocean to Australia, Tasmania, New Zealand, and the southern Pacific islands. From these places it was then easier to continue eastward around the world to return to European ports. Rounding Cape Horn was relatively easy on an eastward voyage, but in the opposite direction, in the face of prevailing stormy westerly winds, was fraught with great danger.

Although the westerly wind belts no longer exert a strong influence over the routes of modern ocean vessels, they are important in long-distance flying. As we have seen, the westerlies are persistent upward through the entire troposphere and locally have powerful jet streams. Flights in the easterly direction require less fuel and a shorter time. On westward flights, strong head winds may eat dangerously into the fuel supply of the plane and may necessitate reduced pay loads.

The *polar easterlies* described in earlier paragraphs, complete the system of surface winds. The concept is greatly oversimplified, if not actually erroneous, for winds in the polar regions take a variety of directions, as dictated by local weather disturbances. Perhaps in Antarctica, where an ice-capped landmass rests squarely upon the pole and is surrounded by a vast oceanic expanse, the outward spiraling flow of polar easterlies from a polar high is a valid concept.

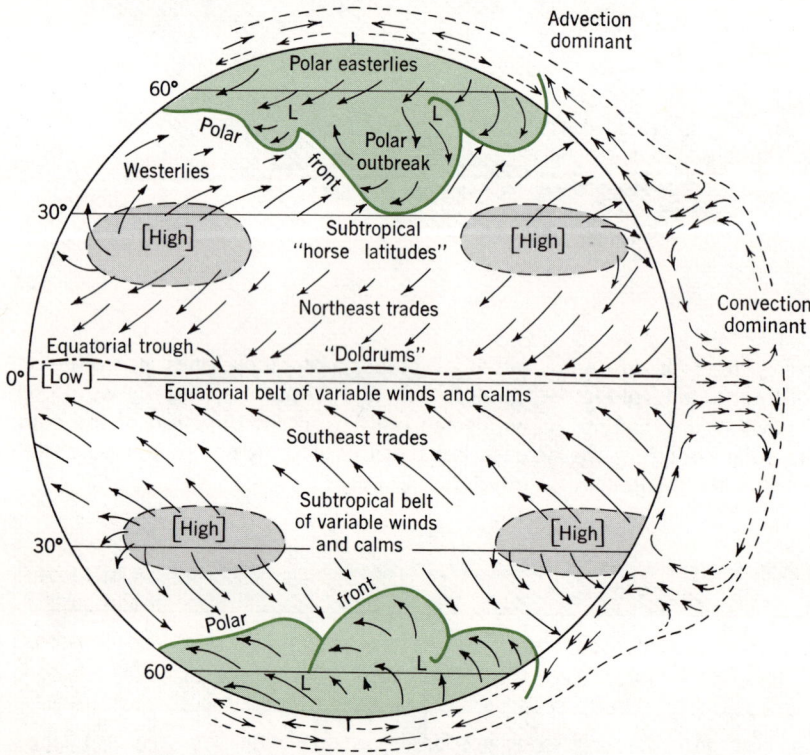

Figure 4.26 Idealized diagram of global surface winds.

104

Monsoon Winds Systems

The powerful seasonal control which Asia and North America exert upon air temperature and pressure requires that these continents must develop wind systems relatively independent of the belted system of earth winds so well illustrated in the Southern Hemisphere.

In summer, southern Asia develops a cyclone into which there is a flow of surface winds. This cyclone is a *heat low*, or *thermal low*, limited to the lower levels of the atmosphere. It shows up nicely on the July pressure maps (Figure 4.22.) From the Indian Ocean and the southwestern Pacific, warm, humid air moves northward and northwestward into Asia, passing over India, Indochina, and China. This air flow constitutes the *summer monsoon*, and is accompanied by heavy rainfall in southeastern Asia.

In winter, north central Asia is dominated by a strong anticyclone, from which there is an outward flow of air, reversing that of the summer monsoon. This circulation shows up well on the January pressure maps (Figure 4.22). Blowing southward and southeastward toward the equatorial oceans, this *winter monsoon* brings dry, clear weather for a period of several months.

North America, being smaller in extent, does not have great seasonal extremes of barometric pressure found in Asia, but there is nevertheless a distinct alternation of pressures and dominant wind directions between winter and summer. Wind records show that in summer there is a prevailing tendency for air originating in the Gulf of Mexico to move northward across the central and eastern part of the United States, whereas in winter there is a prevailing tendency for air to move southward from sources in Canada (Figure 4.22). Strong and persistent westerlies occur in winter over eastern and northern Canada, comparable with winds of the Asiatic winter monsoon. Northern Australia, too, shows a monsoon effect, but being south of the equator it reverses the seasonal timing of Asia.

Prevailing winds are an important environmental factor for the organic world, particularly affecting the land plants. The influence is closely tied in with regimes of precipitation (or lack of precipitation) since the winds represent large-scale movements of air masses that transport water vapor.

Local Winds

A class of local winds known as *drainage winds*, or *katabatic winds*, results from the flow of cold air under the influence of gravity from higher to lower regions. Such cold, dense air may accumulate in winter over a high plateau or high interior valley. When general weather conditions are favorable some of this cold air spills over low divides or through passes to flow out upon adjacent lowlands as a strong, cold wind. On the ice sheets of Greenland and Antarctica, powerful katabatic winds move down the gradient of the ice surface and are funneled through coastal valleys to produce powerful blizzards lasting for days at a time.

Another localized wind type is found in certain highland regions of the world where cold air flows from inland centers of high pressure through gaps in mountain barriers. The *bora* of the northern Adriatic coast and the *mistral* of southern France are well-known examples. In southern California there issues on occasion from the Santa Ana Valley a strong, dry, east wind, the *Santa Ana*, which blows across the coastal lowland. This air is of desert origin and may carry much dust and silt in suspension.

Still other types of local winds, bearing such names as *foehn* and *chinook*, result when strong regional winds passing over a mountain range are forced to descend on the lee side with the result that the air is heated and dried. These winds are explained in Chapter 5.

Wind and Waves

The earth's surface winds, blowing over ocean surfaces, transfer large amounts of momentum from the atmosphere to the oceans. While the major atmospheric circulation systems, such as the Hadley cell and the westerlies, are the primary systems of energy and mass transport powered directly by solar energy, the induced water circulation is largely secondary, forming circuits that draw from and return momentum to the primary circuit.

Two forms of kinetic energy of water motion are induced by winds: wave motion and the down-wind drift of a surface water layer. Air moving over a free water surface sets up a frictional drag, or *skin drag*, tending to pull the water in the same direction as the moving air. Because the air is in turbulent motion, with innumerable small eddies continually forming and dissolving, the drag upon the water surface is not uniformly distributed. Instead it is intensified at some points and reduced at other points. This unequal action of stress sets up ripples which at first are held to small size by the capillary skin of the water. With more intense wind stress, these ripples combine into water waves, which travel in conformity to certain physical laws. Once formed, water waves can be strengthened by the input of energy from direct pressure of wind on the wave surface.

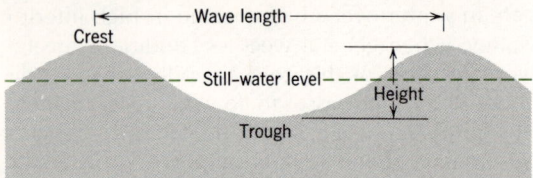

Figure 4.27 Terminology of water waves.

Wind-generated ocean waves belong to a type known as *progressive oscillatory waves*, because the wave form travels through the water and causes an oscillatory water motion. A simple terminology applied to waves is illustrated in Figure 4.27. *Wave height* is the vertical distance between *trough* and *crest*. *Wave length* is the horizontal distance from trough to trough, or crest to crest. The speed at which the wave advances through the water is the *wave velocity* (feet or meters per second). The time elapsed between successive passages of wave crests past a fixed point is the *period* (seconds).

In the progressive oscillatory wave, a tiny particle, such as a drop of water or a small floating object, completes one vertical circle, or *orbit*, with the passage of each wave length (Figure 4.28). Particles move forward on the wave crest, backward in the wave trough. At the sea surface, the orbit is of the same diameter as the wave height, but dies out rapidly with depth. In long, low waves, the water particles return to the same starting point at the completion of each orbit, hence there is no net motion in the direction of the wind.

Only wave energy and wave form are transmitted through the water. One-half of the total energy within the wave moves forward through the water at the speed of the wave itself. In steep, high waves, however, the orbits are not perfect circles. The particle moves just a bit faster forward when on the crest than when it returns in the trough, so that at the end of each circuit the particle has made a slight advance. The result is a very slow surface drift in the direction in which the waves are traveling, at a rate known as the *mass-transport velocity*. Under favorable conditions, the drift may reach a velocity as high as 2 mi (3.5 km) per hour and will tend to raise the water level along a coast against which the waves are breaking. (This motion is not the same as the drift set up by wind friction.)

Far from being simple parallel crests and troughs, ocean waves appear highly irregular in height and form because of interference among several wave trains normally present. These are not only of different periods, but travel in slightly different directions, so as to intersect at many points. Where two wave crests intersect, the wave height is increased, forming a peak. Where two troughs intersect, the depression is accentuated.

Two forms of waves are distinguished by the oceanographer: *wind waves*, which are being actively formed or maintained by the wind; and *swell*, consisting of wind waves that have left the region where they were formed and are gradually dying out in a region of calm or lesser winds. Wind waves grow through two mechanisms (Figure 4.29). First, the direct *push* of wind upon the windward slope of the wave drives it forward, just as with any floating object or sail. Second, the *skin drag* of air flowing over the water surface exerts a pull in the direction of wave motion. Over the wave crest, where drag is strongest, the orbital movement is supplemented, adding energy to the wave. In the trough, which is protected, drag is weaker, hence does not counteract the reverse orbital movement as forcefully as it is assisted on the crests. The result is a steady increase in wave height and length to some maximum point possible under a given wind strength. Surprisingly enough, wind waves commonly reach

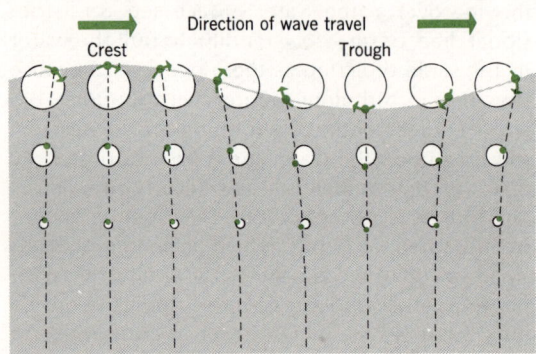

Figure 4.28 Orbital motion in deep-water waves of relatively low height.

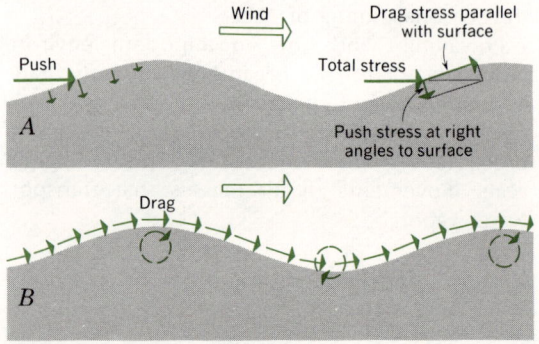

Figure 4.29 Wind waves developed by pressure and surface drag. (From A. N. Strahler, 1971, *The Earth Sciences*, 2nd ed., Harper & Row, New York.)

speeds much faster than the winds that produce and sustain them, a condition that would be impossible without the mechanism of skin drag.

Three factors control the maximum height to which wind waves can grow in deep water. First, wind velocity is obviously a major factor, since this determines the amount of energy that can be supplied. Second, duration of the wind determines whether or not the waves have an opportunity to grow to maximum size. Third, the *fetch*, or expanse of open water available, is important because the waves travel as they grow. If waves are developed in a very large body of water over a period of many hours, so that neither duration nor fetch are limiting factors, the maximum wave height varies as the square of the wind velocity.

As waves continue to grow, they not only increase their speed of travel, but become longer as well. When they have passed beyond the region of strong winds that formed them, waves are transformed into a swell, consisting of very long, low waves of simple form and parallel, even crests. As a rule-of-thumb, for each time that the swell travels a distance in nautical miles equivalent to its length in feet, the swell loses one-third of its height. Swells increase greatly in both length and period as they travel; those that have traveled 3000 to 4000 mi (5000 to 6500 km) may have periods of 15 to 20 seconds, whereas the storm waves at the point of origin may have had periods of 6 to 10 seconds.

Wind-generated waves and swell play an important environmental role in a number of ways and dimensions. First, ocean waves (and waves of lakes as well) represent a flow of kinetic energy that undergoes little loss through friction until the waves reach shallow water. Here the wave energy is transformed into the energy of breakers and currents which not only shape the coastlines, but also distribute sediment and nutrients for plant and animal life of shallow waters. As we found in Chapter 3, orbital motion of waves carries heat down to limited depths to produce a layer of warm surface water (the epilimnion of lakes). Oxygen and carbon dioxide are also carried downward from the surface by the wave-mixing process and are thus made available to animals and plants.

So far as Man is concerned, large storm waves represent an environmental hazard too well known to spell out here in detail. Even the great ocean liners and tankers of our modern day can suffer structural damage or even be broken in two by huge storm waves. The forward water motion associated with storm waves acts with the wind itself to impel ships toward a shore where they can be broken up by intense surf. The spill of oil from such accidents is a matter of great environmental concern. In Chapter 15 we shall refer to the effects of storm energy upon the land itself, causing rapid erosion and a loss of land to the sea.

The Causes of Ocean Currents

The frictional drag of wind over a water surface sets in motion a surface layer, which in turn exerts a drag upon the next layer beneath, and so forth, transmitting the motion to deeper water. The motion dies out rapidly with depth. However, the Coriolis effect acts to turn the flow of water to the right in the Northern Hemisphere. As a result, the direction of motion of the surface water is about 45° with respect to the wind direction. Since the Coriolis force acts upon each successive moving layer of water in turn, deeper layers move at progressively greater angles with respect to the surface layer. The average direction of the entire layer of motion is about 90° to the right of the wind direction (Northern Hemisphere).

Horizontal movement of water can also be induced by differences in sea-water density from place to place in the surface layers. A full explanation of this mechanism is a bit too elaborate to include here. It will be enough to note that where a zone of lower density water adjoins a zone of higher density water, there is set up a surface water slope from the lower-density to the higher-density region. This slope corresponds with a barometric gradient in the case of the atmosphere. Water tends to move down the slope, but is deflected by the Coriolis force until the flow is parallel with the slope, in a manner resembling the geostrophic wind. Because differences in water density can be caused by differences in water temperature, or in water salinity, or in both, the type of current thus produced is termed a *thermohaline current*. While thermohaline currents are locally important, they are secondary to winds as a cause of general oceanic circulation patterns.

Salinity of sea-surface water varies from one latitude belt to the next, depending upon the ratio of evaporation to precipitation, as shown by the meridional profiles in Figure 4.30. Intense evaporation in the dry subtropical high-pressure belts raises salinity to peak values in latitudes 20° to 30°, whereas heavy precipitation in the equatorial trough lowers salinity by dilution. Similarly, the surplus of precipitation in high latitudes reduces salinity, an effect particularly striking over the Southern Ocean at latitude 50° S. In constricted seas and gulfs in the dry subtropical zones, evaporation raises salinity to high values.

Circulation Systems in Atmosphere and Oceans

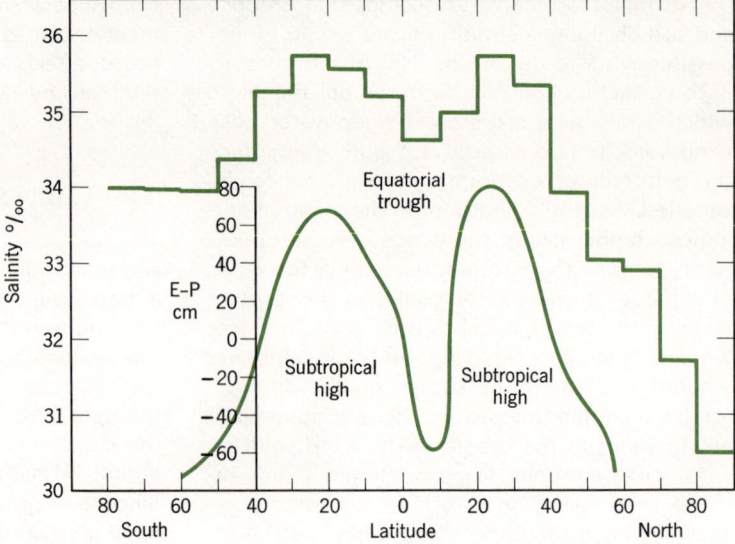

Figure 4.30 Average surface water salinity for 10-degree latitude zones and profile of the ratio of evaporation to precipitation. (Data of Georg Wüst, 1954, 1960.)

For example, the salinity of Mediterranean waters is 38‰, while that of the Red Sea is 40‰ or more. Thermohaline effects are important in those areas where water masses of strongly differing salinities and temperatures are in contact.

The Global Pattern of Ocean Currents

To illustrate the general pattern of surface water circulation imagine an idealized ocean extending across the equator to latitudes of 60° or 70° (Figure 4.31). Perhaps the most outstanding features are the circular movements, or *gyres*, around the subtropical highs, centered about 25° to 30° N and S. These features are prominent on a world map of ocean currents (Figure 4.32). An *equatorial current* marks the belt of the trades. Whereas the trades blow to the southwest and northwest, obliquely across the parallels of latitude, the water movement follows the parallels. This motion confirms the principle that ocean currents trend at an angle of about 45° with the prevailing surface winds, because of the Coriolis force.

A slow eastward movement of water over the zone of the westerly winds is named the *west-wind drift*. It covers a broad belt between 35° and 45° in the Northern Hemisphere, and between 30° or 35° and 70° in the Southern Hemisphere where open ocean occupies the higher latitudes.

The equatorial currents are separated by an *equatorial countercurrent*. This flow is well developed in the Pacific, Atlantic, and Indian oceans (Figure 4.32).

Along the west sides of the oceans in low latitudes the equatorial current turns poleward, forming a warm current paralleling the coast. Examples are the *Gulf Stream* (*Florida stream*), the *Japan current* (*Kuroshio*) and the *Brazil current*, which bring higher than average temperatures along these coasts.

The west-wind drift, upon approaching the east side of the ocean, is deflected both south and

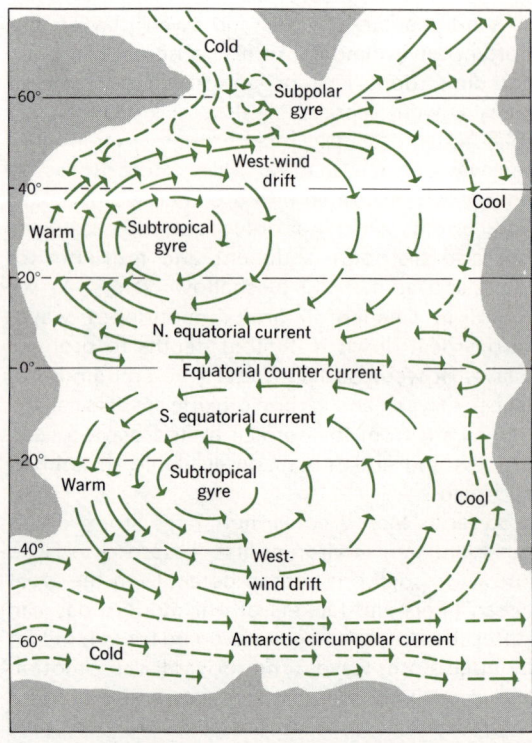

Figure 4.31 Schematic map of ocean currents and gyres.

108

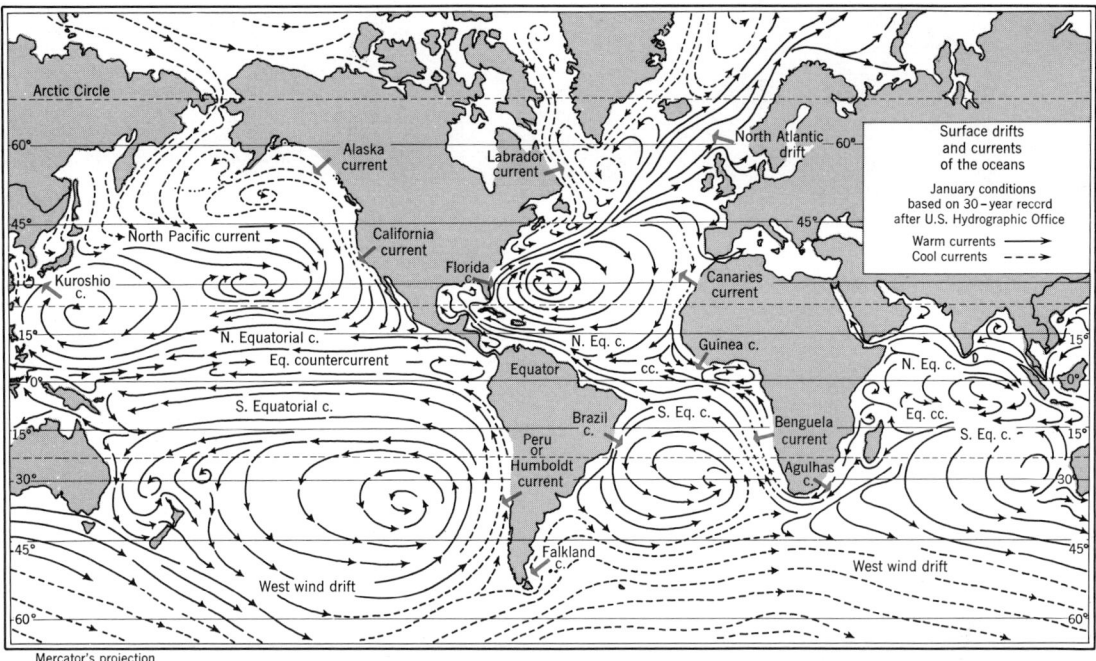

Figure 4.32 Surface drifts and currents of the oceans in January. (Data from U.S. Navy Oceanographic Office.)

north along the coast. The equatorward flow accompanied by upwelling is a cool current. It is well illustrated by the *Humboldt current* (*Peru current*) off the coast of Chile and Peru; by the *Benguela current* off the southwest African coast; by the *California current* off the west coast of the United States; and by the *Canaries current*, off the Spanish and North African coast. Note that this cold water causes a marked equatorward deflection of the isotherms, illustrated in Figure 3.27 by the northward bend in the 70° F (21° C) January isotherm along the east side of the South Pacific and South Atlantic oceans. Examine also the equatorward bend in the 60° F (16° C) isotherm off the California coast in July (Figure 3.28).

In the northeastern Atlantic Ocean, the west-wind drift is deflected poleward as a relatively warm current. This is the *North Atlantic current*, which spreads around the British Isles, into the North Sea, and along the Norwegian coast. The port of Murmansk, on the Arctic Circle, has year-round navigability by way of this coast. Note that in Figure 3.27 isotherms are deflected northward in a great bulge where they cross this current. In winter this effect is much more pronounced than in summer.

In the Northern Hemisphere, where the polar sea is largely landlocked, cold water flows equatorward along the west side of the large straits connecting the Arctic Ocean with the Atlantic basin. Three principal cold currents are the *Kamchatka current*, flowing southward along the Kamchatka Peninsula and Kurile Islands; the *Greenland current*, flowing south along the east Greenland coast through the Denmark Strait; and the *Labrador current*, moving south from the Baffin Bay area through Davis Strait to reach the coasts of Newfoundland, Nova Scotia, and New England.

In both the North Atlantic and North Pacific oceans the Icelandic and Aleutian lows coincide in a very rough way with two centers of counterclockwise circulation involving the cold arctic currents and the west-wind drifts.

The antarctic region has a relatively simple current scheme consisting of a single *Antarctic circumpolar current* moving clockwise around the antarctic continent in latitudes 50° to 65° S, where a continuous expanse of open ocean occurs.

Zones of Convergence and Upwelling

The patterns of surface-water circulation do not show the important vertical motions affecting the full depth of the ocean body, yet these motions are of vital environmental significance. The sink-

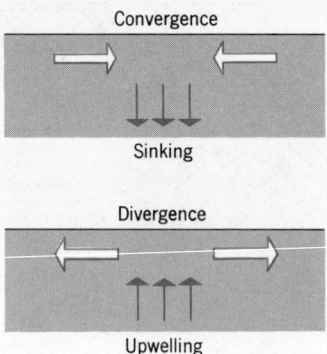

Figure 4.33 Convergence with sinking; divergence with upwelling.

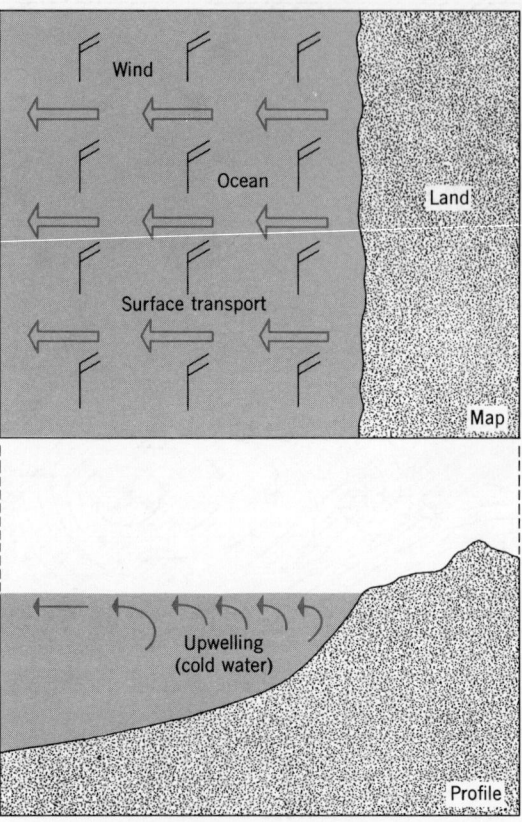

Figure 4.35 Winds paralleling a coast cause surface drift away from the shoreline and induce upwelling.

ing of denser water and the rising, or *upwelling*, of cold water from great depths are an integral part of the oceanic circulation. For lack of space we can't go into a full account of the structure and motions of the ocean throughout its depths.

As Figure 4.33 shows, the convergence of surface water must cause sinking, while the divergence of surface water must cause upwelling. Important zones of convergence and sinking occur in both arctic and antarctic waters, where dense water is formed by intense chilling and sinks to the deep ocean floor. The effect of an increase in salinity by evaporation in the subtropical high pressure belt is to induce a sinking of surface water, resulting in development of the *subtropical convergence*, illustrated in Figure 4.34. Sinking is a mechanism whereby dissolved oxygen can be carried to great depths.

Divergence and upwelling are of great importance to the biosphere because nutrients required by floating marine organisms (plankton) can be continually brought up from great depths. Upwelling of ocean waters off the west coasts of the continents can be explained by prevailing winds having an equatorward component, as illustrated in Figure 4.35. Surface water is dragged away from the coast, so that upwelling occurs and cold water reaches the surface. Upwelling occurs at tropical latitudes in conjunction with the cold equatorward currents already named; for example, the Peru (Humboldt) current, the Benguela current, and the Canaries current.

In the broad sense, the environmental significance of ocean currents lies in the transport of heat from low to high latitudes, an essential part of the earth's heat balance. Mild, poleward-moving air streams, such as prevail over the northeast Atlantic Ocean, drag with them warm surface waters, with the result that both air and water show exceptional warmth for such high latitudes. This effect largely accounts for surprisingly mild winters along western continental coasts as far north as 50° to 60° N latitude. The reverse flow accounts for surprisingly cool year-around temperatures of narrow west-coast belts in subtropical latitudes. With an understanding of the oceanic circulation system we are ready to analyze the total heat balance of the globe.

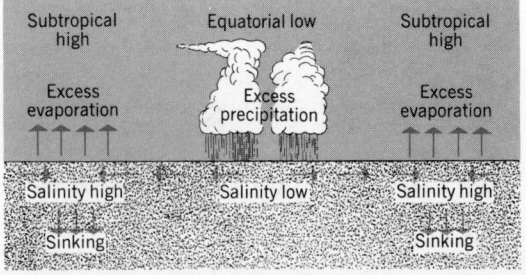

Figure 4.34 Sinking accompanying a high evaporation rate under the subtropical highs.

The Earth's Heat Balance

The heat balance equation developed in Chapter 3 for a unit column of soil or water was the following:

$$R = H + LE + G + F$$

where R is net all-wave radiation at the surface boundary,
H is sensible heat flux through the upper surface,
LE is latent heat flux due to evaporation,
G is net gain or loss of heat by the column,
and F is the horizontal transport of heat in or out of the column.

Let us now extend the unit column to include the total thickness of the atmosphere and oceans. The term F, which was crossed out of the equation for a column of soil, now becomes very important because of the horizontal motions of the air and water. Some other changes will be needed in the terms of the equation. Since there can be no sensible heat flux leaving the atmosphere as a whole (only by radiation can energy pass out into space), we substitute for H a term C, which will represent the net flux of heat horizontally in or out of the column by means of air motion. We will then restrict the definition of F to net flux of heat in or out of the column by horizontal movements of ocean water, e.g., by ocean currents. The term LE will need further modification. Evaporation, represented by E, will be expanded into $(E - r)$, where r is annual precipitation (rain or snow) and represents the mass of water returned to the liquid state from the vapor state. Now $(E - r)$ represents the net gain or loss of water by the unit column of atmosphere. When this term has a positive value, there is an excess of evaporation with a gain in latent heat. A negative value indicates excess precipitation and a loss of latent heat. Finally, we can dispense with the term G, since over long periods of time the unit column cannot gain or lose appreciable amounts of heat or it would not maintain a stable average yearly temperature.

The heat balance equation with terms redefined or eliminated becomes

$$R = C + L(E - r) + F$$

Terms on the right relate to the three meridional heat transport mechanisms: (1) sensible heat carried by air in motion, (2) latent heat carried as atmospheric water vapor, and (3) sensible heat carried by ocean currents.

Average annual values of each of the four terms of the heat balance equation are plotted graphically in Figure 4.36 for 10-degree belts of latitude from pole to pole, in units of kilolangleys per year. The unit column here consists of the entire latitude belt encircling the globe. The existence of a surplus or deficit for each belt is shown by the height of graph column above or below the line of zero value. Where the quantity is a surplus, there will be heat transport into the adjoining latitude belt of lower surplus value. Consequently, it is now possible to draw a second graph, shown in Figure 4.37 in which the actual meridional heat transport rate for the three transport terms, C, $L(E' - r)$, and F, can be plotted as smooth curves. The units used here are kilolangleys per year times 10 to the 19th power (10^{19}). A fourth curve gives the total meridional transport. As a device for graphic presentation, northward heat flow is shown by values increas-

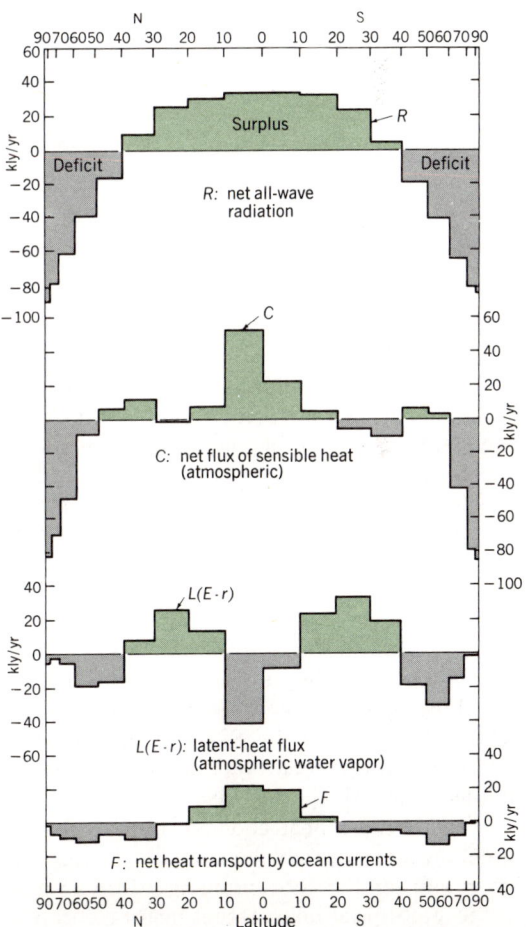

Figure 4.36 Components of the global heat-balance equation. Averages of 10-degree latitude belts are shown on a horizontal scale proportioned to surface area of each belt. Data from W. D. Sellers, 1965, *Physical Climatology*, Univ. of Chicago Press, Chicago, p. 115, Figure 34. (From A. N. Strahler, 1971, *The Earth Sciences*, 2nd ed., Harper & Row, New York.)

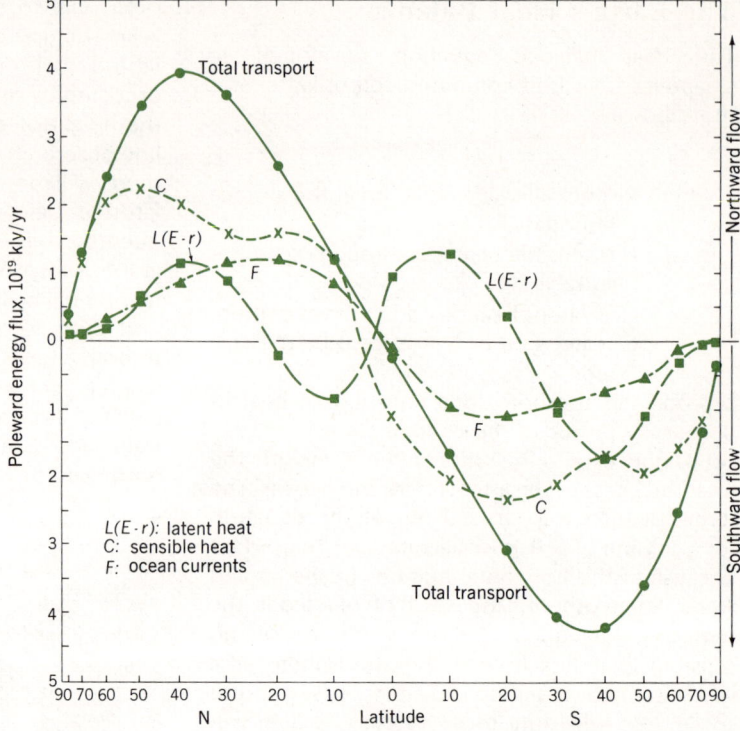

Figure 4.37 Global average heat transport across parallels of latitude. Data source same as in Figure 4.36. (From A. N. Strahler, 1971, *The Earth Sciences*, 2nd ed., Harper & Row, New York.)

ing upward from the zero line, southward flow by values increasing downward from that line.

Now for some interpretation of these graphs, look first at Figure 4.36. The net radiation graph (R) is the same as in Figure 2.14. Note the large energy surplus in middle and low latitudes and the deficit in regions poleward of latitude 40°. The graph for C (sensible heat flux in the atmosphere) has a strong peak over the equatorial belt. Obviously, a great surplus of heat is being generated here; it results both from absorption of radiation and from the release of sensible heat by condensation of water vapor. This zone has been aptly named the "firebox" of the earth's atmosphere. (Notice in the latent heat graph, $L(E − r)$, the corresponding negative values.) Sensible heat flux shows near-zero and negative values in two subtropical belts (20° to 30° latitude). An explanation lies in the great evaporation taking place in this zone, which is a belt of deserts, withdrawing sensible heat and transforming it into latent heat. For proof, look at the latent heat graph and see the corresponding positive values in the subtropical belts. Latent heat taken up by the atmosphere in the subtropical belts is carried equatorward in the Hadley cell circulation, as shown in the transport curve of $L(E − r)$ in Figure 4.37. The gain in heat by ocean water in low latitudes is shown by positive values of F in Figure 4.36; the loss of heat by negative values poleward of 20° N and 30° S. The ocean current transport curve, F, in Figure 4.37 shows that meridional transport by ocean currents is greatest in a broad zone from 10° to 40° latitude, for here the great gyres are most effective in heat exchange. Middle-latitude zones show a net gain of sensible heat of the atmosphere by the poleward movements of air in the westerlies, while the latent heat flux is negative because of heavy condensation occurring in storms, frequent in this latitude zone. In the two polar zones, heat arriving largely in the sensible form (curve C, Figure 4.37) is disposed of by longwave radiation into space.

We have finally been able to combine all of the forms of heat transport into a global heat balance. However, details of the processes of condensation remain to be completed in the next chapter. The heat balance describes only a system of energy flow, but it has obviously been dependent upon transport of matter as well. Consequently, we would need to evaluate the global exchanges of matter in the form of air and water to arrive at a global *mass balance*. The flow of the air itself is not particularly interesting, since it remains a gas. But the flow of water is worthy of close study, since it continually undergoes changes of state. Therefore, the earth's mass balance concerns itself specifically with a *water balance*, but this will not be complete until the gravity flow of water in the lands is explained in Part 3.

The planetary water balance depends upon large-scale condensation of water vapor, for it is by this process that the full circuit of water from ocean to atmosphere and back again to ocean is completed. The cycle is schematically shown in Figure 5.1. So far, we have introduced the process of energy transport as latent heat—the latent heat flux—but only briefly mentioned the process of condensation, whereby latent heat is transformed into sensible heat. Condensation occurs with high intensity in relatively small volumes of air—as compared with the slow evaporation of water into the troposphere over vast surfaces of oceans and lands. The effect of this concentration of condensation is to infuse the atmosphere locally with doses of sensible heat that may be converted into kinetic energy. These events are storms and related weather disturbances in which winds and vertical air motions powered by condensation may be very intense. In many respects this energy release is like a bonfire blazing intensely for a

5

Atmospheric Energy Releases

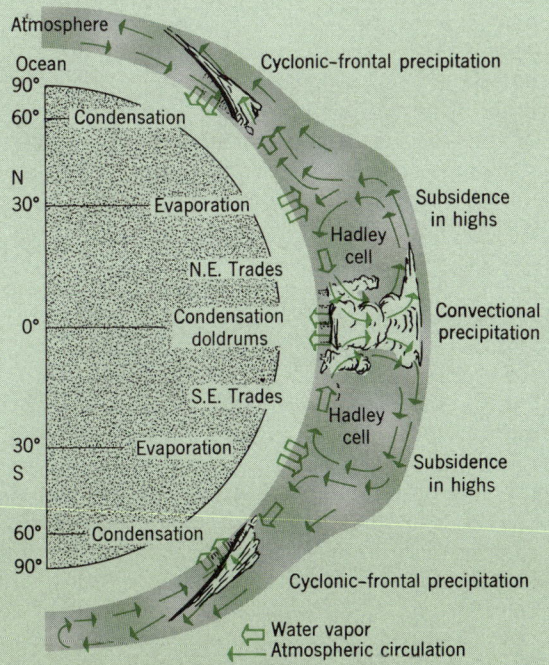

Figure 5.1 Schematic diagram of meridional water transport by circulation of the atmosphere and oceans. (From A. N. Strahler, "The Life Layer," *Jour. of Geography*, vol. 69, no. 2, p. 74, Figure 3. © 1970 by The Journal of Geography; reproduced by permission.)

short period of time until the fuel runs out. Collection of the wood for that fire may have taken a long time and the individual pieces of wood may have been gathered from a large area.

This chapter will be concerned with the causes and effects of condensation of water vapor. Condensation may lead to *precipitation* in the form of rain, snow, hail, or sleet. Not only does precipitation provide fresh water vital to organic processes at the land surfaces, but it may be accompanied by storm winds that act as a strong environmental stress and at times as a real hazard, bringing wholesale loss of life to animals as well as the physical destruction of larger plants.

In studying condensation we shall encounter yet another basic principle of physics of heat, namely, that the expansion of rising air is accompanied by cooling of that air, while contraction of sinking air into a smaller volume is accompanied by heating of the air. This principle is behind the *adiabatic process* and is vital to changes in state of water in the atmosphere. We shall begin, however, by describing the containment of water vapor in the air.

Relative Humidity and Vapor Pressure

The term *humidity* refers generally to the concentration of water vapor in the air. For any specified temperature there is a definite limit to the quantity of moisture that can be held by the air. This limit is known as *saturation*. The proportion of water vapor present relative to the maximum quantity is the *relative humidity,* expressed as a percentage. At the saturation point, relative humidity is 100%; when half of the total possible quantity of vapor is present, relative humidity is 50%, and so on.

A change in relative humidity of the atmosphere can be caused in one of two ways. If an exposed water surface is present, the humidity can be increased by evaporation. This is a slow process, requiring that the water vapor diffuse upward through the air or be carried upward in eddies. The other way is through a change of temperature. Even though no water vapor is added, a lowering of temperature results in a rise of relative humidity. This is automatic and is a logical consequence of the fact that the capacity of the air to hold water vapor has been lowered by cooling; thus the existing amount of vapor represents a higher percentage of the total capacity of the air. Similarly, a rise of air temperature results in decreased relative humidity, even though no water vapor has been taken away. The principle of relative humidity change caused by

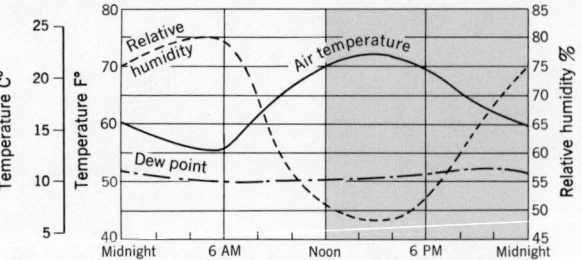

Figure 5.2 Daily cycle of relative humidity, temperature, and dew point for the month of May, Washington, D.C. Data of NOAA, National Weather Service. (From A. N. Strahler, 1971, *The Earth Sciences*, 2nd ed., Harper & Row, New York.)

temperature change is illustrated by a graph of these two properties throughout the day (Figure 5.2). As air temperature rises, relative humidity falls, and vice versa.

A simple example may be given to illustrate these principles. At a certain place the temperature of the air is 60° F (16° C), the relative humidity 50%. Should the air become warmed by the radiant energy from the sun and ground surface to 90° F (32° C), the relative humidity automatically drops to 20%, which is very dry air. Should the air become chilled during the night and its temperature fall to 40° F (5° C), the relative humidity will automatically rise to 100%, the saturation point. Any further cooling will cause condensation of the excess vapor into liquid form. As the air temperature continues to fall, the humidity remains at 100%, but condensation continues. This may take the form of minute droplets of dew or fog. If the temperature falls below freezing, condensation occurs as frost upon exposed surfaces, but continues to occur in suspended liquid droplets well below the freezing point.

The term *dew point* is applied to the critical temperature at which the air is fully saturated, and below which condensation normally occurs. An excellent illustration of condensation due to cooling is seen in summertime, when beads of moisture form on the outside surface of a pitcher filled with ice water. Air immediately adjacent to the cold glass or metal surface is sufficiently chilled to fall below the dew point temperature, causing moisture to condense on the surface of the glass.

In Chapter 1, we found that the weight of a column of atmosphere counterbalances the mercury column in a barometer and that variations in mercury height measure changes in air pressure (See Figure 1.5). When water vapor is added to otherwise pure, dry air, the water molecules diffuse perfectly among the other gas molecules. That part of the total barometric pressure that is due to the water vapor alone is termed the *vapor pressure*. For cold, dry air, the vapor pressure may be as low as 1.7 mb (0.05 in.; 0.013 cm); for very warm, moist air of the equatorial regions, it may be as high as 27 mb (0.80 in.; 2 cm).

Figure 5.3 shows the maximum possible vapor pressure for air of a range of temperatures from very cold to very warm.

Absolute and Specific Humidity

Although relative humidity is an important indicator of the state of water vapor in the air, it is a statement only of the relative quantity with respect to a saturation quantity. The actual quantity of moisture present is denoted by *absolute humidity*, defined as the weight of water vapor contained in a given volume of air. Weight is stated in grams, volume in cubic meters. For any specified air temperature, there is a maximum weight of water vapor that a cubic meter of air can hold (the saturation quantity). Figure 5.3 shows this maximum moisture content of air for a wide range of temperatures.

In a sense, the absolute humidity is the environmental scientist's yardstick of a basic natural resource—water—to be applied from equatorial to polar regions. It is a measure of the quantity of water that can be extraced from the atmosphere as precipitation. Cold air can supply only a small quantity of rain or snow; warm air is capable of supplying huge quantities.

One disadvantage of using absolute humidity in the study of atmospheric moisture is that when air rises or sinks in elevation, it undergoes corresponding volume changes of expansion or compression. Thus the absolute humidity cannot remain a constant figure for the same body of air. The meteorologist therefore makes use of another measure of moisture content, *specific humidity*, which is the ratio of mass of water vapor to mass of moist air (including the water vapor). This ratio is stated in units of grams of water vapor per kilogram of moist air. When a given parcel of air is lifted to higher elevations (without gain or loss of moisture) the specific humidity remains constant, despite volume increase.

Specific humidity is often used to describe the moisture characteristics of a large mass of air. For example, extremely cold, dry air over arctic regions in winter may have a specific humidity of as low as 0.2 grams per kilogram, whereas extremely warm moist air of tropical regions may hold as much as 18 grams per kilogram. The total

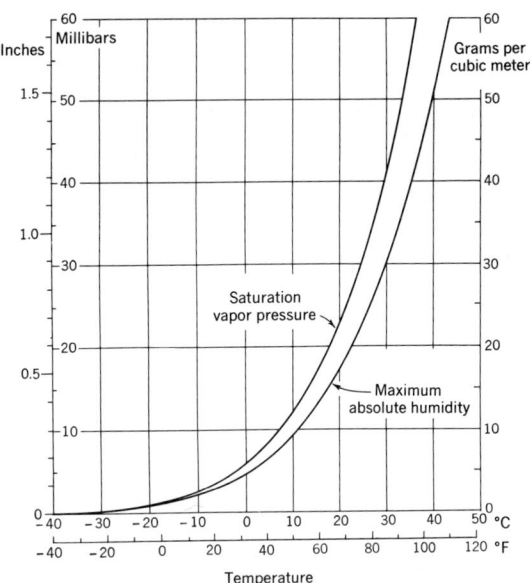

Figure 5.3 Maximum vapor pressure and absolute humidity. (From A. N. Strahler, 1971, *The Earth Sciences*, 2nd ed., Harper & Row, New York.)

natural range on a world-wide basis is such that the largest values of specific humidity are from 100 to 200 times as great as the least.

A measure related to specific humidity is the *mixing ratio*, defined as the ratio of density of water vapor to density of the dry air (not including the water vapor). Numerically, specific humidity and mixing ratio of a given air sample are very nearly the same.

Figure 5.4 is a graph of average relative humidity, vapor pressure, and mixing ratio plotted against latitude. It shows that the warm air of equatorial regions normally holds vastly more water vapor than cold air of arctic and polar regions. Notice the lower prevailing relative humidities over the subtropical belts of high pressure and the higher humidities over the equatorial trough and in middle to high latitudes.

Air Masses

The role of water vapor in atmospheric exchanges is better understood through the concept of the *air mass*, already touched upon in Chapter 4 in an explanation of Rossby waves and the jet stream. A single air mass is a substantial portion of the troposphere, generally covering an area of subcontinental size, that has taken on a distinctive set of properties—primarily a particular combination of temperature and humidity—that are more or less constant in all horizontal directions.

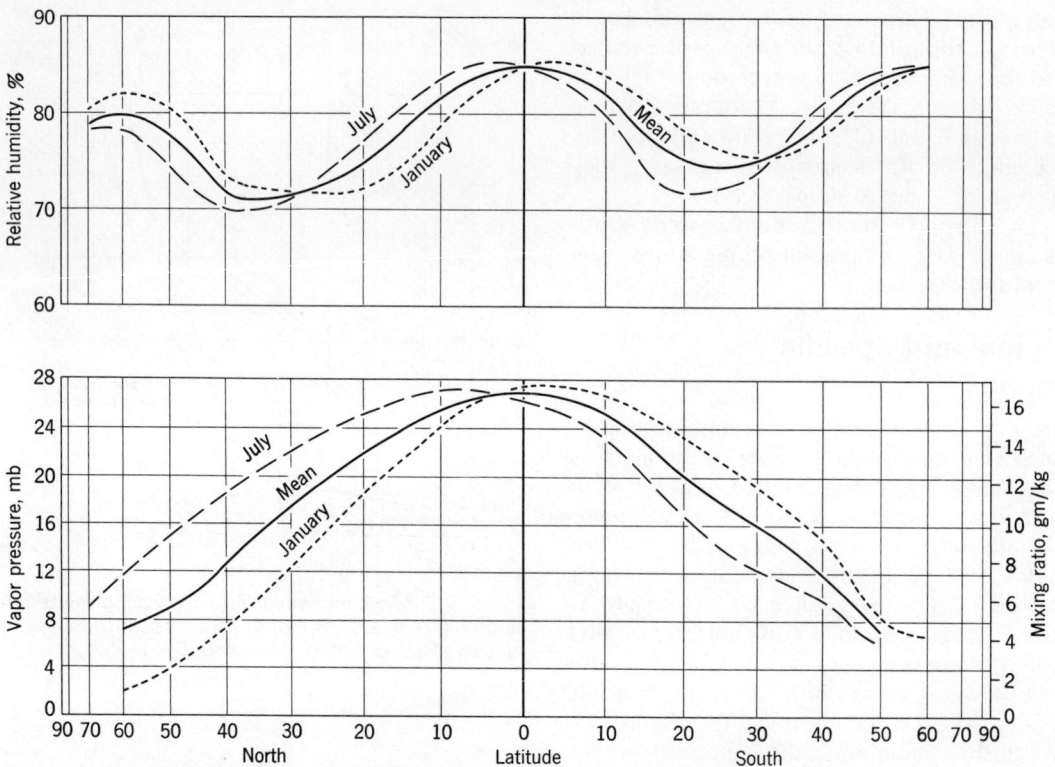

Figure 5.4 Meridional profiles of relative humidity (*above*) and of mixing ratio or vapor pressure (*below*). Values of mixing ratio are numerically similar to values of specific humidity. (After B. Haurwitz and J. M. Austin, 1944, *Climatology*, McGraw-Hill, New York, p. 88, Figure 17, and p. 90, Figure 18.)

Each air mass tends to have a characteristic environmental lapse rate of temperature.

A given air mass may have a rather sharply defined boundary between itself and a neighboring air mass. This discontinuity is termed a *front*. We found an example of a front in the contact between polar and tropical air masses below the axis of a jet stream, as shown in Figure 4.18. This feature was termed the *polar front*; it represents the highest degree of global generalization. Fronts may be nearly vertical, as in the case of air masses having little motion relative to one another; or they may be inclined at an angle not far from the horizontal, in cases where one air mass is sliding over another. A front may be almost stationary with respect to the earth's surface, but nevertheless the adjacent air masses may be in relatively rapid motion with respect to each other along the front, which then represents a *shear zone*.

The properties of an air mass are derived in part from the regions over which it passes. Because the entire troposphere is in more or less continuous motion, the particular air-mass properties at a given place reflect the composite influence of trajectories covering thousands of miles and passing alternately over oceans and continents. This complexity of influences is particularly important in middle and high latitudes in the Northern Hemisphere, within the flow of the global westerlies. However, there are vast tropical and equatorial areas over which air masses reflect quite simply the properties of the ocean and land surfaces over which they move slowly or tend to stagnate. Thus, over a warm equatorial ocean surface the lower levels of the overlying air mass may develop a high water vapor content combined with a steep environmental temperature lapse rate. Over a large tropical desert, slowly subsiding air forms an air mass of high temperatures and low relative humidities. Over cold, snow-covered land surfaces in arctic latitudes in winter, the lower layer of the air mass remains very cold with a very low water-vapor content. Meteorologists have designated as *source regions* those land or ocean surfaces that strongly impress their temperature and moisture characteristics upon overlying air masses moving over them.

Air masses move from one region to another following the patterns of barometric pressure. During such migration, lower levels of the air mass undergo gradual modification, taking up or losing heat to the surface beneath, and perhaps taking up or losing water vapor as well.

TABLE 5.1 Classification of Air Masses

Air Mass	Symbol	Source Region	Temperature
Arctic	A	Arctic ocean and fringing lands	Very cold (winter)
Antarctic	AA	Antarctica	Very cold (winter)
Polar	P	Continents and oceans 50° to 60° N and S	Cold (winter) Cool (summer)
Tropical	T	Continents and oceans 20° to 35° N and S	Warm to very hot
Equatorial	E	Oceans close to equator	Uniformly warm
			Moisture Content
Maritime	m	Oceans	Large, especially if warm
Continental	c	Continents	Small, especially if cold

Air masses are classified according to two categories of generalized source regions: (a) latitudinal position on the globe, which primarily determines thermal properties, and (b) underlying surface, continents or oceans, determining the moisture content. With respect to latitudinal position, five types of air masses are *arctic, antarctic, polar, tropical,* and *equatorial* (Table 5.1).

With respect to type of underlying surface, two further subdivisions are imposed on the types of air masses: *maritime* and *continental* (Table 5.1). By combining types based on latitudinal position with those based on underlying surface a list of six important air masses results; these are listed in Table 5.2. Figure 5.5 shows global distribution of source regions of these air masses. Table 5.2 gives typical values of temperature and specific humidity at the surface, although a wide range in these parameters may be expected, depending upon season. The equatorial air mass, *mE*, holds about 200 times as much water vapor as the extremely cold arctic and antarctic air masses, *cA* and *cAA*. The maritime tropical air mass, *mT*, and maritime equatorial air mass, *mE*, are quite similar in temperature and water vapor content. Both are capable of very heavy yields of precipitation. The continental tropical air mass, *cT*, has its source region over subtropical deserts of the continents. Although typically it has a substantial water vapor content, it has low relative humidity when highly heated during the daytime. The polar maritime air mass, *mP*, with specific humidity of 4 to 6 g/kg

TABLE 5.2 Properties of Typical Air Masses

Air Mass	Symbol	Properties	Temperature °F	°C	Specific Humidity (g/kg)
Continental arctic (and continental antarctic)	cA (cAA)	Very cold, very dry (winter)	−50°	(−46°)	0.1
Continental polar	cP	Cold, dry (winter)	12°	(−11°)	1.4
Maritime polar	mP	Cool, moist (winter)	39°	(4°)	4.4
Continental tropical	cT	Warm, dry	75°	(24°)	11
Maritime tropical	mT	Warm, moist	75°	(24°)	17
Maritime equatorial	mE	Warm, very moist	80°	(27°)	19

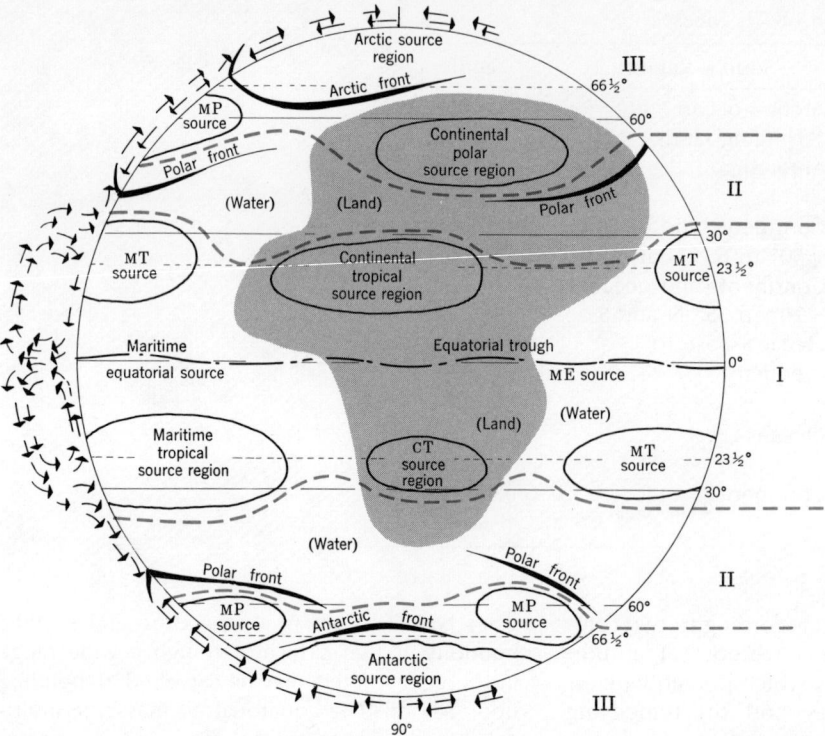

Figure 5.5 Global diagram of air-mass source regions and frontal zones. (After S. Petterssen and others.)

in winter, originates over middle-latitude oceans. Although this quantity of water vapor is not large compared with the tropical air masses, it represents a high relative humidity and can yield substantial precipitation.

Figure 5.5 shows by bold black lines the typical location of the polar front, a zone of interaction between polar and tropical air masses, and an *arctic front*, a zone of interaction between arctic and maritime polar air masses. These frontal zones shift widely depending upon season, and are typically the sites of intense cyclonic storms, described in later paragraphs of this chapter.

Dashed lines in color in Figure 5.5 show the globe divided into three major air-mass regions: *Region I* is a low-latitude region dominated by equatorial and tropical air masses. Polar air masses rarely penetrate into this domain. *Region II* is a middle-latitude region alternately occupied by polar and tropical air masses which interact vigorously. *Region III* is a high-latitude region dominated by polar and arctic air masses. Tropical air masses rarely penetrate into this domain.

Climate, which is the characteristic state of the atmosphere as it generalized through the annual radiation cycle at a given location, reflects the properties of the air masses prevalent from day to day and season to season. If you have a good understanding of the characteristics of air masses, their typical trajectories of motion, their interactions, and the changes that they undergo during migration, you also have at the same time a real understanding of the atmospheric environment of the life layer. This knowledge will allow you to anticipate quite well the climate prevailing at any given place and season on the globe. Anticipation based upon deduction and reason is the thinking person's substitute for an encyclopedic memory for statistics.

Condensation and the Adiabatic Process

Falling rain, snow, sleet, or hail, referred to collectively as *precipitation*, can result only where large masses of air are experiencing a steady drop in temperature below the dew point. This condition cannot be brought about by the simple process of chilling of the air through loss of heat by conduction and by longwave radiation during the night. Instead, it is nesessary that the large mass of air be rising to higher elevations. This statement requires understanding of another principle of physics.

One of the most important laws of meteorology is that rising air experiences a drop in temperature, even though no heat energy is lost to the outside (Figure 5.6). The drop of temperature is a result of the decrease in air pressure at higher elevations, permitting the rising air to expand. Individual molecules of the gas are more widely diffused and do not strike one another so frequently so that the gas has a lower sensible temperature. Temperature change due to volume change only, and without import or export of heat, is described as *adiabatic* change.

When no condensation is occurring, the rate of drop of temperature, termed the *dry adiabatic rate,* is about $5\frac{1}{2}$ F° per 1000 feet of vertical rise of air. In metric units the rate is 1 C° per 100 meters. The dew point also declines with rise of air; the rate is 1 F° per 1000 ft (0.2 C° per 100 m) as shown by a dashed line in Figure 5.6.

When water vapor in the air is condensing, the adiabatic rate is less, on the order of 2 to 3 F° per 1000 ft (0.35 to 0.5 C° per 100 m), owing to the partial counteraction of temperature loss through the liberation of latent heat during the condensation process. This modified rate is referred to as the *wet adiabatic*, or *saturation adiabatic rate* (Figure 5.6). Adiabatic cooling rate should not be confused with the normal environmental lapse rate, explained in Chapter 1. The normal lapse rate applies only to still air whose temperature is measured at successively higher levels.

Clouds and Fog

Precipitation comes from clouds, which are the first evidence of condensation occurring on a large scale within an air mass. Clouds consist of extremely tiny droplets of water, 2 to 40 microns (0.002 to 0.04 mm) in diameter, or minute crystals of ice. These are sustained by the slightest upward movements of air. In order for cloud droplets to form, it is necessary that microscopic dust particles serve as centers, or *nuclei,* of condensation. Dusts with a high affinity for water are abundant throughout the atmosphere.

Where the air temperature is well below freezing, clouds may form of tiny crystals. Water in such minute quantities can remain liquid far below normal freezing temperatures; it is said to be *supercooled.* Thus, water droplets exist at temperatures down to 10° F (−12° C); a mixture of water droplets and ice crystals from 10° to −20° F (−12° to −30° C) or even lower; and predominantly ice crystals below −20° F (−30° C). Supercooled water droplets can be turned rapidly into ice particles by the natural introduction of ice particles falling from higher, colder clouds. This process is known as *seeding.* Below −40° F (−40° C) all of the cloud is ice.

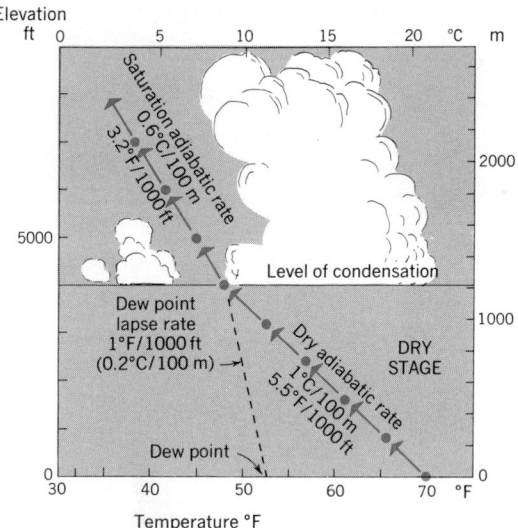

Figure 5.6 Adiabatic changes of temperature in a rising air mass. (From A. N. Strahler, 1971, *The Earth Sciences*, 2nd ed., Harper & Row, New York.)

Clouds appear white when thin or when the sun is shining upon the outer surface. When dense and thick, clouds appear gray or black underneath simply because this is the shaded side.

Cloud types may be classified on the basis of two characteristics: general form and altitude (Figure 5.7). On the basis of form there are two major groups: *stratiform* or layered types, and *cumuliform* or massive, globular types. High clouds, which are *cirrus* and related types, are found in the upper troposphere at the 20,000- to 40,000-foot level (6 to 12 km) (Figure 5.8A). These thin, wispy clouds are composed of ice particles. The banded and wispy types are often associated with the jet stream located near the tropopause. Of the middle-cloud group the blanket-like *altostratus* is important in indicating the presence of a relatively warm, moist air layer sliding over a denser, drier layer beneath. Of the low clouds, the dark, dense *nimbostratus* is a producer of rain or snow and indicates the steady forced ascent of a moist air mass. Figure 5.7 shows schematically the various cloud types in the three height zones, together with their names.

Cumuliform clouds, having small horizontal extent but often a large vertical extent, indicate the rapid spontaneous rise of air in the process of atmospheric convection. Figure 5.9 shows how a small *cumulus* cloud forms within a rising bubble of warm air. When adiabatic cooling reaches the dew point, condensation sets in and the cloud becomes visible. Thus cumuliform clouds typically have flat bases at more or less the same altitude (Figure 5.8B). Rapid vertical ascent of a

Atmospheric Energy Releases

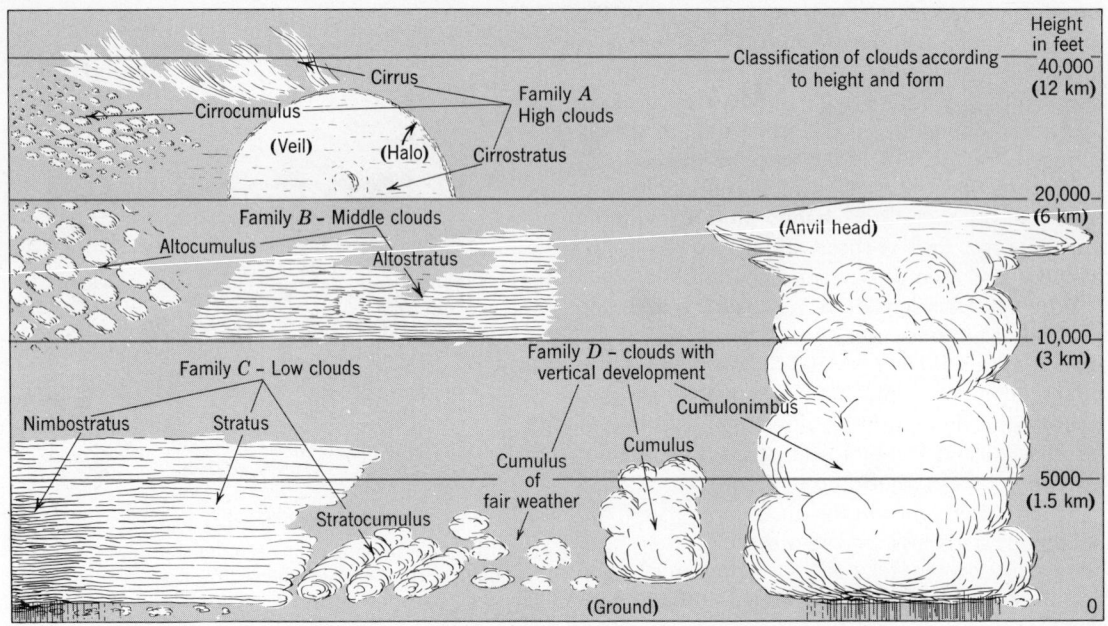

Figure 5.7 Cloud types are grouped into families, according to height range and form.

large body of moist air to high altitudes produces the *cumulonimbus* cloud, which is associated with lightning and torrential rain, and often with hail. Further details of this cloud are given in later paragraphs.

Fog is simply a stratiform cloud lying very close to the ground. One type, known as a *radiation fog*, is formed at night. This type of fog requires still air and clear skies above, so that the nocturnal net radiation loss is large and mixing cannot occur. A temperature inversion then develops. When the basal air temperature falls below the dew point, fog is formed. Another type, *advection fog*, results from the movement of warm, moist air over a cold or snow-covered ground surface. Losing heat to the ground, the air layer undergoes a drop of temperature below the dew point, and condensation sets in. A similar type of advection fog is formed over oceans where air from over a warm current blows across the cold surface of an adjacent cold current. Fogs of the Grand banks off Newfoundland are largely of this origin because here the cold Labrador current comes in contact with warm waters of Gulf Stream origin.

Fog is an environmental hazard, as those who drive the nation's city streets and highways know

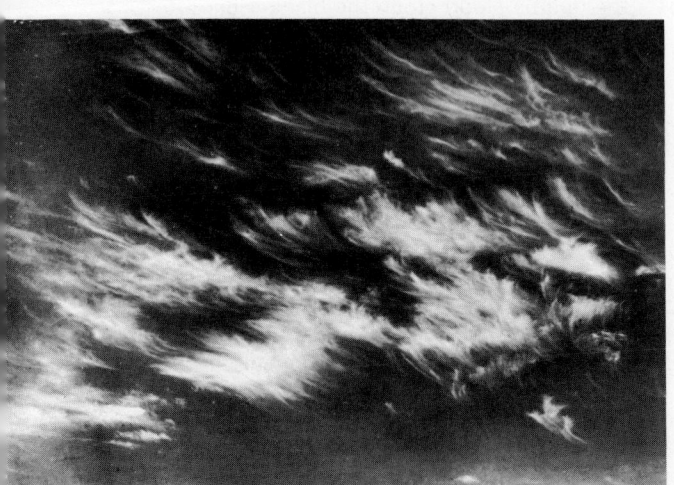 

Figure 5.8 Left: Cirrus in parallel trails, showing rapid air flow in the jet stream. (NOAA, National Weather Service.) Right: Cumulus of fair weather. (NOAA, National Weather Service.)

Forms of Precipitation

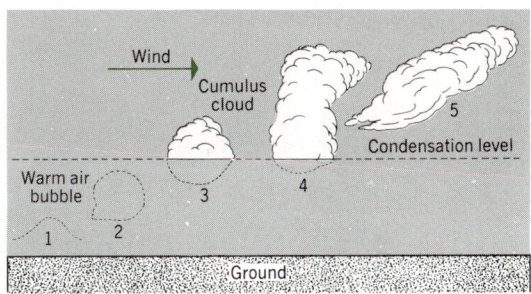

Figure 5.9 Rise of a warm air bubble to produce a small cumulus cloud. (From A. N. Strahler, 1971, *The Earth Sciences*, 2nd ed., Harper & Row, New York.)

Forms of Precipitation

Precipitation results when condensation is occurring rapidly within a cloud. *Rain*, consisting of falling water droplets, is typically formed by melting of snow previously formed in a cloud. In warm, tropical and equatorial regions rain results when cloud droplets in large numbers are caused to coalesce into drops too large to remain suspended in the air. The drops may then grow by colliding with other drops and joining with them to become as large as 0.25 in. (7mm) in diameter; but above this size they are unstable and break into smaller drops. Falling droplets less than 0.22 in (0.5 mm) in diameter make a *drizzle*. Rainfall is measured in terms of the water depth (in., cm) that is received per unit of time.

Ice pellets, often referred to in the United States as *sleet*, are small ice particles produced by freezing of raindrops formed in an upper, warmer layer, but falling through an underlying cold air layer.

Snow consists of masses of crystals of ice, grown directly by condensation from the water vapor of the air, where air temperature is below freezing. Individual snow crystals, which can be carefully caught upon a black surface and examined with a strong magnifying glass, develop in six-sided, flat crystals, or as prisms. They display infinite variations in their beautiful symmetrical patterns (Figure 5.11).

Hail consists of rounded lumps of ice, having an internal structure of concentric layers, much like an onion. Ordinarily the ice is not clear but has a frosted appearance. Hailstones range from 0.2 to 2 in. (0.5 to 5 cm) in diameter. Large hail-

all too well. In marine navigation fog brings grave danger of collision of one vessel with another or with an iceberg, or of running aground. To these classic perils of the sea are now added the perils of aircraft operation in and out of fog-bound landing fields. A large city airport, closed down by fog, incurs enormous losses of revenues due to flight cancellations, to say nothing of loss of productive time to thousands of persons forced to wait in airports or to seek alternative means of transportation.

Frequency of occurrence of dense fog varies greatly with region, Figure 5.10 shows that for the United States, fog incidence is highest in coastal areas, especially adjacent to cold currents (Pacific Coast, New England), over large inland water bodies (Great Lakes), and over mountainous areas in humid climates (Appalachian region). In contrast, dense fogs are rare in interior continental regions, especially in the drier deserts and steppes of the West.

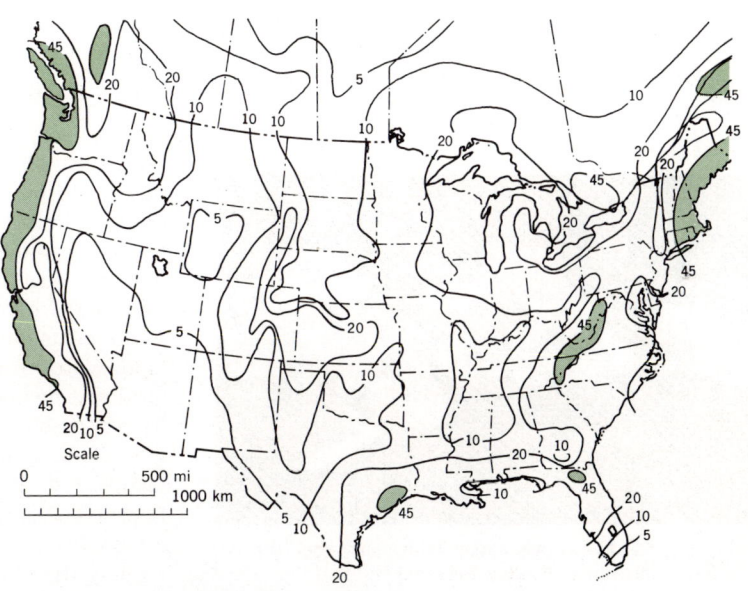

Figure 5.10 Greatly generalized map of the 48 contiguous United States and southern Canada showing number of days annually with dense fog. (Data of J. N. Myers, NOAA, National Weather Service, and Dept. of Mines & Technical Survey, Atlas of Canada.)

Figure 5.11 Individual hexagonal ice crystals such as those shown here are clumped together to form snowflakes. (Photograph by W. A. Bentley, NOAA, National Weather Service.)

Figure 5.13 Heavily coated wires and branches caused heavy damage in eastern New York State in January 1943, as a result of this icing storm. (Courtesy of NOAA, National Weather Service and New York Power and Light Company, Albany, N.Y.)

stones constitute an environmental hazard as they can be extremely destructive to crops and light buildings (Figure 5.12). Hail occurs only from the cumulonimbus cloud, inside of which are extremely strong updrafts of air. Raindrops are carried up to high altitudes, are frozen into ice pellets, then fall again through the cloud. Suspended in powerful updrafts, the hail stone grows by the attachment and freezing of droplets, much as ice accumulates on the leading edge of an airplane wing. Eventually the hailstone escapes from the updrafts and falls to earth.

When rain falls upon a frozen ground surface that is covered by an air layer of below-freezing temperature, the water freezes into clear ice after striking the ground or other surfaces such as trees, houses, or wires (Figure 5.13). The coating of ice that results is called a *glaze*, and an *icing storm* is said to have occurred. Actually no ice falls, so that ice glaze is not a form of precipitation. The phenomenon is called *freezing rain*.

Precipitation takes place only when large bodies of moist air are lifted or otherwise induced to rise rapidly, allowing large-scale condensation to be sustained. So we must next look into the basic conditions under which such large-scale rise of air masses can occur.

Convective Precipitation and Thunderstorms

Convective precipitation occurs within *convection cells*, which consist of bodies of air spontaneously rising rapidly, because that air is less dense than the surrounding air. A simple convection cell is illustrated in Figure 5.9. Unequal heating of the ground occurs under insolation. For example, a bare field over which dry soil is exposed heats more rapidly than surrounding forest land, with the result that a body of air near the ground is heated intensely and expands, reducing its density below that of the surrounding air layer (Step 1 in Figure 5.9). A bubble of warm air detaches itself and rises (Step 2). Adiabatic cooling takes place and condensation sets in, producing a cumulus cloud (Steps 3 and 4). With sufficient cooling the cloud reaches a level of stability; as

Figure 5.12 Hailstones, larger than hens' eggs (*arrow*). (NOAA, National Weather Service.)

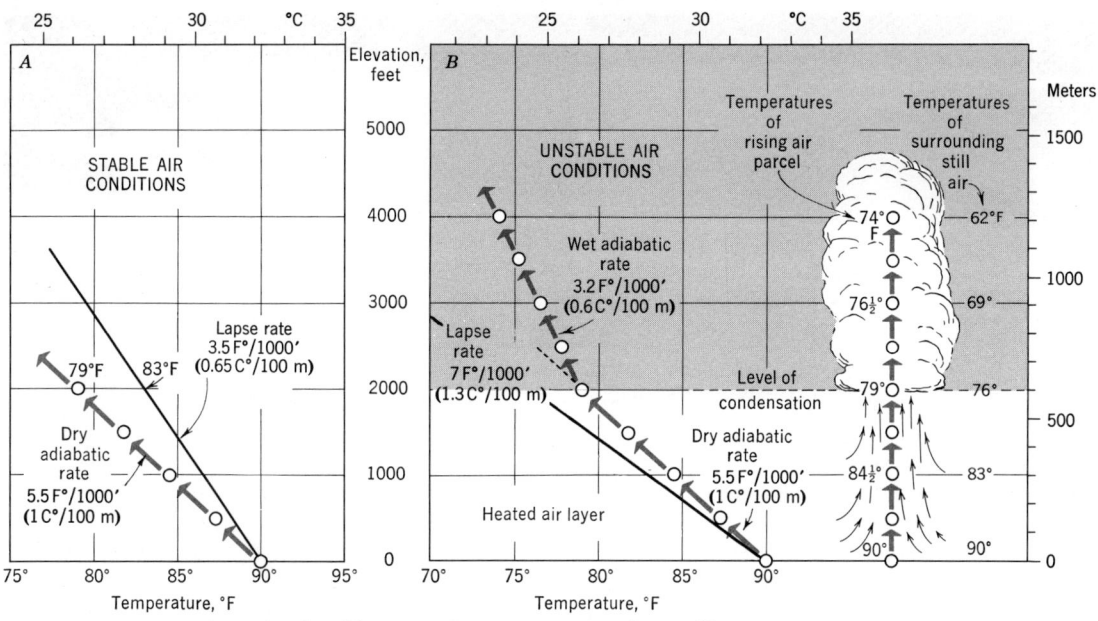

Figure 5.14 A. Forced ascent of stable air. B. Spontaneous rise of unstable air.

it is carried downwind it gradually dissolves (Step 5). Small bubbles of rising warm air are typical of conditions of generally fair weather and are not a cause of precipitation. To understand how cumulonimbus clouds develop and produce heavy rainfall, we must look further into the process of condensation and liberation of latent heat. Figure 5.14 will help to explain why spontaneous rise of air to produce intense convection can take place only when air mass properties are favorable.

Diagram A (*left*) is a plot of altitude against air temperature. The small circles represent a small parcel of air being forced to rise steadily higher, following the same dry adiabatic rate of cooling shown in Figure 5.6. To the right of this line is a solid line showing that the temperature of the undisturbed surrounding air; it decreases upward at the normal environmental lapse rate. Suppose that the air parcel is lifted from a point near the ground, where its temperature is 90° F (32° C). After the air parcel has been carried up 2000 ft (600 m), its temperature has fallen about 11 F° (6 C°) and is now 79° F (26° C); whereas the surrounding air (environment) is cooler by 7 F° (4 C°), and has a temperature of 83° F (28° C). The air parcel would thus be cooler than the environment at 2000 ft (600 m), and if no longer forcibly carried upward, would tend to sink back to the ground. These conditions represent *stable* air, not likely to produce convection cells, because the air would resist lifting.

When the air layer near the ground is excessively heated, the environmental lapse rate is increased. Diagram B of Figure 5.14 shows a *steep* lapse rate of 7 F° per 1000 ft (1.3 C° per 100 m). (As the graph is constructed the steeper lapse rate is represented by a line of lower inclination.) The air parcel near the ground begins to rise spontaneously because it is less dense than air over adjacent, less intensely heated ground areas. Although cooled adiabatically while rising, the air parcel at 1000 ft (300 m) has a temperature of 85° F (29° C), but this is well above the temperature of the surrounding still air. The air parcel, therefore, is lighter than the surrounding air and continues its rise. At 2000 ft (600 m), the dew point is reached and condensation sets in. Now the rising air parcel is cooled at the reduced wet adiabatic rate of 3.2 F° per 1000 ft (0.6 C° per 100 m), because the latent heat liberated in condensation offsets the rate of drop due to expansion. At 3000 ft (900 m), the rising air parcel is still several degrees warmer than the environment, and therefore continues its spontaneous rise.

The air described here as spontaneously rising during condensation is said to be *unstable*. Generally, steep lapse rates in air holding a large quantity of water vapor are associated with instability. In such air the updraft tends to increase in intensity as time goes on, much as a bonfire blazes with increasing ferocity as the updraft draws in greater supplies of oxygen. Of course, at very high altitudes, the bulk of the water vapor

Figure 5.15 A large, anvil-topped cumulonimbus cloud. (Sketched by A. N. Strahler from a photograph by NOAA, National Weather Service.)

will have condensed and fallen as precipitation. With the energy source gone the convection cell weakens and air rise finally ceases.

Recent studies of cumulonimbus clouds show that a single *thunderstorm* consists of individual parts, called *cells*. Air rises within each cell as a succession of bubble-like air bodies, rather than as a continuous updraft from bottom to top. As each bubble rises, air in its wake is brought in from the surrounding region, a process called *entrainment*. Rising air in the thunderstorm cell can reach vertical speeds up to 3000 ft (900 m) per minute. Rapid condensation will be in the form of rain in the lower levels, mixed water and snow at intermediate levels, and snow at high levels. Upon reaching high levels, which may be on the order of 20,000 to 40,000 ft (6 to 12 km), or even higher, the rising speed diminishes and the cloud top is dragged downwind to form an *anvil top* (Figure 5.15). Ice particles falling from the cloud top act as nuclei for condensation at lower levels, a process called *seeding*. The rapid fall of precipitation adjacent to the rising air bubbles exerts a frictional drag upon the air and sets in motion a downdraft. Striking the ground where precipitation is heaviest, this downdraft forms a local squall wind, which is sometimes strong enough to fell trees and do severe structural damage to buildings (Figure 5.16). We have already noted that hail is produced in the updrafts of a thunderstorm and that it can fall with damaging intensity in the thunderstorm downdraft.

Another effect of convection cell activity is to generate *lightning*, one of the environmental hazards that results annually in the death of many persons and livestock, as well as in the setting of forest fires and building fires. (See data in Table 5.2.) Lightning is a great electric arc—a gigantic

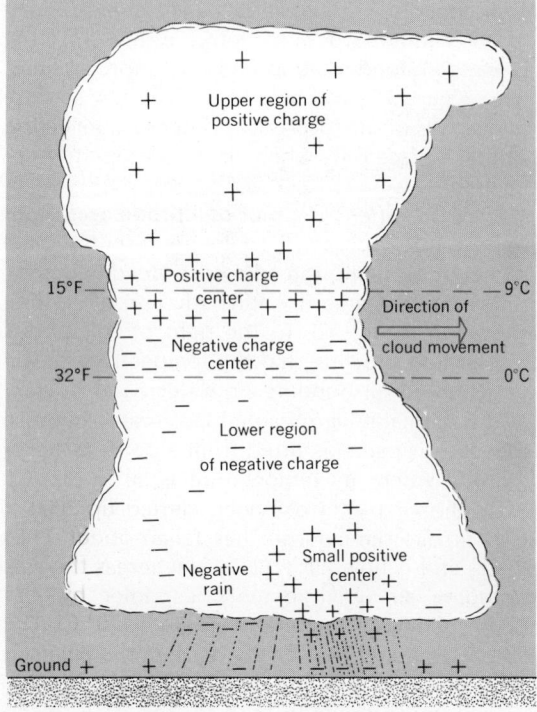

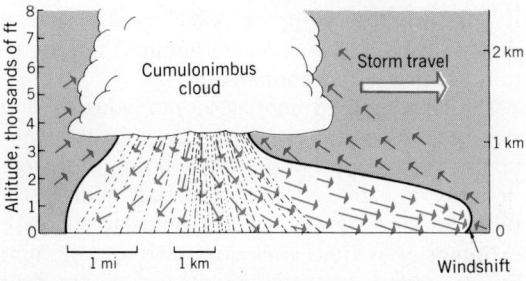

Figure 5.16 Downdraft beneath a thunderstorm. (After H. R. Byers and R. T. Braham, 1949, *The Thunderstorm*, U.S. Govt. Printing Office, Washington, D.C.)

Figure 5.17 Electrical charges within a thunderstorm. (After U.S. Dept. Commerce, 1955, *C.A.A. Technical Manual 104*, U.S. Govt. Printing Office, Washington, D.C., p. 60, Figure 95.)

Orographic Precipitation

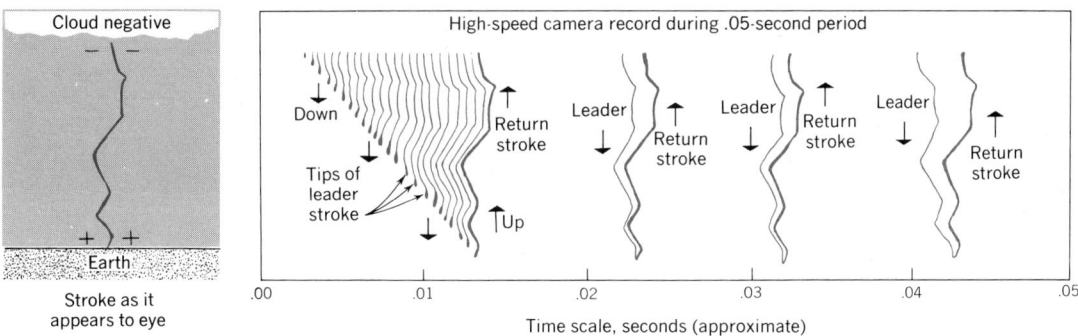

Figure 5.18 A lightning stroke caught by high-speed camera. (After U.S. Dept. Commerce, 1955, *C.A.A. Technical Manual 104*, U.S. Govt. Printing Office, Washington, D.C., p. 62, Figure 96.)

spark—passing between cloud and ground, or between parts of a cloud mass. Although the mechanism of lightning generation is not fully understood, it is known that electrical charges become separated in different cloud levels, as shown in Figure 5.17, so that certain levels have a positive charge while others develop a negative charge. As shown in Figure 5.18, the lower portion of the cloud develops a negative charge; the ground beneath it a positive charge. When sufficient potential has accumulated, an arc passes from cloud to ground, and returns to cloud, with many subsequent alternations of flow until the potential is relieved. During lightning discharge a current of as much as 60,000 to 100,000 amperes may develop. Rapid heating and expansion of the air in the path of the lightning stroke sends out intense sound waves, which we recognize as a thunder clap.

A single large thunderstorm consists of several adjacent cells, each experiencing activity in succession. As a result, the storm can have a long duration and can yield many heavy bursts of rain.

Convective precipitation is characterized by its intensity combined with spotty distribution, so that an area a few miles across may receive 1 to 3 in. (25 to 75 mm) of rainfall in one storm, whereas adjoining areas may receive little or none during the same period. Because spontaneous convection is favored by a warm air mass with a high moisture content, thunderstorms are a dominant form of precipitation in the equatorial zone throughout the year, and in the rainy (monsoon) season of lands in the tropical zone (Figure 5.19). In middle latitudes thunderstorms are largely a summer phenomenon, while in arctic and polar regions their occurrence is rare. Incidence of thunderstorms is far less over mid-oceanic regions in low latitudes than over continents at equivalent latitudes, as Figure 5.19 clearly shows. Evidently terrain features play an important role in causing convective precipitation, and it is to this relationship between land relief and precipitation that we turn next.

Orographic Precipitation

A second precipitation-producing mechanism is described as *orographic*, which means "related to mountains." Air masses moving with prevailing winds may be forced to flow over mountain ranges (Figure 5.20). As the air rises on the windward side of the range, it is cooled at the dry adiabatic rate. Condensation then sets in, and if cooling is sufficient, precipitation will result. After passing over the mountain summit, the air will begin to descend the lee side of the range. Now the air is warmed through the adiabatic process and, having no source from which to draw up moisture, becomes very dry. A belt of dry climate, often called a *rainshadow*, typically exists on the lee side of the range. Several of the important dry deserts of the earth are of this type.

Dry, warm *foehn winds* (Europe) and *chinook winds* (northwestern North America), which occur on the lee side of a mountain range, cause extremely rapid evaporation of snow or soil moisture. These winds result from turbulent mixing of lower and upper air in the lee of the range. The upper air, which has little moisture to begin with, is greatly dried and heated when swept down to low levels.

An excellent illustration of orographic precipitation and rainshadow occurs in the far west of the United States. Prevailing westerly winds bring moist air from the Pacific Ocean over the coast ranges of central and northern California and the great Sierra Nevada range, whose summits rise to 14,000 ft (4000 m) above sea level (Figure 5.21). Heavy rainfall occurs on the windward slopes of these ranges, nourishing rich forests. Passing down the steep eastern face of the Sierras, air

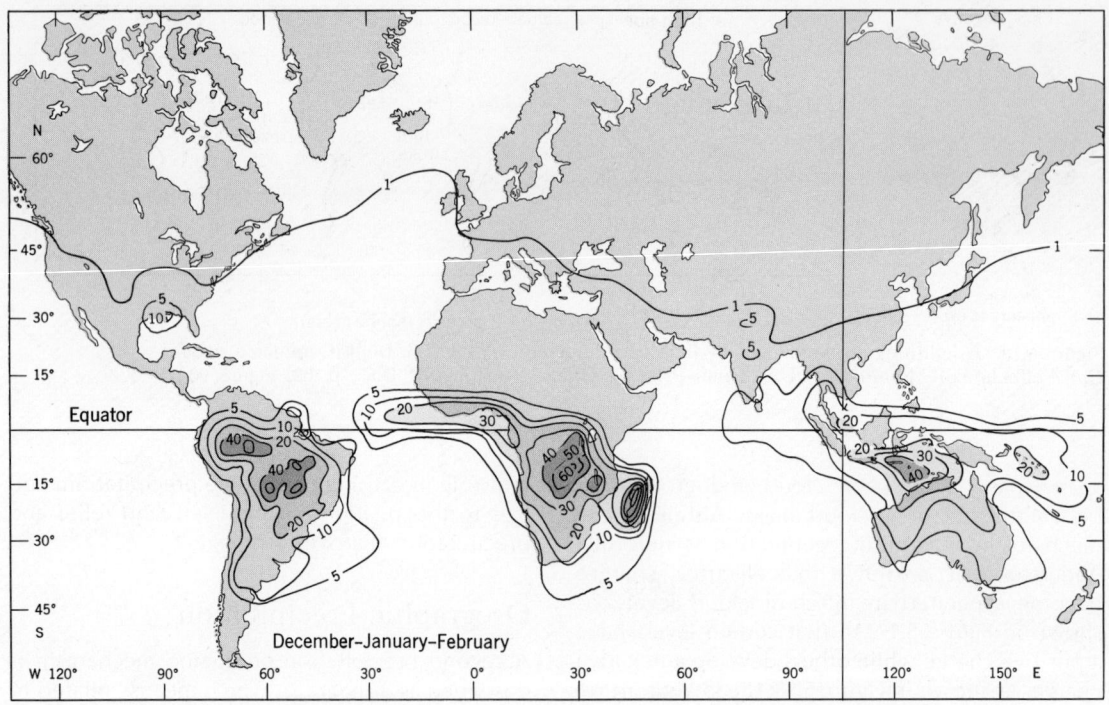
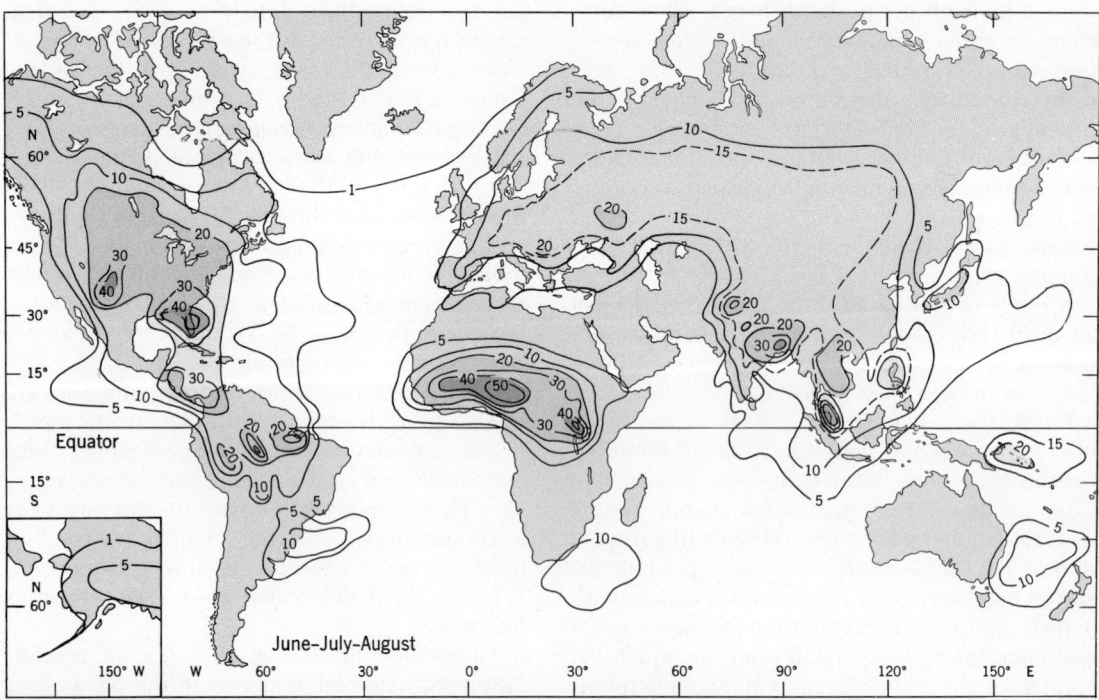

Figure 5.19 World map of frequency of thunderstorms by seasons. Figures are for *thunderstorm days*, or number of days per year on which one or more storms are recorded at an observing station. (Data from U.S. Air Force, 1960, *Handbook of Geophysics*, Macmillan, New York, pp. 9–26 and 9–28.)

Cyclonic and Frontal Precipitation

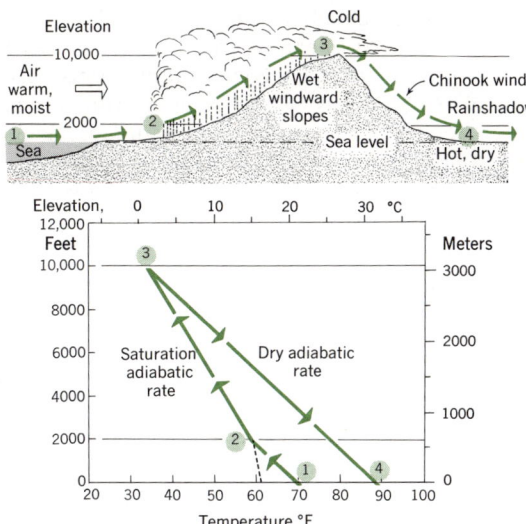

Figure 5.20 Forced ascent of oceanic air masses, producing precipitation and a rainshadow desert. (From A. N. Strahler, 1971, *The Earth Sciences*, 2nd ed., Harper & Row, New York.)

must descend nearly to sea level, even below sea level in Death Valley. The adiabatic heating so caused, and a consequent reduction in humidity, produces part of America's great desert zone, covering a strip of eastern California and all of Nevada.

Much orographic rainfall is actually of the convective type, in that it takes the form of heavy convective showers and thunderstorms. The convection is induced by the forced ascent of unstable air as it passes over the mountain barrier. This effect shows well in the world distribution of thunderstorms, Figure 5.19.

Cyclonic and Frontal Precipitation

The third major class of precipitation includes lift mechanisms in which a denser air mass lying near the surface acts as an obstruction to force the rise of a moisture-ladened air mass of less density, as the two masses interact in a frontal zone. Such frontal precipitation is typically associated with a trough of low pressure or with a center of low barometric pressure—a moving cyclone. Moving cyclones in the troposphere range from comparatively mild disturbances associated with cloudiness, precipitation, and light winds, to intense storms with gale force and stronger winds.

Moving cyclones fall into three general classes. (1) The *extratropical cyclone* is typical of middle and high latitudes. It ranges in severity from a weak disturbance to a powerful storm. (2) The *tropical cyclone* is found in low latitudes over ocean areas. It ranges from a mild disturbance to the terribly destructive *hurricane,* or *typhoon.* (3) The *tornado,* although a very small storm, is an intense cyclonic vortex of enormously powerful winds. It is on a very much smaller scale of magnitude than other types of cyclones and must be treated separately.

About the time of the First World War, Norwegian meteorologist J. Bjerknes presented a new theory to explain extratropical cyclones. He first developed the frontal concept and established the existence of a polar front of interaction between cold polar and warm tropical air. Figure 5.22 shows a detailed map of a section of the polar front; it coincides with a trough of low pressure lying between two anticyclones.

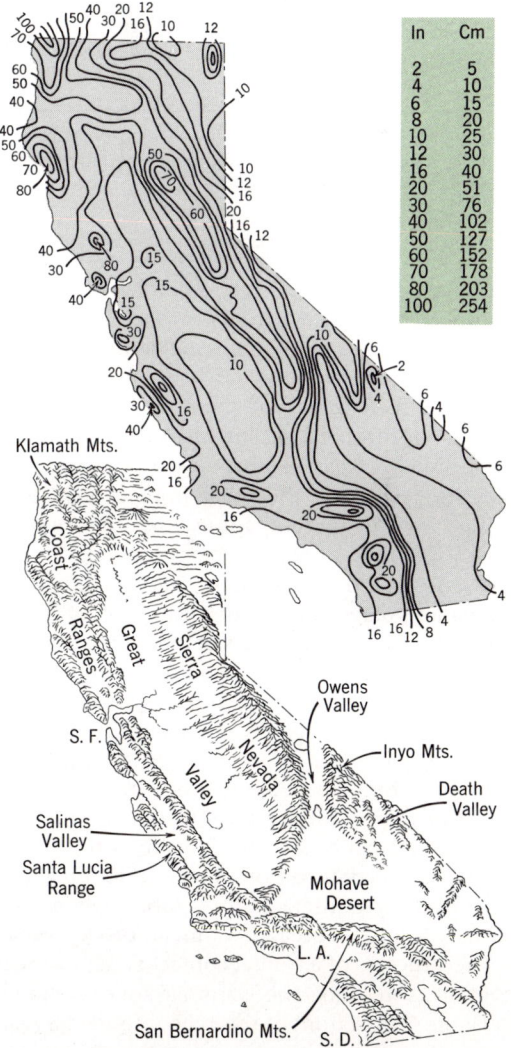

Figure 5.21 The effect of mountain topography on rainfall is well shown by the state of California. Isohyets in inches. (Rainfall map after U.S. Department of Agriculture, *Yearbook of American Agriculture*, 1941.)

127

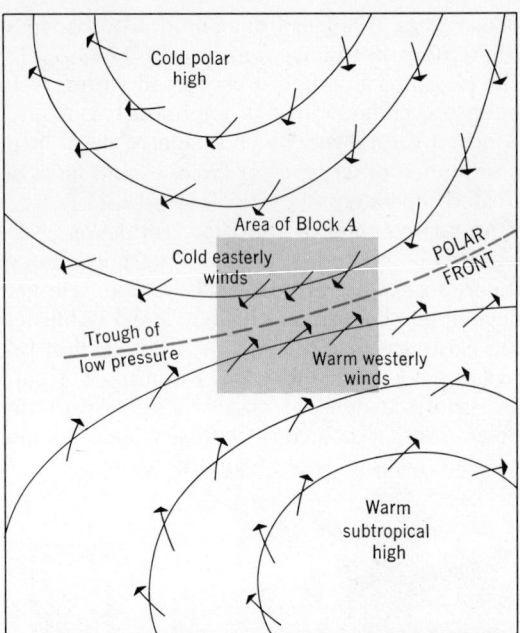

Figure 5.22 The trough between two high-pressure regions is a likely zone for development of a wave cyclone.

figure, being actually of the order of slope of 1 in 40 to 1 in 80 (meaning that the slope rises 1 foot vertically for every 80 feet of horizontal distance). Cold fronts are associated with strong atmospheric disturbance; the warm air being lifted often breaks out in violent thunderstorms. In some instances these storms occur along a line well in advance of the cold front, a *squall line*. Such lines of thunderstorms can be seen on the radar screen (Figure 5.25).

Figure 5.26 illustrates a *warm front* in which warm air is moving into a region blanketed by colder air. Here, again, the cold air mass remains in contact with the ground. The warm air mass is forced to rise, as if ascending a long ramp. Warm fronts have lower slopes than cold fronts, being of the order of 1 in 80 to as low as 1 in 200. Moreover, warm fronts are commonly attended by stable atmospheric conditions and lack the turbulent air motions of the cold front. Of course, if the warm air is unstable, it will develop convection cells and there will be heavy showers and thunderstorms.

Cold fronts normally move along the ground at a faster rate than warm fronts. When both types are in the same neighborhood, as they are in the cyclonic storm, the cold front may overtake the warm front. A curious combination known as an *occluded front* then results (Figure 5.27). The colder air of the fast-moving cold front remains next to the ground, forcing both the warm air and the less cold air to rise over it. The warm air mass is lifted completely free of the ground. The relations between warm, cold, and occluded fronts can now be introduced into the life history of the cyclone.

In Figure 5.23B the wavelike disturbance along the polar front has deepened and intensified. Cold air is now actively pushing southward along a cold front; warm air is actively moving northeastward along a warm front. Each front is convex in the direction of motion. The zone of precipitation is now considerable, but wider along the warm front than along the cold front. In a still later stage the more rapidly moving cold front has reduced the zone of warm air to a narrow sector. In block C, the cold front has overtaken the warm front, producing an occluded front and forcing the warm air mass off the ground, isolating it from the parent region of warm air to the south. The source of moisture and energy thus cut off, the cyclonic storm gradually dies out and the polar front is reestablished as originally as in block D.

The term *front* as used by Bjerknes, was particularly apt because of the resemblance of this feature to the fighting fronts in western Europe, then active. Just as vast armies met along a sharply defined front which moved back and forth, so masses of cold polar air meet in conflict with warm, moist tropical air. Instead of mixing freely, these unlike air masses remain clearly defined, but they interact along the polar front in *wave cyclones* whose structure is quite like the form of steep ocean waves seen in cross section.

A series of individual blocks, Figure 5.23, shows the various stages in the life history of an extratropical cyclone. At the start of the cycle the polar front is simply a smooth boundary along which air of unlike qualities is moving in opposite directions, as shown in Figure 5.22. In Block A of Figure 5.23, the polar front shows a bulge, or *wave*, beginning to form. Cold air is turned in a southerly direction, warm air in a northerly direction, as if each would penetrate the domain of the other. We must now digress from this series of diagrams to consider the interaction occurring when cold air moves into an area of warm air, or vice versa.

The structure of a frontal contact zone in which cold air is invading the warm-air zone is shown in Figure 5.24. A front of this type is termed a *cold front*. The colder air mass, being the denser, remains in contact with the ground and forces the warmer air mass to rise over it. The slope of the cold front surface is greatly exaggerated in the

Long observation on the movements of cyclones has revealed that certain tracks are most commonly followed. Figure 5.28 is a map of the

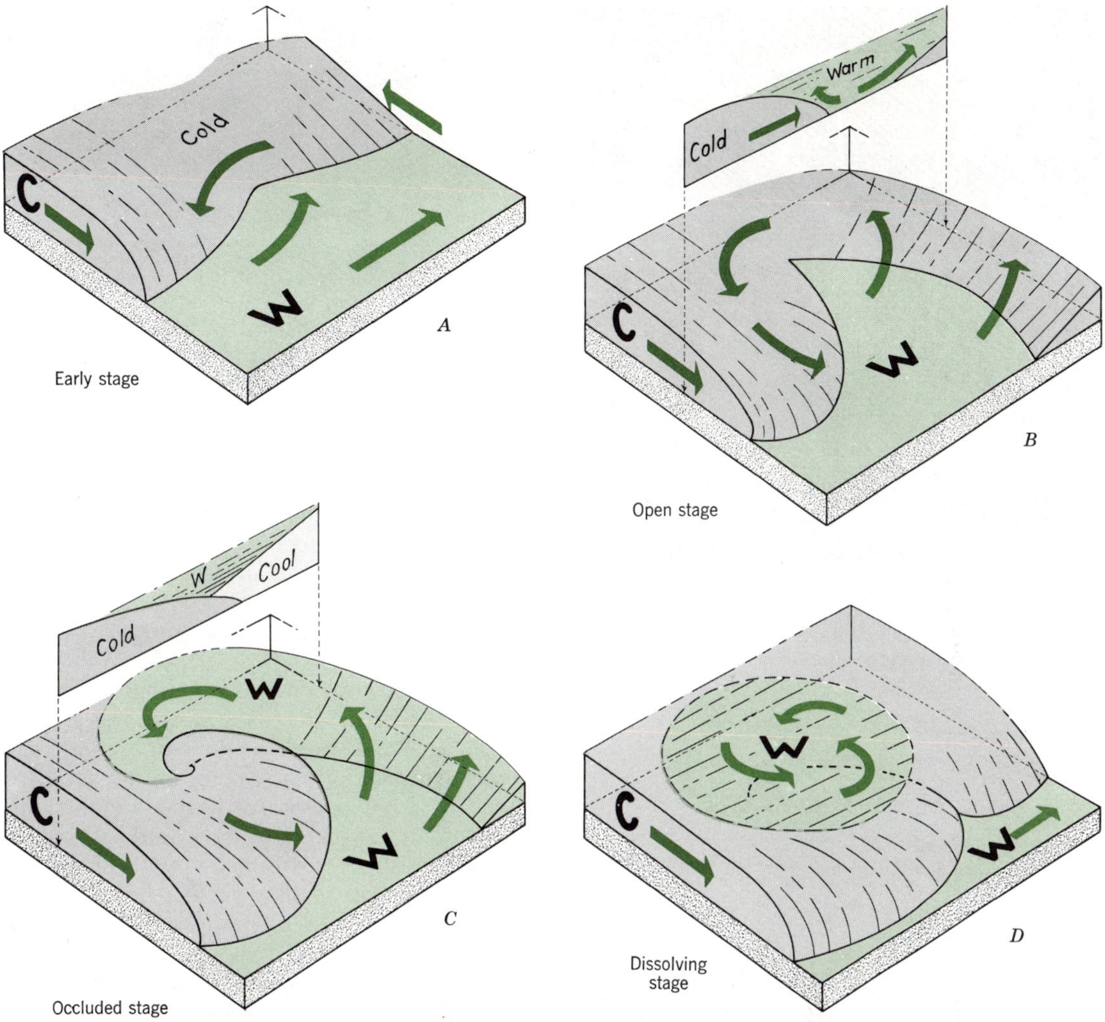

Cyclonic and Frontal Precipitation

Figure 5.23 The development of a middle-latitude cyclone.

United States showing these common paths. Notice that while some cyclonic storms travel across the entire United States from places of origin in the North Pacific, such as the Aleutian low, others originate in the Rocky Mountain region, the central states, or the Gulf Coast. Most tracks converge toward the northeastern United States and pass out into the North Atlantic, where they tend to concentrate in the region of the Icelandic low. A similar concentration occurs in the neighborhood of the Aleutian low. Extratropical cyclones commonly form in succession to travel in a chain across the North Atlantic and North Pacific oceans. Figure 5.29 shows several such *cyclone families*.

In the Southern Hemisphere, storm tracks are more nearly along a single lane, following the parallels of latitude. This appears to be the result of uniform ocean surface throughout the middle

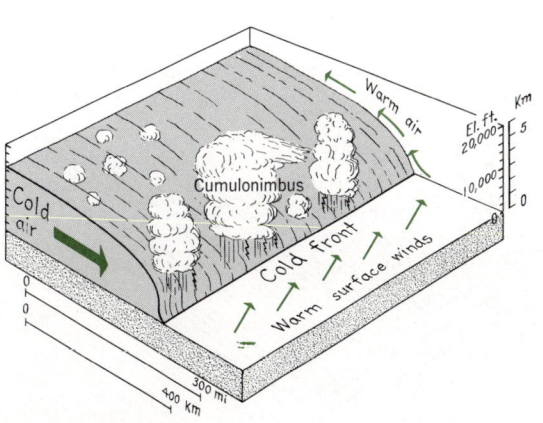

Figure 5.24 A cold front.

Atmospheric Energy Releases

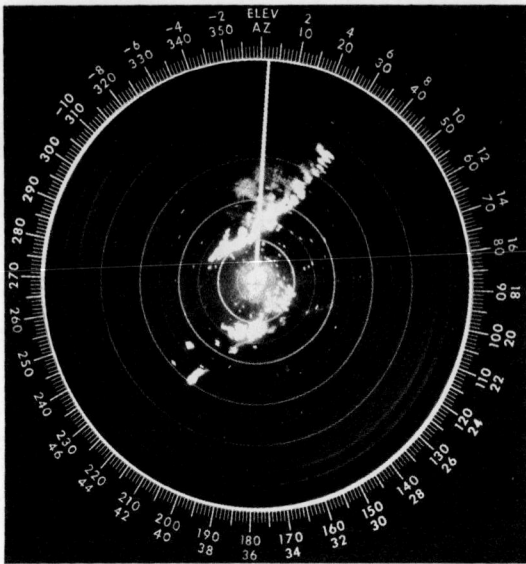

Figure 5.25 Photograph of a radar screen on which lines of thunderstorms show as bright patches. The heavy circles are spaced 50 nautical miles (80 km) apart. (NOAA, National Weather Service.)

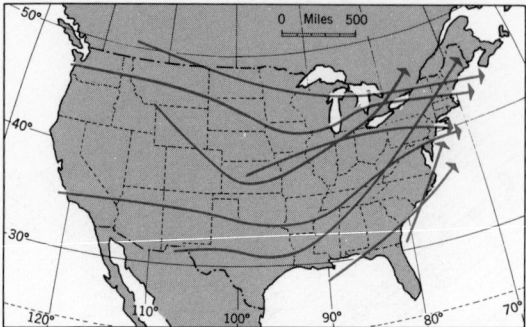

Figure 5.28 Common tracks taken by middle-latitude cyclones passing across the United States. (After Bowie and Weightman.)

latitudes, only the southern tip of South America breaking the monotonous oceanic expanse. Furthermore, the polar-centered antarctic ice sheet provides a centralized source of polar air.

Cyclones and Upper-Air Waves

Wave cyclones of the lower atmosphere are closely tied in with Rossby waves and the jet stream at higher levels in the troposphere. Cyclones typically form at a point on the southern bulge of an upper-air wave and are then intensified and swept along in a northeasterly direction (Northern Hemisphere) by the jet stream. This relationship is shown by a hemispherical weather map (Figure 5.30) in which four families of cyclones are related to four Rossby waves. Wave cyclones are shown on this map by warm and cold fronts pivoting on abrupt V-junctions. The isobars represent upper-air conditions and are essentially also streamlines of west-to-east air flow.

It is evident from Figure 5.30 that wave cyclones and their warm and cold fronts are confined to a fairly shallow lower layer of the troposphere. While a cyclone is formed of surface winds spiralling into the low-pressure center from all points of the compass, the flow of air in the upper troposphere, where the jet stream exists, is smoothly in one direction over the cyclone below it. Similarly, an anticyclone has surface winds spiralling outward in all compass directions, yet the upper-air flow passes smoothly in one direction across the anticyclone. There is obviously coupling of energy between low and high levels. Although oversimplified, the diagrams in Figure 5.31 provide an explanation of the coupling mechanism. The upper diagram is a schematic map, showing a Rossby wave overlying an anticyclone and cyclone in their typical positions. Notice that over the anticyclone the upper-air streamlines of flow

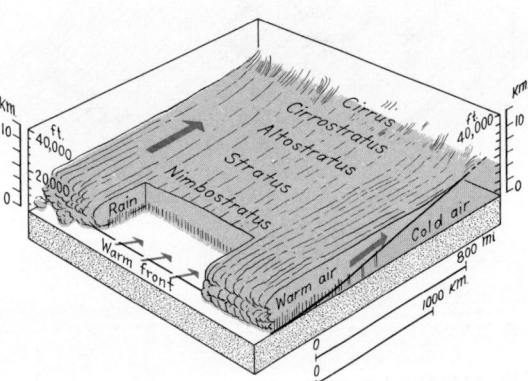

Figure 5.26 A warm front.

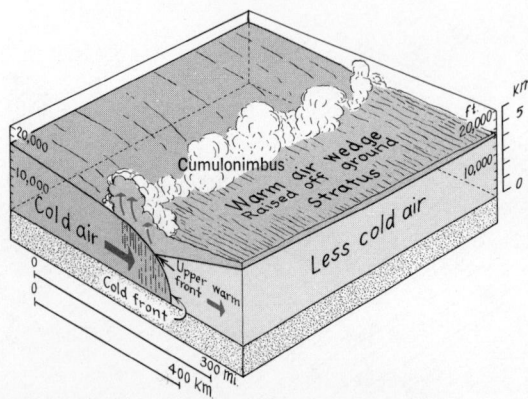

Figure 5.27 An occluded warm front.

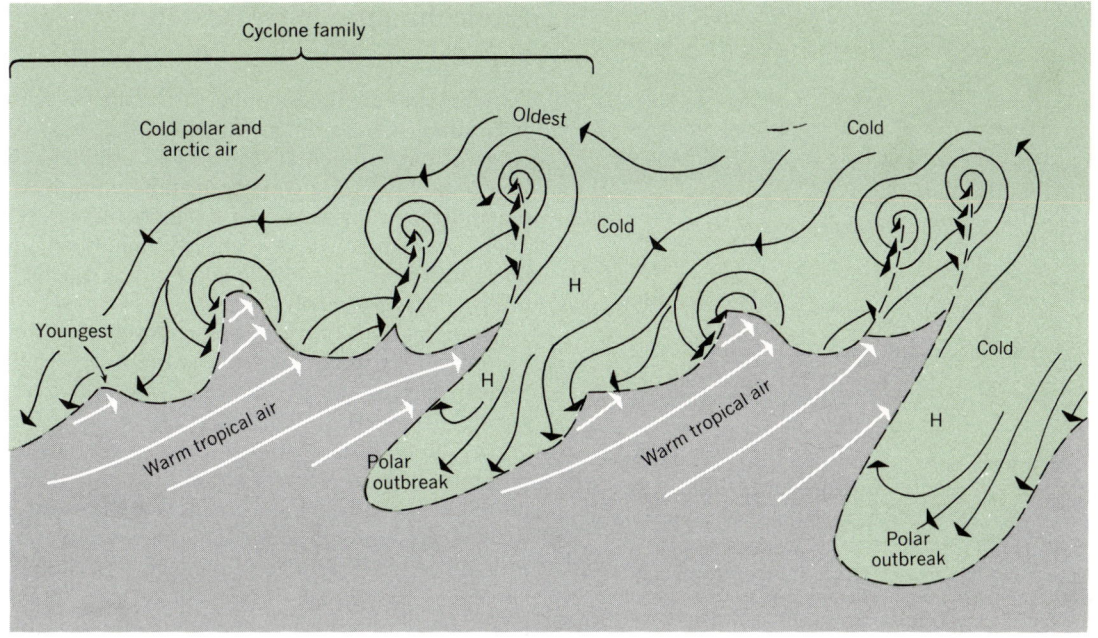

Figure 5.29 Two families of wave cyclones in the Northern Hemisphere, as seen on a schematic weather map. After Bjerknes and Solberg; Petterssen. (From A. N. Strahler, 1971, *The Earth Sciences*, 2nd ed., Harper & Row, New York.)

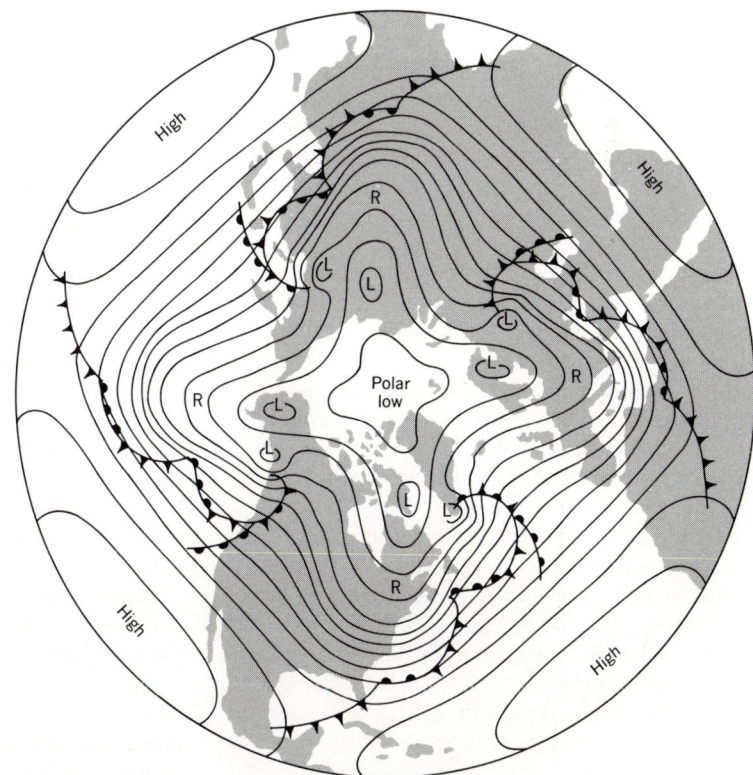

Figure 5.30 Polar map showing four families of wave cyclones. Cyclones are shown by fronts at sea level, whereas upper-air waves are shown by high-level isobars. (Data of Palmén, in S. Petterssen, 1969, *Introduction to Meteorology*, 3rd ed., McGraw-Hill, New York, p. 224, Figure 13.15.)

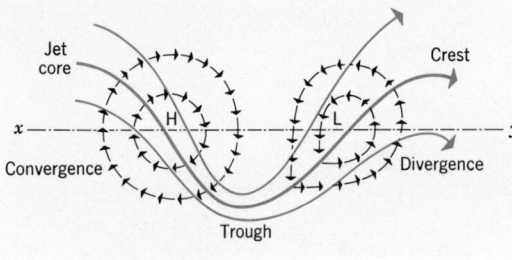

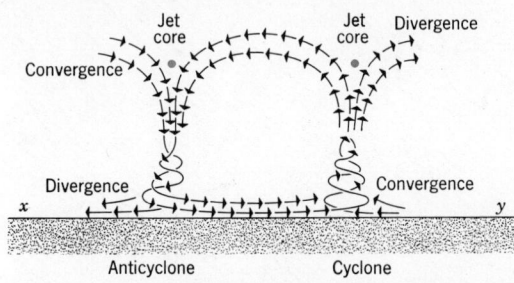

Figure 5.31 Convergence and divergence of upper-air flow related to anticyclone and cyclone, respectively, at low levels. (From A. N. Strahler, 1971, *The Earth Sciences*, 2nd ed., Harper & Row, New York.)

Environmental Stresses and Hazards of Weather Phenomena

Cyclonic and frontal weather disturbances, with associated convectional activity bring a wide range of environmental stresses and problems to bear upon both rural and urban communities. For the most part these are of a much higher frequency and much lesser severity than such intense storms as the tornado and hurricane, which often rate as major disasters.

Torrential rains produced by convective storms bring flooding to city streets and highways when the capacity of storm drains and sewers is overtaxed. Together with reduced visibility such rains are a safety hazard to vehicular traffic, while property damage results from flooding. (River floods are discussed in Chapter 13.) Powerful wind gusts from thunderstorm downdrafts are capable of much structural damage to buildings and trees, while falling tree limbs may break power lines and cause local power failures. Lightning associated with these storms is a major hazard in setting building and forest fires and causing death and injury by electrocution to people and livestock (see Table 5.3). Severe hailstorms are a common cause of crop damage, aggregating a large financial loss each year, and may do structural damage to building roofs and automobiles (see Table 5.3). (More is given on this subject in Chapter 6.)

Snowfall brings a familiar set of problems to residents of urban areas in the northern states. All too frequently, despite massive retaliation by armies of specialized vehicles, snow accumulates faster than it can be removed. Automobiles become stalled and block major arteries; traffic becomes paralyzed, and community activity subsides to await the cessation of snowfall and the signal to start digging out. Airports may be forced to curtail traffic or close down entirely, despite efforts of highly effective ganged snowplows to keep the runways clear. Snow removal, together with spreading of deicing salt and sand, is a major cost item in the budget of many large cities, as it is for county and state governments responsible for keeping highways clear. Several millions of tons of salt are used annually in the United States for this purpose, with deleterious environmental side effects (discussed in Chapter 12).

The *blizzard* is an intense winter storm affecting particularly the northcentral and northeastern United States and adjacent parts of Canada; it is characterized by extremely strong and persistent cold winds accompanying the invasion of polar and arctic air masses. Snow that has already fallen in earlier stages of the storm is carried by the

are converging. Upper-air convergence must be met with subsidence, since that air must escape. As the lower diagram shows in cross-section, the subsiding air feeds into the low-level anticyclone, from which it escapes by surface divergence. On the opposite flank of the upper-air wave the streamlines show a horizontal divergence in the region of the jet stream core. Air must rise from beneath to supply the diverging air motion. The underlying cyclone represents a surface convergence, bringing in air from all sides. The linkage of the two parts of the flow system is completed by surface and upper-level winds, as shown by arrows in the lower diagram.

The upper-air waves, with their divergences and convergences set off, maintain and steer the anticyclones and cyclones at low levels. It is for this reason that upper-air winds are carefully plotted from the data of sounding balloons. From such information the meteorologist can anticipate the formation of cyclones and can predict their path of travel. When upper-air flow is rather uniformly west-to-east and lacking in upper-wave development (a condition known as *zonal flow*) cyclones will be weak and fast-moving because there is little opportunity for exchange of polar and tropical air. When upper-air waves are well developed with strong meridional components of flow, both poleward and equatorward, a sequence of strong cyclones can be anticipated, along with invasions of cold air in strong anticyclones.

wind and accumulates in drifts that bury highways, stranding motorists and threatening death from exposure to severe cold. Death to livestock by starvation on the open range is sometimes a major economic loss following a blizzard.

Associated with a warm front in winter are two major environmental hazards: icing and fog. With warm air aloft, rain falling through a cold air layer beneath to reach frozen ground causes the buildup of ice on tree branches, power lines, walks, and roadways (Figure 5.13). These conditions are extremely hazardous to life and limb, with many accidents occurring to motorists and pedestrians. Power failure can occur on a large scale as wires break under the weight of the ice and under impact of falling tree limbs. As warm moist air behind a warm front passes over cold and often snow-covered ground, a dense advection fog can result. The environmental hazards of fogs have already been discussed in earlier pages of this chapter.

In Chapter 6, under the subject of planned weather modification, we shall look into various attempts that are being made to reduce the severity of hailstorms and to disperse fog.

The Tornado—An Environmental Hazard

Associated with frontal air-mass interactions in middle latitudes is the smallest but most violent of all known storms, the *tornado*. It seems to be a typically American storm, being most frequent and violent in the United States, although occurring in Australia in substantial numbers and reported occasionally in other places in middle latitudes. Storms reported as tornadoes also occur throughout tropical and subtropical zones.

The tornado is a small, intense cyclone in which the air is spiraling at tremendous speed. It appears as a dark *funnel cloud* (Figure 5.32), hanging from a dense cumulonimbus cloud. At its lower end the funnel may be from 300 to 1500 ft (90 to 460 m) in diameter. The funnel appears dark because of the density of condensing moisture, and the presence of dust and debris swept up by the wind.

Wind speeds in a tornado exceed anything known in other storms. Estimates of wind speed run to as high as 250 mi (400 km) per hour. As the tornado moves across the country the funnel writhes and twists. The end of the funnel cloud may alternately sweep the ground, causing complete destruction of anything in its path, then rise in the air to leave the ground below unharmed. Tornado destruction occurs both from the great wind stress and from the sudden reduction of air

Figure 5.32 This funnel cloud was photographed by William L. Males at Cheyenne, Oklahoma, on May 4, 1961. The tornado was less than one mile from the observer when the picture was taken.

pressure in the vortex of the cyclonic spiral. Closed houses literally explode. It is even reported that the corks will pop out of empty bottles, so great is the difference in air pressure.

Tornadoes occur as parts of cumulonimbus clouds, usually those clouds in a squall line that travels in advance of a cold front. They seem to originate where turbulence is greatest. They are commonest in the spring and summer but occur in any month, Figure 5.33. Where maritime polar (*mP*) air lifts warm, moist tropical (*mT*) air on a cold front, conditions may become favorable for tornadoes. They occur in greatest numbers in the central and southeastern states and are rare over mountainous and forested regions. They are almost unknown from the Rocky Mountains westward and are relatively few on the eastern seaboard (Figure 5.34).

Devastation from a tornado is complete within the narrow limits of its path. Only the strongest buildings constructed of concrete and steel can resist major structural damage. Table 5.3 gives data on tornado deaths and property damage. Storm cellars built completely below ground provide satisfactory protection if they can be reached in time. Although a tornado can often be seen or heard approaching, a cold front passing during the hours of darkness, as it often does, may pre-

Atmospheric Energy Releases

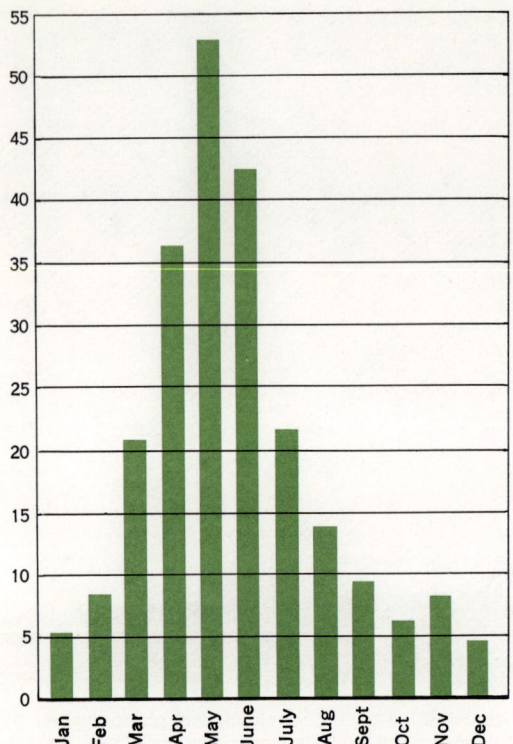

Figure 5.33 Average number of tornadoes reported in each month in the United States for the period 1916–1960. (From NOAA, National Weather Service.)

sent no warning. The National Weather Service maintains a tornado forecasting and warning system. Whenever weather conditions conspire to favor tornado development, the danger area is alerted and systems for observing and reporting a tornado are set in readiness. Communities in the paths of tornadoes may thus be warned in time for inhabitants to take shelter.

Waterspouts are similar in structure to tornadoes but form at sea under cumulonimbus clouds. They are smaller and less powerful than tornadoes. Sea water may be lifted 10 ft (3 m) above the sea surface, and the spray is carried higher. Waterspouts are commonly found in subtropical waters of the Gulf of Mexico and off the southeastern coast of the United States where they result from air turbulence in continental air masses spreading out over the ocean.

Weather Disturbances of Low Latitudes

Close to the equator, in the intertropical convergence zone, the Coriolis force is weak and air masses show little contrast in temperature and humidity. As a result, weather disturbances are weakly defined and difficult to interpret. Wave-depressions develop from time to time and drift slowly westward. These may at times take the form of distinctive low pressure centers with cyclonic inflow of winds, but fronts are not readily discernible.

A disturbance of tropical latitudes (15° to 25° N) is the *monsoon depression*, a weak form of cyclone that typically develops in the rainy monsoon (high-sun) season over southeastern Asia. Figure 5.35 is a weather chart showing two such depressions, each represented by a counter-clockwise vortex in the streamlines of air flow at the 10,000-ft (700-mb) level. Although the barometric pressure in the center of the low is only from 2 to 4 mb less than in the surrounding region, heavy convectional rainfall is set off. West-

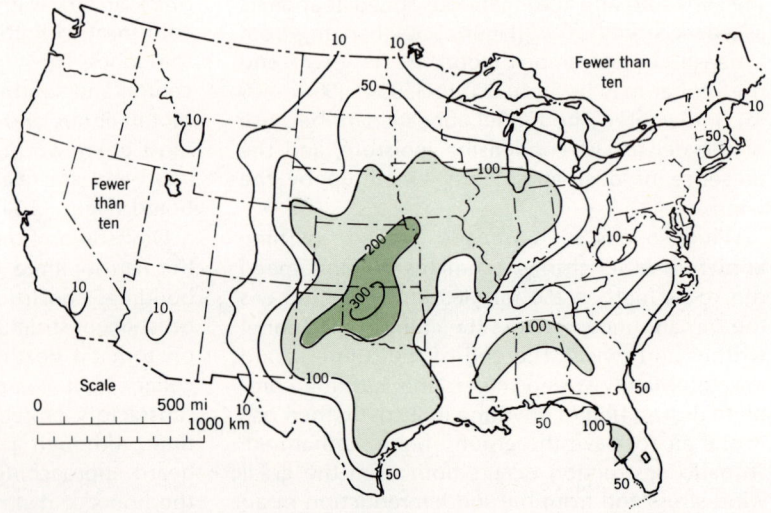

Figure 5.34 Distribution of tornadoes in the United States, 1955 to 1967. Figures tell number of tornadoes observed within 2-degree squares of latitude and longitude. (Data of M. E. Pautz, 1969, ESSA Tech. Memo WBTM FCST 12, Office of Meteorological Operations, Silver Spring, Md.)

The Tropical Cyclone—An Environmental Hazard

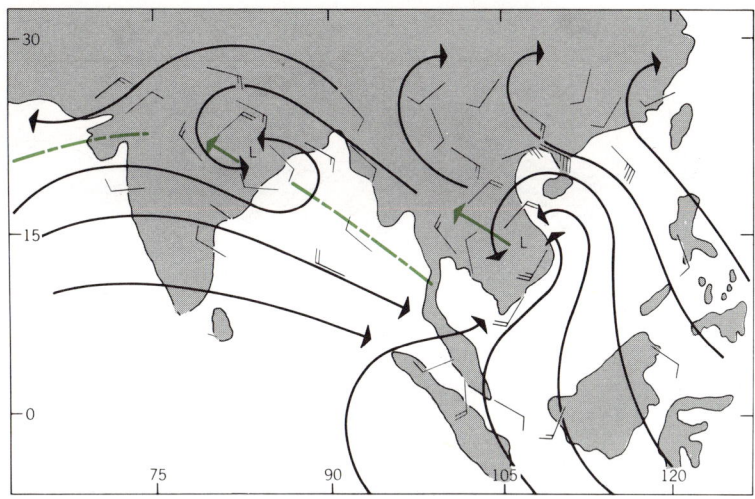

Figure 5.35 Monsoon lows, labeled L, on a weather chart for southeast Asia on a day near the autumnal equinox. Long arrows are streamlines of air flow at 10,000 ft (3 km). Short arrows show surface winds. The dashed line shows position of the equatorial trough. The lows are moving toward the west. (After H. Riehl, 1965, *Introduction to the Atmosphere*, McGraw-Hill, New York, p. 177, Figure 8.5.)

ward motion of these depressions is slow—some 10 to 12 mi (16 to 20 km) per hour. They are evidently steered by the tropical easterly jet stream present overhead at very high levels.

In the tropical zone, or trade-wind belt, latitudes 5° to 30° N and S, much of the precipitation is from *easterly waves* that travel slowly westward across the Atlantic and Pacific oceans. Figure 5.36 is a weather chart showing an easterly wave in the West Indies. The wave is simply a series of indentations in the isobars to form a shallow pressure trough. The wave travels westward, perhaps 200 to 300 mi (325 to 500 km) per day. Air flow tends to converge on the eastern, or rear, side of the wave axis. This causes the moist air to be lifted and to break out into scattered showers and thunderstorms. The rainy period may last for a day or two.

Both monsoon depressions and easterly waves are reinforced and the rainfall increased by the orographic effect of mountainous coastal zones.

The Tropical Cyclone—An Environmental Hazard

One of the most powerful and destructive types of cyclonic storms is the *tropical cyclone*, otherwise known as the *hurricane* or *typhoon*. The storm develops over oceans in latitudes 8° to 15° N and S, but not close to the equator, where the Coriolis force is extremely weak. In many cases an easterly wave simply deepens and intensifies, growing into a deep circular low. High sea-surface temperatures, which exceed 80° F (27° C) in these latitudes, are of basic importance in the environment of storm origin (Figure 5.37). Warming of air at low level creates instability and predisposes toward storm formation. Once formed, the storm moves westward through the trade wind belt. It may then curve northwest and north, finally penetrating well into the belt of westerly winds.

The tropical cyclone is an almost circular storm center of extremely low pressure into which winds are spiraling with great speed, accompanied by very heavy rainfall (Figure 5.38). The diameter of the storm may be 100 to 300 mi (150 to 500 km); the wind velocities range from 75 to 125 mi (120 to 200 km) per hour, sometimes much more; and the barometric pressure in the center commonly falls to 965 mb (28.5 in. 72.4 cm) or lower (Figure 5.39).

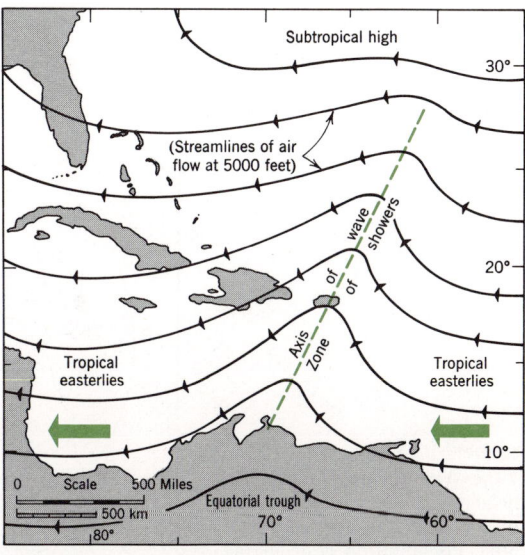

Figure 5.36 An easterly wave passing over the West Indies. (After Riehl.)

135

Atmospheric Energy Releases

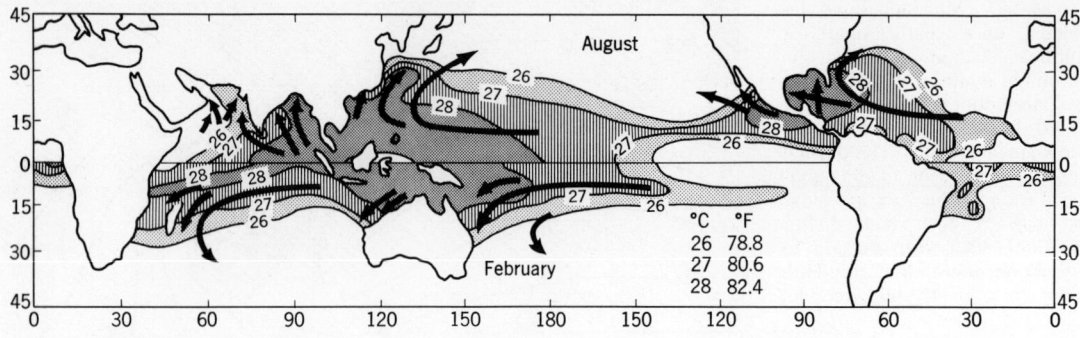

Figure 5.37 Typical paths of tropical cyclones in relation to sea surface temperatures (°C) in summer of the respective hemisphere. (After Palmén, 1948.)

Figure 5.38 is a weather chart of an intense hurricane passing over the eastern tip of Cuba. Winds spiral counterclockwise into the storm center and heavy precipitation occurs in spiral bands (Figure 5.40). These bands are rows of cumulonimbus clouds, seen in cross section in the three-dimensional diagram, Figure 5.41. A cirrus cloud layer spreads downwind at high levels.

Particularly interesting is the presence of a *central eye*, or hole in the clouds, at the storm center. Here air is sinking rapidly and, being adiabatically warmed, is clear and its temperature abnormally high. As the central eye passes over a given location the winds abruptly cease and the sky becomes clear. After the center has passed, perhaps a half hour later, a dense cloud wall arrives and with it a renewal of powerful winds, but these are now reversed in direction.

Tropical cyclones originate as weak lows, and these may intensify into a tropical storm and finally into a full-fledged hurricane or typhoon. Energy is derived from the warm sea surface beneath. For Hurricane Carla (1961) the total latent-heat flux from sea surface to storm was estimated at 21.5×10^{17} calories in a 24-hour period, for an average value of 6×10^{13} calories per second. The

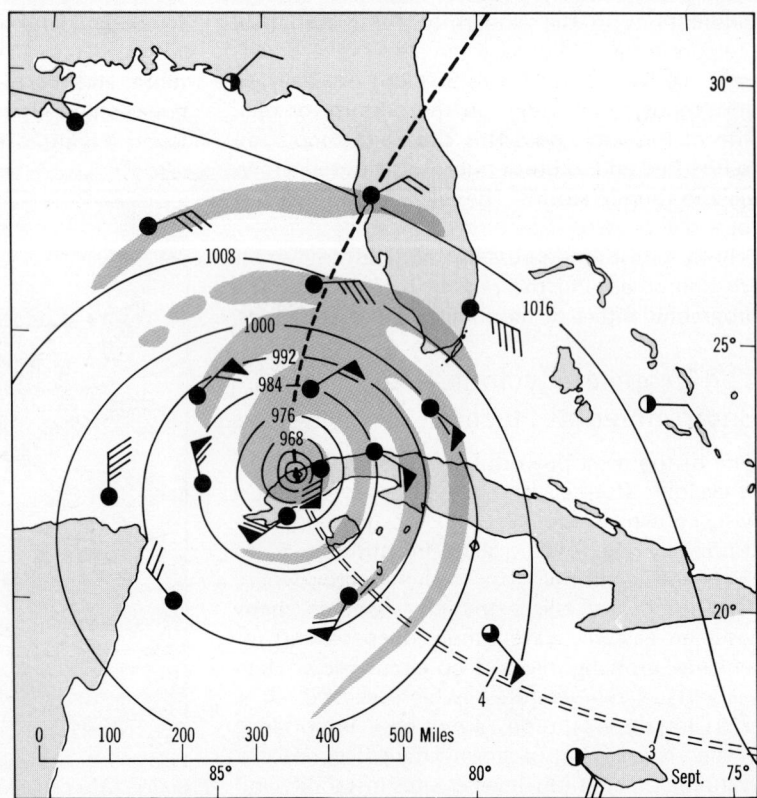

Figure 5.38 A typical hurricane of the West Indies, shown on a surface weather map. Eye of the storm is over the western tip of Cuba. Precipitation is occurring in the shaded areas. Dashed line is storm track.

The Tropical Cyclone—An Environmental Hazard

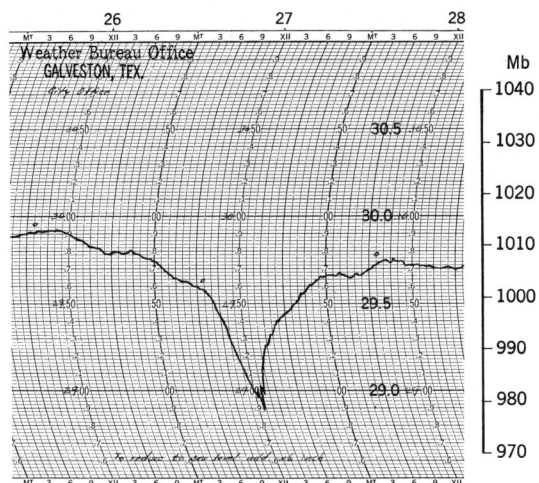

Figure 5.39 Trace sheet from a barograph at Galveston, Texas, during the hurricane of July 27, 1943. (NOAA, National Weather Service.)

sea-surface temperature was lowered by more than 3 F° (1.5 C°) by this great heat flux.

World distribution of tropical cyclones is limited to six regions, all of them over tropical and subtropical oceans (Figure 5.37): (1) West Indies, Gulf of Mexico, and Caribbean Sea; (2) western North Pacific, including the Philippine Islands, China Sea, and Japanese Islands; (3) Arabian Sea and Bay of Bengal; (4) eastern Pacific coastal region off Mexico and Central America; (5) south Indian Ocean, off Madagascar; and (6) western South Pacific, in the region of Samoa and Fiji Islands and the east coast of Australia. These storms are unknown in the South Atlantic, perhaps because sea-surface temperatures do not exceed 80.6° F (27° C). Tropical cyclones never originate over land, although they often penetrate the margins of continents.

Paths, or tracks, of tropical cyclones of the North Atlantic (Figure 5.42) show that most of the storms originate at 10° to 20° latitude, travel westward and northwestward through the trades, then turn northeast at about 30° to 35° latitude into the zone of the westerlies. Here the intensity lessens and the storms change into typical extratropical cyclones. In the trade wind belt the cyclones travel some 6 to 12 mi (10 to 20 km) per hour; in the westerlies, from 20 to 40 mi (30 to 60 km) per hour.

The occurrence of tropical cyclones is restricted to certain seasons of year, depending on the global location of the storm region. Those of the West Indies, and off the western coast of Mexico, occur largely from May through November, with maximum frequency in late summer or early autumn. Those of the western North Pacific, Bay of Bengal, and Arabian Sea are spread widely through the year but are dominant from May through November. Those of the South Pacific and south Indian oceans occur from October through April. Thus, they are restricted to the warm season in each hemisphere.

The environmental importance of tropical cyclones lies in their tremendously destructive effect upon island and coastal habitations (Figure 5.43). Wholesale destruction of cities and their inhabitants has been reported on several occasions. A terrible hurricane which struck Barbados in the West Indies in 1780 is reported to have torn stone buildings from their foundations, destroyed forts, and carried cannon more than a hundred feet from their locations. Trees were torn up and stripped of their bark. More than 6000 persons perished there.

Coastal destruction by storm waves and greatly raised sea level is perhaps the most serious effect of tropical cyclones. Where water level is raised by strong wind pressure, breaking storm waves attack ground ordinarily far inland of the limits of wave action. A sudden rise of water level, known as a *storm surge* may take place as the hurricane moves over a coastline. Ships are lifted bodily and carried inland to become stranded. If high tide accompanies the storm the limits reached by inundation are even higher. The terrible hurricane disaster at Galveston, Texas, in 1900 was wrought largely by a sudden storm surge inundating the low coastal city and drowning about 6000 persons. At the mouth of the Hooghly River on the Bay of Bengal, 300,000 persons died as a result of inundation by a 40-foot (12-meter)

Figure 5.40 Hurricane Gladys, photographed on October 8, 1968, from *Apollo 7* spacecraft at an altitude of about 110 mi (180 km). The storm center was about 150 mi (240 km) southwest of Tampa, Florida. (NASA photograph.)

Atmospheric Energy Releases

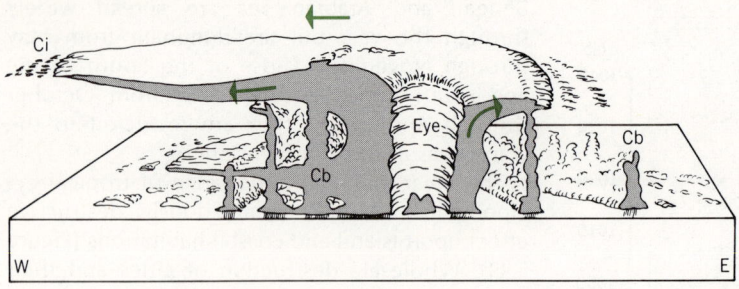

Figure 5.41 Schematic three-dimensional drawing of a typical hurricane. The front section cuts through the eye. Cumulonimbus clouds rise stalklike through dense stratiform cloud layers. Width of diagram is about 600 mi (1000 km); highest clouds are at an altitude often over 30,000 ft (9 km). (Redrawn from R. C. Gentry, 1964, *Weatherwise*, vol. 17, p. 182. Data of NOAA, National Weather Service.)

storm surge which accompanied a severe tropical cyclone in 1737. A similar disaster took place in November, 1970, when a cyclone struck the deltaic coast of East Pakistan and set up a storm surge estimated to have been 25 to 30 ft (8 to 10 m) in height. This wave completely inundated several low-lying islands, drowning tens of thousands of persons in minutes. Total deaths from drowning were estimated at more than 300,000 persons; the ensuing starvation and disease added many more victims. In fact, this event has been labeled as the worst natural disaster of the twentieth century. Low-lying coral atolls of the western Pacific may be entirely swept over by wind-driven sea water, washing away palm trees and houses and drowning the inhabitants.

Of environmental importance, too, is the large quantity of rainfall produced by tropical cyclones. A considerable part of the summer rainfall of certain coastal regions can often be traced to a few such storms. Hurricane Diane, which struck New England in August, 1955, not only caused a loss of 200 lives and heavy damage to property, but also produced unprecedented river floods in Connecticut, Rhode Island, and Massachusetts. Flooding rains accompanying typhoons also occur along the coasts of China and Japan. Destructive mudslides and avalanches are a consequence of such drenching rains falling upon steep mountain slopes.

Table 5.3 gives some comparative statistics on average yearly deaths and property damage caused by four forms of severe weather phenomena described in this chapter. While hurricanes account for the largest dollar value of property damage, their annual death toll is less than that for tornadoes and lightning. Hail is not a producer of fatalities, but results in very heavy property damage, mostly to crops in the Great Plains and Mississippi Valley region.

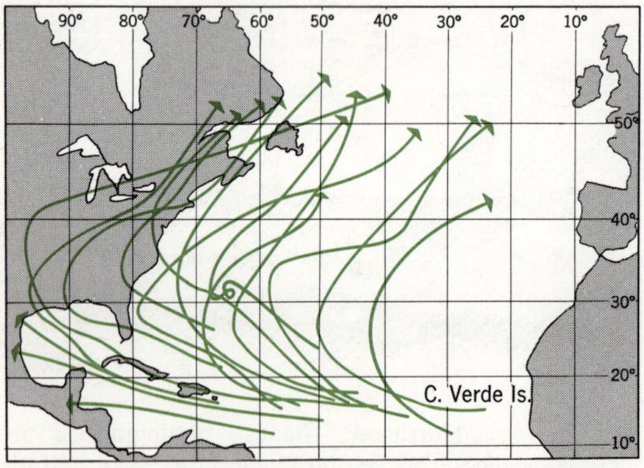

Figure 5.42 Tracks of some typical hurricanes occurring during August. (After U.S. Navy Oceanographic Office.)

Figure 5.43 This devastation along the south coast of Haiti was caused by Hurricane Flora on October 3, 1963. (Miami News photo.)

TABLE 5.3 Losses of Life and Property in the United States Caused by Severe Weather Phenomena

Phenomenon	Average Annual Deaths	Average Annual Property Damage (millions of dollars)
Tornado[a]	125	75
Lightning	150[b]	100[b]
Hail	—	284[c]
Hurricane[a]	75	500

SOURCE: E. Kessler, 1970, *Bull. Amer. Meteorological Soc.,* vol. 51, no. 10, p. 962.
[a] Data of Environmental Data Service, NOAA, for period 1955 to 1969.
[b] Period 1959-1965. Property damage includes building fires set by lightning.
[c] Period 1958-1967. Most is crop damage.

World Precipitation Regions

Patterns of convectional, orographic, and cyclonic precipitation are governed by the source regions and trajectories of air masses and by the unique configurations of continents and their mountain ranges. When we put all of these controls together, some seven world precipitation regions emerge. The strong control which precipitation in turn exerts upon the biosphere, primarily upon the distribution of land plants, makes it important to know about these precipitation regions and their characteristics.

Annual average precipitation is shown on a world map by lines of equal depth of water per year, known as *isohyets* (Figure 5.44). Although this map fails to show seasonal effects, such as a wet season alternating with a dry season, it does reveal the global contrasts in precipitation from place to place.

The seven precipitation regions are as follows:

1. The *wet equatorial belt* of heavy convectional rainfall lies astride the equator. Here rainfall is copious in all, or nearly all, months of the year and totals over 80 in. (200 cm). Cause of this rainfall has been fully covered in earlier pages.
2. *Windward tropical coasts,* or *trade-wind littorals,* having annual rainfall totals of 60 to 80 in. (150-200 cm) or more, extend from near the equator to latitudes 25° to 30° N and S on the eastern continental margins. As we have seen, the orographic mechanism derives rainfall from moist maritime tropical air masses moving with the tropical easterlies and intensified by easterly waves. Good examples of this rainfall regime are along the coastal zones of Central America, northeastern Australia, and the islands of Madagascar, Formosa (Taiwan), and southern Japan. The annual regime of rainfall shows a marked decrease, or even a dry season, associated with the time of low sun.
3. *Tropical deserts* lie approximately astride the tropics of Cancer and Capricorn (23½° N and S). Here annual rainfall is generally less than 10 in. (25 cm), and over large areas is less than 4 in. (10 cm). The dry zones extend westward far out over the adjoining oceans. Dryness is due to persistent subsidence of air in the subtropical high pressure cells.
4. *Middle-latitude deserts* occupy continental interiors in the Northern Hemisphere of both North America and Eurasia in the latitude range 30° to 50°. Here annual precipitation is generally less than 12 in. (30 cm), but with small areas having 4 in. (10 cm) or less. Bordering regions having from 12 to 16 in. (30 to 40 cm) are known as *steppes*. Cause of aridity is complex but is basically due to the isolation of the continental interiors from sources of maritime air masses. The rain-shadow effect with respect to the westerlies is particularly strongly felt in North America, and to a lesser degree in South America. The mountain chains of southern Asia also block off maritime air masses during the wet monsoon.
5. *Humid subtropical* regions are situated on the southeastern sides of North America and Asia, and on the eastern sides of South America and Australia as well. The latitude range is 25° to 40°. These regions, which extend far inland, receive 40 to 60 in. (100 to 150 cm) of precipitation annually; mostly as rain, but some as winter snow. Much of the moisture is derived from maritime tropical (*mT*) air masses moving landward and poleward on the western sides of the oceanic subtropical high pressure cells. Encountering frontal activity over the continents these unstable air masses yield heavy summer rain, most of which is convectional.
6. *Middle-latitude west coasts,* in the latitude range of about 35° to 65° in both hemispheres are narrow zones of heavy precipitation, largely induced by the orographic mechanism acting upon maritime polar (*mP*) air masses in the zone of the prevailing westerlies. Precipitation is from 40 to 80 in. (100 to 200 cm) annually and reaches even

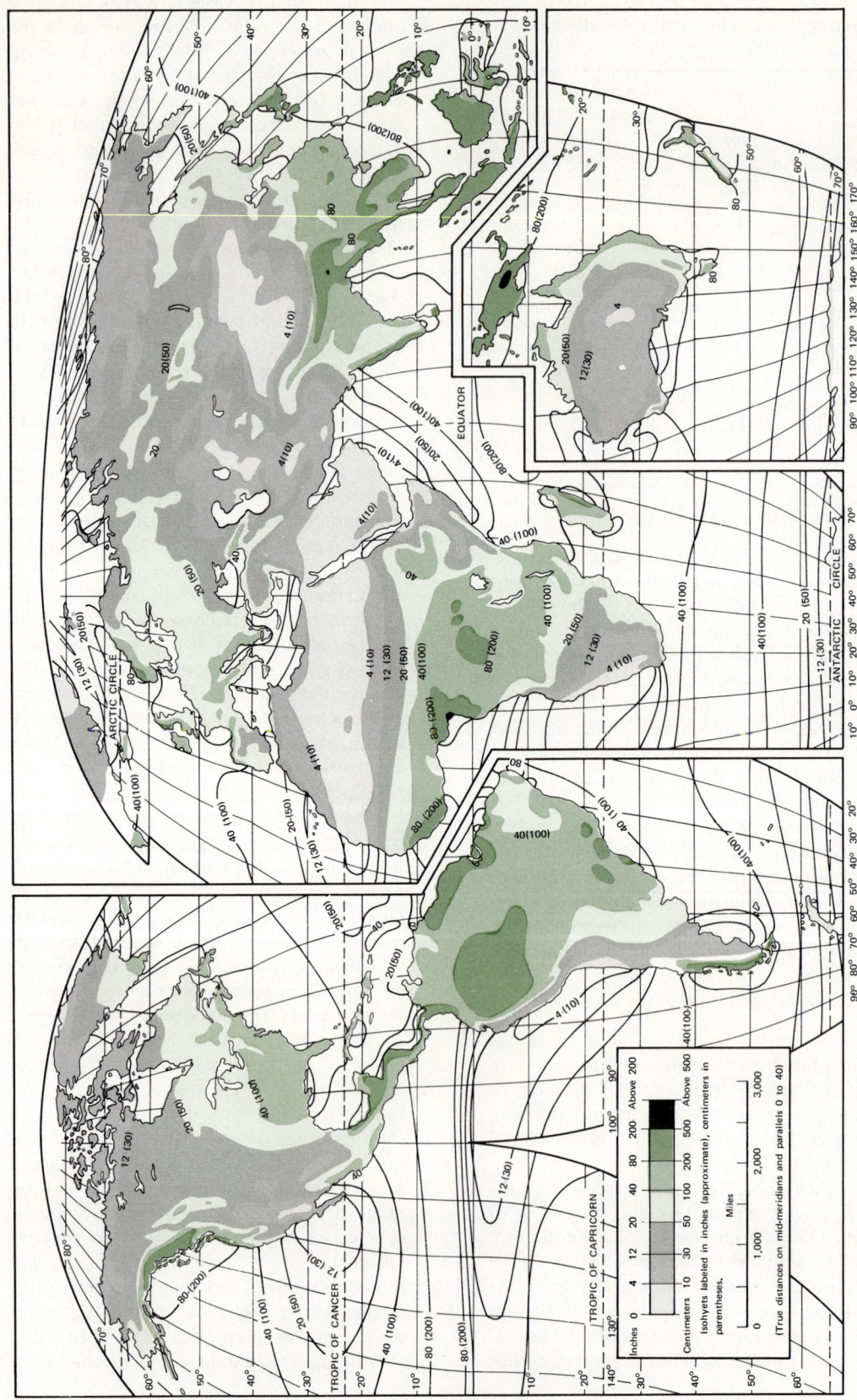

Figure 5.44 World precipitation. Isohyets show mean annual precipitation in inches. (Isohyets modified and simplified from *The Times Atlas*, John Bartholomew, Editor, The Times Publishing Company, Ltd., London, 1958, Volume I, Plate 3). Based on Goode Base Map. Copyright by the University of Chicago. Used by permission of the Department of Geography.

Water Balance of the Atmosphere

higher values where high mountains lie close to the sea. For example, in British Columbia and Patagonia, annual totals locally exceed 120 in. (300 cm). In the higher latitudes much of this precipitation falls as winter snow.

7. *Arctic and polar deserts* comprise the seventh precipitation region. These are cold deserts, in which total precipitation is 12 in. (30 cm) or less, mostly as snow. On the other hand, water is abundantly present as liquid water or ice in the soil, and relative humidity levels may run high.

Water Balance of the Atmosphere

Now that we have analyzed the processes of condensation of water vapor in the atmosphere, we can set up an atmospheric water balance on a global scale. This is a balance of matter, rather than of energy. (Energy was covered in the global radiation and heat balances.) Consequently, we will use as our units the mass of water (kilograms) gained or lost per unit of time (hour, day, month, or year).

For a unit column of the atmosphere, the water balance equation involves the following terms: First, water can enter the column as water vapor, evaporated from land or water surfaces beneath. This term is given the symbol E (kilograms per year). Since transpiration from plants is included, E also stands for evapotranspiration, defined in Chapter 3. Second, there is the horizontal import or export of water vapor into the column by means of the atmospheric circulation; this quantity is designated as c. A positive value of c designates import; a negative value, export. Third, there is loss of water by the process of precipitation (including dew and hoarfrost), designated as P. The increase or decrease of water vapor stored in the column will be designated by g. The water balance equation is then

$$g = E - P - c$$

Now, as an annual average, the change in storage, g, will be practically nil, so that this term can simply be set equal to zero, giving

$$O = E - P - c$$
$$\text{and } c = E - P$$

So we see that the net delivery of water vapor to the column by air circulation, c, is equal to the difference between evaporation (evapotranspiration) and precipitation.

Figure 5.45 is a graph of water vapor flux, c, across the parallels of latitude. Since flux is continuous, a smooth curve has been drawn through points plotted for 10-degree units of latitude. Values above the line are for net northward flux; values below the line for net southward flux. Water vapor is transported in Hadley cell circulation from subtropical high pressure belts toward the equatorial zone, where, as you will recall, it is condensed in convectional precipitation. Also from the Hadley cell, water vapor escapes poleward and is carried in advective circulation of cyclones to polar regions.

Completion of the water balance to include movements of water on lands and in the ocean must wait until Chapter 12 in which terrestrial parts of the hydrologic cycle are explained.

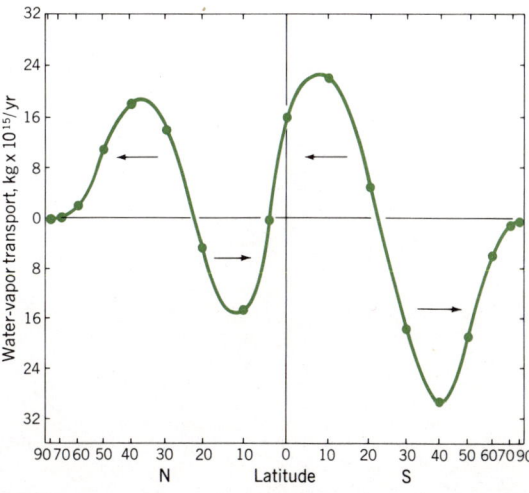

Figure 5.45 Mean annual meridional transport of water vapor. The smooth curve connects points representing the mean value for each 10-degree parallel of latitude. Based on data of W. D. Sellers, 1965, *Physical Climatology*, Univ. of Chicago Press, Chicago, p. 94, Figure 29. (From A. N. Strahler, 1971, *The Earth Sciences*, 2nd ed., Harper & Row, New York.)

Energy Systems Reviewed

The atmosphere and oceans illustrate all three types of energy systems described in the introduction in terms of different time-schedules. First, there is the sun-to-earth radiation system, which, though it shows no signs of running out of fuel, is nevertheless destined ultimately to use up its energy supply and to undergo decay. Second, there are the daily and seasonal rhythms of radiation and heat in response to astronomical controls of unvarying precision. Atmosphere and oceans respond through completely predictable rhythmic changes in the transport and storage of energy and matter. Third, there are irregularly-timed fluctuations in the flow of fluids, due to their characteristic instability. We have in mind the formation and dissolution of upper-air waves and low-level cyclones for which specific long-range prediction is not possible. At an even finer level, in terms of scale of size and difficulty of prediction, are convection cells of thunderstorms and tornadoes.

Energy transport and transformation has been the predominant concern of these early chapters, and we have found that air and water are the forms of matter whose transport provides the pathways of those energy fluxes. Water, in particular, is the key to understanding of the pathways of energy, because water undergoes changes of state and is capable of energy storage and transport in vast quantities as latent heat. Heat and water, which have been traced through many pathways in atmosphere and oceans, are the basic physical environmental controls of life on earth, but this topic remains to be developed in Part 4.

So far we have seen that there are various forms of physical environmental stresses and hazards associated with kinetic energy phenomena of the atmosphere and oceans. Examples were found in storm waves, storm surges, storm winds, hail and lightning. Increasing knowledge of these phenomena will improve predictive processes, which lie in the field of environmental protection. Can Man also take steps to alter the phenomena themselves, so as to ameloriate the intensity of the stress? This question falls within the field of Man-induced environmental changes, and is the subject of the next chapter.

6 Man's Impact upon the Atmosphere

The atmospheric and oceanic circulation systems, which maintain the planetary fluxes of heat and water, operate on such a massive scale that one is at first inclined to doubt that Man's activities could have any measurable effect on the balances of heat and water in the life layer. The enormous energy output of the sun shows no long-range trend of change, and the earth will continue to intercept, absorb, and return that quantity of energy far into the future, subject only to predictable astronomical cycles of variation in distance separating earth and sun.

On the other hand we know that only yesterday, geologically speaking, our planet experienced an ice age during which enormous glacial masses formed on the continents, withdrawing a large quantity of water from the oceans and lowering sea level by more than 300 ft (100 m). Climate zones shifted their positions by hundreds of miles in some areas, causing plant and animal life to make many migrations and adjustments. Primitive Man was on earth during that ice age. His numbers were then so few that he scarcely made any impact upon the biosphere, but perhaps the rigors of the ice age influenced his meteoric evolution as the world-dominating species. We ascribe the profound environmental changes of the ice age to natural causes, and not to Man. This event should make us aware that energy and matter balances of the atmosphere and oceans do fluctuate with time, and that we can expect changes in the future.

In fact, measurements of atmospheric temperature show distinctly that change is in progress in the global heat balance. How much of this measure change is the result of Man's spread over the lands? Agricultural lands have replaced huge areas of forest, converting them permanently to cultivated fields. Industrial countries consume prodigious quantities of hydrocarbon fuels, spewing heat and pollutants into the lower atmosphere. Jet aircraft endlessly weave trails of water vapor, carbon dioxide, and pollutants high in the troposphere and lower stratosphere.

The difficult problem is not to establish that Man's impact upon the atmosphere and oceans is real, for we know it is, but rather to evaluate the quantity of the impact in comparison with natural forces of change. On the one hand some writers have warned us that we can bring on another ice age, or melt off the existing glacial ice and inundate our major cities, or kill off the microscopic plant life of the sea and thus cut off our vital oxygen supply. On the other hand scientists with a deeper understanding of planetary processes are asking us to be very cautious about predicting catastrophe on the basis of current trends in environmental parameters.

This chapter will attempt a sober and cautious evaluation of the effects of Man's activities upon the atmosphere, since we now have at our disposal the basic scientific principles of atmospheric science and of the interactions between atmosphere and oceans, and between atmosphere and lands.

Atmospheric changes induced by Man fall into four categories with respect to basic causes: (1) changes in concentrations of the natural component gases of the lower atmosphere; (2) changes in the water vapor content of the troposphere and stratosphere; (3) alteration of surface characteristics of the lands and oceans in such a way as to change the interaction between the atmosphere and those surfaces; (4) introduction of finely divided solids into the lower atmosphere, along with gases not normally found in substantial amounts in the unpolluted atmosphere.

Carbon Dioxide and Oxygen Levels in the Atmosphere

From the environmental standpoint atmospheric carbon dioxide (CO_2) and free molecular oxygen (O_2) are the most critical of the natural, so-called nonvarying, constituents of pure, dry air (see Figure 1.1). CO_2 is critical because it is an absorber and emitter of longwave radiation as well as a vital compound in plant growth; O_2 because it is chemically very active and plays an essential role in both organic and inorganic processes of the life layer. Although molecular nitrogen (N_2) comprises nearly four-fifths of the bulk of the atmosphere, its inert chemical nature relegates it to a minor environmental role, so far as atmospheric processes are concerned. We shall see later that this huge reservoir of nitrogen is the source of vital nitrogen compounds used in organic processes, but the quantity involved in such uses is quite trivial. Argon (A) is an almost completely inert gas; neither it nor the related inert gases neon, helium, krypton, and xenon, are involved in environmental change. Figure 1.1 lists some very small quantities of molecular hydrogen (H_2), methane (CH_4) and nitrous oxide (N_2O). The last two vary greatly in concentration from place to place in the life layer and are of great environmental importance; they will be referred to again.

Both atmospheric carbon (as CO_2) and O_2 are involved in continuous cycles of interchange between atmosphere and oceans, soil and rock layers, and the biosphere. These cycles are pictured in Figures 19.4 and 19.8. However, you should now examine these diagrams carefully to see how the atmospheric balances of both CO_2 and O_2 are maintained.

As Figure 19.4 shows, carbon is linked with oxygen in CO_2 as a free gas in the atmosphere and in solution in ocean waters and fresh water of lakes, streams, and in soil water. Both land and marine plants withdraw and use CO_2 to create carbohydrate compounds of plant tissues. Animals consume plants, releasing CO_2 back to the atmosphere in the process of biological oxidation (respiration). Decomposition of plant matter also releases CO_2 to the atmosphere. The recycling of CO_2 by organic processes also involves inputs and outputs of energy, and these are explained in Chapter 18.

From the long-range point of view throughout the earth's long history both carbon and oxygen (as CO_2 and H_2O) have been added to the atmosphere by outgassing from the earth's interior. If it had not been for the unceasing work of plants in removing CO_2 and storing it as hydrocarbon compounds in the earth (coal, oil, natural gas), and of marine organisms in removing CO_2 and storing it in sediments as carbonate compounds, the earth would have an atmosphere rich in CO_2, perhaps like that of the planet Venus. Under preindustrial conditions of recent centuries, CO_2 added to the atmosphere by outgassing has been removed and stored at a comparable rate, keeping the atmospheric content fixed at a level of roughly 0.03% by volume, or 300 parts per million (300 ppm). The problem is, of course, that Man has recently begun to extract and burn hydrocarbon fuels at a rapid rate, releasing into the atmosphere the combustion products (principally H_2O and CO_2) and a great deal of heat as well. How does this activity affect our environment?

During the past 110 years there has been an increase in atmospheric CO_2 from about 295 ppm to a 1970 value of about 320 ppm, an increase of about 10%. Moreover, the rate of increase in this period, while slow at first, has become much more rapid toward the end of the period. This increase is shown by the lower curve of Figure 6.1. Now, we also have a fairly good evaluation of the quantity of hydrocarbon fuel burned during the same period, and from this we can calculate the increase in atmospheric CO_2 that would have resulted in the same period, if all of the

Carbon Dioxide and Oxygen Levels in the Atmosphere

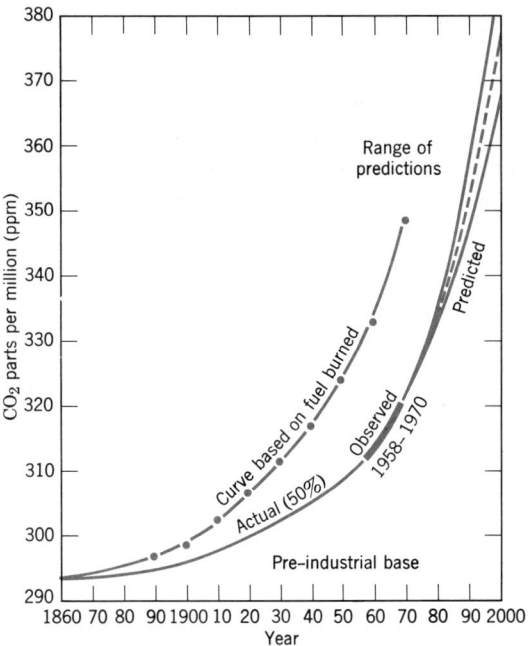

Figure 6.1 Increase in atmospheric carbon dioxide since 1860, with projections to the year 2000. (Data of L. Machta, 1971, as given in *Inadvertent Climate Modification*, SMIC, The MIT Press, Cambridge and London, p. 237, Figure 8.13.)

in part on observations of actual CO_2 values recorded at Mauna Loa, Hawaii, in the period 1958–1970. Projection of this curve into the future shows a CO_2 value of 375 ppm in the year 2000. Other estimates place the 2000 value somewhat higher or lower, as shown by the diverging lines in the upper right-hand part of Figure 6.1.

Let's turn next to consider the environmental effects to be anticipated from an increase of atmospheric CO_2. Because CO_2 is an absorber and emitter of longwave radiation, its presence in larger proportions will raise the level of absorption of both incoming and outgoing radiation, changing the energy balance so as to raise the average level of sensible heat in the atmospheric column. Thus a general air temperature rise is the anticipated result. So, we shall want to look next at available temperature data to see if such an increase has occurred.

Figure 6.2 shows the observed change in mean hemispherical air temperature for approximately the past century. From 1920 to 1940, when fuel consumption was rising rapidly, the average temperature increased by about 0.6 F° (0.4 C°). This relationship follows the predicted pattern. However, since 1940 the graph shows a drop in temperature, despite the rising rate of fuel combustion. Assuming that the atmospheric warming because of increased CO_2 is a valid effect in principle, some other factor, working in the opposite direction, has entered the picture and its cooling effect has outweighed that of warming by CO_2. (This mystery will be explored on later pages.)

When we calculate in terms of complete combustion of all of the estimated world supplies of hydrocarbon fuels, and make the assumption that 50% of the CO_2 thus produced is added to the atmospheric quantity, an average air temperature increase of between 2 and 4 F° (1.1 and 2.2 C°) is indicated, assuming this cause alone is acting. Recent calculations show that as CO_2 increases,

additional CO_2 had remained in the atmosphere. It is estimated that 40 to 50% of the CO_2 produced has remained in the atmosphere. The excess CO_2 has been removed from the atmosphere, most of it going into solution in ocean waters, but some of it perhaps going into increased rate of plant growth. The upper curve of Figure 6.1 shows estimated concentration of CO_2 if none had been absorbed by the oceans. It is based on an assumed absorption of 50% of the industrial production.

It is estimated that 100 billion tons of CO_2 is exchanged annually between atmosphere and a shallow upper layer of the oceans. The extra amount of CO_2 from fuel combustion is therefore a very small part of the total exchange. Given sufficiently long time after the establishment of a constant value of annual input of CO_2 into the atmosphere (should fuel consumption be leveled off), the oceans would absorb the annual increment and place it into storage as carbonate sediment at the same rate; a new steady state would become established. However, the response rate of the oceans is very slow—some thousands of years—so that the immediate prospect is for continued increase in the atmospheric content of CO_2.

The lower curve in Figure 6.1 shows one of several recent calculations of the rising atmospheric CO_2 content from 1860 to 1970. It is based

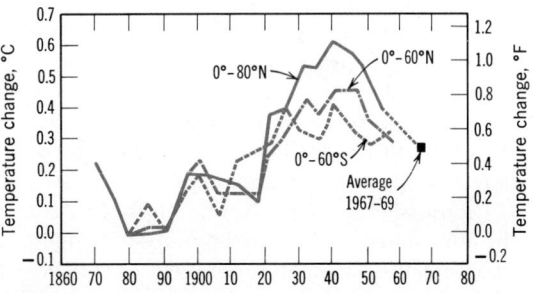

Figure 6.2 Trends in mean annual air temperature for three latitude ranges. (After J. M. Mitchell, Jr., 1970, in *Global Effects of Environmental Pollution*, S. F. Singer, ed., Springer-Verlag, New York, p. 142, Figure 1.)

its temperature-raising influence falls off, so that no really serious run-away heating effect need be feared.

Two feed-back mechanisms may come into play as temperature rises. First, as sea-water temperature increases, more CO_2 is released into the atmosphere. This mechanism would tend to accelerate the air temperature rise. Second, higher temperatures would increase evaporation and raise water vapor content of the atmosphere; in turn increasing cloudiness and the earth's albedo. With more solar radiation turned back into space atmospheric temperatures would be lowered. These two mechanisms tend to counteract each other's effects in a self-regulatory, thermostat-like mechanism, and it is very difficult to predict the outcome.

The possible limit of temperature increase from rising CO_2 is thought to be on the order of 4 F° (2 C°). If a temperature increase actually takes place in this amount, we can anticipate melting of part of the great glacial ice accumulations of Antarctica and Greenland, which in turn will result in a small rise in sea level. One must be very cautious about predicting the extent of ice melting, since the projected temperature increase might not be uniformly felt over the entire globe. Warming of the sea surface beneath maritime air mass sources would increase the water vapor content of those air masses and they would then yield more precipitation. So the result might even be that ice sheets would receive more nourishment and would grow, rather than shrink.

Turning back again to atmospheric oxygen, examine the oxygen cycle as shown in Figure 19.8 Molecular oxygen, O_2, moves in the reverse direction from CO_2 during exchanges between atmosphere and biosphere. Oxygen placed in long-term storage as hydrocarbon compounds in the earth (peat, coal, petroleum) and as carbonate sediments (limestone) balances out the rate at which oxygen (as CO_2) is added to the atmosphere by outgassing. However, Man's large scale combustion of hydrocarbon fuels requires that a large quantity of O_2 be withdrawn from the atmosphere, to be converted with hydrocarbon into CO_2 and H_2O. Combustion releases heat as well. Now the question of environmental significance is: what will be the impact of large-scale withdrawal of free oxygen and addition of heat?

The possibility of a lowering of O_2 content of the atmosphere to levels deleterious or dangerous to animal life has been the subject of concern to a number of environmentalists in recent years. Besides combustion, another source of depletion of atmospheric oxygen has been seen in the possible wholesale destruction of marine plants (phytoplankton) by pollutants. It was argued that the loss of this oxygen source would be a serious threat to animal life, or in any case would have a modifying influence equal to raising the continental life zone to a comparatively high elevation above sea level. However, recent analysis of the oxygen balance and storage capacity have satisfactorily shown that there can be no oxygen depletion by these causes of sufficient magnitude to constitute a serious threat to life.

On the other hand, the increased production of heat by fuel combustion is an environmental impact to be reckoned with. You will recall that this subject was discussed in our introduction to energy systems. At present, the added input of heat from power generation (fuel combustion and nuclear power) is an extremely tiny fraction of the longwave energy radiated from the earth, so that its effect on the planetary heat balance is negligible. We noted that when the existing rates of increase in power generation are projected into the future, an increase of about 2 F° (1 C°) in the earth's mean air temperature from the added heat of combustion might be expected in about 90 years. While the hydrocarbon fuel supply lasts, the warming effect of increased CO_2 can be expected to supplement the direct effect of power-generation heat.

Man-induced Changes in Atmospheric Water Vapor, Clouds, and Precipitation

Like CO_2, water vapor in the atmosphere plays a vital role in the radiation and heat balances through its capacity to absorb and emit longwave radiation. The major difference in roles of these two gases lies in the great place-to-place variation in water vapor content, a fact emphasized in Chapter 5. Figure 5.4 shows this variation with latitude, but since these are average values, the total range is even greater. Should Man's activities increase or reduce the quantity of water vapor present in the atmosphere there should follow a corresponding change in the radiation and heat balances. But, while we can measure CO_2 content and document its steady increase, it is impossible to measure a change in average global atmospheric water vapor content. This difficulty of measurement comes about from the large place-to-place and time-to-time variations in water vapor content. So we will have to approach the problem from a deductive standpoint.

Because the capacity of air to hold water vapor increases with temperature, an overall global rise of air temperature, such as that documented for the period 1920 to 1940, would be expected to result in a larger storage of water vapor in the global atmosphere. It seems reasonable that there would also be increased precipitation, since more water vapor would be available for condensation. While no such increase has been documented, the opposite effect in times past seems to be demonstrated by evidence from conditions prevailing during the last ice age, or Pleistocene glacial epoch (see Chapter 17). When ice sheets had spread widely over the continents, the average air temperature was lowered substantially (as shown by evidence of lowering of the snow line on mountains). The tropical deserts underwent an equatorward expansion, while the wet equatorial belt was correspondingly narrowed. This evidence suggests that reduced air temperature was associated with reduced precipitation generally in low latitudes, and this effect is in line with our deductive reasoning.

Introduction of large amounts of water into the atmosphere by combustion of fuels accompanies the emissions of CO_2 and heat. Does this added water vapor also build up in the atmosphere, as does CO_2? The answer is a probable "no," since water vapor readily returns to the oceans as precipitation by the self-regulating process described in the previous paragraph.

A special case meriting our attention is that of the emission of water vapor, CO_2, heat, and various other substances by jet aircraft, since these emissions occur at high levels in the troposphere, where water vapor content is normally very small. In particular, the SST (supersonic transport) aircraft will fly within the stratosphere, where water vapor content is extremely small, and the low density of the air makes the water vapor and CO_2 emissions comparatively large per unit mass of air, as compared with the lower troposphere. A substantial increase in both of these gases might have a significant warming effect through increased absorption of longwave radiation. Obviously, until large fleets of SST aircraft are in action, this possibility cannot be evaluated, but it has been predicted that the increase in stratospheric water vapor content from this source will prove to be negligible.*

Jet aircraft high in the troposphere produce condensation trails, known as *contrails*, which you have probably seen many times, lacing the blue sky in various directions on a clear day. Perhaps you have noticed that when conditions are right, a contrail spreads laterally and becomes a cirrus cloud. When many such clouds are being produced the cirrus layer may cover much of the sky above you. There has been considerable speculation about the effects of contrail-induced clouds upon the earth's radiation and heat balances, since these clouds are highly reflective. It has been found from data of orbiting earth satellites that the albedo of the planet as a whole depends largely upon its cloud cover. If we increase the albedo by producing contrail clouds, the energy absorbed by the planet will be proportionately reduced and the average temperature must decrease. Again, detection of this effect is not yet possible, but it should be kept in mind in evaluating the results of greatly increased jet aircraft activity in lanes of dense air traffic.

A considerable quantity of water vapor enters the atmosphere from the surfaces of the lands. Actually, this quantity is about one-seventh of that entering from evaporation of ocean surfaces, but it is important in the earth's water balance, treated in Chapter 12. Man's agricultural activities have for centuries been altering the surface characteristics of the lands through deforestation and the maintenance of large areas in crop cultivation. Further landscape alteration through urbanization is adding rapidly to these surface changes. One effect of such surface alterations is to change the rate of evapotranspiration. Complete removal of a dense forest cover will sharply reduce transpiration and thus reduce the quantity of water returning to the atmosphere in vapor form. Just what large-scale effect, if any, surface changes made by Man have had upon the moisture properties of air masses of great vertical extent is uncertain, although we know that climate near the ground is profoundly influenced by agricultural development and urbanization.

There is, however, one form of agricultural activity that may result in measurable changes in precipitation; namely large-scale crop irrigation practiced in semiarid regions. A case in point is found in the High Plains region of Texas, Oklahoma, western Kansas, and western Nebraska. This region is one of sparse grasses and normally has dry soils through the summer; evapotranspiration is then very low. In the last three decades irrigation has been developed throughout this area on a large scale, greatly increasing the evapotranspiration in the summer months. During the past 13-year period increases in July rainfall amounting to 20 to 50% have been recorded over the main irrigated areas. These increases are with respect to the average of the previous 60- to 70-year period. If irrigation is actually responsible for the increased precipitation—and this connection is only hypothesis—the mechanism of cause and effect remains uncertain. A likely possibility is that the added water vapor increases instability

---

*H. E. Landsberg, 1972, personal communication.

of the lower air layer and acts as a triggering mechanism to set off more frequent and more intense convection in the air mass above. This linkage is reasonable since convective precipitation is the dominant rainfall type in summer in this region. However, a much longer period of observation will be needed to establish that a significant precipitation increase has occurred.

Planned Weather Modification

We now come to a related topic, that of Man's deliberate attempts to modify atmospheric processes in order to produce more precipitation, or to ameliorate the severity of storms. Almost everyone knows that throughout human history rainmaking to break severe droughts has been attempted by all sorts of practitioners in many lands. The rainmaker still plies his trade, feeding off of the gullibility of farmers desperate in the need to save their crops from disaster. Early scientific experiments with rainmaking, making use of principles of atmospheric science, were conducted in the 1940s by Irving Langmuir and Vincent Schaefer. These scientists used the principle of *cloud seeding*, mentioned in Chapter 5 as the natural process in cumulonimbus clouds wherein ice crystals falling from the anvil cloud top serve as nuclei of condensation at lower levels. Langmuir and Schaefer released dry-ice (solid CO_2) pellets into a stratus cloud and found that there resulted a rapid growth of ice particles at the expense of liquid particles in a layer of supercooled cloud particles. There followed experimentation by Bernard Vonnegut in which silver iodide smoke was released to provide nuclei. When basic conditions were favorable this method proved capable of intensifying condensation in dense cumulus clouds. The latent heat thus released caused cumulonimbus clouds to form and to yield heavy precipitation. But for many years the efficacy of the method to increase precipitation remained in doubt.

Severe drought in south and central Florida in 1970 and 1971 led to intensification of cloud-seeding experiments begun in 1968 by Joanne Simpson and William L. Woodley of NOAA's Experimental Meteorological Laboratory. By dropping pyrotechnic flares into the tops of individual massive cumulus clouds, these investigators were able to increase rainfall sevenfold. Their latest project has been an attempt to produce mergers between separate clouds, since larger clouds produce disproportionately more rain than smaller ones. Only occasional days are favorable for seeding experiments and the optimum in results is to increase rainfall, but not to break the drought.

Project Skywater, aimed to increase winter snowfall over the San Juan Mountains of southwestern Colorado, was begun in 1969, using the silver iodide method. A snowfall increase of 16% is anticipated. An increase in the snowpack will increase the runoff from mountain streams into the Colorado River system. At the same time, an ecological study will evaluate the effects of both the increased precipitation and the silver iodide fallout upon plants and animals of the mountain region.

Increasing precipitation by seeding methods can have success only where large masses of air have high moisture content and are basically unstable. One situation that seems particularly favorable is that of orographic precipitation. Here moist maritime air masses are brought continually to higher levels over a fixed topographic feature. Seeding of the cloud mass, constantly forming over the barrier, may substantially increase precipitation on windward slopes of the mountain range, augmenting the winter snowpack and increasing the annual runoff of water available for use in irrigation and urban water supplies.

A new twist is added to the field of planned weather modification in the Great Lakes region. Here very heavy snowfalls, totalling as much as 100 in. (2.5 m) per year, occur along the lee shores of the lakes when cold, dry polar air moves southward over the relatively warmer lake surfaces (Figure 6.3). Picking up heat and moisture from the lake surface, the lower air layer becomes unstable. Upon reaching the southern shore, this moisture is deposited in heavy snowfalls in a very narrow zone. Cities such as Buffalo, New York, intensify the snowfall because the city heat and air pollutants reinforce the tendency for air lift and precipitation. It is now proposed that cloud seeding techniques be developed to cause precipitation to occur over a much wider zone, extending farther inland, and thus reduce the snow concentration over the narrow coastal zone.

An attempt to ameliorate the severity of a hurricane by cloud seeding methods was made in August 1969 by scientists of NOAA. The experiment, known as *Project Stormfury*, apparently produced significant results. Five seedings at 2-hour intervals are alleged to have brought about a 30% reduction in wind speeds. One day later the winds had again intensified, but a second multiple seeding program brought another apparent reduction in wind speeds. Seeding would have acted to cause rapid condensation of supercooled liquid particles in the storm, and conse-

Planned Weather Modification

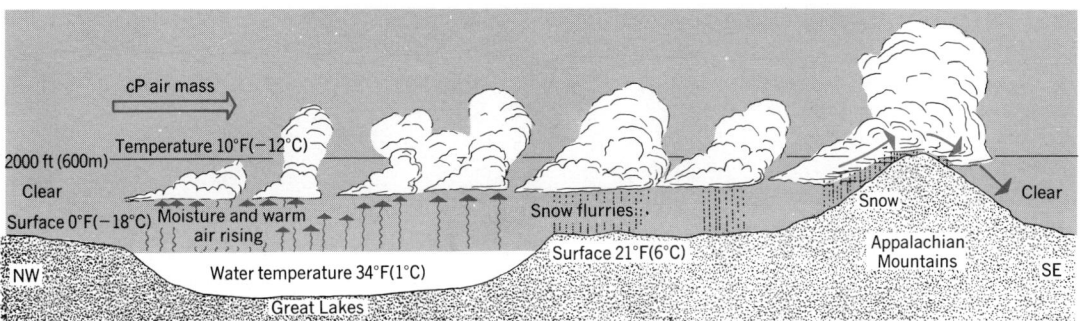

Figure 6.3 Snow flurries produced from water vapor taken up by a polar air mass moving across the Great Lakes. (From A. N. Strahler, 1971, *The Earth Sciences*, Harper & Row, New York.)

quently to quickly drain off the reserves of latent heat. A rise of barometric pressure at the storm center accompanied the reduction in wind speeds. Scientists who conducted the experiment were convinced that hurricane modification was achieved, and that an intensification of research is desirable. Many more trials will be needed to establish efficacy of the treatment.

Yet another aspect of weather modification is that of reducing the severity of hail storms. Annual losses from crop destruction by hail storms are given in Table 5.3 as approaching 300 million dollars. In 1970, insurance companies paid out about 70 million dollars in damages, but uninsured losses ran much higher. Figure 6.4 shows a corn crop destroyed by a single hail storm. Damage to wheat crops is particularly severe in a northsouth belt of the High Plains, running through Nebraska, Kansas, and Oklahoma (Figure 6.5). There occur from four to five hail storms per year at any given place. A much larger region, extending eastward generally from the Rockies to the Ohio Valley, experiences two to four days per year of hail storms at any given place. Much of this larger region is under corn cultivation. A small area of particularly high frequency of hail storms, over eight per year, is situated over the common corners of Wyoming and Nebraska, and a part of northeastern Colorado. This spot is known to atmospheric scientists as *Hail Alley*. A research project begun in 1971 by the National Center for Atmospheric Research (NCAR) at Boulder, Colorado, will undertake to study the structure and dynamics of the isolated cumulonimbus clouds that produce hail in this region. Research will lead to development of cloud-seeding techniques by means of which the severity of a hail storm can be reduced. The principle is that seeding, by supplying vast numbers of nuclei of condensation, can induce the formation of many small ice particles or many small hail stones, rather than the fewer larger stones that would otherwise be formed. The Soviet Union, which has agricultural regions experiencing heavy crop losses through hail, claims to have been successful in developing the seeding technique to reduce hail damage.

Fog dispersal is another form of weather modification that has invited research and experimentation because of its great potential use at airports. Seeding experiments have shown that fog consisting of supercooled droplets can be cleared by seeding, using liquid propane or dry ice (Figure 6.6). Seeding causes rapid transformation of water droplets into ice particles. The very cold fogs to which this method applies are only a small percentage of all fogs that occur in middle and high latitudes. Warm fogs require other methods for dispersal, and these have met with some success.

Figure 6.4 A severe hail storm devastated this corn crop. (NCAR photograph.)

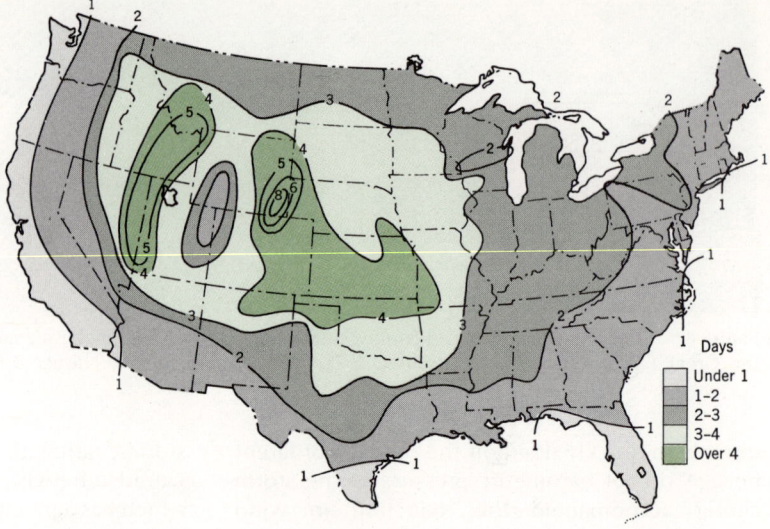

Figure 6.5 Average annual number of days with hail. Map is based on observations at 200 National Weather Service stations in the period 1899–1938. (After U.S. Department of Agriculture.)

It is well to keep in mind that the forms of planned weather modification described here apply to very small areas for short periods of time. No means yet exists to change precipitation appreciably over large areas on a sustained basis.

Urbanization and the Balances of Radiation and Heat

Applying the principles of the radiation and heat balances at the interface of the atmosphere and the solid ground surface, we can anticipate the impact of Man as his cities spread, replacing a richly vegetated countryside with blacktop and concrete. Not only do the thermal properties of the surface change, but the hydrologic factors of evaporation and transpiration are changed as well, altering the water balance itself. The vertical walls of buildings not only add to surfaces of reflection and radiation, but they also change the aerodynamic character of the surface, altering the flow patterns of air and the speed of winds.

In the urban environment the absorption of solar radiation causes higher ground temperatures for two reasons: First, foliage of plants is absent, so that the full quantity of solar energy falls upon the bare ground. Absence of foliage also means absence of transpiration, which elsewhere produces a cooling of the lower air layer through the latent heat flux. A second factor is that roofs and pavements of concrete and asphalt hold no moisture and evaporative cooling does not occur as it would from a moist soil. The thermal effect is that of converting the city into a hot desert. The summer temperature cycle close to the pavement of a city may be almost as extreme as that of the desert floor, illustrated in Figure 3.24. This surface heat is conducted into the ground and stored there. The thermal effects within a city are actually more intense than on a sandy desert floor, because the capacity of solid concrete, stone, or asphalt to conduct and hold heat is greater than that of loose, sandy soil. Because of more rapid conductivity, the solid materials absorb heat to a greater depth than loose dry soil in a given period of heating. An additional thermal factor is that vertical masonry surfaces absorb insolation

Figure 6.6 Airborne cloud seeding cleared the fog from this runway at Elmendorf Air Base, Alaska. Note the dense fog remaining over the far end of the runway. (U.S. Air Force photograph.)

Pollutants in the Atmosphere

or reflect it to the ground and to other vertical surfaces. The absorbed heat is then radiated back into the air between buildings.

As a result of these changes in the radiation and heat balances, the central region of a city typically shows summer air temperatures several degrees higher than for the surrounding suburbs and countryside. Figure 6.7 is a map of the Washington, D.C. area showing isotherms for the afternoon of a typical day in August. The isotherms delineate a *heat island*. You might suppose that at night the city air temperatures would fall below those of the surrounding countryside, since long-wave radiation from bare, dry surfaces would cause rapid heat loss, as it does in the desert. Instead, we find that the heat island persists through the night because of the availability of a large quantity of heat stored in the ground during the daytime hours. In winter additional heat is radiated by walls and roofs of buildings, which conduct heat from the inside. Even in summer, Man adds to the city heat output through use of air conditioners, which expend enormous amounts of energy at a time when the outside air is at its warmest.

A mechanical side effect of the heat island is to increase convectional circulation at night. A general rise of the warm air extends to the upper limit of the polluted air layer. As shown in Figure 6.8, the air then spreads radially outward from the city to reach the surrounding countryside. Cooler air moving into the city at lower levels constitutes a weak *country breeze*. In later pages we will get

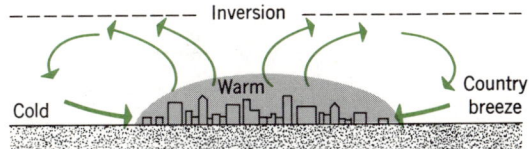

Figure 6.8 Air flow at night within an urban heat island. Dense, cold air from the surrounding countryside flows inward to replace rising, outspreading warm air. (After H. E. Landsberg, 1970, *Science*, vol. 170, p. 1271, Figure 5.)

a better understanding of the relationship of the heat island and its circulation to the domelike structure of the polluted air layer over the city.

The heat-island effect of cities is augmented by the presence of large buildings, which act to reduce wind speeds by as much as 10 to 30% as compared with flat ground surfaces. For example, wind speeds observed in Central Park in New York City averaged from 2.6 to 3.3 mi (4.2 to 5.3 km) per hour less than at La Guardia Airport. These differences existed throughout each quarter of the year and the average reduction was 23%. The Central Park weather station is actually quite far from buildings that surround the park, so that the effect would be even more pronounced among blocks of high buildings.

Pollutants in the Atmosphere

That Man has degraded the quality of the lower atmosphere over densely populated sections of the industrial nations is so obvious to the senses that the statement needs no proof. Industrial activities and related practices of populations in industrialized regions inject into the atmosphere two classes of contaminants or, simply, *pollutants*. First, there are solid and liquid particles, which we shall designate collectively as *particulate matter*. Dusts found in smoke of combustion, as well as droplets naturally occurring as cloud and fog fall into the category of particles. Figure 6.9 gives the scale of sizes of various atmospheric particles. (Note that the word *aerosols* is used in many publications in essentially the same sense as particles, but when properly used *aerosol* describes the entire system of particles of colloidal size, 1 micron or less, together with the gas in which they are suspended.)

Second, there are compounds in the gaseous state that are included in the general terms *chemical pollutants*, in the sense that they are not normally present in measurable quantities in clean air remote from densely populated, industrialized regions. Excess CO_2 produced by combustion is not usually classed as a pollutant. One

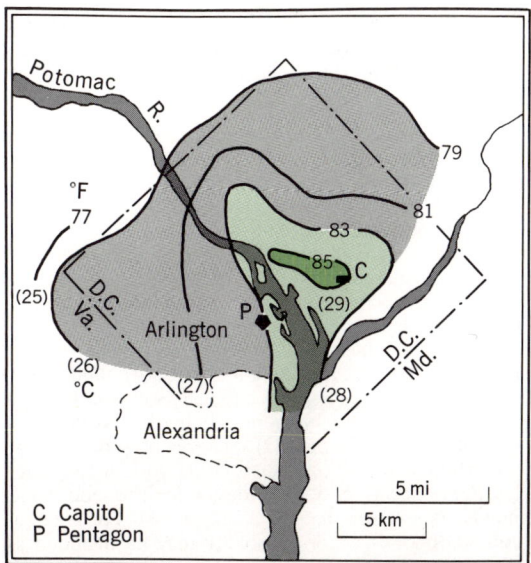

Figure 6.7 Isotherms for 10 P.M. local time show a heat island over Washington, D.C. on a day in early August. (After H. E. Landsberg, 1950, *Weatherwise*, vol. 3, No. 1.)

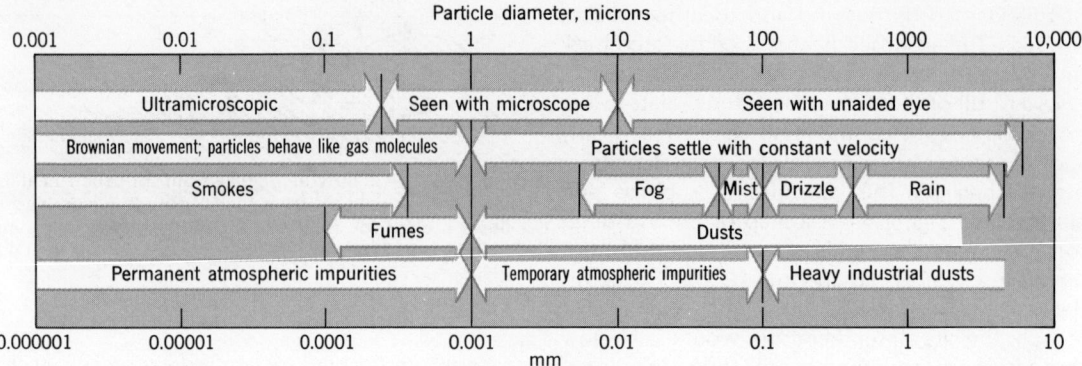

Figure 6.9 Sizes and physical properties of atmospheric particles. (Data from W. G. Frank, American Air Filter Company, Inc., Louisville, Ky. Reference: *Air Conservation*, 1965, Amer. Assoc. Advancement of Sci., Publ. 80, p. 110, Figure 5.)

group of chemical pollutants of industrial and urban areas is of primary origin, e.g., produced directly from a source on the ground. Gases included in this group are *carbon monoxide* (CO), *sulfur dioxide* (SO_2), *oxides of nitrogen* (NO, NO_2, NO_3), and hydrocarbon compounds. However, these chemical pollutants cannot be treated separately from the particulate matter, since they are often combined within a single suspended particle. Certain dusts are said to be *hygroscopic* because they have an affinity for water and easily take on a covering water film (ordinary table salt is an example). The water film in turn absorbs the chemical pollutants. Hygroscopic particles of salt, derived from the sea surface, are normally present in great numbers in the atmosphere.

When particles and chemical pollutants are present in considerable density over an urban area, the resultant mixture is known as *smog*. Just about everyone living in middle latitudes is familiar with smog through its irritating effects upon the eyes and respiratory system and its ability to obscure distant objects. When concentrations of suspended matter are less dense, obscuring visibility of very distant objects, but not otherwise objectionable, the atmospheric condition is referred to as *haze*. Atmospheric haze builds up quite naturally in stagnant air masses as a result of the infusion of various surface materials. Haze is normally present whenever air reaches high relative humidity because water films grow upon hygroscopic nuclei. Nuclei of natural atmospheric haze particles consist of mineral dusts from the soil, crystals of salt blown from the sea surface, hydrocarbon compounds (pollens and terpenes) exuded by plant foliage, and smoke from forest and grass fires. Dusts from volcanoes may, on occasion, add to atmospheric haze.

It's evident at this point that what we are calling atmospheric contaminants, or *pollutants*, are of both natural and man-made origin, and that Man's activities can supplement the quantities of natural pollutants present.* Table 6.1 attempts to illustrate this complexity by listing the primary pollutants according to sources.

Not all man-made pollution comes from the cities. Isolated industrial activities can produce pollutants far from urban areas. Particularly important are smelters and manufacturing plants in small towns and rural areas. Sulfide ores (metals in combination with sulfur compounds) are processed by heating in smelters close to the mine. Here sulfur compounds are sent into the air in enormous concentrations from smokestacks. Fallout over the surrounding area is destructive to vegetation. An example is the Ducktown, Tennessee, smelter located in a richly forested area of the Southern Appalachians. An area of several square miles surrounding the smelter has been denuded of vegetation and resembles a patch of

---

*Despite its wide use, the term *environmental pollution* will rarely be found defined in print. In the light of contexts of its use in current writings, we may define man-made environmental pollution as the introduction directly or indirectly through Man's activities into the atmosphere, hydrosphere, lithosphere, and biosphere of infusions of matter and energy at levels of quantity or intensity appreciably higher than natural levels and usually with undesirable or deleterious effects upon environments of the biosphere. The *pollutant* is the form of energy or matter causing pollution. In *air pollution* the pollutants include gases and particulate solids and liquids of both organic and inorganic chemical classification. *Water pollution* (Chapter 12) includes presence of disease-producing (pathogenic) bacteria and viruses (*biological pollution*) and of undesirable ions and compounds in solution (*chemical pollution*). Presence of suspended solids causing turbidity may be included as forms of water pollution. *Thermal pollution* of air and water, a form of energy infusion, raises the quantity of sensible heat in those fluids to abnormally high levels. *Noise pollution* illustrates energy infusion into the environment by sound-wave transmission.

Pollutants in the Atmosphere

TABLE 6.1 Sources of Primary Atmospheric Pollutants

Pollutants from Natural Sources	Sources of Man-made Pollutants
Volcanic dusts	Fuel combustion (CO_2, SO_2, lead)
Sea salts from breaking waves	Chemical processes
Pollens and terpenes from plants	Nuclear fusion and fission
	Smelting and refining of ores
Aggravated by Man's activities:	Mining, quarrying
Smoke of forest and grass fires	Farming
Blowing dust	
Bacteria, viruses	

SOURCE: Assn. of Amer. Geographers, 1968, *Air Pollution*, Commission on College Geography, Resource Paper No. 2, Figure 3, p. 9.

badlands in the arid lands of the West. Mining and quarrying operations send mineral dusts into the air. For example, asbestos mines (together with asbestos processing and manufacturing plants) send into the air countless thread-like mineral particles, some of which are so small that they can be seen only with the electron microscope. These particles travel widely and are inhaled by humans, lodging permanently in the lung tissue. Nuclear test explosions inject into the atmosphere a wide range of particulates, including many radioactive substances capable of traveling thousands of miles in the atmospheric circulation. Certain of these pollutants belong to a very special and dangerous class, for they are sources of ionizing radiation.

Man-induced forest and grass fires add greatly to smoke palls in certain seasons of the year. Plowing, grazing, and vehicular traffic raise large amounts of mineral dusts from surfaces of deserts and steppes. Bacteria and viruses, which we have not as yet mentioned, are borne aloft in air turbulence when winds blow over contaminated surfaces, such as farmlands, grazing lands, city streets, and waste disposal sites.

Table 6.2 lists major industrial atmospheric pollutants, largely of urban sources, in the United States in 1968. The units are millions of tons per day. Figure 6.10 shows these quantities in percentage form as to both component matter and source. While the figures will increase from year to year, there are some stable relationships that appear in the data. We find that much of the carbon monoxide, half of the hydrocarbons, and about a third of the oxides of nitrogen come from exhausts of gasoline and diesel engines in vehicular traffic. Generation of electricity and various industrial processes contribute most of the sulfur oxides, because the coal and lower-grade fuel oil used for these purposes are comparatively rich in sulfur. These same sources also supply most of the particulate matter. *Fly ash* is the term applied to the coarser grades of soot particles emitted from smoke stacks of generating plants. These particles settle out quite quickly within close range of the source. Very finely divided carbon comprises much of the smoke of combustion and is capable of remaining in suspension almost indefinitely because of its colloidal size. Combustion used for heating is a comparatively minor contributor to pollution, because the higher grades of fuel oil are low in sulfur and are usually efficiently burned. Burning of refuse is also a minor contributor in all categories.

In the smog of cities there are, in addition to the ingredients mentioned above, certain chemical elements contained in particles contributed by exhausts of automobiles and trucks. Included are particles in the size range from 0.02 to 0.06 micron that contain lead, chlorine, and bromine. About half of the particles in automobile exhaust contain lead. Finely ground rubber from automobile tires is another kind of particulate matter contributed to the air over cities.

Primary pollutants are conducted upward from the emission sources by rising air currents that are a part of the normal convective process. The larger particles settle under gravity and return to the surface as *fallout*. Particles too small to settle out are later swept down to earth by precipitation, a process called *washout*. By a combination of fallout and washout the atmosphere tends to be cleansed of pollutants. In the long run a balance is achieved between input and output of

TABLE 6.2 Components of Air Pollution and Their Sources in the United States, 1968

Sources	Contaminants (millions of tons per year)[a]						
	Carbon Monoxide	Particulate Matter	Sulfur Oxides	Hydro-carbons	Nitrogen Oxides	Totals	
Transportation	64	1	1	17	8	91	(83)
Fuel combustion: stationary sources	2	9	24	1	10	46	(42)
Industrial processes	10	7	7	5	0.2	29	(26)
Solid waste disposal	8	1	0.1	2	1	11	(10)
Miscellaneous[b]	17	10	0.6	8	2	37	(34)
Total	100	28	33	32	21	214	(195)
	(91)	(26)	(30)	(29)	(19)		

SOURCE: NAPCA Inventory of Air Pollutant Emissions, 1970. Reference: 1971 Annual Report of the Council on Environmental Quality, U.S. Government Printing Office, Table 1.
NOTE: Figures rounded off.
[a] Metric tons in parentheses.
[b] Largely from forest fires, agricultural burning, and coal waste fires.

pollutants, but there are large fluctuations in the quantities stored in the air at a given time. Pollutants are also eliminated from the air over their source areas by winds which disperse the particles into large volumes of cleaner air in the downwind direction. In middle latitudes the passage of a cold front accompanied by strong winds quickly sweeps away most pollutants from an urban area, but during periods when a stagnant anticyclone is present, the concentrations rise to high values.

In polluted air certain chemical reactions take place among the components injected into the atmosphere, generating a secondary group of pollutants. For example sulfur dioxide (SO_2) may combine with oxygen to produce *sulfur trioxide* (SO_3), which in turn reacts with water suspended droplets to yield *sulfuric acid* (H_2SO_4). This acid is both irritating to organic tissues and corrosive to many inorganic materials. In another typical reaction the action of sunlight upon nitrogen oxides and organic compounds produces *ozone* (O_3) a toxic and destructive gas. Reactions brought about by the presence of sunlight are described as *photochemical*. One toxic product of photochemical action is *ethylene*, produced from hydrocarbon compounds.

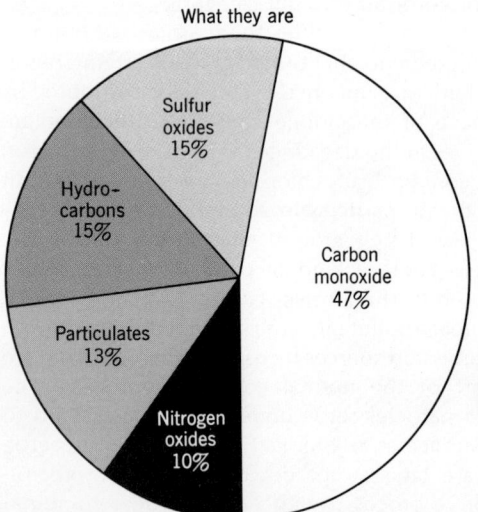
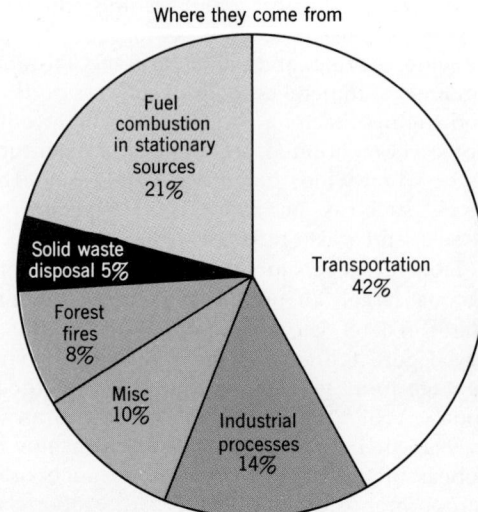

Figure 6.10 Air pollution emissions in the United States, 1968. Percentage by weight. (Data source: National Air Pollution Control Administration, HEW. Reference: 1971 Annual Report of the council on Environmental Quality, U.S. Govt. Printing Office, p. 64.)

Inversion and Smog

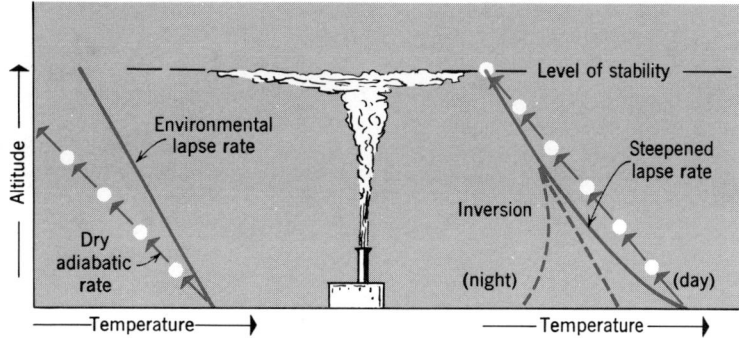

Figure 6.11 Relation of dry adiabatic lapse rate (*circles*) to various environmental lapse rates.

Inversion and Smog

Concentration of pollutants over a source area rises to its highest levels when the vertical mixing (convection) of the air is inhibited by a stable configuration of the vertical temperature profile of the air mass. The principles of stable versus unstable air conditions were covered in Chapter 5 and we shall now apply those principles to the problem of air pollution over cities.

When the normal environmental lapse rate of $3\frac{1}{2}°$ F per 1000 ft (0.6° C per 100 m) is present, there is resistance to mixing by vertical movements, as we have already shown (Figure 6.11, *left*). Air that is forced to rise will cool adiabatically at a rate faster than the normal lapse rate; it will be denser than the surrounding air and will tend to sink back to its original level if allowed to do so. Under these stable conditions, work must be done to lift air to a higher level, and in the absence of a mechanism for performing that work, the air mass remains quiescent.

Consider next, that the environmental lapse rate is steepened by heating of an air layer near the ground, because of excess heat radiated and conducted from hot pavements or roof tops (Figure 6.11, *right*). When the temperature gradient (lapse rate) of the heated air becomes greater than the dry adiabatic rate of 5.5 F° per 1000 ft (1.0 C° per 100 m), a condition of instability exists and a bubble of warm air can begin to rise, like a helium-filled balloon. Assume that the lapse rate lessens with increased altitude, as shown by the curved line in Figure 6.11. Cooled at the dry adiabatic rate, the temperature of the rising bubble then falls faster than does the temperature of the surrounding air. When the bubble has reached an altitude at which its temperature (and therefore also its density) matches that of the surrounding air, it can rise no further and convection ceases.

Now suppose that instead of the bubble of warm air we substitute the hot air from a smokestack (Figure 6.11). The resultant rise follows essentially the same pattern, although initially faster and in the form of a vertical jet. Carrying up with it the pollutants of combustion, the rising hot air gradually cools and reaches a level of stability, where it spreads laterally. Cooling by longwave radiation and eddy mixing with the surrounding air will reinforce the adiabatic cooling, since a truly adiabatic system would not be realistic in nature.

Recall that at night when the air is calm and the sky clear, rapid cooling of the ground surface typically produces a *low-level temperature inversion*, illustrated in Figure 3.24. Such inversions also exist over snow-covered surfaces in winter in polar and arctic air masses and the reversal of the temperature gradient may extend hundreds of feet into the air. A low-level temperature inversion represents an unusually stable air structure. When this type of inversion develops over an urban area conditions are particularly favorable for entrapment of pollutants to the degree that heavy smog or highly toxic fog can develop, as shown in Figure 6.12. The upper limit of the inversion layer coincides with the *cap*, or *lid*, below

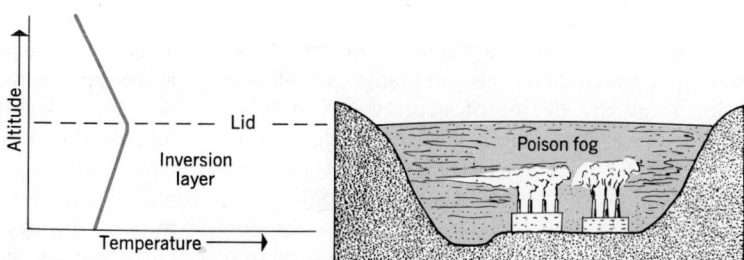

Figure 6.12 A low-level inversion (*left*) was the predisposing condition for poison fog accumulation at Donora, Pennsylvania, in October, 1948. (From A. N. Strahler, 1972, *Planet Earth; Its Physical Systems Through Geologic Time*, Harper & Row, New York.)

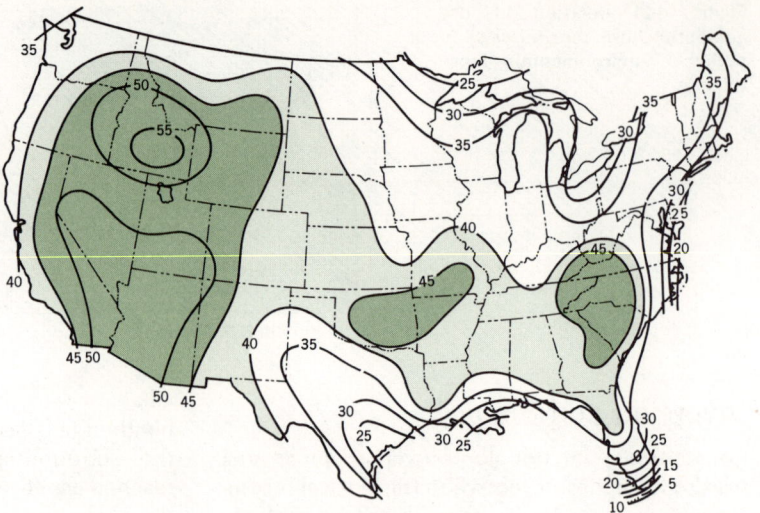

Figure 6.13 Frequency of low-level inversions (under 500 ft; 150 m) in the autumn season. Figures show average percentage of hours of inversion daily. (After D. H. Pack, *Science*, vol. 146, p. 1125. Copyright 1964 by the American Association for the Advancement of Science.)

which pollutants are held. The lid may be situated at a height of perhaps 500 to 1000 ft (150 to 300 m) above the ground. Figure 6.13 shows for the United States the frequency with which low-level inversions may be expected during the fall, which is the season when air mass interactions are least intense generally. Strong temperature inversions conducive to air pollution are also caused in late spring and early summer over coastal cities such as New York, Toronto, and Chicago by cool sea or lake breezes that bring a stable, cool air layer over a narrow coastal zone.

Although situations dangerous to health during prolonged low-level inversion have occurred a number of times over European cities since the industrial revolution began, the first major tragedy of this kind in the United States occurred at Donora, Pennsylvania, in late October of 1948. The city occupies a valley floor hemmed in by steeply rising valley walls which prevent the free mixing of the lower air layer with that of the surrounding region (Figure 6.12). Industrial smoke and gases from factories poured into the inversion layer for five days, increasing the pollution level. Since humidity was high, a poisonous fog formed and began to take its toll. In all, 20 persons died and several thousand persons were made ill before a change in weather pattern dispersed the smog layer.

Observation of smoke plumes from tall stacks of large industrial combustion plants can tell you a lot about the degree of stability or instability of the atmosphere on a given day. Figure 6.14 shows the smooth, level, downwind flow of smoke indicating a high degree of stability with low winds speed; it is described as a *fanning plume*. When air stability is close to neutral and winds are greater than about 20 mi (32 km) per hour, the *coning plume* develops, showing some moderate diffusion and mixing in the downwind direction. Under conditions of instability the *looping plume* develops. The vertical fluctuations are an indication of the large size of turbulent eddies. Typically, since wind speeds and turbulence are higher during the afternoon than at night, a smoke plume will assume the fanning form during the night and early morning hours, but change to the coning or looping forms during the late morning hours. When turbulence is severe, smoke can be carried down to ground levels in the looping plume. As stack height is increased the concentration of smoke reaching the ground is lessened and the horizontal distance from the stack at which the smoke touches ground is lengthened. Under conditions of dead calm with stable air, the smoke from the stack goes straight up and will be seen to spread laterally at a sharply defined level, which coincides with the inversion lid (Figure 6.11).

Related to the low-level inversion, but caused in a somewhat different manner, is the *upper-level inversion*, illustrated in Figure 6.15. Going back to Chapter 5, recall that anticyclones are cells of subsiding air that diverges at low levels (Figure 5.31). Within the center of the cell winds are calm or very gentle. As the air subsides, it is adiabatically warmed, so that the normal temperature lapse rate is displaced to the right in the temperature-altitude graph, as shown in Figure 6.15 by the diagonal arrows. Below the level at which subsidence is occurring the air layer remains stagnant. The temperature curve consequently develops a kink in which a part of the curve shows an inversion. For reasons already

Inversion and Smog

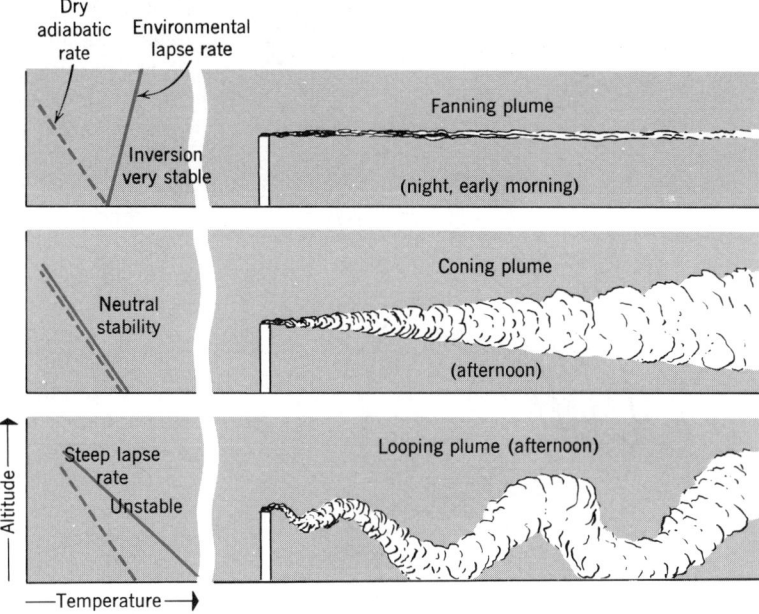

Figure 6.14 Three types of smoke plumes reflecting the degree of stability of the lower air layer. (After E. V. Somers, 1971, in *Air Pollution Control*, Part I, W. Strauss, ed., Wiley-Interscience, New York, p. 25, Figure 13.)

explained, the layer of inverted temperature structure strongly resists mixing and acts as a lid to prevent the continued upward movement and dispersal of pollutants.

Upper-level inversions develop occasionally over various parts of the United States when an anticyclone stagnates for several days at a stretch. Figure 6.16 is a map of the eastern United States showing the frequencies of anticyclonic stagnations, each lasting 4 days or longer. Apparently, stagnations are most frequent over the Southern Appalachians and Piedmont region, but are fairly common in a zone extending northward into the heavily industrialized and populated Atlantic seaboard as far north as New York City.

For the Los Angeles Basin of southern California, and to a lesser degree the San Francisco Bay Area and over other west coasts in these latitudes generally, special climatic conditions produce prolonged upper-air inversions favorable to persistent smog accumulation. The Los Angeles Basin is a low, sloping plain lying between the Pacific Ocean and a massive mountain barrier on the north and east sides. Cool air is carried inland over the basin on weak winds from the south and southwest, but cannot move farther inland because of the mountain barrier. It is a characteristic of these latitudes that the air on the eastern sides of the subtropical high pressure cells is continually subsiding, creating a more-or-less permanent upper-air inversion that dominates the western coasts of the continents and extends far out to sea (Figure 6.17). The effect is particularly marked in the summer season, when the Azores and Bermuda highs are at their largest and strongest. The subsiding air over the Los Angeles Basin is warmed adiabatically as well as heated by direct absorption of solar radiation during the day, so that it is markedly warmer than the cool stagnant

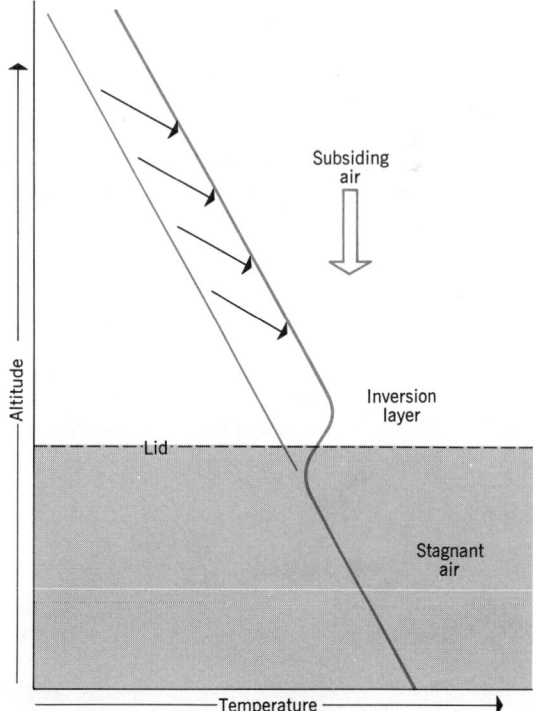

Figure 6.15 An upper-air inversion caused by subsidence. (From A. N. Strahler, 1972, *Planet Earth; Its Physical Systems Through Geologic Time*, Harper & Row, New York.)

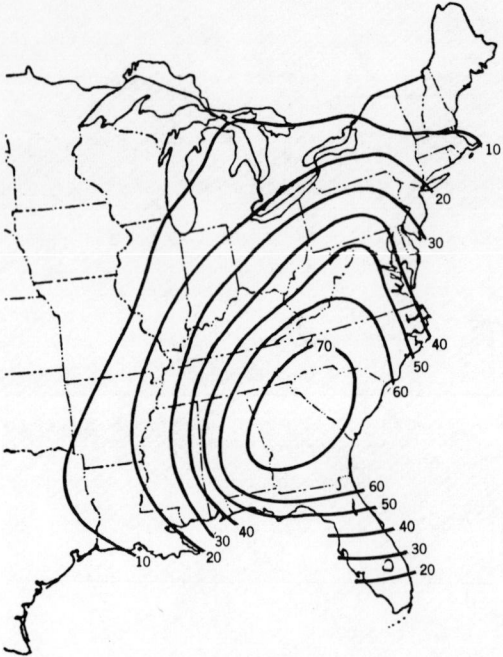

Figure 6.16 Frequency of occurrence of stagnating anticyclones over the eastern United States. Figures tell number of occurrences of stagnation, lasting 4 days or longer, in the period 1936–1960. (After D. H. Pack, *Science*, vol. 146, p. 1126. Copyright 1964 by the American Association for the Advancement of Science.)

air below the inversion lid, which lies at an elevation of about 2000 ft (600 m). Pollutants accumulate in the cool air layer, producing the characteristic smog that first became noticeable for its irritating qualities in the early 1940s. Because water vapor content is generally low, Los Angeles smog is better described as a dense haze than as a fog, for it does not shut out the sun or cause a reduction in visibility to a degree that interferes with vehicular operation or the landing and take-off of aircraft. Close to the coast, true fogs are frequent, but this condition normally is prevalent on dry west coasts adjacent to cold currents. The upper limit of the smog stands out sharply in contrast to the clear air above it, filling the basin like a lake and extending into valleys in the bordering mountains (Figure 6.18).

One important physical effect of urban air pollution is that it reduces visibility and illumination. A smog layer can cut illumination by 10% in summer and 20% in winter. Ultraviolet radiation is absorbed by smog, which at times completely prevents these wavelengths from reaching the ground. Reduced ultraviolet radiation may prove to be of importance in permitting increased bacterial activity at ground level. City smog cuts horizontal visibility to some one-fifth to one-tenth of the distance normal for clean air. Where atmospheric moisture is sufficient, the hygroscopic particles acquire water films and can lead to formation of true fog with near-zero visibility. Over cities, winter fogs are much more frequent than over the surrounding countryside. Coastal airports, such as La Guardia (New York), Newark (New Jersey), and Boston suffer severely from a high incidence of fogs augmented by urban air pollution.

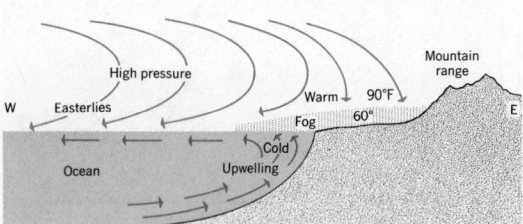

Figure 6.17 Subsidence, temperature inversion, and fog over a desert west coast. Upwelling brings cold water to the surface. (From A. N. Strahler, 1971, *The Earth Sciences*, Harper & Row, New York.)

TABLE 6.3 Climate of a City as Compared with that of the Surrounding Countryside

Radiation	
Total insolation	15 to 20% less
Ultraviolet (winter)	30% less
Ultraviolet (summer)	5% less
Sunshine duration	5 to 15% less
Temperature	
Annual mean	0.9 to 1.8 F° (0.5 to 1.0 C°) higher
Winter minimum	1.8 to 3.6 F° (1.0 to 2.0 C°) higher
Relative humidity	2 to 3% less
Cloudiness	
Cloud cover	5 to 10% more
Fog in winter	100% more
Fog in summer	30% more
Precipitation	
Total quantity	5 to 10% more
Snowfall	5% less
Particulate matter	10 times more
Gaseous pollutants	5 to 25 times more
Wind speed	
Annual mean	20 to 30% lower
Extreme gusts	10 to 20% lower
Calms	5 to 20% more frequent

SOURCE: Data of H. E. Landsberg, 1970, *Meteorological Monographs*, vol. 11, p. 91, Table 1.

Inversion and Smog

Figure 6.18 Smog layer in downtown Los Angeles. Base of the temperature inversion is located immediately above the top of the smog layer, at a height of about 300 ft (100 m) above the ground. (Photograph by Los Angeles County Air Pollution Control.)

A related effect of the urban heat island is the general increase in cloudiness and precipitation over a city, as compared with the surrounding countryside. This increase results from intensified convection generated by heating of the lower air. For example, it has been found that thunderstorms over the city of London produce 30% more rainfall than thunderstorms over the surrounding country. Increased precipitation over an urban area is estimated to average from 5 to 10% over the normal for the region in which it lies.

Table 6.3 summarizes the main climatic differences between a city and the surrounding countryside. Keep in mind that this summary is a generalization applied to highly industrialized nations in middle latitudes, and there are differences among cities with respect to magnitude of a given effect.

As we found in earlier paragraphs, a large city produces a heat island. Within this warm air layer pollutants are trapped beneath an inversion lid. The layer of polluted air takes the form of a broad *pollution dome* centered over the city when winds are very light or near calm. However, when there is general air movement in response to the pressure gradient field, and pollutants are carried far downwind to form a *pollution plume*. Figure 6.19 has two maps showing plumes from the major cities of the Atlantic seaboard. The V-lines show zones of fallout beneath the plumes, while

Figure 6.19 Pollution plumes from cities on the eastern seaboard under conditions of weak, southerly winds (*left*) and strong westerly winds (*right*). The dot at end of each plume represents approximate distance traveled by pollutants at the end of a 24-hour period. (After H. E. Landsberg, 1962, in *Symposium—Air Over Cities*, Sanitary Engineering Center Tech. Report A62-5, Cincinnati, Ohio.)

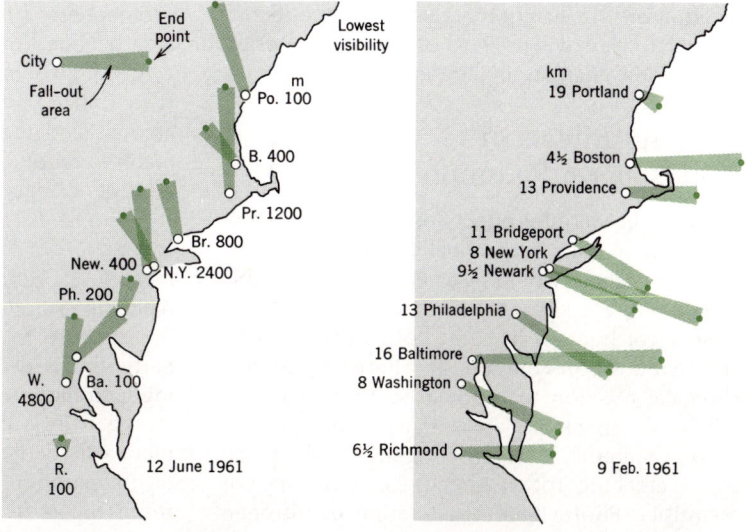

159

the small dot at the open end of the V shows the distance traveled by the air in one day from the source. The left hand map shows the effects of weak southerly winds on a day in June. In this situation pollution from one city affects another and generally the pollutants remain over the land, contaminating suburban and rural areas over a wide zone. The right hand map, for a day in February, shows the effect of strong westerly winds, causing the pollution to be carried directly to sea.

Glacial Ice as a Recorder of Air Pollution

Each snow layer deposited on the surface of an alpine glacier, icecap, or ice sheet holds particulate fallout matter carried in the atmosphere from distant sources. Because glacial ice is usually remote from man-made sources of pollution, the quantity and composition of these particles give us a good overall index of world production of atmospheric pollutants. Older ice layers beneath serve as a guide to the normal, preindustrial levels of pollutants. In snow and ice layers of the Greenland Ice Sheet a strong rise in sulfate concentration since about 1950 is attributed to combustion of fossil fuels, while a marked recent increase in lead is attributed to combustion of leaded gasolines. An annual cycle can be detected, with peak quantities of both sulfate and lead occurring in the winter months. Analysis of mercury in Greenland ice layers shows a doubling in the last two to three decades over the average of previous centuries. The increase in atmospheric mercury is attributed to combustion of fuels, roasting of ores, and various industrial processes using mercury compounds.

Although the studies mentioned above are comparatively new and will require reinforcement by additional research, it appears that the record of the ice layers will be most useful in monitoring world-wide changes in atmospheric pollutants.

Harmful Effects of Atmospheric Pollution

A list of the harmful effects of atmospheric pollutants upon plant and animal life and upon inorganic substances would be a very long one if fully developed. We can only suggest some of these effects. For humans in cities both sulfur dioxide and hydrocarbon compounds, altered by photochemical reaction to produce sulfuric acid and ethylene, respectively, are irritants to the eyes and to the respiratory system. Nitrogen dioxide is also an eye and lung irritant when present in sufficient quantities. Photochemical alteration of nitrogen oxides leads to production of ozone, which acts as irritant in smog and would be lethal if it occurred in large concentrations.

For persons suffering from respiratory ailments, such as bronchitis and emphysema, the breathing of heavily polluted air can bring on disability and even death, as statistics clearly show. During the London fog of December, 1952, the death rate approximately doubled, while an increased death rate over the average rate persisted for weeks after the event. In a more recent London fog, that of December 1962, more than 300 deaths were attributed to the breathing of polluted air. Particularly hard hit were the very young and the old. There is also a suspected linkage between breathing of atmospheric pollutants and lung cancer, since the incidence of that disease is higher in cities than in other areas. It is thought that the accumulation of atmospheric hydrocarbon compounds on lung tissues may predispose to the onset of lung cancer.

Carbon monoxide is a cause of death when inhaled in sufficient quantities. Everyone knows that the carbon monoxide from automobile exhaust will kill in a short time when breathed in a closed garage. Carbon monoxide levels are a general indicator of the degree of air pollution from vehicular exhausts, but concentrations rarely reach sufficient levels in the open air to be a threat to life. Nevertheless, the long-continued inhalation of small amounts of carbon monoxide is suspected of harmful effects, as yet unevaluated.

Ozone in urban smog has a most deleterious effect upon plant tissues and in some cases has caused the death or severe damage of ornamental trees and shrubs. Sulfur dioxide is injurious to certain plants and is a cause of loss of productivity in truck gardens and orchards in polluted air. Atmospheric sulfuric acid in cities has in places largely wiped out lichen growth.

Although secondary in the sense that the loss is in dollars, rather than in lives and health of animals and plants, the deterioration of various materials subjected to polluted air forms an important category of harmful effects. Building stones and masonry are susceptible to the corrosive action of sulfuric acid derived from the atmosphere. Metals, fabrics, leather, rubber, and paint deteriorate and discolor under the impact of exposure to urban air pollutants. Ozone, in particular, is deleterious to natural rubber, causing it to harden and crack. The sulfuric acid produced from sulphur dioxide corrodes exposed metals, particularly steel and copper. Not the least of the economic losses from pollution are simply from soilage of clothing, automobiles, furniture, and interior floors, walls, and ceilings. The cleaning bill totalled for a large city is truly staggering, when calculated to include the labor and cleaning agents expended by householders.

Lead and other toxic metals in the polluted atmosphere are a particular source of concern for human health in the future. The lead bearing particles from auto exhausts tend to concentrate in the grass, leaves, and soil near major highways. There is good reason to suppose that humans ingest lead particles directly from the air and that they may prove to be a health hazard. Although lead poisoning from atmospheric sources has not yet been documented in humans, there is now evidence that it has caused the deaths of animals in city zoos. Tests have shown high levels of lead in the tissues of the dead animals and no source other than atmospheric particles has been found for the ingested lead. Animals in outdoor cages showed higher lead levels than animals kept indoors. These findings are ominous in tone and tend to reinforce the conclusion of the Air Pollution Control Office of the Environmental Protection Agency that atmospheric lead pollution is a possible health hazard. Reduction of lead additives to gasoline is a corrective step having high priority.

Radioactive substances in the atmosphere are a special form of environmental hazard because of the genetic damage that is done to plant and animal tissues exposed to ionizing radiation. This radiation is also important in causing mutations.

TABLE 6.4 Air Pollution Standards—Limits of Concentrations Permitted

Pollutant	Weight of Pollutants per Cubic Meter of Air
Carbon monoxide	10 milligrams maximum 8-hour concentration
Sulfur oxides	80 micrograms annual mean, 365 micrograms maximum 24-hour concentration
Hydrocarbon compounds	125 micrograms maximum 3-hour concentration
Nitrogen oxides	100 micrograms annual mean, 250 micrograms maximum 24-hour concentration
Photochemical oxidants	125 micrograms maximum 1-hour concentration
Particulates	75 micrograms annual mean, 260 micrograms maximum 24-hour concentration

Late in 1970 the Federal Clean Air Act was signed into law, giving the Environmental Protection Agency authority to set national standards for tolerable limits of pollutants in the air. These standards are given in Table 6.4. Included in the act is provision for an emergency alert system that will signal the need to reduce fuel combustion and curtail the use of automobiles when the danger point is reached in a given city. Furthermore, the law sets 1975 as the deadline year by which the automobile industry is to install effective emission control devices on cars.

Global Effects of Particles in the Atmosphere

Let us now investigate the physical effects of increased atmospheric dusts—considered as inert particles—in altering the global radiation and heat balances. As already noted, major natural sources of particles have always existed. One source is smoke from large forest fires, set by lightning. From time to time these conflagrations have produced great smoke palls in the troposphere.

Much more significant have been emissions of particles from volcanoes in eruption; these events must have occurred spasmodically throughout all of geologic time. For example, the Indonesian volcano Krakatoa underwent explosive destruction in 1883, emitting enormous quantities of dust into the upper troposphere and stratosphere, where it was given world-wide distribution in the planetary wind system. Unusually colorful sunsets were observed during ensuing years in the British Isles and Europe; these were favored subjects for landscape artists of the time and became known as the Chelsea Sunsets. Following the Krakatoa eruption, an observatory at sea level at Montpellier, France, recorded short periods of lowered solar radiation intensity, which persisted through 3 years and ranged from 10 to 20% below normal radiation intensity. During the 10-year period following Krakatoa the eruptions of three other volcanoes continued to supply dusts to the upper atmosphere and resulted in additional periods of lowered insolation. Obviously, these dusts absorbed energy and reflected shortwave solar radiation back into the atmosphere, and so reduced the quantity penetrating to the recording instruments at the observatory.

You might guess that the periods of lowered solar radiation would correspond with periods of reduced average air temperature at lower levels, but inspection of the temperature graph, Figure 6.2, does not show this to be the case. No really significant temperature change shows for the decade 1883–1893. Perhaps the quality of the temperature records was inadequate to reveal a temperature increase. There is, however, a significant decline in temperature after 1940, as we have already noted. Since about 1947 there has been a sharp increase in the quantity of dust in the stratosphere because of the eruption of the Icelandic volcano Hekla in that year and a series of eruptions of several other volcanoes in the 1950s and 1960s. However, the preceding three decades had witnessed a low level of volcanic activity. These observations suggest that volcanic dusts are the cause of the temperature decline.

A documented effect of the injection of volcanic dust into the stratosphere by the eruption of Agung in 1963 was a definite warming of the air in the altitude range 10 to 12 mi (16 to 20 km), which is at the tropopause or in the lower stratosphere. In a large part of the latitude belt between about 20° N and 40° S warming at 12 mi (19 km) at the end of one year after the eruption exceeded 7 F° (4 C°) and in certain areas over the equator exceeded 14 F° (8 C°). However, no warming was evident below about 6 mi (10 km). This temperature rise is explained by absorption of solar radiation by the minute dust particles.

On the other hand, because increased stratospheric dust allows less energy to reach lower levels, we can at least propose the hypothesis that the cooling effect of volcanic dusts has more than offset the warming effect of increased CO_2, and has reversed the warming trend that had set in when volcanic activity was unimportant. Climatologists are justifiably cautious about accepting such a simple cause-and-effect relationship, for other natural causes can be invoked for the same temperature changes.

What impact do man-made dusts have upon the radiation and heat balances? Are industrial dusts partly responsible for the air temperature decline since 1940? Snow layers in the Caucasus Mountains, at elevations exceeding 12,000 ft (3.6 km) show a sharp increase in dust content in the period 1930 to 1963. This increase in fallout is attributed to industrial dusts brought from Europe by the prevailing westerly winds. If it be agreed that Man's industrial activities raise the concentration of particles in the lower troposphere, does this activity also add to the dust content as high as the stratosphere?

A tentative answer to this question has been given by scientists of NOAA, based on analysis of radiation data collected at the Mauna Loa Observatory, Hawaii, at an altitude of 11,000 ft (3.4 km). In 1958, a program of solar radiation monitoring was begun under the auspices of the IGY (International Geophysical Year). Data have been collected continuously through 1970. The atmospheric property of *turbidity*, or degree of concentration of suspended particulates, was inferred on the basis of the percentage of solar energy received as compared with the solar constant. Figure 6.20 is a graph showing the percentage of transmission of energy during the period 1958–70. Each dot is the mean value of a month of observations. During the 5-year period 1958–1963 an annual cycle of change with an amplitude of about 1% was detected and attributed to natural causes. One possible cause of such an annual cycle is the seasonal rhythm of photochemical production of particles by oxidation of volatile organic matter emitted by plants; another cause is the seasonal rhythm of change in the planetary atmospheric circulation. But statistical treatment of the data showed no trend of the average either upward or downward in this period, which was one of no significant volcanic activity the world over. Then, in 1963, Mount Agung in Bali erupted and sent large quantities of dust into the stratosphere. As shown in Figure 6.20, there occurred a significant drop in atmospheric transparency to radiation, and this event was followed by a slow rise, still in progress in 1970. Slowness of rise of the curve may be due to the fact that other volcanic eruptions followed in 1965, 1966, and 1968. NOAA scientists who analyzed these radiation data have concluded that for the most part turbidity of the upper atmosphere is dominated by natural causes and does not show any effects of Man's activities.

Man-made dusts in the lower troposphere have the effect of increasing nuclei of condensation and, therefore, of perhaps increasing the global average of cover by lower clouds, which now stands at about 31%. This low cloud cover is largely responsible for determining the earth's average albedo. It has been estimated that should the low cloud cover increase to 36% the planetary albedo would be raised enough to cause a drop of the world average air temperature by 7 F° (4 C°). This drop would be enough to cause significant growth and spread of existing glaciers and ice sheets.*

This forecast has been reinforced by another recent analysis of the effects of increasing turbidity in which it is concluded that if the rate of increase of injection of particles into the atmosphere increases by a factor of from six to eight in the next half-century, the drop of the mean

*Based on data in *First Annual Report of the Council on Environmental Quality,* August, 1970, Chapter 5, p. 97.

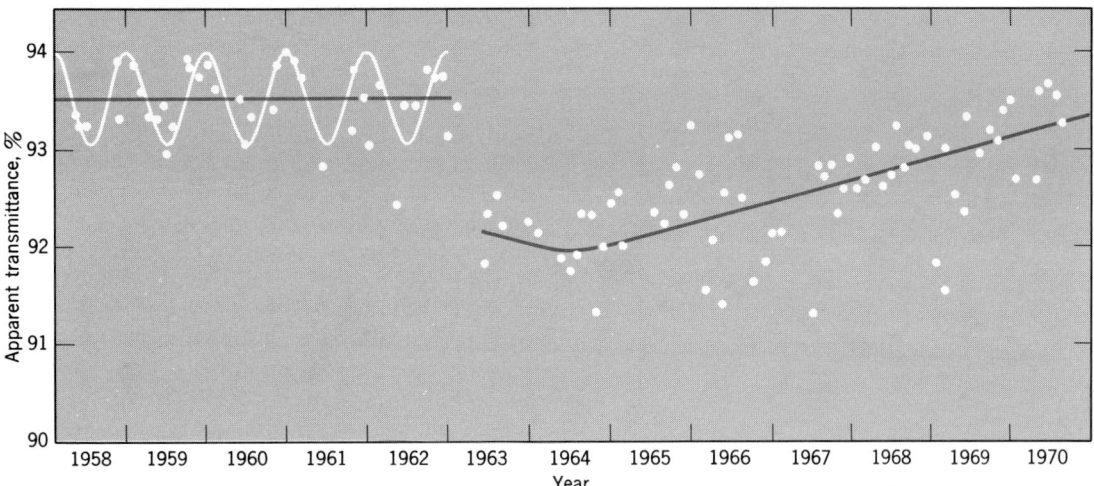

Figure 6.20 Variations in the transparency of the atmosphere to solar radiation as observed at the Mauna Loa Observatory, Hawaii (altitude: 11,000 ft; 3400 m), between 1958 and 1970. Dots are monthly mean values. A solid line shows the inferred annual cycle of change. (After H. T. Ellis and R. F. Pueschel, 1971, Science, vol. 172, p. 846, Figure 1. Data of NOAA.)

average atmospheric temperature may be as much as 6 F° (3.5° C).* Here again, we must be aware of possible feed-back mechanisms that might nullify the change. Lowered temperature would result in less evaporation from ocean surfaces, which in turn would cause decreased cloudiness and a reduction in global albedo. The resulting tendency toward warming of the atmosphere might thus offset the tendency for cooling through increased turbidity.

Our tentative conclusion will be that Man's contributions of atmospheric particles may have far-reaching effects through changes induced in processes of the troposphere, but perhaps little effect upon processes of the stratosphere.

Testimony of the Glacial Ice Layers

Ice layers of the Greenland and Antarctic ice sheets also provide a means of monitoring atmospheric air temperatures and of comparing the modern record with temperatures going back many thousands of years. The method uses analysis of the ratio of an uncommon isotope of oxygen (O^{18}) to the normal oxygen isotope (O^{16}). It is known that the ratio of these isotopes in water varies with water temperature. Consequently, measurement of the oxygen isotope ratio for successive layers of ice and compacted snow reveals changes in the atmospheric temperature in which the snow was first crystallized.

When the method was applied to the Greenland Ice Sheet, there emerged a remarkable record of air temperature fluctuations over the past 800 years (Figure 6.21). While the oxygen-isotope curve is not readily translated into an absolute temperature scale, the correlation with atmospheric temperature is close and gives a qualitative picture. First, we notice that the warming trend of the first half of the present century, illustrated in Figure 6.2 and attributed to increase in carbon dioxide from fuel combustion, is strongly shown by the isotope-ratio curve. Likewise, the reversal of this trend since about 1940 also shows clearly. Now, when we look back into earlier centuries, we find a series of cycles of isotope variation of approximately the same amplitude. Cycles of periods of 78 and 181 years have been derived by analysis of these isotope data. The important point is that air temperature fluctuations of the same order of magnitude that we observe in our century are normal events. What we have ascribed to man-made changes in global atmospheric temperature may, in fact, be caused by variations in solar radiation.

Where Do We Stand?

This chapter has covered a very wide range of effects—real or possible—of Man's activities upon the earth's atmosphere and the global energy balances. It isn't surprising if at this point you feel

*S. I. Rasool and S. H. Schneider, 1971, Science, vol. 173, p. 141.

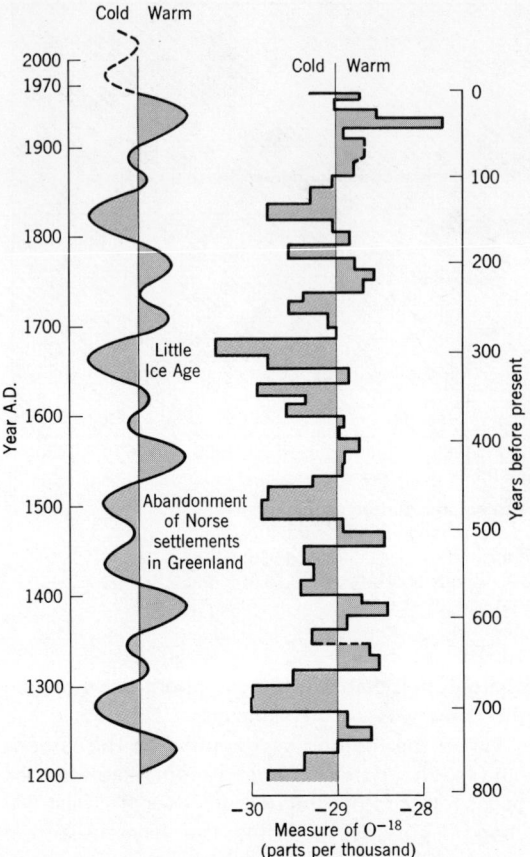

Figure 6.21 Variations in oxygen-isotope ratio within an ice core from the Greenland Ice Sheet. Observed ratios are shown at right for the past 800 years. The smooth curve at left is derived by mathematical analysis (Fourier analysis) to show medium-range trends. (From S. J. Johnsen, et al., 1970, *Nature*, vol. 227, No. 5257, Figure 1.)

confusion and frustration, since a number of potentially important effects have turned out to be postulates for which no firm data are yet available. To make matters worse, opposite effects are superimposed and it is difficult or impossible to decide which is the dominant trend.

Looking back over this chapter we may feel some satisfaction in drawing the following general conclusions: (1) Man definitely has altered the composition of the atmosphere—globally with respect to carbon dioxide released by combustion, and locally over urban areas by means of pollutants introduced by industrial activities. (2) Man has achieved initial successes in increasing precipitation locally and may be on the verge of ameliorating the intensity of certain severe weather phenomena. (3) Man has degraded air quality over cities to a degree that at times constitutes a distinct hazard to human health and in a way that suggests long-term injurious effects as yet to be evaluated. (4) Heat and water balances in the atmosphere over cities have been demonstrably altered through permanent changes in land surface characteristics and the continuous emissions of heat and pollutants. (5) Changes in the global balances of radiation, heat, and water are suggested by rising and falling trends of average world air temperatures, but it is not yet clear to what extent, if any, Man's activities have contributed to these changes. (6) The close association of increases in stratospheric dusts following volcanic eruptions with decreases in intensity of solar radiation has been demonstrated, but there is no evidence that man-made particulates have thus far altered the amounts of radiation received at the top of the troposphere. (7) The world's oxygen supply is not threatened by Man's activities. (8) Prospects for long-term, man-made changes in atmospheric temperatures, bringing serious attendant environmental problems (such as changes in sea level), are real in the light of projected increases in the generation of carbon dioxide and heat and of changes of cloud cover affecting the planetary albedo.

Deterioration of the quality of air over cities is documented and well understood. Here the steps that need to be taken to arrest the deterioration and restore satisfactory quality to the air are for the most part clear-cut. We must reduce emissions of pollutants to a desired level. Implementation of this reduction is the thorny problem, since it can only be done by the expenditure of vast quantities of money and effort that might otherwise go into raising the material standards of society or to alleviating other social ills. This book does not enter into economic, legal, and political aspects of air pollution control. Instead, we have laid the foundation of scientific principles pertinent to the problems and have outlined the nature and extent of the problems.

When it comes to possible long-range effects of Man's activities upon the global balances of radiation, heat, and water, we do not know with certainty what the problems are, or even if they exist. Reason tells us that serious environmental changes of our own making may be in store for future generations. The course of action is clear. We must undertake a massive research program of monitoring of the environmental variables and interpreting the data. Again, the cost of such an effort is high and it is difficult to gain support for projects that deal with rather tenuous propositions, particularly when there are so many urgent problems, such as urban air pollution, that demand direct action now.

PART II

Energy Systems of the Lithosphere

7 Igneous Processes and the Earth's Crust

At first thought, the dense, hard rock that lies beneath the soil layer might seem to play no active role in the environment of life. If you boil a clean piece of fresh granite rock for many hours in pure water, you will find at the end of that time that the water tastes just as fresh as at the start. Evidently the rock is incapable of yielding up chemical ions that might provide nutriment to plants and it does not contaminate the water in which it soaks—or at least our senses fail to detect any degree of dissolution. The same piece of granite, placed out of doors shows no observable change after decades of exposure to sunlight, rain, frost, and winds.

What role, other than that of a stable platform beneath the lands, does the lithosphere play in environmental processes? To get an answer, we must shift gears, so to speak, with respect to the time-scale of events. From one year as the standard unit of time we must jump 4, 5, 6 or more orders of magnitude (powers of ten) to time units in tens and hundreds of thousands of years, and even to millions of years.

Viewed over such spans of time the seemingly inert piece of granite undergoes changes of state and composition; it moves along the pathways of an energy system of geologic scope. If we could see the events of 500 million years taking place deep within the earth compressed into one year, as if by running a movie film made up of individual frames taken one each 100,000 years, we might see a tongue of molten rock push violently upward through the surrounding solid rock, like the rising bubble of warm air in a cumulonimbus cloud. We would see the molten rock cease to rise and then spread out, laterally, suddenly to freeze into immobility as a crystalline solid. Later frames of our time-lapse movie would show the land surface above the frozen granite mass gradually worn away, until the granite body is uncovered and exposed to the atmosphere.

Processes acting within the lithosphere belong to an energy system wholly unconnected to the solar energy system that drives the atmospheric and oceanic circulations. One source of energy of geologic systems is internal; it is energy inherited from the time of formation of the planet. This energy source is *radioactivity*, the spontaneous transformation of mass into energy in certain unstable elements, known as *radioisotopes*. Heat is generated by radioactivity; we call this *radiogenic heat*. Radioisotopes are known to be concentrated in the outermost zone of the solid earth. Here they generate heat to a degree that maintains deep-seated rock close to its melting point. Unequal heating produces differences in rock density, in response to which huge masses of rock tend to rise or sink. Locally, pockets of molten rock are formed and forced up to the earth's surface, making volcanic eruptions and creating new rock before our eyes. A second possible source of heat within the lithosphere is the flexing of the earth by tidal forces as the earth rotates on its axis while deformed by gravitational attraction between earth and moon and between earth and sun. If tidal energy is a source of internal heat, as some scientists have postulated, it too is an energy source inherited from the time of origin of the solar system in the form of kinetic energy of planetary masses in motion.

The internal heat of the earth, whether of radiogenic or tidal origin, acts in various global locations that are quite unrelated to the earth's axis of rotation and to ordered latitudinal belts of heat and atmospheric motion. So we find Mount Erebus, an active volcano of Antarctica, erupting amidst fields of snow and glacial ice. Volcanoes, great alpine mountain chains, and zones of rock rifting make strange patterns over the face of the globe, haphazardly transgressing the belts of temperature, barometric pressure, winds, and precipitation. The North Polar region is occupied by a deep ocean basin; the South Polar region by a continent (Figure 7.1). Turn a globe so that the North Pole faces you, and you will see a great amount of land surrounding the Arctic Ocean. Turn the globe so that the South Pole faces you and you will see mostly ocean surrounding the Antarctic landmass. It is this unbalanced distribution of the continents and oceans, with reference to the axis of earth rotation, that gives a new dimension and variety to global environments.

Already we have seen in Part 1 that the distribution of land and ocean profoundly affects the radiation, heat, and water balances of the earth. Thus Asia produces a monsoon regime; the North Atlantic and North Pacific oceans produce intense, subarctic low-pressure areas as well as strong subtropical cells of high pressure.

But the lithosphere has much greater environmental significance than as a modifier of atmospheric and oceanic processes. The lithosphere is the source of elements and compounds essential to life on earth. These vital substances are released during reactions taking place in the shallow interface between rock and atmosphere in the presence of heat and water. The granite fragment that we thought to be inert and unchanging is ultimately physically disintegrated and chemically decomposed, releasing its tightly bonded atoms to freedom in water solutions in which they can move and be used by plants in the construction of tissues. Granite can furnish plant nutrients such as potassium, calcium, magnesium, and sodium, but only when acted upon by external processes powered by solar energy. Massive granite disintegrates into finely divided particles that we know collectively as the *soil*. Release of large numbers of ions from mineral matter of rock must be preceded by fragmentation of the rock into particles ranging in size from sand through silt to clay composed of particles of colloidal size. This mass of fragments is the parent mineral matter of the soil. Life on the lands could not have been possible without formation of soil; thus one major environmental role of geologic processes has been to furnish the parent matter of soil and with it essential nutrients of plant growth. Well, you might argue that floating plant life of the sea and of lakes and streams—algae, that is—don't need any soil for their growth. While this is true, the algae do need the ions supplied by the breakdown of rock and transported in solution to those water bodies by flowing water.

For modern Man, the lithosphere takes on yet another vital role, that of supplying him with mineral resources without which no industrial development would have been possible. Mineral resources include hydrocarbon fuels, which are Man's principal energy source, and the essential substances of civilization—metals, nonmetallic structural materials, and chemicals for all sorts of industrial processes and synthetic products. In order that you can have a full appreciation of the limitations of these nonrenewable resources, you must know how rocks are formed and how very rare elements dispersed throughout rocks are concentrated into ores.

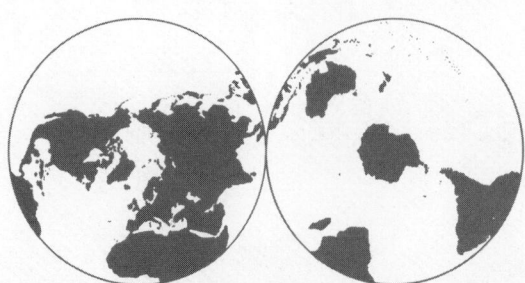

Figure 7.1 Hemispheres—contrasting patterns in land-ocean distribution.

The Earth's Interior

Our solid earth is a nearly spherical body almost 4000 mi (6400 km) in radius. Most of the sphere is included in two major divisions: (1) a *core*, largely of iron and some nickel, with a radius of 2160 mi (3475 km), and (2) a *mantle* of silicate rock 1800 mi (2895 km) in radial thickness (Figure 7.2). The outer part of the iron core is in the liquid state at temperatures in the range of 4000° to 5000° F (2500° to 3000° K) and is under confining pressures from 1.5 to 3 million atmospheres. A relatively small inner portion of the core, with radius of about 780 mi (1255 km) is in the solid state. Here pressures reach 3.5 million atmospheres. Density of the core material is from 10 to 12 times that of water at the earth's surface, e.g., 10 to 12 g/cc. The mantle is in the solid state, although its properties under such high pressures and temperatures are not what we are accustomed to observing in rocks at the surface. Mantle rock is made up of minerals which are silicate compounds (silicate is silicon in combination with oxygen) containing much magnesium and iron. Mantle rock is capable of slow flowage like a very sluggish liquid under unequal stresses, and, in fact, the deeper zones of the mantle cannot break with a sharp fracture as do rocks at the earth's surface. Mantle rock density is on the order of 4.5 to 5.5 g/cc, or about half that of the core. An outermost layer, the earth's *crust*, ranges from 10 to 26 mi (16 to 40 km) in thickness, which is only a very thin, almost insignificant layer in comparison with the combined core and mantle. Yet it is within the crust that all relief features of the earth's surface are contained, and we must give much greater attention to the crust than to the underlying zones.

Most modern theories of the formation of our planet and of the solar system in general call for an originally cold body of dispersed gases and dust at least as large as the radius of the outermost planets. This *solar nebula* underwent some sort of aggregation, or condensation, to form a number of individual masses of greater density, and these then became solid bodies, growing under a steady infall of particles of all sizes through gravitational attraction to become the planets. It is generally supposed that at the time of its formation as a solid body, about 4.5 billion years ago, the earth had not attained a sufficiently high temperature to cause internal melting. The component mineral compounds, largely silicates of magnesium, and iron, were rather uniformly mixed in this primeval solid sphere. Also uniformly distributed throughout the sphere were radioisotopes of the elements uranium, potas-

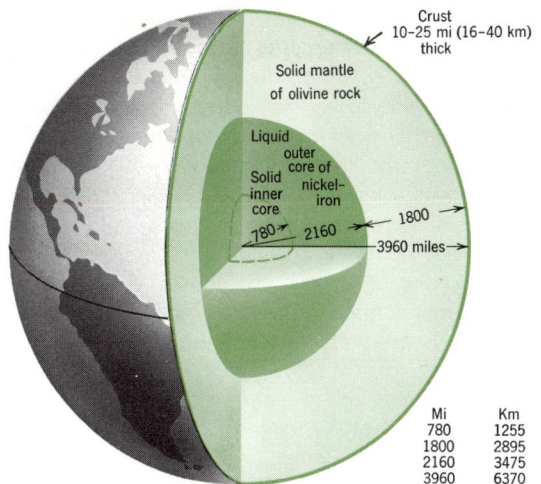

Figure 7.2 Concentric zones of the earth's interior.

sium, and thorium. These produced radiogenic heat and caused a gradual rise in internal earth temperatures, bringing the primary mixture to the melting point from place to place throughout the interior. The molten bodies, being less dense than the surrounding mass, migrated surfaceward where they cooled and solidified. For reasons that we cannot go into here, the result of repeated internal melting and solidification was to allow the denser iron to settle to the earth's center to form the core, while the less dense silicate compounds were segregated into the mantle. At the same time, the radioactive elements were concentrated largely in the upper region of the mantle and in the thin crust. Consequently, radiogenic heat is now produced largely in a shallow outer layer of the earth. If this were not the case, the buildup of radiogenic heat in the deep interior would have subjected the entire earth to repeated episodes of melting and overturn. As it is, the bulk of the earth has been comparatively stable for the past 3 billion years or so with the principal unrest occurring only in the crust and shallow outer portion of the mantle.

While the environmental influence of the core, mantle, and crust upon the biosphere shows no appreciable change over centuries, this structural arrangement is responsible for the production of the earth's magnetic field, a subject covered briefly in Chapter 2. Were it not for a slow rotation of the liquid core with respect to the surrounding mantle, there would be no magnetic field, and no magnetosphere to entrap energetic particles carried in the solar wind or arriving as cosmic radiation. Our planetary magnetic field thus protects the surface environment of life.

In contrast, the moon, with no atmosphere and no appreciable magnetic field, receives the full force of ions of the solar wind. So far, space probes have detected no external magnetic field around either Mars or Venus, our nearest neighbors and most likely planetary contenders for supporting some sort of life. Mars, which is a much smaller planet than earth, probably has no liquid iron core, and this situation would explain the absence of a magnetic field. Venus, which closely matches earth in mass, density, and diameter, may very well have a liquid iron core. But because Venus' rotation is extremely slow, there may not be enough interaction between core and mantle to generate a magnetic field. Here, again, we see the uniqueness of the environment of Planet Earth as compared with the moon and the nearer planets.

The solid earth as a whole consists largely of the elements iron, oxygen, silicon, and magnesium, listed in order of abundance by weight. Table 7.1 shows that these four elements make up about 92% of the earth by weight. Also listed are lesser constituents down to the fifteenth ranking element. Most of the iron and nickel lies in the core, while a large part of the magnesium and a considerable amount of the iron are in the mantle. Because of the density segregation that has taken place, the chemical composition of the thin crust is quite different from the total earth composition.

The Earth's Crust

From the standpoint of environment of life on earth, it is the crust that is really significant, because this paper-thin layer (as compared with the mantle and core) contains the continents and ocean basins and is the source of soil and other sediment vital to life, of salts of the sea, the gases of the atmosphere, and all free water of the oceans, atmosphere, and lands. True, the ultimate source of water and gases delivered to the surface by outgassing and by molten rock may lie deeper than the crust (in the upper mantle, that is), but we will not be far wrong in considering the crust as the direct source of all elements of environmental significance.

The crust, which is a layer of less density than the mantle beneath it, represents only about four-tenths of a percent of the earth's total mass. A study of the principal crustal elements, as listed in order of abundance in Table 7.2, shows some striking differences from the table of element abundances for the earth as a whole (Table 7.1). These crustal figures apply to a layer about 10 miles (17 km) thick. This is approximately the

TABLE 7.1 Average Composition of the Earth

Rank	Element	Symbol	Percent by Weight
1	Iron	Fe	34.6
2	Oxygen	O	29.5
3	Silicon	Si	15.2
4	Magnesium	Mg	12.7
5	Nickel	Ni	2.4
6	Sulfur	S	1.9
7	Calcium	Ca	1.1
8	Aluminum	Al	1.1
9	Sodium	Na	0.57
10	Chromium	Cr	0.26
11	Manganese	Mn	0.22
12	Cobalt	Co	0.13
13	Phosphorus	P	0.10
14	Potassium	K	0.07
15	Titanium	Ti	0.05

(Ranks 1–4 bracketed as 92.0)

SOURCE: Brian Mason, 1966, *Principles of Geochemistry*, 3rd ed., John Wiley & Sons, New York, Tables 3.7 and 2.4.

figure obtained by averaging out the crust, which ranges considerably in thickness and composition from continents to ocean basins. The eight elements listed account for between 98 and 99% of the total crustal composition by weight, and of this amount about half is oxygen. Oxygen is an element of comparatively large atomic size and low atomic weight, so that in terms of volume, oxygen represents nearly 94% and accounts for nearly 63% of the atoms of the crust. Oxygen thus has replaced iron for first place on the list, as compared with the total earth composition (Table 7.1). Silicon takes second place with about 28% by weight, a value about double that for the entire earth. Aluminum has risen from eighth to third place, with an eight-fold increase in abundance. Iron drops to fourth place with only 5%, compared with its leading abundance in the entire earth. There follow four metallic elements known as *bases*: calcium, sodium, potassium, and magnesium. These bases are vital to plant growth and their abundance in the soil is a measure of soil fertility, especially for agricultural crops such as the grains and forage grasses. All four of the bases are of the same order of abundance, from 2 to 4%.

If we were to extend Table 7.2, the ninth-place element would prove to be titanium, which would be followed in order by hydrogen, phosphorus, barium, and strontium. Phosphorus is one of the essential elements in plant growth. As we shall see in Chapter 10, a number of metallic elements that we think of as abundant in manufactured products—copper, lead, zinc, nickel, and

The Silicate Minerals

TABLE 7.2 Abundant Elements in the Earth's Crust

Element	Symbol	Percent by Weight	Percent by Volume	Percent of Atoms Present
Oxygen	O	46.6	93.8	62.6
Silicon	Si	27.7	0.9	21.2
Aluminum	Al	8.1	0.5	6.5
Iron	Fe	5.0	0.4	1.9
Calcium	Ca	3.6	1.0	1.9
Sodium	Na	2.8	1.3	2.6
Potassium	K	2.6	1.8	1.4
Magnesium	Mg	2.1	0.3	1.8

SOURCE: Brian Mason, 1966, *Principles of Geochemistry*, 3rd ed., John Wiley & Sons, New York. Table 3.4, p. 48.
NOTE: Figures have been rounded to nearest one-tenth.

tin—are present only in extremely low percentages and are very scarce elements, indeed. Were it not for the fact that these essential metals have been concentrated into rich ores at a few scattered locations in the crust, the industrial development of civilization could not have progressed beyond a rather primitive iron age.

To proceed further in an investigation of the earth's crust, some basic knowledge of crustal rocks and minerals is essential. After acquiring this knowledge, it will be fairly easy to understand how and why the crust under the continents is different in composition and thickness from the crust under the ocean basins.

The Silicate Minerals

Rock, which is composed of mineral matter in the solid state, comes in a very wide range of compositions, physical characteristics, and ages. A given rock is usually composed of two or more minerals and usually many minerals are present; however, a few rock varieties consist almost entirely of one mineral. Most rock of the earth's crust is extremely old in terms of human standards, the times of formation ranging back many millions of years. But rock is also being formed at this very hour, as a volcano emits lava that solidifies upon contact with the atmosphere. A *mineral* is perhaps easier to define; it is a naturally occurring, inorganic substance usually possessing a definite chemical composition and a characteristic atomic structure. The vast number of known minerals, together with the great number of their combinations in rocks, require that we must generalize and simplify this discussion to a high degree, referring to rocks and minerals in a way meaningful in terms of their environmental properties and value as natural resources.

Most of the rock of the crust is *igneous* in origin, meaning that the crystalline or glassy solid rock has solidified from a high-temperature molten state, or *magma*. Most of the bulk of igneous rock consists of *silicate minerals*, or simply *silicates*, which are compounds containing a combination of silicon (Si) and oxygen (O), usually together with one or more metallic elements. A basic understanding of igneous rocks can be had by considering the various common combinations of only seven silicate minerals or mineral groups; these are shown in Figure 7.3.

Among the commonest minerals of all rock types is *quartz*; its composition is *silicon dioxide* (SiO_2). There follow five mineral groups, collectively forming the *aluminosilicates* because all contain aluminum. Two groups of *feldspars* are set apart: the *potash feldspars* contain potassium (K) as the dominant metallic ion, but sodium (Na) is commonly present in various proportions. The *plagioclase feldspars* form a continuous series, beginning with the *sodic*, or sodium-rich varieties, and grading through with increasing proportions of calcium to the *calcic*, or calcium-rich varieties.

Belonging to the *mica group*, which is familiar because of its property of splitting into very thin, flexible layers, is *biotite*, a dark-colored mica with a complex chemical formula. Potassium, magnesium, and iron are present in biotite, along with some water. The *amphibole group*, of which the common black mineral *hornblende* is a common representative, is a complex aluminosilicate containing calcium, magnesium, iron, and water. Similar in outward appearance and having essentially the same component elements (except for water) is the *pyroxene group*, with the mineral *augite* as a representative. Last on the list is *olivine*, a dense greenish mineral which is a silicate of magnesium and iron, but without aluminum.

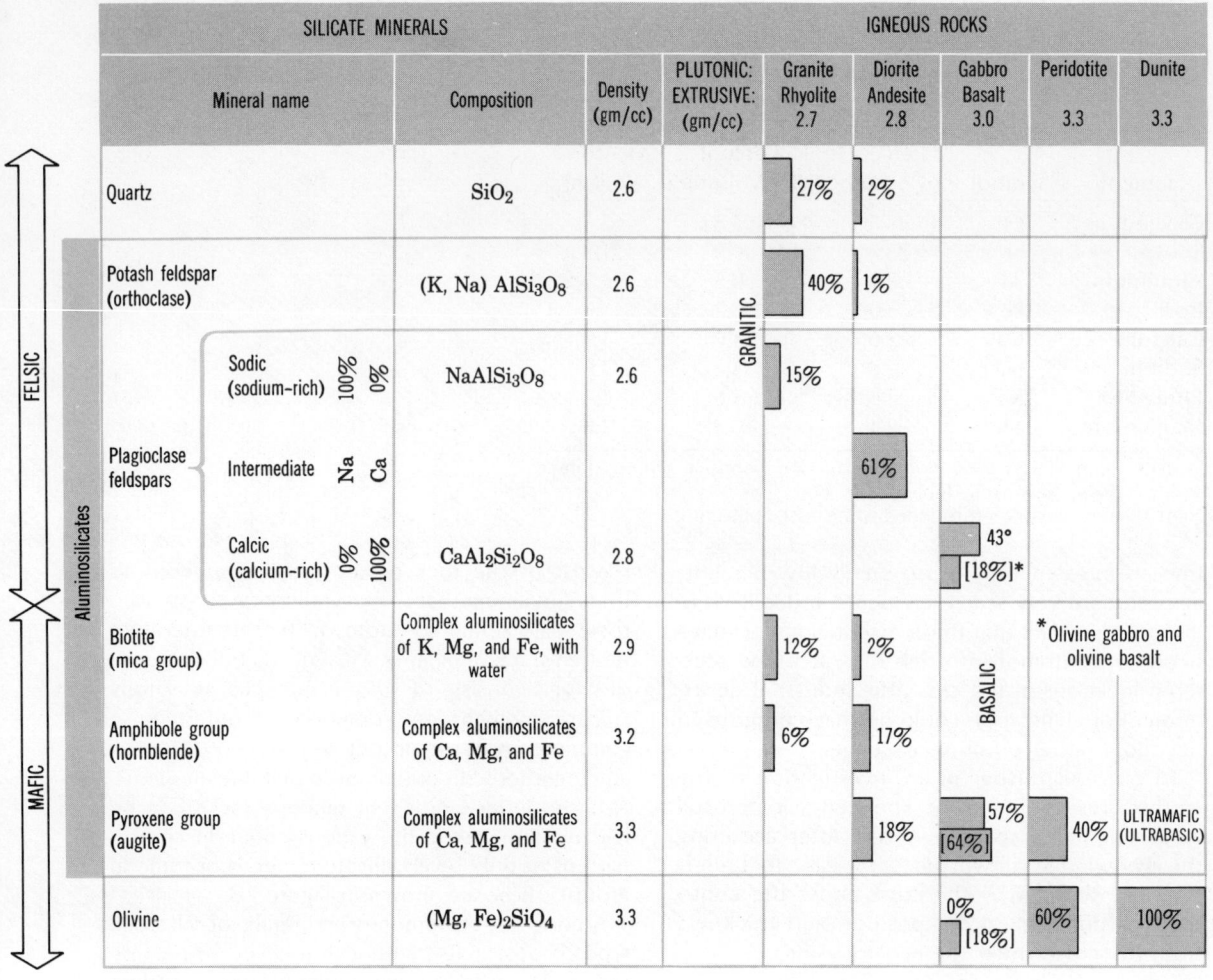

Figure 7.3 A simplified chart of common silicate minerals and abundant igneous rocks. (From A. N. Strahler, 1972, *Planet Earth; Its Physical Systems Through Geologic Time*, Harper & Row, New York.)

Looking down the list of densities given for these silicate minerals you will notice that there is a progressive increase from the least dense, quartz, to the most dense, olivine. This change reflects the decreasing proportion of aluminum and sodium, elements of low atomic weights, and the increasing proportion of calcium and iron, elements of considerably greater atomic weights.

The list as a whole is conveniently divided into two major groups of silicate minerals: a *felsic group*, consisting of quartz and the feldspars; and a *mafic group*, consisting of the silicates rich in magnesium and iron. The coined word felsic is easily recognized as a combination of *fel*, for feldspar, and *si* for silica, while the syllable *ma* in mafic stands for magnesium, the letter *f* for iron (symbol, *Fe*). The felsic minerals are light in color, as well as of comparatively low density; the mafic minerals are dark in color and of comparatively high density.

Two important mafic minerals that are not silicates occur in many igneous rocks. These are *magnetite*, an iron oxide (Fe_3O_4), and *ilmenite*, an oxide of iron and titanium ($FeTiO_3$). Both of these minerals are black and have high densities—4.5 to 5.5 g/cc.

Keep in mind that the minerals listed are important as sources of ions of primary environmental significance in the life layer. Ions in sea water include silicon, calcium, sodium, potassium, and magnesium. The last four named are positive ions in sea salts (see Table 8.4). Bases important in soils include ions of calcium, sodium, potassium, and magnesium; they are also the principal bases found in stream water derived from areas of igneous rocks (see Table 8.5). Oxides of aluminum and iron, derived from chemical alteration of the silicate minerals, are important components of soils and provide essential ores of iron and aluminum. From the geologic standpoint, the silicate minerals can be viewed as the basic materials out of which other rock groups (sedimentary and metamorphic) are created. These are covered in the next two chapters.

Silicate Magmas and Volatiles

About 99% of the bulk of the igneous rocks of the earth's crust consists of the seven silicate minerals or mineral groups listed in Figure 7.3. The remainder consists of minerals of secondary importance in bulk, although their number is very large. Fortunately, the seven silicate minerals or groups combine to form only a dozen or so igneous rock varieties having widespread occurrence. We shall simplify the list to only five representative rock types, with the purpose in mind of illustrating the concepts behind the formation of igneous rocks.

Silicate minerals of the igneous rocks are derived from *silicate magmas* (melts rich in Si and O), formed at considerable depths in the crust or upper mantle under conditions of very high temperatures and pressures. Since these magmas cannot be studied under the conditions of their occurrence at depth, inferences as to their properties depend upon laboratory studies in which minerals are melted and allowed to solidify, and upon direct observations of magmas issuing from active volcanoes. At a depth of several miles silicate magmas probably have temperatures in the range of 900° to 2200° F (500° to 1200° C) and are under pressures on the order of 6 to 12 kilobars, that is, approximately 6000 to 12,000 times as great as atmospheric pressure at sea level.

The presence of substances over and above the elements that comprise the solidified silicate rock is very important in the behavior of magmas. These substances are known as *volatiles*, because they remain in a gaseous or liquid state at much lower temperatures than the silicate compounds. Consequently, some of the volatiles are separated from the magma as it cools and solidifies into a crystalline state. From analysis of gas samples collected from volcanic emissions we know that water is the preponderant constituent of the volatile group. Table 7.3 lists the volatiles found in gases emanating from magma of active Hawaiian volcanoes. For comparison, the table lists the proportions of these volatiles in the atmosphere and hydrosphere. Notice that the abundances of the various constituents are of the same general order of magnitude in both lists, although the ranking is not the same.

As we have noted in earlier chapters, the emanation of volatiles from the earth's interior is called *outgassing* and is the source of free water of the earth's hydrosphere as well as of the atmospheric gases carbon dioxide, nitrogen, argon, and hydrogen. Chlorine and sulfur compounds of sea water have derived their chlorine and sulfur from outgassing. It looks, then, as if silicate magmas with their enclosed volatiles have through geologic time supplied just about all of the essential components of the atmosphere, hydrosphere, and lithosphere. As life evolved, its processes and forms were adapted to the chemical ingredients supplied by fresh magmas and volatiles, and not from an original lithosphere and atmosphere inherited from the time of accretion of the earth (about 4.5 billion years ago). But, as life evolved and spread in the shallow seas and later over the lands, organic processes began to change the composition of atmosphere, the dissolved solids in the oceans, and the mineral deposits on the sea floors. Thus, molecular oxygen (O_2), which is not found in gases of magmas, was released by photosynthesis of plants and eventually built up to a large volume (21%) of the atmos-

TABLE 7.3 Volatiles in Gases of Magmas Compared with Volatiles in Atmosphere and Hydrosphere

	Lava Gases from Mauna Loa and Kilauea Volcanoes (percent by weight)	Volatiles Free in Earth's Atmosphere and Hydrosphere (percent by weight)
Water, H_2O	60	93
Carbon, as CO_2 gas	24	5.1
Sulfur, S_2	13	0.13
Nitrogen, N_2	5.7	0.24
Argon, A	0.3	trace
Chlorine, Cl_2	0.1	1.7
Hydrogen, H_2	0.04	0.07
Fluorine, F_2	—	trace

SOURCE: W. W. Rubey, 1952, *Geological Society of America, Bulletin*, vol. 62, p. 1137.
NOTE: Figures have been rounded off.

phere. The much lower proportion of carbon (as CO_2) in hydrosphere and atmosphere today than in magmatic gases largely reflects the storage of carbon as a mineral compound in solid layers on the bottom of the sea, and to a lesser degree as hydrocarbon compounds in fuels such as coal and petroleum contained in rocks. Sulfur has similarly been removed from the atmosphere and hydrosphere and placed in storage in rocks. The important point is that processes of the life layer have contributed to a sort of inorganic evolution of the earth's crust. This subject will be explored in depth in the next chapter, but the basic concept starts here, with the environmental significance of magmas and their volatile constituents.

The presence of water in silicate magmas is most important in the physical behavior of the magma as it works its way from the place of origin to shallower levels in the crust. Even a small amount of water greatly lowers the minimum temperature at which the magma can remain in the liquid state. Consequently, as the magma rises and becomes progressively cooler, the presence of water allows the magma to reach positions much shallower in depth than would otherwise be possible. Extended fluidity of the magma allows it to reach the surface in places, where it can pour out in enormous volumes in fluid layers and tongues as *lava*.

At a certain critical temperature, in combination with a critical pressure, crystallization begins to take place in a slowly cooling magma. Although all of the eight elements listed in Figure 7.3 will eventually be gathered into compounds as individual crystals of the various silicate minerals, certain minerals are formed first, followed by others on the list. To complicate the picture, a mineral once formed can undergo a change in composition or may be dissolved and reformed to produce another mineral. Change of this sort during magma cooling is referred to as *reaction*. The orderly series of changes is known as the *reaction series*. The crystallization process is made even more complex by the possibility that certain minerals, once crystallized, can be separated from the magma and left behind to form a separate igneous rock. This process, called *fractionation*, changes the chemical composition of the remaining fluid magma and alters the series of reactions that follows, so that another variety of igneous rock is formed.

Let us now follow the typical succession of changes, known as the *Bowen reaction series*, in which a magma containing all eight elements of the common silicate minerals form themselves into silicate minerals as temperature falls. Figure 7.4 shows this series as taking the form of two convergent branches. The lefthand branch involves the mafic minerals; the righthand branch the plagioclase feldspars. Olivine is the first-formed mafic mineral, followed in order by pyroxene, amphibole, and biotite. In this series, each silicate is reformed into the next in a series of discontinuous steps. Within the plagioclase series progression is from calcic to sodic extremes, with a continuous reaction from one end to the other.

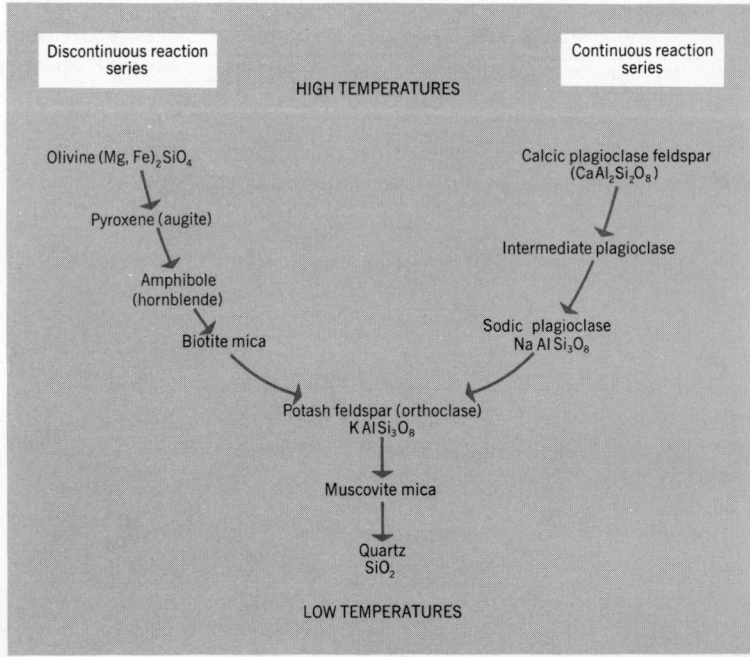

Figure 7.4 The Bowen reaction series.

The Igneous Rocks

Crystallization of potash feldspar then takes place, followed by a light-colored mica, called *muscovite*, and finally quartz.

The Bowen reaction series gives us the key to how various igneous rock types can be formed from a single silicate magma. If fractionation takes place after olivine and pyroxene have crystallized, by the settling of these heavy minerals to the bottom of the magma chamber, these two minerals can form a rock known as *peridotite*, a mixture of crystals of both minerals. If the reaction proceeds through the stage of formation of pyroxene and an intermediate form of plagioclase, the magma can produce a rock known as *gabbro*. The remaining magma is richer in the felsic constituents, aluminum and potassium, but has lost most of the mafic constituents, calcium, iron, and magnesium. When this remaining magma has migrated to a new location, it can crystallize as an aggregation of felsic minerals, principally as quartz and potash feldspar, with minor amounts of biotite and sodic feldspar. The resulting rock is *granite*; it is the principal igneous rock of felsic composition.

Finally, there remains a watery magma rich in silica. Under high pressure this solution leaves the main body of solidifying magma and penetrates the surrounding rock mass in small chambers and narrow passageways to crystallize in the form of *pegmatite* bodies. Pegmatite consists of large crystals of such minerals as quartz, potash feldspar, muscovite, and biotite. Pegmatites contain many other minerals, some of which are extremely rare and provide valuable ores.

Fractionation of magma, leading to the formation of several varieties of igneous rocks is known in general as *magmatic differentiation*. Now that we know how this process operates, it will be used to prepare a related classification of igneous rocks—a classification that makes sense and is not just a collection of names.

Figure 7.5 Granite, a coarse-grained intrusive rock, is made up of tightly interlocking crystals of a few kinds of minerals.

in the crust are described by the adjective *plutonic*. Granite is a typical coarse-grained plutonic rock (Figure 7.5). In contrast, rapid cooling of magma as it comes close to the surface or pours out as lava gives very small mineral crystals, or *fine-grained* texture. The lavas, which are classed as *extrusive* in contrast to the *intrusive* rocks solidified beneath the surface, typically are formed of crystals too small to be distinguished with the unaided eye, or even with aid of a magnifying lens. Rapid cooling of viscous magma yields a natural volcanic glass, known as *obsidian* (Figure 7.6, *right*), while the frothing of a magma as its gases expand near the surface gives a porous rock, known as *scoria* (Figure 7.6, *left*).

To simplify rock classification, we shall recognize five rock varieties according to mineral com-

The Igneous Rocks

Before going further with classification of igneous rocks, it will be necessary to recognize that they are classified not only on the basis of mineral composition, but also upon grades of sizes of the component crystals. The term *texture* applies both to the grain size and to the arrangements of crystals of various sizes.

Generally speaking, very gradual cooling of a magma that lies deep within the crust results in formation of large mineral crystals and results in a rock with *coarse-grained* texture. Huge masses of coarse-grained igneous rock, solidifying at depth

Figure 7.6 A frothy, gaseous lava solidifies into a light, porous scoria (*left*). Rapidly cooled lava may form a dark volcanic glass (*right*).

Igneous Processes and the Earth's Crust

position, and for each of these classes there will be coarse-grained, plutonic varieties as well as fine-grained or glassy extrusive varieties. As shown in the right half of Figure 7.3, there are five named plutonic varieties and three extrusive varieties (the extrusive varieties of the last two on the list are unimportant). This boils down the list of rock names to seven.

Bars of varying width, with attached percentage figures, show the typical mineral compositions of these igneous rocks. To cut through a lot of tedious detail, note simply that granite and its extrusive equivalent, *rhyolite,* are rich in quartz and potash feldspar, with lesser amounts of sodic plagioclase, biotite, and amphibole. *Diorite* and its extrusive equivalent, *andesite,* are almost totally lacking in quartz and potash feldspar, but consist dominantly of intermediate plagioclase and lesser amounts of the mafic minerals. Going progressively in the direction of domination by mafic minerals, we come to *gabbro,* and its lava equivalent, *basalt.* Here the plagioclase feldspar is of the calcic type, making up about half the rock, while pyroxene makes up the other half. In a common variety of gabbro and basalt olivine is also present. The next rock, *peridotite,* is not abundant in the crust, but probably makes up the bulk of the outer mantle. It is composed mostly of pyroxene and olivine. Finally, we list *dunite,* a rock of nearly 100% olivine, as an example of the extreme mafic end of the mineral series. Dunite is a rare rock at the surface, but may comprise the bulk of the lower mantle.

Looking back over the five igneous rock classes defined by composition, granite and diorite and related types rich in felsic minerals can be collectively described as *granitic* rocks; gabbro and basalt and related types as *basaltic* rocks; and the extreme mafic types as *ultramafic* rocks. This generalization will be put to use in understanding the gross composition of the crust and its relation to the underlying mantle.

Perhaps you have noticed that the listing of five rocks by composition in Figure 7.3 is in reverse order of the crystallization order in the Bowen reaction series. However, it is conventional in teaching to start with the felsic end and work down to the mafic end.

Densities of the igneous rocks are proportional to the densities of the component minerals. Thus granite has a density of about 2.7 g/cc; gabbro and basalt, about 3.0; peridotite and dunite, 3.3. Reasoning that geologic processes tend to cause the gross arrangement of igneous rocks to be in layers in order of their densities, granitic rocks will be found in the top layer of the crust, basaltic rocks in the next lower layer, and ultramafic rocks at the bottom. This proves to be a correct deduction as a first order of generalization.

Continents and Ocean Basins

The first-order relief features of the earth are the continents and oceans basins. From a globe or atlas we can compute that about 29% of the globe is land; 71% is ocean. If, however, the seas were to drain away, it would become apparent that broad areas lying close to the continental shores are actually covered by shallow water, less than 600 ft (180 m) deep. From these relatively shallow *continental shelves* the ocean floor drops rapidly to depths of thousands of feet. Figure 7.7 is a map of a portion of the Atlantic continental shelf. Notice how the sea floor descends rapidly from the 100-fathom (600 ft, 180 m) depth contour. This rapid fall-off is the *continental slope.* In a sense, then, the ocean basins are brimful of water and have even spread over the margins of ground that

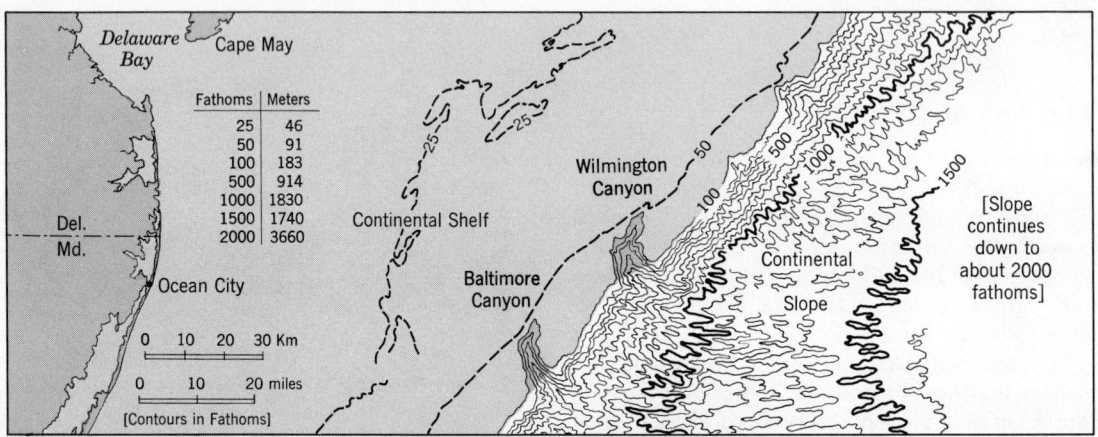

Figure 7.7 Two submarine canyons notch the outer edge of the Atlantic continental shelf off the Delaware-Maryland coast. (After Veatch and Smith.)

Continents and Ocean Basins

would more reasonably be assigned to the continents. If the ocean level were to drop by 600 ft (180 m) the surface area of continents would increase to 35%, the ocean basins decrease to 65%—figures which we may regard as representative of the true relative proportions.

The continental shelves play an important role in the life layer, for in these shallow waters lie most of the world's food resources from the sea. At times when glacial ice reached its maximum spread on lands during the four or more glaciations of the Pleistocene Epoch, sea level fell to a position some 300 to 400 ft (100 to 120 m) below the present level, exposing most of the continental shelves and greatly reducing the extent of shallow waters. That these continental margins are now inundated is most fortunate for mankind, for without the richly productive shelf waters his prospects of feeding a fast-growing world population would be even more bleak than we now find them.

Figure 7.8 shows graphically the percentage distribution of the earth's solid surface area with respect to elevation both above and below sea level. Note that most of the land surface of the continents is less than 3300 ft (1 km) above sea level. There is a rapid drop off from about −3000 to −10,000 ft (−1 to −3 km) until the ocean floor is reached. A predominant part of the ocean floor lies between 10,000 and 20,000 ft (3 and 6 km) below sea level. Disregarding the earth's curvature, the continents can be visualized as platformlike masses; the oceans as broad, flat-floored basins.

To help keep our surface environment in its correct proportions, let us consider the true scale of the earth's landforms in comparison with the earth as a sphere. Most of the relief globes and pictorial relief maps seen commonly in magazines and atlases are greatly exaggerated in vertical scale. For a true-scale profile around the earth we might draw a chalk-line circle 21 ft (6.4 m) in diameter, representing the earth's circumference on a scale of 1 to 2,000,000. A chalk line 3/8 in. (0.15 cm) wide would include within its limits not only the highest point on the earth, Mt. Everest, +29,000 feet (+8840 m) but also the deepest known ocean trenches, below −35,000 feet (−10,700 m).

Figure 7.9 shows profiles correctly curved and scaled to fit a globe whose diameter is 21 ft (6.4 m). The topographic profile is drawn to natural scale, without vertical exaggeration. Although the most imposing landforms of Asia and North America are shown, they seem little more than trivial irregularities on the great global circle.

Enormous as the relief features of the continents and ocean basins may seem to us, they are only the muted surface expression of much greater rock masses that make up the earth's crust. Using our knowledge of igneous rocks and their densities, we can turn to examination of a cross section of the crust, revealing the fundamental geological causes of differences between continents and ocean basins.

The upper diagram of Figure 7.10 shows a continent between two ocean basins. The continental crust consists of an upper granitic layer overlying a lower basaltic layer. The lower diagram of Figure 7.10 shows this structure in more detail. At the base of the basaltic layer the crust gives way abruptly to the underlying mantle, which consists of ultramafic rock. The surface of discontinuity between crust and mantle is formally named the *Mohorovičić discontinuity*, after a seismologist who discovered the change in rock properties by interpreting the records of earthquake waves. At a depth which averages about 25 mi (40 km), the waves show a sudden increase in speed and this change can only mean that the rock abruptly

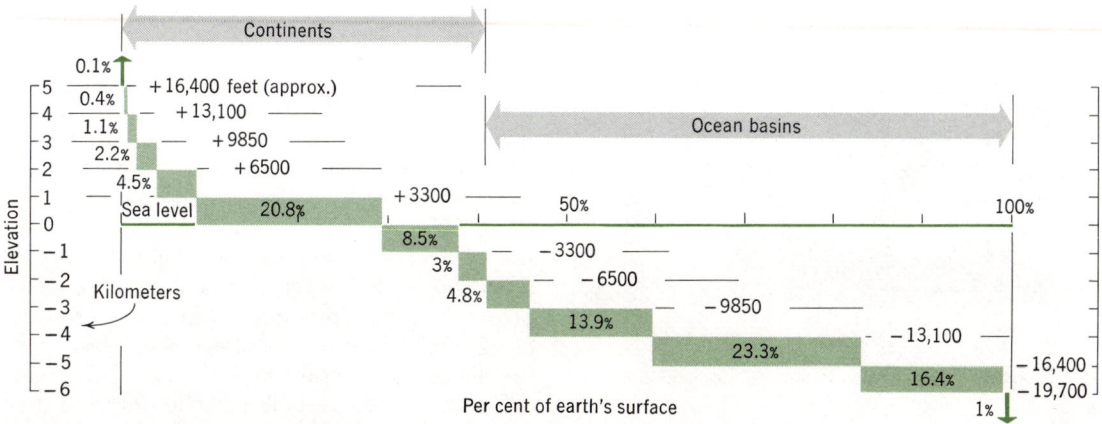

Figure 7.8 Distribution of the earth's solid surface area in successively lower elevation zones.

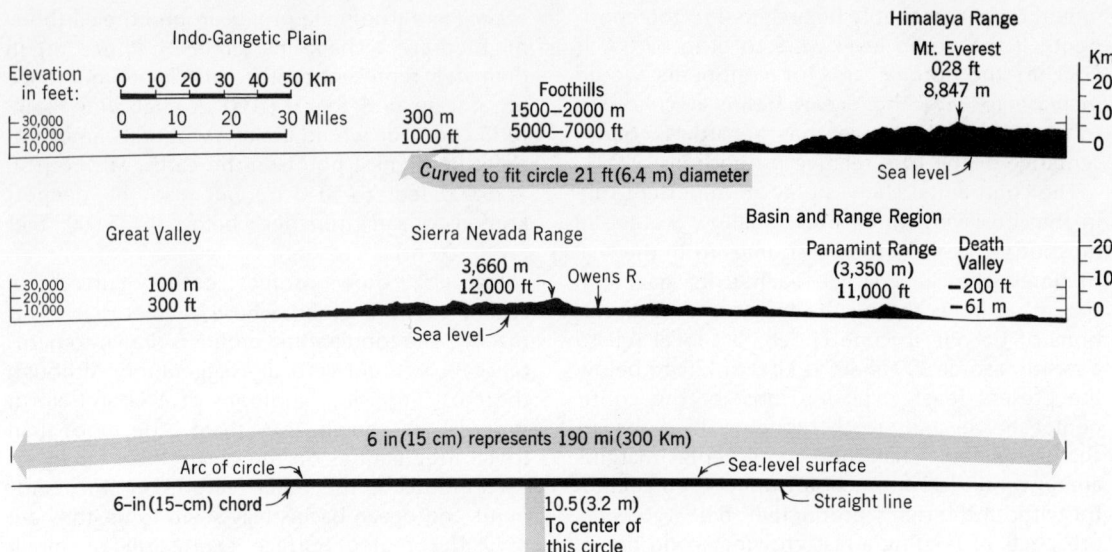

Figure 7.9 These profiles show the earth's great relief features in true scale with sea-level curvature fitted to a globe 21 ft (6.4 m) in diameter.

changes to a denser state. Numbers on the diagram show the typical earthquake wave velocities in the granitic, basaltic, and ultramafic layers. There seems to be only a gradual change from granitic to basaltic rock with increasing depth. The term *Mohorovičić discontinuity* has mercifully been simplified to one of two working terms: *M-discontinuity*, or *Moho*

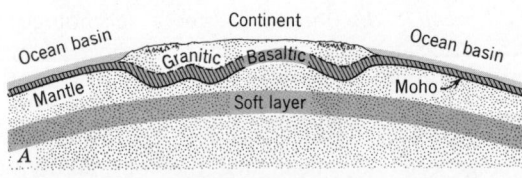

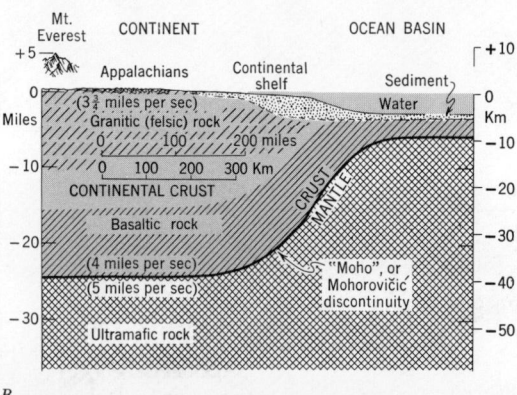

Figure 7.10 The earth's crust is much thicker under continents than beneath the ocean basins.

Beneath the ocean basins, the crust is strikingly different from that of the continents. As Figure 7.10 shows, the oceanic crust is much the thinner, so that the M-discontinuity lies at a depth of only about 7½ mi (12 km). Moreover, the oceanic crust consists of a single basaltic layer about 3 to 4 mi (5 to 6 km) thick. Covering the basaltic layer is a thin layer of sediment, about 0.3 mi (0.5 km) thick, and above this is the ocean water, averaging 2.8 mi (4.5 km) in depth. Where the ocean basin joins the continent the basaltic layer passes down under the continent to form a continuous layer with the basaltic portion of the lower continental crust.

If we were to plot a cross-section of the continental crust passing through a wide range of relief features, from low plains to high mountains, the depth to the M-discontinuity would be greater beneath the mountain ranges than beneath the low plains. In other words, mountains have crustal *roots*. For example, beneath the Pamirs, a great mountain mass of southern Asia, the crust extends down to 46 mi (75 km), the maximum depth to which the M-discontinuity has yet been recorded. Now, a cake of ice floats in water with most of its bulk beneath the water line. Whatever its size, the ice cake will have, say, one-eighth of its volume above the water line and seven-eighths below that line. Consequently, the larger the ice block, the higher it rises above water level and the deeper is its base.

Geologists have established the principle that the crust behaves much like the floating ice, in that it reaches an equilibrium position, as if it were floating upon a yielding plastic mantle of greater density. There is good reason to believe that there exists in the upper mantle, a *soft layer* having little strength. This condition of softness sets in at a depth of about 40 mi (60 km). Strength declines with depth to a minimum in a zone near 125 mi (200 km), below which strength again increases. Loss of strength in this soft zone is caused by accumulated radiogenic heat, which brings the mantle rock close to its melting point. Like iron heated white hot in a furnace, the rock loses strength and will move slowly in response to unequally applied stresses.

The crust in combination with the uppermost mantle comprises a strong, rigid earth shell called the *lithosphere*. (This geological usage is somewhat more restrictive than our previous usage of the word to designate the entire solid earth realm.) Beneath the lithosphere lies the soft, weak layer, or *asthenosphere*. Recognition of these two layers based upon their strength properties is extremely important in understanding modern discoveries of the drifting apart of continents, a subject reserved for the next chapter.

The principle of crustal flotation in equilibrium is known in geological circles as the *principle of isostasy* The word isostasy comes from two Greek words, *isos*, equal, and *stasis*, a standing still. Under the principle of isostasy, the continents, floating in equilibrium over a soft mantle, stand higher and extend deeper because they are thicker and less dense than the oceanic crust, which is not only comparatively thin, but being composed of basalt has an average density substantially greater than that of the granitic rock.

The particular model of isostasy that is generally used today was postulated almost a century ago by Sir George Airy, Astronomer Royal of England. His model is illustrated in Figure 7.11 by a number of prisms of copper immersed in a bath of mercury. Each prism floats at a depth such that the proportion of volume above the liquid level to that lying below the liquid level is the same for all blocks, even though they are of different lengths. Thus the longest block rises highest and extends to greatest depth. The sketch below the block model shows the analogy with continents and ocean basins, and explains the deep crustal root beneath the mountain mass. We will refer to this model of isostasy in later chapters, since it follows that if the upper part of a prism were to be removed, the remaining prism would rise to a new position of equilibrium. Removal of mass from one part of a continent must cause the crust there to rise; while addition of mass to another part of the continent will cause the crust to sink beneath the load.

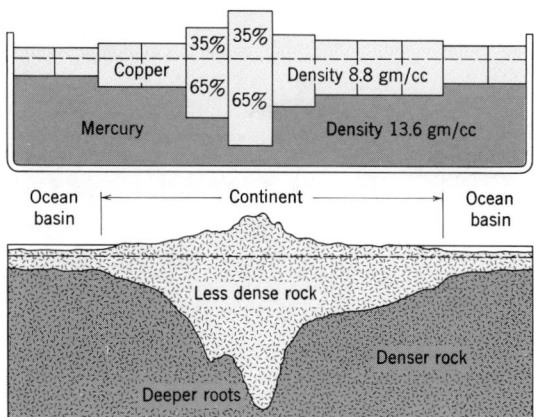

Figure 7.11 Airy model of isostasy. All floating blocks have the same density. (From A. N. Strahler, 1971, *The Earth Sciences*, 2nd ed., Harper & Row, New York.)

Rise of Igneous Rock in the Crust

No igneous rock, or for that matter any terrestrial rock, has been found to exceed an age of about 3.5 billion years. On the other hand, it is fairly certain that the growth of the earth as a planet was completed about 4.5 billion years ago. (Meteorites yield an age of 4.5 billion years.) This leaves a one billion year period of earth history unaccounted for by any tangible record. While geological exploration may yet lead to the discovery of rock somewhat older than any now known, it seems most unlikely that the billion-year gap will ever be filled. The inescapable conclusion is that none of the original surface rock remains in existence; it has all been reworked or replaced by the rise of magma and solidification of newer igneous rock. This conclusion is in line with our previous explanation of the development of core and mantle by episodes of melting and overturn occurring in the early stages of the earth's history.

Plutonic rock, most of it rock of granitic composition, occurs in huge masses called *batholiths* (Figure 7.12). The intruding magma that formed the batholith made its way upward through the surrounding solid rock (called *country rock* in this context) by breaking away and ingesting fragments of the solid rock above it, or by melting the rock above it. At the time of its formation, the batholith lies deep within the crust, covered by a thick layer of the country rock. Only after a long period of crustal uplift and erosional removal of the overlying cover does the batholith appear at the surface (Figure 7.13).

A number of other forms of intrusive igneous rock bodies are illustrated in Figure 7.12. Where the older rock is layered, magma can insinuate itself between layers, lifting the mass above and

Igneous Processes and the Earth's Crust

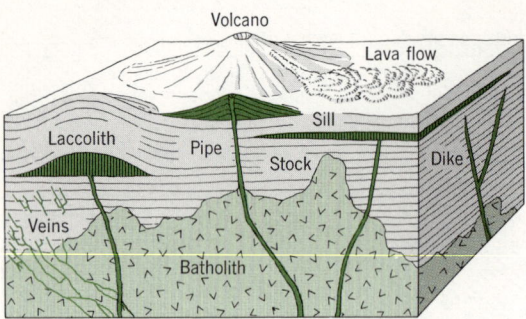

Figure 7.12 Igneous rock bodies.

forming a sheetlike igneous body, a *sill,* or a tack-shaped body, a *laccolith.* Where near-vertical fractures are forced apart by pressure of rising magma, igneous rock solidified in a wall-like mass, called a *dike* (Figure 7.14). The importance of these varied igneous bodies is related to their subsequent expression as landforms, after erosion has removed the enclosing country rock. For example, a dike may later form a rock wall running for miles across the landscape (see Figure 7.23).

Associated with the margins of a batholith are bodies of pegmatite, which have been explained as the final intrusive deposits formed from watery solutions forced out of the crystallizing magma. Figure 7.15 shows how pegatite bodies are related to the batholith in its final stages of solidification. They form in the peripheral region of the batholith itself, as well as in dikes and thinner *veins* in the country rock. The importance of pegmatites lies in their containing ore deposits of great value (Chapter 10).

While the great bulk of basaltic rock is contained in the oceanic crust and in the basaltic lower layer of the continental crust, there has occurred from time to time in geologic history the large-scale rise of basaltic magma within the continental crust. Forced upward through long fractures, called *fissures,* the basaltic magma poured

Figure 7.14 A basaltic dike cutting granite, Cohasset, Massachusetts. (Photograph by John A. Shimer.)

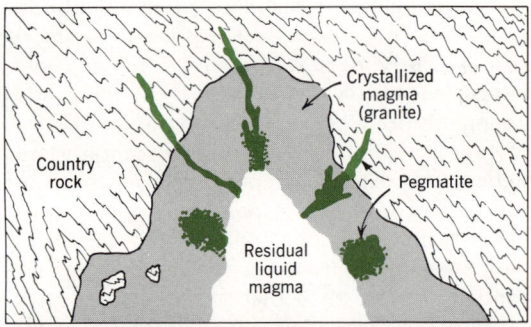

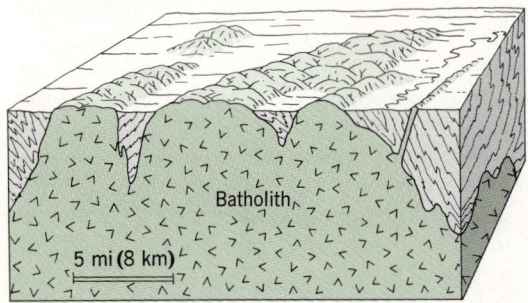

Figure 7.13 Deep-seated igneous rocks appear at the surface only after long-continued erosion has removed thousands of feet of overlying rocks. (After Longwell, Knopf, and Flint.)

Figure 7.15 *Above:* Pegmatite bodies in relation to an intrusive igneous body and the country rock. (From A. N. Strahler, 1972, *Planet Earth; Its Physical Systems Through Geologic Time,* Harper & Row, New York.) *Below:* a pegmatite consisting of tourmaline crystals (black) and of quartz and feldspar crystals (white). (Specimen by courtesy of Ward's Natural Science Establishment, Inc., Rochester, N.Y.)

Volcanoes

out upon the land surface in lava flows that accumulated, one upon the next. These basalt floods have inundated areas on the order of 50,000 to 100,000 sq mi (130,000 to 260,000 sq km), and the total thickness of the flows has reached several thousands of feet. One example is the Columbia Plateau region of Washington, Oregon, and Idaho (Figure 7.16); another is the Deccan Plateau of western peninsular India (Figure 7.17). Environmentally, the plateau basalts are significant, since the basaltic rock has quite different chemical and physical properties from granitic and other kinds of felsic rock that more typically make up the continental surfaces. Mafic mineral constituents of basalt produce distinctive soils rich in calcium, magnesium, and iron. Landforms, too, are distinctive and take the form of platforms (*mesas, buttes*) and steps separated by steep cliffs (Figure 7.18).

Volcanoes

Volcanoes are conical or dome-shaped structures built by the emission of lava and its contained gases from a restricted vent in the earth's surface (Figure 7.12). The magma rises in a narrow pipe-like conduit from a magma reservoir far below. Reaching the surface, igneous material may pour out in tonguelike lava flows, or may be ejected under pressure of confined gases as solid fragments collectively called *ejecta*. Form and dimensions of the volcano are quite varied, depending upon the type of lava and the presence or absence of ejecta.

Volcanoes are built of rocks derived from magmas of one of several types, ranging from granitic to basaltic composition. Here we refer back to the table of igneous minerals and rocks, Figure 7.3. The form of a volcano and the nature of the eruption, whether explosive or quiet, depend upon the type of magma. For purposes of classifying volcanoes, three magma types are recognized:

Magma Composition	Name of Lava	Classification of Lava
Granitic	Rhyolite	Felsic
Intermediate	Andesite	Intermediate
Basaltic	Basalt	Mafic

The important point is that the felsic and intermediate lavas have a high degree of viscosity (property of tackiness, resisting flowage) and hold large amounts of gas under pressure. As a result, these lavas produce explosive eruptions. In contrast, mafic lava is highly fluid (low viscosity) and holds little gas, with the result that the eruptions

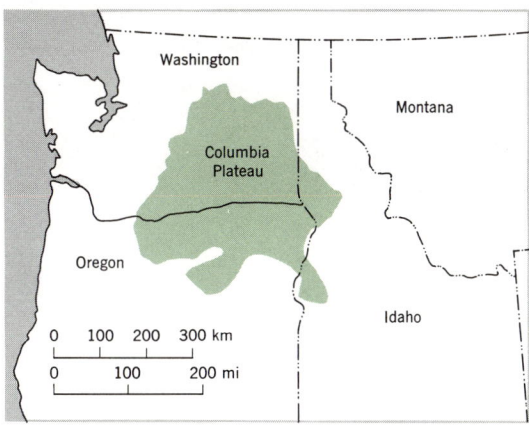

Figure 7.16 Distribution of Columbia Plateau basalts. (From A. N. Strahler, 1971, *The Earth Sciences*, 2nd ed., Harper & Row, New York.)

are quiet and the lava can travel long distances to spread out in thin layers (as in the plateau basalts).

Tall, steep-sided *volcanic cones* are produced by felsic and intermediate lavas. These cones tend to steepen toward the summit, where a comparatively small depresssion, the *crater*, is located. Ejecta in these volcanic eruptions takes the form of fine particles, described as *volcanic ash*, which falls upon the area surrounding the crater and thus contributes to the structure of the cone. The interlayering of ash layers and lava streams produces the *composite volcano*, or *stratovolcano* (Figure 7.19). The world's lofty conical volcanoes, well known for their scenic beauty, are of the

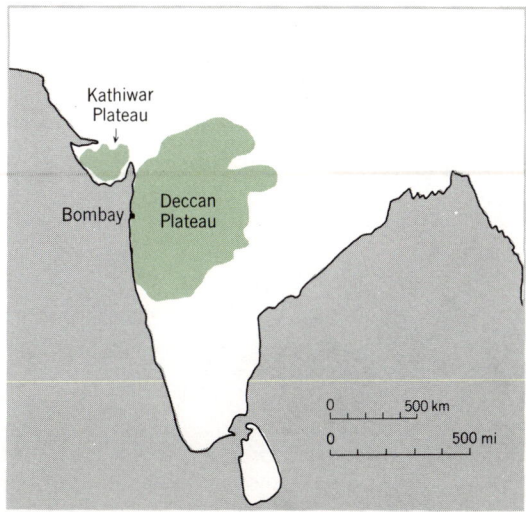

Figure 7.17 Approximate present surface extent of basalts of the Deccan Plateau of India. (From A. N. Strahler, 1971, *The Earth Sciences*, 2nd ed., Harper & Row, New York.)

181

Figure 7.18 Basaltic lava flows of the Columbia Plateau region have been eroded to produce steep cliffs rimming broad, flat-topped mesas. Dry Falls, Grand Coulee, central Washington. (Photograph by John S. Shelton.)

composite type. Examples are Mt. Hood in the Cascade Range, Mt. Fujiyama in Japan, Mt. Mayon in the Philippines (Figure 7.20), and Mt. Shishaldin in the Aleutian Islands. Ejecta also include *volcanic bombs,* solidified masses of lava ranging up to the size of large boulders, that fall close to the crater and roll down the steep slopes (Figure 7.21). Very fine volcanic dust rises high into the troposphere and stratosphere, where it remains suspended for years (Chapter 6).

Another important form of emission from the explosive types of volcanoes is a cloud of incandescent gases and fine ash. Known as a *nuée ardente* (French, glowing cloud), this intensely hot cloud travels rapidly down the flank of the volcanic cone, searing everything in its path. On the island of Martinique, in 1902, a glowing avalanche issued without warning from Mt. Pelée, sweeping down upon St. Pierre, destroying the city and killing all but one of its 30,000 inhabitants.

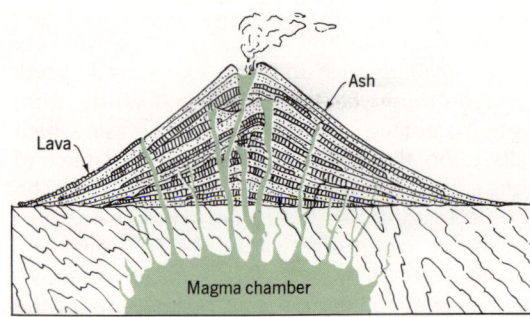

Figure 7.19 An idealized cross section through a composite volcano and the magma chamber beneath. (From A. N. Strahler, 1971, *The Earth Sciences,* 2nd ed., Harper & Row, New York.)

Many of the world's active composite volcanoes lie in a great belt, the *circum-Pacific ring* extending from the Andes in South America, through the Cascades and the Aleutians, into Japan; thence south into the East Indies and New Zealand. (See Figure 9.12.) There is also an important Mediterranean group, which includes active volcanoes of Italy and Sicily. Otherwise, Europe has no active volcanoes.

One of the most catastrophic of natural phenomena is a volcanic explosion so violent as to destroy the entire central portion of the volcano. There remains only a great central depression, termed a *caldera.* Although some of the upper part of the volcano is blown outward in fragments, most of it subsides into the ground beneath the volcano. Although calderas have been formed in historic time, conditions near the volcano do not permit observation of the process. Vast quantities of ash and dust are emitted and fill the atmosphere for many hundreds of square miles around.

Krakatoa, a volcanic island in Indonesia, exploded in 1883, leaving a great caldera. It is estimated that 18 cu mi (75 cu km) of rock disappeared during the explosion. Great seismic sea waves, or *tsunamis,* generated by the explosion killed many thousands of persons living on low

Figure 7.20 Mount Mayon, Philippine Islands. Elevation of the volcano summit is approximately 8100 ft (2400 m). (Photograph by Ed Drews/Photo Researchers.)

coastal areas of Java and Sumatra. Another historic explosion was that of Katmai, on the Alaskan Peninsula, in 1912. A caldera more than 2 mi (3 km) wide and 2000 to 3700 ft (600 to 1100 m) deep was produced at this time. The explosion was heard at Juneau, 750 mi (1200 km) distant, while at Kodiak, 100 mi (160 km) away, the ash formed a layer 10 in. (25 cm) deep.

A classic example of a caldera produced in prehistoric time is Crater Lake, Oregon (Figure 7.22). Mt. Mazama, the former volcano, is estimated to have risen 4000 ft (1200 m) higher than the present rim. Valleys previously cut by streams and glaciers into the flanks of Mt. Mazama were beheaded by the explosive subsidence of the central portion and now form distinctive notches in the rim. Wizard Island, a small volcanic cone with associated flows, has since grown in the floor of the caldera.

After a composite volcano has ceased its activity the erosional action of flowing water carves the cone into a system of radial stream valleys. The crater is breached and disappears, but the magma that solidified in the volcano feeder pipe then begins to come into prominence. Along with a system of radial feeder dikes, the volcanic pipe extends far down into the country rock below the base of the volcano. Ultimately, after erosion has continued for time spans on the order of millions of years, features are brought out into relief, and consist of a central *volcanic neck* and *radial dikes* (Figure 7.23).

A. This distant view of Sakurajima shows the great cauliflower cloud of condensed steam.

C. Reaching the sea, the hot lava makes clouds of steam.

B. A blocky lava flow is advancing slowly over a ground surface littered with volcanic bombs and ash.

D. Volcanic ash has buried this village.

Figure 7.21 Sakurajima, a Japanese volcano, erupted violently in 1914. These pictures show various scenes from the eruption. (Photograph by T. Nakasa.)

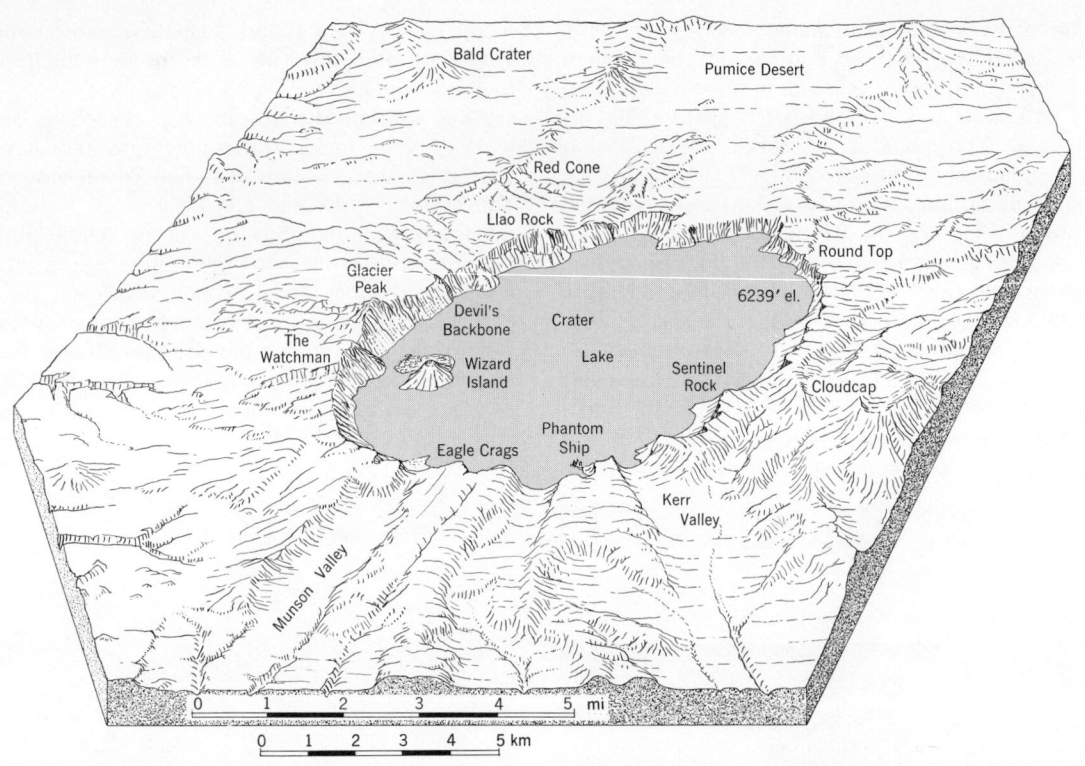

Figure 7.22 Crater Lake, Oregon, is an outstanding illustration of a caldera, now holding a lake. A great composite volcano which existed here was destroyed in prehistoric time by violent explosion. (After E. Raisz.)

Figure 7.23 Shiprock, New Mexico, is a volcanic neck. Radiating from it are dikes. (Spence Air Photos.)

Figure 7.24 Hawaii Volcanoes National Park. (Courtesy of National Park Service, U.S. Department of Interior.) Upper: Air view of Mokuaweoweo, the broad central depression on the summit of Mauna Loa. Lower: A pit crater, Halemaumau, on Mauna Loa, as seen in 1952.

Lava Domes and Shield Volcanoes

The quietly erupting, highly fluid basaltic lavas give rise to a second major group of volcanoes, known as *lava domes*, or *shield volcanoes*. The best examples are from the Hawaiian Islands, which consist entirely of lava domes (Figure 7.24).

Lava domes are characterized by gently rising, smooth slopes which tend to flatten near the top, producing a broad-topped volcano. The Hawaiian domes range to elevations up to 13,000 ft (4000 m) above sea level, but including the basal portion lying below sea level they are more than twice that high. In width they range from 10 to 50 mi (16 to 80 km) at sea level and up to 100 mi (160 km) wide at the submerged base.

Lava domes, as the name implies, are built by repeated outpourings of lava. Explosive behavior and emission of fragments are not important, as they are for composite cones built of acidic magmas. The lava, which in the Hawaiian lava domes is of a dark basaltic type, is highly fluid and travels far down the low slopes, which do not usually exceed 4° or 5°.

Instead of the explosion crater, lava domes have a wide, steep-sided *central depression*, or *sink*, which may be 2 mi (3.2 km) or more wide and several hundred feet deep. These large depressions are a type of caldera produced by subsidence accompanying the removal of molten lava from beneath. Molten basalt is actually seen in the floors of deep *pit craters*, steep-walled depressions 0.25 to 0.5 mi (0.4 to 0.8 km) wide or smaller, which occur on the floor of the sink or elsewhere over the surface of the lava dome (Figure 7.24). Most lava flows issue from cracks, or *fissures*, on the sides of the volcano.

Lava domes of the Hawaiian islands are in various stages of erosion (Figure 7.25). Active volcanoes such as Kilauea and Mauna Loa are in the initial stage and have smooth slopes. Others, such as East Maui, have long been inactive and are partly dissected by deep canyons but still possess sizable parts of the original surface. Still others, such as West Maui, are fully dissected. Rising from the sea are some steep-walled rock masses representing the last vestiges of old domes; and there exist submarine banks some 250 ft (75 m) below sea level, representing the final stage in destruction.

Basaltic Cinder Cones

Very much smaller in size than the composite volcanoes and great basaltic domes are *cinder cones*, consisting only of ejecta. The particles range from *cinder*, which includes gravel-sized and pebble-sized grains, to boulder-sized bombs. Highly gaseous, frothy lava, usually of basaltic composition, is emitted from a small vent and throws out ejecta that pile up around the vent (Figure 7.26). Finer ash settles out as an apron surrounding the cone.

Igneous Processes and the Earth's Crust

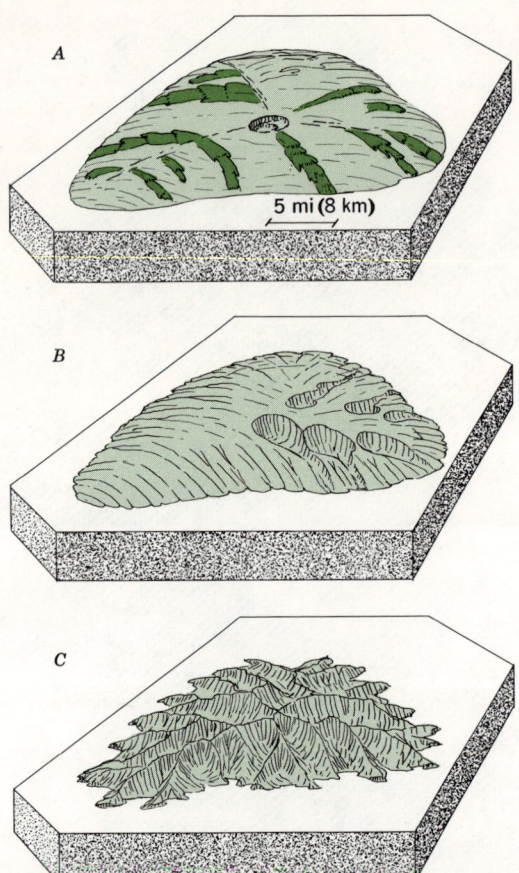

Figure 7.25 Lava domes in various stages of erosion make up the Hawaiian Islands. (Data from Stearns and Macdonald.) *A.* Initial dome with central depression and fresh flows issuing from radial fissure lines. *B.* Early erosion stage with deeply eroded valley heads. *C.* Stage of advanced erosion with steep slopes and great relief.

Cinder cones rarely grow to more than 500 or 1000 ft (150 to 300 m) in height. Growth is rapid. Monte Nuovo, near Naples, Italy, grew to a height of 400 ft (120 m) in the first week of its existence. Paricutin, in Mexico, started as a cinder cone and reached a height of 1000 ft (300 m) in the first three months. The angle of slope of a recently formed cinder cone ranges between 26° and 30°. So loose is the material that it absorbs heavy rain without permitting surface runoff. Erosion is thus delayed until weathering produces a soil which fills the interstices.

Cinder cones normally have proportionately large central craters (Figure 7.26). The rim is often much higher on one side than the other, as the prevailing wind blows the finer cinders and ash to one side of the vent.

Lava flows sometimes issue from the same vent as a cinder cone. They may burst apart the side of the cone but more commonly do not alter its form. Cinder cones may erupt in almost any conceivable topographic location, on ridges, on slopes, and in valleys. Cinder cones usually occur in groups, often many dozens in an area of a few tens of square miles.

Environmental Influences of Igneous Rock and Volcanic Activity

This chapter has dealt with a number of geological phenomena that have environment significance, but in quite different ways. Basically, igneous activity is responsible for the global framework of continents and ocean basins, which is an environmental control of first-order magnitude. Igneous rocks are in themselves of primary importance to the environment of the life layer, since these rocks furnish the soil and sea water with essential ions without which plants could not grow. Both the intrusive and extrusive rocks give rise to characteristic relief features, or landforms, which are passive environmental features of a secondary order of magnitude on the continental surfaces.

The eruptions of volcanoes and lava flows are environmental hazards of the severest sort, often taking a heavy toll of plant and animal life and the works of Man. What natural phenomenon can compare with the Mt. Pelée disaster in which thousands of lives were snuffed out in seconds? Perhaps only an earthquake or storm surge of a tropical cyclone is equally disastrous—you can take your choice. Wholesale loss of life and destruction of towns and cities are frequent in the history of peoples who live near active volcanoes. Loss occurs principally from sweeping clouds of incandescent gases that descend the volcano slopes like great avalanches; from lava flows whose relentless advance engulfs whole cities; from the descent of showers of ash, cinders, and bombs; from violent earthquakes associated with the volcanic activity; and from mudflows of volcanic ash saturated by heavy rain. For habitations along low-lying coasts there is the additional peril of great seismic sea waves, generated elsewhere by explosive destruction of volcanoes.

The surfaces of volcanoes and lava flows remain barren and sterile for long periods after their formation. Certain types of lava surfaces are extremely rough and difficult to traverse; the Spaniards who encountered such terrain in the southwestern United States named it *malpais* (bad

Environmental Influences of Igneous Rock

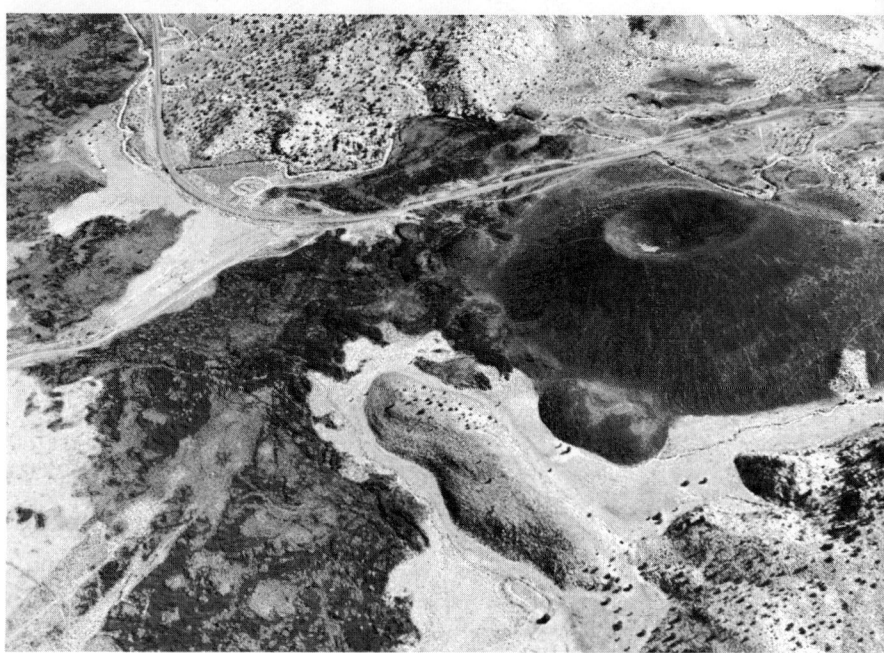

Figure 7.26 Seen in this oblique air view, a fresh cinder cone and its associated basaltic lava flow (*left*) have partially blocked a valley. Dixie State Park, about 17 mi (27 km) northwest of St. George, Utah. (Photograph by Frank Jensen.)

ground). Most volcanic rocks in time produce highly fertile soils that are intensively cultivated.

Volcanic ash may have a remarkably beneficial effect upon productivity of soil where the ash fall is relatively light. The eruption of Sunset Crater, near Flagstaff, Arizona, in 1065 A.D., spread a layer of sandy volcanic ash over the barren reddish soil of the surrounding region and caused it to become highly productive because of the moisture-conserving effect of the ash, which acted as a mulch in the semiarid climate. Because Hopi Indian corn grows well in sand, this development attracted Indians, who settled the area thickly. As the ash was gradually washed off of the slopes by heavy summer rains or blown into thick dunes by wind, the fertility declined and after about 200 years of occupation the region was abandoned to its previous state.

Volcanoes are also a natural resource in terms of recreation and tourism. Few landscapes can rival in beauty the mountainous landscapes of volcanic origin. National parks have been made of Mt. Rainier, Mt. Lassen, and Crater Lake in the Cascade Range, a mountain mass largely of volcanic construction. Hawaii Volcanoes National Park recognizes the natural beauty of Mauna Loa and Kilauea; their displays of molten lava are a living textbook of igneous processes.

Igneous activity is the primary source of practically all of the world's inorganic resources, which range from the atmosphere and oceans, produced by outgassing, to priceless metallic ores. Igneous rock itself is a source of structural materials—building stone and the rock aggregates used in concrete and in foundation layers for highways and buildings. These nonrenewable natural resources are evaluated in Chapter 10.

The internal energy system of the solid earth, powered by heat of radioactivity, is not only the most sluggish of the terrestrial energy systems, but also the least subject to alteration by Man's activities. We cannot even stop the advance of a lava flow, let alone reduce the energy released by a volcanic eruption. Although there is good reason to believe that the production of heat from matter through radioactivity is gradually lessening as eons of time go by, there can be no discernible change in the average energy balance of the system in the span of time that Man has occupied the earth; nor will there be a discernible change throughout the future period in which Man remains on the planet.

8 Rock Alteration and Sediments

The alteration of rocks to soils and sediments is a product of the interaction of two great energy systems of the life layer. One is the solar-powered, external system of energy and matter transformation and transport that maintains the global balances of radiation, heat, and water. The other is the internally powered geologic system of igneous activity with its accompanying crustal movements. At the interface of atmosphere and hydrosphere with lithosphere, oxygen and water are brought into contact with igneous rock. New minerals are formed—compounds that are more stable in the low temperatures and pressures of the surface environment than are the igneous minerals formed at high temperatures and pressures. In this process the solid rock is softened and fragmented, producing loose grains of *sediment*. But the external systems also bring kinetic energy to bear upon the decaying rock surfaces, energy in the form of winds, falling rain and flowing water, slowly grinding ice of glaciers, and the turbulent surf and currents of lake and ocean waters where they meet the land.

Sediment, once formed, undergoes a chemical and physical evolution in place to produce the *true soil* layer, or *solum*. Activities of plants and animals contribute to the development of the soil; they are vital factors in maintaining the stability of the soil as an open system. One of the objectives of Chapter 11 is to outline the basic concepts of soil formation and to recognize several fundamental soil-forming regimes determined by the availability of heat and water in the annual climatic cycle.

But soil does not simply stay in place as a static layer. Soil continually loses some of its mass, the loss being replaced by new raw materials derived from chemical and physical alteration of the underlying rock. Sediment, in the form of both solid mineral particles and ions in solution, is removed from the soil and carried away in the continual flow of surplus fresh water from precipitation. Streams carry the sediment to lower levels and to locations where accumulation is possible. (Wind and glacial ice also transport sediment, but not necessarily to lower elevations or to places suitable for accumulation.) Usually these sites of accumulation are in shallow seas bordering the continent, but they may also be inland seas and lakes. Thick accumulations of sediment may become deeply buried. Over long spans of time the sediments undergo physical or chemical changes, or both, to become compacted and hardened, producing *sedimentary rock*. The process of compaction and hardening is referred to as *lithification*. *Diagenesis* is a more general term for complex changes with time, usually involving chemical change and replacement.

The alteration of igneous rock into sediment and the change of sediments into sedimentary rock is only part of a great cycle of rock transformation. Further steps in this cycle, leading to its completion as a pathway of energy changes by means of work done on mineral matter, remains for exploration in the next chapter.

Sediments and sedimentary rocks contain essential resources for Man's industrial society. First, they provide materials needed to build structures, such as highways and buildings. Second, they supply compounds for industrial processes that yield chemicals, such as fertilizers and acids. Third, they are the source of hydrocarbon compounds—coal, petroleum, and natural gas—that turn the industrial wheels and provide us with transportation and heat. This chapter will develop the basic concepts underlying the sedimentary accumulations of these natural resources. In Chapter 10 we will look into the processes of concentration of useful mineral substances; we will assess the rates of Man's consumption of these resources and contemplate his future courses of action.

Rock Weathering in the Surface Environment

Poets and advertising copy writers assure us that the highly polished granite slab is an everlasting monument in a changing world. Actually, the surface environment is poorly suited to the preservation of a plutonic rock formed under conditions of high pressure and high temperature. Most silicate minerals do not last long, geologically speaking, in the low temperatures and pressures of atmospheric exposure, particularly since free oxygen, carbon dioxide, and water are abundantly available. Rock surfaces are also acted upon by physical forces of disintegration, tending to break up the igneous rock into small fragments and to separate the component minerals, grain from grain. Fragmentation is essential for the chemical reactions of rock alteration, since it results in a great increase in mineral surface area exposed to chemically active solutions.

The term *weathering* is used by geologists for the total process of physical disintegration and chemical decomposition of rock at or near the earth's surface. Weathering is a passive process, in that the products of its action tend to remain in place close to the parent rock. Such residual accumulations of rock alteration comprise the *regolith*. The soil is formed from the surface layer of this body. The removal of weathering products allows more underlying rock to be exposed to the atmosphere. Removal occurs either by *mass wasting*, the spontaneous downhill movement of the weathering products under the force of gravity, or by *erosion*, the application of stress by a moving fluid, which may be wind, running water, glacial ice, or breaking waves and related currents. These active agents of erosion always cause *transportation* of sediment, followed by *deposition* at new sites. The energy of running water in streams and of glaciers is initially potential energy of position; the fluid agent with its contained sediment moves downhill under the influence of gravity, transforming the potential energy to kinetic energy. As for wind and waves, their kinetic energy is traced back to the pressure-gradient force that sets the atmosphere in motion. We shall return in later chapters to mass wasting and the processes of erosion and transportation, with special attention to the landforms that are shaped by removal and accumulation of weathered rock and sediment. Our present concern is with the chemical aspects of rock alteration and with the geologic role of sediment in the cycle of rock transformation.

Mineral Alteration

Mineral alteration, also called *chemical weathering*, includes several types of chemical reactions, all of which take place more-or-less simultaneously. Surface water—whether it be as raindrops, or as water of streams, lakes, or as water held in the soil—is an aqueous solution of the atmospheric gases. The presence of atmospheric gases dissolved in natural water is a matter of great environmental importance, not only because of the inorganic reactions with mineral matter, but also because the presence of two gases in particular—oxygen and carbon dioxide—is vital to the life processes of plants and animals living in water. We shall therefore need to have a short review of the principles of solution of gases in water.

The degree to which a given gas will dissolve in water, i.e., the *solubility* depends upon two variable factors: (1) the temperature of the solution, and (2) pressure of the gas in the atmosphere overlying the water surface. For a given combination of temperature and pressure there is a maximum quantity of the gas that can be dissolved in the water; this is the *saturation* quantity. You might suppose that in pure fresh water exposed to the atmosphere the dissolved atmospheric gases—nitrogen, oxygen, and carbon dioxide—would be present in the same proportions as in the air, namely 78%, 21%, and 0.03%, respectively, of the total quantity of dissolved gas. This guess would prove wrong, however, since each gas has its own particular solubility and its own saturation quantity for a given combination of pressure and temperature. For example, take carbon dioxide by itself. If we place an atmosphere of pure CO_2 over pure water (possible only in a sealed jar in the laboratory) and hold the pressure constant at sea-level atmospheric pressure (1013 mb), we will find a saturation quantity of CO_2 which is temperature-dependent. The table below gives the values for N_2 and O_2, as well as for CO_2. The quantities are in volume of gas per volume of water, both in liters:

Gas	0° C	30° C	100° C
N_2	0.02	0.01	<0.01
O_2	0.05	0.03	0.02
CO_2	1.70	0.7	—

It is obvious that the concentration of CO_2 is many times greater than for either nitrogen or oxygen. Evidently, both the solubility of the gas and its proportion with respect to the other gases in the atmosphere determine its concentration in the fluid solutions.

The saturation quantities of oxygen and carbon dioxide, the vital gases in environmental terms, are as shown in Table 8.1 for a pressure of 1013 mb over a range of temperatures found in natural surface waters. Although the ratio of O_2 to CO_2 in the atmosphere is about 700 to one,

the ratio of the two dissolved gases at 0° C is about 15 to one, reflecting the much greater solubility of CO_2. Notice that the saturation quantity decreases with increasing temperature, but that the rate of decrease is less for O_2 than for CO_2. This fact shows in Figure 8.1 as a steeper slope of the plotted points for CO_2 than for O_2.

Returning now to mineral alteration, the presence of dissolved oxygen in water in contact with mineral surfaces in the soil leads to *oxidation*, which is the combination of oxygen ions with metallic ions, such as calcium, sodium, potassium, magnesium, and iron, abundant in the silicate minerals. At the same time, CO_2 in solution forms a weak acid, *carbonic acid*, capable of reaction with certain minerals. The reaction is given attention in a later section of this chapter. In addition, where decaying vegetation is present, soil water contains complex organic acids, capable of interaction with mineral compounds. Certain common minerals, such as rock salt (NaCl) dissolve directly in water, but simple solution is not particularly effective for the silicate minerals.

TABLE 8.1 Saturation of Dissolved Atmospheric Oxygen and Carbon Dioxide in Pure Fresh Water

Temperature °C	Parts per Million (ppm)	
	O_2	CO_2
0	14.6	1.00
5	12.7	0.83
10	11.3	0.70
15	10.1	0.59
20	9.1	0.51
25	8.3	0.43
30	7.5	0.38

SOURCE: R. E. Coker, 1968, *Streams, Lakes, Ponds*, Harper & Row, New York, p. 24.

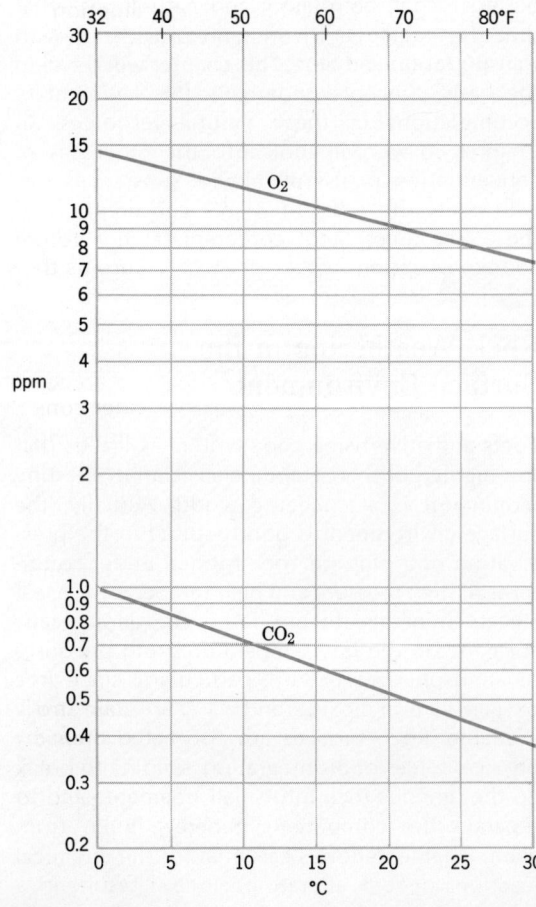

Figure 8.1 Solubility of oxygen and carbon dioxide decrease with increasing water temperature.

Mineral Products of Hydrolysis and Oxidation

Water itself combines with certain mineral compounds in a reaction known as *hydrolysis*. This process is not merely a soaking or wetting of the mineral, but a true chemical change producing a different compound and a different mineral. The reaction is not readily reversible under atmospheric conditions, so that the products of hydrolysis are stable and long-lasting, as are the products of oxidation. In other words, these changes represent a permanent adjustment of mineral matter to a new environment of pressures and temperatures.

Since water is required for mineral alteration, you might think that the rate of rock decay would be directly proportional to the amount of free water available in the rock and soil, and that the dry deserts would be environments of very limited rock decay. To some extent this is a valid conclusion, since polished stone surfaces of the ancient Egyptian monuments are almost perfectly preserved through the centuries in a dry, hot desert climate. The same monuments, taken to middle-latitude cities of humid climate and exposed to the atmosphere undergo rapid disintegration. (Frost may also be a factor.) Laboratory experiments showed that a polished granite surface, exposed to thousands of cycles of intense heating and cooling under dry conditions, produced no alteration of the grain surfaces; the same experiment repeated with water sprays to quench the heat quickly showed the beginnings of change in mineral surfaces. Although mineral alteration is perhaps much slower in dry deserts than in humid lands, there is nevertheless enough water present as water vapor and as dew to allow alteration to proceed, since we observe the decay products in abundance on igneous rock surfaces in most deserts. The effect of cold is quite another matter. When soil water is frozen, chemical reactions are greatly slowed. Regions of perennially frozen soil and rock (permafrost) can be expected to show little chemical decay since seasonal thaw affects only a thin surface layer. Decreasing temperature increases the concentration of dissolved gases in water, as Table 8.1 shows, but this is more than offset by the fact that most chemical reactions take place more rapidly at high temperatures than at low. Consequently, chemical alteration of minerals is most rapid in warm (and moist) climates of low latitudes.

Mineral Products of Hydrolysis and Oxidation

Figure 8.2 shows schematically some important alteration products of the common silicate minerals and mineral groups. Certain of these products are clay minerals. A *clay mineral* is one that has plastic properties when moist, because it consists of minute thin flakes lubricated by layers of water molecules. Potash feldspar ($KAlSi_3O_8$) undergoes hydrolysis to become *kaolinite*, a clay mineral with the composition $Al_2Si_2O_5(OH)_4$ (Figure 8.3). The hydroxyl ion (OH) in the kaolinite formula represents the added water. Notice that potassium (K) is lost in the process; it becomes a free positive ion in the soil water and may be used by plants. Some free silica (SiO_2) is also released. A soft, white mineral with greasy feel, kaolinite becomes plastic when moistened. It is an important ceramic mineral used to make chinaware, porcelain, and tile. Kaolinite can also be derived from the plagioclase feldspars.

Bauxite is an important alteration product of feldspars, occurring typically in warm climates of tropical and equatorial zones where rainfall is abundant year-around or in a rainy season. Bauxite is actually a mixture of minerals, the dominant constituent being *diaspore*, with the formula $Al_2O_3 \cdot 2H_2O$. The *sesquioxide of aluminum*, Al_2O_3, shows that full oxidation has taken place, yielding an unusually stable compound. Unlike kaolinite, which is a true clay with plastic properties, bauxite forms massive rocklike layers below the soil surface. *Laterite* is the name given to these accumulations, which are also rich in oxides of iron. Laterite is almost immune to further chemical change and resists erosion to produce crustlike structures under upland surfaces.

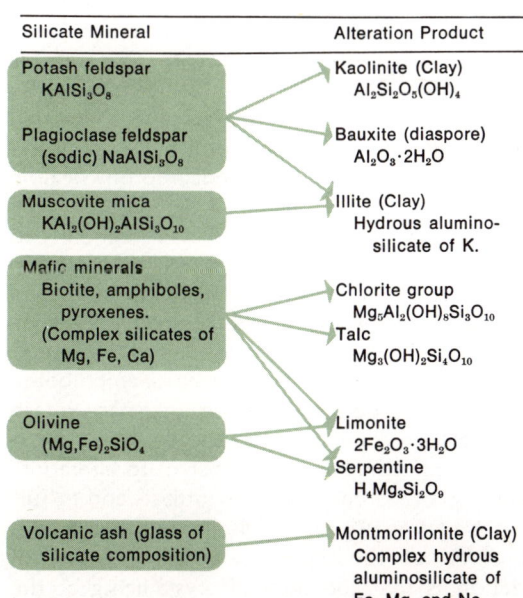

Figure 8.2 Common alteration products of silicate minerals. "Clay" denotes a clay mineral. (From A. N. Strahler, 1971, *The Earth Sciences*, 2nd ed., Harper & Row, New York.)

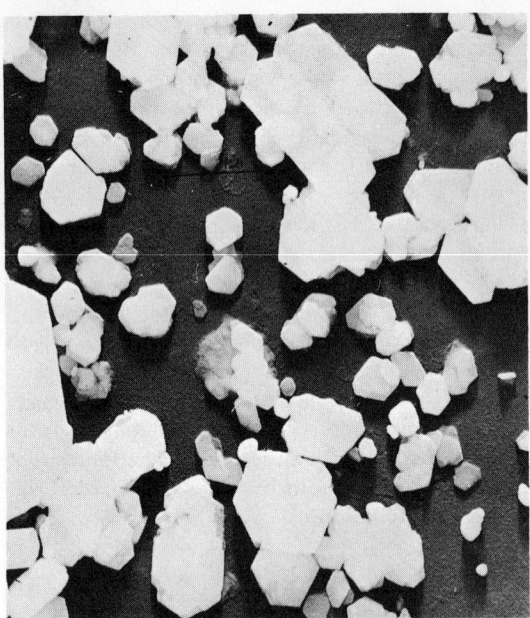

Figure 8.3 Electron microscope photograph of kaolinite crystals magnified about 20,000 times. (Photograph by Paul F. Kerr.)

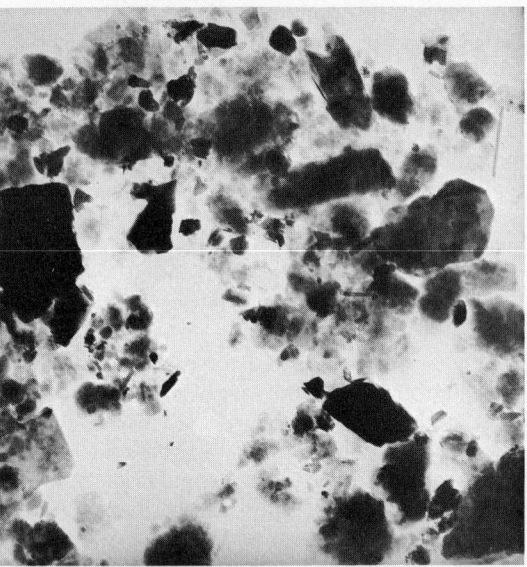

Figure 8.4 Seen here enlarged about 20,000 times are tiny flakes of the clay minerals *illite* (sharp outlines) and *montmorillonite* (fuzzy outlines). These particles have settled from suspension in San Francisco Bay. (Photograph by Harry Gold. Courtesy of R. B. Krone, San Francisco District Corps of Engineers, U.S. Army.)

Laterite is easily cut into blocks, but hardens upon exposure to the drying effects of the free air. It has been extensively used as a building stone in southeast Asia. (Example: the ornate temple ruins of Angkor Wat in Cambodia.) Where bauxite has accumulated in large quantities it is an important ore of aluminum (see Chapter 10).

A second clay mineral is *illite*, formed as an alteration product of feldspars and muscovite mica. Illite is a hydrous aluminosilicate of potassium and is a mineral of great importance in sedimentary rocks. It occurs as minute thin flakes of colloidal dimensions and is carried long distances in streams (Figure 8.4). Skipping down to the last mineral listed in Figure 8.2 we find *montmorillonite*, a common clay mineral (more correctly a group of minerals) derived from alteration of feldspar, or certain of the mafic minerals, or volcanic ash. Fragments of montmorillonite are seen together with illite in Figure 8.4.

Alteration products of mafic minerals include *chlorite* (a mineral group) and *talc*, both hydrous aluminosilicates of magnesium. They are soft, scaly substances. Some of you have experienced the soapy or greasy feel of talcum powder, made from talc, or talc sink tops in older laboratories.

A most important alteration product of the mafic minerals is *limonite*, a hydrous iron compound with the formula $2Fe_2O_3 \cdot H_2O$. *Iron sesquioxide* (Fe_2O_3) in limonite is a stable form of iron oxide and is found widely distributed in rocks and soils. It is closely associated with bauxite in laterites of low latitudes. Limonite supplies the typical reddish to chocolate-brown colors of soils and rocks. Some shallow accumulations of limonite were formerly mined as a source of iron (bog ore).

Another common alteration product of mafic minerals, particularly of olivine, is *serpentine*, a hydrous magnesium silicate with the formula $H_4Mg_3Si_2O_8$. Serpentine has resulted from the alteration of large bodies of peridotite, a rock rich in olivine. One interesting form of serpentine is a type of *asbestos*, which shreds into tiny flexible fibers (see Figure 10.22).

Silicate minerals differ in their susceptibility to alteration; the order in which they fall into sequence is the same as the order of crystallization in the Bowen reaction series (Figure 7.4). Olivine, which crystallizes first, is most susceptible to alteration, followed by the pyroxenes, amphiboles, biotite, and sodic plagioclase feldspar. Potash feldspars are somewhat less susceptible. Muscovite mica is comparatively resistant to alteration, while quartz is immune to hydrolysis and to further oxidation. The principle behind the correlation of order of susceptibility with order of crystallization is that the mineral crystallizing at the highest temperature is most distantly removed from the environment of alteration and is consequently the least stable when exposed to the surface.

As a result of the differing susceptibility of silicate minerals to alteration, the igneous rocks also reflect these differences. The mafic rocks are the more easily decomposed, whereas the felsic rocks are the more resistant. This difference shows in the superior resistance of a felsic rock enclosed in mafic rock (Figure 8.5). Regolith and soil will develop more rapidly on the mafic rock, other conditions being alike.

Size Grades of Sediment Particles

Besides mineral composition, the other essential parameter in classifying soil particles and transported sediments is particle size, including mixtures of sizes, which relate to the general aspect of rock texture. Table 8.2 is an abbreviated list of size grades, giving the names and dimensional ranges of each grade. The full scale, including subdivisions of each grade into coarse, medium, and fine categories, is known as the *Wentworth scale*, and is of standard use in geology. Soil scientists and engineers use somewhat different grade scales. Clay can be subdivided with respect to its behavior. Particles below about 1 micron (0.001 mm) behave as *colloids*, particles that remain indefinitely in aqueous suspension and are responsive to impacts of water molecules so as to engage in *brownian movement*. Since most natural colloidal clays are thin, platelike bodies or irregularly shaped objects, rather than being spherical, grain diameter is not a particularly meaningful term to apply to these very fine grades. Colloidal mineral matter is of vital importance to inorganic physical processes and biological processes of the soil because of the ability of colloids to hold ions, a subject discussed in Chapter 11.

Decreasing particle size brings great increase in surface area of particles contained within a given volume, and when the colloidal size is reached the surface area is truly enormous. Table 8.3 illustrates this point by assuming that we start with a single cube, 1 cm on a size (pebble size); it has a volume of 1 cc and a surface area of 6 sq cm. Consider next that we slice the cube into cubes 0.1 cm on a side (size of coarse sand); yielding us 1000 cubes with a total surface area of 60 sq cm. Subdividing the cube into 10^{12} cubes, each 0.0001 cm on a side (fine clay size) gives a total surface area of 60,000 sq cm. Finally, with subdivision of the original cube into 10^{15} cubes, each 0.000,01 cm (0.1 micron) on a side (colloidal size) yields a total surface area of 600,000 sq cm, which if spread out into a continuous horizontal surface would be an area about 8 m by 8 m.

Figure 8.5 A felsic dike (sharp-cornered blocks in center) in deeply altered mafic rock (rounded masses and rough gray rock surface). Sangre de Cristo Mountains, New Mexico. (Photograph by A. N. Strahler.)

Sediments and Sedimentary Rocks

The environmental importance of sediments is widespread in the life layer. Sediment (including the true soil and the regolith) provides the physical base, or *substrate*, for practically all forms of life except plankton, at least in one phase or another of their life cycles or in food-gathering

TABLE 8.2 Simplified Wentworth Scale

Grade Name	Limits (mm)	(in.)
Boulders		
	256	10
Cobbles		
	64	2.5
Pebbles		
	2	0.08
Sand		(microns)
	0.0625	62
Silt		
	0.004	4
Clay (noncolloidal)		
	0.001	1
Clay (colloidal)		
	(down to 0.001 micron)	

193

TABLE 8.3 Surface Area of Cubes Obtained by Subdividing a One-Centimeter Cube

Cube Dimensions Length of Side (cm)	Grade Equivalent	Number of Particles (per cu cm)	Total Surface Area (sq cm)
1	Pebble	1	6
0.1	Coarse sand	10^3	60
0.0001	Fine clay	10^{12}	60,000
0.000,01 (0.1 micron)	Colloidal clay	10^{15}	600,000 (8 m × 8 m)

practices. Water-saturated sands, silts, clays, and muds form the life environments of thousands of species of aquatic organisms under the shallow seas, in tidal estuaries, in lake bottoms and stream beds, and in swamps and marshes. A principle of paramount importance is that organisms modify the sediment in which they live and feed; they also create sediment through life processes, for example, as shells and skeletons. Consequently, a major class of sediments is organically derived, in contrast to the chemically derived products of rock alteration and the physically derived particles of rock disintegration.

Classification and description of sediments and the sedimentary rocks presents a study in complexity. To help make order out of a large collection of terms, Figure 8.6 attempts to simplify a classification that begins with sediments, lists the mineral components of those sediments, and finally names the derived sedimentary rocks.

Clastic and nonclastic are the two major divisions of sediments. *Clastic sediments* are those derived directly as particles broken from a parent rock source, in contrast to the *nonclastic sediments*, which are of newly created mineral matter precipitated from chemical solutions or from organic activity. The clastic sediments are divided into a *pyroclastic* group and a *detrital* group. Pyroclastic sediment issues directly from volcanoes in the form of cinder, ash, and volcanic dust. These sediments were described in Chapter 7, so that we will give them little further attention here.

The detrital sediments are derived from any one of the three major rock groups—igneous, sedimentary, and metamorphic—which gives a very wide range of parent minerals. One sediment source is from the silicate minerals and the alteration products of those minerals. Since the highly susceptible minerals—mostly mafic—are altered prior to transportation, whereas quartz is immune to such alteration, the most important single component of the detrital sediments generally is quartz (Figure 8.7). Second in abundance are fragments of unaltered fine-grained parent rocks, and these can be of any of the three major rock groups. Feldspar, particularly of the potash and sodic plagioclase types, is third in order of abundance as a detrital mineral. Mica is also present, but in small quantities. Clay minerals, particularly kaolinite, illite, and montmorillonite, are major constituents of the very fine detrital sediments. Then there are also fragments of carbonate rocks which are themselves of nonclastic origin; they will be discussed in later paragraphs.

Also important in the clastic sediments are the *heavy detrital minerals,* of which two examples, magnetite and ilmenite, were mentioned in Chapter 7 as being nonsilicate mafic minerals found commonly in igneous rocks. By "heavy" is meant a mineral of high density. Four important heavy detrital minerals are the following.

	Density (g/cc)	Composition
Magnetite	4.9–5.2	Fe_3O_4
Ilmenite	4.3–5.5	$FeTiO_3$
Zircon	4.4–4.8	$ZrSiO_4$
Garnet group	3.4–4.3	Aluminosilicates of Ca, Mg, Mn, Fe

These minerals are derived from the weathering of igneous, sedimentary and metamorphic rocks. They are extremely resistant to abrasion (in the same range of hardness as quartz) and, like quartz, travel long distances in streams and along beaches. You will see concentrations of these minerals in most beach sands and stream bars. A magnet dragged through the dry sand will accumulate a coating of the magnetite grains. While the heavy detrital minerals are not of direct importance to life processes, they are locally significant in large concentrations as metallic mineral sources. For example, titanium is obtained from ilmenite, and zirconium from zircon. Another example is the mineral *cassiterite,* an oxide of tin (SnO_2), which is mined from concentrations in gravel, where it is known as *stream-tin*.

Sediments and Sedimentary Rocks

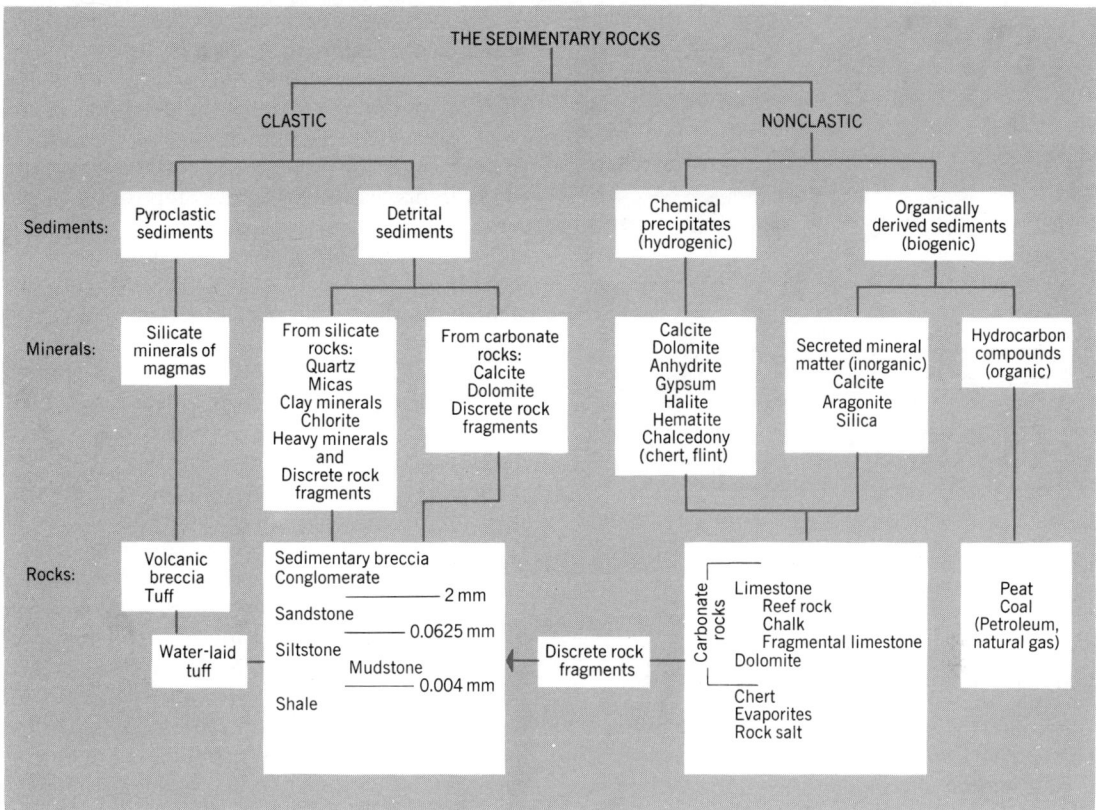

Figure 8.6 Classification of sedimentary rocks according to sediment origin and mineral content. (From A. N. Strahler, 1971, *The Earth Sciences*, 2nd ed., Harper & Row, New York.)

The natural range of particle grades in detrital sediment is a prime determinant of the ease and distance of travel of particles in transport by water currents. Obviously, the finer the particles, the more easily are they held in suspension in the fluid; the coarser particles tend to settle out to the bottom of the fluid layer. In this way a separation of grades, known as *sorting*, occurs and determines the texture of the sediment deposit and of the sedimentary rock derived from that sediment. Colloidal clays do not settle out unless they are made to clot together into larger groups, a process of *flocculation*. In Chapter 14, which explains how streams transport and deposit sediment, we will look further into the process of sorting as it is related to particle sizes.

Compaction and cementation of layers of sediment leads to formation of sedimentary rocks; the principal kinds are listed in Figure 8.6. *Sedimentary breccia* is formed of angular rock fragments, usually in a matrix of finer grade particles; it is not a common rock. A related rock is *volcanic breccia*, formed of lava fragments and ash. *Conglomerate* consists of pebbles or cobbles, usually of a very durable rock, set in a matrix of sand. Conglomerate layers represent lithified beaches or stream-bed deposits. *Sandstone* is formed of the sand grades of sediment, cemented into a solid rock by silica (SiO_2) or calcium carbonate ($CaCO_3$). The sand grains are commonly of quartz, as shown in Figure 8.7, and sometimes of feldspar, but in other cases are sand-sized fragments of fine-grained rock containing several minerals. *Siltstone* and *shale* are lithified layers of silt and clay particles, respectively, while mixtures of silt and clay, often with some sand, produce *mudstone*.

Shale is the most abundant of the sedimentary rocks. It is formed largely of the clay minerals, kaolinite, illite, and montmorillonite. The compaction of the original clay and mud involves a considerable loss of volume as water is driven out.

A characteristic feature of sedimentary rocks is their layered arrangement, the layers being collectively called *strata*. Typically, layers of different textural composition are alternated or interlayered. The planes of separation between layers are known as *bedding planes* (Figure 8.8).

Rock Alteration and Sediments

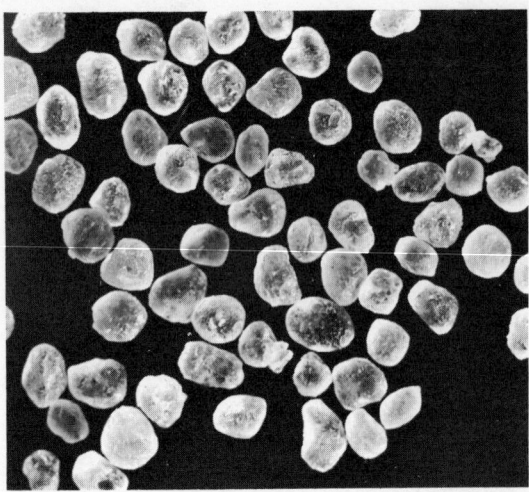

Figure 8.7 Rounded grains of quartz from a sandstone of early Paleozoic age. The grains average about 1 mm (0.04 in.) in diameter. (Photograph by A. McIntyre, Columbia University.)

The Nonclastic Sediments and Sedimentary Rocks

The nonclastic sediments are of particular importance in environmental science because they represent enormous storages of carbon, obtained from atmospheric CO_2 and changed into carbonate and hydrocarbon compounds by both inorganic and organic processes. As Figure 8.6 shows, we recognize two major divisions of the nonclastic sediments: (1) *Chemical precipitates* are compounds precipitated directly from water in which the ions are transported; these sediments are described as *hydrogenic*. (2) *Organically derived sediments* are created by the life processes of plants and animals; these are described by the adjective *biogenic*.

Chemical precipitates are of major importance in sediments of the sea floor, but a second important environment of deposition is in inland salt lakes of desert regions, where evaporation exceeds precipitation by a wide margin. As an aid to understanding hydrogenic sediments of the ocean floors, we look to the composition of sea water in terms of the constituent ions and their concentrations, Table 8.4. The proportions of ions shown in Table 8.4 are maintained in constant proportions through a system of addition and removal of components operating in a steady state. Of the substances listed, chlorine, sulfate radical, carbonic acid radical, and bromine ions and boric acid are volatiles, which we have already explained as introduced into the atmosphere and hydrosphere by outgassing. The positive ions of sodium, magnesium, calcium, and potassium are explained as derived from weathering of igneous rocks. This explanation is verified by examining Table 8.5, showing the abundance of the four principal bases found in stream water.

At first thought, you might reason that the proportions of salts in sea water would reflect closely the proportions in which the elements are introduced by outgassing and by rock weathering. Yet we find instead that chlorine, the most abundant ion of sea water, is only a minor constituent of gases of lavas (Table 7.3); while carbon (as carbonic acid radical) is a very small constituent of sea water but a major constituent (as CO_2) in lava gases. Sodium is second in abundance in sea water, and more abundant than calcium by a factor of about 26; yet in the stream-water tabulation (Table 8.5), calcium is twice as abundant as sodium in areas of varied igneous rocks.

An answer to the problem of proportions of sea-water constituents lies in a factor known as

Figure 8.8 Sedimentary strata exposed in the walls of Grand Canyon, Arizona. This view spans a total thickness of about 3000 ft (1000 m) of Paleozoic sandstones, shales, and limestones. (Photograph by A. N. Strahler.)

The Nonclastic Sediments and Sedimentary Rocks

TABLE 8.4 Principal Constituents of Sea Water

		Concentration (g/kg, parts per thousand)	Percent of Total Salt
Negative ions			
Chlorine	Cl⁻	19.0	55.0
Sulfate radical	SO_4^{-2}	2.5	7.7
Carbonic acid radical	HCO_3^-	0.14 (varies)	0.4
Bromine	Br⁻	0.065	0.2
Positive ions (bases)			
Sodium	Na⁺	10.5	30.6
Magnesium	Mg^{+2}	1.3	3.7
Calcium	Ca^{+2}	0.40	1.2
Potassium	K⁺	0.38	1.1
Neutral			
Boric acid	H_3BO_3	0.024	0.7

SOURCE: D. W. Hood, 1966, *Encyclopedia of Oceanography*, R. W. Fairbridge, ed., Table 1., p. 793.

TABLE 8.5 Abundances of Bases in Stream Water

	Igneous Rock Areas, All Varieties (percent by weight)	Basalt Areas Only (percent by weight)
Calcium	52	73
Magnesium	11	12
Sodium	27	13
Potassium	10	2
	100	100

SOURCE: W. W. Rubey, 1951, *Bull. Geol. Soc. Amer.*, vol. 62, pp. 1121, 1123.

the *residence time*, which is the average time that the constituent remains in solution. Table 8.6 gives residence times of the principal bases and silicon. Sodium heads the list with 260 million years (m.y.); magnesium is next (45 m.y.), calcium and potassium both have short times (8 and 11 m.y.), and silicon a very short time (10,000 years). The concentrations, given in the third column of the table, are in rough agreement with this ordering. Evidently, the sodium ion is comparatively inactive in sea water and maintains a high concentration, whereas calcium is very active and is rapidly precipitated into sediment, keeping the sea-water concentration low. The same can be said of silicon and aluminum (not listed in the tables), which are abundantly supplied from igneous rocks. Knowing these fundamentals, we can predict that the principal nonclastic sediments will consist of calcium in combination with carbon (as carbonate compounds). Magnesium, which is abundantly supplied and has a low resi-

dence time will also be liberally precipitated as a carbonate compound.

The most important hydrogenic and biogenic minerals (hydrocarbon compounds are not included) are listed in Table 8.7, together with their compositions. The first three—calcite, aragonite,

TABLE 8.6 Sea-Water Residence Times of Principal Bases and Silicon

Element	Residence Time in Millions of Years	Concentration in Sea Water
Sodium	260	31%
Magnesium	45	3.7%
Calcium	8	1.2%
Potassium	11	1.1%
Silicon	0.01 (10,000 years)	0.003%

SOURCE: E. D. Goldberg, 1961, *Oceanography*, Mary Sears, ed., A.A.A.S., Washington, D.C., p. 586.

and *dolomite*—are classed as *carbonates*. Calcite is the dominant carbonate mineral and occurs in many forms. Aragonite is of the same composition as calcite, but of different crystal structure; it is secreted in the shells of certain invertebrate animals. Dolomite contains magnesium as well as calcium. All three carbonate minerals are soft substances, as compared with the silicate minerals. The second group, the *evaporites*, are typically formed where sea water is evaporated in shallow bays and gulfs. *Anhydrite* and *gypsum* are composed of calcium sulfate, the latter in combination with water. Gypsum is an economically important mineral used in the manufacture of structural materials. The third evaporite, *halite*, is commonly known as rock salt, with the composition sodium chloride (NaCl). In refined form it is the table salt we use in cooking and flavoring.

Hematite, a sesquioxide of iron (Fe_2O_3) is a common mineral in sedimentary rocks. This compound is described as *ferric iron oxide*, to distinguish it from the *ferrous* form, FeO. Magnetite, a heavy detrital mineral, is a mixture of ferrous and ferric iron oxides. Limonite is the hydrous form of ferric iron oxide. The significance of hematite, the ferric form of oxide, is that the iron holds more oxygen than in the ferrous form, indicating precipitation in an oxygen-rich environment. Hematite is a major ore of iron and we shall refer to it again in Chapter 10. No less important than the carbonates and evaporites is *chalcedony*, the mineral name for silica (SiO_2) in a very fine-grained crystalline form. Chemically, it is not different from the mineral quartz.

The minerals listed in Table 8.7 can be chemically precipitated from sea water, or the water of saline lakes, or can be secreted by organisms to produce the nonclastic sedimentary rocks listed in Figure 8.6. Commonest of the carbonate rocks is *limestone*, of which there are many varieties. One source is in reefs built by corals and algae; another form is *chalk*, made up of the skeletons of a marine form of the algae and of some microscopic animals. Other limestones are formed of shell fragments or other broken carbonate matter. Some limestones are densely crystalline and have undergone recrystallization under pressure. *Dolomite*, a rock of the same name as the mineral that composes it, may have been derived through the alteration of limestone, as magnesium ions of sea water gradually replaced calcium ions. *Chert*, composed of chalcedony, is an important siliceous sedimentary rock. It occurs as nodules in limestone, and in some cases as solid rock layers referred to as *bedded cherts*. Gypsum, anhydrite, and halite are layered rocks of their respective mineral compositions and occur in association with clastic sedimentary rocks.

TABLE 8.7 Common Hydrogenic and Biogenic Minerals

Mineral Name	Composition	Density g/cc
Carbonates		
Calcite	Calcium carbonate $CaCO_3$	2.72
Aragonite	Calcium carbonate $CaCO_3$	2.9–3
Dolomite	Calcium-magnesium carbonate $CaMg(CO_3)_2$	2.9
Evaporites		
Anhydrite	Calcium sulfate $CaSO_4$	2.7–3
Gypsum	Hydrous calcium sulfate $CaSO_4 \cdot 2H_2O$	2
Halite	Sodium chloride NaCl	2.1–2.3
Hematite	Sesquioxide of iron (ferric) Fe_2O_3	4.9–5.3
Chalcedony (chert, flint)	Silica SiO_2	2.6

Chemical Weathering of Carbonate Rocks

Carbonate rocks, after being deposited in shallow seas or in lakes, may become exposed to the atmosphere through crustal uplift and erosion. Because of the presence of dissolved carbon dioxide (CO_2) in all rain water, soil water, and stream water, the carbonate rocks are easily dissolved in a process called *carbonation*.

The solution of CO_2 in water, explained in earlier paragraphs of this chapter, produces a *carbonic acid* solution, which is a mixture of the hydrogen ion (H^+) and the bicarbonate ion (HCO_3^-) in water. The reaction is:

$$HOH + CO_2 \rightleftharpoons HCO_3^- + H^+$$

Water / Carbon Dioxide / Bicarbonate Ion / Hydrogen Ion

Carbonic Acid

This reaction is *reversible*; it is possible for the reaction to go either way. If we dissolve a given quantity of CO_2 in water, some of that CO_2 will react with water to form the bicarbonate ion. An equilibrium is formed, where the concentration of the *reactants* (left side of the reaction) and the *products* (right side of the equation) in the water is stable. If we increase a reactant, CO_2, by dissolving more CO_2 in water, the reaction will go to the right, producing more products. If we add a product, say bicarbonate ion, the reaction will go to the left, producing more reactants.

Suppose now that calcite ($CaCO_3$) is added to the carbonic acid solution. Calcite will slowly dissolve, producing Ca^{++} and CO_3^{-} (carbonate ion).

$$CaCO_3 \rightleftharpoons Ca^{++} + CO_3^{--}$$
Calcite   Calcium ion   Carbonate ion

Although this reaction is reversible, the equilibrium lies far to the left, and dissolution is slow. However, the carbonate ion is unstable, and quickly picks up a hydrogen ion to become a bicarbonate ion:

$$CO_3^{--} + H^+ \rightleftharpoons HCO_3^-$$
Carbonate   Hydrogen   Bicarbonate
Ion   Ion   Ion

This rapid loss of a product pushes the solution reaction leftward, resulting in more rapid solution of $CaCO_3$.

The series of reactions might be diagrammed as follows:

$$CaCO_3 \rightleftharpoons Ca^{++} + CO_3^{--} \quad (1)$$
$$\downarrow$$
$$CO_3^{--} + H^+ \rightleftharpoons HCO_3^- \quad (2)$$
$$\downarrow \quad \downarrow$$
$$CO_2 + HOH \rightleftharpoons H^+ + HCO_3^- \quad (3)$$

At first, many H^+ ions will be present from reaction (3). This high H^+ concentration will force reaction (2) to the right, forming HCO_3^-. Since reaction (2) requires CO_3^{--}, the CO_3^{--} concentration will be reduced, stimulating the dissolution of calcite in reaction (1).

This process, however, is self-defeating. Note that in reaction (2), HCO_3^- is produced. This increased production of HCO_3^- will push reaction (3) to the left, forming CO_2, which will be given off as a gas, and H_2O. Thus the H^+ supply is doubly depleted, first in the conversion of CO_3^{--} to HCO_3^-, and second, by conversion of carbonic acid to CO_2 and water. As a result, $CaCO_3$ dissolution will rapidly stop, and the reactions will come to an equilibrium.

In a landscape where limestone rock is at the surface, however, new supplies of carbonic acid are constantly brought in from precipitation and stream flow in which air and water are mixed so equilibrium is never reached and the rock dissolves rapidly. Further, the reaction products, Ca^{++} and HCO_3^-, are carried downstream and away from the limestone area, also increasing the limestone solution rate. This rapid solution of limestone in areas of strong stream flow results in the formation of caverns and many interesting landforms to be discussed in Chapter 12.

In sea water, however, these three reactions often do reach a near-equilibrium state. Yet we find throughout the geologic column vast accumulations of limestone formed by precipitation in shallow seas. How does such precipitation occur? Recall that the solubility of CO_2 in water is highly temperature dependent. Where cool water is warmed in shallow seas, CO_2 will be given off to the atmosphere. This will reduce H^+ ion concentration, forcing reaction (2) to the left. Since CO_3^{--} concentration will then be reduced, reaction (1) will also proceed to the left, forcing precipitation of $CaCO_3$. The effect will be further enhanced by new supplies of Ca^{++} and HCO_3^- ions received from stream water entering the shallow sea, also acting to force reactions (1) and (2) to the left. These reactions are part of the carbon cycle, discussed further in Chapter 19, in which atmospheric carbon is cycled through air, fresh and sea water, and carbonate rocks.

Hydrocarbon Compounds in Sedimentary Rocks

Hydrocarbon compounds form a second group of biogenic sediments. These organic substances occur both as solids—peat and coal—and as liquids and gases—petroleum and natural gas—but only coal qualifies physically for designation as a rock.

Peat, a soft, fibrous substance of brown to black color, accumulates in a bog environment where the continual presence of water inhibits decay and oxidation of plant remains. A common form of peat is of fresh-water origin and represents the filling of shallow lakes, of which many thousands remained over North America and Europe after recession of the great ice sheets of the Pleistocene Epoch. This peat has been used for centuries as a low-grade fuel. Peat of a different sort is formed in the salt-water environment of tidal marshes (Chapter 15).

At various times and places in the geologic past, conditions were favorable for the large-scale accumulation of plant remains, accompanied by subsidence of the area and burial of the compacted organic matter under thick layers of inorganic sediments. Thus, *coal seams* interbedded

Figure 8.9 A coal seam, 8 ft (2.4 m) thick exposed in a river bank, Dawson County, Montana. (Photograph by M. R. Campbell, U.S. Geological Survey.)

with shale, sandstone, and limestone strata came into existence (Figure 8.9). Groups of strata containing coal seams are referred to as *coal measures;* the individual seams range in thickness from a fraction of an inch to as great as 50 ft (12 m) in the exceptional case. Coal consists mostly of compounds of carbon, hydrogen, and oxygen, and usually contains a small proportion of sulfur compounds. Water and hydrocarbon volatiles make up a large part of lower-grade coals, but fixed carbon is the principal constituent of the highest grades of coal (see Figure 10.8). More information on coals and their geologic occurrence is given in Chapter 10.

We know a great deal about the plants that produced the initial peat deposits, since the leaves, seeds, and stems of these plants occur in abundance as fossils in the coal measures. Land plants evolved in the Devonian Period, which began about 400 million years ago, but forests of sufficient density, growing in vast, fresh-water low-level swamps did not develop until the following Carboniferous Period, 345 to 280 million years ago. As the name implies, this geological period was unparalleled as a producer of coal measures.

From the estimate that 30 ft (9 m) of peat was required to produce 1 ft (0.3 m) of coal, and that this quantity would be produced in a 300-year period, we can conclude that the great 50-ft (15-m) Mammoth Coal Seam of Pennsylvania was formed from 1500 ft (460 m) of peat produced in a continuous 450,000-year period of forest growth! Even if there is a large error in such an estimate, it is obvious that remarkably uniform, or *equable,* environmental conditions must have prevailed during the Carboniferous Period, and at a number of other times in geologic history. At such times the earth's crust must have been extremely quiescent, and sea level extremely stable. A very slow submergence of the peat-forming area is indicated by the typical succession of strata above and below a coal seam, shown schematically in Figure 8.10. Sediment deposition began with nonmarine sandstone, shale, and limestone, in that order, and was followed by the swamp forest and its peat deposit. The sea level was very slowly rising (or the crust was very slowly subsiding), so that the forest was inundated by the ocean. In this shallow water, shale and limestone beds accumulated, but the ocean then retreated and exposed the strata to the atmosphere, allowing a minor amount of erosional removal of the uppermost deposits. This cycle of sedimentation has been named a *cyclothem;* it was repeated over and over during the Carboniferous Period in many parts of the world.

Petroleum, or *crude oil* as the liquid form is often called, includes many hydrocarbon com-

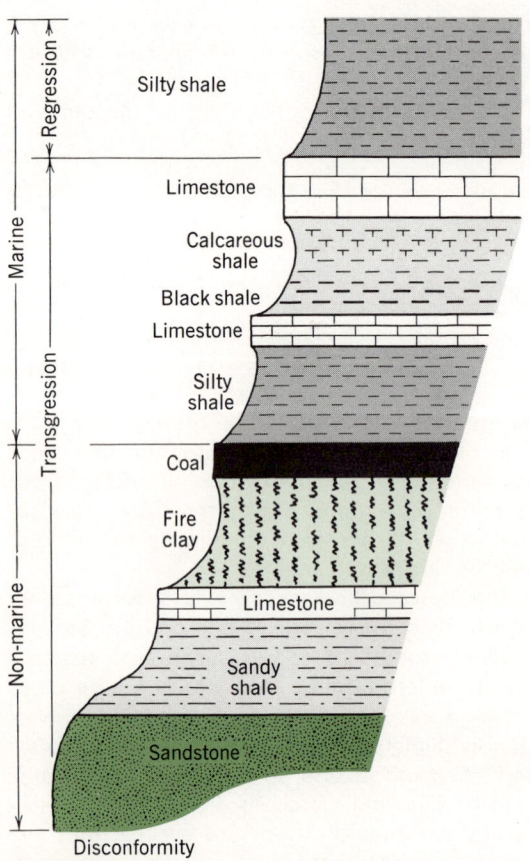

Figure 8.10 Cross section of strata in an idealized cyclothem of Carboniferous age in Illinois. (After J. M. Weller, 1960, *Stratigraphic Principles and Practice,* Harper & Row, New York, p. 372, Figure 145.)

Landforms Developed on Sedimentary Rocks

pounds in the gaseous, fluid, or solid states. Typically, a sample of crude oil might consist of 82% carbon, 15% hydrogen, and 3% oxygen and nitrogen. *Natural gas*, found in close association with accumulations of petroleum, is a mixture of gases. The principal gas is *methane* (*marsh gas*), CH_4, and there are minor amounts of ethane, propane, and butane, all of which are hydrocarbon compounds. Small amounts of carbon dioxide, nitrogen, and oxygen are also present, and sometimes helium. Petroleum and natural gas are not classed as minerals, but they originated as organic compounds in sediments. The occurrence of accumulations of petroleum and natural gas in sedimentary strata is explained in Chapter 10.

While it is generally agreed that petroleum and natural gas are of organic origin, the nature of the process is hypothetical. A favored explanation for petroleum is that the oil originated within microscopic floating marine plants (phytoplankton) such as the *diatoms*. As each diatom died, it released a minute droplet of oil, which became enclosed in muddy bottom sediment. Eventually, the mud became a shale formation, in which the oil was disseminated. We have today *oil shales*, holding petroleum in a dispersed state, and these give support to the organic hypothesis we have stated. However, the occurrence of petroleum in heavy concentrations in porous rock such as sandstone requires that the oil in some manner was forced to migrate from its source region to a reservoir rock. The volatile hydrocarbons and other gases were then segregated to a position above the liquid petroleum.

From the standpoint of Man's energy resources, which is the subject of Chapter 10, the outstanding concept relating to the hydrocarbon compounds within the earth—*fossil fuels* they are collectively called—is that they have required hundreds of millions of years to accumulate, whereas they are being consumed at a prodigious rate by our industrial society. These fuels are *nonrenewable resources*. Once they are gone there will be no more, since the quantity produced in a thousand years by geologic processes is scarcely measurable in comparison to the quantity stored through geologic time. In Chapter 10 we shall attempt to evaluate the extent of these energy resources and predict how long they will last.

Landforms Developed on Sedimentary Rocks

As in the case of the igneous intrusive and volcanic rocks, which are associated with distinctive surface landforms, there are certain landforms uniquely related to the layered structure of sedimentary strata. Large areas of the continental crust are covered by thick sequences of sedimentary-rock layers, which at one time in the geologic past were the bottom deposits of shallow inland seas or were stream deposits spread over vast alluvial plains. When uplifted with little disturbance, strata that were of marine origin became land surfaces, which were at first low, smooth plains. However, as erosion progresses, the strata are carved up by streams into a network of valleys.

The horizontal attitude of the rock layers gives rise to distinctive landforms where the layers are of alternately weak and resistant nature (Figure 8.11). The resistant layers, usually of sandstone and limestone (the latter particularly in arid climates), form *cliffs*. The weak layers, usually of shale or clay, are easily washed away from beneath the lower edges of the resistant layers. This process accentuates the cliffs above and forms smoothly descending slopes at each cliff base. In dry climates, where vegetation is scant and the action of rainwash especially effective, sharply defined topographic forms develop. They comprise what may be described as *scarp-slope-shelf* topography, because the normal sequence of forms is a cliff, or *scarp*, at the base of which is a smooth slope. This in turn flattens out to make a shelf, terminated at the outer edge by the cliff of the next lower set of forms (Figure 8.12). In the walls of the great canyons of the Colorado Plateau region in Colorado, Utah, Arizona, and New Mexico, these forms are wonderfully displayed (Figure 8.13).

In plateau regions underlain by horizontal strata, the erosion processes tend to strip successive layers from the plateau surface. Cliffs, capped by hard rock layers, retreat as near-perpendicular surfaces. When undermined, the rock in the upper cliff face repeatedly breaks away along vertical fractures. Where a cliff has retreated a

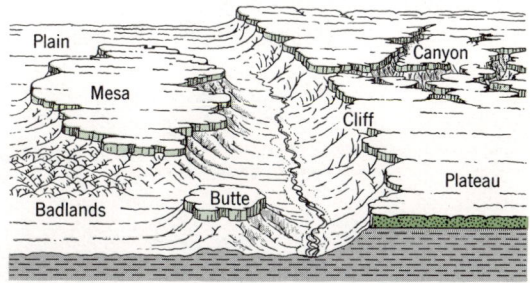

Figure 8.11 In arid climates, a distinctive set of landforms develops in flat-lying sedimentary formations.

Rock Alteration and Sediments

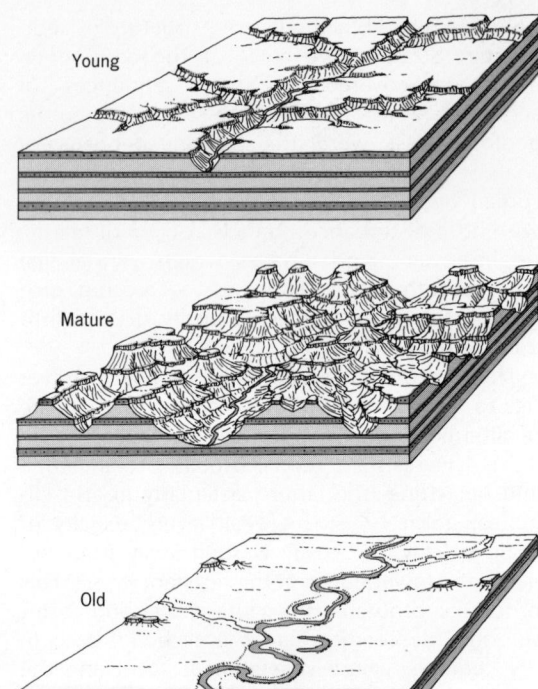

Figure 8.12 Erosional development of horizontal sedimentary strata in arid lands. Upper diagram shows an early erosion stage with narrow canyons and an extensive plateau surface. Lower diagram shows advanced erosion stage. (After E. Raisz.)

Figure 8.13 Below: This panoramic drawing by the noted geologist-artist, W. H. Holmes, published in 1882, shows the Grand Canyon at the mouth of the Toroweap. In this part of the canyon, rarely seen by tourists, a broad bench called the Esplanade is well developed. (From Dutton, *Atlas to accompany Monograph II*, U.S. Geological Survey.)

considerable distance from a canyon, there remains a broad, flat bench which is the exposed surface of the next resistant layer below. Should the entire plateau surface be formed by the complete or almost complete removal of a rock series, leaving a plateau capped by a resistant layer, the plateau is termed a *stripped surface* or *stratum plain*.

In advanced stages of erosion the landscape in an arid region has many *mesas* (Figures 8.11 and 8.12), tabletopped hills or mountains bordered on all sides by cliffs and representing the remnant of a formerly extensive layer of resistant rock. As a mesa is reduced in area by retreat of the cliffs that border it, it maintains its flat top and altitude. Before its complete consumption the final stage is a small, steep-sided hill or peak known as a *butte* (Figure 8.14).

Where extremely weak clays or shales, lacking a protective vegetative cover, are exposed to rainwash and gully erosion in dry regions, a very rugged topography resembling miniature mountains develops. Such areas are termed *badlands* (Figures 8.11, 14.4, and 14.5).

In addition to regions underlain by sedimentary rocks, the regions of horizontal layered rocks may include thick accumulations of lava flows. In some parts of the world, such as the Columbia Plateau region of eastern Washington and Oregon, or the Deccan Plateau of western India, the vast outpourings of highly fluid basalt lavas now cover thousands of square miles and are several thousand feet thick (see Figures 7.16 and 7.17). Interbedded with the lavas are lake and stream deposits of sands, gravels, and clays, and in some cases layers of volcanic ash. Consequently, the structure exhibits alternately weak and resistant layers in which erosion produces landforms very similar to those of sedimentary strata.

Landforms Developed on Sedimentary Rocks

Figure 8.14 This early photograph shows a butte of horizontal red sandstones capped by a gypsum layer, near Cambria, Wyoming. (Photograph by N. H. Darton, U.S. Geological Survey.)

Figure 8.15 Seen from the air, the dissected Allegheny Plateau of West Virginia appears largely forested. Relief of 800 ft (240 m) is here developed on shale strata. (Photograph by J. L. Rich, Courtesy of the *Geographical Review*.)

Generalizations cannot readily be made about the environmental influences of horizontal strata because of the great variations in surface relief that exist. Where the topography is plainlike, and conditions of climate and soil are favorable, agriculture is widely developed. On the high plains of western Kansas and Nebraska, eastern Wyoming and Colorado, New Mexico, and Texas, wheat farming is a predominant activity. Despite comparatively high elevations of 3000 to 5000 ft (900 to 1500 m), this undulating plain is trenched only by a few major through-flowing streams.

Some regions of horizontal strata in the interior United States are hilly to mountainous in relief. An example is found in mountain areas of the Alleghenies or the Cumberland Mountains, where cultivation is limited to a few small tracts, such as the floodplain belts of larger streams, despite the favorable humid climate. The steep mountain slopes are heavily forested (Figure 8.15). In the canyon lands of the Colorado Plateau an extremely low population density exists, not only because the high relief and aridity do not favor agriculture, but also because human access is almost impossible across the network of sheer-walled canyons.

A special, but widespread case of dominance of a landscape by sedimentary strata is the *coastal plain*, a sequence of strata lapping over the margins of a continent. The strata, of comparatively young geologic age, were deposited when the coastal zone was submerged as a continental shelf, and consequently take the form of thin wedges tapering landward to feather-edges, as shown schematically in Figure 8.16A. Upon emerging to become part of the continental surface, the coastal plain is smooth and nearly featureless, sloping gently seaward. However, as erosion proceeds, clay or shale layers are lowered more rapidly than sand or sandstone layers with the result that broad *lowlands* are developed paralleling the shoreline (Figure 8.16B). Between the lowlands are belts of low hills upheld by layers of sand, sandstone, or limestone. These hill belts are known as *cuestas*. In some cases, where the strata are well lithified, the cuesta develops a steep landward face, or *scarp*, contrasting with a gently inclined *backslope* (Figure 8.17).

Splendid examples of coastal plains are present along the Atlantic and Gulf coasts of the United States, in southeastern England, and in the Paris Basin region of north-central France.

The coastal plain of the United States is by far the largest of these, ranging in width from 100 to 300 mi (160 to 500 km) and extending for 2000 mi (3000 km) along the Atlantic and Gulf coasts. The coastal plain starts at Long Island, which is a partly submerged cuesta, and widens rapidly southward so as to include much of New Jersey, Delaware, Maryland, and Virginia (Figure 8.18). Throughout this portion the coastal plain has but one cuesta, that which forms the Atlantic Highlands, Mt. Laurel, Pine Hills, and similar hill groups. The cuesta is underlain by a porous sand formation. An inner lowland is a continuous broad valley developed on a weak clay formation.

In Alabama and Mississippi the coastal plain is more complex. Cuestas and lowlands run in belts roughly parallel with the coast (Figure 8.19). The term *belted coastal plain* is applied to these re-

Rock Alteration and Sediments

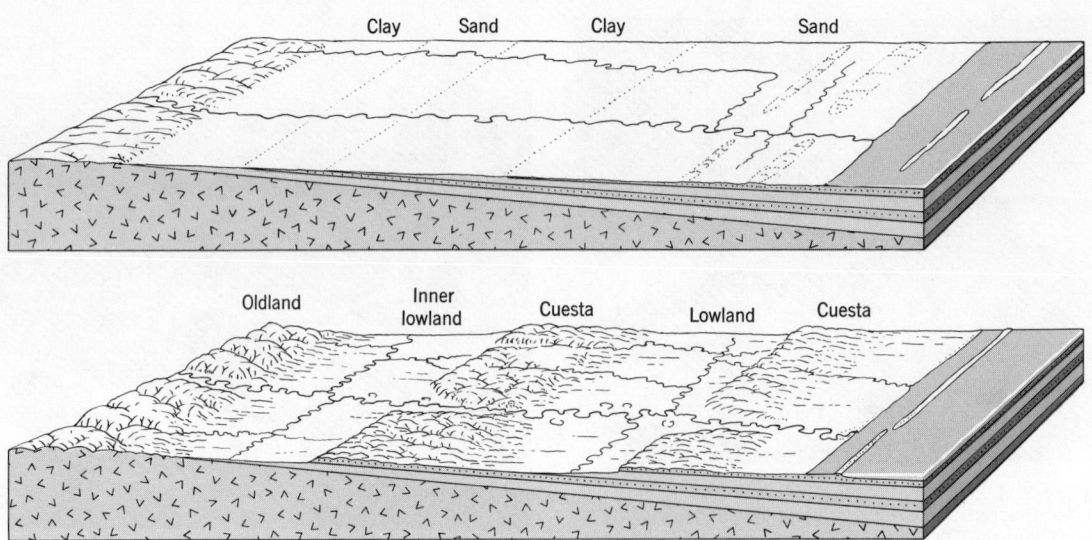

Figure 8.16 Coastal plain development. *Above:* Plain recently emerged from beneath ocean. *Below:* Dissected coastal plain with well-developed cuestas.

gions. The cuestas are underlain by sandy formations, while limestone forms fertile lowlands, such as the Black Belt in Alabama.

The entire southeastern portion of England is a former coastal plain (Figure 8.20). Erosion has developed moderate relief and two bold cuestas dominate the topography. The innermost is of limestone of Jurassic age and is locally named the Cotswold Hills. In England, the term *wold* is applied to a cuesta, *vale* to a lowland. The outer or southeastern cuesta is of white chalk of Cretaceous age and includes the Chiltern Hills. Between cuestas is a lowland in which lie Oxford and Cambridge. An extensive inner lowland runs between the inner cuesta and the old rock masses of Cornwall, Wales, and the Pennine Range. In the inner lowland are the important cities of Bristol, Gloucester, Birmingham, Nottingham, Lincoln, and York, as well as extensive farm lands.

Broad coastal plains, such as those of the eastern United States and southeastern England, have had profound environmental significance throughout the history of western civilization. They have shown intensive agricultural development because of the fertility and easy cultivation of broad lowlands. Cuestas have provided valuable forests. Transportation routes have tended to follow the lowlands and to connect the larger cities located there. For example, important roads and railroads connect New York with Trenton, Philadelphia, Baltimore, and Washington, all of which are situated in an inner lowland (Figure 8.18). Cuesta topography, however, is rarely so rugged as to interfere seriously with the location of communication lines.

The seaward inclination of sedimentary strata in a coastal plain provides a structure favorable to the development of *artesian* water wells. (The

Figure 8.17 This sharply defined cuesta in the Paris basin of northern France has its steep face to the east (*left*), a very gentle slope westward from the crest. (Photograph by Douglas Johnson.)

Landforms Developed on Sedimentary Rocks

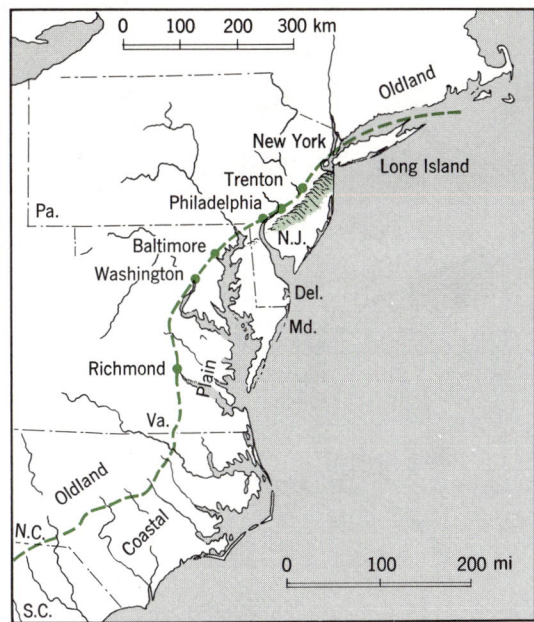

Figure 8.18 The coastal plain of the Atlantic seaboard states shows little cuesta development except in New Jersey. The inner limit of the coastal plain is marked by a series of fall-line cities. (After A. K. Lobeck.)

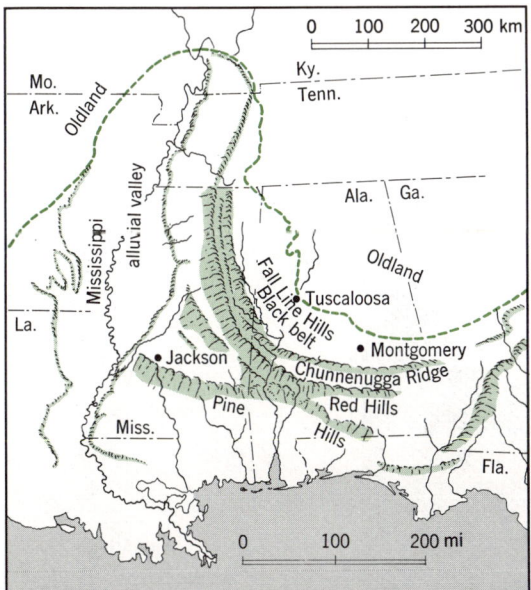

Figure 8.19 The Alabama-Mississippi coastal plain is belted by a series of sandy cuestas and shale or marl lowlands. (After A. K. Lobeck.)

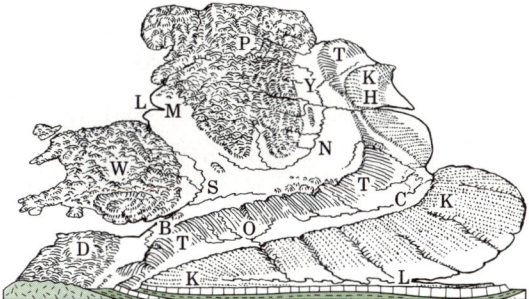

Figure 8.20 Southeastern England is a broadly curved former coastal plain. L = London; K = Cretaceous chalk cuesta; C = Cambridge; O = Oxford; H = Humber River; Y = York; N = Nottingham; S = Severn River; B = Bristol; D = Dartmoor; W = Wales; M = Manchester; L = Liverpool; P = Pennine Range; T = Jurassic limestone cuesta. (After W. M. Davis.)

subject of underground water resources is developed in detail in Chapter 12.) Water penetrates deeply into a sandy cuesta stratum, which is overlain by shales or clays impervious to the flow of underground waters. When a well is drilled into the sand formation considerably seaward of its surface exposure, water under hydraulic pressure reaches the surface (Figure 8.21). Artesian water in large quantities has been available in many parts of the Atlantic and Gulf coastal plains.

Where sedimentary strata have been deformed by upbending, associated with igneous intrusion or other forms of crustal movement, erosion has bared the upturned edges of strata, bringing into relief a number of distinctive landforms (Figure 8.22). Highly resistant layers of sandstone (and of limestone in an arid climate) form narrow, steep-sided ridges known as *hogbacks* (Figure 8.23). Weaker strata, usually shales, form narrow valleys separating the hogbacks. As Figure 8.22 shows, there is a complete series of intergradations from hogbacks, to cuestas, and to mesas and plateaus, as the strata change in attitude from steep inclination to horizontality.

A special case of hogback development is found in a geologic structure known as the *sedimentary dome*, shown in Figure 8.24. A series of sedimentary layers has been lifted, perhaps by igneous intrusion, into a low, flat-topped dome structure (Diagram A). Erosion by streams eventu-

Figure 8.21 An artesian well in gently inclined strata of a coastal plain. Ground-water recharge occurs on sandy cuesta. (After E. Raisz.)

Rock Alteration and Sediments

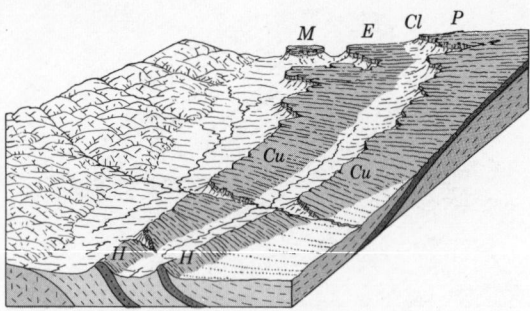

Figure 8.22 Hogbacks gradually merge into cuestas, the cuestas into plateaus and esplanades, where the inclination of the strata becomes less from one place to another. H = hogback ridge; Cu = cuesta; M = mesa; E = esplanade; Cl = cliff; P = plateau. (After W. M. Davis.)

ally removes strata from the central area of the dome (Diagram B). Upturned strata yield hogbacks and narrow valleys that encircle a central core in which older rock, possibly igneous in origin, appears in a central mountainous region.

A classic example of a large and rather complex sedimentary dome is the Black Hills dome of western South Dakota and eastern Wyoming (Figure 8.25). Valleys that encircle this dome are splendid locations for railroads and highways. So, it is natural that towns and cities should have grown in these valleys. One valley in particular, the Red Valley, is continuously developed around the entire dome and has been termed the Race Track because of its shape. It is underlain by a weak shale, which is easily washed away. In the

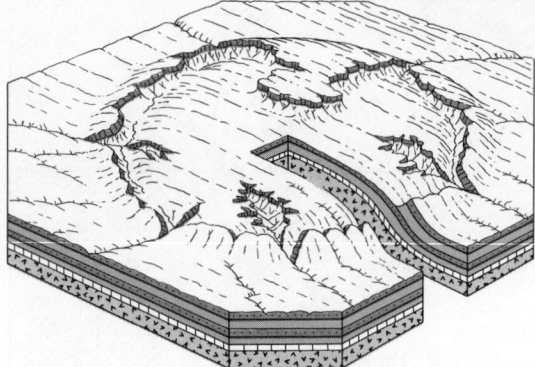

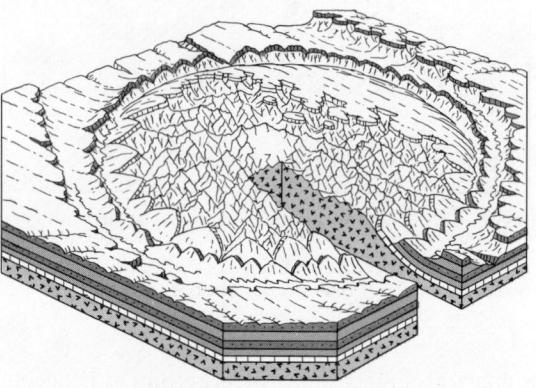

Figure 8.24 Erosion of sedimentary strata from summit of a dome structure. *Above:* Strata partially removed, forming an encircling hogback ridge. *Below:* Strata removed from center of dome, revealing a core of older igneous or metamorphic rock.

Figure 8.23 The Virgin River cuts across a hogback of steeply dipping strata on the flank of the Virgin anticline, southwestern Utah. (Photograph by Frank Jensen.)

Landforms Developed on Sedimentary Rocks

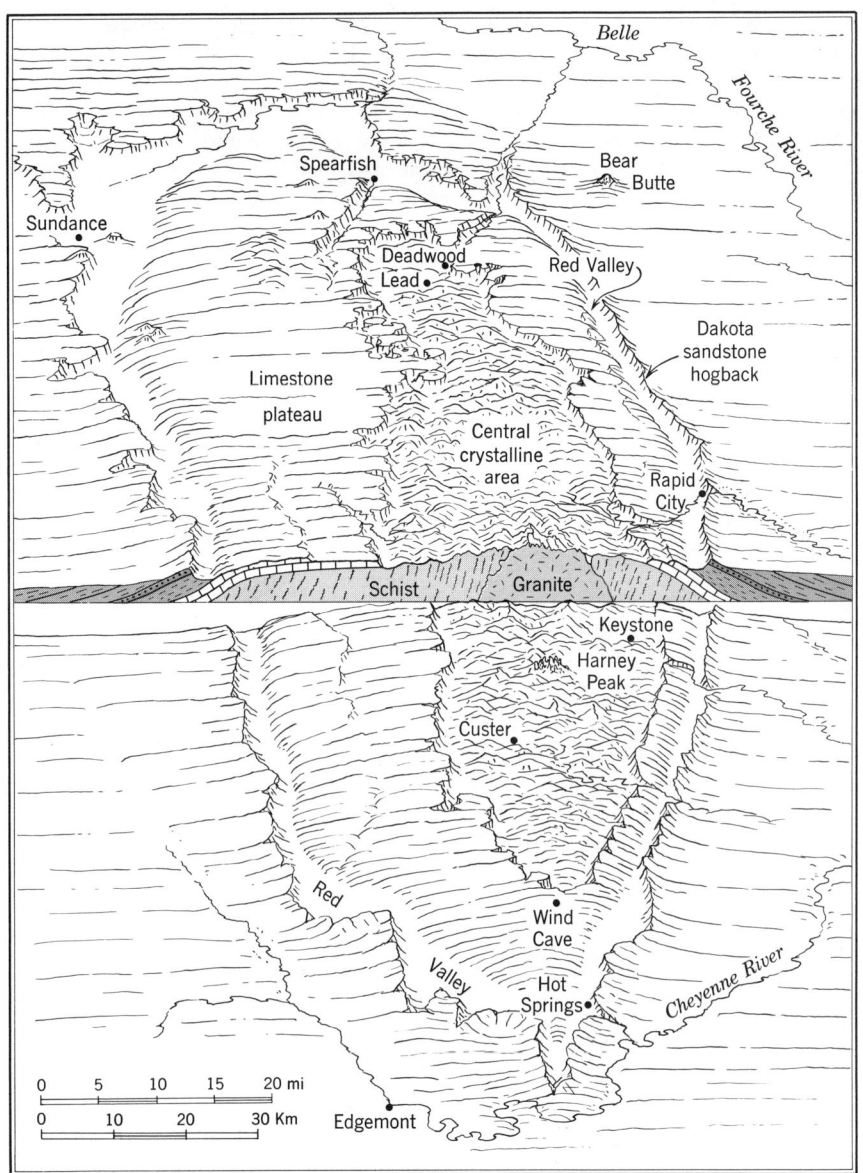

Figure 8.25 The Black Hills consist of a broad, flat-topped dome, deeply eroded to expose a core of igneous and metamorphic rocks.

Red Valley lie such cities as Rapid City, Spearfish, and Sturgis. On the outer side of the Red Valley is a high, sharp hogback of Dakota sandstone, known simply as the Hogback Ridge. It rises some 400 to 500 ft (120 to 150 m) above the level of the Red Valley. Farther out toward the margins of the dome the strata are less steeply inclined and form a series of cuestas. Artesian water is obtained from wells drilled in the surrounding plain.

The eastern central part of the Black Hills consists of a mountainous core of intrusive and other ancient rocks. These mountains are richly forested, whereas the intervening valleys are beautiful open parks. Thus the region is attractive as a summer resort area. Harney Park, elevation 7242 ft (2207 m), is highest of the peaks of the core. In the northern part of the central core, in the vicinity of Lead and Deadwood, are valuable ore deposits. At Lead is the fabulous Homestake Mine, one of the world's richest gold-producing mines.

The western central part of the Black Hills consists of a limestone plateau deeply carved by streams. The original dome had a flattened summit. The limestone plateau represents one of the last remaining sedimentary rock layers to be stripped from the core of the dome.

From these few examples you can appreciate the way in which sedimentary strata play an environmental role through the control which they exert upon landforms. The varied resistance of strata from place to place influences the slope and relief of the land surface. These factors in turn influence the texture and drainage of soils and in this way strongly influence the natural forms of vegetation that occupy those soils. Use of the land by Man as crop lands, grazing lands, and forest lands also follows these original patterns of landforms and soils. Details of the processes of weathering, soil formation, and erosion by running water remain to be filled out in detail in Chapters 11 and 14. Our purpose here is simply to relate geologic controls to surface environments. We shall complete this subject in the next chapter.

Rock Alteration and Sediments in Review

This chapter has followed the mineral products of two interacting energy systems through a succession of changes in physical properties and chemical compositions, and through a succession of translocations beginning on the lands and ending in the oceans. Energy is expended at each step in the alteration and translocation of mineral matter; it is the energy of solar radiation acting through both inorganic and organic systems. The soil layer is an intermediate and temporary resting place for inorganic matter passing from igneous to sedimentary phases of the rock transformation cycle.

Accumulation of sediment is a process of storage of inorganic matter reprocessed from preexisting rock, and also of carbon that was not previously in rock form. This volatile element came directly from the earth's interior as a gas, but was placed into solid form, in part by organic processes, as carbonate minerals in rock. Carbon has also been stored in solid form in vast quantities as hydrocarbon compounds manufactured by plants. In this way Man's supply of available energy has accumulated.

Crustal uplift, followed by erosion, exposes sedimentary rocks to the atmospheric environment. Minerals are recycled—carried as sediment to the sea for deposition as new sedimentary strata. Uplift and exposure of the strata has allowed Man to extract and burn stored hydrocarbon fuels.

In the next chapter we return to the action of the earth's internal energy system of radiogenic heat, as it reprocesses the sediments and creates new rock to build continents.

9 Tectonic Processes and Continental Evolution

Internal earth processes powered by radiogenic heat, and perhaps also by tidal flexing, cause the crust to be severely deformed, as masses of continental dimensions are pulled apart or brought together in head-on collisions. These forms of work done upon the crust are described by the adjective *tectonic*. Preexisting igneous rock and deeply buried sedimentary rocks are caught up in colossal movements and subjected to pressures capable of kneading those rocks as if they were soft dough. One result of tectonic processes is a third major class of rocks—the *metamorphic rocks*. New minerals are formed by the stresses and heat of metamorphism. As in the case of the igneous rocks, the metamorphic rocks gradually become exposed to the surface as erosion of overlying rock occurs. A new phase of rock alteration is then begun and new sediment is produced, leading to formation of more sedimentary rock. One objective of this chapter is to complete the analysis of rock-forming processes as a cyclic phenomenon involving the transformation and transport of mineral matter with an accompanying energy dissipation on an enormous scale.

We are then led to speculate on the origin of continental crust through tectonic processes in which great *lithospheric plates*, massive segments of the rigid lithosphere, are pulled apart or are brought together in compressive action. Tectonic activity causes the earthquake, an environmental hazard feared by Man since time immemorial. As cities grow, the areas of potential death and destruction by earthquakes also increase. We find, too, that certain of Man's activities are setting off earthquakes. This serious environmental problem deserves careful study. Can earthquakes be predicted and even controlled? To complement the classification and description of landforms of volcanic origins, covered in Chapter 7, we will add a description of those landforms produced directly by tectonic movements, for these too have environmental significance in terms of life on the lands. Even more important are erosional landforms controlled by the rock structures of ancient tectonic events.

Because patterns of tectonic activity have been repeated many times over since the earth was formed, the rocks that remain intact from each

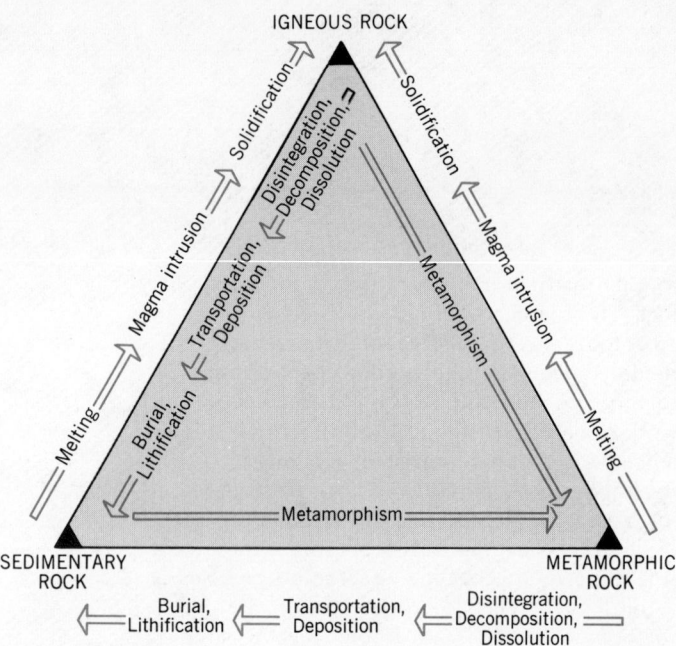

Figure 9.1 Three major rock classes as corners of a triangle. (From A. N. Strahler, 1971, *The Earth Sciences*, 2nd ed., Harper & Row, New York.)

period of activity must be arranged in a time sequence. The geologic time scale and age determination of rocks now become worthwhile topics of study, particularly since life was evolving in an orderly way during much of this time. But to understand how age of rocks is determined, it will be first necessary to study natural radioactivity of certain elements found in rocks. This radioactivity proves to be of great environmental significance, since it is a source of radiation that affects the evolution of life.

The Cycle of Rock Transformation

A triangular diagram, Figure 9.1, shows the three major rock classes with the directions of transformation that are possible. From any one of the three corners change is possible in the direction of the other corners. Labels along the three sides explain the changes that occur in the direction of change.

Since the triangular diagram does not place the changes of rock in an environmental framework, we have set up a different sort of schematic diagram, shown in Figure 9.2. Here we distinguish between a *surface environment* of low pressures and temperatures and a *deep environment* of high pressures and temperatures. We have already encountered both of these environments: the deep environment in a study of igneous rocks and magmas, Chapter 7; the surface environment in the study of rock alteration and sediment deposition, Chapter 8.

Seen in its complete form, the total network of rock changes in response to environmental stress constitutes the *cycle of rock transformation*. The rock cycle is not an energy system in itself; it is the sequence of material products of the interaction of two great global energy systems. The mineral matter provides pathways by which the energies of electromagnetic radiation and radiogenic heat are transformed, stored, and eventually dissipated to outer space.

Sedimentary Accumulations through Geologic Time

Streams transport sediment to the continental margins, where thick accumulations are formed. Returning to the principle of isostasy, explained in Chapter 7, consider what happens when sediment is removed from a mountain mass, transported to adjacent low areas, and deposited there in layers (Figure 9.3). Removal of load from the mountain region results in crustal rise, partially restoring the mass removed. Loading of the crust by sediment deposition causes the crust to sink. A transfer of mass within the soft layer of the mantle (the asthenosphere) is required, as the arrows in the diagram suggest.

Throughout geologic history sediment accumulations have taken place in long, narrow belts where the crust has, through action of deep-seated tectonic forces, been caused to sag slowly in a structure known to geologists as a *geosyncline* (Figure 9.4). Clastic sediment is being

Sedimentary Accumulations through Geologic Time

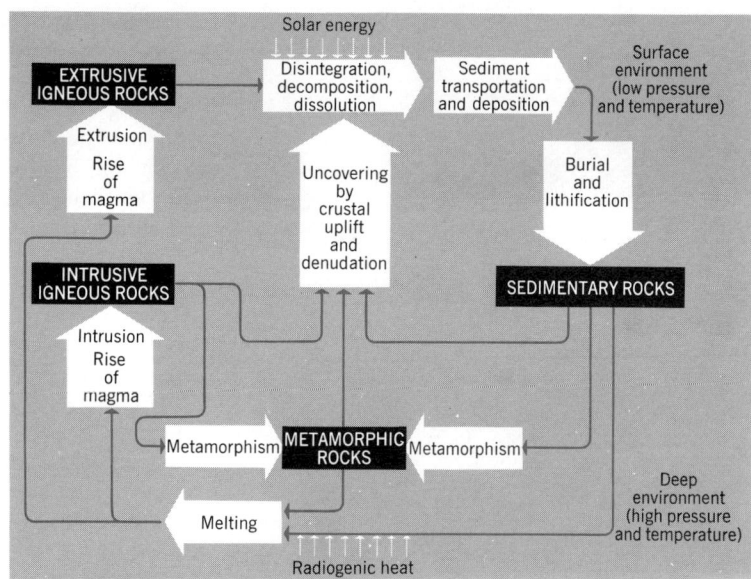

Figure 9.2 Schematic diagram of the cycle of rock transformations. (From A. N. Strahler, 1971, *The Earth Sciences*, 2nd ed., Harper & Row, New York.)

brought from the adjacent continent, and also from an *island arc* of active volcanic growth that typically bounds the geosyncline on the oceanic side. This sediment is spread by currents over the floor of the geosyncline to form clastic sedimentary strata—shales, sandstones, and occasional conglomerate layers. Carbonate sediments are also being precipitated upon the shallow floor of the geosyncline to accumulate as limestone strata. Sediment accumulation keeps pace with slow crustal subsidence, so that the water remains shallow during the life of the geosyncline. Part of the geosynclinal sediment may take the form of large deltas built by rivers draining large continental areas. After some tens of millions of years of sedimentation, the strata of the geosyncline have reached a total thickness that may measure up to tens of thousands of feet in the central axis of the geosyncline, but the sediment mass tapers to a thin wedge at either margin. Length of the geosyncline is on the order of 1000 to 3000 mi (1500 to 5000 km), its width perhaps 300 to 500 mi (500 to 800 km).

A fine example of a geologically young geosyncline, still undergoing sedimentation, is the *Gulf Coast Geosyncline*, which lies beneath the coastal plain of Louisiana and Texas and the con-

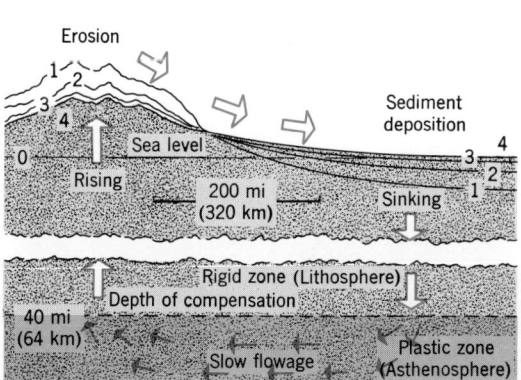

Figure 9.3 Schematic diagram of isostatic rising and sinking of crust accompanying erosion and sediment accumulation. (From A. N. Strahler, 1971, *The Earth Sciences*, 2nd ed., Harper & Row, New York.)

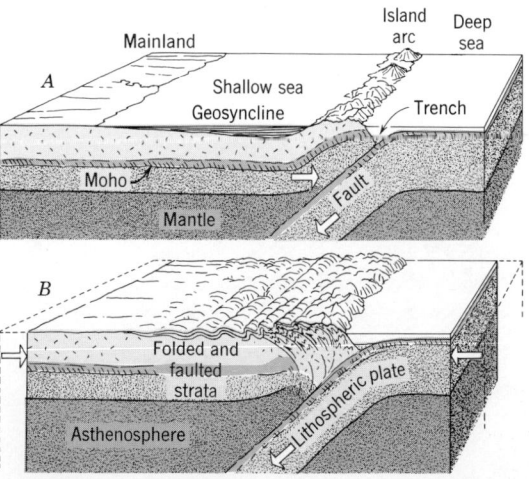

Figure 9.4 Geosyncline bordered by an island arc (*above*) is intensely deformed during orogeny (*below*). (From A. N. Strahler, 1971, *The Earth Sciences*, 2nd ed., Harper & Row, New York.)

Figure 9.5 Complexly folded and faulted strata of Precambrian age, northern Rockies of Glacier National Park, Montana. (Photograph by Chapman, U.S. Geological Survey.)

tinental shelf of the Gulf of Mexico. Sedimentation began in this area as far back as the Mesozoic Era, 150 million years ago, but the thickest accumulations were formed in the Cenozoic Era, spanning the past 65 million years. At its thickest point, approximately beneath the present coastline, the strata of Cenozoic age have a total thickness of over 40,000 ft, or about $7\frac{1}{2}$ mi (12 km)! Incidentally, this huge body of sediments contains a great wealth of petroleum and natural gas, salt, and sulfur.

It is a fact of geologic history, repeated many times over, that the period of geosynclinal deposition is terminated by an intense deformation of the crust, known as an *orogeny* (from the Greek *oros*, mountains, *geneia*, origin). The upper diagram of Figure 9.4 shows what has been happening during the period of geosynclinal deposition. The lithospheric plate beneath the geosyncline has been broken and is bent down under compressive stress. As the edge of the down-bent plate sinks into the asthenosphere there is relative motion of the two lithosphere plates along a surface of separation known as a *fault*, which in this case is inclined downward beneath the continental margin. The fault plane reaches the surface on the oceanward side of the island arc and its location is marked by a deep oceanic *trench*. Seen in terms of the lithospheric plate motions, the geosyncline is a down-bending of the crust induced by pressure of the descending plate upon the continental plate adjacent to it, while the volcanic island arc represents the rise of magma from local pockets of melting along the fault zone. Orogeny, as illustrated by the lower diagram of Figure 9.4, is interpreted as a change in pace of the lithospheric movements. The two plates are suddenly (geologically speaking) forced together, crumpling the geosynclinal belt. Strata that were more-or-less horizontal are thrown into wave-like *folds* (Figure 9.5). The same compression also causes new faults that divide the folded strata into slices, shoving one up over the other. Faults of this type are called *overthrust faults* (Figure 9.6).

Figure 9.7 shows in greater detail the effects of orogeny. Notice the zone of intense folding and overthrusting, grading into a belt of open folds on the continental side. Below the surface, in the zone of most intense deformation, sedimentary rock has been altered under high pressure and temperature into metamorphic rock, while melting of the metamorphic rock has produced a new

Figure 9.6 The Lewis overthrust fault can be seen as a light-colored band (below the arrow) rising gently from left to right along the mountain side. The rock mass above the fault was thrust from left to right and rests upon a base of much younger rock. (Photograph by Douglas Johnson.)

Metamorphic Rocks

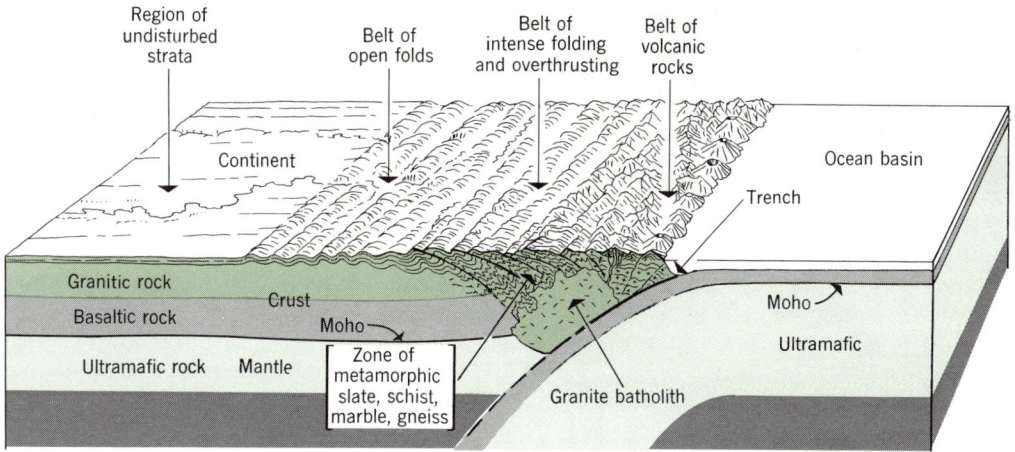

Figure 9.7 Folded and overthrust strata and intrusion of a batholith mark the completion of orogeny. (From A. N. Strahler, 1971, *The Earth Sciences*, 2nd ed., Harper & Row, New York.)

magma and this has become a granite batholith. Here we see the completion of a full cycle of rock transformation, for we are back to igneous rock again. While the origin of granite is still much debated, there is much to recommend the hypothesis that at least some granitic magma comes from melting of preexisting rock, and that its parent rock is largely sedimentary rock, which is felsic in composition.

Metamorphic Rocks

Metamorphic rocks show changes in either structure or in mineral composition, or both. The subject is highly complex and we will not gain very much from the standpoint of environmental science by going into detail about the metamorphic minerals and rocks. Since any variety of igneous or sedimentary rock is subject to metamorphism, the varieties of metamorphic rock are many. We shall mention only a few common metamorphic rocks.

One banded metamorphic rock, known generally as *gneiss*, is derived from the metamorphism of sedimentary rocks (Figure 9.8). The individual layers may be composed largely of quartz or of feldspar. In some banded gneisses, layers are of igneous composition and indicate an injection of granitic magma between sedimentary layers. Intense kneading of clastic sediments, particularly shales, results in the formation of *schist*, a metamorphic rock characterized by irregular partings on surfaces coated with flakes of mica (Figure 9.9). A less intense degree of metamorphism produces *slate* from black or red shales. Sandstone is altered to *quartzite*, an extraordinarily resistant rock. Limestone and dolomite, when subjected to metamorphism undergo a recrystallization of the carbonate minerals to yield *marble*, which is sometimes a white or gray rock of sugary texture.

The environmental significance of metamorphic rocks lies largely in the diverse landforms that result from their weathering and erosion, long after crustal uplift and erosion have bared them to the surface. Figure 9.10 is an idealized diagram showing a typical assemblage of metamorphic rocks with the landforms they produce. Gneiss and schist are typically strong rocks, forming mountains and uplands; slate is of intermediate resistance and forms hill belts; quartzite stands

Figure 9.8 Banded gneiss of Precambrian age, along the east coast of Hudson Bay, Canada. (Photograph G.S.C. No. 125221 by F. C. Taylor, Geological Survey of Canada, Ottawa.)

Tectonic Processes and Continental Evolution

Figure 9.9 This fragment of schist, 6 in. (15 cm) long, has a glistening, undulating surface (*above*) consisting largely of mica flakes. An edgewise view (*below*) of the same specimen shows the wavy foliation planes.

out boldly in narrow ridges; while marble is rapidly dissolved in a humid climate to produce prominent valleys and lowlands. From the standpoint of natural resources, the metamorphic rocks are widely used as structural materials: for example, slate as a roofing material, and gneiss as a building stone. Marble is well known for its ornamental qualities in both interior and exterior surfacing of walls and floors. Certain of the metamorphic minerals are of industrial value.

Plate Tectonics

Now we return to an overview of the global system of tectonics in which the strong, brittle lithosphere moves over a soft asthenosphere, impelled by forces that are little understood, but which we are attributing here basically to accumulation of radiogenic heat in the upper mantle.

Within the past two decades, exploration of the ocean floors and oceanic crust has brought evidence of lithospheric motions on a colossal scale. The new findings constitute a major revolution

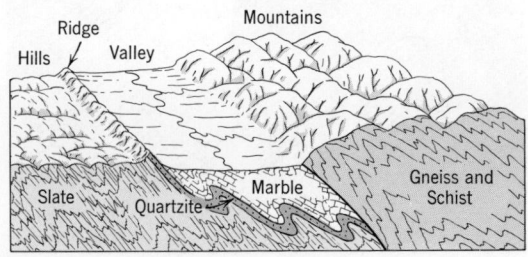

Figure 9.10 Erosion brings out differences in resistance among various types of metamorphic rocks, resulting in belts of hills, mountains, and lowlands.

in geology, and have led to a unified theory of large-scale geological processes. It all began shortly after World War II with discovery of many remarkable features of the ocean floors—features that seemed to have no counterpart on the continents. Foremost of the strange oceanic landforms is the *Mid-Oceanic Ridge*, a narrow submarine mountain range traceable for a distance of about 40,000 mi (64,000 km). It was first mapped in detail in the North Atlantic Ocean, where it is known as the *Mid-Atlantic Ridge* (Figure 9.11). The ridge is a broad belt, rising from low marginal hills to a high, summit zone marked by a narrow cleft termed the *axial rift*. This rift shows as a narrow notch in the profile of Figure 9.11. Subsequent mapping traced the Mid-Oceanic Ridge through the South Atlantic, Indian, and Pacific Oceans (Figure 9.12). It was found that the axial rift is offset at many points by groups of cross fractures, which are a type of fault. The Mid-Oceanic Ridge is a producer of earthquakes and there are volcanic islands at places along its length (for example Iceland).

Although the concentration of tectonic activity along the ridge was obvious for years, the underlying reason for this activity did not become apparent until studies of the magnetic properties of the basaltic rocks on either side of the ridge yielded up evidence that the oceanic crust is in the process of spreading apart on the two sides of the central rift. Although it makes a fascinating story of scientific detection, we can't go into the details here. Suffice it to say that the crustal separation is known to be taking place at the rate of about 1 to 2 in. (2 to 5 cm) per year, as measured from a point in the center of the rift to a point on one side. The rift is occupied by rising basaltic magma as it opens. Thus new oceanic crust is generated constantly.

Further research rapidly led to a general theory of *plate tectonics*, illustrated schematically in Figure 9.13. The entire lithosphere of the earth is considered to be divided up into *lithospheric plates*, of which there are at least six major plates and many smaller blocks within or between the major plates. The Mid-Oceanic Ridge represents the zone of separation of plates that are drifting apart. A lithospheric plate moves slowly over the soft asthenosphere, much as a slab of butter moves over the surface of a warm skillet. Where two lithospheric plates are moving toward each other they are in collision and one of the two plates may bend down to pass beneath the other, as Figure 9.13 shows. This downbending is termed *subduction*. The zones of collision are the orogenic belts already described.

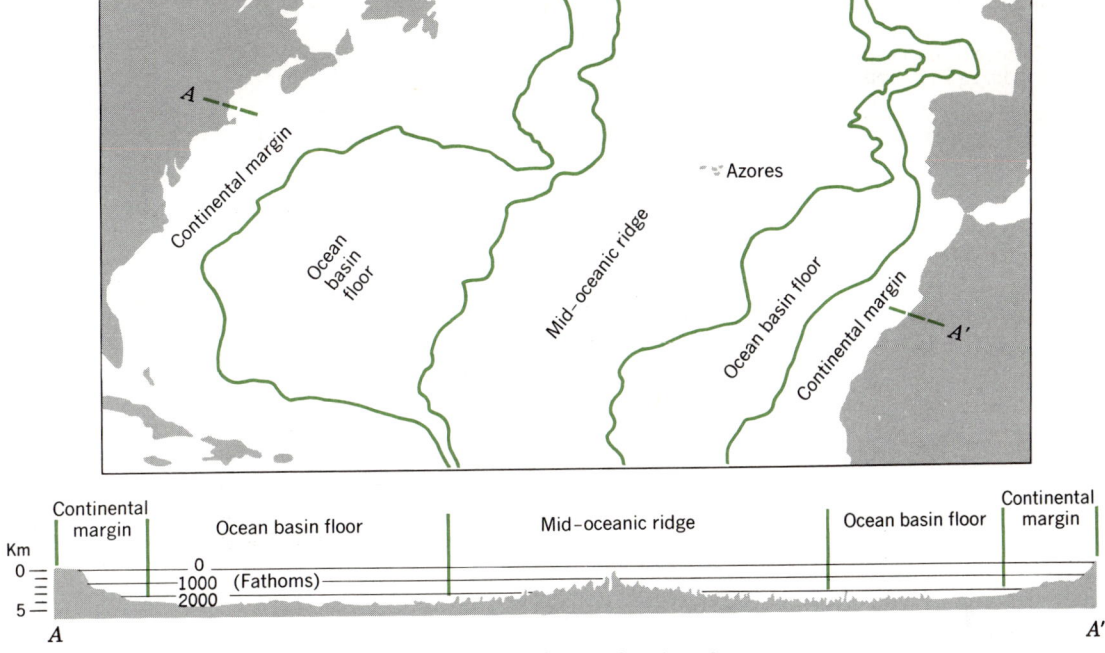

Figure 9.11 Major divisions of the North Atlantic Ocean basin (*above*), and a representative profile from New England to the coast of Africa (*below*). Profile exaggeration is about 40 times. Data of B. C. Heezen, M. Tharp, and M. Ewing, 1959. (From A. N. Strahler, 1971, *The Earth Sciences,* 2nd ed., Harper & Row, New York.)

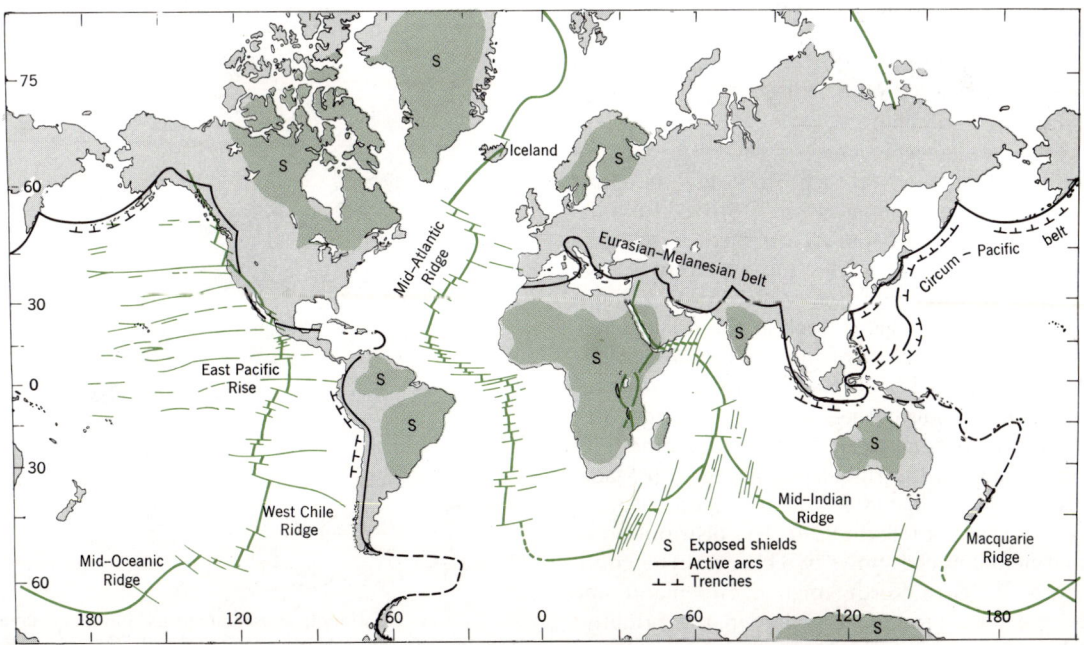

Figure 9.12 Mid-oceanic ridge system. The central rift zone is shown by a bold line; related fracture zones by light lines. (After a map by L. R. Sykes, 1969, *Geophysical Monograph 13,* Amer. Geophys. Union, Washington, D.C., p. 149, Figure 1. Based on data of B. C. Heezen, M. Tharp, H. W. Menard, and others.)

Tectonic Processes and Continental Evolution

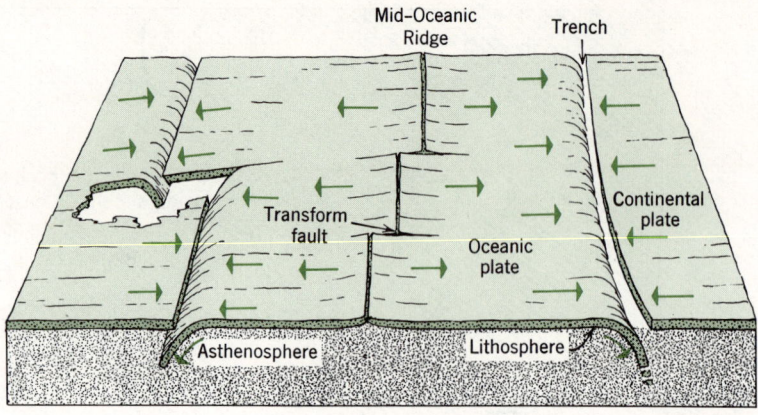

Figure 9.13 Schematic diagram showing major features of plate tectonics. Earth-curvature has been removed. (From A. N. Strahler, 1971, *The Earth Sciences*, 2nd ed., Harper & Row, New York.)

Through geologic time, as lithospheric plates have drifted apart, vast areas of ocean basins with a thin basaltic crust have come into existence. At the same time, the rock cycle of sedimentation, orogeny, and igneous intrusion has gradually created continental crust with its felsic (granitic) upper zone. Thus the continents have evolved and increased in size over a 3- to 4-billion year period. We find today that within the continents there are great patches of extremely ancient rock; most of it is igneous and metamorphic rock. These areas are the *continental shields* (see Figure 9.14). Small core areas within each shield consist of rock from 3.5 to 2.7 billion years (b.y.) in age; these are the *nuclei* of the continents. Nuclei are surrounded by broad zones of somewhat younger rock, 0.8 to 2.7 b.y. in age. All of this rock belongs to the *Precambrian* span of geologic history, a term that designates all rock older than 600 million years.

Since the oceanic crust is produced by upwelling along the Mid-Oceanic Ridge, its age is very young as compared with the continental crust. The oldest dated oceanic crust is in the western part of the Pacific Ocean basin; its age is about 150 million years, which is only about one-fifteenth the age of rock in the nuclei.

As early as the close of the nineteenth century, the proposal had been made that the continents of North and South America, Europe, Africa, Australia, and Antarctica, and the subcontinent of India (south of the Himalayas) along with Madagascar, had originated as a single supercontinent, named *Pangaea* (Figure 9.14). The distribution of continental nuclei shows two clusters. The continental nuclei of North America, Greenland, and Europe were grouped together in the primitive subcontinent of *Laurasia*; the nuclei of South America, Africa, peninsular India, Madagascar, Australia, and Antarctica were grouped together in the primitive subcontinent of *Gondwana*.

Pangaea began to break apart about 200 million years ago. The fragments slowly separated and underwent some rotation as well. Thus the Atlantic Ocean is envisioned as the area opened up

Figure 9.14 Continents reassembled as they may have been prior to the start of continental drift. Oldest shield rocks (older than −1.7 b.y.) shown by dark pattern; rocks ranging from −0.8 to −1.7 b.y., by light pattern. (Redrawn from a map by P. M. Hurley and J. R. Rand, 1969, *Science*, vol. 164, p. 1237, Figure 8.)

Plate Tectonics

by drifting away of the Americas from Africa and Europe. This hypothesis of *continental drift* was generally rejected or skeptically regarded by most of the geological fraternity, particularly the Americans, until about 1960, when the amazing evidence of crustal spreading came to light. Now a majority of geologists have accepted continental drift, much as it was originally outlined, and have fitted it into the global theory of plate tectonics. Figure 9.15 shows postulated stages in the separation of the continents, according to a recent version. (See Table 9.3 for ages of geologic periods.)

Under plate tectonic theory, North and South America are part of the *America Plate*, moving westward with respect to the Mid-Atlantic Ridge. Consequently, the western edge of this plate is being pushed toward adjoining plates on the west. This collision of plates is responsible for the great Cordilleran and Andean mountain chains. A huge submarine trench lies off the South American west coast and another off the Aleutian Island chain of Alaska. These trenches are interpreted as expressions of the subduction of the lithospheric plate to the west, which is being forced into the asthenosphere as the America Plate overrides it.

Looking at the Pacific Ocean as a whole, we find that most of it is a single lithospheric plate

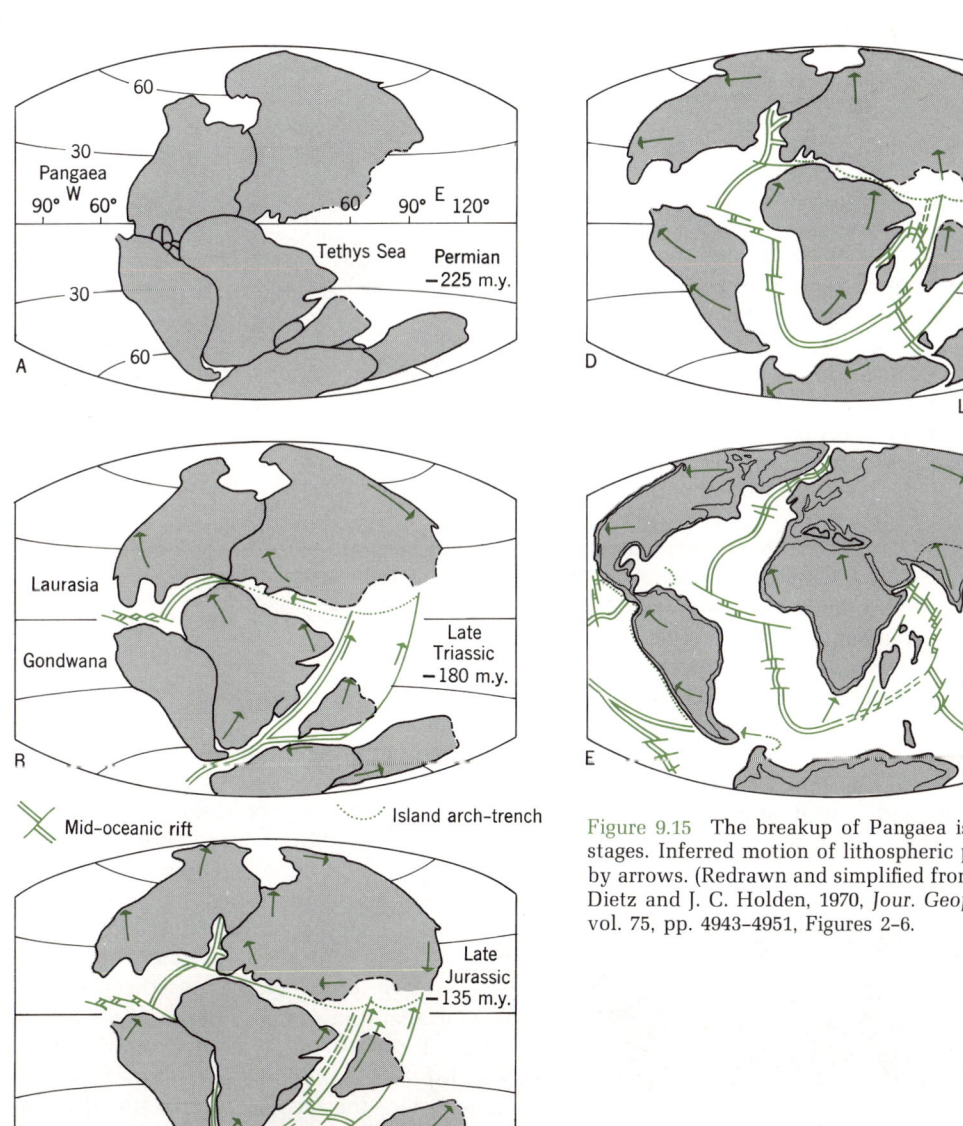

Figure 9.15 The breakup of Pangaea is shown in five stages. Inferred motion of lithospheric plates indicated by arrows. (Redrawn and simplified from maps by R. S. Dietz and J. C. Holden, 1970, *Jour. Geophys. Research*, vol. 75, pp. 4943-4951, Figures 2-6.

capped entirely by oceanic crust. Around the margins of this Pacific Plate is a great tectonic ring—the *Circum-Pacific Belt* of arcs of mountains or volcanic islands, bordered by deep trenches. It is a belt of crustal unrest, manifested by numerous intense earthquakes as well as volcanic activity, for this is a more-or-less continuous line of subduction of lithospheric plates being brought together in collision (Figure 9.12).

Earthquakes—An Environmental Hazard

All of you have read many news accounts about disastrous earthquakes and have seen pictures of their destructive effects. Californians know about severe earthquakes from first-hand experience, but many other areas in North America have experienced earthquakes, and of these a few have been severe. The *earthquake* is a motion of the ground surface, ranging from a faint tremor to a wild motion capable of shaking buildings apart and causing gaping fissures to open up in the ground.

From the energy-systems standpoint, the earthquake is a form of kinetic energy of wave motion transmitted through the surface layer of the earth in widening circles from a point of sudden energy release—the *focus*. Like ripples produced when a pebble is thrown into a quiet pond, the waves travel outward in all directions, gradually losing energy through frictional resistance encountered within the rock that is flexed by the passing waves.

Earthquakes are produced by sudden movements of faults. Figure 9.16 shows two common types of faults as they appear at the surface. The *normal fault* has dominantly vertical motion between adjacent crustal blocks; the *transcurrent fault* has dominantly horizontal motion. Both types generate earthquakes when slippage occurs (Figure 9.17).

We shall not go further into the details of mechanics of faults and how they produce earthquakes. It must be enough to say that rock on

Figure 9.17 Evidences of lateral earth movement accompanying the San Francisco Earthquake of 1906, Marin County, California. (Photographs by G. K. Gilbert, U.S. Geological Survey.) *Above:* A fence offset 8 ft (2.4 m) along the main fault near Woodville. *Below:* A road offset 20 ft (6 m) along the main fault near Point Reyes Station. The shear zone is 60 ft (18 m) wide.

both sides of the fault is slowly deformed over many years as horizontal forces are applied in the movement of lithospheric plates. Potential energy accumulates in the bent rock, just as it does in a bent crossbow. When a critical point is reached, the strain is relieved by slippage on the fault and a large quantity of kinetic energy is instantaneously released in the form of *seismic waves*. Figure 9.18 shows that slow deformation of the rock takes place over many decades. Its release then causes offsetting of features that formerly crossed the fault in straight lines, e.g., a roadway or fence. Faults of this type also show a slow, steady displacement known as *fault creep*, which tends to reduce the accumulation of stored strain.

Earthquake waves travel rapidly through the earth's interior and are received at distant points by delicate sensing instruments (seismographs).

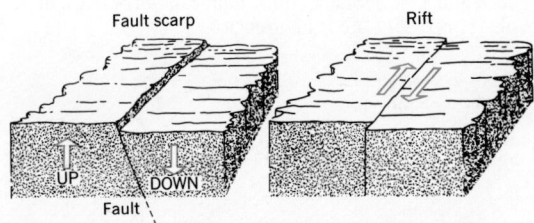

Figure 9.16 Normal fault (left) and transcurrent fault (right). (From A. N. Strahler, 1971, *The Earth Sciences*, 2nd ed., Harper & Row, New York.)

Earthquakes—An Environmental Hazard

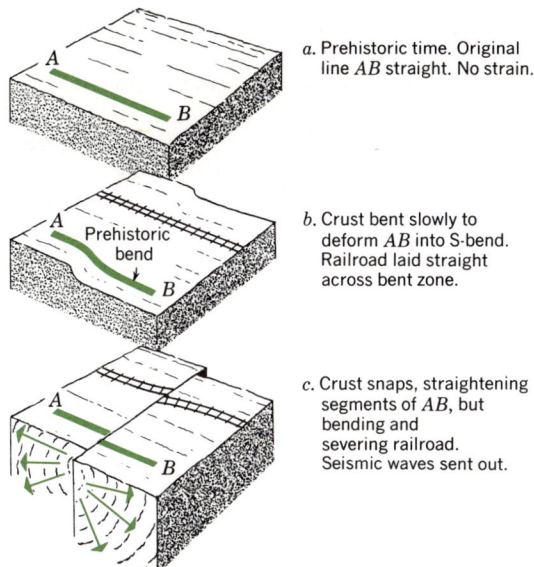

Figure 9.18 Sudden release of elastic strain accumulated by slow bending of rock along a fault. (From A. N. Strahler, 1971, *The Earth Sciences*, 2nd ed., Harper & Row, New York.)

TABLE 9.1 The Richter Scale

Magnitude	Description
0	Smallest quake detectable. Energy released: 3×10^{12} ergs.
2.5–3	If nearby, the quake can be felt. Each year there are about 100,000 quakes of this magnitude.
4.5	Quakes of this magnitude can cause local damage.
5	About equal in quantity of energy to the first atomic bomb, Alamogordo, New Mexico, 1945. (Hiroshima bomb: magnitude 5.7; energy released, 8×10^{20} ergs)
6	Destructive within a limited area. About 100 per year of this magnitude in shallow-focus quakes.
7	Above this magnitude it is a major earthquake and can be recorded over entire earth; 14 per year this great and greater.
7.8	1906 San Francisco Earthquake.
8.4	Close to maximum observed. Examples are Honshu, 1933; Assam, 1950; Alaska, 1964. Energy released: 3×10^{24} ergs.
8.6	Maximum observed between 1900 and 1950. Energy released is 3 million times that of first atomic bomb (see magnitude 5).

Our interest, however, lies in the energy of *surface waves* that travel much as do the bow-waves of a ship moving through calm water. From the environmental standpoint, it is the energy level of the earthquake that counts. We shall be interested in two measures of earthquake energy: first is the quantity of energy released at the focus; second is the intensity of the effect as measured at a given point on the earth's surface at some finite distance from the source. Whereas the earthquake focus lies at varying distances beneath the surface—down to depths as great as 400 mi (600 km)—the source point for measurement of intensity effects is the *epicenter* which lies on the surface directly above the focus.

The *Richter scale* of earthquake magnitudes was devised in 1935 by the distinguished seismologist, Charles F. Richter, to indicate the quantity of energy released by a single earthquake. Scale numbers range from 0 to 9, but since it is a logarithmic scale there is no upper limit except for Nature's own limit of energy release. A value of 8.6 was the largest observed between 1900 and 1950. The data in Table 9.1 suggest the quantities of energy associated with the various numbers of the scale.

It is estimated that the total annual energy release by all earthquakes is about 10^{25} ergs, and most of this is from a small number of earthquakes of magnitude over 7.

To measure the ground-shaking effects of an earthquake an *intensity scale* is used. In the United States the *modified Mercalli scale* is used, as prepared by Richter in 1956. It recognizes 12 levels of intensity, each designated by a Roman numeral. For each level there are certain criteria that can be readily observed by persons experiencing the earthquake. At intensity level IV, for example, hanging objects swing, a vibration like that of a passing truck is felt, standing automobiles rock, and windows and dishes rattle. Degrees of damage to various classes of masonry structures serve as criteria for identifying higher intensity levels (Figure 9.19). For example, at intensity XII, the highest number, damage to manmade structures is nearly total and large masses of rock are displaced.

One of the great earthquakes of recent times was the Good Friday Earthquake of March 27, 1964, with an epicenter located about 75 mi (120 km) from Anchorage, Alaska. Its magnitude was 8.4 to 8.6 on the Richter scale, approaching the maximum known. In Anchorage itself, the

Tectonic Processes and Continental Evolution

Figure 9.19 Destruction by earthquake and fire, San Francisco, California, 1906. View is southwest from the corner of Geary and Mason streets. (Photograph by W. C. Mendenhall, U.S. Geological Survey.)

intensity was judged as VII to VIII on the modified Mercalli scale. Here we come upon an observation of particular interest in connection with earthquakes as environmental hazards—that of *secondary effects*. At Anchorage most of the damage was from secondary effects, since buildings of wooden frame construction often experience little damage where on solid rock areas. Damage was largely from earth movements in weak clays underlying the city (Figure 9.20). These clays developed liquid properties upon being shaken (they are called *quick clays*) and allowed great segments of ground to subside and to pull apart in a succession of steps, tilting and rending houses. Other secondary effects were from the rise of water level and landward movement of large waves, destroying shipping and low-lying structures. The focus of this earthquake lay at a depth of about 20 mi (30 km) below the epicenter and represented slippage along the descending fault plane where the oceanic crust and mantle of the Pacific Plate is slipping beneath the continental plate.

Seismic Sea Waves

Another important secondary effect of a major earthquake, nicely illustrated by the Good Friday Earthquake, is the *seismic sea wave*, or *tsunami*, as it is known to the Japanese. A train of these waves is often generated in the ocean at a point near the earthquake epicenter by a sudden movement of the sea floor. The waves travel over the ocean in ever-widening circles, as shown by the map of wave fronts (Figure 9.21). Notice, for example that the first waves reached the California coast about 9 hours after the time of origin and were impinging upon the coast of Chile some 21 hours later. Speed of travel is on the order of 300 mi (500 km) per hour.

Seismic sea waves are very long—60 to 120 mi (100 to 200 km)—and of very low height—1 to 2 ft (0.3 to 0.6 m). Consequently, they are not perceptible in deep water. However, when a wave arrives at a distant coastline, the effect is to cause a slow rise of water level over a period of 10 to 15 minutes. This rise is reinforced by a favorable configuration of the bottom offshore. Wind-driven waves, superimposed upon the heightened water level allow the surf to attack places inland

Figure 9.20 During the Anchorage, Alaska, earthquake of 1964, unconsolidated sediments underwent slumping and flowage, causing severe property destruction. (U.S. Army Corps of Engineers photograph.)

Figure 9.21 Successive hourly positions of a tsunami wave front, originating in the Gulf of Alaska as a result of the Anchorage, Alaska, earthquake of 1964. Figures give Greenwich Mean Time. (After B. W. Wilson and A. Torum, 1968, U.S. Army Corps of Engineers, *Tech. Memorandum No. 25*, Coastal Engineering Research Center, Washington, D.C., p. 38, Figure 27.)

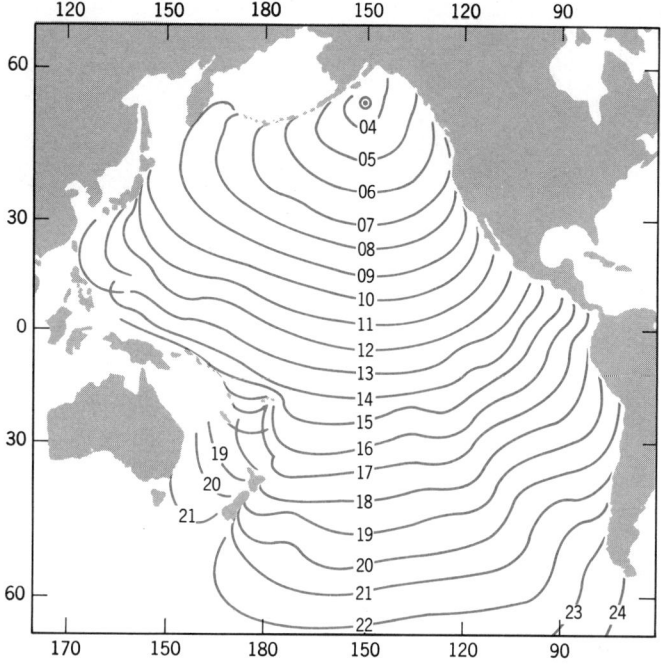

that are normally above the reach of waves. For example, the particularly destructive seismic sea wave of 1933 in the Pacific Ocean caused waves to attack ground as high as 30 ft (9 m) above normal tide level, causing widespread destruction and many deaths by drowning in low-lying coastal areas. It is thought that coastal flooding which occurred in Japan in 1703, with an estimated life loss of 100,000 persons, may have been caused by seismic sea waves.

Because of the far-ranging devastation that seismic sea waves can wreak, a *tsunami warning system* has been established for the Pacific Ocean. When a severe earthquake is recorded by seismographs and its center and time of occurrence have been established, warnings are sent to distant coastal points, using calculated arrival times based upon distance and water depths. Here is an example of Man's attempt to minimize losses of life and property from an environmental hazard (but not to ameliorate its intensity) by application of scientific knowledge and a suitable organizational structure for dissemination of information based on international cooperation.

Earthquake Prediction and Control

We now come to the interesting question of Man's ability (a) to predict the time and place of occurrence of earthquakes, and (b) to reduce the severity of earthquakes. The latter question is also tied in with a third environmental question: Does Man's activity set off earthquakes? Earthquake prediction is still in its very early stages of development, but research shows promise of leading to useful methods. One line of approach uses sensitive instruments located along the fault lines where past earth movements have occurred and new earthquakes are considered likely. For example, such instruments can measure the degree of strain building up in the rock. Through experience with the relationship of past earthquakes to the degree of strain and its rate of buildup, it may be possible to anticipate when an earthquake is imminent. Changes in tilt of the ground are also measured, and these may prove to be indicators of an impending earthquake. A different line of research being explored is that of changes in the earth's magnetic field as indicators of coming earthquakes.

Another type of approach to prediction is based upon the time elapsed since an earthquake-generating slip has occurred along a known active fault. The case of the *San Andreas Fault* of California is interesting in this respect. This fault is one of the transcurrent type and runs for some 600 mi (950 km) from the Salton Basin in southern California in a northwesterly direction to the San Francisco Bay region, beyond which it passes out to sea (Figure 9.22). In terms of plate tectonics, this fault is thought to represent a transverse fracture zone in which the lithospheric plate on the western side is moving to the northwest (oceanward), with respect to the eastern plate (Figure 9.23). Average movement of one

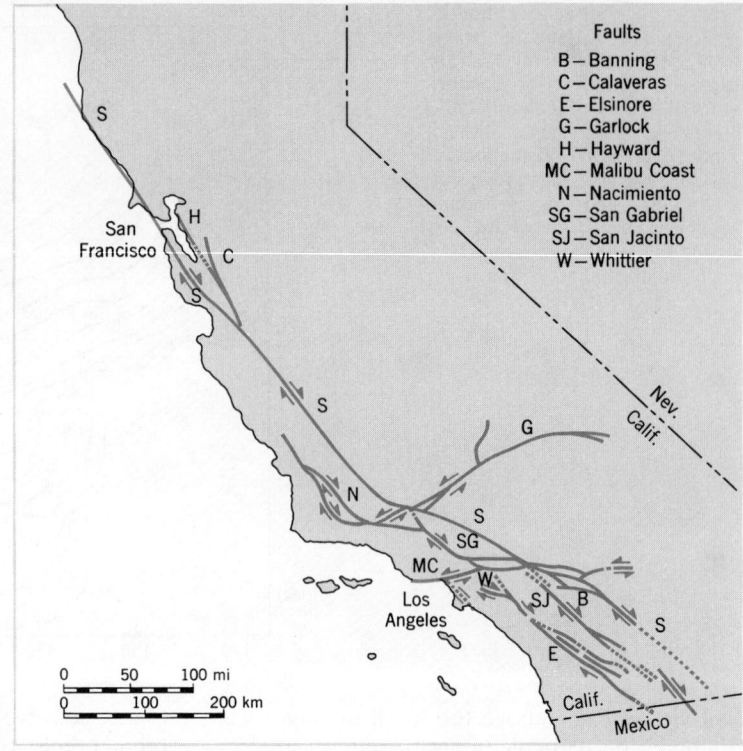

Figure 9.22 Sketch map of the San Andreas fault and associated major faults of California. (Based on data of R. H. Jahns and C. G. Higgins.)

plate with respect to the other is estimated to be on the order of 2 in. (5 cm) per year. Obviously, where a section of the San Andreas Fault has shown no movement for a century the potentiality for earthquake is great (2 in. × 100 years = 200 in.), since movement must occur from time to time. These sections of no known movement are described as *locked sections,* and show no history of seismic activity. They are judged to be the most likely places for major earthquakes. Sections that are seismically active are considered less of a hazard, since strain has been recently relieved, or is being relieved in small fault movements, or as slow fault creep. Figure 9.24 shows the active and locked sections of the San Andreas Fault. Of course, this type of prediction cannot set the time of an earthquake, even within very wide limits.

In a 1971 study of earthquakes of the Aleutian Islands and the southern Alaska coast, Lynn R. Sykes, a Columbia University seismologist, found three important gaps in an otherwise dense chain of earthquake epicenters. In these *seismicity gaps* no major earthquake has been generated in the half-century since 1920. Two of the gaps are off the Alaskan coast; one is in the far western end of the Aleutian chain. One of the two Alaskan gaps is close to epicenters of major earthquakes of 1899 and 1900. All three gaps are considered

Figure 9.23 In this vertical air view, the nearly straight trace of the San Andreas fault contrasts sharply with the sinuous lines of stream channels. San Bernardino County, California. (Photograph by Litton Industries, Aero Service Division.)

Earthquake Prediction and Control

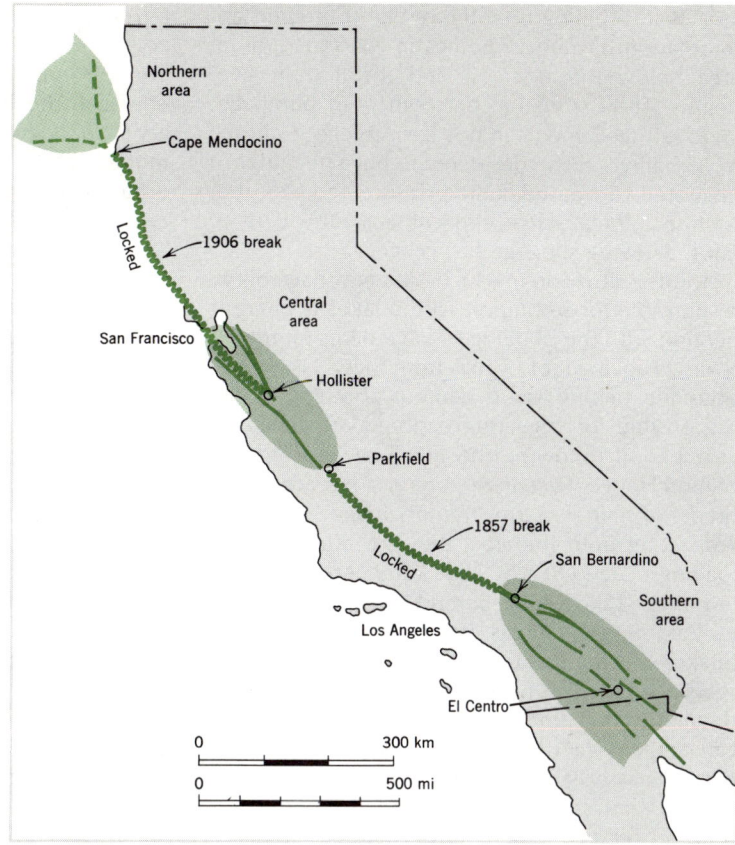

Figure 9.24 Active areas (color) and locked sections of the San Andreas fault. (After C. R. Allen, 1968, in Proc., *Conference on the Geologic Problems of the San Andreas Fault System*, Stanford Univ. Publ., Univ. Ser. Geol. Sci., No. 11, p. 70.)

likely sites for future major earthquakes, with accompanying seismic sea waves. It is recommended that the gaps be intensively monitored by instrumentation capable of detecting events preceding a major shock.

Turning next to earthquake control, we must first see if any activities of Man have set off earthquakes. A case that is now almost a classic is that of the Denver, Colorado, region. Here, near the Rocky Mountain Arsenal, hundreds of earthquakes have been recorded since 1962; they seem to be correlated with pumping of fluids under pressure into a disposal well penetrating to a depth of 12,000 ft (3600 m). As an explanation, it has been proposed that the increased fluid pressure within the rock caused the release of strain already present within the rock; i.e., it had a triggering effect.

Naturally, this hypothesis leads us to wonder if it might not be possible to use engineering methods, such as fluid injections, to induce many small fault movements, and thus to prevent strain buildup to dangerous levels. This possibility has attracted the attention of scientists and they are working on the theoretical concepts. One might also be led to reason that if fluids (natural ground water or petroleum in rock pores) were to be pumped out of the rock near a major fault zone, the tendency for the fault to become locked would be increased, and seismic inactivity be induced. To investigate these possibilities, scientists of the United States Geological Survey undertook a study beginning in 1969 in which the Rangely Oil Field of western Colorado was the guinea pig. They found that the injection of water into deep oil wells (to induce more oil flow) had raised the fluid pressure by as much as 60% above normal. During this period earthquakes were being generated at the rate of 15 to 20 per week from a fault system passing through the oil field. The fluid pressure was then lowered by pumping water out of the same wells for a 6-month period. A dramatic drop in earthquake frequency resulted generally; the number fell to none at all near the wells. Pumping of fluid back into the ground has been resumed to find out if earthquakes will then increase.

It is now suspected that the pumping of fluids into the Inglewood Oil Field to raise the hydrostatic pressure and increase oil recovery was responsible for setting off the earthquake of 1963 which fractured a wall of the Baldwin Hills Reservoir. Water spilling from the reservoir brought an inundation of mud to homes in a 1-square-mile area below the reservoir and resulted in five deaths. The correlation between fluid pumpage and fault movements in this Los Angeles, California, locality is now considered to have been demonstrated. Increased fluid pressure reduces the frictional force across the contact surface of a fault, allowing slippage to occur.

Another situation in which Man may have been responsible for setting off earthquakes is in connection with the building of large dams on major rivers. The load of water from new lakes impounded behind these dams is thought to be responsible for triggering earthquakes. In a 10-year period following the filling of Lake Mead, behind Hoover Dam in Arizona and Nevada, hundreds of minor earth tremors were observed emanating from the area; they are attributed to loading of the crust by lake water. Another case in point is Lake Kariba, behind the Zambezi River in Zambia, which has been generating earthquakes of even greater magnitude.

Several scientists have been concerned with the possibility that underground nuclear explosions can set off significant earthquakes, and that a hazard may exist in this testing activity. Research thus far has shown that an underground blast does set off a large number of small earthquakes close to the site of the blast. Seismic energy of the blast triggers the release of strain along faults in the vicinity, but the radius of the known effects is on the order of 6 to 12 mi (10 to 20 km). These observations have led to the suggestion that underground nuclear blasts can be placed where they will induce strain release and thus prevent buildup of strain to dangerous levels.

Earthquakes and Urban Planning

The toll in human lives and the severe structural effects of the San Fernando earthquake of February 9, 1971, shocked the entire Los Angeles community into renewed awareness of the need for urban planning to minimize or forestall the damaging effects of a major earthquake. Although the earthquake was not in the really severe category by the Richter scale (it measured 6.6, which is moderate in severity), local areas experienced a ground motion of acceleration as high as, or higher than any previously measured in an earthquake. Fortunately, the ground shaking was of brief duration; had it persisted for a longer time the structural damage would have been much more severe than it was. Particularly disconcerting was the collapse of the Olive View Hospital in Sylmar, a new structure supposedly conforming with earthquake-resistant standards. The Veterans Hospital in Sylmar also suffered severe damage and the collapse of several buildings (Figure 9.25). A crack produced in the Van Norman Dam caused authorities to drain that reservoir to prevent dam collapse and disastrous flooding of a densely built-up area. The Sylmar Converter Station, one of the key elements in the electrical power transmission system of the Los Angeles area, was severely damaged. Collapse of a freeway overpass blocked the highway beneath, and freeway pavements were cracked and dislocated (Figure 9.26). Fortunately, the time of the quake was 6 A.M., when most persons were at home and few were traveling the major arteries.

Yet the fault movement that set off the San Fernando earthquake was not on the great San Andreas Fault, but rather from an epicenter some

Figure 9.25 Severe structural damage and collapse of buildings of the Veterans Administration Hospital, Sylmar, Los Angeles County, California, caused by the San Fernando earthquake of February, 1971. (Wide World Photos.)

Figure 9.26 Collapsed pavement and overpass on the Golden State Freeway at the northern end of the San Fernando Valley, California, resulting from the earthquake of February, 1971. (Wide World Photos.)

15 mi (25 km) from that fault, along a system of relatively minor faults. Recall that in this section the San Andreas Fault is locked and is potentially capable of producing an earthquake of far greater intensity than the 1971 San Fernando earthquake, and although the year of this event is not predictable within decades, the progress of urbanization will have greatly expanded the structures and population subject to devastation.

Therefore, soon after the earthquake had occurred, the National Academy of Sciences and the National Academy of Engineering set up a joint panel of experts to study the earthquake effects and to draw up recommendations. The panel concluded: "It is clear that existing building-codes do not provide adequate damage control features. Such codes should be revised."* The panel further recommended that public buildings, such as hospitals, schools, and buildings housing police and fire departments and other emergency services should be so constructed as to withstand the most severe shaking to be anticipated. Fortunately, most school buildings constructed following the Long Beach earthquake of the 1930s showed no structural damage, but many of the older school buildings were rendered unfit for use. Damage from the San Fernando earthquake of 1971 has been estimated at $500 million, but experts think that an earthquake as severe as that of 1906 at San Francisco would cause damage on the order of $20 billion, if it occurred now in a large metropolitan area. Perhaps then, one beneficial effect of the San Fernando earthquake will be that adequate funds will be appropriated for research covering many important aspects of earthquakes and their effects upon the urban environment.

*See *Science*, vol. 172, p. 141.

Landforms of Tectonic Activity

Tectonic activity produces a varied group of landscape features, or primary landforms. Although quickly subdued by weathering and erosion, these *tectonic landforms* are often clearly defined, where faulting has been recently active. The *fault scarp*, a sharply-defined clifflike feature, is illustrated in Figure 9.16. An example of a fresh, recent fault scarp is seen in Figure 9.27. Frequently, the scarps of two normal faults are paired to produce a downfaulted trench—a *graben*—or an uplifted block—a *horst*—both illustrated in Figure 9.28. After fault activity ceases, the fault scarp is eroded

Figure 9.27 Formation of this fresh fault scarp in alluvial materials accompanied the Hebgen Lake earthquake of August 17, 1959, in Gallatin County, Montana. Displacement was about 19 ft (6 m) at the maximum point. Vehicle stands on the upthrown side of the fault. (Photograph by J. G. Stacy, U.S. Geological Survey.)

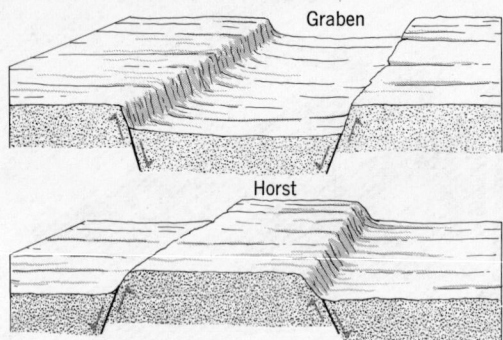

Figure 9.28 Graben and horst.

and subdued (Figure 9.29A). Because the fault plane extends to great depth, its control on erosion rates may persist through millions of years in the form of a *fault-line scarp* (Figure 9.29B). Such ancient scarps are numerous in igneous and metamorphic rocks of the continental interiors (Figure 9.30).

Large-scale faulting produces crustal blocks of mountainous proportions; these are *block mountains* (Figure 9.31). A leading example of a region of block mountains is the Basin-and-Range region of the western United States, largely within the states of Oregon, Nevada, Utah, California, Arizona, and New Mexico. This broad belt of block mountains is interpreted as the result of tectonic plate separation along an extension of the Mid-Oceanic Ridge penetrating the North American continent through the Gulf of California (see Figure 9.12). A related occurrence is the rift valley zone of east Africa, where rifting apart of the crust has produced a number of large grabens.

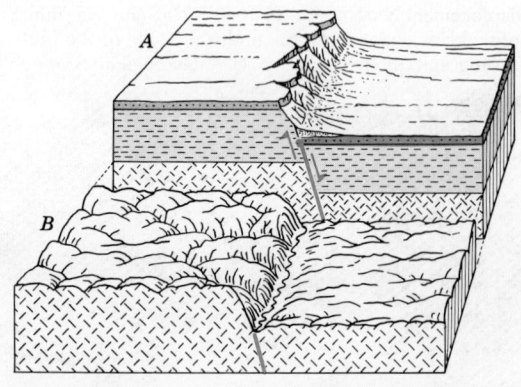

Figure 9.29 A. A fault scarp produced directly by movement on a normal fault. B. A fault-line scarp developed by long-continued erosion on the line of a very old, inactive fault. (From A. N. Strahler, 1971, *The Earth Sciences*, 2nd ed., Harper & Row, New York.)

Figure 9.30 Fault-line scarp, MacDonald Lake, near Great Slave Lake, Northwest Territories, Canada. (Canadian Armed Forces Photograph.)

The environmental significance of a basin-and-range region lies in the mountainous terrain of the uplifted blocks, and in the valley forms of the downfaulted blocks. The mountainous blocks have steep slopes and rocky surfaces; their use lies mainly in forests, grazing lands, and in mineral resources of the exposed rocks. The intermontane valleys are partially filled with sediment derived from the adjacent mountains. Here are found belts of soils favorable to agriculture. If the region is arid, as in much of the Basin-and-Range region of the western United States, irrigation is often available from streams issuing from the adjacent mountain blocks. Because downfaulting is varied in amount from place to place, the intermontane basins may have no drainage outlet and will accumulate shallow lakes. In a desert climate, these basins contain flat central areas known as *playas*, which are the sites of accumulation of evaporites

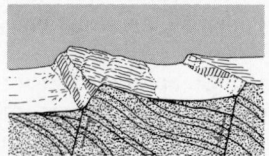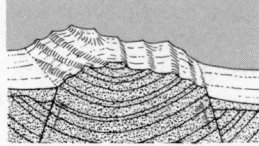

Figure 9.31 Fault block mountains may be of tilted type (*left*) or lifted type (*right*). (After W. M. Davis.)

Landforms of Tectonic Activity

(Chapter 8). Shallow saline lakes may be present, as exemplified by the Great Salt Lake in Utah.

Folding of sedimentary strata during orogeny produces a distinctive group of structures upon which unique landforms are later produced by erosion processes. Erosion of simple, open folds produces a *ridge-and-valley* landscape (Figure 9.32), as weaker formations of shale and limestone are eroded away, leaving hard strata of sandstone and quartzite to stand in bold relief as long, narrow ridges. The term *anticline* is applied to the archlike upwardly convex part of the fold; *syncline* to the adjacent, troughlike downfold (Figure 9.32A). Commonly, the fold axes are inclined; the folds are then described as *plunging folds* (Figure 9.33). Erosion of plunging folds yields ridges of zig-zag pattern. A leading example of ridge-and-valley landforms is found in the Newer Appalachians of the eastern United States within a narrow belt running through the states of Pennsylvania, Maryland, West Virginia, Virginia, Tennessee, and Georgia, and into Alabama. An excellent example of ridges is found in the Harrisburg region of Pennsylvania (Figure 9.34). Of particular interest as an economic resource of the Appalachian folds of Pennsylvania is the region of anthracite coal basins (Figure 9.35). The coal seams lie in deeply down-folded synclines, but the coal-bearing strata have been entirely removed from the remainder of the region by extensive erosion.

The environmental significance of ridge-and-valley belts lies in the arrangement of various sedimentary rock types in long, narrow strips. In humid climates of middle latitudes, such as the

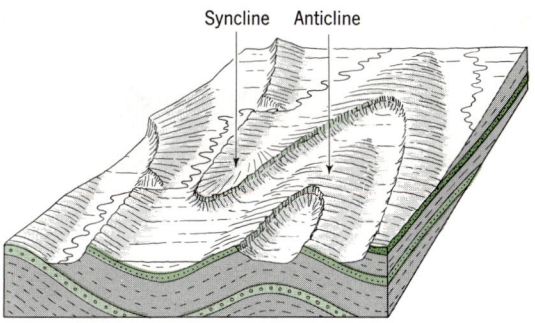

Figure 9.33 Folds with crests that plunge downward give rise to zigzag ridges following erosion.

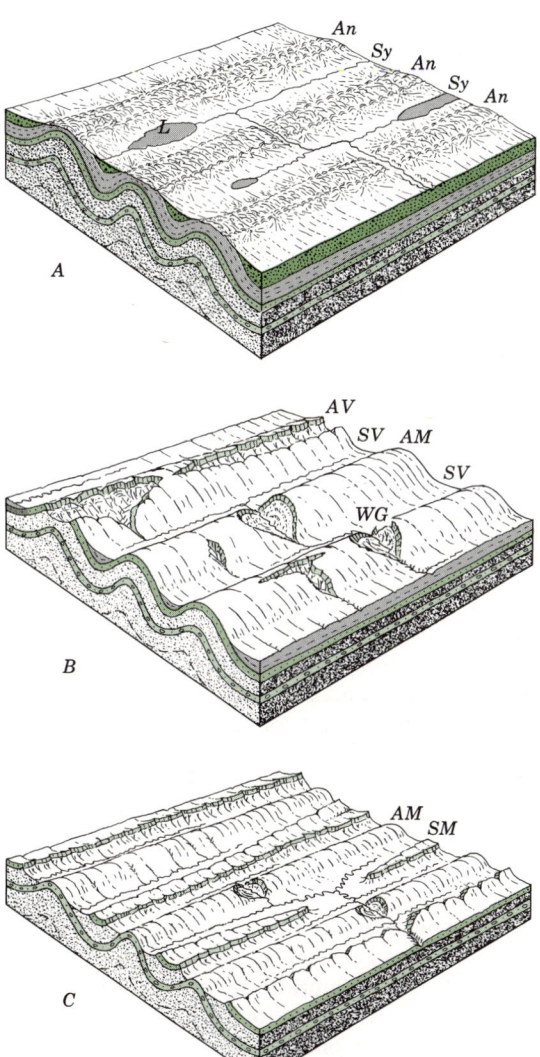

Figure 9.32 Stages in the erosional development of folded strata. A. While folding is still in progress, erosion cuts down the anticlines; alluvium fills the synclines, keeping relief low. An = anticline; Sy = syncline; L = lake. B. Long after folding has ceased, erosion exposes a highly resistant layer of sandstone or quartzite. AV = anticlinal; SV = synclinal valley; WG = water-gap. C. Continued erosion partly removes the resistant formation but reveals another below it. AM = anticlinal mountain; SM = synclinal mountain.

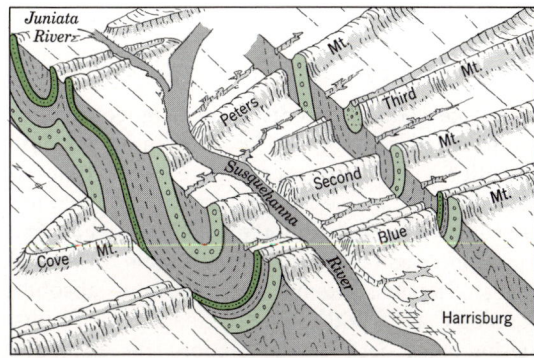

Figure 9.34 A great synclinal fold, involving three resistant quartzite-conglomerate formations and thick intervening shales, has been eroded to form bold ridges through which the Susquehanna River has cut a series of water-gaps. (After A. K. Lobeck.)

227

Tectonic Processes and Continental Evolution

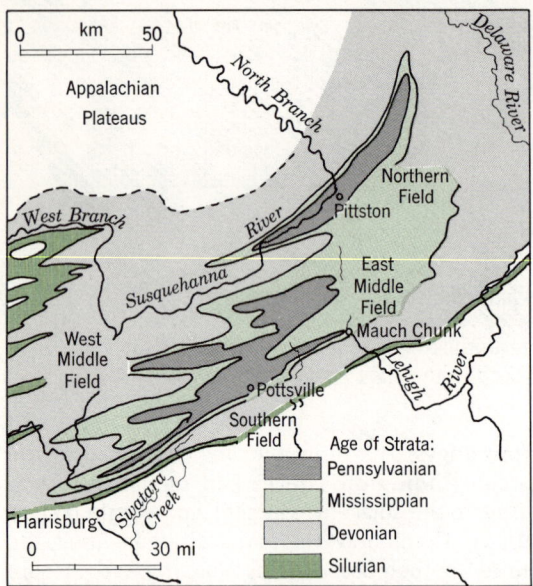

Figure 9.35 Anthracite coal basins of central Pennsylvania correspond with areas of Pennsylvania strata, downfolded into long synclinal troughs.

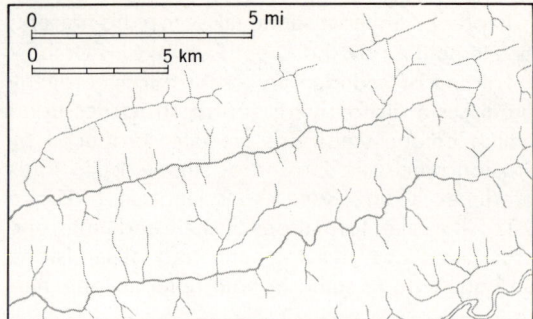

Figure 9.36 A trellis drainage pattern developed on deeply eroded folded strata of the Appalachians.

Appalachian and Ouachita mountains, ridges of sandstone and quartzite, with their steep slopes and rocky soils, are unfit for cultivation but may be valuable for their forests and pastures. In these same regions limestone belts form lowlands with rich soils; shales form hill belts suitable for diversified use as cultivated fields, pastures, and woodlands. The grain of the rock not only controls soils and relief, but imposes its effect upon patterns of streams, which take on a *trellis pattern* (Figure 9.36). Major streams cut across the ridges in narrow *watergaps* (Figures 9.32B and 9.34). Highways and railroads may use these same watergaps, or may require tunnels to pass from one valley to another.

Tectonic Events and Geologic Time

To place geologic events such as igneous intrusion, sedimentation, and orogeny in an absolute position in time, and to correlate events that occurred at widely separated localities, it becomes necessary to establish an *absolute geologic time scale* in years (as distinct from a scale of relative ages). The first attempts to determine absolute geologic age yielded grossly inaccurate results. Using as a basis of calculation the estimated quantity of sodium in the oceans, and dividing that figure by the estimated annual increment of sodium brought to the sea by streams, one late-nineteenth-century analysis gave a figure of about 100 million years (m.y.) for the earth's age since sedimentary processes began. The uncertainty of this figure is evident when we consider (1) that the rate of introduction of sodium into the oceans may have varied greatly in the geologic past, and (2) that great quantities of sodium have been placed in rock storage in sedimentary strata. Another method, used by a long list of geologists between 1860 and 1910, required that the thickness of sedimentary strata of all ages be totalled and divided by an arbitrary average rate of annual accumulation. Naturally, allowance had to be made for periods of nondeposition, and that could be only a blind guess. It is not surprising, then that estimates of the length of recorded geologic time by this method ranged from 20 m.y. to 1.5 b.y.

In the late nineteenth century the English physicist Lord Kelvin made a calculation of the earth's age as based upon rate of contraction and cooling of a molten earth. His figure was a mere 20 to 40 m.y., and this was a source of great disappointment to Charles Darwin, who felt that evolution of the more advanced life forms alone would require some 300 m.y. Since the laws of physics were then unquestioned in validity, there seemed no alternative but to accept Kelvin's estimate. Suddenly, however, the discovery of natural radioactivity by Henri Becquerel in 1896 opened to view a new world of physics, and with it knowledge of an energy source unlike any previously known. In 1898 Marie and Pierre Curie isolated the radioactive element *radium*, ushering in a revolution in theories of the internal sources of earth energy and a key to the age of the earth and its geologic events. John Joly, in 1906 showed that radiogenic heat could power the earth's processes of vulcanism, intrusion, and mountain building, while a year later B. B. Boltwood, a chemist, made the first reliable determinations of rock ages based upon principles of radioactivity.

Spontaneous Radioactive Decay

To understand both the production of radiogenic heat and the determination of absolute ages of rock it will help to have a brief review of some basic principles of nuclear physics and radioactivity.

Spontaneous Radioactive Decay

The key to understanding of radioactivity lies in the internal structure of the atom. The *nucleus*, or dense core of the atom, consists of two kinds of particles: *neutrons* and *protons*. The number of neutrons within a given element is only approximately constant whereas the number of protons is fixed. An example is a radioactive form of the element *uranium* (U), in which the atomic nucleus has 146 neutrons and 92 protons. The total of neutrons and protons is thus 238, a quantity known as the *mass number*; it is designated by a superscript, thus: U^{238} (or simply U-238). The number of protons in the nucleus (92 in uranium), is known as the *atomic number*. While the atomic number is fixed for a given named element, there is a possibility of minor variation in the number of neutrons present. Although in U-238 there are 146 neutrons, there exists another form of uranium with 143 neutrons. The latter form then has a mass number of 235 and is designated as U-235. The different varieties of the same element are known collectively as *isotopes*. Certain isotopes are unstable, with the result that a small part of the nucleus will fly off, resulting in transformation into another element with a different mass number or atomic number, or both. In this process, mass is converted into energy and constitutes the process of *radioactive decay*; it is a spontaneous process which cannot be influenced by external factors.

Radioactive decay gives off both particles and electromagnetic radiation. When the atomic nucleus breaks down an *alpha particle* is emitted; it consists of two neutrons and two protons. Consequently there is a decrease of 4 in the mass number and 2 in the atomic number. A second form of nuclear emission is the *gamma ray*, an energy form similar to that of electromagnetic radiation (Chapter 2). A third form of nuclear emission is the *beta particle*, a high-velocity electron. (A neutrino is also emitted.) Beta emission does not change the mass number, but results in a new element of atomic number 1 greater than before the emission (because a proton has been formed from a neutron). These changes are shown by labeled arrows in Figure 9.37. The three forms of emission are absorbed in the surrounding matter after a short distance of travel and converted to heat. It is possible to calculate exactly the quantity of heat produced by spontaneous decay of a given isotope; this is the basis for calculation of the quantities of radiogenic heat produced by various rock types in the earth's crust and mantle.

Spontaneous radioactive disintegration begins with a parent isotope, such as U-238, and leads

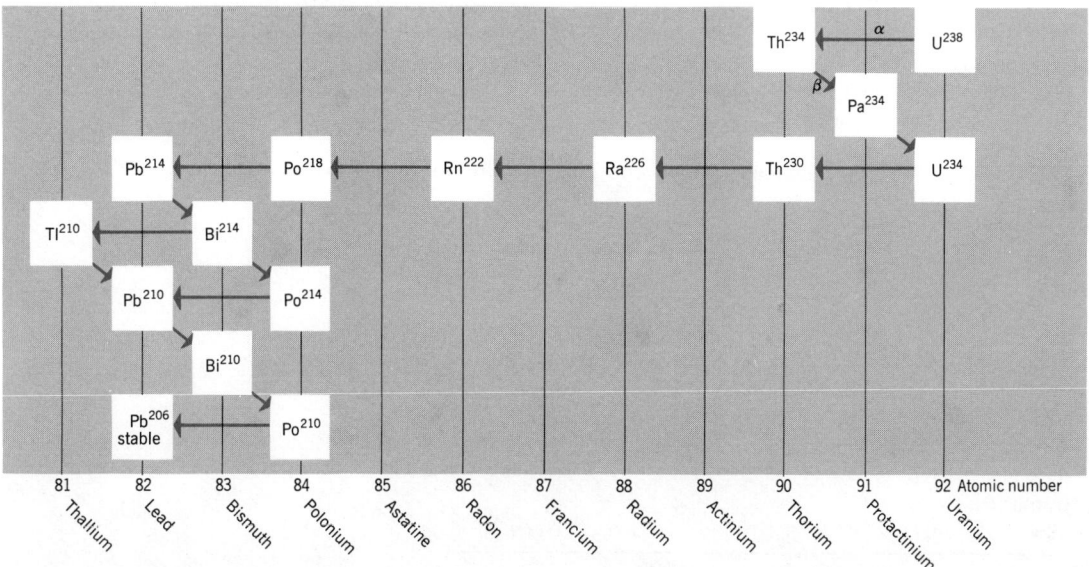

Figure 9.37 Radioactive decay series beginning with U-238 and ending in Pb-206, (After P. M. Hurley, 1959, *How Old Is the Earth?*, Doubleday, Garden City, N.Y., p. 62, Figure 9.)

to the formation of another unstable isotope, called a *daughter product*. As shown in Figure 9.37, the daughter product of U-238 is thorium-234 (Th-234) produced by emission of an alpha particle. One daughter product leads to another in a series that may produce a number of elements in succession, as Figure 9.37 shows. Ultimately, however, the chain ends in a stable isotope, which in this case is *lead-206* (Pb-206). Once the chain of disintegration is complete, the rate of production of heat energy per unit mass of parent isotope becomes constant. In the example, U-238 produces 0.71 calories of heat per year for each gram remaining.

The emission of gamma rays by radioactivity is an environmental phenomenon of great importance to Man and other life forms on earth. This form of radiation, which resembles powerful X rays, is just one form of *ionizing radiation* to which organisms are constantly subjected, and which can produce genetic changes. There are in nature over 320 isotopes of elements, but of these only about 60 are radioactive, i.e., are *radioisotopes*. However, nuclear fission developed by Man has produced over 200 additional radioisotopes, which are in a sense artificial, as they are not found in nature. Recall from Chapter 2 that another source of ionizing radiation is external: it consists of cosmic particles and their byproducts, and energetic particles from the sun entrapped in the magnetosphere. Then there are also man-made sources of ionizing radiation in X rays used in medical diagnosis, and X rays emitted by television picture tubes.

In rocks of the earth's crust, the principal heat-producing isotopes are U-238, U-235, *thorium-232* (Th-232), and *potassium-40* (K-40). These are listed in Table 9.2, together with their stable daughter products. In addition, *rubidium-87* is listed in the table, but is not important as a producer of radiogenic heat. Table 9.3 lists the concentrations of radioisotopes of uranium, thorium, and potassium in granitic, basaltic, and ultramafic rocks, together with the heat production of each element in calories per gram per year. Notice that twice as much heat is generated per unit of mass of granitic rock than for the same mass of basaltic rock, and that the quantity generated by ultramafic rock is extremely small by comparison. These data make quantitative our earlier statement in Chapter 7 that most of the radioisotopes are now concentrated in the upper, granitic crust of the continents. Granitic rocks contribute a major share of the natural ionizing radiation (background radiation) to which all organisms are exposed at the ground surface. This gamma radiation is roughly of equal intensity to that quantity of the ionizing radiation produced by cosmic particles which reaches down to sea level. Although the combined radiation dosage from these two sources is very small and causes no immediate damage to tissues, it is believed to be a major cause of genetic change in organisms.

TABLE 9.2 Important Natural Radioisotopes

Parent Isotope	Stable Daughter Products	Half-Life (b.y.)
Uranium-238	Lead-206, plus helium	4.5
Uranium-235	Lead-207, plus helium	0.71
Thorium-232	Lead-208, plus helium	14
Rubidium-87	Strontium-87	51
Potassium-40	Argon-40, calcium-40	1.3

SOURCE: B. Mason, 1966, *Principles of Geochemistry*, 3rd ed., John Wiley & Sons, New York, p. 9.

TABLE 9.3 Radiosotope Concentrations and Heat Production Rates

	Concentrations			Heat Production (cal/gm/yr)			Total
	U (ppm)	Th (ppm)	K (%)	U	Th	K	
Granitic rock	4	14	3.5	3	3	1	7
Basaltic rock	0.6	2	1.0	1.5	1.5	0.5	3.5
Ultramafic rock	0.015	—	0.011	0.01	0.01	0.001	0.02

SOURCES: B. Mason, 1966, *Principles of Geochemistry*, 3rd ed., John Wiley & Sons, New York, Table 11.1; and P. J. Hurley, 1959, *How Old Is the Earth?*, Doubleday & Co., New York, p. 64.

Radiometric Age Determination

Radioactive decay is used in determining ages of rocks (and of organic substances as well) because the rate of decay of the parent isotope and the rate of accumulation of the stable daughter product can be predicted with great accuracy. This principle is illustrated in Figure 9.38, which shows the time-rate of decay of potassium-40 (K-40) and of the accumulation of its two stable daughter products, calcium-40 (Ca-40) and argon-40 (A-40). The vertical scale on the graph is relative and starts at a given point in time with some given quantity of K-40, represented by unity (1.0). At this initial time there are no daughter products. Once equilibrium has been reached in the decay process, the ratio of decrease in number of parent isotope atoms is constant per unit of time. In this particular case, after 1.31 b.y. have elapsed, the number of K-40 atoms will have been reduced to exactly one-half the starting number. The period 1.31 b.y. is referred to as the *half-life* of K-40. After a second period of 1.31 b.y. has elapsed, the quantity will again have been halved, bringing the remainder to one-quarter (0.25) of the initial quantity. In the meantime, the combined number of atoms of Ca-40 and A-40 will have increased in similar ratio, so that at the end of the first 1.31 b.y. they will equal the number of the remaining K-40 atoms. Note that the ratio between the two daughter products is always the same: 88% is Ca-40 and 12% is A-40. The time-rate of change illustrated by radioactive disintegration is known as a *negative exponential decay*. This type of energy system was briefly explained in our introduction to energy systems. As matter is transformed into energy, following the Einstein equation given in that introduction, the system depletes its own source and the rate of energy production becomes progressively smaller. However, there is no limit to the life of the system, since the quantity of matter remaining can only approach zero (but never reach zero) as time approaches infinity.

Now, let us put these principles to work in devising a method of measuring the absolute age of a rock. When an igneous rock crystallizes to the solid state from a magma, very small amounts of mineral compounds containing radioisotopes become enclosed within the crystal structure of common minerals, or in some cases form discrete crystals of those radioactive compounds. An example of such a mineral is *uraninite* (also called *pitchblende*), which consists mostly of a combination of oxides of uranium and lead. Naturally radioactive uranium occurs as one of three radioisotopes, U-234, U-235, and U-238. Of these, however, U-238 constitutes over 99% of the total abundance of the three combined.

The procedure in determining the age of the rock, i.e., its *radiometric age*, is to measure the ratio between the parent radioisotope and its stable daughter product. For U-238 the stable product is lead-206 (Pb-206). We must also know the half-life of U-238, which is 4.5 b.y. The ratio is then entered into a simple equation, yielding the absolute age in years.* Other series used in determining radiometric ages of rocks are U-235/Pb-207, the potassium-argon series (K-40/A-40), and the rubidium-strontium series (Rb-87/Sr-87). Table 9.2 lists the daughter products and half-lives of these series. Because both of the lead-uranium series are normally present in the same mineral sample, the measurement of one ratio can serve as a cross-check upon the other. Consequently uranium ages are considered accurate to within 2% or less.

The same principles of age determination can be applied to radioisotopes with very short half-lives, and thus determine ages of organic substances containing the isotopes, or to measure the elapsed time between introduction of a sample into a fluid system and its arrival at some distant point (e.g., use as a *tracer*).

Particularly important in environmental science is the use of the radioisotope of carbon, *carbon-14* (C-14, or simply *radiocarbon*), since carbon atoms are included in all hydrocarbon compounds synthesized by plants and carried through the food chain in animals. For example, carbon in calcium-carbonate shells and bones of animals represents the end product of the chain. Substances that can be dated by the radiocarbon method

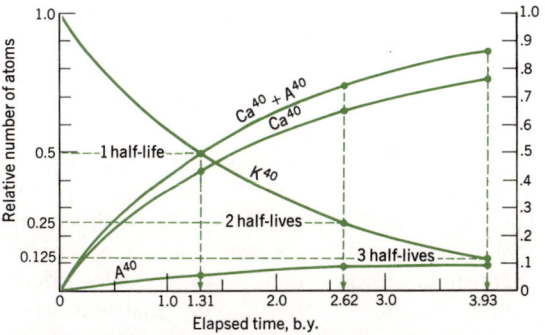

Figure 9.38 Decay and growth curves for K-40, Ca-40, and Ar-40. (After P. M. Hurley, 1959, *How Old Is the Earth?*, Doubleday, Garden City, N.Y., p. 101, Figure 17.)

*For the U-238/Pb-206 series the equation is $t = 6.5 \times 10^9$ $(1 + Pb^{206}/U^{238})$, where t is age in b.y., and the constant 6.5×10^9 is a number derived from the half-life of U-238.

include wood, charcoal, cloth, peat, bones, tusks, and shells. Radiocarbon dating is of great importance in archeology, and in establishing the chronology of events of the Pleistocene Epoch, or Ice Age, since it is useful for ages as far back as about −40,000 years. The method can also be used to date samples of sea water, since CO_2 is held in solution in the water.

Carbon-14 is an isotope of carbon produced by bombardment of atmospheric nitrogen by cosmic radiation. An atom of ordinary nitrogen-14 (N-14) when struck by a neutron, absorbs the neutron and emits a proton, becoming an atom of C-14, and this in turn quickly combines with oxygen to form carbon dioxide (CO_2). The carbon dioxide is diffused through the lower atmosphere and is consumed by plants in photosynthesis. But the C-14 is unstable and decays into N-14. The half-life is about 6000 years (5730 ± 40 yrs). The method of age determination depends upon a constancy of rate of production of C-14 in the upper atmosphere. Thus among the carbon atoms in a sample of organic matter the ratio of C-14 atoms to normal carbon atoms (C-12) will be always the same at the time the plant tissue is synthesized. Thereafter, the proportion of C-14 declines, following the exponential decay curve for its given half-life. The measurement of C-14 present in a sample can thus easily permit calculation of its age, provided, of course, that the original constituents of the sample have remained intact and that there has not been contamination by addition of carbon of more recent age. It is generally recommended that the measured age of the sample be given a range of probable error plus or minus 5%.

Now, it is an interesting fact that radiocarbon ages show some discrepancies with ages determined by counting the annual growth-rings of tree trunks. Tree-ring age determination, or *dendrochronology*, has been carried back to about −8000 years, so that ages based upon the two methods can be compared that far back in time for the same samples. Bristlecone pine trees of California, celebrated as the oldest living organisms on earth, have allowed ages based on tree-ring counts (calendar years) to be compared with C-14 age determinations of the wood itself. When this is done a rather disconcerting discrepancy begins to appear at about −2000 years and to increase on the whole rather steadily to at least −6000 years. In this age range C-14 age runs lower than tree-ring age. For wood dated by C-14 as −4000 years the tree-ring age is about −4600 years, a difference of about 600 years. For a C-14 age of −6000 years the tree-ring age is about −6800 years, or about 800 years older. Small, short-term deviations are also observed in the radiocarbon ages. Causes of the deviations are not well understood, but may be the result of upsets in the carbon cycle of the atmosphere and oceans, or of changes in the flux of cosmic particles, or of variations in intensity of the earth's external magnetic field. Whatever the cause or causes, they suggest past changes in the balances of environmental systems, and as such deserve extended research. One environmental change causing change in radiocarbon ratios of the atmosphere is that of CO_2 added to the atmosphere by combustion of fossil fuels (see Chapter 6). This CO_2 contains practically no radiocarbon because of the great geologic age of the fuel. Consequently ratio of the radiocarbon to normal carbon will be decreased by dilution and leads to the strange effect (*Suess effect*) of making radiocarbon ages of trees grown late in the nineteenth century turn out to be younger than for trees grown in the early twentieth century! Another man-made effect is that of radiocarbon introduced into the atmosphere by atomic bomb tests. At one time this man-made radiocarbon raised total atmospheric radiocarbon to a level 100% above the normal value, and at present it is about 60% above normal. Scientists use this information as a means of calculating the time required for carbon to move through the atmospheric phase of the carbon cycle.

This discussion of radiocarbon has taken us far from the main theme of this chapter, which is on tectonics and crustal evolution through geologic time; it is to the subject of geologic time that we now return.

The Geologic Time Table

The major events of geologic history and the evolution of life forms had been worked out by geologists in considerable detail well before radiometric ages were available, but the chronology was of necessity a relative one. Relative ages of rocks and the division of geologic time into time units was built upon evidence of the sedimentary strata of the continents and the life remains found in them. The study of strata and their history is a branch of geology known as *stratigraphy*.

The *law of stratigraphic succession* was published in 1669 by Nicolaus Steno, a Danish physician. It states that a given layer of sedimentary rock is younger than the layer below it and older than the layer above it. This law is a self-evident statement following from the fact that sediments settling from a fluid medium must accumulate in order from bottom to top. To establish the relative

The Geologic Time Table

TABLE 9.4 Table of Geologic Time

	Era	Period	Duration (m.y.)	Age (m.y.)	Orogenies
Phanerozoic Eon	Cenozoic (65)				Cascadian
				65	
	Mesozoic (160)	Cretaceous	71		Laramian
				136	
		Jurassic	54		Nevadian
				190	
		Triassic	35		
				225	
	Paleozoic (345)	Permian	55		Appalachian (Hercynian)
				280	
		Carboniferous	65		
				345	
		Devonian	50		Acadian (Caledonian)
				395	
		Silurian	35		
				430	
		Ordovician	70		Taconian
				500	
		Cambrian	70		
				570	
			(b.y.)	(b.y.)	
Cryptozoic Eon	Precambrian	Upper Precambrian	0.3–0.4		
				0.9–1.0	Grenville
			0.6–0.8		
		Middle Precambrian		1.6–1.7	Hudsonian
			0.7–0.9		
				2.4–2.5	Kenoran
		Lower Precambrian	0.9–1.0		
		Oldest dated rocks		3.4 ± 0.1	
	Earth accretion completed			4.6–4.7	
	Age of universe			7–9?	

SOURCE: D. Eicher, 1968, *Geologic Time,* Englewood Cliffs, N.J., Prentice-Hall, end paper; M. Kay and E. H. Colbert, 1965, *Stratigraphy and Earth History,* New York, John Wiley & Sons, p. 74.

ages of strata which are found widely separated, with no connection or continuity of strata between (as, for example, on two different continents), the stratigrapher has depended upon *fossils,* the evidences or remains of life forms preserved in sedimentary rocks. Life forms have changed through geologic time by the process of organic evolution. At a given point in time the organisms inhabiting a particular environment of sediment accumulation are represented by a distinctive assemblage of species, known collectively as a *fauna,* or, in the case of plants, a *flora.* Once the succession of faunas in a sedimentary rock sequence is established in a given locality, strata of other localities can be correlated (matched in time) by the presence of similar fossil faunas. It was found that changes in faunas occurred quite suddenly at certain points in the record and it was supposed that these changes could be matched with orogenies that tilted, folded, or faulted all the strata which had accumulated up to that point in time. Using these observations, stratigraphers blocked off the geologic record into time units and gave them names. Table 9.4 is a general table of geologic time giving established names and ages of the most important time units.

Precambrian time, already mentioned in connection with the evolution of the continents, designates all geologic time before the beginning (at about −0.6 b.y.) of the *eras of abundant life.* The Precambrian encompasses about 85% of the entire span of geologic time since the earth was formed, at about −4.6 b.y., Precambrian history is poorly understood. Most of the Precambrian rock is igneous or metamorphic, while the few occurrences of unaltered sedimentary strata contain few, if any, fossils and these show only primitive life forms. It is not surprising, then, that geologists coined the name *Cryptozoic Eon* to designate Precambrian time, for the word *Cryptozoic* is derived from the Greek words *kryptos* (hidden) and *zoo* (life). In contrast, the eras of abundant life that followed are grouped together as the *Phanerozoic Eon*; the title being derived from the Greek word *phaneros*, meaning *visible*. Life forms are indeed conspicuous, numerous, and complex in even the oldest strata of this eon.

The Phanerozoic Eon is divided into three *eras*; their names and root meanings are as follows:

Era	Greek Root	Age (m.y.)	Duration (m.y.)
		(Present)	
Cenozoic	*kainos* (recent)		65
		−65	
Mesozoic	*mesos* (middle)		160
		−225	
Paleozoic	*paleos* (ancient)		345
		−570	
(Precambrian time)			

Thus the *Paleozoic Era* represents the *era of ancient life*, the *Mesozoic Era*, the *era of middle life*; and the *Cenozoic Era*, the *era of recent life*. Major orogenies and major changes in the composition of life forms are associated with the close of each era. The names of these orogenies are shown in Table 9.4.

Each era is in turn subdivided into *periods*. We notice from Table 9.4 that the Paleozoic Era has six periods, averaging about 60 m.y. per period; while the Mesozoic Era has three periods, averaging about 55 m.y. each. In contrast the Cenozoic Era has a duration of only 65 m.y., about the same as one period of the earlier eras. Consequently, Cenozoic time is not now subdivided into periods, but instead into secondary orders of time known as *epochs*. There are 7 epochs within the Cenozoic Era. The first five of these average about 12 m.y. each in duration. The next-to-the-last epoch is the *Pleistocene Epoch*, corresponding with the glaciations of the last ice age; its duration is only about 2 m.y. Finally, the *Holocene Epoch* designates the final bit of time that has elapsed since the last glacial retreat.*

Geologic Systems in Review

This chapter brings to a close a review of the salient features of the solid earth and their role in environmental processes and forms of the life layer. The cycling and recycling of mineral matter in many repetitions of the rock transformation cycle has given the earth its distinctive continental and oceanic environments. Energy for this cycle comes from two sources, the sun and the earth's interior; both sources consist of energy inherited from the time of formation of the solar system. Life forms have evolved with the continents, while organic processes have been active in shaping and modifying the environment.

At the same time that we have studied the mineral substances and their transformations, it has become evident that mineral resources now being consumed at prodigious rates by our industrial society required enormously great spans of time to accumulate; they are both finite and irreplaceable. In the chapter to follow, the earth's mineral and energy resources are evaluated, for availability of these resources will largely determine the future course of society and its institutions. Consumption of the nonrenewable resources has direct impacts upon the environment in many ways, some of which we have already seen in the form of air pollution and inadvertent climate modification (Chapter 6). So we leave the world of geologic time and return to the contemporary scene, in which the passage of even a single decade or year brings significant changes in the relation of Man to his environment.

*An older classification divided the Cenozoic Era into two periods. The first of these was called the *Tertiary Period* and contained the first five epochs; the second was called the *Quaternary Period* and encompassed both the Pleistocene and Holocene epochs.

10 Man's Consumption of Planetary Resources

Repeatedly in the three preceding chapters we have pointed out that Man is rapidly consuming earth resources that required geologic spans of time to be created and stored. So slowly do the geological processes operate that rates of replenishment are infinitesimally small in comparison with the present rates of consumption. The geological resources are therefore finite, and once we know approximately the world extent of a particular resource, we can predict its expiration according to any number of use schedules.*

In the formation of a mineral resource, matter has been concentrated by geologic processes. When these resources are consumed by Man, just the reverse usually occurs—the matter is transformed from a concentrated state to a dispersed state. This pervasive principle—that resource consumption produces dispersion—is a universal fact with which Man will have to deal, sooner or later. For example, coal represents an extremely dense concentration of hydrocarbons in large quantities. Combustion of coal disperses this matter into atmospheric constituents and the stored energy is transformed into heat and dissipated into outer space as longwave radiation. We have no way to reverse this process without expending an even larger amount of energy. In the case of one comparatively rare metal—lead—geologic processes have concentrated the element into rich ores. We disperse much of it into the atmosphere through combustion of leaded gasoline.

On the other hand, a great deal of used lead is recovered and used again—a process known as *recycling*. Aluminum provides an example of a metal of which only a small proportion (about one twelfth) is recycled; the remainder is widely dispersed, as witness the ubiquitous aluminum beer can. Other metals are dispersed in ways we do not see. For example, tungsten, vanadium, and chromium added in small quantities to steel as alloy metals, cannot be recovered without expenditure of considerable additional energy. Another example is silver, dispersed in electronic

*Much of the text of this chapter is taken from *Planet Earth; Its Physical Systems Through Geologic Time*, Harper & Row, Publishers, New York. © 1972 by Arthur N. Strahler. Used by permission of the publisher.

equipment and in photographic film in such ways that it cannot generally be recycled.

Actually, Man himself usually must carry out the final stages of mineral concentration, as in the case of nuclear fuels, copper ores, or most iron ores. While these deposits as mined represent an extraordinary degree of concentration when compared to their average distribution within the earth's crust, they must be concentrated further before they can be used by Man as, for example, pure uranium, copper, or iron. So the principle of concentration and dispersion must be invoked with caution and with due regard for instances when it does not apply.

What connection is there between resource consumption and environment of the life layer? As long as Man existed in a highly dispersed population in preagricultural times he used only renewable resources and those in sufficiently small quantities that most ecosystems remained essentially intact. Thus his food came from the hunting of other animals and from plants in the natural state; his water from surface streams and lakes; his fuel was wood of forest trees; and his clothing was made of plant fibers, fur, and feathers. These resources were renewed seasonally and there was no measurable depletion. As agriculture evolved and animals were domesticated, plant and animal life systems were radically altered in many parts of the continents. One of the first of the nonrenewable resources to be drained was the fertile soil layer, eroded and carried away by streams in areas denuded of protective forest or grass cover. It is thought that the Mediterranean lands, in particular, suffered heavily from soil erosion brought on by overgrazing and trampling of steep slopes. Large areas of northern China underwent devastating soil erosion following the cultivation of a highly erodible soil material (loess). However, an agrarian culture used little in the way of fossil fuels.

Although essentially agrarian cultures, the Roman and earlier civilizations of the Mediterranean lands exploited nonrenewable metallic resources. Important in the expansion of the Roman Empire was the control and extraction of lead, copper, tin, silver, and gold from mines around the Mediterranean Sea and throughout Europe. The depletion of many of these deposits through continuous working coincided with, and may have been to some degree a cause of economic bankruptcy that followed the fall of the Roman Empire.

The coming of the industrial era brought major changes that began to impact the environment in new ways. We have already seen how the burning of hydrocarbon fuels has brought on serious air pollution problems and may be in the process of causing long-term climatic changes. The very process of mining fossil fuels and minerals defaces the land with great scars and pits, destroying ecosystems and bringing on many undesirable side effects such as water pollution and the disturbance of hydrologic systems. As energy sources change over gradually from fossil fuels to nuclear energy, a new set of environmental problems is arising to replace or add to the old ones. Contamination of water and atmosphere by radioactive wastes and excess heat and the threat of accidents within nuclear reactors are already coming to the forefront as environmental problems. Many of the substances produced in industrial processes are rare metals capable of serious environmental pollution, among them compounds of lead and mercury derived from metallic ores and ultimately dispersed into the atmosphere and hydrosphere where they are picked up and concentrated in food chains. As the richer deposits of certain scarce minerals become depleted, poorer grades of deposits are mined and these are removed and processed in much greater volumes. As a result, the devastation of mining extends over increasingly large areas. The impact of consumption of natural resources upon environment is a very real phenomenon, and the magnitude of the impact increases sharply as the expenditure of resources increases. In less than two centuries, Man has replaced a planetary regime of steady states in energy systems with a transient regime in which a host of disturbances in physical and biological systems are interacting rapidly to produce many stresses upon life forms and particularly upon Man himself.

Nonrenewable Earth Resources

Two of the geologic resources falling into the nonrenewable category are discussed in Chapters 11 and 12; these are soils and ground water. Their slow rates of accumulation are only quasi-geologic in time, in comparison with other mineral resources. Nevertheless, the rates of expenditure of both soils and ground water (in arid lands) can vastly exceed the rates of their accumulation, and in this sense they are truly nonrenewable resources.

Nonrenewable earth resources can be grouped about as follows:

1. Soils
2. Ground water (renewable in regions of water surplus)
3. Metalliferous deposits (examples: ores of iron, copper, tin)

Metalliferous Deposits

4. Nonmetallic deposits, including
   a. Structural materials (examples: building stone, gravel, and sand)
   b. Materials used chemically (examples: sulfur, salts)
5. Fossil fuels (coal, petroleum, and natural gas)
6. Nuclear fuels (uranium, thorium)

Notice that the last two groups represent sources of energy, in distinction to the preceding two groups, which are sources of matter (materials). In this brief review, emphasis must be on concepts rather than upon data. Our aim will be to gain insight into the nature and distribution of these natural resources as a guide to broader issues related to planning for the future.

Metalliferous Deposits

Metals occur in economically adequate concentrations as *ores*. An ore is a mineral accumulation that can be extracted at a profit for refinement and industrial use. A number of important metallic elements are listed in Table 10.1, together with the *clarke of abundance*, or percent by weight in the average crustal rock. The *clarke* is a unit named for F. W. Clarke, a geochemist who did extensive research on the chemical composition of the earth's crust. (The clarke is the same quantity given for abundance in Table 7.2.) An important point is that magmas comprise the primary sources of many metals. Our concern is with the natural geological processes of concentration of metallic elements and compounds into ores of various kinds. Whereas aluminum and iron are relatively abundant, most of the essential metals of our industrial civilization are present in extremely small proportions—witness mercury and silver with clarkes of only 0.000008 and 0.000007, respectively.

In a classification of metals by uses, iron stands by itself in the quantities used in production of iron and steel. (Table 10.1 gives annual world consumption.) Related to iron is a group of *ferro-alloy metals,* which are used principally as alloys with iron to create steels with special properties. The ferro-alloys include titanium, manganese, vanadium, chromium, nickel, cobalt, molybdenum, and tungsten, as listed in order of appearance in Table 10.1. Other important metals (*nonferrous*

TABLE 10.1 Selected Metallic Abundances in Average Crustal Rock

Symbol	Element Name	Clarke (percent by weight)	Annual World Consumption (tons)[a]
Al	Aluminum	8.1	6,100,000
Fe	Iron	5.0	310,000,000
Mg	Magnesium	2.1	150,000
Ti	Titanium	0.44	10,000
Mn	Manganese	0.10	6,000,000
V	Vanadium	0.014	7,000
Cr	Chromium	0.010	1,400,000
Ni	Nickel	0.0075	400,000
Zn	Zinc	0.0070	3,800,000
Cu	Copper	0.0055	5,400,000
Co	Cobalt	0.0025	13,000
Pb	Lead	0.0013	2,800,000
Sn	Tin	0.00020	190,000
U	Uranium	0.00018	30,000[b]
Mo	Molybdenum	0.00015	45,000
W	Tungsten	0.00015	30,000
Sb	Antimony	0.00002	60,000
Hg	Mercury	0.000008	9,000
Ag	Silver	0.000007	8,000
Pt	Platinum	0.000001	30
Au	Gold	0.0000004	1,600

SOURCE: Brian Mason, 1966, *Principles of Geochemistry*, 3rd ed., New York, John Wiley & Sons, pp. 45–46, Table 3.3 and Appendix III.
[a] New metal used. Does not include recycling.
[b] As U_3O_8.

metals), standing apart individually with respect to industrial uses, are aluminum, magnesium, zinc, copper, lead, and tin. A minor group listed in Table 10.1 includes antimony, silver, platinum, and gold. Finally, there are metals which are radioactive, including uranium, listed in Table 10.1, thorium, and radium.

From the standpoint of metallic abundances as ores, the clarke is an abstraction of no practical value. Instead, the economic geologist is interested in the proportion of a given metal present in the form of the ore of its usual occurrence, either as an element or a compound. While a few metals, among them gold, silver, platinum, and copper occur as elements (i.e., as *native metals*), most occur as compounds. Oxides and sulfides are the most common forms, but more complex forms are present in many ores. The abundance of an element actually present in an ore is given as a multiplying factor known as the *clarke of concentration*. For example, manganese has a clarke of crustal abundance of 0.1 (Table 10.1). A common ore of manganese is the mineral *pyrolusite*, composition manganese oxide (MnO_2), in which manganese is present in the proportion of 63.2% by weight. The clarke of concentration for pure pyrolusite is therefore 632 (63.2% ÷ 0.1 = 632). A manganese ore containing pyrolusite as the principal mineral would be sufficiently rich for extraction with a concentration clarke of 350, which in this case represents an ore consisting of 35% of the element manganese (0.1 × 350 = 35%). Table 10.2 lists a number of metals with their clarkes of crustal abundance, concentration clarkes, and approximate percentages of the elements required for profitable extraction.

Table 10.3 gives the compositions of important ore minerals of the metals listed in Table 10.1. Not all important minerals are included, but at least one is given for each metal. A knowledge of mineral compositions will also be useful in understanding certain forms of air and water pollution derived from mines and piles of mine wastes (tailings), and from concentrating plants and smelters in which the ores are heated to drive off volatiles.

A notable trend in mineral extraction has been a shift from ores of simple composition to ores of complex composition. Certain ores yield a principal commodity plus one or more byproducts, for example, silver-bearing galena (lead is the principal metal). In certain complex ores each of the constituents is necessary to make the operation profitable.

Principles of magma crystallization, rock metamorphism, mineral alteration, and deposition of hydrogenic sediments explain the origin of metalliferous mineral deposits.

Origin of Metalliferous Ores

One major class of ore deposits is formed within magmas by direct segregation in which mineral grains of greater density settle down through the fluid magma while crystallization is still in progress. Masses or layers of a single mineral can accumulate in this way. One example is *chromite*, the principal ore of chromium with a density

TABLE 10.2 Concentration Clarkes of Ore Bodies for Selected Minerals

Metal	Clarke (percent by weight)	Concentration Clarke Required for Ore Body	Approximate Percent of Metal in Ore Needed for Profitable Extraction
Aluminum	8.13	4	20
Iron	5.00	6	30 (lower possible)
Manganese	0.10	350	35–27
Chromium	0.02	1500	30
Copper	0.007	140	0.8–0.5
Nickel	0.008	175	1.5
Zinc	0.013	300	4[a]
Tin	0.004	250	1
Lead	0.0016	2500	4
Uranium	0.0002	500	0.1

SOURCE: Brian Mason, 1966, *Principles of Geochemistry*, 3rd ed., New York, John Wiley & Sons, p. 50, Table 3.5.
[a] Percent in a multiple-element ore.

of 4.4 g/cc (Table 10.3). Bands of chromite ore are sometimes found near the base of the igneous body. Another example is seen in the nickel ores of Sudbury, Ontario. These sulfides of nickel apparently became segregated from a saucer-shaped magma body and were concentrated in a basal layer (Figure 10.1). *Magnetite* is another ore mineral that has been segregated from a magma to result in an ore body of major importance.

An important process associated with igneous intrusion is *contact metamorphism*, in which the invading magma and the highly active chemical solutions it contains altered the surrounding rock (country rock). In this process, ore minerals were introduced into the country rock in exchange for existing components in the rock.

For example, a limestone layer may have been replaced by iron ore consisting of hematite and magnetite (Figure 10.2). Ores of copper, zinc, and lead have also been produced in this manner. Valuable deposits of nonmetallic minerals have also resulted from contact metamorphism.

TABLE 10.3 Representative Metallic Ore Minerals

Metal	Symbol	Mineral Name	Composition	
Aluminum	Al	Bauxite (not a single mineral)	Hydrous aluminum oxide (complex)	(35–40% Al)
Iron	Fe	Magnetite	$Fe(FeO_2)_2$	(72.4% Fe)
		Hematite	Fe_2O_3	(70.0% Fe)
		Limonite (not a single mineral)	$Fe_2O_3 \cdot nH_2O$	(50–60% Fe)
		Pyrite	FeS_2	(46.6% Fe)
Magnesium	Mg	Magnesite (also from MgCl in brines)	$MgCo_3$	(47.6% MgO)
Titanium	Ti	Rutile	TiO_2	(60% Ti)
		Ilmenite	$FeTiO_3$	
Manganese	Mn	Pyrolusite	MnO_2 (with some water)	(30% Mn)
		Manganite	$Mn_2O_3 \cdot H_2O$	
Vanadium	V	Vanadinite (see also carnotite)	Oxide of lead and vanadium	
Chromium	Cr	Chromite	$FeCr_2O_4$ (with variable Al, Mg, Fe)	
Nickel	Ni	Pentlandite	(Fe, Ni)S	(20–40% Ni)
Zinc	Zn	Sphalerite	(Zn, Fe)S	(67% Zn)
Copper	Cu	Native copper	Cu	
		Chalcopyrite	$CuFeS_2$	(34.5% Cu)
		Chalcocite	Cu_2S	(80% Cu)
Cobalt	Co	Cobaltite	CoAsS	(35.4% Co)
Lead	Pb	Galena	PbS	(86.6% Pb)
Tin	Sn	Cassiterite	SnO_2	(78.6 Sn)
Molybdenum	Mo	Molybdenite	MoS_2	(60% Mo)
Tungsten	W	Wolframite	$(Fe, Mn)WO_4$	
Uranium	U	Pitchblende (uraninite)	Hydrous uranium oxide	
		Carnotite	Hydrous oxides of potassium, vanadium, and uranium	
Antimony	Sb	Stibnite	Sb_2S_3	(71.4% Sb)
Mercury	Hg	Cinnabar	HgS	(86.2% Hg)
Silver	Ag	Native silver	Ag	
		Argentite	Ag_2S	(87.1% Ag)
Gold	Au	Native gold	Au, Ag	
Platinum	Pt	Platinum	Pt (alloyed with Fe and other metals)	

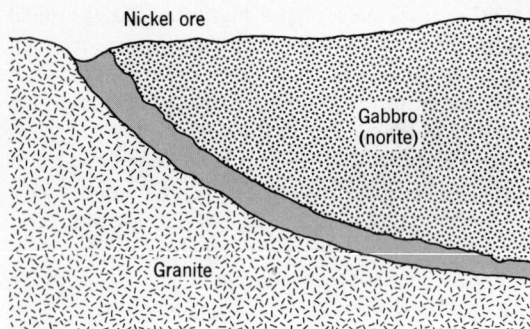

Figure 10.1 Cross section through a nickel ore deposit lying at the base of a body of gabbro and overlying a basement of older granite. (After A. P. Coleman, 1913, Canada Dept. of Mines, *Monograph 170*, p. 34.)

Another type of ore deposit associated with igneous intrusion is found in pegmatite bodies, described in Chapter 7 and illustrated in Figure 7.15. Pegmatite bodies take the form of irregular masses within the parent plutonic rock, and occur as dikes (wall-like bodies), and as veins extending out into the country rock. While the bulk of all pegmatites consists of quartz, feldspar, micas, and other common minerals of granitic rocks, there occur unusual concentrations of rarer minerals, both metallic and nonmetallic. For example, in certain pegmatites of the Black Hills of South Dakota there occur enormous crystals of the mineral *spodumene*, an aluminosilicate of lithium (Figure 10.3). A single crystal of record size measured 47 ft (14m) in length and 6 ft (2m) in diameter; several lesser ones weighed over 30 tons apiece. This and other pegmatite bodies are a principal source of the light metal, *lithium*, which has many important uses in industry. Another metal, *beryllium*, is found in pegmatites in the form of the mineral *beryl*, an aluminosilicate of beryllium. Beryllium is an important component in high-strength alloys of copper, cobalt, nickel, and aluminum. Two other metals found in pegmatites,

Figure 10.3 Large spodumene crystals in the Etta pegmatite, Pennington County, South Dakota. The hammer rests upon a single large crystal of spodumene. (Photograph by J. J. Norton, U.S. Geological Survey.)

both very rare but essential in industry, are *tantalum* and *columbium*. Because of the large size of pegmatite crystals—from several inches to as much as a few feet in diameter—they are important commercial sources of certain common nonmetallic minerals, principally the feldspars, used in manufacture of pottery, tile, porcelain, and glass. Another example is *muscovite mica*, needed in large sheets and plates for electrical insulation and related uses; it occurs as large sheets only in pegmatites.

A fourth type of ore deposit is produced by the effects of high-temperature solutions, known as *hydrothermal solutions*, that leave a magma during the final stages of its crystallization and are deposited in fractures to produce mineral veins. Some veins are sharply defined and evidently represent the filling of open cracks with layers of minerals. Other veins seem to be the result of replacement of the country rock by the hydrothermal solutions. Where veins occur in exceptional thicknesses and numbers, they may constitute a *lode*.

Hydrothermal solutions produce yet another important type of ore accumulation, the *disseminated deposit*, in which the ore is distributed throughout a very large mass of rock. Certain of the great copper deposits are referred to as *porphyry copper* deposits because the ore has in some cases entered a large body of igneous rocks of a texture class known as *porphyry*, which had in some manner been shattered into small joint

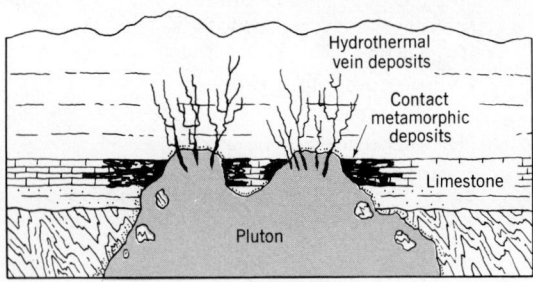

Figure 10.2 Schematic cross section through an intrusive igneous body and the overlying country rock, showing veins and contact metamorphic deposits. (From A. N. Strahler, 1972, *Planet Earth; Its Physical Systems Through Geologic Time*, Harper & Row, New York.)

Figure 10.4 Open-pit copper mine at Bingham Canyon, Utah. (Photograph by courtesy of Kennecott Copper Corporation.)

blocks that permitted entry of the solutions. One of the most celebrated of these is at Bingham Canyon, Utah (Figure 10.4).

Hydrothermal solutions rise toward the surface, making vein deposits in a shallow zone and even emerging as hot springs. Many valuable ores of gold and silver are deposits of the shallow type. Particularly interesting is the occurrence of mercury ore in the form of the mineral *cinnabar* (see Table 10.3) as a shallow hydrothermal deposit. Most renowned are the deposits of the Almaden district in Spain, where mercury has been mined for centuries and has provided most of the world's supply of that metal.

A quite different category of ore deposits embodies the effects of downward moving solutions in the zone of aeration and the ground-water zone (see Chapter 12). Enrichment of mineral deposits to produce ores in this manner is described as a *secondary* process. Consider first a vein containing primary minerals of magmatic origin (Figure 10.5). These minerals, mostly sulfides of copper, lead, zinc, and silver, along with native gold, are originally disseminated through the vein rock and may not exist in concentrations sufficient to qualify as ores. (See Table 10.3 for minerals associated with each metal.) Through long-continued denudation of the region, the ground surface truncates the vein, which was formerly deeply buried. Assuming a humid climate, there will exist a water table and a ground water zone, above which is the zone of aeration (see Chapter 12). Water, arriving as rain or snowmelt, moves down through the zone of aeration. The geologist refers to this water as *meteoric*, which is perfectly acceptable from the standpoint of atmospheric science. The meteoric water becomes a weak acid, since it contains dissolved carbon dioxide (carbonic acid) and will also gain sulfuric acid by reactions involving iron sulfide (mineral *pyrite*, Table 10.3).

The result of downward percolation of meteoric water is to cause three forms of enrichment and thus to yield ore bodies. First, in the zone closest to the surface, as soluble waste minerals are removed, there may accumulate certain insoluble minerals, among them gold and compounds of silver or lead, in sufficient concentration to form an ore. This type of ore deposit is known as a *gossan* (Figure 10.5). Iron oxide and quartz will also accumulate in the gossan. In Colonial times, iron-rich gossans constituted minable iron ores, but they have been exhausted. Leaching of other minerals carries them down into a *zone of oxidation*. In this second zone there

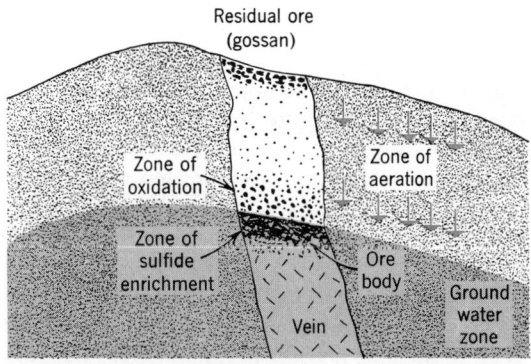

Figure 10.5 Schematic cross section showing secondary ore deposits formed by enrichment of minerals of a vein. (From A. N. Strahler, 1972, *Planet Earth; Its Physical Systems Through Geologic Time*, Harper & Row, New York.)

Figure 10.6 In this oblique air photograph the light-colored patches are strip mines operated during World War II by the Japanese to obtain bauxite. The ore, which averages 7 ft (2 m) thick, was derived by alteration of volcanic rock in a warm, wet equatorial climate. Babelthaup Island, Palau Group (lat. $7\frac{1}{2}°$ N, long. $134\frac{1}{2}°$ E). (Photograph by U.S. Geological Survey.)

may accumulate a number of oxides of zinc, copper, iron, and lead, along with native silver, copper, and gold. A third zone is that of *sulfide enrichment* within the upper part of the groundwater zone, just beneath the water table (Figure 10.5). Sulfides of iron, copper, lead, and zinc may be heavily concentrated in this zone. (Mineral examples are pyrite, chalcopyrite, chalcocite, galena, and sphalerite, described in Table 10.3.) Sulfide enrichment may also affect large primary ore bodies of the disseminated type, such as the porphyry copper of Bingham Canyon, Utah, referred to above. Here, the enriched layer has already been removed and mining has progressed into low-grade primary ore beneath.

Also in the category of secondary ores is *bauxite* (Table 10.3). This principal ore of aluminum accumulates as a near-surface deposit in tropical regions where the soil-forming regime of laterization prevails (see Chapter 12). Bauxite, a mixture of hydrous oxides of aluminum derived from the alteration of aluminosilicate minerals, is practically insoluble under the prevailing climatic conditions and can accumulate indefinitely as the denudation of the land surface progresses. Produced under similar environmental conditions are residual ores of manganese (mineral: *manganite*), and of iron (mineral: *limonite*) (see Table 10.3). The term *laterite* is commonly applied to these residual deposits (Figure 10.6).

Another category of ore deposit is that in which concentration has occurred through fluid agents of transportation: streams and waves. Certain of the insoluble heavy minerals derived from weathering of rock are swept as small fragments into stream channels and carried downvalley with the sand and gravel as bed materials. Because of their greater specific gravity, these minerals become concentrated in layers and lenses of gravel to become *placer deposits*. Native gold is one of the minerals extensively extracted from placer deposits; platinum is another. A third is an oxide of tin, the mineral *cassiterite* (see Table 10.3), which forms important placer deposits. Diamonds, too, are concentrated in placer deposits, as are certain other gem stones. Transported by streams to the ocean, gravels bearing the heavy minerals are spread along the coast in beaches, forming a second type of placer deposits, the *marine placers*.

Finally, we can recognize a group of ore deposits in the hydrogenic category of sediments, explained in Chapter 8. For the most part, sediment deposition is the principal source of nonmetallic mineral deposits, considered below, but some important metalliferous deposits are of this origin. Iron, particularly, occurs as sedimentary ores in enormous quantities. Sedimentary iron ores are oxides of iron—usually *hematite* (see Table 10.3). A particularly striking example is iron ore of the Clinton formation of Silurian age, widespread in the Appalachian region. For reasons not well understood, unusually large quantities of iron oxides, derived by weathering of mafic minerals in rocks exposed in bordering lands, were brought to the sea floor and were precipitated as hematite. Another metal, manganese, has been concentrated by depositional processes into important sedimentary ores.

Metal Demand and Supply

Table 10.4 will give some appreciation of the future demands for the same metals of Table 10.1, as estimated by the U.S. Bureau of Mines for the year 2000. (Primary production is that derived from mining of ores.) The table also shows what part of the 1968 United States demand for primary metals came from domestic sources. This information is highly significant in demonstrating the dependence of a large, heavily industrialized nation upon foreign sources of metals. An approximate factor of increase in demand is given in the last column.

A particularly striking fact shown in Table 10.4 is that a number of the metals for which the factor of increased demand is the largest are those derived largely or entirely from foreign sources. For example, metallic titanium demands are shown to increase by a factor of about 12, but none is now produced in the United States. Aluminum demand will be up seven-fold but we produce only one-tenth of our primary aluminum. A similar situation holds for chromium, nickel, cobalt, and platinum. Our dependence is strong upon foreign supplies of iron, manganese, zinc, lead, tungsten, antimony, silver, and gold. Actually, the only metals which the United States produced in substantial surplus relative to demand in 1968 were uranium and molybdenum. It is predicted by the Bureau of Mines that in the year 2000 domestic production of primary minerals in all categories (including nonmetallic minerals) will supply substantially less of the demand for primary minerals than it does at present.

Of increasing importance in manufacturing today is the secondary production of metals through reprocessing of durable metal goods manufactured in the past one to ten decades. We are not here referring to new scrap metal, derived as cuttings during initial manufacture, but to the old materials from discarded products. Metals can be reclaimed from old scrap by processes of distillation, electrometallurgy, mechanical separation, and chemical processes. As the total output of manufactured goods increases through time, the input of metals from secondary sources will also rise in volume.

Recycling of metals is rising in importance as national mineral resources are becoming depleted at increasing rates and as the grade of ores being mined is declining. Some idea of the importance of recycled nonferrous metals is given by data in Table 10.5. Secondary consumption figures in-

TABLE 10.4 Comparison of United States Primary Metal Demand in 1968 with Projected Demand in the Year 2000

Metal	Units	1968 U.S. Primary Production	1968 U.S. Primary Demand	Projected U.S. Primary Demand in Year 2000	Approximate Factor of Increase
Aluminum	Thousand S.T.[a]	420	3,900	28,000	7
Iron	Million S.T.	56	84	150	2
Magnesium (metal)	Thousand S.T.	98	91	390	4
Magnesium (nonmetal)	Thousand S.T.	1,000	1,100	1,800	1½
Titanium (metal)	Thousand S.T.	0	13	150	12
Titanium (nonmetal)	Thousand S.T.	300	440	1,600	4
Manganese	Thousand S.T.	48	1,100	2,100	2
Vanadium	Short tons	6,100	5,800	31,000	5
Chromium	Thousand S.T.	0	450	1,100	2½
Nickel	Million pounds	30	320	930	3
Zinc	Thousand S.T.	530	1,400	3,000	2
Copper	Thousand S.T.	1,200	1,500	6,400	4
Cobalt	Thousand pounds	1,300	14,000	25,000	2
Lead	Thousand S.T.	360	900	2,000	2
Tin	Thousand S.T.	Almost nil	59	85	1½
Uranium	Short tons	10,000	2,700	64,000	20
Molybdenum	Million pounds	94	56	180	3
Tungsten	Thousand pounds	9,800	16,000	74,000	5
Antimony	Short tons	1,900	21,000	40,000	2
Mercury	Thousand flasks	29	62	130	2
Silver	Million ounces	33	90	210	2
Platinum	Thousand ounces	5	460	1,400	3
Gold	Thousand ounces	1,500	6,600	24,000	4

SOURCE: U.S. Bureau of Mines.
NOTE: Figures rounded to two places.
[a] Short tons.

TABLE 10.5 Comparison of Primary and Secondary United States Consumption of Four Metals in 1967

Metal	Primary Consumption		Secondary Consumption		Total Consumption	
	MST[a]	(%)	MST	(%)	MST	(%)
Copper	1.5	(55)	1.2	(45)	2.7	(100)
Aluminum	3.8	(81)	0.9	(19)	4.7	(100)
Lead	0.75	(58)	0.55	(42)	1.30	(100)
Zinc	1.33	(84)	0.26	(16)	1.59	(100)

SOURCE: U.S. Bureau of Mines.
[a] Millions of short tons.

clude both new and old scrap metal. Comparing secondary consumption with primary consumption for 1967, we find that the secondary quantity of copper is only a little smaller than the primary quantity. A similar ratio prevails for lead consumption. In contrast, secondary consumption values of aluminum and zinc amount to only about one-quarter and one-fifth, respectively, of the primary consumption values.

Since the data of Table 10.5 combine new and old scrap, they conceal marked differences in proportions actually derived from old scrap. *Metal recycling* is defined as the ratio of old scrap metal consumed to total consumption of primary (virgin) metal plus new scrap metal, stated as a percentage. On this basis, recycling during 1968 and 1969 in the United States ran about as follows (listed in order of decreasing percentage):*

Silver	66%
Lead	36%
Copper	25%
Iron	19%
Tin	17%
Mercury	14%
Nickel	7%
Zinc	5%
Aluminum	3%
Cadmium	3%

Silver shows the highest percentage of recycling, while the recovery of lead from old scrap is also comparatively high—three-quarters of this lead comes from plates of discarded batteries. Recovery of copper from old scrap is also of major importance, while in contrast, that of zinc and aluminum is of comparatively minor importance. Recovered metals are largely in the form of alloys. For example, most of the recovered copper is in brass and bronze; most of the recovered lead is antimonial lead from battery plates; most of the recovered zinc is in brass and bronze; practically all of the recovered aluminum is in alloys. These facts indicate that secondary metal sources are not, in general, capable of furnishing substantial quantities of pure metals under prevailing conditions of recovery technology.

Nonmetallic Mineral Deposits

Nonmetallic mineral deposits (not including fossil and nuclear fuels) include such a large and diverse assemblage of substances, and cover such a wide range of uses that it would be impossible to do the subject justice in a few paragraphs. In outline form we offer some examples of these mineral deposits classified by use categories:

Structural materials:
1. Clay: For use in brick, tile, pipe, chinaware, stoneware, porcelain, paper filler, and cement. Examples: kaolin (for china manufacture) from residual deposits produced by weathering of felsic rock; shales, marine and glacio-lacustrine clays for brick and tile.
2. Portland cement: Made by fusion of limestone with clay or blast-furnace slag. Suitable limestone formations and clay sources are widely distributed and are of many geologic ages.
3. Building stone: Many rock varieties are used, including granite, marble, limestone, sandstone. Slate was used widely as a roofing material.
4. Crushed stone: Limestone and *trap rock* (gabbro, basalt) are crushed and graded for aggregate in concrete and in macadam pavements.
5. Sand and gravel: Used in building and paving materials such as mortar and concrete, asphaltic pavements, and base courses under pavements. Sources lie in fluvial and glaciofluvial deposits and in beaches and

*Computations by Charles B. Belt, Jr., 1972, *Geol. Soc. of Amer.* (publication in press); based on data of U.S. Bureau of Mines and American Iron and Steel Institute.

dunes. Specialized sand uses include *molding sands* for metal casting, *glass sand* for manufacture of glass, and *filter sand* for filtering water supplies.
6. Gypsum: (Hydrous calcium sulfate, $CaSO_4 \cdot 2H_2O$) Major use is in calcined form for wallboard and as plaster, and as a retarder in Portland cement. Source is largely in gypsum or anhydrite ($CaSO_4$) beds in sedimentary strata associated with red beds and evaporites.
7. Lime: Calcium oxide (CaO) obtained by heating of limestone, has uses in mortar and plaster, in smelting operations, in paper, and in many chemical processes.
8. Pigments: Compounds of lead, zinc, barium, titanium, and carbon, both manufactured and of natural mineral origin, are widely used in paints.
9. Asphalt: Asphalt occurs naturally, but most is derived from refining of petroleum. It is used in paving, and in roofing materials.
10. Asbestos: Fibrous forms of four silicate minerals, used in manufacture of various fireproofing materials.

Mineral deposits used chemically and in other industrial uses:
1. Sulfur: A major source is free sulfur occurring as beds in sedimentary strata in association with evaporites, but most sulfur is supplied as a byproduct of ore refining and smelting operations. Chief use is for manufacture of sulfuric acid.
2. Salt: Naturally occurring rock salt, or *halite*, is largely sodium chloride (NaCl), but includes small amounts of calcium, magnesium and sulfate. It occurs in salt beds in sedimentary strata, and in salt domes. Major uses include manufacture of sodium salts, chlorine, and hydrochloric acid.
3. Fertilizers: Some natural mineral fertilizers are *phosphate rock*, of sedimentary origin, *potash* derived from rock salt deposits and by treatment of brines, and *nitrates*, occurring as sodium nitrate in deserts (Atacama Desert of Chile).
4. Sodium salts: Found in dry lake beds (playas) of the western United States are various salts of sodium, such as *borax* (sodium borate). These have a wide range of chemical uses. Also important are sodium carbonate and sodium sulfate, found in other dry lake accumulations.
5. Fluorite: The mineral *fluorite* is calcium fluoride (CaF_2). It is found in veins in both sedimentary and igneous rocks. Uses are metallurgical and chemical, e.g., to make hydrofluoric acid.
6. Barite: *Barite* is barium sulfate ($BaSO_4$) and occurs as a mineral in sedimentary and other rocks. It is used as a filler in many manufactured substances, and as a source of barium salts required in chemical manufacture.
7. Abrasives: A wide variety of minerals and rocks have been used as abrasives and polishing agents. Examples are seen in garnet, used in abrasive paper or cloth, and diamond, for facing many kinds of drilling, cutting, and grinding tools.

The above list is by no means complete, and it can serve only to give an appreciation of the strong dependence of industry and agriculture upon mineral deposits and the substances manufactured from them.

Mineral Resource Depletion

The impact of Man upon all forms of mineral resources of the continents is admirably summarized in a statement written by a distinguished economic geologist, Thomas S. Lovering, in a report by the *Committee on Resources and Man* of the National Academy of Sciences—National Research Council:

> The total volume of workable mineral deposits is an insignificant fraction of 1 percent of the earth's crust, and each deposit represents some geological accident in the remote past. Deposits must be mined where they occur—often far from centers of consumption. Each deposit also has its limits; if worked long enough it must sooner or later be exhausted. No second crop will materialize. Rich mineral deposits are a nation's most valuable but ephemeral material possession—its quick assets. Continued extraction of ore, moreover, leads, eventually, to increasing costs as the material mined comes from greater and greater depths or as grade decreases, although improved technology and economics of scale sometimes allow deposits to be worked, temporarily, at decreased costs. Yet industry requires increasing tonnage and variety of mineral raw materials; and although many substances now deemed essential have understudies that can play their parts adequately technology has found no satisfactory substitutes for others.*

*From T. S. Lovering, "Mineral Resources from the Land," Chapter 6, p. 110, of *Resources and Man*, copyright © 1969 by the National Academy of Sciences, W. H. Freeman and Company, Publishers. Reproduced by permission of the National Academy of Sciences.

Mineral Resources from the Sea

If the prospect of eventually running out of various mineral resources from the lands seems all too real, we may want to turn to consider possible substitutions of mineral resources from the sea.* Sea water has always been available as a resource, and it has long provided the bulk of the world's supply of magnesium and bromium, as well as much of the sodium chloride. The list of elements present in sea water includes most of the known elements and, despite their small concentrations, these are potential supplies for future development. It is thought that sodium, sulfur, potassium, and iodine lie in the category of recoverable elements. It is, however, beyond reason to hope for extraction of ferrous metals (principally iron) and the ferro-alloy metals in significant quantities to provide substitutes for ore deposits of the continents.

The continental margins, with their shallow continental shelves and shallow inland seas, are already being exploited for mineral production, as witness the working of placer deposits of platinum, gold, and tin in shallow waters. The petroleum resources of the North American continental shelf are already under development along the Gulf coast; zones of potential development are believed to exist on the shelf off the Atlantic coast as well. Possibility exists of finding and using mineral deposits of continental crystalline rocks submerged to shallow depths, although this has not yet happened.

Exploration of the deep ocean floor as a source of minerals is still in an early stage, but already the layer of manganese nodules found in parts of all of the oceans is regarded by some as a major future source of manganese, along with a number of metals in lesser quantities. Presence of substantial amounts of silica with the manganese oxide may render the nodules unfit for exploitation of manganese by present extraction methods, but this does not rule out the possibility of future use.

In reviewing the overall prospects of mineral resources from the oceans and ocean basins we are only being realistic in concluding that contributions from sea water itself are limited only to a few substances, that most of the contributions of the sea floor will be from shallow continental shelves where petroleum and natural gas are the major resources, and that prospects of substantial mineral contributions from the deep ocean floor

*Based on data of Preston Cloud, "Mineral Resources from the Sea," Chapter 7, pp. 135-55 of *Resources and Man*, 1969, National Academy of Sciences. San Francisco, W. H. Freeman and Company.

are rather poor at this time. In the light of these conclusions, the need for conservation and careful planning for the use of mineral resources of the lands becomes all the more evident.

Fossil Fuels—Geologic and Development Aspects

The fossil fuels, or fossil hydrocarbon compounds, were described in Chapter 8 as biogenic sedimentary accumulations. Whereas the various forms of peat and coal are remains of larger land plants, petroleum and natural gas are thought to have been derived originally from microscopic marine plants (plankton) and to have been originally disseminated through thick shales. We shall now treat some of the geological aspects of these fossil fuels and their development as natural resources.

Coal is classified into three types, representing a developmental sequence. *Lignite*, or brown coal, is soft and has a woody texture (Figure 10.7A). It represents an intermediate stage between peat and true coal. Further compaction resulting from deep burial resulted in the transformation of lignite into *bituminous coal*, often called *soft coal*. In areas where the crust was compressed and folded by mountain-making forces, bituminous coal was further changed, becoming *anthracite*. Whereas bituminous coal typically breaks into prismatic fragments, anthracite exhibits a glassy (conchoidal) type of fracture (Figure 10.7B,C).

The coals consist largely of the elements carbon, hydrogen, and oxygen, with small amounts of sulfur also present. Inorganic impurities are also present and are designated as *ash*, the noncombustible residue after coal is completely burned. For purposes of energy analysis as fuels, the contents of lignite and coal are given in terms of percentages of fixed carbon, volatiles, and water. Figure 10.8 shows typical analyses of samples of lignite, bituminous coal, and anthracite. In the transition from lignite to bituminous coal a large quantity of water is driven off. In the transition from bituminous coal to anthracite most of the volatiles are driven off, with the result that anthracite is composed almost entirely of fixed carbon. Coals are evaluated, or ranked, in order of the *fuel ratio*, which is the ratio of fixed carbon to volatile matter. Volatiles burn in the form of gas and give a long, smoky flame, whereas fixed carbon produces a short, hot, smokeless flame that is steady. The lower-ranking coals—lignite and *subbituminous coal* (intermediate between lignite and bituminous coal)—not only have low heating value, but are subject to spon-

Figure 10.7 Specimens of coals. A. Lignite from North Dakota. (Photograph by M. E. Strahler.) B. Bituminous coal, Virginia. (Photograph by J. B. Eby, U.S. Geological Survey.) C. Anthracite from Pennsylvania. (Photograph by M. E. Strahler.)

taneous combustion. Lignite readily disintegrates (*slakes*) after drying in air. Volatiles in bituminous coal are a source of gas used as a fuel, and of *coke*, the fixed carbon remaining after heating has driven off the volatiles.

Coal is found in all ages of sedimentary strata following the Devonian Period, when land plants evolved. The Carboniferous Period is particularly noted for coals of that age having worldwide distribution. Coals of Permian age are also widespread. The Triassic and Jurassic strata contain coals, but of limited world distribution. Cretaceous coals are second in importance only to those of the Carboniferous Period. Most of the world's lignite is from Cenozoic strata, but some high-ranking coals were also produced in that era.

Figure 10.9 shows coal fields of the 48 contiguous United States differentiated as to rank of coal. This map has significance beyond the subject of resource distribution, since the environmental impact of mining will be felt in the areas shown as coal fields. The most important producer is the Appalachian Field, which includes some 70,000 square miles (180,000 sq km) of both flat-lying and folded strata of Carboniferous and Permian age. The Anthracite Field of northeastern Pennsylvania is very small in area, but has been a major producer. (A map of the synclines in which the anthracite occurs is given in Figure 9.35.) The Interior Fields of bituminous coal give particularly important production in parts of Indiana, Illinois, and Kentucky. Here the strata are of Carboniferous age and lie nearly horizontal. The Rocky Mountain Fields contain bituminous coals and lignite of Cretaceous age, while a large reserve of Cenozoic lignite occurs in the Dakotas and eastern Montana. A large area of lignite in Cenozoic strata underlies the Gulf Coastal Plain. There are minor coal fields in the Pacific Northwest. Alaska has large coal reserves of Cenozoic age.

Estimates of the world's total minable coal resources are shown in Table 10.6. These figures are based on the assumption that only 50% of all existing coal is minable, since there is a lower

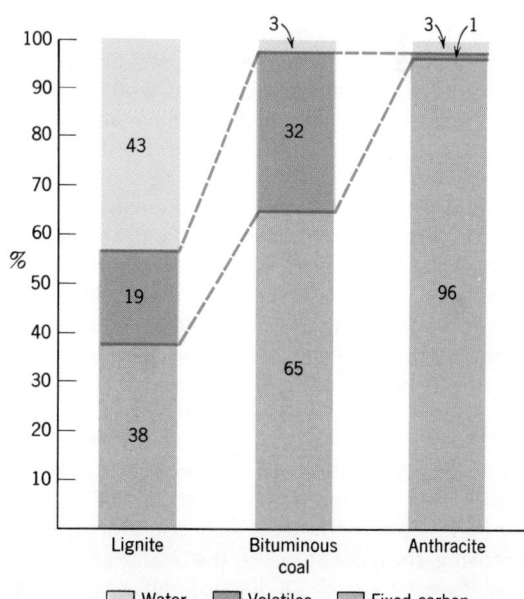

Figure 10.8 Percentage by weight of three variable constituents of representative coals. (From A. N. Strahler, 1971, *The Earth Sciences*, 2nd ed., Harper & Row, New York.)

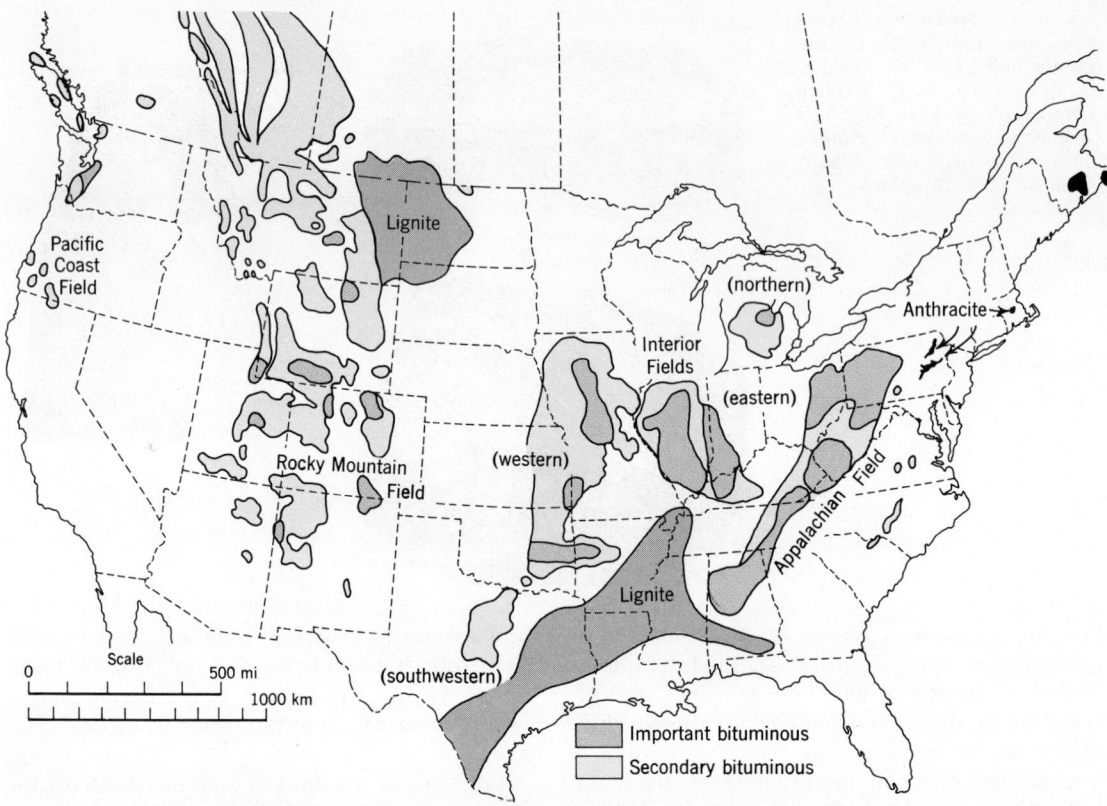

Figure 10.9 Generalized map of coal fields of the United States and southern Canada. (After U.S. Geological Survey, Canada Dept. of Mines & Technical Survey; and A. M. Bateman, 1950, *Economic Mineral Deposits*, 2nd ed., John Wiley & Sons, New York, p. 645, Figure 16.3.)

limit to the thickness of coal seams that can be mined. The statistics are not only of great importance in terms of world energy resources, but have environmental significance as well, because the combustion of this fuel will have an impact upon the energy systems of the atmosphere and hydrosphere.

The mining of coal has a strong environmental impact in many ways upon the land and water of the coal fields, and upon the miners themselves. We can only understand these effects if we know something of the manner in which coal is mined. For deep-lying coal seams, vertical shafts are driven down from the surface to reach the coal, which is mined by extension of horizontal *drifts* and rooms into the face of the seam. In mountainous terrain drifts can be driven directly into the seam where it is exposed on the mountain side. Two methods are in common use in removing the coal. One is the *room-and-pillar system*, in which about half of the coal is left behind in supporting pillars. Figure 10.10 shows the plan of such a mine. A second method is the *long-wall system* in which all of the coal seam is removed centripetally inward from a large circumference, leaving a large central block to support the mine shaft. The roof above the excavated area is allowed to settle as coal is removed.

TABLE 10.6 Estimated World Coal Resources

Continent or Country	Coal Reserves in Billions of Metric Tons
U.S.S.R. in Asia and Europe	4,310
Asia exclusive of U.S.S.R.	681
United States	1,486
North America, exclusive of United States	601
Western Europe	377
Africa	109
Central and South America	14
Oceania, including Australia	59
Total	7,640

SOURCE: Data of Paul Averitt, U.S. Geological Survey, as given by M. K. Hubbert in *Scientific American*, 1971, vol. 224, no. 3, p. 64.

Fossil Fuels

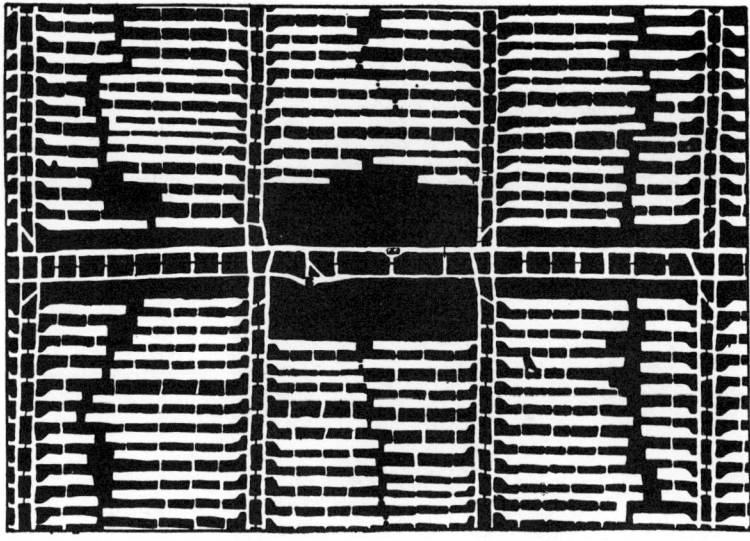

Figure 10.10 Plan (map) of a room-and-pillar mine showing rooms (white areas) and coal (black). About half of the coal remains to support the roof. (Illinois Geological Survey, Bull. 56, p. 47, Figure 17.)

Where the coal seams lie close to the surface or actually outcrop along hillsides the *strip mining* method is used. Here, earthmoving equipment removes the covering strata (*overburden*) to bare the coal, which is lifted out by power shovels. There are two kinds of strip mining, each adapted to the given relationship between ground surface and coal seam. *Area strip mining* is used in regions of nearly flat land surface under which the coal seam lies horizontally (Figure 10.11A). After the first trench is made and the coal removed, a parallel trench is made, the overburden of which is piled as a spoil ridge into the first trench. Thus the entire seam is gradually uncovered and there remains a series of parallel spoil ridges (Figure 10.12). In this connection, it is interesting from the aspect of environmental impact to know that phosphate beds are mined extensively by the area strip mining method in Florida, and that the method is also used for mining clay layers. The *contour strip mining* method is used where a coal seam outcrops along a steep hillside (Figure 10.11B). The coal is uncovered as far back into the hillside as possible and the overburden dumped on the downhill side. There results a bench bounded on one side

Figure 10.12 Area strip mining in Ohio County, Kentucky. Dragline in background is removing overburden and piling it at the right. Loader in foreground is removing the exposed coal. (TVA photograph.)

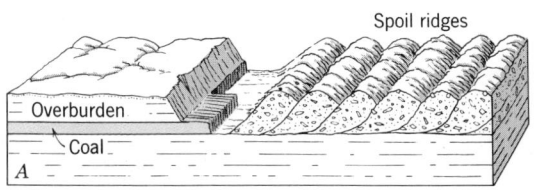

Figure 10.11 A. Area strip mining. B. Contour strip mining.

Figure 10.13 Contour strip mining in Wise County, Virginia. (Photograph by Kenneth Murray, Kingsport, Tennessee.)

TABLE 10.7 Element Composition of Typical Petroleum

Element	Percentage by Weight		
	Crude Oil	Asphalt	Natural Gas
Carbon	82–87	80–85	65–80
Hydrogen	12–15	8.5–11	1–25
Sulfur	0.1–5.5	2–8	trace–0.2
Nitrogen	0.1–1.5	0–2	1–15
Oxygen	0.1–4.5	—	—

SOURCE: A. I. Levorsen, 1967, *Geology of Petroleum*, 2nd ed., W. H. Freeman and Company, San Francisco and London, p. 177, Table 5.5.
NOTE: Figures rounded to the nearest one-half percent.

by a freshly cut rock wall and upon the other by a ridge of loose spoil with a steep outer slope leading down into the valley bottom. The benches form sinuous patterns following the plan of the outcrop (Figure 10.13). Strip mining is carried to depths as great as 100 feet below the surface. Associated with contour strip mining is *auger mining* in which enormous auger drills are run horizontally into the exposed face of the coal seam after the initial strip mining is completed. Augers with cutting heads several feet in diameter are used and are capable of penetrating as far as 200 feet into the seam.

The general term *petroleum* spans the range from crude oil to natural gas in the one direction, and to *asphalt* and related semisolid hydrocarbon substances in the other. Table 10.7 gives the range of carbon, hydrogen, sulfur, nitrogen, and oxygen in typical analyses of the three forms of petroleum. Crude oil in the natural state is a mixture of a large number of hydrocarbon compounds. Since more than 200 compounds have been isolated and analyzed in crude oil, and the range and abundance of hydrocarbon compounds differs greatly from one oil field to another, we cannot enter very deeply into the subject here.

Crude oils differ in terms of the relative abundances of four hydrocarbon series: paraffins, naphthenes, asphalts, and aromatics. Of these, only the first three series are dominant in petroleum. Generally speaking, the paraffin series is the most abundant of hydrocarbons in both liquid petroleum and natural gas. Crude oil is described as *paraffin-base* when paraffins are dominant; it is of low density and typically yields good lubricants and a large proportion of kerosene. An example is the paraffin-base crude oil of the Pennsylvania fields. *Asphalt-base* oil has a high density and is referred to as a *heavy* oil; its primary yield is in the form of fuel oils.

Petroleum and natural gas have been concentrated into accumulations known as *oil pools*, a misnomer since the oil and gas does not occupy a single large rock cavity. Instead, the oil and gas are held in pore spaces in a *reservoir rock*. Usually this is a porous sand or sandstone, but certain other rocks, such as porous carbonates, can serve

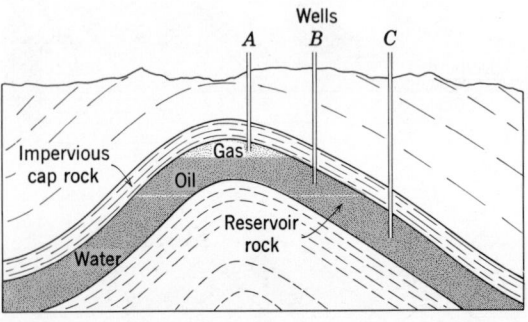

Figure 10.14 In this idealized cross section of an oil-bearing anticline or dome, Well A yields gas, Well B, oil, and Well C, water. The caprock is a shale formation; the reservoir rock is sandstone. (From A. N. Strahler, 1972, *Planet Earth; Its Physical Systems Through Geologic Time*, Harper & Row, New York.)

Fossil Fuels

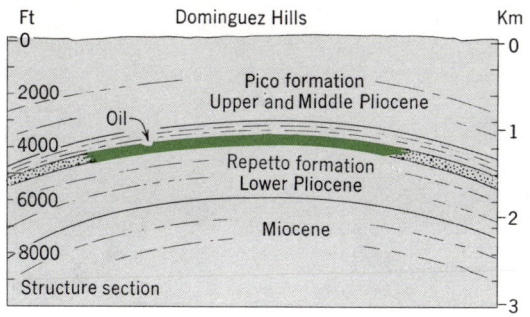

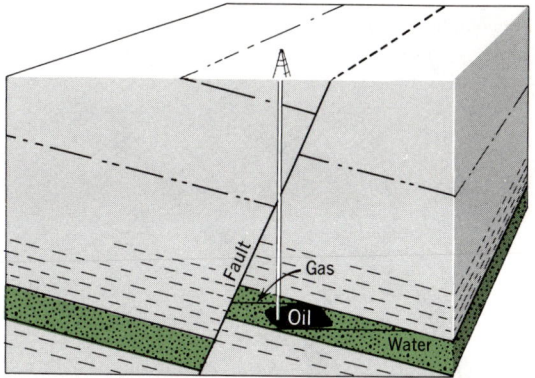

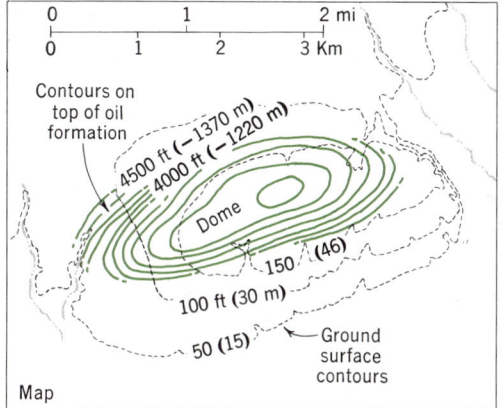

Figure 10.15 The Dominguez Hills, a low dome in an early stage of erosion, has beneath it a valuable oil pool. (After H. W. Hoots and U.S. Geological Survey.)

Figure 10.16 An oil pool has accumulated in the permeable sandstone beds and is prevented from escaping by the impermeable shales faulted against the edge of the sandstone layer.

as reservoirs. One of the simplest arrangements of rock strata holding oil and gas is the *dome* or *anticline,* an up-arching of strata shown in cross-section in Figure 10.14. The sandstone layer holds natural gas, petroleum, and water in layered order from top to bottom, while an impervious shale formation serves as a *cap rock.*

While a few oil pools occur in structures as simple as that shown in Figure 10.14, most are more complex. A typical stratigraphic dome containing a valuable oil accumulation is illustrated in Figure 10.15. At the ground surface this structure is expressed as an elliptical area of low hills, showing that the up-arching occurred quite recently. Many dome structures of this type have been important oil producers in the Rocky Mountain region.

Other arrangements and structures of sedimentary strata are capable of holding large quantities of oil and gas; these structures are known generally as *traps*. One type of trap is the *fault trap,* illustrated in Figure 10.16. The fault has displaced sedimentary strata and petroleum has migrated up the inclined sandstone formation to accumulate against the barrier formed by shale on the opposite side of the fault surface. Another general class of trap is the *stratigraphic trap.* One trap of this class is illustrated in Figure 10.17. The sand formations become thinner from right to left as the elevation of the formation rises. The formation is said to *pinch out.* Petroleum accumulates in the highest part of the sand layer, held within impervious shales above and below. The pinch-out trap is particularly important in strata of the Gulf Coast region (described as a geosyncline in Chapter 9). Figure 10.18 shows the oil pools associated with

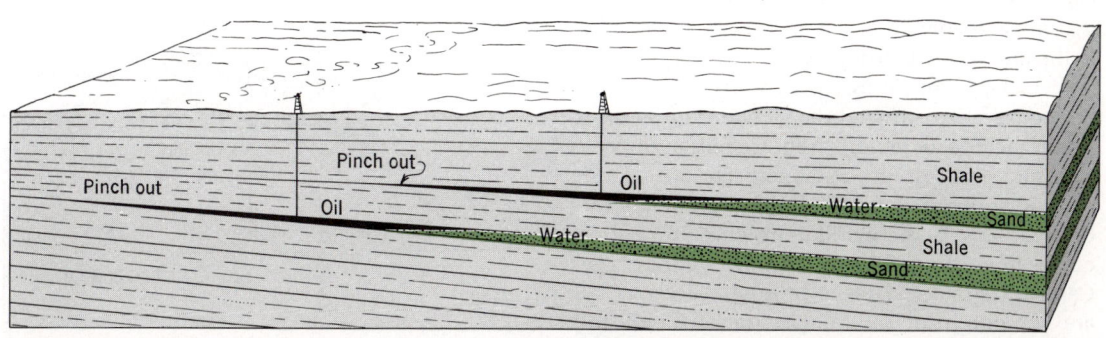

Figure 10.17 Oil pools can form in the fringes of sand formations which pinch out in the direction of rise.

251

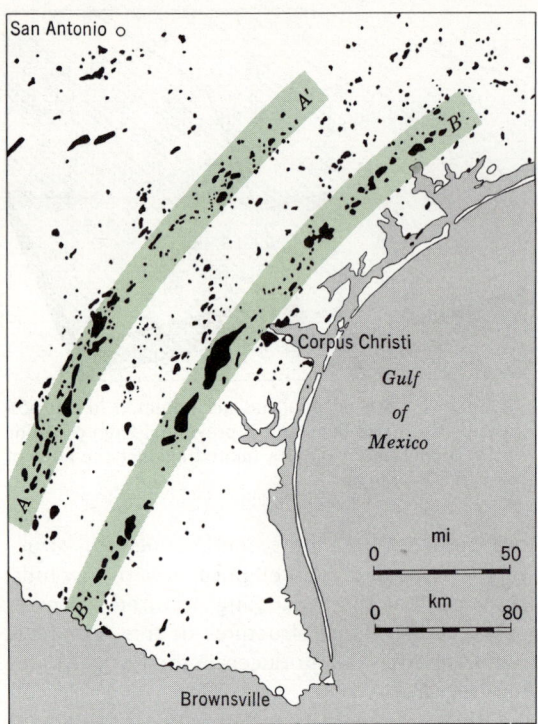

Figure 10.18 Two zones of oil pools on up-dip pinch-outs of sands of Eocene age (*AA'*) and Oligocene age (*BB'*). (After A. I. Levorsen.)

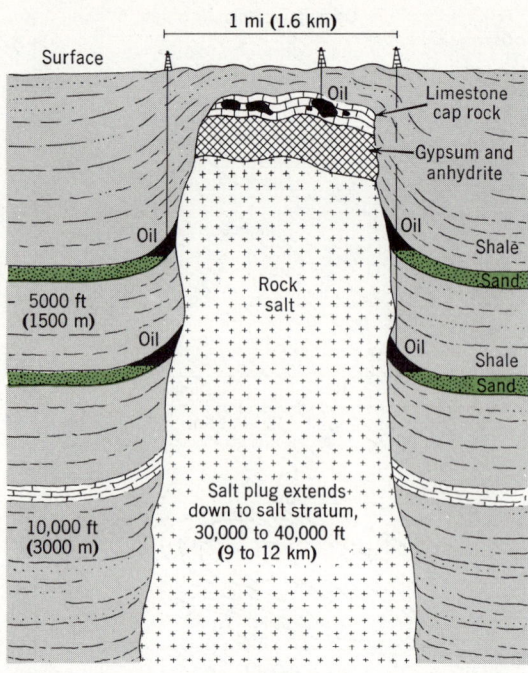

Figure 10.19 Idealized structure section of a salt dome.

two pinch-out structures in sand formations of Eocene and Oligocene ages under the coastal plain of Texas.

Another type of oil pool, abundant in the thick geosynclinal strata of the Gulf Coast, is associated with *salt domes* or *salt plugs* (Figure 10.19). These strange, stalklike bodies of rock salt project upward through thousands of feet of strata. Apparently they were forced up by slow plastic flowage from thick salt formations lying in deep lower layers. Surrounding strata are sharply bent up and faulted against the side of the salt plug, making traps for petroleum. Salt plugs commonly have a cap rock of limestone resting upon a plate of gypsum and anhydrite. Oil may collect in cavities in the limestone. Distribution of salt domes of the Gulf Coast is shown in Figure 10.20.

Geologic ages of strata in which petroleum is found include all periods from Cambrian to present, but strata of the Paleozoic Era are of oil-bearing importance only in the United States and Canada. Most of the world petroleum production comes from strata of Cretaceous and younger age. Cenozoic strata of Eocene through Miocene age are the dominant oil sources for oil fields the world over, including those of South America, Europe, the Middle East, and Indonesia.

In this brief discussion we cannot even begin to go into the world distribution of known petroleum deposits. The important point is that petroleum reserves seem to be heavily concentrated in a few world regions. Some two-thirds of the world reserves lie in two regions that are antipodal with respect to one another (meaning that they lie at diametrically opposite points on the globe). These *oil poles* are the Gulf of Mexico-

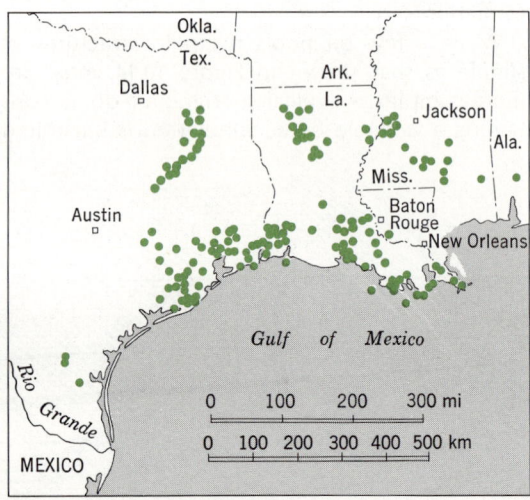

Figure 10.20 Distribution of salt domes of the Gulf Coast region is indicated by dots. (After K. K. Landes.)

TABLE 10.8 Estimated World Petroleum Resources

Region	Billions of Barrels
United States	200
Canada	95
Latin America	225
Europe	20
Africa	250
Middle East	600
Far East	200
U.S.S.R. and China	500
Total	2,100

SOURCE: Data by W. P. Ryman, Standard Oil Company of New Jersey, as given by M. K. Hubbert in *Scientific American*, 1971, vol. 224, no. 3, p. 65.

Caribbean region and the Middle East region. Of the two poles, that of the Middle East has by far the greater reserves. Table 10.8 shows one estimate of the total world petroleum resources by regions. The total quantity, about 2000 billion barrels, is one of the highest yet made. Other estimates run as low as 1350 billion barrels. The newly discovered Alaska oil field at Prudhoe Bay may be estimated to contain about 10 billion barrels, and future discoveries in that area may raise reserves in that area to a figure as high as 30 to 50 billion barrels. In terms of world reserves, or even of total U.S. reserves, this quantity is not large.

Additional petroleum resources are disseminated through large volumes of sedimentary rock as *tar sands* and as *oil shales*. The significance of these reserves in terms of the total hydrocarbon energy resource is evaluated at a later point in this chapter. A significant fact that will emerge is that the world resources of coal and lignite are vastly greater than those of petroleum in all its forms, when evaluated in terms of recoverable energy.

Environmental Impact of Mineral Extraction and Processing

The environment of the life layer is affected adversely in many varied ways by removal of mineral deposits and fossil fuels and by the processing of those substances. Then, too, there are direct threats upon Man's safety and health as he engages in these activities. Listed below in outline form are many of these important environmental effects.

1. *Scarification of the land; waste (spoil) accumulation* by
    a. Open-pit mining of ores (Chapter 11)
    b. Quarries for extraction of structural materials, such as limestone, granite, gypsum, crushed rock, glass sands (Chapter 11)
    c. Borrow pits, sand and gravel pits, clay pits, phosphate pits (Chapter 11)
    d. Dredging and hydraulic mining of stream gravels (Chapter 14)
2. *Mass movements of rock and overburden caused by mining*
    a. Subsidence of land surface due to collapse of mines and withdrawal of petroleum (Chapter 11)
    b. Sliding and flowage of soil and bedrock due to removal of support in mining operations and to instability of spoil accumulations (Chapter 11)
3. *Hydrologic effects upon runoff, ground water, and streams*
    a. Sedimentation of stream channels induced by exposure of soil and rock and by spoil accumulations (Chapter 14)
    b. Water pollution from mines and spoil accumulations, including acid mine drainage (Chapter 12)
4. *Air pollution* by
    a. Smelting and treatment of ores, especially sulfide ores
    b. Combustion of fossil fuels, particularly through emissions of sulfides and carbon dioxide (Chapter 6)
5. *Ocean pollution* by
    a. Petroleum hydrocarbons
    b. Metals
    c. Synthetic industrial compounds
6. *Health hazards associated with mining*, including
    a. Death and injury within mines due to accidents
    b. Respiratory disorders and diseases related to mining and mineral processing
    c. Ionizing radiation exposure

Of the topics listed above, two are not covered in other chapters and will be briefly treated here. One is the physical aspect of pollution of the oceans by oil. The other is health hazards associated with mining.

Marine Oil Pollution

Everyone is familiar with two major marine oil-pollution events of recent years. One was the *Torrey Canyon* disaster of 1967 off the coast of Cornwall, England. A huge oil tanker of that name ran aground and broke apart, releasing 100,000 metric tons of oil with lethal effects upon ecosystems of the shore zone. Another was the Santa

Barbara accident, beginning in 1969, in which crude oil leakage from an offshore well produced a large oil slick that spread to the shoreline, polluting beaches and damaging marine life of the coast. Altogether some 10,000 metric tons of oil were introduced into the ocean from the Santa Barbara accident. Great as these quantities of oil may seem, they are only a very small proportion of the total quantity of oil introduced into the oceans annually from all sources.

Evidence of widespread marine oil pollution comes from collection of floating oil-tar lumps over wide reaches of the oceans. These lumps represent the nonvolatile residues of crude oil spilled in oil transportation. Marine scientists pick up the oil-tar lumps in a type of net (neuston net) towed behind a vessel to pick up zooplankton (microscopic floating animals) from the surface water layer. On one cruise of the Woods Hole Oceanographic Institution's Research Vessel *Chain* oil-tar lumps as large as 3 in. (5 cm) in diameter were picked up in the Sargasso Sea of the subtropical Atlantic Ocean. Within two to four hours of towing the nets became so encrusted with tar that they had to be cleaned with solvent. Data from similar plankton tows in the northwestern Atlantic Ocean and the Mediterranean Sea showed tar concentrations of 1 mg/m^2 for the North Atlantic Ocean and 20 mg/m^2 for the Mediterranean Sea. Thor Heyerdahl reported that during his 1970 voyage across the Atlantic Ocean in the papyrus vessel, *Ra*, pollution by tarlike and asphaltlike lumps was visible during 6 of the 52 days the trip lasted. While vast areas of the oceans remain unsampled, it is obvious that marine oil pollution exists on a large scale.

Sources of marine oil pollution are varied (see Table 10.9). A major source is that connected with oil transportation on the open ocean and in estuaries and ports where oil is transferred. In addition to major oil spills from collision and breakup of grounded tankers there is a continual infusion of oil from tank and bilge cleaning operations and leakage of lubricating oil from propeller shaft bearings. It has been estimated that the total annual contribution of oil to the oceans from transport sources alone runs to at least 1 million metric tons and may even be double that amount. The lower value represents one-tenth of one percent (0.1%) of all oil transported by water. (About 60% of all oil produced is transported by water.)

An additional infusion of crude oil occurs through natural and accidental seepages from the ocean floor, as in the case of the Santa Barbara accident. Tankers sunk during World War II pose a continual threat of oil leakage. Another major source is oil released from the lands in the form of refinery wastes, unburned fuels, and used lubricants carried to the oceans by streams and by ocean-outfall sewage systems (Table 10.9). Yet another source of marine oil pollution (not listed in Table 10.9) is the direct fallout of petroleum hydrocarbon particles from the atmosphere from automobile exhausts and various industrial activities. Altogether the various sources listed in Table 10.9 contribute about 2 million metric tons per year of oil and oil products to the oceans, as calculated for the year 1969. This figure represents about 0.1% of the 1969 world petroleum production of 1.8 billion metric tons.

Figure 10.21 shows schematically the conditions under which offshore oil drilling can lead to seepage of crude oil from the ocean floor. It is somewhat similar to the type of structure that resulted in the Santa Barbara accident. The drill hole has passed through a fault zone along which the rock has been crushed and rendered permeable to fluid flow. Once the oil pool is tapped, oil under pressure escapes from the hole into the fault zone and makes its way to the ocean floor. Natural seepages of the same type may occur along active fault zones which cut through oil reservoir structures.

Long-term physical effects of marine oil pollution have not yet been detected, but the qualita-

TABLE 10.9 Marine Oil Pollution Sources, 1969

Source	Thousands of Metric Tons per Year	Percent of Total
Tankers in normal operations		
Controlled	30	1.4
Uncontrolled	500	24.0
Other ships (bilge pumping)	500	24.0
Offshore oil production (normal operations)	100	4.8
Accidental spills		
Ships	100	4.8
Nonships	100	4.8
Refineries	300	14.4
Rivers carrying automobile and industrial hydrocarbons	450	21.6
Total	2,080	100.0

SOURCE: *Man's Impact on the Global Environment*, 1970, SCEP report, The MIT Press, Cambridge, Mass., p. 267, Table 5.10.

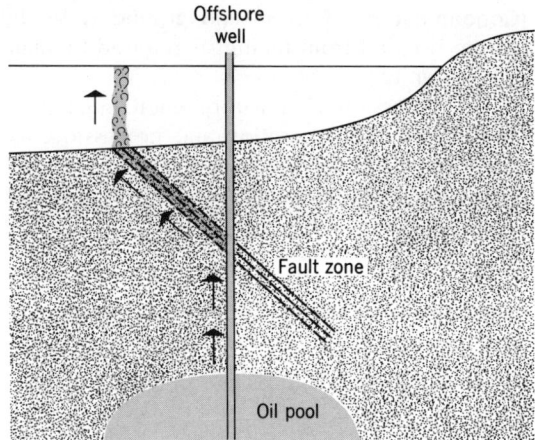

Figure 10.21 Schematic diagram of petroleum seepage along a fault zone penetrated by an offshore well.

tive nature of the effects can be predicted in a general way. Assuming that there comes to exist over significant areas and for long enough spans of time an oil cover on the sea surface, the effects upon the global heat and water balances would be felt in two ways. First, inhibition of normal evaporation would alter the latent-heat flux term of the heat and water balances. Second, the presence of oil might reduce the intensity of momentum transfer from winds to the sea surface and thereby reduce rates of mass water transport and the intensity of wave energy reaching the shores. Reduced wave size would in turn reduce the extent of mixing in the surface layer and alter both the thermal structure and oxygen content of that layer. As yet, none of these and other predictable physical effects has been noted or even inferred to have taken place. However, we should remain alert to the possibilities of such effects in the future, particularly in small, largely landlocked seas and estuaries where oil transport by water and shore-based industrial activities are concentrated. Special attention should also be given to arctic and polar waters because bacterial oxidation of hydrocarbons is extremely slow at low temperatures and oil spilled into the ocean may last as long as 50 years.

Although technology of oil-spill removal and treatment is advancing rapidly, the increase in petroleum transport and in the size of tank vessels will tend to increase the frequency and size of major spills. However, removal and treatment methods cannot be effective against the dispersed sources of oil during transport, the land-based sources, and atmospheric fallout, all of which can be expected to increase in proportion to the consumption of hydrocarbon fuels. International control and regulation of water transport to reduce oil spillage is difficult to implement because of the distribution of ship registry among many nations and the practical impossibility of effective patrolling of vessels on the high seas.

Health Hazards Associated with Mining

That mining is a hazardous occupation is amply borne out by statistics. Coal mining, in particular, has chalked up a terrifying record of death and injury from explosions, fires, and cave-ins. Since reliable records have been kept, over 80,000 men have lost their lives in coal mine accidents in the United States alone. Another extreme hazard lies in inhalation of dusts produced in mines, and these effects are felt to a lesser degree in the environs of mines and mineral processing plants. Among coal miners the prevalent *black lung* disease, *pneumoconiosis,* is caused by inhalation of coal dust; it leads to disability ranging from slight to severe and often results in death because no cure is known. A recent estimate places at 4000 the annual death toll attributable to respiratory diseases contracted in coal mining. Among large groups of coal miners examined recently in the United States from 30 to 60% within each group were found to have black lung disease. Another respiratory disease produced by mine dusts is *silicosis,* a lung fibrosis caused by accumulation of silica particles inhaled in close proximity to rock drilling in mines and quarries in silicate rocks.

Yet another health hazard is found in the mining, processing, and handling of asbestos, one variety of which is the fibrous form of the mineral serpentine (see Chapter 8). Asbestos of this type occurs in commercially useful quantities in metamorphic rocks. As Figure 10.22 shows, asbestos

Figure 10.22 A specimen of one form of asbestos. The mineral is a fibrous variety of serpentine. (Photograph by courtesy of Ward's Natural Science Establishment, Rochester, N.Y.)

fibers are extremely fine and flexible. Long fibers can be woven into cloth, while short fibers are matted into boards and used in other ways in fire-resistant and heat-insulating applications. Individual filaments of the mineral are as thin as 0.0004 in. (10 microns) in diameter. Broken filaments are carried in suspension as dust particles and lodge in the lung, giving rise to a disease known as *asbestosis,* a form of pneumoconiosis. Among workers in asbestos mines and manufacturing plants asbestosis is a prevalent disease and its occurrence extends to workers in the building trades who handle asbestos products. Asbestos can be found in dusts of all large industrial cities and the particles are present in the lungs of city dwellers. Asbestosis is known to lead to certain rare forms of cancer, and may prove to be widely implicated in lung diseases generally.

The mining and milling of uranium ores has brought into importance a comparatively new environmental hazard, that of ionizing radiation. (See Chapters 2 and 9 for background information.) Within uranium mines, radiation doses can exceed 5600 mrem/yr, a quantity some 40 to 70 times the average background radiation dose from all sources. Besides the direct gamma radiation from the uranium ore there is a radiation hazard in the inhalation of *radon* gas. (Radon is a radioactive daughter product of uranium decay; see Figure 9.37.) The gas rapidly produces other radioactive daughter products and these lodge in the lungs. Miners of pitchblende (an ore of uranium and radium) in Germany and Czechoslovakia were the first groups to show the effects of this form of ionizing radiation. Some 50% of these miners ultimately died of lung cancer, attributed to the radiation exposure. The lesson went unheeded, however, for when uranium mining in the United States boomed in the period of World War II and after, standards of individual exposure and mine ventilation were set too low. In a study group of about 3400 uranium miners 46 had died of lung cancer by the year 1967, and it has been forecast that of the 6000-odd men who have been uranium miners some 600 to 1100 will ultimately be dead of lung cancer because of exposure to radiation. Tailings (rock wastes) from uranium mines are another source of ionizing radiation for they contain radium-226 and other radioisotopes. Radium is particularly toxic in stream waters derived from leaching of the tailings, while radon gas, the daughter product, can accumulate to hazardous levels in buildings resting on tailings. In recent years dangerous radon gas levels have been discovered or suspected in homes in Uravan and Grand Junction, Colorado.

(Contamination of rivers and ground water by radium leached from tailings is referred to again in Chapter 12.)

This discussion of environmental impacts of mineral and fuel extraction and processing, together with discussions elsewhere in the book, is far from being a complete coverage in detail, but it at least gives some idea of the scope of environmental effects of industrialization. Most of the observed detrimental effects could be eliminated or greatly reduced by application of appropriate regulations and known techniques. Unfortunately, these measures can be extremely costly and their implementation is not easy to achieve.

Sources of Energy

Before looking into the sources of energy that are derived from the solid earth, let us review the full picture of world energy resources to gain a better perspective in terms of natural physical systems.* Sources of energy are found in both sustained-yield and exhaustible categories. A sustained-yield source is one that undergoes no appreciable diminution of energy supply during the period of projected use. Consider first the sustained-yield sources.

Solar energy has been described in some detail in Chapter 1, dealing with the radiation balance. Stated in terms of power, solar radiation intercepted by one hemisphere (or rather, the area of a circle equal in diameter to the earth and presented at right angles to the sun's rays) is calculated to be about 100,000 times as great as the total existing electric power generating capacity. The problem is, of course, that solar radiation derived from a large receiving area must be concentrated into a very small distribution center. To produce power equivalent to that of a large generating plant (e.g., 1000 megawatts capacity) would require at an average location a collecting surface of about 16 sq mi (42 sq km), represented by a square measuring 4 mi (6.5 km) on a side. While there seems to be no technological barrier to building such a plant, the cost at present is far too high to make this energy source a practical one. However, the source will always be available for future use.

Water power under gravity flow is a second source of sustained energy and has been devel-

---

*Data in the remainder of this chapter have been drawn largely from M. K. Hubbert, "Energy Resources," Chapter 8 of *Resources and Man,* National Academy of Sciences. Published by W. H. Freeman and Company, San Francisco.

oped to a point just over one-quarter of its estimated ultimate maximum capacity in the United States. In 1965 water power supplied about 4% of the total energy production of the United States (see Figure 10.24). Since we have a good knowledge of stream runoff and stream profiles, the estimate of maximum capacity is probably not much in error. For the world as a whole, present development is estimated to be about one-nineteenth of the ultimate maximum capacity. Potential power is particularly great in South America and Africa, where coal is in very short supply. A serious defect in such calculations is one referred to in Chapter 14—the loss of capacity of artificial reservoirs through sedimentation. Most large reservoirs behind big dams have an estimated useful life of a century or two at most. Perhaps, after all, water power is not in reality to be categorized as a "sustained" source of energy.

Tidal power is a third sustained-yield energy source. (The nature of tidal energy is discussed in Chapter 15.) To utilize this power a bay subject to a large range of tide is selected for development. Narrowing of the connection between bay and open ocean intensifies the differences of water level that are developed during the rise and fall of tide. A strong hydraulic current is produced and alternates in direction of flow every 12½ hours (for a semidiurnal tidal cycle). The flow is used to drive turbines and electrical generators, with a maximum efficiency of about 20 to 25%. Assessment of the world total of annual energy potentially available by exploitation of all suitable sites comes to only 1% of the energy potentially available through water-power development.

Yet another sustained-yield energy source is classified as *geothermal*; it uses heat within the earth at points of locally high concentration—e.g., hot springs, fumaroles, and active volcanoes. Wells drilled at such places yield superheated steam, which can be used to power an electric generating plant (Figure 10.23). Electric power is presently being generated from a number of geothermal fields, the most important being located in Italy, New Zealand, and the United States. Table 10.10 gives the extent of world power development from geothermal sources in the year 1969. Planned development and development under construction will in the near future add at least 850,000 kw to the total given in Table 10.10, so will about double the 1969 figure. Of this added power, some 600,000 kw will come from The Geysers field in California. Since 1969 the production at Wairakei, New Zealand, has risen to 290,000 kw. An estimate made in 1965 by Donald E. White of the United States Geological

TABLE 10.10 Geothermal Power Development

Country	Name of Field	1969 Capacity (thousands of kilowatts)
Iceland	Hveragerdi	17
Italy	Larderello	365
	Monte Amiata	25
Japan	Matsukawa	20
	Otake	13
Mexico	Pathé	4
New Zealand	Wairakei	160
	Kawerau	10
United States	The Geysers	83
U.S.S.R.	Pauzhetsk	3
	Total	700

SOURCE: J. B. Koenig, 1971, *Geotimes*, vol. 16, no. 3, p. 12.

Survey shows that geothermal power production can be increased to an amount at least 10 times larger than in 1965 and can be sustained for at least 50 years. This total capacity is about the same as the estimated value for potential tidal power development; both being very small fractions of existing energy requirements.

Energy from Fossil Fuels

So we return to the principal sources of energy expended in the past century. These are the fossil fuels: coal and lignite, petroleum (as crude oil),

Figure 10.23 This group of wells in The Geysers geothermal field of California produces about 400 tons of steam per hour. Condensation plumes are generated during blow-out procedures to remove subterranean debris. (Photograph by Pacific Gas and Electric Company.)

and natural gas. Figure 10.24 is a graph showing the production of energy (scaled in units of heat) from fossil fuels plus water power, since 1900. Water power has, in this 65-year period, amounted to about 3 to 4% of the yearly production of energy, so that only a small allowance needs to be made for its inclusion. The relative contributions of the several sources of energy over the past 65 years are also shown in Figure 10.24. The two graphs tell us that while the total production of energy has increased over five-fold since 1900, the contributions of the several sources have changed markedly in ratio. Coal and lignite together have been reduced to less than half of the starting percentage, while anthracite has declined to almost nothing. Both petroleum and natural gas have increased in proportion in the same period, but of the two, natural gas has greatly expanded its ratio to become about equal with petroleum.

Since the quantity of stored hydrocarbons in the earth's crust is finite, and the rate of geologic production and accumulation of new hydrocarbons is immeasurably small in comparison with the rate of their consumption, the ultimate exhaustion of this energy source is inescapable. When this event will happen is, however, a very difficult thing to predict, since we have to project into the future two independent curves. First is the rate of production, which at present is increasing by about 6% per year for petroleum (crude-oil). So far, discoveries of new oil reserves have more than kept pace with production, so there has been a moderate increase of known reserves. In terms of time, we are running about 12 years ahead on discoveries of new oil reserves in the United States. But in due time the rate of increase in new discoveries must slow down and then begin to decline. The reserves will then dwindle and eventually be entirely used, after

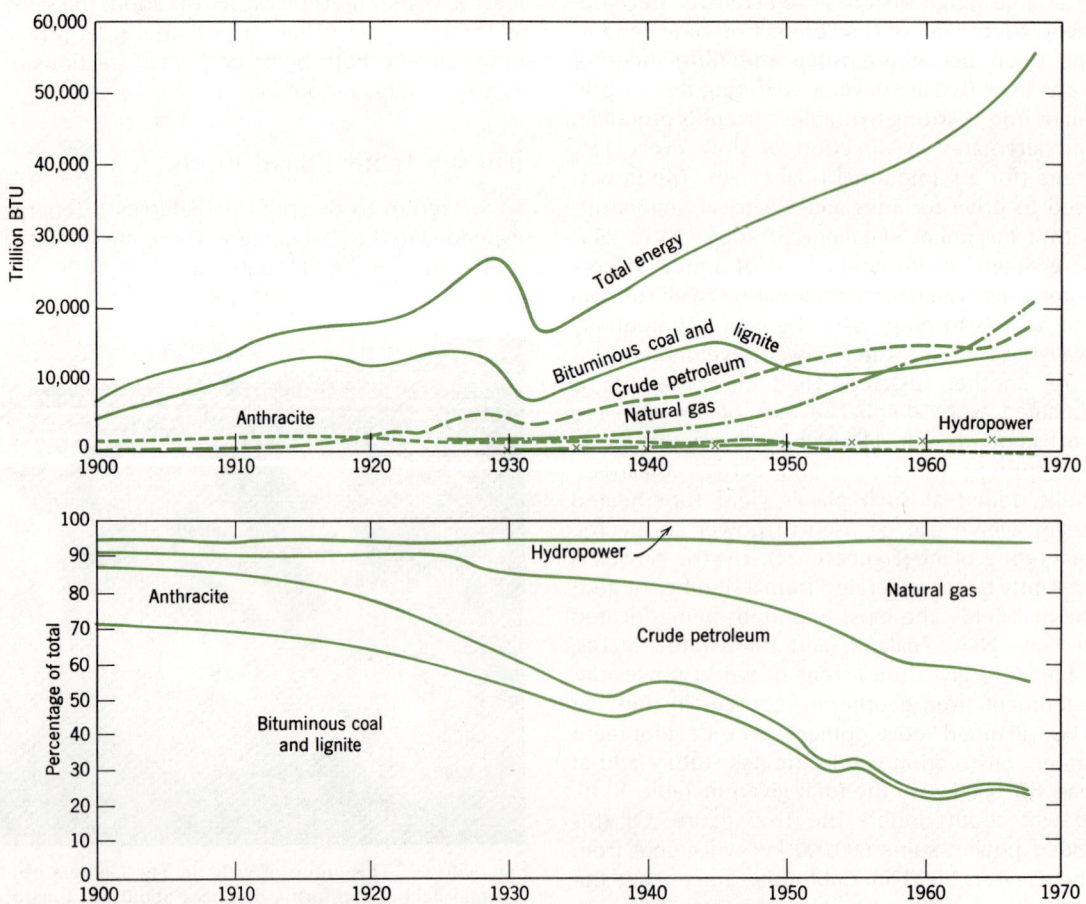

Figure 10.24 Changes in production of fossil-fuel energy resources and hydroelectric power, 1900 to 1968. Quantity of nuclear power is too small to show on this scale. Data of U.S. Bureau of Mines, Minerals Yearbook. (From A. N. Strahler, 1972, *Planet Earth; Its Physical Systems Through Geologic Time*, Harper & Row, New York.)

which point production itself must begin a decline and will ultimately approach zero.

The complete cycle of petroleum production is envisaged as a symmetrical, bell-shaped curve (Figure 10.25). Two estimates are shown, each based upon a different value for the ultimate amount of liquid petroleum produced. Peak of production is shown as occurring in either 1990 or 2000; with decline to a rate equal to that of 1927 being arrived at in either 2050 or 2070.

Estimates of the world resource of natural gas are more difficult to arrive at, but a similar bell-shaped curve of production is anticipated on roughly the same time schedule. Are there other petroleum sources that can be developed? One possibility is the use of *heavy oil* enclosed in sands (oil sands) and not as yet exploited. Estimated reserves of these heavy-oil accumulations show that they are an important fraction as compared with petroleum reserves and development work is presently under way to begin production.

Another possibility is the use of *oil shales* as hydrocarbon sources. Although the hydrocarbon is in solid form, disseminated throughout the shale formation, treatment can ultimately yield petroleum in an amount estimated to range from 10 to 100 gallons of oil per ton of shale. Known resources from oil shales are thought to be comparable to the ultimate world capacity of petroleum (about 2500 billion barrels of oil), although less than a tenth of that amount is considered recoverable under present conditions.

What of the future of our coal resources? The picture is quite different from that of petroleum and natural gas. Estimates of coal resources are regarded as quite realistic, since the existence and thickness of coal seams can be directly sampled from borings. The estimate given in Table 10.7 of total world coal that can be mined comes to about 7600 billion (7.6×10^{12}) metric tons, of which more than half lies in Asia and the European Soviet Union, and about one-third in North America. (The figure includes coal already mined, but this is only a small fraction of the total.) Projected curves of coal production indicate a peak around the year 2100 to 2200, which is a century or two beyond the peak for petroleum. Furthermore, the present rate of production (about equal to that of petroleum) is only a small fraction of the peak value. The conclusion must be that coal will become our main hydrocarbon energy source by the year 2050 or thereabouts and will continue in that role thereafter until it, too, is exhausted.

Nuclear Energy as a Resource

The controlled release of energy from concentrated radioactive isotopes can be achieved through one of two processes: *fission* and *fusion*. Atomic fission makes use of uranium-235. The fission of one gram of this substance yields an amount of heat equivalent to the combustion of about 3 metric tons of coal or about 14 barrels of crude oil. While uranium-235 is the basic fissionable material, it is a rare isotope of a very rare element (see Table 10.1) and the world supply would be rapidly exhausted if nothing else were used. It is, however, possible to induce fission in other isotopes, notably other isotopes of uranium and isotopes of plutonium and thorium.

Of particular interest in the development of future energy sources is the process of *breeding*, carried out in the *breeder reactor*; it burns uranium or plutonium in such a way as to convert thorium-232 or uranium-238 into fissionable materials, which then become the new fuels. The importance of the breeder reactor lies in the fact that it produces more new fuel than it burns and that it makes use of the naturally more abundant isotopes. A breeder can double the original quantity of fuel within 6 to 20 years time. Moreover,

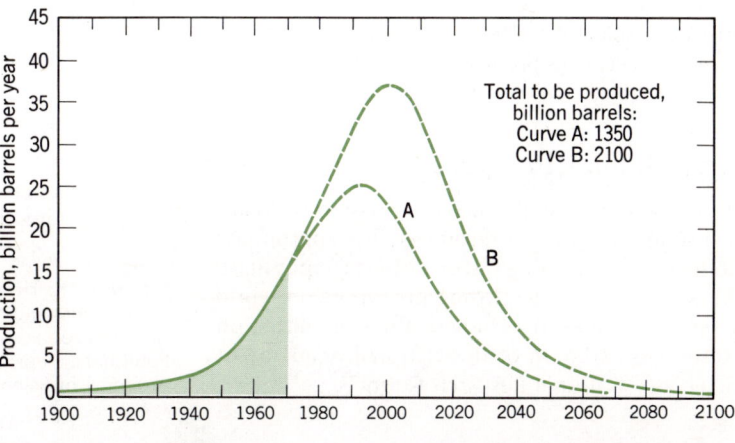

Figure 10.25 World crude-oil production projected into the future according to two estimates of the total ultimate production. (After M. K. Hubbert, 1969, in *Resources and Man*, Freeman, San Francisco, p. 196, Figure 8.23.)

while the existing nuclear reactors make use of only between 30 and 50% of mined uranium, the breeder reactor is capable of using more than 70%, with a consequent great reduction in rate of depletion of uranium resources.

Great concern has been expressed by scientists over the necessity to develop breeder reactors to conserve uranium. If such development is successful, low-grade deposits of uranium can be exploited, making available a source of energy judged to range from hundreds to thousands of times greater than all reserves of fossil fuels.

Energy from fusion depends upon fusing isotopes of hydrogen into helium with a consequent large release of energy. As everyone knows, the explosive release of enormous quantities of energy has been achieved through the hydrogen bomb (thermonuclear bomb). As yet, controlled release of energy through hydrogen fusion has not been achieved, although research is in progress. In theory, the quantities of energy available through fusion could exceed that of fossil fuels by a factor ranging up to hundreds of thousands.

Nuclear energy development is accompanied by a host of difficulties and attendant environmental problems. One is the generation of unwanted heat that raises water temperatures of rivers or lakes into which it is discharged. This activity, which is one of several forms of industrial thermal pollution, is the subject of debate (see Chapter 13). Of more far-reaching importance is the environmental problem of disposal of radioactive wastes from chemical plants which process nuclear fuels and the possibility of accidental release of radioactive substances from reactors in nuclear power plants. Consideration of these problems brings into play all of the concepts of systems of water circulation within the hydrosphere, including runoff, ground-water movement, and oceanic circulation. In addition, atmospheric circulation systems may become involved, as in the case of fallout of radioactive particulates released into the atmosphere by nuclear explosions. The relation of these problems to the earth's physical systems of energy and mass exchange should be kept in mind.

The Human Prospect

To bring to a close this brief discussion of Man's consumption of earth resources, it is appropriate to read the words of a distinguished geophysicist, M. King Hubbert, in closing his report on world energy resources in a study by the Committee on Resources and Man of the National Academy of Sciences—National Research Council:

To sustain a high-energy-dependent world culture for a period much longer than a few centuries requires, therefore, a reliable source of energy of appropriate magnitude. The largest and most obvious of such sources is solar radiation, the continuance of which at close to present rates may be relied upon for millions of years into the future. The energy from solar radiation, with the exception of that fraction manifested as water power, does not offer much promise as a means of large-scale power production, although future technology may circumvent this difficulty. This leaves us with nuclear energy as our only remaining energy source of requisite magnitude. Although the earth's resources of uranium and thorium, and of deuterium, are finite and therefore exhaustible, the magnitudes of these resources in terms of their potential energy contents are so large that with breeder and fusion reactors they should be able to supply the power requirements of an industrialized world society for some millenia. In this case, the limits to the growth of industrial activity would not be imposed by a scarcity of energy resources, but by the limitations of area and of the other natural resources of a finite earth. It now appears that the period of rapid population and industrial growth that has prevailed during the last few centuries, instead of being the normal order of things and capable of continuance into the indefinite future, is actually one of the most abnormal phases of human history. It represents only a brief transitional episode between two very much longer periods, each characterized by rates of change so slow as to be regarded essentially as a period of nongrowth. It is paradoxical that although the forthcoming period of nongrowth poses no insuperable physical or biological problems, it will entail a fundamental revision of those aspects of our current economic and social thinking which stem from the assumption that the growth rates which have characterized this temporary period can be permanent.*

Whether you share this scientist's views or not, his long look into the future of Man on earth must receive your serious attention as you ponder our best course in the light of our knowledge of the planetary energy systems.

*From M. K. Hubbert, "Energy Resources," pp. 238-39 of *Resources and Man,* copyright © 1969 by the National Academy of Sciences, W. H. Freeman and Company, Publishers. Reproduced by permission of the National Academy of Sciences.

PART III

Energy Systems at the Fluid-Solid Interface

11 Wasting of the Continental Surfaces

In this third part of our study of environmental science we will build further on the concepts developed in Parts 1 and 2. Our attention now focuses more sharply on the shallow life layer itself, for here the externally acting, solar-driven energy systems of the atmosphere and oceans mesh with the internally driven geologic system that has created and raised the continental masses, allowing their varied rock types to become exposed to the surface environment. The interaction of these two great planetary systems has already been examined in Chapter 8 with respect to the chemical alteration of rock and the production of sediment, a process essential to the rock transformation cycle and to growth of the continental crust. Our interest in that chapter was centered on the broader geologic aspects of rock alteration, whereas in this part of the book we will look at the role of rock alteration and associated external processes in shaping the surface of the lands and in the production of soil. Landforms, described in earlier chapters as the distinctive configurations of the land surface—mountains, hills, valleys, plains, plateaus, and the like—make up the varied habitats of plants and animals, while the soil layer on those landforms supplies the nourishment to sustain life forms. Landforms are environmentally significant because they influence the place-to-place variation in such ecological factors as water availability and exposure to radiant solar energy. Through varying degrees of inclination of the ground surface, i.e., the property of *slope*, landforms directly influence hydrologic and soil-forming processes, which in turn act as direct controls on life forms.

Geomorphology and Hydrology

Study of the life layer of the continents brings together two major branches of the earth sciences: geomorphology and hydrology. *Geomorphology*—the study of landforms, including their history and processes of origin—deals largely with the action of *fluid agents* which erode, transport, and deposit mineral and organic matter. These fluid agents are (a) *running water* in surface and underground flow systems, (b) *waves* and *cur-*

rents, acting in oceans and lakes, (c) *glacial ice,* moving sluggishly in great masses, and (d) *wind* blowing over the ground. Of these four agents, three are forms of water. Consequently the science of *hydrology,* which is a study of the complex flow paths of water on and beneath the land surfaces, is inseparably interwoven with geomorphology. One might be not far wrong in saying simply that the hydrologist is preoccupied with "where water goes"; the geomorphologist with "what water does." Hydrology concerns itself with the hydrologic cycle in an attempt to calculate the water balance and to quantify flow rates and energy transformations related to the water itself. Geomorphology concerns itself with geologic work that the water in motion performs on the land. Quantification of geomorphic systems is far more difficult than for hydrologic systems, since many forms of energy expenditure take place simultaneously in shaping of landforms; moreover, their effects are often difficult to refer to a time scale.

Weathering and Mass Wasting

Before we turn to the action of fluids in motion as agents of landform development, we must consider certain important processes that do not involve the direct stress of fluid flow. One such group of forces includes those set up in mineral solids by changes in volume, strength, and internal structure, caused by such changing physical parameters as temperature and water content. Another force to consider is that of gravity acting directly upon soil and rock. In the introductory chapter on energy systems we emphasized the role of gravity as a pervasive environmental factor. All processes of the life layer take place in the earth's gravity field and all particles of matter tend to respond to gravity.

Those processes whereby rock and soil are physically disintegrated and chemically decomposed during exposure to atmospheric influences in the surface zone are referred to collectively as *weathering processes.* In addition to the irreversible chemical and physical changes in mineral matter that result from weathering, there is continued agitation and volumetric change in soil because of changes in temperature and water content. These daily and seasonal rhythms of change continue endlessly.

The spontaneous downward movement of soil and rock under the influence of gravity (but without the dynamic action of moving fluids) is included under the general term *mass wasting.* Movement to lower levels takes place when the internal strength of a mass of soil or rock declines to a critical point below which the force of gravity cannot be resisted. This failure of strength with respect to the ever-present force of gravity takes many forms and scales, and we shall see that Man is a major agent in causing several forms of mass wasting.

This chapter deals with processes and forms of weathering and mass wasting. Weathering prepares the parent matter of the soil both physically and chemically for complex soil processes. All of these processes act simultaneously in a given region and their forms and structures evolve concurrently. Consequently, a great deal of interaction takes place.

Bedrock, Soil, and Overburden

Examination of a freshly cut cliff, such as that in a new highway cut or quarry wall, will reveal several kinds of earth materials, shown in Figure 11.1. Solid, hard rock which is still in place and relatively unchanged is called *bedrock.* It grades upward into a zone where the rock has become decayed and has disintegrated into clay, silt, and sand particles. This may be called the *residual overburden,* or *regolith.* At the top is a layer of *true soil,* or *solum,* often called *topsoil* by farmers and gardeners. Over the soil may be a protective layer of grass, shrubs, or trees. Figure 11.2 shows in greater detail the upward transition from bedrock to regolith and to true soil. Figure 8.5 shows a typical exposure of regolith.

One or more of these zones may be missing. Sometimes everything is stripped off down to the bedrock, which then appears at the surface as an *outcrop* (Figure 11.1). Sometimes following cultivation or forest fires the true soil only is eroded away, leaving exposed the regolith, which is infertile and may become scored by deep gullies (see Figure 14.7). The thickness of soil and regolith is quite variable. Although the true soil is rarely more than a few feet thick, the regolith of decayed and fragmented rock may extend down tens or even hundreds of feet. Formation of regolith is greatly aided by the presence of innumerable bedrock cracks termed *joints* (Figure 11.1), along which water can move easily to promote rock decay.

Another variety of overburden that may be found covering the bedrock is *transported overburden.* It consists of such materials as stream-laid gravels and sands, floodplain silts, clays of lake bottoms, beach and dune sands, or rubble left by a melting glacier. All types have in common a history of having been transported by streams, ice, waves, or wind. Whereas residual overburden, or regolith, formed in place by disintegration of bedrock below, is of local origin, transported overburden consists of rock and mineral varieties

Geometry of Rock Breakup

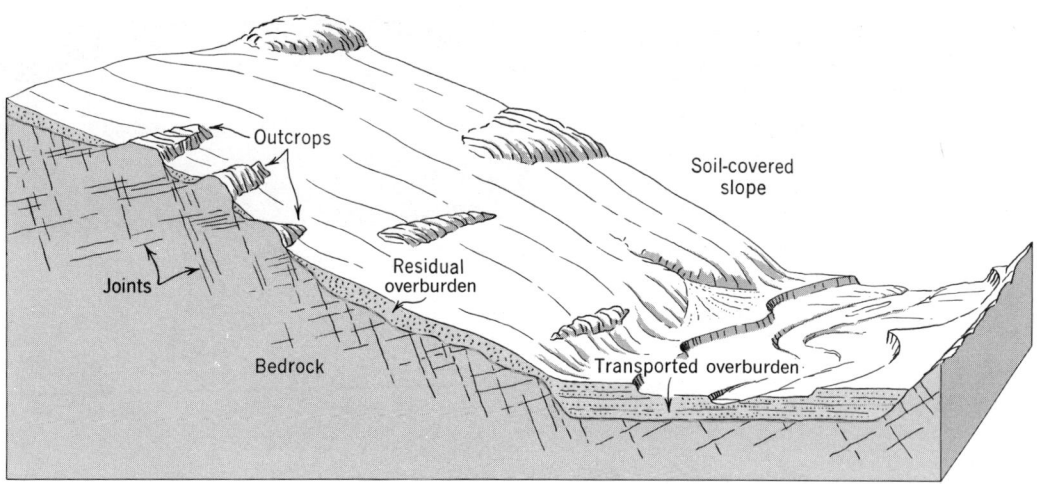

Figure 11.1 Residual and transported overburden.

from distant sources and may be quite unlike the underlying minerals and rocks. Figure 11.1 shows stream valley deposits, called *alluvium*, which would be designated as transported overburden in contrast to residual overburden of the adjacent hill slope. Once deposited, transported overburden may remain undisturbed for many thousands of years, in which case a soil layer is formed in its uppermost layer.

Geometry of Rock Breakup

Before examining weathering processes, it is useful to introduce four terms applied to the geometrical forms in which bedrock breaks into smaller pieces. At this point we are not considering the possible forces involved, but only the shapes of the rock fragments.

Rocks composed of rather coarse mineral grains (intrusive igneous rocks of granitoid texture and coarse clastic sedimentary rocks) typically fall apart grain by grain, a form of breakup termed *granular disintegration* (Figure 11.3). The product is a gravel or sand in which each grain consists of a single mineral particle separated from its fellows along the original crystal or grain boundaries. *Exfoliation* is the formation of curved rock shells which separate in succession from the original rock mass, leaving behind successively smaller spheroidal bodies (see Figure 11.14). This type of breakup is also called *spalling*.

Where a rock has well-developed sets of joints produced previously by orogenic pressures or by shrinkage during cooling from a magma, the common form of breakup is by *block separation* (Figure 11.3). Obviously, comparatively weak forces can separate such blocks, whereas much greater forces are required to make fresh fractures

through solid rock. In sedimentary rocks the planes of stratification, or bedding planes, comprise one set of planes of weakness commonly cutting at right angles to the joints. Figure 11.9 shows joint blocks being separated by weathering forces. Of course, it is quite likely that a single, solid joint block will later break up either by granular disintegration or by exfoliation.

Shattering is the disintegration of rock along new surfaces of breakage in otherwise massive, strong rock, to produce highly angular pieces with sharp corners and edges (Figure 11.3). The surface of fracture may pass between individual mineral crystals or grains, or may cut through them. Blocks seen in Figure 11.4 are joint blocks, many of which have been shattered into smaller pieces.

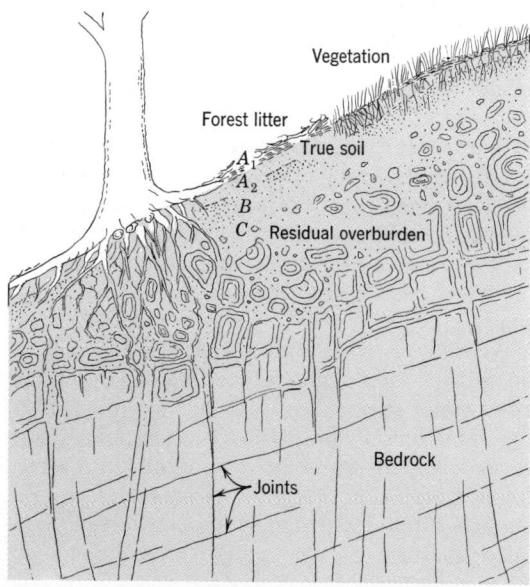

Figure 11.2 Bedrock, residual overburden, and soil.

Wasting of the Continental Surfaces

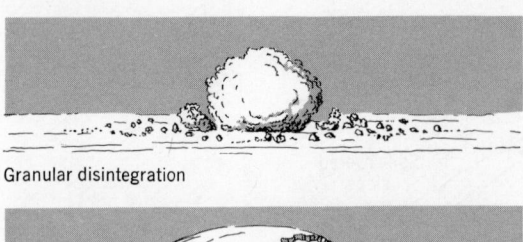
Granular disintegration

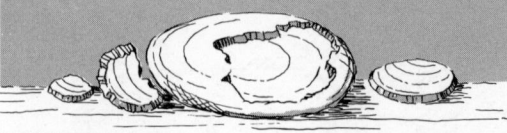
Exfoliation

Block separation

Shattering

Figure 11.3 Rock breakup takes various forms.

Figure 11.4 A felsenmeer atop Medicine Bow Peak, Snowy Range, Wyoming, at 12,000 ft (3650 m) elevation. The rock is quartzite. (Photograph by A. N. Strahler.)

Physical Weathering Processes and Forms

The *physical*, or *mechanical*, processes of weathering produce fine particles from massive rock by the exertion of stresses sufficient to fracture the rock, but do not change its chemical composition. One of the most important physical weathering processes in cold climates is *frost action*, the repeated growth and melting of ice crystals in the pore spaces of soil and in rock fractures. As water in joints freezes, it forms needlelike ice crystals extending across the openings. As these ice needles grow, they exert a powerful force against the confining walls and can easily pry apart the joint blocks. When soil water freezes, it tends to form ice layers parallel with the ground surface, *heaving* the soil upward in an uneven manner.

Freezing water strongly affects soil and rock in all middle- and high-latitude regions having a cold winter season, but its effects are most striking in high mountains, above the timberline, and in arctic regions. Here the separation and shattering of joint blocks may produce extensive ground surfaces littered with angular blocks (Figure 11.4). Such a surface is called a *felsenmeer* (rock sea), or *boulder field*.

Frost action on cliffs of bare rock in high mountains and in arctic regions detaches rock fragments that fall to the cliff base. Where production of fragments is rapid they accumulate to form *talus slopes* (also called *scree slopes*). Most cliffs are notched by narrow ravines which funnel the fragments into individual tracks, so as to produce conelike talus bodies arranged side by side

Figure 11.5 Talus cones at the base of a frost-shattered cirque headwall. Moraine Lake in the Canadian Rockies. (Photograph by Ray Atkeson.)

Physical Weathering Processes and Forms

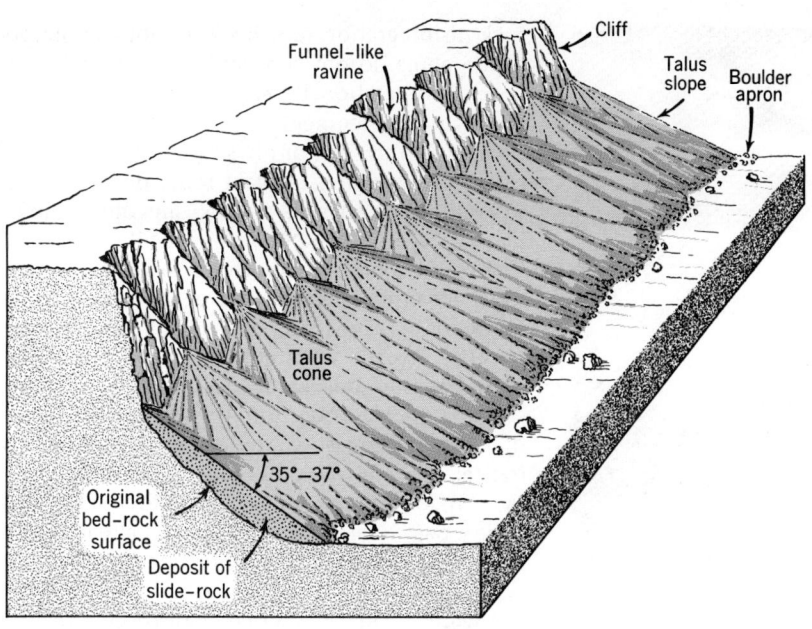

Figure 11.6 Idealized diagram of talus cones formed at the base of a cliff, which might be 200 to 500 ft (60 to 150 m) high.

along the cliff (Figure 11.5). Where a large range of sizes of particles is supplied, the larger pieces, by reason of their greater momentum and ease of rolling, travel to the base of the cone, whereas the smaller grains lodge in the apex. This mechanism tends to sort the fragments by size, progressively finer from base to apex (Figure 11.6).

Most fresh talus slopes are unstable, so that the disturbance created by walking across the slope, or dropping of a large rock fragment from the cliff above, will easily set off a sliding of the surface layer of particles. The upper limiting angle to which coarse, hard, well-sorted rock fragments will stand is termed the *angle of repose*, and is typically an angle of about 33° to 36° with respect to the horizontal. Other examples of this critical angle of slope are seen in the leeward surfaces (slip faces) of sand dunes and on the side slopes of small volcanic cones.

Closely related to the growth of ice crystals is the weathering process of rock disintegration by growth of salt crystals. This process operates extensively in dry climates and is responsible for many of the niches, shallow caves, rock arches, and pits seen in sandstone formations. During long drought periods, ground water is drawn to the surface of the rock by capillary force. As evaporation of the water takes place in the porous outer zone of the sandstone, tiny crystals of salts are left behind. Salts involved in this process are commonly gypsum, calcite, soda niter (sodium nitrate), and hydrous sulfates of sodium and magnesium. The growth force of these crystals is capable of producing granular disintegration of the sandstone, which crumbles into a sand and is swept away by wind and rain. Especially susceptible are zones of rock lying close to the base of a cliff, for there the ground water tends to seep outward, perhaps prevented from further downward percolation by impervious layers below (Figure 11.7). In the southwestern United States, many of the deep niches formed in this manner were occupied by Indians, whose cliff dwellings obtained protection from the elements as well as safety from armed attack (Figure 11.8).

Salt crystallization also acts adversely upon masonry buildings and highways. Brick and concrete in contact with moist soil are highly suscep-

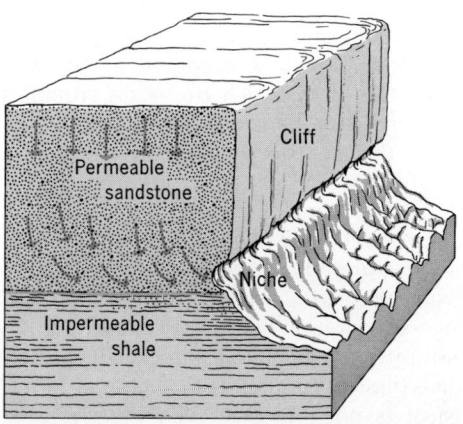

Figure 11.7 Seepage of water from the cliff base localizes development of niches through rock weathering.

267

Figure 11.8 White House Ruin occupies a deep niche in the sandstone wall of Canyon de Chelly, Arizona. (Photograph by Mark A. Melton.)

tible to granular disintegration from this cause. The salt crystals will be seen as a soft, white, fibrous layer on basement floors and walls. Salts are commonly hydrous calcium sulfate (gypsum), sodium sulfate, or magnesium sulfate. Man has added to these destructive effects by spreading deicing salts on streets and highways. Sodium chloride (rock salt, or halite), widely used for this purpose, is particularly destructive to concrete pavements and walks, curbstones, and other exposed masonry structures.

An important but little appreciated process of physical weathering is the swelling and shrinking of soils as the particles of fine silt and clay absorb or give up soil water in alternate periods of rain and drought. Shrinkage forms soil cracks in dry periods, making the infiltration of rainfall much more rapid in early stages of an ensuing rain. In clay-rich sedimentary rocks such as shales, the swelling accompanying water absorption is largely responsible for a spontaneous breakup known as *slaking*, in which the shale crumbles into small chips or pencil-like fragments when exposed to the air. Low grade coals and lignite are strongly affected by slaking.

Most rock-forming minerals expand when heated and contract when cooled. Where rock surfaces are exposed daily to the intense heating of the sun alternating with nightly cooling, the resulting expansion and contraction exerts powerful disruptive forces upon the rock. Given sufficient time (tens of thousands of such daily alternations), it is possible that even the strongest rocks may develop fractures. The importance of temperature change alone as a natural agent of rock disintegration remains in doubt. Laboratory experiments do not confirm its efficacy, whereas rock disintegration by crystallization of ice and salts is fully documented.

Finally, in this list of physical weathering processes, the wedging of plant roots deserves consideration as a possible mechanism whereby joint blocks may be separated. We have all seen at one time or another a tree whose lower trunk and roots are firmly wedged between two great joint blocks of massive rock (Figure 11.9). Whether the tree has actually been able to spread the blocks farther apart or has merely occupied the available space is open to question. However, it is certain that osmotic pressure exerted by growth of tiny rootlets in joint fractures must be of great importance in loosening countless small rock scales and grains, particularly when a rock has already been softened by chemical decay or fractured by frost action.

A widespread process of rock disruption related to physical weathering results from *unloading*, the relief of confining pressure, as rock is brought nearer to the earth's surface through the erosional removal of overlying rock. Rock formed at great depth beneath the earth's surface (particularly igneous and metamorphic rock) is in a slightly contracted state because of the confining pressure of overlying rock. Rock at depth may also be in a state of strain acquired during mountain-making crustal deformations. On being brought to the surface, the rock expands slightly in volume and, in so doing, thick shells of rock break free from the parent mass below. The new surfaces of fracture are a form of jointing termed *sheeting*

Figure 11.9 Jointing in sandstone resembles pavement blocks at Artists View, Catskill Mountains, New York. (Photograph by A. N. Strahler.)

Forms Produced by Chemical Weathering

structure and show best in massive rocks such as granite and marble, because in a closely jointed rock the expansion would be taken up among the blocks. The rock sheets or shells produced by unloading generally parallel the ground surface and therefore tend to be inclined toward valley bottoms. On granite coasts the shells are found to incline seaward at all points along the shore. Sheeting structure is well developed in granite quarries, where it greatly facilitates the removal of rock (Figure 11.10). Individual sheets may break free and arch upward with explosive violence. A similar phenomenon in deep mines and tunnels is the breaking away of ceilings. Known as "popping rock," this spontaneous rock disintegration is a hazard to miners.

Where sheeting structure has formed over the top of a single large body of massive rock, an *exfoliation dome* is produced (Figure 11.11). In the Yosemite Valley region, California, where domes are spectacularly displayed, the individual rock shells may be as thick as 50 ft (15 m).

The energy source for disruption of rock by spontaneous expansion during unloading is that of gravity, where the rock was compressed by the overlying load of rock, or it may be internal energy of radiogenic origin where the compression was caused by tectonic processes.

For most of the physical weathering processes described here the basic energy source is that of solar radiation. Heat energy for the evaporation of water during drying out of soil and rock comes from solar radiation. This heat is required to increase the kinetic energy of water molecules and enable them to overcome the attractive forces by

Figure 11.11 North Dome and Basket Dome, Yosemite National Park, California. (Photograph by Douglas Johnson.)

which they are held to one another and to soil colloids. As water is lost, colloidal particles exert their attractive forces more intensively upon one another and pull the mass more tightly together, with resulting shrinkage and hardening of the soil. Plant growth utilizes solar energy, and a by-product is osmotic pressure in rootlets which exerts disruptive force upon the enclosing soil or rock.

The evaporation of water from soil solutions, requiring solar energy, causes the formation of salt crystals whose growth can exert a mechanical force on soil and rock particles. Even the action of ice crystal growth in moving mineral particles is powered by the sun. As freezing proceeds at the rock-ice-water interface, water molecules moving into the ice crystal lattice release latent energy. Most of this energy is released as heat, but some is used to do work in moving the rock particle. Because this energy was originally stored during melting, work done by freezing water is solar powered.

Forms Produced by Chemical Weathering

Chemical weathering processes were discussed in detail in Chapter 8 under the synonymous title of *mineral alteration*. Recall that the dominant processes of chemical change affecting silicate minerals of igneous rocks (and also metamorphic

Figure 11.10 Sheeting of granite, a large-scale form of exfoliation, facilitates quarrying operations. (Photograph by courtesy of Smith Quarry Division of Rock of Ages Corporation, Barre, Vermont.)

rocks with similar silicate compositions) are oxidation and hydrolysis. Carbonate minerals (calcite, dolomite) respond readily to reaction with carbonic acid. On the other hand, quartz is a highly stable mineral, although under favorable climatic conditions it is susceptible to removal in solution by surface waters. In Chapter 8 it was further pointed out that the clastic sedimentary rocks are composed of minerals that are highly stable in the surface environment. The clay minerals in particular undergo no further chemical change when reexposed at the surface in outcrops of sedimentary rock. In contrast limestone and dolomite are quickly redissolved when exposed in a humid climate and the bicarbonate ions are carried away to the sea for recycling as new carbonate sediments. Certain of the evaporite minerals, such as halite, are easily dissolved in surface water. Our interest here is to point out some of the surface forms and products of chemical weathering.

An important natural function of decomposition by hydrolysis and oxidation is to change a very strong rock into a very weak surface layer, allowing erosion processes to operate with great effectiveness, should the regolith be exposed directly to attack through destruction of the protective vegetative cover. Weakness of the regolith also makes it susceptible to natural forms of mass wasting, a topic discussed in later paragraphs of this chapter.

In warm, humid climates of tropical and equatorial zones hydrolysis and oxidation can result in the decay of igneous and metamorphic rocks to depths as much as 100 to 300 ft (30 to 90 m). Geologists who first studied this deep rock decay in the Southern Appalachians termed the rotted layer *saprolite* (literally "rotten rock"). To the civil engineer, such occurrences of deep weathering are of major importance in construction of highways, dams, or other heavy structures. Advantageous as is the property of softness of the saprolite, so as to be removable by power shovels with little blasting, there is serious danger in the weakness of the material in bearing heavy loads, as well as undesirable plastic properties because of a high content of clay minerals.

The hydrolysis of granite, with accompanying granular disintegration and some exfoliation of thin scales, produces many interesting boulder and pinnacle forms by rounding of angular joint blocks (Figures 11.12 and 11.13). These forms are particularly conspicuous in arid regions because of the absence of any thick cover of soil and vegetation. There is ample moisture in most deserts for hydrolysis to act, given sufficient time. The products of granular disintegration form a coarse desert gravel known as *grus*, which consists largely of grains of quartz and partially decomposed feldspars. Hydrolysis in fine-grained basic igneous rocks, such as basalt, commonly gives small-scale exfoliation of a type called *spheroidal weathering* (Figure 11.14).

The solution of limestone produces interesting surface forms, many of them of such small dimension as to be classed as *microrelief features*. Outcrops of limestone typically show cupping, rilling, grooving, and fluting in intricate designs (Figure 11.15). In a few limited regions of the world the scale of deep grooves and high wall-

Figure 11.12 Egg-shaped granite boulders are produced from joint blocks by granular disintegration in a semiarid climate near Prescott, Arizona. (Photograph by A. N. Strahler.)

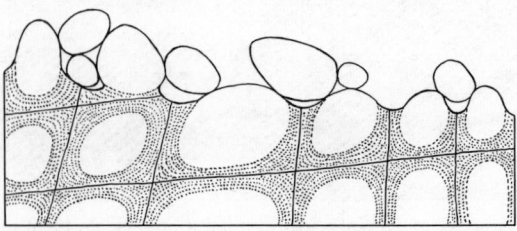

Figure 11.13 Stages in the development of egg-shaped boulders from rectangular joint blocks. (After W. M. Davis.)

Forms Produced by Chemical Weathering

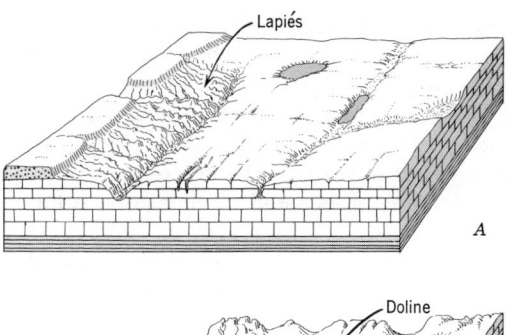

Figure 11.14 Spheroidal weathering, shown here, has produced many thin concentric shells in a basaltic igneous rock. Lucchetti Dam, Puerto Rico. (Photograph by C. A. Kaye, U.S. Geological Survey.)

like rock fins reaches proportions that impede passage of men and animals. A classic example is seen in the Dalmatian coastal region of Yugoslavia, described as a *karst landscape*. Here the large grooves are referred to as *lapiés* (Figure 11.16A). Associated with these surface rugosities are numerous deep conical depressions, known as *dolines* in the Dalmatian karst region, and as *sinkholes*, or *swallow holes*, in the English-speaking lands (Figure 11.16B and Figure 11.17). Sinkholes lead down into limestone caverns, seen in cross-section in Figure 11.16. As caverns are en-

Figure 11.16 Stages of evolution of a karst landscape show increased relief and cavern development. (Drawn by E. Raisz.)

Figure 11.15 Solution rills in limestone, west of Las Vegas, Nevada. Scale is indicated by pocket knife in center. (Photograph by John S. Shelton.)

Figure 11.17 Outcrops of horizontal limestone strata show in the walls of this deep sinkhole on the Kaibab Plateau of northern Arizona. (Photograph by A. N. Strahler.)

larged and deepened, surface collapse occurs on a large scale, leaving rock arches, shown in Figure 11.16C. In this advanced stage of karst development surface streams occupy valleys floored by insoluble shales (labeled *polje*). Other well-known karst regions or karstlike regions displaying landforms dominated by limestone solution are the Mammoth Cave region of Kentucky, the Causses Region of France, and parts of Cuba and Puerto Rico. Each region has unique environmental qualities, determined by properties of the limestone, by climate, by elevation and relief, and by relationship of the limestone formations to noncarbonate rocks. (The origin of limestone caverns is discussed in Chapter 12.)

The Dynamic Soil Layer

It would be a mistake to think of the soil as a lifeless, residual layer, which has somehow accumulated over a long period of time and which merely holds a supply of things necessary for plant growth. The soil is a dynamic layer in which many complex chemical, physical, and biological activities are going on constantly. Far from being a static, lifeless zone, it is an active open system having inputs and outputs of energy and matter. Through time soils become adjusted to prevailing conditions of climate and plant cover and will change internally when those controlling conditions change.

The true soil, or solum usually has developed distinctive layers, known as *horizons*. These horizons are suggested by the letters A, B, and C in Figure 11.2. Beneath the true soil is the *parent matter*, which may be altered rock, such as the regolith overlying an igneous rock, or sediment previously deposited after transportation by streams, wind, ice, or waves and currents. The true soil contains organic matter, whereas the underlying parent matter usually does not.

Soil is made up of matter existing in three states: solid, liquid, and gaseous. For plant growth a proper balance of all three states of matter is necessary.

The solid portion of soil is both inorganic and organic. The inorganic, or mineral, particles give a soil the main part of its weight and volume. The organic solids consist of both living and dead plant and animal materials, most being plant roots, fungi, bacteria, worms, insects, and rodents. Colloidal particles of organic matter share with inorganic colloidal particles an important function in soil chemistry.

The liquid portion of soil, the *soil solution*, is a complex chemical solution necessary for many important activities that go on in the soil. Soil without water cannot have these chemical reactions, nor can it support life. Gases in the open pore spaces of the soil form the third essential component. They are principally the gases of the atmosphere, together with gases liberated by biological and chemical activity in the soil.

For an understanding of soils, we must collect some information about (1) the physical-chemical properties and materials of soils, and (2) the processes that make and maintain soils.

Physical Properties of Soils

Although not a dynamic factor in itself, *soil color* is perhaps the characteristic that is first noticed about a soil. Color can tell much about how a soil is formed and what it is made of. Soil horizons are usually distinguishable by color differences. One sequence of colors ranges from white, through brown, to black as a result of an increasing content of *humus* which is finely divided, partially decomposed organic matter.

Reds and yellows are common colors in soils and are the result of small quantities of iron compounds. The red color is particularly associated with sesquioxide of iron (the mineral hematite, Fe_2O_3), whereas the yellow color may indicate the presence of limonite. Red color indicates that the soil is well drained with ample exposure to free atmospheric oxygen, but locally the color may be derived from a red source rock containing small amounts of hematite.

Grayish and bluish colors in soils of humid climates often mean the presence of reduced iron compounds (such as FeO) and indicate poor drainage or bog conditions with a deficiency of oxygen. Grayish soils in dry climates mean a meager amount of humus; a white color may be a result of the deposit of salts in the soil. Although some recently formed soils retain the color of the parent matter or bedrock, the color of fully developed soils is usually independent of what lies beneath.

Soil texture, a major characteristic of the soil, refers to particle sizes composing the soil. Grade sizes used by the agricultural soil scientist differ in minor details from the Wentworth scale (Table 8.2), but these are not particularly important here.

The U.S. Department of Agriculture has set up standard definitions of soil-texture classes in which the proportions of sand, silt, and clay are given in percentages. Rather than attempt to list these classes and give the limiting percentages for each, the information is given in a triangular diagram (Figure 11.18), which enables the percentages of all three components to be shown simultaneously. The corners of the triangle represent

Chemical Properties of Soils

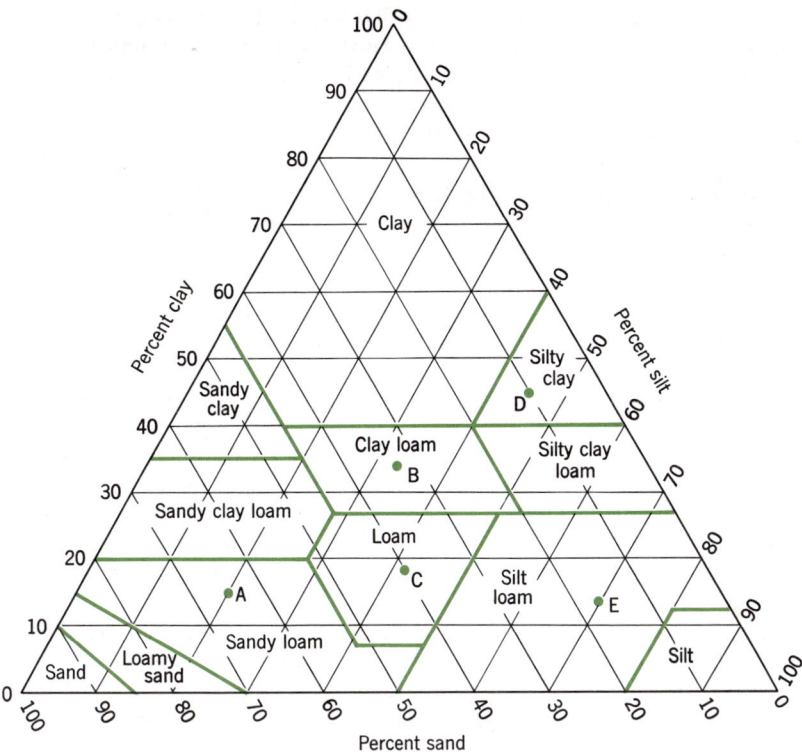

Figure 11.18 Texture classes shown as areas bounded by heavy lines on a triangular graph. (From U.S. Department of Agriculture and Millar, Turk, and Foth, *Fundamentals of Soil Science*, John Wiley & Sons, New York.)

100% of each of the three grades of particles—sand, silt, or clay. The word *loam* refers to a mixture in which no one of the three grades dominates over the other two. Loams therefore appear in the central region of the triangle. A particular soil whose components give it a position at Point A in the triangle has 65% sand, 20% silt, and 15% clay; it falls into a texture class known as *sandy loam*. Another soil, whose texture is represented by Point B has $33\frac{1}{3}$% sand, $33\frac{1}{3}$% silt, and $33\frac{1}{3}$% clay; it falls into the class of a *clay loam*.

Texture is important because it largely determines the water retention and transmission properties of the soil. Sand may drain too rapidly; in a clay soil the individual pore spaces are too small for adequate drainage. Where clay and silt proportions are high, root penetration is difficult. From the agricultural standpoint, the loam textures are best for plant growth.

Chemical Properties of Soils

By far the most important constituents of the soil are the colloids, which may be of either inorganic matter (clay minerals) or organic matter (humus colloids). While the silt and sand particles are largely inert and act more or less as a skeleton for the soil, the colloids make up the active fraction because of their high surface area and chemical activity. You can picture a clay particle as a thin, platelike body, shown in Figure 11.19. The molecular structure of clay minerals is such that oxygen atoms, which are relatively negatively charged, are nearest the upper and lower surfaces. This means that positive ions, or *cations*, will be attracted to the clay particle surface and held there by electrostatic attraction. Ions of hydrogen (H^+), sodium (Na^+), potassium (K^+), calcium (Ca^{++}) and magnesium (Mg^{++}) are commonly present in soil solutions and all are found on clay particle surfaces. In many soil reactions, these cations replace one another in the process of *cation exchange*.

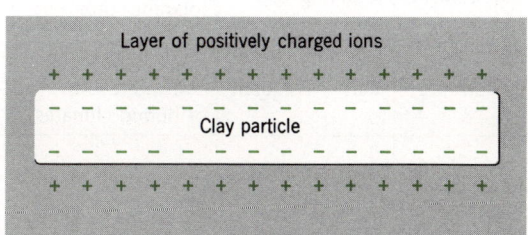

Figure 11.19 A colloidal particle with negative surface charges and a layer of positively charged ions.

273

Cation exchange is governed by a replacement order, indicating which ion is capable of replacing another. This is a kind of seniority system in which the ion of a given rank can take over the position of ions of lower seniority. Hydrogen can displace any of the metallic ions, so it occupies the top position on the list. In order of replacement ability there follow cations of strontium, barium, calcium, magnesium, potassium, sodium, and lithium. One application of this principle is that where acids are formed in the soil by decay of plant matter, hydrogen ions are in abundant supply and will replace the common base cations, which are then carried out of the soil by surplus water of rainfall. As a result these plant nutrients are lost and a soil of low fertility with respect to certain plants, such as the grasses, results.

The capacity of a given quantity of soil for cation exchange is called the *cation-exchange capacity*, and is a general indicator of the degree of chemical activity of a soil. Capacity is indicated in a unit known as the *milliequivalent* which is a measure of the ratio of weight of ions to weight of soil. The exact definition of this unit is not important here, but its relative magnitude is, so we find that the exchange capacities of various soil colloids are as follows:

Organic colloids	200
Montmorillonite	100
Illite	30
Kaolinite	8

The numbers tell us that organic colloids are a major contributor to total cation exchange of a given soil, and this should serve as a warning that practices which reduce the humus content of the soil will seriously decrease its chemical activity and hence also its ability to hold plant nutrients. These are ions of calcium, magnesium, potassium, and sodium, and are known as *bases*, or *base cations*.

The hydrogen ion, although a cation, is not a base. The extent to which the hydrogen ion is held by soil colloids is a measure of *soil acidity*. On the other hand, the abundance of the important base cations largely determines the soil *alkalinity*. Since the hydrogen ion can replace any of the base cations in the layer of positively charged particles surrounding the soil colloid particle, there is a wide range in soils from those that are strongly acid to those that are strongly alkaline. This range is measured in terms of the *pH* of the soil solution.* A pH value of 7.0 is neutral in this scale; values below 5.0 represent a strongly acid soil solution; values above 10.0 represent a strongly alkaline soil solution. Table 11.1 shows a classification of soils according to acidity and alkalinity. For agricultural soils, this quality is very important, since certain crops such as the grains, require near-neutral values of pH and cannot thrive on acid soils unless the acidity is counteracted by application of lime. Plants differ considerably in their preference for soil acidity or alkalinity, and this is an important factor in the distribution of plant types.

*The term pH is a measure of the concentration of hydrogen ions; it is the logarithm to the base 10 of the reciprocal of the weight in grams of hydrogen ions per liter of water. Consequently, the smaller the pH number, the greater is the hydrogen ion concentration.

TABLE 11.1 Soil Acidity and Alkalinity

pH	4.0	4.5	5.0	5.5	6.0	6.5	6.7	7.0	8.0	9.0	10.0	11.0
Acidity	Very strongly acid		Strongly acid	Moderately acid	Slightly acid			Neutral	Weakly alkaline	Alkaline	Strongly alkaline	Excessively alkaline
Lime requirements	Lime needed except for crops requiring acid soil		Lime needed for all but acid-tolerant crops		Lime generally not required			No lime needed				
Occurrence	Rare	Frequent	Very common in cultivated soils of humid climates						Common in sub-humid and arid climates		Limited areas in deserts	
Pedogenic regime	Podzolization		Gleization		Laterization				Calcification		Salinization	

SOURCE: C. E. Millar, L. M. Turk, and H. D. Foth, 1958, *Fundamentals of Soil Science*, 3rd ed., John Wiley & Sons, New York, Chart 4.

Soil colloids also play an important role in determining the water content of the soil. Water molecules are attracted to the charged surface of the colloidal particle and form a sort of atmosphere of water molecules; those nearest the colloid surface are tightly held; those more distant are more loosely held. This phenomenon is referred to as *adsorption*. Consequently, while colloid-rich soils will give up a great deal of water by evaporation in dry periods, it is impossible under normal atmospheric temperatures for all of the adsorbed water to be driven off. Colloids allow the soil to remain moist under arid conditions that would render a pure silt or sand almost devoid of water. (We will find in Chapter 12, in the discussion of soil water and ground water, that water is also held in soils in the form of capillary films.)

Soil Forming Processes

The processes that develop and maintain a soil in equilibrium with its environment are extremely complex, so that we can only look into a few of the more important relationships. The characteristic set of horizons, making up the *soil profile* requires a long time for development, a fact of great environmental importance because Man's agricultural and industrial activities using vast energy sources, can destroy in a short time a delicate soil profile that took centuries to form. Starting with a new layer of parent matter—for example, the deposits of a river flood, or of a glacier—in which organic matter is lacking and no horizons exist, evolution to a stable profile configuration may take one to two centuries to develop under the most favorable of conditions, but the figure is probably better estimated in terms of thousands of years. An estimate of the rate of development of soils of wet equatorial and tropical climate zones places a time span of 20,000 to 100,000 years on the production of 1 meter (3 ft) of soil from the parent granitic rock. Over large parts of the continental surfaces in low latitudes soils have an age of 1 to 6 million years, but we do not know what part of this period is required to bring the profile to a stable, or equilibrium state. The important point is that the true soil, or solum, is a nonrenewable resource in terms of agricultural production. Once the natural soil is degraded or destroyed, the loss in terms of useful plant production is permanent. In Chapter 14 we will investigate the physical destruction of the soil layer by water erosion induced by Man. The natural soil can also be degraded chemically through removal of the nutrient bases, and these changes also lead to changes in physical constitution of the soil.

The role of climate upon soil development is paramount, because climate not only acts directly upon the soil profile, but it is also the primary control of plant and animal activity, which in turn is a major agent in forming the soil body and its horizons. The two essential ingredients of climate as it affects soils are water availability and heat. Water availability reduces down to a question of whether the water balance of the soil is one of a water surplus or a water deficit. At the one extreme is a climate in which all months show a water surplus, at the other a climate in which all months show a severe water deficit. We shall go into the matter of the soil-water balance in Chapter 12, since it is vital to flow of streams and the occurrence of ground water. Availability of water largely determines soil chemistry through its effects upon the cation exchange in soil colloids, and at the same time is a strong control over the rate of plant production, thereby affecting organic content and processes in the soil. Heat, the second factor in soil development, has its strongest influence through plants, since heat is a prime factor in determining the rates of production and destruction of organic matter. Before turning to an analysis of how the world pattern of climate dictates a world pattern of several distinctive soil groups, we need to devote some attention to the interaction of climate, organisms, and the soil.

For purposes of this discussion consider all terrestrial plants to fall into two groups: the *macroflora* (trees, shrubs, and herbs) and *microflora* (bacteria and fungi). We shall find in Chapter 20 that the macroflora are the *producers* of organic matter—carbohydrate compounds—whereas the microflora are the *consumers* of organic matter.

Grasses and trees require somewhat different chemical substances for growth. Trees, particularly the conifers, use little calcium and magnesium. Hence they thrive well in soils from which these substances have been leached away and which are usually acid. Grasses and small grains (wheat, oats, barley) need abundant calcium and magnesium and do well in soils of the semiarid and subhumid lands. For grasses to grow well in acid soils, calcium must be added to the soil in the form of lime or crushed limestone. Plants tend to maintain the fertility of soil by bringing the base cations (calcium, magnesium, potassium) from lower layers of the soil into the plant stems and leaves, then releasing them to the soil surfaces as the plant decomposes. This natural recycling of matter is a key process in maintaining the soil system in an equilibrium state.

Fallen leaves and stems—plant litter—are the source of humus, the finely divided organic matter of the soil. The process of humus development, or *humification*, is essentially the slow oxidation of the vegetative matter. Organic acids, formed during humification, aid in decomposing the minerals of the parent soil material. The hydrogen ions of the acid solution tend to replace the base cations of potassium, calcium, magnesium, and sodium, which are removed by leaching as excess soil water percolates down through the soil layer. Soils of the cold humid climates are therefore deficient in the bases and are consequently of low fertility for crop farming.

Turning now to the microflora, we find that bacteria consume humus, oxidizing the organic compounds and releasing carbon dioxide. In cold climates bacterial growth is slow, with the result that humus accumulates on and in the soil. Soils of the subarctic climates (the tundra regions) have much undecomposed organic matter, which locally forms layers of peat, but in humid tropical and equatorial climates, bacterial action is intense and all plant litter is rapidly oxidized by the bacteria. Here humus content of the soil is low. The organic acids formed by humus are therefore also lacking, and certain bases such as aluminum, iron, and manganese accumulate in a large proportion relative to silica. In this way the fundamental differences in soils of cold and warm climates can be traced back to intensity of bacterial activity.

Figure 11.20 shows the relation between mean annual air temperature and the relative rate of production and destruction of organic matter by macroflora and microflora, respectively. In the colder climate production exceeds destruction and humus accumulates. Above a mean annual temperature of 77° F (25° C) destruction exceeds production and humus is absent. Note, however, that under anaerobic conditions of swamps, consumption rates are low, even at high temperatures, so that organic matter accumulates in the swamp environment in all wet climates, whether cold or warm.

Another function of some bacteria and other soil organisms (algae) is to take gaseous molecular nitrogen from the air and convert it into a chemical form that can be used by plants. This process is known as *nitrogen fixation*. One kind of bacteria (*Rhizobium*) lives in the root nodules of leguminous plants and there fixes nitrogen beneficial to the plant host.

The influence of animals in the soil is largely mechanical, but nevertheless important. Earthworms are a particularly important agent in humid regions. They not only continually rework the soil by burrowing, but also change the texture and chemical composition of the soil as it passes through their digestive systems. Ants and termites bring large quantities of soil from lower horizons to the surface. Such burrowing animals as prairie dogs, gophers, ground squirrels, moles, and field mice disturb and rearrange the soil. Digging of burrows brings soil of lower horizons to the surface; collapse of burrows carries surface soil into lower horizons. In a sense, then, activities of animals within the soil are one of the various physical weathering processes that continue to agitate the surface layer, long after it has achieved a stable configuration with respect to climate and organic processes.

Since water availability is, together with heat, the principal determining factor of the character of the soil and its plant cover, we can expect to find great place-to-place differences in the soil profile depending upon the soil water balance that prevails. These differences are discussed in Chapter 12, in connection with the soil-water budget. We shall find that both the soil-water balance and soil-forming processes are linked together in a world pattern of distinctive regimes, making for great environmental diversity within the life layer.

Mass Wasting

Everywhere on the earth's surface, gravity pulls continually downward on all materials. Bedrock is usually so strong and well supported that it remains fixed in place, but should a mountain slope become too steep through removal of rock at the base, bedrock masses break free, falling or sliding to new positions of rest. In cases where huge masses of bedrock are involved, the result may be catastrophic in loss to life and property in towns and villages in the path of the slide. As such, the phenomenon is a major form of environmental hazard in mountainous regions. Soil and overburden, being poorly held together, are

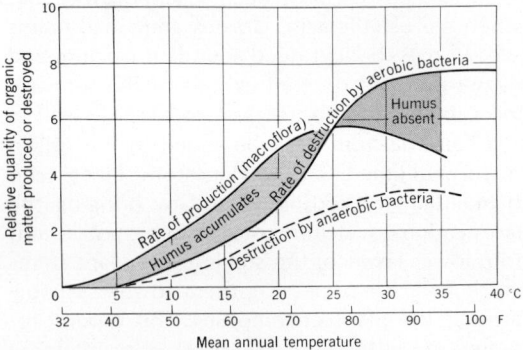

Figure 11.20 Relative rates of production and destruction of organic matter as related to mean annual temperature. Data of M. W. Senstius. (From A. N. Strahler, 1971, *The Earth Sciences*, 2nd ed., Harper & Row, New York.)

much more susceptible to gravity movements. There is abundant evidence that on most slopes at least a small amount of downhill movement is going on at all times. Much of this is imperceptible, but sometimes the overburden slides or flows rapidly.

Taken altogether, the various kinds of downslope movements occurring under the pull of gravity, which we have collectively termed mass wasting, constitute an important process in slope wasting and denudation of the continental surfaces. Man's activities in moving enormous volumes of rock and soil on construction sites of dams, canals, highways, and buildings has added to the natural forms of mass wasting.

Soil Creep

On almost any moderately steep, soil-covered slope, some evidence may be found of extremely slow downslope movement of soil and regolith, a process called *soil creep*. Figure 11.21 shows some of the evidence that the process is going on. Joint blocks of distinctive rock types are found moved far downslope from the outcrop. In some layered rocks such as shales or slates, edges of the strata seem to bend in the downhill direction. This is not true plastic bending, but is the result of slight movement on many small joint cracks (Figure 11.22). Fence posts and telephone poles lean downslope and even shift measurably out of line. Retaining walls of road cuts lean and break outward under pressure of soil creep from above.

What causes soil creep? Heating and cooling of the soil, growth of frost needles, alternate drying and wetting of the soil, trampling and burrowing by animals, and shaking by earthquakes all produce some disturbance of the soil and mantle. Because gravity exerts a downhill pull on

Figure 11.22 Slow creep has caused this downhill bending of steeply dipping sandstone layers. (Photograph by courtesy of Ward's Natural Science Establishment.)

every such rearrangement that takes place, the particles are urged progressively downslope.

Creep affects rock masses enclosed in the soil or lying upon bare bedrock. Huge boulders which have gradually crept down a mountain side in large numbers may accumulate at the mountain base to produce a boulder field containing blocks the size of a house. Creep also affects the rock fragments in a talus slope, causing the angle of the talus surface gradually to become flatter.

Earthflow

In humid climate regions, where slopes are steep, a mass of water-saturated soil, overburden, or weak clay or shale layers may slide downslope during a period of a few hours in the form of an *earthflow*. Figure 11.23 is a sketch of an earthflow showing how the material slumps away from the top, leaving a steplike terrace bounded by arcuate scarps, and flows down to form a bulging toe.

Shallow earthflows, affecting only the soil and overburden, are common on sod-covered and

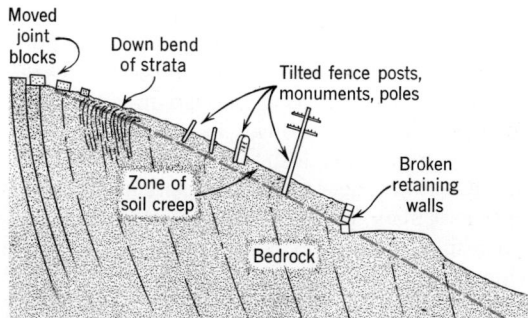

Figure 11.21 Evidences of the slow, downhill creep of soil and weathered overburden. (After C. F. S. Sharpe.)

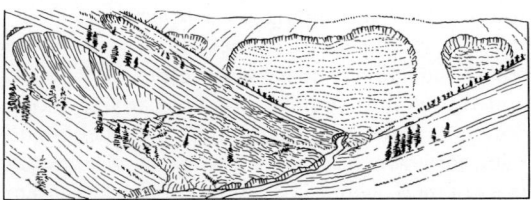

Figure 11.23 Earthflows in a mountainous region. (After W. M. Davis.)

Figure 11.24 This earthflow near Snowmass, Colorado, involved the rapid movement of clay-rich glacial moraine, weakened by water saturation. (Photograph by Mark A. Melton.)

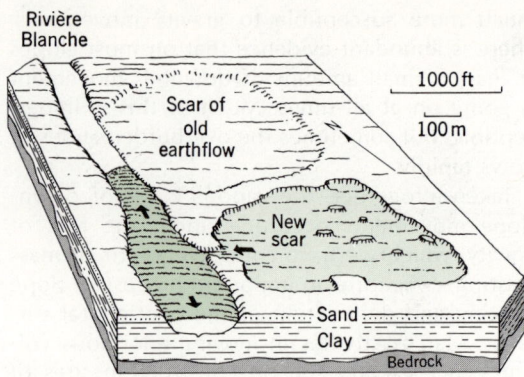

Figure 11.25 Block diagram of the 1898 earthflow near St. Thuribe, Quebec. (After C. F. S. Sharpe, 1938, *Landslides and Related Phenomena*, Columbia Univ. Press, New York, p. 51, Figure 7.)

forested slopes that have been saturated by heavy rains. An earthflow may affect a few square yards, or it may cover an area of several acres (Figure 11.24). If the bedrock of a mountainous region is rich in clay (shale or deeply weathered igneous rocks), earthflow sometimes involves millions of tons of bedrock, moving by plastic flowage like a great mass of thick mud.

Earthflows are a common cause of blockage of highways and railroad lines, usually during periods of heavy rains, but as the rate of flowage is slow, the flows are not often a threat to life. However, property damage to buildings, pavements, and utility lines is often large where construction has taken place on unstable soil slopes.

One special form of earthflow has proved to be a major environmental hazard in parts of Norway and along the St. Lawrence River and its tributaries in Quebec Province of Canada. In both areas the flowage involves horizontally layered clays, sands, and silts of Pleistocene age that form low, flat-topped terraces adjacent to rivers or lakes. Over a large area, that may be 2000 to 3000 ft (600 to 900 m) across, a layer of silt and sand 20 to 40 ft (6 to 12 m) thick begins to move toward the river, sliding on a layer of soft clay that has spontaneously turned into a near-liquid state. The moving mass also settles downward and breaks into steplike masses. Carrying along houses or farms, the layer ultimately reaches the river, into which it pours as a great, disordered mass of mud. Figure 11.25 shows in block diagram form an earthflow of this type that occurred in 1898 in Quebec, along the Rivière Blanche. Beyond is the scar of a much older earthflow of the same type. The Rivière Blanche earthflow involved about 3.5 million cubic yards (3 million cu m) of material and required three to four hours to move into the river through a narrow, bottleneck passage. Disastrous earthflows have occurred a number of times since the occupation of Quebec by Europeans. A particularly spectacular example was the Nicolet earthflow of 1955 which carried a large chunk of the town into the Nicolet River (Figure 11.26). Fortunately only three lives were lost, but the damage to buildings and a bridge ran into the millions of dollars.

Mudflow

One of the most spectacular forms of mass wasting and one that is potentially a serious environmental hazard, is the *mudflow*, a mud stream of fluid consistency which pours down canyons in mountainous regions (Figure 11.27). In deserts, where vegetation does not protect the mountain soils, local convective storms produce rain much faster than it can be absorbed by the soil. As the water runs down the slopes it forms a thin mud, which flows down to the canyon floors. Following stream courses, the mud continues to flow until it becomes so thickened that it must stop. Great boulders are carried along, buoyed up in the mud. Roads, bridges, and houses in the canyon floor are engulfed and destroyed. If the mudflow emerges from the canyon and spreads across a piedmont plain, property damage and loss of life can result, because in desert regions the plains lying at the foot of a mountain range which supplies irrigation water may be densely populated.

Mudflows also occur on the slopes of erupting volcanoes. Freshly fallen volcanic ash and dust is

Landslide

Figure 11.26 A portion of the city of Nicolet, Quebec, was carried into the channel of the Nicolet River by an earthflow that occurred in November, 1955. The flow moved downvalley (*right*), passing beneath the bridge. (Photograph by Raymond Drouin.)

turned into mud by heavy rains and flows down the slopes of the volcano. Herculaneum, a city at the base of Mt. Vesuvius, was destroyed by a mudflow during the eruption of 79 A.D., when the neighboring city of Pompeii was buried under volcanic ash.

Mudflows show varying degrees of consistency, from a mixture about like the concrete that emerges from a mixing truck to thinner consistencies that are little different than in turbid stream floods. The watery type of mudflow is commonly called a *debris flood* in the western United States, and particularly in southern California, where it occurs commonly and with disastrous effects (discussed in later paragraphs of this chapter).

Landslide

Landslide is the rapid sliding of large masses of rock with little or no flowage of the materials as in the previous types. Two basic forms of landslide are (a) *rockslide* in which the bedrock mass slips on a relatively flat inclined rock plane, such as a fault or bedding plane, and (b) *slump*, in which there is backward rotation on a curved up-concave slip plane (Figure 11.28).

Wherever steep mountain slopes occur, there is a possibility of large, disastrous rockslides. In Switzerland, Norway, or the Canadian Rockies, for example, villages built on the floors of steep-sided valleys have been destroyed and their inhabitants killed by the sliding of millions of cubic yards of rock, set loose without any warning.

The Turtle Mountain landslide of 1903 in Alberta, shown in Figure 11.29 involved the sliding of an enormous mass of limestone, its volume estimated at 35 million cubic yards (27 million cu m), through a descent of 3000 ft (900 m). The debris buried a part of the town of Frank, with a loss of 70 persons. A similar disaster occurred

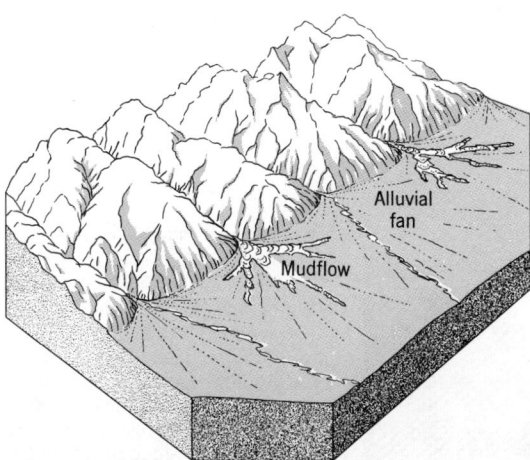

Figure 11.27 Thin streamlike mudflows occasionally issue from canyon mouths in arid regions, spreading out upon the piedmont alluvial fan slopes.

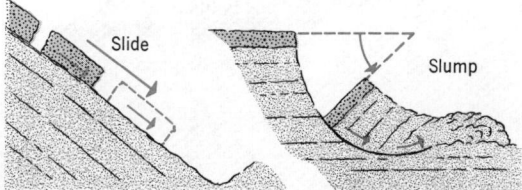

Figure 11.28 Landslides may involve slip on a nearly plane surface (*left*) or slump with rotation on a curved plane (*right*).

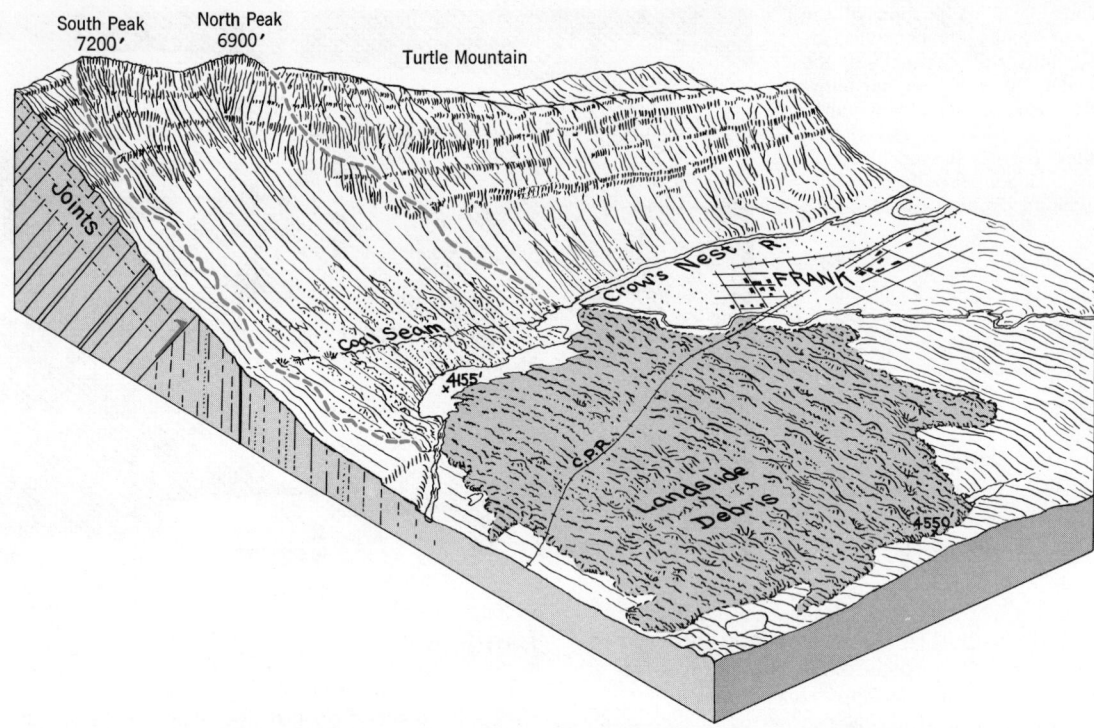

Figure 11.29 A classic example of a great, disastrous landslide is the Turtle Mountain slide, which took place at Frank, Alberta, in 1903. A huge mass of limestone slid from the face of Turtle Mountain between South and North peaks, descended to the valley, then continued up the low slope of the opposite valley side until it came to rest as a great sheet of bouldery rock debris. (After Canadian Geological Survey, Dept. of Mines.)

in 1959 in Montana when a severe earthquake (Hebgen Lake earthquake, Richter scale 7.1) caused an entire mountainside to slide into the Madison River gorge, killing 27 persons (Figure 11.30). The Madison Slide, as it is known, formed a debris dam over 200 ft (60 m) high and produced a new lake. Volume of this slide was about the same as in the Turtle Mountain slide. The lake has now been made permanent by construction of a protected spillway. Severe earthquakes in moun-

Figure 11.30 The Madison Slide, southwestern Montana, photographed shortly after it occurred. Water of the Madison River, impounded by the slide, is rising to form a new lake, named Earthquake Lake. Within three weeks the lake was nearly 200 ft (60 m) deep. (Photograph by N. R. Farbman, *Life Magazine*, © 1959, Time Inc.)

tainous regions are a major immediate cause of landslides, earthflows, and other forms of gravity displacements.

Aside from occasional great catastrophes, rockslides have rather limited environmental influence because of their sporadic occurrence in thinly populated mountainous regions. Small slides may, however, repeatedly block or break an important mountain highway or railway line.

The second form of landslide produces *slump blocks*, great masses of bedrock or overburden that slide downward from a cliff, at the same time rotating backward on a horizontal axis (Figure 11.31). Wherever massive sedimentary strata, usually sandstones or limestones, or lava beds, rest upon weak clay or shale formations, a steep cliff tends to be formed by erosion. As the weak rock is eroded from the cliff base, the cap rock is undermined. When a point of failure is reached, a large block breaks off, sliding down and tilting back along a curving plane of slip. Slump blocks may be as much as 1 to 2 mi (1.5 to 3 km) long and 500 ft (800 m) thick. A single block appears as a ridge at the base of the cliff. A closed depression or lake basin may lie between the block and the cliff.

Slumping commonly occurs on a small scale wherever weak overburden, such as floodplain alluvium or a glacial deposit, is cut away, and it is seen in caving river banks or along a sea cliff.

Most rapid of all mass wasting processes is *rockfall*, the free falling or rolling of single masses of rock from a steep cliff. Individual fragments may be as small as a boulder, or as large as a city block, depending upon the overall scale of the cliff and the manner in which the rock breaks up. Large blocks disintegrate upon falling, strewing the slope below with rubble and leaving a conspicuous scar on the upper cliff face.

A related phenomenon of high, alpine mountain chains, where glacial erosion has produced extremely steep valley gradients and where large quantities of glacial rock rubble (moraine) and relict glaciers are perched precariously at high positions, is the *alpine debris avalanche*. This sudden rolling of a mixture of rock waste and glacial ice can produce a tongue of debris traveling down-valley at a speed little less than that of a freely falling body. A recent disaster of this type occurred in 1970 in the high Andes of Peru. A severe earthquake (magnitude 7.7 on the Richter scale), which caused widespread death and destruction, set off the fall of a large snow cornice from a high peak, Huascarán. After a free fall of 3000 ft (900 m) the snow mass was partially melted by impact and incorporated a great quantity of loose rock to become a debris avalanche (Figure 11.32). Traveling down-valley at a speed calculated to have reached 300 mi (480 km) per hour, the avalanche wiped out the town of Yungay and several smaller villages. The death toll, which included earthquake casualties, was estimated in the thousands.

Figure 11.32 Vertical air photograph showing the lower portion of the Huascarán, Peru, debris avalanche. The main tongue of the debris (*far right*) turned and followed the valley of the Santa River (*below*). A secondary tongue (*center*) spread over the alluvial cone surface, burying much of the town of Yungay. (NASA photo.)

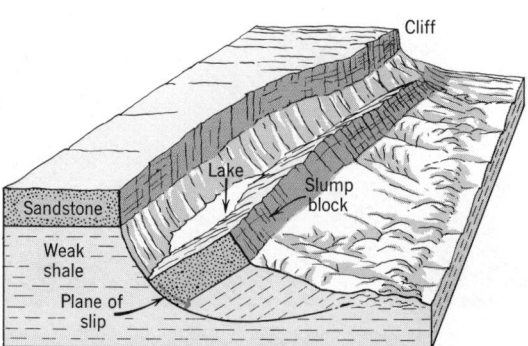

Figure 11.31 Slump blocks rotate backwards as they slide down from a cliff.

Surface Processes and Forms of the Arctic Environment

Permafrost regions of the arctic land fringes of North America, Greenland, and Eurasia are subjected to a special set of physical weathering and mass wasting processes; these give rise to a unique assemblage of surface forms, mostly in the category of microrelief features. Since the arctic land surface underlain by permafrost is highly sensitive to disturbance by man-made activities, an understanding of the environmental factors will go far to assure wise planning for the extended occupation and economic development of arctic regions.

Recall from the discussion on arctic permaforst in Chapter 3 that poleward of the 60th parallel all moisture in the soil, overburden, and bedrock is solidly frozen to depths of many feet. Only a shallow surface layer experiences thawing in the short summer season. The shallow zone of seasonal thaw and freeze is called the active zone.

A distinctive feature of the arctic lands is the development of *patterned ground,* a general term for the occurrence of a pattern of nested polygons in the soil or unconsolidated overburden. In fine-grained material such as floodplain silts, ice has been segregated into *ice-wedges* perpendicular to the surface (Figure 3.7). Seen from above, the ice wedges form a polygonal network, which is one of the forms of patterned ground (Figure 11.33). Where the surface layer is composed of coarse clastic materials—pebbles, cobbles, and boulders—the larger fragments are concentrated by repeated alternations of freeze and thaw of soil ice into wedges forming a polygonal pattern. The term *stone polygons* (also *stone rings, stone nets*) is given to these features (Figure 11.34). Where a flat upland surface grades into a marginal hillslope, the polygons are drawn out by downslope creep into elliptical forms and then into parallel *stone stripes* (Figure 11.35). Similar features will be found at high elevations in the alpine environment above timberline over a wide range of latitudes.

Because the seasonal thaw affects only a shallow layer, soil moisture reaches the saturation level and the water cannot escape through the impermeable frozen layer below. The thawed layer is then in a weakened and plastic condition and is highly susceptible to mass gravity wasting where the surface has an appreciable inclination. Under a protective sod of tundra vegetation the saturated soil moves unevenly downslope to produce bulges. (Figure 11.36). The process is called *solifluction* from the Latin words for "soil" and "to flow." In addition to the *solifluction lobes* shown in Figure 11.36, there result *solifluction terraces* which give a distinctive stepped pattern to large expanses of mountainsides.

Figure 11.33 Ice-wedge polygons in fine-textured floodplain silts near Barrow, Alaska.. Dark areas within polygons are lakes. In the middle distance is a meandering river channel. (Photograph by courtesy of R. K. Haugen, U.S. Army Cold Regions Research & Engineering Laboratory.)

Figure 11.34 Stone rings near Thule, Greenland. Notice that a set of smaller, secondary rings has begun to form within the floor of the larger ring in the foreground. (Photograph by A. E. Corte, Geology Department, Universidad Nacional del Sur, Bahia Blanca, Argentina.)

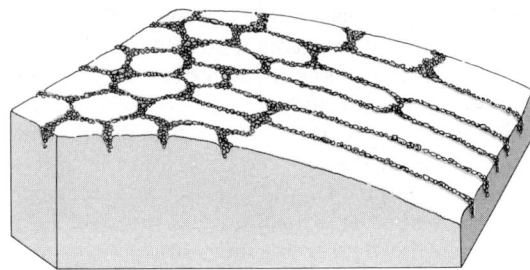

Figure 11.35 Sketch of stone polygons on upland, grading into stone stripes on adjacent slope. (After C. F. S. Sharpe. From A. N. Strahler, 1971, *The Earth Sciences*, 2nd ed., Harper & Row, New York.)

Environmental degradation of permafrost regions arises from man-made surface changes, usually related to the destruction or removal of an insulating surface cover, which may consist of a moss or peat layer in combination with living plants of the tundra or arctic forest. When this layer is scraped off, the summer thaw is extended to a greater depth, with the result that ice wedges and other ice bodies melt in the summer and waste downward. This activity is called *thermal erosion*. Meltwater mixes with silt to form mud, which is then eroded and transported by water streams, with destructive effects that are evident in Figure 11.37.

The consequences of disturbance of permafrost terrain became evident in World War II, when military bases, airfields, and highways were hurriedly constructed without regard for maintenance of the natural protective surface insulation. In extreme cases, scraped areas turned into mud-filled depressions and even into small lakes which expanded in area with successive seasons of thaw, engulfing nearby buildings. Engineering practices now call for placing buildings on piles with an insulating air space below, or for the deposition of an insulating pad of coarse gravel over the surface prior to construction. Steam and hot water lines are placed above ground to prevent thaw of the permafrost layer.

Another serious engineering problem of arctic regions is in the behavior of streams in winter. As the surfaces of streams and springs freeze over, the water beneath bursts out from place to place, freezing into huge accumulations of ice. Highways are thus made impassable.

The lessons of superimposing Man's technological ways upon a highly sensitive natural environment were learned the hard way—by encountering unpleasant and costly effects that were not anticipated. Yet even now a new threat has emerged in form of the proposed Trans-Alaska Pipeline, which will carry hot oil from the northern shores of Alaska, across a permafrost land-

Figure 11.36 Solifluction lobes cover this Alaskan mountain slope in the tundra climate region. (Photograph by P. S. Smith, U.S. Geological Survey.)

Figure 11.37 After one season of thaw this vehicular winter trail through the Alaskan arctic forest had suffered severe thermal and water erosion. (Photograph by courtesy of R. K. Haugen, U.S. Army Cold Regions Research & Engineering Laboratory.)

scape to the southern coast. Effects of this pipeline upon permafrost and other elements of the environment have been heavily debated and are the subject of intensive investigation. In addition to the prospects of thaw of the permafrost layer by heat of the pipeline there is the possibility of environmental damage from spills due to pipe breakage and of the effect of the pipeline as a barrier to animal migration.

Man as an Agent of Mass Wasting and Land Scarification

Because Man is now possessed of enormous machine power and explosives, he is capable of moving from one place to another vast quantities of soil, overburden, and bedrock for two basic purposes: first, to extract mineral resources; second, to reorganize terrain into suitable configurations for highway grades, airfields, building foundations, dams, canals, and various other large structures. Both activities involve removal of earth materials, which destroys entirely the preexisting ecosystems and habitats of plants and animals, and building up of new land upon adjacent surfaces, using those same earth materials, a process that also destroys by burial the preexisting ecosystems and habitats. What distinguishes man-made forms of mass wasting from the natural forms described in earlier paragraphs is that Man can use machinery to raise earth materials against the force of gravity. Also, Man's use of explosives in blasting can bring to bear in a rock mass disruptive forces that exceed by many orders of magnitude the forces of physical (mechanical) weathering. Rock breakup by shattering, the fourth of the forms of rock breakup illustrated in Figure 11.3, can be viewed as predominantly a man-made phenomenon.

Scarification is a general term for excavations and other land disturbances produced for purposes of extracting mineral resources; it includes concomitant accumulation of waste matter (spoil, tailings). Among the forms of scarification are open-pit mines, strip mines, quarries for structural materials, borrow pits along highway grades, sand and gravel pits, clay pits, phosphate pits, scars from hydraulic mining, and stream gravel deposits reworked by dredging. Open-pit mining of low-grade copper ores is illustrated in Figure 10.4. Strip mining for coal is explained in Chapter 10 and illustrated in Figures 10.11, 10.12, and 10.13. Hydraulic mining and dredging are discussed in Chapter 14.

Scarification is on the increase both because demands for coal to meet energy requirements are on the rise and because of increased demands for industrial minerals used in manufacturing and construction. At the same time, as the richer and more readily available mineral deposits are consumed, industry turns to poorer grades of ores and to less easily accessible coal deposits, with the result that the rate of scarification is further increased. Table 11.2 gives the surface areas disturbed by mineral extraction up to the year 1965. Although the figures are now out of date, they give a correct idea of the magnitude and relative importance of disturbance in each of several categories. Of the total area disturbed by strip mining, most is accounted for in the following state subtotals:

State	Area in Thousands of Acres
Alabama	50
Illinois	130
Indiana	95
Kansas	46
Kentucky	120
Missouri	32
Ohio	210
Pennsylvania	300
Tennessee	30
Virginia	30
West Virginia	190

Reference to the map of United States coal fields, Figure 10.9, will help to explain the concentration of operations in these 12 states. The U.S. Department of Agriculture estimated that as of the start of 1965 about one-third of the acreage shown in Table 11.2 does not require reclamation because it has been restored by nature or by Man's effort.

Without question, the environmental devastation produced by strip mining exceeds in quantity and intensity any of the other varied forms of man-made land destruction. Not surprisingly, there is sharp conflict between conservation and industrial groups and within legislative and administration groups in the government as to what course of action should be taken to minimize environmental damage. At the one extreme are conservationists who advocate immediate cessation of strip mining. A bill introduced into Congress in 1971 (Hechler bill) asks for a ban on all coal strip mining. Other bills seek to restrict and control strip mining and to require restorative measures in which environments favorable to plant and animal life are artificially reconstructed upon the devastated areas. Reclamation is expensive and poses many problems of implementation. Perhaps the most serious basic dilemma in the question of environmental control related to strip

TABLE 11.2 Land Disturbance by Mineral Extraction through 1964

Product	Thousands of Acres Disturbed
Coal (mostly as strip mining)	1,300
Clay	110
Stone	240
Sand and gravel	820
Gold (mostly in California)	200
Phosphate rock (mostly in Florida and Tennessee)	180
Iron ore (mostly in Alabama and Minnesota)	160
All others	160
Total	3,200
(about 5,000 sq mi)	

SOURCE: Data from U.S. Department of Agriculture and U.S. Department of Interior.
NOTE: Figures are rounded off.

mining in the United States is that this form of mining furnishes over 45% of the total national coal production. To close down strip mines would result in serious disruption of the nation's electrical power system, since coal combustion furnishes about one-half of the energy for generating of electricity. Nevertheless, in England and France total environmental rehabilitation is practiced for both coal and iron strip mining, while regulations include preservation of roads, hedges, and forests. Environmental degradation associated with strip mining includes water pollution and the upset of stream activities. These topics are discussed in Chapters 12 and 14.

Man's activities induce mass wasting in forms ranging from mudflow and earthflow to rockslide and slump. These activities include (a) piling up of waste soil and rock into unstable accumulations that fail spontaneously, and (b) removal of support by undermining natural masses of soil, overburden, and rock. Referring again to strip mining, the spoil bank produced by contour strip mining is unstable and a constant threat to the lower slope and valley bottom below. When saturated by heavy rains and melting snows the spoil generates earthflows and mudflows that descend upon houses, roads, and forest. The spoil also supplies sediment that clogs stream channels far down the valleys (Chapter 14).

At Aberfan, Wales, there occurred a major disaster when a hill 600 ft (180 m) high, built of rock waste (culm) from a nearby coal mine spontaneously began to move as an earthflow. The pile had been constructed on a steep hill slope and upon a spring line, as well, making a potentially unstable configuration. The debris tongue overwhelmed part of the town below, destroying a school and taking over 150 lives (Figure 11.38). Phenomena of this type are often called "mudslides" in the news media.

In Los Angeles County, California, real estate development has been carried out on very steep hillsides and mountainsides by the process of

Figure 11.38 Debris flow at Aberfan, Wales, sketched from a photograph. (From A. N. Strahler, 1972, Planet Earth; Its Physical Systems Through Geologic Time, Harper & Row, New York.)

bulldozing roads and homesites out of the deep regolith. The excavated regolith is pushed out into adjacent embankments where its instability poses a threat to slopes and stream channels below. When saturated by heavy winter rains, these embankments can give way, producing earthflows, mudflows, and debris floods that travel far down the canyon floors and can spread out upon the piedmont surfaces (alluvial fans), burying streets and yards in bouldery mud. Many debris floods of this area are, however, produced by heavy rains falling upon undisturbed mountain slopes that have been denuded by fire of vegetative cover in the preceding dry summer (Figure 11.39). Many of these fires are set by Man, whether inadvertently or deliberately. Disturbance of slopes by construction practices is simply an added source of debris, and serves to enhance an already important environmental hazard.

Some large earthflows and landslides have been induced by removal of supporting rock and steepening of unstable slopes during excavation for canals and dams. A classic example is from the digging of the Panama Canal, which suffered from large earthflows of clay-rich rocks. These flows filled the canal at certain points and continued to move long after the initial masses of debris were removed. In connection with the excavation of Coulee Dam on the Columbia River deposits of older river silts moved as earth flows into excavations made for the base of the dam. In these same materials a large slump movement affected 2 million cubic yards (1.5 million cu m) of silt at the downstream toe of the dam. To stabilize these mass movements, elaborate water drainage systems and even freezing of the entrapped water were required. The list of examples of major and minor occurrences of a wide range of mass wasting phenomena induced by engineering works is almost endless. The important engineering field of *soil mechanics* has developed in part from the need to understand and forestall mass movements resulting from instability produced during construction projects of various sorts. Similarly, a large segment of the field of *engineering geology* is devoted to studies of the conditions of rock and geologic structure that affect strength and stability of dam foundations, canal walls, tunnel walls, highway excavations, bridge pier footings, and many other forms of heavy construction.

Examples of both large and small earthflows induced or aggravated by Man's activities are found in the Palos Verdes Hills of Los Angeles County, California. These movements occur in shales that tend to become plastic when water is added. The upper part of the earthflow undergoes a slump motion with backward rotation of the down-sinking mass. The interior and lower parts of the mass move by slow flowage and a toe of extruded flowage material may be formed. Largest of the earthflows in this area is the Portuguese Bend "landslide" which affected an area of 300 to 400 acres. The total motion over a three-year period was about 70 ft (20 m). Damage to residential and other structures totalled some $10 million. The most interesting observation, from our point of view in assessing the impact of Man on the environment, is that the slide has been attributed by geologists to infiltration of water from cesspools and from irrigation water applied to lawns and gardens. A discharge of over 30,000 gallons of water per day from some 150 homes is believed to have sufficiently weakened the shale beneath to start and sustain the flowage.

Man-Induced Land Subsidence

Yet another category of land disturbance of man-made origin is *subsidence,* a sinking of the ground following the removal of solid mineral matter or fluids (water, oil) from the rock beneath. Subsidence over coal mines is a serious hazard to habitations built on the land surface above. One example comes from the city of Scranton, Pennsylvania. Here the collapse of abandoned anthracite mine workings has repeatedly caused settling and fracturing of the ground, damaging streets and houses. Near Hanna, Wyoming, the cave-in of abandoned shallow coal mines has produced many deep pits, while

Figure 11.39 This layer of debris followed a flood in January, 1952. Mouth of Benedict Canyon, California. (Photograph by U.S. Army Corps of Engineers.)

Figure 11.40 Contours superimposed on this air view of the Long Beach, California, harbor area show the amount of subsidence in feet in the period 1928–1960. The photograph was taken in 1971, by which time all subsidence had been halted by brine injection. (Port of Long Beach photo.)

underground burning of the coal seam has added to progressive collapse.

Perhaps the best-known example of land subsidence caused by withdrawal of subterranean fluids from an oil field is that at Long Beach, California. Here the Wilmington oil field lies within the harbor area of the city (Figure 11.40). Oil production began in 1936 and was developed over an area of about 10 sq mi (26 sq km). Subsidence began shortly after the field was opened and has continued at a rate ranging from 0.5 to 2.0 ft (0.15 to 0.6 m) per year in the center of the affected area. Figure 11.40 shows by contour lines the total subsidence by 1962, which was just over 27 ft (8 m) in the center. Subsidence is attributed to the lowering of fluid pressure, permitting the sedimentary particles to pack more closely and reduce rock volume in strata far beneath the surface. Because the subsidence brought a large land area below sea level, protective dikes and levees were built to keep out the sea. The subsidence also did much costly damage to the oil wells. In view of the prospect of continued subsidence, remedial measures have been taken in the form of water injection into deep wells to raise the underground fluid pressure. This costly procedure was successful in reversing subsidence in the southern part of the field. (Subsidence of the land surface caused by withdrawal of ground water is discussed in Chapter 12.)

In this chapter we have compared natural processes of wasting of the continental surfaces with Man-induced changes of a similar nature. While the processes of weathering and soil formation are for the most part slow-acting and produce effects that are visible only when accumulated over centuries, the natural mass-wasting process includes catastrophic events. Indeed, some of the large and rapid mass gravity movments dwarf anything that Man has accomplished in equal time by earth-moving with machinery and explosives. But Man works constantly with enormous quantities of energy at his disposal, and the cumulative environmental damage and destruction continues to mount, almost unchecked. Only now are the conflicts of interest beginning to emerge and to be squarely faced in designing and implementing environmental protection measures that will control the spread of scarification and its many harmful side effects.

The global water balance, for which we analyzed the atmospheric phases in Chapter 5, is completed by flow of water on the lands. The *hydrologic cycle* traces all pathways of flow water in its vapor, liquid, and solid states (Figure 12.1). This cycle consists of a rather complex set of interconnected loops involving water transport in atmospheric and oceanic circulation systems and the flow of water from higher to lower levels on the continents under the force of gravity. It is with the terrestrial pathways of flow that we are concerned in this and the following two chapters.

Consider first the energy pathways and transformations in the hydrologic cycle. Precipitation brings water as rain or snow to an elevated position above sea level, giving each unit mass of water a quantity of potential energy proportional to its vertical distance above sea level. Work done in raising the water mass to its elevated position on the land requires the expenditure of solar energy, first to supply the latent heat of evaporation of that water from the ocean surfaces, then to generate the atmospheric motions by which the moist air masses are carried over the continents. During condensation the latent heat is returned to the atmosphere, cancelling out the energy used in evaporation. In addition, the fall of raindrops, snow flakes, hail, or sleet from cloud to ground transforms potential energy to kinetic energy, and this is largely dissipated in friction due to air resistance, eventually leaving the system by longwave radiation. So we see that of the total energy required to power the ocean-to-land mass transport within the hydrologic cycle, only a fraction remains as potential energy of position on the elevated land surface. The gravity flow of water on the lands transforms potential energy to kinetic energy of moving fluids in the form of surface streams, underground water, and glacial ice. (Glacial ice behaves as a sluggish fluid when it has accumulated in large masses.) The kinetic energy of water motion is dissipated by frictional resistance, both within the water and along its solid mineral boundaries of rock and soil. Heat thus generated is dissipated to the atmosphere and eventually leaves the planet.

There is however, an added ingredient of mass transport and energy dissipation connected with

12

Subsurface Water of the Lands

The World Water Balance

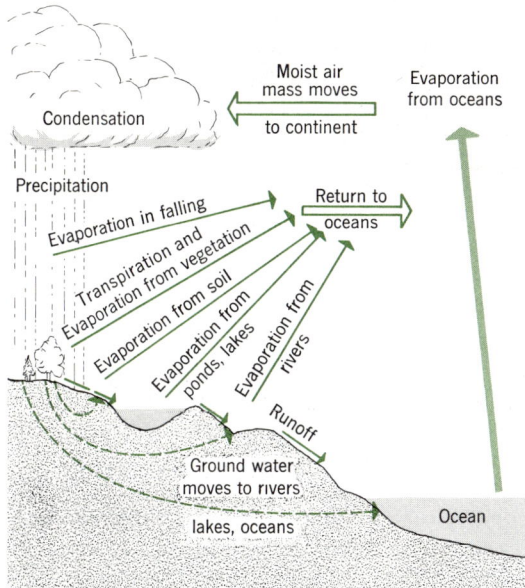

Figure 12.1 The hydrologic cycle.

The World Water Balance

The pictorial diagram of the hydrologic cycle given in Figure 12.1 can be quantified for the earth as a whole. Figure 12.2 is a mass-flow diagram relating the principal pathways of the water circuit. We can start with the oceans, which are the basic reservoir of free water. Evaporation from the ocean surfaces totals about 109,000 cu mi per year. (Metric equivalents shown in table in Figure 12.2.) At the same time, evaporation from soil, plants, and water surfaces of the continents totals about 15,000 cu mi. Thus the total evaporation term is 124,000 cu mi; it represents the quantity of water that must be returned annually to the liquid or solid state. Precipitation is unevenly divided between continents and oceans: 26,000 cu mi is received by the land surfaces; 98,000 cu mi by the ocean surfaces. Notice that the continents receive about 11,000 cu mi more water as precipitation than they lose by evaporation. This excess quantity flows over or under the ground surface to reach the sea; it is collectively termed *runoff*.

Using the same terms as in the water-balance equation of the atmosphere (Chapter 5), we can now state the global water balance, as follows:

$$P = E + G + R$$

where P is precipitation,
       E is evaporation,
       G is net gain or loss of water in the system, a storage term,
and R is runoff (positive when out of the continents, negative when into the oceans).

All terms have the dimensions of mass per unit

the gravity flow of water on the lands—the transport of mineral and organic matter within the water itself. Weathering processes, which we investigated in the previous chapter, produce free mineral particles, ranging in size from ions to grains of rock as large as cobbles and boulders. These particles, together with bits of organic matter, are entrained in the flowing water of streams and in slowly moving glaciers. Even the water that moves beneath the ground surface carries ions in solution. As these solids move from higher to lower levels, potential energy of position is transformed into kinetic energy, and we must include this quantity in the calculation of the energy balance of the hydrologic cycle. Kinetic energy of motion of both the fluid medium and the rock particles it carries is partially expended in breaking free other particles from the enclosing rock and soil surfaces. This is the geological process of erosion, which we shall examine in detail in Chapter 14. Rock attained its elevation above sea level from geologic work done by internal earth forces, which we have already attributed to inequalities in production of radiogenic heat. Without this geologic activity there would be no surface gradients on which water could run. Instead, the solid earth surface would long ago have been completely inundated beneath a world ocean and we would by now have no terrestrial environments. Using this overview of energy transformations of the hydrologic cycle as a background, let us next evaluate the mass balance of the global water-transport system.

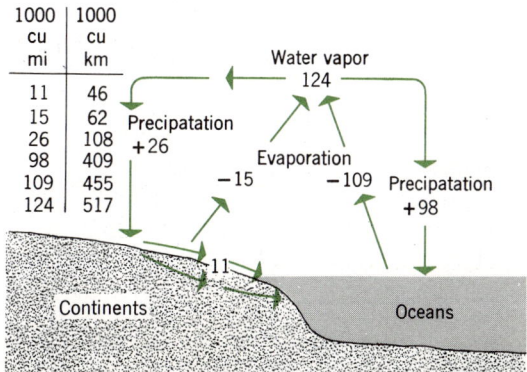

Figure 12.2 The global water balance. Figures give average annual water flux in and out of world land areas and world oceans. (Data of M. I. Budyko as given in *Inadvertent Climate Modification*, 1971, Report of the Study of Man's Impact on Climate (SMIC), MIT Press, Cambridge, Mass., p. 97, Table 5.3, p. 98, Table 5.4.)

time (for example, metric tons per day), or volume per unit time (for example, cubic miles per year). When applied over the span of a year, and averaged over many years, the storage term G can be neglected, since the system is essentially closed so far as matter is concerned. The quantities of water in storage in the atmosphere, on the lands, and in the oceans will remain about constant from year to year.

The equation then simplifies to:
$$P = E + R$$
Using the figures given above, for the continents
$$26{,}000 = 15{,}000 + 11{,}000,$$
and for the oceans
$$98{,}000 = 109{,}000 - 11{,}000.$$
For the globe as a whole, combining continents and ocean basins, the runoff terms cancel out:
$$26{,}000 + 98{,}000 = 15{,}000 + 109{,}000.$$
$$124{,}000 = 124{,}000$$

From a global water balance in terms of water volume per unit of time (cu mi/yr; cu km/yr), we now turn to water balance data in terms of water depth per unit of time (in./yr; cm/yr). Whereas the first set of data tell us the total quantity of water transported into or out of a specified area in one year; the second set will tell us the intensity of water flow, independent of total surface area and total water quantity. (The distinction between *total flow rate* and *flow intensity rate* is explained in Appendix I.)

Table 12.1 gives water-balance figures in flow intensity units for oceans and lands. Data were originally stated in metric units (cm/yr); we have converted these into English units (in./year) and rounded them off. The English units are less sensitive to small differences and for this reason the figures for Asia do not equate. For the oceans, it is necessary to take into account the flow of water by ocean currents from one ocean basin to another. This quantity is combined with the runoff term (R) in the water balance equation. Data for the land areas are interesting in that they reflect the dominant climatic influence. For example, South America has an enormous area within the very wet equatorial zone where precipitation is large and evaporation is small. As a result, runoff for the whole continent is exceptionally large. In contrast, Australia is dominated in area by subtropical desert, with the result that evaporation comes close to equalling precipitation and runoff is very small. Figures for Africa, Asia, Europe, and North America are quite alike, showing that areas of extremes of wet and dry climate tend to cancel out, giving averages similar to one another and to the world average for land areas. The three oceans show far less range in precipitation and evaporation than the land areas.

The water balance can be estimated for latitude belts of 10-degree width to reveal the response to the latitudinal changes in radiation and heat balances from equatorial zone to polar zones. Figure 12.3 shows the average annual values of precipitation, evaporation, and runoff. Again units are those of flow intensity—water depth per year. The runoff term, R, in this case includes import or export of water in or out of the belt by ocean

TABLE 12.1 Water Balance of the Oceans and Lands

	Precipitation		Evaporation		Runoff		Inflow (−) or Outflow (+)	
	(in./yr)	(cm/yr)	(in./yr)	(cm/yr)	(in./yr)	(cm/yr)	(in./yr)	(cm/yr)
Oceans								
Atlantic	35	(89)	49	(124)	−9	(−23)	−5	(−12)
Indian	46	(117)	52	(132)	−3	(−8)	−3	(−7)
Pacific	52	(133)	52	(132)	−3	(−7)	+3	(+8)
World ocean	45	(114)	50	(126)	−5	(−12)	0	(0)
Lands								
Africa	27	(69)	17	(43)	10	(26)		
Asia	24	(60)	12	(31)	11	(29)		
Australia	19	(47)	17	(42)	2	(5)		
Europe	25	(64)	15	(39)	10	(25)		
North America	26	(66)	13	(32)	13	(34)		
South America	64	(163)	28	(70)	36	(93)		
Total land areas	29	(73)	17	(42)	12	(31)		

SOURCE: M. I. Budyko, 1971, *Climate and Life*, Hydrological Publ. Co., Leningrad, as reproduced in *Inadvertent Climate Modification*, Report of the Study of Man's Impact on Climate (SMIC), 1971, The MIT Press, Cambridge, Mass., pp. 97–98.

The World Water Balance

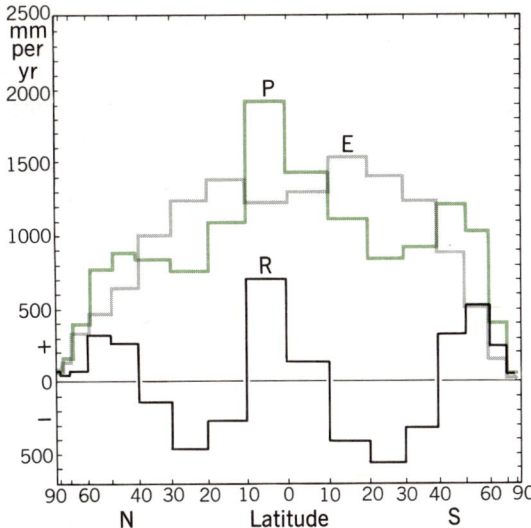

Figure 12.3 The water balance for 10-degree latitude zones. Data from several sources compiled by W. D. Sellers, 1965, *Physical Climatology*, Univ. of Chicago Press, Chicago, p. 5, Table 1. (From A. N. Strahler, 1971, *The Earth Sciences*, 2nd ed., Harper & Row, New York.)

currents as well as by stream flow. Notice the equatorial zone of water surplus with positive values of R, in contrast to the subtropical belts, with an excess of evaporation. The runoff term, R, which is here negative in sign, represents the importation of water by ocean currents to furnish the quantity needed for evaporation. Water surpluses occur poleward of the 40th parallel but the values of all three terms decline rapidly at high latitudes, going nearly to zero at the poles. This graph should be compared with Figures 4.36, 4.37, and 5.45. Notice that the precipitation surpluses are sustained by importation of water vapor, evaporation surpluses by export of water vapor (Figure 5.45).

Runoff, as used in the global water-balance equation, has two basic forms of occurrence: *surface water*, flowing exposed or impounded upon the land, or *subsurface water*, occupying openings in the soil, overburden, or bedrock. That subsurface water which is held in the soil within a few feet of the surface is termed *soil water*, and is the particular concern of the climatologist, ecologist, soil scientist, and agronomist; it is a vital ingredient of the environment of the life layer. Water which is held in the openings of the bedrock or deep within thick layers of transported overburden is referred to as *ground water*. It is studied by the geologist, who is concerned with the storage and flow of this water in various kinds of rocks and natural rock structures. Surface water in streams, lakes, and glaciers and various aspects of ground-water are studied by hydrologists.

Great inequalities exist in global amounts of water stored in the gaseous, liquid, and solid states. Table 12.2 gives a breakdown of storage quantities. Water of the oceans constitutes over 97% of the total, as we would expect. Next comes water in storage in glaciers, a little over 2%. The remaining quantity, about one-third of 1%, is mostly held as subsurface water, so that surface water in lakes and streams is a very small quantity, indeed. Yet it is this surface water together with the very small quantity of soil water that sustains all life of the lands. Indeed, some environ-

TABLE 12.2 Distribution of the World's Water Resources

Location	Surface Area (sq mi)	Surface Area (sq km)	Water Volume (cu mi)	Water Volume (cu km)	Percent of Total
Surface water					
Fresh-water lakes	330,000	860,000	30,000	125,000	0.009
Saline lakes and inland seas	270,000	700,000	25,000	104,000	0.008
Average in stream channels	—	—	300	1,250	0.0001
Subsurface water	50,000,000	130,000,000			
Soil moisture and intermediate-zone (vadose) water			16,000	67,000	0.005
Ground water within 0.5 mi (0.8 km) depth			1,000,000	4,170,000	0.31
Ground water, deep-lying			1,000,000	4,170,000	0.31
Total liquid water in land areas			2,070,000	8,637,000	0.635
Icecaps and glaciers	6,900,000	18,000,000	7,000,000	29,200,000	2.15
Atmosphere	197,000,000	510,000,000	3,100	13,000	0.001
World ocean	139,500,000	360,000,000	317,000,000	1,322,000,000	97.2
Totals (rounded)			326,000,000	1,360,000,000	100

SOURCE: Dr. Raymond L. Nace, U.S. Geological Survey, 1964.

mentalists consider that fresh surface water will prove to be the limiting factor in the capacity of our planet to support the rapidly expanding human population. The quantity of water vapor held in the atmosphere is also very small—only about ten times greater than that held in stream channels.

Long-term changes in water-balance quantities are associated with atmospheric environmental changes, discussed in some detail in Chapter 6. For example, atmospheric cooling on a global scale would bring a reduction of water-vapor storage and would reduce precipitation and runoff generally. But at the same time, a greater proportion of that precipitation would be in the form of snow, so that the storage in ice accumulations would rise and the storage in ocean waters would fall. These changes describe the changing water balance associated with onset of an *ice age*, or *glaciation*, and constitute a major environmental change already experienced by our planet at least four times in the past 2 to 3 million years. We shall return to this subject in Chapter 17.

Interception and Infiltration of Precipitation

Turning now to important details of the hydrologic cycle on the lands, let us follow the progress of precipitation through its alternative pathways. Imagine that a hillside slope which has been thoroughly drained of moisture in a period of drought is subjected to a period of rain. If dense vegetation such as forest is present, much of the rain at the beginning is held in droplets on the leaves and plant stems, a process termed *interception* (Figure 12.4). This water may be returned directly to the atmosphere by evaporation, so that if the rainfall is brief, little water reaches the ground.

When rain continues to fall beyond the limits of interception, and reaches the soil surface, it enters the soil as *infiltration*. Most soil surfaces in their undisturbed, natural states are capable of absorbing completely by infiltration the water from light or moderate rains. Such soils have natural passageways between poorly fitting soil particles, as well as larger openings, such as earth cracks resulting from soil drying, borings of worms and animals, cavities left from decay of plant roots, or openings made by heaving and collapse of soil as frost crystals alternately grow and melt. A mat of decaying leaves and stems breaks the force of falling drops and helps to keep these openings clear. Eventually, however, the soil passages are sealed or obstructed, dropping the infiltration rate to a low value. Any excess precipitation now remains on the ground surface, first accumulating in small puddles or pools which occupy natural hollows in the rough ground surface or are held behind tiny check dams formed by fallen leaves and twigs (Figure 12.4). *Surface detention* is the term applied to the holding of water on a slope by such small natural containers. Assuming that the rain continues to fall with sufficient intensity, water then overflows from hollow to hollow, becoming *overland flow*, a form of runoff.

At this point we must make a decision as to whether to follow the surface or the subsurface pathways of the hydrologic cycle. We shall chose the subsurface alternative, leaving the surface-water pathways for consideration in the next chapter.

Relationships among precipitation, infiltration, and runoff for a unit area of ground surface are shown in Figure 12.5. We can imagine the soil surface to be a fine sieve which receives rainfall of a given intensity and is capable of transmitting water downward at a given rate. Rainfall intensity is usually stated for short time periods, e.g., in./hr, or in./10-min. Infiltration is also stated as depth per unit of time. When rainfall intensity exceeds infiltration rate, the excess water leaves the surface as overland flow. Overland flow may also be stated in intensity units, as water depth per unit of time. The following equation then applies:

$$P - I = R_o$$

where P is precipitation rate (intensity),
I is infiltration rate,
and R_o is overland flow,

all stated as in./hr, or cm/hr. Evaporation is assumed to be zero during the rainfall period. Ob-

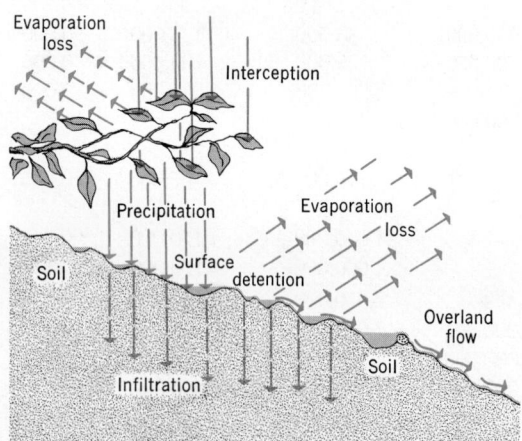

Figure 12.4 Interception, detention, and overland flow.

Evaporation and Transpiration of Soil Moisture

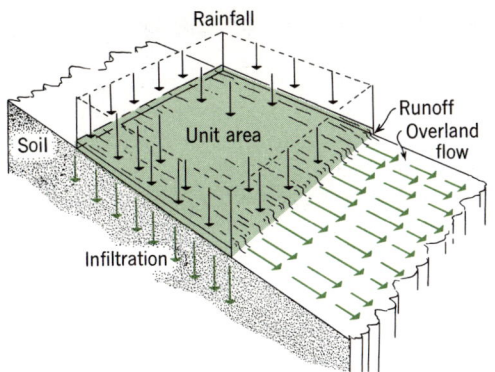

Figure 12.5 Rainfall on a unit area of ground surface is disposed of by infiltration and by runoff as overland flow. (From A. N. Strahler, 1971, *The Earth Sciences*, 2nd ed., Harper & Row, New York.)

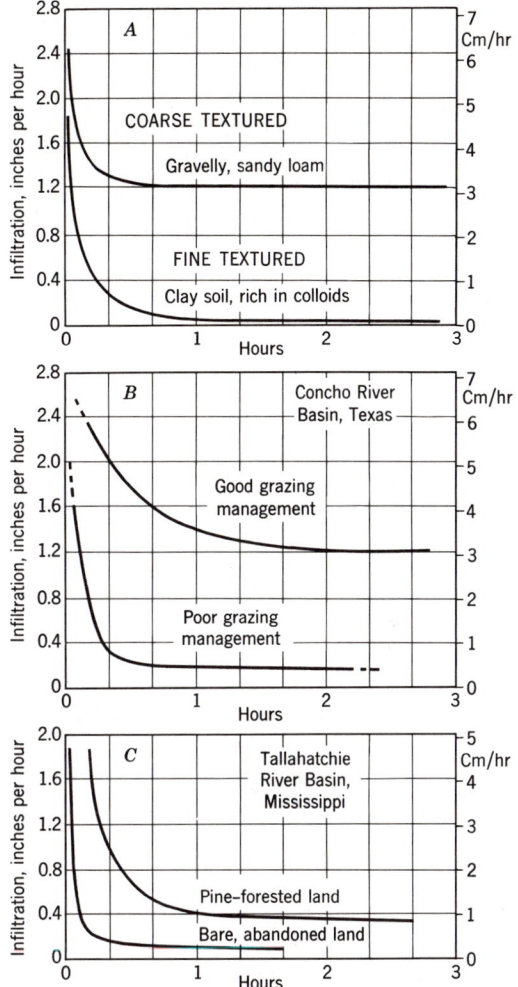

Figure 12.6 Infiltration rates vary greatly according to soil texture and land use. (Data from Sherman and Musgrave; Foster.)

viously infiltration rate will equal precipitation rate until the limit of the infiltration rate, or *infiltration capacity*, is reached.

Now, it is an important fact about soils that their infiltration capacity is usually great at the start of a rain which has been preceded by a dry spell, but drops rapidly as the rain continues to fall and to soak into the soil. After several hours the soil's infiltration capacity becomes almost constant. The reason for the high starting value and its rapid drop is, of course, that the soil openings rapidly become clogged by particles brought from above, or tend to close up as the colloidal clays take up water and swell. From this effect we can easily reason that a sandy soil with little or no clay will not suffer so great a drop in infiltration capacity, but will continue to let the water through indefinitely at a generous rate. In contrast, the clay-rich soil is quickly sealed to the point that it allows only a very slow rate of infiltration. This principle is illustrated by the graph in Figure 12.6A showing the infiltration curves of two soils, one sandy, one rich in clay.

It also follows that a sandy soil may be able to infiltrate even a heavy, long-continued rain without any surface runoff occurring, whereas the clay soil must divert much of the rain into overland flow. Many forms of artificial disturbance of soils tend to decrease the infiltration capacity and to increase the amount of surface runoff (Figure 12.6B, C). Cultivation tends to leave the soil exposed so that raindrop impact quickly seals the soil pores. Fires, by destroying the protective vegetation and surface litter, also expose the soil to raindrop impact. Trampling by livestock will tamp the porous soil into a dense, hard layer. It is little wonder, then, that Man has, through his farming and grazing practices, radically changed the original proportions of infiltration to runoff. As a result of reduced infiltration severe erosion damage has occurred. At the same time, there has been a decrease in the reserves of soil moisture, which might otherwise sustain plant growth in droughts. (More on this subject in Chapters 13 and 14.)

Evaporation and Transpiration of Soil Moisture

Between periods of rain, water held in the soil as *soil moisture* is gradually given up by a twofold drying process: (1) direct evaporation; (2) withdrawal by plants. Evaporation into the open air occurs at the soil surface and drying progresses downward. Air also enters the soil freely and may actually be forced alternately in and out of the soil by atmospheric pressure changes. Even if the soil did not "breathe" in this way, there would

Subsurface Water of the Lands

be a slow diffusion of water vapor surfaceward through the open soil pores. Ordinarily only the first foot (30 cm) of soil is dried by evaporation in a single dry season, but in the prolonged drought of deserts, drying will extend to depths of many feet. Plants draw the soil moisture into their systems through vast networks of tiny rootlets. This water, after being carried upward through the trunk and branches into the leaves, is discharged in the form of water vapor, through specialized leaf pores into the atmosphere, a process termed *transpiration*.

In studies of climatology and hydrology it is convenient to use the term *evapotranspiration* to cover the combined moisture loss from direct evaporation and the transpiration of plants. The rate of evapotranspiration slows down as soil moisture supply becomes depleted during a dry summer period because plants employ various devices to reduce transpiration. In general, the less moisture remaining, the slower is the loss through evapotranspiration.

Figure 12.7 shows diagrammatically the various terms explained up to this point and serves to give a more detailed picture of that part of the hydrologic cycle involving the soil. As the plus signs show, the *soil-moisture zone* gains water through precipitation and infiltration. As the minus signs show, the soil loses water through transpiration, evaporation, and overland flow, and by *gravity percolation* downward through the soil to the ground-water zone below. Between the soil-moisture zone and the ground-water zone is an *intermediate zone*, holding moisture but at a depth too great to be returned to the surface by evapotranspiration.

Moisture Capacity of the Soil

When infiltration occurs during heavy and prolonged rains (or when a snow cover is melting) the water is drawn downward by gravity through the soil pores, wetting successively lower layers. Soon the soil openings are filled with water moving downward, except for some air entrapped in the form of bubbles. Then the percolation continues downward into the bedrock. Suppose now that the rain stops and a period of several days of dry weather follows. The excess soil water continues to drain downward, but some water clings to the soil particles and completely resists the pull of gravity through the force of *capillary tension*. We are all familiar with the way in which a water droplet seems to be enclosed in a "skin" of surface molecules, drawing the droplet together into a spheroidal shape, so that it clings to the side of a glass without flowing down. Similarly, tiny films of water adhere to the soil grains, particularly at the points of grain contacts, and will stay until disposed of by evaporation or by absorption into plant rootlets.

When a soil has first been saturated by water, then allowed to drain under gravity until no more water moves downward, the soil is said to be holding its *field capacity* of water. This drainage takes no more than two or three days for most soils; most is drained out within one day. Field capacity is measured in units of water depth, usually inches or centimeters, just as with precipitation. This means that for a given cube of soil, say 12 in. on a side (1 cu ft), if we were to extract all of the field moisture, it might form a layer of water 3-in. deep in a pan 1-ft square. This would

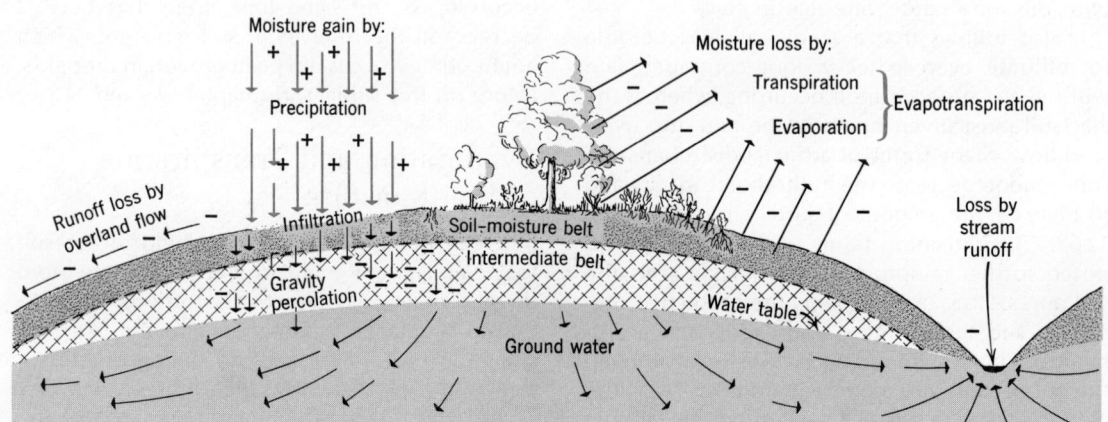

Figure 12.7 The soil-moisture zone occupies an important position in the water balance of the lands.

The Soil-Moisture Cycle

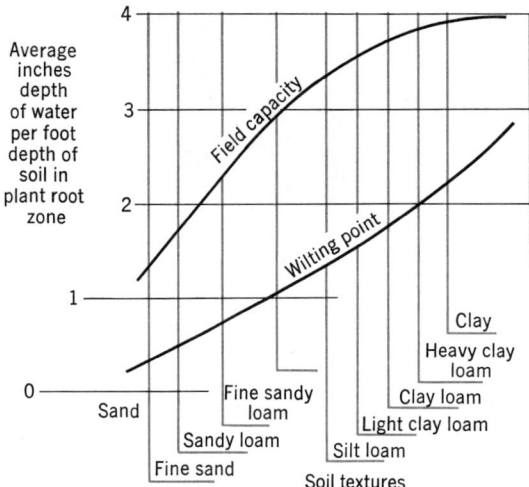

Figure 12.8 Field capacity and the wilting point. (After Smith and Ruhe, *Yearbook of Agriculture*, 1955.)

be equivalent to complete absorption of a 3-in. rainfall by a completely dry 12-in. layer of soil.

Field capacity of a given soil depends largely on its texture. Sandy soil has a very low field capacity; clay soil has high field capacity. This effect is shown in Figure 12.8, a graph in which field capacity is plotted against soil texture, from coarse to fine. (Refer to Figure 11.18 for explanation of terms.) It should also be noted that sandy soils reach their field capacity very quickly, both because of the ease with which the water penetrates and the low quantity required. Clay soils take long rain periods to reach field capacity because the infiltration is slow and the total quantity required to be absorbed is great.

Also in use as a measure of soil moisture is the *wilting point*. This is the quantity of soil moisture below which plants will be unable to extract further moisture from the soil and the foliage will wilt. As Figure 12.8 shows, the wilting point also depends upon particle size.

The Soil-Moisture Cycle

Equipped with the foregoing explanations of processes and terms relating to water gains and losses in the soil, we can turn next to consider the annual water budget of the soil, involving principles of great importance not only in plant ecology and agriculture, but in the further study of ground water, surface runoff, stream flow, and therefore of the erosion of land slopes.

Figure 12.9 shows the annual cycle of soil moisture for the year 1944 at an agricultural experiment station in Coshocton, Ohio. If we follow the changes in this example, the cycle it shows can be considered generally representative of conditions in humid, middle-latitude climates where there is a strong temperature contrast between winter and summer. Let us start with the early spring (March). At this time the evaporation rate is low, because of the low rate of energy input. The abundance of melting snows and rains has restored the soil moisture to a surplus quantity. For two months the quantity of water percolating through the soil and entering the ground water keeps the soil pores nearly filled with water. This is the time of year when one encounters soft,

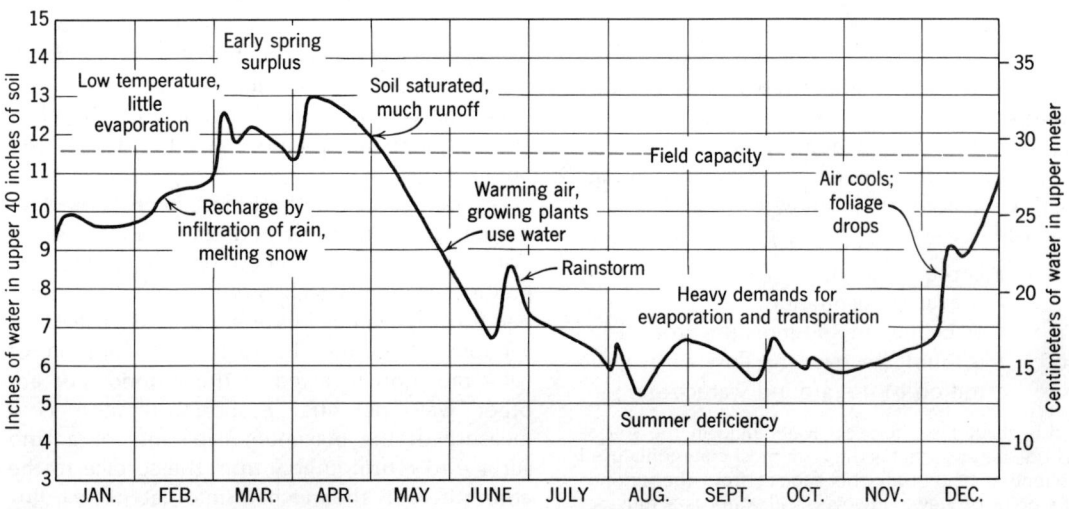

Figure 12.9 Annual cycle of soil moisture at Coshocton, Ohio. (After Thornthwaite and Mather, *The Water Balance*, 1955.)

Subsurface Water of the Lands

muddy ground conditions, whether driving on dirt roads or walking across country. This, too, is the season when runoff is heavy and major floods may be expected on larger streams and rivers. In terms of the soil-moisture budget, a *water surplus* exists.

By May, increasing energy of solar radiation, increasing evaporation, and the rapid growth of plant foliage—bringing on heavy transpiration—have reduced the soil moisture to a quantity below field capacity, although it may be restored temporarily by unusually heavy rains in some years. By midsummer, a state of *moisture deficiency* exists in the water budget. Even the occasional heavy thunderstorm rains of summer cannot restore the water lost by steady and heavy evapotranspiration. Small springs and streams dry up, the soil becomes firm and dry as soil colloids contract. By November (and sometimes as early as late September), however, the soil moisture again begins to increase. This is because the plants go into a dormant state, sharply reducing transpiration losses, while, at the same time, falling air temperatures reduce evaporation. By late winter, usually in February at this location, the field capacity of the soil is again restored.

The Soil-Moisture Budget

From the foregoing example of soil-moisture changes throughout a single year's time, we go on to a more generalized concept of the soil-moisture budget.* Let us first return to the basic water-balance equation, modifying it to apply specifically to the soil-moisture zone. This zone has a depth equal to the lower limit to which plants send their roots; it is the storage zone for moisture available to plants. Figure 12.10 is a schematic diagram of the terms in the moisture-balance equation. The soil column is assumed to have a unit cross sectional area, and we can therefore use flow units of depth per unit time (e.g., mm/day or mm/month), as in previous statements of the general water-balance equation. The equation is as follows:

$$P = E + G + R$$

where *P* is precipitation,
    E is evapotranspiration,
    G is change in soil-moisture storage,
and *R* is runoff (by overland flow, or by infiltration to the ground water zone.)

*In this chapter we refer to water held in the soil as *moisture*, because that is the term most commonly used by scientists who study this subject from the agronomist's point of view. However, *soil water* is a fully acceptable alternative term for *soil moisture*.

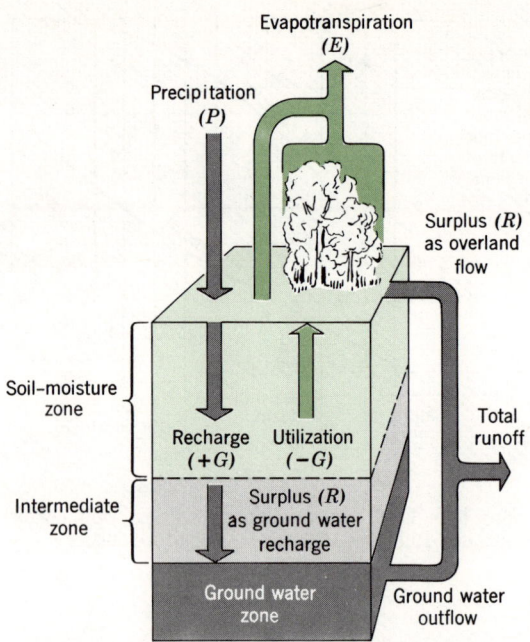

Figure 12.10 Schematic diagram of the soil-water balance in a soil column of unit cross section.

To proceed further, we must recognize two ways to define evapotranspiration. First is *actual evapotranspiration*, which is the true or real rate of water-vapor return to the atmosphere from the ground and its plant cover. Second is *potential evapotranspiration*, representing the water vapor flux under an ideal set of conditions. One condition is that there be present a complete (or closed) cover of homogeneous vegetation consisting of fresh green leaves, and no bare ground exposed through that cover. The leaf cover is assumed to have a uniform height above ground—whether the plants be trees, shrubs, or grasses. A second condition is that there be an adequate water supply, such that the field capacity of the soil is maintained at all times. This condition can be fulfilled naturally by abundant and frequent precipitation, or artificially by frequent irrigation.

Several rigorous methods have been devised for estimating the monthly potential evapotranspiration at any given location on the globe. The method used in this chapter is based on air temperature, as well as latitude and date. The latter variables determine intensity and duration of solar radiation received at the ground. Put another way, potential evapotranspiration is a measure of the maximum capability of a land surface to return energy from the surface to the atmosphere by the mechanism of latent heat flux under the defined conditions.

The Soil-Moisture Budget

Beyond this point, we shall use the symbol E_p to designate potential evapotranspiration; the symbol E_a to designate actual evapotranspiration. Note that when P exceeds E_p or when the soil is at field capacity, E_p and E_a will have the same numerical value, but at all other times E_a will be a smaller value than E_p.

Suppose, now, that mean daily values of precipitation, P, and both forms of evapotranspiration, E_p and E_a, are plotted on a graph to show a full year's cycle for a place in the Northern Hemisphere (Figure 12.11). For the sake of simplicity we have used smooth curves, rather than a sequence of 365 points, to depict the changes in the three variable quantities. We shall assume that precipitation is uniformly distributed throughout the entire year, and shows as a horizontal line on the graph. On the other hand, potential evapotranspiration, E_p, is assumed to show a strong annual cycle with a low value in winter and a high value in summer, as is typical in middle latitudes. Consider two idealized cases. First (as shown in Diagram A), full irrigation is supplied during the summer, so that E_a and E_p remain identical in value. During the period when P is greater than E_p, there is a *water surplus* and runoff (R) is generated. In the second case (shown in Diagram B), no irrigation water is available in the summer. When E_p exceeds P the plants must draw upon moisture stored in the soil in an attempt to sustain as rapid growth as possible at all times. There is, of course, a continuing supply of precipitation constituting an input of moisture into the soil zone, but the plants are capable of using more than this amount. Under these circumstances, the curve of actual evapotranspiration, E_a, drops off to a lower value than the curve of E_p. As we have previously stated, it is assumed that the rate at which soil moisture is depleted falls off at a rate proportional to the quantity of soil moisture remaining. The curve of E_a falls off rapidly at first then begins to level off. The store of soil moisture is never fully depleted, but may approach a negligible quantity after a long season of depletion. Daily change in soil-moisture storage, G, is represented by the vertical distance between the curves of P and E_a (G has a negative sign).

The difference between the daily values of E_p and of E_a (or $E_p - E_a$), when accumulated from the time of onset of soil-moisture depletion, until the depletion period is ended, is referred to as the *soil-moisture deficiency* (D); it is stated in units of water depth (mm or in.). (In other words, D is a quantity of water, rather than a flow rate of water.) The soil-moisture deficiency is repre-

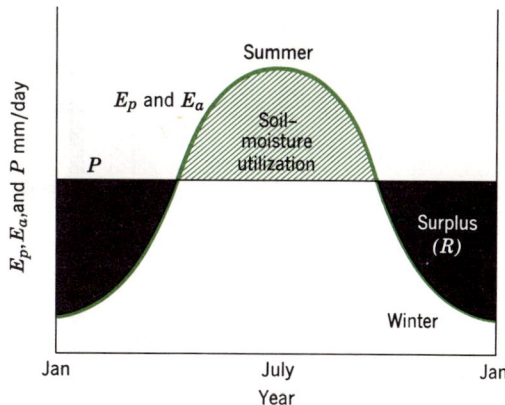

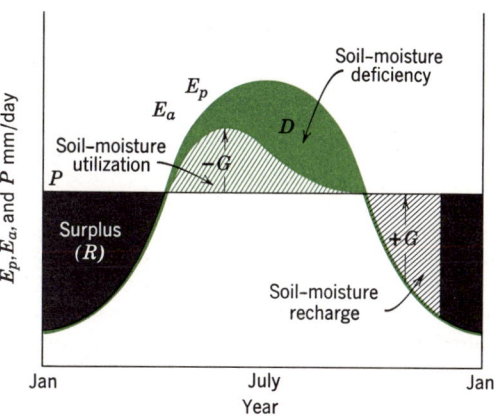

Figure 12.11 Schematic graphs of the soil-moisture budget.

sented graphically in Diagram B as the area between the curves of E_p and E_a.

Once the period of soil-moisture deficiency has ended (crossing of the curves E_p and P), there must follow a period of *soil-moisture recharge*, in which soil moisture is restored to the field capacity. All P in excess of E_p now goes into recharge. The daily rate of increase in soil-moisture storage is represented by the term G with a positive sign. When recharge is complete, a period of *water surplus* sets in and runoff (including ground water recharge or overland flow, or both) takes place. The period of soil-moisture recharge is shown in Diagram B by a distinctive pattern between the curves of P and E_a. This area is equal to the area on the same graph representing soil-moisture utilization.

We are now ready to turn to the practical application of the soil-moisture balance equation to the simplified data of an observing station. Monthly, rather than daily, means of P are used,

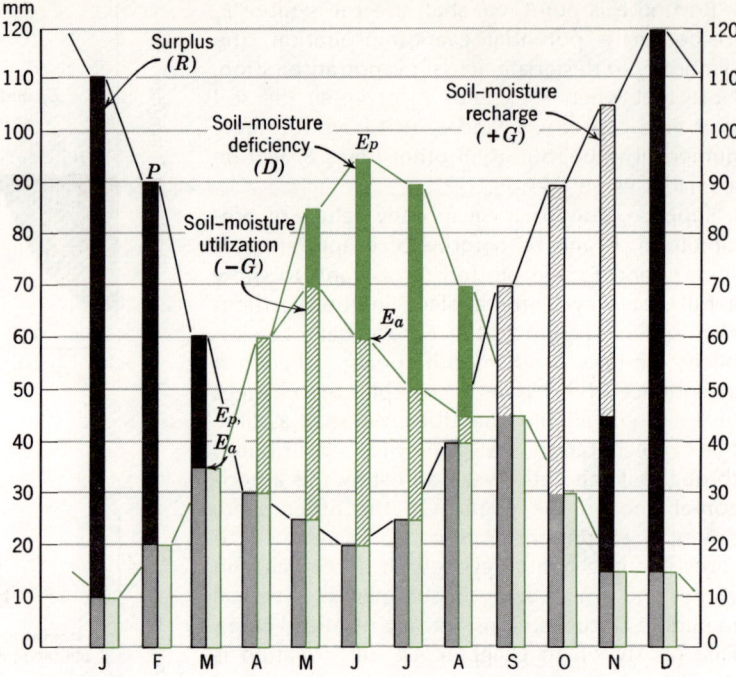

Figure 12.12 A model soil-moisture budget.

and these represent long-term averages. Figure 12.12 is an idealized yearly moisture budget. Each month is represented by two vertical bars, whose heights are proportional to flow rate in units of millimeters per month. Each bar is divided up into segments showing the several terms of the water-balance equation. The left-hand bar shows P, part of which matches an equal quantity of moisture lost by evapotranspiration, and part of which may represent recharge ($+G$) or surplus (R). The right-hand bar shows E_p, part of which may represent moisture deficiency (D) or soil-moisture utilization ($-G$).

We can now assign numerical values to each term of the water-balance equation for each month, as shown in Table 12.3 (all values in this example are multiples of 5 mm).

By means of simple arithmetic, we have calculated the water balance for each month singly and for the year as a whole. Tallied separately at the right are the monthly differences between E_p and E_a, giving a total soil-moisture deficiency, D, of 125 mm for the year. This is the quantity of water that would have to be supplied by irrigation to sustain the full value of E_p. The importance of the water-balance calculation in estimating the need

TABLE 12.3 A Model Soil-Moisture Budget

Equation:	$P = E_a$	$+G$	$+R$	E_p	$(E_p - E_a) = D$
January	110 = 10		+100	10	0
February	90 = 20		+ 70	20	0
March	60 = 35		+ 25	35	0
April	30 = 60	−30		60	5
May	25 = 70	−45		85	15
June	20 = 60	−40		95	40
July	25 = 50	−25		90	40
August	40 = 45	− 5		70	25
September	70 = 45	+25		45	0
October	90 = 30	+60		30	0
November	105 = 15	+60	+ 30	15	0
December	120 = 15		+105	15	0
Totals	785 = 455	−145 +145	+330	570	125
	785 = 785				

for irrigation of crops should be obvious. We have also been able to calculate the annual runoff, 330 mm, and can use this information to estimate the recharge of ground water and the runoff into streams. In this way an assessment of the water resource potential of a region can be made and is a vital consideration in planning regional economic development and resource management.

Soil-Moisture Regimes

C. Warren Thornthwaite, a distinguished climatologist who was concerned with practical problems of crop irrigation, developed the foregoing principles of the soil-moisture balance and proposed a system of classification of world climates based on those principles. Associates and students of Thornthwaite extended his work and collected hydrologic data for a vast network of observing stations in all land areas of the globe. Moisture deficiencies were computed on the basis of a field capacity of 300 mm, considered as the best single representative value for general use.

Figure 12.13 is a map of the United States showing isopleths of equal total annual potential evapotranspiration. Notice that the highest values occur over the desert basins of the Southwest. Here insolation is intense and air temperatures are generally high. High values also prevail over a belt bordering the Gulf Coast from Texas to Florida. As you would expect, values fall steadily as one progresses northward because of lowered total annual insolation and colder mean annual air temperatures. The world range in values of total annual E_p is much larger than this map indicates. Over large areas of the southern Sahara Desert, for example, the values are over 1500 mm/yr., while values of 1400 to 1750 mm/yr are prevalent over equatorial Africa. In arctic latitudes annual E_p is reduced in some localities to as little as 200 to 250 mm/yr. Not only does the annual total E_p grade from high values in low latitudes to low values in high latitudes, but the pattern of seasonal distribution changes radically. As with monthly values of net radiation (Figure 2.15), monthly E_p values in the equatorial zone show nearly uniform high values throughout the year. With increasing latitude an annual cycle becomes stronger, and, where ground is frozen and plants are dormant in a winter season, there are several consecutive months of virtually zero E_p. These global trends will show conspicuously in the examples to follow (see Figure 12.14).

Rather than attempt to set up a world-wide system of climates we shall simply offer some examples of the wide global range in *soil-moisture regimes*. By "regime" we mean "prevail-

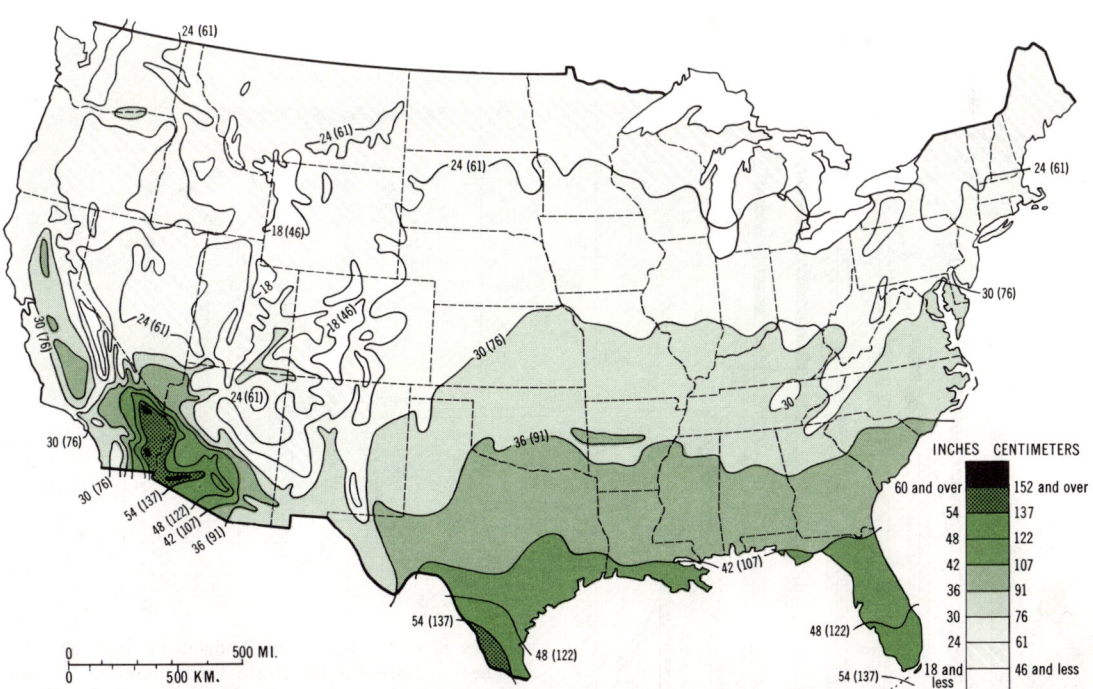

Figure 12.13 Average annual potential evapotranspiration in the United States. (Courtesy of *The Geographical Review*, vol. 38, 1948, Copyrighted by the American Geographical Society, New York.)

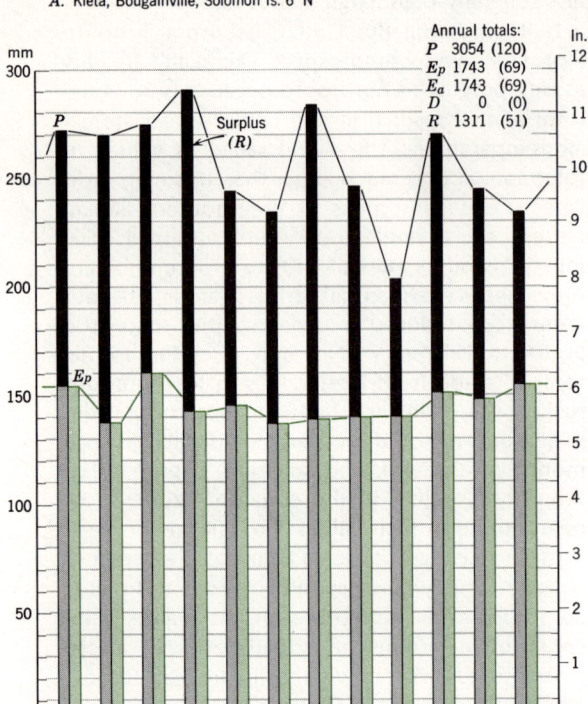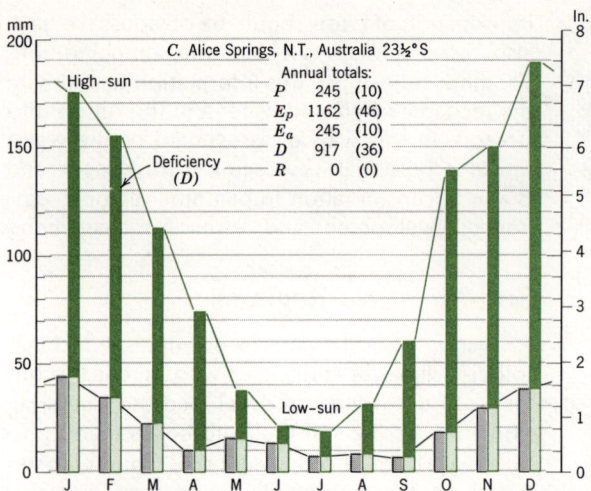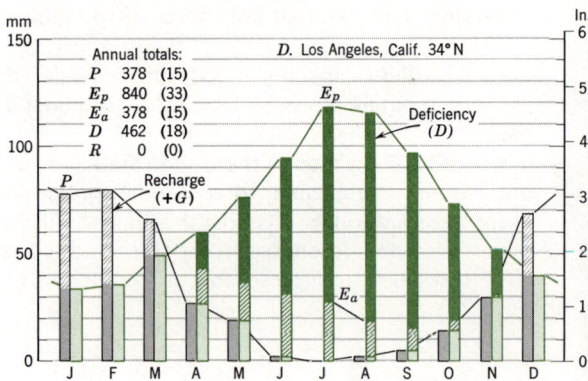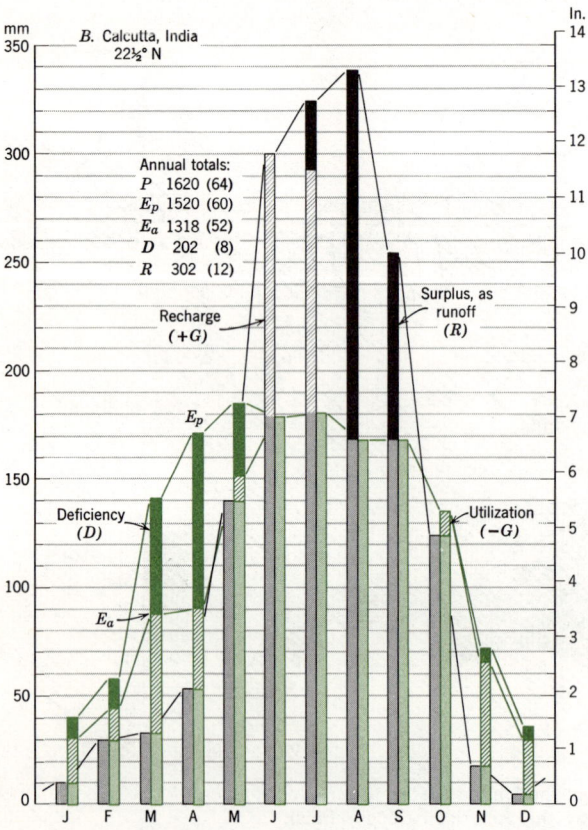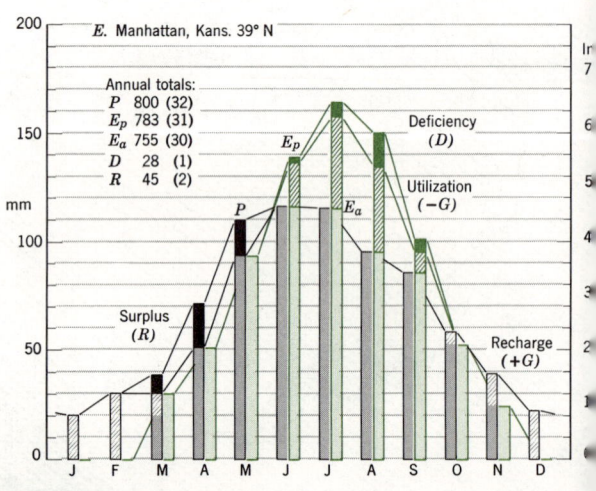

Figure 12.14 Soil-moisture budgets for eight stations. Inches in parentheses. (Data from *Average Climatic Water Balance Data of the Continents*, C. W. Thornthwaite Associates, Laboratory of Climatology, Publ. in Climatology, 1962–1965, Centerton, New Jersey.)

Soil-Moisture Regimes

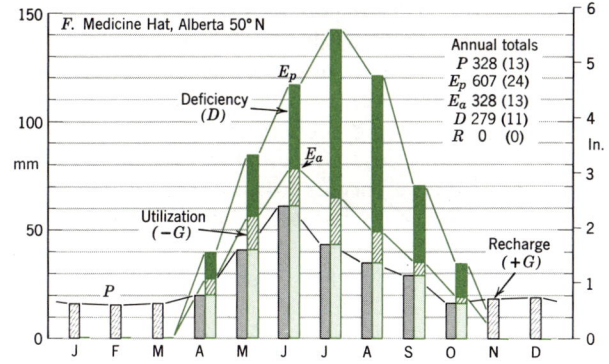

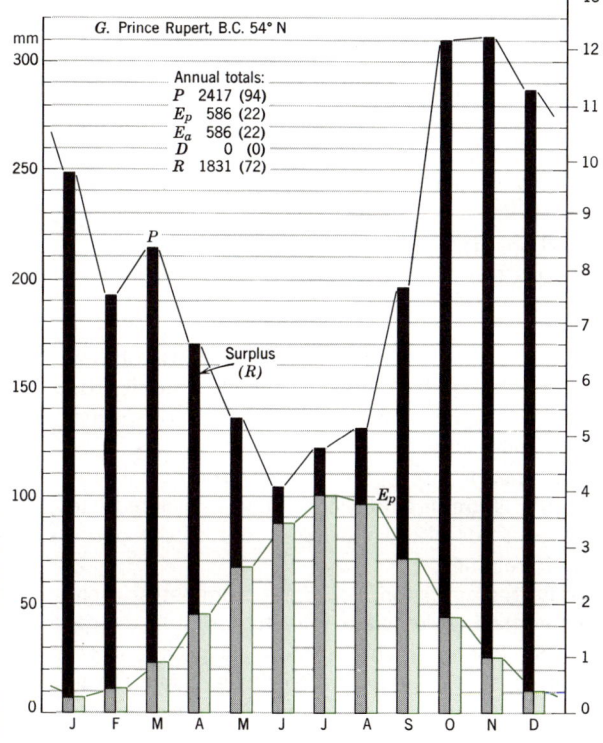

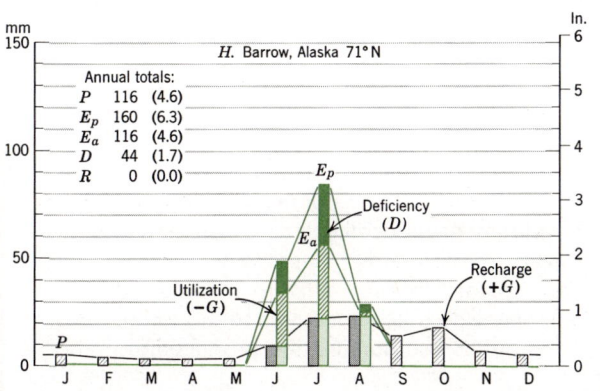

ing tendency," or "dominant pattern" with respect to the annual rhythms of change in soil-moisture deficiencies and water surpluses. Figure 12.14 is a collection of moisture-balance graphs for stations showing a wide range of regimes; they are arranged in order of increasing latitude. Let us start with the *Mediterranean regime*, illustrated by the soil-moisture budget for Los Angeles, California (Graph D). The regime name derives from its prevalence in lands bordering the Mediterranean Sea, but it is found generally on western continental coasts in the latitude range 30° to 50°. Notice that the graph for Los Angeles is very similar to the idealized example we used to show calculation of the terms of the moisture-balance equation (Figure 12.12). This regime is one of a large soil-moisture deficiency. Because P declines to low values at the very time when E_p rises to its maximum, the large summer deficiency is accentuated. Here is a region of dry climate in which crop irrigation will be extremely valuable; in fact, summer crop cultivation will not be possible here without irrigation. The native vegetation must be adapted to a long, hot summer with little or no rainfall and a long period of severely depleted soil moisture. The winter excess of P over E_p, on the other hand, gives a period of soil moisture recharge, but it is insufficient to generate a surplus.

An example of a regime with an enormous water surplus is the *equatorial regime*, illustrated in Graph A, showing data of Bougainville, a Pacific island of the Solomon group, lying close to the equator. Both P and E_p are remarkably uniform throughout the year, but P is much the greater in every month. As a result there is a tremendously large annual total water surplus. Soil moisture storage remains at capacity—300 mm—in all months. Plant growth can proceed at the maximum rate in all months; the equatorial rainforest is the typical natural vegetation. Stream flow in this regime is copious throughout the year and leads to the possibility of development of hydroelectric power where mountainous relief is present and a demand for power exists.

The *tropical wet-dry regime* is illustrated by Graph B, showing data for Calcutta, India, lying close to the Tropic of Cancer. Here is a regime of great moisture contrasts. Precipitation is slight in the low-sun months (November–February) but rises to huge values in the rainy season (June–September). The annual curve of E_p also shows a strong seasonality, being moderately high in coincidence with the rainy season. As a result, there is a large water surplus (June–September) but a severe soil-moisture deficiency preceding the rains (March–May). Agriculture depends upon these monsoon rains; should they not materialize, the moisture deficiency can extend through the year, bringing crop failure and famine.

The *tropical desert regime* is illustrated by data for Alice Springs which lies in the heart of the Great Australian Desert (Graph C). While P is measurable in small amounts in all months, the values of E_p are always the larger. Consequently a soil-moisture deficiency prevails throughout all months of the year and racks up a very large total. Soil moisture storage is close to zero at all times. It is easy to understand why vegetation is sparse in a desert such as this. Notice that, because this station is in the Southern Hemisphere, the annual cycle seems inverted in phase. Actually the cycles of P and E_p, which are closely in phase, have their peak when the sun is highest in the sky.

Passing up Los Angeles, Graph D, which we have already analyzed, we arrive in middle latitudes. Graph E shows a station in the heart of the United States and illustrates the *continental humid regime*. The annual cycles of both P and E_p are closely in phase at Manhattan, Kansas; both show a strong summer maximum. However values of E_p are the larger from June through October, so that a small soil-moisture deficiency develops in summer. But, because the deficiency does not make serious inroads into the soil-moisture storage (300 mm), crops thrive without irrigation and yields are high in most years. In winter, E_p falls to zero in three months (December–February) and recharge sets in. By spring water surplus is produced, but it is not a large amount.

We now travel north, still in the heart of the North American continent, to reach Medicine Hat, Alberta, a station in the Great Plains region of Canada (Graph F). The regime is much like that of Manhattan, Kansas, but P is much less than E_p in summer months, generating a large total soil-moisture deficiency. We can call this pattern the *continental arid regime*. Recharge in winter is insufficient to restore the field capacity of the soil, so that no surplus occurs.

At about the same latitude as Medicine Hat, but located on the Pacific Coast, is a station showing a wet variety of the Mediterranean regime (Graph G). Although the regime at Prince Rupert, British Columbia, may seem totally unlike that for Los Angeles (Graph D), the annual cycles of P and E_p show the same out-of-phase relationship. Notice the reduced values of P in the summer. However, since P is always larger than E_p, a surplus occurs in every month and yields a total annual surplus—even greater than for Bougainville (Graph A). This huge water surplus might be put to Man's use both for hydroelectric power and to be transported to regions of water deficiency (Medicine Hat, for example).

Finally, we travel to Barrow, Alaska, a station on the shores of the Arctic Ocean, latitude 71° N, lying poleward of the Arctic Circle (Graph H). Because the ground is frozen from September through May, no E_p occurs for nine consecutive months. However, E_p rises to a sharp peak in summer, and even exceeds P to the extent that a soil-moisture deficiency develops in three months.

This very brief investigation of only a few examples of soil-moisture budgets can give you a better appreciation of the great global range in qualities of the environments of the life layer. Considerations of soil-moisture deficiency and water surplus are vital, not only in understanding the adaptation of plants and animals to their environments, but also in understanding the opportunities and limits of Man's use of plant resources. Intelligent planning for environmental management depends heavily upon an understanding of the water balance.

Pedogenic Regimes and the Soil-Moisture Budget

Because the characteristics of the soil profile in a given region are adapted to availability of heat and water and to organic activity that is also dependent upon heat and water, it is not surprising that there are soil-forming regimes, here called *pedogenic regimes,* corresponding with the soil-moisture regimes and their associated climates. Figure 12.15 illustrates four of the pedogenic regimes.

The regime of *podzolization* dominates in climates having sufficient cold to inhibit bacterial action, but sufficient moisture to permit larger green plants (macroflora) to thrive. Such conditions exist only in middle and high latitudes, and at high altitudes. Favorable climates range from the cool marine-type climates of west coasts to interior continental climates with a great annual temperature range and very cold winters. The soil-moisture regime shows a marked water surplus in winter and spring, while the summer shows either a small moisture deficiency or none at all.

Figure 12.16 shows the major world regions where podzolization is dominant. In its extreme development podzolization is associated with forests of coniferous trees (spruce, fir, pine). These plants do not use large amounts of bases, and hence do not restore them to the soil surface. The result is that humic acids, produced from the

Pedogenic Regimes and the Soil-Moisture Budget

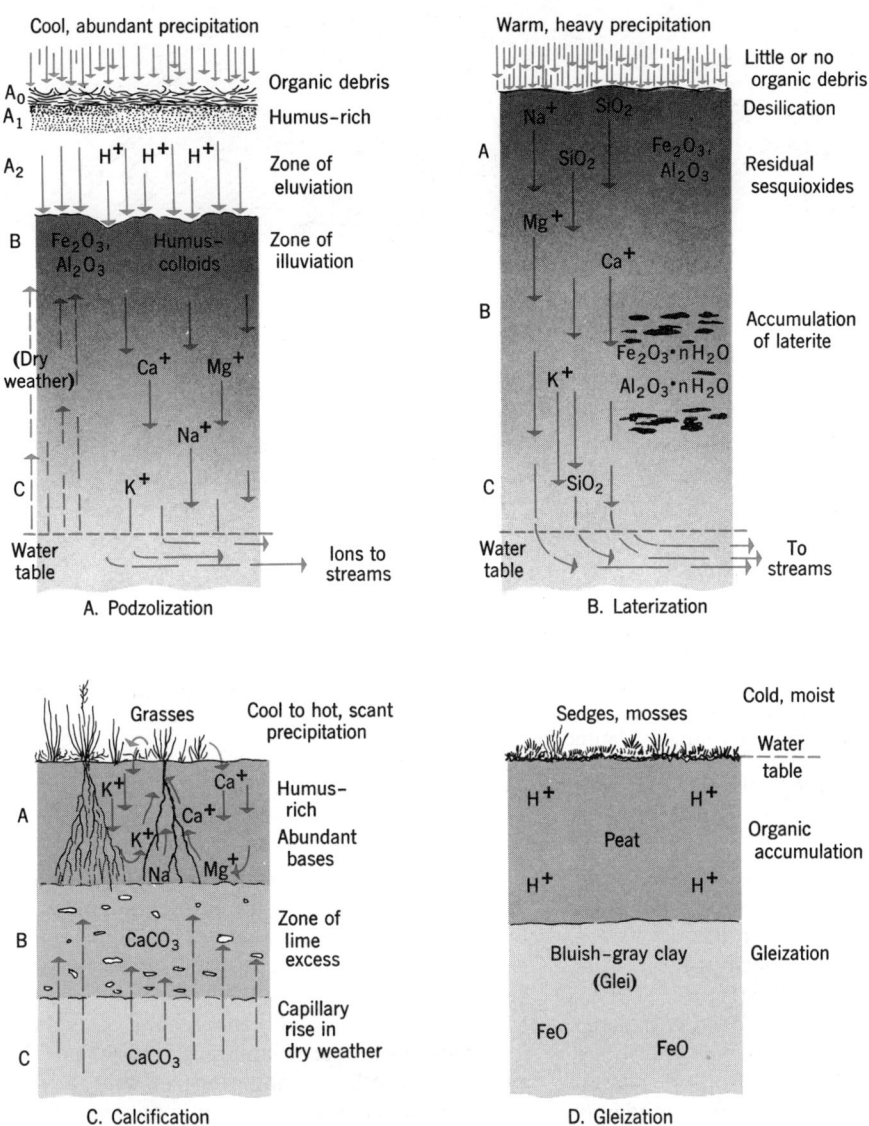

Figure 12.15 Soil profile development under four pedogenic regimes.

abundant plant litter and humus, leach the upper soil strongly of bases, colloids, and the oxides of iron and aluminum, leaving a characteristic ash-gray A_2 soil horizon composed largely of silica (SiO_2) (Figure 12.15A). Colloids, humus, and oxides of iron carried out of the A_2 horizon accumulate in the B horizon, which may be dark in color, dense in structure, and in some cases hardened to rocklike consistency.

The process of downward movement of ions and colloids is termed *eluviation;* the zone of that name is identical with the A_2 horizon. Accumulation of these same materials is referred to as *illuviation;* the zone of that name is identical with the B horizon (Figure 12.14A).

Soils produced under the regime of podzolization go by the name of *podzols,* a Russian word connoting "ash soil." A typical podzol soil profile is shown in Figure 12.17A). Podzols and related podzolic soils underlie much of those areas of North America and Eurasia that were subjected to continental glaciation in the Pleistocene Epoch, so that most of the parent matter is glacially deposited and was uncovered only from 12,000 to 15,000 years ago.

The pedogenic regime of *laterization* is in some respects a warm-climate relative of podzolization, in that both are associated with climates of ample precipitation and with forests. Laterization takes place in warm climates having a large water sur-

303

Subsurface Water of the Lands

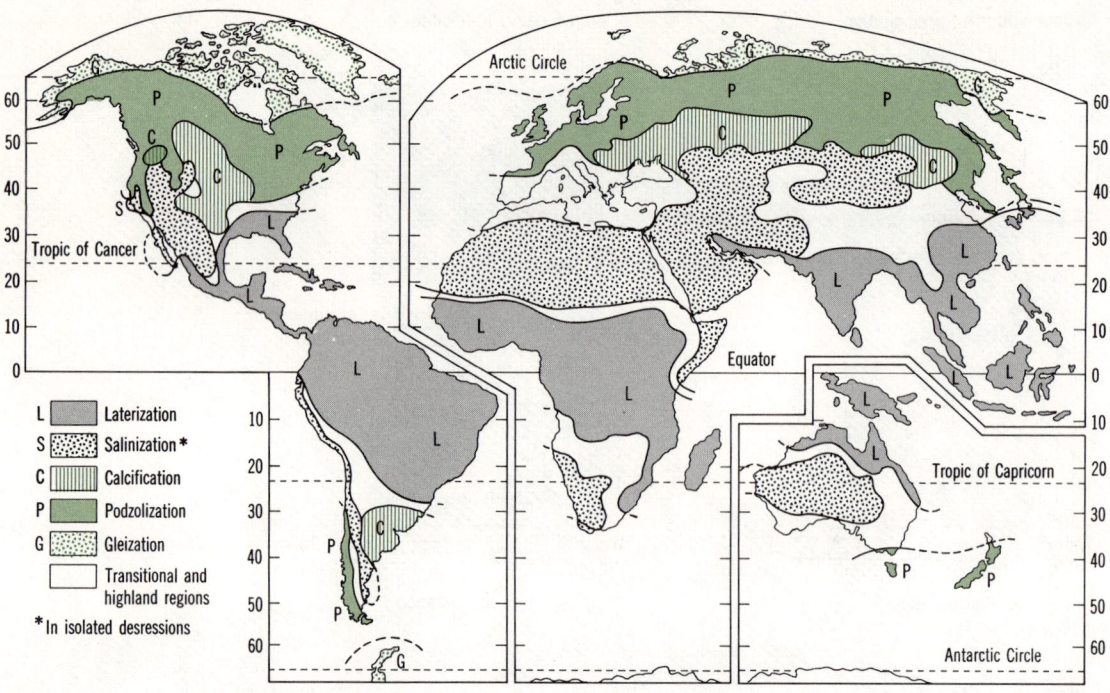

Figure 12.16 Generalization of distribution of the pedogenic regimes. The map should be considered as a schematic diagram in terms of latitude and continental position. Major highland areas are left blank. (Data of A. C. Orvedal, U.S. Department of Agriculture. Based on Goode Base Map. Copyright by the University of Chicago. Used by permission of the University of Chicago Press.)

plus either seasonally or year-around. These climates occupy the equatorial zone and those parts of the tropical zones having a monsoon season with abundant rainfall. A typical soil-moisture budget is that of Bougainville (Figure 12.14A). Monsoon climates of southea Asia, with large water surpluses in the rainy monsoon season are also included (Figure 12.14B). Tropical coastal zones in the trade-wind belt also show the regime of laterization (Figure 12.16).

A high mean annual temperature and a lack of severe winter season permit sustained bacterial action which destroys plant litter as rapidly as it is produced. Consequently little or no humus is found upon or in the soil (Figure 12.15B). In the absence of humic acids the sesquioxides of iron and aluminum are insoluble and accumulate in the soil as yellow and red clays, nodules, and rocklike layers (laterite). Silica, on the other hand, is leached out of the soil and disposed of eventually by stream flow in the process of *desilication*. No distinctive soil horizons are developed. In the absence of silicate colloids (true clay minerals) the soil tends to be firm and porous rather than sticky and plastic, and will transmit water readily. Laterization results in very low soil fertility because bases are not held in the soil and humus is lacking.

Soils produced by laterization are known generally as *latosols*. They are easily recognized by their deep red and reddish brown colors. The general pattern of world distribution is shown in Figure 12.16.

Calcification is a pedogenic regime of climates in which potential evapotranspiration on the average exceeds precipitation so that a soil-moisture deficiency prevails in one season and little or no surplus is developed (see Figure 12.14F). These climates control the semiarid steppes and range widely in latitude from tropical to middle latitude zones (Figure 12.16). Downward movement of surplus water is not enough to leach out the bases, so that calcium and magnesium ions remain in the soil (Figure 12.15C). Grasses, which use these bases, restore them to the soil surface. Colloids remain essentially in place and are not leached out, but are in a dense (flocculated) state and hold the soil into aggregate structures. Calcium carbonate, brought upward by capillary water films and evaporated in the season of moisture deficiency, is precipitated in the B horizon of the soil in the form of nodules, slabs, and even dense stony layers. These rocklike accumulations are referred to as *calcrete*, or locally in the southwestern United States as *caliche*. Microbial activity is restricted and humus may be abundantly distributed throughout the A and B horizons. Humus occurs in progressively smaller amounts as one traces the soil into climate zones of increasing aridity. Calcification is characteristically associated with grasslands.

In middle-latitude regions in which annual evapotranspiration is only slightly in excess of annual precipitation—we can call these *subhumid* climates—calcification has produced a type of soil named the *chernozem* (Russian) or *black earths*. The soil-moisture budget for Manhattan, Kansas (Figure 12.14E) is representative for this regime. Notice that water surplus is small and that there is a summer moisture deficiency. A typical chernozem profile is shown in Figure 12.17B). The chernozem regions are noted for their high productivity of small grains—wheat, oats, barley, and rye.

The pedogenic regime of *gleization* is characteristic of poorly drained environments under a cool or cold climate. Gleization is thus associated with the subarctic lands, or tundra (Figure 12.16). Annual totals of both precipitation and evapotranspiration are small and soil-moisture storage remains adequate in the short summer, despite depletion (Figure 12.14H). Gleization is also effective in bog environments of middle-latitude climates with cold winters. Low temperatures permit heavy accumulations of organic matter to form a surface layer of peaty material (Figure 12.15D). Beneath this is the *glei horizon*, a thick layer of compact, sticky, structureless clay of bluish gray color. In bogs the glei horizon lies generally within the zone of ground-water saturation; consequently the iron is in a partially reduced condition and imparts the bluish gray color.

Finally, there is the pedogenic regime of *salinization*, or accumulation of highly soluble salts in the soil. Salinization is associated with the desert climate of a severe soil-moisture deficiency and takes place in poorly drained locations where occasional surface runoff derived from outside the region can evaporate. Such locations are typically low-lying valley floors, flats, and basins in the continental interiors, and coastal flats in arid climates (Figure 12.16). Sulfates and chlorides of calcium and sodium are common salts in such soils. A representative soil-moisture budget is that of Alice Springs (Figure 12.14C).

The significance of the pedogenic regimes and their associated varieties of soils looms so large in the environment of the life layer on the lands that it would be hard to overemphasize by any words we could muster. Animal life of the lands depends for sustenance upon the plants, which are the primary organic producers, and there would be no primary production on the lands without the soil. A knowledge of world distribu-

Figure 12.17 Left: Podzol soil profile developed on a sandy glacial deposit in Maine. Right: Chernozem soil profile developed on glacial deposits in North Dakota. (Photograph by C. E. Kellogg and W. M. Johnson, Division of Soil Survey, U.S. Dept. of Agriculture.)

tion of the pedogenic regimes and their associated soil-moisture budgets enables us to evaluate the world land environments with respect to food resources. When we do this, we find that the world's crop production potential is severely limited, not only in terms of available water and heat and suitable soil chemistry, but also by the inhospitality of enormous areas of mountain lands unfit for crop cultivation because of their steep, rocky slopes. Other extensive areas are waterlogged swamps and bogs, difficult to use without elaborate drainage works.

Ground Water

Water that is drawn downward by gravity through the soil zone to lower levels becomes part of the ground-water body; the geometrical relationships are shown in Figure 12.18. Strictly speaking, ground water is that part of subsurface water which fully saturates the pore spaces of the rock or its overburden and which behaves in response to gravitational force. The ground water occupies the *zone of saturation*. Above it is the *zone of*

Subsurface Water of the Lands

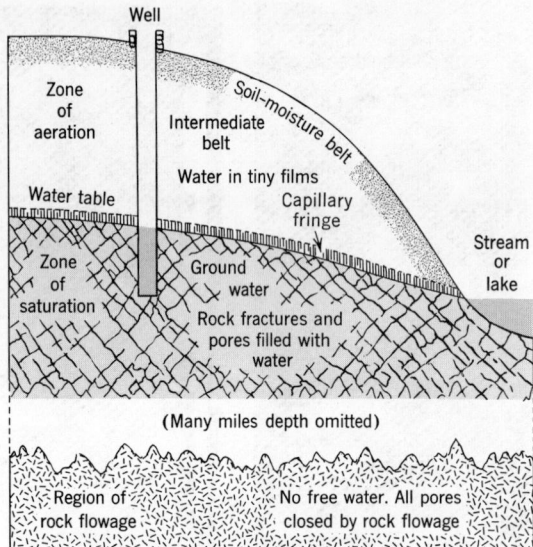

Figure 12.18 Schematic diagram of zones of subsurface water. (After Ackerman, Colman, and Ogrosky.)

aeration, in which water does not fully saturate the pores. We have seen that the soil-moisture belt is the uppermost layer of the zone of aeration and that moisture is held in this belt by capillary force in tiny films adhering to the soil particles. A similar condition prevails through the underlying intermediate belt. The sole basis for distinguishing these two belts is that the soil-moisture belt represents a shallow zone of moisture usable by plants, whereas the intermediate belt is too deep for capillary water to be returned to the atmosphere by either direct evaporation or transpiration. The depth of the zone of aeration may be very shallow or missing, when the ground water is close to the surface in low, flat regions, or up to several hundred feet thick in hilly or mountainous regions with a deep water table.

At the base of the zone of aeration is the *capillary fringe*, a thin layer in which the water has been drawn upward from the ground-water body through capillary force. The action is much like the rise of kerosene in a lamp wick, or of water in a blotter whose edge is immersed. Water in the capillary fringe largely fills the soil pores, hence, is continuous with the ground-water body. Thickness of the capillary fringe depends on the soil texture, because capillary rise is higher when the openings are smaller. Thus, in a silty material the capillary fringe may be 2 ft (0.6 m) thick, but only a fraction of an inch (1 cm) thick in coarse sand or fine gravel with large pore spaces.

Ground water in the zone of saturation moves under the force of gravity and therefore its upper surface, the *water table*, would tend to become a horizontal surface, just as with a free-water body such as a lake. But because water moves very slowly through the rock, the water table actually maintains a sloping surface. A difference in water-table level, or *head*, is built up and maintained between areas of high elevation and those of low elevation.

If wells are numerous in an area, the position of the water table can be mapped in detail by plotting the water heights and noting the trends in elevation from one well to the other. When this is done it is usually seen that the water table is highest under the highest areas of ground surface, namely, hilltops and divides, but descends toward the valleys where it may appear at the surface close to streams, lakes, or marshes (Figure 12.19). The reason for such a configuration of the water table is that water percolating down through the zone of aeration tends to raise the water table, whereas seepage into streams, swamps, and lakes tends to draw off ground water and to lower its level.

The ground-water body receives its replenishment, or *recharge*, by gravity movement of surplus soil water that percolates down through the intermediate zone. Upon reaching the water table and entering the zone of saturation, water movement follows a different set of physical laws, those that apply to movement through a porous medium.

Contrary to what the average person might predict, all of the ground water does not move directly from divides to the lines of seepage by flow close to the top of the water table. If such were the case, the lower parts of the ground-water body would be stagnant. Certain geological phenomena, such as the cementation of rocks and the transfer of dissolved mineral matter from place to place, would not take place without some ground-water flow, even though extremely slow.

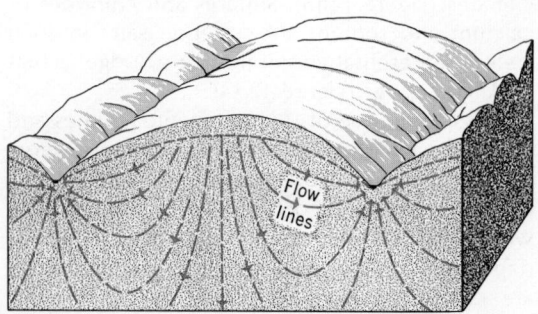

Figure 12.19 Theoretical paths of ground-water movement between valleys. (After M. K. Hubbert.)

Ground Water

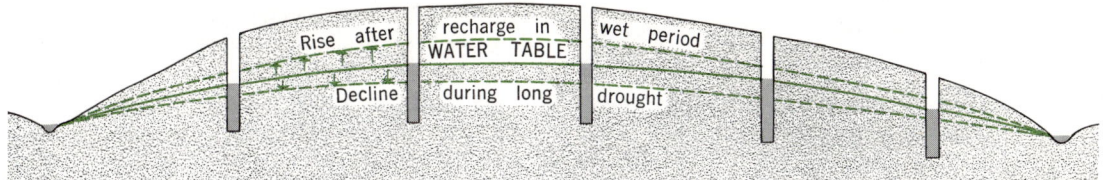

Figure 12.20 The water table rises and falls according to natural variations in annual recharge.

Figure 12.19 shows the theoretical paths of flow of ground water as calculated by use of basic principles of fluid mechanics. Water follows paths curved concavely upward. Water entering the slope midway between divide and stream flows rather directly toward the stream valley in shallow paths. Close to the divide point on the water table, however, the flow lines go almost straight down to great depths in the earth from which they recurve upward to points under the streams. Progress along these deep paths would be incredibly slow; that near the surface would be faster. The most rapid flow is encountered close to the line of discharge in the stream, where the arrows are shown to converge.

The subsurface phase of the hydrologic cycle is completed when the ground water emerges along lines or zones where the water table intersects the ground surface. Such places are the channels of streams and the floors of marshes and lakes. By slow seepage and spring flow the water must emerge in sufficient quantity to balance that which enters the ground-water table by percolation through the zone of aeration.

So we see that the ground-water flow comprises an open system of mass transport which tends to achieve a steady state of operation in which the rate at which water enters the system is balanced against the rate at which it leaves the system. When this balance exists the geometry of the water table and flow paths have a stable configuration. Like other natural open systems, the ground-water flow system is subject to natural seasonal rhythms and to variations in input following longer cycles; it is also sensitive to disturbances caused by Man's activity in changing the rate of recharge by surface modifications and by withdrawing ground water from wells.

In climate regimes showing a season of large soil-water surplus alternating with a season of soil-water deficiency (either a winter season or a dry season), the recharge of ground water is seasonal and the water table rises and falls accordingly (Figure 12.20). Figure 12.21 is a type of *hydrograph* (water graph) in which the elevation of the water table in an observation well is plotted against time. Water level is measured once each month and the plotted points connected by a line. In this climate regime there is a large water surplus in the winter, starting usually in November and ending in April. Correspondingly, the water table usually rises from December through May. The summer is a period of soil-water deficiency, so that recharge is largely cut off. Since the ground water is continually in motion, the water table typically subsides from May or June through November. In this example the annual fluctuation of water-table level is on the order of 2 to 3 ft (0.6 to 1.0 m).

During a period of drought years, recharge is reduced and the trend of water-table levels is downward. Figure 12.21 shows the effects of a drought period in the mid 1960s. The water table reached its lowest point in 1966, after which a return to normal precipitation brought water levels up to average. When water is withdrawn by pumping for Man's use, and a large part of what is consumed is not returned to the ground-water body, the water table undergoes a similar decline.

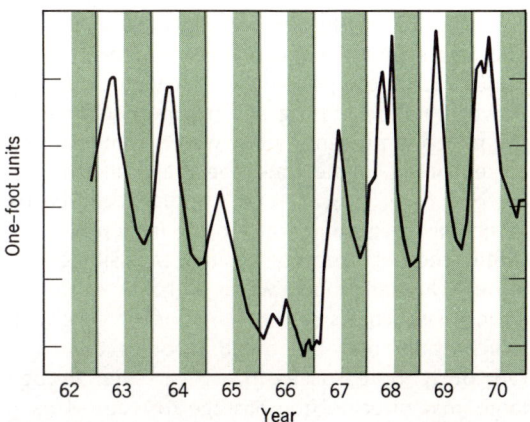

Figure 12.21 Hydrographs of an observation well on Cape Cod., Massachusetts, showing the characteristic annual cycle of rise and fall of the water table. (Data of U.S. Geological Survey.)

Pore Spaces in the Ground-Water Zone

Ground water can saturate a great variety of geological materials ranging from relatively soft overburden of both residual and transported types, to hard bedrock of any origin. The term *porosity* refers to the total volume of pore space present in a given volume of rock. The amount of water that can be held in storage in a rock is measured by its porosity. A knowledge of rocks enables us to understand something of the variation in porosity that might be expected in rocks. Among the sedimentary rocks, the coarse-grained clastic rocks, such as sandstone or conglomerate, can have large porosity. Similarly, any transported overburden consisting of sands and gravels laid down by streams or shore currents will have large porosity. Shale, because of its dense compaction, has relatively low porosity, but soft clay or mud, by contrast, has high porosity even though the pores are extremely tiny. Limestones may have large openings, such as caverns, produced by solution of the rock. Scoriaceous (frothy) lavas commonly have great porosity.

Dense, massive rocks, such as the igneous and metamorphic types, have little or no pore space between the individual mineral crystals, but the presence of numerous fractures, such as joints and faults, offers a large number of interconnected openings in which water can be stored and through which it can move. In sedimentary strata the bedding planes may provide additional openings.

How far down into the earth does the ground-water zone extend? No single depth can be stated in answer to this question, but it is certain from experience with deep wells that water becomes very scanty at depths of more than 2 mi (3.2 km). Furthermore, geologists have evidence that at depths greater than about 10 mi (16 km), rock is under such great pressure that it yields by slow flowage, tightly closing any natural openings in the rock and preventing any water from entering or remaining in the rock. We may call this the *region of rock flowage* and say that it limits the extent of the ground water body in depth.

Still another property besides porosity determines whether ground water will move through a rock mass. This is *permeability*, or the ease with which water may be forced through the rock. Obviously, pore spaces in a rock, even though large, may offer no free passage to water if each pore is sealed off from its neighbors by mineral matter. Thus the degree to which openings are interconnected is important in determining permeability. Second, the size of the openings exerts a strong influence upon rate of flow. Coarse sands or gravels permit rapid flow and are therefore rated as highly permeable materials. The microscopic pores of a clay impede flow of water so effectively that clays and rocks containing much clay are commonly designated as *impermeable* (not permeable), at least for practical purposes of obtaining useful flow of water from them. Although intrusive igneous rocks are highly impermeable where massive and not decomposed, they may actually constitute permeable bodies if broken by numerous, closely set systems of joints and fractures.

Aquifers, Aquicludes, Perched Water Tables, and Artesian Wells

Where rock layers lie nearly horizontal, or in gently inclined positions, the ground water relations may be quite different from the simple pattern illustrated in Figure 12.19 in which completely uniform geologic materials were assumed to exist. Suppose, for example, that the region is one of sedimentary strata with beds of sandstone alternating with beds of shale (Figure 12.22). Sandstone is commonly both porous and permeable, providing a large ground-water storage reservoir through which water may move easily. Such a rock body is termed an *aquifer*. By contrast, a shale bed with low permeability virtually prevents flow of ground water and is called an *aquiclude* In the particular case shown in Figure 12.22, a thin bed of shale has effectively blocked the downward percolation of water to the main water table below, creating a *perched water table*, separated from the main water table by a zone of aeration. Where the perched water table meets the valley side a *seep* or *spring* (that is, a slow flow of water emerging from the ground) is formed.

Most natural springs are mere trickles of water, unseen and unnoticed under a cover of dense

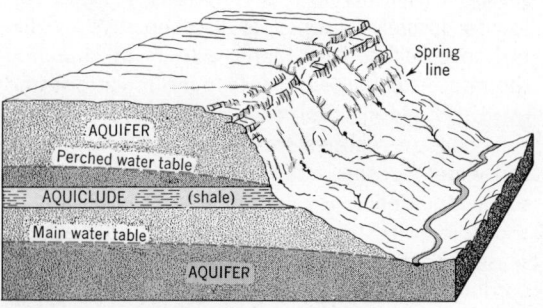

Figure 12.22 A perched water table.

Relation of Fresh to Salt Ground Water

Figure 12.23 Thousand Springs, Idaho, emerges from the north side of the Snake River Canyon, opposite the mouth of the Salmon River. The spring extends for 0.5 mi (0.8 km) and issues from a scoriaceous basalt layer with a nearly constant discharge of 500 cfs (14 cms). (Photograph by I. C. Russell, U.S. Geological Survey.)

vegetation. A few springs, however, discharge enormous volumes of water where an unusually good aquifer, abundantly fed from a large source area, is exposed in a deep canyon (Figure 12.23).

Where strata are inclined a favorable situation may exist for development of an *artesian* spring or well, one in which the water flows upward toward the surface through its own pressure. In Figure 12.24 we see a highly diagrammatic representation of such conditions, the vertical exaggeration being very great merely to show the principle. The eroded edge of a sandstone aquifer is exposed to intake of water at a high position. Water entering here passes deep underground to a position below the valley floor, at which point the water is under a strong pressure, or head, from the weight of the overlying water. This pressure is sufficient to force water up to the surface in a well drilled down through the impervious shale layer into the aquifer. Similar flow as an artesian spring may occur naturally if there are faults in the strata which permit water to seep upward through the shale layer.

Relation of Fresh Ground Water to Salt Ground Water

Permeable rock beneath the ocean floors is saturated with salt water. This zone of salt ground water comes in contact with the fresh ground-water body beneath coastlines, where an equilibrium relationship is developed. The principles are illustrated by conditions beneath an island or a narrow peninsula jutting into the sea. Figure 12.25 shows the relationships in a schematic way. The body of fresh ground water takes the shape of gigantic lens with convex faces, except that the upper surface has only a broad curvature whereas the lower surface, in contact with the salt ground water, bulges deeply downward. Because fresh water is less dense than salt water, we can think of this fresh-water lens as floating upon the salt water, pushing it down much as the hull of an ocean liner pushes aside the surrounding water. The ratio of densities of fresh water to salt water is about as 40 to 41, so, if the water table is, say, 10 feet above sea level, the bottom of the fresh water lens will be located 400 feet below sea level, or forty times as deep as the water table is high with respect to sea level.

The fresh ground water extends seaward some distance beyond the shoreline. Although the salt ground water is stagnant, the fresh water travels in the curved paths shown by arrows in Figure 12.25. Ground water is discharged by seepage through the ocean bottom close to shore. The effect of this discharge in bays and tidal estuaries is to cause some measure of natural dilution of the salt water, creating a favorable environment for certain forms of marine life.

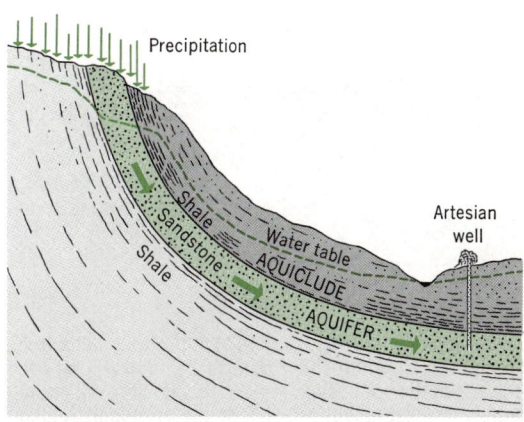

Figure 12.24 Geological conditions for artesian flow.

309

Subsurface Water of the Lands

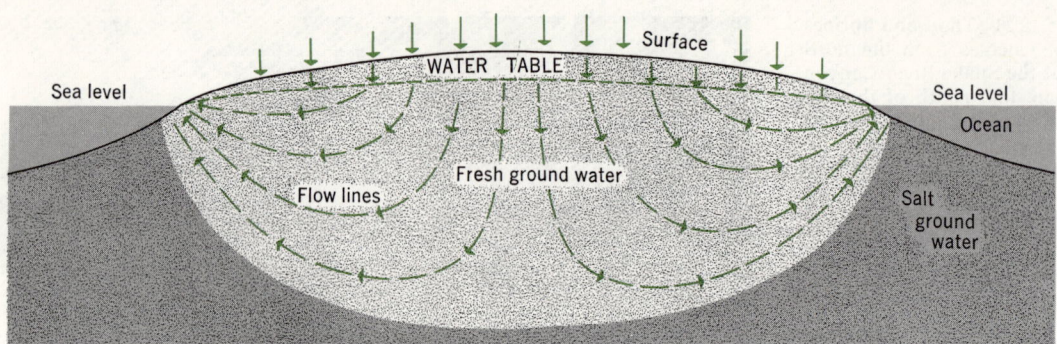

Figure 12.25 Lens of fresh ground-water displacing salt ground-water on an island or peninsula. (After G. Parker.)

Limestone Caverns

Limestone caverns consist of interconnected subterranean cavities in the form of passageways and rooms. Within a single cavern system the total length of such openings can aggregate many miles. Figure 12.26 is a detailed map of passageways in a portion of one cavern. The rectangular patterns suggests control of passageways by sets of joints in the limestone.

The consensus of opinion among geologists is that most cavern systems were opened out in the ground-water zone. Figure 12.27 suggests how caverns may develop. In the upper diagram the process of carbonation (action of carbonic acid upon calcium carbonate) is shown to be particularly concentrated just below the water table. (Refer to Chapter 8 for background.) Products of solution are carried along the ground-water flow paths to emerge in streams and leave the region in stream flow. In a later stage, shown in the lower diagram, the stream has deepened its valley and the water table has been correspondingly lowered to a new position. The cavern system previously excavated is now in the zone of aeration. Evaporation of percolating gravity water on exposed rock surfaces in the caverns now begins the deposition of carbonate matter, known as *travertine*. Encrustations of travertine take many beautiful forms—stalactites, stalagmites, columns, drip curtains, and terraces—that make many caverns tourist attractions. Above the cavern system will often be found a karst landscape (see Chapter 11).

In their natural state, caverns comprise one of the natural habitats of a few animal species. Bats inhabit many caverns and make nightly flights out of and into the cave openings. Their accumulated excrement, known as *guano*, is rich in nitrates and has been used in the manufacture of fertilizers and explosives. Bat guano was taken from Mammoth Cave for making gunpowder during the war of 1812. Much more recently a valuable guano deposit in a limestone cavern in the wall of the Grand Canyon of the Colorado River was mined and the guano lifted to the canyon rim by cable car.

Throughout Man's early development, caves were an important habitation. Now we find the skeletal remains of these people, together with their implements and cave drawings, preserved through the centuries in caves in many parts of the world. Today, with increasing destructiveness of weapons of warfare, caverns are achieving more importance as possible sites for command and control installations, for storage of valuable materials, as emergency shelters, and as factories for important types of production.

Karst regions, underlain by extensive cavern systems, present certain environmental problems related to construction engineering. As we noted in Chapter 11, collapse of the ground is a process that occurs naturally in karst regions, as caverns

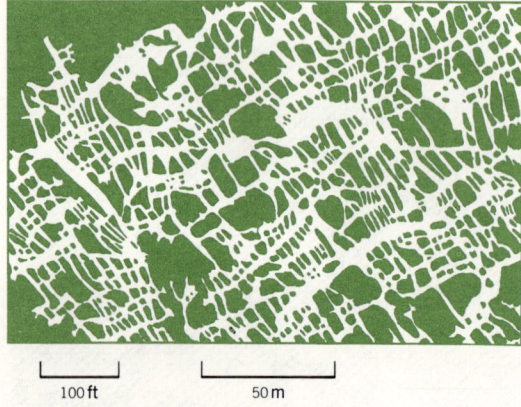

Figure 12.26 Map of a part of Anvil Cave, Alabama. Passageways (white) show control by intersecting sets of joints in the rock layer. (By courtesy of W. W. Varnedoe and the Huntsville Grotto of the National Speleological Society.)

Utilization of Ground Water as a Resource

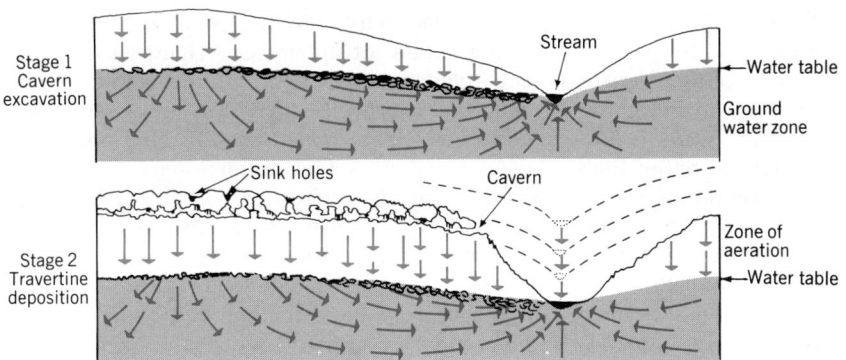

Figure 12.27 Cavern development in the ground-water zone, followed by deposition in zone of aeration. (From A. N. Strahler, 1971, *The Earth Sciences*, 2nd ed., Harper & Row, New York.)

are slowly enlarged and support is removed from the overlying rock layer. It is not surprising, then, that many instances have been recorded of destruction and damage to buildings and highways through sudden collapse of the bedrock beneath to form a new sinkhole. Construction of large concrete dams across valleys underlain by cavernous limestone or dolomite may involve serious problems, not only of collapse of the bedrock under heavy load, but also of leakage of water from the reservoir through subterranean solution passages beneath the dam and in the valley walls. In such localities the dam site must be thoroughly explored by numerous drill holes. If cavities are found, it may be possible to seal them by *grouting*, the pouring of concrete through bore holes to fill the cavities, but such procedures can be extremely costly.

Utilization of Ground Water as a Resource

Since before the dawn of civilization, Man has dug shallow wells to reach the water table and withdraw ground water for domestic and agricultural purposes. However, as long as his energy resources were limited to power provided by his own muscles or those of draft animals, or by windmills, the amount of ground water he could divert from natural recharge was usually of little consequence to the operation of the natural ground-water system. Discharge of ground water through surface seepages was sufficient to maintain the flow of streams in the summer and to sustain the water levels of ponds, marshes, and bogs. With the invention of powerful pumps utilizing electricity generated from the combustion of fossil fuels, Man's industrial society could meet heavy water demands of an expanding population. Enormous quantities of ground water were pumped for irrigation to increase food production. Not only has Man seriously depleted this natural resource, but his disturbances to the ground water system have had a number of deleterious effects upon the environment.

In agricultural lands of the semiarid and desert climates, heavy dependence is placed upon irrigation water from pumped wells, because the major river systems have already been fully developed for irrigation from surface supplies. Wells can be drilled within the limits of a given agricultural or industrial property, and so can provide immediate supplies of water without need to construct expensive canals or aqueducts. A few of the hydraulic principles of water wells are treated here to aid you in understanding the basis of complex economic, legal, and environmental problems arising from ground-water development and use.

Formerly the small well needed to supply domestic and livestock needs of a home or farmstead was actually dug by hand as a large cylindrical hole, lined with masonry where required. By contrast, the modern well put down to supply irrigation and industrial water is drilled by powerful machinery which may bore a hole 12 to 16 in. (30 to 40 cm) or more in diameter to depths of 1000 ft (300 m) or more, although much smaller-scaled wells and well-boring machines suffice for domestic purposes. Drilled wells are sealed off by metal casings which exclude impure near-surface water and prevent clogging of the tube by caving of the walls. Near the lower end of the hole, where it enters the aquifer, the casing is perforated so as to admit the water through a considerable surface area. Rate of flow of a well or spring is stated in units of gallons or liters per minute or per day. The yields of single wells range from as low as a few gallons per day in a domestic well to many millions of gallons per day for large, deep industrial or irrigation wells.

High-capacity pumps can easily bring water to the surface more rapidly than it can enter the well, so that the delivery of ground water is limited by the properties of the aquifer rather than by the mechanical equipment. The rate at which water can enter the well depends on the permeability of the aquifer, which limits the rate of flow of water through the aquifer from the surrounding area. Flow of ground water is extremely slow, in any case, compared to flow of streams. It is estimated that ground water may move at a speed of 5 ft (1.5 m) per day through a formation in which wells of good yield are developed, that in exceptional cases of coarse gravels the velocity may reach 30 to 60 ft (10 to 20 m) per day. In dense clays and shales the rate may be immeasurably slow.

When rate of pumping of the well exceeds the rate at which water can enter, the level of water in the well drops and the surrounding water table is lowered in the shape of a conical surface, termed the *cone of depression*, the height of which is termed the *drawdown* (Figure 12.28). By producing a steeper gradient of the water table, the flow of ground water toward the well is also increased, so that the well will yield more water. This effect holds only for a limited amount of drawdown, beyond which the yield fails to increase. The cone of depression may extend as far out as 10 mi (16 km) or more from a well where very heavy pumping is continued. Where many wells are in operation, their intersecting cones produce a general lowering of the water table.

Depletion often greatly exceeds the rate at which the ground water of the area is recharged

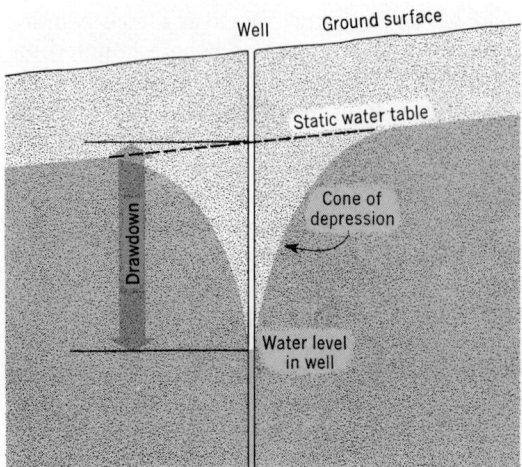

Figure 12.28 Drawdown of water table and cone of depression produced by pumping of a well.

by percolation from infiltration of precipitation or from the beds of streams (see Figure 14.43). In an arid region, much of the ground water for irrigation is from wells driven into thick sands and gravels which are lowland deposits of transported overburden of a type termed *alluvium*. (These features are described in Chapter 14.) Recharge of such deposits depends on the seasonal flows of water from streams heading high in adjacent mountain ranges. Where such highly permeable materials exist, the extraction of ground water by pumping can greatly exceed the recharge by stream flow. Cones of depression deepen and widen; deeper wells and more powerful pumps are then required. Overdrafts of water accumulate and the result is exhaustion of a natural resource not renewable except by long lapses of time. In humid areas where annual rainfall is copious—from 30 to 50 in. (75 to 125 cm) annually—natural recharge is by general percolation over the ground area surrounding the well. Here the prospects of achieving a balance of recharge and withdrawal are highly favorable through the control of pumping and the return of waste waters or stream waters to the ground-water table by means of *recharge wells* in which water flows down, rather than up.

Environmental Effects of Overdraft of Ground Water

Sustained withdrawal of ground water at a rate substantially greater than the natural recharge rate produces a number of undesirable side effects upon the environment. Sometimes these effects are slow in coming, but by the time they are recognized it may be too late to correct the damage. Because the drawdown from a well spreads to the limits of an aquifer, a general lowering of the water table will be felt over a large area after many wells have been in operation for a period of years. For example, in Nassau County on western Long Island, New York, a survey of water levels in wells over a 50-sq mi (130-sq km) area showed that between 1953 and 1966 the water table had declined an average of 10 ft (3 m), and locally as much as 15 ft (5 m). Most of the decline is attributed to the introduction of sanitary sewer systems which collect and treat used water. The effluent is discharged directly into the ocean (e.g., *ocean outfall*). Formerly in this area sewage was disposed of by domestic septic tanks, which return much of the used water directly to the ground-water body over the area of consumption. In this way a recycling of water was carried out and the natural recharge was not substantially lowered. However, with the intro-

Environmental Effects of Overdraft of Ground Water

duction of the sewerage system with ocean outfall, a large quantity of used ground water was not returned to the ground-water body, with the result that water-table levels steadily declined. Some of the decline has been attributed to pumping of wells in adjacent Queens County, which lies to the west.

You may well wonder what is harmful about a decline in water-table levels. After all, you may say, the ground water is far below the surface and is not used anyway by plants. One harmful effect will be upon ponds, lakes, and bogs that normally stand at the level of the surrounding water table. These water bodies can be thought of as surface exposures of the water table itself (Figure 12.29). Thousands of such water-table ponds and bogs exist in the northern United States within the area formerly covered by the Pleistocene ice sheets. Although the pond water levels rise in the spring and decline in late summer and fall, along with the water table itself, the pond level remains on the average more or less constant from year to year and from decade to decade. As a result, these aquatic habitats have come to sustain stable communities of plants and animals. If the water table is severly lowered by pumping of ground water, as shown in Figure 12.29, these water bodies will cease to exist and the ecosystems within them will be destroyed. There will, of course, be esthetic losses as well and these cannot easily be measured.

Water table decline also affects stream flow adversely. In humid regions the year-around flow of many smaller streams is sustained by ground-water seepage (Chapter 13). If the water table falls permanently below stream level, the stream will contain water only following heavy rain or rapid snowmelt, when water reaches the stream directly by overland flow. Again, as in the case of the ponds, the ecosystem of the permanent stream is destroyed. Reduced stream flow has other undesirable effects farther down valley. One effect is to allow pollutants in stream water to be more heavily concentrated; another is to permit the water of estuaries, into which streams discharge fresh water, to become more saline. The upstream invasion of salt water may be enough to reach and contaminate municipal water supply intakes.

Serious depletion of flow from a large stream can take place where a well of large capacity is placed near the stream, as shown in Figure 12.30. The drawdown cone intersects the stream channel and water travels from stream bed to well. Since the stream water is filtered in transit, this arrangement may improve the quality of the well water in addition to augmenting the quantity available.

In coastal regions, fresh ground-water supplies are vulnerable to contamination by *salt water intrusion*. As shown in Figure 12.29, pumping of a well has not only caused the water table to decline, but has also caused the interface between salt ground-water and fresh ground-water to move inland as an invading wedge. Ultimately, the salt water is drawn into the well and the water is no longer fit for consumption. Notice that in Figure 12.29 a small coastal well also suffered salt contamination. In the case of an island or narrow

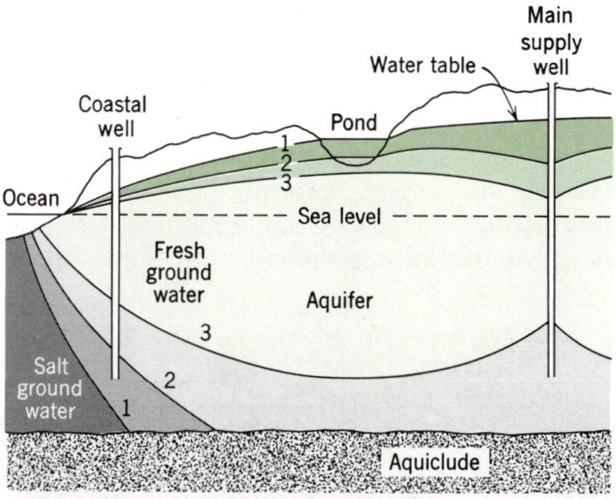

Figure 12.29 Schematic cross section showing lowering of water table and pond levels, and intrusion of salt ground water, as a result of ground-water withdrawal. (Vertical scale is greatly exaggerated above sea level.)

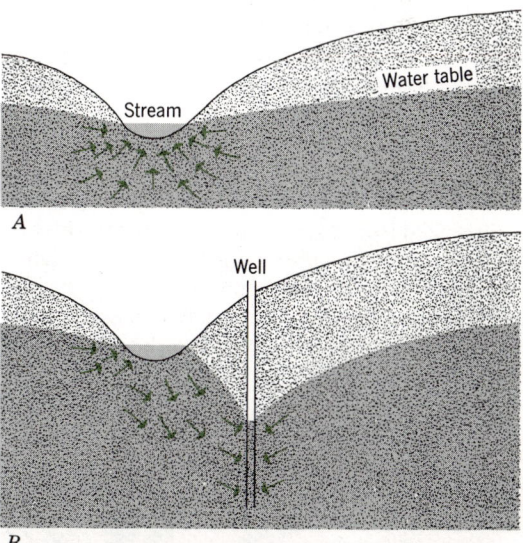

Figure 12.30 Diversion of stream flow into a nearby well. (After D. K. Todd. From A. N. Strahler, 1972, *Planet Earth; Its Physical Systems Through Geologic Time*, Harper & Row, New York.)

peninsula, where salt ground water lies beneath a fresh-water lens, salt water is easily drawn directly upward to contaminate a well. A case in point is the town water supply of Provincetown, Massachusetts, at the tip of Cape Cod. The water has become brackish during summer periods of heavy pumping. If pumping is stopped after salt contamination has occurred the salt water will be gradually pushed back to its original limits. To hasten this process and to prevent further salt intrusion, fresh water can be pumped down into recharge wells located between the contaminated well and the shore.

Another serious side effect of excessive ground-water withdrawal is that of subsidence of the ground surface, a phenomenon already discussed in Chapter 12 in connection with pumping of crude oil. The major examples of subsidence occur over aquifers of the artesian type. Several localities have been affected in California, where ground water has been pumped from basins filled with alluvial (stream- and lake-deposited) sediments. Water-table levels in these basins have dropped over 100 ft (30 m), with a maximum drop of 400 to 500 ft (120 to 150 m) being recorded in one locality in the San Joaquin Valley. Here the maximum ground subsidence has been about 10 ft (3 m) in a 35-year period, and has caused damage to wells in the area. Another important area of ground subsidence accompanying water withdrawal is beneath Houston, Texas, where the ground has dropped from 1 to 3 ft (0.3 to 1.0 m in a metropolitan area 30 mi (50 km) across. Damage has resulted to buildings, pavements, airport runways, and flood-control works. Displacement has been observed on a number of minor faults in the Houston area and this activity is attributed to a triggering effect of water withdrawal from the aquifer.

Perhaps the most celebrated case of ground subsidence is that affecting Mexico City. Carefully measured ground subsidence between 1891 and 1959 ranged from 13 to 23 ft (4 to 7 m). Subsidence began at a much earlier date as a result of withdrawal of ground water from an artesian aquifer system beneath the city and has caused many serious engineering problems. Clay beds overlying the aquifer have contracted greatly in volume as water has been drained out. To combat the ground subsidence, recharge wells have been drilled to inject water into the aquifer. In addition, new water supplies from sources outside the city area have been developed to replace local ground-water use.

Pollution of Ground Water

Disposal of solid wastes poses a major environmental problem in the United States because our advanced industrial economy is an endless source of garbage and trash. Little or no effort is being made to reclaim and recycle such materials as glass, metals, paper, and wood. Durable goods of metallic construction add great bulk to the waste, as do scrapped construction materials. Traditionally, these waste products have been trucked to the town or city dump, and there burned in continually smoldering fires that emit foul smoke and gases. The partially consumed residual waste is then buried under earth. In recent years, a major effort has been made to improve solid-waste disposal methods. One method is high-temperature incineration. Another is the *sanitary land-fill* method in which waste is not allowed to burn, but is continually covered by protective overburden, usually sand, silt, or clay available on the land-fill site. The waste is thus buried in the zone of aeration and is subject to reaction with percolating rainwater infiltrating the ground surface. This water picks up a wide variety of ions from the waste body and carries these down as *leachate* to the water table. Once in the water table, the leachate follows the flow paths of the ground water.

As shown in Figure 12.31, there normally develops a *mound* in the water table beneath the disposal site. Loose soil of the disposal area facilitates infiltration of precipitation, while lack of vegetation reduces the evapotranspiration. Consequently, the recharge here is greater than elsewhere and the mound is maintained. After leachate has moved vertically down by gravity percolation to the water-table mound, it moves radially outward from the mound to surrounding

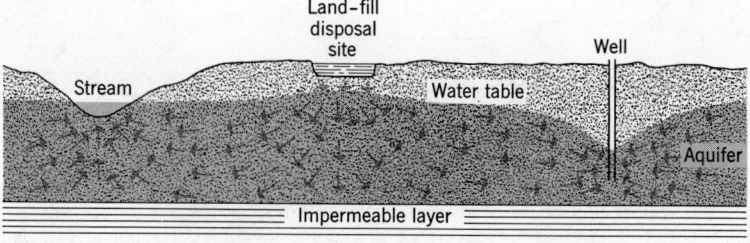

Figure 12.31 Leachate from a waste disposal site moving toward a supply well (right) and a stream (left). (From A. N. Strahler, 1972, Planet Earth; Its Physical Systems Through Geologic Time, Harper & Row, New York.)

lower points on the water table. As shown in Figure 12.31, a supply well with its cone of depression draws ground water from the surrounding area. Linkage between outward flow from the waste disposal site and inward flow to the well can bring leachate into the well, polluting the ground-water supply.

An important step in guarding against this form of pollution is to place a *monitor well* (or several monitor wells) on a line between the disposal site and the well. Chemical tests for presence of leachate are made regularly, while the slope of the water table can also be determined. Movement of leachate toward the supply well may be blocked by placement of a recharge well (or wells), building a fresh-water accumulation (actually an inverted cone), which will oppose the movement of the leachate.

Pollution of supply wells by partially treated effluent infiltrating the ground at sewage disposal plants can occur in a basically similar manner.

It is not necessary for a ground-water mound to be present for pollutants to travel to distant points. Where the water table has a pronounced slope, as it generally does everywhere except near the summit of a broad ground-water divide, leachate or any pollutant introduced at a given point migrates as a *pollution plume* along the flow paths of the ground water. Figure 12.32 is a map on which are drawn lines of equal elevation of the water table, labeled in feet above sea level. As the numbers on these lines show, the water table is sloping from north to south across the area of the map. An ocean or lake shoreline lies to the south. Flow of the ground water, as seen from above, is along the direction of the arrows, which cross the ground-water contour lines at right angles.

On the left side of the map we find that a municipal well, W, is located at an inland position high on the water table. The municipal sewage or solid-waste disposal site, D, is located to shoreward at lower elevation on the water table. The pollution plume from the disposal site will move shoreward away from the well to be discharged by seepage into water offshore. Although the well is safely situated, there may be a problem of pollution in coastal waters immediately offshore.

On the right side of the diagram we see that the disposal site, D, is located at an inland position high on the water table. A municipal well has been placed at a shoreward position and at a lower level on the water table, along the line of flow of ground water from the point D. A pollution plume from the disposal site will travel along the indicated flow path and may in time enter the field of influence of the well. Monitor wells (M), placed along the line of ground-water movement could detect the presence of pollutants in the ground water long in advance of arrival at the well. Because of the very slow rate of movement of ground water, some years might elapse before these substances would reach the well.

Pollution of ground-water supplies can take place rapidly in limestone regions where extensive cavern systems are below the water table. Movement of ground water is rapid through large passageways, so that liquid wastes discharged into sinkholes can travel freely to distant points where water is being withdrawn from water-table ponds or supply wells. The peninsula of Florida, which has large areas underlain by cavernous limestone and a high water table, has presented a number of problem localities of this type. Detailed studies by ground-water geologists can reveal the presence of such pollution hazards and are essential in planning for water resource development and waste disposal systems.

Yet another potential source of pollution of ground-water supplies is from highways and streets, through spillage of chemicals and from deicing salt applied during the winter months. Spillage of large volumes of liquids from tank trucks and tank cars as a result of highway traffic and railroad accidents poses a serious threat because a large slug of pollutant can be injected into the ground-water recharge system. The commonest pollutants to be feared are automobile fuel and heating oil, but many toxic industrial

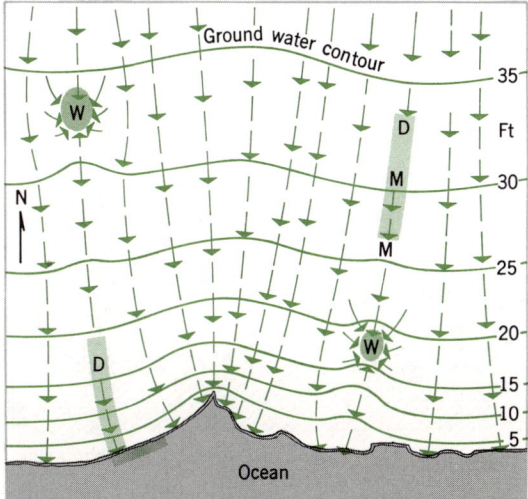

Figure 12.32 Idealized map showing contours on the water table and ground-water flow paths leading toward a shoreline. Pollution plumes follow the flow paths.

chemicals are also transported in tank trucks and cars. Leakage of fuels from underground storage tanks used in all gasoline stations, is a related source of possible pollution.

Richard R. Parizek evaluates the problem as follows: "The potential danger of releasing large volumes of these materials in surface-water or ground-water reservoirs used for drinking purposes is obvious. When released into surface-water reservoirs, chemicals derived from accidental spills may be adequately diluted in some cases or monitored as they are carried away from the region. Where they enter the ground-water reservoir, their presence may not be known in advance of damaging a water supply, the rate and direction of movement may not be known, and they may not be flushed from or diluted within aquifers to tolerable limits for years to come."*

In recent years, the great increase in use of sodium chloride and calcium chloride as deicing salts on roads and highways has led to an increasing incidence of contamination of ground-water supplies. In some cases wells have had to be drilled deeper and cased to greater depth to reach uncontaminated ground water. In other cases new wells have had to be drilled at greater distance from highways, and at state expense. Obviously, planning for future water supply development should take into account the possibility of highway pollution.

Effects of Engineering Works upon Ground-Water Systems

Major engineering works, including dams, canals, and highway cuts, in many instances cause significant changes in the water table and the movement of ground water. These effects can be predicted in advance and should be taken into account in assessing the total environmental impact of a new project.

Where a major dam is built and a large reservoir of water impounded behind it, water will percolate from the reservoir into the surrounding rock, raising the water table in the surrounding region. As a result, adjacent low-lying areas may become saturated wetlands with damaging effect upon agricultural land, towns, and highways. In some instances, however, the rise of water table brings water within range of plant roots, with a beneficial effect upon croplands.

*R. P. Parizek, 1971, "Impact of Highways on the Hydrogeologic Environment," *Environmental Geomorphology*, D. R. Coates, ed., Publications in Geomorphology, State University of New York, Binghamton, N.Y., p. 188.

Major excavations of overburden and bedrock often penetrate well below the normal water-table and expose the ground-water zone, with resulting major changes in ground-water flow. Figure 12.33 shows a deep highway cut in cross section. What was formerly a region of ground-water recharge now becomes a discharge zone. Frequently springs will be seen emerging from the walls of the cut. The water table is drawn down on both sides, with the result that nearby shallow wells may go dry. Flow lines now emerge at the base and floor of the cut. This water seepage must be disposed of by adequate drains. If the material exposed in the walls of the cut is a clay-rich rock or overburden, it may be subject to failure by earthflow and slump because of seepage forces of the emerging water.

A sea-level canal, such as that which was planned for Florida and since abandoned, poses problems of salt-water contamination of adjacent aquifers. The denser salt water enters the ground-water zone beneath the canal, forming a salt ground-water wedge that moves out beneath fresh ground-water. Contamination of nearby wells follows according to the principles already explained for coastal zones. Use of a system of locks that shut out salt water in damaging quantities and maintain fresh water in the canal is one possible answer to the problems of a sea-level canal. If the fresh-water level is maintained several feet above sea level, the hydraulic pressure will prevent salt water from moving inland.

Related to environmental problems of engineering works is the problem of *acid mine drainage*, in which ground water issuing from abandoned coal mines is contaminated with sulfuric acid derived from sulfur compounds in coal. This problem is discussed in Chapter 13 under the subject of pollution of stream water.

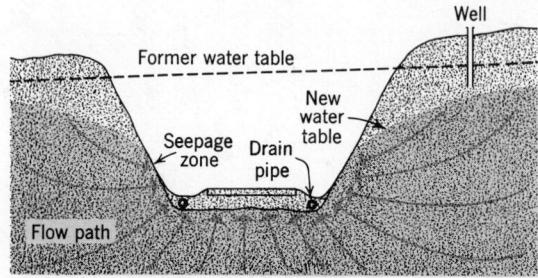

Figure 12.33 Cross section through a deep highway cut, showing drawdown of the water table and modified ground-water flow paths bringing water seepages into the cut. (After R. R. Parizek, 1971, in *Environmental Geomorphology*, D. R. Coates, ed., State Univ. of New York, Binghamton, N.Y., p. 157, Figure 2.)

Disposal of Radioactive Wastes into the Ground

The serious environmental problem of storage and disposal of radioactive wastes was mentioned in Chapter 10 in evaluating the future of world energy resources. An adequate understanding of the problem and its ramifications will require a substantial background of information on the categories of such wastes with regard to both place of origin and level of radioactivity. Only a very brief sketch is possible here to preface this discussion of disposal of wastes into the ground and their possible contamination of the groundwater body.

Radioactive wastes are produced in three basic phases of activity. First is the mining, milling, and fuel fabrication of natural radioisotopes; these are principally isotopes of uranium, thorium, radon, and radium. (Refer to Chapter 9 for background.) This source of wastes involves natural radioactivity. A second category of wastes comes from *fission products*, produced during the irradiation of nuclear fuels in reactors. Chemical processing of these irradiated nuclear fuels is a source of highly dangerous radioactive wastes. At present this activity is done solely by Atomic Energy Commission (AEC) plants. A third class of wastes comes from *activation products*, which are radioisotopes produced by exposure of nonfuel material to radiation of fuels. The coolant liquids used in nuclear power plants may be activated, along with structural materials surrounding the fuels. In addition, many radioisotopes are created for specialized uses in industry, scientific research, and medicine. An example is cobalt-60 used in cancer treatment. These radioisotopes eventually require disposal.

With respect to levels of radioactivity, wastes are classified as low-level, intermediate-level, and high-level. *Low-level* liquid wastes generally can be treated to remove radioisotopes, diluted, and released into the environment; i.e., poured into the ground or into streams. *Intermediate-level* wastes cannot be diluted sufficiently to be released but can be processed in a variety of ways for safe disposal or storage. Treatment of low-level solid wastes consists of containing them indefinitely in storage. For example, at the Oak Ridge National Laboratory, Tennessee, the AEC practices shallow land burial of waste containers in unlined pits and trenches (Figure 12.34). For some wastes of higher activity levels, burial in concrete lined wells is required. Storage at the surface is also practiced to allow radioactivity to decay to lower levels. At the Hanford Plant in

Figure 12.34 Burial of containers of radioactive wastes in a trench, Oak Ridge National Laboratory, Tennessee. (A.E.C. photograph.)

Washington some intermediate-level liquid wastes are discharged into trenches and allowed to infiltrate the zone of aeration, which is here 200 to 600 ft (60 to 180 m) thick (Figure 12.35). Mineral colloids permanently hold the radioactive ions and other chemicals. Retention zones for various components are shown in Figure 12.35. Depth of penetration is monitored with a system of wells. Another method of intermediate-level waste disposal under study is to inject the waste in a cement mixture into a deep well drilled in a shale formation. The mixture then solidifies in place in fractures in the rock.

High-level liquid wastes, produced in processing of nuclear fuels, present the greatest environmental hazard, particularly since they contain long-lived radioisotopes (strontium-90 and cesium-137) that will not decay to safe levels for hundreds of years. The production of heat by these wastes presents a serious problem. As shown in Figure 12.36, total heat production drops rapidly with time but strontium-90 and cesium-137 heat production diminishes very slowly and is ultimately the principal heat producer. The high-level liquid wastes are contained in elaborate, concrete-encased steel tanks set in massive retaining basins. It is of course, absolutely essential that these storage facilities function perfectly for centuries. The tanks will require periodic replacement.

Subsurface Water of the Lands

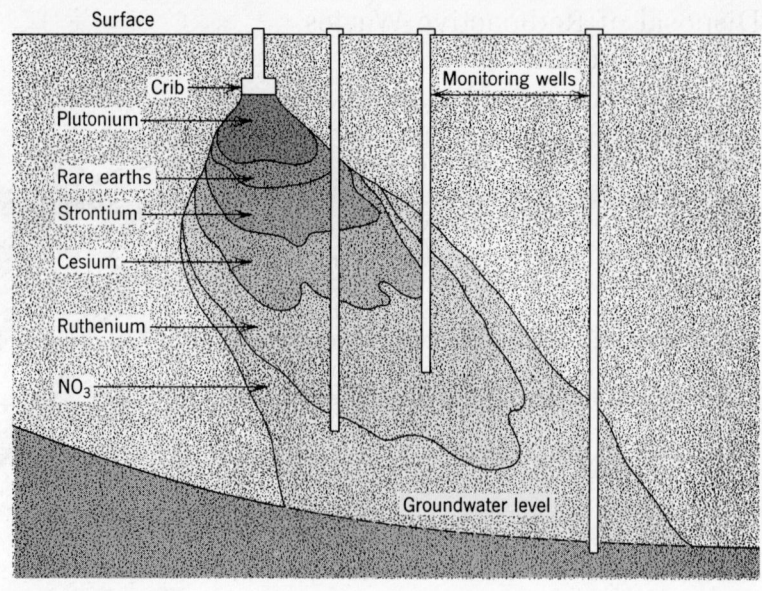

Figure 12.35 Cross section showing retention zones of various types of radioactive liquid wastes allowed to percolate downward through the zone of aeration. Depth to water table varies from 200 to 600 ft (60 to 180 m). (After C. H. Fox, 1969, *Radioactive Wastes*, U.S. Atomic Energy Commission, Understanding the Atom Series, p. 21, Figure 10.)

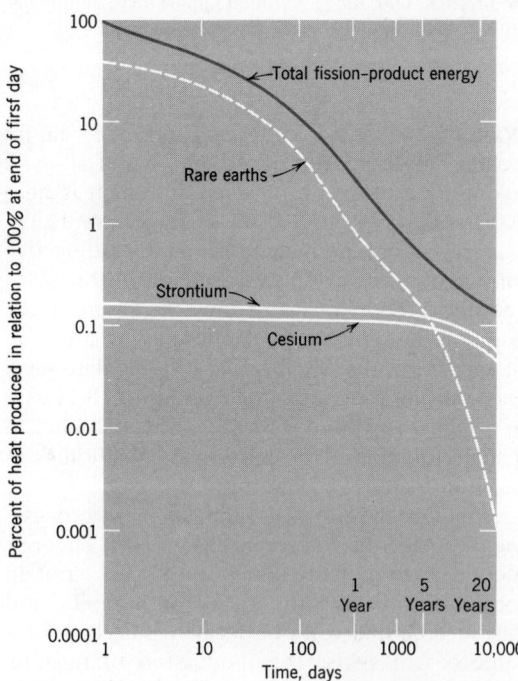

Figure 12.36 Decline with time in heat production from various high-level radioactive wastes produced during fuel processing. Scales are logarithmic. (After C. H. Fox, 1969, *Radioactive Wastes*, U.S. Atomic Energy Commission, Understanding the Atom Series, p. 24, Figure 12.)

Research is now directed to conversion of the high-level liquid wastes to solid form for safer permanent containment. Interest is turning to storage of these solid wastes in underground locations. One proposal is to store the containers in thick layers of rock salt (halite) in sedimentary strata. Salt has a favorable quality of closing fractures by mineral flowage, should ruptures occur. Experimental procedures have been carried out (Figure 12.37). Plans by the AEC to initiate rock-salt storage in an abandoned salt mine in Lyons, Kansas, have met with strong opposition from geologists who feel that studies are not yet adequate to demonstrate safety of the method. Depth of the mine is only about 1000 ft (300 m) and leakage would result in ground water contamination. Other possibilities under study include storage in bedrock tunnels in dense crystalline bedrock. In Chapter 13 we will take up the subject of environmental impact of nuclear waste disposal upon surface waters.

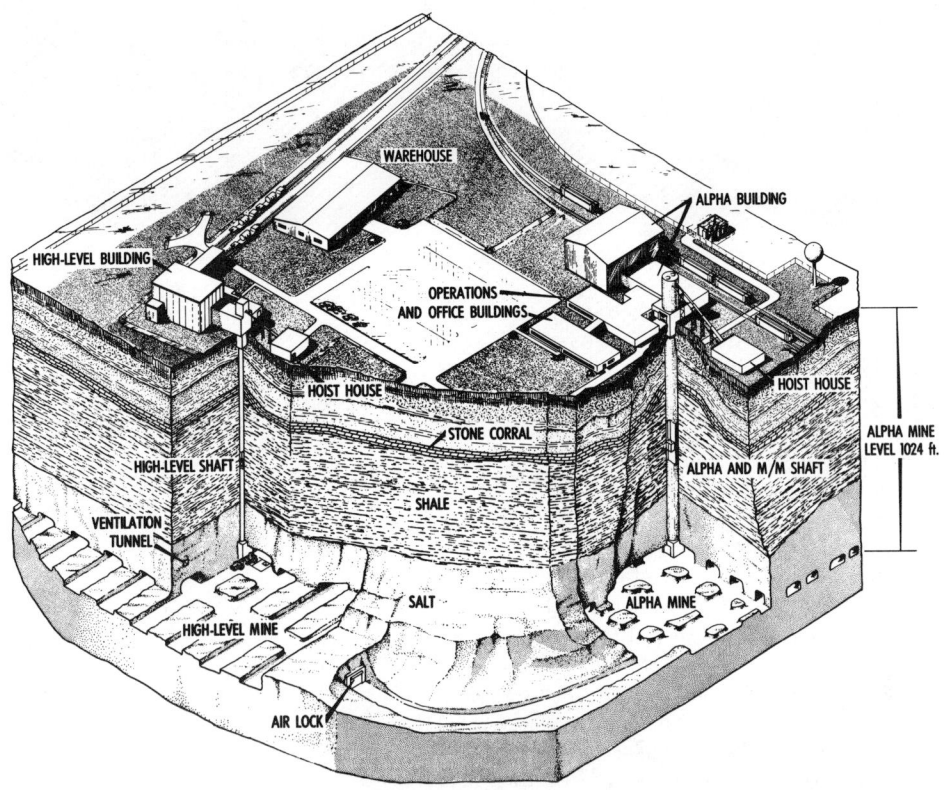

Figure 12.37 Above: Cut-away diagram showing proposed waste-disposal system in a salt mine at Lyons, Kansas. (A.E.C. illustration.) Right: In an experiment conducted by the Oak Ridge National Laboratory, cannisters containing radioactive wastes are placed in holes 12 ft (3.7 m) deep, lined with stainless steel. Location is seen in lower-left part of diagram above. (A.E.C. photograph.)

13
Hydrology of Streams and Lakes

In this chapter we return to the surface-water flow paths of the hydrologic cycle, tracing the movement of water on the lands by overland flow, into streams and lakes, and eventually reaching the sea. As we explained in some detail in opening paragraphs of the preceding chapter, surface flow of both water and ice (as glaciers) belongs to the class of irreversible (one-way) systems in which the potential energy of position is converted into kinetic energy of motion. Because of frictional resistance, kinetic energy is transformed into sensible heat, which is subsequently conducted or radiated out of the system. An equivalent amount of energy is dissipated into outer space to maintain the earth's heat balance.

In this chapter we examine the flow of water itself—largely from the hydrologist's point of view—leaving geological considerations for the next chapter. Streams, lakes, bogs, and marshes are highly varied, specialized habitats of plants and animals; their ecosystems are particularly sensitive to changes induced by Man in the water balance and in water chemistry. Not only does our industrial society make radical physical changes in water flow by construction of engineering works (dams, irrigation systems, canals, dredged channels), but we also pollute and contaminate our surface waters with a large variety of wastes. Some of these wastes are in the form of ions in solution. The subject of quality of water is appropriate to discuss in this chapter, and we shall want to study the sources and quantities of ions introduced into the natural flow systems by Man, comparing the data with natural conditions. Introduction of mineral and organic sediment into surface waters as a result of land disturbances is a suitable topic for the next chapter.

Surface water is also a basic natural resource essential to Man in his varied and intense agricultural and industrial activities. Runoff held in reservoirs behind dams provides water supplies for great urban centers, such as New York City and Los Angeles; diverted from large rivers, it provides irrigation water for highly productive lowlands in arid lands, such as the Imperial Valley of California and the Nile Valley of Egypt. To these uses of runoff are added hydroelectric power, where the grade of a river is steep; or routes of inland navigation, where the grade is gentle. Hydropower as an energy resource has been reviewed in Chapter 10.

Unlike ground water, which represents a large water-storage body that can be withdrawn much as a mineral is mined for exploitation, fresh surface-water in the liquid state is stored only in small quantities. (An exception is the Great Lakes system.) Referring back to Table 12.2, note that the quantity of available ground water is about thirty times as large as that stored in fresh-water lakes, while the water held in streams is only about one-hundreth that in lakes. Both surface-water and ground-water resources are renewable, but because of small natural storage capacities surface water can be drawn only at a rate comparable with its annual renewal through precipitation. Dams are built to develop useful storage capacity for surplus runoff that would otherwise escape to the sea, but once the reservoir has been filled, water use must be scaled to match the natural supply rate averaged over the year. Development of surface-water supplies brings on many environmental changes, both physical and biological, and these must be taken into account in planning future development.

Surface-water flow also exerts its stress upon Man and his structures through natural events. These are river floods that inundate low-lying land surfaces and are capable of wholesale destruction of life and property. Many great river floods have qualified as major disasters. Consequently, we shall need to make a study of floods and their control by engineering works.

Runoff and Natural Drainage Systems

As defined in Chapter 12, runoff consists of all surface water flow, both over the ground surface as overland flow and in streams as *channel flow*. Runoff may be derived directly from excess precipitation which cannot infiltrate the soil, or it may originate as the outflow of ground water along lines where the water table intersects the earth's surface.

Seeking to escape to progressively lower levels and eventually to the sea, runoff becomes organized into *drainage systems*, which are more-or-less pear-shaped areas bounded by divides, within which ground slopes and branching stream networks are adjusted to dispose as efficiently as possible of the runoff and its contained load of mineral particles. Most drainage systems possess a constricted exit, normally the mouth of the master stream, where it meets a large body of water. Thus, a drainage system is a converging mechanism for funneling and integrating the weaker and more diffuse forms of runoff into progressively deeper and more intense paths of activity.

Progress of Overland Flow

Consider now a hillside slope which is receiving moderate to heavy rainfall over a period of several hours. After the first hour or so, the infiltration rate has fallen to a sustained rate less than the precipitation rate, producing overland flow in one of several forms. It may be a continuous thin film, called *sheet flow*, where the soil or rock surface is extremely smooth, or a series of tiny rivulets connecting one water-filled hollow with another, where the ground is rough or pitted. On a grass-covered slope, overland flow is subdivided into countless tiny threads of water, passing around the stems. Even in a heavy and prolonged rain, overland flow in full progress on a sloping lawn may not be visible to the casual observer. On heavily forested slopes bearing a thick mat of decaying leaves and many fallen branches and tree trunks, overland flow may pass almost entirely concealed beneath this cover.

Overland flow is measured in inches or centimeters of water per hour, just as for precipitation and infiltration. Therefore, the simple equation given in Chapter 12 can be rewritten to express the rate at which overland flow will be produced by a given unit of ground surface as follows:

$$R_o = P - I$$

where R_o is rate of production of overland flow,
P is precipitation intensity,
and I is infiltration rate.

For example, if the rate of infiltration became constant at a value of 0.4 in./hr, and the rate of rainfall was a steady 0.6 in./hr (a heavy rain), the runoff would be produced at a rate of 0.2 in./hr, assuming none to be returned to the atmosphere by evaporation.

Because any given unit square of ground on a hillside must receive the overland flow from the entire strip of ground of that width lying upslope of it, we may expect the rate of discharge (volume of water passing across a given line in a given unit of time) to increase in direct proportion to the length of the total path of the flow. Depth of the flowing layer might therefore be expected to increase the farther downslope it progresses, but this increase may be small because the flow velocity will also be increasing down the slope. Figure 13.1 shows heavy runoff at the base of a long slope, where the accumulated overland flow has converged into broad shallow streams spreading across the slope.

Figure 13.1 Overland flow running down an 8% slope following a heavy thunderstorm. The ditch in the foreground receives the runoff and conducts it away as channel flow. (Soil Conservation Service photograph.)

At the base of a hill slope, overland flow is disposed of by passing into a stream channel or lake, or by sinking into the ground, should a highly permeable layer of sand or gravel be encountered.

Stream Channels

The channel of a stream is a long, narrow trough, shaped by the forces of flowing water to be most effective in moving the quantities of water and sediment supplied from the drainage basin. Channels may be so narrow that a person can jump across them, or as wide as 1 mile (1.5 km) for great rivers such as the Mississippi (see Figure 14.42). Taking the entire range of natural channel widths as between 1 foot and 1 mile, a 5000-fold difference in size can exist.

Hydraulic engineers who must measure stream dimensions and flow rates have adopted a set of terms to describe channel geometry (Figure 13.2). *Depth*, in feet or meters, is measured at any specified point in the stream as the vertical distance from surface to bed. *Width* is the distance across the stream from one water's edge to the other. *Cross-sectional area*, A, is the area in square feet or square meters of a vertical slice across the stream at any specified place. *Wetted perimeter*, P, is the length of the line of contact between the water and the channel, as measured from the cross section. An important characteristic of streams is the *hydraulic radius*, R, which is defined as the cross-sectional area, A, divided by the wetted perimeter, P, or $R = A/P$.

Another important ratio expressing channel geometry is *form ratio*, defined as depth, d, divided by width, w, or d/w. Form ratio is commonly stated as a simple fraction such as 1/100 or 1:100, meaning that the stream channel is 100 times as wide as it is deep. Finally, a most important measure is *slope*, S (or *gradient*), which is the angle between the water surface and the horizontal plane. Slope can be stated in feet per mile or meters per kilometer. Thus a slope of 5 feet per mile means that the stream surface undergoes a vertical drop of 5 feet for each mile of horizontal distance downstream. Slope can also be given in terms of *per cent grade*, a common practice in engineering. A grade of 3%, or 0.03, means that the stream drops 3 feet for every 100 feet of horizontal distance.

Stream Flow

Gravity acts upon the water of a stream to exert pressure against the confining banks and bed of its channel. A small part of the gravitational force is aimed downstream parallel with the surface and bed, causing flow. Resisting the downstream flow is the force of resistance, or friction, within the water and between the water and the floor and sides of the channel. As a result, water close to the bed and banks moves slowly; that in the deepest and most centrally located zone flows fastest. Figure 13.2 indicates by dotted lines the manner in which flow takes place, or the *velocity distribution*. We can imagine that each arrow traces a given drop of water and that we observe its subsequent positions at equal time intervals. The single line of highest velocity is located in mid-stream, if the channel is straight and symmetrical, but about one-third of the distance down from surface to bed.

The above statements about velocity need to be qualified. Actually, in all but the most sluggish streams, the water is affected by *turbulence*, a

Stream Velocity and Energy

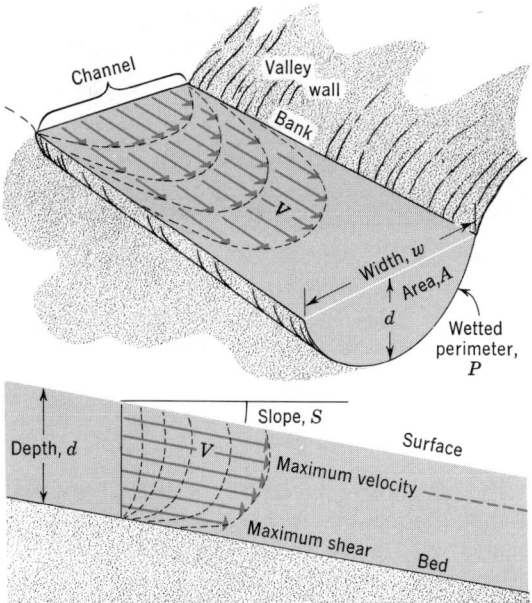

Figure 13.2 Hydraulic geometry of a stream in its channel.

system of innumerable eddies that are continually forming and dissolving. Therefore, a particular molecule of water, if we could keep track of it, would actually describe a highly irregular, corkscrew path as it is swept downstream. Motions would include upward, downward, sideward, and backward directions. Turbulence in streams is extremely important because of the upward elements of flow that lift and support fine particles of sediment. The murky, turbid appearance of streams in flood is ample evidence of turbulence, without which sediment would remain near the bed. Only if we measure the water velocity at a certain fixed point for a long period of time, say several minutes, will the average motion at that point be downstream and in a line parallel with the surface and bed. It is such average values that are shown by the arrows in Figure 13.2.

Because the average velocity at a given point in a stream differs greatly according to whether it is being measured close to the banks and bed, or out in the middle line, a single figure, the *mean velocity*, is computed for the entire cross section to express the activity of the stream as a whole. Mean velocity in streams is commonly equal to about six-tenths of the maximum velocity, but depends on the relative depth of the stream.

The last and most important measure of stream flow is *discharge*, Q, defined as the volume of water passing through a given cross section of the stream in a given unit of time. Commonly, discharge is stated in cubic feet per second, abbreviated to *cfs*. Sometimes the hydraulic engineer simply states this quantity as *second feet*. In metric units discharge is stated in cubic meters per second (cms). Discharge may be obtained by taking the mean velocity $\bar{V}$, and multiplying it by cross-sectional area, A. This relationship is stated by the important equation, $Q = A\bar{V}$.

If no water enters or leaves the banks and bed of a stream channel, the discharge (Q), must be the same at all cross-sections along its length for a stream in steady state of flow. A particular segment, or *reach*, of a stream channel is a true open system and when steady state is achieved the rate at which matter enters the system must be equalled by the rate at which matter leaves the system. It follows that the product of cross-sectional area and mean velocity must also be the same at all channel cross sections. It is for this reason that engineers refer to the equation $Q = A\bar{V}$ as the *equation of continuity*. However, either the area, A, or mean velocity, $\bar{V}$, can vary from section to section along the channel, in such a manner that the product of the two terms remains constant. As Figure 13.3 shows, that portion of the channel in which water is flowing swiftly in rapids has a small area of cross-section; that portion where the flow velocity is small has a large cross-section.

Stream Velocity and Energy

As water moves in a stream channel, resistance to flow is set up by a physical property of the water known as the *viscosity*, a measure of the ease of flow of a fluid under a given stress. Viscosity of water is less than for a heavy motor oil, while that of molasses is greater than for the oil. As the water moves in the channel, we can visualize infinitesimally thin layers of water molecules moving over one another, a type of motion known as *fluid shear*. Viscosity is a measure of resistance to shear. Work must be done to overcome viscosity. A water layer close to the solid channel boundary also encounters frictional resistance. As the stream flows downgrade, potential energy of elevation is transformed into kinetic energy of motion. If there were no resistance to flow (i.e., zero viscosity), the water body of the stream would accelerate as if it were an object falling in a vacuum. Instead, as velocity increases, resistance to flow increases, withdrawing more of the energy in friction. The result is that a *terminal velocity* is very quickly attained and maintained; the mean velocity becomes a constant. Sensible heat generated by the resistance is conducted through the channel walls and upper stream surface and is lost at a rate equal to its production

Hydrology of Streams and Lakes

rate. We have thus completed a tracing of the full pathway of energy in the flowing stream system.

Stream channels differ in the amount of resistance offered to flow. Where the hydraulic radius, R, is a small ratio, meaning that the stream is wide, but shallow in depth, resistance is large because of the relatively large boundary surface area of stream channel in proportion to the cross-sectional area. A narrow, deep channel offers less resistance. (The optimum channel section would be a semicircle.) However, streams are often required to have broad, shallow channels because of the load of mineral matter that must be carried, or because the banks are weak and will not hold a steep attitude. Resistance also varies according to the minor irregularities of the bed. Flow over coarse cobbles and boulders offers more resistance than flow over a smooth bed of fine sand or clay. Vegetation, where present, is also a source of resistance to flow.

Velocity, Depth, and Slope

It is intuitively obvious that water will flow faster in a channel of steep gradient than in one of low gradient, since the component of gravity acting parallel with the bed is larger for the steeper grade. It is also reasonable to conclude that a deeper stream will flow faster than one of less depth, on the same gradient, since the motion between successive water layers is accumulated upward from the bed and outward from the banks. For nearly two centuries hydraulic engineers have tried to refine an empirical equation relating velocity, slope (gradient), and depth in streams. A French engineer was the first to obtain the following working equation, named after him as the *Chezy equation*:

$$\bar{V} = C\sqrt{R \cdot S}$$

where $\bar{V}$ is mean velocity,
R is hydraulic radius,
S is slope, as percent grade,
and C is a numerical constant.

The equation can be rewritten as
$$V = C \cdot \sqrt{R} \cdot \sqrt{S}$$
or, $V = C \cdot R^{1/2} \cdot S^{1/2}$

Stated in words, the equation reads "mean velocity varies directly as the square-root of the hydraulic radius times the square root of the slope." Hydraulic radius is numerically about equivalent to mean stream depth. The Chezy equation has been refined by various investigators but not greatly changed. One of the more commonly used modifications (Manning equation) substitutes the two-thirds power of the hydraulic radius ($R^{2/3}$) for the square root.

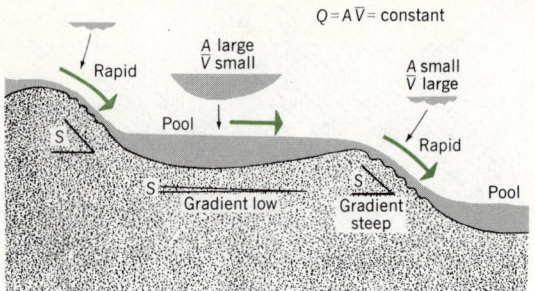

Figure 13.3 Schematic diagram of relationships among cross-sectional area (A), mean velocity ($\bar{V}$), and gradient (S). (From A. N. Strahler, 1971, *The Earth Sciences*, 2nd ed., Harper & Row, New York.)

The practical significance of the Chezy and Manning equations is that they tell us approximately how much the velocity of a stream will be increased when the gradient increases, as for example where the stream passes over a series of rapids. As shown in Figure 13.3, greater slope steepness over the rapid requires that velocity be increased. In the pool between rapids, where the gradient is low, velocity is decreased. We shall find that these basic relationships are of importance in predicting specific environmental impacts of man-made structures modifying stream channels. Increases in velocity can cause increased erosion of the stream bed and banks and thus increase the stream's load of mineral particles. These effects are discussed in the next chapter.

Drainage Systems

Examine a water-drainage system from the standpoint of an open energy system having a definite boundary through which matter enters, and an exit point where matter leaves the system. Figure 13.4 is a map of a real drainage system. The surface boundary is a line known as the *basin perimeter*, or *drainage divide*; it can be traced from the exit point of the principal stream channel, where divide elevation is lowest, to a point of maximum elevation at the opposite end of the basin, then back again to the exit point. The basin is typically elliptical or pear-shaped in outline. All precipitation falling on the area inside the divide must enter the flow system. Let us assume that the underlying rock mass is homogeneous, so that there will be a ground-water divide beneath the surface divide. In that case precipitation infiltrating the surface passes down to the water table and moves toward the axis of the basin where it emerges in streams that intersect the water table. (Various geological structures can exist that will

Drainage Systems

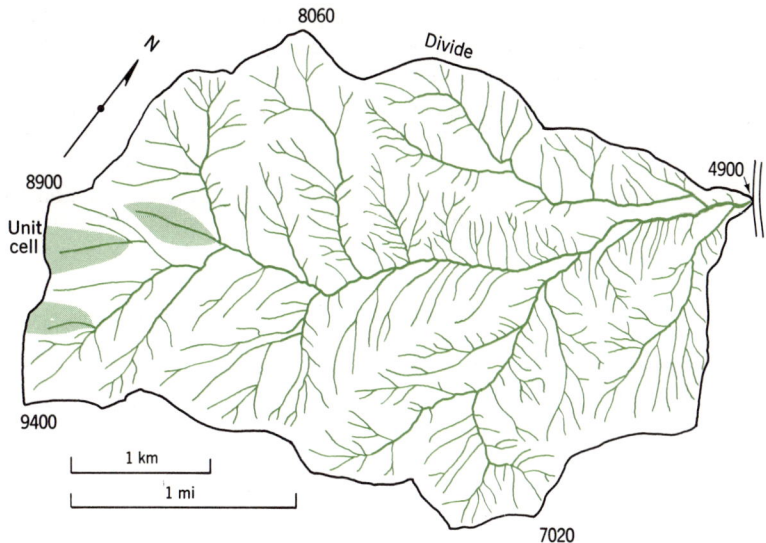

Figure 13.4 Channel network of the basin of Pole Canyon, Spanish Fork Peak Quadrangle, Utah. Data of U.S. Geological Survey and Mark A. Melton. (From A. N. Strahler, 1972, *Planet Earth; Its Physical Systems Through Geologic Time*, Harper & Row, New York.)

cause ground water to move independently of the surface configuration of the basin.)

The drainage basin is dominated by a branching channel system (Figure 13.4). The adjective *dendritic,* which means a branching figure resembling a tree, is applied to such a channel system. Overland flow originating near the divide makes its way along a surface path of steepest possible grade to reach the nearest stream channel (Figure 13.5). The ground surfaces between divides and channels are referred to as *valley-side slopes.* Overland flow on these slopes provides a channel with its sustenance. For a given set of environmental conditions of soil, bed rock, and vegetation, under a given soil-water budget, a particular quantity of surface area is needed to maintain a permanent channel.

The *unit cell* of the drainage system is defined by the smallest, or fingertip channel and includes all of the ground surface contributing overland flow to that channel. Examples of unit cells are outlined in Figure 13.4. Fingertip channels join with one another to produce channels of larger carrying capacity, and these in turn join into larger channels, eventually converging to form the main trunk. However, fingertip channels can enter any of the larger channels directly from unit cells adjacent to those channels, as clearly shown in Figure 13.4. The dendritic channel system must dispose of all water that enters it by overland flow and ground water seepage, but it must also dispose of all mineral waste entrained in that water. There must be a built-in geometrical compromise between two aspects of system efficiency. On the one hand, the shortest distance between a point on the divide and the exit point will have the steepest average gradient and will minimize flow resistance for a single channel. On the other hand, one large stream channel is more efficient (i.e., offers less resistance in ratio to its discharge) than several small channels carrying the same flow. The dendritic pattern of channels represents the most effective compromise between length of flow path and stream discharge, while at the same time carrying out water and mineral waste supplied by the drainage basin surface. Adjustment of the system has taken place over a long time span of land erosion. In Chapter 14 we shall examine the erosion process and seek further insight into the attainment of an equilibrium configuration of land surface and the channel network. We will find there an explanation of the fact that as the stream channels of the network increase in size, and hence in water discharge, they also decrease in gradient. Therefore, the channel gradients will be steepest for the fingertip channels and least for the main trunk channel at the exit point.

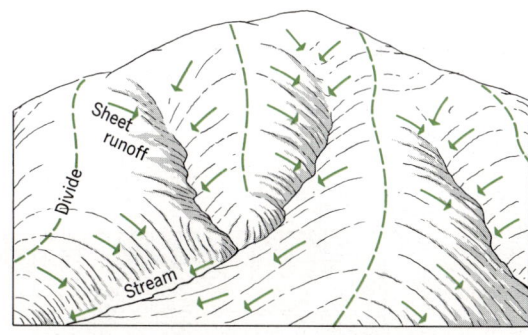

Figure 13.5 Overland flow from slopes in the headwater area of a stream system supplies water and rock debris to the smallest elements of the channel network.

Using the Chezy equation as a basis for reasoning, you might be led to think that as channel slopes (gradients) decrease downstream in a drainage basin, the mean velocity of flow will become less, since velocity varies about as square root of channel slope. On the other hand, you could reason that as channel cross-section increases in the downstream direction depth will also increase, and greater depth will cause an increase in mean velocity. Actually, observations show that while the two factors tend to offset one another, depth increase is somewhat more effective than slope decrease, with the result that most streams show a slight downstream increase in mean velocity.

Stream Gauging

To assess and manage our natural resources in the form of surface waters it is necessary to measure stream flow at key points in the stream networks of the nation. To this end an important activity of the U.S. Geological Survey is the measurement, or *gauging*, of stream flow in the United States. In cooperation with states and municipalities this organization maintains over 6000 river-measurement stations on principal streams and their tributaries. The discharge data are published by the Geological Survey in a series of *Water-Supply Papers*. Information on daily discharge and flood discharges is essential for planning the management of surface waters as well as for design of flood-protection structures and for the prediction of floods as they progress down a river system. Study of gauge records can also lead to assessment of the environmental impact of Man's activities on delicately adjusted drainage systems.

A stream gauging station requires a device for measuring the height of the water surface, or *stage* of the stream. Simplest to install is a *staff gauge*, which is simply a graduated stick permanently attached to a post or bridge pier. This must be read directly by an observer whenever the stage is to be recorded. More useful is an automatic-recording gauge, which is mounted in a *stilling tower* built beside the river bank (Figure 13.6). The tower is simply a hollow masonry shaft into which water enters through a pipe at the base. By means of a float connected by wire to a recording mechanism above, a continuous ink-line record of the stream stage is made on a graph paper attached to a slowly rotating drum.

To measure stream discharge, it is necessary to determine both the area of cross section of the stream and the mean velocity. This requires that a *current meter* (Figure 13.7) be lowered into the stream at closely spaced intervals so that the velocity can be read at a large number of points evenly distributed in a grid pattern through the stream's cross section (Figure 13.6). A bridge often serves as a convenient means of crossing over the stream; otherwise a cable car or small boat is used. The current meter has a set of revolving cups whose rate of turning is proportional to current velocity. The Price current meter, pictured in Figure 13.7 is in general use by the Geological Survey and will measure velocities from 0.2 to 20 ft (0.06 to 6 m) per second. As the velocities are being measured from point to point, a profile of the river bed is also made by sounding the depth. The cross-sectional area is measured from the profile. Mean velocity is computed by summing all individual velocity readings and dividing by the number of readings. Discharge can then be computed using the formula, $Q = A\bar{V}$.

In practice, the number of velocity readings at each point of sounding is reduced to two: one at 0.2 the depth; one at 0.8 the depth. The average of these two readings gives a close approximation to the true mean velocity. For shallow streams, a single velocity reading at 0.6 the depth suffices at each point of sounding.

Because of the time and labor required to measure discharge repeatedly by current readings, it is practical to take instead only a limited set of such measurements over a wide range of discharges. From these measured discharges is constructed a *rating curve*, or *stage-discharge curve*, permitting discharge to be estimated directly from gauge height. For the sample curve in Figure

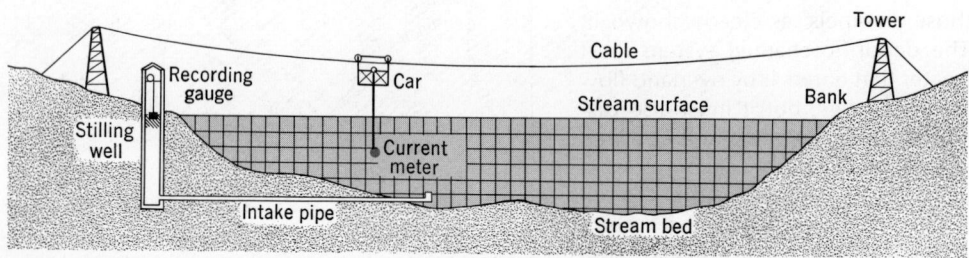

Figure 13.6 Idealized diagram of stream gauging installation.

13.8, eight points of discharge were actually measured by the current-meter method. These were plotted on the graph against gauge height in feet. A smooth curve was then drawn through the points. Thus, if the gauge height is known to be 20 ft, we can estimate that the discharge is occurring at a rate of about 16,500 cfs. Rating curves enable estimates of total discharge to be computed from stage records alone, despite wide fluctuations in stage.

A single rating curve may be useful only for limited periods of time because of changes in form of the river channel. Such changes may take place by channel erosion in floods. The rating curve is therefore recomputed and corrected as necessity demands.

Relation of Discharge to Basin Area

Common sense tells us that the average discharge of a stream increases with increasing drainage basin area. It remains to be determined what mathematical model applies to such an increase.

Figure 13.9 shows the observed relationship of *average discharge*, $\bar{Q}$, to drainage area, A, for the Potomac River basin. Each point represents one gauge. Obviously, gauges in the headwater regions show as points at the lower left; those far downstream lie at the upper right. Logarithmic scales are used on this graph. While the individual points show marked departures from the fitted straight line, the trend is obvious. Because the fitted straight line runs at 45° across the graph, it can be said that the discharge increases in direct proportion to the area.

One practical use of an established mathematical relationship between stream discharge and basin area is that it enables the hydrologist to estimate mean discharge at any point in the system by measuring the watershed area lying above that point. Such knowledge would be essential in designing hydraulic structures, such as dams, bridges, and irrigation diversions. Moreover, if man-made changes in a particular area should alter the soil-water budget, the effects may be recognized by a marked departure of data of a given gauge from the normal trend line.

Stream Flow and Precipitation

By studying the records of stream discharge in relation to precipitation on a given watershed, the hydrologist has developed a set of basic principles applying to the variations in stream discharge with different lengths and intensities of storms and with different sizes of watersheds.

Figure 13.7 In gauging large rivers, the current meter is lowered on a cable by a power winch. Earphones, connected by wires to the meter, receive a series of clicks whose frequency indicates water velocity. (U.S. Geological Survey photograph.)

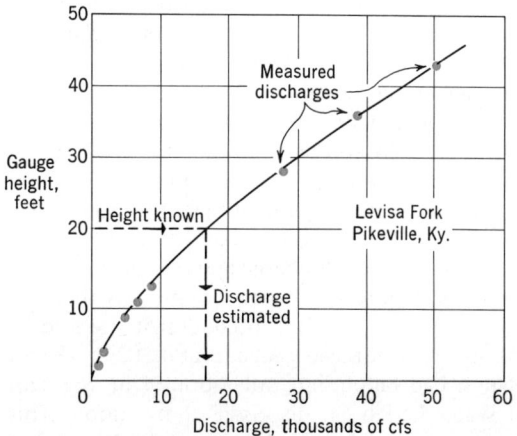

Figure 13.8 This rating curve applied to Levisa Fork, Kentucky, during the period October 1945 to January 1946. (After W. G. Hoyt and W. B. Langbein, 1955, *Floods*, Princeton Univ. Press, Princeton, N.J., p. 69, Figure 23.)

Hydrology of Streams and Lakes

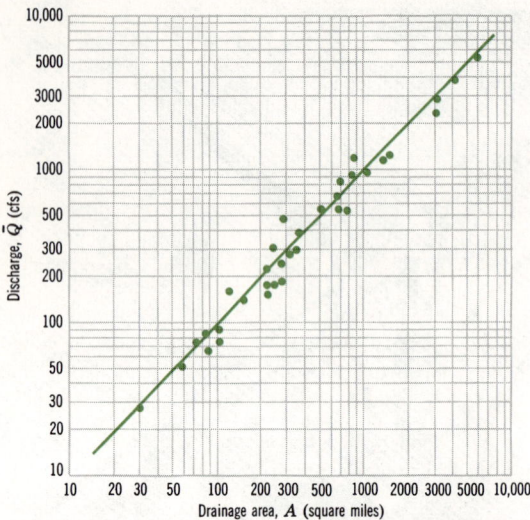

Figure 13.9 Relation of mean stream discharge to drainage area for all gauging stations in the Potomac River basin. Each point represents a gauge. (Data from J. T. Hack, 1957, U.S. Geological Survey, *Professional Paper 294-B*, p. 54, Figure 15.)

response of runoff is rapid. Response of runoff to precipitation is stated in terms of the *lag time*, measured as time difference between center of mass of precipitation and center of mass of runoff. In Figure 13.10 these centers of mass are labeled *CMP* and *CMR* respectively; the lag time is about 12 minutes. Let us now look at the hydrographs of larger areas over longer periods of time to see the effect of watershed size and storm duration on stream flow.

Figure 13.11 shows the hydrograph of Sugar Creek, Ohio, with a watershed area of 310 sq mi (805 sq km). Sugar Creek basin, a part of the much larger Muskingum River watershed, is outlined in Figure 13.12, a map showing by isohyetal lines the rainfall during the 12-hour storm of August 6 and 7, 1935, for which the hydrograph was constructed. Over the area of Sugar Creek, the average total rainfall was 6.3 in. (16 cm) for the entire storm, but the total quantity discharged by Sugar Creek was only 3 in. (7.5 cm). This means that 3.3 in. (8.5 cm), or more than half of the rainfall, was

Consider first a very small watershed, just over one acre (0.4 hectare) in area, upon which a heavy rain fell in a total period of about an hour. Figure 13.10 is a graph showing what happened to the water from beginning to end of the storm. Rainfall was measured with the rain gauge and is shown in terms of the intensity of rainfall, or quantity falling in each 5- or 10-minute period. Rain began at 4:21 P.M. and was extremely heavy for nearly 40 minutes, after which it let up rapidly and ceased entirely by 5:40. Discharge, measured at the outlet point of the small watershed, is shown in Figure 13.10 by a smooth curve scaled in cubic feet per second. Because of the high initial infiltration capacity of the soil, all of the rain was at first absorbed by the soil or was detained in surface irregularities. About 6 minutes after the rain began, discharge set in and rose rapidly for a half hour, reached the peak just after 4:50, then declined again and became zero by 5:50 P.M.

The lower part of Figure 13.10 is another form of representation in which the quantities of rainfall and runoff are accumulated from beginning to end. Here both discharge and rainfall are scaled in terms of inches of water depth, so that the values can be directly subtracted. At the end of the storm, about 5:40 P.M., a total of 1.2 in. (3 cm) of rain had fallen, but only about 1 in. (2.5 cm) of water had been disposed of by runoff. This leaves 0.2 in. (0.5 cm) of loss through combined evaporation and infiltration. An important principle of this water graph, or *hydrograph*, as it is generally called, is that for a small watershed, the

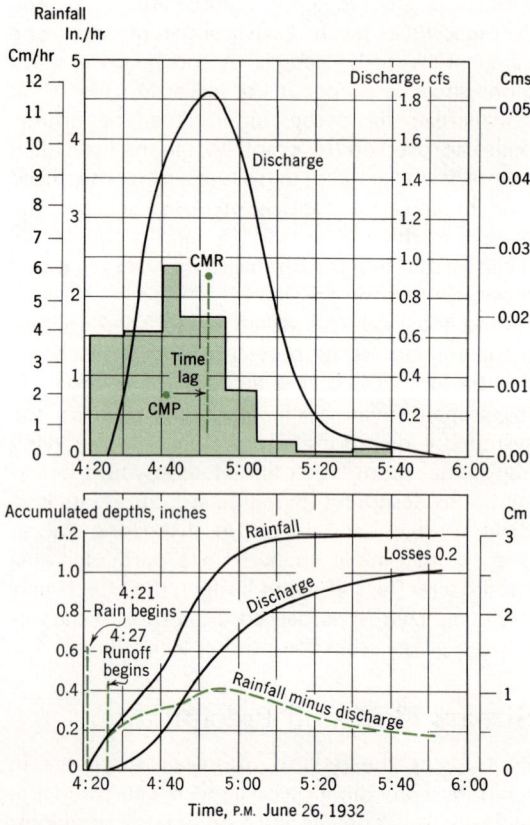

Figure 13.10 Hydrograph of a very small drainage area, about one acre (0.4 hectare), near Hays, Kansas, during a rainstorm in June. (Data of E. E. Foster, 1949, *Rainfall and Runoff*, Macmillan, New York, p. 306, Figure 114.)

Base Flow and Surface-Water Flow

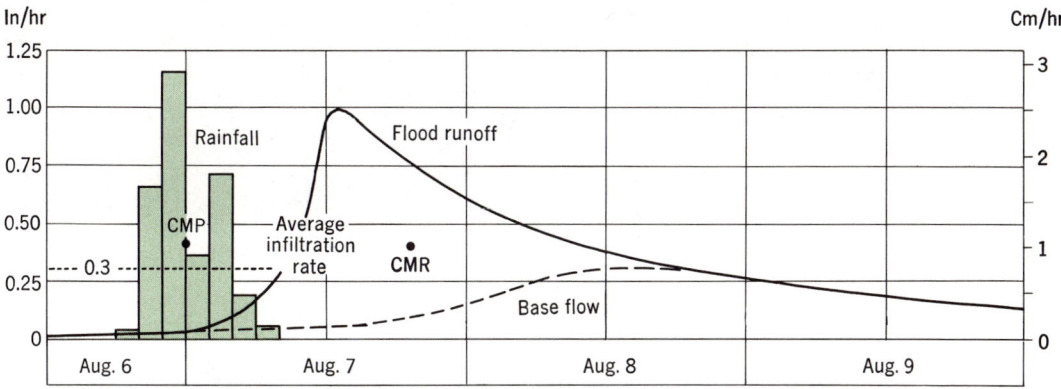

Figure 13.11 Hydrograph of Sugar Creek, Ohio, for four days during and after a heavy rainstorm. (Data from W. G. Hoyt and W. R. Langbein, 1955, *Floods*, Princeton Univ. Press, Princeton, N.J., p. 45, Figure 13.)

retained on the watershed, having infiltrated to become part of the soil and ground water, or had evaporated.

Studying the rainfall and runoff graphs in Figure 13.11, we see that prior to the onset of the storm, Sugar Creek was carrying a small discharge. This was being supplied by the seepage of ground water into the channel and is termed *base flow*. After the heavy rainfall began, several hours elapsed before the stream gauge at the basin mouth began to show a rise in discharge. This time lag indicates that the branching system of channels was acting as a temporary reservoir, receiving inflow more rapidly than it could be passed down the channel system to the stream gauge. The term *channel storage* is applied to runoff delayed in this manner during the early period of a storm.

The peak of flow in Sugar Creek was reached almost 18 hours after the rain began, or about 6 hours after the cessation of rainfall—a vastly greater delay than was observed in the 1-acre plot (Figure 13.10). Lag time, as previously defined, is approximately 18 hrs. (Center of mass of runoff is based on area under runoff curve minus base flow area.) Observe also that the rate of decline in discharge was much slower than the rate of rise. In general, then, the larger a watershed, the larger is the lag time.* Because much rainfall had entered the ground and had reached the water table, a slow but distinct rise is seen in the amount of discharge contributed by base flow.

Base Flow and Surface-Water Flow

In regions of humid climates, where the water table is high and normally intersects the important stream channels, the hydrographs of larger streams will show clearly the effects of two sources of water: (a) base flow, and (b) surface-water flow, derived as overland flow. Figure 13.13 is a hydrograph of the Chattahoochee River, Georgia, a large river draining a watershed of some 3350 sq mi (8700 sq km) much of it in the humid southern Appalachian Mountains. The sharp, abrupt fluctuations in discharge are produced by surface flow following rain periods of one to three days duration. These are each similar

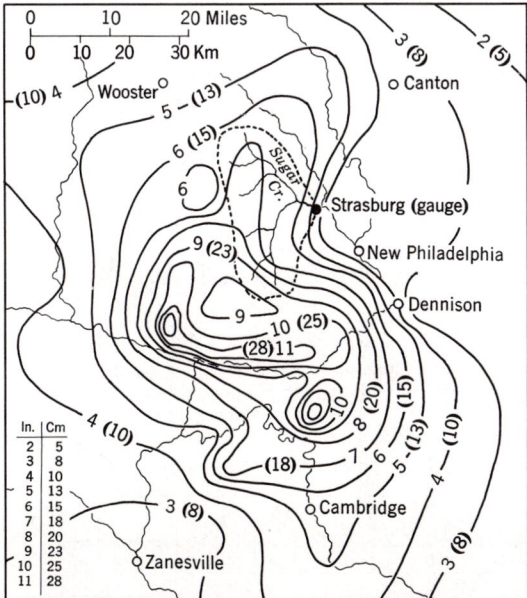

Figure 13.12 Total rainfall in inches from the rainstorm of August 6–7 on the Sugar Creek, Ohio, watershed. Figures in parentheses are centimeters. (Same data source as Figure 13.11.)

*Hydrologists have found that lag time is directly proportional to basin length (long dimension of basin) and inversely proportional to the square root of the mean basin gradient.

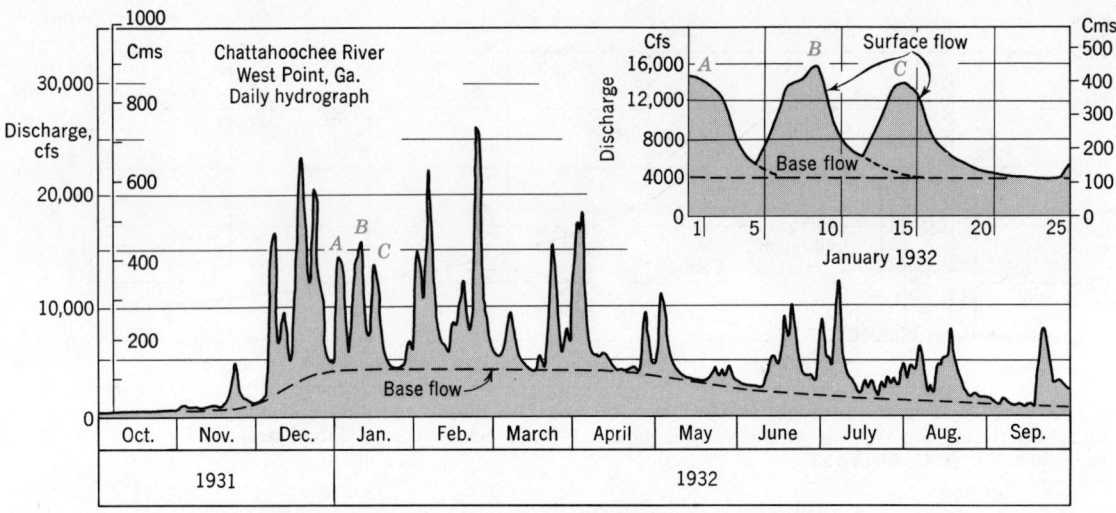

Figure 13.13 Flow peaks of the Chattahoochee River, Georgia. (After E. E. Foster, 1949, *Rainfall and Runoff*, Macmillan, New York, p. 303, Figure 111, p. 304, Figure 112.)

to the hydrograph of Figure 13.11, except that they are here shown much compressed by the time scale.

After each rain period, the discharge falls off rapidly, but if another storm occurs within a few days, the discharge rises to another peak. The enlarged inset graph, showing details of the month of January, shows how this effect occurs. Where a long period intervenes between storms, the discharge falls to a low value, the base flow, where it levels off. Throughout the year the base flow, which represents ground-water seepage into the stream, undergoes a marked annual cycle. During the period of ground-water recharge (winter and early spring), water-table levels are raised and the rate of inflow into streams is increased. For the Chattahoochee River, the rate of base flow during January, February, March, and April holds uniform at about 4000 cfs (110 cms). As the heavy evapotranspiration losses of spring reduce soil water and therefore cut off the recharge of ground water by downward percolation (see Chapter 12), the base flow falls steadily. The decline continues through the summer, reaching by the end of October a low of about 1000 cfs (30 cms) supplied entirely from base flow.

With a knowledge of soil-water budget regimes and rock types to guide our reasoning, it may be guessed that base flow will be important in regions of ample, well-distributed soil-water surplus, but unimportant or absent in semiarid regions having a major soil-water deficiency.

Figure 13.14 shows comparative hydrographs for 1-year periods for three watersheds of approximately the same areas. That of Ecofina Creek, Florida, is unusual in showing a large proportion of base flow and lack of strong discharge peaks. The explanation may lie partly in the low relief and gentle slopes of the watershed, but more particularly in the presence of cavernous limestone beneath the surface. Most of the excess

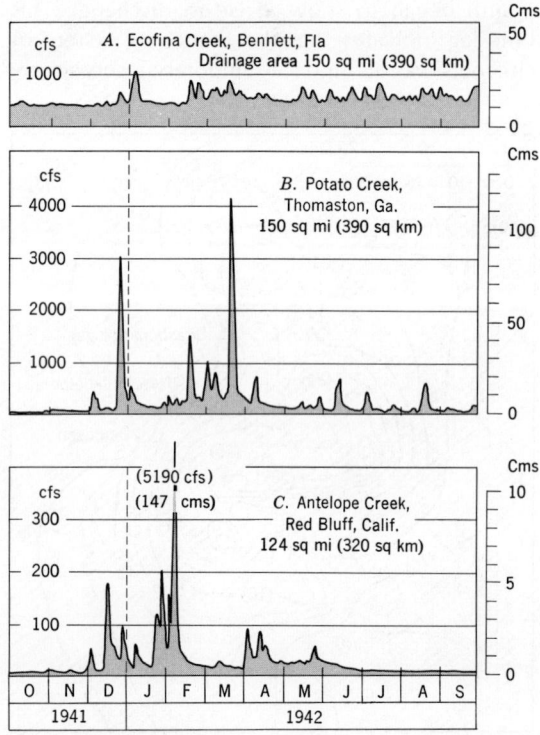

Figure 13.14 Hydrographs of three streams differing in flow characteristics because of climate, relief, and rock type. (After E. E. Foster, 1949, *Rainfall and Runoff*, Macmillan, New York, p. 300.)

River Floods—An Environmental Hazard

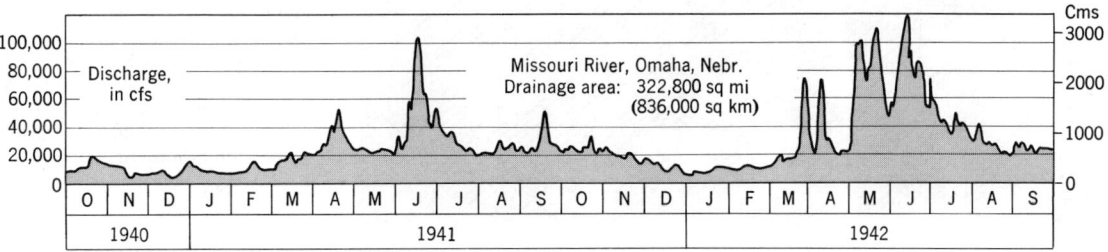

Figure 13.15 Hydrograph of the Missouri River at Omaha, Nebraska. (After E. E. Foster, 1949, *Rainfall and Runoff*, Macmillan, New York, p. 301.)

rainfall enters the ground-water system and is discharged copiously through solution passages. Potato Creek, Georgia, in a region of steep slopes, has occasional extreme peak flows that occur in winter and early spring when there is a soil-water surplus and the proportion of surface runoff is high. Antelope Creek, gauged at Red Bluff, California, in the northern part of the state, likewise shows the extreme discharges typical of mountain watersheds, but the complete absence of peaks from June through October reflects the long summer drought of the Mediterranean-type climate regime during which only base flow can be maintained (compare with Figure 12.14E).

Finally, examine the hydrograph of the Missouri River at Omaha, Nebraska, from October 1940, to September 1942 (Figure 13.15). This great river, draining 322,800 sq mi (840,000 sq km) of watershed, is a major tributary of the Mississippi River. Notice that the discharge, ranging from 10,000 to over 100,000 cfs (280 to 2800 cms), is many times greater than the discharges of smaller streams considered thus far. High rates of flow are chiefly from snowmelt, which occurs on the High Plains in spring and in the Rocky Mountains headwater areas in early summer. This source explains the sudden high discharges from April through June. During midwinter, when soil moisture is frozen and total precipitation small over the watershed as a whole, the discharge rises little above the base flow. Ground-water recharge occurring in the spring raises summer levels of base flow to about 20,000 cfs (570 cms), or two to three times the winter base flow.

River Floods—An Environmental Hazard

In our modern day of newspapers and television, everyone has seen enough pictures of river floods to have a good idea of the appearance of river flood waters and the havoc wrought by their erosive power and by the silt and clay that they leave behind. Nevertheless, even the hydraulic engineer may not be fully satisfied that he can exactly define the term *flood*. Perhaps it is enough to say that a condition of flood exists when discharge of a river cannot be accommodated within the margins of its normal channel, so that the water spreads over adjoining ground upon which crops or forests are able to flourish. Most of our larger streams have a *floodplain,* a belt of low flat ground bordering the channel on one or both sides inundated by stream waters about once a year or every other year, at the season when abundant supplies of surface water combine with effects of a high water-table and a soil-water surplus to supply more runoff than can stay within the heavily scoured troughlike channel (Figure 13.16). Such annual or biennial inundation

Figure 13.16 The Wabash River in flood near Delphi, Indiana, February, 1954. An ice dam clogs the river channel, while lines of trees mark the crest of the bordering natural levees. The floodplain itself is inundated on both sides of the channel and reaches to the base of the bluff, at left. (U.P.I. Telephoto.)

Figure 13.17 The city of Hartford was partly inundated by the great Connecticut River flood of March 1936. The river channel is to the left, its banks marked by a line of trees. (Official Photograph, 8th Photo Section, A.C., U.S. Army.)

is considered a flood, even though its occurrence is expected and does not prevent the cultivation of crops after the flood has subsided, nor does it interfere with the growth of dense forests which are widely distributed over low, marshy floodplains in all humid regions of the world.

Still higher discharges of water, the rare and disastrous floods which may occur as infrequently as several decades or longer, inundate ground lying above the floodplain, principally affecting broad steplike expanses of ground known as *terraces* (Figure 13.17). (Floodplains and terraces as landforms are explained in Chapter 14.)

For practical purposes, the National Weather Service of NOAA, which provides a flood-warning service, designates a particular stage or gauge height at a given place as the *flood stage*, implying that the critical level has been reached above which overbank flooding may be expected to set in. Immediately at or below flood stage, the river may be described as being in the *bankfull stage*, the flow being entirely within the limits of the heavily scoured channel.

Downstream Progress of a Flood Wave

The rise of a river stage to its maximum height, or *crest*, followed by a gradual lowering of stage, is termed the *flood wave*. The flood wave is simply a large-sized rise and fall of river discharge of the type already analyzed in earlier paragraphs, and follows the same principles. Figure 13.18A shows the downstream progress of a flood on the Chattooga-Savannah river system. In the Chat-

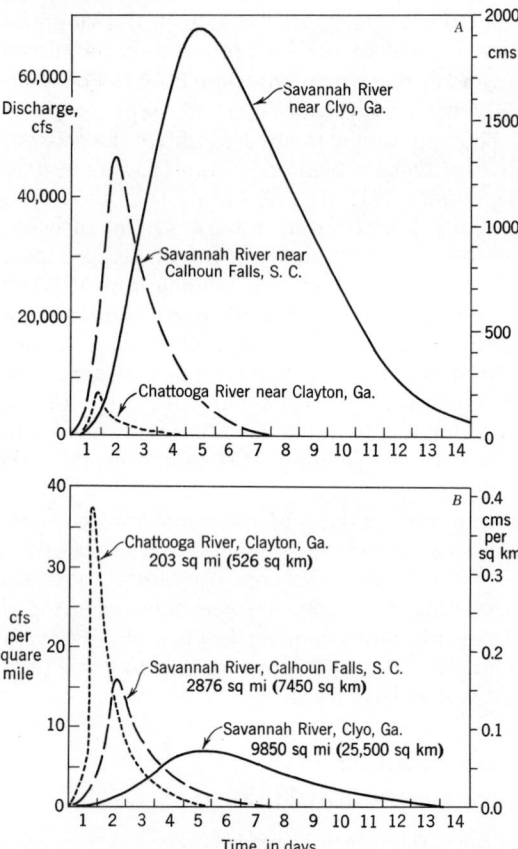

Figure 13.18 Downstream progress of a flood wave on the Savannah River in South Carolina and Georgia. (Data from W. G. Hoyt and W. B. Langbein, 1955, *Floods*, Princeton University Press, Princeton, N.J., p. 39, Figure 8.)

tooga River near Clayton, Georgia, the flood crest was quickly reached—one day after the storm—and quickly subsided. On the Savannah River, 65 mi (105 km) downstream at Calhoun Falls, South Carolina, the peak flow occurred a day later, but the discharge was very much larger because of the larger area of watershed involved. Downstream another 95 mi (153 km), near Clyo, Georgia, the Savannah River crested five days after the initial storm with a discharge of over 60,000 cfs (1700 cms). This set of three hydrographs shows that (a) the lag time in occurrence of the crest increases downstream, (b) the entire period of rise and fall of flood wave becomes longer downstream, and (c) the discharge increases greatly downstream as watershed area increases.

Figure 13.18B is a somewhat different presentation of the same flood data, in that the discharge is given in terms of a common unit of area, the square mile, thus eliminating the effect of increase in discharge downstream and showing us only the shape or form of the flood crest. In other words, the lag time and sharpness of peaking of the flood wave are emphasized without respect to the total discharges involved.

Flood Prediction

The National Weather Service operates a River and Flood Forecasting Service through 85 selected offices located at strategic points along major river systems of the United States. Each office issues river and flood forecasts to the communities within the associated district, which is laid out to cover one or more large watersheds. Flood warnings are publicized by every possible means. Close cooperation is maintained with such agencies as the American Red Cross, the U.S. Army Corps of Engineers, and the U.S. Coast Guard, in order to plan evacuation of threatened areas, and the removal or protection of vulnerable property.

Long and intensive study of stream flow data enables the National Weather Service to prepare graphs of flood stages telling the likelihood of occurrence of given stages of high water for each month of the year. Figure 13.19 shows expectancy graphs for four selected stations. The meaning of the strange-looking bar symbols is explained in the key.

The Mississippi River at Vicksburg illustrates a great river responding largely to spring floods so as to yield a simple annual cycle. The Colorado River at Austin, Texas, is chosen to illustrate a river draining largely semiarid plains. Summer floods are produced directly by torrential rains from invading moist tropical air masses. Floods of the late summer and fall are often attributable to tropical storms (hurricanes) moving inland from the Gulf of Mexico. The Sacramento River at Red Bluff, California, has a winter flood season when rains are heavy, but a sharp dip to low stages in late summer, which is the very dry period for the California coastal belt. The flood expectancy graph for the Connectict River at Hartford shows two seasons of floods. The more reliable of the two is the early spring, when snowmelt is rapid over the mountainous New England terrain; the second, in the fall, when rare but heavy rainstorms, some of hurricane origin, bring exceptional high stages. Thus the exceptional maximum flood stage of the month of September was set by the hurricane of September 21 to 23, 1938, which added an enormous quantity of runoff to channels already carrying bankfull flow from heavy rains of September 18 to 20.

Engineering Practices in Floodplain Management

In the face of repeated disastrous floods, vast sums of money have been spent on a wide variety of engineering measures intended to reduce flood crests and to retain flood waters within specified bounds. Two physical approaches to achieve these ends are: (a) to detain and delay runoff by various means on the ground surfaces and in smaller tributaries of the watershed; (b) to modify the lower reaches of the river and its floodplain where inundation is expected.

The first form of flood reduction is aimed at treatment of watershed slopes, usually by reforestation or planting of other vegetative cover so as to increase the amount of infiltration and reduce the rate of production of overland flow. This type of treatment, together with construction of many small flood-storage dams in the valley bottoms, may greatly reduce the flood crests and allow the discharge to pass into the main stream over a longer period of time. These measures are closely tied in with measures to control soil erosion (Chapter 14).

Under the second type of flood management, designed to protect the floodplain areas directly, two quite different theories can be practiced. First, the building of *levees,* or *dikes,* parallel with the river channel on both sides can function to contain the overbank flow and prevent inundation of the adjacent floodplain (Figure 13.20). Such levees are broad embankments built of earth and must be designed with great care, not only to possess the physical resistance to water pressures, but must be high enough to contain the flood discharge; otherwise they will be breached

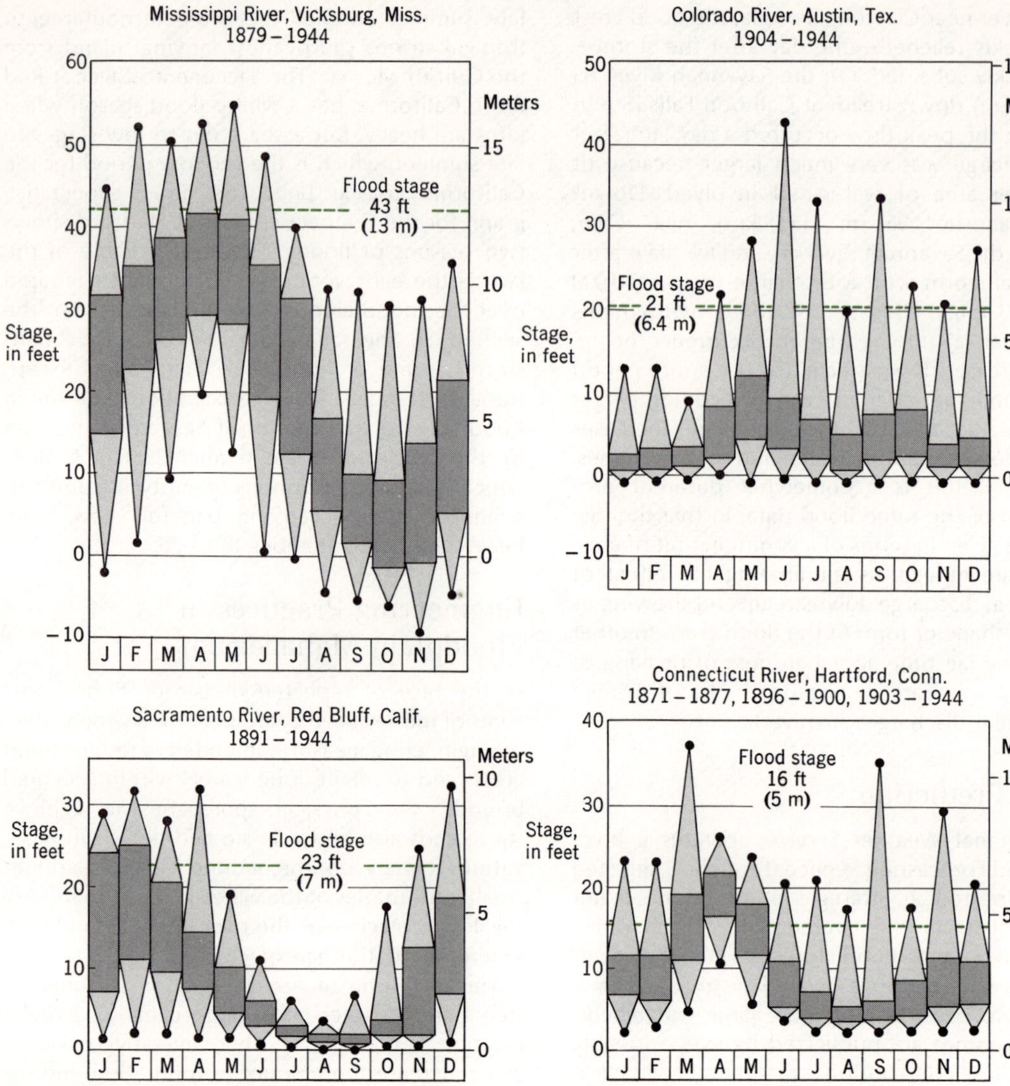

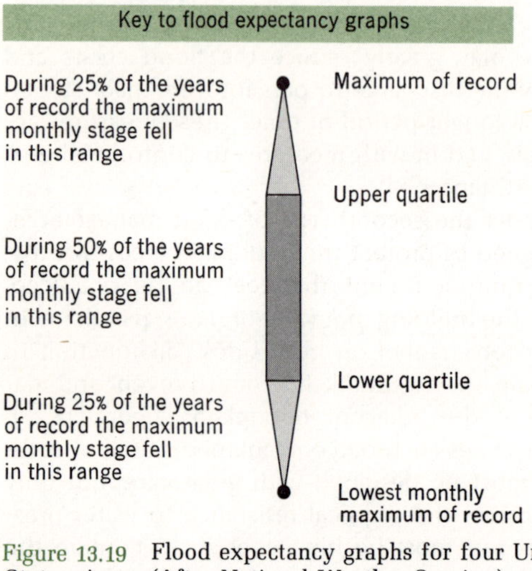

Hydrology of Streams and Lakes

Figure 13.19 Flood expectancy graphs for four United States rivers. (After National Weather Service.)

rapidly by great gaps, termed *crevasses,* at the points where water spills over (Figure 13.21).

Under the control of the Mississippi River Commission, which began in 1879, a vast system of levees was built along the Mississippi River in the expectation of containing floods. Figure 13.20 shows such a levee during the flood of 1903, when it was necessary in Louisiana to add to the top of the levee by means of planks and sand-filled bags for a distance of 71 mi (114 km) to prevent overflow. Levees have been continuously improved and now total more than 2500 mi (4000 km) in length and in places are 30 ft (10m) high.

Because a levee system will prove inadequate for very high floods it is desirable to create *floodways* in the back-swamp zones of the floodplain. These floodways are zones limited by levees and connected by spillways to the main levee system. When a certain critical flood stage is reached, the

334

Figure 13.20 This old photograph shows the artificial levee of the Mississippi River near Greenville, Mississippi, during the great flood of March 1903. A crevasse, or break, at the distant point, X, is discharging flood water into the lower floodplain on the left. (Mississippi River Commission photograph.)

main levees can be artificially breached to divert discharge into the floodways. In other instances, a sill is provided in the main levee and will permit overflow as soon as the sill level is topped. In the lower Mississippi River valley a distributary channel, the Atchafalaya River, has been developed as a floodway leading to the Gulf of Mexico. Above New Orleans the Bonnet Carré spillway provides for diversion of flood discharge into Lake Pontchartrain. Yet another engineering device is the *retarding basin*, produced by construction of a low dam across the river. A series of such basins can retard the flood flow and reduce the flood crest.

The second theory, practiced in recent decades on the Mississippi River by the U.S. Army Corps of Engineers, is to shorten the river course by cutting channels directly across the great meander loops to provide a more direct river flow. These changes are discussed in further detail in Chapter 14. Shortening has the effect of increasing the river slope, which in turn increases the mean velocity. Greater velocity enables a given flood discharge to be moved through a channel of smaller cross-sectional area; the flood stage is correspondingly reduced. Channel improvement had a measurable effect in reducing flood crests along the lower Mississippi.

Management of floodplains can also make use of nonstructural alternatives to elaborate protective engineering works. For example, zoning regulations can be implemented to recognize the reality that inundation will occur in certain low floodplain areas. Such high-risk land can be kept free of construction of homes and factories. Nationally-funded flood insurance is another important aspect of flood protection. The engineering methods we have described represent only one aspect of the complex subject of environmental management of floodplains.

Hydrologic Effects of Urbanization

In the previous chapter we examined certain effects of urbanization upon the ground-water body. An example, the lowering of the water table on Long Island by introduction of sewer systems with ocean outfall, showed how one phase of the hydrologic cycle is thrown out of balance. We will now consider how urbanization affects the surface runoff pathway in the hydrologic cycle.

Hydrologic characteristics of a watershed are altered in two ways by urbanization. First, an

Figure 13.21 This air view, taken in April 1952, shows a break in the artificial levee adjacent to the Missouri River in western Iowa. Water is spilling from the high river level at right to the lower floodplain level at left. (Photograph by Forsythe, U.S. Department of Agriculture.)

increasing percentage of the surface is rendered impervious to infiltration by construction of roofs, driveways, walks, pavements, and parking lots. It has been estimated that in residential areas, for a lot size of 15,000 sq ft (0.34 acre; 1400 sq m) the impervious area amounts to about 25%, whereas for a lot size of 6000 sq ft it rises to about 80%. As you would anticipate, an increase in proportion of impervious surface reduces infiltration and increases overland flow generally from the urbanized area. In addition to increasing flood peaks during heavy storms, there is a reduction of recharge to the ground water body beneath, and this reduction in turn decreases the base flow contribution to channels in the area. Thus the full range of stream discharges, from low stages in dry periods to flood stages is made much greater by urbanization.

A second change caused by urbanization is the introduction of storm sewers that allow storm runoff from paved areas to be taken directly to stream channels for discharge. Runoff travel time to channels is thus being shortened at the same time that the proportion of runoff is being increased by expansion of impervious surfaces. The two changes together conspire to reduce the lag

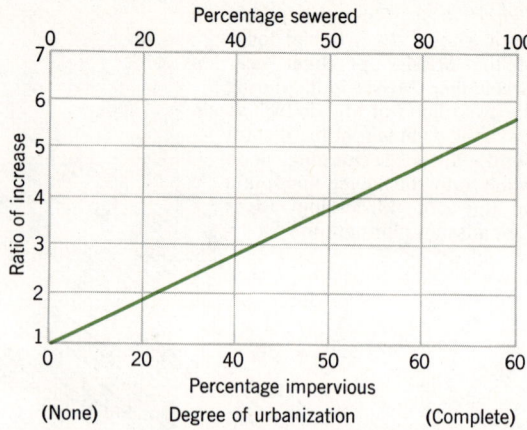

Figure 13.23 Increase in frequency of overbank flows with increase in degree of urbanization. Data are adjusted to a drainage area of 1 square mile. (After L. B. Leopold, 1968, U.S. Geological Survey, Circular 554.)

time, as shown by the schematic hydrographs in Figure 13.22. Figure 13.23 shows how combined effect of increase in sewered area and impervious area is to increase the frequency of occurrence of overbank floods as compared with frequency that existed prior to urbanization. The ratio given on the vertical axis is the increase in yearly number of overbank flows based on a drainage area of 1 square mile. Thus, for 50% impervious area and 50% sewered area the number of overbank flows yearly has increased by a factor of about $3\frac{1}{2}$ times.

Increase in flood peaks results in inundation of areas previously above flood limits except during rare flood events, with attendant water damage to properties adjacent to the channel. If the stream should be free to adjust its dimensions by deepening and widening its channel to accommodate higher flood peaks there would be a serious problem arising from the resulting load of sediment, a factor that is discussed in Chapter 14.

One partial solution to the problem created by storm sewering is to return storm runoff to the ground water body by means of infiltrating basins. This program has been adopted on Long Island, where infiltration rates are high in sandy glacial materials. Another method of disposal of storm runoff is by recharge wells. At Orlando, Florida, storm runoff enters wells that penetrate cavernous limestone. The capacity of the system to absorb runoff without clogging appears to be adequate. In Fresno, California, a large number of gravel-packed wells of 30-in. (76-cm) diameter receive runoff from streets. The system has proved successful in disposing of storm drainage.

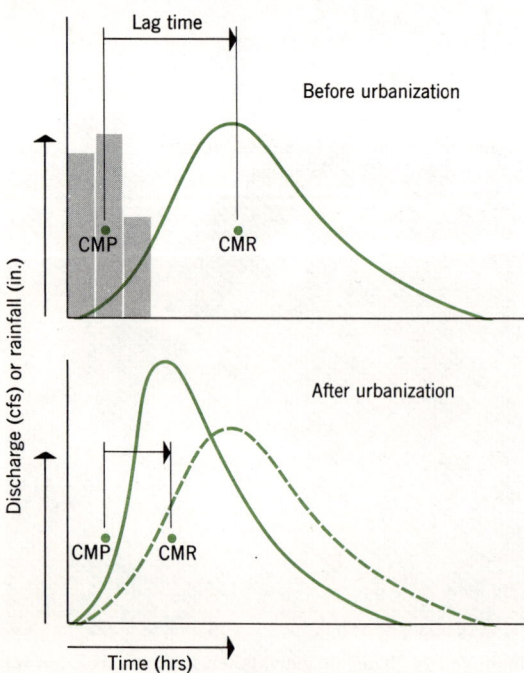

Figure 13.22 Schematic hydrographs showing effect of urbanization as reducing lag time and increasing peak discharge. Points CMP and CMR are centers of mass of rainfall and runoff, respectively, as in Figure 13.10. (After L. B. Leopold, 1968, U.S. Geological Survey, Circular 554.)

Lakes as Physical Features of the Environment

Use of the term *lake* to include all forms of standing water covers a very wide range of kinds of water bodies, having in common only the requirement that they have an upper water surface exposed to the atmosphere and no appreciable gradient with respect to a level surface of reference. Ponds (small, usually shallow water bodies), marshes, and swamps with standing water can be included. Lake water may be fresh or saline, and we may have some difficulty in deciding whether a body of salt water adjacent to the open ocean is to be classed as a lake or an extension of the sea. A practical criterion rules that a coastal water body is not a lake if it is subject to influx of salt water from the ocean. Lake surfaces may, however, lie below sea level, an example being the Dead Sea with surface elevation of -1300 ft (-396 m). The largest of all lakes, the Caspian Sea, has a surface elevation of -80 ft (-25 m). Significantly, both of these large below-sea-level lakes are saline.

Basins occupied by lakes show a wide range of origins as well as a vast range in dimensions. Basins are created by geologic processes and it should not be surprising that there are lakes produced by every category of geologic process: tectonic and volcanic activity; weathering and mass wasting; erosion and deposition by streams, glaciers, waves and currents, and winds. Using these geologic processes as the primary basis of classification of lake basins, we list below some of the important types:

Process	Type of Lake Basin
Tectonic	Examples: Lake Victoria (downwarping) and Lake Nyasa (graben faulting) in East Africa.
Volcanic	Craters of extinct or dormant volcanoes and calderas. Example: Crater Lake, Oregon. Valleys blocked by lava flows. Example: Lake Kivu, East Africa.
Mass wasting	Valleys blocked by landslides. Example: Slide Lake, Madison River, Montana (see Chapter 11).
Chemical weathering	Limestone solution lakes. Example: Deep Lake, Florida
Stream action	Floodplain lakes occupying sections of abandoned river channels, commonly *oxbow lakes*. Countless examples on floodplains of all major rivers of low gradient (see Chapter 14).
	Basins formed by blockage of a valley by alluvial fans (see Chapter 14). Numerous examples of playa lakes in Basin-and-Range region of southwestern United States.
Glacial action	Basins eroded in bedrock by alpine glaciers or blocked by morainal debris in cirques and troughs (see Chapter 17).
	Basins eroded by continental ice sheets or blocked by moraines of those ice sheets. Examples: Great Lakes, Finger Lakes of New York (see Chapter 17). Ice-block collapse cavities in glacial deposits (moraines, outwash plains). Countless examples in glaciated northcentral and northeastern United States and southern Canada.
Shoreline processes	Basins formed behind barrier beaches and bars, or behind coral reefs, and subsequently uplifted by positive crustal movements. Example: Okefenokee Swamp, Georgia and Florida, a depression landward of an uplifted barrier beach.
Wind action	Basins excavated by wind scour or blocked by dune accumulations or both (see Chapter 16). Examples: Lakes in Sand Hills Region of Nebraska; Moses Lake, Washington.
Organic processes (nongeologic)	Basins dammed by accumulation of plant matter. Example: mountain ponds behind beaver dams.

To the above classes we add man-made lakes. These are reservoirs behind dams, or they may be lakes occupying abandoned mine pits and quarries. Most man-made lakes occupy stream-eroded valleys, with the result that many of these reservoirs take on the dendritic branching pattern of the preexisting channel and valley system. As such, they are easily identified from the air.

Since the kinds of lake basins are so numerous and varied and their occurrence covers such a wide range of climate regimes, the environmental role of lakes as habitats of aquatic life is also highly varied.

An important point about lakes in general is that they are for the most part short lived features in terms of geologic time. Lakes disappear from the scene by one of two processes, or a combination of both. First, lakes that have stream channel outlets will be gradually drained as the outlet channels are eroded to lower levels. The principle is illustrated by the catastrophic breaching of landslide dams, such as that of the Madison slide described in Chapter 11. Where a strong bedrock threshold underlies the outlet erosion will be slow, but nevertheless certain. Second, lakes accumulate inorganic sediment carried by streams entering the lake and organic matter produced by plants within the lake.

Lakes also disappear by excessive evaporation accompanying climatic changes. Many former lakes of the southwestern United States flourished in moister periods of glacial advance during the Pleistocene Epoch, but today are greatly shrunken or have disappeared entirely under the present arid regime. This subject is developed further in Chapter 17. A special case of lake disappearance is that of the lowering of the water table by excessive withdrawal of ground water (Chapter 12).

Water Balance of Lakes and Reservoirs

In this chapter our theme is a hydrologic one, related to the surface flow of water in the hydrologic cycle. Consequently, it is appropriate here to look into the water balance of lakes. A lake represents a simple open system with respect to the mass balance of the water itself. A lake has an upper surface boundary exposed to the atmosphere and a lower surface boundary in contact with a solid mineral surface. Water may enter and leave through both of these boundary surfaces. Incoming streams and overland flow from ground surfaces draining into the lake represent point and line sources of water input, while an overflow channel represents an output point. Consequently, we can set up a fairly simple water-balance equation (Figure 13.24). Let the letter I stand for input of water; O, for output of water. The units used may be volume or mass of water per unit time. Units of volume per unit of time are identical with discharge, Q, in stream flow and are the more useful choice, since input and output from stream channels will be gauged in terms of discharge. By means of subscript letters we can differentiate the sources of input and output as follows: r represents runoff by stream flow and overland flow; g represents subsurface flow as ground-water movement; p represents direct fall of precipitation upon the lake surface; e represents evaporation from the free-water surface.

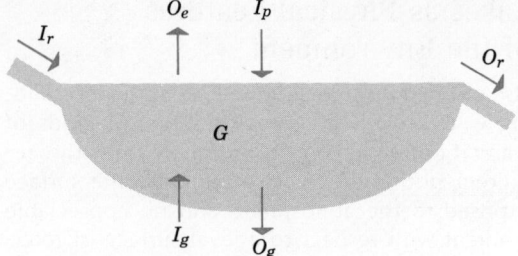

Figure 13.24 Schematic diagram of the water balance of a lake. (Terms are defined in text.)

Finally, let G stand for the net change in storage of water in the lake. The equation can now be written as

$$G = (I_r + I_p + I_g) - (O_r + O_e + O_g)$$
$$\text{Input} \qquad \text{Output}$$

When the system is in steady state, there will be no change in storage, so that the term G becomes zero and drops out. Then

$$I_r + I_p + I_g = O_r + O_e + O_g$$
$$\text{Input} \qquad \text{Output}$$

Actual measurement of each term for a given lake is a most difficult technical problem. Stream gauges at input and output points can give reasonably reliable data on the major elements of surface flow in the system but the increment from direct overland flow is not subject to direct measurement. Groundwater flow is next-to-impossible to measure directly and must be estimated from a knowledge of local gradients of the water table and permeability of the aquifer. Precipitation can be directly measured with a rain gauge. Evaporation from the free-water surface is not easy to measure, since it depends upon water temperature, atmospheric temperature and humidity, barometric pressure, and wind speed. Formulas taking into account these variables are available for estimating evaporation from lakes.

Direct measurement of evaporation from a free-water surface makes use of the *evaporating pan*, which is simply a circular container 4 to 6 ft (1.2 to 1.8 m) in diameter and 10 in. (25 cm) or more in depth. Water is added as required and the amount of surface lowering due to evaporation measured with a gauge (Figure 13.25). Evaporation data are collected by the National Weather Service at many observing stations. The evaporation from the pan is generally greater than from a lake or reservoir, so that the pan readings require correction by a reduction factor ranging from 60% to 80%. Figure 13.26 is a map of the United States showing the average annual evaporation from a free-water surface as measured by

Water Balance of Lakes and Reservoirs

Figure 13.25 An evaporating pan with anemometer at side. Notice a thermometer immersed in the water (*upper left*). The cylindrical device at near edge of the pan is a hook gauge to measure height of water level. (NOAA National Weather Service photograph.)

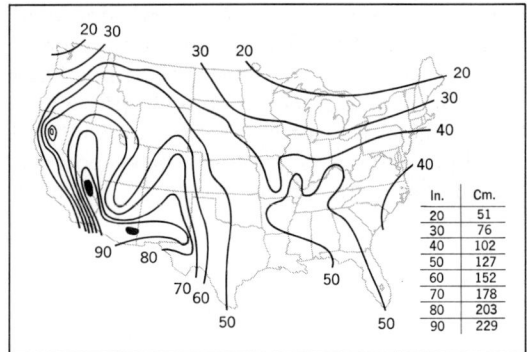

Figure 13.26 Annual evaporation from a free-water surface in the United States. (After Mead.)

the pan method. Notice the high values in the hot desert regions of the southwestern United States in contrast to low values in the cool northeastern region. Listed below are some representative figures giving annual total evaporation for actual reservoir surfaces:

	Inches	Cm
Gardiner, Maine	24	61
Birmingham, Alabama	43	109
El Paso, Texas	71	180
Tucson, Arizona	60	152
San Juan, Puerto Rico	55	140
Gatun, Canal Zone	48	122
Atbara, Sudan	124	315

Lakes with no outlet are characteristic of arid regions in which there is a yearly strong water deficiency. The evaporation term, O_e, represents the dominant or entire output of the system and, of course, the term O_r is zero and drops out entirely. Since dissolved solids are brought into the lake by streams—usually ones that head in distant highlands where a water surplus exists—and there is no surface outlet, the solids accumulate with resultant increase in salinity of the water. Eventually, salinity levels reach a point where salts are precipitated in the solid state as evaporites (Chapter 8).

Evaporation control is a subject of major importance in conserving the water supplies in a man-made reservoir, particularly where the reservoir is situated in a region of arid climate having a large annual water deficit. This situation occurs where an exotic river is dammed. (An *exotic river* is one that is sustained in its flow across an arid region through runoff derived from a distant region of water surplus.) The Colorado River in Arizona is a good example. A large reservoir, such as Lake Mead behind Hoover Dam, presents an enormous water surface exposed to intense evaporation. Since the input of stream water is finite, a reservoir may be designed with such a large capacity that it will never completely fill, because there is a point at which annual evaporation equals the annual input.

Evaporation from reservoirs in arid lands not only constitutes a loss of water resources, but has the effect of raising the salinity of water released into river flow below the reservoir. For example, salinity of the Colorado River below Hoover Dam is about 700 ppm, but increases to 900 ppm at Yuma, where water is diverted into the All-American Canal for use in the Imperial Valley of California. Increase in salinity results largely from evaporation in Lakes Mohave and Havasu, which are desert reservoirs lying between Hoover Dam and Yuma. Such increases in salinity can have adverse effects upon croplands using river water for irrigation in downstream areas. Evapotranspiration losses from the irrigated surfaces further raise the salinity of soil water and ultimately the soil is ruined for crop cultivation. Increase in salt content in soil water is referred to as *salinization*; it has proved a costly penalty for development of irrigation projects in arid lands. The prevention or cure for salinization is to flush down soil salts into lower levels by use of more water, a process that requires much greater water use than for crop growth alone. Furthermore, when salts are flushed down to lower levels, they may enter the ground water body and contaminate it with toxic ions, such as nitrate.

Reduction of evaporation from reservoir surfaces has been approached through the application of a synthetic chemical compound that will spread over the surface as a film of monomolecular thickness, cutting down on the water evaporation. *Hexadecanol*, a fatty alcohol, is one such compound that has been used. One problem is that the chemical must be nontoxic to plant and animal life. Another is that losses by evaporation and solution of the chemical require that it be continually replaced. On large reservoirs the film is easily broken up by winds and rain. Nevertheless, experimental applications have achieved a reduction of evaporation of about 10% on large reservoirs to as high as 25% on small ponds on the order of 0.1 to 1.0 acre (0.04 to 0.4 hectare). The saving in water must be weighed against the cost of the method.

Dissolved Solids in Surface Waters

Runoff from the lands to the oceans carries important quantities of dissolved inorganic matter in the form of ions. We discussed this subject in Chapter 8 in connection with the origin of salts in sea water and found that ions of calcium, magnesium, sodium, and potassium are normally present in stream waters (Table 8.5). These base cations were originally derived from breakdown of silicate minerals of igneous rocks, but they are recycled through complex pathways. Ions in stream water and in lakes serve as tracers of the flow paths of runoff, and as such are an appropriate topic to place in this chapter. Because various ions enter surface waters with the precipitation itself, we must first investigate the chemical composition of rainwater, including melted snow.

We may like to think of rain that falls in sparsely settled humid regions, far from urban sources of pollution and desert sources of dust, as being "pure" water. In fact, however, all rainfall, wherever it occurs, carries down with it a variety of ions, some introduced into the atmosphere from the sea surface, some from land surfaces undisturbed by Man, and some from man-made sources.

Sea salts enter the atmosphere as minute droplets detached from wave crests. These droplets evaporate, leaving salt residues which are carried upward in turbulent winds. Sea salts are distributed throughout the entire troposphere and are a major form of nuclei of condensation (Chapter 5). When carried down to earth as rain, sea salts contribute all of the ions that are present in sea water, but as you can see from the data of Table 8.4, most of the ions contributed will be chlorine (Cl^-) and sodium (Na^+). Ions of magnesium (Mg^{++}), sulfate (SO_4^{--}), calcium (Ca^{++}), and potassium (K^+), will be contributed in minor amounts.

The contribution to rainwater of ions from sea water will, of course, be much higher along the oceanic coasts than far inland, because local sea spray is incorporated into rain. Table 13.1 shows this effect quite well. Rainwater sampled at two coastal stations, Cape Hatteras, North Carolina, and Brownsville, Texas, show high levels of both sodium and chloride ions, in contrast to low values for the two inland stations. However, the concentration of these ions in the rainwater of Cape Hatteras is less than one-third as great as at Brownsville, largely because of dilution in the much larger quantity of rain that falls annually at Cape Hatteras.

Mineral dusts lifted from the ground into the atmosphere and carried upward in turbulent wind account for much of the potassium, calcium, and magnesium found in rainwater. (Table 13.1). (Magnesium content is not shown in the table.) Calcium content is relatively high at Brownsville and Ely, Nevada, both located in regions of semiarid climate, and reflects the large amount of

TABLE 13.1 Average Chemical Composition of Rainwater at Selected Localities

Place	Distance from Ocean (mi)	(km)	Mean Annual Precipitation (in.)	(cm)	Ions Present (ppm)						
					Na^+	K^+	Ca^{++}	Cl^-	SO_4^{--}	NO_3	NH_4^+
Cape Hatteras, N. C.	0	(0)	54	(140)	4.5	0.2	0.4	6.5	0.9	1.0	0.1
Brownsville, Tex.	1	(1.6)	25	(64)	22.3	1.0	6.5	22.0	5.3	1.8	0.3
Ely, Nev.	410	(660)	15	(38)	0.7	0.1	3.8	0.3	1.0	0.8	0.4
Columbia, Mo.	650	(1050)	40	(100)	0.3	0.3	2.2	0.1	1.2	3.8	0.4

SOURCE: C. E. Junge and R. T. Werby, 1958, *Journal of Meteorology*, vol. 15, pp. 417–425.
NOTE: Figures have been rounded off to one decimal place.

calcium carbonate (calcite) and calcium sulfate (gypsum) in soils and alluvial sediment of dry lands.

Sulfate ion (SO_4^{--}) in rainwater comes largely from sulfate particles and gaseous sulfur compounds (largely H_2S and SO_2) injected into the atmosphere by fossil fuel combustion, forest fires, volcanoes, and biological activity (Chapter 6). Locally, sulfate may be high because of abundance of calcium sulfate (gypsum) in soil dusts. This effect may explain the high sulfate concentration observed at Brownsville (Table 13.1). Nitrate ions (NO_3^-) and ammonium ions (NH_4^+) are produced from gaseous forms of nitrogen introduced into the atmosphere from a variety of sources, including combustion of fuels, decay of organic matter, and fertilizers. It has often been stated that nitrate is produced by lightning in thunderstorms, but recent research suggests that this source is unimportant. Phosphate ions (PO_4^{-3}) are also present in rainwater, but in much smaller amounts than ammonium and nitrate. Hydrogen ions (H^-) and bicarbonate ions (HCO_3^+) are important constituents of rainwater, but are not included in our discussion; they are products of the solution of atmospheric carbon dioxide in rainwater, as explained in Chapter 8.

Upon entering the soil, ions of rainwater are selectively removed or detained, resulting in a different array of ion concentrations in soil water. Nitrate in normal concentrations is largely removed by plants, yielding bicarbonate ions in the following reaction:

$$H_2O + CO_2 + NO_3^- \rightarrow HCO_3^- + \text{organic nitrogen}$$

Concentrations of sulfate and chloride ions are not appreciably changed in the soil. The base cations may be absorbed by soil colloids in the cation exchange process, explained in Chapter 11. However, unless the soil chemistry is undergoing a change, we may expect that on the average the soil is not experiencing a net gain or loss of base cations. If so, the principal base cations (Na^+, K^+, Ca^{++}, Mg^{++}) arriving in rainwater will be discharged into surface runoff and to ground water without significant depletion. In addition, decomposition of minerals of the soil parent matter and bedrock will gradually release various ions and silica (SiO_2) to runoff. The actual chemical composition of stream water and ground water thus varies considerably according to the type of rock present beneath the soil. Stream water derived from areas underlain by felsic igneous and metamorphic rocks has a high silica and calcium content, whereas limestone bedrock yields water rich in both calcium and carbonate ions, but low in silica. Water derived from bedrock rich in gypsum (hydrous calcium sulfate) shows a very high concentration of sulfate ion as compared with water from other rock types; the same is true of water from some shale areas.

Based on extensive chemical analyses made by the U.S. Geological Survey, concentrations of ions in major rivers of the United States are known in considerable detail. Table 13.2 lists dissolved solids in three major U.S. rivers, selected to span a wide range of environments, and so to display

TABLE 13.2 Principal Ions in Major United States Rivers

Ion		Savannah River, Georgia (ppm)	(%)	Mississippi River, Minneapolis, Minnesota (ppm)	(%)	Colorado River, Grand Canyon, Arizona (ppm)	(%)
Anions							
Chlorine	Cl^-	2½	5	2	½	90	10
Sulfate	SO_4^{--}	3	6	18	6	320	36
Bicarbonate	HCO_3^-	22	45	190	65	200	22
Nitrate	NO_3^-	⅓	1	2	1	5	½
Cations							
Sodium	Na^+	3½	7	10	3½	120	13
Potassium	K^+	1	2½			6	1
Calcium	Ca^{++}	4	9	40	14	94	10
Magnesium	Mg^{++}	1⅓	3	14	5	34	4
Total dissolved solids[a]		50	100	290	100	900	100

SOURCE: U.S. Geological Survey, Professional Papers 236, 638, 889-E.
NOTE: Figures rounded off.
[a] Total includes silica (SiO_2) and iron (Fe).

a wide range in relative abundance of the constituents as well as in the total concentration of dissolved solids in the water. These data have environmental significance in a number of ways. First, the ions furnish nutriment to aquatic plants at the bottom of the food chain and are required for tissues of aquatic animals as well. Excess quantities of certain of the ions, for example nitrate, introduced by Man's activities, may act as an abnormally large nutrient supply. Second, by comparing natural abundance of ions in stream flow with abundances in rainwater, we can find out what ions are exchanged within the soil. Downstream changes in ion concentrations will allow us to calculate the input of various chemical pollutants produced in industrial processes and in urbanized areas.

Although we shall need to be very cautious about reading into the figures of Table 13.2 a variety of unwarranted interpretations, certain points are worth nothing. First the total concentration of dissolved solids spans a wide range from a low value for the Savannah River, where a large water surplus is produced and temperatures are warm much of the year, through an intermediate value for the Mississippi River in an interior region of small water surplus and low winter temperatures, to a high value for the Colorado River draining a largely semiarid region with prevailing water deficit.

In a region of water surplus ion concentrations are low because of dilution; in a region of water deficiency, evaporation results in ion concentration. On the other hand, the bicarbonate ion is relatively more abundant in the humid environment than the arid because of the greater action of carbonic acid upon carbonate rocks. The drainage basin of the Colorado River contains vast areas of sedimentary rocks, including strata bearing gypsum (calcium sulfate) and halite (NaCl). Thus geological characteristics of the region may account for the large proportions of sulfate, chloride, and sodium ions in Colorado River water. Notice that chlorine is very low in relative concentration in the upper Mississippi River. Assuming that most of the chlorine of river waters is brought into the system in rainwater (from sea salts carried inland), the remoteness of the Mississippi River watershed from the ocean may explain the low value.

The sulfate ion is derived not only from solution of gypsum, but also from breakdown of iron sulfide, occurring commonly as the mineral pyrite in sedimentary rocks. As we shall find in later paragraphs, abnormally high proportions of the sulfate ion may indicate contamination from mining and industrial processes.

Phosphorus, as phosphate ion (PO_4^{-3}), is not listed in Table 13.2 because its concentration is normally very small—about one-twentieth that of nitrate. Phosphorus is essential to plant growth and is readily taken up by microscopic plants of lakes and streams. High phosphate concentrations are indicative of man-made pollution from fertilizers and detergents.

Effects of Urbanization on Quality of Surface Waters

Urbanization not only influences flow characteristics of runoff, but causes changes in water quality, as seen in changes in concentrations and proportions of the eight common ions listed in Table 13.2. Although research on the nature and extent of these changes is yet in its exploratory stages, a preliminary study carried out by Professor Wallace S. Broecker and associates of the Lamont-Doherty Geological Observatory of Columbia University has given us some highly significant information about the problem. The study area is a suburban region in northern Bergen County, New Jersey, and southern Rockland County, New York. The area lies immediately northwest of New York City, on the opposite side of the Hudson River. The study area includes a stream system and lakes with a watershed area about 17 mi (27 km) long and 9 mi (14 km) wide. Runoff leaves the area through a single trunk stream, the Hackensack River.

As a first step in tracing ions in stream flow, the composition of rainwater was determined. Anions were found present in the following typical proportions:

Anion	ppm
Cl^-	13
SO_4^{--}	40
HCO_3^-	Negligible
NO_3^-	25

Chlorine ions in rainwater are from sea salts suspended in the atmosphere, but another possible source is from crystals of deicing salts swept into the air from streets and highways. Sulfate and nitrate ions come from atmospheric pollutants typical of any large industrial city (see Chapter 6). This origin is confirmed by the fact that the concentration of these ions diminishes with increasing distance from the city.

Once the rainwater comes in contact with the soil, important changes in ion composition take place, as already explained. The resulting total ion composition of surplus soil water, after modifica-

TABLE 13.3 Calculated Ion Inputs into Bergen County Surface Waters

	Cl^-	SO_4^{--}	HCO_3^-	NO_3^-	Na^+	K^+	Ca^{++}	Mg^{++}
1. Residual rain modified by soil contact	15	44	30	0.5	7	1	14	6
2. Deicing salt	50	——	——	——	30	——	1	——
3. Lawn lime	——	——	20	——	——	——	6	1
4. Lawn fertilizer	——	10	——	15	——	3	——	1
5. Subtotals (75%)	65	54	50	15	37	4	21	7
6. Sewage effluent (25%)	160	160	210	(20)	150	20	40	16
7. Totals (weighted averages)	90	72	94	16	68	8	26	9

SOURCE: W. S. Broecker, 1970, Lamont-Doherty Geological Observatory of Columbia University, New York. Used by permission.
NOTE: Figures give concentrations in parts per million (ppm).

tion in the soil, is given in the top line of Table 13.3, which estimates the inputs of ions into runoff as it passes through the suburban system.

The second line of Table 13.3 shows the large input of chlorine and sodium ions calculated from known use of deicing salts on roadways and highways during the winter. A minor contribution of calcium ions in this category comes from calcium chloride used for deicing. The third line of the table shows the calculated input of bicarbonate, calcium, and magnesium ions from lawn lime, largely pulverized limestone of calcium-carbonate composition (mineral calcite). Calculated input from lawn fertilizers, line 4, increases the concentration of sulfate, nitrate, potassium, and magnesium ions. Line 5 gives subtotals for the first four lines.

Sewage effluent is next taken into consideration. About five-eights of this effluent is partially treated sewage from a single plant located on the trunk stream below the exit point from the study area. The remainder is released from individual septic tanks through cesspools and leaching fields. Line 6 of Table 13.3 shows the calculated concentration of ions within the effluent. Since the volume of effluent is estimated to comprise about 25% of the total runoff, the effluent values are combined with the subtotals on line 5 by means of a weighted average (75% to 25%); the final concentrations are given in line 7. While all concentrations are substantially greater than in soil water, the 30-fold increase of nitrate is especially striking, in view of the nutrient value of nitrogen to aquatic plants. Because the data are from a preliminary phase of the Columbia research project they should be treated as tentative. However, they give a good idea of the relative importance and orders of magnitude of input of the common ions into surface runoff as a result of Man's urbanization of the environment. The study illustrates a valuable application of the science of *geochemistry* to environmental problems.

Other Forms of Chemical Pollution of Surface-Water Supplies

In addition to pollution of streams and lakes by deicing salt, lawn conditioners, and sewage effluent in urban and suburban areas, major sources of pollution are associated with mining and processing of mineral deposits. In addition, there is thermal pollution from discharge of coolant waters from nuclear generating plants, and the possibility of contamination from radioisotopes produced in those same forms of activity described in Chapter 12 as sources of ground-water contamination.

Taking the last possibility first, as a follow-up of the problems of nuclear waste disposal covered in Chapter 12, we had noted that low-level wastes can be treated and diluted to a degree that reduces background radiation to safe levels and permits the wastes to be passed into the environment, meaning surface water flow systems or into the ocean directly.

The AEC has undertaken studies to evaluate the effects of low-level waste disposal upon the environment into which the diluted liquids are dispersed. One study area is the Clinch-Tennessee River system below Oak Ridge, Tennessee. Physical, chemical, and biological effects of the radioactivity are evaluated. Water-sampling stations extend as far as 125 mi (38 km) downstream from the Oak Ridge plant.

A particular form of chemical pollution of surface water goes under the name of *acid mine drainage* and is an important form of environmental degradation in parts of Appalachia where abandoned coal mines and strip mine workings are concentrated. Ground water emerging from abandoned mines, as well as soil water percolating through strip-mine spoil banks, is charged with sulfuric acid and various salts of metals, particularly of iron. The sulfuric acid is formed by reaction of water with iron sulfides, particularly the mineral pyrite (Fe_2S), which is a common constituent of coal seams. Acid of this origin in stream waters can have adverse effects upon animal life. In sufficient concentrations it is lethal to certain species of fish and has at times caused massive fish kills. The acid waters also cause corrosion to boats and piers. Government sources have estimated that nearly 6000 mi (10,000 km) of streams in this country, together with almost 40 sq mi (100 sq km) of reservoirs and lakes are seriously affected by acid mine drainage. The acid may be gradually neutralized by reaction with carbonate rocks, but other chemical pollutants may persist. One particularly undesirable byproduct of acid mine drainage is precipitation of iron to form slimy red and yellow deposits in stream channels.

Sulfuric acid may also be produced from drainage of mines from which sulfide ores are being extracted, and from tailings produced in plants where the ores are processed. Such chemical pollution also includes salts of various toxic metals, among them zinc, lead, arsenic, copper, and aluminum. We have already taken note in Chapter 12 of the chemical pollution of streams by radium derived from the tailings of uranium ore processing plants.

Chemical pollution by direct disposal into streams and lakes of wastes generated in industrial plants is a process that requires no elaborate physical explanation. It is a phenomenon well known to the general public and can be seen first-hand in almost any industrial community in the United States. Direct outfall of sewage, whether raw (untreated) or partially treated, is another form of direct pollution of streams and lakes that does not need to be elaborated upon, for it, too, is a commonplace phenomenon for all to see (and smell).

Presence of sulfur in stream water is an important index of industrial pollution generally, since it is produced in large amounts in the combustion of fossil fuels used in industrial energy conversion plants. Sulfur emitted into the atmosphere as SO_2 is oxidized into SO_4 and returned to the ground by washout and fallout where it enters the runoff system. Sewage effluent and fertilizers also contribute much of the sulfur to runoff.

A recent study of runoff data on a global scale by Robert A. Berner provides some revealing statistics on the industrial component of sulfur in the total flow of sulfur from land to the sea. Table 13.4 shows results of this study for five continents (Australia is omitted because of the small quantities it produces.) The chlorine ion serves as a reference quantity, since it is not appreciably increased by industrialization. In South Africa, South America, and Asia, which have low levels of industrial activity, chlorine and sulfate ion concentrations are about equal. In contrast, Europe and North America have much higher concentrations of sulfate ion than of chlorine ion. The excess sulfate concentration is attributed to

TABLE 13.4 Sulfate Ion Flow from Continents to Oceans

Continent	Ion Concentration (ppm)			Flow to Oceans (millions of tons per year)	
	Total Cl^-	Total SO_4^{--}	Pollutant SO_4^{--}	Natural SO_4^{--}	Pollutant SO_4^{--}
Europe	6.9	24	17	17	45
North America	8.0	20	12	37	55
South America	4.9	4.8	0	39	0
Africa	12.1	13.5	0	81	0
Asia	8.7	8.4	0	94	0
			Total	268	100

SOURCES: R. A. Berner, 1971, *Jour. Geophys. Research*, vol. 76, p. 6598, Table 1. Data for chlorine and sulfate concentrations from D. A. Livingstone, 1963, U.S. Geological Survey, Professional Paper 440-G.

industrial sources (third column, labeled "Pollutant SO_4"). The table then gives total yearly mass rates of sulfate discharge as contributed by the natural environment and by pollutant sulfate. Roughly one-third of the world discharge of sulfate ion from lands into oceans is apparently a result of Man's activities.

Toxic metals, among them mercury, and products of detergent wastes—for example phosphates—along with pesticides and a host of other chemicals are introduced into streams and lakes in quantities that are locally damaging or lethal to plant and animal communities. In addition, sewage introduces live bacteria and viruses that are classed as biological pollutants; these pose a threat to health of Man and animals.

Thermal Pollution of Surface Waters

Thermal pollution is a term applied generally to the discharge of heat into the environment from combustion of fuels and from nuclear energy conversion into electric power. We have covered thermal pollution of the atmosphere and its effects in Chapter 6. Thermal pollution of water is different in its evironmental effects because it takes the form of heavy discharges of heated water locally into streams, estuaries, and lakes. The thermal environmental impact may thus be quite drastic in a small area.

Nuclear power plants require the flow of large quantities of water for cooling. Other industrial processes also rely upon water as a coolant. Large air conditioning plants also use water to cool the condensing coils. A cooling system may be designed to recycle the coolant, causing the heat to be dissipated into the atmosphere, the principle being the same as the water-cooled automobile engine and most small air conditioners. In this case the heat load is placed on the atmosphere by radiation and conduction. Evaporation of water may be used as a means of cooling in recycling systems, but this requires an input of water to replace that lost by evaporation. A third method is to withdraw cold water from a lake or stream or from the ground water body, pass it through the cooling system, and discharge the heated water into the same source body. For example a large air-conditioning plant may use cold ground water and dispose of the heated water into recharge wells. Commonly, such disposal by well recharge is required by law to conserve ground water.

Nuclear power plants have large cooling requirements, with the result that they may be sited near a large body of cold water. For example, a proposed nuclear generating plant on Lake Cayuga, a deep finger lake in New York State, will require the pumping of 750 million gallons (2.85×10^9 liters) per day from the cold lower depths (hypolimnion) of the lake, where water temperature is on the order of 43° F (6° C). This water will be discharged into the surface layer of the lake at a temperature of about 70° F (21° C). Environmentalists fear that this infusion of warm water may delay the annual fall cooling of the surface layer (epilimion), which in turn will delay the annual overturn of lake water and lengthen the period of time that the lower water remains stagnant (see Chapter 3). It is also feared that the bringing to the surface of nutrients present in the deep water will increase plant growth in the surface layer and lead to an oxygen deficiency.

The general physical effects of abnormal increase of water temperature in lakes and streams relate to the solubility of gases and the rates at which chemical reactions take place. As explained in Chapter 8, the concentration of atmospheric gases dissolved in water decreases as temperature increases (see Table 8.1). Thermal pollution thus tends to reduce the quantity of oxygen available to decompose organic matter. At the same time, the higher temperature accelerates the rates of certain chemical reactions, such as oxidation, making greater demands for dissolved oxygen at a time when it is available in lesser quantities.

It is appropriate here to point out that the forms of pollution we have listed here often do severe damage to the esthetic qualities of landscapes of streams and lakes. You must decide what value to place upon these intangible resources and be prepared to treat esthetic degradation on equal terms with economic losses incurred by deterioration of water quality.

Our National Water Resources

We close this chapter with an overview of our national surface- and ground-water resources and a look into the future. Two sets of figures make up the data. First is *withdrawal*, representing the quantity of water actually taken from surface-water and ground-water sources. Much of this water is returned after use to those same sources, so that the quantity of water actually lost to the atmosphere through evaporation and transpiration is a second quantity, the *consumptive use*. Note that withdrawal of ground water in many areas of the country greatly exceeds the rate of recharge, and thus represents a net decrease in the ground-water storage. We did not allow for this depletion in the water-balance equation given in Chapter 12.

Hydrology of Streams and Lakes

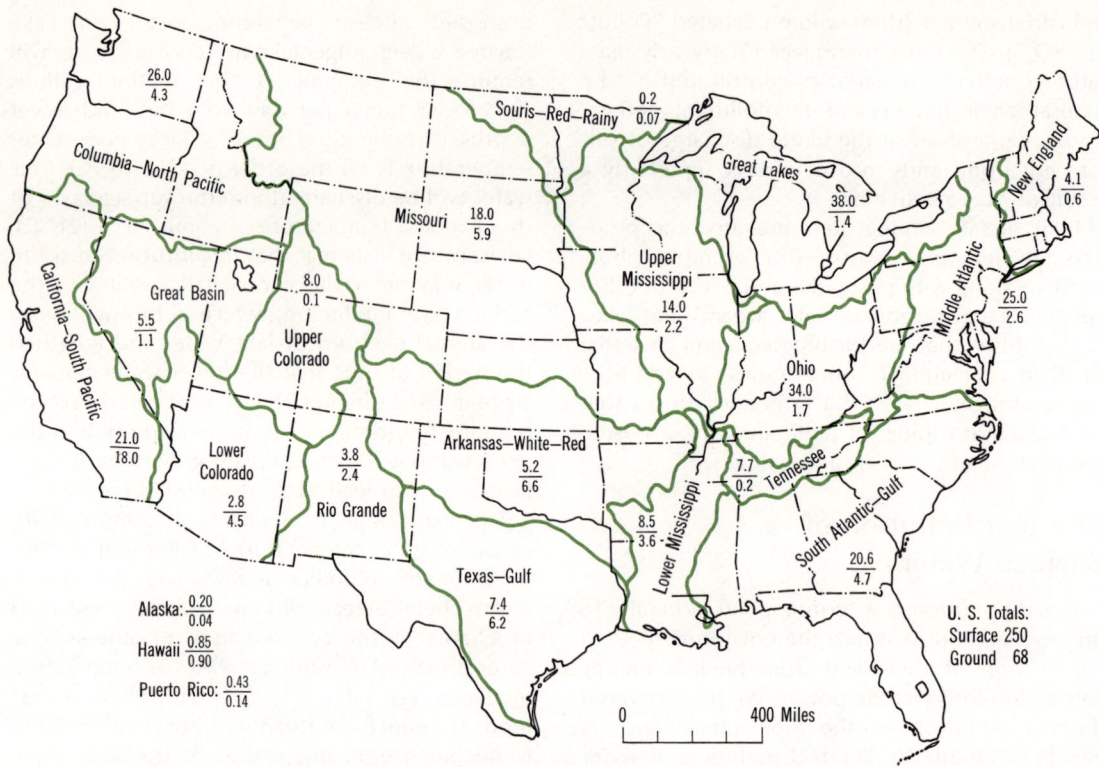

Figure 13.27 Surface-water (*upper number*) and ground-water withdrawal (*lower number*) for water-use regions of the United States in 1970. Units are billions of gallons per day. Water withdrawn for hydroelectric power is not included. (Data of C. R. Murray and E. B. Reeves, 1972, Geological Survey, Circular 676, Table 17, p. 36.)

Figure 13.27 shows average daily United States water withdrawal for the year 1970. The nation is divided up into *water-use regions,* named on the map. Within each region, figures give daily water withdrawal from surface-water (upper figure) and ground-water sources (lower figure). Water used for hydroelectric power is not included. The ratio of the two figures within each region is highly revealing as to water-source differences across the nation. For the eastern states and the Pacific northwest surface water is overwhelmingly the dominant source of water withdrawal. However, in the southwest the balance shifts in favor of ground water. Table 13.5 summarizes total United States water withdrawal and consumptive use for both public water supplies and rural supplies. Rural supplies are used for domestic and livestock needs. Irrigation use is not included, nor is self-supported industrial use (i.e., industries having their own water supplies). Public water supplies use surface water in twice the amount of ground water; of the total only about 20% is lost in consumptive use. Rural supplies use ground water in ratio of four times as much as surface water and the consumptive use is proportionately very high—about 75%.

When we turn to examine water use by industry and irrigation, we find that the total United States rates of withdrawal in both categories are much greater than that for public supplies. Figure 13.28 shows withdrawals for public supplies (upper number), self-supplied industrial use (middle number), and irrigation (lower number), by the same water-use regions as in Figure 13.27. As we would expect, irrigation withdrawal is very small in the humid eastern half of the nation, but large in the western states. Of the self-supplied industrial withdrawal of fresh water, by far the greatest amount is used by electric utilities. Table 13.6 shows industrial withdrawal and consumptive use of fresh water in 1970. Since most of this water is used for driving turbines and for cooling, the proportion of return to source is very large, and the total consumptive use is only about 3% of withdrawal.

Figure 13.29 shows estimated future withdrawal for 1980 and 2000 as compared with 1960. We can expect withdrawals to increase only moderately for public supplies and for combined irrigation and rural use, in contrast to an enormous increase in the industrial requirement. Figure 13.30 shows consumptive use projected to 1980 and A.D. 2000

Figure 13.28 Water withdrawal for public supplies (*upper number*), self-supplied industrial use (*middle number*), and irrigation use (*lower number*) in the United States in 1970. Units are billions of gallons per day. (Data of C. R. Murray and E. B. Reeves, 1972, U.S. Geological Survey, Circular 676.)

in the same use categories as in Figure 13.29. Public supplies will hold to a constant percentage of consumptive use while that use is about doubled in the year 2000. Industrial consumptive use will increase enormously, but the percentage with respect to withdrawal will remain small—under 4%. Consumptive use in the rural and irrigation category (mostly irrigation) is so much larger than the other two categories that it requires a different scale on the graph. Consumptive use in irrigation is a very large percentage of withdrawal, as we would expect, and this percentage will rise to nearly 70% by the year 2000, as consumptive use is about doubled over the 1960 figure.

Assuming that the above estimates are fairly accurate, will we have enough fresh water to meet future demands? To get at the answer we may go back to the water-balance equation and find the mean annual runoff leaving the boundaries of the United States. The assumption will be that we cannot count on removing any ground water from storage. Consequently all ground-water withdrawal must equate to annual ground-water recharge. In steady state of the hydrologic cycle, the amount of water annually added to surface water by seepage from the water table into streams and lakes will equal the annual recharge. This being the case, we need to deal only with the annual runoff, which is estimated to be 480 cu mi (2000 cu km) per year. This figure converts to about 1400 billion gallons (b.g.) (5300 billion liters) per day. Compare this figure with the total estimated water withdrawal in the year 2000, which is about 900 b.g. (3400 b.l.) per day. Assuming that the consumptive use in A.D. 2000 will average 20% of the total withdrawal, we will be actually consuming (losing to the atmosphere) 180 b.g. (580 b.l.) per day, which is only a small fraction of the runoff. Recycling procedures can be developed to reduce even further the percentage of runoff required to be withdrawn.

This analysis suggests that for the nation as a whole there is an adequate fresh water supply for the forseeable future. For individual regions of the country, however, we can anticipate severe stortages that must be remedied by transfer of water from regions of surplus. The enormous concentration of persons and industry into a small area, as for example Los Angeles County and the New York City region, require development of surface water facilities in distant uplands, but it is only by continual and costly expansion of these

TABLE 13.5 United States Water Use from Public and Rural Supplies in 1970

	Withdrawal (billion gallons per day)	Consumptive Use (billion gallons per day)	Percent of Withdrawal
Public water supplies			
Surface water	18.0 (68)		
Ground water	9.4 (36)		
Total	27.4 (104)	5.9 (22)	21.5%
Rural water supplies			
Surface water	0.9 (3)		
Ground water	3.6 (14)		
Total	4.5 (17)	3.4 (13)	75.6%

SOURCE: C. R. Murray and E. B. Reeves, 1972, "Estimated Use of Water in the United States in 1970," U.S. Geological Survey Circular 676.
NOTE: Figures in parentheses give billions of liters per day.

TABLE 13.6 United States Industrial and Irrigation Water Use in 1970

	Withdrawal (billion gallons per day)	Consumptive Use (billion gallons per day)	Percent of Withdrawal
Electric utilities			
Surface water	120.0 (456)		
Ground water	1.4 (5)		
Subtotal	121.4 (461)	0.8 (3)	0.6%
Other industry			
Surface water	31.0 (118)		
Ground water	8.0 (30)		
Subtotal	39.0 (148)	4.1 (16)	10.5%
Total industrial use	160.4 (609)	4.9 (19)	3.1%
Irrigation			
Surface water	81.0 (308)		
Ground water	45.0 (171)		
Total irrigation use	126.0 (479)	95.0 (361)[a]	75.4%

SOURCE: Same data source as Table 13.5.
NOTE: Figures in parentheses give billions of liters per day.
[a] Includes conveyance losses by evapotranspiration.

facilities that shortages can be avoided. In other words, the real problem of supply is not in the total quantity of fresh water available but in its distribution. Compounding distribution problems are those of water pollution. If we continue to degrade the quality of our fresh water supplies we will need to resort increasingly to expensive water treatment procedures necessary to keep the water usable.

For those areas experiencing or anticipating water shortages, certain alternatives exist to going farther afield in search of new supplies. One is to reduce per-capita water use by increasing the water cost to the consumer, with the anticipation that wasteful uses will be curtailed. Another means of saving water is to reduce leakage from municipal distribution systems. Leakage is estimated to run from 10 to 30% in many city water systems.

For arid and semiarid lands the maintenance of existing irrigation systems and their further expansion by reaching out to new and more distant runoff sources has been questioned in recent years. The grounds for such questioning are that

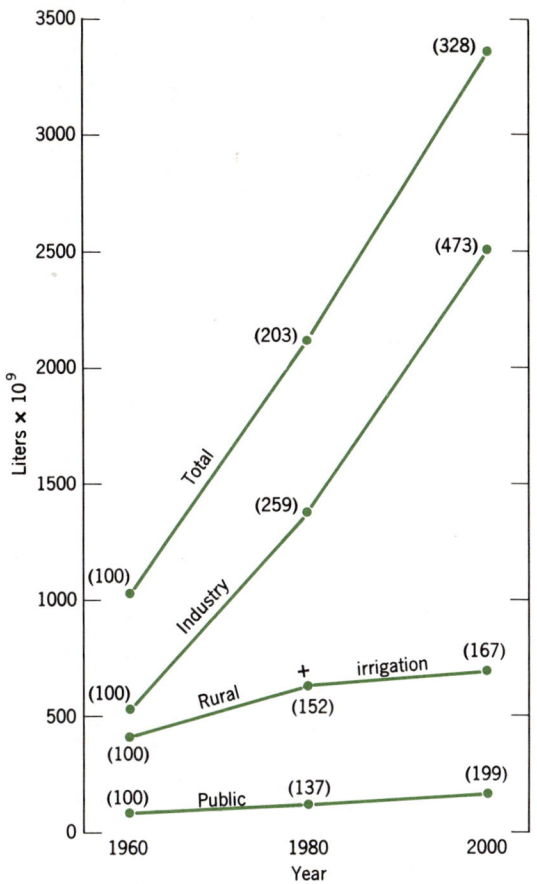

Figure 13.29 Estimated future water withdrawal in the United States. The figures in parentheses give percentage increase over 1960 values. (Data of C. R. Murray, 1968, U.S. Geological Survey, Circular 556; A. M. Piper, 1965, U.S. Geological Survey Water-Supply Paper 1797.)

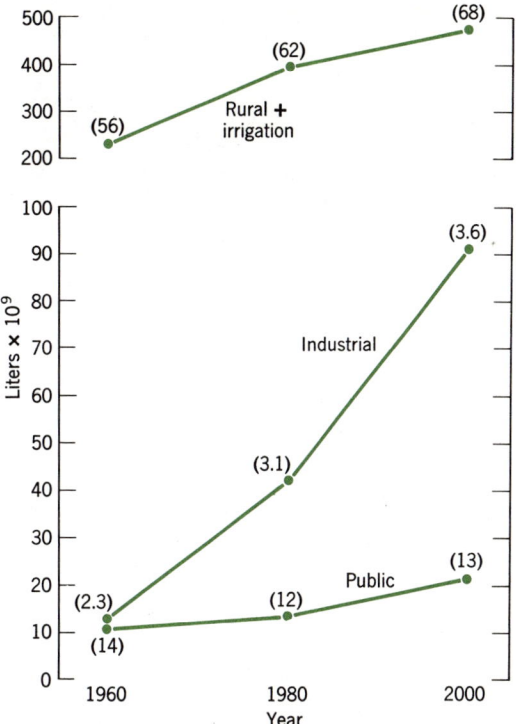

Figure 13.30 Estimated future water consumption in the United States. Figures in parentheses give percentage of estimated withdrawal. (Same data sources as Figure 13.29.)

most of the crops grown on these irrigated lands are low-value types (such as feed grains and forage crops) and can be grown in humid regions at less cost. As already stated, consumptive water use is extremely high in crop irrigation in hot, dry climates. It has been estimated that for the state of Arizona, 90% of the water use is for irrigation but that agriculture accounts for only about 10% of the total value of the state's economic production. Salinization of irrigated soils, explained earlier in this chapter, can increase further the demand for water merely to sustain productivity.

While we have only touched upon a number of alternatives to increasing the transport of water in meeting increasing demands, it should be obvious that water management involves a broad spectrum of economic, political, and cultural factors. We have dealt only with the basic hydrologic and geologic factors and with some facts of present-day nationwide water use.

All forms of terrestrial life evolved throughout the Phanerozoic Eon on continental surfaces undergoing *denudation*, the total action of all processes of weathering, mass wasting, and erosion by running water, glaciers, winds, waves, and currents. Denudation is an overall lowering of the land surface tending toward the reduction of the continents to nearly featureless sea-level surfaces, and ultimately, through wave action, to submarine surfaces. We have already suggested the concept that denudation, if not repeatedly counteracted by tectonic activity throughout geologic time, would have eliminated all terrestrial environments. The important point that emerges as we look back through geologic time is that terrestrial life environments have been in constant change, even as plants and animals have undergone their evolutionary development. The varied denudation processes have produced, maintained, and changed a wide variety of landforms, which have been the habitats for those evolving life forms. In turn, the life forms have become adapted to those habitats and have diversified to a degree that matches the diversity of the landforms themselves. In this chapter we will be concerned with those landforms shaped by runoff in the form of overland flow and stream flow. This study is the geologic aspect of the gravity flow system of water on the lands as distinguished from the hydrologic aspect covered in the previous chapter.

Initial and Sequential Landforms

Because the configuration of continental surfaces reflects the balance of power, so to speak, between internal earth forces, acting through volcanic and tectonic processes, and external forces acting through the agents of denudation, landforms in general fall into two basic categories.

Landforms produced directly by vulcanism and tectonic activity are *initial landforms* (Figure 14.1). We have investigated these landforms in Chapters 7 and 9. They include volcanoes and lava flows, fault blocks and rift valleys, and fold mountains elevated in active zones of contact between lithospheric plates. The energy for lifting molten rock and rigid crustal masses to produce the initial landforms has an internal heat source, which we

14
Fluvial
Processes
and
Landforms

Fluvial Processes and Landforms

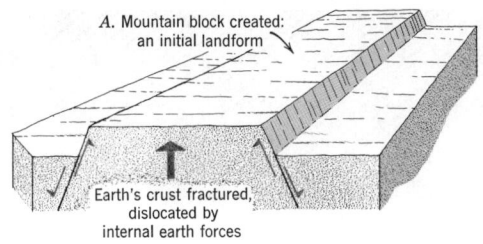

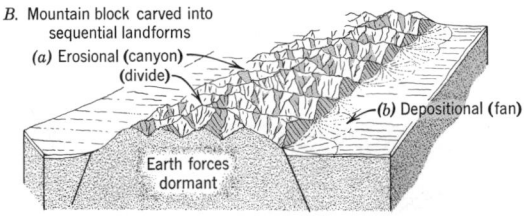

Figure 14.1 Initial and sequential landforms.

have ascribed primarily to radioactivity. The body of rock lying above sea level and exposed to denudation is the *available landmass*.

Landforms shaped by processes and agents of denudation belong to the class of *sequential landforms*, meaning that they follow in sequence after the initial landforms are created and the landmass has been raised to its elevated position. As shown in Figure 14.1, a single uplifted fault block, itself an initial landform, is set upon by agents of denudation and carved up into a large number of sequential landforms.

Fluvial Processes and Landforms

Landforms created by running water are conveniently described as *fluvial landforms* to distinguish them from landforms made by weathering, mass wasting, glacial ice, wind, and waves. Fluvial landforms are shaped by the *fluvial processes* of overland flow and channel flow. Fluvial landforms and fluvial processes dominate the continental land surfaces the world over. Under what are typically prevailing environmental conditions of the geologic regimen, glacial ice is present only in comparatively small global areas located in the polar zones and in high mountains. Landforms made by wind action occupy only trivially small parts of the continental surfaces; while landforms made by waves and currents must necessarily be restricted to a very narrow contact zone between oceans and continents. The last statement is not intended to underplay the importance of specialized life habitats of dunes, shorelines, and estuaries, but simply to emphasize that in terms of areal extent the fluvial landforms are dominant in the environment of terrestrial life and are the major source areas of Man's food resources

through the practice of agriculture. Almost all lands in crop cultivation and almost all grazing lands have been shaped by fluvial processes. It is true that those parts of our continent which have recently emerged from beneath glacial ice, or were belts of active sand dunes in Pleistocene time, may derive their landform configurations from nonfluvial processes, but they are now under the fluvial regime and are undergoing change by fluvial processes. Volcanic landforms, such as the vast lava fields of the Snake River Plain of Idaho, are also examples of landforms of nonfluvial origin, but, again, they are now under the fluvial regime and are rapidly being changed by fluvial processes, as witnessed by incision of the great winding gorge of the Snake River.

Fluvial processes engage in the geological activities of erosion, transportation, and deposition. Consequently, there are two major groups of fluvial landforms: erosional landforms and depositional landforms (Figure 14.2). Where rock is eroded away by fluvial agents, valleys are formed. Between the valleys are ridges, hills, or mountain summits representing unconsumed parts of the landmass. All such sequential landforms shaped by progressive removal of the bedrock mass are designated *erosional landforms*.

Rock and soil fragments that are removed from the parent mass are transported by the fluid agent and deposited elsewhere to make an entirely different set of surface features, the *depositional landforms*. Figure 14.2 illustrates the two groups of landforms. The ravine, canyon, peak, spur, and col are erosional landforms; the fan, built of rock fragments below the mouth of the ravine, is a depositional landform. The floodplain, built of material transported by a stream, is also a depositional landform.

A depositional landform, once created, may in turn be eroded, with the result that a new generation of erosional landforms is created. An example is shown in Figure 14.2 where the bluff, an erosional landform, has been carved out of the fan, a depositional landform. Second generation

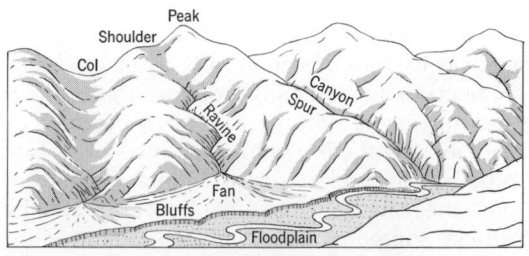

Figure 14.2 Erosional and depositional landforms.

erosional landforms are abundant in regions of fluvial denudation, since fluvial activity tends to be cyclic, with a depositional phase alternating with an erosional phase at the same site.

Normal and Accelerated Slope Erosion

Fluvial processes begin their action at the divides of drainage basins, for it is here that the input of water into the system is derived from excess precipitation unable to infiltrate the soil surface as fast as it arrives. Overland flow, by exerting a dragging force over the soil surface, picks up particles of mineral matter ranging in size from fine colloidal clay to coarse sand or gravel, depending on the speed of the flow and the degree to which the particles are bound by plant rootlets or held down by a mat of leaves. Added to this solid matter is dissolved mineral matter in the form of ions produced by acid reactions or direct solution. Such slow removal of soil is part of the natural geological process of landmass denudation and is both inevitable and universal. Under stable, natural conditions, the erosion rate in a humid climate is slow enough that a soil with distinct horizons is formed and maintained, enabling plant communities to maintain themselves in a stable equilibrium. Soil scientists refer to this state of activity as the *geologic norm*.

By contrast, the rate of soil erosion may be enormously speeded up through Man's activities or rare natural events to result in a state of *accelerated erosion*, removing the soil much faster than it can be formed. This condition comes about most commonly from a forced change in the plant cover and physical state of the ground surface and uppermost soil horizons. Destruction of vegetation by clearing of land for cultivation, or by forest fires, directly causes great changes in the relative proportions of infiltration to runoff. Interception of rain by foliage is ended; protection afforded by a ground cover of fallen leaves and stems is removed. Consequently the rain falls directly upon the mineral soil.

Direct force of falling drops (Figure 14.3) causes a geyserlike splashing in which soil particles are lifted and then dropped into new positions, a process termed *splash erosion*. It is estimated that a violent rainstorm has the ability to disturb as much as 100 tons of soil per acre (225 metric tons per hectare). On a sloping ground surface, splash erosion tends to shift the soil slowly downhill. A more important effect is to cause the soil surface to become much less able to infiltrate water because the natural soil openings become sealed by particles shifted by raindrop splash. Reduced in-

Figure 14.3 A large raindrop (*above*) lands on a wet soil surface, producing a miniature crater (*below*). Grains of clay and silt are thrown into the air and the soil surface is disturbed. (Official U.S. Navy photograph.)

filtration permits a much greater proportion of overland flow to occur from rain of given intensity and duration. The depth and velocity of overland flow then increase greatly, intensifying the rate of soil removal.

Another effect of destruction of vegetation is to reduce greatly the resistance of the ground surface to the force of erosion under overland flow. On a slope covered by grass sod, even a deep layer of overland flow causes little soil erosion because the energy of the moving water is dissipated in friction with the grass stems, which are tough and elastic. Similarly on a heavily forested slope, countless check dams made by leaves, twigs, roots, and fallen tree trunks take up the force of overland flow. Without such vegetative cover the eroding force is applied directly to the bare soil surface, easily dislodging the grains and sweeping them downslope.

Summarizing these things, we note that the eroding ability of overland flow is directly proportional to the rate of precipitation and length of slope, but inversely proportional to both the infiltration capacity of the soil and the resistance of the surface. To complete this equation, we need only to add the effect of the steepness of ground slope. Obviously, the steeper the slope of ground, the faster is the flow and the more intense the erosion. We therefore add that the eroding capacity of overland flow increases directly with angle of slope. As the slope angle approaches the vertical, however, erosion will become less intense from overland flow because the ground surface intercepts much less of the vertically falling rain.

We can get an appreciation of the contrast between normal and accelerated erosion rates by comparing the quantity of sediment derived from cultivated surfaces with that derived from naturally forested or reforested surfaces within a given region in which climate, soil, and topography are fairly uniform. *Sediment yield* is a technical term for the quantity of sediment removed by overland flow from a unit area of ground surface in a given unit of time. Sediment yield is usually stated in tons per acre, or metric tons per hectare. Sediment concentration is determined for water samples of the storm runoff and, knowing also the water discharge, total sediment leaving the watershed can be estimated. This total quantity is then divided by the surface area to give sediment yield, which measures erosion intensity.

Table 14.1 gives data of annual average runoff by overland flow and sediment yield from several types of upland surface in northern Mississippi. Notice that both surface runoff and sediment yield decrease greatly with increasing effectiveness of the protective vegetative cover. Sediment yield from cultivated land undergoing accelerated erosion is over twenty times greater than from pasture and about one thousand times greater than from pine plantation land. The reforested land has a sediment yield rate representing the geologic norm of soil erosion for this region; it is about the same as for mature pine and hardwood forests that have not previously experienced cultivation.

The distinction between normal and accelerated slope erosion applies to regions in which the soil-moisture balance shows an annual surplus, or only a moderate seasonal deficiency. Under a regime of rather severe summer water deficiency, such as that shown in the soil-mositure budget of Los Angeles, California (Figure 12.13E), or of Ankara, Turkey (Figure 12.13F), the natural plant cover, though sparse and providing rather poor ground cover of plant litter, is strong enough that the geologic norm of erosion can be sustained. However, in these environments the sparse grasses and scattered shrubs or trees provide only minimal protection to the soil surface. Consequently, the natural equilibrium is highly sensitive to upset accompanying depletion of plant cover by fires or the grazing of herds of domesticated animals. These sensitive, marginal environments require cautious use, for they lack the potential to recover rapidly from accelerated erosion, once it has set in.

Erosion at a very high rate by overland flow is actually a natural geological process in certain

TABLE 14.1 Runoff and Sediment Yield from Various Types of Upland Surfaces in Northern Mississippi

Land Use or Cover Type	Average Annual Rainfall (in.)	(cm)	Average Annual Runoff (in.)	(cm)	Average Annual Sediment Yield (tons per acre)	(metric tons per hectare)
Open land						
Cultivated	52	(132)	16	(40)	22	(50)
Pasture	51	(129)	15	(38)	1.6	(36)
Forest land						
Abandoned fields	51	(129)	7	(18)	0.13	(0.29)
Depleted hardwoods	51	(129)	5	(13)	0.10	(0.22)
Pine plantations	54	(137)	1	(2.5)	0.02	(0.045)
Mature pine-hardwoods	51	(129)	9[a]	(23)	0.02	(0.045)

SOURCE: S. J. Ursic, 1965, Paper No. 6, *Proceedings of the Federal Inter-Agency Sedimentation Conference, 1963*, U.S. Department of Agriculture, Misc. Publ. No. 970, U.S. Government Printing Office, Washington, D.C., p. 49, Table 2.
[a] This forest type occurs on sandy soils underlain by dense clay that inhibits deep infiltration. Runoff is therefore high and cannot be compared with other runoff values.

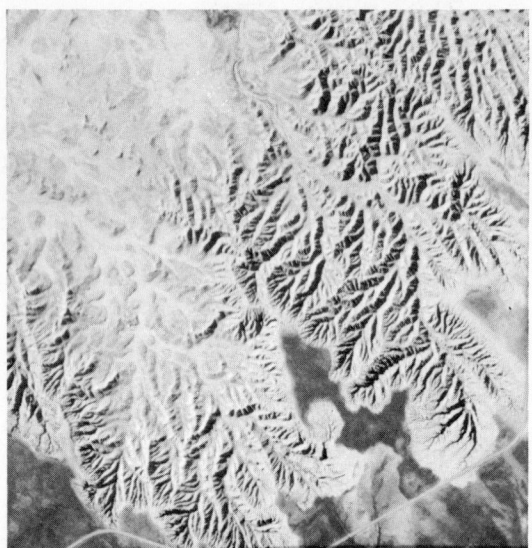

Figure 14.4 Vertical air photograph of an area of one square mile in the Big Badlands of South Dakota. (U.S. Dept. of Agriculture.)

Figure 14.5 Badlands, such as these in the Petrified Forest National Monument, Arizona, are like miniature mountain topography on bare clay formations. (Photograph by B. Mears, Jr.)

favorable localities in semiarid and arid lands; it takes the form of *badlands*. Two well-known areas of badlands are the Big Badlands of South Dakota, along the White River (Figure 14.4), and the Painted Desert and related areas of northern Arizona (Figure 14.5). These localities are underlain by clay formations that are almost impervious to infiltration of precipitation when in the moist state and are also easily eroded by overland flow. Erosion rates are too fast to permit plants to take hold and no true soil can develop. A maze of small stream channels is developed and valley-side slopes are very steep. Badlands such as these are self-sustaining and have been in existence at one place or another on continents throughout much of geologic time.

Forms of Accelerated Soil Erosion

Regions of substantial soil-water surplus, for which the natural plant cover is forest or prairie grasslands, experience accelerated soil erosion when Man expends enough energy to remove the plant cover and keep the land barren by annual cultivation or by allowing intensive overgrazing and trampling by livestock. With fossil fuels to power his machines of plant and soil destruction, Man has easily overwhelmed the restorative forces of nature over vast expanses of continental surfaces. We now turn to consider the consequences of these activities.

When a plot of ground is first cleared of forest and plowed for cultivation, little erosion will occur until the action of rain splash has broken down the soil aggregates and sealed the larger opening. Following this, overland flow begins to remove the soil in rather uniform thin layers, a process termed *sheet erosion*. Because of seasonal cultivation, the effects of sheet erosion are often little noticed until the upper horizons of the soil (*A* and *B* horizons) are removed or greatly thinned. Reaching the base of the slope, where the angle of surface is rapidly reduced to meet the valley bottom, soil particles come to rest and accumulate in a thickening layer termed *colluvium*, or simply *slope wash*. This deposit, too, has a sheetlike distribution and may be little noticed, except where it can be seen that fence posts or tree trunks are being slowly buried.

Material that continues to be carried by overland flow to reach a stream in the valley axis is then carried farther down valley and may be built up into layers on the valley floor, where it becomes *alluvium*, a word applied generally to any stream-laid deposits. As used by the agricultural soil scientist the process of accumulation of colluvium and alluvium constitutes *valley sedimentation*. In many ways, sedimentation at the base of slopes and in valley bottoms is a process equally serious to erosion from the agricultural standpoint, because it results in burial of soil horizons under relatively infertile, sandy layers

Forms of Accelerated Soil Erosion

and may choke the valleys of small streams, causing the water to flood broadly over the valley bottoms.

Where slopes are exceptionally steep and runoff from storms is exceptionally heavy, sheet erosion progresses into a more intense activity, that of *rill erosion,* or *rilling* (Figure 14.6), in which innumerable, closely spaced channels, called *shoestring rills,* are scored into the soil and subsoil. If these rills are not destroyed by soil tillage, they may soon begin to integrate into still larger channels, termed *gullies.* This transformation comes about as the more active rills deepen more rapidly than their neighbors and incorporate the adjacent drainage areas. Erosive action thus is concentrated into a few large channels which can deepen into steep-walled, canyonlike trenches whose upper ends grow progressively upslope (Figure 14.7).

Ultimately, a rugged, barren topography, resembling the badland forms of the arid climates, may result from accelerated soil erosion allowed to proceed unchecked.

As we pointed out in Chapter 10, the natural soil with its organic-colloid and base-cation content and its well-developed horizons is a nonrenewable natural resource. The rate of soil formation is extremely slow in comparison with the rate of its destruction when accelerated erosion has set in and is allowed to go unchecked. Soil erosion as a potentially disastrous form of environmental degradation was brought to public attention decades ago in the United States.

Major efforts to affect soil conservation measures were initiated in the late 1920s, in large part through the effective crusading of H. H. Bennett, a pioneer figure in the study of soil erosion. Congress in 1930 established ten erosion experiment stations in which research was begun on ways to control accelerated erosion. By 1933 the Soil Erosion Service had been established as an agency under the Department of the Interior and began demonstration projects in erosion control. Then in 1935 the Soil Conservation Service was set up as a permanent bureau under the Department of Agriculture and an era of revolutionary developments in soil management followed. Although tremendous progress has been made in introducing better land management methods and in creating a general awareness of the need for soil and water conservation practices, a great deal remains to be done to reduce soil losses to the minimum practicable level.

A first step in implementing soil conservation practices is the classification of agricultural and forest lands according to *land capabilities.* The Soil Conservation Service recognizes eight classes

Figure 14.6 Shoestring rills on a barren slope. (Soil Conservation Service photograph.)

of land, ranging from land so nearly level as to be largely immune from excessive erosion (Class 1) to land so steep and vulnerable to erosion as to be unsuited to any productive agricultural use. Such surfaces are relegated to wildlife and recreation urposes (Figure 14.8). For intermediate classes the degree of susceptibility to erosion is stated in terms of management practices that are a requirement of agricultural use.

A number of common practices effective in soil erosion control are the following: *Contouring* is a general term for plowing, planting, cultivating,

Figure 14.7 This great gully, eroded into deeply weathered overburden, was typical of certain parts of the Piedmont region of South Carolina and Georgia before remedial measures were applied. (Soil Conservation Service photograph.)

Figure 14.8 Seven out of the eight land capability classes are seen in this single photograph. Class I presents practically no erosion hazard. Classes VII and VIII are steep slopes highly susceptible to damaging erosion. (Soil Conservation Service photograph.)

and furrowing along the natural contour lines of sloping ground (Figure 14.9). Because the crop rows or belts, and the minor ridges their cultivation produces are oriented at right angles to the downslope lines on which overland flow would normally move, the effect of contouring is to provide increased surface detention that allows greater infiltration. Contouring also creates obstacles that greatly reduce the velocity of downhill flow which would otherwise occur on a smooth slope. A second practice is that of *crop rotation*, in which a given soil belt is planted in a different crop in each year of a cycle of rotation that includes legumes to increase nitrogen content of the soil, a grass crop to improve tilth (ease of cultivation), and a clean-cultivated crop such as corn or tobacco.

Strip cropping is the planting of different crops side by side in narrow, parallel belts. When practiced as contouring, strip cropping allows the close-growing crops to catch soil particles entrained by runoff from upslope strips of clean-cultivated crops. Strip cropping is also practiced to inhibit soil loss through wind action. *Stubble mulching* is the practice of leaving the base of the plant with its root system intact and allowing plant leaves and stems to remain on the ground. As a result, erosion and soil-moisture evaporation are reduced in the season following harvest.

Terracing, a form of contour development, is the construction of a system of ditches and embankments along the contour so as to direct overland flow on a low-gradient path to the edges of the field, where it is drained off in controlled runoff channels. Terracing is an ancient agricultural practice that has been carried to elaborate levels of effectiveness in mountain and hill lands of steep slopes throughout southern and eastern Asia. The terraces are nearly flat and are bounded by perpendicular steps constructed as retaining walls. Figure 14.10 shows terracing in a region of China underlain by highly erodible wind-blown silt (loess). The area had been deeply gullied prior to being reclaimed by terracing.

Gully control follows various practices. Immediate measures consist of construction of check dams that trap sediment. Plantings of vines, grasses, and shrubs can quickly provide a protective cover over the bare gully walls and bottom. A system of permanent dams may be installed and the gully walls and head graded to a lower slope that can be stabilized by grass or other dense-growing plants.

Soil erosion control by the practices listed above also provides an important measure of water conservation (by increased infiltration) and flood control (by reducing and delaying flow peaks). For this reason the conservation districts laid out by the Department of Agriculture combine soil and water conservation. We will return in later pages of this chapter to the relationship between these watershed controls and the regulation of stream channels that must receive the water and sediment entrained by overland flow.

Stream Erosion

Erosion by a stream is the progressive removal of mineral material from the floor and sides of the channel, whether this be carved in bedrock, or in residual or transported overburden. Streams erode in various ways, depending on the nature of the channel materials and the tools with which the current is armed. The force of the flowing water alone, exerting impact and a dragging action upon the bed, can erode poorly consolidated materials such as gravel, sand, silt, and clay, a process termed *hydraulic action*. Where rock particles carried by the swift current strike against bedrock channel walls, mineral grains and chips of rock are detached. The rolling of cobbles and boulders over the stream bed will further crush and grind smaller grains to produce an assortment of grain sizes. These processes of mechanical wear

Figure 14.9 Contour cultivation near Edson, Kansas. Crops are grown in broad contour belts. Shallow terrace channels between crop belts conduct the overland flow on a gentle gradient. Slopes are in Classes II and III. (Soil Conservation Service photograph.)

are combined under the term *abrasion*, which is the principal means of erosion in bedrock too strong to be affected by simple hydraulic action (Figure 14.11). Finally, the chemical processes of rock weathering—acid reactions and solution—are effective in removal of rock from the stream channel; a process called *corrosion*. Effects of corrosion are most marked in limestone, which is a hard rock not easily carved by abrasion, but yielding readily to the action of carbonic acid in solution in the stream water.

The hydraulic action of flood waters is capable of excavating enormous quantities of unconsolidated materials in a short time (Figure 14.12). Not only is the channel greatly deepened in flood, but the undermining of the banks causes huge masses of silt, sand, and gravel to slump into the river where the particles are quickly separated. This process, known as *bank caving*, is an important source of sediment during high river stages, and is associated with rapid sidewise shifts in channel position on the outsides of river bends. Typically, a large stream occupies a channel carved in alluvium deposited by the stream itself in earlier phases of activity. Thus most bank caving involves alluvium and represents a recycling of stream-transported materials.

Stream Transportation and Load

Mineral matter carried by a stream constitutes the *load*. We have already investigated in Chapter 13 the load of dissolved solids, consisting of ions in solution, present in all natural streams. Although the dissolved solids give a slightly increased density to the stream water, their physical effect upon the stream is negligible.

Clay, silt, and sometimes fine sand are carried in *suspension*, that is, held up in the water by the upward elements of flow in turbulent eddies in

Figure 14.10 Deeply eroded wind-blown silt (loess) in Shansi Province, China (lat. 38° N, long. 112° W). Steep slopes have been terraced for cultivation, thereby arresting the extension of gully heads. Nevertheless, this region leads the world in sediment yield (see Table 14.2). (U.S. Geological Survey photograph.)

Figure 14.11 These potholes have been carved in granite in the channel of the James River, Henrico County, Virginia. (Photograph by C. K. Wentworth, U.S. Geological Survey.)

the stream. This fraction of the transported matter is termed the *suspended load.* Sand, gravel, and still larger fragments move as *bed load* close to the channel floor by rolling or sliding and an occasional low leap.

The load carried by a stream varies enormously in the total quantity present and the size of the fragments, depending on the discharge and stage of the river. In flood, when velocities of 20 ft (6 m) per second or more are produced in large rivers, the water is turbid with suspended load. Even large boulders may be moving over the stream bed, if the channel gradient is steep.

Figure 14.12 A river in flood eroded this huge trench at Cavendish, Vermont, in November 1927. An area 1 mi (1.6 km) wide and 3 mi (4.8 km) long, once occupied by eight farms, was cut away by the flood waters. Damage was great because the material consisted of sand and gravel which offered little resistance. (Wide World Photos.)

Stream load is measured in units of mass carried through a given cross section per unit of time; it is commonly stated as tons (English or metric) per day or per year. Because the velocity and turbulence of a stream increase greatly as its depth and discharge increase in flood stage, most of the sediment transport takes place in high stages. Figure 14.13 gives an example of the increase in suspended load with discharge, as measured at a single gauging station. The scales are logarithmic and show that the load in flood stage may be over 10,000 times as great as in the lowest stages. From the inclination of the line fitted to the plotted points it can be seen that a 10-fold increase in discharge brings a 100-fold increase in suspended load. This load may be derived from overland flow on the watershed surfaces during heavy rains, or from bank erosion.

The great rivers of the world show an enormous range in quantity of suspended sediment load and in sediment yield. Data for seven selected major

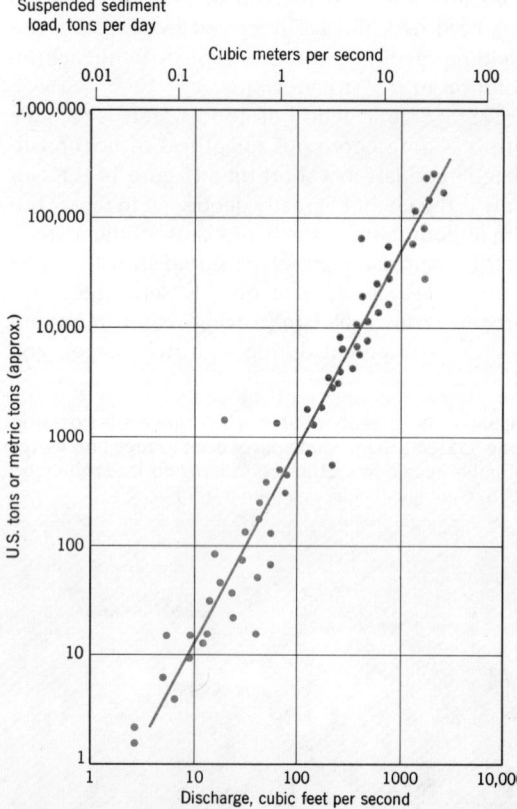

Figure 14.13 Increase of suspended sediment load with increase in discharge. Both scales are logarithmic. The dots show how individual observations are scattered about the line of best fit. Data are for the Powder River, Arvada, Wyoming. (After L. B. Leopold and T. Maddock, 1953, U.S. Geol. Survey, Professional Paper 252, p. 20, Figure 13.)

rivers (Table 14.2) reveal some interesting relationships among load, climate, and land surface properties. The Yellow River (Hwang Ho) of China heads the world list in annual suspended sediment load, while its sediment yield is one of the highest known for a large river basin. The explanation lies in a high soil-erosion rate on intensively cultivated upland surfaces of wind-deposited silt (loess) in Shensi and Shansi provinces (see Chapter 16 and Figure 14.10). Much of the drainage area is in a semiarid climate with dry winters; vegetation is sparse and the runoff from heavy summer rains entrains a large amount of sediment. The Ganges River derives its heavy sediment load from steep mountain slopes of the Himalayas and from intensively cultivated lowlands, all subjected to torrential rains of the tropical wet monsoon season. The Colorado River represents an exotic stream, deriving its runoff largely from snowmelt and precipitation on high mountain watersheds of the Rockies, but most of its suspended sediment is from tributaries in the semiarid plateau lands through which it passes. The sediment load of the Mississippi River comes largely from subhumid and semiarid grassland watersheds of its great western tributary, the Missouri River. The Amazon River, a colossus in discharge and basin area, has a very low sediment yield, because like the Congo River, much of its basin lies within a wet equatorial climate where the land surface bears a highly protective rainforest. The Yenisei River of Siberia has a remarkably low sediment load and sediment yield for its vast drainage area, most of which is in the needleleaf forest, or taiga, of the subarctic and high middle-latitude zones.

Although it is difficult to assess the importance of man-induced soil disturbance upon the sediment load of major rivers, most investigators are generally agreed that cultivation has greatly increased the sediment load of rivers of eastern and southeastern Asia, Europe, and North America. The increase due to Man's activities is thought to be greater by a factor of 2½ than the geologic norm for the entire world land area. For the more strongly affected river basins the factor may be ten or more times larger than the geologic norm.

The sediment load carried by a large river is of considerable importance in planning for construction of large storage dams and in the construction of canal systems for irrigation. Sediment will be trapped in the reservoir behind a dam, eventually filling the entire basin and ending the useful life of the reservoir as a storage body. At the same time, depriving the river of its sediment in the lower course below the dam may cause

TABLE 14.2 Suspended Sediment Loads and Sediment Yields of Selected Large Rivers

River	Drainage Area	Average Discharge	Average Annual Sediment Load	Average Annual Sediment Yields
	(thousands of sq mi)	(thousands of cu ft/sec)	(thousands of tons)	(tons/sq mi)
Yellow (Hwang Ho), China	280	53	2,100,000	7,500
Ganges, India	370	410	1,600,000	4,000
Colorado, U.S.A.	250	5.5	150,000	1,100
Mississippi, U.S.A.	1,200	630	340,000	280
Amazon, Brazil	2,400	6,400	400,000	170
Congo, Congo	1,500	1,400	71,000	46
Yenisei, U.S.S.R.	950	600	12,000	12
	(sq km $\times$ 10^3)	(cu m/sec)	(metric tons $\times$ 10^3)	(metric tons/sq km)
Yellow (Hwang Ho), China	715	1.6	1,900,000	2,600
Ganges, India	960	12	1,500,000	1,400
Colorado, U.S.A.	640	0.17	140,000	380
Mississippi, U.S.A.	3,200	19	310,000	97
Amazon, Brazil	6,100	190	360,000	60
Congo, Congo	4,000	42	65,000	16
Yenisei, U.S.S.R.	2,500	18	11,000	4

SOURCE: J. N. Holeman, 1968, "The Sediment Yield of Major Rivers of the World," *Water Resources Research*, vol. 4, no. 4, pp. 737-747.
NOTE: Data rounded to two digits.

serious upsets in river activity. Resulting deep scour of the bed and lowering of river level may render irrigation systems inoperable. In designing for canal systems, the forms of artificial channels must be adjusted to the size and quantity of sediment carried by the water, otherwise obstruction by deposition or abnormal scour may follow.

How Channels Change in Flood

We tend to think of a river in flood as changing largely through increase in height of water surface, which causes channel overflow and inundation of the adjoining floodplain. Because of the turbidity of the water we cannot see the changes taking place on the stream bed, but these can be determined by sounding the river depth during stream-gauging measurements (Chapter 13). Figure 14.14 shows how a river channel changes its configuration with rising and falling stages. At first the bed may be built up by large amounts of bed load supplied to the stream during the first phase of heavy runoff. This phase is soon reversed, however, and the bed is actively deepened by scour as stream stage rises. Thus, in the period of highest stage, the river bed is typically at its lowest elevation. When the discharge then starts to decline, the level of the stream's surface drops and the bed is built back up by the deposition of bed load. In the example shown in Figure 14.14, about 10 ft (3 m) of thickness of alluvium was *reworked*, that is, moved about in the complete cycle of rising and falling stages.

Alternate deepening by scour and shallowing by deposition of load are responses to changes in the stream's ability to transport its load. The maximum bed load of debris that can be carried by a stream is a measure of the stream's *bed-load capacity* (also called *tractive capacity*).

Where a stream is flowing in a channel of hard bedrock, it may not be able to pick up enough bed materials to supply its full capacity for bed load. Such conditions exist in streams occupying deep gorges and having steep gradients, so that when flood occurs, the channel cannot be quickly deepened in response. In an *alluvial river,* one in which the channel is carved in alluvium, thick layers of silt, sand, gravel, and boulders underlie the channel. Under these conditions the rising river easily picks up and sets in motion all of the material that it is capable of moving. In other words, the increasing capacity of the stream for bed load is immediately and fully satisfied.

Capacity for bed load increases sharply with the stream's velocity, because the swifter the current the more intense is the turbulence and the stronger is the dragging force upon the bed. Bed-load capacity goes up about as the third to fourth power of the velocity. Thus, if a stream's velocity is doubled in flood, its ability to transport bed load is increased from eight to sixteen times. It is small wonder, then, that most of the conspicuous changes in the channel of a stream, such as sidewise shifting of the course, occur in flood stage, with very few important changes occurring in low-water stages.

When the flood crest has passed and the discharge begins to decrease, the stream's capacity to transport load also declines. Therefore, some of the particles that are in motion must come to rest on the bed in the form of sand and gravel bars. First the largest boulders and cobblestones will cease to roll, then the pebbles and gravel, then the sand. At even lower velocities fine sand

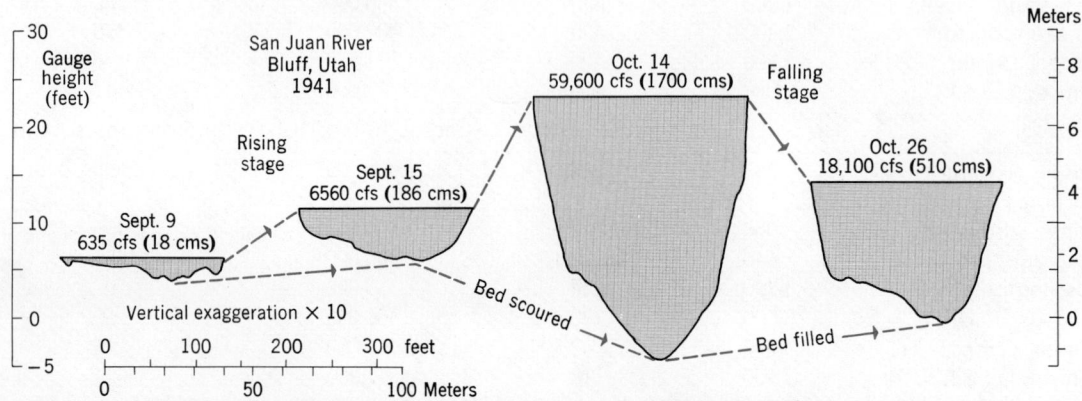

Figure 14.14 Changes in channel form of the San Juan River near Bluff, Utah, during the progress of a flood. (Based on data of L. B. Leopold and T. Maddock, 1953, U.S. Geol. Survey Professional Paper 252, p. 32, Figure 22.)

and silt carried in suspension can no longer be sustained, and settle to the bed. Clay particles continue far downstream with the flood wave. In this way the stream adjusts to its falling capacity. When restored to low stage the water may become quite clear, with only a few grains of sand rolling along the bed where the current threads are fastest.

Stream Gradation

A stream system, fully developed within its drainage basin of contributing valley-side slopes, as described in Chapter 13, has undergone a long period of adjustment of its geometry so that it can discharge through the trunk exit not only the surplus water produced by the basin but also the solid load with which the channels are supplied. A purely hydraulic system could operate without a gradient, because accumulated surplus water can generate its own surface slope and is capable of flowage on a horizontal surface. However, the transport of bed load requires a gradient, and it is in response to this requirement that a stream channel system has adjusted its gradient and achieved an average steady state of operation, year in and year out, and from decade to decade. In this condition the fluvial system is then said to be *graded,* and to have achieved an *equilibrium* state of operation.

It will be helpful in developing the concept of a graded stream system to investigate the changes that will take place along a stretch of stream that is initially poorly adjusted to the transport of its load. Such an *ungraded* channel is illustrated in Figure 14.15. The initial profile is imagined to be brought about by crustal uplift in a series of fault steps, bringing up a surface that was formerly beneath the ocean and exposing it to fluvial processes for the first time. Overland flow collects in shallow depressions, which fill and overflow from higher to lower levels. In this way a through channel originates and begins to conduct runoff to the sea. Figure 14.16 illustrates the gradation process in a series of block diagrams. Over *falls* and *rapids,* which are simply steep-gradient portions of the channel, flow velocity is greatly increased and abrasion of bedrock is most intense (block A). As a result, the falls are cut back and the rapids trenched, while the ponded reaches are filled by sediment and are lowered in level as the outlets are cut down. In time the lakes disappear and the falls are transformed into rapids, which in turn are reduced to a gradient more closely approximating the average gradient of the entire stream (block B). At the same time, branches of the stream system are being extended into the landmass, carving out a drainage basin and transforming the initial landscape into a fluvial landform system. In these early stages of gradation and extension, the capacity of the stream exceeds the load supplied to it, so that little or no alluvium accumulates in the channels. Abrasion continues to lower the elevation of the channels, with the result that they come to occupy steep-walled *gorges* or *canyons* (Figure 14.17). Weathering and mass wasting of these rock walls contributes an increasing supply of rock debris to the channels. Debris from ground surfaces contributing overland flow to the newly developed branches is also on the increase.

We can anticipate that the gradual decrease in the stream's capacity for bed load, resulting from the gradual reduction in channel gradient, will be converging upon the increasing load with which it is supplied, so that there will come a point in time at which the supply of load exactly matches the stream's capacity for transport. It is at this point that the stream has achieved the graded condition and possesses an *equilibrium profile.*

It is important to understand that the balance between load and a stream's capacity exists only as an average condition over periods of many years. As already explained, streams scour their channels in flood and deposit load when in falling stage. Thus in terms of conditions of the moment,

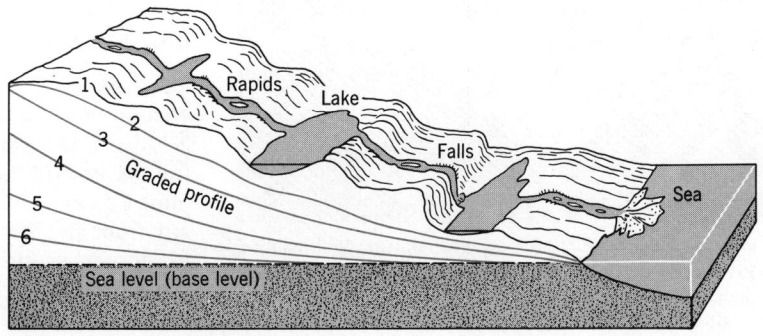

Figure 14.15 Schematic diagram of gradation of a stream, originally consisting of a succession of lakes, falls, and rapids. (From A. N. Strahler, 1972, *Planet Earth; Its Physical Systems Through Geologic Time,* Harper & Row, New York.)

Fluvial Processes and Landforms

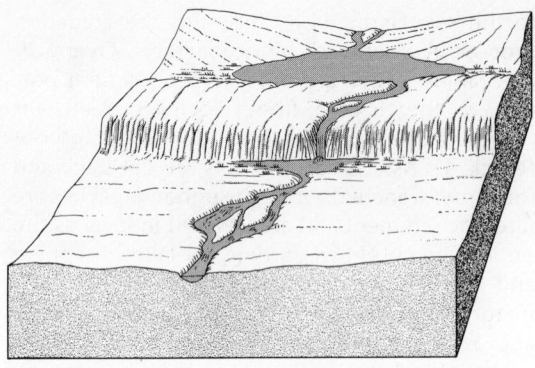

A. Stream established upon a land surface dominated by landforms of recent tectonic activity.

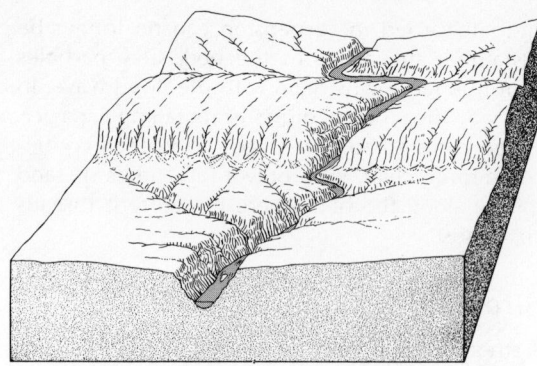

B. Gradation in progress; lakes and marshes drained; deepening gorge; tributary valleys extending.

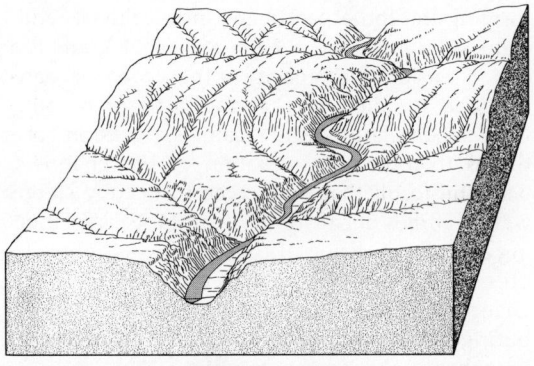

C. Graded profile attained; beginning of floodplain development; valley widening in progress.

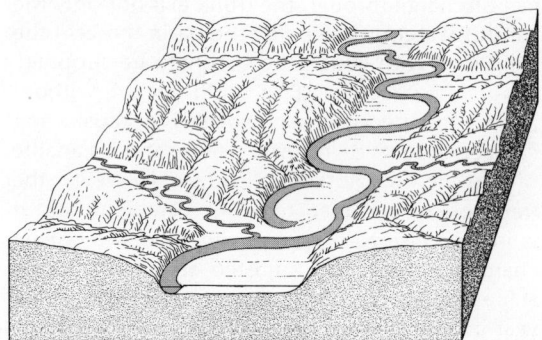

D. Floodplain widened to accommodate meanders; floodplains extended up tributary valleys.

Figure 14.16 Evolution of a stream and its valley. (After E. Raisz.)

Figure 14.17 The Grand Canyon of the Yellowstone River, viewed from over Inspiration Point, illustrates a deep canyon carved into the surface of a lava plateau. (U.S. Army Air Service Photograph.)

a stream is rarely in equilibrium; but over long periods of time, the graded stream maintains its level by restoring those channel deposits temporarily removed by the excessive energy of flood flows.

Having attained this state of balance, the stream continues to cut sidewise on the outsides of banks. The lateral cutting does not appreciably alter the gradient and therefore does not materially affect the equilibrium (Figure 14.16C).

Environmental Significance of Gorges and Waterfalls

Deep gorges and canyons of major rivers exert environmental controls in a variety of ways. There is little or no room for roads or railroads between the stream and the valley sides so that road beds must be cut or blasted at great expense and hazard from the sheer rock walls. Maintenance is expensive because of flooding or undercutting by the stream and the sliding and falling of rock, which can wipe out or damage the road bed. Yet

Environmental Significance of Gorges and Waterfalls

Figure 14.18 The Royal gorge of the Arkansas River in the Colorado Rockies illustrates the canyon of a river with a steep gradient. Seen above is a suspension bridge, 1053 ft (321 m) above river level. (Photograph by Josef Muench.)

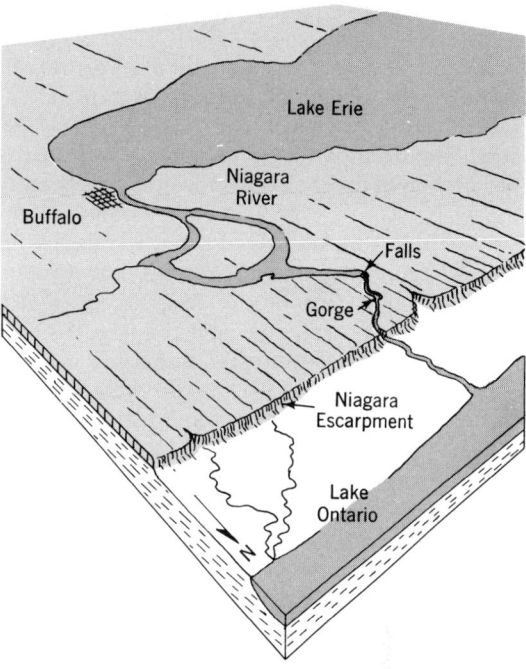

Figure 14.19 A bird's-eye view of the Niagara River with its falls and gorge carved in strata of the Niagara Escarpment. View is toward the southwest from a point over Lake Ontario. Redrawn from a sketch by G. K. Gilbert, 1896. (From A. N. Strahler, 1971, The Earth Sciences, 2nd ed., Harper & Row, New York.)

a gorge may afford the only feasible passage through a mountain range. The Royal Gorge of the Arkansas River, in the Rocky Mountain Front Range of southern Colorado, is a striking example (Figure 14.18). This illustration also brings out the point that a deep canyon is a barrier to movement across it and may require construction of expensive bridges.

Another environmental consideration is that a river with a steep gradient is not navigable without locks, even though it might otherwise have a sufficient discharge.

Although small waterfalls are common features of alpine mountains carved by glacial erosion (Chapter 16), large waterfalls on major rivers are comparatively rare the world over. New river channels resulting from flow diversions caused by ice sheets of the Pleistocene Epoch provide one class of falls and rapids of large discharge. Certainly the preeminent example is Niagara Falls (Figure 14.19). Overflow of Lake Erie into Lake Ontario happened to be situated over a gently inclined layer of limestone, beneath which lies easily eroded shale. As the detailed diagram shows (Figure 14.20), the fall is maintained by continual undermining of the limestone (Lockport dolomite) by erosion in the plunge-pool at the base of the fall. In this way the falls have retreated about 6.5 mi (10.4 km) since they were formed some 12,000 to 13,000 years ago at a point on the Niagara Escarpment, leaving behind a

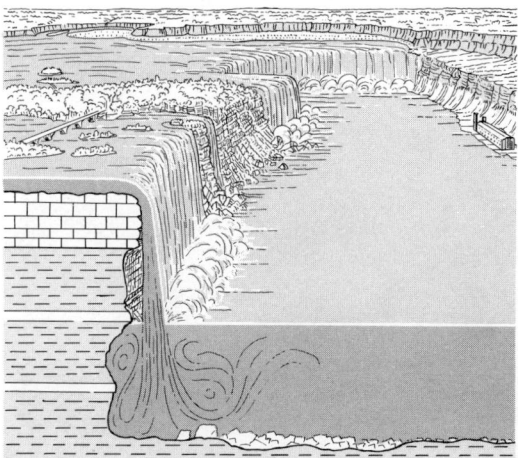

Figure 14.20 Niagara Falls is formed where the river passes over the eroded edge of a massive limestone layer. Continual undermining of weak shales at the base keeps the fall steep. (After G. K. Gilbert and E. Raisz.)

spectacular gorge. Present rate of recession of the Canadian Falls is some 4 to 6 ft (1 to 2 m) per year but has varied considerably from century to century. The height of the falls is now 170 ft (52 m) and its discharge about 200,000 cfs (17,000 cms). The drop of Niagra Falls is utilized for the production of hydroelectric power by the Niagara Power Project, in which water is withdrawn upstream from the falls and carried in tunnels to generating plants located 4 miles downstream from the falls. Capability of this project is 2400 megawatts of power, making it the largest single producer in the Western Hemisphere.

Entirely different as a class are waterfalls are those formed by recent tectonic activity. In the rift valley region of East Africa, crustal blocks have been dislocated by recent faulting, creating lake basins and discontinuities in the gradients of major rivers, and giving rise to falls and rapids. An example is Murchison Falls on the upper (White) Nile River near the north end of a graben in which Lake Albert is situated. The height of this fall is 130 ft (40 m). Victoria Falls on the Zambezi River, height 355 ft (108 m), owes its drop to erosion of weak rock along a fault zone (Figure 14.21).

Most large rivers of steep gradient do not, however, possess falls, and so it is necessary to build dams in order to create artificially the vertical drop necessary for turbine operation. An example is the Hoover Dam, behind which lies Lake Hoover (formerly Lake Mead) occupying the canyon of the Colorado River. With a dam height of 726 ft (220 m) the generating plant of Hoover Dam is capable of producing 1345 megawatts of power, about half as much as the Niagara Project.

We should not lose sight of the esthetic and recreational values of gorges, rapids, and waterfalls of major rivers. The Grand Canyon of the Colorado River, probably more than any single product of fluvial processes, epitomizes the scenic value of a great river gorge. Against the advantages in obtaining hydroelectric power and fresh water for urban supplies and irrigation by construction of large dams we must weigh the permanent loss of some large segments of our finest natural scenery, along with the destruction of ecosystems adapted to the river environment. It is small wonder, then, that new dam projects are meeting with stiff opposition from concerned citizen groups who fear that the harmful environmental impacts of such structures far outweigh their future benefits. We shall return to consider more of these environmental effects later in this chapter.

The Graded Stream System and Its Profile

The first indication that a stream has attained a graded condition is the beginning of floodplain development. On the outside of a bend the channel shifts laterally into a curve of larger radius and thus undercuts the valley wall. On the inside of the bend alluvium accumulates in the form of a *point-bar* deposit. Widening of the point-bar deposit produces a crescentic element of low ground, which is the first stage in floodplain development (Figures 14.22 and 14.23A). This stage is also illustrated in (Figure 14.16C). As lateral cutting continues, the floodplain strips are widened and the channel develops sinuous bends, termed *meanders* (Figure 14.23B,C). The

Figure 14.21 This air view of Victoria Falls of the Zambezi River shows that the river has excavated a long cleft in the bedrock, probably along a fault zone. (Photographer not known.)

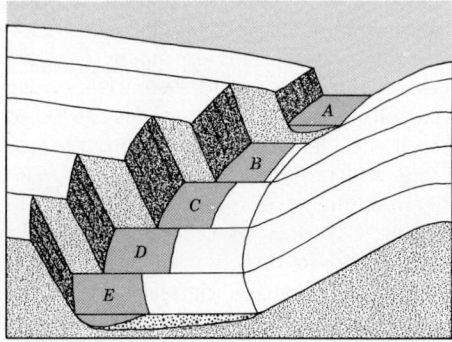

Figure 14.22 A graded stream cuts sidewise against the outside of a bend, leaving a floodplain belt on the inside. (After W. M. Davis.)

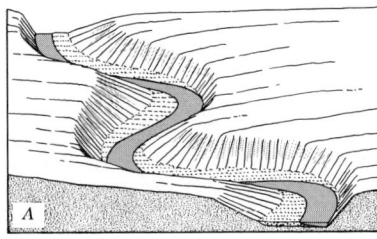
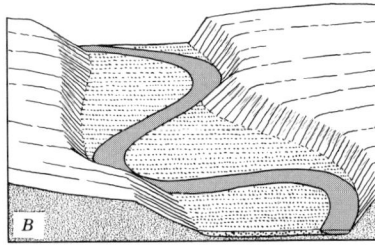
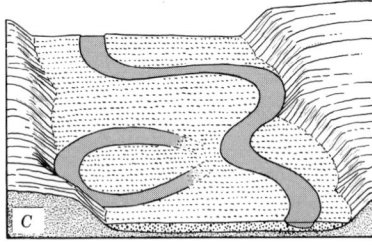

Figure 14.23 Widening of a valley by lateral cutting of a graded stream permits the free growth and cutoff of meander bends. (After W. M. Davis.)

floodplain is now a continuous belt between steep valley walls, while the meanders are well developed. The meander loops are free to grow and occlude to produce *cutoffs,* leaving crescentic *oxbow lakes* (Figure 14.16D).

Because floodplain development reduces the frequency with which channel scour attacks and undermines a given stretch of the adjacent valley wall, weathering, mass wasting, and overland flow act to reduce the steepness of the valley-side slopes (Figure 14.24). As a result, in a humid climate the gorge-like aspect of the valley gradually disappears and gives way to an open valley with soil-covered slopes that may be protected by a dense plant cover.

The profile of a graded stream, which we have referred to as an equilibrium profile, when plotted on a graph of elevation versus distance, will be found to be upwardly concave, so that the gradient diminishes from head to mouth (Figure 14.25). The explanation of diminishing gradient in the downstream direction lies in increased efficiency of a stream as its cross section becomes larger. Given two channels of identical cross-sectional form (and therefore with the same form ratio), the larger channel has proportionately less wetted perimeter in relation to its cross-sectional area than does the smaller stream. As a result, the larger stream expends a smaller proportion of its total energy in friction with the channel. In compensation, the larger stream has adjusted its gradient to a lower value and is able to perform its function of transport of water and load on that lower gradient. The principle of downstream decrease in gradient of a stream because of increasing efficiency is often referred to as *Gilbert's law of declivities,* after the American geologist Grove Karl Gilbert, who in 1877 offered the explanation. (Gilbert used the word *declivity* for gradient.)

We should not expect the profile of the graded stream to be a smooth curve, since the joining of large tributaries causes abrupt increases in discharge and load from point to point along the main stream. Ideally, at each major junction the profile should show an abrupt decrease in gradient, reflecting the abrupt increase in stream efficiency. In actuality, then, the profile of a graded stream is *segmented,* with the segments forming a general upward concavity of the total profile.

Another factor tending to a downstream decrease in stream gradient is average particle size of the bed load. When particle size decreases in the downstream direction, as has been measured in many streams, a lesser gradient suffices to carry the finer particles as bed load than is required by an equivalent load of larger particles.

The profiles of most large streams show irregularities of profile that reflect differences in rock type from region to region. These irregularities are typically found in the upper reaches of the stream system. Where the stream crosses a belt of weaker rock, such as shale, its profile will have a less-than-average gradient; where it crosses a

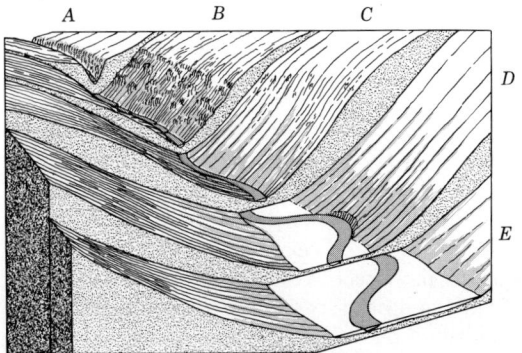

Figure 14.24 Following stream gradation the valley walls become more gentle in slope and the bedrock is covered by soil and weathered rock. (After W. M. Davis.)

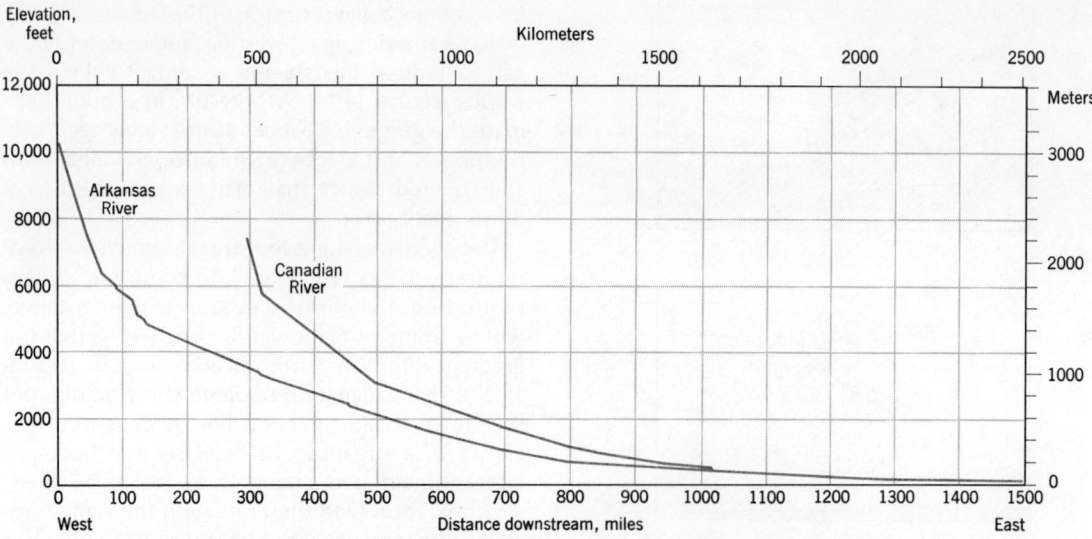

Figure 14.25 Longitudinal profiles of the Arkansas and Canadian rivers. The middle and lower parts of the profiles are for the most part smoothly graded, whereas the poorly graded upper parts reflect rock inequalities and glacial modifications within the Rocky Mountains. (From Gannett, U.S. Geological Survey.)

belt of resistant rock such as granite or gneiss, the gradient is steeper than average for that section of the profile. Where the profile is developed throughout in easily eroded alluvial materials, we usually find a close approach to the ideal conditions based on downstream increase in discharge and load.

Goal of Fluvial Denudation as a Geologic Process

With the concept of a graded stream in mind, this is a good place to expand our thinking to take in some of the broader aspects of fluvial denudation in its geologic context. Consider a landmass made up of drainage basins and their branching stream systems. Assume that the streams have become graded and rock debris is being transported out of each drainage basin by the trunk streams at the same average rate as the debris is being contributed from the land surfaces within the basin. Obviously, this equilibrium cannot be sustained, since the production of debris must lower the land surface generally, and must decrease the available landmass lying above sea level. If so, the average elevation of the land surface is steadily declining and this decline must be reflected in a reduction in the average gradients of all streams. In essence, the open system is consuming its own quantity of mass in storage and at the same time is experiencing a reduction in potential energy of the system through lowering of surface elevation.

To describe the ensuing events briefly, as time passes the streams and valley-side slopes of the drainage basins must undergo gradual and continual change to lower gradients. Rate of expenditure of energy within the system also undergoes a gradual decline. The decreases in both surface elevation and intensity of energy transformation follow the negative exponential decay curve with respect to time. In theory, the ultimate goal of the denudation process is a reduction of the landmass to a featureless plain at sea level. In this process, sea level projected beneath the landmass represents the lower limiting level, or *base level*, of the fluvial denudation. But, because rate of land surface reduction becomes progressively slower, the land surface approaches the base level surface of zero elevation as time approaches infinity. Under this program, the ultimate goal can never be reached. Instead, after the passage of some millions of years, the land surface is reduced to an undulating surface of low elevation, referred to by geologists as a *peneplain*. This term was coined by an American, W. M. Davis, who combined the words *penultimate* and *plain*.

Production of a peneplain requires a high degree of crustal and sea-level stability for a period of many millions of years. Evidence of peneplains having been produced in the geologic past is well established by finding of erosion surfaces of low relief preserved beneath sedimentary strata of younger geologic age. To find a good example of a peneplain in process of development at the present time is not easy, since the continents reflect widespread crustal uplift by tectonic activity in late Cenozoic time. One region that has been cited as a possible example of contemporary peneplain is the Amazon-Orinoco basin of South America.

Peneplains that have been uplifted after their development, and are now high-standing land surfaces, are numerous. These regions are characterized by an upland surface of uniform elevation trenched by stream valleys that are graded with respect to a lower baselevel. Figure 14.26 shows a landscape interpreted by geologists as an uplifted peneplain deeply trenched by a stream valley. Notice how the gently undulating upland surface, used for agriculture, contrasts with the steep canyon walls.

In the United States, a good example of an uplifted peneplain is the Piedmont Upland lying east of the Blue Ridge mountains in Maryland, Virginia, the Carolinas, and Georgia. This rolling upland is underlain by igneous and metamorphic rocks. Here and there harder rock masses that were not reduced to the general level of the peneplain stand as isolated mountains. An example is Stone Mountain, near Atlanta, Georgia, which rises several hundred feet above the surrounding Piedmont surface (Figure 14.27).

The environmental significance of peneplains such as the Piedmont Upland lies in the thick soils they have developed over millions of years. In a warm, humid climate the soil is underlain by thick saprolite (Chapter 11). Because of the gentle land surface gradients, uplands are intensively cultivated, with the result that accelerated soil erosion can develop with devastating effects, including deep gullying of the saprolite.

Fluvial denudation systems have been operating continuously throughout all recorded geologic time. Huge volumes of sediment have been carried to the sea and deposited, mostly in marginal accumulation zones—geosynclines—with only a small part reaching the deep ocean floors. Perhaps we can gain insight into the total significance of this process by estimating current rates of mass transport and extrapolating these data over long spans of time.

Sheldon Judson, a geologist, estimates that the current rate of mass transport of sediment from continents to oceans by streams is a little over 9 billion metric tons per year. Added to this quantity is a much smaller transport of sediment by glacial ice and winds—perhaps somewhere between 0.2 and 0.5 billion tons per year. The total transport is therefore estimated at about 10 billion (10^{10}) tons per year. Geologists estimate that the total quantity of sedimentary rock present in the earth's crust is just under 2×10^{18} metric tons.

Figure 14.26 The gently rolling upland surface in the distance is part of the St. John peneplain; its elevation is about 2000 ft (600 m). The peneplain is deeply trenched by Canyon de San Cristobal (*foreground*), carved by the Rio Usabon. (Photograph by R. P. Briggs, U.S. Geological Survey.)

Dividing by 10^{10} gives a figure of 2×10^8 years, or 200 million years, for the production of all existing sediment by denudation. This length of time is equal to that elapsed since the late Triassic Period, whereas the oldest known sedimentary

Figure 14.27 Stone Mountain, on the Piedmont upland near Atlanta, Georgia, is a striking residual mass of hard rock about 1.5 mi (2.4 km) long and rising 650 ft (193 m) above the surrounding Piedmont peneplain surface. The rock is a light-gray granite, almost entirely free of joints, and has been rounded into a smooth dome by weathering processes. (Photograph by U.S. Army Air Service.)

rocks (now metamorphosed) are of Precambrian age and are about 3 billion years (b.y.) in age. Where has most of the sediment produced since −3 b.y. gone? One answer is that it has been recycled through melting into igneous rock, and then by alteration of igneous minerals, into new sediments. If so, and if our calculations are correct, about 25 such cycles have occurred in earth history. Even if the estimates of sediment transport are too large by a factor of two or even four, a great deal of recycling may have taken place.

Readjustments of Stream Grade

A graded stream, delicately adjusted to its environment of supply of water and rock waste from upstream sources, is highly sensitive to changes in those controlling parameters. Changes in climate and in surface characteristics of the watershed bring changes in discharge and load at downstream points, and these changes in turn require channel readjustments.

Consider first the effect of an increase in bed load beyond the capacity of the stream. At the point on a channel where the excess load is introduced, the coarse sediment accumulates on the stream bed in the form of bars of sand, gravel, and pebbles (see Table 8.2). These deposits raise the elevation of the stream bed, a process called *aggradation*. As more bed materials accumulate the stream channel gradient is increased, and the increased flow velocity enables bed materials to be dragged downstream and spread over the channel floor at progressively more distant downstream reaches. But the building up of the channel also reduces the channel gradient upstream from the place where excess load is entering, decreasing the stream capacity in that reach. As a result, bed materials accumulate in the upstream direction, as well, and the effects of aggradation progress headward in the system.

Aggradation typically changes the channel cross section from one of narrow and deep form (large form ratio, d/w) to a wide, shallow cross section (small form ratio). Because bars are continually being formed, the flow is divided into multiple threads and these rejoin and subdivide repeatedly to give a typically *braided* form to the channel (Figure 14.28). The coarse channel deposits spread across the former floodplain, burying fine-textured alluvium with coarse material.

How is aggradation induced in a stream system by natural processes? Figure 14.28 illustrates one natural cause of aggradation that has been of major importance in stream systems of North America and Eurasia during the Pleistocene Epoch. Advance of a valley glacier has resulted

Figure 14.28 The braided stream in the foreground is aggrading the floor of a glacial trough. A shrunken glacier in the distance provides the meltwater and debris, Peters Creek, Chugach Mountains, Alaska. (Photograph by Steve McCutcheon, Alaska Pictorial Service.)

in the input of a large quantity of coarse rock debris at the head of the valley. (Glacier action is explained in Chapter 17.) During summer periods of rapid ice-melting the stream is supplied with a greater quantity of bed load than it can transport, so that aggradation spreads far down the valley. Valley aggradation was widespread in a broad zone marginal to the great ice sheets of the Pleistocene Epoch, and the accumulated alluvium filled valleys to depths of many tens of feet. Figure 14.29A shows a valley filled in this manner by an aggrading stream. The case could represent any one of a large number of valleys in New England or the Middle West.

Suppose, next, that the source of bed load is cut off or greatly diminished. In the case illustrated in Figure 14.29, the ice sheets have disappeared from the headwater areas, and with them the supplies of coarse rock debris. Reforestation of the landscape has restored a protective cover to valley-side and hill slopes of the region, holding back coarse mineral particles from entrainment in overland flow. Now the streams have copious water discharges but little bed load. In other words, they are operating below capacity. The result is channel scour and deepening. The channel form becomes deeper and narrower. Gradually, the stream profile is lowered in elevation, a process of *degradation*. Because the stream is very close to being in the graded condition at all times, its dominant activity is lateral (sidewise) cutting by growth of meander bends, as shown in block B of Figure 14.29. The valley alluvium is gradually excavated and carried downstream, but it cannot all be removed because the channel enounters in many places hard bedrock that lies beneath the alluvium. Consequently, as shown in block C of Figure 14.29, there remain step-like surfaces on both sides of the valley. The treads of these steps are *alluvial terraces*. The terraces are bounded by steep *terrace scarps*, carved by the stream in the form of arcs concave toward the channel.

Alluvial terraces have always attracted occupation by Man because of their advantages over both the valley-bottom floodplain, which is subject to annual flooding, and the hill slopes beyond, which may be too steep and rocky to cultivate. Terraces, on the other hand, were easily tilled and made prime agricultural land. Towns were easily constructed on the flat ground of a terrace, and roads and railroads were easily run along the terrace surfaces parallel with the river. Consequently, control of cultural patterns by alluvial terraces is seen widespread throughout the central and eastern United States. (Figure 14.30).

A graded stream experiences a major change in activity when the crust beneath it is raised with respect to sea level. Such broad-scale upwarping of the continental crust is referred to as *epeirogenic movement* in contrast to orogenic movement, which causes surface dislocation by faulting and folding. During and following upwarping, the stream undergoes channel degradation in an attempt to reestablish grade at a lower level. This process, termed *rejuvenation*, begins as a series of rapids at the stream's mouth, where the water passes from the former mouth down to the lowered sea level. The rapids quickly shift upstream, and soon the entire stream valley is being trenched to form a new valley (Figure 14.31).

If rejuvenation occurs when a stream has already developed a floodplain, the effect is to give a steep-walled inner gorge, on either side of which lies the former floodplain, now a flat terrace high above river level (Figure 14.32). This

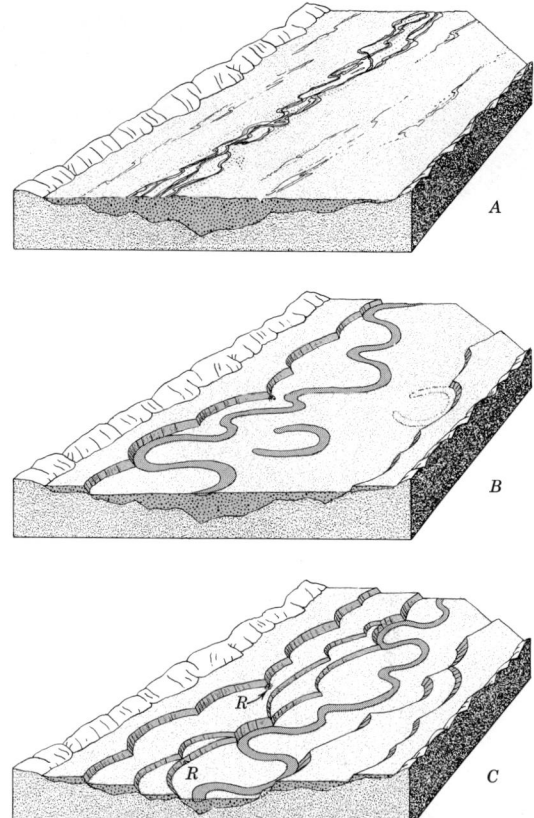

Figure 14.29 Alluvial terraces form when a graded stream slowly cuts away the alluvial fill in its valley. At points labeled R, outcrops of hard rock protect the adjacent terraces from further undermining.

feature is called a *rock terrace* to distinguish it from an alluvial terrace.

Where the graded stream had developed meanders on a broad floodplain, rejuvenation causes the meanders to become impressed into the bedrock and give the inner gorge a meandering pattern. These sinuous bends are termed *entrenched meanders* to distinguish them from the floodplain meanders of an alluvial river (Figure 14.33).

Although entrenched meanders are not free to shift about as floodplain meanders, they can enlarge slowly so as to produce cutoffs. Cutoff of an entrenched meander leaves a high, round hill surrounded on three sides by the deep abandoned river channel and on the fourth by the shortened river course (see rear part of Figure 14.33). As you might guess, these hills formed ideal natural fortifications. Many European fortresses of the Middle Ages were built on such cutoff meander spurs. A good example is Verdun, near the Meuse River.

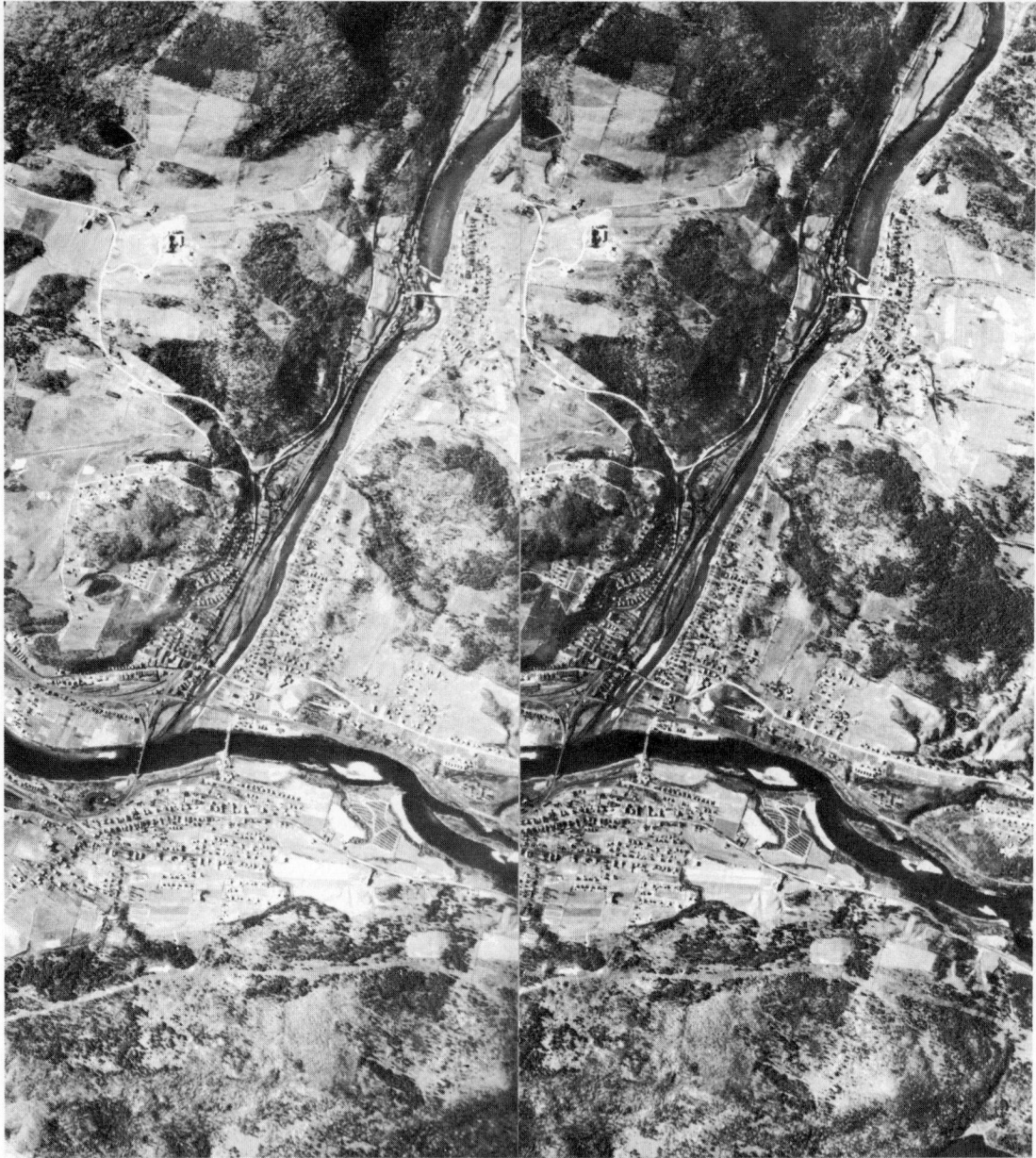

Figure 14.30 Vertical air photographs (stereopair) showing terraces of the Connecticut and White rivers at White River, Vermont. Note the occupation of the multiple terrace treads by fields, roads, and houses. Wooded slopes on older crystalline rocks are largely unoccupied. To obtain three-dimensions, view with stereoscope. (U.S. Geological Survey photographs.)

Under unusual circumstances, where the bed rock includes a strong, massive sandstone formation meander cutoff leaves a *natural bridge*, formed by the narrow meander neck (Figure 14.33). One well-known example is Rainbow Bridge at Navajo Mountain, in southeastern Utah; other fine examples can be seen in natural Bridges National Monument at White Canyon in San Juan County, Utah.

Entrenched meanders do not offer ideal locations for railroads and highways, but in a few instances they have been the best available choices for arteries of travel. This point is well illustrated by the Moselle River, whose winding entrenched meanders through the Ardennes mountain upland of Belgium and Western Germany have been utilized (Figure 14.34). Engineers have cut tunnels through the narrow necks.

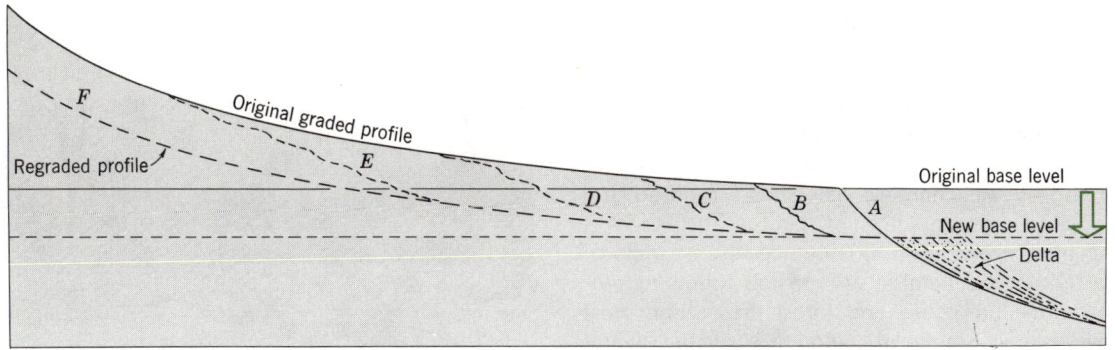

Figure 14.31 A drop in base level brings on rejuvenation and regrading of the stream profile, starting at point A and progressing upstream. (From A. N. Strahler, 1971, *The Earth Sciences*, 2nd ed., Harper & Row, New York.)

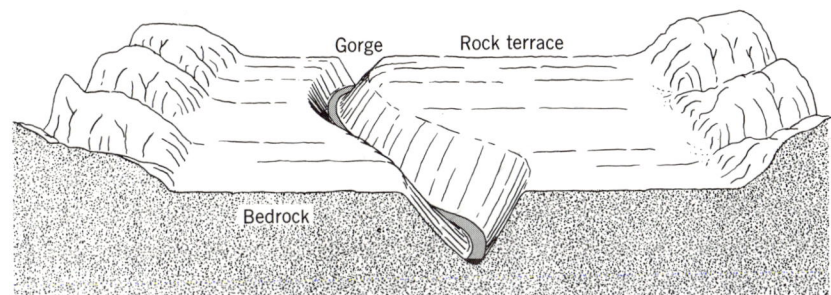

Figure 14.32 Following rejuvenation, a winding gorge has been carved into a former floodplain, which has become a high rock terrace. (From A. N. Strahler, 1971, *The Earth Sciences*, 2nd ed., Harper & Row, New York.)

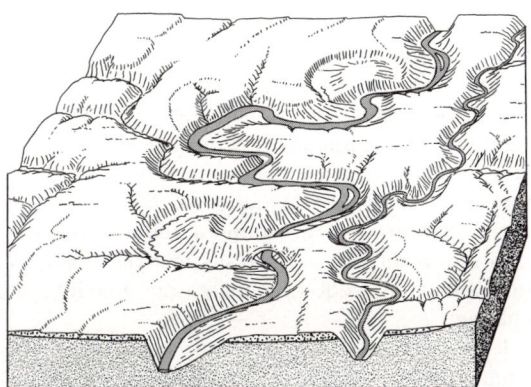

Figure 14.33 Rejuvenation of a meandering stream has produced entrenched meanders. One meander neck has been cut through, forming a natural bridge.

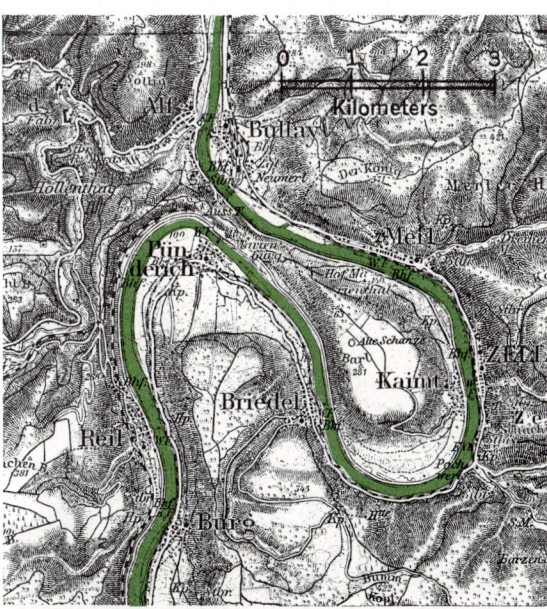

Figure 14.34 Entrenched meanders of the Moselle River in the Eifel district of Western Germany. The railroad following the river bank passes by tunnel through the narrow meander neck. (Portion of German 1:100,000 topographic map, 1890.)

Aggradation and Sedimentation Induced by Man

With an understanding of the principles of stream aggradation and degradation, it is possible to interpret and predict the impact of Man's activities upon stream channels. The most common and obvious consequence of land disturbance is channel aggradation; it is induced by a variety of activities. Accelerated soil erosion following cultivation, lumbering, and forest fires is the most widespread source of sediment for valley aggradation. On the other hand, this form of aggradation may be less conspicuous than other forms, since the land disturbance involves only the uppermost soil horizons. The results are typically seen in a gradual accumulation of sandy colluvium and alluvium in the smaller valleys, burying the finer-textured soils and lowering the agricultural quality of the valley bottoms. The silts, clays, and organic particles (humus) suspended in the runoff are carried far down valley to distant sites of deposition. In localities of particularly severe soil erosion, such as that which affected the Piedmont Upland of the southeastern states and was manifested in deep gullying, aggradation was rapid in valley bottoms, inundating the surface with sediment and destroying the surface for productive use. Filling of reservoirs by sediment is a major consequence of this type of accelerated valley sedimentation.

Mining operations have been the cause of extreme aggradation of channels in many places. One particularly important example was the consequence of hydraulic mining of gold-bearing gravels in the Sierra Nevada range of California. Along the walls of valleys high in the range are thick gravels containing the gold particles. These gravels were carved out by mine operators using powerful water jets. The gravel was passed through sluices to trap the gold. The gravel was then swept into the nearest stream channel and carried downstream. As aggradation extended to channels in the lowlands of the Great Valley, flooding of agricultural lands began to set in and rapidly reached a point that required preventative action. As a result, legislation was passed to regulate hydraulic mining, requiring that the gravel be trapped and retained at the site.

Aggradation of channels has also been a serious form of environmental degradation in coal-mining regions, along with water pollution (acid mine drainage) described in Chapter 13. Throughout the Appalachian coal fields, channel aggradation is widespread because of the huge supplies of coarse sediment from mine wastes (Figure 14.35). Strip mining has enormously increased the

Figure 14.35 Valley bottom in Kentucky choked with coarse debris from a strip mine area. The natural channel has been completely buried under the rising alluvium. (Photograph by W. M. Spaulding, Jr., Fisheries and Wildlife, U.S. Department of the Interior.)

aggradation of valley bottoms because of the vast surfaces of broken rock available to entrainment by runoff.

Urbanization and highway construction are also major sources of excessive sediment, causing channel aggradation. Major earth-moving projects are involved in creating highway grades and preparing sites for industrial plants and housing developments. While these surfaces are eventually stabilized, they are vulnerable to erosion for periods ranging from months to years. The regrading involved in these projects often diverts overland flow into different flow paths, further upsetting the regimen of streams in the area.

Mining, urbanization, and highway construction not only cause drastic increases in bed load, which cause channel aggradation close to the source, but also increase the suspended load of the same streams. Suspended load travels downstream and is eventually deposited in lakes, reservoirs, and estuaries far from the source areas. This sediment is particularly damaging to the bottom environments of aquatic life. Moreover, the accumulation of fine sediment reduces the capacity of reservoirs and results in rapid filling of tidal estuaries, requiring increased dredging of channels.

It has been estimated that the yield of sediment from strip-mined areas is as much as one thousand times as large as from the same land surfaces in a natural condition. This ratio is seen in an estimate that in strip-mined areas of Kentucky the spoil banks yielded some 27,000 tons of sediment per square mile in a 4-year period while the yield from undisturbed forested areas was only 25 tons per square mile.

The effect of urbanization upon suspended sediment yield is well illustrated by data from the Baltimore-Washington region, where expanding suburbs are replacing agricultural lands or lands that were previously in agricultural use but have lain abandoned for decades. Portions of the developing area that lie within the Piedmont region are rolling surfaces underlain by thick residual soils and saprolite. The climate is one of ample rainfall (42 in./yr; 110 cm/yr) well-distributed throughout the year, but with high-intensity rains more common in the summer. In wooded rural areas the sediment yield is on the order of 200 to 300 tons/sq mi/yr, while for areas being farmed the yield is on the order of 500 tons. On abandoned farm lands the yield has dropped well below the level of cultivated land. In contrast, sediment yield from construction sites for housing developments and industrial parks has been measured as ranging from 1000 to over 100,000 tons/sq mi/yr. The lower figures apply to larger watersheds for which the proportion of disturbed area is small; the high figures to small watersheds largely occupied by the construction site. Sediment concentrations are high in streams throughout the year in this region.

Suspended sediment is a form of water pollution and may make water unfit for use in municipal water supplies and for certain industrial uses. Turbidity of reservoirs and lakes also represents a deterioration of esthetic qualities of those water bodies with resultant loss of recreational value.

While accelerated sediment production will always be associated with urbanization, much can be done to reduce the concentrations and effects. Regulation includes limitation of duration of bare surfaces to exposure and the immediate application of erosion reduction measures, such as mulching and planting of graded surfaces, and the construction of sedimentation basins to trap coarser sediment.

Environmental Effects of Large Dams

Large dams, such as those mentioned on earlier pages of this chapter and in Chapter 13 introduce major side effects into the river channel far upstream and downstream from the dam and its reservoir. We have already considered in Chapter 13 the effects of reservoir evaporation upon quality of water.

Let us take up first the upstream changes brought about by a large dam. Bedload and most if not all of the suspended load brought into the reservoir comes to rest in the standing water. This activity is a form of delta-building, explained in Chapter 15. The delta surface constitutes a horizontal extension of the graded stream channel, as illustrated in Figure 14.36. However, because a gradient is required to move bedload across the deposit, aggradation accompanies lengthening of the channel. The aggradation is propagated upstream as a thickening wedge of coarse alluvium, which buries fine-grained floodplain deposits. A possible example is the valley of the Rio Grande,

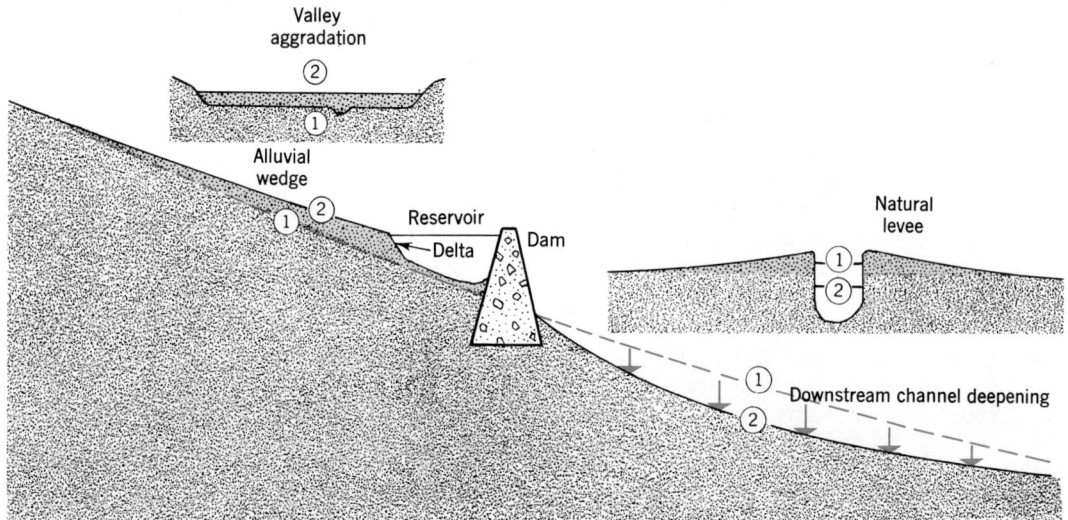

Figure 14.36 Schematic profile and cross sections of a river showing upstream and downstream effects of a dam and reservoir. (From A. N. Strahler, 1972, *Planet Earth; Its Physical Systems Through Geologic Time*, Harper & Row, New York.)

upstream from the Elephant Butte Reservoir in New Mexico. In a 30-year period after completion of the reservoir aggradation had reached a thickness of 10 ft (3 m) at the head of the reservoir. The village of San Marcial was literally buried in these deposits. At the city of Albuquerque, 100 mi (160 km) upstream from the reservoir, aggradation in the same period had reached a depth of 4 ft (1.2 m).* The process will continue indefinitely, since the delta deposit in the reservoir will continue to be extended. Regrading of the channel must accompany this extension in order to maintain a gradient on which to transport the load. Ultimately, however, after a reservoir is completely filled and the stream itself begins to flow over the dam, a fixed grade will eventually be reestablished.

In the downstream direction another set of changes takes place. Below the dam, water without load is released in large, but controlled discharge; it flows out over a channel previously adjusted for the transport of a large quantity of coarse bed load. To satisfy its capacity for bed material, the stream scours its bed and lowers the channel to a new gradient. Below Hoover Dam on the Colorado River the scour ceased only when a residual layer of boulders prevented further lowering. This phenomenon is described by engineers as *armoring* of the bed. Farther downstream a new equilibrium was established. At Yuma, 350 mi (560 km) below Hoover Dam, permanent channel changes consisted of a lowering of average position of the stream bed by about 10 ft (3 m). Stream depth has about doubled, but the stream surface in high stages is only a little lower than before, because of downcutting. At Yuma the stream load has been reduced to about $\frac{1}{15}$ of its original value, and this comes from tributary streams and as direct contributions from bank caving. Because of lowering of the stream bed, the channel is set lower with respect to the crest level of its banks (natural levees) and the result has been to render unworkable an irrigation system of gravity flow previously developed along reaches of the river that have a floodplain. Instead, costly pumping must be used to lift the water for irrigation. Channel degradation below dams has been documented for many streams, with average lowering of the bed on the order of 1 to 2 ft (0.3 to 0.6 m) for such rivers as the Missouri, Red, Canadian, Platte, and Rio Grande.

*Aggradation of the Rio Grande channel had been in progress at San Marcial for two decades prior to filling of the reservoir. This sediment was coming in large part from tributaries heading in areas experiencing accelerated erosion. The stated effects of the dam are therefore not clearly demonstrated.

Environments of Alluvial Rivers and Their Floodplains

An *alluvial river* is one that flows upon a thick accumulation of alluvial deposits constructed by the river itself in earlier stages of its activity. Depth of alluvium is at least as great as the maximum depth to which the channel is scoured in flood, and may be much deeper. A characteristic of an alluvial river is that it experiences overbank floods with a frequency ranging between annual and biennial occurrence during the season of large water surplus over the watershed. This surplus may occur in a spring period of rapid snowmelt or in a winter rainy season in middle latitude regions, or during the wet monsoon season in tropical zones (see Chapter 13 and Figure 13.19). Overbank flooding of an alluvial river normally inundates part or all of a floodplain that is bounded on either side by rising *bluffs*. These bluffs consist of nonalluvial materials, or in some cases of older alluvium.

Typical landforms of an alluvial river and its floodplain are illustrated in Figure 14.37. Dominating the floodplain is the meandering river channel itself, and abandoned reaches of former channel, as well (Figure 14.38). the cutoff of a meander is quickly followed by deposition of silt and sand across the ends of the abandoned channel, producing an oxbow lake. The oxbow lake is gradually filled in with fine sediment brought in during high floods and with organic matter produced by aquatic plants. Eventually the oxbows are converted into swamps, but their identity is retained indefinitely.

The growth of alluvial meanders leaves distinctive marks upon the floodplain. As shown in Figure 14.39, the channel in a meander bend is deep

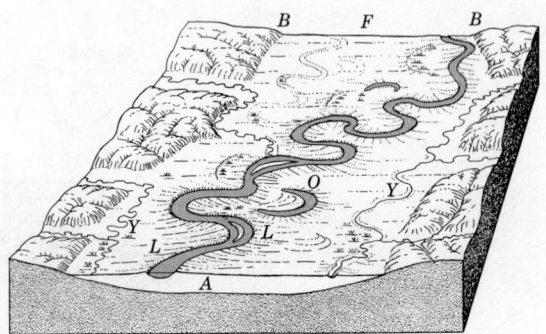

Figure 14.37 Landforms of an alluvial river floodplain with freely developed meanders. L = levees; O = oxbow lake; Y = yazoo stream; A = alluvium; B = bluffs; F = floodplain.

close to the outside or *undercut* bank, which yields by caving and allows the bend to grow in radius. Here the channel has its greatest depth, forming a *pool*. As flow passes from one bend to the next the threads of swiftest current cross the channel diagonally in a zone known as the *crossing*. This element of the channel, which is shallow and has many shifting bars, constitutes a *riffle*. Thus pools and riffles occur in alternation, corresponding with each meander bend. The meander bend not only grows laterally, but also shifts slowly down-valley in a migratory movement known as *down-valley sweep*. Combined effect of meander growth and sweep gives to the point-bar deposits nested arcuate patterns consisting of *bars* (embankments of bed material) and *swales* (troughs between bars). *Bar-and-swale topography* thus produced is clearly visible in Figure 14.38.

During periods of overbank flooding, when the entire floodplain is inundated, water spreads from the main channel over adjacent floodplain deposits. As the current rapidly slackens, sand and silt are deposited in a zone adjacent to the channel; these sediments form *overbank deposits*. The result is an accumulation known as a *natural levee*. Because deposition is heavier closest to the channel and decreases with distance away from the channel, the levee surface slopes away from the channel (Figure 14.40). Referring back to Figure 13.6, notice that the higher ground of the natural levees is revealed by a line of trees on either side of the channel. Between the levees and the bluffs is lower ground, which is often poorly drained and constitutes the *back-swamp* (Figure 14.40).

Overbank flooding results not only in the deposition of a thin layer of silt upon the floodplain, but brings an infusion of dissolved mineral substances which saturate the soil. As a result of base cation infusions, floodplain soils retain their remarkable fertility in regions of soil-water surplus from which these bases are normally leached out under the pedogenic regimes of podzolization and laterization (Chapter 11).

One effect of the growth of natural levees is to inhibit the free flow of tributary streams that enter upon the flood-plain. Typically, the tributary stream flows down-valley in the back-swamp zone to a point where the main channel impinges upon the floodplain bluffs. A stream whose junction is deferred in this manner is called a *yazoo stream*, after the Yazoo (Tallahatchie) River, a tributary of the Mississippi River (Figure 14.37).

Alluvial rivers attracted Man's habitations long before the dawn of recorded history. Early civilizations arose in the period 4000 to 2000 B.C. in

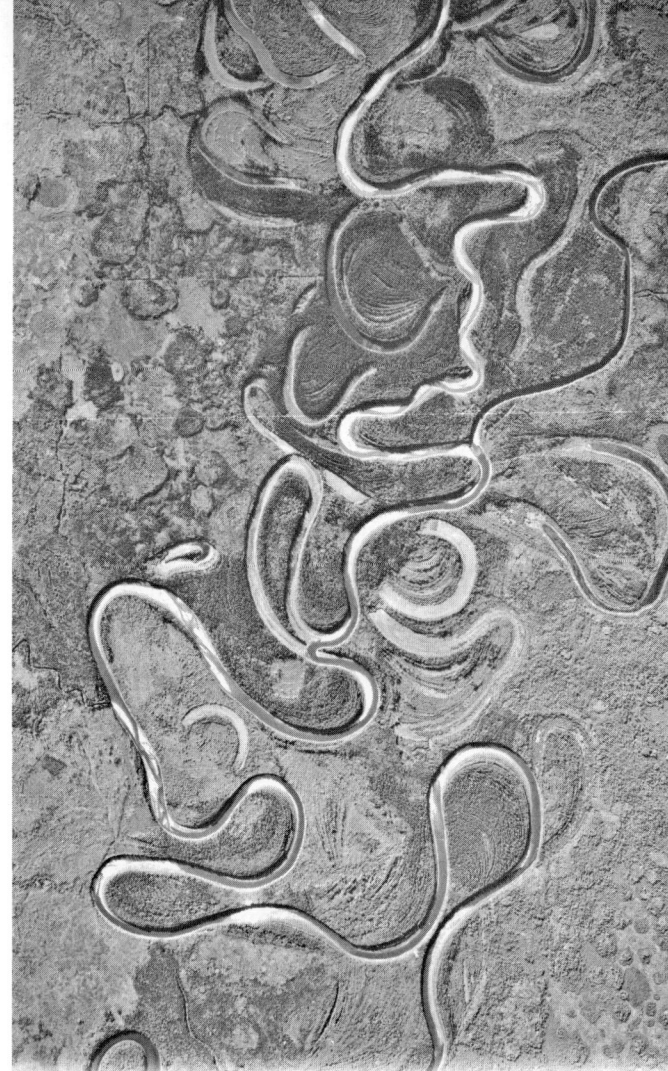

Figure 14.38 This vertical air photograph, taken from an altitude of about 20,000 ft (6100 m) shows meanders, cutoffs, oxbow lakes and swamps, and floodplain of the Hay River, Alberta (lat. 58° 55′ N, long. 118° 10′ W). (National Air Photo Library, Surveys and Mapping Branch, Department of Energy, Mines and Resources, Canada.)

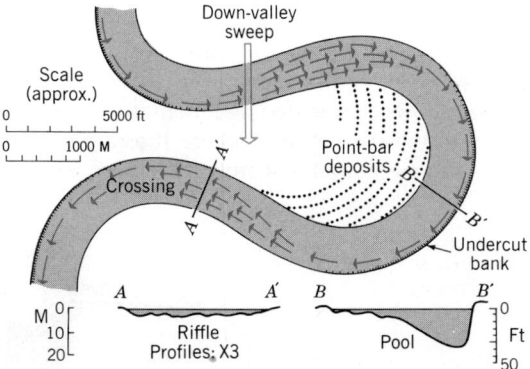

Figure 14.39 An idealized map and cross-profiles of a meander bend of a large alluvial river, such as the lower Mississippi River. Small arrows show position of the swiftest current. (From A. N. Strahler, 1971, *The Earth Sciences*, 2nd ed., Harper & Row, New York.)

alluvial valleys of the Nile River in Egypt, the Tigris and Euphrates Rivers of Mesopotamia, the Indus River of what is now West Pakistan, and the Yellow River of China. Fertile and easily cultivated alluvial soils, situated close to rivers of reliable flow, from which irrigation water was easily lifted or diverted, led to intense utilization of these fluvial zones. (With respect to human culture, the alluvial river and its floodplain are sometimes referred to as the *riverine environment*.) Today about half of the world's population lives in southern and southeastern Asia; the bulk of these persons are small farmers cultivating alluvial soils of seven great river floodplains.

To the agricultural advantages of the alluvial zone are added the value of the river itself as an artery of transportation. Navigability of large alluvial rivers led to growth of towns and cities, many situated at the outsides of meander bends, where deep water lies close to the bank (see Greenville, Mississippi, Figure 14.41), or at points on the floodplain bluffs where the river is close by (examples: Memphis, Tennessee, and Vicksburg, Mississippi).

Effects of Regulation of Alluvial Rivers

Methods practiced in attempts to prevent overbank flooding along alluvial rivers have consisted of two basic procedures (Chapter 13): (a) construction of levees and (b) shortening of the river by artificial cutoffs. Success of a levee system also has some undesirable side effects upon the environment of the floodplain. If overbank flooding is effectively prevented year after year, normal sedimentation in oxbows, abandoned channel reaches, and back-swamps is cut off. Not only is a normal process of upbuilding thus prevented, but the nutrient base cations needed by plants are no longer added annually to the floodplain surface, as we have already noted. As to the main river channel, confinement of all flow between levees allows suspended sediment to travel directly to the river mouth, where the rate of offshore sedimentation is sharply increased. Bottom environments are thus changed, often with deleterious effects upon shellfish and other bottom-dwelling organisms of shallow coastal waters.

A program of artificial cutoffs was begun by the U.S. Army Corps of Engineers in the early 1930s after that organization took over river control from the Mississippi River Commission. Figure 14.41 shows a series of these cutoffs about two years after completion. Reduction in channel length steepened the average channel gradient and caused channel deepening, which in turn allowed flood discharges to pass through without rising to levels that would overtop the artificial levees.

Unfortunately, nature does not passively accept such artificial channel control, but works instead to restore the meander bends. Only by extremely expensive control works can the return to a serpentine course be prevented. It is interesting to ponder upon the point that for over a century prior to Man's interference, the total length of channel of the Mississippi River between Cairo and Baton Rouge—some 850 mi (1370 km)—remained about constant despite the growth and cutoff of many meander bends. Figure 14.42 is a set of four channel maps superimposed to show changes in one stretch of the Mississippi River between 1820 and 1932, during which period the river was allowed to change its form without appreciable restriction. Notice meander cutoff and growth, and a general down-valley sweep. Clearly, the river was compensating for cutoffs and other forms of channel shortening by an equal average rate of channel lengthening.

Stream Channelization

Related to flood-regulation practiced on large alluvial rivers, but on a much smaller scale, is a form of environmental modification applied to smaller streams and their floodplains. Under the general term of *stream channelization*, modification consists of straightening, widening, and deepening of channels in order to prevent seasonal overbank flooding and to provide permanent drainage to water-saturated bog soils of the

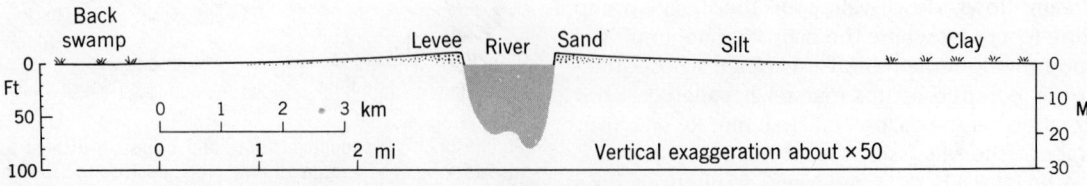

Figure 14.40 Profile across the Mississippi River showing the natural levees flanking the channel. Notice the great vertical exaggeration of the profile. (From A. N. Strahler, 1971, *The Earth Sciences*, 2nd ed., Harper & Row, New York.)

Stream Channelization

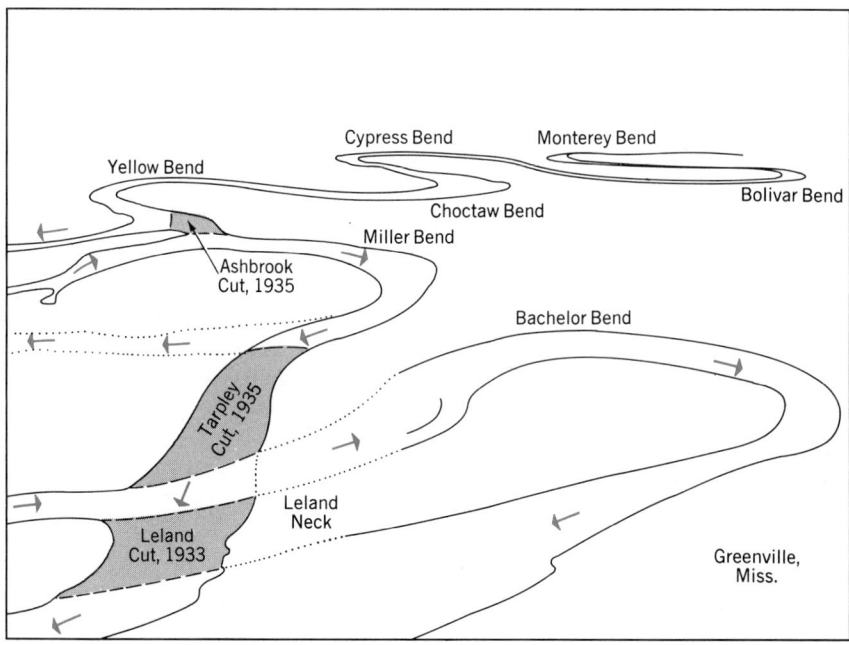

Figure 14.41 In the oblique air photograph, looking north, we see three artificial meander cutoffs of the Mississippi River. The photograph was taken in 1937. The diagram below gives the names and dates of each cutoff. White patches in the photograph are sediment plugs blocking the ends of Batchelor Bend, in which the river formerly flowed past the city of Greenville, Mississippi (*lower right*). (Photograph by War Department, Corps of Engineers. Diagram from A. N. Strahler, 1971, *The Earth Sciences*, 2nd ed., Harper & Row, New York.)

floodplain. Meanders are eliminated in the process. Forest is cleared from the floodplain and channel banks. The land thus drained is placed in cultivation.

Stream channelization carried out by the U.S. Department of Agriculture, Soil Conservation Service, largely since the mid-1950s, has altered more than 8000 mi (13,000 km) of channelways of smaller streams not affected by flood regulation activities of the army Corps of Engineers. A large area of land has thus been brought under cultivation. On the other hand, economic benefits

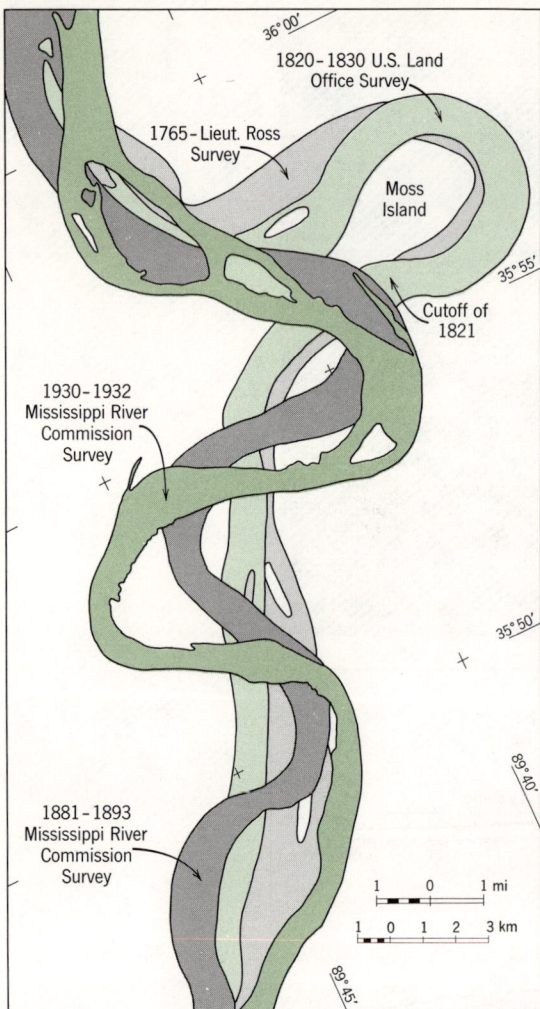

Figure 14.42 A series of four surveys of the Mississippi River shows considerable changes in the position of the channel and the form of the meander bends. Note that one meander cutoff has occurred (1821) and new bends are being formed. (After U.S. Army Corps of Engineers.)

Fluvial Processes and Forms of Arid Environments

Much of the world's land surface lies in a regime of sustained potential water deficit. Potential evapotranspiration greatly outweighs precipitation in these warm deserts. Once soil moisture has been depleted evapotranspiration stops, and when precipitation does again occur, it is barely enough to restore the soil water to its field capacity. as a result, surplus water is generally not available over upland surfaces for percolation to a ground-water body, and consequently base flow is not found. Stream channels are therefore normally dry except when fed by direct surface flow immediately following high-intensity rainstorms. (These streams are described as *ephemeral*.) Yet, on such occasions overland flow can entrain much debris and can fill channels to bankfull stage with a raging torrent. On steep gradients these desert floods perform a great deal of erosion and transportation in a short period of flow.

Furthermore, desert precipitation, being largely from convective cells, is highly localized. A torrential rain in one small watershed may be entirely lacking in another watershed a mile or two distant. Added to this localization of precipitation is the orographic effect, in which mountain ranges promote convection and trap copious rainfall in contrast to low-lying, intermontane basins and plains. Channel flow derived from mountain watersheds typically decreases downstream through direct evaporation and the infiltration of water into highly permeable beds of coarse alluvium. The desert stream is typically *influent,* losing discharge by percolation; whereas the stream of humid lands is typically *effluent,* gaining discharge by base flow (Figure 14.43). In alluvial basins of arid lands, a ground-water body is built up and recharged by influent streams

of increased crop production have been offset in part by economic losses through reductions in local populations of fish and waterfowl. Severe damage or even total destruction of the ecosystem of the natural channel and its adjacent wetlands results from channelization. Increasing awareness of environmental degradation accompanying channelization has brought increasing opposition to the activity. Conservation-minded members of the Department of Interior, particularly those concerned with fish, wildlife, and parks, have voiced opposition to further alteration of wetlands by channelization and related engineering activities.

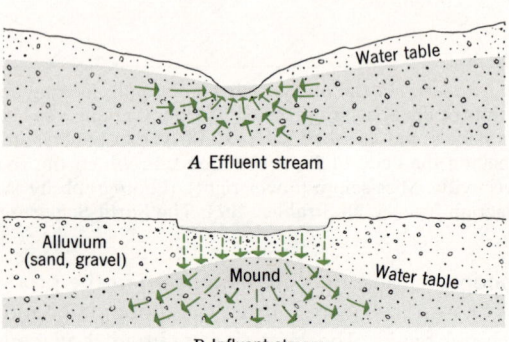

Figure 14.43 Effluent and influent streams. (From A. N. Strahler, 1971, *The Earth Sciences*, 2nd ed., Harper & Row, New York.)

Fluvial Processes of Arid Environments

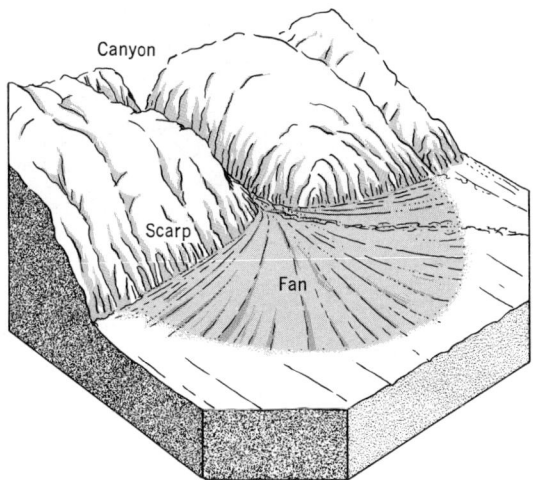

Figure 14.44 A simple alluvial fan.

deriving flow from mountain watershed. A water-table mound forms beneath the channel. Ground-water reserves of alluvial basins of California, referred to in Chapter 12, are of this type.

The combination of stream flow issuing from mountain canyons and carrying a heavy load of coarse debris, together with the loss of discharge through evaporation and influent seepage, causes the growth of a distinctive landform, the *alluvial fan* (Figure 14.44). Its capacity reduced by diminishing discharge, the stream aggrades its channel after it emerges from the canyon. Aggradation in turn causes sidewise shifting of the channel, but the narrow canyon mouth, lined with resistant bedrock, acts as a fixed point on the channel. as a result, the shifting braided stream sweeps in a radial manner and builds a conical deposit of alluvium. Since the larger particles of the bed load come to rest first, while the finer ones travel further, the fan shows a size gradation from coarse to fine, and a decrease in gradient, from the fan

Figure 14.45 A great alluvial fan in Death Valley, built of debris swept out of a large canyon. Notice that the main stream issuing from the canyon has regraded its profile to a lower gradient, trenching the head of the fan and simultaneously building newer alluvial deposits (*lower left*). Older remnant fan surfaces are being eroded by many narrow, subparallel streams. (Copyrighted Spence Air Photos.)

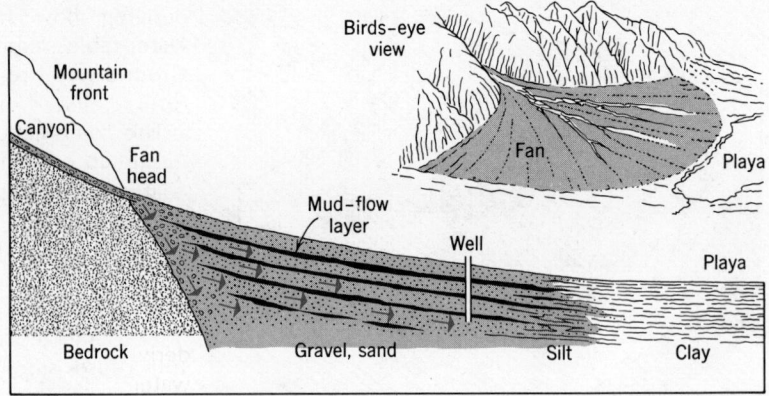

Figure 14.46 Idealized cross section of a complex alluvial fan showing the movement of ground water from fan head through aquifers of gravel and sand. (From A. N. Strahler, 1972, *Planet Earth; Its Physical Systems Through Geologic Time*, Harper & Row, New York.)

apex to its outer periphery. Over long periods of time, where great mountain blocks rise high above low basins, large and complex alluvial fans have accumulated and may have radii several miles long (Figure 14.45).

Complex desert fans also include *mudflows*, mud streams which are interbedded with the channel deposits (Figure 11.27). As a result, the structure of such a fan is favorable to the accumulation of ground water under artesian pressure (Figure 14.46). Water infiltrating the fan apex makes its way down the inclined alluvial beds, which are the aquifers in the system, and is held under pressure under mudflow layers, which are

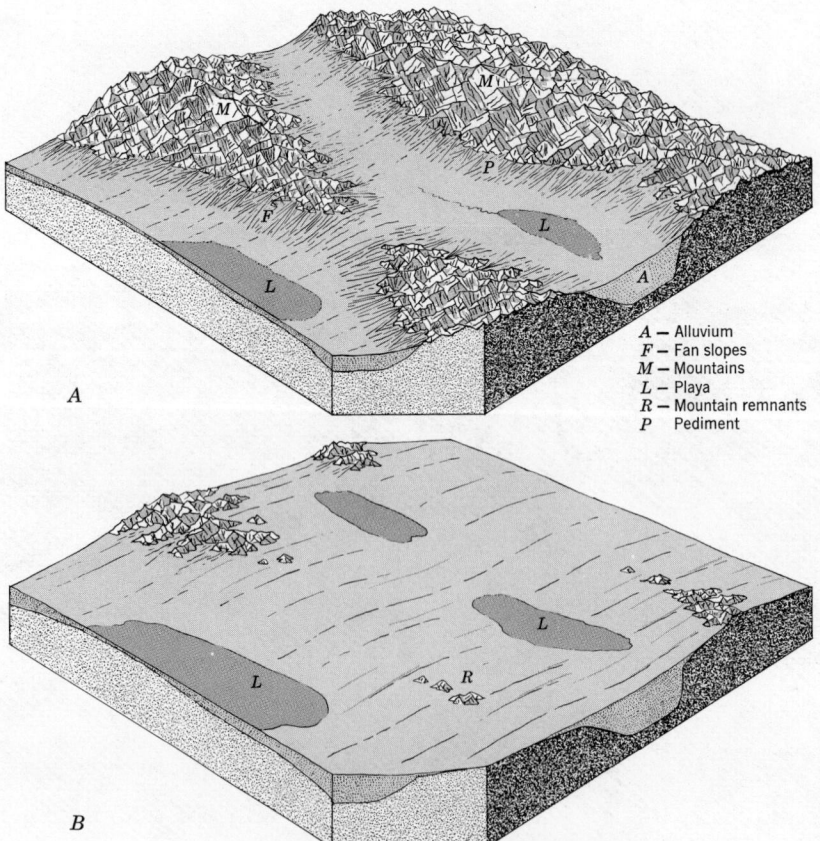

Figure 14.47 Idealized diagrams of landforms of the mountainous deserts of the southwestern United States. *A.* Stage of rapid filling of tectonic basins with debris from high, rugged mountain blocks. *B.* Advanced stage with small mountain remnants and broad playa and fan slopes.

Fluvial Processes of Arid Environments

Figure 14.48 This air view of Death Valley, California, shows a desert landscape comparable to that shown in Figure 14.47A. (Copyrighted Spence Air Photos.)

the aquicludes. Alluvial fans are the dominant class of ground-water reservoirs in the southwestern United States. Sustained heavy pumping of these reserves for irrigation has lowered the water table severely in many fan areas. Rate of recharge is extremely slow in comparison. However, efforts are made to increase this recharge by means of waterspreading structures and infiltrating basins on the fan surfaces.

Closely related in outward form to the alluvial fan is a desert land surface known as a *pediment*. Unlike the fan, however, the pediment is carved from bedrock by water erosion and forms a rock surface sloping gently from the steep mountain base toward the basin, where it passes down under a cover of alluvium (Figure 14.47). An explanation of the process of pediment formation is beyond the scope of this chapter, since it is complex and there are conflicting hypotheses of origin.

In the centers of desert basins lie the saline lakes and dry lake basins referred to in Chapter 13. Accumulation of fine sediment and precipitated salts produces an extremely flat land surface, referred to in the southwestern United States and in Mexico as a *playa*. Where an evaporite layer forms the surface the term *salt-flat* is applied. In other cases shallow water stands in the basin as a true salt lake.

Figure 14.47 shows in perspective the major landform assemblages of a mountainous desert region. Figure 14.48 is an air photograph of just such a landscape—the Death Valley region of southeastern California. The three environmental zones are: (1) rugged mountain masses dissected into branching canyons with steep, rocky walls; (2) a piedmont zone of alluvial fans or a pediment; and (3) the playa occupying the central part of the basin. It should be obvious that fluvial processes in such a region are limited to local transport of detritus from a mountain range to the nearest adjacent basin, which receives all of the sediment and must be gradually filling as the mountains are diminishing in elevation. Since there is no outflow to the sea, the concept of a base level of denudation has no meaning, and a peneplain does not represent the penultimate stage of denudation, as it does in the humid environment. Each arid basin becomes a closed system so far as mass transport is involved. Only the hydrologic system is open, with water entering as precipitation and leaving as evaporation.

Wind-generated water waves on oceans and lakes constitute a secondary kinetic energy system, driven by the primary system of atmospheric circulation. In Chapter 4 we investigated the transfer of momentum and energy from winds to the ocean surface, resulting in both progressive water waves and a drift of the surface water layer. Winds result from the pressure-gradient force, itself derived from inequalities in barometric pressure that are basically the result of non-uniform heating of the global atmosphere under solar radiation. Kinetic energy gained by air in horizontal motion is dissipated in frictional resistance, both within the moving air (because, like water, air has viscosity and resists shearing) and along the contact surface, or interface, with water or the solid land. Recall that surface friction slows the speed of winds close to the ground and causes their direction to be modified with respect to the geostrophic wind direction at high levels (Chapter 4, Figure 4.11).

The energy lost from winds by surface friction can be disposed of in two ways. First is by transporting momentum downward to the water layer beneath and thus transferring energy from the air to the water. This transferred kinetic energy takes the form of progressive oscillatory water waves and surface water motion. Over a land surface, a similar energy transfer takes the form of entrainment and downwind motion of solid particles (sand, dust, or snow) and the flexing of plant foliage and stems. Second, wind blowing over a solid surface of high strength and rigidity, such as bare rock surfaces, glacial ice, or the buildings and pavements of a city causes no observable motion but also dissipates energy. Although in theory these solids are elastically bent under wind stress the amount of distortion is not measurable. Instead, intensified shearing occurs within the wind near the surface and there develop intense eddies within the lower air layer. These shearing and eddying motions transform kinetic energy into heat energy, which in turn is radiated or conducted out of the system. Recall from Chapter 4 that frictional loss of energy of surface winds is least over smooth surfaces, but greatest over a rough terrain surface. Vegetation of a forest withdraws a comparatively large quantity of mo-

15
Waves, Currents, and Coastal Landforms

mentum from wind, with the result that an air layer near the ground may be nearly stagnant.

As wind generated waves enter shallow water they lose energy by frictional resistance with the solid bottom and by impacts with shore features. Secondary surges and currents are produced, and these in turn encounter frictional resistance. These water motions are capable of performing the geological roles of erosion, transportation, and deposition. It is with the geological processes and landforms of wave and current action that this chapter is concerned.

We shall need to bring into this chapter another energy system—one totally different in action from that of wind-driven waves and currents. This newcomer to our thinking is the *tide* of the oceans, a rhythmic rise and fall of ocean level caused by mutual gravitational attraction of earth and moon and of earth and sun. Tides are an expression of the kinetic energy of planetary bodies in motions of rotation on their axes and of revolution in orbits, one body around the other. The kinetic energy is inherited from the time of origin of the solar system, and in this respect the tidal energy system has something in common with the sun's internal energy system. Both originated at about the same time and both are running down, if considered from the point of view of geologic time. However, in spans of time as short as centuries, the energy budgets of both systems show no appreciable change. Like the solar constant of electromagnetic radiation, the energy of tides is almost perfectly reliable.

Tides produce horizontal water motions in the form of *tidal currents*, and these are superimposed upon the motions generated by shoaling waves. The resulting water motions are highly complex and subject to great variations in directions and velocities. The life environments of shallow ocean waters are strongly influenced by the total patterns and rhythms of water motions, not only because of the direct physical effects of the water motion, but also because those motions transport nutrients and wastes, as well as the sediment of which bottom environments are formed. Water salinity, a critical factor in environment of marine life, depends upon mixing of fresh water of continental runoff with salt water of the open oceans. The motions of water caused by waves, wave-induced currents, and tidal currents determine the mixing ratios of fresh and salt water.

As in the case of the gravity-flow systems of running water on the lands, Man is a potent agent of change in coastal environments maintained by action of waves, currents, and tides. Armed with powerful machines run by fossil fuels, Man can make radical alterations in the landforms of shorelines and estuaries. Certain of these natural environments are maintained in a stable state by a very delicate balance of forces and can easily be disrupted, with severe impact upon their ecosystems. We shall want to investigate these sensitive areas where environmental changes have occurred, and try to assess Man's role in implementing the change as distinguished from similar natural changes that have taken place repeatedly in the past as responses to fluctuations in climatic and hydrologic parameters. To make such assessments requires a good working knowledge of fundamental principles of wave action and tides.

First, however, let us agree on the meaning of two important terms: shoreline and coast. The *shoreline* is the line of contact of water surface with the land. Because this line fluctuates constantly with the changing water level of breaking waves and tides, we need to expand the term to include the entire surface over which the water line sweeps; this surface is the *shore* or *shore zone*. A *coast* is a zone that includes not only the shore, but also a shallow water zone adjoining it and a belt of land above the limit of water action that is influenced by marine processes. Consequently, cliffs and dunes bordering the shore are part of the coast.

Shoaling Waves and Breakers

Most shore zones have a fairly smooth, sloping bottom extending offshore into deeper water. As a train of waves enters progressively shallower water depths there comes a point at which the orbital motion of the waves encounters interference with the bottom. As a general rule this critical depth is about equal to one-half of the wave length (Figure 15.1). Beyond this point frictional resistance with the bottom causes the wave orbits to become distorted into elliptical figures (Figure 15.2). As each wave crest passes a given reference line on the bottom, the forward drag of water moves sand grains shoreward; with each passing wave trough, sand is dragged seaward. The resulting back-and-forth motion causes *sand ripples* to be formed. Ripples of this particular kind are called *oscillation ripples*, since they are formed by an alternating current (Figure 15.3).

As the waves continue to travel shoreward, the wave length decreases, while the wave height increases, as shown in Figure 15.1. Consequently, the wave is steepened, and becomes unstable. Rather suddenly the crest of the wave moves forward and the wave is transformed into a *breaker*, which then collapses (Figure 15.4). The

Waves, Currents, and Coastal Landforms

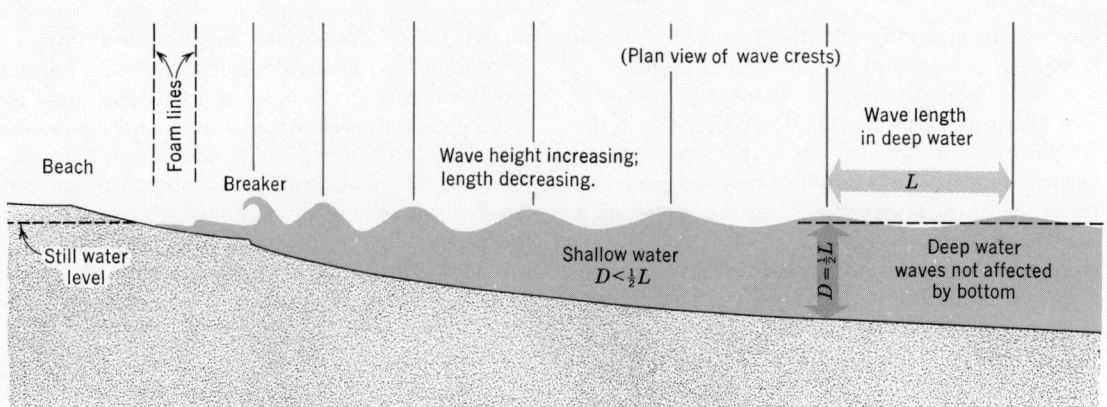

Figure 15.1 Waves entering shallow water increase in height and steepness until the breaking point is reached.

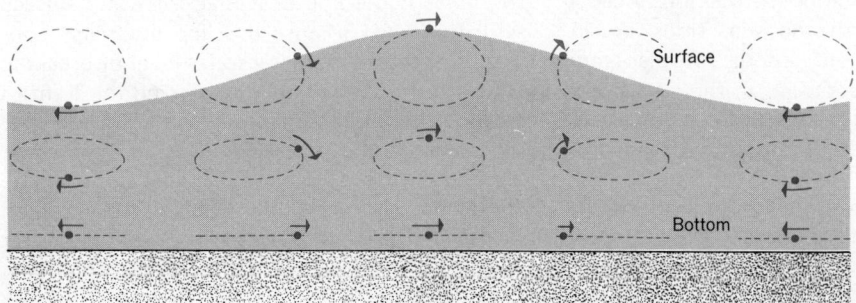

Figure 15.2 Elliptical wave orbits in shallow water produce an oscillating current drag on the bottom. (From A. N. Strahler, 1971, *The Earth Sciences*, 2nd ed., Harper & Row, New York.)

turbulent water mass then rides up the beach as the *swash,* or *uprush*. This powerful surge causes a landward movement of sand and gravel on the beach. When the force of the swash has been spent against the slope of the beach, the return flow, or *backwash,* pours down the beach, but much disappears by infiltration into the permeable beach sand. Sands and gravels are swept seaward by the backwash.

The *plunging* type of breaker is illustrated in Figure 15.1 and 15.4; it is sometimes referred to as a *combing wave,* because it develops a hollow cylindrical form immediately before collapse. The plunging breaker is ideally illustrated by the breaking of a long, even swell on an otherwise calm ocean surface. Another form, the *spilling* breaker, becomes unstable far from shore and continually loses height by a forward flow of turbulent water down the forward slope of the wave.

There are a number of significant similarities between the action of breaking waves with the forms they produce and the action of a stream with the channel forms it produces. For one point,

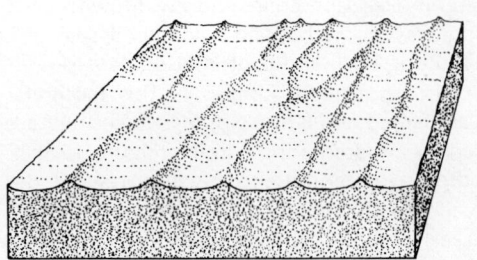

Figure 15.3 Oscillation ripples in sand. (From A. N. Strahler, *A Geologist's View of Cape Cod.* Copyright © 1966 by Arthur N. Strahler. Reproduced by permission of Doubleday & Co., Inc.)

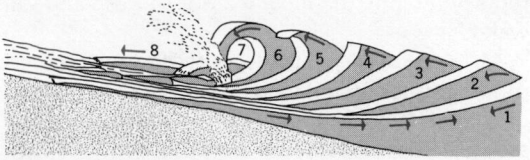

Figure 15.4 A breaking wave. (After W. M. Davis.)

the energy of waves extends through a great range from that of small waves in periods of relative calm to that of great storm waves. As in the case of a stream in flood, the ability of storm waves to transport sediment and modify the profile of the shore is great, whereas, like a stream in low stage, small waves tend to build sediment accumulations and to restore the masses of sediment removed in a previous storm.

Analogous with streams that occupy bedrock channels and have irregularities of gradient are those shore zones in which bedrock is exposed to direct attack by waves. On the other hand, analogous with alluvial stream channels, are those shorelines on which waves act upon thick accumulations of sands and gravels, which we can call collectively *beach deposits*. Both the stream and the breaking waves tend to shape their respective alluvial or beach deposits into equilibrium forms, adjusted to the energy level present in the system. It seems reasonable, then, that there is an *equilibrium shore profile*, analogous with the equilibrium profile of a graded stream. Moreover, there must be early stages in the adjustment of a shoreline when the bedrock forms dominate, just as in the case of an ungraded stream channel with its gorge, falls, and rapids. Using this concept, let us consider first the ungraded shore profile carved into bedrock.

Wave Erosion and Marine Cliffs

Consider a coast that has experienced recent changes of sea level with respect to the land, such that the crust has subsided or the water level has risen, or both. This event is referred to as *coastal submergence*, since it submerges a belt of land that was formerly above sea level, and the new shoreline comes to rest against what was formerly a set of rather steep hill slopes shaped by fluvial denudation or by glacial action.

Diagram A of Figure 15.5 illustrates the new shoreline created by rapid submergence. With fairly deep water offshore, waves break upon the shoreline with little loss of energy. At first the hydraulic action of the waves will make little headway upon hard bedrock, such as granite or a massive limestone, but gradually a small cliff, or *nip*, is carved at the water line. As this notch deepens, two shoreline elements emerge: a perpendicular *marine cliff* and a sloping *abrasion platform* (Diagram B of Figure 15.5). Rock fragments, torn from the cliff by storm swash, along with fragments derived by weathering of the higher sections of the cliff, accumulate as a litter of pebbles, cobbles, and boulders at the foot of the cliff. Collectively called *shingle*, these frag-

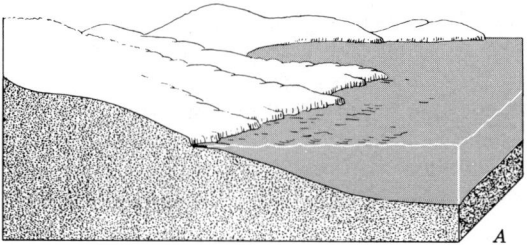

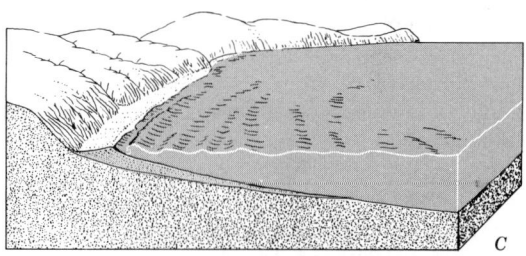

Figure 15.5 Evolution of a marine cliff. A. Appearance of a nip. B. Intensive erosion has produced a stack (S), an arch (A), a notch (N), a cave (C), a crevice (R), and an abrasion platform (P). The beach (B) is thin. C. Later stage in which an equilibrium profile is maintained in a thick beach deposit. (From A. N. Strahler, 1971, *The Earth Sciences*, 2nd ed., Harper & Row, New York.)

ments provide tools of abrasion for carving of a *notch* at the base of the cliff and for widening the abrasion platform. Irregularities in rock resistance are expressed in irregularities in the cliff. Weak rock zones are excavated into *crevices* and *sea caves*. Rock remnants rise above the abrasion platform as *stacks*. In rare cases the waves cut through a rock promontory to produce a rock *arch* (Diagram B of Figure 15.5).

Many thousands of miles of cliffed marine shorelines exist over the globe today; they reflect the rather radical changes of land and sea levels typical of the most recent events of geologic history. Depression of the continental crust under the load of ice sheets, and the rise of sea level accompanying melting of those ice sheets have been major causes of coastal submergence. Essentially similar results with respect to the conditions favorable to cliffs of marine erosion are found where recent tectonic activity has been expressed in a rapid rise of a coast, often by upfaulting of crustal blocks. Construction of volcanic cones and lava flows also favors the development of marine cliffs. Thus marine cliffs are extensive along the Pacific Ocean shorelines and reflect late Cenozoic tectonic and volcanic processes around the circum-Pacific belt of lithospheric plate activity. (Figure 15.6).

When the earth's crust rises abruptly through fault-block movements a marine cliff and its platform may be suddenly raised far above the limits of wave action. The abrasion platform now becomes a *marine terrace* and is acted upon by mass wasting and fluvial erosion (Figure 15.7). A second marine cliff and platform are now carved at the new shoreline; but this in turn may be elevated by another episode of faulting. In this way, some coasts have come to show a succession of marine terraces, rising like a flight of broad stairs. Examples occur at numerous places along the Pacific Coast of the United States. The terraces provide sites for cities and highways on what might otherwise be precipitous mountain slopes descending to the sea.

The marine cliff-abrasion platform type of shoreline represents a unique habitat for life forms; it is an environment of vigorous surf action with a dominantly hard rock surface and little or no fine-grained sediment. Marine animals and plants cling to rock surfaces or take refuge in rock-lined pools. Various marine mammals—for example, seals and sea lions—live on such rocky shores, while the marine cliff may support large numbers of shore birds that feed upon the marine life and nest in crevices in the cliffs. Although few beaches suitable for bathing and surfing will be found along such cliffed shorelines, the spectacular scenery of the coast constitutes a natural resource of unmatched recreational value. Acadia National Park in Maine has a strikingly beautiful cliffed granite coast displaying many of the forms of bedrock erosion by waves, and protected from disfigurement by summer homes, motels, and restaurants at the top of the cliff. Point Lobos in the city of San Francisco, in contrast, stands as a monument to environmental disfigurement by the lowest level of commercialism, and there are many more. Somewhere in between in the scale of degradation is the famed Pebble Beach Country Club with its golf course situated on a marine terrace at the brink of a marine cliff at Monterey, California.

In a few places along the coasts of North America and northern Europe marine cliffs are being rapidly eroded back in weak glacial deposits. In contrast to the extremely slow progress of erosion in hard varieties of bed rock, these cliffs of glacial materials are easily undermined and undergo rapid retreat. For example, the Atlantic Ocean (eastern) shore of Cape Cod has a 15-mile (24-km) stretch of marine cliff ranging in height from 60 to 170 ft (18 to 52 m) carved into unconsolidated glacial sands and gravels (Figure 15.8). This feature is better described as a *marine scarp*, since the loose sand holds a slope angle of about 30° to 35° from the horizontal and is by no means a sheer wall. Here the rate of shoreline retreat has averaged nearly 3 ft (1 m) per year for the past century and a single winter storm will cut the cliff back a distance of several feet. Naturally, there is no rock abrasion platform along this shoreline, but instead a beach of sand and pebbles derived from the glacial deposit.

Figure 15.6 Marine cliffs bordered by a broad abrasion platform. A pocket beach lies at lower left. Pacific coast, south of Cape Flattery, Washington. (Photographer not known.)

Figure 15.7 Elevated marine terraces on the western slope of San Clemente Island, off the southern California coast. More than twenty terraces have been identified in this series; the highest has an elevation of about 1300 ft (400 m). (Photograph by John S. Shelton.)

Wave Refraction and Its Effects

The flow of energy in waves impinging on a shoreline would be uniform only along a straight shoreline having a perfectly uniform profile of seaward slope at all points. This degree of perfection is rarely, if ever, found in nature. Instead, many shorelines, particularly those of the rocky, cliffed type we have just described, are irregular in plan (i.e., as seen on a map, or when looking straight down from an airplane). The seaward-projecting points, or *promontories*, alternate with landward indentations, or *bays;* we call this an *embayed coast*. Typically, the bottom configuration reflects the pattern of promontories and bays, so that water depth is greater along the axis of the bay than along the axis of the promontory. This relationship is to be expected if the bay represents a partially submerged former stream valley or glacial trough.

Change in direction of travel of water waves in response to change in bottom configuration is referred to as *wave refraction*. When undergoing refraction, the wave crest appears to become bent as seen from above. Figure 15.9 shows a train of uniform waves approaching a shoreline with bays and promontories. Successive positions of a wave are indicated by the lines numbered 1, 2, 3, etc. In deep water the wave fronts are parallel. As the shore is neared, the retarding influence of shallow water is felt first in the areas in front of the promontories. Shallowing of water reduces speed of

Figure 15.8 A marine scarp, 70 ft (21 m) high, eroded by winter storm waves in unconsolidated glacial outwash sands. Sand continually rolls down the scarp in dry weather, producing a series of talus cones. Rate of scarp retreat averages about 3 ft (1 m) per year. (Photograph by Harold L. R. Cooper, Cape Cod Photos.)

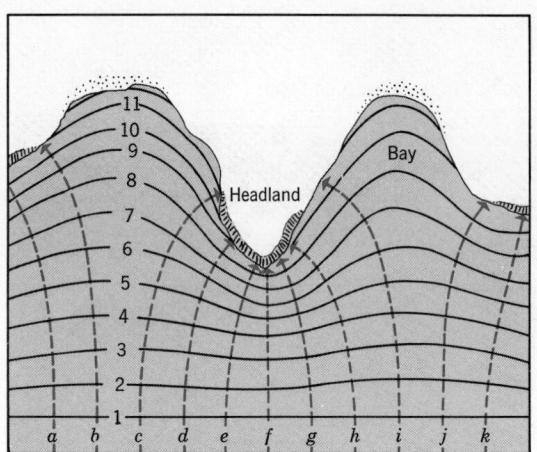

Figure 15.9 Wave refraction along an embayed coastline.

Littoral Drift and Its Beach Forms

Beach deposits consist of hard mineral grains ranging in size from fine sand, through pebbles, to large cobbles. (The grade size limits for sand, pebbles, and cobbles are given in Table 8.2.) Because of the intense water turbulence in the surf zone, silt and clay cannot accumulate as beach deposits, but are instead carried in suspension into deeper offshore waters where they can settle to the bottom and remain. As a result, beach deposits are usually rather well graded in terms of particle size. Sand and pebbles are easily dragged and rolled by the swash and backwash and by bottom currents. Cobbles are less easily moved, but are shifted about by storm swash; they tend to be pushed landward to become stranded in positions well above the limits of all but the most energetic waves and become shingle beaches (Figure 15.10). The slope of the beach in the zone of swash and backwash is quite closely adjusted to the average size of the particles comprising the beach deposit. Fine sand has a low angle of slope, coarse sand a higher slope, and cobbles a very steep slope. With increasing coarseness, a greater proportion of the swash disappears by infiltration. On a shingle beach composed of cobbles all of the swash may sink into the deposit, yielding no backwash at all, and consequently the particles can only be pushed landward but not dragged seaward.

In an idealized situation in which waves approach a straight shoreline, their crests parallel

wave travel at those places, but in the deeper water in front of the bays the retarding action has not yet occurred. Consequently, the wave front is bent, or *refracted*, in rough conformity with the shoreline. The wave will break first upon the promontory and on the bay head last, as indicated in Figure 15.9.

Particularly important in understanding the development of embayed shorelines is the distribution of wave energy along the shore. On Figure 15.9, dashed lines (lettered *a*, *b*, *c*, *d*, etc.) divide the wave at position 1 into equal parts, which may be taken to include equal amounts of kinetic energy traveling forward with the wave. Along the headlands the energy becomes concentrated into a short piece of shoreline; along the bays it is spread over a much greater length of shoreline. Consequently, the breaking waves act as powerful erosional agents on the promontories, but are relatively weak and ineffective at the bay heads. The important principle is that promontories are rapidly eroded back, whereas the bays experience little or no erosion. Thus in time wave action acts to produce a simple shoreline as an ultimate form. Over short distances this simplified shoreline may be nearly a straight line, but more typically shows a curvature in an arc of large radius. Diagram C of Figure 15.5 shows a late stage in development of an embayed coast. Notice that the promontories have been cut away, leaving a simple shoreline following a broad sweeping curve.

Simplification of the shoreline minimizes the extent of wave refraction and tends to equalize the distribution of wave energy along the shoreline. In some respects, this evolutionary process is analogous to the gradation of a stream profile, in which a nearly uniform gradient is attained by elimination of rapids.

Figure 15.10 A multiple-crested shingle beach, Smith Cove, Guysboro, Nova Scotia. The beach forms a crescent between rocky headlands. (Photograph by Maurice L. Schwartz.)

Littoral Drift and Its Beach Forms

with that line, the wave breaks at the same instant at all points and the swash rides up the beach at right angles to the shoreline. The backwash returns along the same line. Consequently particles move up and down the beach slope along a fixed line. They may, of course, travel farther landward than seaward and thus creep up the beach; or they may travel farther seaward then landward with each stroke, and thus creep down the beach and out into deeper water. The landward or seaward shift of sand is a normal process and the direction of net mass movement depends upon the form and size of the incoming waves. Generally speaking, long, low waves (such as those of a swell) cause sand to creep shoreward, building up the beach; whereas short, steep waves (produced in local storms) tend to transport sand seaward and thus to cut back the beach. Engineers refer to out-building of the beach as *progradation;* to cutting-back as *retrogradation.*

Let us now turn to processes whereby sediment can be moved along the shoreline. Movement parallel with the shore is referred to generally as *sediment drift.* Referring back to the subject of wave refraction, consider what happens when a train of waves approaches a straight shoreline obliquely, as shown in Figure 15.11. The wave crests are uniformly bent so that the crests tend to become more nearly parallel with the shore. However, the obliquity of approach persists as the wave breaks, so that the swash is directed obliquely up the slope of the beach, as shown in Figure 15.12. As a result the sand, pebbles, and cobbles are moved obliquely up the slope. After the swash has spent its energy, the backwash flows down the slope of the beach, being controlled by the pull of gravity which moves it in the most direct downhill direction. The particles are therefore dragged directly seaward and come to rest at positions to one side of the starting points. Because, on a particular day, wave fronts approach consistently from the same direction

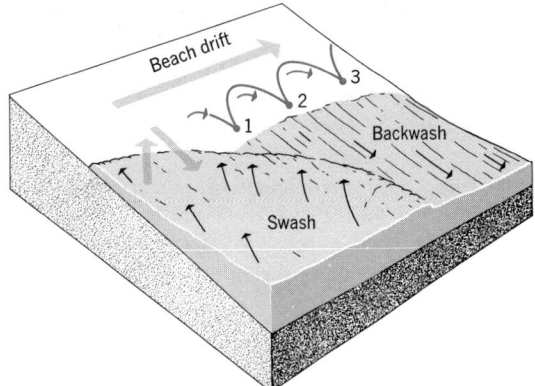

Figure 15.12 Beach drift of sediment caused by swash of obliquely approaching waves.

this movement is repeated many times. Individual rock particles thus travel a considerable distance along the shore. Multiplied many thousands of times to include the numberless particles of the beach, this form of mass transport, called *beach drift,* is a major process in shoreline development.

It is a rare occasion to find that beach drift is not taking place, in one direction or the other, along a marine shoreline. Usually, a given stretch of shoreline is subjected to a dominant direction of wave approach throughout a given season of the year, or throughout the entire year. Consequently, beach drift can be assigned a single direction of net transport as the seasonal or yearly average.

A process related to beach drifting is *longshore drift.* When waves approach a shoreline under the influence of strong winds, the water level is slightly raised near shore by a slow shoreward drift of water. There is thus an excess of water pushed shoreward, which must escape. A *longshore current* is set up parallel to shore in a direction away from the wind (Figure 15.13). When wave and wind conditions are favorable, this cur-

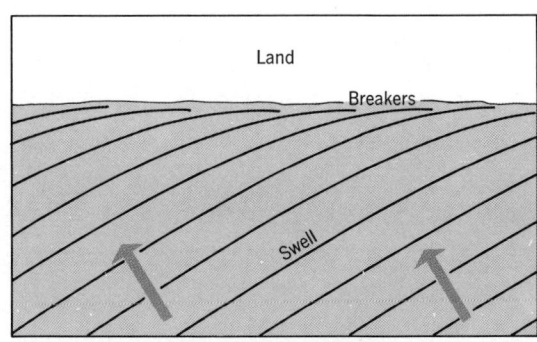

Figure 15.11 Wave refraction along a straight shoreline.

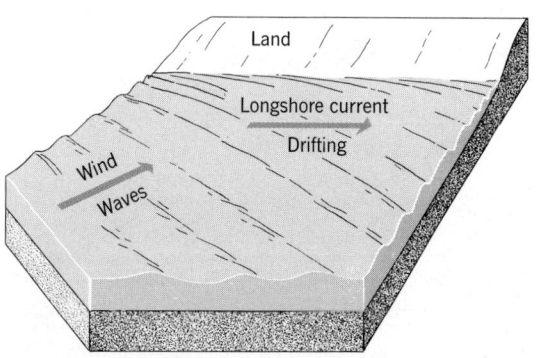

Figure 15.13 Longshore current drifting.

rent is capable of moving sand along the bottom in the breaker zone in a direction parallel to the shore.

Both beach drifting and longshore drifting move particles in the same direction for a given set of onshore winds and oblique wave fronts and therefore supplement each other's influence in sediment transportation. The combined transport by beach and longshore drift is termed *littoral drift*. Let us now apply the principle to the evolution of beach deposits.

Going back to the embayed shoreline, and analyzing the oblique approach of breaking waves with respect to the shoreline, we find that littoral drift will occur along the sides of the bays, with the result that sediment generated by intensified wave attack on the promontory is carried toward the bay head, where wave action is weak (Figure 15.14A). Deposition of sediment forms a *bay-head beach*, which because of its shape, is known as a *crescentic beach*, or because of its isolation, a *pocket beach*. An example is shown in Figure 15.6. Typically, pocket beaches are composed of coarse sediment, in the size range of pebbles to cobbles.

Littoral drift along a straight section of shore is illustrated in Figure 15.14B. Where an embayment occurs, drift continues along the line of the straight shore, with the result that an embankment of sediment is constructed along that line. This narrow beach deposit, extending out into open waters, is called a *sandspit*, or simply a *spit*. Through wave refraction sediment is carried around the spit end, which develops a landward curvature. The spit is then described as *recurved* (Figure 15.15).

Figure 15.15 A recurved sandspit with sand dunes, Georgia Strait, British Columbia. (Photograph by D. D. Rahm. © by McGraw-Hill Book Co. Used by permission.)

Construction of a spit completely across a bay results in a *baymouth bar,* which cuts off the bay from action of waves of the open ocean (Figure 15.16). In some instances the bar has an *inlet,* which is a narrow gap through which water can flow alternately landward and seaward in response to tidal rise and fall of water level. Figure 15.17 shows an island tied to the mainland by two bars; each of which is designated as a *tombolo*.

Where littoral drift converges from opposite directions upon a given point on a shoreline sediment accumulates in the form of a *cuspate bar* (Figure 15.18). Continued progradation occurs in the form of a succession of *beach ridges*, separated from each other by belts of low, marshy ground called *swales*. In time a *cuspate foreland*

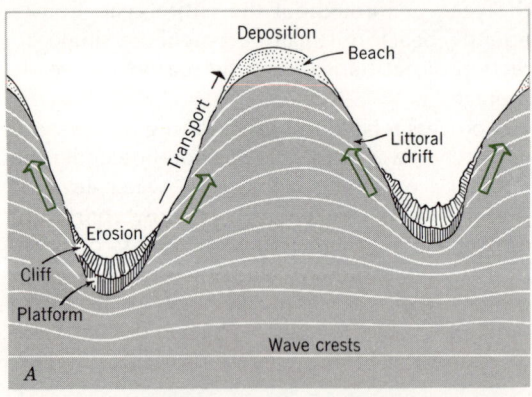
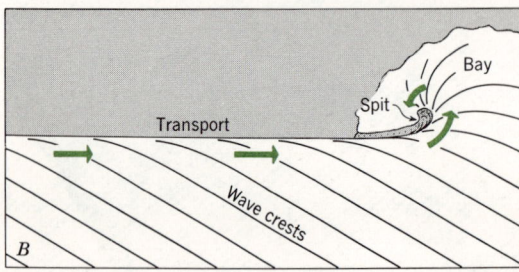
Figure 15.14 A. Littoral drift on an embayed coast. Compare with Figure 15.9. B. Littoral drift and sandspit growth along a straight shoreline. (From A. N. Strahler, 1971, The Earth Sciences, 2nd ed., Harper & Row, New York.)

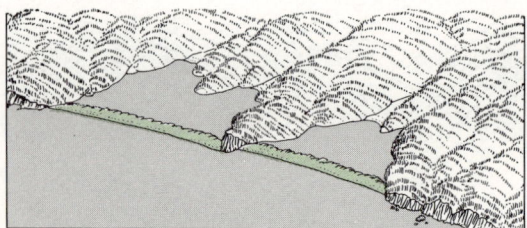
Figure 15.16 Baymouth bars, sealing off two bays. (After W. M. Davis.)

The Equilibrium Beach Profile

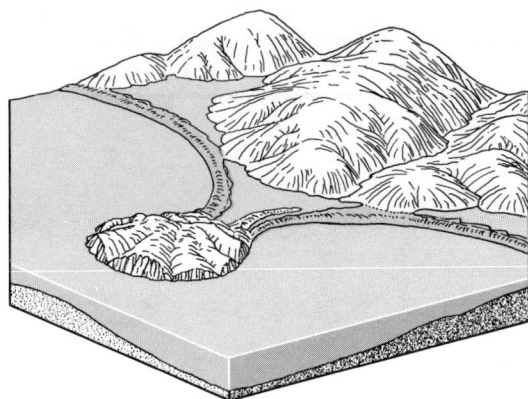

Figure 15.17 Two tombolos connecting an island to the mainland. (After W. M. Davis)

is built (Figure 15.19). Where the beach ridges are formed of sand, they will be acted upon by winds and changed into coastal dunes, but the broad outlines of the beach ridges will persist, as on the Provincelands of Cape Cod (Figure 16.21).

The Equilibrium Beach Profile

A sand beach with its bordering shallow-water zone of breaking waves, on the one side, and a zone of wind action with dune development forming a border on the landward side, represents a succession of unique life habitats in which each assemblage of plant and animal forms is adapted to a different environment. Figure 15.20 is an idealized profile across a typical beach developed by exposure to waves of the open ocean in a middle-latitude location experiencing a strong contrast between wave action of summer and winter seasons. The profile shows summer conditions in which waves are of low height and comparatively low levels of energy. During the summer, progradation takes place, building a *summer berm*, which is a benchlike structure. A higher *winter berm* lies behind the summer berm. The

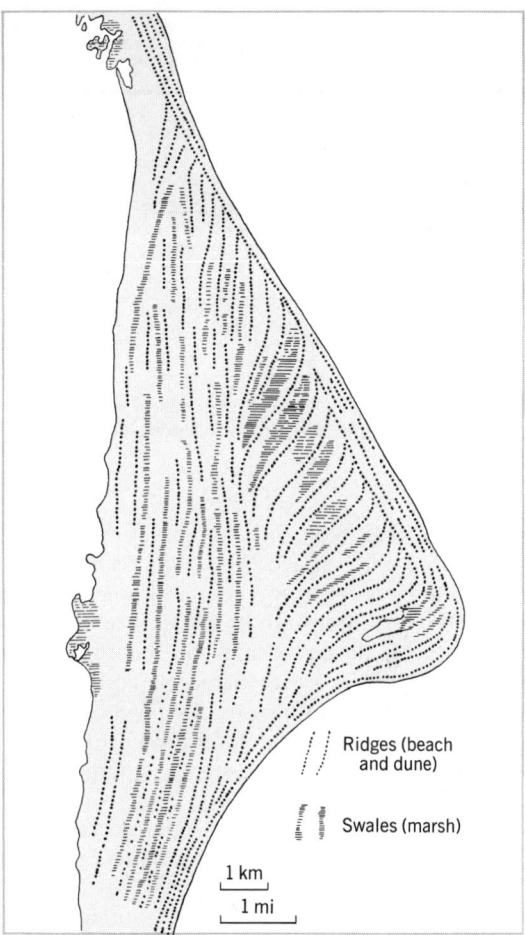

Figure 15.19 Sketch map of Cape Canaveral (renamed Cape Kennedy), Florida, as it was in 1910, before man-induced modification. Ridges near the shore (right) are beach ridges; those farther inland are dune ridges built upon older beach ridges. (After Douglas Johnson, 1919. From A. N. Strahler, 1917, *The Earth Sciences*, 2nd ed., Harper & Row, New York.)

term *foreshore* is applied to the sloping beach face in the zone of swash and backwash. Beneath the breaker zone is a low underwater bar, called an *offshore bar*, since it lies in the *offshore* region of the beach, in the zone of shoaling waves and below the level of mean low tide. During the winter, storm waves of high energy levels will cut away the summer berm and the offshore bar may be moved seaward into deeper water. Actually, there are many variations in beach-profile forms from place to place and season to season, depending upon wave form and energy and upon the composition of the beach.

If there is a valid concept of an equilibrium profile of a graded stream in alluvium, we might expect to find an analogous concept in the beach profile, since both are shaped by water move-

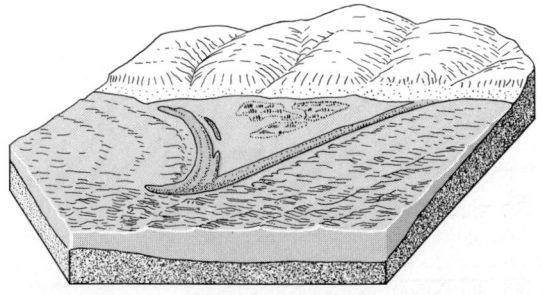

Figure 15.18 A cuspate bar, enclosing a triangular lagoon. (After W. M. Davis.)

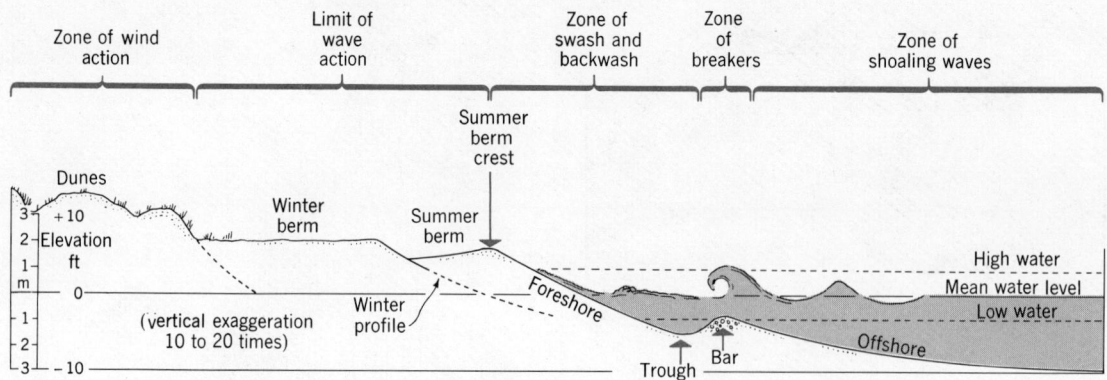

Figure 15.20 Typical forms and zones of a sand beach in the middle-latitude zone. (From A. N. Strahler, 1971, *The Earth Sciences*, 2nd ed., Harper & Row, New York.)

ments that drag sediment along the contact of a fluid with a solid surface. We can find some close similarities (but also important differences) between bed-load transport by a stream and littoral drift along a beach. Rising and falling stream stages are analogous with increasing and decreasing wave sizes; in both cases, the level of energy being expended changes through a wide range and the geometry of the system is appropriately adjusted.

Figure 15.21 is a schematic diagram of an equilibrium shore profile, including a narrow slice of beach to provide an open system having side boundaries as well as an upper surface. The beach-sand body available for reworking by waves is called the *surf lens*. Energy enters the system with shoaling waves; their outer limit of action upon bottom sediment in the offshore zone defines the *surf base*. In the zone of shoaling waves, oscillatory drag upon the bottom can move sediment landward or seaward, expending energy in friction. In the breaker zone a similar alternating current motion expends wave energy. When the wave characteristics of height, length, and energy remain constant for a sufficiently long period (a few days), the beach profile is adjusted to be neither built up (prograded) nor cut back (retrograded), while the entire quantity of input energy is dissipated in friction within the water and along its lower boundary.

Now, we must also introduce the action of littoral drift, shown by opposed arrows parallel with the shoreline. In the equilibrium state the quantity of sediment entering the system from one side boundary must exactly equal the quantity leaving the opposite boundary. It is obvious that if a greater quantity of sediment enters than leaves, progradation will result, the profile will be built seaward, and multiple beach ridges will be formed as shown in Figure 15.22A. If more sediment leaves than enters, the beach undergoes retrogradation and the profile is moved landward (Figure 15.22B.)

The importance of investigating the beach profile in terms of concepts of open systems and the steady state lies in its usefulness in predicting the changes that will occur when Man interferes with natural shoreline processes by adding or removing sediment sources through construction of engineering works.

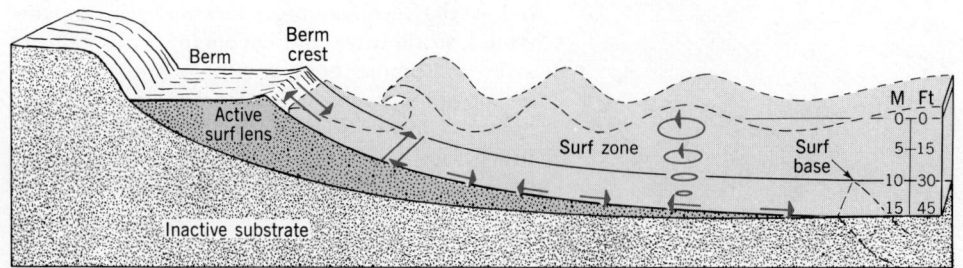

Figure 15.21 Idealized diagram of the equilibrium shore profile developed on a surf lens. Wave size is exaggerated. (From A. N. Strahler, 1971, *The Earth Sciences*, 2nd ed., Harper & Row, New York.)

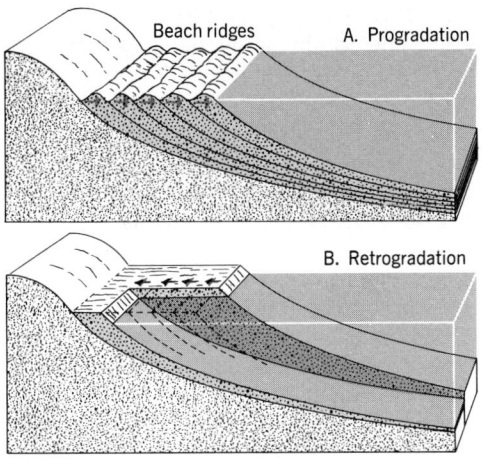

Figure 15.22 Progradation and retrogradation.

Figure 15.23 Storm waves breaking against a coast underlain by weak sand quickly undermined this shore h at Seabright, New Jersey. The barrier of wooden pilings (*right*) proved ineffective in preventing cutting back of the cliff. This event occurred early in the present century, before this shoreline was adequately protected by massive engineering structures. (Photograph by Douglas Johnson.)

Man's Impact on Shorelines

We have already referred to the natural cutting-back, or retrogradation, of land under the attack of storm waves along promontories and other exposed stretches of coast. For the most part, promontories such as those pictured in Figures 15.5 and 15.6 are supported by hard rock—igneous, metamorphic, and certain massive sedimentary types—and show no appreciable change in a century's time or more. Comparison photographs taken over such time spans often show no more than the fall of a few precariously situated joint blocks from the marine cliff. Changes this slow are measured in geological time and cause no problem to the works of Man.

In contrast, marine scarps carved in unconsolidated materials—glacial deposits, geologically recent sedimentary strata, stream alluvium, or former beach and dune deposits—yield rapidly under the impact of storm waves. Consider the case already cited—that of the outer shore of Cape Cod—in which measured retreat has averaged close to 3 ft (1 m) per year over more than a century. Loss of 300 feet of land in a century is perhaps not in itself a large amount, even along some 10 to 20 mi (16 to 32 km) of coast. However, coasts of this type are usually favored with excellent bathing beaches, because of the abundant quantities of loose sediment available to waves. As a result, along many such coasts resort communities have grown up at the very brink of the marine scarp. Costly homes, beach clubs, restaurants, and hotels are crowded together, as close as possible to the sea. Here, even a few yards of retrogradation threaten property destruction, and if it is not prevented, the threat is realized (Figure 15.23). A lesser problem lies in depletion of a sand beach, leaving a narrowed zone of pebbles and cobbles, with resultant loss of recreational usage.

Engineering structures designed to meet wave attack head-on are not only extremely expensive, but also prone to failure. The impacting pressure of breaking storm waves upon a vertical wall runs up into the thousands of pounds per square foot. Breakwaters must, of course be constructed and maintained to protect important harbors, and it is from such structures that we gain an appreciation of the force of the battering-ram action of storm waves. For example, in Amsterdam Harbor, a block of stone weighing 20 tons (18 metric tons) was lifted vertically 12 ft (4 m) by breaking storm waves and came to rest on a pier nearly 5 ft (1.5 m) above the high-water mark. At Cherbourg, France, rock masses weighing as much as 3.5 tons (3.2 metric tons) were thrown over a sea wall 20 ft (6 m) high by storm waves. To protect retrograding scarps, various designs of sea walls are used. They are subject to collapse as beach deposits are undermined from beneath the base of the structure, and to flank attack where neighboring properties are not equally well fortified.

If direct resistance to frontal attack is not always the best answer to coastal erosion, what else can be done? In some circumstances a successful strategy is to install structures that will cause progradation, building a broad beach with an ample berm. The principle here is that the excess energy of storm waves will be dissipated in reworking the beach deposits. Cutting back of the

Figure 15.24 A system of groins for trapping of beach sand, Willoughby Spit, Virginia. Littoral drift is from lower left to upper right. (Photograph by Department of the Army, Corps of Engineers.)

berm in a single storm will (hopefully) be restored by berm-building between storms. Progradation requires that sediment moving as littoral drift be trapped by the placement of baffles across the path of mass transport. To accomplish this result, groins are installed at close intervals along the beach. A *groin* is simply a wall or embankment built at right angles to the shoreline (Figure 15.24); it may be constructed of huge rock masses (*rip-rap*), of concrete, or of wooden pilings. Figure 15.25 shows the shoreline changes induced by groins. Sand accumulates on the updrift side of the groin, developing a curved shoreline. On the downdrift side of the groin the beach will be depleted because of the cutting off of the normal supply of drift sand. The result may be harmful retrogradation and cutting back of the scarp. For this reason groins must be closely spaced so that the trapping effect of one groin will extend to the next. Ideally, when the groins have trapped the maximum quantity of sediment, beach drift will be restored to its original rate for the shoreline as a whole. However, there have been many instances of damaging retrogradation induced by groin construction on the updrift side of an unprotected shoreline.

More serious and permanent effects often accompany the construction of long *jetties* to maintain a navigable inlet in a bar connecting a bay with the open sea (Figure 15.26). Tidal currents then maintain a deep channel. These jetties, which are walls similar in plan to groins, cause sediment to accumulate on the updrift side, but result in serious beach depletion on the downdrift side. The effects of a long jetty can extend far along the shoreline (Figure 15.27).

In some instances the source of beach sand is from the mouth of a river. Construction of dams on the river may drastically reduce the sediment load and therefore also cut off the source of sand for littoral drift. Retrogradation may then occur on a long stretch of shoreline.

Barrier Island Coasts

In striking contrast to cliffed coasts are coasts along which the land surface is a coastal plain; i.e., a low plain descending almost imperceptibly to the sea and extending seaward as a shallow continental shelf. Under these conditions it is

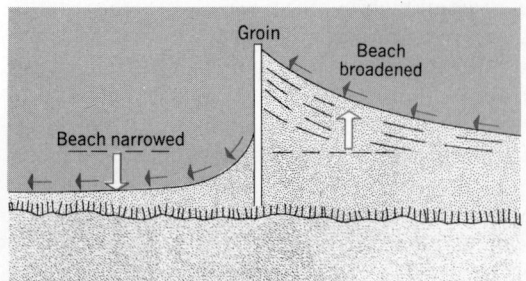

Figure 15.25 Effects of groin construction upon a sand beach. (From A. N. Strahler, 1972, *Planet Earth; Its Physical Systems Through Geologic Time*, Harper & Row, New York.)

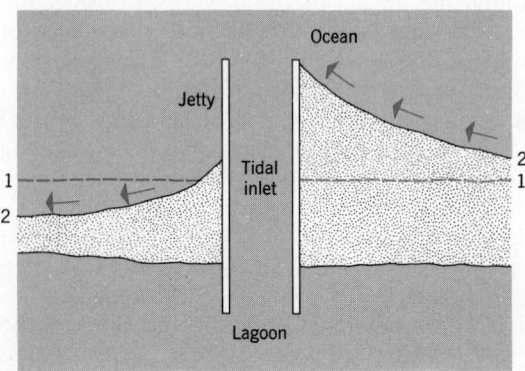

Figure 15.26 Effects of twin jetties upon beach on either side of an inlet. (1) Original shoreline; (2) modified shoreline. (From A. N. Strahler, 1972, *Planet Earth; Its Physical Systems Through Geologic Time*, Harper & Row, New York.)

Figure 15.27 Jetties at Cold Spring Inlet, New Jersey, mark the entrance to Cape May Harbor. Beach sand has accumulated on the updrift side (*right*) of the inlet, but the beach has been depleted on the downdrift side (*left*). (Photograph by Department of the Army, Corps of Engineers.)

Figure 15.28 shows inferred history of development of a barrier island on the Gulf Coast of Texas. Some 5000 years ago sea level stood much lower than today because a great deal of ocean water remained locked in storage in the wasting ice sheets (see Chapter 16). Streams had previously carved valleys in the exposed surface of the continental shelf at a time when the sea level was lowest. As sea level rose and the shoreline shifted landward, waves scoured the shallow bottom and pushed the sediment into a primitive barrier island. As water level continued to rise, the barrier was built higher by wave action and became a complex belt of beach ridges, one upon the other. During this process the broad, shallow lagoon accumulated muds from streams and tidal currents, and these have now partially filled the lagoon to form tide-level flats. Sand dunes derived from the beach ridges encroached upon the tidal flats.

To understand the processes and forms of sediment deposition in shallow lagoons and bays, we will need to investigate the ocean tides and weave the rhythmic variations in water level and currents into the processes of stream and wave action that also contribute to the evolution of this distinctive coastal environment.

typical to find that a beach deposit has formed a narrow offshore strip of land, called a *barrier island* (or *barrier beach*), separating the open ocean from a shallow water body called a *lagoon*. Similar, but less extensive barrier beaches are also built across wide, shallow bays.

Deltas and Delta Environments

Fluvial and marine processes act in concert to form a unique coastal environment, the *delta*, which is any accumulation of sediment formed in standing water of a lake or the ocean by depo-

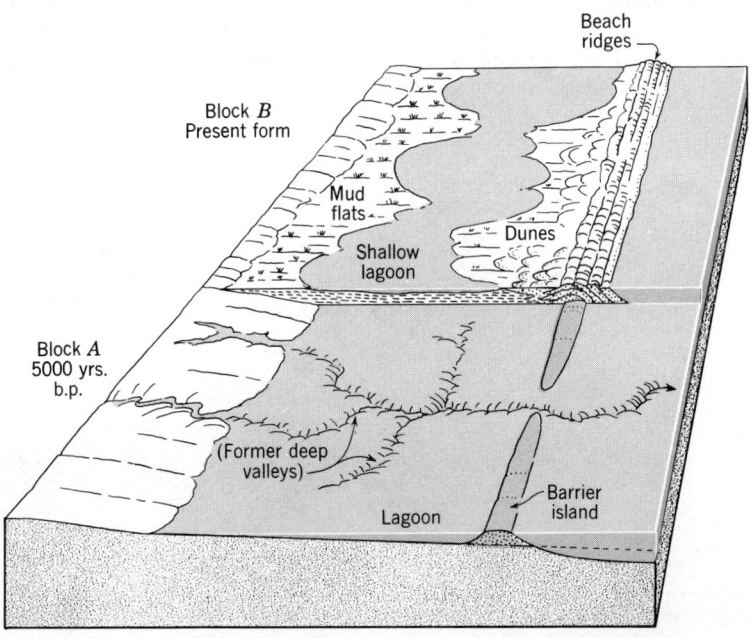

Figure 15.28 Upbuilding of a barrier island during postglacial rise in sea level is an essential part of the history of the Texas Gulf Coast. (Based on data of H. N. Fisk. From A. N. Strahler, 1971, *The Earth Sciences*, 2nd ed., Harper & Row, New York.)

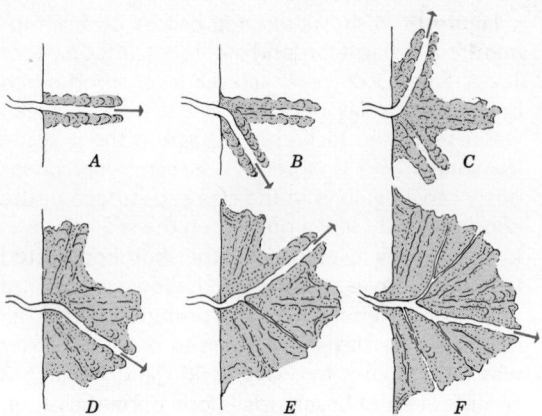

Figure 15.29 Stages in formation of a simple delta built into a small lake. (After G. K. Gilbert.)

sition from the mouth of a stream. Action of waves and shore currents goes on simultaneously with fluvial deposition, with the result that most deltas are partly shaped by marine processes and contain varying proportions of beach deposits.

The simplest delta form, which can be found in lakes receiving streams carrying a substantial proportion of sand and coarser particles as bed load, is illustrated in a series of diagrams in Figure 15.29. Entering the still water as a jet, the stream flow is quickly slowed and deposits its bed load as a lengthening ramp of sediment with two lateral embankments of finer sediment. Because of aggradation of the newly extended channel, the stream breaks through the embankment and extends a new sediment arm. This process is repeated many times and results in a fanlike delta. Each channel branch is a *distributary,* and two or more may be carrying flow simultaneously. Figure 15.30 shows the internal structure of a simple delta. Bed load coming to rest on the steep delta front builds *foreset beds* of sand and pebbles, and beyond these are thin layers formed of silt and clay that has settled from suspension, the *bottom-set beds.* As the delta is expanded the aggrading stream adds *topset beds* of coarse bed material.

Large deltas built into the ocean show varying degrees of control by wave and current action, as well as constraints to growth by preexisting shoreline and bottom configurations. Figure 15.31 shows four types of delta outlines. The Nile Delta, whose triangular outline led to use of the Greek letter *delta* for the landform, exhibits the fan outline. Littoral drift of sediment from the distributary mouths has given rise to barrrier beach deposits making an arcuate shoreline behind which are shallow lagoons. The Tiber delta, a small deposit, has been strongly shaped by wave action. Littoral drift carries sediment away from the mouth to form two arcuate beach deposits, so that the delta is described as *cuspate* (toothlike). The delta of the Seine River represents an *estuarine delta,* built within the confines of a narrow estuary resulting from submergence of a river floodplain. Postglacial rise of sea level has produced many such drowned valley mouths and the estuarine delta is a common type.

The modern Mississippi River delta (Figure 15.31*B*) exhibits the *bird-foot* outline formed by narrow distributary fingers built out into open water. Distributary ends are here called *passes.* Figure 15.32 shows the structure and distribution of sediment in this delta. Beneath each distributary is a sand deposit, the *bar finger,* while clays form the deposit between distributaries. Subsidence of the delta has rendered much of the area tidal marsh, with only the natural levees

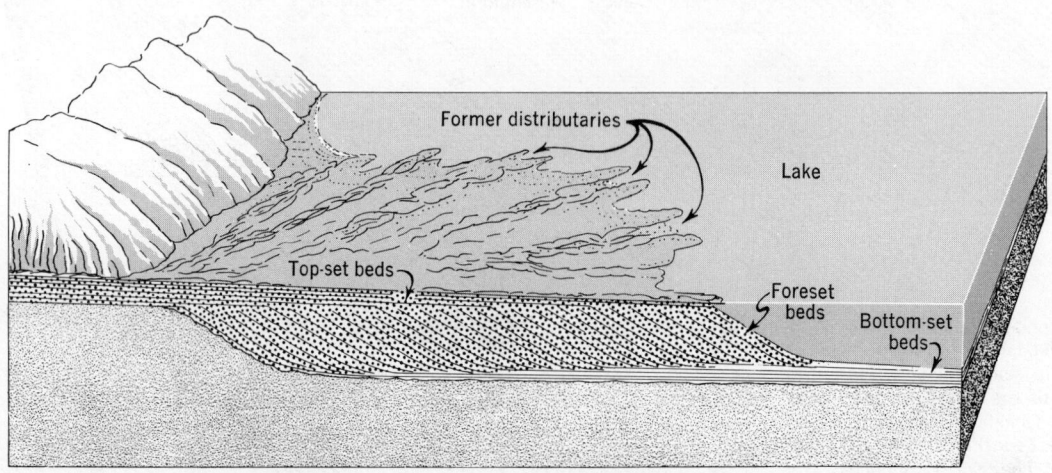

Figure 15.30 Structure of a simple delta built into a lake. (After G. K. Gilbert.)

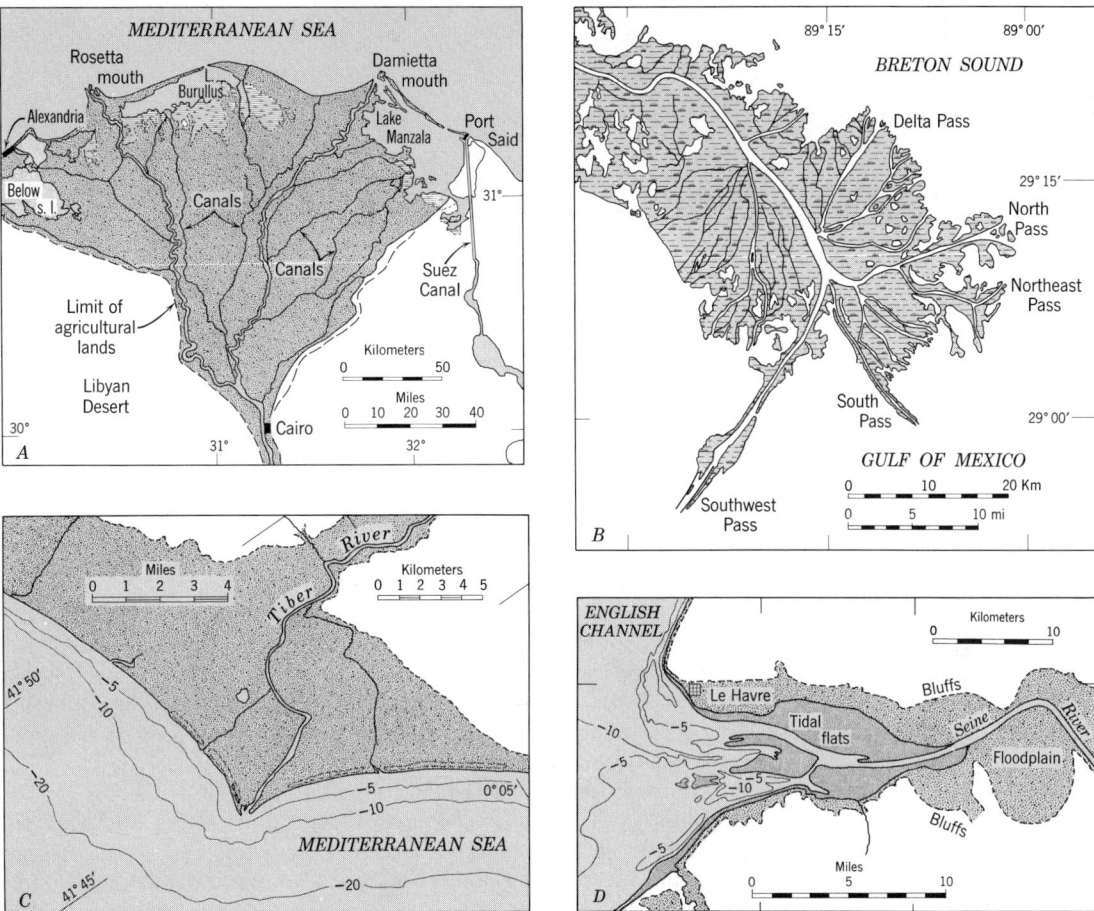

Figure 15.31 Delta outlines. A. The Nile delta has an arcuate shoreline and is triangular in plan. B. The Mississippi delta is of the branching, bird-foot type with long passes. C. The Tiber delta on the Italian coast is pointed, or cuspate, because of strong wave and current action. D. The Seine delta is filling in a narrow estuary.

rising above tide level. As we noted in Chapter 14, the Mississippi River carries about 90% of its total load in suspension. Consequently, the delta is largely built of silts and clays. An important process in allowing clay particles to settle to the bottom is *flocculation,* the clotting together of the particles to form larger aggregates with much faster settling rates. Flocculation of colloids results from mixing of fresh water with salt water. Dissolved salts form an *electrolyte* solution which alters the electrical charges of the colloidal particles so that they attract, rather than repel, one another. Flocculation is a vital process for the coastal environment, for without it suspended clays would disperse widely through the oceans and would not accumulate close to stream mouths.

Throughout its 5000-year history, the postglacial Mississippi River delta has shifted its location several times, with the result that there is a *deltaic plain* many times larger in extent than the modern bird-foot delta. Figure 15.33 is a map showing this deltaic plain with its intricate lacing of abandoned distributary channels. Subsidence due to compaction of the sediment has allowed a large part of the outer deltaic plain to become submerged beneath shallow water.

Deltas of large rivers have been of great environmental importance to Man from earliest historical times because their extensive flat areas and rich alluvial soils support dense agricultural populations. Important coastal cities, linking ocean and river traffic, are situated on or near deltas; for example, Alexandria on the Nile, Calcutta on the Ganges-Brahmaputra, Rotterdam on the Rhine, Shanghai on the Yangtze, Marseilles on the Rhone, Venice on the Po, Bangkok on the Chao Phraya, Phnom Penh on the Mekong, and New Orleans on the Mississippi. Delta growth is often rapid, ranging from about 10 ft (3 m) per year for

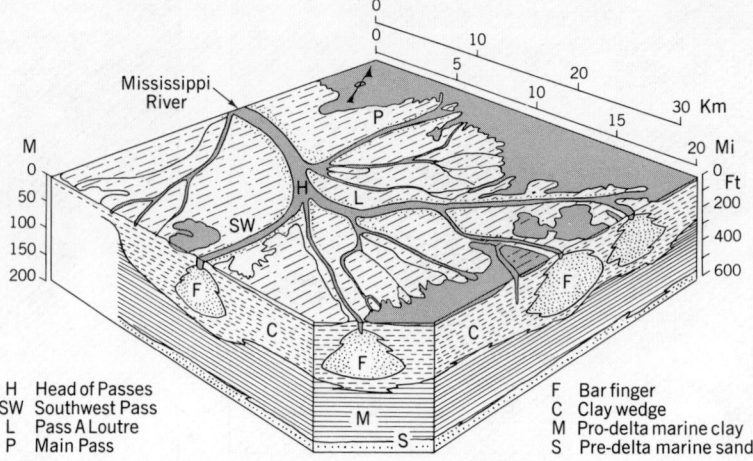

Figure 15.32 Idealized diagram of structure and composition of the modern bird-foot delta of the Mississippi River. (Simplified from H. N. Fisk, E. McFarlan, Jr., C. R. Kolb, and L. J. Wibert, Jr., 1954, *Jour. of Sedimentary Petrology*, vol. 24, p. 77, Figure 1.)

H   Head of Passes
SW Southwest Pass
L   Pass A Loutre
P   Main Pass

F   Bar finger
C   Clay wedge
M   Pro-delta marine clay
S   Pre-delta marine sand

the Nile to 200 ft (60 m) per year for the Po and Mississippi rivers. Thus, some cities and towns that were at river mouths several hundred years ago are today several miles inland. An important engineering problem is to keep an open channel for ocean-going vessels which have to enter the delta distributaries to reach port. The natural passes of the Mississippi River delta have been extended by the construction of jetties, between which the narrowed stream is forced to move faster, thereby scouring a deep channel.

Deltas and deltaic plains are particularly vulnerable to inundation by storm surges produced when tropical storms move toward the coast. (This phenomenon is explained in Chapter 5.) Deltaic plains of the Mississippi River and of the Ganges-Brahmaputra Rivers have suffered heavily from the onslaught of such storms. Dense population of the agriculturally productive deltaic lands bordering the Bay of Bengal account for extremely large casualty figures from these great inundations.

Coastal waters adjacent to large deltas comprise a special environment of life forms. These waters share the general characteristics of all shallow coastal waters and estuaries, but with the added

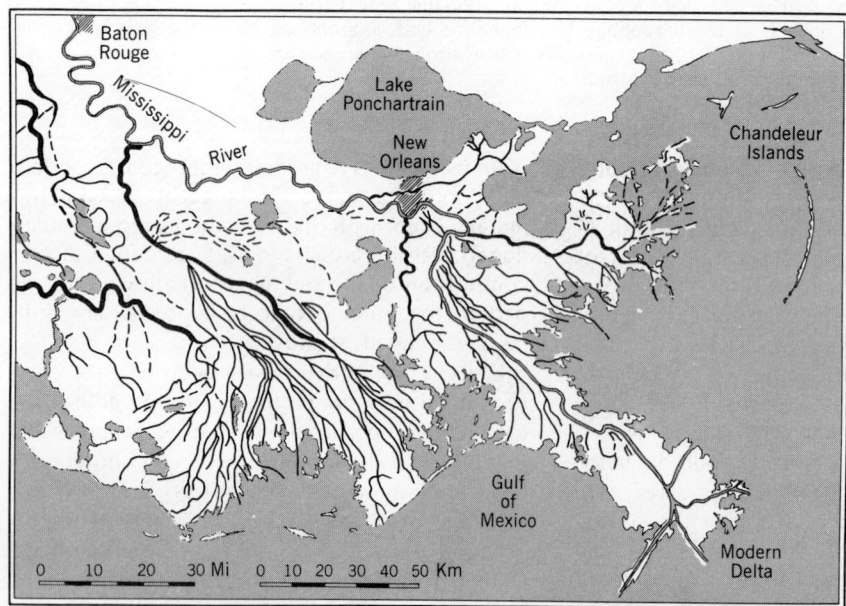

Figure 15.33 Deltaic plain of the Mississippi River, formed during the last 5000 years, has a large number of abandoned distributary channels, shown in bold lines on this map. (Redrawn and simplified from a map by C. R. Kolb and J. R. Van Lopik, 1966, in *Deltas in Their Geologic Framework*, M. L. Shirley, ed., Houston Geol. Soc., Houston, Texas, p. 22, Figure 2.)

factors of large contributions of fine sediment from the river mouths, along with abundance of nutrient ions essential to growth processes. By the same token, the deltaic marine environment is highly vulnerable to environmental changes of the river system. Pollutants arrive from numerous upstream sources and these may be toxic to organisms, or may promote eutrophication. If the river discharge is reduced by upstream water withdrawals, salinity of coastal waters near the river mouth can be drastically altered.

Trapping of sediment behind dams reduces the rate of sedimentation off the river mouth and may allow wave action to do extensive damage through retrogradation of barrier beaches. It is said that the reduction of sediment load of the Nile River, due to construction of the Aswan High Dam, has already caused rapid retrogradation of beaches along the Mediterranean shoreline of the delta, since river sand is no longer supplied for distribution by littoral drift.

Ocean Tides

Most marine coasts are to some degree influenced by the *tide,* or rhythmic rise and fall of sea level under the influence of changing attractive forces of moon and sun upon the rotating earth. Where tides are great, the effects of changing water level and of currents set in motion are of major importance in shaping coastal landforms.

Without going into the causes of ocean tides and their many variations, some of the important principles of tidal action can be understood by considering the common type of *tide curve* in which the cycle of rise and fall is *semidaily* (*semidiurnal*), taking approximately 12½ hours to complete. If we make half-hourly observations of the position of water level against a measuring stick, or *tide staff,* attached to a pier or sea wall, we can plot the changes of water level and thus draw the tide curve. Figure 15.34 is a tide curve for Boston Harbor covering a day's time. The water reached its maximum height, or *high water,* at the 12-ft (3.7-m) mark on the tide staff, then fell to its minimum height, or *low water,* occurring about 6¼ hours later. A second high water occurred about 12½ hours after the previous high water, completing a single semidaily tide cycle.

The *range of tide,* or difference between heights of successive high and low waters, is about 9 ft (2.7 m) for the example shown in Figure 15.34. The tide range and the form of the tide curve vary throughout the lunar month of about 29½ days. Twice during the month the range is greater than average, comprising *spring tides;* one week later the range is less than average, com-

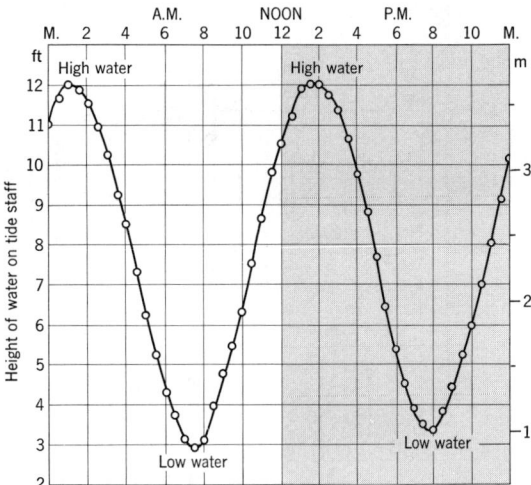

Figure 15.34 Height of water at Boston Harbor measured every half hour. (After H. A. Marmer.)

prising *neap tides* (Figure 15.35). This monthly variation is caused by the changing relative positions of sun and moon, both of which exert tide-raising forces. When moon and sun are on a common line with the earth (*conjunction* and *opposition*) the two tide-raising forces are combined, resulting in spring tides. When a line from moon to earth forms a right angle with a line from sun to earth (*quadrature*), the sun's force is subtracted from that of the moon, giving neap tides.

Also of major importance in determining the range of tide is the changing distance from moon to earth. Tide range is greater when the moon is at its closest position (*perigee*) and less when at its most distant position (*apogee*). Because the period from one perigee to the next is only 27½ days, while that from one conjunction to the next is 29½ days, the effect of perigee at times reinforces the spring-tide effect and at other times tends to nullify the neap-tide effect.

Besides the rhythmic astronomical controls that govern the tide range throughout the month and year, the average range at a given coastal point is determined by geometrical properties of the ocean body, with the result that the range differs greatly from place to place along shorelines over the globe. The Mediterranean Sea has a scarcely perceptible tide range and the phenomenon was not familiar to early inhabitants of the Mediterranean lands, except through information imported from the lands bordering the English Channel and North Sea, where the average tide range runs as high as 10 to 15 ft (3 to 6 m). Along the United States Gulf Coast, the tide range is only 1 to 2 ft (0.3 to 0.6 m), whereas along the New England coast it is on the order of 8 to 20 ft

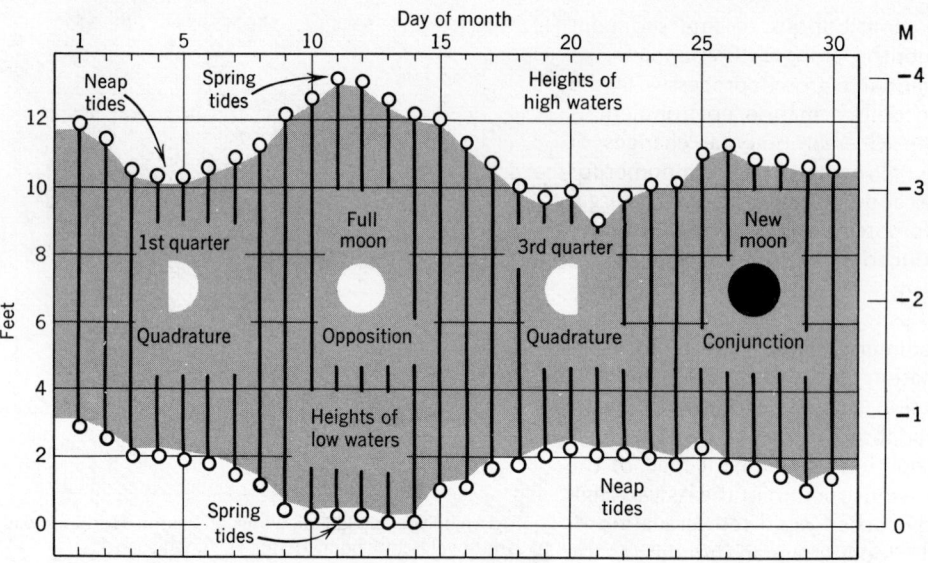

Figure 15.35 Neap and spring tides. (After H. A. Marmer.)

(2.4 to 6 m) and even greater along the New Brunswick–Nova Scotia coast where the configuration of the Bay of Fundy is favorable to producing a great range (Figure 15.36).

The significance of tide range is considerable in the shore environment. Where a large range occurs wave action is effective many feet above mean tide level. Beach ridges and berms can be built high above the mean sea level. The *intertidal zone*, which constitutes the *littoral environment* is extremely narrow where tide range is very small, but makes a very broad belt where tide range is high and the bottom slope gentle. Furthermore, the velocity and power of tidal currents, discussed below, depend upon tide range.

Tidal Currents

The rising tide sets in motion in bays and estuaries landward and seaward currents of water. The relationships between tidal-current speed and the tide curve are shown in Figure 15.37. When the tide begins to fall, an *ebb current* sets in, reaching maximum speed about at *midtide*. Flow ceases about the time when the tide is at its lowest points, a condition known as *slack water*. As the tide begins to rise, a landward current; the *flood current*, begins to flow and gains in strength to attain a maximum speed about at midtide. Note that the ebb current is stronger than the flood current, a condition explained by the fact that a large river contributes a considerable discharge of fresh water from the land which must escape to the sea. This stream discharge augments the ebb current but opposes the flood current.

In a comparatively deep, narrow estuary having no constrictions, the tide crest and trough travel landward as a true wave form, the speed being controlled by water depth. This tide form is known as a *river tide*. In the case of the Hudson River, which in its lower portion is a true arm of the sea and not a terrestrial river, the tide wave travels 130 mi (210 km) from the entrance to New

Figure 15.36 At low tide along the shores of the Bay of Fundy shad are being removed from nets completely inundated in the previous high water period. (Photograph by National Film Board, Ottawa, Canada.)

Tidal Currents

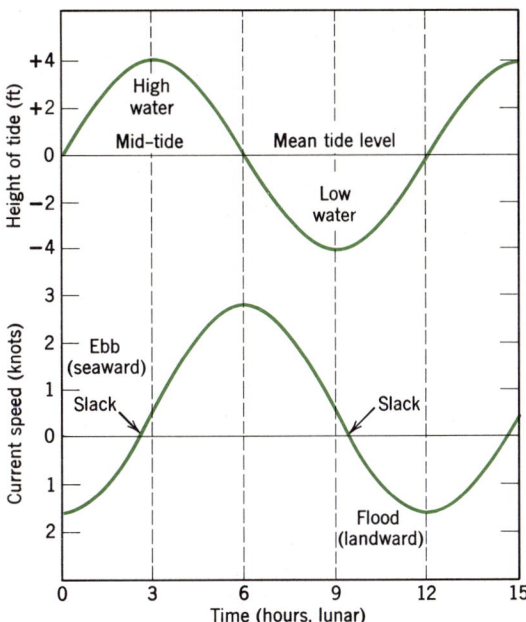

Figure 15.37 Ebb and flood currents in relation to the tide curve.

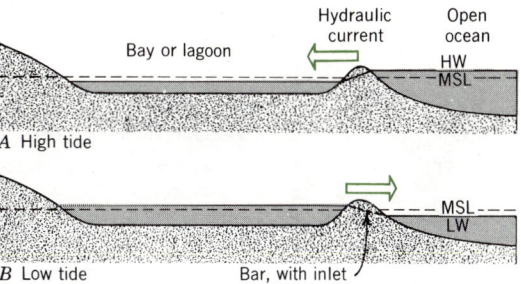

Figure 15.39 Alternating hydraulic currents through the inlet of a bar. (From A. N. Strahler, 1971, *The Earth Sciences*, 2nd ed., Harper & Row, New York.)

York Harbor. In this distance the tide range decreases from 4.4 ft (1.3 m) at Fort Hamilton to 3.0 ft (0.9 m) at Troy, New York, located at the head of tidal water. One important environmental activity of river tides is to transport pollutants, such as sewage and industrial waste products, to the open sea. This transport results from the fact that a particle moves farther seaward with the ebb current than it moves inland with the flood current. Thus, while the path alternates in direction, the net movement is seaward.

When salt water enters a tidal river, the denser salt water remains close to the bottom, while the fresh water lies over it. With flood tide the salt water moves upriver as a *salt wedge*, while on the ebb current fresh water flows seaward over the salt wedge (Figure 15.38). For a river of large discharge, such as the Mississippi, the contact between salt and fresh water in the river mouth is sharply defined, with little mixing, but in other cases mixing occurs between the two layers and salt water dilutes the fresh water above it.

A second variety of tidal current is described as *hydraulic* and operates on a different principle. Favorable conditions for hydraulic currents exist where a bay is cut off from the open ocean except for a narrow inlet in the bar or barrier beach. Figure 15.39 shows how the rise of tide in the open ocean sets up a hydraulic gradient, causing a strong flow through the inlet. As tide level falls in the open ocean the gradient is reversed and a strong current flows oceanward. Hydraulic currents have great erosional and transportational power. The strongly scoured bed typically has large *sand ripples*, showing that the bottom sediment is constantly being reworked (Figure 15.40).

Along barrier islands and barrier beaches, *tidal inlets* occur at intervals. These gaps are initially produced by the swash of storm waves breaking over the barrier at high tide. Hydraulic tidal flow thereafter maintains the inlet. Sediment arriving at the inlet by littoral drift is carried through the inlet by the flood current and deposited in quiet water of the lagoon to form a *tidal delta* (Figure 15.41). Sandy sediment spreads over muds of the lagoon floor. Some inlets do not persist, but are sealed over by sediment carried as littoral drift. Where an inlet persists for a long time on a shore subject to strong littoral drift of sediment, the inlet changes form in such a way that the beach on the updrift side of the inlet comes to overlap the beach on the downdrift side, as shown in Figure 15.42. Gradually the inlet migrates along the shore in the direction of shore drift.

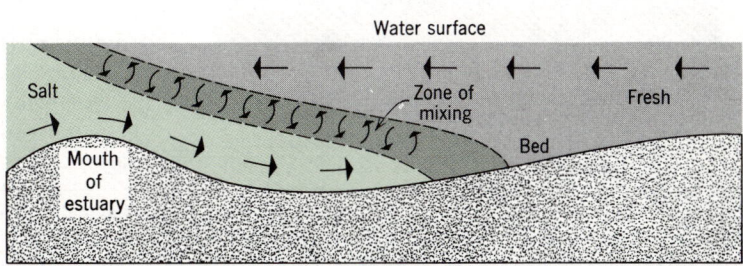

Figure 15.38 A wedge of salt water beneath fresh water in a tidal estuary.

Figure 15.40 These large sand ripples were produced by a strong ebb tide current in the Avon Estuary, Nova Scotia. Current direction is indicated by the arrow. (Photograph by Canada Department of Mines and Technical Survey, Ottawa.)

Tidal Flats and Marshes

Tidal currents carry fine silt and clay in suspension, derived from streams which enter the bays or from offshore bottom muds agitated by storm wave action. This fine sediment tends to become clotted by flocculation into small aggregates where fresh water mixes with salt water. The sediment then settles to the floors of the bays and estuaries where it accumulates in layers and gradually fills the bays. A high proportion of locally derived organic matter is normally present in such sediments, and they can be called *organic-rich muds*.

In time, tidal sediments fill the bays and produce *mud flats*, which are barren expanses of mud exposed at low tide but covered at high tide (Figure 15.43). Next, there takes hold upon the mud flat a growth of salt-tolerant plants (such as the genus *Spartina*). The plant stems entrap more sediment and the flat is built up to approximately the level of high tide, becoming a *salt marsh* (Figure 15.44). The uppermost layer of the salt marsh is composed largely of organic matter and is one of the forms of peat. Salt-marsh peat is resilient and tough in structure, capable of effectively resisting wave and current action. Tidal currents maintain their flow through the salt marsh by means of a highly complex network of sinuous *tidal streams* in which the water alternately flows seaward and landward (Figure 15.45).

The Estuarine Environment

The term *estuary* is in wide use among marine ecologists and oceanographers, but when we read their scientific publications we find that the term has been applied to coastal water bodies spanning a very wide range in physical and biological properties and environments. One carefully considered definition is as follows: An estuary is a semienclosed coastal body of water which has a free connection with the open sea and within which sea water is measurably diluted with fresh water derived from land drainage.* Open bays and gulfs having full ocean salinity are excluded from the definition. The connection of an estuary with the sea allows salt water to enter as a tidal current.

From the physical standpoint, estuaries fall into four major classes. One of these includes the *drowned river valleys*, mentioned earlier in the case of the estuarine delta (see Figure 15.31*D*). Estuaries of this class are found along coasts that

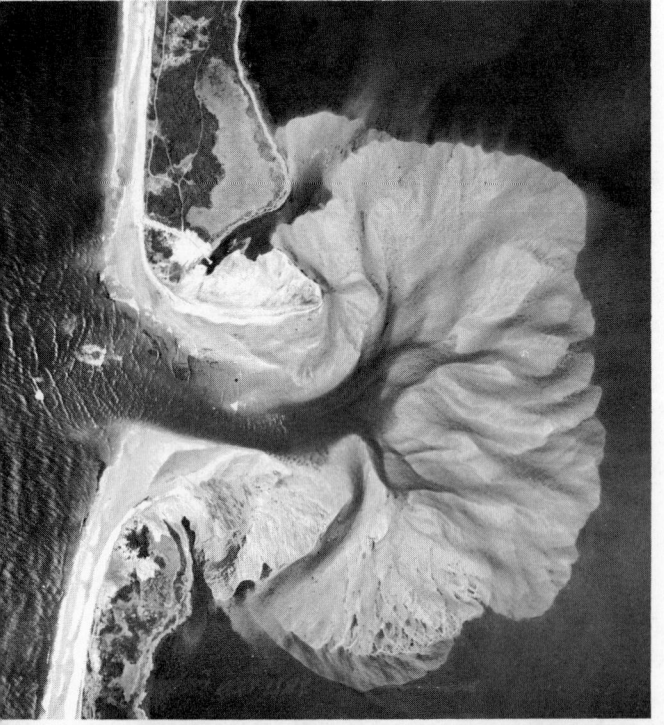

Figure 15.41 East Moriches Inlet was cut through Fire Island, a barrier island off the Long Island shoreline, during a severe storm in March 1931. This aerial photograph, taken a few days after the breach occurred, shows the underwater tidal delta being built out into the lagoon (right) by currents. The entire area shown is about 1 mi (1.6 km) long. North is to the right; the open Atlantic Ocean on the left. (U.S. Army Air Forces Photograph.)

*Donald W. Pritchard, in *Estuaries,* 1967, G. H. Lauff, ed., Publication No. 83, Amer. Assn. Advancement of Science, Washington, D.C., p. 3.

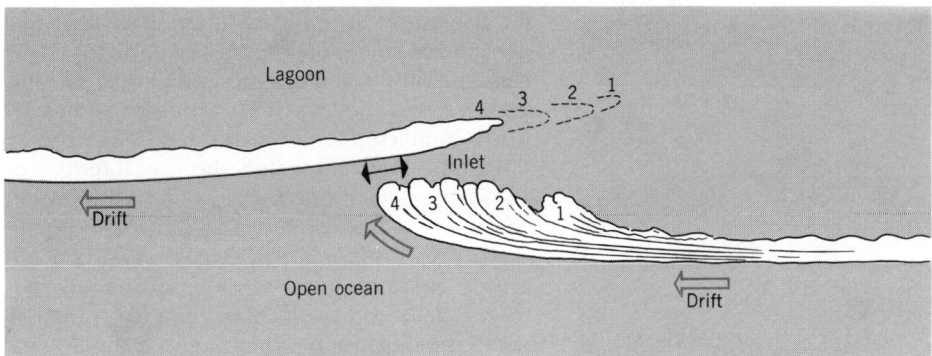

Figure 15.42 Because of littoral drift, right to left, an inlet in a barrier beach has migrated toward the left. (From A. N. Strahler, 1971, *The Earth Science*, 2nd ed., Harper & Row, New York.)

have experienced recent submergence, either by crustal subsidence or by postglacial rise of sea level or by a combination of both. An excellent example is Chesapeake Bay, whose dendritic pattern of branches immediately suggests partial inundation of a dendritic stream-valley system. Each tip of an estuary branch represents a point of input of fresh water from runoff. The Hudson River, referred to in connection with river tides, is another example, but is almost free of branches since it occupies a rock trench carved in a previous episode of stream rejuvenation (Chapter 14). (In some usages, any stream mouth where fresh water meets salt water is an estuary.)

Bar-built estuaries are a second general class. We have already covered the origin of such estuaries in explaining the development of barrier islands and of barrier beaches. The lagoon, referred to earlier as a shallow water body between the barrier island and the inner shoreline is therefore one form of a bar-built estuary. The *fiord*, an arm of the sea occupying a deeply carved former glacial trough, is another class of estuary, physically very different from the estuaries formed by drowning of shallow valley systems and by bar-building (see Figure 17.8D). On the other hand, the Hudson estuary shows the effects of deep scour by ice of the Pleistocene ice sheets, and in some places is very fiord-like in cross section. Another class of estuaries results from

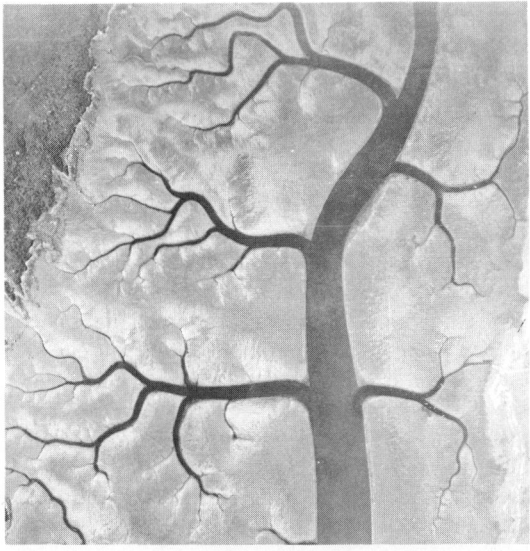

Figure 15.43 Viewed from the air at low tide these tidal mud flats near Yarmouth, Nova Scotia, show a well-adjusted branching system of tidal streams. The area shown is 1.5 mi (2.4 km) wide. (Canadian Armed Forces Photograph.)

Figure 15.44 Salt marsh at South Wellfleet, Massachusetts. Late winter stubble and dead leaves of salt-marsh grass (*Spartina*) cover the peat layer. A small tidal channel lies drained empty at low tide. (Photograph by A. N. Strahler.)

Figure 15.45 This broad tidal salt-marsh along the east coast of Florida is laced with serpentine tidal channels. (Aerial photograph by Laurence Lowry.)

down-faulting of small sections of the crust. These *tectonic estuaries* are not common, but San Francisco Bay is an outstanding example.

Estuaries also can be classified in terms of the balance between fresh and salt water contributions from land and ocean and the relative importance of evaporation. In the *positive* estuary, input of fresh water from runoff and direct precipitation exceeds evaporation, so that salt water is diluted by fresh. In the *negative* estuary, evaporation of ocean water exceeds fresh-water input with the result that the estuarine water develops a salinity greater than that of sea water, a condition known as *hypersalinity*. An example of a negative, or *hypersaline*, estuary is Laguna Madre, a lagoon lying between a 100-mi (160-km) barrier beach and the mainland off the south coast of Texas (see Figure 10.18). A small section of this estuary is shown in Figure 15.28.

Under the broad definition we have accepted for an estuary a wide range of environments and life habitats is included, with the result that few generalized statements apply to all estuaries. Perhaps the only really common properties of estuaries are that they are sea-level water bodies directly connected with the sea and influenced by tides, and that their waters show an interface or a salinity gradient between fresh terrestrial water and water of normal ocean salinity.

Estuaries form one of the most important but at the same time most sensitive of the environments of life on earth. Many types of estuaries rank as among the most productive environments of organic matter to be found on earth. Ecosystems of estuaries are extraordinarily elaborate and include all elements of the food chain. Plant nutrients coming into the estuary from the land allow marine plants, such as the algae, to flourish and provide food for marine animals, some of which are bottom dwelling invertebrates—shellfish, worms, crustaceans. Fish of many species spawn in estuaries, which nourish the young before they migrate to the open ocean. Many species of waterfowl are adapted to feed upon both plants and animals of the estuaries. Our concern here has been mostly with physical aspects of the estuarine environment and with physical changes induced by Man.

Unfortunately, from the standpoint of ecosystems of estuaries, human population is heavily concentrated close to estuaries, many of which are adapted as harbors of major port cities. One cannot list the major coastal cities of the United States and Europe without naming many linked to an important estuary: New York City on the Hudson, Philadelphia on the Delaware, Baltimore on Chesapeake Bay, Washington on the Potomac, Mobile on the Mobile, Houston on Galveston Bay, San Francisco on San Francisco Bay, Seattle on Puget Sound, London on the Thames, Hamburg on the Elbe—and so on.

Man's impact upon the marine life of estuaries takes a number of basic forms. One is through the alteration of water salinities and salinity gradients, either by withdrawal of fresh river water at upstream points, or by engineering changes in the natural systems of tidal inlets and channels. Patterns of water circulation are also altered by these same engineering works. Pollution is a second major category of environmental change. Pollution is heavy from both municipal sources, as sewage and solid wastes, and from industrial sources in the form of toxic substances. Of the latter, petroleum and its derived products are among the most serious from the ecological standpoint, because of the tens of thousands of complex organic compounds found in these pollutants, many are toxic to organisms and many do not break down chemically into harmless or nutrient compounds. Urbanization can result in large increases in suspended sediment of streams, as explained in Chapter 14. When this sediment reaches the head of an estuary, sedimentation is rapidly increased and the bottom environment drastically altered. In an opposite manner, building of large dams withholds suspended sediment, and the lack of a normal input of sediment into an estuary may induce bottom scour by tidal currents. Yet another form of estuarine pollution is that of heating of water—thermal pollution—by discharge of coolant water from electricity generating plants, including nuclear plants. (We have discussed this subject in Chapter 13.)

From this rather general discussion of estuaries and their environmental problems, we turn to specific estuarine types—tidal flat and the salt marsh—and to the problems originating from their utilization by Man.

Reclamation of Tidal Lands

Under pressures of expanding populations in need of more food, agricultural lands in Europe, the British Isles, and to a lesser degree the New World, have been expanded at the expense of the sea by *reclamation*—literally the reclaiming of shallow tidal lands that had fallen victim to the rising postglacial sea levels and to crustal subsidence. In general, reclamation applies to two types of estuarine environments. First is the mud-flat environment—shallow, but open water in which expanses of mud are exposed at low tide. These sediments are largely inorganic silts, brought by streams emptying into broad estuaries (e.g., deltaic sediments) or by tidal currents deriving sediment from wave erosion of exposed cliffs and scarps. The second environment is that of the salt marsh, with its peat layer, already filling to a greater or lesser degree what were previously open tidal estuaries.

The greatest reclamation region has been that of the Low Countries along the North Sea, where extensive areas have been reclaimed in Belgium and the Netherlands (Figure 15.46). A large area of coastal tidelands, ranging in width from 25 to 40 mi (40 to 65 km) and bordered by a belt of coastal sand dunes, occupies the North Sea coast of the Netherlands. Reclamation began as early as A.D. 900 by the building of dikes, earth embankments made of indigenous sediment. With exclusion of sea water, many large lakes were formed, and removal of peat led to their enlargement. In the seventeenth century these lake basins were drained by pumpage. The method was to cut off a small area by dikes and pump out the water through use of windmills. The area of drained land, much of it lying more than 8 ft (2.5 m) below sea level, thus consists of diked land parcels, called *polders,* within an elaborate network of drainage canals. Development of this period reclaimed over 90 sq mi (235 sq km) of land. By 1920 the country had some 1000 mi (1600 km) of sea dikes and the reclaimed area totaled about 330 sq mi (860 sq km).

The Zuider Zee, a large shallow estuary lying east and north of Amsterdam was converted into a fresh-water body by construction of a 20-mi (32-km) dam (Figure 15.46), completed in 1932. Chlorinity in this water body dropped rapidly and by 1937 all salinity was excluded except near the locks by which the new lake, Lake Ijssel, (Ijsselmeer) is drained into the Wadden Zee. Salt water organisms rapidly died out in the new lake and were replaced by fresh-water forms, save for the eel, which thrives in the new lake. Reclamation of Lake Ijssel is taking place in sections (Figure 15.47). When complete, the new land will total about 860 sq mi (2300 sq km), leaving a much smaller lake for drainage of water from the Ijssel River (a distributary of the lower Rhine).

A second area of new reclamation in the Netherlands affects four estuaries lying to the southwest of Rotterdam. These elongate coastal water bodies are the drowned mouths of the Rhine, Maas (Meuse), and Scheldt Rivers, and the land itself represents deltaic deposits of these rivers. Under the name of the Delta Plan, reclamation has begun with the construction of three large primary dams across the mouths of the estuaries (Figure 15.46) and a number of secondary dams. The dammed areas will become fresh-water lakes, and these in turn will be reclaimed by dike-building and pumping. As in the case of the Zuider Zee project, the change from saline to fresh water will eliminate salt-water organisms and these will be replaced by fresh-water organisms. Economic losses to the mussel and oyster industry will be large.

A second area of major reclamation is along the east coast of England. The largest single area is that of the Fens, a marshy lowland lying inland of the Wash, a conspicuous coastal embayment between Lincolnshire and Norfolk. In postglacial time the interior portion of the Fens accumulated peat up to thicknesses of 20 ft (6 m), while the seaward portion accumulated silty sediment. Reclamation was accomplished in the seventeenth century by extensive ditching and construction of embankments. Since then, the surface of peat lands has been lowered many feet in places by a combination of removal and burning of the peat, by its natural oxidation under exposure to

Waves, Currents, and Coastal Landforms

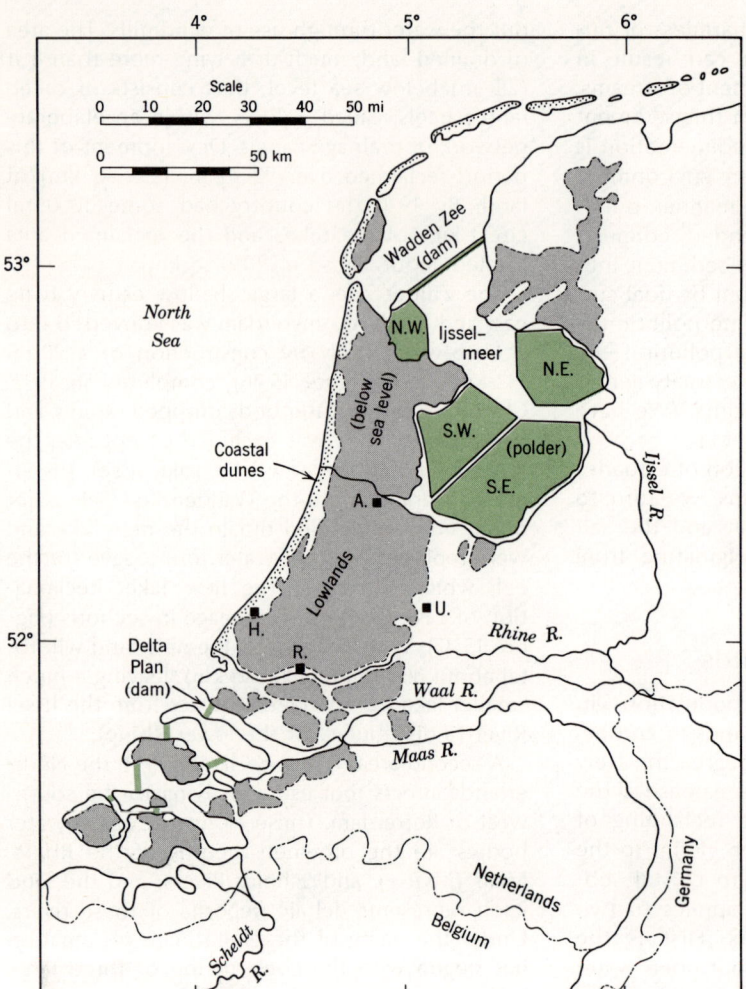

Figure 15.46 Map of Netherlands coastal lowlands and deltaic plains. Areas below sea level are shaded. Zuider Zee and Delta Plan reclamation works, including dams and polders, are shown in color. (Data of Zuider Zee Board and Geological Survey of Netherlands.)

the atmosphere, and by shrinkage from dehydration. One of the remarkable exhibits of this land-lowering is a 30-ft (9-m) iron post that in 1848 had been driven through the peat into clays below (Figure 15.48). The post now protrudes about 14 ft (4 m) above the ground surface and the rate of protrusion is about 1 inch per 2 to 3 years (0.8 to 1.4 cm per year).

Reclaimed tidal marshes, or *fenlands,* occupy many localities of the English coast from the Humber River estuary on the north to the Thames estuary on the south. These lands lie below sea level and are highly susceptible to inundation by salt water during storms that allow the ocean level to overtop the various dune ridges, beach ridges, and embankments separating the reclaimed land from the sea. A particularly devastating North Sea storm, which also severely damaged reclaimed areas of the Netherlands, occurred in January, 1953. A combination of extremely high winds, a spring tide, and a storm surge resulted in flooding of over 300 sq mi (780

sq km) of fenlands and a loss of over 300 lives.

Flooding of fenlands and polders with salt water does serious damage to croplands. The sodium ion of sea salt (NaCl) replaces the calcium ion in the soil, with a resulting deterioration in the physical state of the soil (the *tilth,* or ease of cultivation is destroyed). The salt itself is toxic to most crops, so that flushing out of salts is required. Exclusion of salts by normal rains takes place in time, but from 1 to 5 years may elapse before crops can be grown. Applications of gypsum (calcium sulfate) accelerate the replacement of sodium with calcium.

Flooding of the fenlands and polders by winter storms is a long-standing environmental problem bringing together processes and principles of many sciences, including meteorology, oceanography, geomorphology, civil engineering, and soil science. The subject makes a good example of the interdisciplinary nature of environmental science. The case we have described shows that Man, in his drive for more food to meet the needs

Reclamation of Tidal Lands

of an increasing population has radically altered the natural environment and ecosystems of large areas, and at the same time has aggravated the natural hazards of coastal flooding by his land-use practices.

Ultimately, the limits of land reclamation will be reached, while at the same time a sizeable fraction of the estuarine environment will be permanently destroyed, with consequent losses in food resources of the sea. Careful planning will be required if reclamation is to provide more gain than it loses in terms of natural resources. Fortunately, reclamation of tidal lands in North America, while long practiced, never reached the proportions found in the Old World. Tidal lands are now being increasingly protected by strict legislation governing the uses of wetlands generally. In the United States pressures upon tidal lands are predominantly from urbanization and industry, rather than from agriculture. In all urban seaboard areas tidal mud flats and salt marshes are disappearing under expanding earth fills upon

Figure 15.47 The Zuider Zee reclamation project, Netherlands. (Photographs by courtesy of Netherlands Information Service.) Right: Construction of an enclosing dike within Lake Ijssel, the first step in creating the Southern Flevoland Polder. Below: New polder land adjacent to the city of Medemblick, in the northwestern part of Lake Ijssel. A large pumping station, adjacent to the enclosing dike at the right, removes water from canals draining the diked area.

Waves, Currents, and Coastal Landforms

Figure 15.48 Subsidence of peat of the English fenlands is documented by gradual emergence of this iron post which in 1848 was driven through 22 ft (7 m) of peat into firm clay beneath. The horizontal marks and dates show that almost 11 ft (3.4 m) of subsidence had occurred by 1932, the year the photograph was taken. (Potograph by Major Gordon Fowler.)

which industrial installations, housing developments, highways, and airfields are built. Although the organic content of these sediments poses difficulties in terms of foundation engineering (the runways of Laguardia Airport in New York have been subsiding continuously for years), these problems can be met in a number of ways and are not a limiting factor in use of tidal lands.

Coral-Reef Coasts

Coral reefs are unique from any shoreline forms mentioned thus far, because they represent mineral matter accumulated in the surf zone through organic processes. The reef-forming organisms are *corals*, which secrete lime to form their skeletons, and *algae*, plants that also make limy encrustations. Corals are colonial types of animals, that is, they occur in large colonies of individuals. As coral colonies die, new ones are built upon them, thus developing a coral limestone made up of the strongly cemented limy skeletons. Coral fragments torn free by wave attack and pulverized may be deposited to form beaches, spits, and bars, which later are cemented into a limestone.

Coral-reef shorelines occur in warm, tropical and equatorial zones between the latitude limits 30° N and 25° S. Water temperatures above 68° F (20° C) are necessary for dense reef coral growth. Furthermore, reef corals live near the water surface, down to limiting depths of about 200 ft (60 m). Water must be free of suspended sediment and well aerated for vigorous coral growth; so that corals thrive in positions exposed to wave attack from the open sea. Because muddy water prevents coral growth, reefs are missing opposite the mouths of turbid streams. Coral reefs are remarkably flat on top (Figure 15.49) and have a surface level approximately equal to the upper one-third mark of the range of tide. Thus they are exposed at low tide and covered at high tide.

Three general types of coral reefs may be recognized: (1) fringing reefs, (2) barrier reefs, and (3) atolls. *Fringing reefs* are built as platforms attached to shore (Figure 15.50). They are widest in front of headlands where wave attack is strongest, and the corals receive clean water with abundant nutrients. Fringing reefs are usually absent near the mouths and deltas of streams, where the water is muddy. This is a fact of great environmental importance where the problem is to find reef-free places for landing of ships. Fringing reefs may be from 0.25 to 1.5 mi (0.4 to 2.5 km) wide, depending on exposure to surf and the length of time that the reef has been developing.

Barrier reefs lie out from shore and are separated from the mainland by a lagoon which may

Figure 15.49 A fringing reef on the south coast of Java forms a broad bench between surf zone (*left*) and a white coral-sand beach. Inland is rainforest. (Photograph by Luchtvaart-Afdeeling, Bandung.)

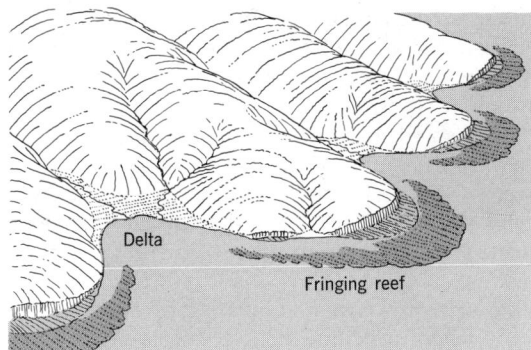

Figure 15.50 Fringing reefs are widest in front of headlands and may be absent near the mouths of streams. (After W. M. Davis.)

range from 0.5 to 10 mi (2.5 to 16 km) or more in width (Figure 15.51). The reef itself may be from 20 to 3000 ft (6 to 900 m) wide. The lagoon is shallow and flat-floored, usually 120 to 240 ft (35 to 75 m) deep. There are, however, many towerlike columns of coral in the lagoon. *Passes*, which occur at intervals in barrier reefs, are narrow gaps through which excess water from breaking waves is returned from the lagoon to the open sea. They sometimes occur opposite deltas on the mainland shore, because of the inhibiting effect of turbid water on coral growth. Passes are of environmental importance because they provide the only means of entrance by ship into the lagoon.

Atolls are more or less circular coral reefs enclosing a lagoon, but without any land inside (Figure 15.52). In all other respects they are similar to barrier reefs. On large atolls, parts of the reef have been built up by wave action and wind to form low island chains, connected by the reef. A cross section of an atoll shows that the lagoon is flat-floored and shallow, and that the outer slopes are steep, often descending thousands of feet to great ocean depths.

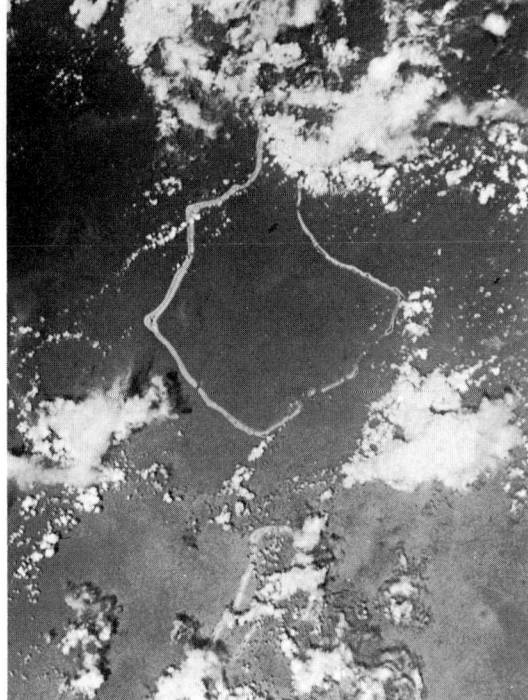

Figure 15.52 Photographed from an altitude of about 150 mi (240 km) by astronauts aboard *Gemini V* spacecraft, Rongelap Atoll appears as a closed loop among cloud patches. Marshall Islands, Pacific Ocean. (NASA photograph.)

Several plausible theories have been advanced for the origin of atolls and barrier reefs. To explain each one and discuss the advantages and disadvantages of each would take many pages. One interesting theory of origin, which has been popular since it was first outlined by a great scientist, Charles Darwin, in 1842, may be called the *subsidence theory* (Figure 15.53). He supposed that small islands, such as volcanoes, slowly subsided in a general downwarping of the earth's crust over

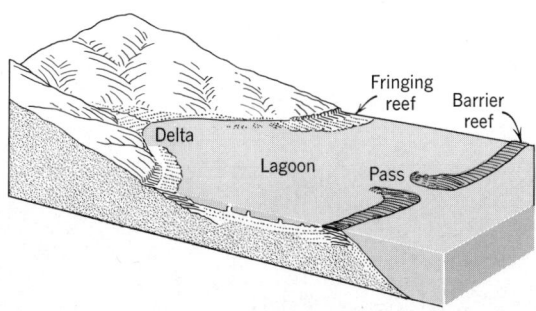

Figure 15.51 A barrier reef is separated from the mainland by a shallow lagoon. (After W. M. Davis.)

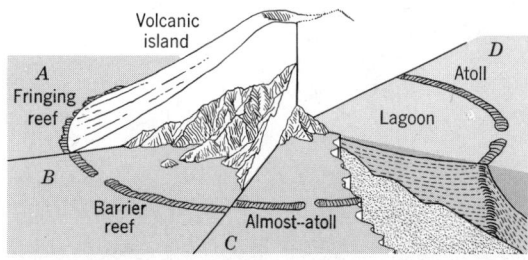

Figure 15.53 The subsidence theory of barrier-reef and atoll development is shown in four stages, beginning with a fringing reef attached to a volcanic island and ending with a circular reef. (After W. M. Davis.)

parts of the ocean basin. Coral reefs, which were originally fringing reefs attached to the island shores, continued to build upward as the island subsided. Thus the area of the island shrank and a lagoon formed, creating a barrier reef. Finally the island sank out of sight, but the reef persisted, maintained at sea level by vigorous coral and algal growth.

The environments of atoll islands are unique in some respects. First, there is no rock other than coral limestone, composed of calcium carbonate. This means that trees requiring other minerals, such as silica, cannot be cultivated without the aid of fertilizers or some outside source of rock from a larger island composed of volcanic or other igneous rock. The cocoanut palm tree is native to atoll islands because it thrives on brackish water, and the seed, or palm nut, is distributed widely by floating from one island to another. Native inhabitants cultivated the cocoanut palm to provide food, clothing, fibers, and building materials. Fresh water is scarce on small atoll islands because there is not enough surface area for the collection of rainfall, and the land is so low that a high water table of fresh water is not present to supply springs, streams, and wells. Rainfall must be caught in open vessels or catchment basins and carefully conserved. Fish and other marine animals are an important part of the human diet on atoll islands. Calm waters of the lagoon make a good place for fishing and for beaching canoes. Coral islands of the western Pacific stand in continual danger of devastation by tropical cyclones (typhoons). Breaking waves wash over the low-lying ground, sweeping away palm trees and houses and drowning the inhabitants. There is no high ground for refuge. In the same way, great seismic sea waves of unpredictable occurrence may inundate atoll islands.

Adjacent to shallow coastal waters that are turbid with deltaic sediment the coral reef is replaced by the *mangrove coast* dominated by growth of mangrove plants.

Wind Action and Coastal Landforms

Transportation and deposition of sand by wind is an important process in shaping coastal landforms. We have made a number of references in this chapter to coastal sand dunes derived from beach sand. In the next chapter we will investigate the transport of sand by wind and the shaping of dune forms in order to complete the linkage between wind action and wave action in controlling coastal environments.

The transfer of momentum and kinetic energy from winds to the land surfaces was reviewed in opening paragraphs of Chapter 15. We pointed out that wind expends energy in boundary friction and eddy turbulence upon solid surfaces of rock or plant structures. Ordinarily, wind strengths are not sufficient to dislodge mineral matter from the surfaces of tightly knit rock or of moist, clay-rich soils, or of soils bound by a dense plant cover. Consequently, the action of wind in eroding and transporting sediment is limited to those surfaces where small mineral and organic particles are in the loose state. While such areas are typically limited to deserts and semiarid lands (steppes), an exception is the coastal environment, where beaches provide abundant supplies of loose sand, even where the climate is one of a substantial soil-water surplus and the natural land surface inland from the coast is well protected by a cover of trees, shrubs, or grasses.

Landforms shaped and sustained by wind erosion and deposition represent distinctive life environments, often highly specialized with respect to the assemblages of animals and plants they support. Because, in climates with barely sufficient soil water, there is a contest between wind action and the growth of plants that tend to stabilize landforms and protect them from wind action, we will find precarious balances in the ecosystems of certain marginal areas. These balances are not only altered by natural changes in climate, but are easily upset by Man's activities, and often with serious consequences. To understand these environmental changes, we will need to acquire a working knowledge of the physical processes of wind action upon the land surfaces.

Erosion by Wind

Wind performs two kinds of erosional work. Loose particles lying upon the ground surface may be lifted into the air or rolled along the ground. This process is *deflation*. Where the wind drives sand and silt particles against an exposed rock or soil surface, causing it to be worn away by the impact of the particles, the process is one of *abrasion*. Whereas abrasion requires cutting tools carried by the wind; deflation is accomplished by air currents alone.

16

Wind Action and Dune Landscapes

Deflation acts wherever the ground surface is thoroughly dried out and is littered with loose particles derived by rock weathering or previously deposited by running water, glaciers, or waves. Dry river courses, beaches, and areas of recently formed glacial deposits are highly susceptible to deflation. In dry climates, the entire ground surface is subject to deflation because the soil or rock is everywhere bare of plant cover.

Wind is selective in its deflational action. The finest particles, those of clay and silt size grades are lifted most easily and raised high into the air. Sand grains are moved only by moderately strong winds and tend to travel close to the ground. Gravel fragments and rounded pebbles up to 2 or 3 in. (5 to 8 cm) in diameter may be rolled over flat ground by strong winds but do not travel far. They become easily lodged in hollows or between their fellows. Consequently, where a mixture of sizes of particles is present on the ground, the finer sizes are removed while the coarser particles remain behind—a winnowing action.

The principal landform produced by deflation is a shallow depression termed a *blowout*, or *deflation hollow*. This depression may be from a few yards to a mile or more in diameter, but is usually only a few feet deep. Blowouts form in plains regions in dry climate where the natural plant cover consists of short grasses or small shrubs. Any small depression in the surface of the plain, particularly where the grass cover is broken through, may develop into a blowout. Rains fill the depression, creating a shallow pond or lake. As the water evaporates the mud bottom dries out and cracks, forming small scales or pellets of dried mud which are lifted out by the wind. In grazing lands, large hoofed animals can trample the margins of the depression into a mass of mud, breaking down the protective grass-root structure and facilitating removal when dry. Consequently, the depression is progressively enlarged (Figure 16.1). Blowouts are also found on rock surfaces where the rock is being disintegrated by weathering.

In the great deserts of southeastern California, Arizona, and New Mexico, the floors of intermontane basins are subject to deflation. Areas of silty soils bordering playas have in some places been reduced by deflation as much as several feet over areas of many square miles.

Where deflation has been active on a ground surface littered with loose fragments of a wide range of sizes, the pebbles that remain behind tend to accumulate until they cover the entire surface (Figure 16.2). By rolling or jostling about as the fine particles are blown away, the pebbles become closely fitted together, forming a *desert pavement*. In Arabia and North Africa such a pebble-covered surface is called a *reg*. Desert pavements develop readily on surfaces of alluvial fans and stream terraces, no longer acted upon by flowing water. The precipitation of calcium carbonate, gypsum, and other salts near the surface, as ground water is drawn to the surface and evaporated in dry weather, tends to cement the pebbles together, forming a highly effective protection against further deflation.

Sandblast action of wind against exposed rock surfaces is limited to the basal few feet of a cliff, hill, or other rock mass rising above a relatively flat plain, because sand grains do not rise high

Figure 16.1 A blowout hollow on the plains of Nebraska. A remnant column of the original soil provides a natural yardstick for the depth of material removed by deflation. (Photograph by N. H. Darton, U.S. Geological Survey.)

Figure 16.2 This desert pavement of quartzite fragments was formed by action of both wind and water on the surface of an alluvial fan in the desert of southeastern California. Fragments range in size from 1 to 12 in. (2.5 to 33 cm). Silt underlies the layer of stones. (Photograph by C. S. Denny, U.S. Geological Survey.)

into the air. Wind abrasion produces small pits, grooves, and hollows in the rock. Where a small rock mass projects above the plain it may be cut away at the base to make a *pedestal rock*, delicately balanced upon a thin stem. Most pedestal rocks, or *mushroom rocks*, are, however, produced by weathering processes.

On plains subjected to blowing sand, the bases of electric power and telephone poles may require protective coverings of sheet steel or heaped boulders to prevent their being cut through by abrasion. Automobile glass can be frosted rapidly by sandblast and the body finish badly etched at the same time.

Dust Storms and Sand Storms

In dry seasons over plains regions, strong, turbulent winds lift great quantities of fine dust into the air, forming a dense, high cloud called a *dust storm*.* The dust storm is generated where ground surfaces have been stripped of protective vegetative cover by cultivation or grazing, or where they naturally carry little or no vegetation cover because of aridity of the climate. A dust storm approaches as a great cloud extending from the ground surface to heights of several thousand feet (Figure 16.3). Within the dust cloud deep gloom or even total darkness prevails, visibility is cut to a few yards, and a fine choking dust penetrates everywhere.

It has been estimated that as much as 4000 tons of dust may be suspended in a cubic mile of air (875 metric tons per cubic kilometer). On this basis, a dust storm 300 mi (500 km) in diameter might be carrying more than 100 million tons (90 million metric tons) of dust—enough to make a hill 100 ft (30 km) high and 2 mi (3 km) across the base.† A region that supplied the dust for thousands of such storms would thus lose a considerable mass over a span of thousands of years.

The true desert *sandstorm* is a low cloud of moving sand that rises usually only a few inches and at most 6 ft (2 m) above the ground. It consists of sand particles driven by a strong wind. Those who have experienced sandstorms report that a man standing upright may have his head and shoulders entirely above the limits of the sand cloud. The reason why the sand does not

Figure 16.3 Front of an approaching dust storm, Coconino Plateau, Arizona. (Photograph by D. L. Babenroth.)

rise higher is that the individual particles are engaged in a leaping motion, termed *saltation* (Figure 16.4). Grains describe a curved path of travel and strike the ground with considerable force but at a low angle. The impact causes the grain to rebound into the air. The phenomenon of saltation explains why the erosional effect of blown sand is concentrated on surfaces exposed less than a foot or two (0.3 to 0.6 m) above the flat ground surface.

The impact of leaping sand grains causes other grains to move slowly forward, a process known as *surface creep*. Because the energy of a leaping sand grain is capable of moving an impacted grain as much as six times its own diameter, or two hundred times its own weight, saltation also causes downwind creep of particles of coarse sand and pebble grades. However, the large grains moving by surface creep travel much slower than sand grains in saltation, with the result that a sorting action takes place. The coarse sand and pebbles remain behind and ultimately accumulate to form protective sheets that prevent the finer sand beneath from being removed.

Another effect of saltation and surface creep is to produce *sand ripples*, which are small ridges and troughs resembling sand ripples produced by water (Figure 16.5).

Recall from Chapter 14 that the transporting ability of a stream increases very greatly with an increase in discharge during rising flood stages. We estimated that the capacity of a stream to

---

*As used here, the term *dust* refers to mineral particles in the size range 0.001 to 0.03 mm, which spans the range from medium clay to medium silt. Coarse silt often behaves in a manner similar to find sand and is not carried high in suspension, but is nevertheless present in lower layers of the dust cloud.

†A. K. Lobeck, *Geomorphology*, McGraw-Hill Book Co., New York, 1939, p. 380.

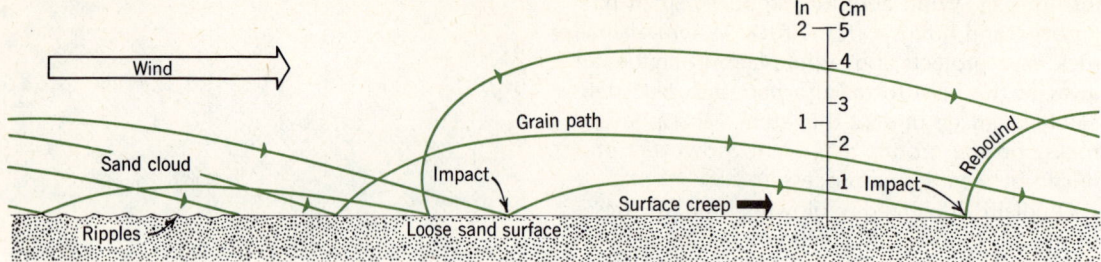

Figure 16.4 Sand particles travel in a series of long leaps. (After R. A. Bagnold.)

transport bed load varies about as the third to fourth power of mean velocity. A similar relationship exists between the rate of downwind mass transport of sand and wind speed. Figure 16.6 is a graph showing transport rate in relationship to average wind speed, as measured at a height of 1 m (3.3 ft) above the ground surface. From the inclination of the straight line on the graph, which uses logarithmic scales on both axes, we can conclude that mass rate of sand transport increases about as the third power (cube) of wind speed. In practical terms, for example, a wind of 35 mi (55 km) per hour can move in one day about the same quantity of sand that a wind of half that speed can move in three weeks. An understanding of this relationship can help us in anticipating the effectiveness of winds from different quarters of the compass at different seasons, and is an aid in planning measures for the control of sand movement.

Sand Dunes

A *dune* is any accumulation of sand transported and shaped by the wind. Dunes may be active, or *live,* when bare of vegetation and constantly changing form under wind currents. They may be inactive, or *fixed,* dunes, covered by vegetation that has taken root and serves to prevent further shifting of the sand. One estimate places the total world area of sand dunes at 5 million sq mi (13 million sq km), which is nearly twice the area of the contiguous 48 United States.

Dune sand is most commonly composed of the mineral quartz, which lacks natural parting (cleavage) and has a high hardness rating. Quartz also is highly resistant to chemical alteration or solution, with the result that it is continually recycled in sediments. Most dune sands are therefore derived from stream-deposited alluvium, from beaches, and from sandstones which were

Figure 16.5 Rippled surface of a sand dune, Quatif, Saudi Arabia. (photograph by courtesy of Arabian American Oil Company.)

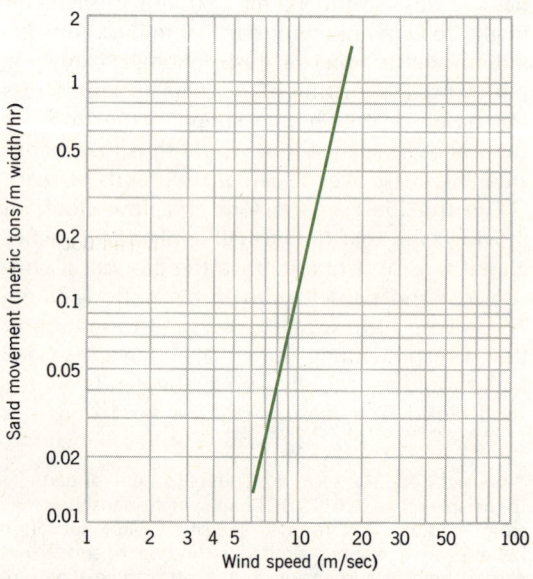

Figure 16.6 Increase in rate of mass transport of sand with increase in wind speed. Both scales are logarithmic. (Based on data of R. A. Bagnold.)

of similar origin. Dune sand close to a fluviatile or beach source may contain minor quantities of many other rock forming minerals. The quartz grains become beautifully rounded and have a frosted surface. Figure 8.7 shows such spherical grains, derived from a sandstone formed of dune sand of Cambrian age. Sorting of dune sands is exceptionally good; almost all grains fall in the size range of 0.1 to 1 mm, and the predominant size is medium sand (0.25 to 0.5 mm). As a result, clean dune sand has a number of industrial uses, among them being sand filters for water purification. The porosity and permeability of such sand is high, so that the rate of infiltration of precipitation into dune surfaces is high and sustained. As a result, dunes rarely show any evidences of erosion by overland flow (e.g., shoestring rills or gullies). Dune sand and sandstones of dune sand generally make excellent ground-water aquifers.

In a few localities dunes are composed of minerals other than quartz and its related rock-forming minerals. Examples are shell fragments (from beaches), particles of volcanic ash, gypsum, and various heavy minerals. Dunes of White Sands National Monument in New Mexico are composed of gypsum and are glaringly bright in full sunlight. Coarse silt and very fine sand also accumulate as dunes, or dunelike mounds and drifts, where these grades constitute the parent material.

A characteristic feature of dunes of free sand, devoid of vegetation, is the formation of the *slip face*, a steep sand slope on the lee side of a dune, inclined downward in the direction of sand transport. Figure 16.7 shows how a slip face develops from a smoothly rounded dune. Sand is removed from the upwind side of the dune and carried in saltation to be deposited on the lee side, which becomes steeper in angle. When the lee slope becomes sufficiently steep, grains in saltation leap beyond a sharp crest and fall to rest upon the surface beyond. When this slope reaches an upper limiting angle, about 35° from the horizontal, the sand is unstable and begins to slide. Sliding affects a surface layer of the sand, which moves down to the dune base. Following sliding, the upper slip face is reduced to a stable slope, but as this steepens again from continued saltation, sliding again occurs. In this way the dune advances and its slip face buries the ground surface ahead of it.

Dunes of Free Sand

Dunes of free sand, having slip faces, range from simple to complex forms, depending upon sand supply and the variations or alternations of wind directions that shape them. Perhaps the simplest type is the *barchan*, or *crescentic dune*, an isolated body of sand moving over a ground surface that is immobilized because it consists of bare rock or pebbles (Figure 16.8). Seen from above, the outline of a barchan resembles an ellipse from which a semicircular bite has been removed (Figure 16.9). The slip face is concavely bowed in the downwind direction. The leading ends of the dune are softly rounded. From these ends, sand moving off the dune in saltation produces *spray zones* of lighter color. However, within the arc of the slip face, the ground surface is free of sand in saltation and appears darker in color. Barchans grow to heights of 30 to 100 ft (9 to 30 m) and have widths up to 1200 ft (360 m). Downwind movement may average several inches per year, but individual dunes have been recorded as traversing distances of 30 to 50 ft (9 to 15 m) in a year's time.

The barchan may originate as a sand drift in the lee of some obstacle, such as a small hill, rock, or clump of brush. Once a sufficient mass of sand has formed it begins to move downwind, taking the crescent form. Consequently, barchans are commonly arranged in chains extending downwind from the source drifts.

Where sand is so abundant that it completely covers the immobile ground beneath, dunes take the form of wavelike ridges separated by troughlike furrows. The dunes are called *transverse dunes* because their crests trend at right angles to direction of the prevailing wind (Figure 16.10). The entire area may be called a *sand sea*, for it resembles a storm-tossed sea suddenly frozen to immobility. The term *erg*, referring to any large

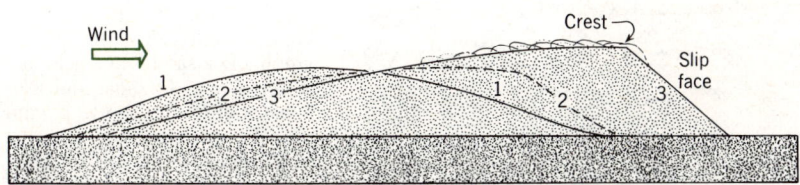

Figure 16.7 Development of a dune slip face. (After R. A. Bagnold. From A. N. Strahler, 1971, *The Earth Sciences*, 2nd ed., Harper & Row, New York.)

Wind Action and Dune Landscapes

Figure 16.8 Barchan dunes at Biggs, Oregon. (Photograph by G. K. Gilbert, U.S. Geological Survey.)

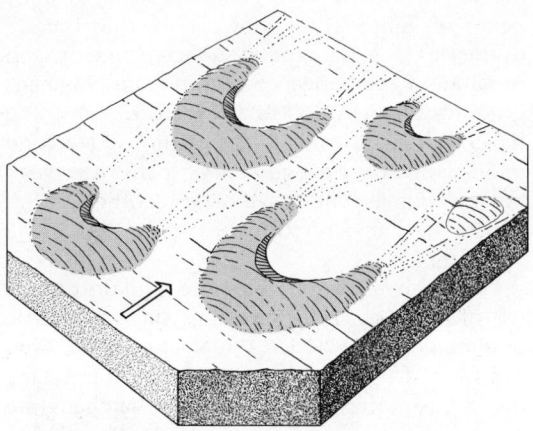

Figure 16.9 Sketch of barchan dunes. Arrow shows prevailing wind direction.

expanse of dunes in the Sahara Desert, has been adopted for this type of landscape. Individual sand ridges have sharp crests and are asymmetrical, the gentle slope being on the windward, the slip faces on the lee side. Deep depressions lie between the dune ridges. Sand seas require huge quantities of sand, often derived from weathering of a sandstone formation underlying the ground surface, or from adjacent alluvial plains. Still other transverse dune belts form adjacent to beaches which supply abundant sand and have strong onshore winds (see Figure 16.15).

In the vast deserts of North Africa, Arabia, and southern Iran are large, complex dune forms of free sand not represented in North America. One of these is the *seif dune,* or *sword dune,* a huge tapering sand ridge whose crestline rises and falls in alternate peaks and saddles and whose side

Figure 16.10 This air photograph of a sand-dune field between Yuma, Arizona, and Calexico, California, shows a sand sea of transverse dunes in the background and a field of crescentic barchan dunes in the foreground. (Copyrighted Spence Air Photos.)

slopes are indented by crescentic slip faces. Seif dunes may be a few hundred feet high and tens of miles long. Another Saharan type is the *star dune*, *pyramidal dune*, or *heaped dune*, a great hill of sand whose base resembles a many-pointed star in plan (Figure 16.11). Radial ridges of sand rise toward the dune center, culminating in sharp peaks as high as 300 ft (100 m) or more above the base. Star dunes seem to remain fixed in position for centuries and can serve as reliable landmarks for desert travelers.

Longitudinally oriented with respect to the wind, but not a true dune, is the *sand drift*, which is a long, tapering sharp-crested ridge of free sand extending downwind from some topographic obstacle, such as a hill that might rise above a desert plain (Figure 16.12). Sand moving in saltation passes over or around the obstacle, lodging to the leeward and gradually building the drift until the zone of quiet air is filled. Additional sand is carried along the length of the drift, extending it far downwind.

Phytogenic Dunes

Certain forms of dunes develop in the presence of a partial cover of plants; they are classed as *phytogenic dunes*. Areas favorable for phytogenic dunes are coasts in a wide range of climates and inland regions of semiarid climate (steppes) which typically bear sparse, short grasses and scattered small shrubs. A characteristic form element of the phytogenic dune is its association with a deflation hollow, or blowout. To the lee of this depression, which is the source of sand, lies a dune ridge convexly bowed toward the downwind direction (curvature opposite to the barchan). These relationships are seen in the *coastal blowout dune* shown in Figure 16.13A. In an earlier stage sand derived from a beach had been built up into a low ridge of irregularly shaped hills and depressions, known generally as *foredunes* (see Figure 16.19). At a point where the cover of beachgrass and associated plants was broken down, deflation set in, producing a blowout depression. The growing dune ridge expanded its arc and moved slowly landward, producing the horseshoe form shown in the diagram. In many cases the forward slope of the dune ridge develops a slip face of free sand, and this advances upon adjacent forest, burying and killing the trees (Figure 16.14). Notice that the dune crest is equal in height to tree-top elevation.

In some cases, a coastal blowout dune continues to advance landward, with the result that the sides become drawn out into two long, almost-parallel ridges, as shown in diagram C of

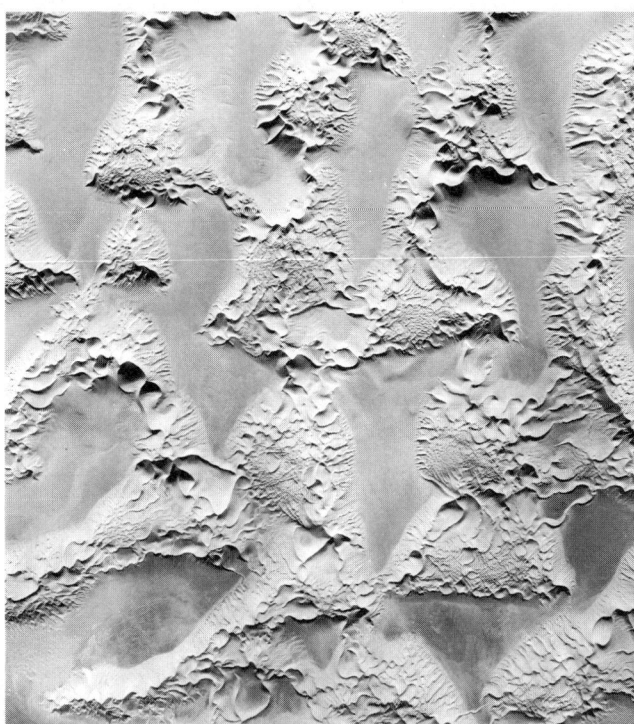

Figure 16.11 Seen from an altitude of 6 mi (10 km), these sand dunes of the Libyan desert appear as irregular patches which rise to star-shaped central peaks 300 to 600 ft (90 to 180 m) higher than the intervening flat ground. Width of the photograph represents about 7 mi (11 km). (Photograph of Aero Service Corporation, Litton Industries.)

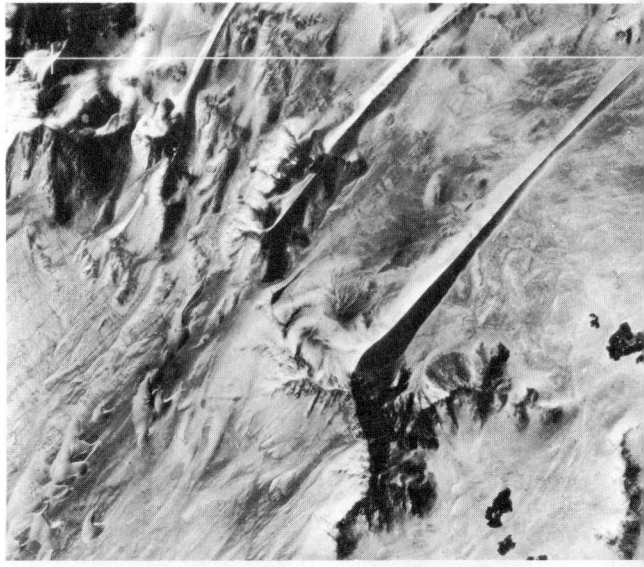

Figure 16.12 Longitudinal sand drifts appear on this air photograph as sharp-crested bladelike streamers of sand drawn out to the lee of a hill. Several barchan dunes are present in the lower left-hand corner. Width of area shown is about 0.75 mi (1.2 km). Chao-Viru-Moche area, Peru. (Courtesy Ministerio de Fomento.)

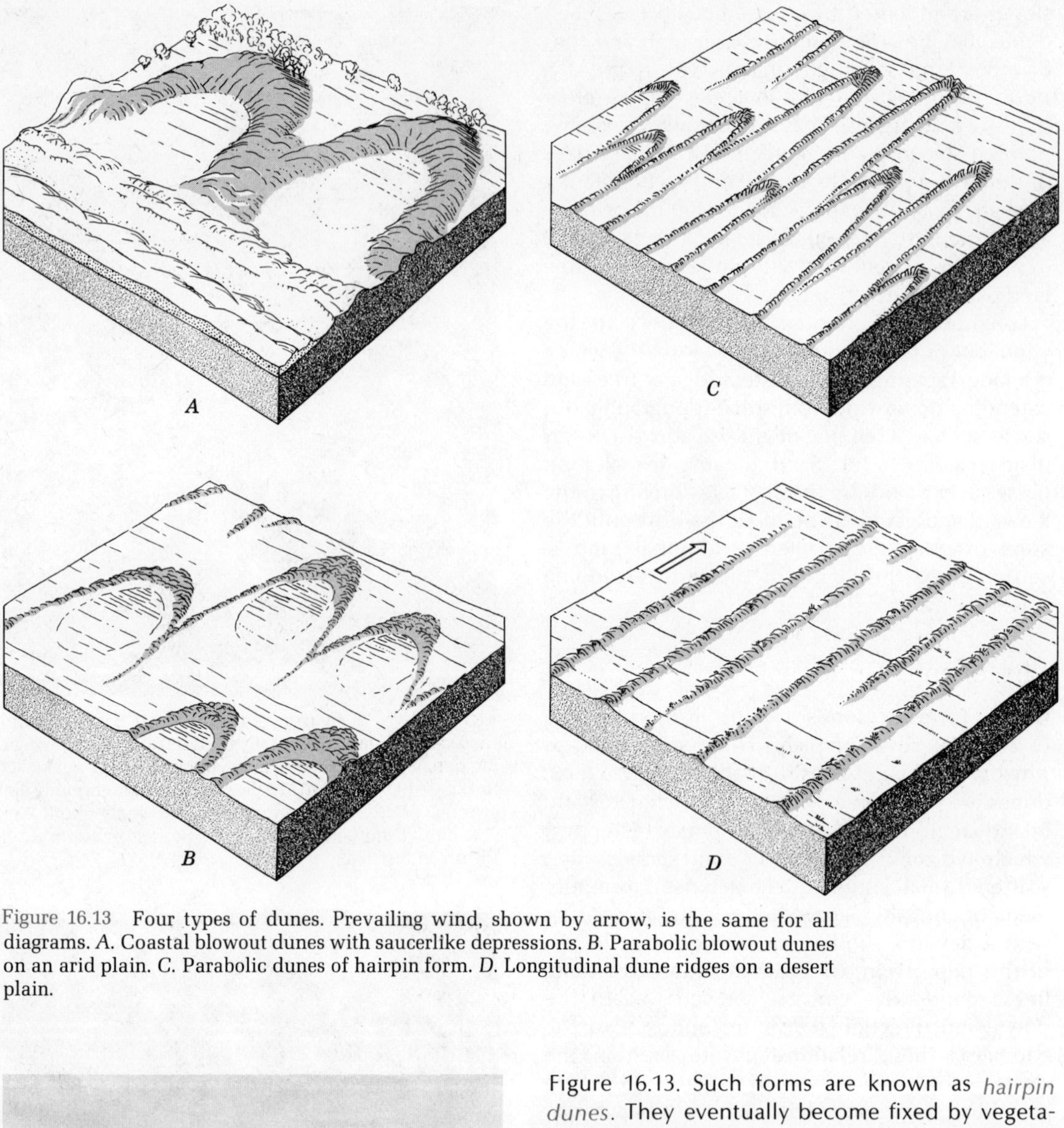

Figure 16.13 Four types of dunes. Prevailing wind, shown by arrow, is the same for all diagrams. A. Coastal blowout dunes with saucerlike depressions. B. Parabolic blowout dunes on an arid plain. C. Parabolic dunes of hairpin form. D. Longitudinal dune ridges on a desert plain.

Figure 16.14 Slip face of a coastal blowout dune advancing over a forest. Cape Henry, Virginia. (Photograph by Douglas Johnson.)

Figure 16.13. Such forms are known as *hairpin dunes*. They eventually become fixed by vegetation and cease to move. Figure 16.15 shows several fixed hairpin dunes of coastal origin. They are being overridden by fresh waves of transverse dunes. Close to the beach one or two blowout dunes can be identified. Notice that these dunes are formed from sand of a bayhead beach, brought to this location by littoral drift from cliffed headlands on either side. Coastal dunes are most abundant adjacent to prograding shorelines.

In both coastal and inland areas, but perhaps primarily in the latter, we find a phytogenic dune whose curve resembles the mathematical curve form of the parabola. Called *parabolic dunes*, these low ridges, bearing grass or shrub vegetation, represent sand (or, in some areas, coarse silt) derived from a shallow blowout depression (Figure 16.13C). Parabolic dunes rarely develop slip

Phytogenic Dunes

Figure 16.15 The arrows on this photograph point to elongate blowout dunes of hairpin form, which once advanced from the beach and have since become stabilized by vegetation. Active transverse dunes are overriding the blowout dunes in a fresh wave, San Luis Obispo Bay, California. (Spence Air Photos.)

faces, but if the plant cover is depleted, they may be transformed into barchans. Transitional forms are known. Parabolic dunes are widespread over the higher, more arid western parts of the Great Plains region of the United States and on elevated plateaus of the Southwest. Their patterns are clearly seen from the air, particularly when drifted snow brings their forms into conspicuous relief.

A possibly related dune form, found in both middle-latitude and tropical deserts, is the *longitudinal sand ridge*, shown in Figure 16.13D. This low ridge, which may bear a sparse plant cover, or none at all, may run for many miles parallel with others. Prevailing wind direction is in the line of the ridges. In some places longitudinal dune ridges appear to be merely extended hairpin dunes, but in the central desert of Australia they are a unique dune form occupying a vast land area (Figure 16.16). Here ridges average 30 to 50 ft (10 to 15 m) in height, are spaced 0.25 to 1.5 mi (0.4 to 2.4 km) apart, and run in continuous length as much as 25 to 50 mi (40 to 80 km). The ridges themselves are largely devoid of plants, but sparse vegetation occupies the intervening low ground.

Many dune areas that were active under conditions immediately following recession of the great Pleistocene ice sheets became stabilized by development of plant cover as climate ameliorated. As plants gained hold, the dunes underwent a

Figure 16.16 This oblique air photograph shows longitudinal sand-dune ridges reaching as far as the eye can see. Simpson Desert, southeast of Alice Springs, Australia. (Photograph by George Silk, *Life Magazine*, © Time Inc.)

phytogenic phase in which sharp crests and steep slip faces were obliterated. Ultimately, practically all sand movement by wind was halted, and the dunes became fixed. A striking example is the Sand Hills region of Nebraska, an area occupying about one-quarter of that state (see Figure 16.24). Dunes that were active in late Pleistocene time in this area are now fixed beneath a grass cover that has been given over to cattle ranching (Figure 16.17). Water-table ponds in the deeper depressions supply fresh water. The grass cover is easily broken, with rapid development of blowouts.

Man as an Agent in Inducing Deflation and Dune Activity

Cultivation of vast areas of plains under a climatic regime of substantial seasonal water deficiency, where only sparse grasses are normally sustained, is a practice inviting deflation of soil surfaces. Much of the Great Plains region, including all or parts of the states of New Mexico, Texas, Oklahoma, Kansas, Colorado, Nebraska, and the Dakotas is such a marginal region and has in past centuries experienced many dust storms gener-

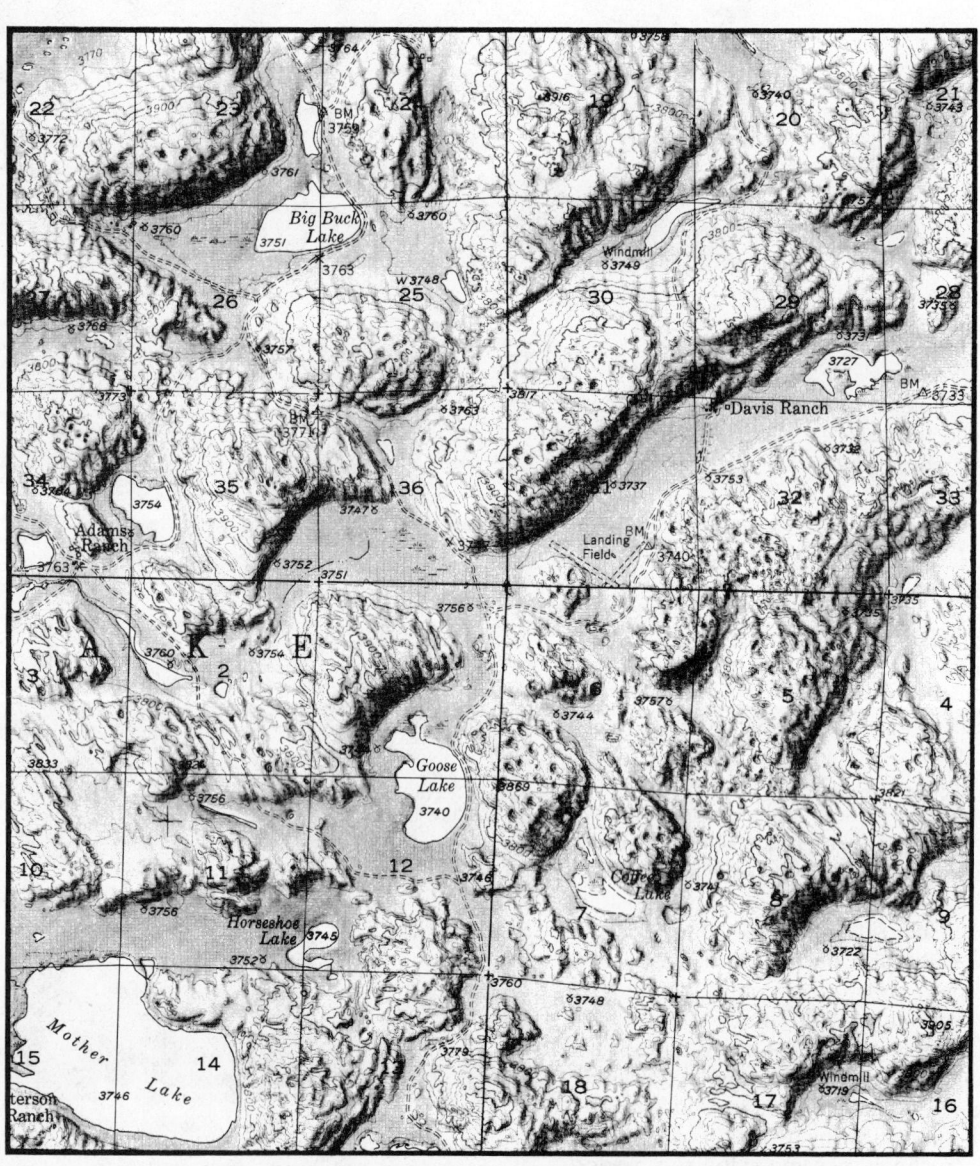

Figure 16.17 Fixed dunes of the Sand Hills Region, Nebraska. Steeper southeast-facing slopes suggest former slip faces and a prevailing northwesterly wind. Lakes represent level of water table. (Ashby, Nebraska, Quadrangle, U.S. Geological Survey.)

ated by turbulent winds. Strong cold fronts frequently sweep over this area, lifting dust high into the troposphere at times when soil moisture is low. However, deflation and soil drifting reached disastrous proportions during a series of drought years in the middle 1930s, following a great expansion of wheat cultivation. These former grasslands are underlain by humus-rich brown and chestnut soils. During the drought a sequence of exceptionally intense dust storms occurred. Within their formidable black clouds visibility declined to nighttime darkness, even at noonday, and the choking dust penetrated into tightly closed buildings, making breathing difficult. The area affected became known as the *Dust Bowl*. Many inches of topsoil were removed from fields and transported out of the region as suspended dust, while the coarser silt and sand particles accumulated in drifts along fence lines and around buildings (Figure 16.18). The combination of environmental degradation and repeated crop failures caused widespread abandonment of farms and a general exodus of farm families. Well known through popular novels and screen plays is the migration of the impoverished "Okies" westward to new homes in California.

Among scientists who have studied the Dust Bowl phenomenon, there is a difference of opinion as to how great a role soil cultivation and livestock grazing played in augmenting deflation. The drought was a natural event over which Man had no control, but it seems reasonable that the natural grassland would have sustained far less soil loss and drifting, had it not been destroyed by the plow.

Although Man cannot prevent cyclic occurrences of drought over the Great Plains, measures can be taken to minimize the deflation and soil drifting occurring in periods of dry soil conditions. Improved farming practices include use of *listed furrows* (deeply carved furrows) that act as traps to soil movement. Stubble mulching, mentioned in Chapter 14, will reduce deflation when land is lying fallow. Tree belts may have significant effect in reducing the intensity of wind stress at ground level.

Man's activities in the very dry, hot deserts have contributed measurably to raising of dust clouds. In the desert of northwest India and West Pakistan (the Thar Desert bordering the Indus River) the continued trampling of fine-textured soils by hooves of grazing animals, and by human feet as well, produces a blanket of dusty hot air that hangs over the region for long periods and extends to a height of 30,000 ft (9 km). In other deserts, such as those of North Africa and the southwestern United States, ground surfaces in the natural state contribute comparatively little dust because of the presence of desert pavements and sheets of coarse sand from which fines have already been winnowed. This protective layer is easily destroyed by wheeled vehicles, exposing finer textured materials and allowing deflation to raise dust clouds. It is said that the disturbance of large expanses of North African desert by tank battles during World War II caused great dust clouds, and that dust from this source was traced as far away as the Caribbean region.

Initiation of rapid sand transport by wind on dunes previously partially fixed or fully fixed by plant cover comes about readily by manmade activities and can lead to locally disastrous consequences. Foredunes bordering a beach are a particularly important example. Here beachgrasses and other small plants, sparse as they seem to be, act as baffles to trap sand in saltation moving landward from the adjacent berms (Figure 16.19). As a result, the foredune ridge is built up as a barrier rising many feet above high tide level. For example, dune summits of the Landes coast of France reach elevations of 250 to 300 ft (80 to 90 m) and span a belt of 2 to 6 mi (3 to 10 km) wide.

The swash of storm waves, acting at high water of tide and under conditions of raised water level due to lowered barometric pressure and the onshore drift of surface water, cuts away the upper part of the beach berm and the dune barrier is

Figure 16.18 A typical scene in the Dust Bowl during the late 1930s. Drifts of sand and silt have accumulated around abandoned farm buildings and a fence in Dallam County, Texas. (Photograph by Soil Conservation Service, U.S. Department of Agriculture.)

Figure 16.19 Foredunes protected by beach grass, Provincelands of Cape Cod, Massachusetts, (Photograph by A. N. Strahler.)

attacked. Undermining of the dunes supplies sand to the swash and this is spread over a wide area of beach slope. Wave energy is then largely expended in the riding of water up a long, sloping ramp, and the force of storm attack is dissipated. Between storms the berm is rebuilt and in due time the dune ridge is also restored if plants are maintained. In this way the foredunes form a protective barrier for tidal lands lying on the landward side.

If, now, the plant cover of the dune ridge is depleted, by vehicular and foot traffic, or by bulldozing of sites for approach roads and buildings, a blowout will rapidly develop. The cavity thus formed may extend as a trench across the dune ridge. With the onset of a storm with high water levels, swash is funneled through the gap and spreads out upon the tidal marsh or tidal lagoon behind the ridge. Sand swept through the gap is spread over the tidal deposits, a phenomenon called *overwash*. A new tidal inlet may be created in this way, but few of these persist.

For many coastal communities of the eastern United States seaboard, the breaching of a dune ridge with its accompanying overwash brings a certain measure of environmental damage to the tidal marsh or estuary, and there may be extensive property damage as well. However, this effect is minor compared with the effects of breaching of dune barriers along the North Sea Coast, which are part of the system to exclude sea water from reclaimed polders and fenlands. Protection of the dunes of the Netherlands coast assumes vital importance in view of the loss of life and property that a series of storm breaches can bring.

The Low Countries are protected from the North Sea by a sand barrier built in postglacial time by northeastward littoral drift from a point near Calais, France, where the marine cliff of chalk ends. Building of dune ridges progressed concomitantly with lengthening and widening of the beach-sand barrier. Older dune ridges, those lying nearest the mainland, have bases now situated 10 to 12 ft (3 to 4 m) below mean sea level, showing that coastal submergence has been in progress as the barrier has been built up. Figure 16.20 is a cross section of the dune barrier north of Haarlem (refer also to map, Figure 15.46). The sand barrier thus protects polder lands that lie below sea level. The dune surfaces are stabilized with forest and beachgrass.

Following the early period of barrier growth, the past 1800 years witnessed a retrogradation of the Netherlands outer shoreline. It is estimated that from 2 to 6 mi (3 to 10 km) of shoreline retreat occurred during this period. The retrogradation was effectively halted by extensive planting of a species of beachgrass (*Arundo arenacea*) that develops long and deep roots. An area 200 mi (320 km) long and from 0.3 to 3 mi (0.5 to 5 km) was planted by hand with this grass in rows about 2 ft (0.6 m) apart. Dunes farther inland were reforested. It is small wonder that the dune vegetation is preserved with greatest care!

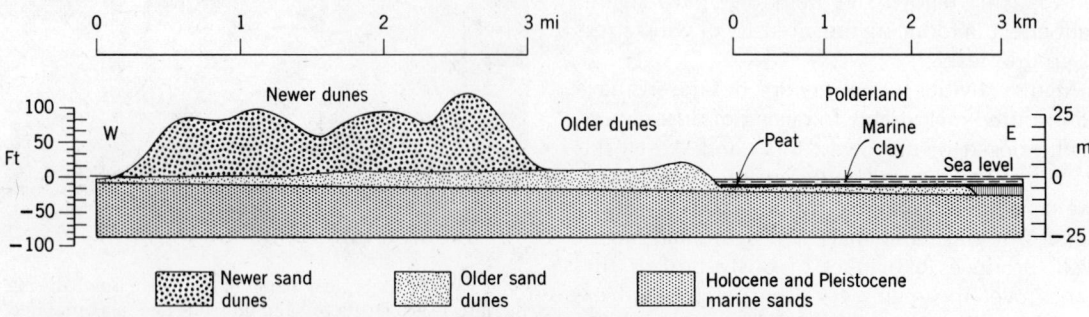

Figure 16.20 Cross section through the great dune barrier of the Netherlands coast in vicinity of Haarlem. Notice that the surface of the polderland is somewhat below mean tide level. (After P. Tesch, Geological Survey of Netherlands.)

Man as an Agent in Inducing Deflation

A second category of environmental damage related to dunes is the rapid downwind movement of sand when the dune status is changed from one of fixed or plant-controlled forms to that of live dunes of free sand. When plant cover is depleted, saltation rapidly reshapes the dunes to produce crests and slip faces. These free sand slopes advance upon forests (Figure 16.14), roads, buildings, and agricultural lands. In the Landes region of coastal dunes on the southwestern coast of France, landward dune advance has overwhelmed houses and churches and even caused entire towns to be abandoned.

A striking case of Man's interference with a dune environment is that of the Provincelands of Cape Cod, located at the northern tip of that peninsula, making up the fist in the armlike outline of the Cape (Figure 16.21). The Provincelands has been constructed of beach sand carried by northwestward littoral drift from a fast-eroding marine scarp in glacial deposits of the arm of the Cape. (See Chapter 15 and Figure 15.8.) The structure consists of a succession of beach ridges, and these have been modified in form and increased in height by dune-building. When the first settlers arrived in Provincetown, a city now occupying the south shore of the Provincelands, the dunes were naturally stabilized by grasses and other small plants covering the dune summits and by pitch pine and other forest trees on lower surfaces and in low swales between dune ridges; although dunes were probably active then, as now, on ridges close to the northern shore. Inhabitants grazed their livestock on the dune summits and rapidly cut the forests for fuel, with the result that the dunes were activated and began to move southward. By 1725, Provincetown was being overwhelmed by drifting sand. Some buildings were partially buried and some had to be abandoned. Sand was carted away from the streets in large volumes. By the early 1800s the major dune ridges were moving southward at a rate estimated to be 90 ft (27 m) per year. Fortunately, beachgrass plantings, begun in 1825, and the rigorous enforcement of laws forbidding grazing and tree-cutting, resulted finally in stabilization of all but the northernmost dunes. Even today, high slip faces advance upon a major highway and into Pilgrim Lake (Figure 16.21). The area is today a part of the Cape Cod National Seashore. Authorities have made extensive new plantings of dunegrass and have minimized vehicular traffic in an attempt to bring further control to sand movement.

Another example of effective dune control is that of the Warrenton dune locality situated at the mouth of the Columbia River in Oregon. Active dunes were threatening to overwhelm agricultural lands, military installations, highways, towns, and expensive resort homes. Sand reaching the mouth of the Columbia River was accumu-

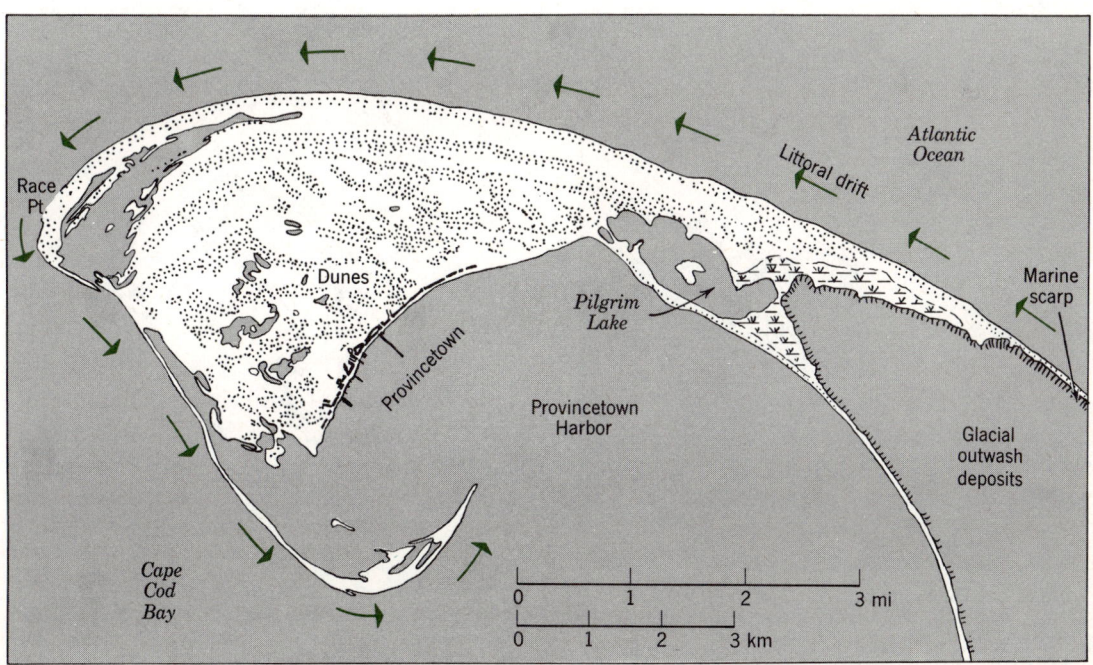

Figure 16.21 Sketch map of the Provincelands of Cape Cod, Massachusetts, as mapped in 1887. Dunes shown in stippled pattern. (Data of U.S. Geological Survey.)

Wind Action and Dune Landscapes

lating in such a way as to threaten prevention of movement of ocean-going vessels through the river mouth. Dune-control work, started in 1935, consisted of plantings made in two stages. The first stage used American and European beachgrasses and American dunegrass. Fertilizer was applied. The second stage consisted of plantings of permanent plant species, including grasses, hairy vetch, and purple beachpea. Careful maintenance of the cover has since kept the dunes stabilized.

Loess—A Natural Resource

In various parts of the world throughout middle-latitude zones the parent matter of rich agricultural soils is a layer of wind-transported sediment called *loess*. This German-derived word has given us a painful problem in pronunciation. Some say it to rhyme with "thus"; others say "less" or form two syllables ("low-us"); and purists insist on pronouncing an umlaut ö as in the original German, löss. Loess is usually unstratified (lacks layers) but commonly exhibits natural vertical parting (cleavage) that allows it to stand in vertical faces, prisms, and pillars, even though the material is soft enough to be cut with a knife or shovel. In color, loess is usually yellowish or tawny orange, and sometimes brown. Particles composing loess fall mostly in the size range 0.005 to 0.06 mm, which spans the category of silt in the Wentworth scale (Table 8.2). Five to ten percent of the particles may fall into the coarse clay grade; about the same proportion may be very fine sand. Although in some areas loess has stratification and is interpreted as being deposited in water, most upland deposits are considered to have been transported in suspension by wind, with some of the coarser fraction arriving by saltation.

Loess varies in mineral composition from one locality to another, but most is composed dominantly of quartz grains, which are mostly angular from fresh breakage. Other common silicate minerals are usually present, while calcite is a common ingredient and in some deposits constitutes as much as 40% by weight.

Strong evidence favors the source of North American and European loess in partially dry, braided stream channels and aggrading floodplains. These channels carried meltwater from the margins of stagnant and wasting ice sheets. During the summer strong winds deflated the alluvial surfaces, carrying the silt in suspension and depositing it over the uplands between valleys. This process can be seen in action today in summer in alluvial valleys in Alaska (Figure 16.22). When

Figure 16.22 Dust clouds raised by strong winds blowing over vegetation-free bars of the braided channel of the Delta River, central Alaska. Silt carried in this way builds modern loess deposits upon adjacent upland surfaces. (Photograph by U.S. Navy, from T. L. Péwé, 1951, *Jour. of Geology*, vol. 59, p. 400.)

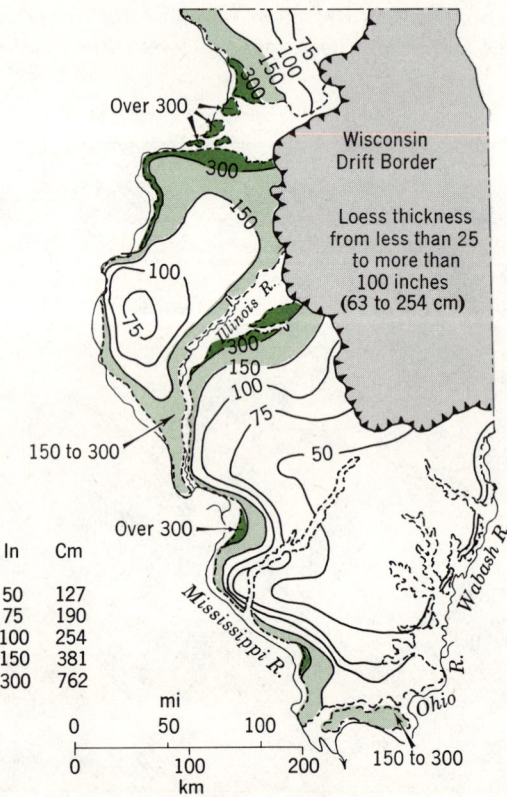

Figure 16.23 Loess thickness (inches) in Illinois. (Based on a map by R. F. Flint.)

we examine a map of thickness of loess in Illinois (Figure 16.23), we see that the deposit is thickest immediately to the east of each of the river valleys, and that this is the leeward position with respect to prevailing westerlies. A map of distribution of loess over the central United States (Figure 16.24) shows a general correspondence with the Mississippi, Missouri, and Platte rivers and their tributaries. A profile drawn from west to east from the bluffs of the Mississippi River floodplain in the state of Mississippi shows the loess to be in the form of a wedge, thickest at the bluffs and tapering rapidly eastward (Figures 16.25 and 16.26). The presence in abundance of fossil shells of air-breathing snails throughout loess of the central United States is evidence that the loess was accumulating while providing a habitat for the snails.

While the loess of the central states attains a maximum thickness of about 100 ft (30 m) in Kansas, thicknesses up to 200 ft (60 m) are measured in central Alaska. European loess is particularly important in eastern Rumania and the Ukraine region of the U.S.S.R., where it forms a continuous blanket 30 to 50 ft (10 to 15 m) thick over thousands of square miles. Loess of northern China is commonly over 100 ft (30 m) thick and reaches a maximum depth of around 300 ft (90 m) (Figure 16.27; see also Figure 14.10). This loess was brought as dust from deserts of central Asia by strong westerly and northwesterly winds of the spring season, and its deposition continues today. Other loess areas are found in New Zealand and Argentina.

The importance of loess in world agricultural resources cannot be easily overestimated. Where occurring in a subhumid climate favoring the pedogenic regime of calcification, loess plains have developed rich, black soils especially suited to cultivation of grains. The highly productive plains of southern Russia, the Argentine Pampa, and the rich grain region of north China are underlain by loess. In the United States, corn is extensively cultivated on the loess plains in those states, such as Iowa and Illinois, where rainfall is sufficient; wheat is grown farther west on loess plains of Kansas and Nebraska and in the Palouse region of eastern Washington.

Because loess forms vertical walls along valley sides and is able to resist sliding or flowage, but at the same time is easily dug into, it has been widely used for cave dwellings both in China and in Central Europe. In China, old trails and roads in the loess have become deeply sunken into the ground as a result of the pulverization of the loess of the road bed and its removal by wind and water (Figure 16.27).

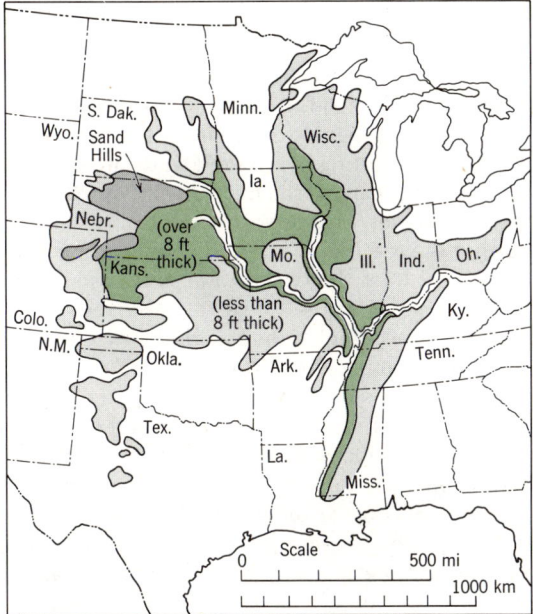

Figure 16.24 Map of loess distribution in the central United States. (Data from Map of Pleistocene Eolian Deposits of the United States, Geol. Soc. of Amer., 1952.)

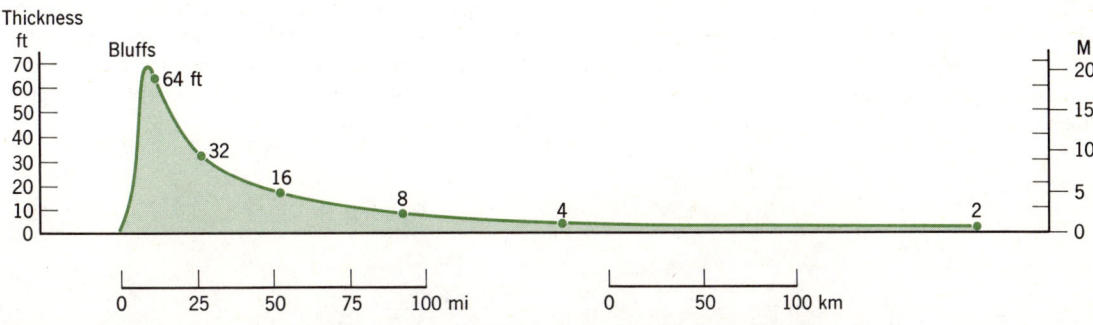

Figure 16.25 Schematic cross section of loess thickness along a west-to-east line through uplands east of the lower Mississippi River alluvial plain. (Same data source as Figure 16.24.)

Wind Action and Dune Landscapes

Figure 16.26 This perpendicular road cut in loess south of Vicksburg is typical of thick glacial loess accumulations on the eastern bluffs of the Mississippi River. (Photograph by Orlo Childs.)

Figure 16.27 Road sunken deeply into loess, Shensi, China. (Photograph by Frederick G. Clapp, courtesy of *The American Geographical Society*.)

Although remarkably stable in the natural state, loess when reworked by machinery or trampling of livestock, so as to break down the original cohesion, yields readily to accelerated soil erosion with the development of deep gullying. Deep gullies can be seen in the loess pictured in Figure 14.10. In loess of the lower Mississippi valley region, in the states of Mississippi and Tennessee, erosion has been a serious problem on cultivated loess lands. The lower graph of Figure 12.6 shows the effects upon infiltration rates of deforestation and abandonment of loess areas. The rapid drop in infiltration capacity to low values causes overland flow to be produced with comparatively light-intensity rains, with resulting severe sheet erosion. However, all loess regions are potential problem areas with regard to accelerated erosion and ensuing sedimentation of stream valleys.

Wind as an Environmental Agent in the Life Layer

This chapter and the one before it have had much in common through the mutual mechanism of transfer of momentum and kinetic energy from the wind to a liquid or solid surface over which it blows. While wind-driven waves are the dominant agent in shaping shorelines, the action of wind in transporting beach sand landward to form coastal dunes can exert a measure of control over breaking waves, because sediment carried inland beyond the reach of waves of normal energy subtracts mass from the beach zone, and in so doing slows progradation of beaches. On the other hand, dune accumulations absorb the exceptional high-energy impact of storm waves and so inhibit retrogradation and the overwash of barrier beaches. Both breaking waves and wind can carry sediment upgrade to higher positions, something that gravity flow of runoff cannot accomplish.

One final process remains to be examined on the list of active agents of erosion, transportation, and deposition: glacial ice. One unique feature of the action of glaciers is that they are quite insensitive to the activities of Man. This quality probably derives from their vast bulk and sluggish motion. No one has yet tried to build a dam to stop the flow of a great glacier, as we have done so many times to halt the flow of a large river. So we turn from considering processes and forms that are extremely sensitive to the impact of Man to a geologic process that imposes its environmental qualities upon Man and accepts no degree of control in return.

17 Glacier Systems and the Pleistocene Epoch

Glacial ice as a geologic agent has played a dominant role in shaping landforms of large areas in middle-latitude and subarctic zones. Glacial ice also exists today in two great accumulations of continental dimensions and in many smaller masses in high mountains. Therefore, glacial ice is an environmental agent of the present, as well as of the past, and is itself a landform. Glacial ice of Greenland and Antarctica strongly influences the radiation and heat balances of the globe. Moreover, these enormous ice accumulations represent water in storage in the solid state, and constitute a major component of the global water balance. Changes in ice storage can have profound effects upon the position of sea level with respect to the continents. We have already seen that the coastal environments of today have evolved with a rising sea level following melting of ice sheets of the last ice advance in the Pleistocene Epoch, or Ice Age. When we examine the evidence of former extent of those great ice sheets, we need to keep in mind that the evolution of modern Man as an animal species occurred during a series of climatic changes which placed many forms of environmental stress upon all terrestrial plants and animals in the middle-latitude zone, and had important effects extending into tropical zones and even equatorial zones, as well.

As we noted in closing the last chapter, glacial ice dictates its own behavior and accepts no appreciable modification or interference by Man, unless it be through subtle, long-range changes in climate parameters. Consequently, in this chapter there will be no section devoted to Man's impact upon glacial systems. Instead we will devote the chapter to an inquiry into glaciers as hydrologic systems and as geologic agents of landform development. A brief review of Pleistocene history is then in order and will place the tiny time-fragment of the Holocene, or Recent, Epoch in its proper perspective.

Glaciers

Most of us know ice only as a brittle, crystalline solid because we are accustomed to seeing it only in small quantities. Where a great thickness of ice exists, let us say 200 to 300 ft (60 to 90 m) or more, the ice at the bottom behaves as a plastic material and will slowly flow in such a way as to spread out the mass over a larger area, or to cause it to move downhill, as the case may be. This behavior characterizes *glaciers*, which may be defined as any large natural accumulations of land ice affected by present or past motion.

Conditions requisite to the accumulation of glacial ice are simply that snowfall of the winter shall, on the average, exceed the amount of melting and evaporation of snow that occurs in summer. (The term *ablation* is used by glaciologists to include both evaporation and melting of snow and ice.) Thus, each year a layer of snow is added to what has already accumulated. As the snow compacts, by surface melting and refreezing, it turns into a granular ice, then is compressed by overlying layers into hard crystalline ice. When the ice becomes so thick that the lower layers become plastic, outward or downhill flow commences, and an active glacier has come into being.

At sufficiently high altitudes, whether in high or low latitudes, glaciers form both because air temperature is low and mountains receive heavy orographic precipitation (Chapter 5). Glaciers that form in high mountains are characteristically long and narrow because they occupy previously formed valleys and bring the plastic ice from small collecting grounds high upon the range down to lower elevations, and consequently warmer temperatures, where the ice disappears by ablation (Figure 17.1). Such *alpine glaciers*, or *valley glaciers*, are distinctive types.

In arctic and polar regions, prevailing temperatures are low enough that ice can accumulate over broad areas, wherever uplands exist to intercept heavy snowfall. As a result, areas of many thousands of square miles may become buried under gigantic plates of ice whose thickness may reach several thousand feet. The term *icecap* is usually applied to an ice plate limited to high mountain and plateau areas. During glacial periods an icecap spreads over surrounding lowlands, enveloping all landforms it encounters and ceasing its spread only when the rate of ablation at its outer edge balances the rate at which it is spreading. This extensive type of ice mass is called a *continental glacier* or *ice sheet*.

Figure 17.2 is a schematic drawing comparing an alpine glacier with an icecap. Both represent open systems of gravity flow of matter. Whereas the alpine glacier has a sloping floor, as does a water stream, the icecap may have a horizontal floor, or even an upwardly concave (saucer-shaped) floor. Flowage of an icecap is induced by surface gradient of the ice. A simple analog is a spoonful of pancake batter poured on a flat skillet. As more batter is added at the center the pancake increases in size. For both alpine glaciers and icecaps, matter enters the systems as snow, possessing potential energy of position. As flow takes place, potential energy is transformed to kinetic energy of motion, and this in turn is dissipated in friction within the ice and in contact with the rock floor beneath it.

Figure 17.1 The Eklutna Glacier, Chugach Mountains, Alaska, seen from the air. A deeply crevassed ablation zone with a conspicuous medial moraine (*foreground*) contrasts with the smooth-surfaced firn zone (*background*). (Photograph by Steve McCutcheon, Alaska Pictorial Service.)

Alpine Glaciers

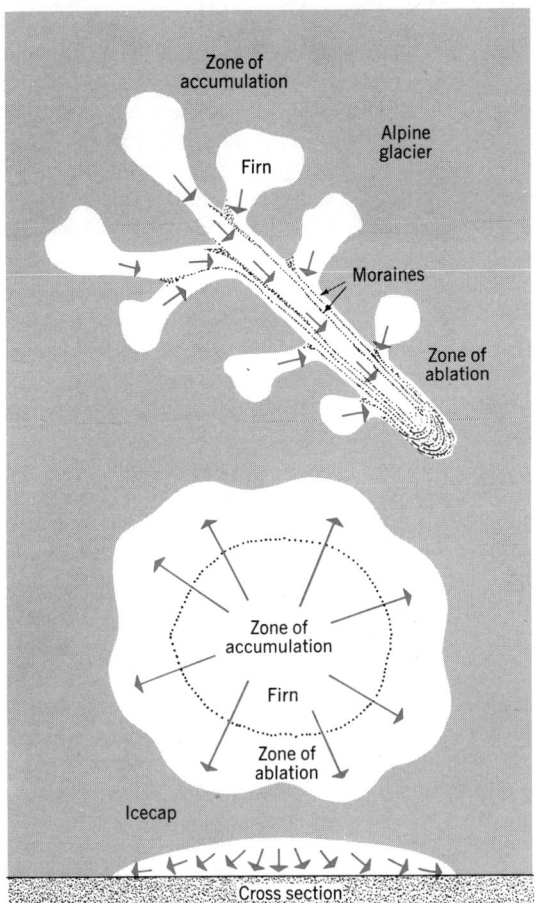

Figure 17.2 Schematic maps and a cross section of an alpine glacier and an icecap. (From A. N. Strahler, 1972, *Planet Earth; Its Physical Systems Through Geologic Time*, Harper & Row, New York.)

Further details of an alpine glacier as an open energy system are shown in Figure 17.3. Input of matter takes place within the *zone of accumulation*. Layers of snow in stages of transformation to ice comprise the *firn*. Through flowage, indicated by arrows, the ice moves both down-valley and into lower depths within the glacier. Upon entering the *zone of ablation*, flow lines turn surfaceward. Here ice is lost through the upper boundary by evaporation and from the lower end, or *terminus*, by melting. Because of ablation, the lower glacier surface is deeply pitted and furrowed, in contrast to the smooth, white surface of the firn. The lower limit of the firn, or *firn line*, shows in Figure 17.1, although it is not a sharp line.

Glacier equilibrium, a steady state, is achieved when the rate of mass input in the firn zone is balanced by the rate of mass output in the ablation zone. Velocity of flow within the ice is adjusted to transport the mass through the system and the geometry of the glacier remains constant with time. By geometry we mean both volume and cross-sectional form of the glacier. In an equilibrium state the glacier terminus neither advances nor recedes.

A change in glacier equilibrium is caused by a change in the net input of snow in the accumulation zone, as averaged over several year's time. If there is an increase in input, the glacier thickens and flow velocity is increased. The glacier terminus then advances and the area of ablation is increased. Also, as the terminus reaches lower elevations, warmer air temperatures are encountered and ablation is accelerated. When ablation

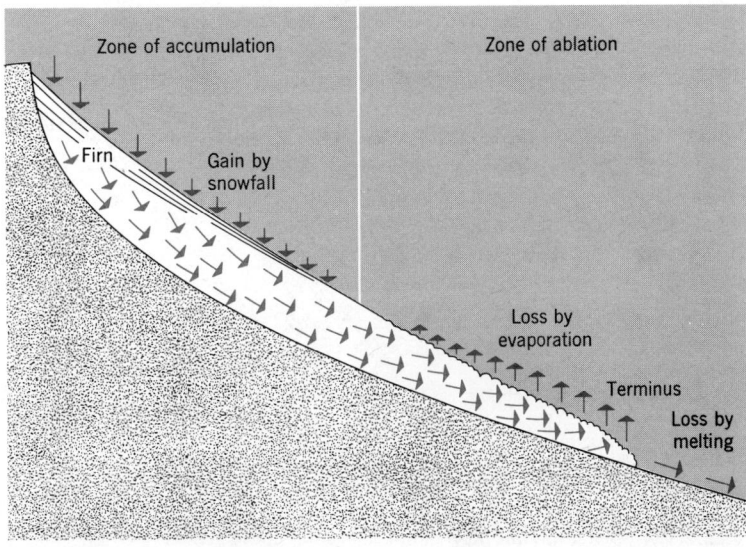

Figure 17.3 An idealized longitudinal cross section through an alpine glacier, showing how a water balance is achieved. (From A. N. Strahler, 1972, *Planet Earth; Its Physical Systems Through Geologic Time*, Harper & Row, New York.)

rate again matches accumulation rate within the system, a new steady state is achieved and the terminus is again stabilized. Should the net snow accumulation be reduced, the glacier will be reduced in thickness and will flow more slowly. As a result, ablation exceeds the rate at which ice is supplied by flowage and the terminus recedes to a higher position. In time a new steady is achieved and the terminus is again stabilized. A similar analysis of mass transport rates and steady state can be applied to the icecap.

Alpine glaciers are rather sensitive indicators of changes in climatic parameters that control both net accumulation and ablation. Consequently, many modern glaciers are kept under surveillance by repeated topographical surveys, to determine their rates of advance or retreat and thus to draw inferences as to climatic changes occurring over decades of time.

Figure 17.4 illustrates a number of details of form and structure of alpine glaciers. The center illustration is of a simple glacier occupying a steeply inclined valley bottom between steep rock walls. Snow is collected at the upper end in a bowl-shaped depression, the *cirque*. The smooth firn field is slightly concave up in profile (upper right). An abrupt steepening of grade of the floor comprises a *rock step,* over which the rate of ice flow is accelerated and produces deep *crevasses* (gaping fractures) which form an *ice fall*. In the zone of ablation, old ice is exposed at the glacier surface, which is extremely rough and deeply crevassed. The glacier terminus or *snout* is heavily charged with rock debris. The lower part of the glacier is usually of upwardly convex cross-profile, the center being higher than the sides (upper right).

The uppermost layer of a glacier, perhaps 200 ft (60 m) in thickness, is brittle and fractures readily into crevasses, whereas the ice beneath behaves as a plastic substance and moves by slow flowage (lower left). If one were to place a line of stakes across the glacier surface, the glacier flow would gradually deform the line into a para-

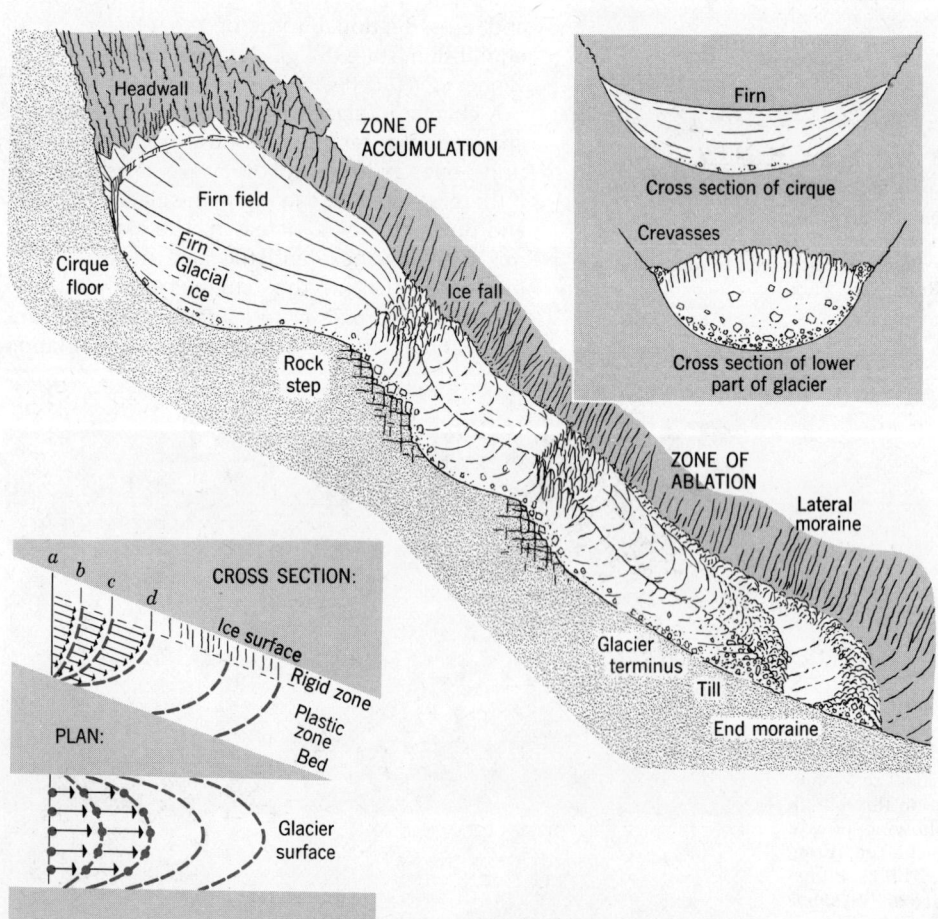

Figure 17.4 Structure and flowage of a simple alpine glacier.

bolic curve, indicating that rate of movement is fastest in the center and diminishes toward the sides.

Rate of flowage of both alpine and continental glaciers is very slow, amounting to a few inches per day for large ice sheets and for the more sluggish valley glaciers, up to several feet per day for an active valley glacier. In addition to slow, continuous flowage, some glaciers have been observed to have fast surges of motion of short duration.

Glacier flow is of the *laminar* (*streamline*) type, lacking in turbulent eddies that characterize flow of water or air. Consequently, rock debris dragged out upon the ice surface at junctions of ice streams is carried down-valley in parallel lines, called *medial moraines* (Figure 17.2). A prominent medial moraine shows on the surface of the glacier pictured in Figure 17.1; its source can be traced to the junction point between two major glacier branches.

Glacial Erosion

Most glacial ice is heavily charged with rock fragments, ranging from pulverized rock (*rock flour*) to huge angular boulders. This material is derived from the rock floor upon which the ice moves, and in alpine glaciers, from material that slides or falls from valley walls. Glaciers are capable of great erosive work, both by *abrasion*, erosion caused by ice-held rock fragments that scrape and grind against the bedrock, and by *plucking,* in which the moving ice lifts out blocks of bedrock that have been loosened by freezing of water in joint fractures. The process of plucking out of joint blocks is illustrated in Figure 17.4, at the rock steps.

The debris entrained in the ice must eventually be left stranded at the outer edge or lower end of a glacier when the ice is dissipated. Thus there are two glacial activities to consider: erosion and deposition. Both result in distinctive landforms, which in some cases are further differentiated according to the type of glacier, whether alpine or continental.

Landforms Made by Alpine Glaciers

Landforms made by alpine glaciers can be studied by a pair of diagrams (Figure 17.5), in which a mountainous region is modified by glaciers, after which the glaciers disappear and the remaining landforms are exposed to view.

An early stage of glaciation is shown at the right hand side of block *A,* where snow is collecting and cirques are being carved by the outward motion of the ice and by intensive frost shattering of the rock near the firn margins. Glaciers have filled the valleys and are integrated into a system of tributaries that feed a trunk glacier, just as in a stream system. Glaciers are, of course, enormously thicker than streams, because the extremely slow rate of ice motion requires a great cross section if a glacier is to maintain a discharge equivalent to a swiftly flowing stream. Tributary glaciers join the main glacier with smooth, accordant junctions, but, as we shall see later, the bottoms of their channels are quite discordant in level.

Intense freezing and thawing of meltwater from snows lodged in crevices high upon the walls of a cirque shatters the bare rock into angular fragments, which fall or creep down as talus upon the snowfield and are incorporated into the glacier. Frost shattering also affects the rock walls against which the ice rests. The cirques thus grow steadily larger. Their rough, steep walls eventually intersect from opposite sides, producing a jagged, knifelike ridge, called an *arête.* Where three or more cirques grow together, a sharp-pointed peak is formed by the intersection of the arêtes. The name *horn* is applied to such peaks in the Swiss Alps (Figure 17.6). One of the best known is the striking Matterhorn. Where the intersection of opposed cirques has been carried further, a pass or notch, called a *col* is formed.

Glacier flow constantly deepens and widens its channel so that after the ice has finally disappeared there remains a deep, steep-walled *glacial trough,* characterized by a relatively straight or direct course and by the U-shape of its transverse profile (block *B*, Figure 17.5). Tributary glaciers also carve U-shaped troughs, but they are smaller in cross section, with floors lying high above the floor level of the main trough, so are called *hanging troughs.* Streams, which later occupy the abandoned trough systems, form scenic waterfalls and cascades where they pass down from the lip of a hanging trough to the floor of the main trough. These streams quickly cut a small V-shaped notch in the trough bottom.

Valley spurs that formerly extended down to the main stream before glaciation occurred have been beveled off by ice abrasion and are termed *truncated spurs* (block *A*). Under a glacier the bedrock is not always evenly excavated, so that the floors of troughs and cirques may contain *rock basins* and *rock steps.* Cirques and upper parts of troughs later are occupied by lakes, called *tarns* (see Figure 11.5). The major troughs fre-

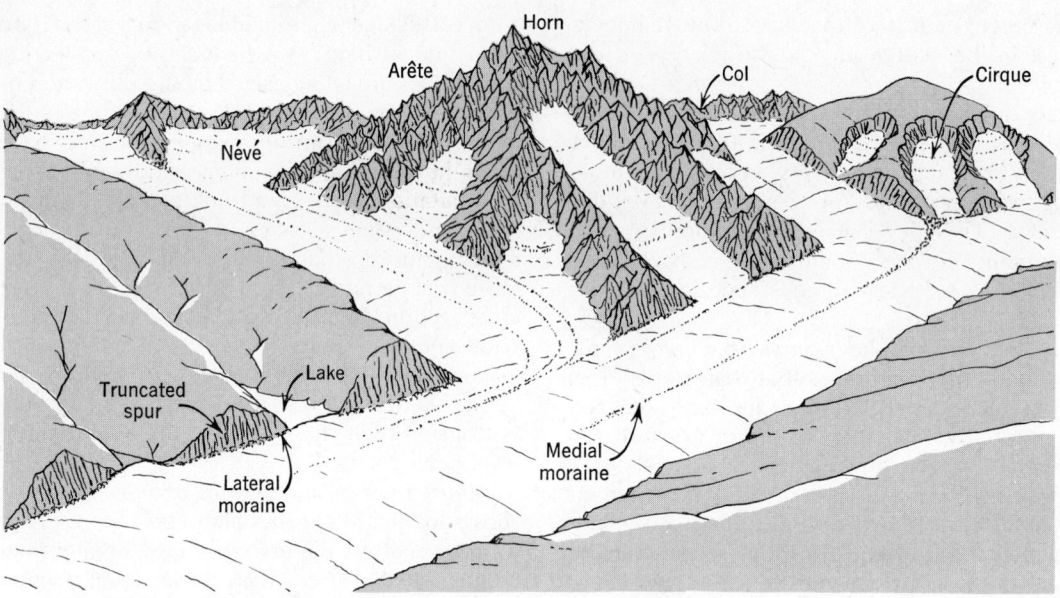

A. Phase of maximum glacial activity.

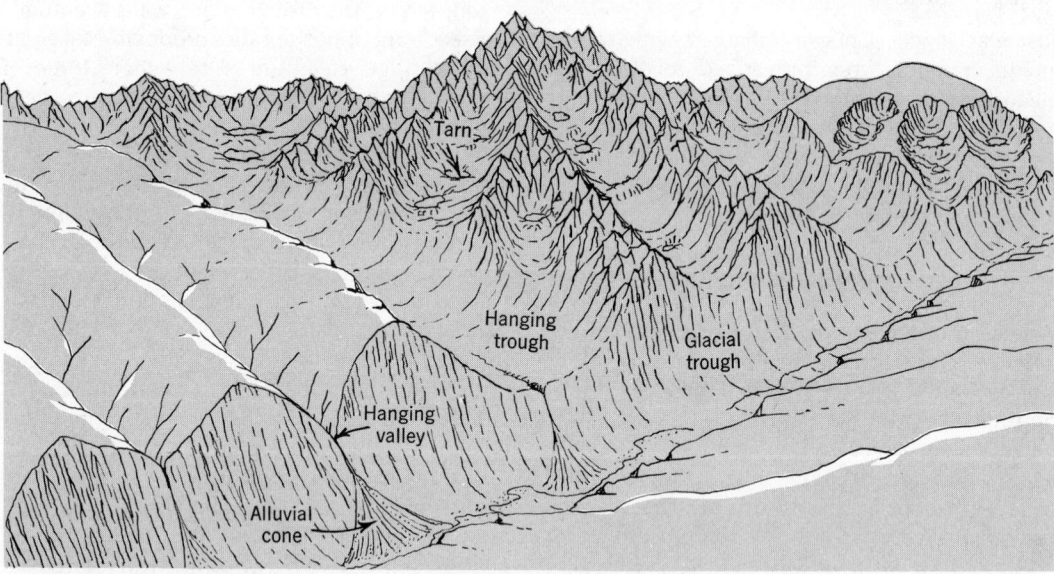

B. Postglacial landscape devoid of glacial ice.

Figure 17.5 Landforms produced by alpine glaciation. (After W. M. Davis and A. K. Lobeck.)

quently contain large, elongate *trough lakes*, sometimes referred to as *finger lakes*. Landslides are a common phenomenon in troughs because glaciation leaves oversteepened and unsupported rock walls. In glaciated countries such as Switzerland and Norway slides are a major type of natural disaster, because most towns and cities lie in the trough floors where they are readily destroyed by rock avalanches and landslides (Chapter 11).

Debris may be carried by an alpine glacier within the ice, or it may be dragged along between the ice and the valley wall as a *lateral moraine* (Figures 17.4 and 17.5A). Where two ice streams join, this marginal debris is dragged along to form a medial moraine. At the terminus of a glacier debris accumulates in a heap known as a *terminal moraine*, or *end moraine*. This heap is usually in the form of a curved embankment lying

Environmental Aspects of Alpine Landscapes

Figure 17.6 The Swiss Alps appear from the air as a sea of sharp arêtes and toothlike horns. In the foreground is a cirque. (Swissair Photo.)

Figure 17.7 Thick morainal deposits at the lower end of the remnant of the Black Glacier, Bishop Range, Selkirk Mountains, British Columbia. Talus cones line the base of the trough wall. (Photograph by H. Palmer, Geological Survey of Canada.)

across the valley floor and bending up-valley along each wall of the trough to merge with the lateral moraines (Figure 17.4). As the end of the glacier wastes back, a succession of such moraines is left behind (Figure 17.7).

Glacial Troughs and Fiords

Many large glacial troughs now are nearly flat-floored because aggrading streams that issued from the receding ice front were heavily laden with rock fragments. Figure 17.8 shows a comparison between a trough with little or no fill and another with alluvial-filled bottom. The deposit of alluvium extending down-valley from a melting glacier is the *valley train*. (Refer to Figure 14.28 showing valley-train deposition in progress by a meltwater stream.)

Where the floor of a trough open to the sea lies below sea level, the sea water enters as the ice front recedes, producing a narrow estuary known as a *fiord* (Figure 17.8D). Fiords may originate either by submergence of the coast or by glacial erosion to a depth below sea level. Fiord excavation below sea level occurs because ice is of such a density that, when floating, from three-fourths to nine-tenths of its mass lies below water level. Therefore, a glacier several hundred feet thick can erode to considerable depth below sea level.

Fiords are observed to be opening up today along the Alaskan coast, where some glaciers are melting back rapidly and the fiord waters are being extended along the troughs. Fiords are found largely along mountainous coasts in latitudes 50° to 70° N and S, because orographic precipitation of snow was particularly heavy where maritime polar air masses carried in the westerlies encountered coastal mountain ranges. These same mountains are also high-precipitation zones today (Chapter 5, Figure 5.44).

Environmental Aspects of Alpine Landscapes

The ruggedness of fully glaciated mountains such as the Alps, Pyrenees, Himalayas, Andes, or Sierra Nevada makes for sparseness of population and difficulty of access. Land above timber line has little use except for summer pasture and the extraction of such minerals as may occur in the rocks. Locally, recreational uses—mountain climbing and winter sports—are intensively developed and are rapidly becoming a major economic asset of the alpine landscape.

Below timber line are rich forests. U-shaped glacial troughs provide broad, accessible strips of land at relatively low levels. These are utilized for town sites, for pasture, and as arteries of transportation. In the Italian Alps several great flat-floored glacial troughs extend from the heart of the Alps southward to the plain of northern Italy. These are important environmental controls because they provide smooth and easy access into

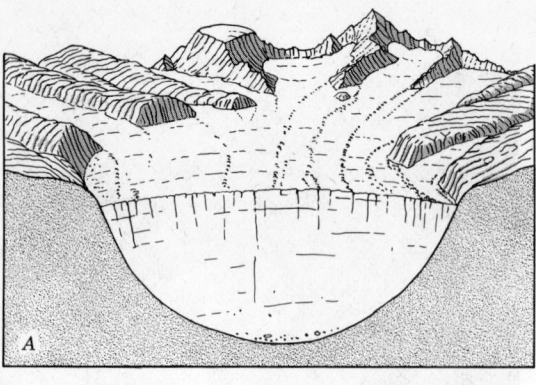
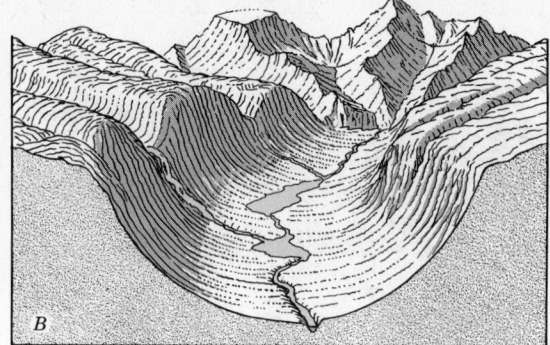
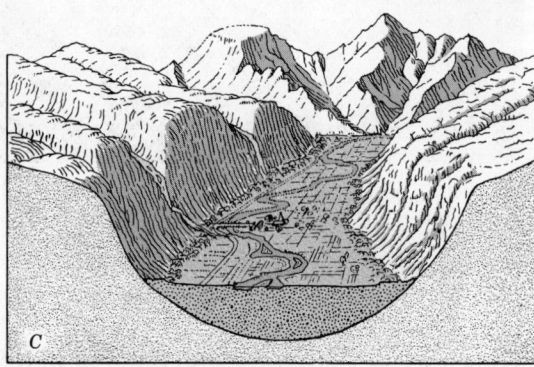

Figure 17.8 Development of a glacial trough. (After E. Raisz.) A. During maximum glaciation, the U-shaped trough is filled by ice to the level of the small tributaries. B. After glaciation, the trough floor may be occupied by a stream and lakes. C. If the main stream has a heavy load, it may fill the trough floor with alluvium. D. Should the glacial trough have been deepened below sea level, it will be occupied by an arm of the sea, or fiord.

the heart of the Alps and to the principal Alpine passes. The Brenner Pass lies at the head of a magnificent trough of this type, the Adige River valley.

The steep-walled troughs of alpine ranges contain many waterfalls and rapids readily adapted to production of hydroelectric power.

Alpine landscapes are not easily subject to degradation by Man's activities, but a notable exception is surface disfigurement by roads, truck trails, mine pits, and waste slopes created by mining operations. An additional problem connected with mining is the disturbance of ecosystems of the alpine and subalpine life zones. Between glacial cirques and troughs in the Rocky Mountains there are broad upland expanses of preglacial surface (seen in the upper right and lower left of block B of Figure 17.5). These surfaces of moderate slope have thin, rocky soils and support a sparse plant cover. Erosion by overland flow quickly follows ground disturbance and removal of vegetation. In the state of Montana, this environmental zone is being degraded by mining activities. As a result, local environmentalists have been challenging the rights of mining companies to operate in higher parts of the Beartooth Mountains, where recent exploration in search of low-grade nickel and copper ores has led to accelerated erosion and disturbance of the subalpine ecosystem.

Ice Sheets of the Present

Two enormous present-day accumulations of glacial ice are the Greenland and Antarctic ice sheets. These are plates of ice several thousand feet thick in the central areas, resting upon landmasses of subcontinental size. The Greenland Ice Sheet has an area of 670,000 sq mi (1,740,000 sq km), about the same as the area of Alaska, and occupies about seven-eighths of the entire island of Greenland (Figure 17.9). Only a narrow, mountainous coastal strip of land is exposed.

The Antarctic Ice Sheet covers about 5 million sq mi (13 million sq km), compared with 3.6 million sq mi for the 50 United States, and in places

Ice Sheets of the Present

spreads out into the ocean to form floating ice shelves (Figure 17.10). One significant point of difference between these two ice sheets is their position with reference to the poles. Whereas the Antarctic Ice Sheet rests almost squarely upon the south pole, the Greenland Ice Sheet is considerably offset from the north pole, with its center about at 75° N lat. This position indicates a fundamental principle; that a large area of high land is essential to the accumulation of a great ice sheet. A substantial portion of Siberia extends poleward of 75° N lat, but today has no ice sheet. Here precipitation is not adequate, because there is insufficient orographic effect and no large influx of moist maritime air masses, such as Greenland receives from the adjacent North Atlantic. No land exists near the north pole so that ice accumulation there is restricted to a thin layer of floating sea ice.

Contours drawn upon the surface of the Greenland Ice Sheet show that it is in the form of a very broad, smooth dome. From a high point of about 10,000 ft (3000 m) elevation east of the center there is a gradual slope outwards in all directions. The rock floor under the ice lies close to sea level under the central region, but is higher near the edges. Accumulating snows add layer upon layer of firn to the surface, while at great depth the plastic ice slowly flows outward toward the edges.

At the outer edge of the sheet the ice thins down to a few hundreds of feet. Continual loss through ablation keeps the position of the ice margin relatively steady where it is bordered by a coastal belt of land. Elsewhere the ice extends in long tongues, called *outlet glaciers*, to reach the sea at the heads of fiords. From the floating glacier edge huge masses of ice break off and drift out to open sea with tidal currents. The breakup of the ice front is known as *calving* and is brought about by strains caused by the rise and fall of tide level as well as by the undercutting and melting at and below the water line. The calving of floating glacier fronts is an extremely rapid process compared to ablation of ice fronts on land. Consequently, ice sheets are limited in their seaward extent and rarely extend far into the ocean beyond the limits of the shallow continental shelves.

Ice thickness in Antarctica is even greater than that of Greenland. For example, on Marie Byrd Land a thickness of 13,000 ft (4000 m) was measured, the rock floor lying 6500 ft (2000 m) below sea level. Over a large central area where the ice dome has its summit the ice is over 10,000 ft (3000 m) thick and the rock floor is near or below

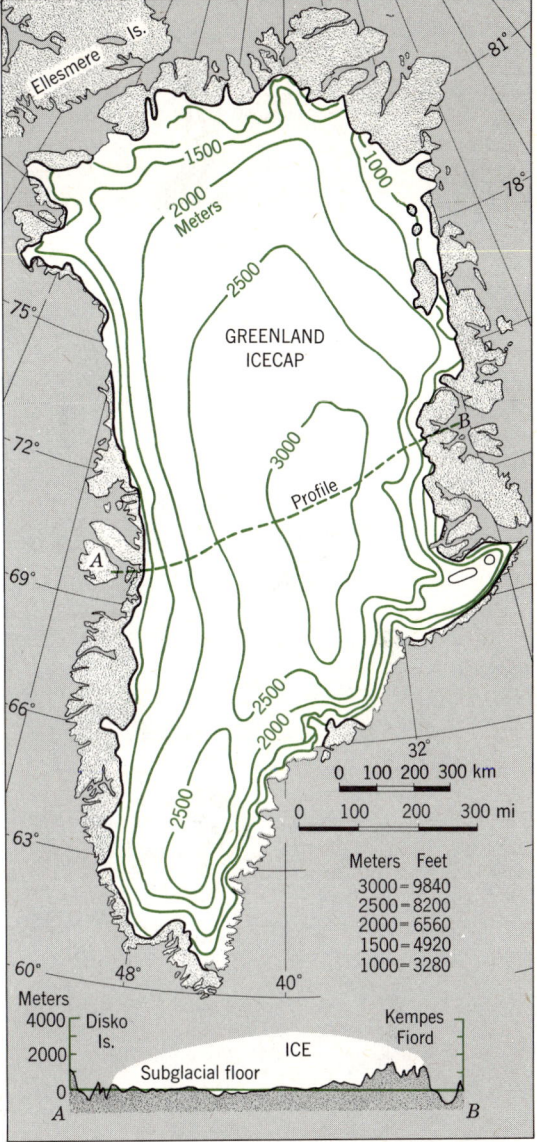

Figure 17.9 Generalized map of Greenland. (After R. F. Flint, 1957, *Glacial and Pleistocene Geology*, John Wiley & Sons, New York.)

sea level. Much of the ice moves toward the coast in large outlet glaciers (Figure 17.11).

An important glacial feature of Antarctica is the presence of great coastal plates of floating glacial ice, termed *ice shelves* (Figure 17.12). The largest of these is the Ross Ice Shelf with an area of about 200,000 sq mi (520,000 sq km) and a surface elevation averaging about 225 ft (70 m) above the sea (Figure 17.10). Ice shelves are fed by the ice sheet, but also gain new ice through the accumulation of snow.

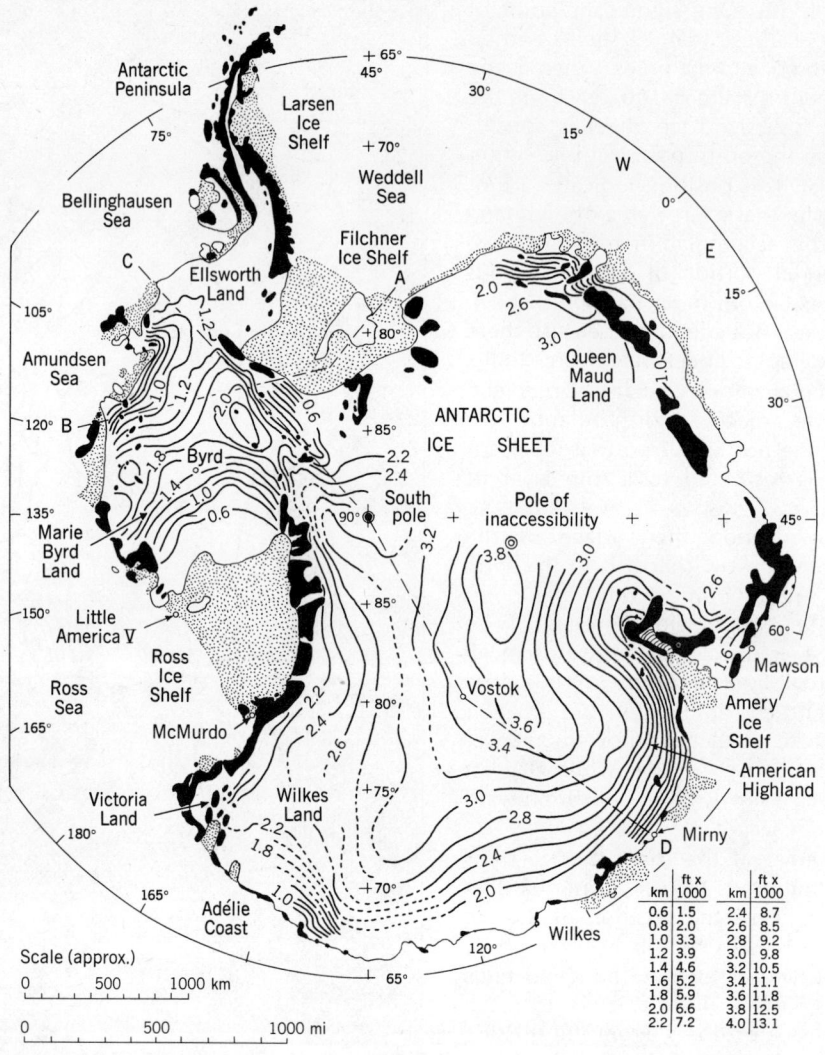

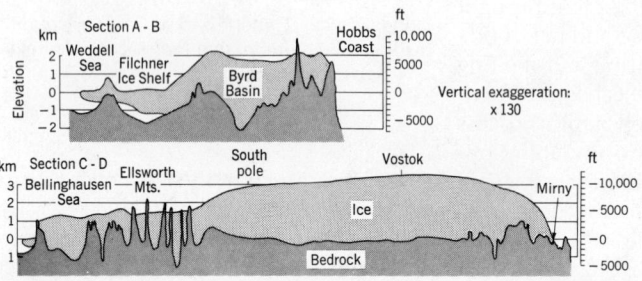

Figure 17.10 Ice surface elevations are shown on this map by contours with an interval of 0.2 km (660 ft). Areas of exposed bedrock are shown in black; ice shelves by a stippled pattern. Data from C. R. Bentley, 1962, *Geophysical Monograph No. 7*, Amer. Geophys. Union, Washington, D.C., p. 14, Figure 2. Cross sections adapted from Amer. Geog. Soc., 1964, *Antarctic Map Folio Series*, Folio 2, Plate 2. (From A. N. Strahler, 1972, *The Earth Sciences*, 2nd. ed., Harper & Row, New York.)

Figure 17.11 In this oblique air view toward the head of the Shackleton Glacier we see in the distance the polar ice plateau. Queen Maude Mountains, Antarctica, lat. 85° S, long, 177° W. (U.S. Geological Survey photograph.)

Estimated rate of outward motion of the Antarctic Ice Sheet surface is between 80 and 165 ft (25 to 50 m) per year near the periphery, but is much faster in outlet glaciers, which average around 1300 ft (400 m) per year. Outward movement of ice shelves is even more rapid—some 3300 to 4000 ft (1000 to 1200 m) per year.

The Water Balance of Antarctica

Because of the great volume of ice held in the Antarctic Ice Sheet, the water balance of that continent is important to the global environment. A careful documentation of mean sea level shows that it is undergoing a very slow rise that has averaged about 1.2 mm (0.05 in.) per year since about 1900. The trend is observed at coastal points in all four oceans and indicates either (a) an increase in water volume of the oceans through melting of glacial ice, or (b) an increase in holding capacity of the ocean basins caused by tectonic activity. (Increase in water temperature is also a possible contributing factor.) The Greenland Ice Sheet appears to be very close to an exact water balance at the present time. So we ask, is the Antarctic Ice Sheet losing or gaining ice volume?

According to one careful estimate* average annual net accumulation for the entire Antarctic ice mass is equal to 6 in. (15 cm) of water depth. As shown in Table 17.1, this amount is equivalent to about 1900 billion metric tons per year. Water loss, by ablation which is largely through melting and calving of outlet glaciers and ice shelves, is estimated to total about 1660 billion metric tons per year. Subtraction of the two figures gives a net annual gain of 240 billion metric tons per year. Even if the result has a substantial error, its positive sign fails to support the hypothesis of sea-level rise from release of Antarctic glacial ice in storage. If the rise is a result of water contributed from alpine glaciers and small icecaps, they would have to be wasting away at a rate such that they will all disappear in 500 years.

TABLE 17.1 The Water Balance of Antarctica

	Billions of Metric Tons per Year
Accumulation (P)	1900
Ablation losses ($E + R$) from	
Ice-sheet surface	10
Ice-sheet contact with ocean	50
Outlet glacier margins	520
Shelf-ice margins	880
Bottom melting beneath shelves	200
Total ablation losses	1660
Net accumulation (G)	240

SOURCE: F. Loewe, 1967, "The Water Budget in Antarctica," Proc. Symposium on Pacific-Antarctic Sciences, Tokyo, Japan, JARE Sci. Reports, Special Issue No. 1, pp. 101–110.

*F. Loewe. See reference in Table 17.1.

Figure 17.12 An ice cliff 50 to 150 ft (15 to 45 m) high marks the edge of the Ross Ice Shelf, Antarctica. (Official U.S. Coast Guard Photograph.)

The Surface Environment of Ice Sheets

The surfaces of the Greenland and Antarctic ice sheets constitute a unique environment, one almost totally devoid of life forms because of intense cold, the unavailability of liquid water, and the lack of mineral soil. Albedo of the white snow surface is extremely high, so that a large percentage of incident shortwave solar radiation is reflected. An air layer close to the ground becomes intensely chilled, resulting in a strong temperature inversion. Added to surface properties is the high altitude of the interior plateau region, much of it 8000 ft (2500 m) and higher. Lowered density of the air mass at this altitude facilitates longwave radiation losses into space.

Severity of ice-sheet climate is evident from the air temperature records of stations on both Greenland and Antarctica (Figure 17.13). On the polar ice plateau of Antarctica the mean temperature of all months is far below 0° F (−18° C). For the interior Greenland ice plateau, average monthly air temperature in summer does not rise above the freezing point. During the long south polar winter (6 months of no-sun at 90° S latitude) temperatures continue to fall from month to month through the polar night as the negative radiation balance becomes more severe. The coldest time of the year is in August and September, which represents a time lag of two to three months after solstice.

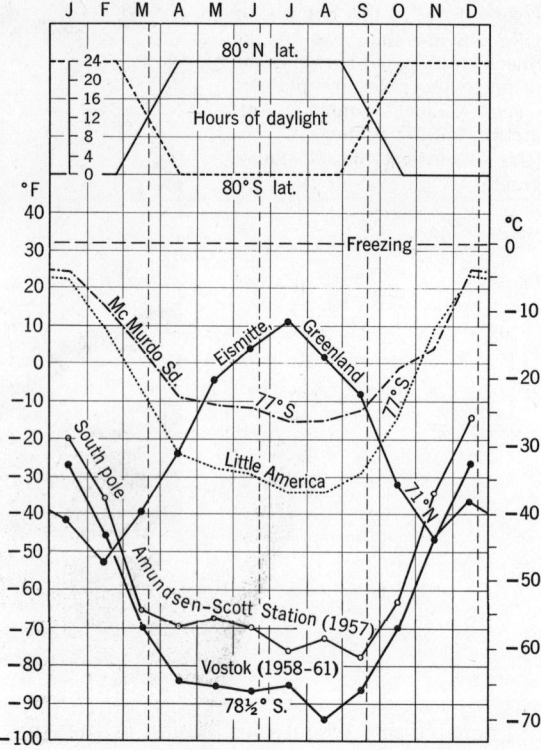

Figure 17.13 Monthly mean air temperatures for five ice sheet stations. (Data from Trewartha; *I.G.Y. Bulletin*; and *Weatherwise*, vol. 16, 1963.)

Besides low temperatures, intense winds blow for long periods at a time over the ice sheets. One effect of strong winds is to cause abrasion and drifting of hard-packed snow to produce a furrowed ice surface—the forms are called *sastrugi* (Figure 17.14). Another environmental hazard of the ice sheets, and also of any snow-covered area at high latitude is the *whiteout*, an optical phenomenon in which an observer cannot discern horizon, shadows, or clouds. The surroundings give a uniform white glow and the observer loses his sense of orientation. Blowing snow may, or may not, be present during a whiteout.

Figure 17.14 Sastrugi (wind eroded furrows) on the Greenland Ice Sheet, 5 mi (8 km) from the ice margin, lat. 76° N. (Photograph by L. H. Nobles.)

Ice Sheets of the Pleistocene Epoch

As if the vast ice sheets of Greenland and Antarctica do not seem fantastic enough, geologists have brought to light abundant and convincing evidence that much of North America and Europe and parts of northern Asia and southern South America were covered by enormous ice sheets in the Pleistocene Epoch. This ice age ended 10,000 to 15,000 years ago with the rapid wasting away of the ice sheets. Consequently, landforms made by the last ice advance and recession are very little modified by erosional agents.

Figures 17.15 and 17.16 show the extent to which North America and Europe were covered at the maximum known spread of the last advance of the ice. Over North America, the dominant ice body was the *Laurentide Ice Sheet*, centered about over Hudson Bay. It began as an icecap over the Labrador Highlands, which enlarged and spread south, west, and northwest to reach the base of the Cordilleran ranges on the west and the arctic islands on the north. Over the Cordilleran ranges coalescent icecaps and alpine glaciers formed a single ice body reaching down to the Pacific coast on the west and to the mountain foothills on the east. In the United States, all the land lying north of the Missouri and Ohio rivers was covered, as well as northern Pennsylvania and all of New York and New England.

In Europe, the *Scandinavian Ice Sheet* centered upon the Baltic Sea, covering the Scandinavian countries and spreading as far south as central Germany. The British Isles were almost covered by an icecap that had several centers on highland areas and spread outward to coalesce with the Scandinavian ice sheet. The Alps at the same time were heavily inundated by enlarged alpine glaciers, fused into a single icecap. All high mountain areas of the world underwent greatly intensified alpine glaciations at the time of maximum ice-sheet advance. Today, only small remnant alpine glaciers exist as vestiges of these great valley glaciers. In less favorable mountain regions no glaciers remain.

Proof of the former great extent of ice sheets has been carefully accumulated since the middle of the nineteenth century when the great naturalist Louis Agassiz first announced the bold theory. In general, the evidence of past glaciation lies in the recognition throughout North America and Europe of landforms identical with those now seen near the margins of the Greenland ice sheet and other glaciers. Although Agassiz's pronouncement was greeted with much skepticism a century ago, continental glaciation of the Pleistocene Epoch is now universally accepted among scientists. Moreover, careful study of the deposits left by the ice has led to the knowledge that not one

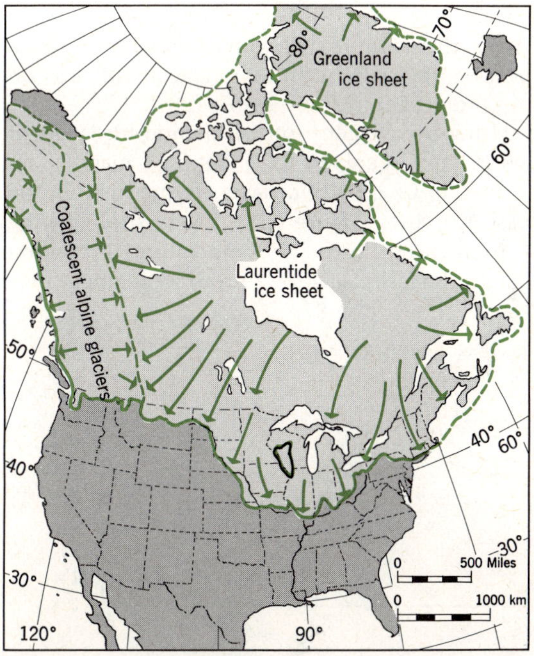

Figure 17.15 Pleistocene ice sheets of North America at their maximum spread reached as far south as the present Ohio and Missouri rivers. (After R. F. Flint.)

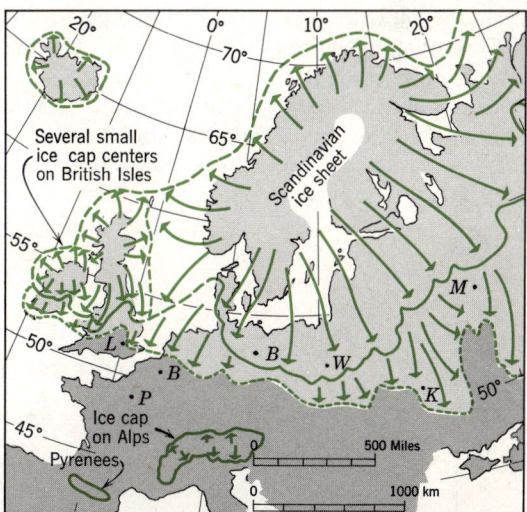

Figure 17.16 The Scandinavian ice sheet dominated northern Europe during the Pleistocene glaciations. Solid line shows limits of ice in the last glacial stage; dotted line on land shows maximum extent at any time. (After R. F. Flint.)

but at least four, and perhaps six major advances and retreats occurred, spaced over a total period of about a million years. Each major ice advance is designated a *glaciation;* the preceding or following event of ice disappearance is an *interglaciation*

Causes of Glaciations

Although we can't go into much detail on merits and shortcomings of several contending hypotheses purporting to explain repeated glaciations and interglaciations, a few general statements are of environmental interest because they bring up certain of the same mechanisms of global climatic change we examined in Chapter 6. The difference between this and our earlier discussion is one of time scale, since each episode of glaciation persisted many thousands of years.

Consider first that glaciation requires a substantial decrease in average global air and sea-surface temperatures, enough to reduce ablation rates appreciably and so allow ice to accumulate. Second, there must be an adequate supply of snowfall over highland areas serving as the sites of development of the initial icecaps. A basic counteracting mechanism exists here, because lowered air and sea surface temperatures result in reduced evaporation and reduced moisture-holding capacity of air masses.

Evidence of lowered global air temperatures during glaciations comes from both terrestrial and marine sources. One line of evidence is the lowered elevation of the *snowline,* or lower limit of snow accumulations lasting through the entire year. The snowline is essentially equated to the firn line of alpine glaciers. Figure 17.17 is a meridional profile along the Cordilleran ranges and Andes from 65° N to 60° S latitude, showing altitude of the snowline of today, as compared with that of a glaciation. These snowlines are smoothed lines, evening out local differences in altitude due to varying local influences. In North America, the amount of lowering was about 6500 ft (2000 m); in the Andes at low latitudes, about 2500 ft (800 m). Similar figures apply on other continents. The lowering of snowline is interpreted as representing a drop of mean annual global air temperature of between 9 and 13 F° (5 and 7 C°).

Very important to terrestrial life forms and their ecosystems is the consideration that a lowering of the snowline must have been accompanied by a lowering of all life zones stratified according to altitude. This change would have required plants and animals to migrate to lower elevations or to lower latitudes to remain in the same thermal environments. A reverse effect would be caused by the rising of snowline during deglaciation (*Deglaciation* is the transition from glaciation to interglaciation.)

Another major source of evidence is from deep-sea sediments and the microorganisms they contain. Shells of such tiny planktonic animals as *Globigerina,* a group of calcareous foraminifera living in shallow marine waters, sink to the ocean floor and are incorporated as microfossils in the sediment. When cores of sediment are brought to the surface, the changing proportions of species from layer to layer can be measured and inferences derived as to the changing water temperatures. A more quantitative method is to evaluate the oxygen-isotope ratios within the microfossil shells of such sediments and translate the information into a water-temperature scale. When this is done, the water-temperature range between glaciations and interglaciations runs on the order of 9 to 11 F° (5 to 6 C°).

Using the oxygen-isotope method for layers of ice in the Greenland Ice Sheet, in a manner described in Chapter 6 for layers formed within the past 800 years (Figure 6.21), a very strong swing is found from lower air temperatures prevailing during the last glaciation to warmer temperatures

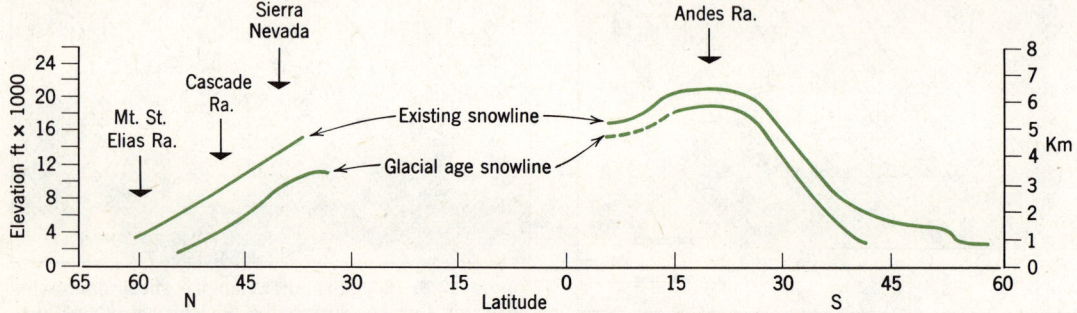

Figure 17.17 Generalized meridional profile of the present snow line and lowered Pleistocene snow line of maximum glaciations. (Data of R. F. Flint, 1957, *Glacial and Pleistocene Geology,* John Wiley & Sons, New York, p. 47, Figure 4.1.)

Causes of Glaciations

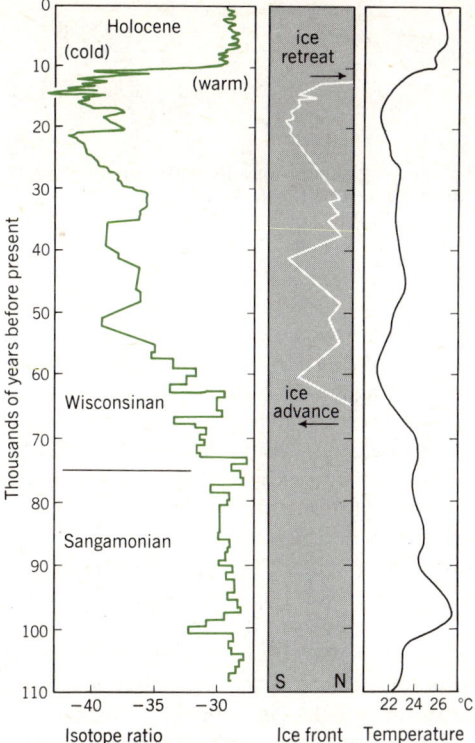

Figure 17.18 Oxygen-isotope ratios (*left*) measured from an ice core from the Greenland Ice Sheet are compared with the record of ice advances and retreats in the Great Lakes Area (*center*). Curve at right shows temperature fluctuations interpreted from oxygen-isotope measurements of sediment in deep-sea cores. (Data of W. Daansgaard, and others, 1969, *Science*, vol. 166, p. 380, Figure 5.)

following rapid deglaciation (Figure 17.18). This graph shows an air-temperature drop of about 9 F° (5 C°) during the last deglaciation.

An important basic factor contributing to onset of glaciation, whatever the primary mechanism of control, is the widespread development of alpine mountain systems and high plateaus over the continents in late Cenozoic time. These features could act as orographic traps for snowfall and also permit the snow accumulations to be retained at high elevations where air temperatures are prevailingly low. Once a large icecap had developed, it became its own orographic trap for continued precipitation.

One group of hypotheses of glaciation invokes a reduction in solar energy received by the earth. It has been proposed that glaciations occur when radiant output of the sun is reduced; in other words, that the value of the solar constant is diminished. A reduced planetary temperature would follow. Actually, no evidence has been found for past reductions in solar output on a scale sufficient to cause glaciation. On the contrary, evidence obtained from lunar rocks suggests a more-or-less constant solar output for a period much longer than the Pleistocene Epoch.

The quantity of solar radiation intercepted by the earth varies with cyclic changes in distance between earth and sun. The eccentricity of the earth's orbit causes the distance to vary yearly from a minimum at the closest approach (perihelion, January 3) to a maximum at the most distant point (aphelion, July 4). This changing distance has the effect of giving the south polar region a slightly larger input of solar energy during the summer of that hemisphere than the north polar region receives during the summer of its hemisphere. Without going into details, it can be added that there is a cycle of variation in eccentricity of the earth's orbit, and this results in a cycle of changing distance with a period of about 90,000 years in which the distance can deviate as much as about 5% from the mean value. Another astronomical cycle varies the inclination of the earth's axis with respect to the ecliptic plane with a period of 40,000 years. When these astronomical cycles are combined, a curve of varying solar radiation can be computed for a selected latitude, such as 45° N or 55° N and shows appreciable departures from the average value. There are a number of scientists who consider these changes in insolation at middle latitudes as sufficient to bring on a glaciation, or a deglaciation. The astronomical hypotheses are strongly debated.

Solar energy is invoked as a causative agent in glaciation in yet another way. It has been postulated that increased quantities of volcanic dust in the atmosphere might bring on a glaciation because more solar energy would be reflected back into space, permitting less to enter the atmosphere. Along with the reduced air temperature would be the increase in numbers of dust particles to serve as nuclei for the condensation of moisture, thus favoring increased precipitation. The role of volcanic dusts in varying radiation received at tropospheric levels has been discussed in Chapter 6. Evidence is lacking of intensified vulcanism in phase with glaciations.

An important and widely held theory attributes glaciation to a reduction of the carbon dioxide content of the atmosphere. The role of carbon dioxide in absorbing longwave radiation and thus warming the atmosphere has been explained in Chapter 6. It is estimated that, if the carbon dioxide content of the atmosphere, which is now about 0.03% by volume, were reduced by half that amount, the earth's average surface temperature would drop about 7 F° (4 C°).

Glacier Systems and the Pleistocene Epoch

Another group of theories proposes shifts in the positions of the continents with respect to the poles, bringing various landmasses into polar positions favorable for the growth of ice sheets. Still another theory requires changes in oceanic currents, specifically the diversion or blocking of such warm currents as the Gulf Stream, which would have brought colder climates to the subarctic regions.

Major glaciations have occurred at a number of points in geologic time, but the reasons are not known and the schedule has been episodic. Durations of glaciations have doubtless been extremely small in comparison with nonglacial times. Earth history of the last 1 to 3 million years of geologic time can therefore be considered as highly atypical from the standpoint of the environments of life on earth and organic evolution.

Erosion by Ice Sheets

Like alpine glaciers, ice sheets are effective eroding agents. The slowly moving ice can scrape and grind away much solid bedrock, leaving behind smoothly rounded rock masses bearing countless minute abrasion marks. Scratches, or *striations*, trend in the general direction of ice movement (Figure 17.19), but variations in ice direction from time to time often result in intersecting lines.

Where a strong, sharp-pointed piece of rock was held by the ice and dragged over the bedrock surface, there resulted a series of curved cracks fitted together along the line of ice movement. These *chatter marks* and closely related *crescentic gouges* whose curvature is the opposite, are good

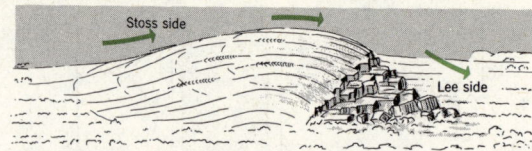

Figure 17.20 A glacially abraded rock knob. (From A. N. Strahler, 1971, *The Earth Sciences*, 2nd ed., Harper & Row, New York.)

indicators of the direction of ice movement (Figure 17.19). Some very hard rocks have acquired highly polished surfaces from the rubbing of fine clay particles against the rock. The evidences of ice erosion described here are common throughout the northeastern United States. They may be seen on almost any exposed hard rock surface.

A common landform shaped by ice abrasion is a knob of solid bedrock that has been shaped by the moving ice (Figure 17.20). One side, that from which the ice was approaching, is characteristically smoothly rounded and shows a striated and grooved surface. This is termed the *stoss* side. The other, or *lee* side, where the ice plucked out angular joint blocks, is irregular, blocky, and steeper than the stoss side. The quaint term *roches mountonnées* has long been applied by glaciologists to such glaciated rock knobs.

Vastly more important than the minor abrasion forms are enormous excavations that the ice sheets made in some localities where the bedrock is weak and the ice current was accentuated by the presence of a valley paralleling the direction of ice flow. Under such conditions the ice sheet behaved much as a valley glacier, scooping out a deep, U-shaped trough. The Finger Lakes of western New York State are fine examples. Here

Figure 17.19 Glacial striations and fracture marks, mostly crescentic gouges, cover the smoothly rounded surface of this rock knob. These marks were made by the East Twin Glacier, Alaska. The ice moved in a direction away from the photographer. (Photograph by Maynard Miller.)

Figure 17.21 Seen from the air, this esker in the Canadian shield area appears as a narrow embankment crossing the terrain of glacially eroded lake basins. (Photograph by Canadian Department of Mines, Geological Survey.)

a set of former stream valleys lay parallel to southward spread of the ice, which scooped out a series of deep troughs. Blocked at the north ends by glacial debris the basins now hold elongated lakes. Many hundreds of lake basins were created by ice action all over the glaciated portions of North America and Europe. Countless small lakes of Minnesota, Canada, and Finland occupy rock basins scooped out by ice action (Figure 17.21). Irregular debris deposits left by the ice are also important to causing lake basins.

Deposits Left by Ice Sheets

The term *glacial drift* has long been applied to include all varieties of rock debris deposited in close association with glaciers. Drift is of two major types: (1) *Stratified drift* consists of layers of sorted and stratified clays, silts, sands, or gravels deposited by meltwater streams or in bodies of standing water adjacent to the ice. (2) *Till* is a heterogeneous mixture of rock fragments ranging in size from clay to boulders and is deposited directly from the ice without water transport. The moraine of a valley glacier, seen in Figure 17.7, is composed largely of till, whereas the valley train, seen in Figure 14.28, is composed of stratified drift.

Over those parts of the United States formerly covered by Pleistocene ice sheets, glacial drift averages from 20 ft (6 m) thick over mountainous terrain such as New England, to 50 ft (15 m) and more over the lowlands of the north-central United States. Over Iowa, drift is from 150 to 200 ft (45 to 60 m) thick; over Illinois, it averages more than 100 ft (30 m) thick. Locally, where deep stream valleys existed prior to glacial advance, as in Ohio, drift may be several hundred feet thick.

To understand the form and composition of deposits left by ice sheets, we need first to consider the conditions prevailing at the time of existence of the ice, as shown in Figure 17.22. Block A shows a region partly covered by an ice sheet with a stationary front edge. This condition occurs when the rate of ice ablation balances the amount of ice brought forward by spreading of the ice sheet. Any increase in ice movement would cause the ice to shove forward to cover more ground; an increase in the rate of wasting would cause the edge to recede and the ice surface to become lowered. Although the Pleistocene ice fronts did advance and recede in many minor and major fluctuations, there were long periods when the front was essentially stable. This condition is represented in Block A.

The transportational work of an ice sheet resembles that of a great conveyor belt. Anything carried on the belt is dumped off at the end and if not constantly removed will pile up in increasing quantity. Rock fragments brought within the ice are deposited at the forward margin as the ice evaporates or melts. There is no possibility of return transportation.

Glacial till that accumulates at the immediate ice edge forms a rubbly heap of irregular thickness, the *terminal moraine*. After the ice has disappeared, as in diagram B, the moraine appears as a belt of knobby hills interspersed with basin-like hollows, some of which hold small lakes. The term *knob-and-kettle* is often applied to such morainal belts (Figure 17.23). Terminal moraines tend to form curving patterns, the convex form of curvature directed southward and indicating that the ice advanced as a series of *lobes* (Figure 17.24). Where two lobes came together, the moraines curved back and fused together into a single moraine pointed northward. This is termed an *interlobate moraine* (Figure 17.22, block B). In its general recession accompanying deglaciation, the ice front paused for some time along a number of lines, causing morainal belts similar to the terminal moraine belt to be formed. These belts, known as *recessional moraines* (Figures 17.22 and 17.24), run roughly parallel with the terminal moraine but are often thin and discontinuous.

Block A of Figure 17.22 shows a smooth, sloping plain lying in front of the ice margin. This is the *outwash plain*, formed of stratified drift left by aggrading streams issuing from the ice. Their deposits are broad alluvial fans upon which were spread layer upon layer of sands and gravels. The adjective *glaciofluvial* is applied to stream-laid stratified drift. Where outwash accumulated around isolated ice blocks left behind in a previous episode of ice advance and recession, the blocks later melted away, leaving kettles (Figure 17.22). The outwash plain is then described as a *pitted* plain.

Large streams issue from tunnels in the ice, particularly when the ice for many miles back from the front has become stagnant, without forward movement. Tunnels then develop throughout the ice mass, serving to carry off the meltwater. After the ice has gone (block B, Figure 17.22) the outwash plain remains in its original form, but may be bounded on the iceward side by a steep slope which is the mold of the ice against which the outwash was built. Such a slope is called an *ice-contact slope*. Farther back, behind the terminal moraine, the position of a former ice tunnel is marked by a long, sinuous ridge known as an *esker*. The esker is the deposit of sand, pebbles, and cobbles formerly laid upon the floor of the ice tunnel. Because ice formed

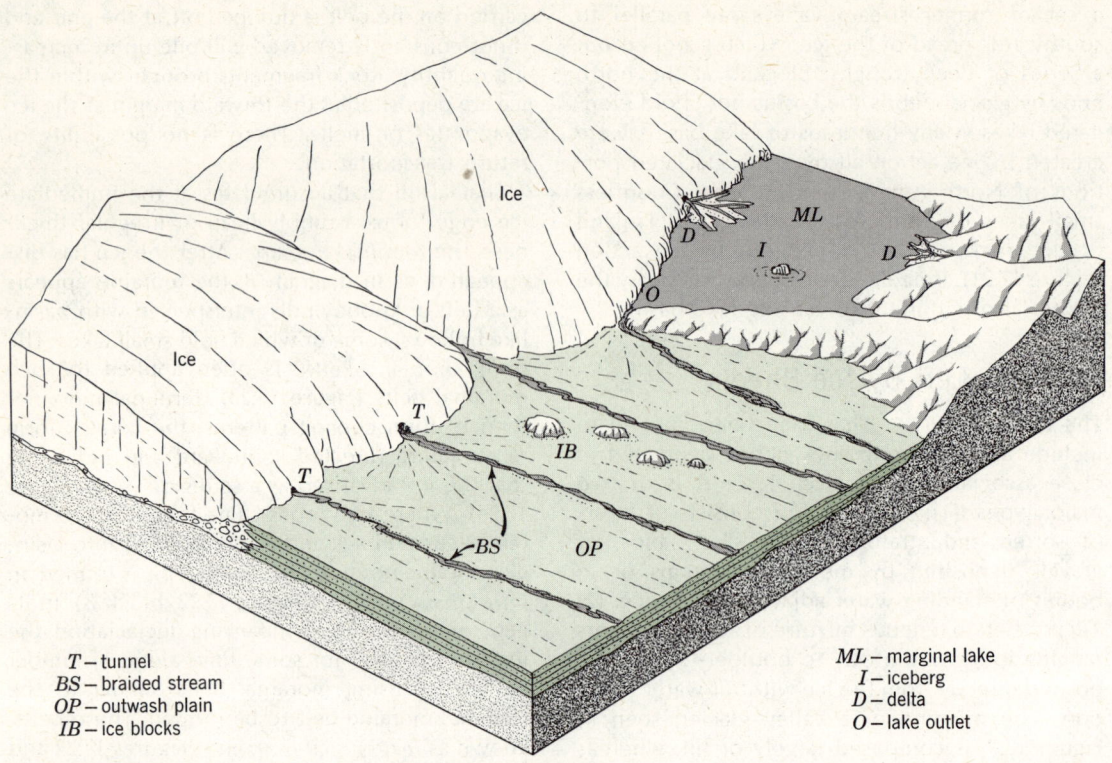

T — tunnel
BS — braided stream
OP — outwash plain
IB — ice blocks

ML — marginal lake
I — iceberg
D — delta
O — lake outlet

A. With the ice front stabilized and the ice in a wasting, stagnant condition, various depositional features are built by meltwater.

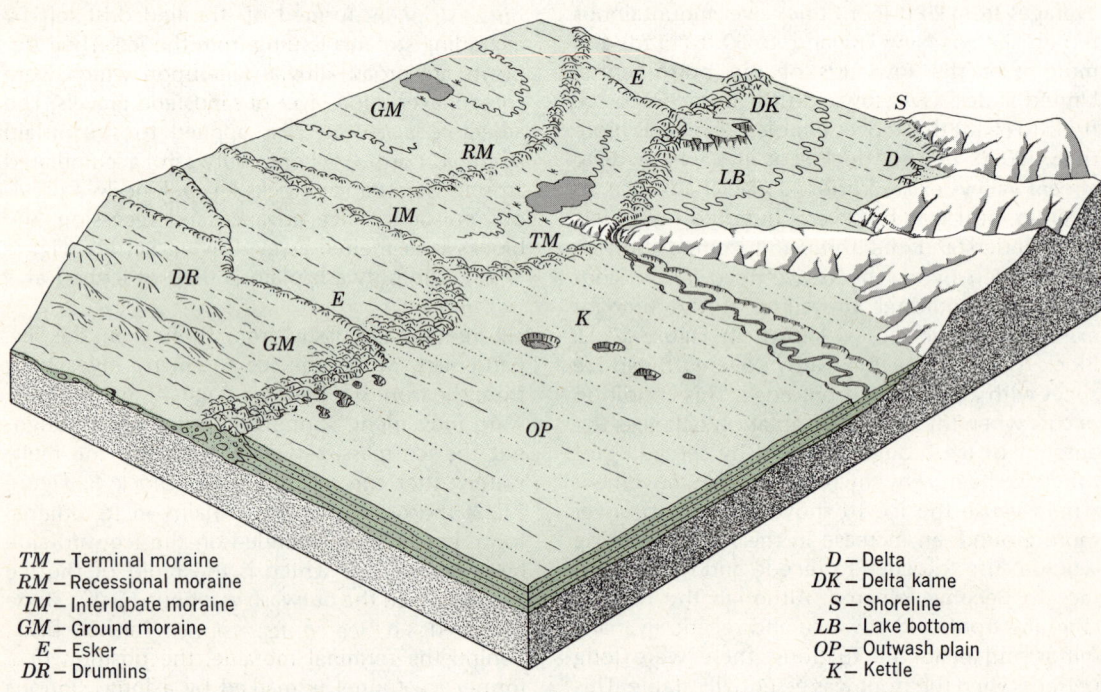

TM — Terminal moraine
RM — Recessional moraine
IM — Interlobate moraine
GM — Ground moraine
E — Esker
DR — Drumlins

D — Delta
DK — Delta kame
S — Shoreline
LB — Lake bottom
OP — Outwash plain
K — Kettle

B. After the ice has wasted completely away, a variety of new landforms made under the ice is exposed to view.

Figure 17.22 Marginal landforms of continental glaciers.

Deposits Left by Ice Sheets

Figure 17.23 Rugged topography of small knobs and kettles characterizes this interlobate moraine northeast of Elkhart Lake, Sheboygan County, Wisconsin. (Photograph by W. C. Alden, U.S. Geological Survey.)

Figure 17.25 This esker has developed a cover of soil and vegetation, concealing the coarse gravel that lies within it. Dodge County, Wisconsin. (Photograph by W. C. Alden, U.S. Geological Survey.)

the sides and roof of the tunnel, its disappearance left merely the stream-bed deposit, which now forms a ridge (Figure 17.25). Eskers are often many miles long; in parts of the Ungava Peninsula of Canada, some are more than 150 mi (240 km) long. Some have branches just as streams do.

Another curious glacial landform is the *drumlin*, a smoothly rounded, oval hill resembling the bowl of an inverted teaspoon. It consists of glacial till (Figure 17.26). Drumlins invariably lie in a zone behind the terminal or recessional moraine. They commonly occur in groups or swarms, which may number in the hundreds. The long axis of each drumlin parallels the direction of ice movement so that the drumlins point toward the terminal moraines and serve as indicators of direction of ice movement. From a study of the composition and structure of drumlins, it has been generally agreed that they were formed under moving ice by a plastering action in which layer upon layer of bouldery clay was spread upon the drumlin. This accumulation would have been possible only if the basal ice were so heavily choked with debris

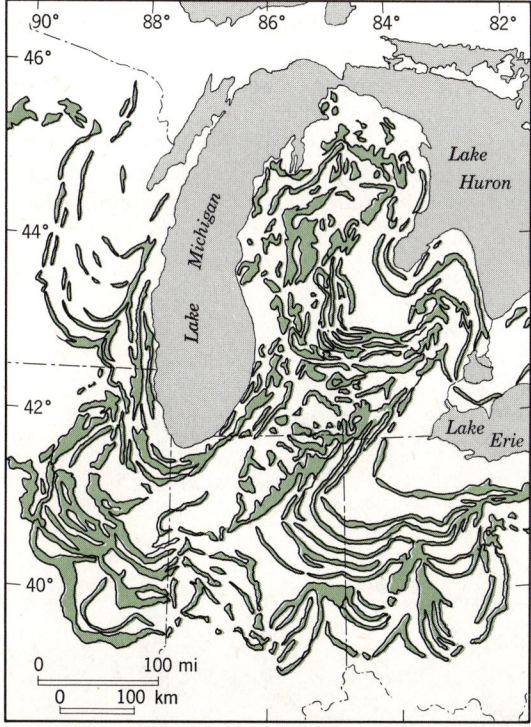

Figure 17.24 Moraine belts of the north-central United States have a festooned pattern left by ice lobes (After R. F. Flint and others, *Glacial Map of North America*, 1945.)

Figure 17.26 This small drumlin, located south of Sodus, New York, shows a tapered form from upper right to lower left, indicating that the ice moved in that direction (north to south). (Photograph by Ward's Natural Science Establishment, Inc., Rochester, N.Y.)

Glacier Systems and the Pleistocene Epoch

that the excess had to be left behind. Furthermore, some sort of knob or surface irregularity may have been required to start the plastering action and localize its occurrence.

Between moraines, the surface left by the ice is usually overspread by a cover of glacial till known as *ground moraine*. This cover is often inconspicuous because it forms no prominent or recognizable topographic features. Even so, the ground moraine may be thick and can obscure or entirely bury hills and valleys that existed before glaciation. Where smoothly spread, the ground moraine forms a level *till plain*, but this feature is likely to be found only in regions already fairly flat to start with. In more hilly and mountainous regions, such as New England, the preglacial valleys and hills retain their same general outlines despite glaciation.

Deposits Built into Standing Water

Where the general land slope is toward the front of an ice sheet, a natural topographic basin is formed between the ice front and the rising ground. Valleys that may have opened out northward are blocked by ice. Under such conditions, *marginal glacial lakes* form along the ice front (Figure 17.22, block A). These lakes overflow along the lowest available channel, which lies between the ice and the ground slope or over some low pass along a divide. Into marginal lakes streams of meltwater from the ice build *glacial deltas*, similar in most respects to deltas formed by any stream flowing into a lake. Streams from the land also build deltas into the lake. When the ice has disappeared the lake drains away, exposing the bottom upon which layers of fine clay and silt have been laid. These fine-grained sediments, which have settled out from suspension in turbid lake waters, are called *glaciolacustrine* sediments and are a variety of stratified drift. The layers are commonly of banded appearance, with alternating dark and light layers, termed *varves*. Glacial lake plains are extremely flat, with meandering streams and extensive areas of marshland.

Deltas, built with a flat top at what was formerly the lake level, are now curiously isolated, flat-topped landforms known as *delta kames*. Delta and stream channel deposits built between a stagnant ice mass and the wall of a valley become *kame terraces*, whose steep scarps are ice-contact slopes (Figure 17.27). Kame terraces are difficult to distinguish from the uppermost member of a series of alluvial terraces, but most kames have undrained depressions or pits produced by the melting of enclosed ice blocks. Built of very well-washed and sorted sands and gravels, kames commonly show the steeply dipping foreset beds characteristic of deltas (Figure 17.28).

Environmental and Resource Aspects of Glacial Deposits

Because much of Europe and North America was glaciated by the Pleistocene ice sheets, landforms associated with the ice are of fundamental environmental importance and the deposits constitute a natural resource as well. Agricultural influences of glaciation are both favorable and unfavorable, depending on preglacial topography and whether the ice eroded or deposited heavily.

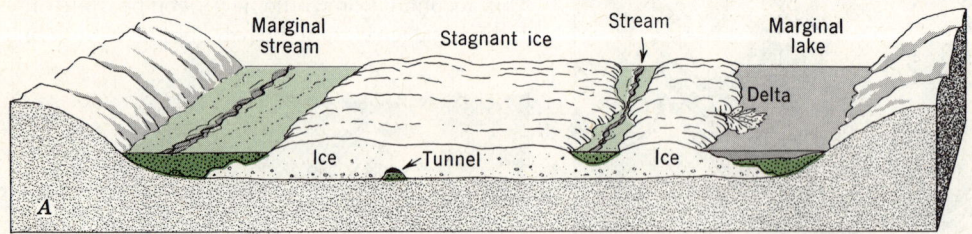

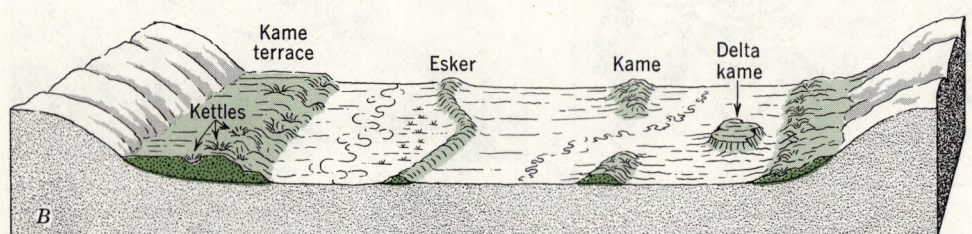

Figure 17.27 Kames may originate as stream or lake deposits between a stagnant ice mass and the valley sides. (After R. F. Flint.)

Environmental Aspects of Glacial Deposits

Figure 17.28 These cross-bedded, sorted sands were laid down in a glacial delta near North Haven, Connecticut. (Photograph by R. J. Lougee.)

most productive agricultural land in the world. In this class belong the prairie lands of Indiana, Illinois, Iowa, Nebraska, and Minnesota. We must not lose sight of the fact that in these areas upland loess forms a blanket over clay-rich till and sandy outwash. Exposed glacial drift would be a poor parent base for soil.

Glaciofluvial deposits are of great economic value. The sands and gravels of outwash plains, kames, and eskers provide the aggregate necessary for concrete and the base courses beneath highway pavements (Figure 17.29). The purest sands may be used for molds needed for metal castings.

Glaciofluvial deposits, where they are thick, form excellent aquifers and are a major source of ground water supplies. Deep accumulations of stratified sands in preglacial valleys are capable of yielding ground water in quantities sufficient for municipal and industrial uses. Water development of this type is widespread in Ohio, Pennsylvania, and New York. But, as we pointed out in Chapter 12, consumptive use of ground water must be kept to a low level by recycling of used water through recharge wells. Fortunately, most of the glaciated areas of the Northern Hemisphere are today in climatic regimes showing a large water surplus. Consequently, natural recharge rates are high and surface water sources are available for additional artificial recharge.

In hilly or mountainous regions, such as New England, the glacial till is thinly distributed and extremely stony. Podzolic soils which developed on glacial deposits of the northern United States and Canada are acid and of low fertility. Extensive bogs, unsuited to agriculture unless transformed by water drainage systems, are another unfavorable element. Early settlers found cultivation difficult because of countless boulders and cobbles in the soil. Till accumulations on steep mountain slopes are subject to mass movements in the form of earth flows and debris avalanches. Clays in the till become weakened upon absorbing water from melting snows and spring rains. Where slopes have been oversteepened by excavation for highways, movement of till is a common phenomenon. Along morainal belts the steep slopes, irregularity of knob-and-kettle topography, and abundance of boulders conspired to prevent crop cultivation but invited use as pasture. These same features, however, make morainal belts extremely desirable as suburban residential areas. Pleasing landscapes of hills, depressions, and small lakes make ideal locations for large estates.

Extensive till plains, outwash plains, and lake plains, on the other hand, comprise some of the

Figure 17.29 Thick layers of outwash sands and gravels such as these on the north shore of Long Island were, excavated in great quantities for use in highway and building construction. The dark layer at the top is a bed of glacial till, left by a glacial advance. Boulders in the foreground are glacial erratics that have rolled down from the till bed. (Photograph taken about 1920 by A. K. Lobeck.)

447

The Pleistocene Epoch

Because the Pleistocene Epoch witnessed the evolution of Man during wide climatic fluctuations between glaciations and interglaciations, a brief review of the salient historical features of this epoch will give you a better insight into the sudden rise and spread of Man to a position of world dominance in the very brief Holocene Epoch in which we now live.

Recall that in Chapter 9, in an explanation of the geologic time table, we said in a footnote that an older system of time units placed the Pleistocene and Holocene epochs into a single time unit, the Quaternary Period. Although obsolete, the term Quaternary continues in use because of convenience in referring to both epochs together. Actually, we may now be in an interglaciation, with another glaciation yet to come, and there is no way to evaluate our position in geologic time.

The beginning point of the Pleistocene Epoch, that is, the time boundary between Pliocene and Pleistocene epochs, is established by geologists on the basis of a change in composition of microfossil faunas in marine sediments. This is a distinction of no special meaning in environmental science, but places the start of the Pleistocene at about -2 m.y. (2 million years before present). From deep-sea sediment cores we learn that substantial icecaps and ice sheets had existed in high latitudes long before the start of the Pleistocene, suggesting that global climatic cooling was begun late in the Pliocene Epoch. Antarctica, because of its polar position, would have had its ice sheet as early as perhaps -4 m.y., or perhaps even earlier. From this source, ice-rafted rock debris was carried out into the Southern Ocean to be released and settle to the bottom as sedimentary layers. Microfossils seem to indicate that a general cooling of the global oceans occurred at about -2 m.y., and this change is associated with the first of several glaciations that saw extensive spread of ice sheets into middle latitudes.

For several decades geologists have recognized in North America four glaciations and three interglaciations on the basis of successive deposits of sediment and organic materials. A similar history was also derived for European glaciations. The names of these events are shown in order in Table 17.2. However, there is evidence in Europe and the British Isles of two earlier cold periods, so that the number of glaciations may prove to be six, rather than four, and there may have been even more than that number.

Unraveling the history of glaciations and interglaciations has been a complex process, accompanied by many controversies. Because ice-sheet advance and recession were evidently not synchronous over all parts of even one continent, the correlation of events is extremely difficult. One form of evidence comes from interpretation of layered deposits (science of *stratigraphy*). The till of one glaciation is typically followed by a loess layer associated with deglaciation, then by formation of an ancient soil (*paleosol*) and by deposition of organic matter such as bog peat indicative of an interglaciation with its mild climate. Older tills show varying degrees of chemical alteration by weathering. Landforms of earlier glaciations, where not buried under new glacial deposits, show increasing degrees of modification by mass wasting and fluvial erosion with increasing age.

Deep-sea sediments also furnish a stratigraphic record with fossils which can be interpreted in terms of sea-water temperature fluctuations associated with glaciations and interglaciations. Ideally, the chronology of the deep-sea sediments should correlate with the stratigraphic record on the lands, but not necessarily in close phase. The radiocarbon method of age determination, described in Chapter 9, has been a mainstay in establishing ages of deposits and events back to about $-40,000$ years, while paleomagnetic data (epochs of magnetic polarity reversals) have allowed absolute age dating of marine sediments to be carried back through the entire Pleistocene Epoch. Even so, no really firm timetable has yet been established to the satisfaction of a large majority of research workers.

For the four glaciations of North America, glacial deposits have been identified and mapped to the general satisfaction of most students of the Pleistocene (Figure 17.30). The oldest deposits,

TABLE 17.2 North American and European Glaciations

Glaciations		Interglaciations
North America	(Alps)	
Wisconsinan	(Würm)	
		Sangamonian
Illinoian	(Riss)	
		Yarmouthian
Kansan	(Mindel)	
		Aftonian
Nebraskan	(Günz)	
?		
?		
?		

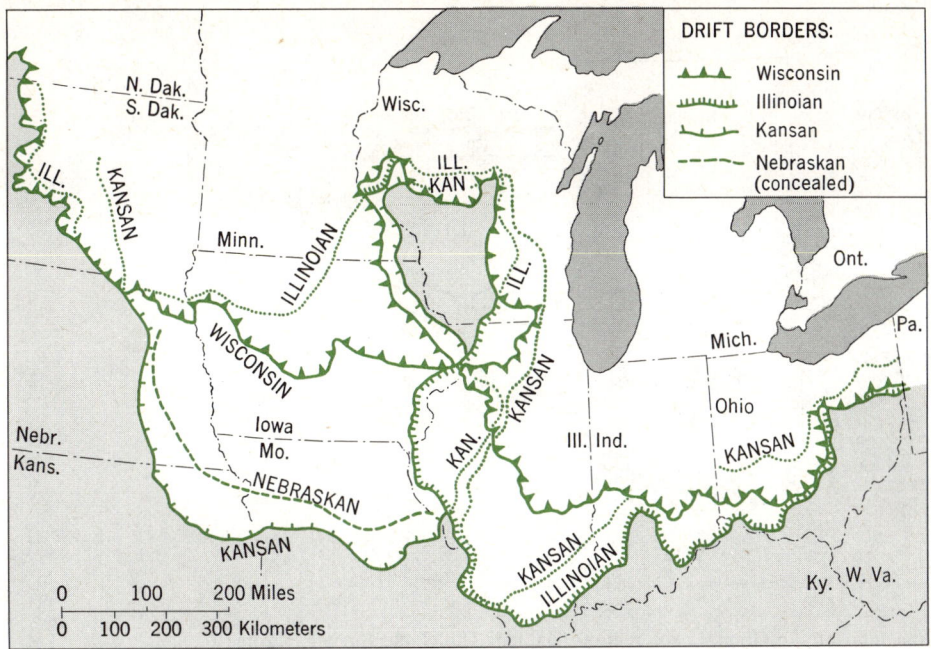

Figure 17.30 Limits reached by ice sheet in each of four Pleistocene glaciations. (After R. F. Flint, 1957, *Glacial and Pleistocene Geology*, John Wiley & Sons, New York, p. 338, Figure 20.1.)

those of the *Nebraskan glaciation*, are found only in Nebraska, Kansas, Iowa, and Missouri. They lie buried beneath deposits of the next advance, the *Kansan glaciation*, which extended farther south in those same states. The two glacial deposits are separated by materials of the *Aftonian interglaciation*. Kansan glacial deposits can be found beneath earlier deposits in Illinois, Indiana, and Ohio. The ice advance of the *Illinoian glaciation* followed the *Yarmouthian interglaciation* and reached a maximum southward limit in Illinois, Indiana, and Ohio. There followed the *Sangamonian interglaciation*. The final advance, that of the *Wisconsinan glaciation* reached limits beyond earlier glaciations only in the Dakotas and Nebraska in the central United States. Notice on Figure 17.30 a remarkable area in southwestern Wisconsin that completely escaped overriding by glacial ice. This *Driftless Area* seems to have been protected by highlands to the north, which caused the ice to diverge around it in lateral lobes. The Driftless Area is interesting because it preserves soils and landforms of pre-Pleistocene age.

In the eastern United States, Wisconsinan ice spread beyond all earlier advances along the coast eastward from New York City and far out upon the Continental Shelf. In Europe, ice of the Wisconsinan glaciation (*Würm glaciation*) nowhere overspread the limits of earlier glaciations.

As to dating of glaciations and interglaciations in terms of years before present, we are still very much up in the air for events older than Wisconsinan. Referring back to Figure 17.18, evidence of oxygen-isotope ratios of the Greenland ice shows a change from warm conditions of the Sangamonian interglaciation to cold conditions of the Wisconsinan glaciation occurring in the time range of $-70,000$ to $-80,000$ years. This information correlates rather well with a value of $-75,000$ years assigned by many geologists to the earliest ice advances of the Wisconsinan glaciation into the middle-western states. For older events, the wide disparity of chronologies worked out by different investigators makes it unwise to even attempt to give figures in years. All are agreed on a rapid world-wide close of the Wisconsinan glaciation by deglaciation starting 10,000 to 12,000 years ago, with rapid warming and ensuing ice-sheet recession.

History of the Great Lakes

The Great Lakes of the United States and Canada are a natural phenomenon unique on the globe and at the same time dominating the environment of a large industrialized and densely populated region of both nations (Figure 17.31). While we don't need to go through the history of these lakes in detail, a brief sketch of their evolution

Glacier Systems and the Pleistocene Epoch

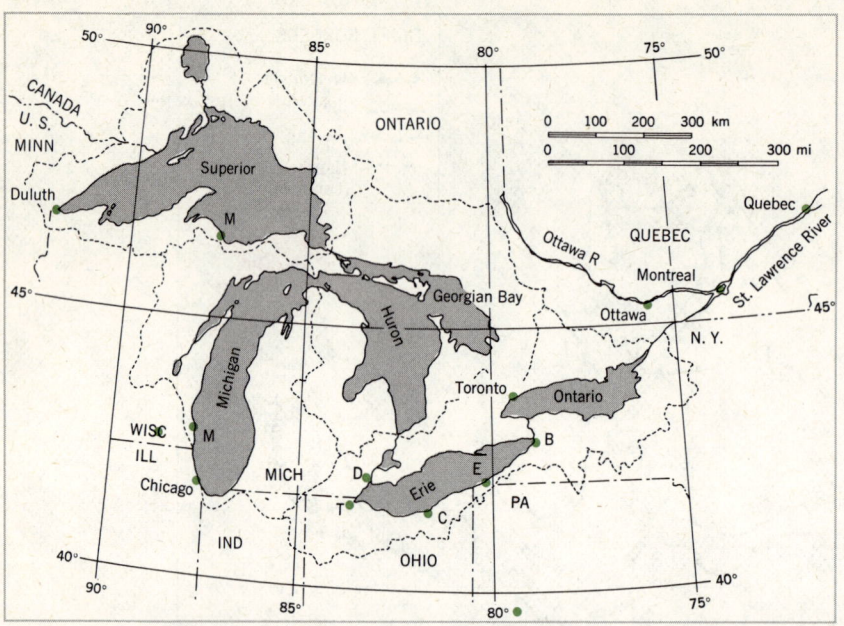

Figure 17.31 Outline map of the Great Lakes region. (Data of U.S. Lake Survey. (From A. N. Strahler, 1971, *The Earth Sciences*, 2nd ed., Harper & Row, New York.)

will give a broader understanding of the environmental problems they now present.

Ranked in order of surface area, the five lakes stand as follows:

	sq mi	sq km
Superior (2)	31,800	82,400
Huron (5)	23,000	59,600
Michigan (6)	22,400	58,000
Erie (12)	9,900	26,000
Ontario (14)	7,500	19,400

Figures in parentheses show ranking among all world lakes in area. Only the Caspian Sea is larger than Superior (by a factor of five). Figure 17.32 is a schematic profile and cross section of the lakes giving surface elevations and depths. Four of the five lakes have bottom depths well below sea level. In contrast, Erie is shallow, a factor that has been important in allowing its advanced pollution and eutrophication.

In preglacial time there existed lowlands where the lakes now stand. These lowlands were occupied by major streams and had been gradually opened out in zones of weaker sedimentary rocks by fluvial denudation. Little is known of pre-Wisconsinan history of the lake area, but the repeated ice advances of the four glaciations had extensively and deeply eroded weak rocks of the former lowlands and had carried the debris south

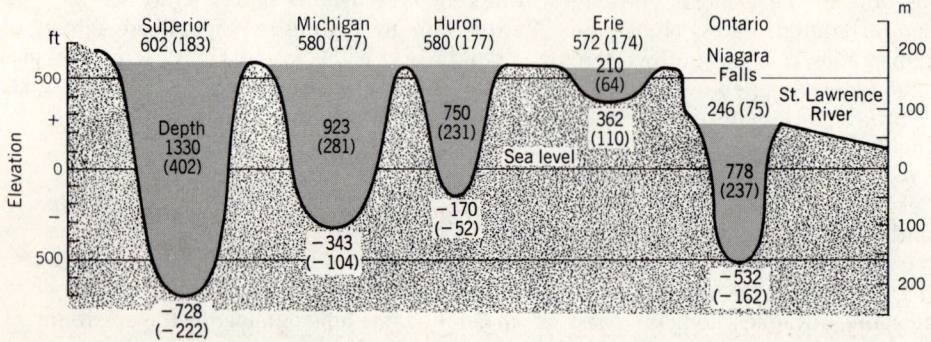

Figure 17.32 Generalized profile and cross section of the Great Lakes. (Data of U.S. Army Corps of Engineers. From A. N. Strahler, 1971, *The Earth Sciences*, 2nd ed., Harper & Row, New York.)

History of the Great Lakes

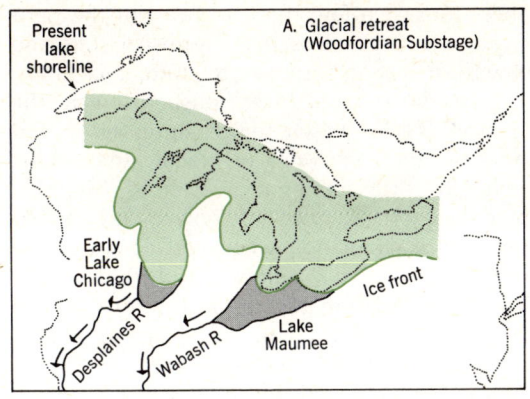
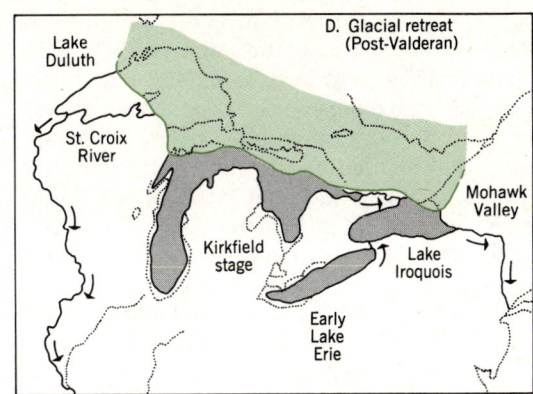
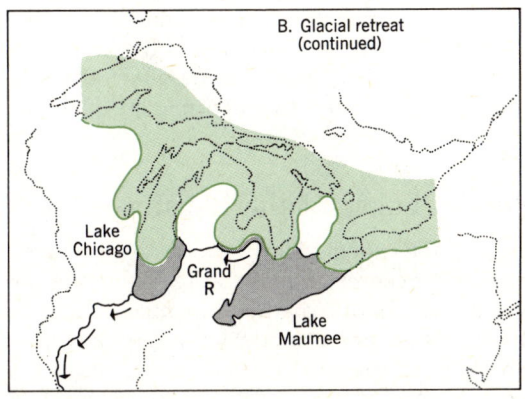
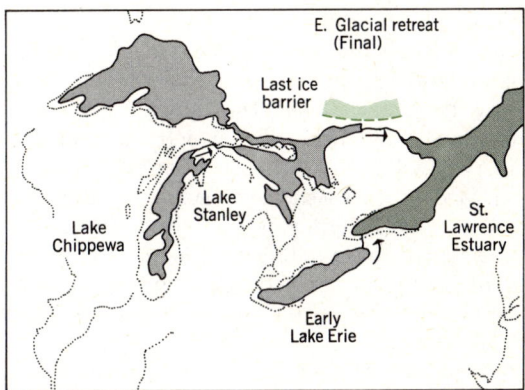
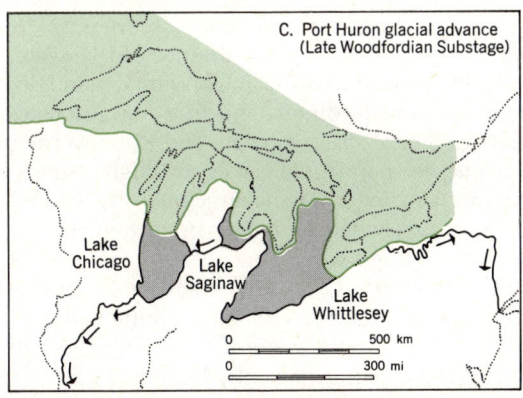
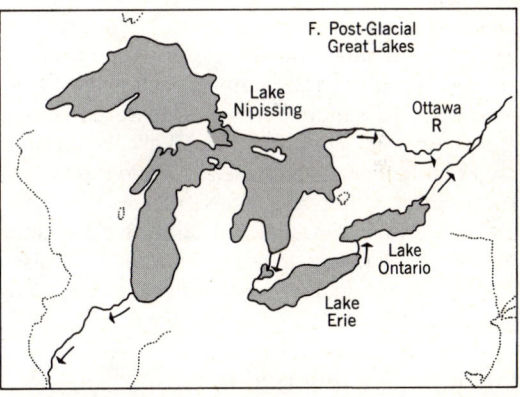

Figure 17.33   Six stages in evolution of the Great Lakes. (After J. L. Hough, 1958, *Geology of the Great Lakes*, Univ. of Illinois Press, Urbana, Illinois, pp. 284–296, Figures 54, 56, 60, 69, 73, and 74. From A. N. Strahler, 1971, *The Earth Sciences*, 2nd ed., Harper & Row, New York.)

to form the great moraine systems bordering the lakes on the south (Figure 17.24).

Stages in Great Lakes evolution are shown in Figure 17.33. These maps are simplified and only a few representative stages are shown. We intend that you grasp the nature of the evolutionary changes, rather than try to name all of the features and remember them in order. Map *A* shows the earliest lakes beginning to form as the ice front receded. Lakes Chicago and Maumee were marginal glacial lakes, ponded between the ice front and higher, moraine-enclosed ground to the south. Both lakes overflowed southward by streams draining into the Mississippi River system. Map *B* shows continued retreat and the diversion of one lake into another by a marginal stream following the ice front. Map *C* catches the action at a point when drainage was established eastward along the ice front to enter the Hudson River system, by way of the Mohawk valley. In map D, final ice recession was under way and part of ancestral Lake Superior had opened up, draining south into the Mississippi system, while the other lakes drained east to the Hudson system.

An important complicating factor, which we do not have space to explain here, was that while the ice sheet was in its recessional phase, with occasional pulses of readvance, the crust beneath the area was rising because of unloading of ice weight, following the principle of isostasy. Crustal rise was progressively greater toward the north, and produced a southward tilt, much as if one were to raise one edge of a table so as to tilt all objects resting on the table.

Map E shows the very last of the ice disappearing in Ontario and opening an outlet along what is now the Ottawa River. This outlet led directly into the St. Lawrence valley, which was then an estuary of salt water, because of crustal depression. Map F shows a stage of maximum extent of the Great Lakes. Further crustal tilting caused the Ottawa River outlet to be abandoned, and lowering of lake levels caused abandonment of the drainage of Lake Michigan into the Mississippi system.

One result of this complex history of changing lake levels and areal extents is that today there are broad marginal zones of lake plains along the shores of Lakes Michigan, Huron, Erie, and Ontario. These plains have successions of shoreline features, including low, sandy beach ridges, but with large areas underlain by fine-textured lacustrine sediments. They are intensively developed as agricultural lands and have absorbed the urban expansion of major lake cities such as Chicago, Toledo, Detroit, Cleveland, Toronto, Buffalo, and Erie. Niagara Falls, discussed in Chapter 14, is also a salient natural feature formed during evolution of the Great Lakes.

Periglacial Pleistocene Environments

The southward spread of ice sheets brought with it an advance zone of frost-controlled climate similar to that found today in coastal lands bordering the Greenland Ice Sheet, on arctic islands of northern Canada, and along the Alaskan arctic coast. The term *periglacial* describes this environmental zone, which would have had a tundra landscape with some areas of barren soils and, elsewhere, a tundra plant cover. Ground ice formed ice-wedges and frost polygons, while spontaneous flowage of saturated sediments occurred in summers as mudflows and solifluction lobes. Today we find these features of the periglacial environment as relict forms in soils and alluvium in a zone bordering the former ice limits.

Other climate zones were pushed toward the equator, as lowered air temperatures set in and shifts occurred in zones of global atmospheric circulation. The upper-air westerlies extended their zone of influence into lower latitudes, as Hadley cell circulation weakened and the subtropical high-pressure belt migrated equatorward. With this migration the tropical deserts moved into lower latitudes and may have produced desert conditions as far equatorward as 10° to 15° latitude. With lowered atmospheric temperatures, it seems likely that resultant reduction in atmospheric moisture would have reduced convectional activity and precipitation in the equatorial belt. All of these climatic shifts affected plants and animals, forcing them to migrate with the shifting climate zones in order to survive.

Pluvial Lakes of the Pleistocene

Among the nonglacial phenomena associated with Pleistocene glaciations were changes in the water balances of closed intermontane basins of the southwestern United States, largely within what is today the arid Basin-and-Range region. As we explained in Chapter 13, on the subject of water balance of lakes, the great excess of potential evaporation over precipitation in hot, dry basins results today in total absence of lake water in most of these basins, in occasional stands of shallow water in others, and in a few instances such as Great Salt Lake, Utah, and Pyramid Lake, Nevada, in permanent lakes of high salinity. Runoff reaching these lakes comes by way of streams receiving discharge from neighboring mountain ranges where a water surplus occurs at high elevations. Obviously, there is a delicate equilibrium among evaporation, influent runoff, and storage in the water-balance equation of those closed basins presently holding water.

During Pleistocene glaciations the water balance changed in favor of small net surpluses, with the result that water occupied a large number of the intermontane basins, bringing into existence a large number of *pluvial lakes*. The word *pluvial* suggests an increase in precipitation during glaciations as the cause of the lakes. Evaporation would also have been less under a regime of lower air temperatures. We know that alpine glaciers of certain of the neighboring higher ranges, such as the Wasatch Mountains and Sierra Nevada, made major advances during glaciations to reach low altitudes, showing the effects of greater net accumulation and reduced ablation.

Pluvial Lakes of the Pleistocene

Figure 17.34 is a map showing pluvial lakes as they were during maximum extent during the Wisconsinan glaciation. Altogether, there were about 120 pluvial lakes in existence then. Some overflowed into others and probably held fresh water at times. Largest of the pluvial lakes was *Lake Bonneville,* an expansion of the present-day Great Salt Lake in Utah. It reached an areal extent of 20,000 sq mi (52,000 sq km), about the same as Lake Michigan, and for a time overflowed northward into the Snake River. Its maximum depth was 1000 ft (330 m). Abandoned shorelines of Lake Bonneville can be seen today along the mountain slopes against which the lake waters rested.

Expansion, contraction, and changing salinities of the pluvial lakes constituted great swings in environmental conditions affecting ecosystems of the basins. A remarkable example of adaptation of animals to changing environments is seen in the case of the desert pupfish (*Cyprinodon*). There are today some twenty populations of these

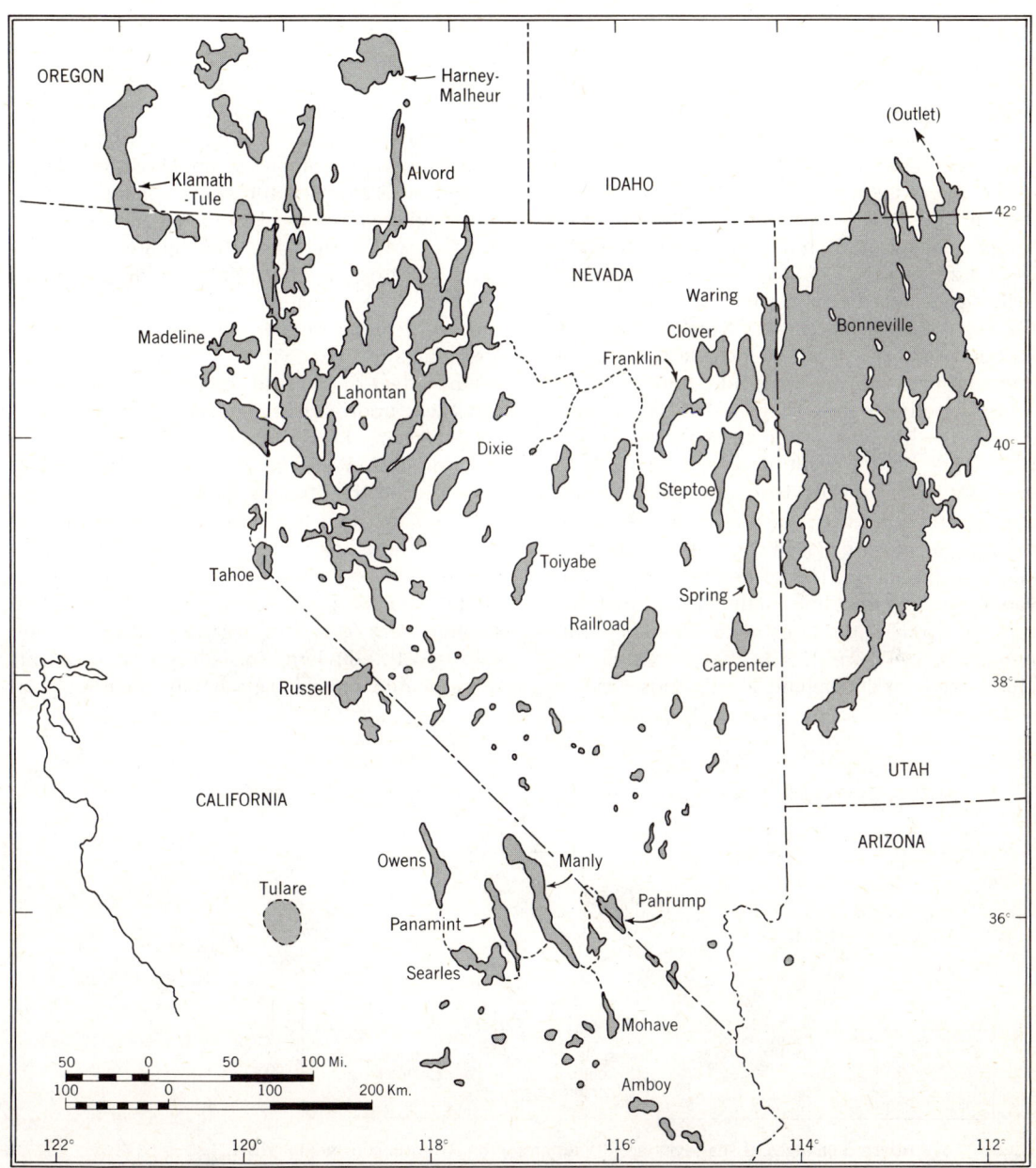

Figure 17.34 Pluvial lakes of the western United States. The dotted lines are overflow channels. (Based on a map by R. F. Flint, 1957, *Glacial and Pleistocene Geology*, John Wiley & Sons, New York, p. 227, Figure 13.2.)

tiny fish surviving in isolated spring-fed streams and tiny ponds in Death Valley, California. This tectonic basin, which lies below sea level and is one of the hottest surface environments on earth, was occupied by pluvial Lake Manly (Figure 17.34). As lake waters disappeared, the fish were forced into a few remaining spring localities and became isolated from one another. Their tolerance to a wide range of temperatures is quite phenomenal. Blue-green algae provide the fish with food.

Pleistocene Changes of Sea Level

At many points in earlier chapters we have referred to changes in sea level associated with glaciations. We know that the volume of ice held on Antarctica is such, that if it all melted, together with all other glacial ice, there would result a sea level rise of about 200 ft (60 m). Imagine the large coastal zones that would be inundated by such a change, if it were to occur in the near future. Major coastal urban centers of North America and Europe would be drowned and a large land area removed from agricultural production.

On the other hand, the growth of Pleistocene ice sheets withdrew vast quantities of water from the oceans, with a resulting decline in sea level. We now have good evidence on which to base a reconstruction of the sea level changes accompanying the closing glacial event within the Wisconsinan glaciation. This evidence comes from a variety of organic and sedimentary materials brought up from the floor of the Atlantic continental shelf (Figure 17.35). These materials include shells and coralline algae whose growth habitats are known to lie close to sea level. Salt-marsh peat is also a valuable indicator of sea level. The samples can be dated by the radiocarbon method. Plotting the present depths of these samples against age, a generalized curve of sea level can be drawn. About 35,000 years before present, sea level stood near its present position, for this was a time of ice recession within the Wisconsinan glaciation—a sort of mini-interglaciation (Farmdalian substage). As the final ice advance set in, sea level declined and reached a low point of about −400 ft (−125 m) at about −15,000 years. At this time a broad zone of the continental shelf was exposed and the shoreline lay some 60 to 125 mi (100 to 200 km) east of its present position. Remains of fresh-water plants show that this exposed shelf was a richly vegetated landscape; animal remains show that it supported land animals, such as elephants (mastodons and mammoths). Although glacial ice stood not far away at the time, the climate was not as severe as one might suppose, being essentially like that of the subarctic lands of Canada, and at times not much different than that of northern New England today.

One effect of lowered sea level during glaciations was upon streams entering the sea. All of these streams were extended in length to reach the more distant shoreline. Lowered stream baselevel caused the streams to degrade their channels and to carve trenches into the continental shelves. Trenching and valley widening also progressed far upstream into what is now the continental mainland. As sea level rose, the same streams were forced to aggrade and to fill their valleys with alluvium. This succession of events was repeated with each glaciation and interglacia-

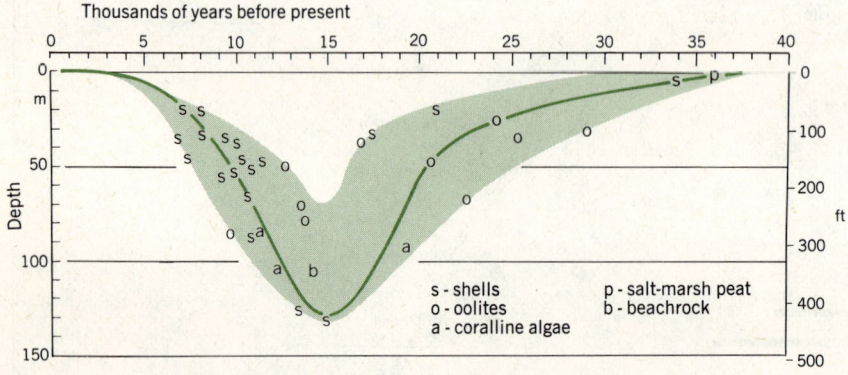

Figure 17.35 Inferred changes of sea level (*solid line*) along the Atlantic continental shelf, based on radiometric ages of samples of various kinds of materials. The shaded zone shows depth limits of samples. (Data of J. D. Millman and K. O. Emery, 1968, *Science*, vol. 162, p. 1122, Figure 1. From A. N. Strahler, 1971, *The Earth Sciences*, 2nd ed., Harper & Row, New York.)

Mammals of the Pleistocene

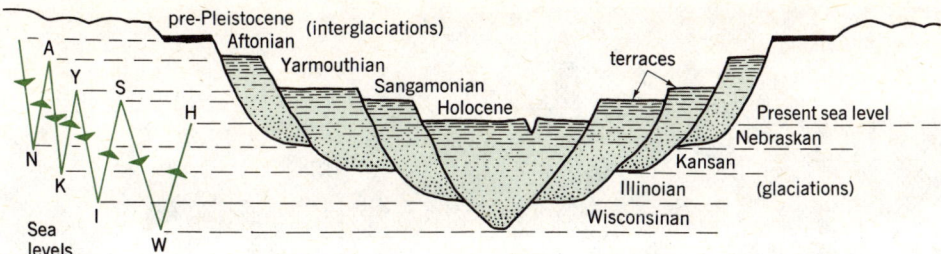

Figure 17.36 Schematic diagram of development of paired alluvial terraces during oscillations of sea level, superimposed on a general crustal rise. Actual deposits would be fragmentary. (From A. N. Strahler, 1971, *The Earth Sciences*, 2nd ed., Harper & Row, New York.)

tion, but with an added effect—that of gradually rising continental crust. The combined effect is illustrated schematically in Figure 17.36. Each successive trenching was carried to a lower level, but each filling ceased at a lower level than the previous filling. The result is a set of paired and nested alluvial terraces. Remember that this diagram is highly idealized and that you would not expect to find a complete succession of paired terraces in any one locality. Nevertheless, major streams draining the Atlantic and Gulf coasts of North America show at least partial development of terraces of this type. (Aggradation and degradation in headwater areas near the ice margin showed a different pattern of events, as explained in Chapter 14.)

Mammals of the Pleistocene

The plant life of the Pleistocene Epoch outwardly differed little in composition and appearance from that of the present for equivalent climates. Nevertheless, a visitor going back into even late Wisconsinan time would be startled to see species of mammals quite different in appearance from those found today in North America and Europe. Throughout the Old World, migrations were continuously possible during the entire Pleistocene Epoch between Africa and Eurasia, allowing African mammals to migrate into Europe during the mild interglaciations. Man, or *Homo sapiens*, was evolving rapidly in the Old World in constant association with the mammals. However, conditions were somewhat different in the New World, for Man had not yet reached the Americas by way of a land bridge that connected Siberia to Alaska in Pleistocene time. Moreover, South America had been separated from North America until late Pliocene time, so that different mammalian assemblages had evolved in the two continents. When a connection was established in the Pleistocene, a mixing of grossly unlike faunas took place. For example, the ground sloths and huge animals called glyptodonts reached North America and were quite unlike any mammals found in Eurasia.

Among the most interesting of Pleistocene mammals of North America were members of the elephant group—mastodons and mammoths (Figure 17.37). Woolly mammoths had arrived from Asia by way of the Bering Strait land bridge (Figure 17.38). They had massive, flat-crowned teeth adapted to grinding of coarse grasses. They lived, as did the muskox and woolly rhinoceras, in a rigorous periglacial climate near the ice sheet margin (Figure 17.38).

Other grazing mammals present in North America in the Pleistocene Epoch were the horse, bison, camel, and peccary (wild pig) (Figure 17.37). Carnivores were well represented by the saber-tooth tiger (Figure 17.39), a now-extinct species with a pair of enormous teeth used for stabbing the prey, and the dire wolf. Remains of these and other extinct Pleistocene mammals are found in abundance in asphalt pits at LaBrea, California.

From the environmental standpoint, it is most interesting to know that few mammalian extinctions in North America took place in the Pleistocene Epoch until near or just after the close of the Wisconsinan glaciation, about −10,000 years. By this time Man had arrived from Asia by way of a land connection in the Bering Straits and had made his way down to middle latitudes as the ice front receded and avenues of approach were opened up. Evidences of Man's hunting prowess are found in spear points lodged in remains of mastodons, and it has been postulated that the mastodons and mammoths were made extinct through hunting by early Man. A great increase in brain capacity marks the emergence of Man from common ancestry with the apes. His intelli-

Figure 17.37 Reconstruction of a summer scene along the Missouri River in late Pleistocene time. *Left page:* mastodon. *Right page:* royal bison, horse. (Painting by C. R. Knight, courtesy of the American Museum of Natural History.)

gent use of weapons, tools, and trapping devices gave him an important advantage over the other mammals, large and powerful as they were. If the hypothesis of mammalian extinction by early Man is valid, it is an ominous precursor to a long list of animal extinctions caused by Man as he has expanded his populations and increased the efficiency of means of killing.

Homo sapiens, the species that includes all living races of humans, appeared on the scene early in middle Pleistocene time, some 500,000 years ago. Space does not permit us to give details of the evolution of *Homo sapiens*. A number of different varieties of this species, formerly considered as separate species, appeared in Europe. Neanderthal Man, who inhabited Europe from about −100,000 to −40,000 years (late Sangamonian and early Wisconsinan) was heavy-boned, short, and stocky (Figure 17.40). He was able to make good stone tools, such as axes, scrapers, and points, and was a capable hunter. Neanderthal Man is associated with the Middle Paleolithic culture, or "Old Stone Age."

At about −35,000 years, Neanderthal Man was replaced in Europe by a more advanced race of *Homo sapiens*, called Cro-Magnon Man. Cro-Magnon Man had a high forehead and small lower jaw, but a prominent chin, in contrast with the low forehead and massive jutting jaws of Neanderthal Man. Cro-Magnon Man stood erect, his skull resting directly over the top of the spine. Although he belonged to the final stages of the Paleolithic culture, the quality of his finely-chipped stone implements was very high and, moreover, he was able to shape ivory into tools, weapons, and ornaments. Paleolithic culture was

Figure 17.38 Restoration of a winter scene in Europe in Pleistocene time showing the woolly mammoth (*left*) and the woolly rhinoceras (*right*). (From a painting by C. R. Knight, courtesy of the Field Museum of Natural History.)

Holocene Environments

succeeded by the *Neolithic culture*, starting at about −10,000 years in the Near East. Newer races of this culture learned to make pottery and to domesticate animals, and then turned to agriculture. The Age of Metals began about −5000 years (3000 B.C.), and soon thereafter came the dawn of recorded history.

Our comprehension of Man's interaction with his environment would be greatly increased by a study of those phases of human history dealing with agricultural expansion and land use generally, with population growth and migrations, and with exploitation of mineral resources. Unfortunately, we don't have space in this book to review even the highlights of Man's history from primitive cultures of early Holocene time to the industrial societies of today.

Holocene Environments

With onset of the Holocene Epoch, about −10,000 years, rapid warming of ocean surface temperatures set in and continental climate zones shifted rapidly poleward. Soil-forming processes began to act upon new parent matter of glacial deposits in middle latitudes. Plants became reestablished in glaciated areas in a succession of climate stages. First of these was known as the *Boreal stage*. Boreal refers to the present subarctic

Figure 17.39 Reconstruction of the saber-tooth tiger (*Smilodon*) from remains found in the La Brea, California, asphalt pits. The specimen is about 3 ft (1 m) high. The lower jaws could open more widely, allowing the upper teeth to be used for stabbing prey. (Sketched from a photograph prepared by Carl O. Dunbar from models by R. S. Lull in the Yale Peabody Museum.)

Figure 17.40 A reconstruction of Neanderthal Man. (Courtesy of the Field Museum of Natural History.)

regions where needle-leaf forests dominate the vegetation. The history of climate and vegetation throughout the Holocene time has been interpreted through a study of spores and pollens found in layered order from bottom to top in postglacial bogs. This study is called *palynology*. Plant species can be identified and ages of samples can be determined by the radiocarbon method. Interpretation of pollens indicates that the Boreal stage in middle latitude had a vegetation similar to that now found in the subarctic zone; a dominant tree was spruce.

There followed a general warming of climate until the *Atlantic* climatic stage was reached about −8000 years. Lasting for about 3000 years, the Atlantic stage had average air temperatures somewhat higher than those of today—perhaps on the order of 4.5 F° (2.5 C°) higher. We call such a period a *climatic optimum* with reference to the middle-latitude zone of North America and Europe. There followed a period of temperatures below average, the *Subboreal* climatic stage, in which alpine glaciers showed a period of readvance. In this stage, which spanned the age range −5000 to −2000 years, sea level had reached a position close to that of the present, and coastal submergence of the continents was largely completed.

The past 2000 years, from the time of Christ to the present, shows climatic cycles on a finer scale than those we have described as Holocene climatic stages. This refinement in detail of climatic fluctuations is a consequence of the availability of historical records and of more detailed evidence generally. Referring back to the record of temperature fluctuations of the past 800 years derived from ice layers of the Greenland Ice Sheet, Figure 6.21, you will see that the larger temperature departures have been of about the same order of magnitude during the entire period. Particularly interesting is a colder period of the early nineteenth century. Evidence from glaciers shows that there was a "Little Ice Age," beginning in the eighteenth century and persisting until near the close of the nineteenth century. During this time valley glaciers made new advances to lower levels. In the process the ice overrode forests and so left a mark of its maximum extent.

The Lesson of Pleistocene Environmental Change

Perhaps the salient concept to emerge from a study of the glaciations and interglaciations of the Pleistocene Epoch is that environmental changes of a large order of magnitude occurred on a global scale with no causation by Man. Even the climatic cycles of the Holocene Epoch, which are minor in comparison with those of the Pleistocene, span a range of average air temperatures much larger than the changes recorded since the combustion of fossil fuels began in large quantities about 1900. In view of this evidence, are we really in a strong position to associate global atmospheric changes of the past half century with Man's activities in a cause-and-effect relationship? In the judgment of many investigators, it is far too soon to assess the true extent of Man's impact on climate. This uncertainty should not lead us to ignore the long-range effect of heat and pollutants we are now injecting into the atmosphere at an increasing rate. Rather, the uncertainty tells us that sustained research with greatly improved monitoring facilities must be carried on for a long time to come.

The realities of environmental degradation and looming shortages of natural resources are nevertheless all too obvious to ignore. Air and water pollution, soil erosion and sedimentation, and scarification of the land are severe measurable environmental changes caused by Man. The review of basic principles of environmental science will serve its purpose if it trains us to separate fact from hypothesis and to evaluate each new piece of evidence with the caution it deserves.

Our chapters on environmental geoscience have stressed the physical processes and products of the life layer, with only brief references to the biological processes and products of those same environments. Yet organisms play a major role in the cycling and recycling of energy and matter between themselves and their inorganic surroundings within the earth's varied ecosystems. The remaining chapters of this book investigate organic processes and the pathways of energy and matter of the biosphere, so that we can complete our study of environmental science.

PART IV

Energy Systems of the Biosphere

Past chapters have dealt with the energy systems of the fluid and solid earth layers. We have seen that all energy systems have certain components and processes in common: energy inputs, energy storage and transformations, and energy outputs, as well as associated transport and storage of matter. Although living systems are considerably more complex than the physical systems which have been examined thus far, the interactions of the life layer can be placed in a similar framework. This framework is drawn from the science of *ecology*, broadly defined as the study of organism-environment relationships. The *ecosystem*, the basic unit of ecological study, constitutes the sum of all organisms, energy flows, and inorganic substances and structures affecting organisms within a set of boundaries. Like other energy systems, the ecosystem is delicately adjusted and balanced, and resembles an open system in the equilibrium state.

Ecosystems are somewhat different from physical energy systems in that the pathways of energy transport and transformation which characterize them are very delicate and sensitive, since they depend for their nature on life processes. Thus, while the global pattern of energy distribution which produces our weather and climate requires vast amounts of energy or material input for its modification by Man, we find that Man can disrupt ecosystems with little input of materials or energy. The accumulation of toxic quantities of DDT by organisms from trace concentrations within their environments is an example of how a very small amount of material input can have widespread effects on an ecosystem. In part, such difficulties arise because Man's knowledge of the movements of material and energy within ecosystems is so fragmentary, a result of the great complexity of biological systems in general.

Although we will begin this chapter by introducing basic concepts about ecosystems, our main topic will be a detailed survey of the organic world as seen by the biologist, from cell to organism. The next chapter, Chapter 19, deals with how organisms become adapted to their environments through natural selection and will review the great impact of life on the earth's surface through geologic time. Chapters 20 and 21 present

18

Life on Earth

the ecosystem as a biological energy system and show how energy and matter move through the ecosystem under the influence of both physical and biological processes. In Chapter 22, a new dimension is added: time. Here we deal with how populations and communities change in response to such biotic pressures as competition between species, or predation of one species by another, or to pressures created by a changing physical environment. The final chapters of Part IV survey aquatic and terrestrial ecosystems and include discussions of Man's impact on a wide range of global environments.

The Ecosystem and the Food Web

As an example of an ecosystem, consider a salt marsh (Figure 18.1). A variety of organisms are present: algae and aquatic plants, microorganisms, insects, snails, and crayfish, as well as larger organisms such as fishes, birds, shrews, mice, and rats. Inorganic components will be found as well: water, air, clay particles and organic sediment, inorganic nutrients, trace elements, and light energy. Energy transformations in the ecosystem occur by means of a *food chain,* or *food web*. Photosynthetic organisms (the plants and algae of Figure 18.1) are the *primary producers,* using light energy to convert carbon dioxide and water to carbohydrates (long chains of sugar molecules) and eventually to other biochemical molecules needed for the support of life. These organisms form the base of the food web. At the next level are the *primary consumers* (the snails, insects, and fishes) who live by feeding on the producers, metabolizing their chemical constituents in the process of *respiration*. At a still higher level are the *secondary consumers* (the mammals and birds), who feed on the primary consumers. In many ecosystems, still higher levels of feeding are evident (the marsh hawks and owls). Feeding on dead organisms from all levels are the *decomposers,* which are mostly microorganisms and bacteria feeding on detritus (decaying organic matter).

Thus, the food web is really an energy flow diagram, tracing the path of solar energy through the ecosystem. Solar energy is stored by photosynthetic organisms in the chemical products of photosynthesis. As these organisms are consumed, chemical energy is released in metabolism and used to power new biochemical reactions, which again produce stored chemical energy in the bodies of the consumers. At each transformation, however, energy is lost as waste heat. In addition, much of the energy input to each organism must be used in respiration simply for bodily maintenance and cannot be stored for use by other organisms higher up in the food web. This means that generally both the numbers of organisms and their total amount of living tissue, or *biomass,* must decrease drastically as one goes up the food chain. The concept of the ecosystem as an energy system will be more fully discussed in Chapter 20.

Because feeding, or obtaining the necessary energy to support life, is so important in the structure of ecosystems, ecologists have developed a set of terms to describe the feeding patterns of organisms, shown in the table below. The *autotrophic* (self-feeding) organisms depend on external inorganic energy sources, utilizing light or stored chemical energy to support their life processes. Technically, organisms which photosynthesize should be termed *photoautotrophic,* but most ecologists simply refer to them as *autotrophs* because *chemoautotrophs* (users of energy stored in inorganic compounds) are rare. Unless *chemoautotrophic* is specifically referred to, we will use autotroph to refer to any photosynthetic organism. The *heterotrophic* (diverse-feeding) organisms depend on other organisms for their energy source. Since heterotrophic nutrition simply rearranges existing food supplies, it

Autotrophic (self-feeding)	Photoautotrophic	Uses light energy through photosynthesis
	Chemoautotrophic	Uses energy stored in inorganic compounds as an energy source
Heterotrophic (diverse-feeding)	Holotrophic	Ingests bulk food
	Parasitic	Lives in or on the host, absorbing nutrients directly or feeding on host tissues
	Saprotrophic	Obtains energy from dead and decaying organisms

The Diversity of Life on Earth

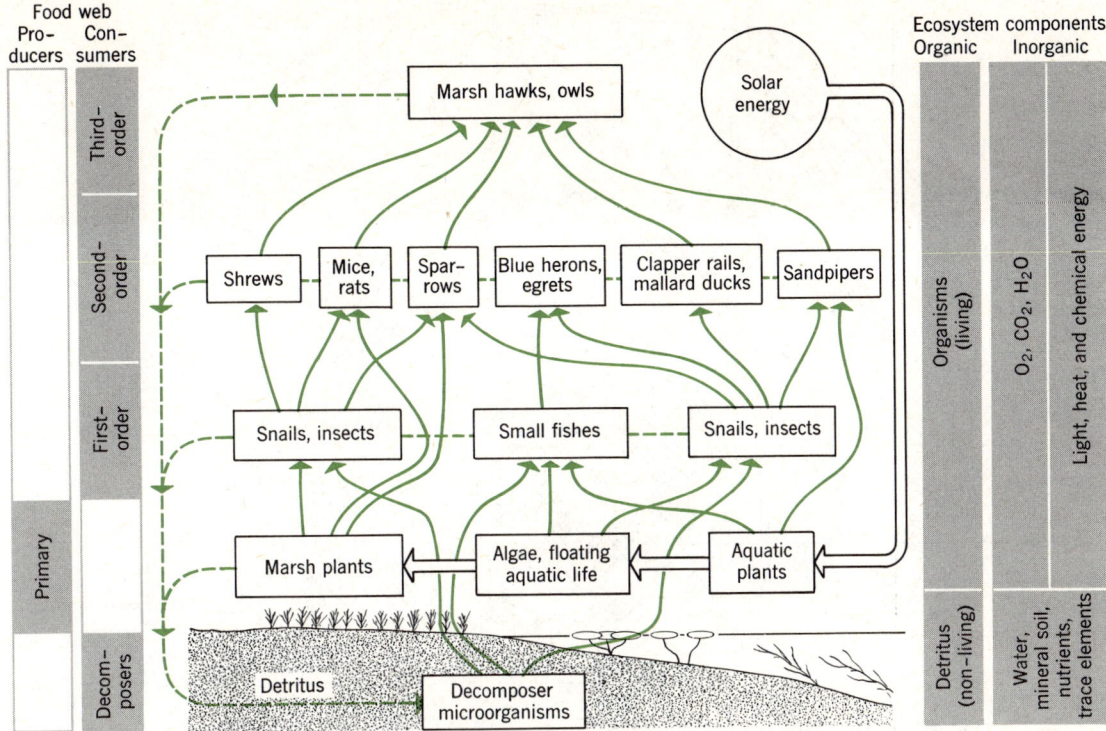

Figure 18.1 Diagram of a salt marsh ecosystem in winter. Arrows show how energy flows from the sun to producers, consumers, and decomposers. (Food web after R. L. Smith, 1966, *Ecology and Field Biology*, Harper & Row, New York, p. 30, Figure 3.1.)

is limited by the amount of autotrophic production. In most ecosystems, the autotrophs are plants while the heterotrophs are animals.

The Diversity of Life on Earth

The basic principles of ecosystems and food webs given in the previous section are necessary for our next topic: the organisms themselves, which serve as the organic components of the ecosystem and the links in the food chain. An astonishing number of organisms exist on earth, each adapted to the environment around it. So far, about 37,500 species of microorganisms, 341,000 species of plants, and 2,158,000 species of animals, including some 800,000 insect species, have been described and identified. In spite of several centuries of study, many organisms remain unclassified. Estimates suggest that species of plants will probably total to about 540,000, of which nearly two-thirds have been identified. Of some 2,000,000 estimated insect species, slightly more than one-third have been described and classified.

The realm of life on earth is usually divided into two parts: *plants*, which are autotrophic and immobile or passively mobile, and *animals*, which are heterotrophic and mobile. Exceptions to this quick and easy subdivision, however, are many. Some forms of unicellular life possess the biochemical mechanisms for photosynthesis, are mobile, and will ingest organic particles, thus behaving as animals as well. Still other forms of unicellular organisms are chemoautotrophs, and are thus independent of solar energy. A third example are the saprotrophic fungi, which are usually classified as plants, yet do not carry out photosynthesis. Mature sponges feed holotrophically, yet are fixed in position like plants.

These exceptions, and others like them, have lead many biologists to reject the simple plant-animal breakdown. Instead, most recognize four great groups of organisms: *Monera, Protista, Metaphyta,* and *Metazoa*. These groups are further subdivided into *phyla* (singular, *phylum*), which represent major evolutionary groups descended from a common ancestral stock. In classifying the plants, botanists have dropped the term *phylum* and use *division* as its approximate equivalent. The great groups, divisions and phyla, along with some environmental characteristics, examples and comments, are shown in Table 18.1. Before we can describe the classification, however, a quick review of the cell and its parts will be necessary.

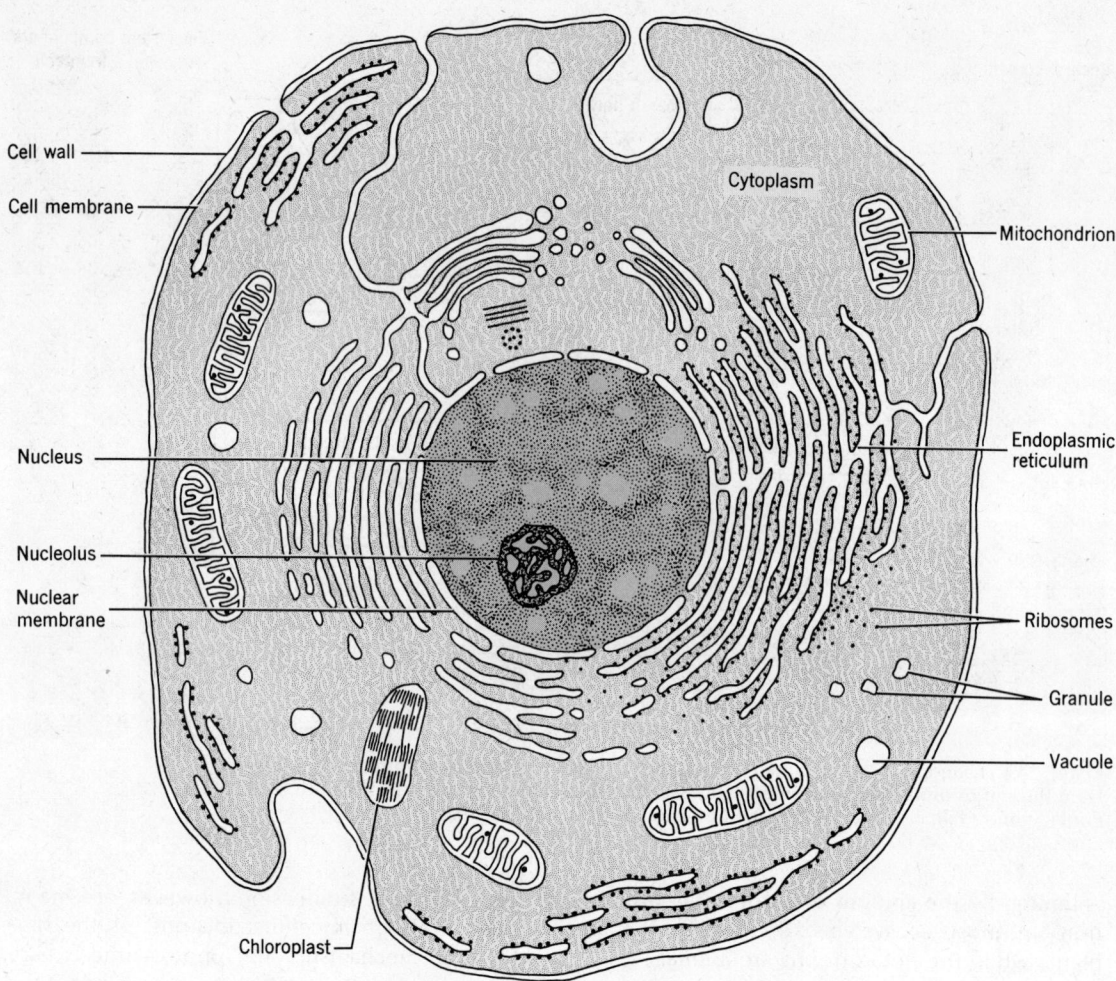

Figure 18.2 Parts of the cell. The composite diagram shows some of the more important cellular structures of different types of cells. For clarity, structures are shown only on a portion of the cell in the diagram; most structures, for example, the endoplasmic reticulum and ribosomes, are actually present throughout the cytoplasm.

The Cell

Almost all forms of life on earth have one thing in common: the *cell* (Figure 18.2). The cell consists of a *cell membrane*, enclosing the *cytoplasm*, and a *nuclear membrane*, which encloses the *nucleus*. In addition, plant cells possess a supporting *cell wall*, usually built of cellulose, which is formed by the linking of many sugar molecules into chains, or *polymers*. The cell wall is synthesized on the outside of the cell membrane once cellular enlargement has ceased. Within the cytoplasm are a number of *organelles* (literally, "little organs") which have specific purposes in the cell's biochemical machinery. *Mitochondria* (singular, *mitochondrion*) (Figure 18.3) are capsules of membranes on whose surfaces are located enzymes and other biochemical molecules. *Enzymes* are large biochemical molecules (macromolecules) which catalyze cellular reactions. The mitochondrion is believed to be the site of *respiration*, the process of breaking down fuel molecules (usually sugars) into CO_2 and H_2O. Energy obtained this way is stored in high-energy molecules of *ATP* (adenosine triphosphate) and other molecules which act as energy sources for most other cellular activities. (We shall return to a more detailed discussion of respiration in Chapter 20.

Another important organelle in plant cells is the *chloroplast* (Figure 18.4). Like the mitochondrion, the chloroplast is a membrane package, but is many times larger and is quite different in internal structure. Several different types of chlorophyll are found in the chloroplast, along with other

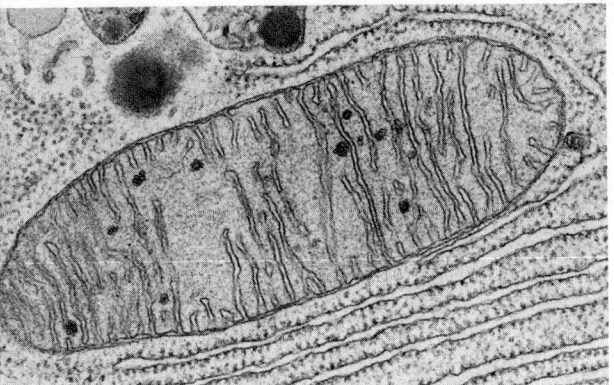

Figure 18.3 A mitochondrion from a bat pancreas cell. The linear structures within the mitochondrion are the membranes on which the enzymes are located. Magnification near 40,000×. (Courtesy of Dr. Keith R. Porter.)

enzymes which participate in *photosynthesis*, the construction of food molecules from carbon dioxide and water with an input of light energy. (photosynthesis will also be discussed in Chapter 20.)

The *endoplasmic reticulum* is a network of membranes, present throughout the cell, which channels the movement of biochemical molecules and also serves to support and partition the cell and its components. The *ribosomes*, which resemble granules under the electron microscope, are attached to the endoplasmic reticulum, and function as part of the cell's protein synthesizing machinery. Some other structures, with which you may be familiar, are shown in Figure 18.2. These structures include *granules* of solid matter and *vacuoles* of liquid droplets, which serve as storage centers for food, wastes, or pigments.

Within the nucleus resides *DNA* (deoxyribonucleic acid) (Figure 19.1), the genetic material of the cell. Through a series of complicated biochemical pathways, which will be discussed in more detail in the following chapter, DNA controls the chemical composition of proteins and enzymes. Since enzymes control almost all aspects of biochemical activity, the DNA indirectly controls nearly all cellular functions. Chemical agents or radiation can disrupt the DNA molecule in the process of *mutation*, altering the behavior of the cell which contains it. DNA and its alteration by mutation are important topics in evolution, the subject of Chapter 19.

Viruses

In a position intermediate between living organisms and nonliving organic molecules are the *viruses* (Figure 18.5). These entities are particles

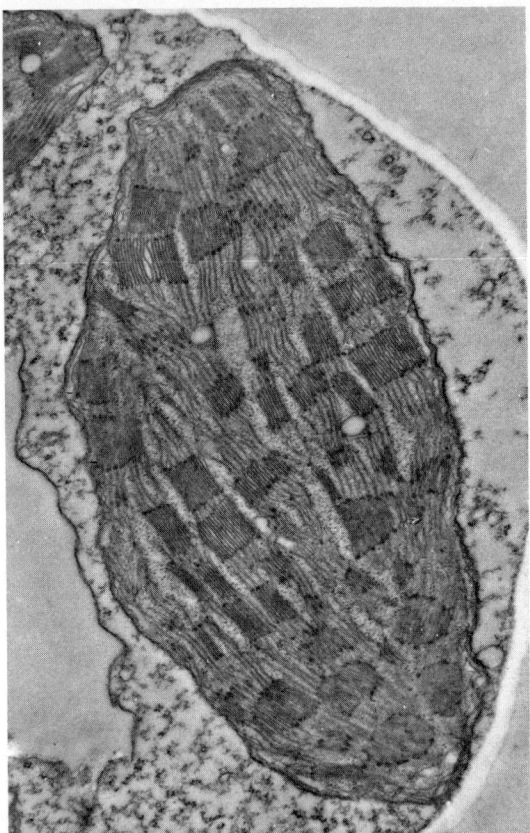

Figure 18.4 Electron micrograph of a chloroplast. Chlorophyll and associated molecules are located on the surfaces of the membrane stacks within the chloroplast. (Courtesy of Dr. A. E. Vatter.)

composed of nucleic acids (DNA and/or its relative, ribonucleic acid, RNA) surrounded by a coat of protein. When a virus particle comes in contact with a cell, its protein coat in some way opens the cell membrane and the nucleic acid contents of the viral body pass into the cell. The viral nucleic acids act in a fashion similar to the cell's own nucleic acids, and soon new proteins are synthesized to the specifications of the viral nucleic acids. These new compounds, however, subvert the cell's normal biochemical manufacturing apparatus, with the result that the cell begins to synthesize new virus particles. Deprived of much of its metabolic capabilities, the cell either dies or ruptures from the pressure of the many virus particles within it, releasing new virus particles to infect other cells. Existing outside of the cell, however, viruses are molecular compounds without life capabilities—they do not ingest organic foods, or photosynthesize, or reproduce. Therefore, virus particles are not living in the usual sense of the word.

Figure 18.5 Electron micrographs of virus particles. *Above:* The T5 bacteriophage showing polyhedral "heads" and short "tail" fibers. (Courtesy of Dr. Robley C. Williams.) *Right:* Crystal of tobacco necrosis virus, constructed of individual virus particles. (Courtesy of Dr. Ralph W. G. Wyckoff.)

Monera

The Monera (Table 18.1) include the most primitive of organisms. Members of the Monera are basically unicellular, although they may group together in *colonies*. The Monera lack a nuclear membrane, endoplasmic reticulum, chloroplasts, and vacuoles. The two phyla included in the Monera are Schizomycota (*bacteria*) and Cyanophyta (*blue-green algae*). The bacteria, the smallest of organisms, are commonly in one of three shapes: spherical, rod-shaped, or spiralled. Bacteria are found in almost all environments and utilize almost any energy source. Bacteria include photoautotrophs, chemoautotrophs, and heterotrophs, as well as saprotrophic and parasitic forms. Bacteria which are parasitic (derive nourishment from a living organism) and cause disease are *pathogenic*. An example is *Pneumococcus*, which causes pneumonia. Still other bacteria have the ability to use nitrogen (N_2) from the atmosphere. These bacteria, called *nitrogen-fixing* bacteria, perform the valuable service of converting atmospheric nitrogen, which is relatively inert, to forms usable by plants. These forms range from ammonium compounds (containing NH_4^+ to nitrites and nitrates (NO_2^- and NO_3^-). Still other bacteria utilize sulfur or other elements instead of oxygen in respiration, and are therefore chemotrophic. Nitrogen- and sulfur-using bacteria play important roles in the nitrogen and sulfur cycles, and we shall return to them in Chapter 21. Although photosynthetic bacteria do not contain chloroplasts, they do contain chlorophyll, in the form of *bacteriochlorophyll*, which is dispersed throughout the cell as tiny granules.

The blue-green algae contain another form of chlorophyll, *chlorophyll A*, which is also found in higher plants. These organisms contain other pigments which are believed to aid in cellular utilization of light energy, and which give blue-green algae colors from black to blue, green, yellow, and red. Blue-green algae are found in almost all aquatic and moist terrestrial environments. Some species have become accustomed to extreme temperatures; blue-green algae are found on the surfaces of glaciers as well as in thermal springs, where the water temperature may be close to the boiling point. Many members of the blue-green algae can fix atmosphere nitrogen in a fashion similar to bacteria.

Protista

The Protista (common name, *protists*) are a group of organisms with cellular characteristics more like their advanced relatives, the true plants and animals, but which lack much intercellular organization and specialization. Unlike the Monera, the Protista possess a more or less complete set of cellular parts, including nuclei, nuclear membranes, and organelles. Protists have adapted to their environment in four basic patterns (Figure 18.6). A protist in the *flagellate* state is endowed with a *flagellum*, (plural *flagella*), a threadlike strand of contractible fibers which

Protista

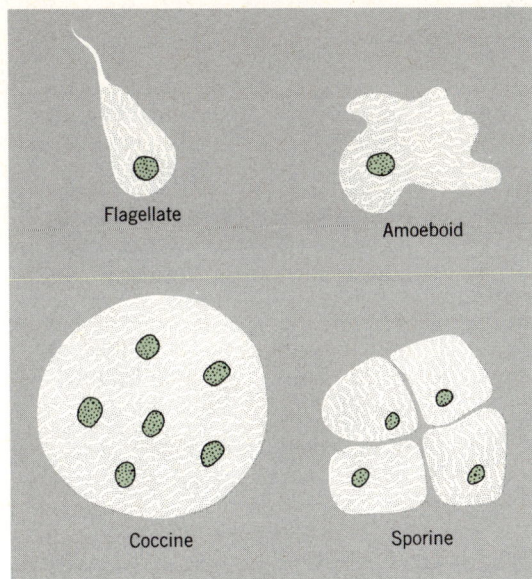

Figure 18.6 Four forms of protists.

provides mobility for food-seeking behavior. Protists in the *amoeboid* state move by extending *pseudopodia* (false feet), or projections of the cytoplasm, which then pull the rest of the cell along behind them. In many protists the two states are interconvertible. Motile protists are typically holotrophic (bulk feeding), enveloping and ingesting organic particles or smaller organisms as their energy source. Besides the amoeboid state, protists may exist in one of two other states, described as sporine and coccine. *Sporine* protists group together in colonies, forming strands, globs or sheets. Each colony is composed of many separate cells, each with its own cell membrane (and perhaps cell wall) and nucleus. *Coccine* protists also group together, but lack dividing cell membranes or walls. A coccine organism, therefore, consists of many nuclei inside one very large cell. Most coccine organisms, however, do form individual cells in reproduction, producing uninucleate spore cells which can later exhibit nuclear divisions of their own to form a new coccine organism.

The Protista can be subdivided into groups of divisions referred to by some more familiar terms: the algae, fungi, and protozoa. The *algae* are a large and important group of organisms, including seven groups of protists (Table 18.1). (The moneran division Cyanophyta is usually included in the term *algae*, although its members are much more primitive than those of the other phyla of algae.) Algae exhibit all protistan forms from unicellular flagellate and amoeboid types to more complex sporine and coccine types with stemlike and leaflike parts. Unicellular algae commonly have nonphotosynthetic counterparts which are otherwise almost identical but lack chloroplasts and are holotrophic. Some sporine types of brown and red algae are large organisms indeed. For example, kelp, a brown alga, sometimes attains a length of 200 ft (60 m) (Figure 18.7).

Figure 18.7 Two forms of algae. *Below*: Two species of marine algae. The large flat blades are of kelp, *Laminara sp.* The finer textured alga with small bladders within the blades is rockweed, *Fucus* sp. (R. H. N. Ailles.) *Right*: An algae bloom filling a small pond. (Grant Heilman.)

TABLE 18.1 Life on Earth

	Phylum or Division	Common Name	Number of Species	Environment	Size
Monera	Schizomycota	Bacteria	2,000	Aqueous and moist terrestrial	Unicellular to multicellular colonies
	Cyanophyta	Blue-green algae	2,500	Aqueous and moist terrestrial	Unicellular to multicullular colonies
Protista	Chlorophyta	Green algae	6,000	Aqueous and moist terrestrial	Unicellular to multicellular colonies
	Charophyta	Stoneworts	250	Fresh water	Small
	Euglenophyta	Euglenas	350	Aqueous	Unicellular
	Chrysophyta	Golden-brown algae	6,000	Aqueous	Unicellular to multicellular colonies
	Pyrrophyta	Fire algae	1,000	Aqueous	Mostly unicellular
	Phaeophyta	Brown algae	1,000	Mostly marine	Mostly multicellular; some very large
	Rhodophyta	Red algae	3,000	Marine	Mostly multicellular
	Phycomycota Ascomycota Basidiomycota	Fungi	200,000	Almost all terrestrial; a few aquatic	Small
	Myxomycota	Slime molds	500	Aquatic and moist terrestrial	Multicellular or multinuclear colonies
	Protozoa	Protozoa	30,000	Aquatic and moist terrestrial	Mostly unicellular
Metaphyta	Bryophyta	Mosses, liverworts	25,000	Moist terrestrial; a few fresh water	Small
	Tracheophytes[a]	Vascular plants	260,000	Aquatic and terrestrial	Small to large
Metazoa[b]	Porifera	Sponges	5,000	Aquatic	Small
	Cnidaria	Coelenterates	10,000	Mostly marine, some fresh water	Small (a few medium to large)

468

Food Source	Example	Comments
Diverse: autotrophic, heterotrophic	*Pneumococcus*	Present in almost all environments; parasitic forms can cause disease. Important ecological role as decomposers.
Photosynthetic		Large group with primitive cellular characteristics; present in wide range of environments. Ecologically important as reef-builders and nitrogen-fixers.
Photosynthetic	*Protococcus* (common on tree bark), *Spirogyra*, *Ulva* (sea lettuce)	Large, diverse group.
Photosynthetic		A small group of advanced algae adapted to a specific environment.
Autotrophic; a few are holotrophic	*Euglena*	Many of these organisms are both holotrophic and saprotrophic.
Autotrophic, holotrophic	Diatoms	One class, diatoms, are the most abundant single group of floating organisms; provide base of ocean food chain.
Autotrophic, holotrophic	Dinoflagellates—*Gymnodinium*	Dinoflagellate blooms may cause "red tides."
Photosynthetic	Rockweed, kelp	Some members provide an important food source.
Photosynthetic	Irish moss	Some members are of economic value for providing thickening agents, as in ice cream, toothpastes; some ecologically important as reef-builders.
Saprotrophic, symbiotic	Mushroom, puffball, truffle	Some are edible delicacies; others deadly poisonous.
Saprotrophic, holotrophic		Many of these organisms are little known.
Holotrophic, parasitic, some saprotrophic	Amoeba, *Trypanosoma* (sleeping sickness), *Plasmodium* (malaria), *Paramecium*	Includes many important disease-producing organisms.
Photosynthetic	*Sphagnum* (peat moss)	Play important role as colonizers in bogs, rocky places.
Photosynthetic, a few saprotrophic, parasitic	Ferns, conifers, flowering plants and trees, grasses	Most common plants; many used as food source or for wood.
Holotrophic		Natural sponges have economic value.
Holotrophic	Corals, jellyfish	Most reef-builders fall in this phylum.

Life on Earth

TABLE 18.1 (Continued)

	Phylum or Division	Common Name	Number of Species	Environment	Size
Metazoa[b]	Platyhelminthes	Flatworms	10,000	Aquatic, some terrestrial	Small to medium
	Aschelminthes	Sac worms	15,000	Aquatic and terrestrial	Small
	Ectoprocta	Moss animals	5,000	Aquatic	Multicellular, colonial
	Mollusca	Mollusks	50,000	Aquatic and terrestrial	Small to large
	Annelida	Segmented worms	15,000	Aquatic and terrestrial	Small
	Arthropoda	Insects and relatives	1,000,000	Aquatic and terrestrial	Small
	Echinodermata	Spiny-skinned animals	6,000	Marine	Small
	Chordata	Chordates	50,000	Aquatic and terrestrial	Small to large

SOURCES: Compiled from P. B. Weisz, 1969, *Elements of Biology*, 3rd ed., McGraw-Hill Book Co., New York, pp. 107–215; C. J. Alexopoulos, 1962, *Introductory Mycology*, 2nd ed., John Wiley & Sons, New York, pp. 34–35; G. H. Orians, 1969, *The Study of Life*, Allyn and Bacon, Boston, pp. 140–237; and H. C. Bold, 1965, *The Plant Kingdom*, Prentice-Hall, Englewood Cliffs, N.J., p. 3.

[a] Informal grouping, including nine divisions as shown in Table 18.2.
[b] Only phyla with substantial numbers of species are shown for the Metazoa.

Algae are an important food source for many organisms. As *primary producers*, they are at the bottom of the oceanic food chain and thus directly or indirectly support most higher forms of life in the seas. Under appropriate conditions, algae are very efficient producers indeed—tropical coral reefs where algae abound rank with the most productive environments known. The larger algae serve as a food source for Man. Much of the diet of the peoples of Japan, China, and the Pacific islands is supplemented by algae, which contain about 50% carbohydrate. Algae can also provide significant amounts of iodine, potassium, vitamins, and mineral salts.

Excessive amounts of algae can cause serious water quality problems. Where nutrients such as phosphates and nitrates are in abundant supply along with high light intensities and warm water, algae can multiply rapidly, producing a *bloom* (Figure 18.7). A bloom can have devastating effects on a water body through oxygen depletion. Although the algae produce oxygen through photosynthesis during the daylight hours, they consume oxygen in respiration at night, creating an oxygen demand on the water. An even greater oxygen demand is created by the microorganisms which decompose the remains of dead algae, increasing their numbers as their food source increases. As a result, the oxygen concentration in a water body experiencing an algae bloom may fall low enough to be lethal to fish and other forms of aquatic life. Once the kill begins, the situation is only aggravated, for the remains of the fish and other dead organisms merely add to the organic material available for decomposition. Undoubtedly, many algal blooms are related to Man's enrichment of nutrients in water bodies, a subject discussed in Chapter 23.

As noted in Table 18.1 the three *fungi* (divisions Phycomycota, Ascomycota, and Basidiomycota) are *saprotrophs*, consuming dead organic matter as their energy source. Most of the time, fungi exist in the form of *hyphae*, thin multicellular strands which interlace and intertwine throughout the food source. Under appropriate conditions, many hyphae grow together upward and produce a

Protista

Food Source	Example	Comments
Holotrophic, parasitic	*Planaria,* liver flukes, tapeworms	Includes parasites harmful to Man.
Holotrophic, parasitic	Rotifers, nematodes, trichina worms, hookworms	Nematodes are very numerous, second only to insects; includes harmful parasites.
Holotrophic	Bryozoans	Many with calcareous exoskeletons; living on seaweeds, wharf pilings.
Holotrophic	Snails, clams, squids	Many members economically important for food.
Holotrophic	Earthworms, leeches	Earthworms ecologically important as decomposers, soil mixers.
Holotrophic	Insects, centipedes, spiders, ticks, crustaceans	Largest phylum in numbers and species; marine arthropods important as food for Man; control of insect populations a major human problem.
Holotrophic	Starfish, sea urchins	Many are important reef predators.
Holotrophic	Vertebrates, fish, reptiles, birds, mammals, Man	Includes Man as well as most animals of importance to Man.

fruiting body which releases spores (Figure 18.8). This fruiting body is most familiar to us—examples are the mushroom and puffball—and is considered a gourmet delicacy. Although some forms of fungi are extensively cultivated and have

Figure 18.8 These mushrooms are the fruiting bodies of saprotrophic fungi. *Left:* The edible common morel, *Morchella esculenta.* (Jesse Lunger/National Audubon Society.) *Right:* The poisonous fly amanita, *Amanita muscaria.* (Leonard Lee Rue III/National Audubon Society.)

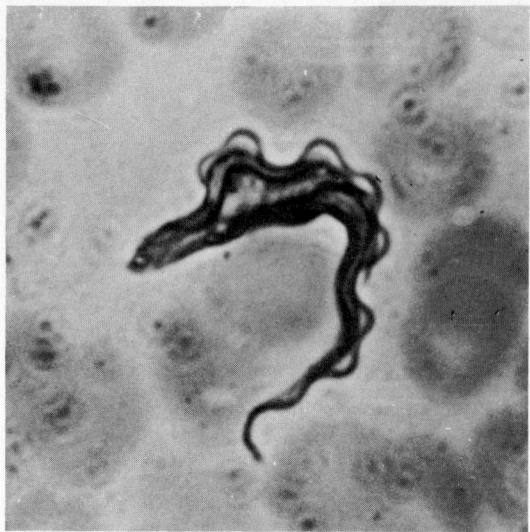

A. The flagellate protozoan, *Trypanosoma*. (Eric V. Gravé.)

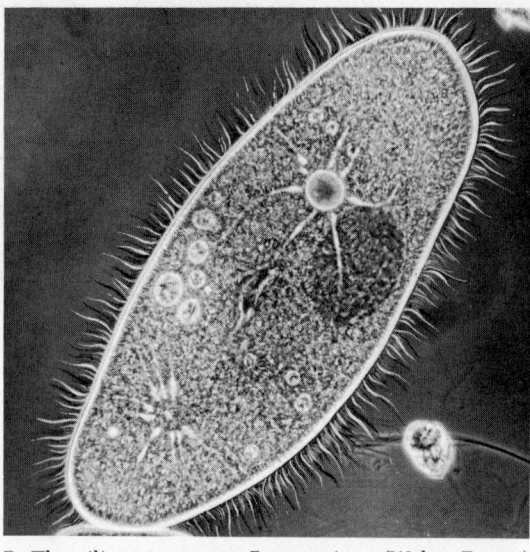

B. The ciliate protozoan *Paramecium*. (Walter Dawn.)

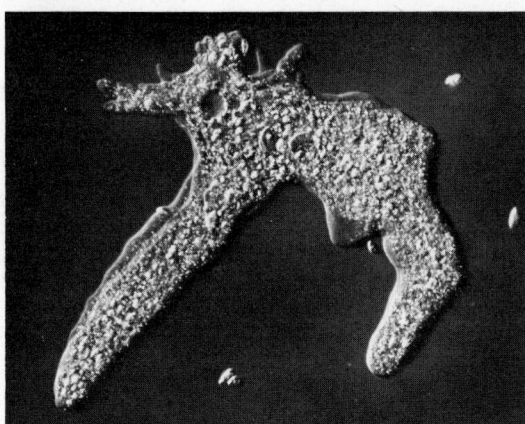

C. The amoeboid protozoan, *Amoeba proteus*. (Carolina Biological Supply Company.)

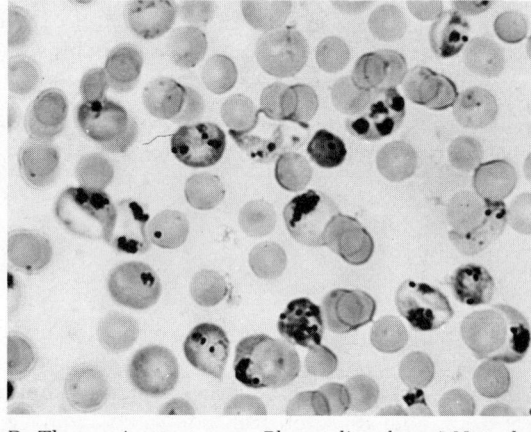

D. The coccine protozoan, *Plasmodium bergei*. Note the many multinucleate cells. (Carolina Biological Supply Company.)

Figure 18.9   Some representative protozoans.

a high economic value, they are low in food value, containing mostly water.

Sometimes included with the fungi are the *slime molds* (division Myxomycota). These interesting organisms consist of many cells aggregated together as a superamoeba which may have a surface area of several square feet. The slime molds are mostly terrestrial, preferring moist woodlands where they creep over rotting logs and vegetation, ingesting microorganisms and organic food particles. Some slime molds are coccine, existing without intercell membranes or walls, while others are sporine and retain cellular boundaries.

The Protozoa (common name, *protozoa*) are microorganisms which lack chlorophyll and cell walls of cellulose, and thus resemble true animals more than plants. While lacking chloroplasts, protozoa often exhibit a highly developed, well-organized cell structure. No sporine protozoa are known, and unicellular forms predominate. Most protozoa are free-living in moist environments and feed holotrophically, although some species are saprotrophic or parasitic. Four types of protozoa are recognized (Figure 18.9): flagellate, ciliate, amoeboid, and coccine. *Flagellate* protozoa possess flagella used both for locomotion and for creating water currents to bring food toward the organism. Most flagellate protozoa are holotrophic and free-living; flagellate parasitic and pathogenic protozoa seem to be derived from free-living types. As an example, the flagellate *Trypanosoma* (Figure 18.9A) is a parasite, causing sleeping sickness in Man.

Ciliate protozoa possess *cilia*, hairlike projections which are used like flagella for both movement and feeding and are often arranged in rows, bands, or spirals. All ciliates are multinuclear, often possessing numerous, small micronuclei and large macronuclei. The common *Paramecium* (Figure 18.9B) is an example.

The *amoeboid* protozoa are characterized by pseudopodia and amoeboid movement. The simple *Amoeba* (Figure 18.9C) is an example. Many organisms in this group possess a siliceous or calcareous protective external lattice through which the pseudopodia project. Examples are *Radiolaria* and its relatives, possessing delicate siliceous shells (Figure 18.10), and the *Foraminifera*, possessing calcareous shells. Over the span of geologic time, foraminiferan shells can accumulate on the ocean floor and eventually produce thick beds of limestone which are often interbedded with chert or flint derived from the siliceous shells of radiolarians.

The *coccine* protozoa are all parasitic. Most exist as multinucleate forms until reproduction, whereupon they synthesize cell membranes and divide into individual cells, each of which is a spore capable of producing nuclear divisions and a return to the coccine state. The best known of the coccine types is *Plasmodium* (Figure 18.9D), which produces malaria in Man and in other animals as well.

Just as the algae serve as producers within the biosphere, so the bacteria, fungi, and protozoa serve as *decomposers*, ingesting organic matter and releasing CO_2 and water, the end products of respiration. Thus the decomposers complete the cycle by which carbon dioxide is converted to organic matter, requiring light energy, then reconverted to carbon dioxide again, releasing chemical energy. (The carbon cycle is discussed further in Chapter 21.)

Metaphyta

The Metaphyta (Table 18.1) include the organisms we commonly refer to as true plants. The true plants are divided into two groups—one, the *Bryophyta*, whose members lack a vascular system for the internal transport of fluids, and another, the *tracheophytes*, whose members possess such a vascular system.

The Bryophyta (*bryophytes*) include the *mosses*, *liverworts*, and *hornworts*. Most bryophytes are terrestrial, usually preferring the moister environments. A few live in fresh water; none are marine. The mosses are probably the most familiar (Figure 18.11). These small plants possess a thin, multicellular filament which creeps along the ground and puts forth stems, leaves, and *rhizoids*. Rhizoids are multicellular projections serving the same function as true roots. The

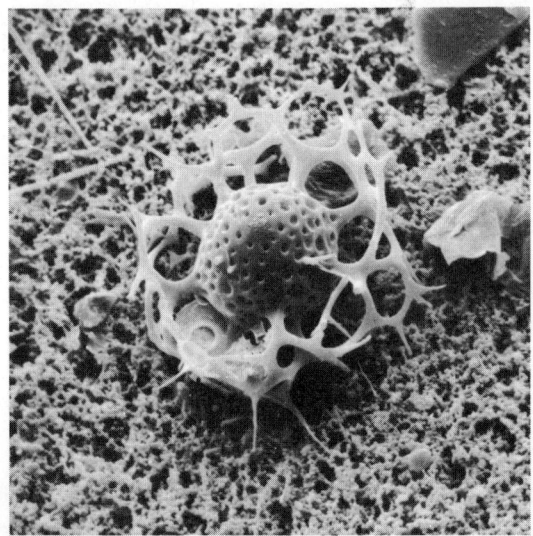

Figure 18.10 Two forms of radiolarians. These specimens were taken from plankton nets towed within the upper 650 ft (200 m) of the North Atlantic Ocean. Magnifications are about ×900 (left) and ×600 (right). (Courtesy of Allen W. H. Bé, Lamont-Doherty Geological Observatory of Columbia University.)

Figure 18.11 A moss of the genus *Sphagnum*, which is characteristic of northern bogs. The stalked fruiting bodies silhouetted against the hand release spores to propagate the species. (Grant Heilman.)

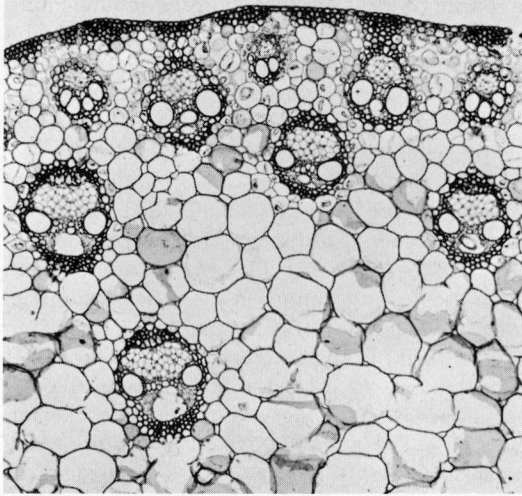

Figure 18.12 Vascular bundles within the corn stem, *Zea mays*. The 2–4 large vessels are the xylem; the smaller open network of vessels is the phloem. (Copyright by General Biological Supply House, Inc.)

liverworts and hornworts have small, flattened bodies which are several cells thick. Upper cells are green and photosynthesize, while lower cells put forth rhizoids, similar to those of the mosses.

The *tracheophytes* are a group of organisms which includes most familiar plants. The group is very large; only the phylum *Arthropoda*, which includes the insects, has more species. Tracheophytes all have one thing in common: they possess *tracheary tissue*, layers or bundles of specialized cells which form a *vascular system* for the transport of water and nutrients throughout the organism. There are two types of tracheary tissue: *xylem*, which conducts water and inorganic nutrients from the roots upward, and *phloem*, which conducts organic nutrients both up and down (Figure 18.12). Undoubtedly this vascular system has contributed much to the success of the tracheophytes on the earth's land surfaces, for it has

TABLE 18.2 Divisions of the Tracheophytes

		Division		Common Name	Number of Species	Environment
	Fern Allies	Psilophyta		Psilopsids	4	Terrestrial
		Microphyllophtya		Club mosses	1,000	Terrestrial
		Antherophyta		Horsetails	25	Terrestrial
Higher Plants	Ferns	Pterophyta		Ferns	9,500	Terrestrial, aquatic
	Gymnosperms	Cycadophyta		Cycads	100	Terrestrial
		Ginkgophyta		Ginkgoes	1	Terrestrial
		Coniferophyta		Conifers	550	Terrestrial
		Gnetophyta		(no inclusive name)	71	Terrestrial, aquatic
	Angiosperms	Anthophyta	Dicots	Dicots	200,000	Terrestrial, aquatic
			Monocots	Monocots	50,000	Terrestrial, aquatic

SOURCE: After H. C. Bold, 1964, *The Plant Kingdom,* Prentice-Hall, Englewood Cliffs, N.J., p. 3.

permitted the diversification of plant structure into specialized leaves, stems, and roots, each with its own function of photosynthesis, transport and support, and absorption and food storage.

Table 18.2 presents the divisions within the tracheophytes. The divisions Psilophyta, Microphyllophyta, and Antherophyta are sometimes lumped together as the *fern allies*, for they share some of the characteristics of the ferns. As the table shows, these subphyla have few species; they are, for the most part, evolutionary relicts whose earlier relatives were much more common in previous geologic eras.

The *pteropsids*, or *higher plants*, include the ferns, angiosperms, and gymnosperms. The higher plants are distinguished by two characteristics of their vascular system: (1) they possess a branched vascular system within the leaf, and (2) where the vascular system enters a leaf, they possess *leaf gaps* in the vascular cylinder of the stem. For purposes of discussion, we will divide the higher plants into four major groups (Table 18.2), the ferns, the conifers, the monocots, and the dicots.

The *ferns* (division Pterophyta) are the seedless vascular plants. They are quite diverse in form and habitat, ranging from the large tree ferns (Figure 18.13) found in tropical rainforests to a few tiny, free-floating aquatic species, to *epiphytic* types which grow on trees or rocks without actually rooting in the soil. The vast majority prefer the moist, shaded environments where we have come to expect them (Figure 18.14). Most ferns have a perennial *rhizome*, or underground stem, which creeps along the forest floor and puts forth leaves

Figure 18.13 Large tree ferns in an Australian eucalyptus forest. The largest fern, just to the right of center, is about 25 ft (8 m) tall. (Courtesy of Australian News and Information Bureau.)

Size	Energy Source	Examples
Medium	Photosynthetic	*Psilotum*
Medium	Photosynthetic	*Lycopodium* (ground pine)
Medium	Photosynthetic	Horsetails
Medium to large	Photosynthetic	Bracken
Large	Photosynthetic	Cycads
Large	Photosynthetic	Ginkgo tree
Medium to large	Photosynthetic	Pine, spruce, fir, redwood
Medium to large	Photosynthetic	(no common example)
Medium to large	Photosynthetic, some saprotrophic, parasitic	Woody plants, trees, daisy, rose
Medium to large	Photosynthetic	Iris, daffodils, lilies, palms

Figure 18.14 The common bracken fern, *Pteridium aquilinium*. The rhizome is a thick woody structure, here concealed by fine roots, to which the stems of the fronds are anchored. (Grant Heilman.)

and roots. The fern leaves, or *fronds*, grow as tight spirals in the bud (Figure 18.15). This gives the expanding leaves in the spring the appearance of "fiddle-heads." In some areas, particularly New England, fiddle-heads of certain species are eaten as a spring vegetable delicacy.

Figure 18.15 Immature fronds of the cinnamon fern, *Osmunda cinnamomea*. Because of their coiled appearance, such immature fronds are referred to as "fiddle-heads." (Jack Dermid.)

The ferns and fern allies reproduce through the dissemination of *spores*—single cells which divide to form a small, intermediate organism, called a *gametophyte*, which lives independently from its parent. The gametophyte then develops reproductive organs, producing eggs or sperms. In some species, male and female gametophytes bear different organs; in others, a single gametophyte bears organs of both kinds. After egg and sperm unite, the young plant is nourished by the gametophyte at first, but soon outstrips it and develops into a mature adult.

In contrast to the ferns, the advanced tracheophytes reproduce by *seeds*. The seed contains an *embryo*, which develops from a fertilized egg and is a collection of cells differentiated into primordial tissues and organs. Surrounding the embryo is a layer of nutritive tissue and a protective coating called the *integument*, or *seed coat*. Biologists divide the seed-bearing plants into two groups—those which bear the seeds openly and those whose seeds are enclosed within a special organ. The first group is the *gymnosperms*, which means "with naked seeds," and is appropriate since the seed develops without a covering. The second group is the *angiosperms*, or flowering plants. In the angiosperms, the female egg is surrounded by an organ called the *ovary*, a capsule of tissue which shields the egg and developing embryo and forms the outer part of the fruit; therefore, angiosperms have enclosed seeds.

The gymnosperms are woody perennials. *Wood* is the name given to accumulations of xylem cells whose cellulose walls have been impregnated with *lignin*, an organic compound which reinforces the cellulose fibers and may act as a preservative to prevent rot. For wood to accumulate, xylem must keep forming throughout the life of the plant; this xylem is *secondary xylem*, since it is produced after the plant stem has ceased its vertical growth and elongation. Secondary xylem is produced by a layer of constantly dividing cells, the *vascular cambium*, which lies between the phloem and xylem layers (Figure 18.16).

The gymnosperms include four divisions, the first of which is the *Cycadophyta*, or cycads. The *cycads* are a small group of plants which resemble tree ferns or palms and are restricted to tropical and equatorial environments. They are particularly slow-growing and long-lived, and bear seeds inside of cones (Figure 18.17) which may be 2 to 3 ft (1 m) in length, 1 ft (0.3 m) in diameter, and weigh over 70 lbs (30 kg)! Also included in the gymnosperms is the *ginkgo* (division *Ginkgophyta*), a tree with fanlike leaves. The modern ginkgo tree, which is commonly planted in cities because of its high tolerance for fallout particles produced by combustion, has changed little in appearance since its ancestors evolved in the

Figure 18.17 A cycad, *Dioon spinulosum*, from Mexico. Note the huge seed-bearing cone. (Courtesy of Dr. T. Elliot Weier.)

Permian Period, some 230 million years ago. It is sufficiently unique to require its own, separate division, the *Ginkgophyta*.

The *conifers* (division *Coniferophyta*) include such familiar trees as pine, spruce, and fir (Figure 18.18). These plants dominate vast areas of the Northern Hemisphere and provide much of our supply of wood for construction and fiber for kraft paper and pulp. The largest known organisms are conifers—the giant California redwood, *Sequoia sempervirens*, attains a height of 340 feet (104 m). The more familiar conifers are characterized by needle leaves (pine or spruce), but other conifers have awl-shaped leaves (juniper or cedar) or even large, broad leaves (some South American conifers). Conifers may be *evergreen* (leaves remaining several years) or *deciduous* (leaves dropping in a cold or dry season). Examples of deciduous conifers are larch and bald cypress. The remaining division of gymnosperms, the Gnetophyta, is a small, heterogeneous group of plants which are mostly of evolutionary interest.

The *angiosperms* (division *Anthophyta*), are divided into two groups: the dicots and the monocots, based on the number of seed leaves, or

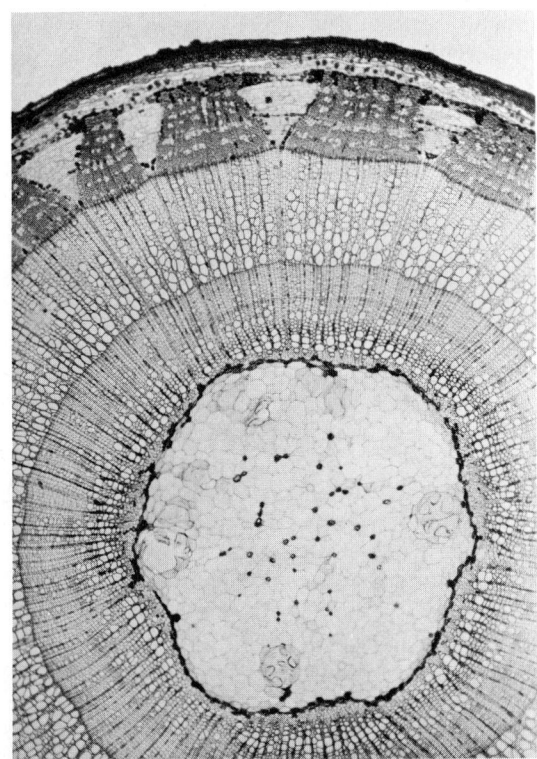

Figure 18.16 Cross section of a basswood stem. The large cells at the center are the pith. The two inner rings are the secondary xylem tissues. Beyond these rings is the layer of vascular cambium, which produces by constant division new layers of xylem on the inside and phloem on the outside.

Life on Earth

Figure 18.18 A mixed stand of white pine, Douglas fir, and Engelmann spruce, Cascade Range, Washington. (U.S. Forest Service photograph.)

leaves are usually net-veined, and flower parts and fruit chambers are often in fives or fours, or their multiples.

The monocots have only a single cotyledon attached to their seeds (Figure 18.19), usually have parallel-veined rather than net-veined leaves, and have flower parts and fruit chambers in threes or sixes. No monocots develop secondary xylem; consequently, none are truly woody, although some possess woody fibers scattered throughout the stem. Large monocots, like the banana plant and the palms (Figure 18.20) resemble trees but are morphologically just overgrown herbs. Examples of monocots include: grasses, sedges, Easter lily, palms, banana, iris, orchid, daffodil, onion, asparagus, jack-in-the-pulpit. The grass family is economically the most important, for it includes the grains and cereals on which most of mankind and his meat-producing livestock depend for existence. Wheat, corn, rice, barley, rye, oats, sorghum, millet, and wild rice are all economically important grasses.

Metazoa

The Metazoa, or true animals, resemble their distant relatives, the Metaphyta, in that they possess a complicated multicellular construction of tissues differentiated into organs and organ systems. The animals, however, lack cell walls or coverings and possess only cell membranes. Animals are often characterized by larval stages, temporary but stable forms which precede the adult stage, when reproduction occurs. Because they lack a photosynthetic mechanism, animals are heterotrophic, living as saprotrophs, parasites, or holotrophs. Most ingest bulk food, requiring the development of a digestive, or alimentary, system.

Animals are also characterized by motion. Most commonly, animals are mobile, moving their bodies from one place to another in search of food, to elude predators, or find mates. Some animals are sessile, or fixed, during one stage of their life cycle (all these are marine organisms), and most of these use flagellant or muscular motion to create a current of water to bring in food. Food-seeking in a mobile animal also requires a sensory system to help the organism discriminate between food and nonfood, and to guide food-seeking motion in general. The nervous system allows coordination of muscular motion as well as other bodily activities.

Lacking cell walls for support of masses of tissue, animals rely on a skeletal system which is either calcareous or siliceous. Besides the familiar endoskeleton of bone and cartilage, some animals

cotyledons, borne by the plant embryo (Figure 18.19). A bean seed or a peanut is an example of a dicot seed. In the peanut, the cotyledons are the two halves of the nut; the embryo is the tiny structure at the base of the cotyledons. The cotyledons are usually thick, green, and fleshy. They function (1) as absorptive organs to take up nutrients from seed tissues, (2) as storage organs for the nourishment of the developing plant after gemination, and (3) as photosynthetic organs which help support the plant before leaves emerge.

The dicots are the more numerous group. Many dicots are woody, possessing secondary xylem and phloem. Plants which lack secondary xylem growth are herbs, or have a herbaceous habit. Some examples of dicots are: apple, pear, peach, tomato, bean, pea, daisy, milkweed, ash, elm, maple, oak, poplar, holly, aspen, alder, willow. Dicots are recognized by a number of characteristics in addition to the number of cotyledons:

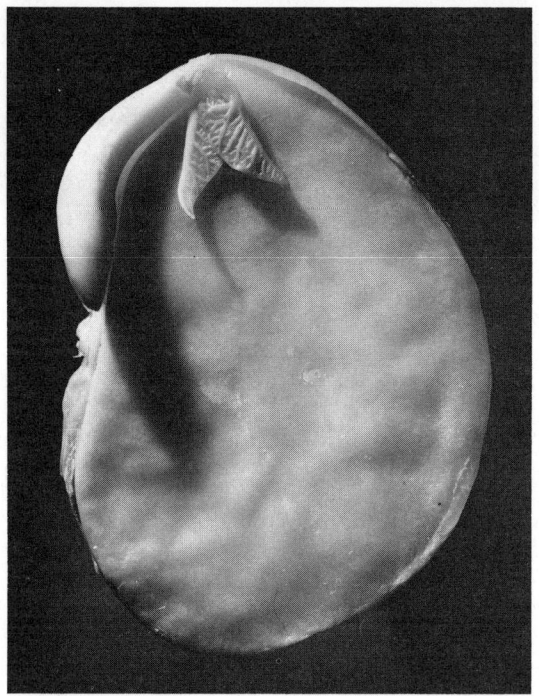
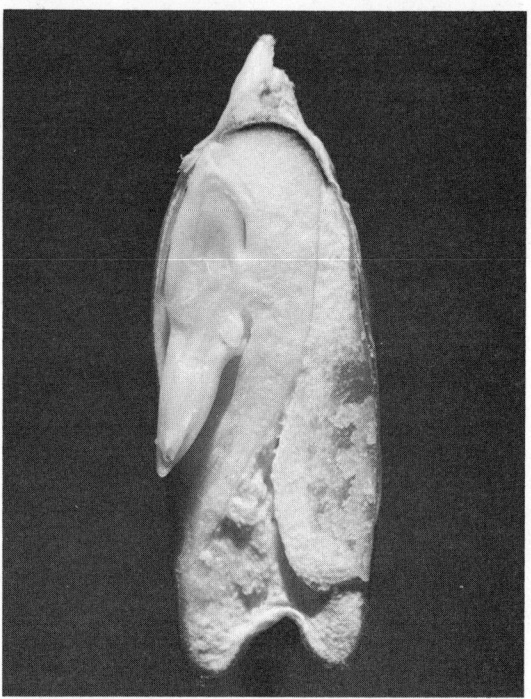

A. A lima bean (dicot) seed. This longitudinal section shows one of the two cotyledons. Attached at the upper left is the young embryo with its first pair of leaves.

Figure 18.19 Dicot and monocot seeds. (Photos by Grant Heilman.)

B. A corn (monocot) seed. The plant embryo points downward on the upper left side of the seed. Next to it, running from the top to lower left, is the single cotyledon, here cut lengthwise. The remainder of the seed is occupied by endosperm, a nutritive tissue not directly part of the embryo.

have instead an *exoskeleton*, a hard external covering often jointed in some way to facilitate movement.

Accumulation of cells into tissues and organs requires that each cell which is no longer in contact with the external environment be given a supply of nutrients and oxygen and have its waste products removed. These are the functions of the *circulatory system*, which maintains an appropriate fluid environment for each cell. Attached to this system must be the organs which exchange fluids with the external environment (gills, lungs, kidneys). In addition, the *reproductive system* must not only provide eggs and sperm, but also in many cases some measure of protection for the delicate embryo whose bodily systems are in the process of developing.

A complete and thorough examination of all the Metazoa is more properly the subject of a biology text than a text in environmental science. Still, it will be useful if you are acquainted with the more important phyla of animals (as listed in Table 18.1) as a background for the studies of ecosystems in further chapters. Although a dis-

Figure 18.20 A stand of cocoanut palms, *Cocos nucifera*. Severe wave action on this beach in the Solomon Islands is undercutting the trees. (American Museum of Natural History.)

Figure 18.21 The breadcrumb sponge, *Halichondria panicea*. Water circulates in through tiny pores and flows out through the central hole, or osculum, which is readily visible. (Runk-Schoenberger/Grant Heilman.)

Figure 18.22 Polyps of the ringed sea anemone. (Robert C. Hermes/National Audubon Society.)

cussion of the various forms of animal life is incomplete without some essentials of biological structure and function, we will instead emphasize the environmental characteristics of the major groups.

The *sponges* (Figure 18.21) are perhaps the simplest of animals. Sponge larvae are typically free-swimming, flagellate, globose, and two cell layers thick. After finding a suitable spot, the larvae metamorphose into the more familiar adult form. Sponges are basically hollow sacs perforated by tiny pores (hence the name of the phylum, Porifera). *Collar cells*, on the inside of the sac, possess flagella which produce a current for feeding by drawing water in through the pores and out of a central hole, or *osculum*. The sac may be quite simple or composed of many interconnecting chambers. Inside the sponges' body, the interior layer of cells secretes needlelike *spicules*, which may be calcareous, siliceous, or organic, and give the sponge its characteristic form.

The phylum Cnidaria, the *coelenterates*, contains aquatic organisms including jellyfish, sea anemones, and corals. The coelenterates exist in two forms: as medusas and as polyps. *Polyps* (Figure 18.22) are sessile forms in which the organism may grow and branch, superficially much like plants. At the end of each branch is a stinging tentacle which is released by contact and entangles passing prey. The external layer of polyp cells often secretes an exoskeleton, which may be thin and organic or massive and calcareous. The exoskeletons of dead coral polyps form extensive deposits of *coral reefs*. *Medusas* (Figure 18.23) are free-swimming forms which develop masses of jelly and tentacles used to trap prey. Jellyfishes are medusa forms of coelenterates. The life cycle

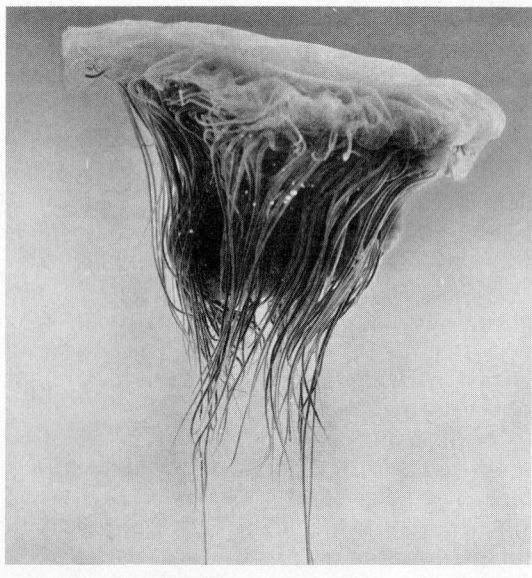

Figure 18.23 Medusa form of the lion's mane jellyfish. (Robert C. Hermes/National Audubon Society.)

A. Free-living planaria, *Dugesia dorotocephala*. (Runk-Schoenberger/Grant Heilman.)

Figure 18.24 Representative flatworms.

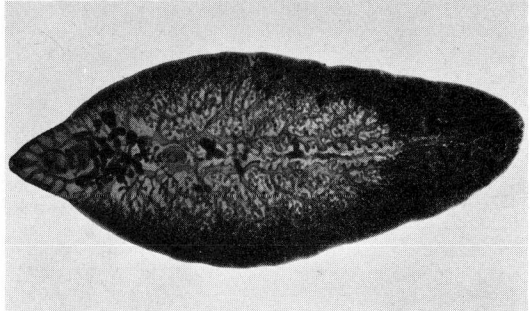

B. A parasitic liver fluke, *Fasciola hepatica*. (Runk-Schoenberger/Grant Heilman.)

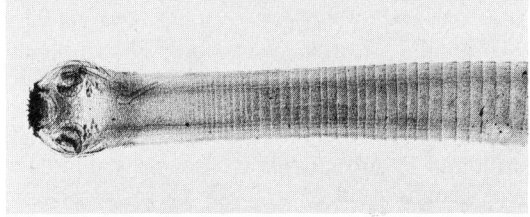

C. The head portion of the tapeworm, *Taenia scolex*. (Carolina Biological Supply Company.)

of coelenterates often includes both polyp and medusa phases. Like the sponges, the coelenterates develop from a two-layered swimming larva. The coelenterates also possess a rudimentary nervous system, and some exhibit muscle cells or contractile fibers used for motion.

The *flatworms* (phylum Platyhelminthes) include the free-living planarians as well as the parasitic flukes and tapeworms (Figure 18.24). These organisms typically have flattened bodies encased by a muscular sheath, as well as a nervous system for muscle control. The *planarians*, which are the basic types, are free-living and holotrophic. The *flukes* are parasites, often exhibiting a complex life cycle with intermediate hosts. They possess hooks or suckers as attachment devices and lack well-developed sensory organs; they feed by eating host tissues. The *tapeworms* are also parasites, obtaining nourishment by absorbing fluids from the digestive systems of their hosts. Although the parasitic habit of tapeworms has led to the degeneration of many organ systems, the reproductive system is particularly well developed. The tapeworms and flukes both include species parasitic in Man which present severe health hazards in many areas of the world. The fluke produces in Man the dreaded and debilitating tropical disease *schistosomiasis* (bilharziasis), in which fresh-water snails are intermediate hosts. Building of dams and irrigation systems on the Nile produced waterways in which snails breed, spreading the disease widely to inhabitants of the Nile floodplain. It is estimated that by 1971 about half of the population of Egypt had acquired the disease.

The two important groups within the Aschelminthes are the rotifers and the roundworms (Figure 18.25). The *rotifers* are small aquatic organisms which are free-living and holotrophic. They are biologically interesting because some species lack males altogether, being propagated by female eggs without benefit of fertilization. The *roundworms*, or *nematodes*, are very numerous organisms; although many types have already been identified, the nematodes have been estimated to contain on the order of 500,000 species. They resemble tiny worms and have well developed muscular and alimentary systems. Many species have a parasitic phase involving a particular plant or animal host. Man harbors some fifty kinds of parasitic roundworms, including *Trichina*, which causes trichinosis and enters the body through ingestion of undercooked pork. Other human parasites are *hookworms*, which burrow through the skin, and *filaria* worms, which block lymph vessels and cause the swelling disease, *elephantiasis*.

The Ectoprocta, or *bryozoans*, are small marine or fresh-water filter-feeding organisms exhibiting a mouth surrounded by ciliated tentacles. Marine types secrete calcareous exoskeletons, forming boxes or short tubes for protection or support. Marine bryozoans are common on rocks, seaweeds, and wharf pilings.

The *mollusks* (phylum Mollusca) include the snails (*gastropods*), clams (*pelecypods*), and

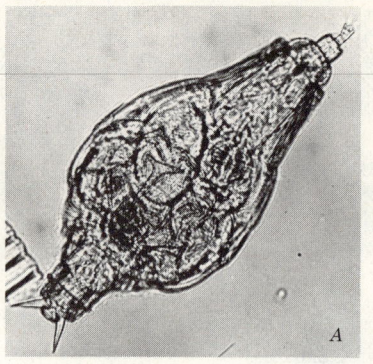
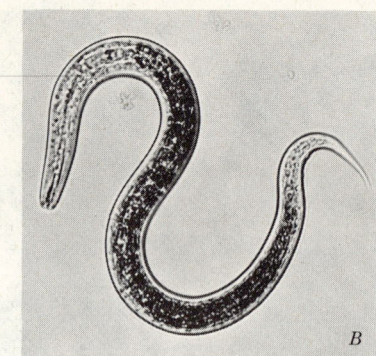

Figure 18.25 Representative sac worms.

A. A rotifer of the genus *Habrotrocha*. See also Figure 23.5A. (Walter Dawn.)

B. A nematode—this one is a species of round worm. (Runk-Schoenberger/Grant Heilman.)

squids (cephalopods) as major groups. Most are characterized by a free-swimming larva called a trochophore. In the *snails,* body development occurs unevenly, and the organ systems are spiralled as a result. This uneven growth extends to the formation of the calcareous, spiralled shell exhibited by most snails. *Clams* are semisedentary, gaining food by straining pieces of organic matter from the stream of water passing constantly through the gills. The mouth opens into the gill chamber and food-rich mucus is swept in by cilia. The clam possesses a muscular foot used for burrowing; some clams also exhibit long tubular siphons which project into clear water, leaving the clam sheltered in the mud below. Some pelecypods are sessile; examples are the oyster and the giant clam. *Squids* (Figure 18.26) are unlike the clams and snails in that they possess only a small remnant of a shell which is enclosed within the soft tissues and is used for body support. The squid's ten tentacles (eight short and two long) trap its prey, which is drawn to the mouth where horny jaws rend it apart. Giant squids may reach a length of 60 ft (18 m), and are the largest invertebrates. Similar to the squids are the octopuses, which have no shell remnant and only eight tentacles.

The Annelida (segmented worms) all have a common body structure of variable numbers of more or less similar segments (Figure 18.27). Typically, the first segment contains the sensory organs and central nerve ganglia; the second segment contains the mouth and associated parts. Other segments contain body muscles and the intestine. The terminal segment contains the anus. Lungs or gills are absent; instead annelids exchange fluids through a capillary circulatory system in flesh outside the muscles. Earthworms and leeches are examples of annelids.

The Arthropoda (arthropods) is the largest phylum in the world of life on earth. It includes such familiar organisms as crabs, lobsters, spiders, centipedes, beetles, bees, flies, scorpions, silkworms, butterflies, and moths. The arthropods are related to the annelids in that they, too, have segmented bodies. In the arthropods, however, the segments have become highly specialized; often segments are united into a compound structure or are present in the embryo or larva only, degenerating in the adult. The arthropod body is covered by an exoskeleton of chitin, a tough, fibrous, impervious material. The exoskeleton is typically divided into head, thorax, and abdomen. Much of the circulatory system of arthropods is open; a few well-developed vessels and a heart serve to channel and facilitate movement of blood freely throughout the body. Lungs or gills serve for fluid exchange in many arthropods; still others possess tracheal systems of ducts leading from interior parts to external pores which serve for respiration.

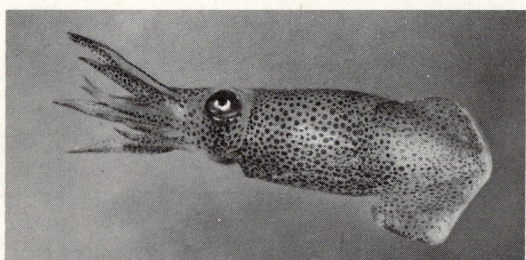

Figure 18.26 A small squid, a member of the phylum *Mollusca.* (Robert C. Hermes/National Audubon Society.)

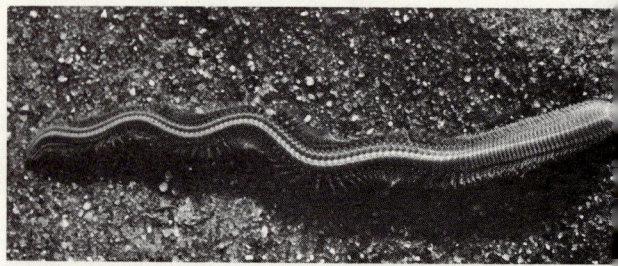

Figure 18.27 A representative annelid, the sandworm (clam worm) *Nereis virens.* (Gordon Smith/National Audubon Society.)

A. The familiar lobster, *Homarus americanus*. (Hal Harrison/Grant Heilman.)

B. The white shrimp, *Penaeus setiferus*. (John H. Gerard/National Audubon Society.)

C. The rock barnacle, *Balanus balanoides*. The feathery appendages are actually legs modified for food gathering. (Runk-Schoenberger/Grant Heilman.)

Figure 18.28 Representative crustaceans.

Many terrestrial arthropods have developed wings and are capable of flight—a mechanism which has lent considerable success to the group as a whole.

The arthropods can be divided into seven groups (Table 18.3). We will discuss two of the larger groups: the crustaceans and the insects. The crustaceans (Figure 18.28) are the dominant arthropods in the sea and include lobsters, crabs, barnacles, shrimps, prawns, and crayfish. The marine forms are largely scavengers, feeding on dead and decaying organic matter. The marine crustaceans are used by Man as an important food source. Although crustaceans on the whole are beneficial to Man, a few are detrimental—the barnacle (Figure 18.28C), as an example, weakens wooden dock pilings and boat hulls. Some fresh water crustaceans serve as intermediate hosts for parasitic flatworms or roundworms. The close relationship of crustaceans to insects often renders them particularly sensitive to pesticides; spraying programs can backfire when toxic compounds are carried into estuaries where crustaceans may spend a good part of their juvenile life.

Insects (Figure 18.29) are virtually all terrestrial, although a good many spend part of their life cycle in fresh water. Salt waters, on the other hand, are almost completely devoid of insect life. The insect realm, then, consists mainly of the land and air. Insects are very abundant for a number of reasons: (1) they are mostly primary consumers, depending on the plant world for most of their food; (2) their small size and rapid movement allows escape from predators; (3) they reproduce rapidly and can increase their numbers quickly to take advantage of an overabundant food supply; (4) their bodies are particularly well adapted to conserving water, an undependable and sometimes scarce environmental ingredient; (5) the mode of flight, exhibited by many insects, allows them to search out new food supplies and evade predators.

TABLE 18.3 Major Groups of Arthropoda

Group	Number of Species	Environment
Horseshoe crabs	5	Marine
Sea spiders	500	Marine
Spiders	35,000	Terrestrial
Crustaceans	30,000	Aquatic; a few terrestrial
Centipedes	3,000	Terrestrial
Millipedes	10,000	Terrestrial
Insects	800,000	Terrestrial; some in fresh water; rarely marine

A. The seventeen-year locust, *Magicicada septendecim*. This is the voracious species responsible for the locust plague. (Grant Heilman.)

Figure 18.29 Some common insects.

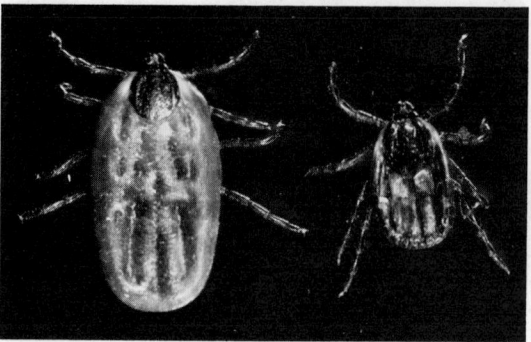

B. The brown dog tick, *Rhipicephalus sanguineus*. These insects are exoparasites; the one on the left is gorged with blood. (Grant Heilman.)

C. A male mosquito of the genus *Anopheles*. The female of the species is an important disease carrier. (Julius Weber.)

Insects are indispensable to terrestrial life as we know it. Many flowering plants depend on insects for their fertilization. Insects are the major food source for many fresh water fish and birds. Silk, honey, and shellac are examples of insect products of importance to Man. The extensive changes Man has made on the face of the earth have greatly affected the insect population. In clearing land for agriculture, Man has destroyed many forest insects and removed their natural habitats. In the process, he has benefited those insects who feed on his crops and domestic animals. Perhaps one of the greatest threats to mankind is the disease potential harbored by insect populations. Many parasitic and pathogenic lower organisms spend part of their life cycle in insects, some relying on biting insects for their entrance to the body. A catalog of insect-transmitted diseases includes malaria, yellow fever, typhus, sleeping sickness, encephalitis, Rocky Mountain spotted fever, parrot fever, and even the deadly black plague. Man is not the only organism affected by insect-borne diseases—many plant and animal pathogens are transmitted by insects as well.

Crop damage and the threat of disease from large insect populations has led Man on a long search for ways of reducing insect populations. One of the most successful methods has been large-scale use of pesticides, toxic chemicals which kill insects and other pests. As an example, the threat of malaria has been greatly reduced by widespread application of the pesticide DDT, killing the common biting mosquito which transmits the disease. Such widespread applications of toxic compounds are not without side effects on other organisms, Man included. Problems induced by pesticides are covered in Chapter 21.

The Echinodermata, or *spiny-skinned animals*, are all marine organisms. Instead of an exoskeleton, these organisms possess an endoskeleton composed of cartilage. The spines possessed by many echinoderms are outgrowths of the endoskeleton. Examples of echinoderms are the starfish, sea urchin, sea cucumber, and sand dollar. Movement in many species is accomplished by a unique water-vascular system, where sea water under the pressure of contractile muscle sacs extends *tube feet*, e.g., the "suckers" of the starfish. Exchange of gases is accomplished in *skin gills*, where fluids circulating in the body cavity come in close contact with surrounding salt water. The echinoderms are primarily carnivorous predators. Starfishes often feed on clams by enveloping the clam with their rays and pulling upon the shell halves until the clam tires and the shell opens.

Vertebrates

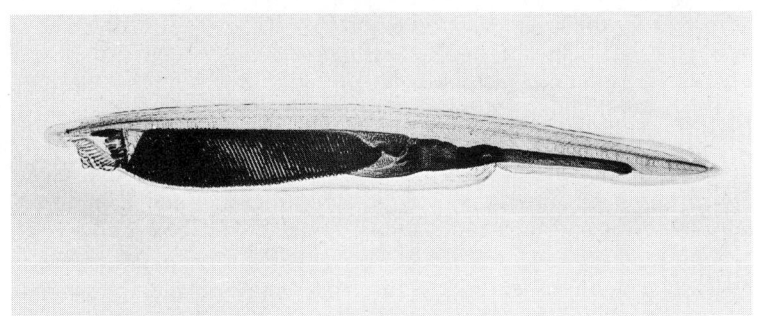

Figure 18.30 The lancet, *Amphioxus*. This primitive chordate is a marine filter-feeder. The nerve cord and notochord are visible together as the dark line in the head and tail regions. (Ward's Natural Science Establishment, Inc.)

The *chordates*, including Man and most of the animals he depends on for food and labor, all possess during at least one stage of development a *notochord*, a length of jointed bone or cartilage running down the center of the back. In the vertebrates, a *vertebral column* of bone surrounds the notochord, which may or may not be present in the adult form. Primitive chordata include the *tunicates*, which are sessile, filter-feeding marine animals in the adult stage; they are recognizable as chordates only in their larval form. Another primitive chordate is *Amphioxus*, the lancet (Figure 18.30), representing a second division of chordates. This marine animal is also a filter-feeder, living in coastal sands.

Vertebrates

If we exclude the insects, the *vertebrates*, the third division of the chordates, are the most numerous animal group, dominating most of the land, sea, and air around us. Development of the air-breathing lung has allowed the vertebrates to escape dependence on an aqueous environment and, along with the insects, to dominate the land. As might be expected for such a large and diverse group, vertebrates are at all levels above the bottom of the food chain, from primary consumers to the top-of-the-chain predators.

The vertebrates can be divided into seven groups (Table 18.4). The *jawless fishes* include lampreys (Figure 18.31), aquatic parasites possessing a suckerlike mouth with which they attach to host fishes, obtaining sustenance by bloodsucking. The *cartilage fishes* include sharks, rays, and skates; their skeleton is permanently cartilaginous. The *bony fishes* include most familiar food and sport fishes. The *amphibians* have evolved from primitive lung-breathing fish, and most are adapted to a terrestrial environment in at least one stage of their life cycle. Amphibian eggs and larvae develop in water, in contrast to those of *reptiles*, which have evolved the shelled land egg. Although many reptiles are aquatic, all must come

TABLE 18.4 Divisions of the Vertebrates

Group	Number of Species	Environment
Jawless fishes	50	Fresh water and marine
Cartilage fishes	600	Marine
Bony fishes	25,000	Fresh water and marine
Amphibians	3,000	Fresh water and marine
Reptiles	6,000	Terrestrial, some mostly aquatic
Birds	10,000	Terrestrial
Mammals	5,000	Terrestrial, aquatic

SOURCE: Data from P. B. Weisz, 1969, *Elements of Biology*, McGraw-Hill Book Co., New York, p. 205.

Figure 18.31 The sea lamprey, *Petromyzon marinus*. Note the sharp teeth and suckers in the mouth of this blood-sucking exoparasite. (Runk-Schoenberger/Grant Heilman.)

ashore to lay their eggs. Reproduction by eggs is also a characteristic of *birds*. Birds possess remarkable adaptations to their habit of flight, including hollow bones, feathers, and a highly efficient lung-circulatory system. The large, flightless birds of today are species which have lost the power of flight, rather than evolutionary bridges between birds and their reptilian ancestors.

The *mammals,* with two exceptions, all exhibit development of the embryo within the female parent, and all possess mammary glands for the suckling of young. Mammals occupy many diverse habitats. Whales (cetaceans) have mastered the aquatic environment; grazing animals (ungulates) are adapted to a life of running on the plains and feeding on vegetation. The rodents and rabbits also are adapted to primary consumption of vegetation, but make their homes in burrows. The primates have developed a two-legged walk and considerable manual dexterity, well suited to their life in the forests.

The mammals include the one species which has assumed almost complete dominance of the terrestrial environment—Man. His ability to reason, communicate, and employ the natural materials of the earth to his own ends has allowed him to escape the bounds of the food chain and the biological limitations of his body, thus living as an organism apart from, not a part of, nature. As humans and their activities multiply, more and more of the earth's surface is converted from primary habitats to secondary ones, causing great shifts in plants and animals long adjusted to pre-existing environments. For this reason, since the Ice Age ended, no agent has had a greater impact on the natural world than Man.

The world of life on earth is incredibly varied in form, size, and mode of existence. From the ocean's abyssal depths to the land's loftiest mountains, life in one form or another has sought out every habitat and become adjusted to just about all earthly environments. How has life gained this astonishing diversity? Through the process of *evolution*, the subject of this chapter, the environment itself has acted on organisms to create this diversity of life forms, even as organic processes have been a prime factor in shaping that environment.

You are probably all familiar with the name of Charles Darwin, whose monumental work, *The Origin of Species by Means of Natural Selection*, was published in 1859. Through exhaustive studies, Darwin showed that all life possessed *variation*, the differences which arise between parent and offspring. Thus, individuals breeding with their kind have the capacity to produce individuals which are different from either parent. The environment acts on variation in organisms, Darwin observed, in much the same way as a plant or animal breeder does, selecting for propagation only those individuals with the best qualities, those best adjusted to their environment. This survival and reproduction of the fittest was termed *natural selection* by Darwin. Darwin saw that variation could, when acted upon by natural selection through geologic time, bring about the formation of new species whose individuals differed greatly from their ancestors. Thus Darwin viewed the formation of new species as a product of variation acted upon by natural selection.

The weakness in Darwin's theory lay not in the origin of species as products of natural selection, but in the process of variation. He was at a loss to explain why variation occurred, and simply accepted it as a natural and automatic property of life. We know now that variation results from the interaction of two sources: *mutation,* which alters the genetic material and consequently the biochemical adjustment of organisms, and *recombination,* the pairing of old and new genetic material in offspring in unique assortments. As you know (Chapter 18), the genetic material in the cell is DNA (deoxyribonucleic acid), contained within the *chromosomes,* linear bodies within the nu-

19

The Evolution of Life Forms

cleus which are visible during cellular reproduction. Both mutation and recombination are products of the molecular structure of DNA and its behavior at reproduction. To understand these two processes fully, however, we must first discuss the genetic basis for life itself.

The Genetic Code

DNA, the genetic material of the cell, controls cellular activities mainly by specifying how enzymes are to be synthesized. You should recall from Chapter 18 that *enzymes* are complicated proteins which catalyze biochemical reactions, and that the sequence of nucleotides in the DNA molecule controls protein synthesis. Nucleotides are molecules composed of two parts: an organic *base* and a *sugar* molecule (Figure 19.1). There are four different bases in the nucleotides which form DNA: *adenine* (A), *thymine* (T), *guanine* (G), and *cytosine* (C). The sugar is *deoxyribose*, one of the many different sugars and sugar-related compounds synthesized by living cells. In DNA, nucleotides are linked into a chain by phosphate molecules attached to the deoxyribose portions of the nucleotides (Figure 19.1). In addition, adjacent nucleotides are linked by *hydrogen bonds* between base pairs. Hydrogen bonds are not ordinary chemical bonds, but result because the nitrogen and oxygen atoms of the bases are both negatively charged and are attracted by the small, positively charged hydrogen ion which they end up sharing.

The molecular configuration of the bases dictates that adenine must pair with thymine (A-T), and guanine must pair with cytosine (C-G). However, the pairs may be freely reversed (T-A, G-C). At any position on the DNA molecule, any pair may occur and may be in either order. Thus, there are four possibilities: A-T, T-A, G-C, and C-G. If we consider two adjacent positions, there are 4^2 or 16 possible arrangements. If we consider three positions, there are 4^3 or 64 possible arrangements, and so forth.

Enzymes, like all proteins, are composed of chains of subunits, the *amino acids*, which are strung together in linear fashion. There are twenty-three different amino acids found in living cells. By careful experimentation, molecular biologists have learned that a sequence of three nu-

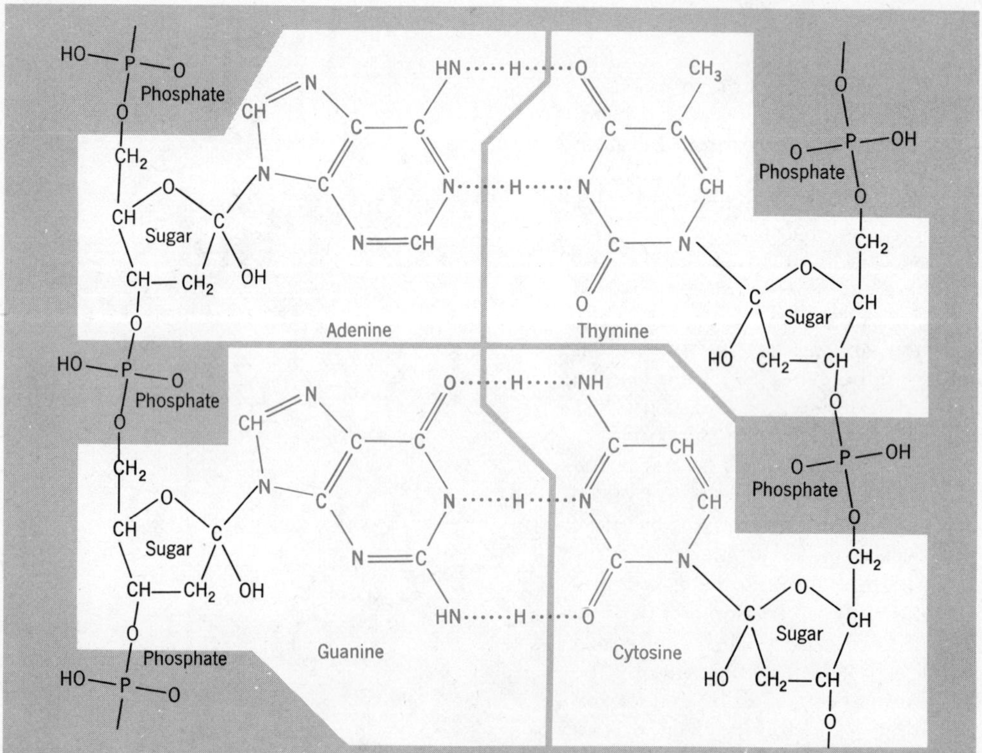

Figure 19.1 The structure of DNA. Each nucleotide (structure enclosed by gray band) is composed of a base (color) and a sugar (black). Nucleotides are joined by hydrogen bonds (dotted lines).

cleotides, called a *triplet*, codes for each of twenty of the twenty-three amino acids. Yet there are sixty-four possible sequences for the bases in a triplet, and only twenty coded amino acids. As it turns out, the coding is *redundant*, meaning that several different triplets will specify the same amino acid. In fact, as many as six different triplets can each specify the same amino acid. Of the twenty coded amino acids, only two are coded by a single, unique triplet sequence. In addition to triplets coding for amino acids, two other triplets are needed: one to start and one to end each protein chain. The list of triplets and the amino acids they correspond to is called the *genetic code*.

Thus the DNA molecule can be thought of as a series of groups of amino acid–specifying codes used for protein synthesis. Each group possess a starter code, then a code for each of the amino acids in the protein, and a terminator. The actual synthesis of proteins is quite complicated, and involves two forms of ribonucleic acid, designated RNA,* and several enzymes. Protein synthesis occurs outside the nucleus on the surface of the ribosomes.

Genes

A *gene* is composed of a series of protein-specifying groups of triplets which control some particular aspect of cellular biochemistry. A gene, then, might consist of a length of DNA which codes for an enzyme needed for the cellular breakdown of sugar molecules, or for an enzyme used in the synthesis of fats for food storage. Or, a gene may control the synthesis of a pigment which colors the eye of the embryo, or perhaps controls an enzyme affecting the growth and differentiation of a limb or wing. Because cells, tissues, and organs must respond to food inputs, or to demands for energy outputs, or to a developmental need for a particular biochemical molecule, enzymes cannot be constantly synthesized. There must exist a regulatory mechanism which controls gene action. As it turns out, molecular biologists believe that DNA also codes for synthesis of regulatory molecules which, through *suppression* or *induction*, act to turn off and on genes or sets of genes.

Although genes control almost all cellular biochemistry, we are most familiar with genes which control morphology, the form or shape of the organism, which is readily determined on examination. Early experiments by Gregor Mendel, the father of genetics, showed that many genes exist in two forms: dominant and recessive. Consider the gene which controls eye color in Man. Because all cells receive two sets of chromosomes, each has two genes controlling eye color. In Man, these genes come in two forms, or *alleles:* one which produces brown eyes, and another which produces blue eyes. The brown allele is a *dominant* type; when both the blue and brown alleles are given to the same offspring, the brown allele will dominate and the offspring will have brown eyes. Blue eyes will only arise when two blue alleles are present. Thus the blue allele is *recessive*, and yields to the brown allele when it is present. Alleles are not always dominant or recessive; some alleles may be neutral, producing an intermediate effect when both are present. For example, in some snapdragon plants there are two alleles for flower color—one which produces red flowers, the other, white. When both are present, they are both active and the flower is pink.

Mutations and Mutagenic Agents

Any change in the structure of DNA is a *mutation*. Since the sequence of bases in the DNA molecule determines the sequence of amino acids in proteins, any change in that sequence will usually produce a change in the protein which is synthesized. (Recall that since an amino acid can be specified by as many as six different triplets, a simple substitution of one base for another might not produce a different protein.) If the DNA strand is broken and a short piece is lost before the ends reunite, the loss will act to remove some amino acids from the protein being synthesized, and will probably also throw the remaining bases in the gene out of sequence. This type of mutation is called a *deletion*. In an analogous fashion, a small portion of the DNA strand may be reduplicated by accident during cell division. This type of mutation is called a *duplication*. Deletions and duplications are only two examples of the many types of mutations known to affect the genetic character of organisms.

The chromosome breaks which produce mutations are induced by mutagenic agents, which are ionizing radiation, ultraviolet light, chemical mutagens, and others. *Ionizing radiation* includes a number of fast-moving charged and uncharged particles which interact with other atoms to produce ions. Recall from Chapter 9 that α particles are helium nuclei, consisting of two protons and two neutrons, and are therefore positively charged. On the other hand, β particles are electrons, and so are negatively charged. As these

---

*RNA is similar to DNA except that the base uracil replaces thymine, and a different sugar molecule, ribose, is used in the nucleotides.

fast-moving particles pass near other atoms, they rip off electrons, producing ionized atoms along their path. These ions are unstable and react quickly with other nearby atoms to form new compounds. Thus, ionizing radiation is capable of breaking the chemical bonds in DNA and permanently altering DNA structure.

Other forms of particulate ionizing radiation are *cosmic particles* (single protons, or hydrogen nuclei) and *neutrons* (Chapter 2). These particles are not normal products of radioactive decay, and are emitted from atomic nuclei only after collision with other particles or from thermonuclear reactions of stars. The molecules of the atmosphere intercept most galactic and solar cosmic particles before they reach the earth, thus shielding life on earth from dangerous amounts of ionizing radiation. Therefore intensity of ionizing radiation of extraterrestrial origin increases with altitude above the earth's surface.

In contrast to particulate ionizing radiation, *X-rays* and *gamma rays* are forms of electromagnetic radiation, possessing no mass and moving at the speed of light. These bundles of energy collide with atoms, increasing atomic energy to the point where electrons are released (producing secondary beta radiation), again forming ions and breaking chemical bonds. *Ultraviolet radiation* (UV), another form of electromagnetic radiation, does not produce ions, but is absorbed directly by atoms and molecules, raising their energies sufficiently to break chemical bonds.

Chemical mutagens react with DNA or other biochemical molecules to produce chromosome breakage, inducing mutation. Many different mutagens are used in the laboratory for studies in genetics; one which may be familiar is mustard gas, which was used in chemical warfare during the First World War. More recently, the halucinogenic drug LSD has been implicated as a mutagen which causes chromosome breakage in experimental animals, but the evidence for damage in Man is not complete and conclusions remain controversial.

Heat can also induce mutations; certain types of mutations in fruit flies are much more readily formed at higher temperatures. Another mutagenic factor is aging. Aged seeds and pollen from plants show higher incidences of mutation than seeds and pollen which are not aged.

Although mutations are produced when DNA structure is altered, the alteration need not be produced directly by the mutagenic agent. Rather, many cellular biochemical mechanisms are required for maintenance and replication of DNA, and mutagenic agents can act through these mechanisms as well.

Effects of Radiation on Genes and Chromosomes

The effects of radiation on organisms can be conveniently separated into two types: somatic effects and genetic effects. *Somatic effects* are produced on the body as a whole. High levels of radiation burn and cause extensive cellular damage or death; lower doses may predispose organisms to cancer or leukemia. Table 19.1 presents some effects of high radiation doses on Man. While somatic effects can be important health problems for existing populations, we will be more concerned here with *genetic effects*, which are passed on to future generations. The dosages are in *rems*, radiation dose units which express a certain probability of biological damage. For comparison, background radiation is about 0.1 rem per year.

The assessment of the genetic effects of exposure to mutagenic agents is difficult, for the mutations produced may not be readily obvious. As an example, one mutation might result in the loss of ability to synthesize a particular amino acid. If the organism is not raised in an environment where that particular amino acid is deficient, the mutation may go unnoticed in the laboratory or be unimportant to natural selection processes affecting that organism's evolution. Figure 19.2 presents the mutational effects of X-ray dosage on *Aspergillus terreus,* a fungus. The graph clearly shows an important point: there is no radiation threshold below which mutations are not produced. In other words, even the smallest X-ray

TABLE 19.1 Somatic Effects of Radiation in Man

Dose (in rems)	Somatic Effect
100	Reduced life expectancy from cancer and leukemia; temporary to permanent sterility.
200	Death within months for 10% of those exposed; 90% survive.
700	Death within months for 90% of those exposed; 10% survive.
1,000	Death within days.
10,000	Death within hours.
100,000	Death within minutes.

SOURCE: After Earl Cook, in W. W. Murdoch, ed., 1971, *Environment,* Sinauer Associates, Inc., Stamford, Conn., p. 257.

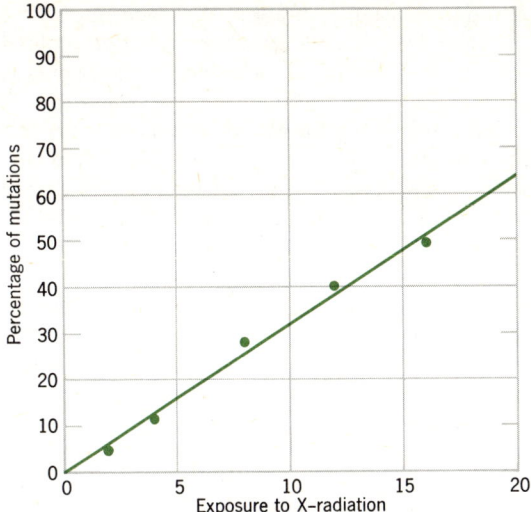

Figure 19.2 Increase in mutation frequency with increasing dosage of X-radiation for the fungus *Aspergillus terreus*. Abscissa in minutes of exposure to 8500 r/min. (Data of Swanson, Hollaender, and Kaufman, 1948, *Genetics*, vol. 33, p. 431.)

Species and Populations

Thus far we have discussed mutations and how they arise. Mutations occur in individuals, but have an impact on species. For our purposes, a *species* (plural, *species*) consists of all individuals capable of interbreeding to produce fertile offspring. A *genus* (plural, *genera*) is a collection of closely related species. Each species has a *scientific name*, composed of a generic name and a specific name in combination. Thus red oak, a common tree, is *Quercus rubra;* the related white oak is *Quercus alba.* Man's scientific name is *Homo sapiens,* which means "wise man."

Although the true test of the species is the ability of all of its individuals to reproduce with one another, this criterion is not always easily applied. Instead, species are usually defined by morphology. The *phenotype* of an individual is the morphological expression of his gene set, or *genotype*, and includes all the physical aspects of his structure which are readily perceivable. Species, then, are usually defined by a characteristic phenotype, or range of phenotypes.

Although in theory all organisms of a particular species are able to reproduce with one another, they rarely do. Instead, they reproduce within local *populations*. As an example, the small brown salamander *Desmognathus wrightii* is found only on mountain peaks in the southern Appalachians.* Colonies on nearby mountains are totally isolated from one another, so that each peak has a separate population whose genotypes do not mix with those of other populations. Yet the phenotypes of individuals from all the populations are sufficiently similar to consider all as a single species.

Mutations in Populations

As we have seen, random mutations arise in populations exposed to mutagenic agents. The organisms which possess these mutations are acted upon by the environment in the process of *natural*, or *environmental, selection.* In competition with one another and with organisms of different species, only the better adapted tend to survive to reproduce their genotypes. The environment, then, acts to sift the mutations which are constantly occurring, eliminating those which are detrimental to the organism's survival

When mutations arise, they are distributed through the population by *recombination*. If a mutation is beneficial, that is, has a positive survival value, the organisms which bear it will have a slightly better chance of growing, maturing and reproducing, insuring that the mutant allele will become more widely distributed through the

dose will produce some genetic damage. In addition, the radiation effects are cumulative from generation to generation, for once DNA is altered, that alteration is passed on to future generations.

While the effects of high radiation dosages on laboratory populations of organisms have been clearly demonstrated in years of research, these data must be compared with normal mutation rates. At all places on the earth's surface, some ionizing radiation is present—this constitutes normal *background radiation*. Ionizing radiation is greater at higher altitudes and near granite bodies containing radioactive isotopes of uranium, thorium, and potassium. This background radiation is one of a number of factors which contribute to the *normal mutation rate*, which differs from organism to organism and from gene to gene. There are several methods of expressing the normal mutation rate. One is by the half-life of the gene, i.e., the time required for the gene to have a 50% chance of mutating. The half-life of a gene, then, is analogous to the half-life of a radioactive isotope. As an example, the half-life of a gene in *Drosophila melanogaster,* a common species of fruit fly, is about 10^3 to 10^4 years, whereas in Man it is about 10^6 years. The difference is related to the much shorter time between generations of fruit flies. When reproduction time is taken into account, the mutation rate may be expressed by the number of mutants per gamete produced. In fruit flies, about one gamete in twenty contains a mutation; in Man the rate is about one gamete in ten!

*Example by courtesy of Professor Theodore H. Eaton.

population. This process will be slower or faster depending on the survival value of the new allele. As an example, imagine a mutation arising in a population of one million annual plants. Suppose the new allele produced by mutation is such that those who possess it will have a 1% better chance of surviving than those without it. In other words, if 100 plants possess the allele in one generation and pass it to all their offspring, 101 of their offspring will (on the average) survive to reproduce the next generation. The time required for such a mutation to be spread to all one million in the population will be quite short, about 1,820 years.

Recall that the half-life of a gene is the time required for the gene to have a 50% chance of being mutated to form a new allele. Putting this another way, by the end of the half-life period about half the genes will mutate, although not all the mutations will be beneficial and preserved through natural selection. A reasonable half-life for the genotype of such an annual plant might be 10,000 to 100,000 years. By comparing this value with the 1,820-year figure above, you can appreciate just how changeable species populations really are.

Isolation and Evolution

Let us return to the brown salamander restricted to peaks in the southern Applachians. How did such an unusual pattern of distribution arise? The species' occurrence on mountain summits shows that it is well adapted to cooler environments. During the Pleistocene Epoch, when temperatures were lower, the species was probably at home everywhere in the Appalachians. With the climatic warming which followed the retreat of the ice sheets, only those individuals in the cooler environments (the mountain peaks) were able to survive. These *endemic* populations (that is, those which are restricted to small areas without genetic intercommunication) have existed for at least 8,000 years, but so far mutations have not been sufficient to differentiate the phenotypes from one another. If the present pattern persists, however, it will be only a question of time before different, randomly occurring mutations in each population will change them sufficiently to be considered new species. This example illustrates the importance of *geographic isolation* in the formation of new species.

Another important mechanism in the evolution of new species is *genetic isolation*. Two populations are near one another; occasionally genes are exchanged between them, but for the most part the populations are genetically independent. Through constant mutation, the *gene pools* (the set of all available alleles) of the two populations change slightly. A mutation occurs, then spreads throughout one population. The mutation is lethal when combined with an allele from the other population. This mutation genetically isolates the two populations, for they can no longer interbreed. The two gene pools are thus forced to go their separate ways, and as random mutations occur, the populations will undergo *genetic drift*, and soon will be sufficiently different to be considered separate species.

Four factors determine the rate of species evolution. The mutation rate is the basic control and is dependent on the exposure of the breeding organisms to mutagenic agents and the sensitivities of the genotypes to these agents. Population size will also be important; the more DNA present in the gene pool, the more chance that mutations will arise. Generation time, the length of the reproductive cycle, will determine just how fast new mutations can spread throughout the population as a whole. And last, mode of reproduction is important; if self-fertilization is possible, as it is in many plants and lower animals, the spread of mutations through a population will be slowed considerably.

The second effect, that of population size, probably explains why the brown salamander populations discussed earlier have not diverged in the 8,000 or more years in which they have been separated. The populations are so small that only a few mutations are gained each year, and as a result the genotypes are relatively stable.

The Evolution of Life

As we have seen, evolution of species occurs through the action of natural selection on variation in populations. All of the diverse life forms now present on earth are thought to have arisen from a few types of primordial organisms existing near the dawn of geologic time. Through constant mutations, DNA has been altered time and time again to produce the myriad genotypes of billions of organisms. In the following pages we will trace the path of organic evolution through geologic time from the simplest of organisms to the most complex. Along the way we will note two important principles. First, when environments change, species change. Therefore, geologic periods encompassing changes in the distributions of land and sea as well as climate will also show the greatest changes in life forms. Second, life itself modifies the environment around it. Thus, precipitation of carbonates by organisms has helped to decrease the quantity of atmospheric carbon dioxide, while oxygen production by autotrophs has helped to raise the atmospheric content of free oxygen and thus paved the way for the evolution of the many heterotrophic life forms we see today.

The pattern which life has followed in evolution from its first primitive forms to the great diversity we have noted in Chapter 18 is revealed by the study of *fossils*, which are the remains or traces of life structures preserved in sedimentary rock. Fossils may occur in many forms. Impressions of soft-bodied organisms on bedding plane surfaces, buried carbonate skeletons which have become replaced by silica, imprints of leaves and stems of plants within coal seams, and even microscopic structures such as indidivual algae or pollen are all fossils found in the geologic record. Fossils usually occur in distinctive assemblages from one or a few sedimentary beds of a certain age. The term *fauna* is used to denote an assemblage of animal species; *flora* similarly denotes an assemblage of plant species.

Species respond to changing environments through two forms of evolutionary changes. When environments change, older life forms give rise through evolution to newer ones which are better adapted to the new environments. One parent class may expand, producing many new genera and species; this process is called *radiation*. Environmental change also produces *extinctions*, in which species or whole lines of species do not evolve rapidly enough to keep pace, and so die out, never again to appear in the geologic record.

The course of evolution, then, follows a branching pattern similar to a tree. Lines of evolution diverge from a central stem or stems and branch time and time again. Not all branches survive, however; these are the extinctions which are so common through geologic time. Just as the branches of a tree do not first diverge and then reunite, neither do species reunite, for their different genotypes keep them incompatible.

Organisms from different lines of descent, however, can come to resemble one another in the process of *convergent evolution*. When natural selection favors organisms with a particular mode or form of life, many different genetic lines will be steered toward that mode or form. Porpoises and sharks resemble each other superficially, and both are carnivores at the top of the oceanic food chain. Yet biologically, the shark is one of the most primitive of fishes, while the porpoise is a mammal which has returned to the sea after a long period of terrestrial evolution. Thus these two organisms have obtained their resemblance through convergent evolution.

Throughout our discussion of life's evolution through geologic time, we shall see abundant examples of radiations, extinctions, and convergent evolution.

Biogenesis

At the time of its formation about 4.7 billion years ago (-4.7 b.y.), the earth must have presented a hostile environment indeed for life as we know it. With a very thin atmosphere, temperatures must have fluctuated greatly from day to day and season to season. Ultraviolet and cosmic radiation must have been intense, too intense for present biological processes to proceed unimpeded. During the first billion years, the earth's environments slowly came to resemble those of the present in certain respects. Recall from Chapters 7, 8, and 9 that by episodes of radiogenic heating, the earth became differentiated into crust, mantle, and core. The less dense, felsic igneous rocks formed the continental shields, and the denser mafic rocks formed the ocean basins. The earth's magnetic field was established, providing a partial shield from the solar wind and from harmful cosmic radiation. At the same time, an atmosphere and hydrosphere were accumulating through outgassing.

An examination of Table 7.3 shows what this early atmosphere must have been like. Water (H_2O), carbon dioxide (CO_2), sulfur (S_2), and nitrogen (N_2) dominated this early atmosphere. Some researchers believe that methane (CH_4) was an important component as well. Ammonia (NH_3) may also have been present in small quantities along with the minor gases argon, chlorine, hydrogen, and fluorine listed in the table. A point of major importance is that free molecular oxygen (O_2) was largely absent. While the dissociation of water into H_2 and O_2 by ultraviolet light provided some oxygen and ozone within this early atmosphere, the amount was very small, probably well below 0.001% of the present atmospheric level. Further, any free oxygen or ozone would combine rapidly with surface minerals in weathering and be removed from the atmosphere. Thus, the oxygen on which almost all present living organisms depend was missing.

It was in just such a hostile environment that *biogenesis*, the origin of life, occurred. Scientists have shown through many experiments that electrical discharges, radiation, and other energy sources could have acted on the components of the early atmosphere and hydrosphere to produce almost all of the basic biological molecules necessary for life. Lack of oxygen was essential to this process, for in the presence of oxygen most of these molecules are highly unstable and are quickly oxidized.

Thus life probably had its origins in an organic "soup," a thin, dilute solution of organic compounds in water. Undoubtedly the first life forms were chemotrophs, obtaining energy from organic molecules within the soup. As water molecules were the only protection available against lethal ultraviolet radiation, life would have been possible only in the oceans, at depths below 33 ft (10 m) according to present estimates. These first life forms may have been bacterialike organisms, for even today, we find that many bacteria are highly resistant to ultraviolet radiation.

At some time after biogenesis, primitive photosynthetic organisms evolved. These early forms were probably monerans similar to the photosynthetic bacteria or blue-green algae. Recall from Chapter 18 that the photosynthetic bacteria possess bacteriochlorophyll, which is distributed throughout the cell in granules, and which differs from the chlorophyll in blue-green and protistan algae. While the photosynthetic bacteria do use sunlight to synthesize organic compounds, they do not evolve oxygen. Most use *hydrogen sulfide* (H_2S) in photosynthesis, liberating free sulfur (S) or other sulfur compounds. The blue-green algae, in contrast to the photosynthetic bacteria, employ H_2O, a much more common molecule, and release O_2.

While some investigators have suggested that the blue-green algae followed and evolved from the photosynthetic bacteria, the earliest fossil evidence from Swaziland, South Africa, seems to show both forms existing simultaneously in rocks of middle Precambrian time. The fossils which are considered to represent bacteria and blue-green algae are spheroidal or filamentous microstructures within black chert layers of the Fig Tree Series and Onerwacht Series (Figure 19.3). These rocks date from − 3.2 b.y. to − 3.0 b.y., that is, about 1.5 b.y. following the earth's formation. In addition to microstructures, chemical evidence indicating photosynthetic life has also been obtained from the Fig Tree Series. Unique molecules have been found in the rocks which are believed to be the remaining products of organic metabolism and imply the presence of chlorophyll. The Soudan Iron Formation in northeastern Minnesota contains similar, but less convincing, forms of evidence, and dates as old as −2.7 to −3.0 b.y.

The evolution of the blue-green algae, with their oxygen-releasing photosynthetic mechanism, produced new environments in middle and late Precambrian time. The oxygen released into

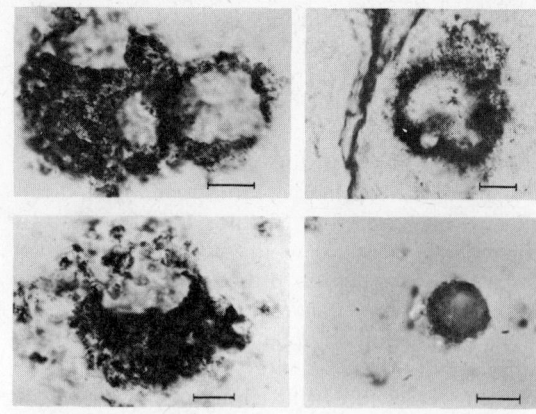

A. Microstructures believed to represent primitive algal-like forms. From the Onerwacht Series of Eastern Transvaal, South Africa, dated at more than 3.2 billion years. (Photo by B. Nagy and L. A. Nagy, 1969, *Nature*, vol. 223, p. 1227.)

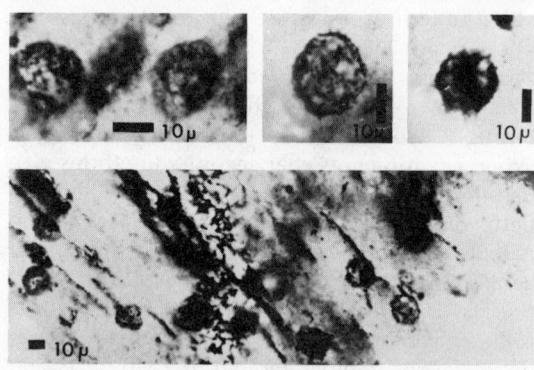

B. Microstructures from the Fig Tree Series, near Barberton, South Africa, dated at more than 3.1 billion years old, which are believed to be algal-like structures. (Photo by Dr. E. S. Barghoorn; see J. W. Schopf and E. S. Barghoorn, 1967, *Science*, vol. 156, p. 509, Figures 1–4.)

Figure 19.3 Photos of microstructures within rocks; these probably represent the earth's oldest fossils. Length of the black bar on the photos is 10 microns (0.0004 in.; 0.01 mm).

the surrounding sea water by their activities was highly reactive and readily combined with surrounding molecules. Reaction was rapid with ferrous iron oxide (FeO), washed into the shallow seas by stream action, producing ferric iron oxide (Fe_2O_3), which precipitated as magnetite or hematite on the sea floor. This phenomenon led to the formation of extensive deposits of *banded iron formations* during the interval from −3.0 b.y. to −2.0 b.y. Today these formations are a major source of iron ore. The banding of the ores suggests that precipitation occurred in cycles, perhaps related to growth cycles of the algae.

Biogenesis

Acceptable evidence of algal activity in such environments comes from fossils in the Gunflint Chert, a series of chert layers in banded iron formations in Ontario and Minnesota. The chert layers show fossil microorganisms which are interpreted as algal forms in unicellular and sporine configurations (Figure 19.4). The chert has an age of at least -1.7 b.y. and perhaps -1.9 b.y.

During the billion or so years that these banded iron formations were being formed, little oxygen would have been released to the atmosphere. However, by about -2.0 b.y. to -1.8 b.y., atmospheric oxygen had probably reached 0.1% of the present level. At this concentration, oxidation of ferrous iron began to proceed on land as well as in the sea. The result was the formation of redbeds, sedimentary rocks composed of particles whose surfaces are covered with a thin, red-purple stain of ferric iron. Redbeds began to be deposited extensively after about -1.8 b.y.

As the oxygen content of the atmosphere was increasing, the carbon dioxide content was decreasing. We have seen in Chapter 8 how inorganic precipitation of carbonates proceeds in shallow, warm seas receiving ample supplies of calcium and magnesium ions derived from terrestrial rock weathering. This process produced substantial accumulations of limestones and dolomites during the period of redbed accumulation in middle to late Precambrian time. Here again, increasing oxygen played a role, speeding the weathering of terrestrial rocks to release the needed calcium and magnesium ions.

A. A stromatolite dome from the Paradise Creek limestone formation, northwest Queensland. The dome is approximately 10 ft (3 m) high, and the rock is dated as 1.6 billion years old.

B. Top views of fossil stromatolites from Dolomite Series, Boetsap, South Africa. These fossils are dated as 2.0 billion years old.

Figure 19.5 Two views of stromatolites of late Precambrian age. (Photographs courtesy of Dr. Preston Cloud.)

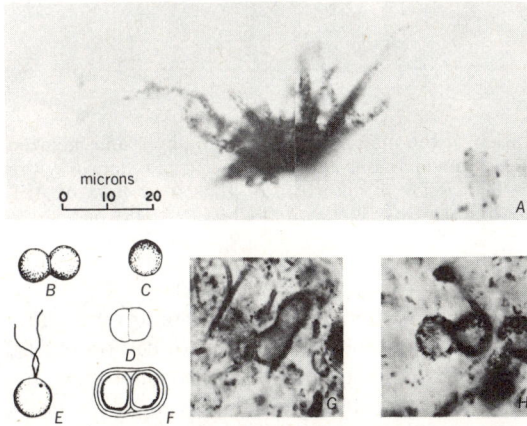

Figure 19.4 Fossil algae from the Gunflint Chert, north shore of Lake Superior, Ontario. A. An algal colony with radiating filaments. B, C, and D, drawings of single cells and cell division of modern blue-green algae; E and F, cell division of living forms of dinoflagellates; G and H are believed to show fossil algae in the process of cell division. (Photographs by courtesy of Dr. Preston Cloud.)

Organic precipitation was also important in the formation of these carbonate rocks, for some of the earliest forms of algae were capable of calcium carbonate precipitation. Living in matlike colonies, these algae produced layered domes and pillars, called stromatolites (Figure 19.5). Although rare, stromatolites are known today from sites in Australia and the West Indies, where they

The Evolution of Life Forms

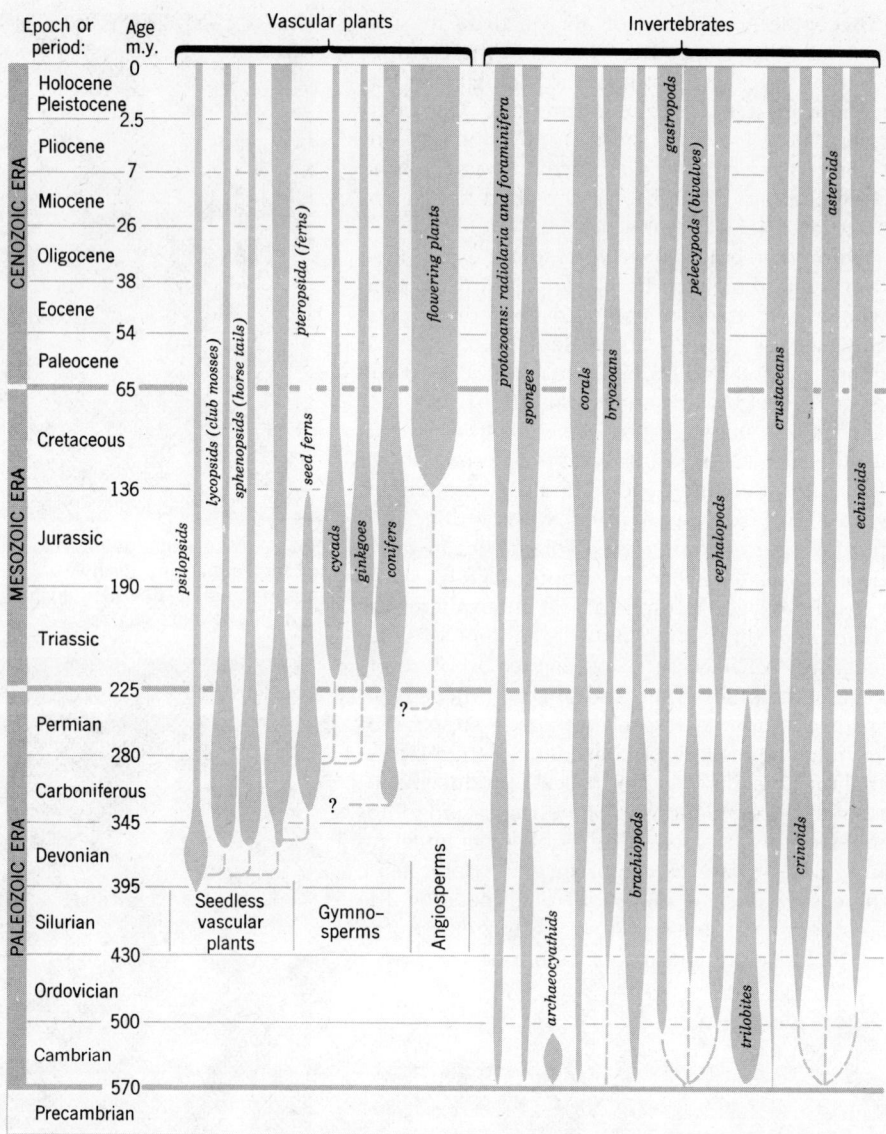

Figure 19.6 Summary chart of evolution of the major groups of metaphyta and metazoa. Widening and narrowing of bands suggest the increases and decreases in abundances within groups, but are not drawn to scale. Time scale is not uniform. (Based on data of A. L. McAlester, 1968, *The History of Life*, Prentice-Hall, Englewood Cliffs, N.J., 151 pp.)

form in intertidal and shallow water zones. Stromatolites are found as early as −2.7 b.y. in rocks of middle and late Precambrian age.

The net effect of this organic and inorganic precipitation was to reduce greatly the quantity of CO_2 in the atmosphere, and by late Precambrian time carbon dioxide concentration had probably reached a value somewhere near that of the present (about 0.03% by volume). The volume of CO_2 stored in rock at that time would have been very large, since at present about 2×10^{16} metric tons (2.2×10^{16} short tons) of carbon is stored in sedimentary rocks. This represents about five hundred times the amount of carbon in CO_2 presently circulating in the atmosphere and hydrosphere. At about this time, too, the salinity, pH, and temperature of the oceans reached values near those of the present. The persistence of organisms which are sensitive to changes in these factors from late Precambrian time to the present helps confirm this supposition.

By the close of Precambrian time, at about −0.7 b.y. to −0.6 b.y., atmospheric oxygen probably reached levels of 1% of the present value. Scien-

Life in the Paleozoic Era

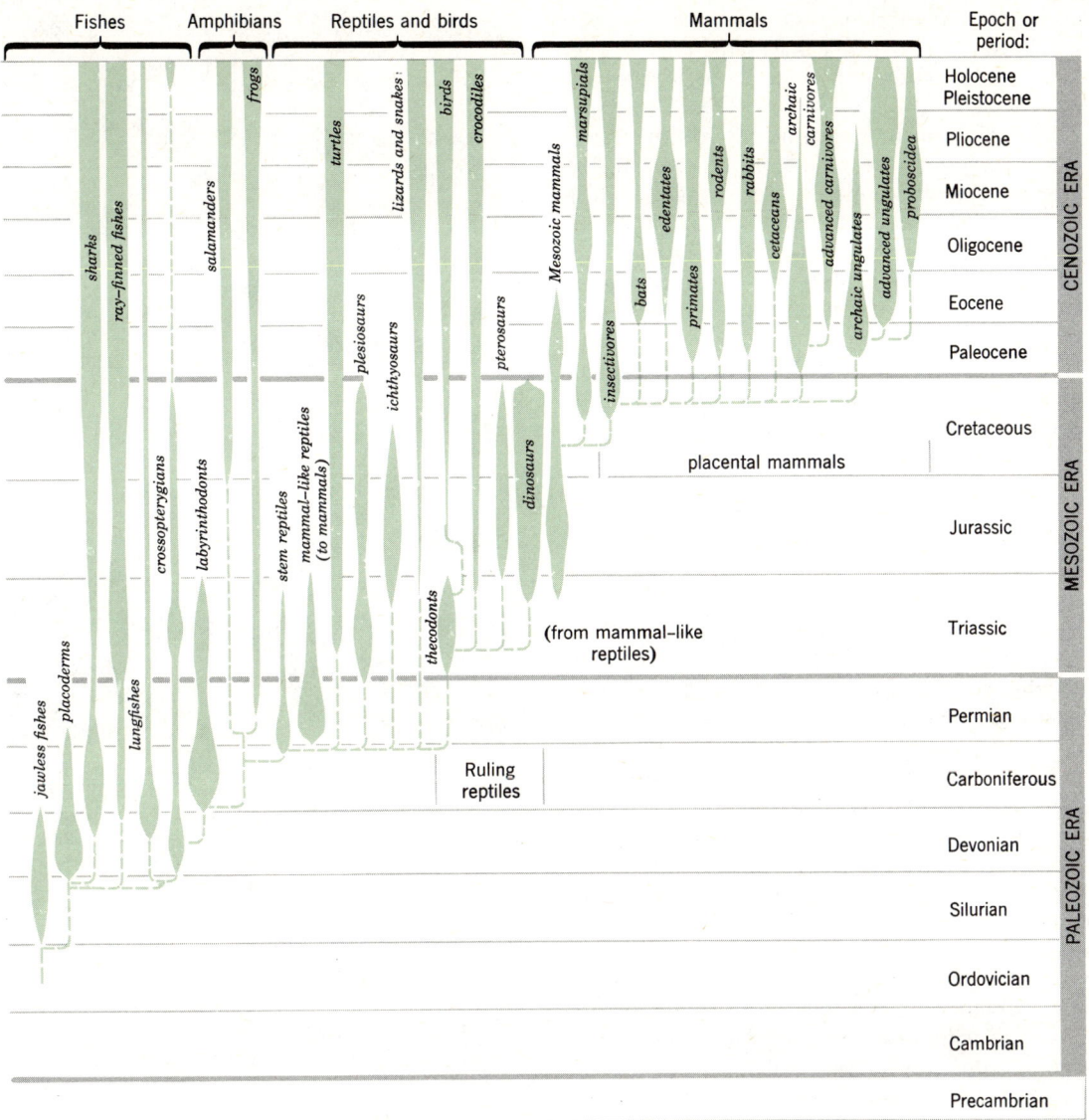

tists remain unsure whether photosynthesis was primarily responsible for the increase. Photo-dissociation of water to H_2 and O_2 through the long Precambrian interval may also have been an important oxygen-forming process. In any case, ozone (O_3) was now formed from O_2 in sufficient quantity in the upper atmosphere to shield the earth from most ultraviolet radiation, and the small amounts of radiation which did penetrate were stopped by a few centimeters of water. Under these conditions, even the shallowest of marine environments could support life. Thus the stage was set for the great proliferation of life forms in the Paleozoic Era.

Life in the Paleozoic Era

From Cambrian time to the present, an abundant fossil record describes the changes in life which occurred through evolution in changing environments. Figure 19.6 summarizes the evolution of the major groups of metaphyta and metazoa through the Paleozoic, Mesozoic, and Cenozoic Eras. Paleozoic life was first restricted to marine environments, but then, in response to increasing oxygen in the atmosphere, life conquered the land as well. During the Mesozoic Era, evolution continued and reptiles and spore-bearing plants came to dominate terrestrial environments. As an offshoot of early reptilian life, mammals evolved and began their spectacular radiation in the Cenozoic to assume their key role in present ecosystems.

The earliest of the Paleozoic periods, the Cambrian, begins with the rapid appearance of highly developed metazoan marine life. Sponges, corals, brachiopods, annelids, and other metazoan forms suddenly appear in the stratigraphic column. The sudden appearance of well-developed life forms at the base of the Cambrian sequence has long puzzled geologists and evolutionary biologists, for there is almost no hint of their slow evolution from preexisting forms. For some time it was thought that the earliest animals had soft bodies and were therefore not well enough preserved to appear in Precambrian rocks. Yet there are abundant Precambrian shales capable of preserving impressions of soft-bodied organisms.

The explanation for this puzzle may lie in the existence of evolutionary pockets in the sea where algae supplied abundant oxygen, but sediments and currents contributed little oxidizable material. Such pockets may actually have been density-stratified layers between 165 and 500 ft (50 and 150 m) within the open ocean. In these oxygenic environments, animals gradually evolved from oxygen-producing algae which had lost their ability to photosynthesize. The evolving Metazoa would have had to lack skeletal materials in order to remain afloat within the layers. In death, those soft-bodied organisms escaping bacterial decomposition would sink to the deep ocean bottom, where the very slow rate of sedimentary accumulation would keep them in contact with ocean water, reducing their chances for preservation. Such a mode of evolution would explain the lack of early metazoan fossils. Then, as the oceans and atmosphere accumulated free oxygen in late Paleozoic time, these early metazoans proliferated rapidly, producing the abundant fossil record of the Cambrian Period.

Still another, simpler mechanism to explain this rapid metazoan evolution is an evolutionary spurt produced by a change in the mutation rate. Such a change might be produced by an increase in ultraviolet or ionizing radiation during a period of intense solar flares.

The earliest metozoan fossils are from late Precambrian rocks of Australia. These fossils, together comprising the *Ediacara fauna*, include about 1,000 specimens (Figure 19.7). Some of the forms observed are probably early jellyfishes and annelids; still other forms have no known affinities. These organisms apparently were soft-bodied, lacking skeletons, and are preserved as

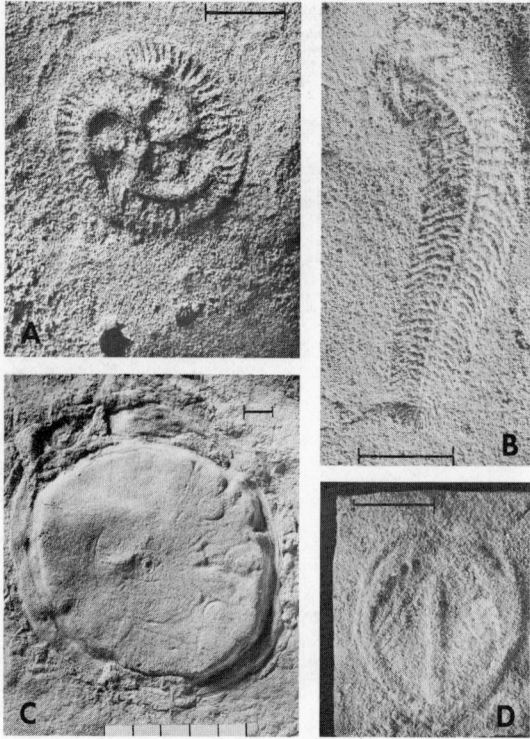

Figure 19.7 Late Precambrian fossils from the Ediacara fauna of the Pound Sandstone, South Australia. (Photographs by courtesy of M. F. Glaessner, University of Adelaide, Australia.) *A.* Circular organism, probably with coiled arms. *B.* Form resembling a burrowing worm. *C.* Fossil medusa form of a jellyfish. *D.* A shield-shaped form, probably a crustacean.

casts, or imprints, on the bedding planes of sandstone layers.

Trilobites are by far the most distinctive Cambrian fossils. These early arthropods were crustaceans, having a hard, three-lobed, segmented shell (Figure 19.8); they were bottom-dwellers, or

Figure 19.8 A trilobite of Cambrian age from the Grand Canyon, shown approximately at natural size. (Department of the Interior, National Park Service.)

Life in the Paleozoic Era

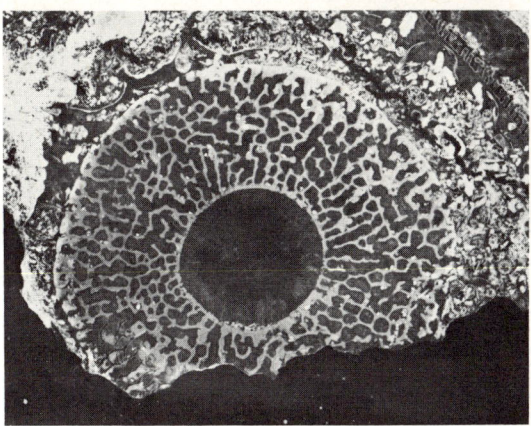

Figure 19.9 A fossil archaeocyathid of the genus *Flindersicyathus*. From northern British Columbia. (Courtesy of Dr. R. C. Handfield.)

benthic organisms. Whereas trilobites are now extinct, their descendants are the horseshoe crabs, which are common in coastal waters. Another important group were the brachiopods. These organisms resembled modern clams (pelecypods), but were constructed quite differently. You may note from Figure 19.6 that one group of invertebrate animals, the archeocyathids, is restricted solely to the Cambrian Period. These organisms (Figure 19.9) possessed a coiled shell resembling a cornucopia. Because of their restriction to Cambrian time, the archeocyathids serve for geologists as index fossils which can be used to date the rocks in which they occur. Other Cambrian invertebrates include sponges, annelids and coelenterates, as well as crustaceans other than trilobites. Figure 19.10 presents a reconstruction of a shallow ocean environment during Cambrian time.

Beginning in the Ordovician Period, radiations produced several new groups of organisms. The crinoids, asteroids, and echinoids appeared suddenly, as did the gastropods (snails) and cephalopods (squids); they were joined in the mid-Ordovician by the pelecypods (clams). The bryozoans, unknown from the Cambrian Period, evolved, and the corals radiated rapidly to form many new species. As these groups waxed, so the trilobites waned. These trends continued into the Silurian Period. Figure 19.11 shows how the ocean floor might have looked in Silurian time.

Throughout the span of 175 million years (175 m.y.) from the beginning of the Cambrian to the close of the Silurian, atmospheric oxygen was increasing. By late Silurian time, it may have reached a value of 10% of the present atmospheric level. With this much oxygen in the atmosphere, ozone production would be sufficient to reduce ultraviolet radiation at the earth's surface to levels compatible with terrestrial life. Thus we see in the remainder of the Paleozoic Era the conquering of terrestrial environments, first by plants, then by animals.

The transition from a salt water environment to a terrestrial one would not have been easy for life. Delicate cellular systems had to be protected from desiccation, and fluids had to be conserved. Oxygen was no longer in a dilute aqueous medium, but was much more concentrated and had to be absorbed from a gaseous atmosphere directly. Gametes could no longer be simply released into water, but needed to be protected and conveyed in new ways. New systems of cellular

Figure 19.10 A reconstruction of the sea floor in middle Cambrian time, based on the fauna of the Burgess shale, eastern British Columbia. Sponges are on the far left and right; trilobites and arthropods on the left and in center; segmented worms on the ocean floor; a jellyfish left of the center. (Photograph by courtesy of the Smithsonian Institution.)

Figure 19.11 Reconstruction of a Silurian coral reef, based on a fossil assemblage from Illinois. The stalked, sessile organisms are crinoids; the larger, head-like masses are corals. Bottom-dwelling organisms include brachiopods, clams, cephalopods, and trilobites. (Photograph by courtesy of the Field Museum of Natural History.)

support were required in a new environment where the buoyancy of the surrounding medium was now greatly reduced.

Undoubtedly life made the transition to land through fresh water. The green algae (chlorophyta) may have been first to evolve species whose metabolism was capable of living in a nonsaline environment, for their unique photosynthetic system is preserved largely unmodified in the terrestrial vascular plants. Present-day bryophytes (mosses and liverworts) may be similar to the plants which made the transition from water to land. Adapted to moist, shady environments, the bryophytes lack a vascular system and so depend on cellular conduction for the transport of fluids. Their male gametes are free-swimming sperms, depending on water films for motility.

The first terrestrial vascular plants to arise (the psilopsids) were indeed primitive. Photosynthesis was carried out in the stem tissues or in scalelike appendages, for these plants lacked leaves. Roots were also absent, but underground stem structures served a similar function. Club mosses, horsetails, and ferns, arising in mid-Devonian time, presumably evolved from these primitive ancestors. By the Carboniferous Period, these forms of vascular plants came to dominate the terrestrial landscape. Many of these were large, treelike organisms; as explained in Chapter 8, forest growth was heavy in lowland swamps and coastal plains, and in many places sedimentation accompanying inundation by shallow seas preserved layers of dead vegetation, producing what are now extensive coal beds. Figure 19.12 presents a reconstruction of a Carboniferous forest.

Figure 19.12 Reconstruction of a Carboniferous forest. F, seed-ferns; S, sphenopsid; L, lycopsid. (Photograph by courtesy of Illinois State Museum.)

Life in the Paleozoic Era

As terrestrial plant life continued to evolve in the Carboniferous Period, a new line of organisms arose. These were the *seed ferns*, which reproduced by seeds rather than by spores. The seeds were borne on the leaves, often on the undersides or at the tips, but in many other arrangements as well. Now completely extinct, the seed ferns gave rise to the conifers (probably in the early Carboniferous) and, toward the close of the Paleozoic, to the cycads and gingkoes. Through a separate line of development, the angiosperms (flowering plants) also may have evolved from the seed ferns before the seed ferns became extinct in the mid-Mesozoic.

In the meantime, animal evolution had proceeded as well. During the Silurian Period, fishes arose in the sea. These early vertebrates lacked jaws, and presumably survived by filter-feeding on the ocean floor. Among earlier forms were lobe-finned fishes (*crossopterygians*); they developed lungs and acquired stubby legs to form the class of early amphibians, the *labyrinthodonts*, which was common in Carboniferous swamp forests (Figure 19.13). The arthropods also radiated new groups: the insects, with their tough protective covering and advantages of flight, and the arachnids (spiders and scorpions) which preyed upon the insects. Gastropods also underwent radiation to produce many species of land-dwelling snails.

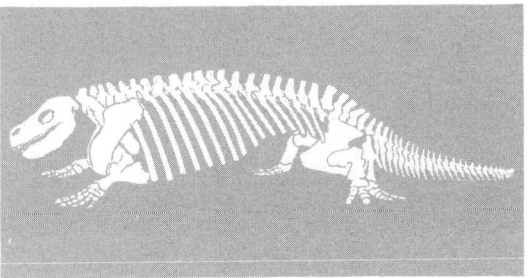

Figure 19.14 The skeleton of a Permian cotylosaur, reconstructed from a Texas fossil. The animal, a member of the extinct genus *Diadectes*, was about 6 ft (2 m) long. (From E. H. Colbert, 1969, *Evolution of the Vertebrates*, John Wiley & Sons, New York, Figure 42B.

Reptiles arose from the labyrinthodont amphibians in mid-Carboniferous time. These new animals laid their eggs on land and were thus free from dependence on an aqueous environment. These earliest reptiles were *cotylosaurs*; they attained several feet in length, and resembled alligators with thick and heavy limbs (Figure 19.14). The cotylosaurs are also referred to as *stem reptiles*, for they served as the ancestral stock for the dinosaurs as well as the mammallike reptiles which arose in the late Permian Period.

The Permian Period marked the end of the Paleozoic Era. Beginning in the upper Carbon-

Figure 19.13 Early vertebrates make the transition from water to land. *Left:* Devonian crossopterygians—lobe-finned fishes—embarking onto the terrestrial environment. *Right:* The later labyrinthodonts, which were true amphibians. (Painted by F. L. Jacques under the direction of W. K. Gregory; American Museum of Natural History.)

iferous, widespread orogeny occurred, resulting in folding and faulting of Paleozoic strata in many areas. Parts of the continents stood at high elevations, increasing the area and volume of sediments which characterize Permian time.

Climatic change was also great. While evidence points to the Carboniferous climate as mild and frost-free, the Permian exhibits more variation. Recall from Chapter 9 that at this time the continents were still united into one or two large masses (Gondwana and Laurasia). During the Permian, extensive glaciation occurred, centered on southern Gondwanaland; the deposits of this glaciation today contribute to the confirmation of continental drift theory. Warm, shallow seas and swamp forests, although reduced in extent, did provide environmental continuity for many Carboniferous organisms.

The transition from the Paleozoic Era to Mesozoic Era was not without change in marine environments. Trilobites continued their decline through the Permian, and became extinct by lowest Triassic. The abundance of brachiopods was greatly reduced, and from the Mesozoic to the present, these organisms have played a minor role in invertebrate faunas. Several dominant orders of Paleozoic corals, bryozoans, and crinoids also failed to survive the transition. Following these extinctions, radiations occurred in many groups. The foraminferans, following their rise in the Carboniferous and decline in the Permian, expanded once again. Other major radiations occurred in the sponges, corals, bryozoans, bivalves, crustaceans, and echinoids. Figure 19.6 shows a small decline for many of these invertebrate groups in the Permian, but since the figure represents numbers of groups, it does not show the extinctions and radiations. In reality, about half the invertebrate families became extinct during the transition from Paleozoic to Mesozoic eras. Similar extinctions occurred in the vertebrates and vascular plants. Thus was the stage set for a great radiation of life forms when environmental stability returned in the Mesozoic Era.

Life in the Mesozoic Era

The Mesozoic Era, the era of middle life, has been referred to as the *Age of Reptiles*, for in this interval a great radiation of reptilian life occurred, culminating in the dinosaurs which ruled wet floodplains and deltas during the Jurassic and Cretaceous Periods. As the Paleozoic Era closed, stem reptiles and mammallike reptiles had radiated from the labyrinthodonts. While the stem reptiles and mammallike reptiles themselves both became extinct at the close of the Triassic Period, their various branches had produced the Mesozoic mammals as well as the seven major groups of reptiles which were common throughout the rest of the Mesozoic (Figure 19.6).

By the middle Triassic, the stem reptiles had radiated into five major groups (Figure 19.6). The *turtles* had hard, bony shells and were similar to those of the present. The *lizards* arose in late Triassic time, and the *snakes* are presumed to have evolved from them. The *plesiosaurs* and *ichthyosaurs* were aquatic reptiles, occupying an ecological niche similar to that of the cetaceans (whales and porpoises) of today. Both these groups were extinct by the beginning of the Cenozoic Era. The stem reptiles also produced the *thecodonts* (Figure 19.15). These reptiles were small and light, and relied on swiftness in predation. Some thecodonts were *bipedal* (two-legged), running more or less upright. Their forelegs, freed from walking, were able to evolve into limbs used for clawing, holding, and grasping.

The thecodonts were the ancestors of the *ruling reptiles*, including the crocodiles, pterosaurs, and dinosaurs which dominated the late Mesozoic. The *pterosaurs* (Figure 19.16), batlike reptiles with large wing spans, probably achieved flight by gliding and soaring. The *crocodiles* were similar to the present reptiles inhabiting aquatic environments. The *dinosaurs* radiated into six major groups, only one of which (the *theropods*) was carnivorous. One example of a theropod is *Tyrannosaurus rex*, which was the largest terrestrial

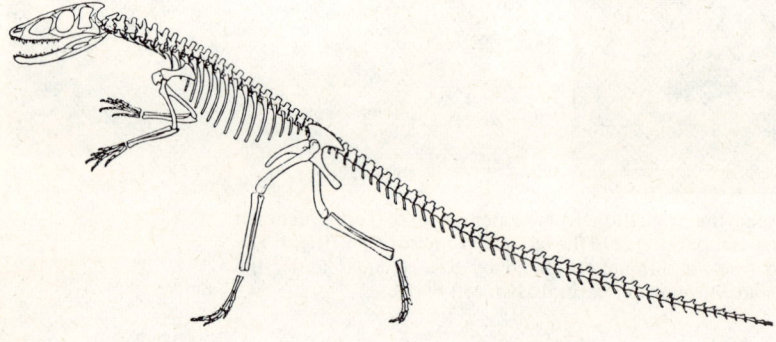

Figure 19.15 Drawing of the skeleton of a representative thecodont. The skeleton is about 4 ft (1.2 m) long. (From E. H. Colbert, 1969, *Evolution of the Vertebrates*, John Wiley & Sons, New York, Figure 55.)

Life in the Mesozoic Era

Figure 19.16 A reconstruction of pterosaurs of the Cretaceous period, shown in a habitat of marine cliffs. (Mural painting by Constantin Astori, American Museum of Natural History.)

predator of all time (Figure 19.17). Probably the best known of the dinosaurs, thanks to advertising by a major oil company, is the *Brontosaurus*. A member of the suaropod group, this huge beast was a placid vegetarian.

Evolution of the vascular plants also proceeded during the Mesozoic Era. Beginning in the Devonian Period, at about their time of evolution, conifers slowly became more and more important in the Paleozoic forests, replacing the seed ferns and primitive arborescent plants. Rapid evolution in the Permian Period made the conifers the forest dominants as the Mesozoic Era began. The conifers, along with the cycads and gingkoes, continued to dominate the land plants until mid-Cretaceous, when the rise of the flowering plants (angiosperms) is recorded.

The rise of the angiosperms in the late Mesozoic presents an evolutionary puzzle, for the derivation of the flowering plants from earlier forms is unclear. A few fossils of Triassic and Jurassic age may represent the ancestors of this largest and most diverse group of plants, but the exact evolutionary pattern of the group remains uncertain for lack of a well-documented fossil record. Perhaps angiosperm ancestors were restricted to upland environments, where erosion was dominant over sedimentation; perhaps their ancestors were exceedingly rare. In any case, angiosperms radiated rapidly in late Cretaceous time, and by the end of the era had displaced the gymnosperms to become the dominant land plants.

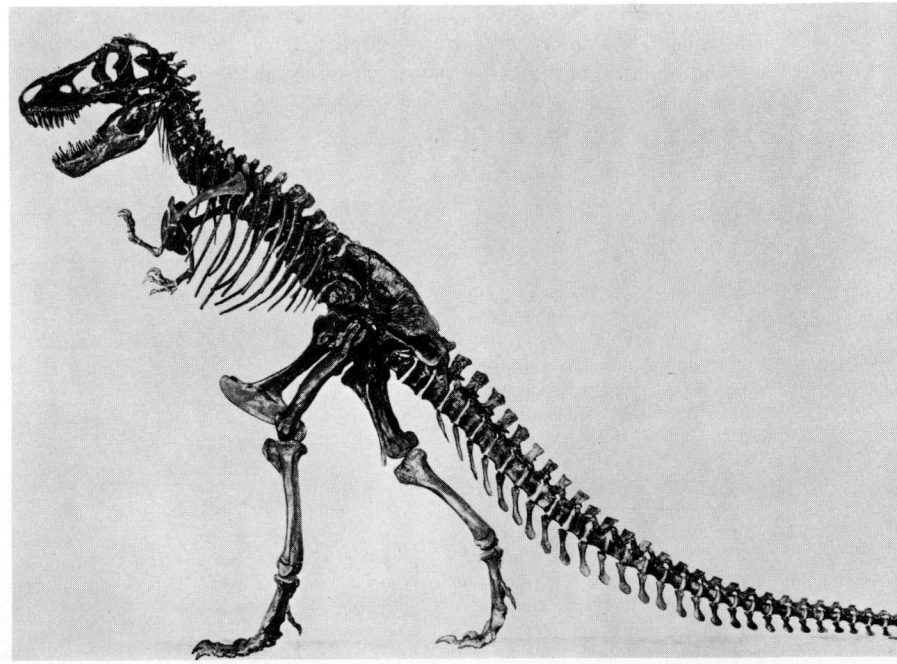

Figure 19.17 A reconstructed skeleton of *Tyrannosaurus rex*, measuring about 50 ft (15 m) in length, standing about 18 ft (5.5 m) high. (American Museum of Natural History.)

Life in the Cenozoic Era

It was not until the Cenozoic Era that life on the earth's surface took on the character that it has today. During this last era, two groups gained ascendancy in terrestrial environments—the mammals, and the angiosperms.

By the close of Cretaceous time, the angiosperms had replaced the gymnosperms as the dominant land plants, and by the early Cenozoic, most present plant genera had evolved. The climate in the early Cenozoic was warm and mild over most of the continents, and fossil assemblages from the Paleocene and Ecocene are comprised largely of broad-leaved plants which at the present time have tropical and subtropical affinities. Deciduous trees were common as far north as Alaska. Climatic cooling proceeded throughout the Tertiary, and by the end of the Pliocene, most plants had migrated southward to somewhere near their present positions.

During the Pleistocene glaciations, many plants were forced still further south; the north-south ranges of the Rockies and Appalachians provided convenient paths in North America for those plants better adapted to cold. During each interglacial period, these plants returned to the north in response to climatic warming to conditions which were perhaps slightly warmer than present. The result of these forced migrations was a great number of species extinctions and some radiations; in fact, it is the glaciations of the Pleistocene which have made the middle-latitude flora what it is today.

We have already noted the presence of early mammals during the middle and upper Mesozoic. These early mammals were quite small and resembled shrews, probably subsisting on seeds and insects (Figure 19.18). Two lines of mammals rose to importance following the reptilian extinctions at the close of the Mesozoic—the marsupials and placental mammals.

The *marsupial mammals* are most familiar to us as the Australian kangaroos, wallabys, and koala "bears." These mammals give birth to partly developed young which are reared in skin pouches where suckling nipples are exposed to provide nourishment. The marsupials are a very important group in the animal evolution of Australia and South America.

In *placental mammals*, the young remain in the mother's body for a longer time and are nourished by the placenta, an organ within the uterus which assimilates nutrients and oxygen from the mother's blood directly. This has the evolutionary advantage of allowing the young to be supported for a longer time in the safety of the mother's body.

The placental mammals never reached Australia, except for a few bats and rodents, and consequently the marsupials radiated into herbivorous and carnivorous forms strikingly similar to the placental mammals which evolved on the continents of North America, Europe, Asia, and Africa. Marsupial mammals are also important in South America; but an intermittent Cenozoic connection with North America mixed the faunas of the two hemispheres at intervals, resulting in quite a different evolutionary pattern. Along with the marsupials, the placental insectivores (shrews, moles, hedgehogs, and others) apparently arose in the middle and upper Cretaceous, and had radiated by Paleocene time to form the principal orders of placental mammals known today. Throughout the Cenozoic, these orders have increased in size and diversity (Figure 19.6). Some

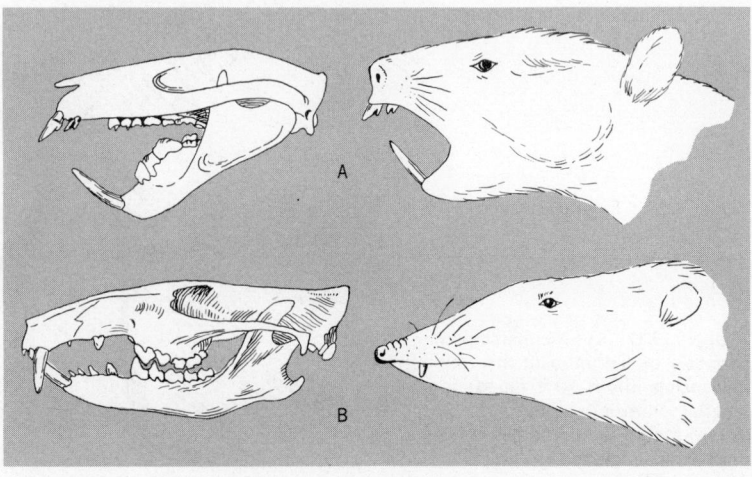

Figure 19.18 Drawings of skulls and reconstructed heads of early mammals. A. A multituberculate mammal of Jurassic age. B. A small insectivore from the Cretaceous period. (After C. O. Dunbar and K. M. Waage, 1969, *Historical Geology*, 3rd ed., John Wiley & Sons, New York, p. 365, Figure 15.20 and p. 395, Figure 16.24. From A. N. Strahler, 1972, *Planet Earth: Its Physical Systems through Geologic Time*, 2nd ed., Harper & Row, New York, p. 321, Figure 14.8.)

The Evolution of Man

reached their peak in mid-Cenozoic time (elephants, horses and their relatives, sloths and armadillos, and others), while others are still expanding at present. The Pleistocene glaciations had much less effect on the mammals than on the angiosperms, probably because the forced migrations were easy for mobile animals.

The Evolution of Man

One of the earlier mammalian orders to arise from insectivores was the *primates*, of which Man is a member. Primate evolution is diagrammed in Figure 19.19. The early primates were *prosimians* (or "premonkey" primates), represented today by the lemurs, lorises, and tarsiers, small tree-dwelling mammals of Africa and Asia. These early ancestors of Man had developed long, dextrous fingers on their feet for grasping tree limbs, as well as forward-set eyes to give binocular vision for perceiving depth.

In Oligocene time, prosimians gave rise to three groups comprising the *anthropoids:* the New World monkeys, the Old World monkeys, and the *hominoids*, including apes and men. By Miocene time, the hominoids were represented by two genera, *Pliopithecus* and *Dryopithecus*. *Pliopithecus* is the ancestor of the gibbon, while *Dryopithecus* is Man's ancestor. There were several species of *Dryopithecus*, each differing in size and distribution. Whether *Dryopithecus* was principally a tree-dweller or had adapted to ground feeding is still uncertain. In any case, by late Miocene time, *Dryopithecus* had given rise to *Ramapithecus*, a direct ancestor of Man. Unfortunately, little is known of *Ramapithecus*, for only jawbones have been found thus far. The pattern of dentition (the shape and arrangement of teeth), however, is strikingly Manlike. *Ramapithecus* is one of the three genera in Man's family, the *Hominidae*, and is thus a *hominid*. The other hominids are *Australopithecus* and, of course, *Homo*.

Of these last two, *Australopithecus* (Figure 19.20) is the older. Known from the early Pleistocene, the genus *Australopithecus* contained two species, one the size of a gorilla, the other about as large as a chimpanzee. These two species were both upright ground-dwellers; they fashioned and used crude stone tools. By middle Pleistocene, a species of hominid had arisen that was so similar to Man that it is placed in the same genus: *Homo erectus* (Figure 19.21). This Man ranged widely throughout Africa and Asia, and is found fossilized in deposits ranging in age from 700,000 to 200,000 years. Like *Australopithecus, Homo erectus* used stone tools, principally hand axes fashioned from stones by flaking and chipping. By about 500,000 years ago, however, *Homo sapiens*, evolved from *H. erectus*, had arrived on the scene.

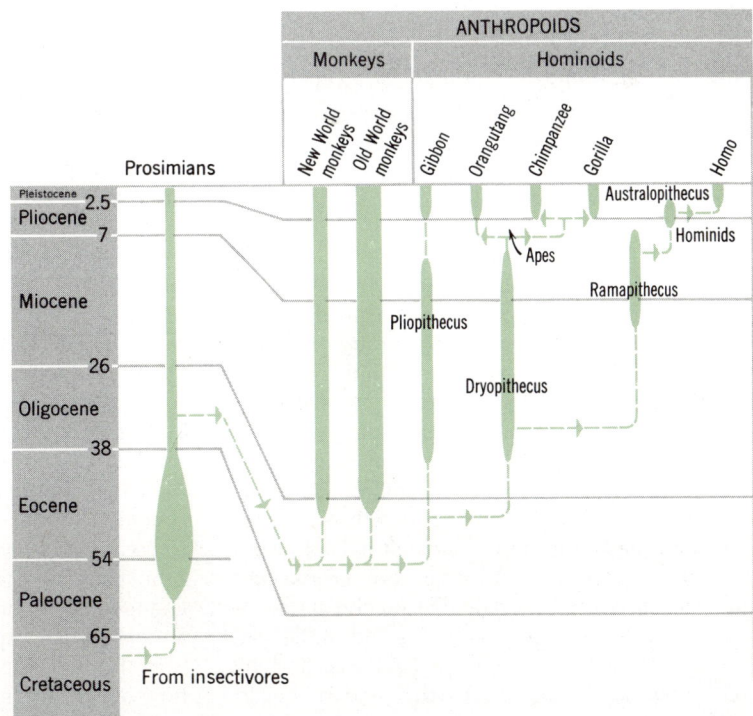

Figure 19.19 Evolutionary diagram for the primates. (Data from A. L. McAlester, 1968, *The History of Life*, Prentice-Hall, Englewood Cliffs, N.J., pp. 130–131, Figure 7.1. From A. N. Strahler, 1971, *The Earth Sciences*, 2nd ed., Harper & Row, New York, p. 543, Figure 30.26.)

The Evolution of Life Forms

Figure 19.20 Individuals of *Australopithecus* as they may have appeared in their eastern African habitat during early Pleistocene time. (Reproduced by permission from J. Augusta and Z. Burian, *Prehistoric Man*, Artia, Prague.)

The transition from *Australopithecus* to *Homo erectus* to *H. sapiens* is characterized mainly by an increase in brain size, shown below.

Hominid	Brain Size (cc)
Australopithecus	600–700
Homo erectus	900–1100
Homo sapiens	1400–1600

The skulls of these hominids, as well as that of Neanderthal man, a heavy-boned race of *H. sapiens*, are shown in Figure 19.22. Thus the brain size of Man more than doubled through his evolution in the Pleistocene Epoch. With the increase in brain size came the development of human culture. As Man developed tool- and weapon-making to a fine art, he became more and more successful at hunting and gathering food, and his numbers increased. By about 15,000 years ago, Man had made the important cultural advances of making pottery, cultivating root crops, and domesticating animals. Freed from the limits of a food supply gained by hunting and gathering, Man developed cities and civilizations, and by 5,000 years ago, learned to write; and so began recorded history.

Man and Evolution

It is appropriate to close our discussion of evolution with Man, for many of the principles we have discussed in this chapter have important implications for the human species. While some of these topics will be discussed in greater detail in later chapters, a preview should increase your understanding of Man's important role in the evolutionary process.

Man acts as an evolutionary agent by greatly modifying the earth's environments. As an example, consider agriculture. Clearing vast areas and raising a relatively small number of plant species has opened the door to exploitation by animals adapted to feed on them. By removing preexisting food sources, Man has simultaneously acted to greatly limit the number of organisms depending on the natural food sources which were replaced. Man's agricultural activities, then, favor a few species at the expense of others. Damming rivers to create lakes, polluting air and water, and many other human activities can also be viewed as changes of environments to which plants and animals must respond.

In the latter part of this chapter we have seen how environmental change leads to radiations and extinctions. A crafty and efficient predator,

Figure 19.21 Individuals of *Homo erectus*, shown in a Far Eastern tropical forest habitat, as they might have existed during middle Pleistocene time. (Reproduced by permission from J. Augusta and Z. Burian, *Prehistoric Man*, Artia, Prague.)

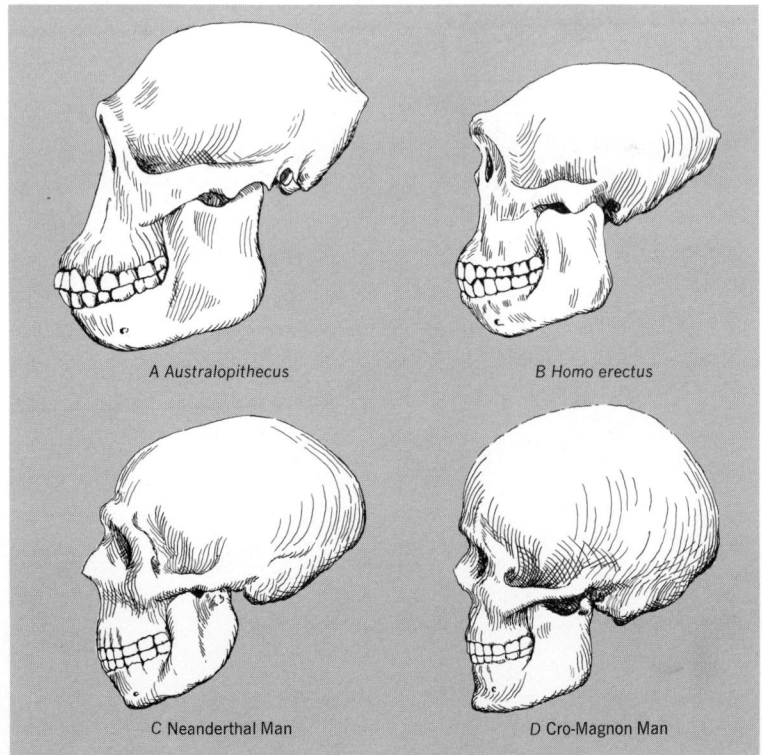

Figure 19.22 Sketches of fossil skulls of Pleistocene hominids. (Drawn from photographs by the American Museum of Natural History. From A. N. Strahler, 1972, *Planet Earth: Its Physical Systems through Geologic Time*, Harper & Row, New York, p. 335, Figure 14.20.)

Man has already produced many extinctions or near extinctions in his food-gathering activities. The passenger pigeon, now totally extinct, and the American bison, reduced to a population of a few thousand, are familiar examples. Many biologists believe that most species of whales have been so reduced in number that their continued survival is no longer possible. And every year the list of endangered species grows as Man's impact on the environment increases.

Since man-induced change is so rapid, we might expect that extinctions will be much more common than radiations. For radiations, we must turn to groups which evolve rapidly. Recall that the rate of evolution depends on (1) the mutation rate, (2) the size of the gene pool, (3) the time between generations, and (4) the mode of reproduction. These factors act to favor three groups, the insects, the annual weeds, and the bacteria, mostly because of their large numbers and rapid generation times. Since the discovery of modern chemical insecticides in the Second World War, insects preying on crops and other economically important plants and animals have been deluged with vast quantities of toxic agents. Under such severe selection pressure, only the more resistant strains have survived, producing an evolutionary race between insects and industrial chemists developing new pesticides. Annual weeds have come to dominate wide areas where Man has cleared or cultivated the land, and like insects, have evolved strains which make the most of man-influenced environments. Mutations in microorganisms have similarly occurred, and now many pathogenic bacteria and viruses have also evolved strains resistant to common antibiotics. Rather than forming a biological conspiracy, these new forms are the result of evolutionary mechanisms which have been active since the dawn of life on earth.

On the other hand, Man's applications of evolutionary principles have proved exceedingly beneficial to the human race. Largely by inducing mutations, crop plants with higher yields and resistance to diseases have been bred, greatly increasing Man's food supply. Cattle, fowl, and swine have similarly been bred to produce genetic lines which mature faster and have more usable meat. It is largely these agricultural advances which have thus far allowed the world's food supply to keep pace with its rising human population.

Changing environments to affect natural selection is only one of the many facets of man's environmental impact. In order to appreciate some of the other biological changes induced by human activity, we must first gain some basic knowledge about how energy and matter move through ecosystems—and these topics are the subjects of the two following chapters.

20 Energy Flow in Organisms and Ecosystems

For continued survival, organisms and ecosystems require constant inputs of energy. Recall from our chapter on energy systems that entropy measures the state of disorder of a system, and that entropy tends to a maximum when a system is closed. In other words, a system will spontaneously lose energy as heat, and the matter contained in the system will degrade to its simplest states, if it is left without new inputs of energy. Thus the prevailing tendency in the physical world will be toward the lowest energy level. When an organism is viewed as an energy system, it is indeed characterized by a very high state of order and organization, particularly organization on the biochemical level and spatial organization on the cellular level and above. Therefore, the maintenance of organic life requires a continual energy input, for it runs counter to the physical law of spontaneous increase in entropy.

Energy once received by an organism or ecosystem passes through a series of transformations, mostly from one form of chemical energy to another. With each transformation, however, some energy will be converted to heat and ultimately radiated away, lost to the surroundings. Thus energy movement through an organism or ecosystem can be conceived of as a one-way flow which diminishes in amount with each transformation until it all has been lost as waste heat.

In addition to sustained energy inputs, organisms require matter inputs. Since individual organisms are continually losing matter to their surroundings, inputs of matter will be constantly needed for growth and maintenance. Flow of matter differs from flow of energy in that matter is not lost to the system at each transformation. Even the matter which leaves one organism may enter another and be reused completely. Thus matter in ecosystems tends to be used and reused, or *cycled*. A fundamental difference, then, in the movement of energy and matter through ecosystems is that energy flows through ecosystems whereas matter cycles within ecosystems. Cycling of materials will be dealt with in Chapter 21.

Organisms differ in the types of energy and matter needed to sustain them. The autotrophs require only the simplest forms of energy and matter; these are light, water, carbon dioxide, and oxygen, as well as some sixteen inorganic nutrients. Their biochemical systems are capable of producing enzymes and other macromolecules entirely from these inorganic substances. Heterotrophs, on the other hand, have lost through evolution the ability to utilize light energy directly and instead depend on chemical energy in the form of reduced carbon compounds of organic origin. In addition, heterotrophs may require specific molecules which they are no longer able to synthesize for themselves—for example, most heterotrophs cannot synthesize all the amino acids and must obtain those they lack from organic sources in their diet. Thus we might divide energy and matter inputs to organisms into two subgroups each: *inorganic inputs*, which will be more important for autotrophs, and *organic inputs*, which will be more important for heterotrophs. This same division holds, of course, for outputs as well.

Variation in Input Rates

The amount of energy and matter available to an organism varies considerably through time and from place to place. When an input rate is reduced, either the output rate must be reduced or storage within the organism must be utilized. When an input rate is increased, output rates or storage, or both, must be increased. This observation can be placed in mathematical terms:

Input rate = Storage rate + Output rate

which is simply the *budget equation* discussed in the introductory chapter on energy systems. Each term represents time rate of change for the whole system. For example, the input, storage, and output of energy all can be stated in units of kilocalories per day.

The variation in light energy as it affects autotrophs provides one example of energy input variation. During the day, energy inputs in the form of light are high, and so considerable amounts of carbohydrate—a chemical compound of carbon, oxygen, and hydrogen—are manufactured by photosynthesis and are accumulated in storage. During the night, light inputs are reduced to zero, but energy consumption continues as biochemical processes proceed. The energy lost in metabolism must be replaced by withdrawals of chemical energy in the form of stored carbohydrate. Thus autotrophs solve the problem of daily energy input variation by maintaining an energy storage reservoir sufficient to last the night. Hibernating animals, another example, have a similar problem—seasonal lack of food supply—but have evolved a different adaptation. They reduce energy output drastically by entering hibernation, in which life processes are greatly slowed. In this fashion, the energy storage they have built up during the favorable season lasts for the duration of the unfavorable season because it is depleted so slowly. Thus organisms can cope with input variations by changing storage and/or output rates.

Many organisms possess the ability to increase or reduce input rates as well. Broad-leaved plants may turn their leaves to face the sun fully or may tilt them at an oblique angle to reduce the amount of intercepted solar radiation. And, obviously, much of the behavior of animals is conditioned to providing suitable input rates of chemical energy as organic foods.

Tolerance Ranges and Ecotypes

The ability of organisms to control input, storage, and output rates is limited, however. If input is too slow and output cannot be reduced, storage may suffice for a time, but ultimately damage to the organism will result if such an unfavorable balance of input and output rates is sustained. The same result occurs if input is too fast and output cannot be raised—unwanted storage will occur which can interfere with normally functioning biochemical processes. Therefore, we can define a *tolerance range* for each input to an organism—an upper and lower rate of input of energy or matter above or below which that organism's life processes are impaired.

As an example of a tolerance range, consider the effect of heat on the hatching of brook trout eggs. Since the eggs are small, little heat storage within the egg is possible—thus, heat input must equal heat output. At high levels of heat input and output (high temperatures), energy absorbed by *macromolecules* (largely complicated molecules such as enzymes) on which biochemical processes depend, will be released by overspeeding the reactions or by actual molecular breakdown, and thus the delicate life systems will be damaged. At low levels of heat input and output, energy absorbed by macromolecules will be insufficient for reactions to proceed normally and injury will also result. By experiment we can determine that the eggs require a heat input rate which results from a surrounding water temperature between 0° and 10° C (32° and 50° F), the *minimum* and *maximum* for heat content in brook trout eggs.

Naturally, different species will have different tolerance ranges, minimums, maximums, and optimums. Frog eggs (*Rana pipiens*) will develop between 0° and 30° C (32° and 86° F), and have an optimum for growth at 22° C (70° F). Thus frog eggs are much more tolerant of variations in heat flow and heat content than are brook trout eggs. Ecologists refer to species which have broad heat content tolerance ranges as *eurythermal;* the opposite term is *stenothermal,* meaning with a narrow heat content tolerance range. In the examples above, trout eggs are stenothermal, while frog eggs are eurythermal. The prefixes *eury-* and *steno-* may be used with other suffixes. An example is *euryhaline,* which describes a species adjusted to broad variations of salinity.

Not only do tolerances vary from one species to another; tolerances also will vary from one stage in the life cycle to the next. As a general rule, the youngest stages or the actively reproducing stages of an organism's life cycle have the narrowest tolerances and therefore are most sensitive to environmental changes. As an example, adult crabs are often found in brackish water or fresh water with a high chloride level. Yet their larvae are much less tolerant of low salt levels and cannot survive in brackish or fresh water. Thus a much broader range of environments is open to the adult crab than to its juvenile offspring. The reproductive process in many adult organisms is also quite sensitive to environmental variation. Often an adult will not reproduce, even though its needs are minimally satisfied—plants found in extreme environments will often be healthy enough, but may only rarely flower and set seed.

Even within species there will be a considerable variation in the response to different input levels of environmental factors. Figure 20.1 shows how the heights of the perennial herb *Achillea lanulosa,* or yarrow (Figure 20.2), vary along a west-to-east transect of central California and Nevada,

Figure 20.2 Woolly yarrow, *Achillea lanulosa.* This species exhibits ecotypic differentiation, as shown in Figure 20.1. (American Museum of Natural History.)

across the Sierras and into the Great Basin. The heights of the plants decrease systematically with elevation, an effect we might at first think is attributable to cooler temperatures at higher elevations. However, the heights are not of individuals measured in the field, but are of plants started from seed taken from each elevation and grown

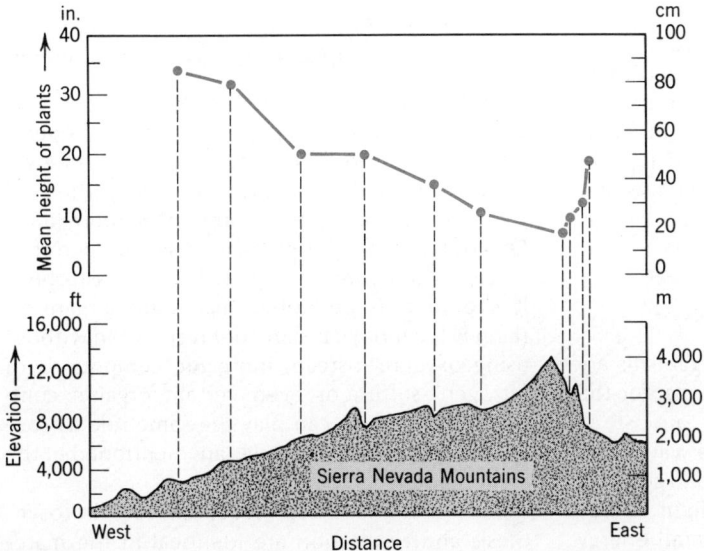

Figure 20.1 Heights of *Achillea lanulosa* plants grown from seeds collected at ten locations in the Sierra Nevada Mountains. (Data of J. Clausen, D. D. Keck, and W. M. Hiesey, 1948. After E. J. Kormondy, 1969, *Concepts of Ecology,* Prentice-Hall, Englewood Cliffs, N.J., p. 145, Figure 5.20.)

under uniform conditions in a greenhouse. Thus the heights are not the immediate result of environmental differences, but are produced by genetic differences alone. Because interbreeding is rare between the populations of *Achillea* which were collected, each population has gone its own way, evolutionarily speaking. Since each population is slightly different ecologically from the next, these populations are referred to as *ecotypes*. Because ecotypes can interbreed, they are not recognized as different species, although they may be distinguished as subspecies or varieties by taxonomists. Most widely distributed species have many ecotypes, each grading into the next.

Multiple Inputs and Limiting Factors

A final point to be made before turning to a systematic coverage of energy inputs in this chapter and of matter inputs in the next is that many biological processes of organisms and ecosystems require *multiple inputs*. Thus, as we shall see in the following pages, light, carbon dioxide, and water are needed as inputs for photosynthesis. Each factor must be present for the process of photosynthesis to proceed. If one factor is present only in minimal quantities, the entire process will be limited. Ecologists use the term *limiting factor* to describe an energy or matter input whose low rate of input holds an array of reactions or processes in check.

That limiting factors can control the growth of crops was first observed by Justis Leibig in 1840 and is known as *Leibig's Law of the Minimum*. Leibig's law states that the growth of a crop is held in check by the nutrient which is scarcest. Thus, yields of corn may be significantly improved by addition of nitrate fertilizer where nitrogen is limiting, or by addition of phosphate fertilizer when phosphorus is limiting. When Leibig's law is applied beyond nutrients to more general factors, we find that plant growth and numbers can be limited by low energy inputs of light or heat. Similarly, the size of animal populations may be, and usually is, limited by organic energy inputs.

Light as an Energy Input

Light energy inputs to autotrophic organisms are of the highest importance in ecosystems, for the products of photosynthesis support almost all earthly life. As we have seen, light is a wave form of energy and can be readily converted to sensible heat by absorption. In this process, light energy is absorbed by a molecule whose overall energy level increases. At a slightly later time, this energy is suddenly released as the molecular energy level falls. For molecules at temperatures near that of the earth's surface, this energy is usually released as heat radiation. Thus light energy interacts with molecules, is stored momentarily, then released in the form of heat. In the case of autotrophs, however, this light energy becomes trapped and can be passed through a complicated biochemical chain to reside eventually in stable chemical storage, a process called *photosynthesis*.

Photosynthesis and Respiration

The overall effect of photosynthesis is to unite the hydrogen atoms of water with the atoms of carbon dioxide to form carbohydrate and release oxygen. *Carbohydrate* is the name given to a class of molecules with a three-to-six carbon chain along which each carbon also combines with water, as H^+ and OH^-. The overall reaction for photosynthesis can be expressed as follows:

$$H_2O + CO_2 + \text{Light energy} \xrightarrow{\text{Chlorophyll}} -HCOH- + O_2$$

The symbol —HCOH— stands for one of the units in the carbon chain which are linked to form the carbohydrate.

While this reaction may seem simple as presented above, it is actually a very complex process involving two forms of *chlorophyll*, a green pigment particularly adapted to absorbing light energy and releasing it in a usable form.

Respiration is the complimentary process to photosynthesis in which carbohydrate is broken down and combined with oxygen to yield carbon dioxide and water. The overall reaction is

$$-HCOH- + O_2 \longrightarrow CO_2 + H_2O + \text{Chemical energy}$$

As in the case of photosynthesis, the actual reactions are far from simple. The chemical energy released is stored in many types of energy-carrying molecules and used for later synthesis of proteins, fats, and other forms of carbohydrate. Of course, the energy transfer is not perfectly efficient, and some energy is lost as waste heat. It should also be noted that some organisms (largely bacteria) can carry out respiration without using oxygen. Instead, inorganic compounds of nitrogen, sulfur, or even certain organic compounds of carbon can play the same role as oxygen. Some specific examples are mentioned in the next chapter.

Note that the overall reactions for photosynthesis and respiration are identical in the mate-

Response of Plants to Varying Light Intensities

rials involved, but are reversed. Because respiration and photosynthesis are complimentary processes which are always occurring when a plant is exposed to light, we must distinguish between gross photosynthesis and net photosynthesis. *Gross* photosynthesis is the total amount of carbohydrate produced by photosynthesis; *net* photosynthesis is the amount of carbohydrate remaining after respiration has broken down sufficient carbohydrate to power the plant. Stated as an equation,

Net photosynthesis
 = Gross photosynthesis − Respiration

Since both photosynthesis and respiration occur in the same cell, gross photosynthesis cannot be readily measured. Instead, we will deal with net photosynthesis in further discussions. In most cases, respiration will be held constant, so that use of the net instead of the gross will show the same trends.

Figure 20.3 presents a model curve for the response of photosynthesis in a plant to increasing light intensities. Net photosynthesis is measured by the CO_2 uptake rate. When light is absent, respiration will liberate CO_2 and the CO_2 uptake rate will be negative (and will therefore be an output). At a certain low light intensity (*A*), photosynthesis will equal respiration and CO_2 uptake will be exactly zero. This is the *compensation point*, for here photosynthesis is exactly compensating for respiration. At low to moderate light intensities, net photosynthesis increases linearly (in direct proportion) with light intensity. This portion of the curve is termed the *linear range* (*B*). Since each additional increment of light produces a corresponding constant increment in photosynthetic rate, in this range light must be limiting to photosynthesis and therefore must be controlling the photosynthetic rate.

At some point (*C*), photosynthesis ceases to increase. This point is the *saturation intensity*, for here photosynthesis is saturated with respect to light. Beyond this saturation intensity, net photosynthesis is limited by a necessary factor other than light. Returning to the photosynthetic equation, note that it requires multiple inputs—water, CO_2, and light. Therefore, one of these multiple inputs besides light has probably become limiting. In most cases, the limiting factor will be CO_2. Since this gas is present in such low concentrations in the atmosphere (0.03%), the plant simply cannot take in CO_2 fast enough to accomodate increasing levels of synthesis. This deficiency produces the flat upper part of the curve. At very high light intensities, net photosynthesis declines (*D*) because light produces heating effects which increase respiration, therefore decreasing the net CO_2 uptake. In addition, very high light intensities may destroy chlorophyll and other molecules important to photosynthesis, reducing the photosynthetic rate.

Response of Plants to Varying Light Intensities

Genetic differences between plant species can restrict them to shady or sunny environments through adaptation to low and high light input rates. Plants which do best in sun are *heliophytes* Figure 20.4*A*); those adapted to shade are *sciophytes* (Figure 20.4*B*). Foresters also classify tree species as tolerant or intolerant. An *intolerant* species cannot grow from seed in the shade. Many conifers are intolerant, requiring the strong sunlight of open areas for seedling growth. *Tolerant* species are those which can germinate and grow in the shade of the forest understory. Tolerant saplings remain in the understory for many years; when an opening occurs in the forest canopy, they are released and quickly grow up to fill the opening. These physiological differences between species are important in determining the course of plant succession (discussed in Chapter 22), the orderly series of changes which the vegetation in an area follows after clearing, burning, blowdown, or other disruption.

Table 20.1 gives the compensation points for several common tree species. The data are obtained from seedlings rather than mature trees.

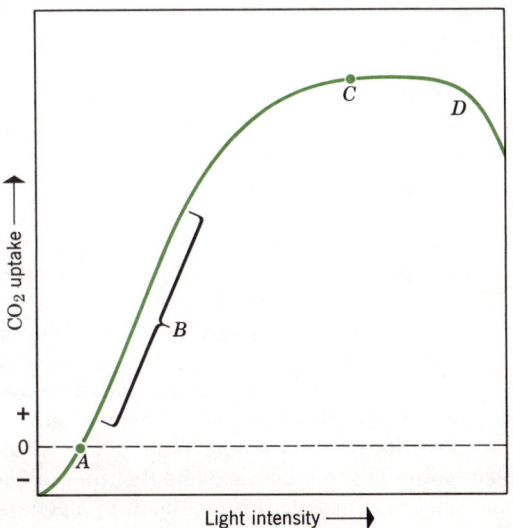

Figure 20.3 Model curve for the response of photosynthesis in a plant to increasing light intensities. (After A. C. Leopold, 1964, *Plant Growth and Development*, McGraw-Hill, New York, p. 21, Figure 2.10.)

A. The heliophyte Queen Anne's lace, *Daucus carota*, dominating an abandoned field.

B. The sciophyte bunchberry, *Cornus canadensis*, rampant in the dense shade of this northern hardwood-fir forest.

Figure 20.4 Heliophytes and sciophytes. (Photographs by Grant Heilman.)

There is a wide range of variation—about tenfold between ponderosa pine and sugar maple. These values also illustrate tolerance and intolerance. In general, the trees with higher compensation points are intolerant; those with low compensation points are tolerant and thus better able to survive shaded conditions. In addition, multicellular plants have a higher compensation point than do unicellular photosynthetic organisms such as algae. This effect arises because not all the cells in a multicellular plant can engage in photosynthesis, and their respiration must be supported by the photosynthetic cells.

Figure 20.5 shows net photosynthesis rates for a single oat leaf and for a whole oat plant. Net rate of photosynthesis is expressed as CO_2 uptake per hour for each 50 cm^2 of leaf area. Note that the single leaf reaches the saturation intensity at about 600 foot-candles of light intensity. (For comparison, bright summer sunlight has an intensity of about 10,000 foot-candles.) At this same light intensity, the whole plant has a lower photosynthetic rate. This effect occurs because some of the leaves are shaded by others and therefore photosynthesize more slowly. At higher light intensities, the photosynthetic rate of the whole plant continues to increase while that for a single leaf remains constant. This fact leads to a general observation true of almost all broad-leaved plants: the plant as a whole operates more efficiently at higher light input levels than does a single leaf.

TABLE 20.1 Compensation Points of Tree Seedlings

Tree Species	Compensation Point[a]
Ponderosa pine	31
Northern white cedar	19
Larch	18
Red oak	14
Engelmann spruce	11
Eastern white pine	10
Eastern hemlock	8 1/2
Beech	7 1/2
Sugar maple	3 1/2

SOURCE: Data from G. P. Burns, 1923, *Ecology*, vol. 4. (Data rounded.)
[a] Expressed as a percentage of full winter sunlight in Maine.

Photoperiodism

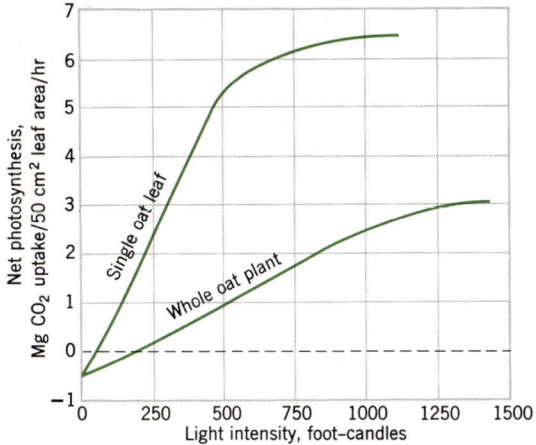

Figure 20.5 Response of photosynthesis in single oat leaf and whole oat plant to increasing light intensity. (Data of Boysen-Jensen, in A. C. Leopold, 1964, *Plant Growth and Development*, McGraw-Hill, New York, p. 22, Figure 2.11.)

Experiments have also shown that leaves from the same plant can become adapted to differing light input rates. Figure 20.6 shows two curves for leaves of a beech tree: one for leaves growing in the shade at the bottom of the tree, the other for leaves growing at the top of the tree and exposed to full sunlight. At low light intensities, the shade-grown leaves photosynthesize faster. As light input rates increase, however, these leaves are soon saturated. Beyond this point, the sun-grown leaves have a higher photosynthetic rate. Thus, photosynthesis in an individual leaf is often adjusted to local conditions, and therefore the plant as a whole system photosynthesizes more efficiently.

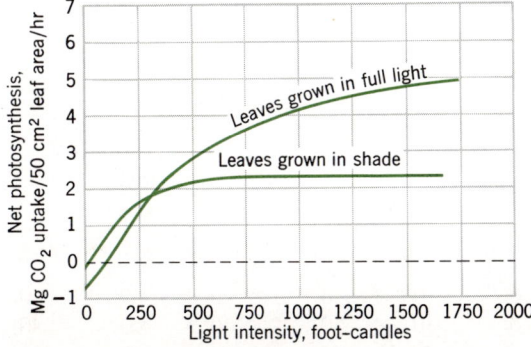

Figure 20.6 Variation of photosynthesis with light intensity for shade-grown and sun-grown beech leaves. (Data of Boysen-Jensen and Müller, in A. C. Leopold, 1964, *Plant Growth and Development*, McGraw-Hill, New York, p. 22, Figure 2.12.)

Photoperiodism

Besides providing the energy necessary to support ecosystems, light has another important function—it provides cues to organisms which regulate growth and behavioral activities. In this way, a small amount of light input can act as a *trigger* for complex behavioral responses. Insolation varies with the season of the year and the latitude, as shown in radiation graphs (Figures 2.6 and 2.7). Day length, or *photoperiod*, acts to signal activities ranging from dormancy to reproduction in both plants and animals. This response of organisms to day (or night) length is *photoperiodism*.

Figure 20.7 shows how day length varies with latitude and season. At the spring and fall equinoxes, the day and night are of approximately equal length at all latitudes. At any other time of the year at locations away from the equator, day and night lengths show seasonal variation, increasing with latitude. The variation is greatest in extreme northern or southern latitudes; above the arctic and antarctic circles, day or night lasts a full 24 hours at the solstice, while at the poles the sun is above or below the horizon continuously for six months.

As an example of the significance of photoperiodism, consider the flowering of a plant. If flowering is too early, a late frost can damage the flowers; if it is too late, an early frost may kill the fruit before it has had a chance to mature completely. Sometimes timing is critical to ensure that pollinating birds or insects are abundant; in other cases, small plants may have to flower and begin ripening fruit early in the spring before they are shaded out by larger plants. While temperature is sometimes used by the plant as a cue to flowering, there will be great variations from year to year. One factor which remains unchanging, however, is day length, and indeed, most of the plants of the middle-latitude zone depend on day length to regulate flowering. This dependence of flowering on photoperiod has led plant physiologists to classify plants as short-day or long-day plants. A *long-day* plant requires a day length longer than a certain critical length for flowering; a *short-day* plant requires a day length shorter than a certain critical length to flower. Those plants whose flowering occurs independent of day length are *day-neutral* plants.

Table 20.2 shows how three populations of alpine sorrel (*Oxyria digyna*, Figure 20.8) respond to different photoperiods. Alpine sorrel is a small, perennial, long-day herb which is adapted to alpine (mountain) and arctic environments, and is found scattered throughout the North American Cordillera as isolated populations on mountain

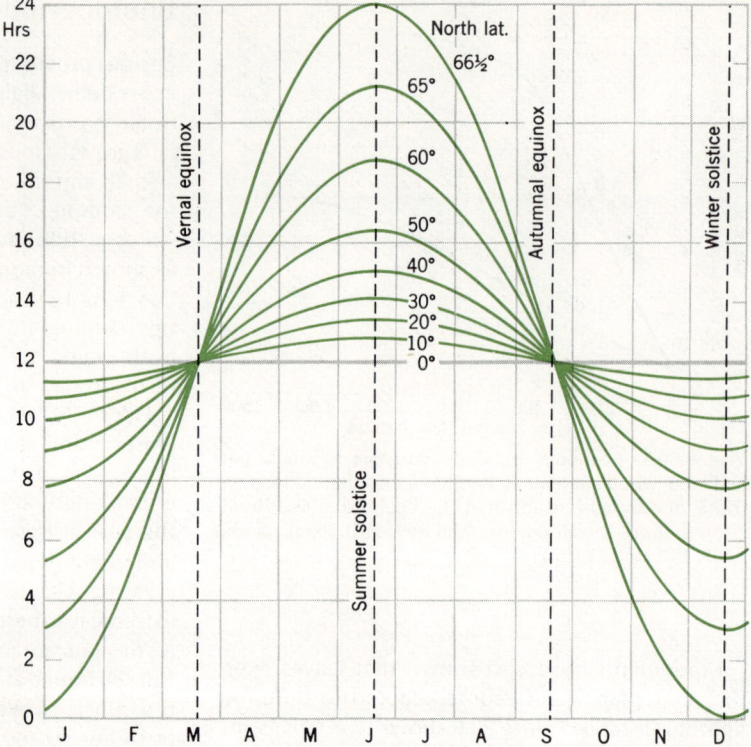

Figure 20.7 Duration of the day at various latitudes throughout the year. Vertical scale gives number of hours the sun is above the horizon. (Not corrected for refraction of sun's rays by the atmosphere.)

summits. The populations used in the table are grown from seed collected at three locations: Elephant's Back Mountain, California, in the Sierra Nevada; Logan Pass, Montana, in the Rockies; and Sagavanirktok River, Alaska. In an interesting experiment, ecologists grew the plants in environmental chambers, varying only the length of day in each. Three chambers were used, with photoperiods of 12, 15, and 24 hours. As each week passed, the number of plants which had flowered were counted, and expressed as a percentage of the number of plants of each population within each chamber. These percentages are shown in the table.

The results of the experiment showed that each population of alpine sorrel is adjusted to the photoperiod of the region where the samples were collected. At Elephant's Back, the normal summer photoperiod is between 13 and 15 hours; thus no plants from this location flowered in the 12-hour environmental chamber, but all flowered in the 15- and 24-hour chambers. At Logan Pass the normal summer photoperiod is 16 hours; thus none flowered in the 12-hour chamber, and all flowered in the 24-hour chamber. In the 15-hour chamber, where the photoperiod was 1 hour less than the normal one, about half the plants flowered, showing that some variation in day length

TABLE 20.2 Flowering of Alpine Sorrel Plants Collected from Three Locations

Week	Chamber:	Elephant's Back, California			Logan Pass, Montana			Sagavanirktok River, Alaska		
		12-hr	15-hr	24-hr	12-hr	15-hr	24-hr	12-hr	15-hr	24-hr
1		0	0	0	0	0	0	0	0	0
5		0	80	60	0	10	10	0	0	40
10		0	100	80	0	20	90	0	0	100
15		0	100	100	0	50	100	0	0	100

SOURCE: Data from H. A. Mooney and W. D. Billings, 1961, *Ecological Monographs*, vol. 31, pp. 1–29.

Figure 20.8 The alpine sorrel, *Oxyria digyna*, in one of its native habitats, the Beartooth Mountains, Wyoming. (Courtesy of Dr. W. D. Billings.)

Biological Clocks

Light also influences animal behavior. Many animals possess *biological clocks*—inborn timing mechanisms regulating behavior which are triggered or updated by external rhythms, including tides, lunar cycles, and variations in photoperiod. Of interest here are the light-regulated rhythms, which are largely of two types: *circadian rhythms*, centered on a daily cycle and cued by day and night alternation, and *annual rhythms*, centered on an annual cycle and cued by photoperiod.

An example of a circadian rhythm is found in the activity patterns of the North American flying squirrel (Figure 20.10). This small mammal is perhaps the most common squirrel in our woodlands, but because of its nocturnal habit, it is rarely seen by most people. In its natural setting, the flying squirrel begins its activity at dusk; with the coming of dawn it returns to its nest in a hole within a tree. If the squirrel is brought into the laboratory, however, and subjected to conditions of constant darkness, its activity pattern still fits a diurnal pattern, with the greatest activity occurring at night. This behavior seems to occur with-

requirement for flowering exists within the Logan Pass population. Plants from Sagavanirktok River, Alaska, where the summer day is 24 hours long, flowered only in the 24-hour chamber. Thus these data illustrate the dependence of flowering in alpine sorrel on photoperiod. The data also show that alpine sorrel is differentiated into ecotypes. Although all the plants are similar enough to be placed in the same species, their ecological characteristics are different, as shown by these critical experiments on their photoperiod.

In addition to flowering, photoperiod can control other plant functions. Initiation of dormancy and breaking of dormancy, time of leaf fall, initiation and growth of tubers, pigmentation, and nutrient requirements have all been observed as properties or processes which are affected by photoperiod in a variety of plants. The biochemical mechanisms responsible for photoperiodism are largely unknown.

Of course, photoperiod is only one of many cues which can initiate developmental processes. Spring garden bulbs are mostly dependent on temperature for flowering. Precipitation can also trigger plant growth—you are probably all familiar with the massive blooming of desert plants following a chance thundershower (Figure 20.9). By and large, however, photoperiod is the dominant timer for most plant developmental processes.

Figure 20.9 Blooming of the desert. Following winter rains, the Mojave Desert is carpeted with brilliant spring flowers. (Richard Weymouth Brooks.)

Figure 20.10 The North American flying squirrel. The flaps of skin between this small mammal's legs and body are used to guide its leaps from one branch to another. (Photograph by Leonard Lee Rue III/National Audubon Society.)

out any known natural cues, and is therefore assumed to be an internal, or *endogenous*, rhythm. Just whether such rhythms are completely internal or are actually somehow cued by variations in air pressure, tidal gravitational field, magnetic field, or by other factors difficult to control in the laboratory, is a point still open to scientific debate. The circadian rhythm of the flying squirrel does not always follow the clock accurately; its biological clock tends to run slightly fast, and thus the normal cycle is slightly shorter than 24 hours. Individual squirrels, though, vary substantially in their daily cycle—variations observed are between about $22\frac{1}{2}$ hours and $24\frac{1}{2}$ hours. After a period of darkness lasting several days, the squirrel may get out of phase with the natural cycle; however, once he is returned to a natural light-dark cycle, the squirrel quickly becomes synchronized again. Thus the natural diurnal cycle of light and dark acts to "set the clock," keeping physiological activity attuned to biological activity.

Annual rhythms in animals are similar to photoperiodic responses in plants. Both regulate critical functions of growth and reproduction, and both are cued by photoperiod. The storing of nuts by the flying squirrel in late fall and early winter presents an example of annual rhythms affected by day length. In nature, storing begins in early October and increases to a peak in January. In a laboratory experiment, squirrels kept during the winter at a constant 15-hour photoperiod (equivalent to midsummer) showed only weak food-storage activity. A reduction of the photoperiod to 13 hours in mid-December produced a sharp upswing in food-storing behavior which continued through the winter. Another group of squirrels was subjected to a mid-November photoperiod in mid-October; food storage activity quickly followed the mid-November pattern.

Just as in plants, the annual rhythm has a purpose—it makes sure that young are born when food is abundant, or that birds migrate at suitable times to take advantage of far-distant food supplies, or that fish swim upstream to spawn when the rivers are running high. As with photoperiodism, the underlying biological mechanisms for circadian and annual rhythms remain unknown.

Heat as an Energy Input

The continuous input of heat energy to organisms from their surroundings maintains an internal environment suitable for biochemical reactions which sustain life. An equation for the heat balance is somewhat different than that of other environmental factors:

Input rate + Internal heat generation rate
$$= \text{Storage rate} + \text{Output rate}$$

Since waste heat is produced with each biochemical transformation within an organism, internal heat generation must be added to the equation. Of course, this extra input simply represents the release as heat of chemical energy which has been stored earlier. In some cases, the internal heat term can be fairly large; warm-blooded animals adjust rates of metabolism to balance variations in heat inputs and thus achieve a constant internal temperature. These animals are also highly capable of varying heat output rates by such mechanisms as sweating (which increases heat output) or by a covering of fur and feathers (which decreases heat output). The storage term in the equation must of necessity be small because it depends mostly on the mass of the organism. The larger the organism, the more heat is stored within it, but even the largest of organisms will cool quickly when heat input is greatly reduced, and degradation of biochemical life systems will eventually result. We should note here that input need not only be in the form of longwave radiation; the absorption of incident shortwave (light) energy of the solar radiation spectrum results in its conversion to sensible heat, providing a supplement to longwave inputs from the surrounding air, and from adjacent solid or liquid surfaces.

The flow of heat energy into and out of an organism is shown in Figure 20.11. Two types of energy impinge onto the animal—shortwave radiation directly from the sun or indirectly from the sky, and longwave energy from the earth, atmosphere, and objects surrounding it. Heat energy is lost by longwave radiation of the animal, by con-

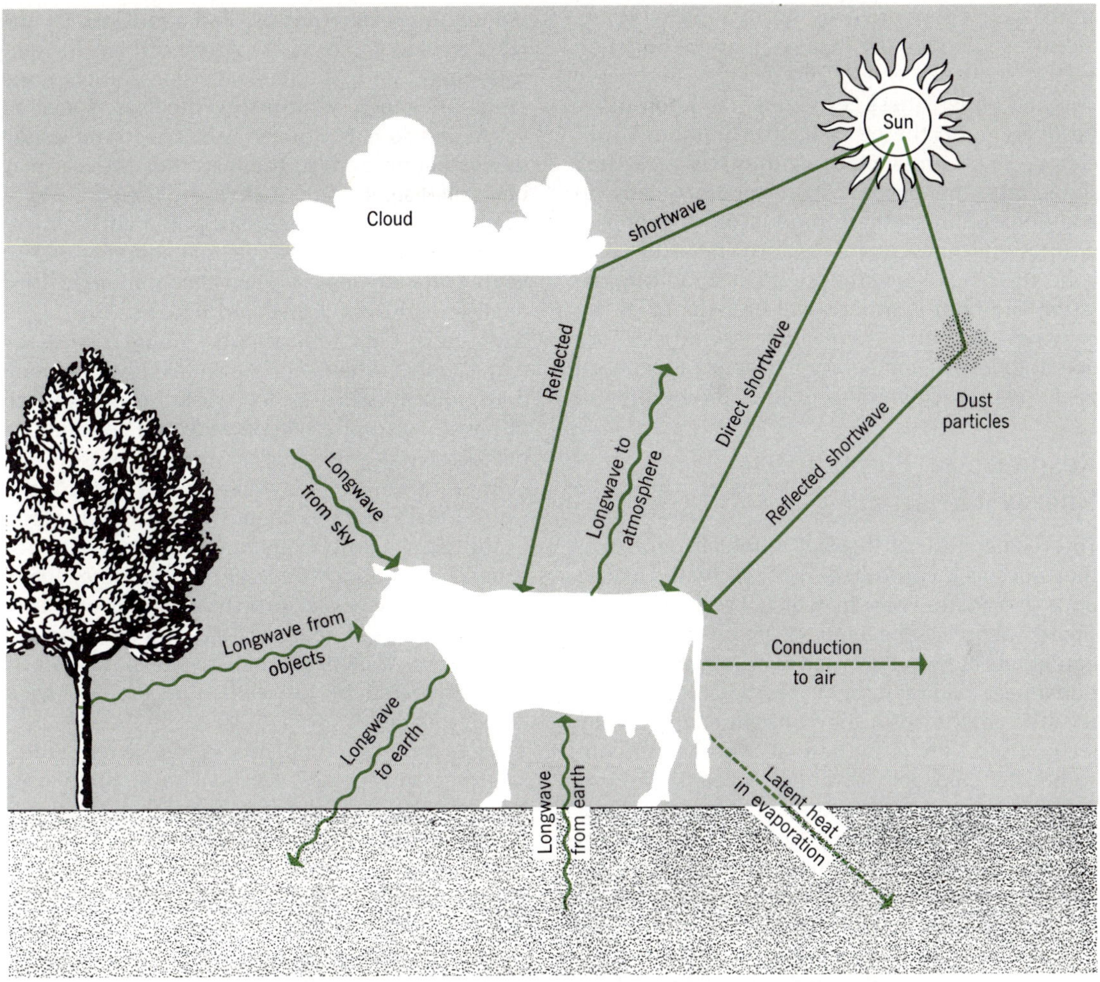

Heat as an Energy Input

Figure 20.11 Radiation energy flows to and from an animal. (After D. M. Gates, 1962, *Energy Exchange in the Biosphere*, Harper & Row, New York, p. 13.)

duction of its heat directly to the surrounding air, and by latent heat in the evaporation of water from its body surface. Both the wind speed and relative humidity of the air will affect losses by conduction and evaporation. The gains and losses of heat energy must equal each other, since energy cannot be destroyed. Thus a *heat balance* between an organism and its environment is maintained.

The heat balance between an organism and its environment determines the *internal temperature* of an organism, which reflects its heat content. In organisms whose tissue temperatures are not controlled by internal metabolism, temperature has a great effect by controlling the rates of biochemical reactions on which life is dependent. To illustrate this point, let us return for a moment to photosynthesis and respiration.

Figure 20.12 shows how gross and net photosynthesis, as well as respiration, vary with temper-

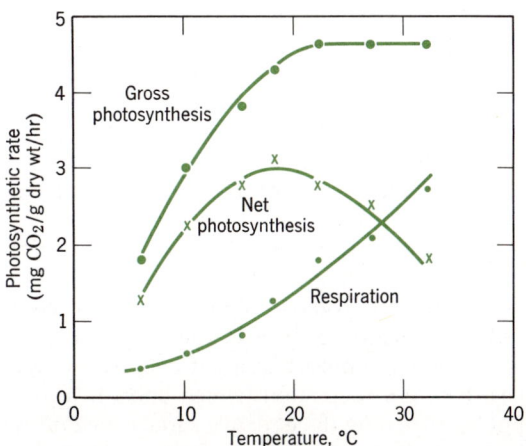

Figure 20.12 Variation in respiration and gross and net photosynthesis with temperature. *Sphagnum* moss in 750 foot-candle illumination. (Data of Stofelt, 1937, presented in A. C. Leopold, 1964, *Plant Growth and Development*, McGraw-Hill, New York, p. 31, Figure 2.26.)

ature. As temperature rises, gross photosynthesis increases more rapidly than respiration, but soon reaches a plateau where CO_2 or light is probably limiting under these experimental conditions. Respiration, on the other hand, continues to increase. Since gross photosynthesis is now at a stable rate, the increase in respiration takes up additional photosynthetic products and net photosynthesis declines. At very high temperatures (beyond the range shown in the figure), both respiration and photosynthesis will be reduced as the enzymes on which these processes depend are denatured. Thus, temperature changes affect each of the two biochemical processes differently.

Response of Plants to Heat Energy Variation

The temperature of the leaf in which photosynthesis occurs is controlled by the heat load placed upon it and the way in which the heat is dissipated. Whereas the temperature of a leaf should equilibrate at that of the surrounding air, an additional heat load (and higher temperature) usually results from the absorption of light energy by the leaf. Absorption is highest at wavelengths absorbed by chlorophylls, but significant amounts of radiation in other portions of the solar spectrum are also absorbed. Naturally, the angle of incidence of sunlight on the leaf is important; an orientation of just 10° from the direct line of radiation is sufficient to reduce absorption by 15%. If the leaf is tilted greater than 70°, absorption becomes negligible and the inflow of heat to the leaf is controlled only by the surrounding air. Some plants growing in strong sunlight will exhibit leaves tilted away from the horizontal, a mechanism presumably developed to cut light absorption and heating. Still other plants have ruffled or crinkled leaves which represent similar adaptations. Another method of reducing this energy input is to increase reflection. The glassy or shimmery appearance of desert plants represents increased reflection produced by scales or waxy secretions.

Heat output from a leaf occurs by conduction to the surrounding air, by longwave radiation, and, through transpiration, in the form of latent heat. Conduction of heat to the surrounding air is probably the most important cooling method. Leaf hairs are adaptations which boost heat output by increasing the effective surface area of the leaf and roughening the leaf surface, inducing turbulence and better mixing of the air layers nearest the leaf surface. Transpiration may account for as much as one-third of the leaf's heat loss; when leaf pores close to conserve water, leaf temperatures rise quickly. Radiation loss by the leaf results when it is warmer than its surroundings. At night under clear skies, much of the radiation energy emitted by the leaf is lost to space and so is no longer available for reflection or reradiation back to the leaf; thus the leaf may be cooled significantly below the temperature of the surrounding air. Smudge pots lighted in orange groves when a freeze is imminent help to keep radiation losses down and minimize frost damage to leaves, fruits, and flowers.

Whereas high temperatures injure tissues by denaturing proteins and enzymes, low temperatures injure tissues by ice crystal growth which physically disrupts cell structure and contents. In plants, crystal growth occurs mostly outside the cells, between the cell walls. If freezing is slow, water passes more easily out from the cytoplasm to the growing intercellular crystals, and damage is minimized. Rapid freezing, on the other hand, encourages ice crystal growth within the cell itself and so is much more damaging. Slow thawing will also minimize damage because the water can be reaccomodated by the cell more easily than if thawing is rapid.

Exposure of young plants to low temperatures can, in many species, *harden* the plant to frost injuries. Although the biochemical mechanisms of hardening are little understood, hardened plants usually have a thicker sap, thicker cell walls, and fewer starch granules within the cells. The thickening of the sap seems to slow water movements in and out of the cell; the thicker cell walls resist penetration by ice crystals; and reducing the number of starch granules removes these sources of mechanical abrasion which are moved about as internal ice crystals grow.

Response to temperature variation, or *thermoperiodicity,* can also play an important role in an organism's growth and development. Many plants reach optimum growth rates only when exposed to lower night temperatures in alternation with higher day temperatures. Often a difference of only a few degrees can result in significant growth rate increases. In addition, temperature requirements can vary from one stage of the life cycle to the next. Figure 20.13 shows optimum temperature plotted against age for one variety of tulip. To initiate the earliest developmental stage of flower buds within the bulb, warm temperatures in the preceeding fall are required. Further development of the buds proceeds best at lower temperatures, but warmer temperatures are again required for bud and stem elongation. Thus, a warm period is needed for the initial development of buds; following this warm period, the plant is adjusted to the cold temperatures which gradu-

Response of Animals to Temperature Variation

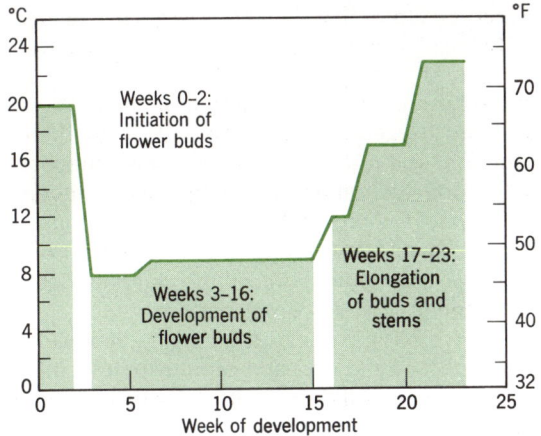

Figure 20.13 Optimal temperature for each week of development in bulbs of the Copeland tulip. (Data of Hartsema, et al., 1930; after A. C. Leopold, 1964, McGraw-Hill, New York, p. 381, Figure 22.20.)

ally rise as winter wanes. As spring progresses, increasing temperatures become favorable for bud and stem elongation, and finally the flower buds burst, sending forth their familiar blooms.

In many plants, dormancy is regulated by temperature variation. Typically the short days of late summer or fall will initiate dormancy of buds, but a winter period of low temperature is necessary if the buds are to break dormancy in the following spring. In addition, seeds of many plants require a low temperature period of several weeks or months for germination. These mechanisms insure that the delicate shoots and leaves of new growth will not emerge under hostile conditions.

Response of Animals to Temperature Variation

The effects on animals of variations in heat input rates are moderated by their physiology and by their ability to seek sheltered environments. Most animals lack a physiological mechanism for internal temperature regulation. These animals, including reptiles, invertebrates, fishes, and amphibians, are *poikilotherms*; their body temperatures are *poikilothermal*, passively following the environment. With a few exceptions (notably fish and some social insects), poikilotherms are active only during the warmer parts of the year. They survive the cold weather of the middle-latitude zone winter by becoming dormant. The smaller invertebrates may use this dormant period as a convenient time to progress from one stage of a life cycle to the next. Some vertebrates enter a state of suspended animation, or *hibernation*, in which metabolic processes virtually halt and body temperatures closely parallel those of the surroundings. (Summer dormancy is *estivation*.) Most hibernators seek out burrows, or nests, or other environments where winter temperatures do not reach extremes or fluctuate rapidly. As we have seen in Chapter 3, annual range of soil temperatures is greatly reduced below the uppermost layers (Figures 3.5 and 3.6); thus soil burrows are particularly suited to hibernation.

The range of temperatures which poikilotherms can endure actively is quite large; only below about 43° F (8° C) and above 108° F (42° C) do they become inactive. Since their body temperatures must follow their environment, their physiological demands must fluctuate as metabolism proceeds rapidly or slowly. For example, bass require only 10% of the oxygen at 41° F (5° C) than at 84° F (29° C). On a daily basis, many poikilotherms regulate their activity cycle to correspond with favorable temperature periods. Some desert dwellers, for example, are inactive during the day, but emerge at night or near dawn or dusk to carry out normal activities.

In contrast to the poikilotherms are the *homeotherms*, those animals whose tissues are maintained at a constant temperature by internal metabolism. This group includes the birds and mammals. Homeotherms possess a variety of adaptations to influence heat output rates. Fur, hair, and feathers are the most obvious examples— these materials act as insulation by trapping dead air spaces next to the skin surface, reducing conductive heat loss to the surrounding air or water. A thick layer of fat will also provide excellent insulation. Other adaptations act to increase heat outputs—for example, sweating or panting capitalizes on the high latent heat of vaporization of water to remove heat. Heat loss is also facilitated by exposing vascularized areas to the cooler surroundings. The seal's flippers and bird's feet serve this function. Most of these structural adaptations can be regulated to control heat output rates from moment to moment or season to season.

Another interesting point is that homeothermal characteristics were evolved independently in two different vertebrate groups—the birds and the mammals. Recall from Chapter 18 that both groups radiated from the reptilian vertebrates during the later Mesozoic Era. The aquisition of homeothermal characteristics by birds and mammals, then, presents another example of parallel evolution.

Hibernation is a characteristic of some homeotherms as well as most poikilotherms. However, the onset of hibernation in the two groups is quite different. A poikilotherm's heartbeat and respiration rate slow as the surroundings cool and tissue temperatures descend, whereas in homeothermal hibernators heart and respiration rates drop suddenly, and only then do tissue temperatures fall. In homeotherms, the onset of hibernation is sometimes preceeded by short initiation periods resembling sleep. During these daily periods, tissue temperatures fall progressively lower until they are close to the external environment. These short periods apparently serve to precondition the brain and body for the long, cold sleep to come. During hibernation itself, respiration may fall to below one breath per minute; heart rate may range around 2–3 beats per minute. Body temperatures will follow the environment for the most part, but below about 37° F (3° C), metabolism increases to keep body temperatures from descending further. Emergence from hibernation is often very rapid; the body temperature of the arctic ground squirrel, for example, rises from 37° F (3° C) to 90° F (32° C) in the space of three hours.

External temperature serves to regulate internal metabolism and therefore food intake. Temperature change has opposite effects on poikilotherms and homeotherms. In cold weather, metabolism slows for poikilotherms and so food intake declines. For homeotherms, cold periods are times of higher food intake because heat output is much greater and metabolism must increase to keep internal temperatures constant. In warm weather, the appetite of the poikilotherm must increase while that of the homeotherm decreases. Since most animal activities are oriented to food gathering, temperature strongly affects this aspect of animal behavior.

Chemical Energy as an Energy Input

All heterotrophic organisms ingest chemical energy as a power source. This energy is derived from organic compounds of carbon, hydrogen, and oxygen which have been previously synthesized by other organisms. The energy of organic compounds is usually released by combining them with oxygen in the process of *oxidation*. You are probably most familiar with oxidation in *burning*, the uncontrolled reaction of substances with atmospheric oxygen. Although oxidation by organisms may produce the same products as burning, it is a controlled reaction and the energy of organic compounds is released slowly. This form of oxidation is *biological oxidation*.

Biological oxidation involves breaking the molecular bonds between atoms of organic molecules and inserting oxygen atoms until the stable end products of CO_2 and H_2O are formed. Thus, a sugar molecule, $C_6H_{12}O_6$, is oxidized until all six carbon atoms have been converted to CO_2 and all twelve hydrogen atoms have been converted to H_2O. At that point, no further energy may be derived from the end products, CO_2 and H_2O, for no more oxygen atoms may be added to them. Therefore, the fewer the oxygen atoms contained in the original compound, the greater is the energy capable of release by oxidation. Compounds with relatively few oxygen atoms are described as *reduced*; the more reduced the compound, the more energy it yields on oxidation.

The two basic types of food-source molecules are *lipids* (fats and oils) and carbohydrates. Lipids are long chains of carbon atoms with hydrogen atoms attached to the carbons, as shown in Figure 20.14A. Most biological lipids terminate in an *organic acid* configuration, —COOH, and are termed *fatty acids*. The oxygens in the organic acid portion of the molecule are the only ones present; therefore each carbon in the lipid chain can yield one CO_2 and one H_2O molecule. Thus lipids are highly reduced and yield much energy in oxidation.

In contrast to the lipids are the carbohydrates, which we have seen before as the products of photosynthesis. A typical carbohydrate is a six-carbon sugar molecule, shown in Figure 20.14B. Note that each carbon atom already has at least one oxygen atom attached, and therefore is less reduced than the carbon of a lipid molecule. Each carbon of the main body of the carbohydrate molecule yields one CO_2 molecule and only one-half of a water molecule on oxidation. As a result, carbohydrates yield less energy on complete oxidation than do lipids.

Proteins also present an energy source to organisms. As we have seen, proteins are made up of amino acids strung together in chains. Amino acids vary in the amount of oxygen atoms they

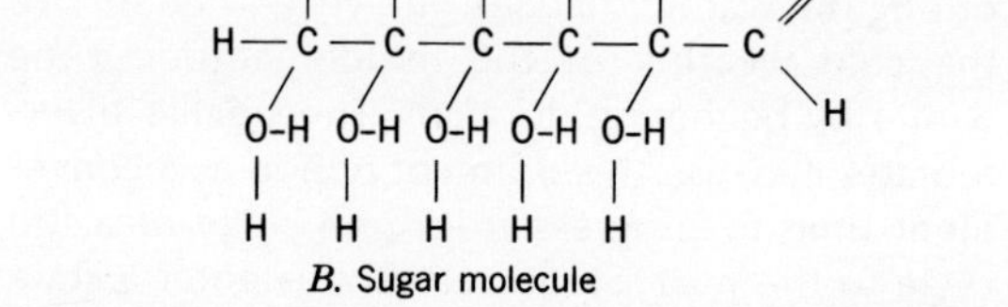

Figure 20.14 Chemical structures of lipid and sugar molecules.

contain per carbon molecule. In general, they contain somewhat fewer than one oxygen per carbon atom, and so are somewhat more reduced than are carbohydrates. Proteins, then, represent an energy source intermediate in yield between carbohydrates and lipids.

Heat energy released by the oxidation of a variety of different compounds is shown below:

Substance	Composition	Heat of Combustion (approx., kcal/gm)
Hydrogen gas	H_2	34
Carbon (example: anthracite coal)	C	8
Methane	CH_4	13
Petroleum	Complex hydrocarbons	10
Lipid	$C_{47}H_{104}O_6$	9.5
Protein	Complex amino acid chain	5.5
Glucose (six-carbon sugar)	$C_6H_{12}O_6$	4
Wood	Mostly carbohydrate	4
Plant tissue: terrestrial plants		4.5
Animal tissue: vertebrates		5.5

The values are obtained by rapid oxidation in a closed container known as a *bomb calorimeter*, under controlled laboratory conditions. It was shown in the 1880s by Max Rubner, a German physiologist, that when proteins, carbohydrates, and fats are oxidized in the animal body by slow normal processes of respiration, the heat equivalents of those substances are almost identical with values obtained by the bomb calorimeter. The value shown for protein is somewhat low, however, for some of the energy is released by oxidizing the amino group, $-NH_2$, of the amino acids to yield NO_2 and water. In organic metabolism, this group is normally not oxidized, and therefore protein yields only about 4 kcal/gm in biological oxidation.

Notice that of lipids, proteins, and carbohydrates, the lipids yield most energy per unit of weight. Thus energy stored as body fats and oils occupies much less bulk than an equivalent quantity of energy stored in the form of either carbohydrate or protein. Fat is therefore used preferentially in energy storage by animals, which must move their biomass from one place to another. The energy value shown in the table for animal tissue, higher than that shown for plant tissue, reflects this difference.

Primary Productivity

Organisms which obtain their energy supply by ingesting organic compounds are ultimately dependent on the conversion of light energy to chemical energy in the synthesis of carbohydrate by autotrophs. From the energy obtained by breaking down carbohydrates in respiration, plants synthesize lipids and proteins as well. Ecologists use the term *primary production* to identify the amount of chemical energy of all forms produced by autotrophs from inputs of light and inorganic matter. Of interest to us is *primary productivity*, the rate at which chemical energy is produced.*

Table 20.3 gives some typical values of primary productivity for different ecosystems, expressed as kcal/m^2/day. In areas of low biological activity, such as deserts or arctic tundra areas, primary productivity may be as little as one one-hundredth of that in such rich and diverse ecosystems as tropical rainforests or coral reefs. Note, however, that primary productivity is expressed on an areal basis, and low productivity does not necessarily mean that the plants themselves are less productive—just that there are fewer plants.

The table also includes *efficiency of primary production* in each ecosystem. In general, an *efficiency* of any operation of an organism or ecosystem is defined as the ratio between two energy flows. In this case, the efficiency of primary production is defined as the ratio between the flow of energy stored by photosynthesis in carbohydrate and the flow of light energy which the autotroph intercepts. The efficiency of primary production ranges from a few hundredths of a percent to as much as 3.5% for a tropical rainforest. Thus, ecosystems are capable of converting and storing at best only a few percent of available light energy.

It is important to note that primary productivity is a gross rather than a net figure; it ignores the rate at which chemical energy is consumed in respiration. Whereas the word *respiration* has been used earlier in this chapter to refer simply

---

*This section and successive sections of this chapter draw information from H. T. Odum, 1971, *Environment, Power and Society*, Wiley-Interscience, New York, Chapter 3, and E. P. Odum, 1971, *Fundamentals of Ecology*, 3rd ed., W. B. Saunders Co., Philadelphia, Chapter 3.

TABLE 20.3 Primary Productivity in Different Ecosystems

Ecosystem	Primary Productivity (kcal/m²/day)	Efficiency of Primary Production (percent)
Natural ecosystems		
Subtropical deep ocean	2.9	0.09
Desert	0.4	0.05
Arctic tundra	1.8	0.08
Coral reef	39–151	2.4
Tropical marine meadow	20–144	2.0
Tropical rainforest	131	3.5
Fertilized ecosystems		
Algal culture	72	3.0
Sugar cane field	74	1.8
Tropical forest plantation	28	0.7

SOURCE: Data from several authors, compiled by H. T. Odum, 1971, *Environment, Power and Society,* Wiley-Interscience, New York, p. 83.

to the breakdown of carbohydrate, the word will be used in the rest of this chapter in an ecological (rather than physiological) sense to include all energy-yielding breakdown processes as well as all the processes of synthesis which produce the different molecular components necessary to support the life system.

When the rate at which concurrent respiration consumes stored energy is subtracted from gross primary productivity, *net primary productivity* remains. Net primary productivity is equivalent to growth, for it represents new material added to the organism. An annual crop, such as corn or soybeans, has a high net primary productivity. The net production can be harvested and used to sustain Man or the animals on which he feeds. On the other hand, a tree in a mature forest may have a low net primary productivity, consuming stored energy at night and during the winter to yield little net gain on an annual basis.

The amount of energy stored within an ecosystem is measured by its *biomass,* the mass of living matter in a unit area. Table 20.4 presents some typical values of biomass for the primary producers (autotrophs) of several different ecosystems. Note the great range possible among ecosystems—the biomass of primary producers in the tropical rainforest is 10,000 times that in the English Channel. Biomass need not be expressed as mass in grams; it can be converted to energy if we have some idea of the energy values stored in each gram of biomass. For example, a gram of pure carbohydrate is equivalent to about 4 kcal of energy; a terrestrial plant yields about 4.5 kcal per gram (dry weight).

The Autotroph as an Energy System

Figure 20.15A illustrates the flow of light energy into an autotroph and its conversion to chemical energy through photosynthesis. The diagram uses three symbols: a circle represents an energy source; a triangle represents an energy conversion process; and a roofed bin with a round bottom (a storehouse) represents a reservoir of stored energy. The arrows connecting the symbols represent energy flows. Because this diagram resembles an electrical circuit with current flows, the symbol used for ground in such circuits is used here to represent an *energy sink* into which energy flows and is lost from the system.

The flow of light energy from the sun, shown as F_1 in the diagram, provides the power source

TABLE 20.4 Autotrophic Biomass in Different Ecosystems

Ecosystem	Biomass, (gm dry weight/m²)
Ocean water, English Channel	4
Lake, Wisconsin	96
Old field, Georgia, U.S.A.	500
Coral reef, Eniwetok	703
Tropical rainforest, Panama	40,000

SOURCE: Data from several sources, compiled by E. P. Odum, 1971, *Fundamentals of Ecology,* W. B. Saunders Co., Philadelphia, p. 80.

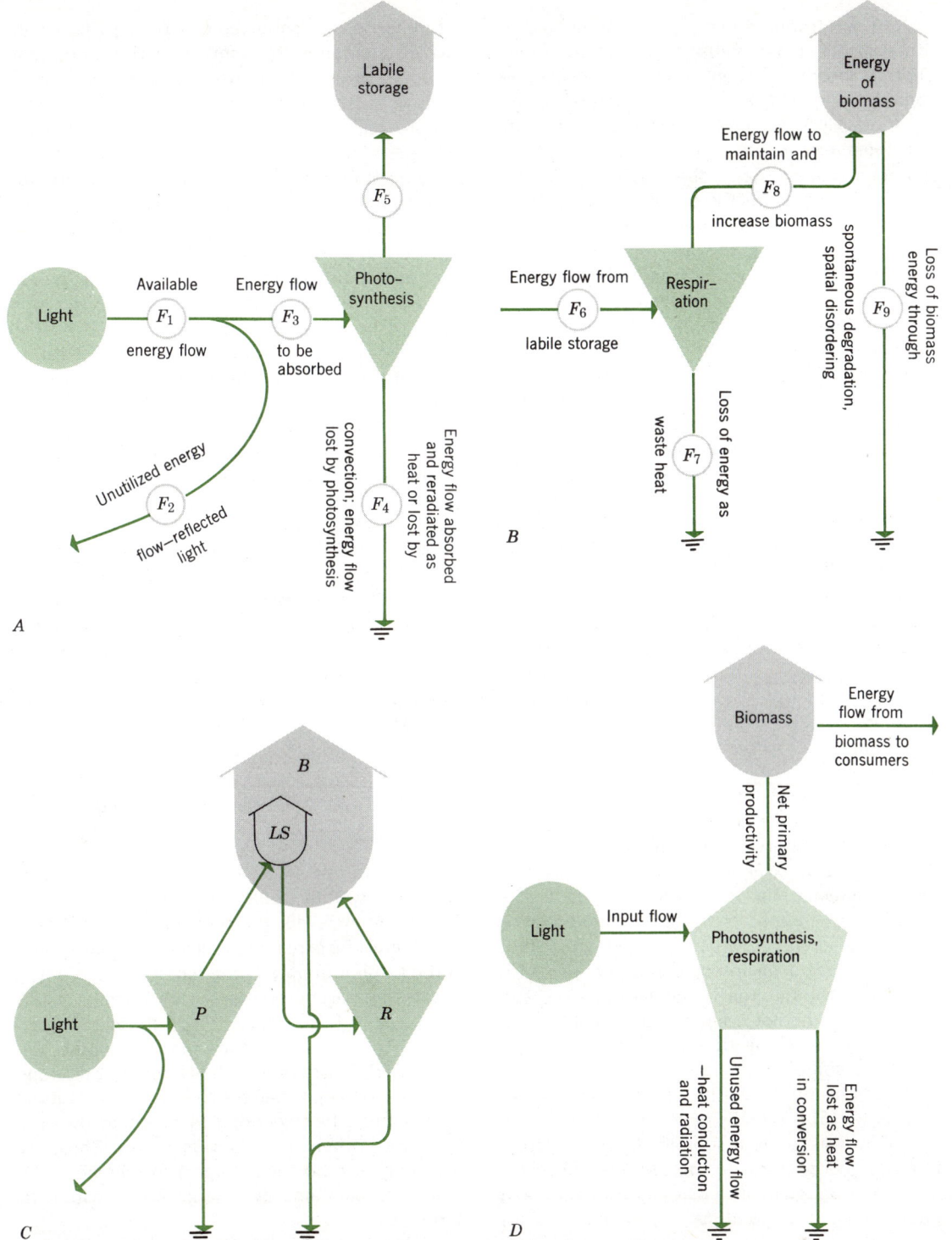

Figure 20.15 Energy flow diagrams for an autotroph. A. Photosynthesis. B. Respiration. C. Combined diagram of photosynthesis and respiration. D. Simplified representation of the autotroph as an energy flow system. (Symbols modified from H. T. Odum, 1971, *Environment, Power, and Society*, Wiley-Interscience, New York, pp. 38–39.)

for the autotroph. However, only about half of this available energy is absorbed in a unit of time. Of this absorbed energy flow (F_3), most is absorbed by molecules other than chlorophylls and so is quickly reradiated away as heat (F_4). The remainder can actually be used in the photosynthetic process. Since this energy conversion process is not perfectly efficient, some energy flow is lost during photosynthesis, contributing additional energy to the flow F_4. Some of the absorbed energy, however, is stored as chemical energy in the formation of carbohydrate. The diagram shows this energy as *labile storage*, storage in a form which can be readily broken down. This flow into labile storage (F_5) is the primary productivity.

The flow of energy through the respiration process and its use to maintain biomass are illustrated in Figure 20.15B. The power source in the autotroph is the chemical energy in labile storage produced by photosynthesis; the flow of energy from this source (F_6) is partially lost by inefficiencies of respiration (F_7), but the remainder (F_8) is used to maintain and increase the total energy stored in the biomass. This maintenance is necessary because the biomass is spontaneously losing energy. As we have noted in the introductory chapter on energy systems and in the opening paragraph of this chapter, in the absence of energy input, the matter of an energy system will degrade to a state of lowest energy and maximum disorder. This loss of energy is shown in the diagram as F_9. Thus the energy flow F_8 acts to maintain the energy stored within the biomass. When this energy flow exceeds the value necessary only for maintenance, the total biomass is increased, resulting in growth. Net primary productivity, then, is the energy flow being stored within biomass in excess of maintenance requirements. In Diagram B, the difference between flows F_8 and F_9 represents the net primary productivity.

When we combine the two diagrams (Figure 20.15C), a complete energy flow diagram for an autotroph results. Since the labile storage of carbohydrate (LS) is actually part of the biomass (B), it is shown as being a subportion of the larger biomass energy reservoir. In addition, all losses of energy through inefficiencies of respiration and photosynthesis as well as spontaneous losses within the biomass are combined into one sink. Diagram D presents a simplification of Diagram C. The autotroph is represented by two parts; a pentagon, symbolizing the internal energy conversion processes of photosynthesis and respiration, and a storehouse, symbolizing the energy stored within its biomass. The pentagon and the storehouse are connected by an energy flow path which combines the sum of all the inputs and outputs to the storehouse shown in Diagram C. This flow, then, represents the net flow of energy to storage, and it is therefore the net primary productivity.

From the viewpoint of photosynthesis the autotroph has one input—solar energy—and three outputs. The first output is a flow of energy, unused in photosynthesis, which is released to the surroundings—in this case it is largely reradiated solar energy. The second output is also a flow of heat energy—the energy lost as heat in the conversion by photosynthesis of light to chemical energy and the energy lost by respiration in converting one form of chemical energy to another. Also included here are the spontaneous losses produced by degradation of the biomass. Like the first output, this energy loss occurs by heat conduction and longwave radiation to the surroundings; thus both outputs are shown as being received by the energy sink. The third output is a flow of energy from storage in biomass to consuming organisms which are further up the food chain and which ingest the biomass of the autotroph for their own energy source. Note that the first and second outputs enter into the thermal budget of the autotroph, so that energy unused by photosynthesis may well enter into other metabolic reactions.

Maintenance of Energy Storage and Its Efficiency

Because the chemical energy in storage as biomass will spontaneously degrade to products of lower energy levels, the continual work of respiration must be expended to maintain that storage. If the respiration rate is compared to the biomass (expressed as energy, not weight), there results a measure of the efficiency of storage maintenance. This measure is the *Schrödinger ratio*, defined as R/B, where R is the respiration rate and B is the biomass. When the R/B ratio is high, more energy must be expended to maintain the biomass than when the R/B ratio is low. Thus, the Schrödinger ratio for an organism tells how efficiently it maintains the energy stored within its biomass.

Ecologists have made the interesting observation that the Schrödinger ratio decreases with the increasing size of the organism. Thus, larger organisms can maintain their biomass with less energy on a per-unit-of-weight basis and are therefore more efficient maintainers of storage. This relation is shown in Figure 20.16. Size of organism is

Energy Flow through a Heterotroph

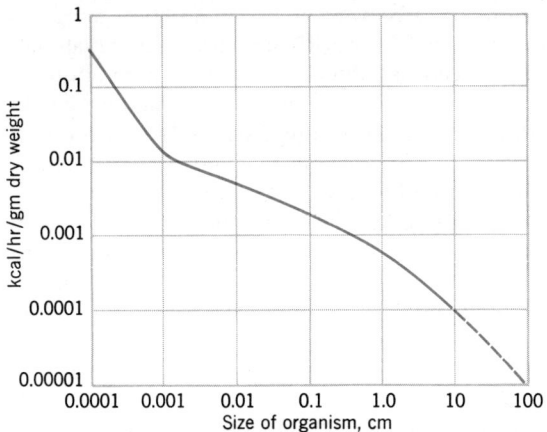

Figure 20.16 Respiratory metabolic rate as a function of size of organism. (After H. T. Odum, 1971, *Environment, Power, and Society*, Wiley-Interscience, New York, p. 75, Figure 3.6b. Based on data of K. E. Zeuthen.)

In almost all ecosystems, chemical energy stored by autotrophs is utilized by heterotrophs higher up the food chain. The intake of chemical energy and its conversion to useful forms by a heterotroph is the process of *assimilation*. Assimilation in most heterotrophs is a two-step process. First, large molecules are broken down into smaller components which will be used for later synthesis. Proteins are divided into their component amino acids; large molecules of fat are cleaved into smaller, more easily handled fatty acids, and complex carbohydrates are broken down into simple sugars. This process is termed *degradation* Since degradation produces simpler products of lower stored energy value, it proceeds more or less without energy input, and in fact liberates energy which is lost as heat. In organisms with intestinal cavities, resident bacteria usually aid in the degradation process by breaking down molecules which the host organism cannot. When the bacteria die, their bodies constitute a supply of degradable matter which, in part, has been obtained from the original unusable materials.

The second step in assimilation is the *transportation* of the products of degradation to cells throughout the organism where they are used as energy sources or as building blocks for the organism's own proteins, fats, and carbohydrates. While some transportation occurs as simple diffusion between cells, the movement of many substances across cell membranes requires energy. In a heterotrophic organism with a circulatory system, energy is also required for intercellular transportation of degradation components in the circulatory fluids.

Energy flow through the assimilation process of a heterotroph is diagrammed in Figure 20.17A. The similarity of this diagram to Figure 20.15A shows that assimilation can be thought of as the counterpart to photosynthesis in an autotroph. Just as a portion of the incoming light energy is unutilized by photosynthesis and reradiated as heat, so a portion (F_2) of the total chemical energy ingested (F_1) by the heterotroph is unused and excreted as fecal matter. And, both processes lose some incoming energy (F_3) as they work and are therefore less than 100% efficient. The product of both processes is a flow of energy (F_4) into labile storage; in the autotroph, this flow is the gross primary productivity, and in the heterotroph it is the *gross rate of assimilation*. The labile storage produced by the assimilation process is, however, somewhat different from that produced by photosynthesis. Assimilation produces not only usable

plotted on the horizontal axis; on the vertical axis is plotted the amount of energy needed to sustain 1 gram (dry weight) of biomass of organism. This quantity is not strictly the Schrödinger ratio, for the Schrödinger is expressed on a caloric-ratio basis rather than a weight-ratio basis. However, the weight-ratio basis has a similar meaning. The declining shape of the curve shows that larger organisms need less energy per unit of biomass to maintain themselves than do smaller organisms over the same time period.

Perhaps an example will clarify this relationship further. A large tree with a dry weight of 300 kg will require about 3 kcal/hr to sustain it, assuming that its storage maintenance efficiency is roughly equivalent to that of a spherical organism of 100 cm diameter. That same 3 kcal/hr, however, will support only 175 gm (dry weight) of algae, assuming that they are about 0.1 mm in diameter. Thus, 175 gm of algae require about as much energy input per unit time as 300,000 gm of tree, on a dry weight basis.

Suppose, now, that we are comparing a forest ecosystem with a marine ecosystem. In the forest, the primary producers will be large organisms (trees); in the marine ecosystem, the primary producers will be small organisms (algae and other phytoplankton). If the energy inputs to both ecosystems are equal, the total biomass of the phytoplankton will be much smaller than the total biomass of the forest trees, because the algae and phytoplankton are much less efficient at maintaining biomass. This effect, in part, explains the wide range in biomass values shown in Table 20.4.

Energy Flow in Organisms and Ecosystems

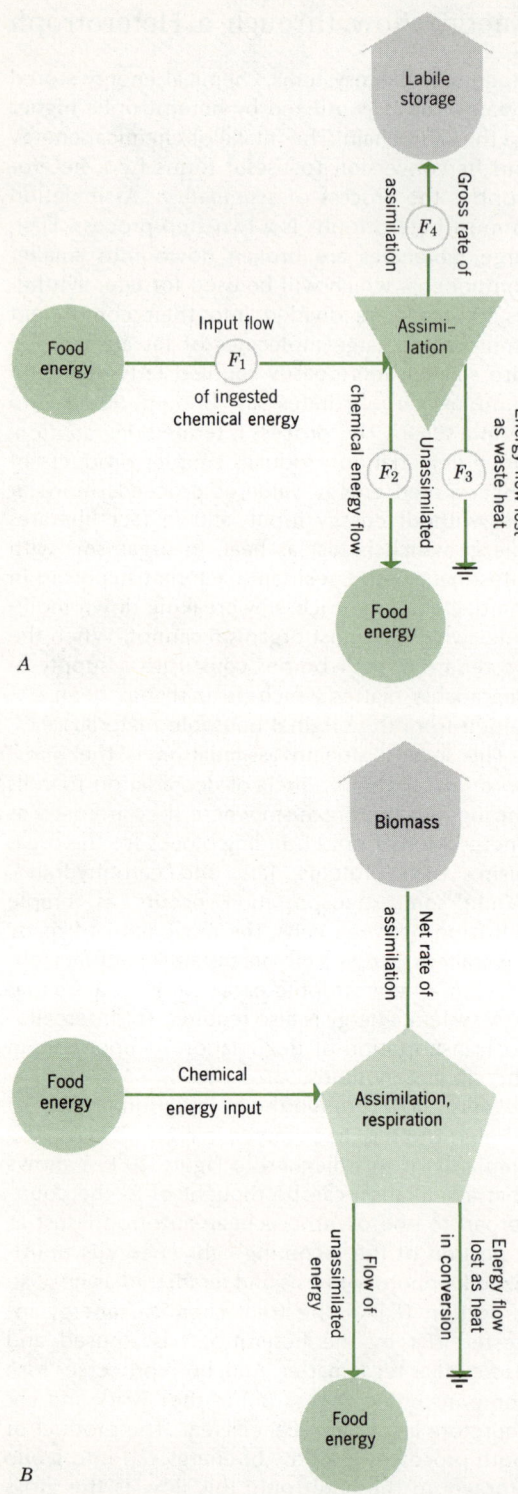

Figure 20.17 Energy flow for a heterotroph. *A*. Assimilation. *B*. Simplified representation of the heterotroph as an energy flow system. (Symbols modified from H. T. Odum, 1971, *Environment, Power, and Society*, Wiley-Interscience, New York, pp. 38–39.)

forms of carbohydrate, but in addition produces amino acids, fatty acids, and many other materials which can be either degraded by respiration for their energy content or used as building blocks to be assembled through synthesis into the biochemical components needed to maintain a functioning biomass.

Figure 20.17*B* diagrams the flow of energy through the heterotroph in a fashion similar to Figure 20.15*D*, which applies to an autotroph. The differences are as follows. First, the input energy as well as the unused energy output differ in type between the two: for autotrophs, they are light and heat respectively, and for heterotrophs both are chemical energy. Second, the flow of energy to storage within the biomass is referred to by different terms: in the autotroph it is net primary productivity, whereas in the heterotroph it is *net rate of assimilation*. The production of biomass by the heterotroph uses chemical energy which has been previously stored by autotrophs, and is therefore a *secondary production*. Since secondary production adds no new energy to an ecosystem, it is not really production in the sense used for autotrophic production; therefore the term rate of assimilation is preferable.

Table 20.5 gives some efficiencies for energy flows within heterotrophs. The table shows that a 50% efficiency for ingestion into assimilation is probably a reasonable figure for heterotrophs. The efficiencies which depend on growth, or net rate of assimilation, vary widely, however. Where management is for growth, as much as 50% of the ingested energy may be stored in new biomass. Growth efficiencies considering assimilation alone may range to near 75%. Under the natural conditions of elephants grazing on the range, however, as little as 1.5% may be stored. In most natural populations of heterotrophs, a 10% efficiency for ingestion to new growth is probably representative.

Energy Flow along the Food Chain

The autotrophs, as primary producers, trap the energy of the sun and make it available to support consuming organisms at upper trophic levels. Since energy is lost at each processing step up the food chain, the total number of steps is limited. In general, anywhere from 10% to 50% of the energy in storage at a particular trophic level can be passed up the chain. The higher values are usually characteristic of the higher trophic levels, but some primary consumer populations, notably some species of zooplankton, also show high efficiency.

Figure 20.18*A* presents a schematic diagram of energy flow in an ecosystem. The symbols previ-

TABLE 20.5 Selected Efficiencies for Heterotrophs

Heterotroph	Assimilation/ Ingestion ratio (percent)	Growth/ Assimilation ratio (percent)	Growth/ Ingestion ratio (percent)
Managed			
Steers being fattened	66	74	49
Marine zooplankton culture	59	57	34
Natural			
Elephants on grazing range	32	1.5	0.5

SOURCE: From data of several authors compiled by H. T. Odum, 1971, *Environment, Power and Society*, Wiley-Interscience, New York, p. 91.

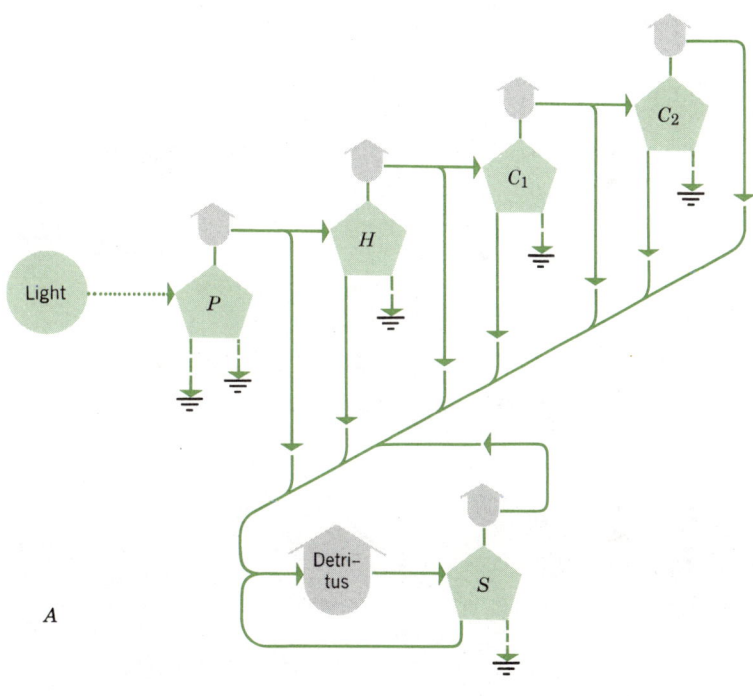

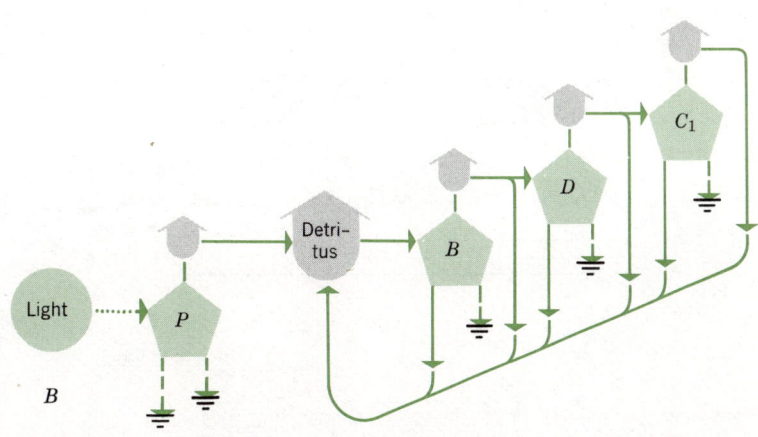

Figure 20.18 Energy flows through two types of ecosystems. A. Ecosystem with grazing food chain. B. Ecosystem with detritus food chain.

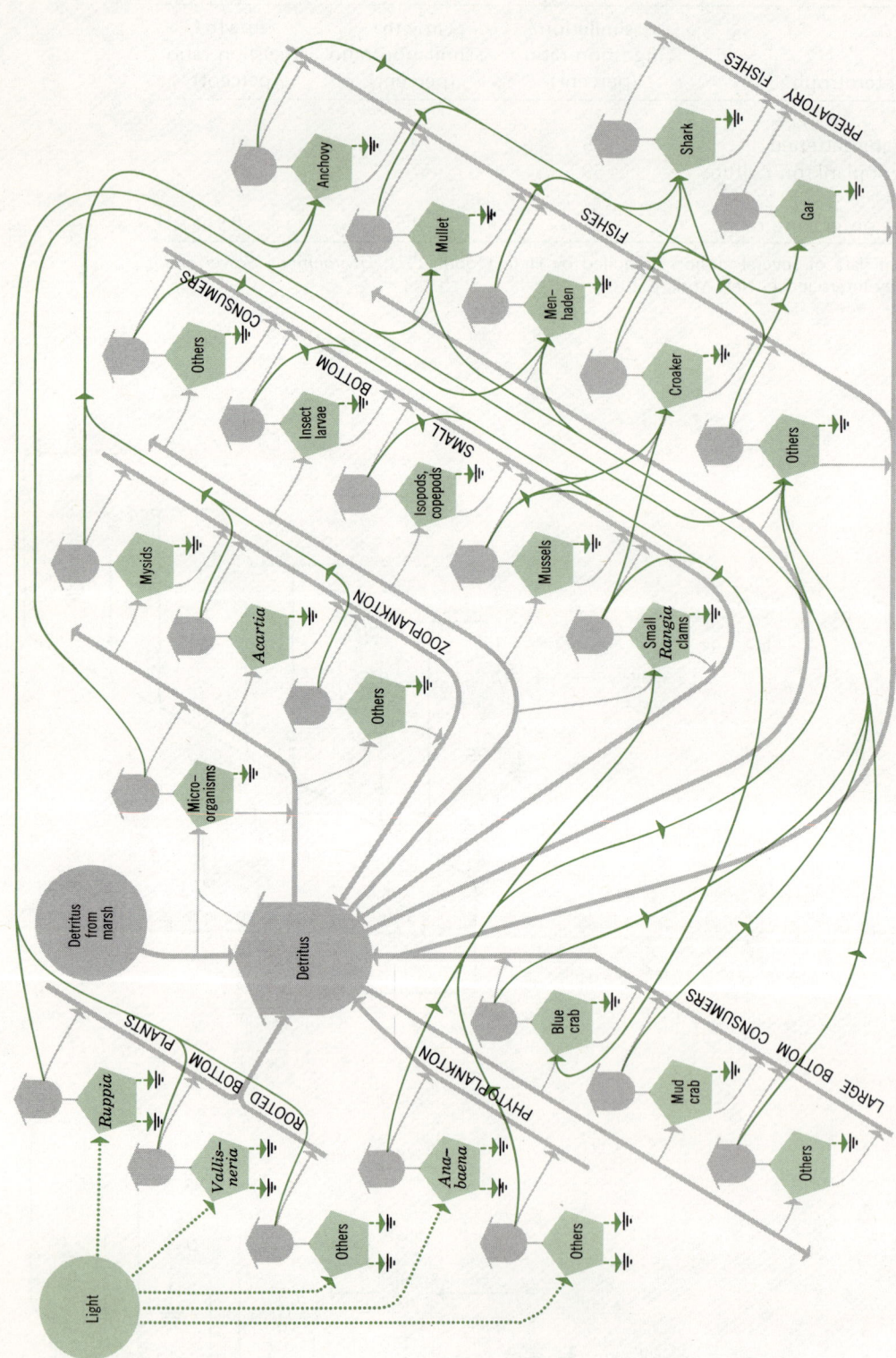

Figure 20.19 Energy flow for a detritus food web in the Lake Pontchartrain estuary, Louisiana. Note that many of the larger organisms feed at several different trophic levels. Solid color lines indicate grazing energy flows; dashed color lines, heat flows; dotted color lines, light energy flows. Chemical energy flows of detritus in black or as large shaded arrows. (After R. M. Darnell, 1961, *Ecology*, vol. 42, pp. 555, 567, and 553–568 *passim*; H. T. Odum, 1971, *Environment, Power and Society*, Wiley-Interscience, New York, p. 64, Figure 3.2b.)

ously used to represent an individual autotroph or heterotroph are now used to represent all autotrophs or heterotrophs feeding at the same level. In addition, energy flows are distinguished by type: a dotted line indicates light, a dashed line, heat, and a solid line, chemical energy. Energy enters the ecosystem as light, and is absorbed by the primary producers (P). Part of the energy is lost as reradiated heat. If gross production exceeds respiration, the biomass of the primary producers increases.

This biomass supports the upper trophic levels of herbivores (H) and carnivores (C_1, C_2) by serving as their energy input. Energy leaves each level (1) as biomass lost to consumers above it, (2) as biomass of dead individuals or parts of individuals, and (3) as ingested but unassimilated food energy in excreted materials. Energy lost in (2) and (3) above contributes to the reservoir of decaying organic matter, or *detritus*, which is acted upon by bacteria and other saprotrophic organisms (S). This simple model assumes that energy in storage does not leave the ecosystem; therefore all incoming solar energy which is trapped and stored by autotrophs is eventually released as waste heat in the processes of assimilation and respiration by heterotrophs.

The food chain shown in Figure 20.18A is a *grazing food chain*, in which the biomass of the primary producers is directly consumed by organisms at upper levels. It is probably a good model for most natural terrestrial ecosystems. Another type of food chain, the *detritus food chain*, is diagrammed in Figure 20.18B. In the detritus food chain, the dead biomass of the primary producers yields a reservoir of detritus, which is acted upon by bacteria (B). Detritivores (D) ingest particles of detritus, gaining energy largely from digestion of the energy-rich bacteria. The detritivores are in turn suitable food for various carnivores (C_1), and so supply energy for the upper levels of the food chain (which may be present but are omitted from Figure 20.18B). Many aquatic ecosystems, and particularly those of estuaries, have well-developed detritus food chains.

Figure 20.19 diagrams energy flow within the ecosystem of Pontchartrain Estuary, Louisiana. The food web shown here is based on organic detritus derived from two sources; bottom plants and phytoplankton. The aquatic plants *Ruppia* and *Vallisneria*, and the phytoplankton *Anabaena* are the principal primary producers; the less important plant species are shown as "others." The energy contained within the dead plant material enters the reservoir of organic detritus, where it is acted upon by bacteria and other unicellular organisms. The bacteria-rich detritus supports three types of zooplankton, five types of small bottom consumers, and three types of large bottom animals. The five types of fishes are supported by energy obtained at several levels. As an example, the menhaden grazes the phytoplankton, but also consumes organic detritus directly. The croaker as well relies on organic detritus as part of its diet, but also feeds on the small bottom consumers. Other fishes may depend on energy obtained from many sources at multiple levels. These feeding patterns show that it is difficult to classify one species as belonging to a particular trophic level; when a complex ecosystem is simplified to a diagram like Figure 20.18, a single species may need to be represented by symbols at several levels.

The fact that energy flow decreases up the food chain or food web of an ecosystem has important implications. First, the number of levels (counting autotrophs as the first level) is ultimately limited, usually to four, but sometimes to five, levels. An ecosystem is actually an energy system distributed over space, and as we go up the food chain, less and less energy is available in a unit of area. Consequently, organisms feeding at the highest levels must range over a wide area—an insect larva grazing on vegetation may find all the food he needs within a few square meters of leaf area, but an eagle or mountain lion must command a vast territory of many square miles to find enough large organisms to sustain itself. Eventually there reaches a point where the available energy is spread so thin that more work must be done to obtain it than is gained by it. At this point, no organisms can be supported and the limits of the food chain are reached.

Second, if an organism feeds at more than one level, a larger population can be supported by feeding at the lower level than at the upper one. This fact results because only a portion of the available energy can be passed up grazing food chains without damaging the supporting level through excessive grazing. If Man depends on meat for his sustenance, he is operating at the third level; if he is a vegetarian, he can operate at the second level. Thus more energy is available to vegetarians than meat-eaters, and if Man continues his population increase, more and more humans may have to be supported at the plant-eating level.

Pyramids of Energy Flow, Biomass, and Numbers

One way to represent the flow of energy up the trophic levels of the food chain is with a *pyramid diagram*, shown in Figure 20.20A. Energy con-

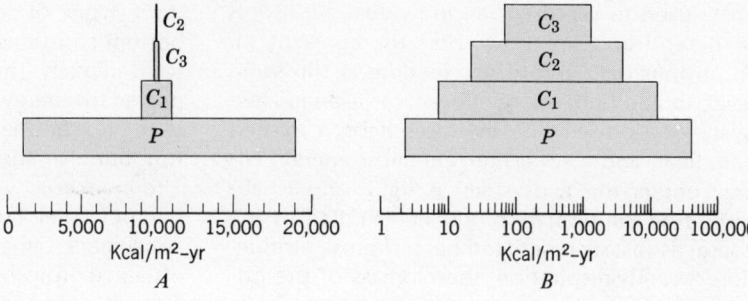

Figure 20.20 Energy flow for a theoretical ecosystem which fixes 2.0 kcal/m²-hr as average primary production, and passes ten percent of the energy flowing through each trophic level on to the next. P, producers; C_1, C_2 and C_3, first-, second- and third-order consumers. Energy flow through saprotrophs is omitted. A. Energy flow plotted linearly. B. Energy flow plotted on a logarithmic scale.

verted by autotrophs forms the bar at the base of the pyramid; atop that are bars whose length represents energy flows through higher levels. Since the size of energy flow decreases with each step to about one-tenth that of the preceding level, the top and bottom values may be as far apart as four orders of magnitude. Figure 20.20B shows the same pyramid diagrammed by logarithmic value. Thus, the length of each bar represents the order of magnitude (power of ten) of the value. The result is the *log pyramid*, which we will use throughout the rest of this discussion.

Just as energy flow decreases up the food chain, so biomass would be expected to decrease. However, smaller organisms are less efficient at maintaining biomass than are larger organisms, as shown in Figure 20.16. Therefore, if the organisms are very small, a much smaller biomass will process the same energy input. This means that a *biomass pyramid*, constructed in a fashion similar to an energy flow pyramid, need not have a pyramidal shape.

Figure 20.21 shows biomass pyramids which result when it is assumed that (1) size decreases with level or (2) size increases with level. The first case is probably representative of a forest ecosystem (large trees, small deer, smaller wolves), while the second provides an approximation to a deep water marine ecosystem (phytoplankton, fingerling fishes, large fishes). The pyramids agree fairly well with our impressions of these ecosystems—in the forest the trees are the most obvious feature, while in the open ocean small fishes seem most plentiful.

Another factor which sometimes interests ecologists is the number of organisms at each trophic level, and it is possible to construct a *numbers pyramid* for an ecosystem. However, the numbers of organisms will depend on the size of the organism and its efficiency at maintaining biomass. Figure 20.22 shows two numbers pyramids based on the assumptions for the biomass pyramids of the preceding paragraphs. Numbers pyramids are easy to obtain by sampling and counting, but they tell very little about the energy dynamics of an ecosystem. Only when factors of size, metabolic efficiency, and differences in biomass are known can energy flow characteristics be inferred from numbers pyramids.

Agricultural Ecosystems

The principles of energy utilization and flow in natural ecosystems which have been discussed above also apply to agricultural ecosystems. Important differences exist, however, between natural ecosystems and those which are highly managed for agriculture. The first major difference is the reliance of agricultural ecosystems on inputs of energy which are ultimately derived from fossil fuels. The most obvious of these inputs is the fuel which runs the machinery used to plant, cultivate, and harvest the crops. Fossil fuels also power such activities as breeding plants which have higher yields and are resistant to disease; developing new chemicals to combat insect pests; and transporting crops to distant sources of consumption, thus enabling large areas of similar climate

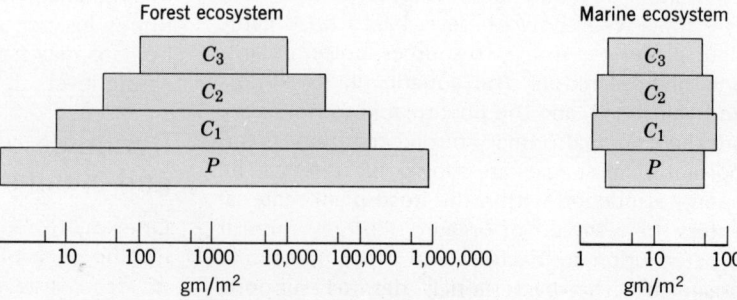

Figure 20.21 Biomass per unit area for two theoretical ecosystems utilizing energy as in Figure 20.20. In the forest ecosystem, size decreases with trophic level, whereas in the marine ecosystem, size increases with trophic level.

Agricultural Ecosystems

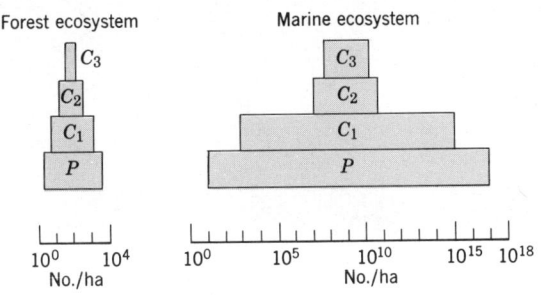

Figure 20.22 Number of individuals per hectare in two theoretical ecosystems as described in Figures 20.20 and 20.21.

and soils to be used for the same crop. In these and many other ways, the high yields obtained today are brought about only at the cost of a significant energy subsidy derived from fossil fuels.

From the standpoint of ecosystem structure and function, agricultural ecosystems are very simple, usually comprised of one genetic strain of one species. Such ecosystems are overly sensitive to attacks by one or two well-adapted insects which can multiply very rapidly to take advantage of an abundant food source. Thus, pesticides are constantly needed to reduce insect populations. Weeds, too, are a problem—adapted to rapid growth on disturbed soil in sunny environments, these plants can divert much of the productivity to undesirable forms. Chemical pesticides and herbicides are often the solution to these problems. Whereas we will discuss the specific effects of these compounds on natural ecosystems in later chapters, we note here that they require inputs of energy derived from fossil fuels for their synthesis. In addition, the chemical starting materials and reagents themselves are often derived from hydrocarbon fractions. Thus, these agricultural chemicals usually owe a double debt to fossil fuels.

Application of agricultural chemicals is one of the ways in which Man uses energy inputs to increase net primary productivity. Large increases in productivity are also produced by application of nutrient elements and compounds—usually of nitrogen and phosphorus—which are in short supply in the normal soil. However, the normal nutrient cycles of these elements, in which they are returned to the soil following the death of the plants which concentrate them, are interrupted by harvesting the crop for consumption at a distant location. Thus, nutrients must be added each year in the form of fertilizers, which are synthesized or mined only with considerable energy input from fossil fuels.

The inputs of energy which Man adds to managed ecosystems in the form of agricultural chemicals and fertilizers, as well as those in the form of the work of farm machinery, have acted to boost greatly the net primary productivity of the land. Table 20.6 presents some examples of how agricultural yields have responded to these inputs, which are all ultimately made possible by fossil fuels. Thus, Man has acted to raise the net primary productivity of his agricultural ecosystems by more than five-fold through the use of energy contained in fossil fuels.

Diverting solar energy to his own service through management of ecosystems is only one example of the great changes Man has produced in natural environments. Other changes are concerned with the movement of matter through ecosystems, influenced not only by the agricultural demands mentioned above, but also through many other types of impact. These changes, and their implications for Man, are the subject of the next chapter.

TABLE 20.6 Crop Productivity and Efficiency with and without Fossil Fuel Energy Subsidy

Crop	Net Productivity, as Harvested (kcal/m²/day)	Efficiency of Primary Production (percent)
Without fossil fuel energy inputs		
Grain, Africa, 1936	0.72	0.02
All U.S. farms, 1880	1.28	0.03
With fossil fuel energy inputs		
Grain, North American average, 1960	5	0.12
Rice, U.S., 1964	10	0.25

SOURCE: Data of several sources, compiled in H. T. Odum, 1971, *Environment, Power and Society*, Wiley-Interscience, New York, p. 116, Table 4.1.

21 Cycling of Materials in Organisms and Ecosystems

Matter in organisms and ecosystems serves two functions: first, matter serves as a vehicle for storage of energy in chemical or potential (gravitational) form; and second, matter serves as a physical framework to support the biochemical activities of life. Thus, life is not possible without molecules which intercept and transform energy from one form to another and without molecules which contain and provide the physical and chemical environment necessary to energy transport and transformation. As molecules are formed and reformed by chemical and biochemical reactions within an ecosystem, the atoms which compose them are not changed or lost in the same fashion that energy is lost to an ecosystem as heat within an energy transformation. Thus, matter can be conserved within an ecosystem, and atoms and molecules can be used and reused, or *cycled* within ecosystems.

Atoms and molecules move through ecosystems under the influence of both physical and biological processes. The pathways of a particular type of matter through the earth's ecosystem comprise a *material cycle*.* In this chapter, five major material cycles are presented: water, carbon, nitrogen, oxygen, and sulfur cycles. These materials serve valuable functions in organisms and ecosystems, and our discussion of cycling will show not just how and where these materials move, but why they are important to organisms and ecosystems.

Ecologists recognize two types of material cycles—gaseous and sedimentary. In the *sedimentary cycle*, the compound or element is released from rock by weathering, then follows the movement of running water either in solution or as sediment to the sea, where eventual precipitation, sedimentation, and diagenesis convert it to rock. When the rock is uplifted and exposed to weathering, the cycle is completed. In a *gaseous cycle*, a shortcut is provided—the element or compound can be converted to a gaseous form, diffuse through the atmosphere, and thus arrive over land or sea, to be reused by the biosphere, in a much shorter time. By and large, more is known about gaseous cycles than sedimentary cycles, and in

---

*Sometimes referred to as a *biogeochemical cycle*, or *nutrient cycle*.

Composition of the Biosphere

addition, the primary constituents of living matter—carbon, hydrogen, oxygen, and nitrogen—all move through gaseous cycles. For this reason, we will examine the major gaseous cycles in depth.

The major features of a material cycle are diagrammed in Figure 21.1. Any area or location of concentration of a material is a *pool*. There are two types of pools: *active pools*, where materials are in forms and places easily accessible to life processes, and *storage pools*, where materials are more or less inaccessible to life. A system of pathways of material flows connects the various active and storage pools within the cycle. Pathways between active pools are usually controlled by life processes, whereas pathways between storage pools are usually controlled by physical processes. The magnitudes of the total storage and total active pools can be very different. In many cases, the active pools are much smaller than storage pools, and materials move more rapidly between active pools than between storage pools or in and out of storage. Taking an example from the carbon cycle, photosynthesis and respiration will cycle all the CO_2 in the atmosphere (active pool) through plants in about ten years, but it may be many millions of years before the carbonate sediments (storage pool) now forming as rock will be uplifted and decomposed to release CO_2.

In comparing the movements of different materials through ecosystems, a variety of units have been commonly used. To help compare one cycle to the next, a universal unit is used in this chapter. This unit is not based on weight or volume, but on the number of molecules or atoms actually in circulation. The unit is the *mole*, or *gram-molecular weight*, and should be familiar to all students with a background in chemistry.* Each mole contains 6.023×10^{23} molecules; but since a mole is still a relatively small quantity (1 mole of H_2O = 18 g, 1 mole of CO_2 = 44 g), we will use one trillion moles (10^{12} moles) as the basic unit of quantity of matter.

*The weight of a mole of any chemical compound is found by adding the atomic weights of each of its consitituents. For example, water is a compound of two hydrogen atoms (weighing 1 atomic weight unit each) and one oxygen atom (weighing 16 atomic weight units). Its total molecular weight is 18 (2 + 16), and therefore one mole of water weighs 18 grams. Carbon dioxide is a compound of one carbon atom (12 atomic weight units) and two oxygen atoms (16 atomic weight units each), and therefore 1 mole of CO_2 weighs 44 grams (12 + 32). If the compound is a gas, one mole of the gas will occupy a volume close to 22.4 liters at 0° C and 1 atmosphere pressure. For atoms, rather than molecules, a mole of atoms is simply equal to the atomic weight of the atom expressed in grams (H = 1, C = 12, O = 16). A mole of water contains 2 moles of hydrogen and 1 mole of oxygen.

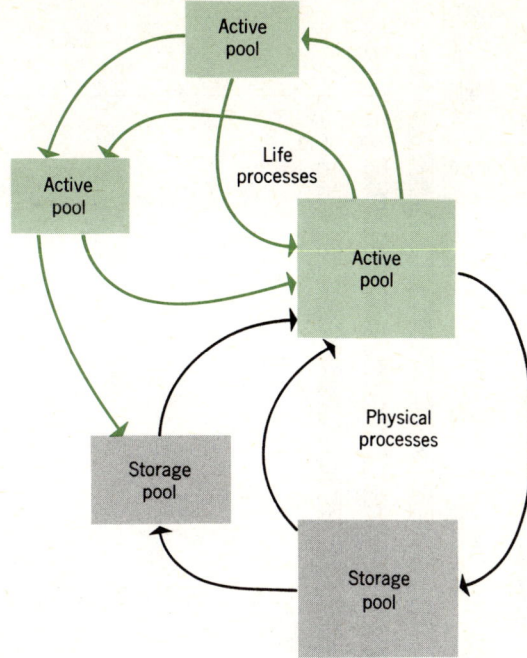

Figure 21.1 General features of a material cycle.

Composition of the Biosphere

The elemental composition of the materials comprising the biosphere is given in Table 21.1 and presented graphically in Figure 21.2; amounts are given in moles/hectare. Because of the wide variation between the largest and smallest amounts, lengths of the bars in Figure 21.2 are scaled in

TABLE 21.1 Elemental Composition of Materials Comprising the Biosphere

Element	Symbol	Quantity (moles/hectare)
Hydrogen	H	13,100
Carbon	C	6,560
Oxygen	O	6,540
Nitrogen	N	71.6
Calcium	Ca	18.9
Potassium	K	11.7
Silicon	Si	8.62
Magnesium	Mg	8.18
Sulfur	S	4.44
Aluminum	Al	4.12
Phosphorus	P	3.35
Chlorine	Cl	2.80
Sodium	Na	1.65
Iron	Fe	1.38
Manganese	Mn	0.765

SOURCE: Data from E. S. Deevey, Jr., 1970, *Scientific American*, vol. 223, no. 3, p. 150.

Cycling of Materials in Organisms and Ecosystems

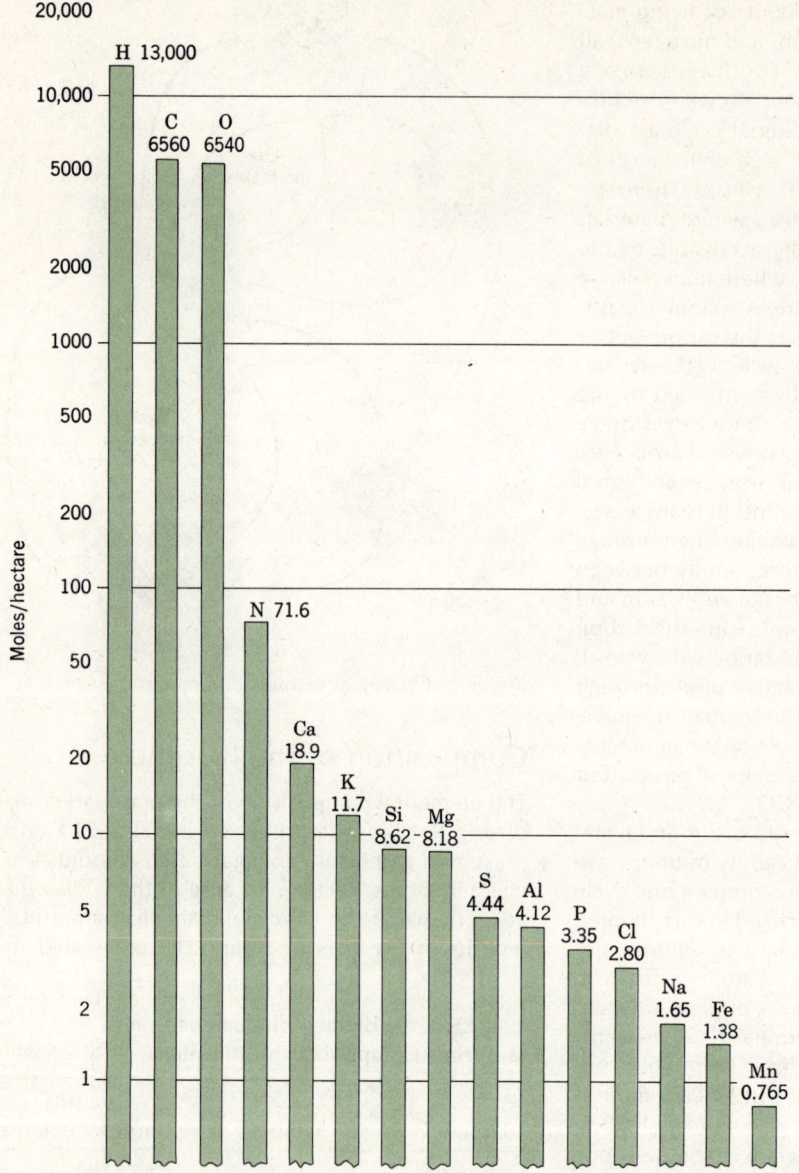

Figure 21.2 Average composition of living matter, expressed in moles per hectare. (Calculated from E. S. Deevey, Jr., 1970, *Scientific American*, vol. 223, no. 3, p. 151.)

proportion to the logarithm of the value; each vertical division represents an order of magnitude. The principal components of living matter are *carbon*, *hydrogen*, and *oxygen*. These three elements are in correct proportion for the compound —CHOH—, which, as we have seen in Chapter 20, is the basic building block of carbohydrate. This proportion occurs because most of the world's biomass is in the wood of its forests, and wood is largely composed of chains of carbohydrate molecules.

Following the three leading elements rank twelve additional elements, including the six most often referred to as the *macronutrients* (Na, Ca, Mg, S, and P). Of these, *nitrogen* is most abundant, for it is contained within each amino acid building block of protein (Chapter 19). The macronutrients are all required in substantial quantities for organic life to proceed. *Phosphorus* is used in the skeletal structure of vertebrates, but is also vital to processes of biochemical synthesis at the cellular level in both plants and animals. *Calcium* and *potassium* both play a role in regulating osmotic pressures in cells; in addition, calcium is precipitated as carbonate or phosphate compounds in external and internal, skeletons, and is particularly important for mollusks and vertebrates. *Magnesium* is an important constituent of enzymes as well as of such macromolecules as chlorophyll. *Sulfur* is a component of proteins and is necessary in some enzymes as well. *Sodium* and *chlorine* are required by animals in fair amounts, but seem unused by all but a few species of plants. These elements are obtained by animals as ions in salt and help produce gastric secretions, regulate the acid-base balance in the body, and promote proper osmotic balance.

The *micronutrients* are elements required by organisms only in small amounts; they are, however, just as vital to the survival of many species as are the macronutrients. Because micronutrients are needed only in trace concentrations, they are sometimes referred to as *trace elements*. A list of micronutrients includes such familiar elements as copper (Cu), iron (Fe), chlorine (Cl), manganese (Mn), and zinc (Zn) as well as the less familiar elements boron (B), molybdenum (Mo), cobalt (Co), and vanadium (V). Most of these elements play specialized roles with enzymes, vitamins, and other molecules involved in biochemical reactions ranging from photosynthesis to nitrogen metabolism.

A comparison of Table 21.1 with Table 7.1 reveals some interesting differences between the composition of living matter and that of the whole earth. Of the three elements most abundant in living matter, only one—oxygen—is present on the list of the fifteen most abundant in the earth. Both carbon and hydrogen are minor constituents of the earth, but are concentrated in living matter. Nitrogen, another element required in large amounts, is also absent from the list in Table 7.1. This concentration of carbon, hydrogen, oxygen, and nitrogen in the biosphere is no accident—these are the lighter, more volatile elements, and, because of the many organic and inorganic compounds they readily form, are most conveniently used for the transfer and transformation of biochemical energy. In addition to their utility in biochemical processes, these elements are widely available at the earth's surface—CO_2, O_2, H in H_2O, N in N_2 and other compounds have been concentrated within the atmosphere and hydrosphere because of their volatile nature as liquids and gases. Thus, the chemistry of life processes has evolved to use these available materials and take advantage of their reactive characteristics.

Oxidation and Reduction

Intimately involved with the movements of materials are chemical and biological reactions which convert one form of material to another. These reactions are typically *oxidations* or *reductions*. As we have seen in Chapter 20, compounds containing many hydrogen atoms and few oxygen atoms are reduced, while compounds containing more oxygen atoms than hydrogen atoms are oxidized. From a chemical viewpoint, however, we can define *oxidation* of an atom as the loss of electrons from an atom, and *reduction* as gain of electrons by an atom.* The reaction of atomic hydrogen and atomic oxygen to form water provides an example.

$$O + 2H \longrightarrow H_2O$$

Hydrogen and oxygen vary in their ability to hold electrons: hydrogen holds its single electron quite loosely, whereas oxygen holds its eight electrons tightly. Chemists measure the ability of an atom to attract electrons by its *electronegativity*; oxygen has the highest electronegativity of all the elements except fluorine. This high electronegativity of oxygen means that when the molecular bond is formed between hydrogen and oxygen, oxygen will tend to hold to the bond very tightly the electron which hydrogen contributes. The effect is to cause hydrogen to lose its electron, or be oxidized, and for oxygen to gain an electron, or be reduced. Note that both oxidation and reduction will always occur in the same reaction, for if one atom loses electrons (is oxidized), some other atom must gain them (be reduced).

Another example is the reaction of methane with oxygen to form carbon dioxide and water:

$$CH_4 + 4O \longrightarrow CO_2 + H_2O$$

Since carbon has a higher electronegativity than hydrogen, the electrons contributed by hydrogen to the methane molecule are held by the carbon atom. After the reaction, however, the electrons in the CO_2 molecule are held by the oxygen atoms, since oxygen is more electronegative than carbon. Considering carbon only, carbon has lost electrons in the reaction, and has therefore been oxidized. The oxygen atoms, on the other hand, have gained electrons from carbon, and are therefore reduced. For the hydrogen atoms, little has changed; their electrons were held by carbon in the methane molecule, and are now held by oxygen in the water molecule. Since the hydrogen

*This relation can be conveniently remembered by the phrase *LEO* (*L*oss of *E*lectrons = *O*xidation), the lion, roars *GER* (*G*ain of *E*lectrons = *R*eduction).

atoms have neither gained nor lost electrons through the reaction, they have been neither oxidized nor reduced.

The importance of oxidation and reduction reactions lies in their energy relationships. Removal of an electron requires energy, just as energy must be used to separate two magnets. However, when the electron is replaced, that energy can be recovered, just as the movement of one magnet toward another can be harnessed to do work. Consider the transfer of an electron from carbon to oxygen. Since carbon holds the electron loosely, only a small amount of energy is needed to remove it. When the electron is conveyed to an oxygen atom, however, it is attracted very strongly and a large amount of energy is released. Since the energy put in is smaller than the energy given out, the net effect is to release energy. In respiration, this energy is harnessed to drive metabolic processes.

We are now in a position to appreciate the fundamental role of oxygen to organisms: it is readily available in its molecular (unreduced) state, and the conveying of electrons to it yields so much energy that there is almost always enough left over to extract for biochemical work. Almost any loss of electrons (oxidation) by carbon, nitrogen, or sulfur will yield energy if those electrons are passed on to oxygen. Thus, the main biological role of oxygen is to serve as an *electron acceptor*

These energy relationships are also important for the inorganic reactions involved in material cycling. Because so much energy is released by adding electrons to oxygen (reducing it), almost all elements will add oxygen and lose electrons (be oxidized) spontaneously. Thus, reduced compounds such as NH_2 (ammonia), H_2S (hydrogen sulfide), or CH_4 (methane) will be converted to NO_2, SO_2, and CO_2 under the right conditions and, in the presence of oxygen at the life layer, will tend to be converted inorganically to oxidized forms. This fact also means that if an organism requires an element in reduced form, such as the carbon in proteins or lipids, or the nitrogen in ammino acids, it will most likely have to spend energy to obtain it, since it will probably be available only in oxidized form. In the examination of the reactions and movements of materials in cycles, we will encounter many important examples of both biological and chemical oxidations and reductions.

The Water Cycle

The water, or hydrologic, cycle has been discussed in detail in Chapter 12, so that only a rapid review is necessary. Of all the material cycles, it is the most important in the biosphere, for it carries along with it the materials moving through many other cycles. It serves as a reservoir medium for many materials and carries carbon, oxygen, nitrogen, sulfur, phosphorus, and many other biologically important elements in solution from the land to the sea. Not only is water important as a carrier of solutes and suspended matter, but it is also important in its own right, for it is a vital component of living cells and takes part in biochemical reactions.

Table 21.2 presents a summary of the active and storage pools in the water cycle. Some 97% of all water is present in the six active pools listed, and of these one is many times larger than the others—the world ocean. The remaining five active pools, including the fresh waters on land and water in the air, account for less than $1/1000$ of the total water in active pools, yet support the greater part of the biomass. Only about $2/74$, or less than 3%, of the world's water is inaccessible to life, residing in the storage pools of icecaps, glaciers, and ground water. This small portion, however, is still some 100 times larger than the amount in terrestrial active pools.

Figure 21.3 shows these pools and in addition provides some values for the movements of water between pools. Although rounding errors may prevent an exact mathematical balance in the figure, somewhat more than 200 million × 10^{12} moles of water is in circulation each year as measured by total precipitation or total evapora-

TABLE 21.2 Active and Storage Pools within the Water Cycle

	Pool Capacity (10^{12} moles)
Active pools	
Atmosphere	720,000
Soil moisture	3,700,000
Stream channels	69,000
Fresh water lakes	6,900,000
Saline lakes and inland seas	5,800,000
Oceans	74,000,000,000
Total active pools	74,017,189,000
Rounded total	(74,000,000,000)
Storage pools	
Icecaps and glaciers	1,600,000,000
Ground water	460,000,000
Total storage pools	2,060,000,000
Rounded total	(2,100,000,000)

SOURCE: Data converted from Table 12.2, from Dr. Raymond L. Nace, U.S. Geological Survey.

The Carbon Cycle

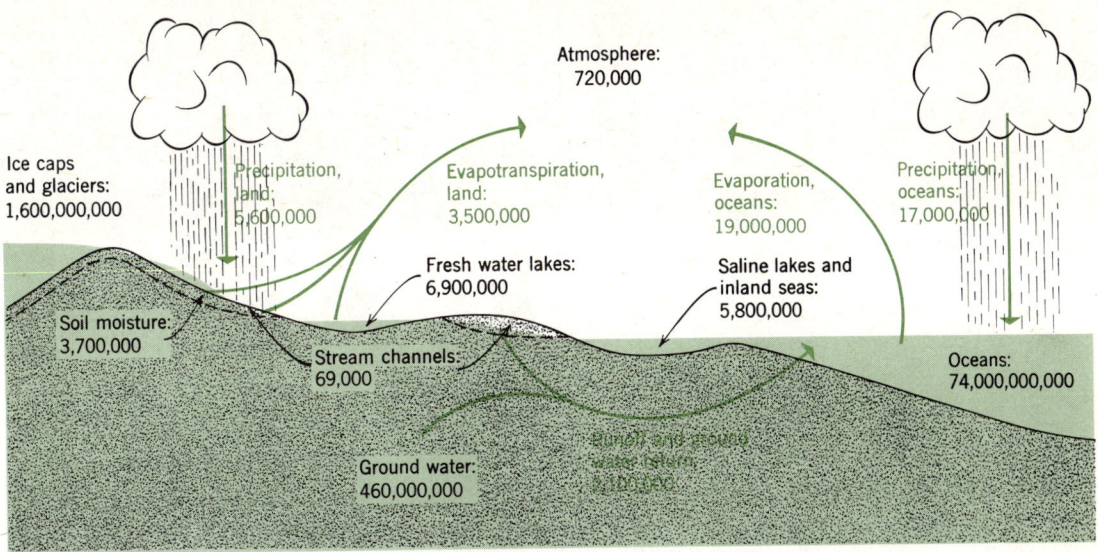

Figure 21.3 The water cycle. Pool capacities (black) in units of 10^{12} moles per yr. (Data of R. L. Nace, converted from Table 12.2 and rounded to two significant figures.)

tion. At this rate, the water within the atmosphere is completely replaced every two weeks or so. However, a complete cycle of all the water in the oceans—through the atmosphere and return to oceans—requires a much greater time, probably somewhere near 4,000 years. The water cycle, then, might be characterized as having one very large active pool (the oceans) which feeds a few very small active pools (water in the air and on land) which support most of the world's biomass.

The Biological Role of Water

Water is perhaps the most important inorganic compound in the biosphere, for it has conditioned the ecology and evolution of terrestrial organisms to an extent far greater than any other compound. Water is used by organisms in at least three ways: (1) as a reactant in biochemical reactions which supplies hydrogen (H^+) and hydroxyl (OH^-) ions (e.g., photosynthesis, Chapter 20); (2) as a cellular medium holding biochemical molecules in solution; and (3) as a circulatory medium in multicellular organisms transporting nutrient ions and molecules to cells and carrying away the waste products of metabolism.

Adaptations of organisms to variation in water supply are remarkably varied and effective; they have been produced by evolutionary processes acting through a pattern of excesses and deficiencies determined by the water cycle. As we found in Chapter 12, it is the balance between precipitation, evaporation, runoff, and infiltration which determines the availability of water to terrestrial organisms at a particular point in time or space. This balance is, in turn, affected by organisms—mainly the plant cover. Through transpiration, plants return much of the soil water to the atmosphere. By obstructing overland flow and increasing soil porosity, plants reduce runoff and increase infiltration. But when we note that over half the global land surface is tundra, desert, mountain, or icecap contributing little or no water vapor to the atmosphere through low evaporation or transpiration, we can see that the major pattern of variation in moisture from place to place is still determined by the overall dynamics of the atmosphere and oceans. Although the movements of water by biological processes are vital from the point of view of organic life, their effects are small compared to those of the physical processes which control the major features of the water cycle.

The Carbon Cycle

The movements of carbon through the life layer are of great importance, for all life is composed of carbon compounds of one form or another. Figure 21.4 presents the carbon cycle in diagrammatic form; Table 21.3 lists the active and storage pools for carbon within the ecosystem of the earth. Carbon is available for life processes in two forms: as CO_2, and as organic carbon. As you know from Chapter 1, CO_2 is a minor constituent of the atmosphere, representing about 1/30 of 1%

539

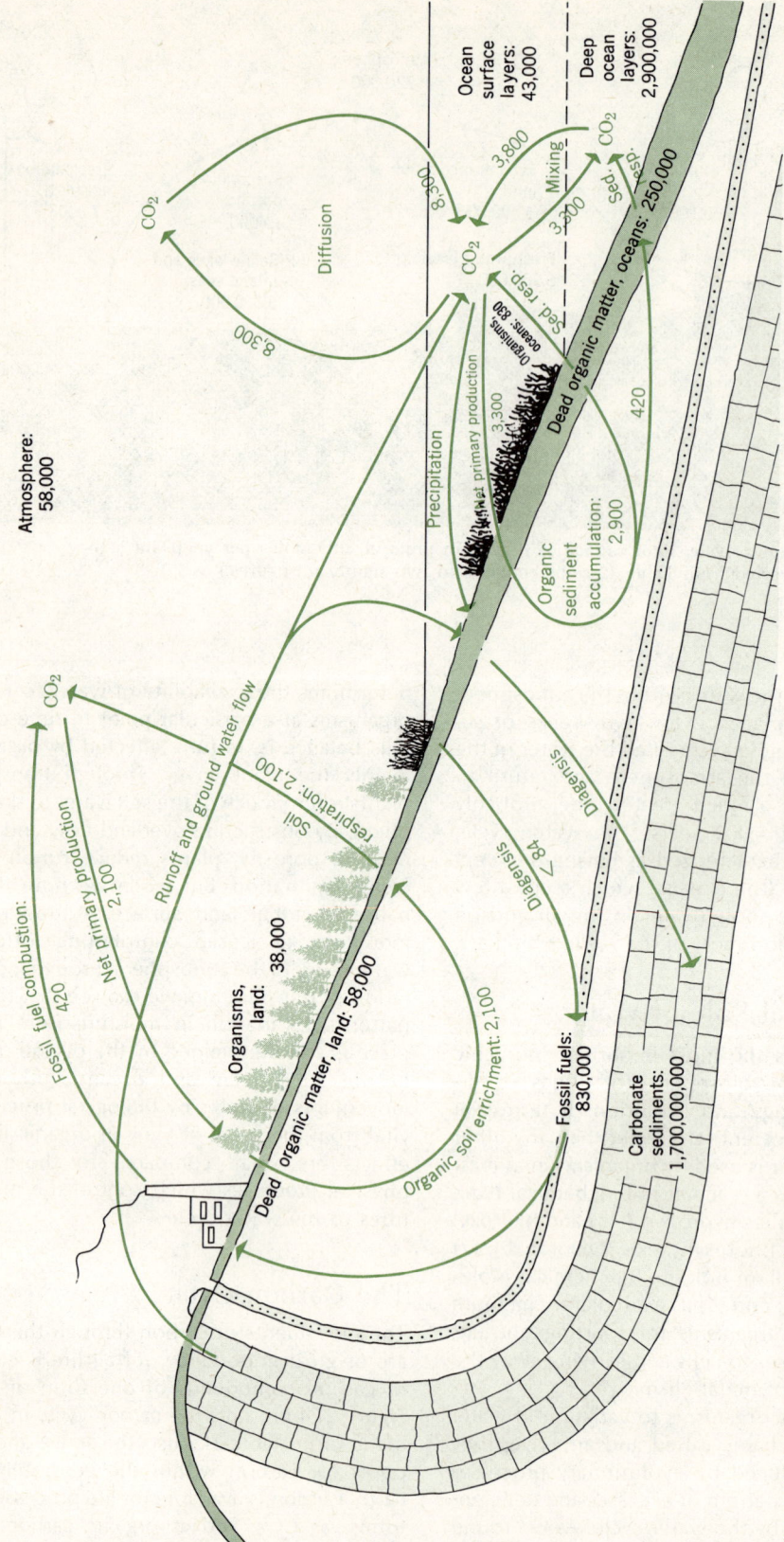

Figure 21.4 The carbon cycle. Pool capacities (black) in units of 10^{12} moles. Flows between pools (color) in units of 10^{12} moles per yr. (Data calculated from B. Bolin, 1970, *Scientific American*, vol. 223, p. 130, and rounded to two significant figures.)

The Carbon Cycle

TABLE 21.3 Active and Storage Pools within the Carbon Cycle

	Pool Capacity (10^{12} moles)
Active pools	
As CO_2 or $CO_3^=$ in solution	
Atmosphere	58,000
Ocean surface layers	43,000
Deep ocean layers	2,900,000
As organic carbon	
Land organisms	38,000
Decaying organic matter, land	58,000
Marine organisms	830
Decaying organic matter, oceans	250,000
Total active pools	3,347,830
Rounded total	(3,300,000)
Storage pools	
As carbonate	
Carbonate sediments	1,700,000,000
As organic carbon	
Fossil fuels	830,000
Total storage pools	1,700,830,000
Rounded total	(1,700,000,000)

SOURCE: Data calculated from Bert Bolin, 1970, "The Carbon Cycle," *Scientific American*, vol. 223, no. 3, p. 130.
NOTE: Values rounded to two significant figures.

(300 ppm) of the total atmospheric volume. As a result, the atmospheric reservoir of CO_2 is relatively small. A much larger amount of CO_2 is dissolved in the water of the world oceans—about $2,900,000 \times 10^{12}$ moles, compared to a figure of $58,000 \times 10^{12}$ moles for the atmosphere; in fact, CO_2 concentrations in water are normally more than 100 times greater than concentrations in air. Most of this marine carbon is in the form of carbonate ion ($CO_3^=$), but since carbonate and CO_2 readily interchange through the reactions described in Chapter 8, CO_2 is always available. The higher concentrations of CO_2 in surface waters means that CO_2 is more readily accessible to aquatic organisms than to terrestrial ones, and this fact also implies that life processes of terrestrial autotrophs may be limited by the small amounts of CO_2 in the atmosphere.

Carbon is also available to organisms in a reduced form—as carbohydrate, fat, protein, or their degradation products, produced by metabolic processes. This reduced carbon contains the energy which powers the heterotrophic portion of ecosystems, and the total active carbon in reduced form is about the same as that in the active form of CO_2. By far the greatest amount of carbon lies in storage as carbonate sediments; in fact, the amount in storage as carbonate is some five hundred times that in active pools. The organic carbon in storage within fossil fuels is only about double the total amount of organic carbon contained within active pools (Table 21.3).

The dynamics of the carbon cycle are presented in Figure 21.4. Intimately involved with the movements of carbon through the life layer is biochemical oxidation and reduction. Through photosynthesis, CO_2 is reduced to organic carbon having the general formula —CHOH—; through respiration, organic carbon is oxidized to CO_2, yielding energy. Involved in these transformations are changes of state—from gas to solid to gas. Thus, carbon moves rapidly from place to place through diffusion as CO_2 gas, but as an organic solid it moves more slowly, if at all, and is usually assisted by running water. For this reason, organic carbon is sometimes referred to as *fixed* carbon—carbon which is fixed into solid carbohydrate.

Our discussion of the movements of carbon through the carbon cycle begins with the fixation and reduction of carbon by land plants. The annual fixation rate is simply the net primary productivity (Chapter 20) expressed in moles of carbon converted rather than in kilocalories of energy stored, and amounts to some $2,100 \times 10^{12}$ moles of carbon. This figure is about 5% of the total terrestrial organic carbon. As we have seen in Chapter 20, all of this fixed carbon will ultimately be consumed by heterotrophic soil bacteria and will eventually return to the atmosphere for recycling as CO_2. Complete oxidation in the soil, however, takes a long period of time; in tropical forests it may require only a few decades, whereas several hundred years may be required in soils of boreal forests.

Within the oceans the total net primary production is greater—about $3,300 \times 10^{12}$ moles per year, or one and one-half times the figure for land. This production comes initially from CO_2 dissolved in surface layers. Most of the organic matter produced is rapidly oxidized to CO_2 by heterotrophs and returned to the surface layers, but about one-eighth moves from the surface layers to the deep ocean floor where it is oxidized much more slowly. Because the solubility of CO_2 in water increases with increasing pressure, the deep ocean layers contain a much greater amount of dissolved CO_2 and $CO_3^=$ than do the surface layers. Vertical diffusion does occur, however; about $3,800 \times 10^{12}$ moles of carbon per year interchanges between the deep and surface layers. At this rate, the CO_2 in deep ocean layers is completely changed about once every 1,000 years or so. The interchange between ocean surface layers and the atmosphere is more rapid—about

Cycling of Materials in Organisms and Ecosystems

$8,300 \times 10^{12}$ moles are interchanged each year, enough to replace the CO_2 in surface layers about every 6 years. Put another way, all of the CO_2 in the atmosphere passes through the oceans once every 7 to 8 years, on the average.

Of the carbon in circulation in active pools, an increment is added each year through precipitation and accumulation to sedimentary deposits, which eventually become lithified into carbonate rocks and fossil fuels. Little is known of the exact quantity which goes into storage each year, but it is probably very small indeed. Carbon returns from storage in sediments as carbonate rocks which are brought to the earth's surface by tectonic forces and which weather to yield carbonate ions ($CO_3^=$) in chemical equilibrium with CO_2 (see Chapter 8). Fossil fuel deposits yield their carbon through mining and combustion for power generation at a much more rapid rate—about 420×10^{12} moles per year, a rate about equal to one-fifth of the net terrestrial fixation. This rapid conversion of carbon from a storage to active form has produced a significant increase in the amount of CO_2 in the atmosphere during the last 100 years or so. The full implications and climatic effects of rising atmospheric CO_2 levels are discussed in Chapter 6.

Organisms and CO_2 Levels

The low level of CO_2 in the atmosphere—320 ppm—has significant implications for the use of carbon dioxide by plants. Figure 21.5 shows how varied CO_2 levels affect photosynthesis in wheat seedlings. In this experiment, wheat seedlings were grown under four different light intensities. At each light intensity level, the CO_2 content of the air was increased, and the effect on the rate of photosynthesis was observed. The rapid rise of photosynthesis with increasing CO_2 concentrations occurs because CO_2 is limiting at first and as more CO_2 becomes available, the photosynthetic process can be speeded up. After a certain point, however, the curves level off, for then light is limiting. At each illumination level, a higher saturation intensity is reached. Note that the curves level off at values for CO_2 nearly double that in the atmosphere. This fact shows that under normal atmospheric conditions, photosynthesis is limited by the CO_2 concentration in the air. Light is not normally limiting, for intensity levels in full sunlight (10,000 to 12,000 foot-candles) are even greater than those shown in the figure.

The fact that CO_2 levels normally inhibit the photosynthesis of plants is of great interest. First, it suggests that excess quantities of CO_2 introduced into the atmosphere by burning of fossil fuels can stimulate photosynthesis, and therefore, stimulate removal of CO_2 by building up the world biomass. The plants, then can serve as a regulatory mechanism to remove some of the excess CO_2. Another process reduces the excess atmospheric CO_2—solution of CO_2 in the world oceans. These two processes combined have probably reduced the increase of CO_2 concentrations produced by burning by about one-half (Chapter 6 has provided more details). Second, the limiting of photosynthesis by low CO_2 levels has suggested to some students of life history that

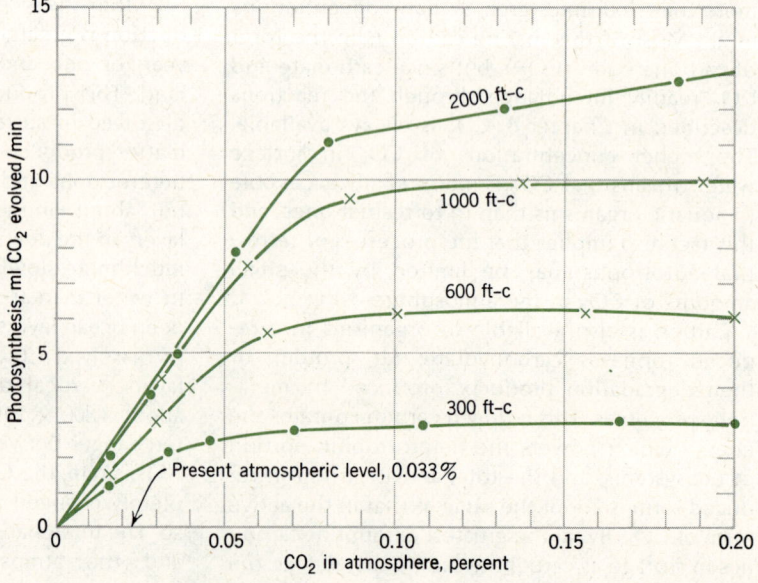

Figure 21.5 Dependence of photosynthesis on carbon dioxide concentration. The curves are for wheat seedlings exposed to four different levels of illumination. (Data of W. H. Hoover, et al., as presented in A. C. Leopold, 1964, *Plant Growth and Development*, McGraw-Hill, New York, p. 26, Figure 2.19.)

the mechanisms of photosynthesis evolved under conditions of much higher CO_2 concentrations, and that photosynthesis through geologic time has been a major factor in contributing to the present high O_2 and low CO_2 content of the atmosphere. This hypothesis has been examined more fully in Chapter 19.

Besides the fact that CO_2 concentrations are normally limiting to photosynthesis, another important implication of the low level of CO_2 in the atmosphere is that removal of CO_2 from the atmosphere by photosynthesis can lower CO_2 concentrations locally. Because of the very low levels at which CO_2 is normally present, small withdrawals by photosynthesis can have a significant impact. Figure 21.6 shows how CO_2 levels vary during a 24-hour cycle at three elevations within a forest. A significant reduction in CO_2 concentration occurs during the day at all levels, reaching lowest values in the afternoon. At night, CO_2 levels rise as the plants respire but do not photosynthesize. Note also that levels are always highest near the ground, an effect produced by the continuous action of soil bacteria degrading organic compounds to produce CO_2. Variation in CO_2 levels can also be noted on a seasonal basis. At latitudes north of about 30°, the summer-winter oscillation of growth rate can produce a decrease of as much as 20 ppm (about 6%) in mean CO_2 concentrations from April to September. Levels of CO_2 rise because soil and plant respiration exceed photosynthesis. This annual variation is most pronounced at the earth's surface. At higher elevations in the troposphere, the variation is smaller, but can even be observed in the lower stratosphere, where it amounts to about 2 ppm decrease between April and September.

In vertebrates and insects, CO_2 acts as a regulator for respiratory activity. An increased CO_2 level of air taken into the body will produce an increased respiration rate, while a decrease will produce slower breathing. Regulation appears to be keyed to levels of CO_2 in the circulatory fluid; in Man, this level ranges at about 5% of the dissolved blood gases.

Forms of Nitrogen

The most commonly encountered form of nitrogen within organisms is the *amino group*, —NH_2, which is a component of amino acids. Where it exists by itself, the amino group is in the form of *ammonia*, NH_3, or *ammonium ion*, NH_4^+. Lacking chemical bonds to oxygen, nitrogen in the forms of amino, ammonia, or ammonium is in its most highly reduced state. The more oxidized states of nitrogen occur in the *nitrate ion*, NO_3^-, and in the slightly less oxidized *nitrite ion*, NO_2^-. Inorganic nitrogen, in the soil or dissolved in water, is usually in these two forms. Somewhat less oxidized is *nitrous oxide*, N_2O, which is produced by a few organisms. For the earth as a whole, by far the vast bulk of nitrogen is in *molecular nitrogen*, N_2, which constitutes about 78% of the air by volume. Molecular nitrogen is considered a neutral form since it contains neither oxygen nor hydrogen. In addition, the gases *nitrogen dioxide* (NO_2) and *nitric oxide* (NO) are also present in the atmosphere in small quantities, and are largely pollutants produced by human activities. In order from most oxidized to most reduced, then, the common forms of nitrogen are: nitrate ion (NO_3^-), nitrogen dioxide (NO_2), nitrite ion (NO_2^-), nitric oxide (NO), nitrous oxide (N_2O), molecular nitrogen (N_2), and ammonia (NH_3).

Biochemical Reactions of Nitrogen

Nitrogen is converted from one form to another by the biochemical processes summarized in Table 21.4. The first three reactions are reductions, involving the loss of oxygen and addition of hydrogen to nitrogen. *Nitrate assimilation* is the reduction of nitrate or nitrite ion to ammonia. Atmospheric ammonia can be assimilated directly

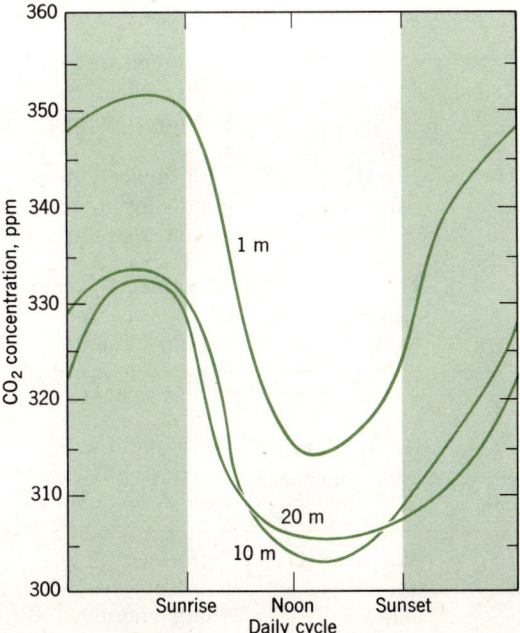

Figure 21.6 Variations in CO_2 concentration through the daily cycle in a forest. A curve is shown for three elevations: 1 m, 10 m (at about the middle of the canopy), and 20 m. (Data from B. Bolin, 1970, *Scientific American*, vol. 233, no. 3, p. 127.)

by plants, but since most available nitrogen is in the soil as nitrate or nitrite ion, plants and soil bacteria have therefore evolved the mechanism of nitrate assimilation to convert nitrate or nitrite to ammonia. The reaction is not a spontaneous one—it requires the input of biochemical energy in one form or another to proceed, and also involves several steps. The higher animals lack this ability to use nitrate or nitrite and must receive nitrogen as the amino group in amino acids.

The second and third reactions in Table 21.4 act together to control the amount of nitrogen available to living organisms; nitrogen fixation converts inert molecular nitrogen from the atmosphere into the usable form of ammonia, whereas denitrification converts usable nitrates or nitrites back to molecular nitrogen. *Nitrogen fixation* is the conversion of molecular nitrogen (N_2) to the usable form of ammonia. From there, the nitrogen may be incorporated into amino acids or other biochemical compounds which can be cycled and recycled through an ecosystem. Only microorganisms possess this ability to fix nitrogen, and we can classify them into two types: free-living nitrogen fixers, and bacteria which fix nitrogen through a symbiotic relationship with higher plants.

The *free living nitrogen fixers* include various types of aerobic soil bacteria (*Azotobacter*), anaerobic soil bacteria (*Clostridium*), photosynthetic bacteria (*Rhodospirillum*), and blue-green algae. Present in nearly all terrestrial environments (and some aquatic environments as well), these organisms normally can fix as much as 5 to 6 kg of nitrogen per hectare per year (4.4 to 5.3 lbs/acre/yr) although in most areas the value is lower—2 to 3 kg/hectare/yr (1.8 to 2.7 lbs/acre/yr). In tropical grasslands, it may be much higher. *Symbiotic nitrogen* fixers are bacteria of the genus *Rhizobium* and are associated with some 190 species of trees and shrubs as well as almost all members of the legume family. The latter is important for such agricultural crops as clover, alfalfa, soybeans, peas, beans, and peanuts. Trees and shrubs associated with nitrogen fixers include such common groups as the buckthorns (*Ceanothus* spp.) and alders (*Alnus* spp.) as well as such primitive plants as cycads and the gingko. All these plants fix nitrogen only when their root cells become infected with the bacterium *Rhizobium*. Bacterial action is concentrated in *root nodules* (Figure 21.7) or swellings which are produced by the joint action of plant and bacterium. The relationship seems truly symbiotic, with

TABLE 21.4 Reactions of Nitrogen in the Nitrogen Cycle

Process	Reaction	Comments
Reduction Nitrate assimilation	NO_3^- or NO_2^- ⟶ NH_3 Nitrate ion / Nitrite ion / Ammonia	Performed by plants and soil bacteria. Requires energy.
Nitrogen fixation	N_2 + H_2 ⟶ NH_3 or NH_4^+ Molecular nitrogen / Molecular hydrogen / Ammonia / Ammonium ion	Performed by nitrogen-fixing organisms. Requires energy.
Denitrification	NO_3^- or NO_2^- ⟶ N_2 or N_2O Nitrate ion / Nitrite ion / Molecular nitrogen / Nitrous oxide	Performed by anaerobic soil bacteria. Requires energy.
Ammonification	$\underset{\text{Glycine, an amino acid}}{\text{COOH—CH}_2\text{—NH}_2}$ + $\underset{\text{Molecular oxygen}}{O_2}$ ⟶ $\underset{\text{Carbon dioxide}}{CO_2}$ + $\underset{\text{Water}}{H_2O}$ + $\underset{\text{Ammonia}}{NH_3}$	Performed by decomposing bacteria. Yields energy.
Oxidation Nitrification	NH_3 or NH_4^+ + O_2 ⟶ NO_2^- + H_2O Ammonia / Ammonium ion / Molecular oxygen / Nitrite ion / Water	Performed by *Nitrosomonas*. Yields energy.
	NO_2^- + O_2 ⟶ NO_3^- Nitrite ion / Molecular oxygen / Nitrate ion	Performed by *Nitrobacter*. Yields energy.

The Nitrogen Cycle

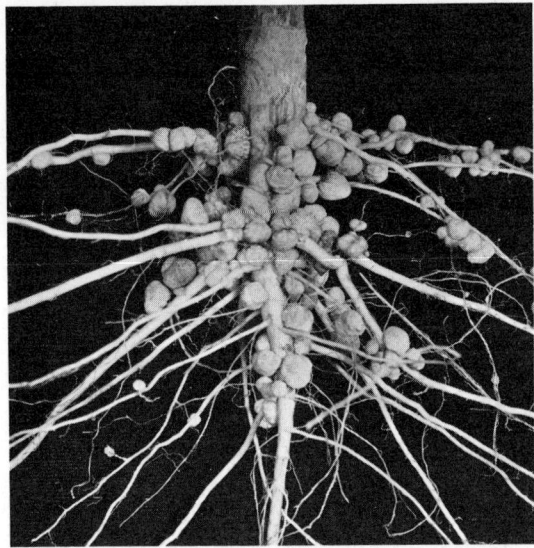

Figure 21.7 Root nodules on the soybean, *Glycine max*. The nodules contain bacteria of the genus *Rhizobium*, and convert atmospheric nitrogen to forms usable by the plant. (Courtesy of the Nitragin Co., Inc.)

Rhizobium supplying the nitrogen in the form of ammonia or amino acids, and the plant supplying the nutrients and organic compounds necessary for bacterial growth. When grown separately, neither plant nor bacterium can fix nitrogen.

Acting together, the bacterium and plant can supply large quantities of fixed nitrogen—as much as 350 kg/hectare/yr (310 lbs/acre/yr) can be added if the crop is returned to the soil. This figure may be many times the amount ordinarily supplied by nitrogen fixers. Many symbiotic nitrogen fixers have been shown to excrete ammonia and some amino acids directly into the soil. This direct addition to the soil is, of course, above and beyond the addition which occurs when the plant dies and its nitrogen-containing compounds are decomposed by microorganisms in the soil.

Denitrification is the name given to the biochemical process which converts nitrate or nitrite to molecular nitrogen or nitrous oxide. Because both forms are stable gases, they leave the soil and enter the atmospheric storage pool and are lost, at least temporarily, to the biosphere. Denitrification occurs only under anaerobic conditions and is performed by a variety of soil bacteria. Because this reaction requires energy, the question arises as to why it is performed at all, since it does not yield a usable form of nitrogen. The reason is that nitrate or nitrite ions in these bacteria can serve the same function as molecular oxygen under aerobic conditions—that of electron acceptors. In this way, organic compounds can be oxidized (and nitrogen reduced) in the absence of molecular oxygen to power the life processes.

Because reaction of organic compounds with O_2 yields somewhat more energy than reaction with NO_3^- or NO_2^-, denitrifying bacteria use oxygen when it is available. As a result, denitrification is restricted to such anaerobic environments as swamps and bogs. Significant amounts of denitrification also occur in permanently saturated soils, where aerobic bacteria quickly consume oxygen dissolved in soil water, producing an anaerobic environment suitable for denitrification.

Ammonification is the biochemical reaction in which amino acids are broken down to release energy. Table 21.4 shows the reaction of a typical amino acid, *glycine*, with oxygen to produce CO_2, H_2O, and NH_3. The reaction yields energy, but the energy yield is obtained from the conversion of organic carbon to CO_2 rather than from the conversion of amino nitrogen to ammonia. Because the starting amino form and the finishing ammonia form are equivalent to one another, the nitrogen is neither oxidized nor reduced. Ammonification is exploited by decomposer microorganisms, and is a one-way reaction; it cannot be reversed to synthesize amino acids from CO_2, H_2O, and NH_3.

Nitrification is the biochemical process in which ammonia is oxidized to nitrate, yielding energy (Table 21.4). The reaction is performed in two steps, each by a different group of bacteria. The *Nitrosomonas* group reacts ammonia or ammonium ion with oxygen to yield nitrite ion and water; the *Nitrobacter* group oxidizes nitrite ion to nitrate ion. Both groups are chemoautotrophic and utilize ammonia or nitrite as their sole energy source. The wide-spread occurrence of these two groups ensures that most soil nitrogen will be in the form of nitrate rather than ammonia, for as NH_3 is produced by nitrogen fixation or ammonification it is rapidly oxidized to NO_3^- by these organisms. Fortunately, nitrogen in this form can be used by almost all plants and microorganisms through the process of nitrate assimilation described above. For most agricultural uses, nitrate is preferred because it is more stable and, compared to ammonia, is much less toxic in high concentrations.

The Nitrogen Cycle

Table 21.5 summarizes the capacities of active and storage pools for nitrogen in the life layer. Within the active pools, nitrogen exists in two forms: as *inorganic nitrogen* (nitrate and nitrite) and as *organic nitrogen* (amino acids and other nitrogen-containing organic compounds). By far the

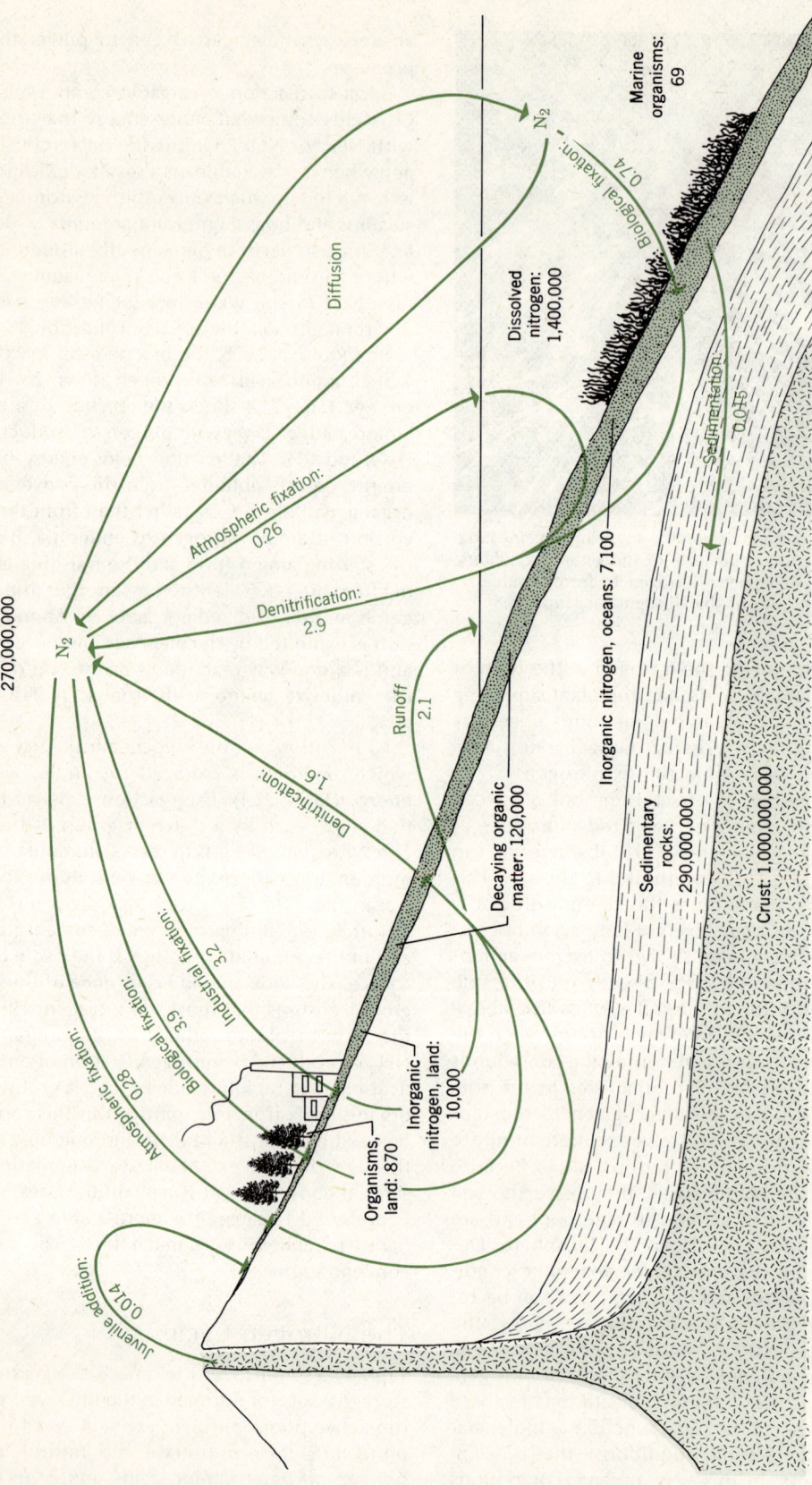

Figure 21.8 The nitrogen cycle. Pool capacities (black) in units of 10^{12} moles. Flows between pools (color) in units of 10^{12} moles per yr. (Data calculated from C. C. Delwiche, 1970, *Scientific American*, vol. 223, no. 3, p. 140, and rounded to two significant figures.)

The Nitrogen Cycle

TABLE 21.5 Active and Storage Pools within the Nitrogen Cycle

	Pool Capacity (10^{12} moles)
Active pools	
As organic nitrogen	
Land organisms	870
Marine organisms	69
Decaying organic matter	120,000
As inorganic nitrogen	
Soils	10,000
Ocean sediments	7,100
Total active pools	138,039
Rounded total	(140,000)
Storage pools	
As N_2	
Atmosphere	270,000,000
Oceans	1,400,000
As rock-forming minerals	
Sedimentary rocks	29,000,000
Crustal rocks	1,000,000,000
Total storage pools	1,561,400,000
Rounded total	(1,600,000,000)

SOURCE: Data calculated from C. C. Delwiche, 1970, "The Nitrogen Cycle," *Scientific American*, vol. 223, no. 3, p. 140.
NOTE: Values rounded to two significant figures.

largest active pool of nitrogen is that in decaying organic matter, amounting to some $120,000 \times 10^{12}$ moles. The total mass of living matter contains only about 1% as much nitrogen. Inorganic nitrogen in soils and marine sediments is also only a fraction (about one-seventh) of the amount in decaying organic matter. Nitrogen in storage pools is in two forms: as relatively inert molecular nitrogen (N_2); and as nitrogen in rock-forming minerals. Of the two forms, more than five times more nitrogen is stored within crustal and sedimentary rocks than is stored in the atmosphere. A point of interest is the great difference between the sizes of the active and storage pools in the nitrogen cycle. Even if the nitrogen in rock-forming minerals is excluded on the grounds that atmospheric nitrogen supplies almost all the active nitrogen through nitrogen fixation, the active pools still total to less than 1/200 of the atmospheric storage pool.

Figure 21.8 pictures the dynamics of the nitrogen cycle. Three processes of nitrogen fixation are shown: biological fixation (discussed earlier), industrial fixation, and atmospheric fixation. Of the three, biological fixation is greatest, totaling nearly 3.9×10^{12} moles/yr. *Industrial fixation*, amounting to about 3.2, includes production of nitrogen fertilizer (about 2.1) and oxidation of nitrogen during fossil fuel combustion (about 1.1). Nitrogen fertilizer is manufactured using the *Haber process*, in which N_2 and H_2 are combined catalytically at temperatures near 500° C and pressures near 200 atmospheres to yield ammonia. The ammonia can then be oxidized to nitric acid (HNO_3), which is reacted with more ammonia to produce ammonium nitrate (NH_4NO_3), a common form of nitrogen fertilizer. Or, the ammonia can be reacted with CO_2 to yield urea (NH_2—CO—NH_2), which is another widely used fertilizer.

Nitrogen fixation also occurs in the combustion of fossil fuels, when N_2 and O_2 are brought together at high temperatures. Besides the carbon dioxide and water produced by the fuel itself, significant quantities of NO (nitric oxide) and NO_2 (nitrogen dioxide) gases are often produced as an undesirable by-product. Released to the atmosphere, these two gases soon combine with H_2O vapor to yield HNO_2 (nitrous acid) and HNO_3 (nitric acid), which reach the soil or ocean surface through precipitation and supply NO_2^- and NO_3^- ions to the biosphere. Nitrogen oxides are primarily produced by two sources—industrial and electrical power and steam plants, and internal combustion engines. Nitrogen oxides are important in their own right as pollutants (discussed in Chapter 6), but our concern here is with the amount of nitrogen so oxidized—about 1.1×10^{12} moles/yr, estimated in 1968. This means that all of industrial fixation, including fertilizer manufacture and oxidation by combustion, converts about 85% as much nitrogen to usable forms as do all nitrogen-fixing organisms.

Minor amounts of nitrogen are also added to active pools by atmospheric fixation and juvenile addition. *Atmospheric fixation* occurs only during lightning strikes, when the high temperatures and pressures created by the arc act to permit oxidation of N_2 to NO and NO_2, which eventually reach the earth as nitrate and nitrite. This process only contributes a small amount of fixed nitrogen each year, however. Also making a minor contribution is the outgassing of NO and NO_2 by volcanoes, a process termed *juvenile addition*.

Of primary importance in our analysis of the nitrogen cycle is the fact that fixation is far exceeding denitrification at the present time, and thus usable nitrogen is accumulating in the life layer. This excess of fixation is produced almost entirely by human activities—in the industrial fixation of nitrogen and in the widespread cultivation of legumes. Most of this excess nitrogen is carried off into rivers and lakes and ultimately reaches the ocean. One of our major water pollution problems arises because nitrogen is limiting in many aquatic environments. Augmentation of nitrogen inputs through Man's activities has permitted increased growth of algae and other phytoplankton, often to the detriment of such desirable aquatic organisms as finfish and shellfish. Thus, increased nitrogen runoff has greatly boosted productivity in many aquatic environments and has contributed to the process of *eutrophication*, (aging of water bodies accompanied by reduction of oxygen content, Chapter 23). This problem will be accentuated in years to come, for industrial fixation of nitrogen in fertilizer manufacture is doubling about every 6 years at present. Just what impact such large amounts of nitrogen reaching the sea will have on the earth's ecosystem remains uncertain. Hopefully, increased denitrification will ameliorate some of the effects, but denitrification by marine organisms is a little understood process. At any rate, Man's careful management of his impact on the nitrogen cycle may very well be a critical problem in the near future.

Forms and Reactions of Sulfur

Like nitrogen, sulfur is present within the life layer in a wide variety of forms. *Sulfur dioxide* (SO_2) and *hydrogen sulfide* (H_2S) are the gaseous forms of primary importance. *Sulfate ion* (SO_4^-) is the common form in water and soils. Within organisms, sulfur is a constituent of three of the twenty-three naturally occurring amino acids.

Sulfur undergoes a number of inorganic and organic (biological) reactions which convert it from one form to another. The inorganic reactions of sulfur important to the sulfur cycle are shown in simplified form in Table 21.6. These are all energy-yielding oxidations, for inorganic reactions require energy and do not take place spontaneously except under special conditions. Much of the sulfur in the atmosphere is in the form of SO_2, which is the starting point for reactions (2), (3), and (4). The SO_2 arises by direct discharge from fossil fuel burning, or is formed by the oxidation of H_2S, shown in reaction (1). In reaction (1), pathways are shown utilizing oxygen in any

TABLE 21.6 Inorganic Reactions of the Sulfur Cycle

	Reaction	Comments
Gaseous		
(1)	$H_2S + \{O, O_2, O_3\} \longrightarrow SO_2$ Hydrogen sulfide + Any form of oxygen → Sulfur dioxide	Normally a slow reaction, but will be rapid in the presence of aerosols or water droplets.
(2)	$SO_2 + O + m \longrightarrow SO_3 + m$ Sulfur dioxide + Atomic oxygen + Any molecule → Sulfur trioxide + Any molecule $SO_3 + H_2O \longrightarrow H_2SO_4$ Sulfur trioxide + Water → Sulfuric acid	This gaseous reaction requires the simultaneous collision of SO_2, O, and any other molecule, m, which serves to carry off excess energy. Proceeds slowly. Addition of water to [Oxidation of] SO_3 is rapid, however. [as in reaction (1).]
(3)	SO_2 + Photochemical smog $\longrightarrow SO_3$ Sulfur dioxide → Sulfur trioxide $SO_3 + H_2O \longrightarrow H_2SO_4$ Sulfur trioxide + Water → Sulfuric acid	The oxidizing agents in photochemical smog make this reaction rapid in polluted atmospheres; SO_3 is highly reactive and will quickly form H_2SO_4.
Aqueous		
(4)	$SO_2 + H_2O \xrightarrow{\text{Trace metal salts}} SO_3^=, HSO_3^-, H^+$ Sulfur dioxide + Water → Sulfite ion, Bisulfite ion, Hydrogen ion $SO_3^=, HSO_3^- + O_2 \longrightarrow SO_4^=, HSO_4^-$ Sulfite ion, Bisulfite ion + Dissolved oxygen → Sulfate ion, Bisulfate ion	Reaction requires dissolved trace metal salts as catalysts; proceeds slowly in fresh water where few salts may be present.

of its three atmospheric forms: O, O_2, and O_3. The reaction is fairly slow, however, except where surfaces are available on which the reaction can occur. Such surfaces include aerosols and water droplets.

Reactions (2) and (3) are oxidations of SO_2 to H_2SO_4. Reaction (2) requires that three different gaseous molecules or atoms all collide at once: SO_2, O (atomic oxygen), and any other molecule. In the reaction, SO_2 combines with O to yield the gas SO_3, *sulfur trioxide;* the other molecule carries off excess energy, which must be removed to keep SO_3 from splitting back to SO_2 and O. The sulfur trioxide molecules thus formed rapidly combine with water vapor to produce H_2SO_4 molecules, which are accumulated in water droplets because of their high solubility and yield *sulfuric acid*, a mixture of H^+, HSO_4^- (*bisulfate ion*), and $SO_4^=$ ions in aqueous solution. This process is not very rapid in the troposphere, but within the ozone layer, where atomic oxygen released by the splitting of O_3 to O_2 and O is widely available, it produces a layer rich in H_2SO_4, or sulfate particles, at an elevation of about 18 km (59,000 ft). Reaction (3) occurs under conditions of photochemical smog (Chapter 6). This smog contains powerful oxidizing agents which convert SO_2 to SO_3; the latter soon combines with water to produce sulfuric acid. Because the reaction requires photochemical smog, it is important only within polluted areas; but, on the other hand, SO_2 discharges are likely to be greatest in these areas because of the concentration of fossil fuel burning facilities.

Reaction (4) is normally the principal pathway for oxidation of SO_2 to sulfate. The reaction occurs in water droplets in the air or within bodies of both fresh and salt water. In the reaction, dissolved SO_2 adds water to yield sulfurous acid, which is then oxidized to sulfuric acid by dissolved oxygen. Trace metal salts act as catalysts to speed the reaction, so that the rate of SO_2 oxidation is greater in salt water than in fresh.

The biological reactions important to the sulfur cycle are shown in Table 21.7. Reactions (1) and (2) are reductions. The first is the mechanism by which sulfur is assimilated by plants and microorganisms. Sulfate ion is absorbed from the soil and reduced by metabolic processes, ultimately to be incorporated as the sulfhydryl group or a related form in proteins and other biochemical molecules. Animals lack the capacity to perform this reaction, and must receive sulfur in prereduced form, usually within one of the sulfur-containing amino acids. Reaction (2) is carried out largely at depth in the sea by *Desulfovibrio* bacteria. Under anaerobic conditions, these bacteria reduce the sulfate in order to oxidize organic compounds to yield energy. This reaction is analogous to denitrification except that the end product is not free sulfur, but hydrogen sulfide, which escapes as a gas and replenishes sulfur in the atmosphere lost by precipitation.

Reaction (3) is of particular interest, for it is performed by photosynthetic bacteria. Recall from Chapter 20 that photosynthesis involves the reduction of CO_2 to organic carbon (—CHOH—), and the simultaneous oxidation of water to molecular oxygen. Photosynthetic bacteria do not use water, H_2O, but instead use its sulfur analog, H_2S, as an electron-accepting compound. Just as O_2 is formed in the process, so the photosynthetic

TABLE 21.7 Biological Reactions of the Sulfur Cycle

	Reaction	Comments
Reductions		
(1)	$SO_4^= \longrightarrow$ —SH (Sulfate ion → Sulfhydryl group)	Normal method of assimilating sulfur by plants and microorganisms. Requires metabolic energy.
(2)	$SO_4^= \longrightarrow H_2S$ (Sulfate ion → Hydrogen sulfide)	Performed by *Desulfovibrio* bacteria under anaerobic conditions in order to oxidize organic compounds for metabolic energy.
Oxidations		
(3)	$H_2S \longrightarrow S$ (Hydrogen sulfide → Free sulfur)	Performed by photosynthetic bacteria, including members of the *Chlorobacteriaceae* and *Thiorhodaceae* groups under anaerobic conditions. H_2S serves the role of H_2O in aerobic photosynthesis.
(4)	H_2S, S, $SO_3^= \longrightarrow SO_4^=$ (Hydrogen ion, Free sulfur, Sulfite ion → Sulfate ion)	Performed by aerobic chemoautotrophic sulfur bacteria. Yields energy which supports metabolism (*Thiobacillus* group).

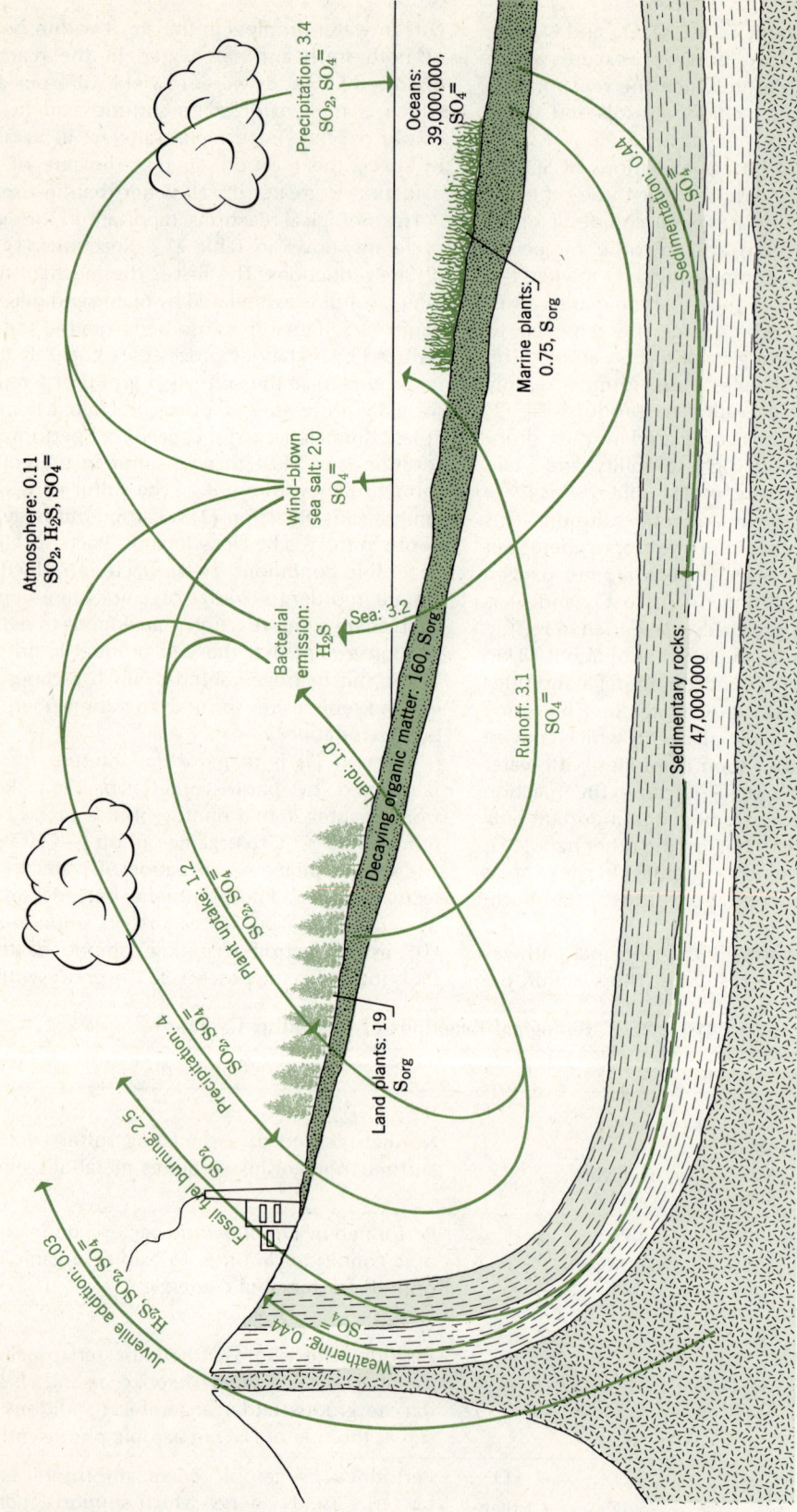

Figure 21.9 The sulfur cycle. Pool capacities (black) in units of 10^{12} moles. Flows between pools (color) in units of 10^{12} moles per yr. (Pool capacities calculated from W. W. Rubey, 1951, Geological Society of America Bulletin, vol. 62, p. 1120, and E. Eriksson, 1960, Tellus, vol. 12, p. 95, Figure 7.11. Flows calculated from W. W. Kellogg, et al., 1972, Science, vol. 175, p. 594, Figure 2. Values rounded to two significant figures.)

bacteria produce free sulfur. The Chlorobacteriaceae (green photosynthetic bacteria) and the Thiorhodaceae (purple photosynthetic bacteria) are two important groups which carry out this reaction. They are commonly found on tidal mud flats where conditions are anaerobic but weak light is still available for photosynthesis.

Reaction (4) is the oxidation of reduced forms of sulfur to sulfate. Since the reaction yields energy, it is employed by aerobic chemoautotrophic bacteria to power their metabolism. It is analogous to the nitrification reactions performed by *Nitrosomonas* and *Nitrobacter*, and is carried out by bacteria of the *Thiobacillus* group. Of these, one organism is of particular interest: *Thiobacillus thiooxidans*, which oxidizes inorganic sulfur to sulfuric acid by the following reaction

$$S + 1\tfrac{1}{2}O_2 + H_2O \longrightarrow H_2SO_4$$

The organism is noteworthy for its ability to withstand highly acid conditions, and survives at pH values as low as 1.0 (about equivalent to laboratory acid).

The Sulfur Cycle

Figure 21.9 shows the dynamics of the sulfur cycle. About $47,000,000 \times 10^{12}$ moles of sulfur is contained within the storage pool of sedimentary rocks. This figure includes the fossil fuels, which may have a sulfur content as high as 6%. Of the active pools, the oceans contain the largest amount, about $39,000,000 \times 10^{12}$ moles. Most of this is in the form of sulfate ion. Decaying organic matter contains only a small fraction of that amount of sulfur: 160×10^{12} moles, which is in organic form as a constituent of proteins and other organic molecules. Land plants contain some 19×10^{12} moles of sulfur, which is about 25 times the 0.75×10^{12} moles contained in marine plants. The active sulfur pool in the atmosphere is smallest of all—about 0.1×10^{12} moles—and includes sulfur as SO_2, H_2S, and $SO_4^=$. The small size of this pool compared to the much larger yearly flows in and out of it shows that sulfur turnover in the atmosphere is very rapid indeed—on the order of 4 to 5 days.

There are four major flows of sulfur to the atmosphere: bacterial emission (4.2×10^{12} moles/yr), fossil fuel burning (2.5), the blowing of sea salts (2.0), and juvenile addition (only 0.03). Most of the sulfur in the form of SO_2 or H_2S is converted to $SO_4^=$ ion by the inorganic reactions of Table 21.6, and is eventually dissolved in water droplets. Precipitation returns some of this sulfur to the oceans, but a large portion (4.0) falls out over land, reversing the seaward flow of sulfate (3.1) in ground and surface water. Direct land plant uptake also removes a substantial portion of sulfur (1.2) from the atmosphere, converting it to biologically useful forms.

In addition to movements through the atmosphere, sulfur moves through the lithosphere as sulfate and becomes incorporated into sedimentary rocks under sea. As sulfate-bearing rock weathers, $SO_4^=$ is released to the terrestrial portion of the biosphere. Movement through this portion of the cycle is slow, however, adding an annual increment of only 0.44×10^{12} or about 8%, to the total terrestrial gain.

A careful analysis of the flows in Figure 21.9 shows that the amount of sulfur is increasing on land (at a rate of about 1.5×10^{12} moles/yr) and in the oceans (about 1.0). This gain is created by the burning of fossil fuel and comes about at the expense of lithospheric sulfur. What are the implications of such a gain? Effects in the ocean should be small because of the vast size of the active pool of dissolved marine sulfate. However, effects on land may be quite different. Especially important is the fact that the increased sulfur is gained mostly in the form of sulfuric acid in rain water. As explained in Chapter 11, the pH of a solution reflects the concentration of hydrogen ions within it (Table 11.1); values below 7.0 indicate acidity, values above 7.0 indicate alkalinity. Rainwater is normally somewhat acid (pH 5 to 6) because of the solution of CO_2 in rainwater to yield carbonic acid.

In recent years, water chemists have noted a dramatic lowering of the pH of rain in northwestern Europe and the northeastern United States to values ranging roughly between pH 3 and 5. Because pH numbers are on a logarithmic scale, these values mean that rain in these areas is now often 100 to 1,000 times more acid than previously. Figure 21.10 shows four maps of pH values for rainwater in northwestern Europe for the years 1956, 1959, 1961, and 1966. The maps show both a dramatic lowering of pH levels and a widespread increase of areas receiving significantly acid precipitation. Although such complete data are not available for the United States, the annual pH of rain at three stations in the Finger Lakes region of New York state and a fourth at Hubbard Brook, New Hampshire, ranged between 3.91 and 4.03 for the year 1970-71. Although a large part of this excess acidity is due to oxidation of SO_2, nitrogen dioxide conversion to nitric acid in rainwater may also be a significant factor in lowering pH.

What effects can be expected from such a change in the quality of rain? In a recent study

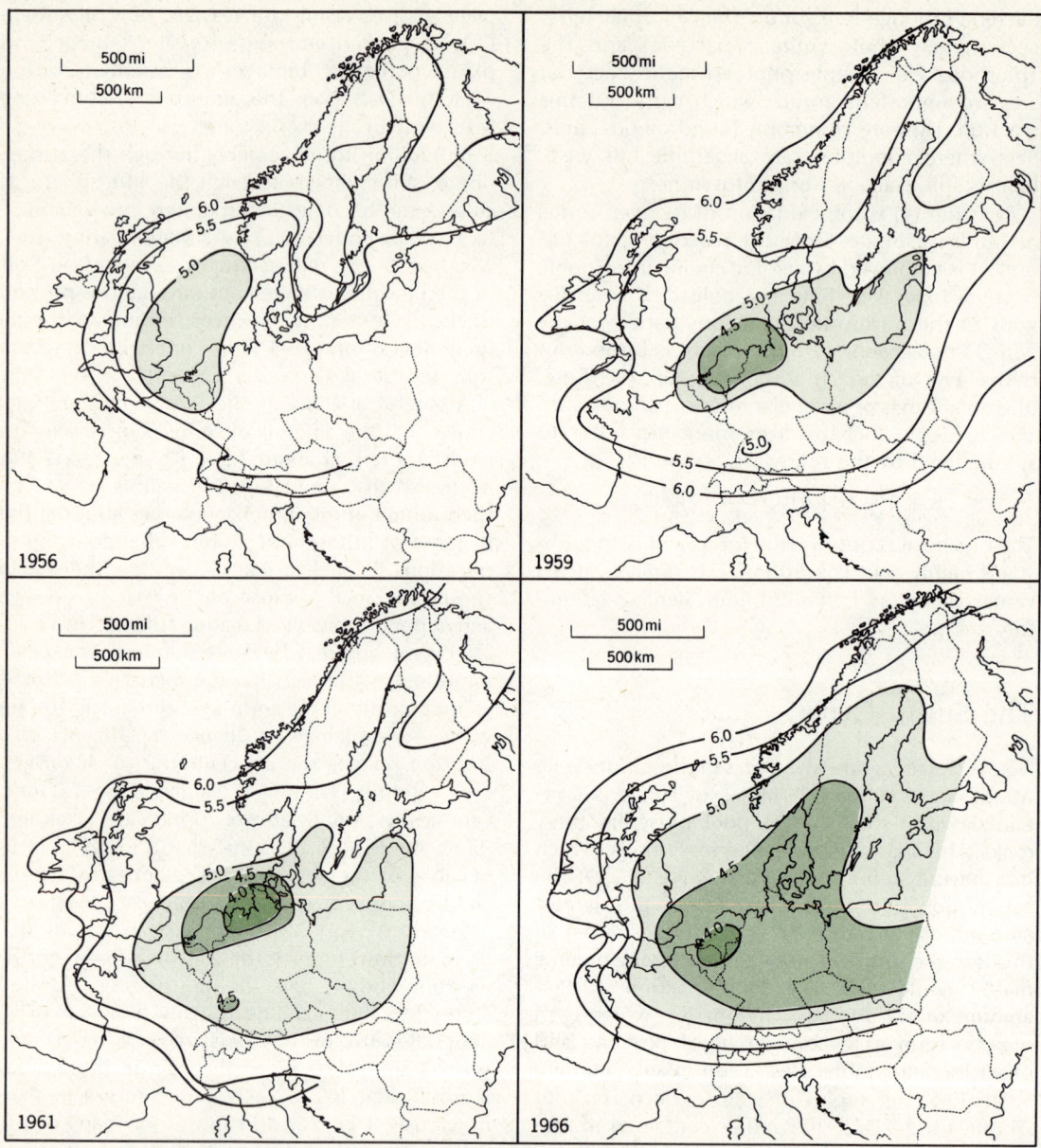

Figure 21.10 Rainwater acidity in northern Europe in the years 1956, 1959, 1961, and 1966. Figures give pH values. (Data of S. Oden, 1972; after G. E. Likens et al., *Environment*, vol. 14, no. 2, p. 36, Figure 1.)

which gathered the data presented above, a team of ecologists concluded:

> The ecological effects of this change are as yet unknown, but potentially they are manifold and very complex. Effects may range from changes in leaching rates of nutrients from plant foliage, changes in leaching rates of soil nutrients, acidification of lakes and rivers, effects on metabolism of organisms, and corrosion of structures.*

The third effect, acidification of lakes and rivers, has already been noted in southern Norway, where lowered pH in certain streams has eliminated salmon runs by inhibiting egg development. Whether similar effects have been created in North America is uncertain, but hopefully by switching to low sulfur fuels and by reducing nitrogen oxides from auto emission this startling trend will be reversed.

*G. E. Likens, F. H. Bormann, and N. M. Johnson, 1972, *Environment*, vol. 14, no. 2, p. 37–38.

The Oxygen Cycle

Water, molecular oxygen, and carbon dioxide are the three inorganic forms of oxygen most important to life processes. The size of oxygen pools is stated in Table 21.8. By far the greatest part of this oxygen, almost 70 billion $\times 10^{12}$ moles is in the form of water in the oceans. Molecular oxygen in the atmosphere is the next largest active pool, comprising some 150 million $\times 10^{12}$ moles, followed by oxygen in molecules of fresh water, in dissolved CO_2, and in molecules of soil water. Only a relatively small amount of oxygen is tied up in the molecules of living and dead organic matter. In the storage pools, the largest quantity is in oxides of crustal rock. Storages in ground water and carbonate sediments are smaller.

TABLE 21.8 Active and Storage Pools within the Oxygen Cycle

	Pool Capacity (10^{12} moles)
Active pools	
As O_2	
Atmosphere	150,000,000
Oceans	210,000
As CO_2	
Atmosphere	120,000
Oceans	5,900,000
As H_2O	
Atmosphere	700,000
Oceans	70,000,000,000
Soil moisture	3,700,000
Surface water, fresh	6,900,000
As organic oxygen	
Land organisms	38,000
Marine organisms	1,830
Organic matter, soils	58,000
Organic matter, marine sediments	250,000
Total active pools	70,167,876,830
Rounded total	(70,000,000,000)
Storage pools	
As H_2O	
Ground water	460,000,000
As carbonates	
Sedimentary rocks	5,100,000
As rock oxides	
Crustal rocks	7,700,000,000
Total storage pools	8,165,100,000
Rounded total	(8,200,000,000)

SOURCE: Data calculated from Table 19.1 and 19.2, and from W. W. Rubey, 1951. The geologic history of sea water, Geol. Soc. Amer., Bull., vol. 62, pp. 1120-1121.
NOTE: Values rounded to two significant figures.

The movements of oxygen are shown in Figure 21.11. Because oxygen is so highly reactive and takes part in so many reactions, flow values beyond those given under discussions of the carbon, nitrogen, and sulfur cycles are difficult to determine and so are not listed. In terms of simple movement of oxygen-containing compounds, flows of water in the hydrologic cycle are probably the most important. In terms of reactions, the biological interconversion of H_2O, CO_2, and O_2 by photosynthesis and respiration probably proceeds at the greatest rate. Chemical reactions also occur. Silicate minerals in rocks brought to the surface are weathered by hydrolysis and oxidation after exposure to molecular oxygen and water. These processes remove oxygen from availability to the actions of organisms. Most of this mineral oxidation and hydrolysis occurs on land, where oxygen is abundant. In anaerobic environments, which usually are found under water, oxygen depletion and acid accumulation will in some places produce *reducing conditions,* which may reverse a portion of this mineral oxidation. Also, during lithification of sediments, water may be removed from hydrolyzed clay minerals and released to ground-water storage.

The needs and uses of oxygen by organisms have been described in earlier chapters; however, Man's impact on the oxygen cycle remains to be considered. Man is reducing the amount of oxygen in the air by (1) removal of oxygen by fossil fuel burning, (2) clearing and draining land, speeding the oxidation of soils and soil organic matter, and (3) reducing photosynthesis by clearing forests for agriculture and by paving and covering previously productive surfaces. This latter effect can be appreciated by the fact that every six months a land area about the size of Rhode Island is covered by new construction in the United States alone. In fact, calculations have shown that the oxygen consumed in the continental U.S. in 1966 exceeded the amount produced by nearly 70%. Fortunately, a recent analysis shows that the oxygen pool is so large that Man's impact or potential impact is very small, at least at this time.

Sedimentary Cycles

Unlike the gaseous cycles, sedimentary cycles are essentially one-way flows. Potassium, magnesium, calcium, phosphorus, chlorine, iron, sodium, and other elements are released by rock weathering on the land, then are carried to the sea, where sedimentation returns them to storage in the lithosphere. Because these elements do not form gaseous compounds at normal temperatures and

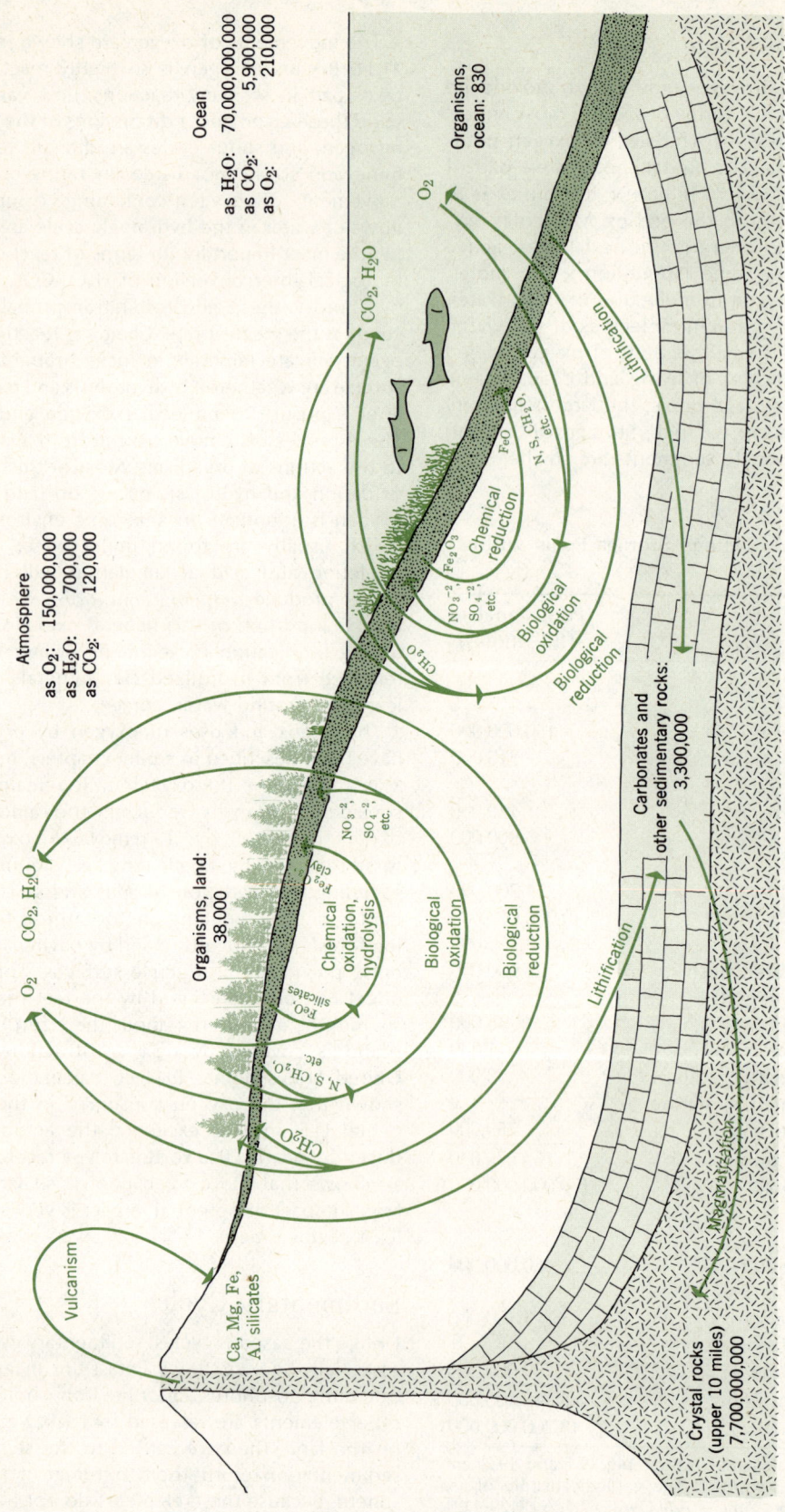

Figure 21.11 The oxygen cycle. Pool capacities (black) in units of 10^{12} moles. (Data calculated from sources of Figures 21.3, 21.4, 21.8 and 21.9. Values rounded to two significant figures.)

Sedimentary Cycles

pressures, they cannot return to the atmosphere and then to the land except by the blowing of sea salts over continents. The potential, then, exists for these elements to be in short supply.

In Chapter 13 we gained some insight into the flow of common ions in stream water. Tables 13.1 and 13.2 gave examples of ion concentrations in rainwater and stream water in a variety of environments. This discussion was followed by an example of the introduction of ions into stream water by Man's activities in a suburban region. In returning to this topic, we shall be focusing attention upon the role of organisms in taking up ions from surface water, utilizing them in growth processes, and returning them to the surface water flow paths.

Table 21.9 shows the input and output budget of nutritive ions for a small watershed at Hubbard Brook, New Hampshire. Inputs come from rain, and outputs are in the form of runoff. Of interest is the net export, shown in the last column. Note that the forest in the watershed is accumulating sulfate, potassium, ammonium, and nitrate, while it is losing silica, sodium, chloride, magnesium, and calcium. The gains represent the incorporation of ions into new plant material as the forest grows. Because the input figures do not include the release of elements from weathering within the watershed, the losses are probably not from the plants or organic matter, but instead are derived from the rocks. Note also that the two materials lost in largest amounts (Na^+ and SiO_2) are nonessential to plants. The data show how organisms tend to accumulate elements and compounds of value to them, and thus slow the movement of these materials to the sea as part of the sedimentary cycle.

When Table 7.2, listing the eight most abundant elements in the earth's crust, is compared to Table 21.1, which expresses the average composition of living matter, only three of the macronutrients listed in Table 21.1 do not rank in the table: phosphorus, chlorine, and manganese. Of these, phosphorus is most critical for organisms, since it is used by all organisms as an energy carrier—for example, in ATP (Chapter 19)—and is required in the largest amount. Chlorine is the major constituent of sea salt in precipitation, and so is likely to be available in spite of its terrestrial scarcity. Manganese is rarer, but as a constituent of enzymes is needed in much smaller quantities than phosphorus or chlorine, for it is a micronutrient. It would appear, then, that phosphorus is most likely to be in short supply.

Lack of phosphorus inhibits the development of many organisms, and is, in most cases, the cause of soil infertility. Phosphorus is soluble only in acid solutions or under reducing conditions, and so tends to remain in the form of calcium phosphate or ferric phosphate in soils. Typical concentrations of phosphate in natural surface waters may be on the order of a few ppm, and commonly algae and other phytoplankton in fresh water are limited by low phosphorus concentrations. When phosphate levels in lakes are increased by addition of sewage effluent, often the result is a greatly increased productivity, yielding an algae bloom. Or, if nitrogen now becomes limiting because of an ample supply of phosphate, blue-green algae may be favored, because they possess their own mechanism for fixing atmospheric nitrogen. In any case, these changes are part of the process of eutrophication, discussed in Chapter 23.

TABLE 21.9 Input and Output of Nutritive Ions in a Small Watershed

Ion	Input (moles/hectare)	Output (moles/hectare)	Net Export (moles/hectare)
NH_4^+	116	17	−99
NO_3^-	108	77	−31
K^+	46	28	−18
$SO_4^=$	312	306	−6
Ca^{++}	70	75	+5
Mg^{++}	45	74	+29
Cl^-	79	116	+37
Na^+	91	182	+91
SiO_2	32	348	+316

SOURCE: Data calculated from F. H. Borman, et al., as shown in E. S. Deevey, Jr., 1970, "Mineral Cycles," *Sci. American*, vol. 223, no. 3, p. 155.

The DDT Cycle

The cyclic movement of matter through ecosystems applies not only to such naturally occurring compounds as nitrates and phosphates, but also to Man-made compounds which are absorbed by organisms. DDT is an example of a synthetic compound which moves through the food web and has produced a significant impact on the biosphere.

DDT is one of many toxic chemical compounds, some simple and some complex, classed as *pesticides*, substances used to reduce the numbers of an unwanted plant or animal population. Before the widespread advances in chemical technology accomplished during and after World War II, two types of insecticides were in use: (1) *inorganic compounds*, including compounds of arsenic, sulfur, copper, or cyanide, and (2) complex organic compounds, often referred to as *botanicals*, which were obtained from plant tissues. Use of these compounds was not widespread, and except in local areas, impact was minimal. In the postwar era, however, many new organic compounds were developed with toxic properties suitable to their use as biocides of one sort or another. One large class of pesticides thus developed is the *chlorinated hydrocarbons*, of which DDT is a member.

Besides their apparent low toxicity to Man, the chlorinated hydrocarbons have two general characteristics which make them desirable as insecticides: first, they are toxic to a large range of organisms and therefore can control many pests at once, and second, they are relatively long-lasting and therefore need to be applied less frequently. Unfortunately, these characteristics also make the chlorinated hydrocarbons all the more dangerous to natural ecosystems. The danger lies in their *biological magnification* within the food chain, a phenomenon produced by the high solubility of these compounds in fats and oils and their storage within the fatty tissues of organisms.

Table 21.10 presents some concentrations of DDT in the body tissues of organisms sampled from a salt marsh in Great South Bay, on the southern shore of Long Island, New York. DDT enters the marsh by aerial spraying for mosquito control. Because it is highly insoluble in water, most of the DDT remains in the marsh soil or settles to be incorporated in the bottom sediments of channels and pools. A very small amount, however, does dissolve in the water. The algae, phytoplankton, and decomposer bacteria of the marsh are thus exposed to very low concentrations of DDT in their environment. Because DDT is much more readily soluble in fats and oils, it tends to accumulate in the bodies of these tiny

TABLE 21.10 DDT Residues from an Estuary at the Eastern End of Great South Bay, Long Island, New York

Sample	DDT Residues (ppm)
Water	0.00005
Plankton, mostly zooplankton	0.04
Shrimp	0.16
Atlantic silverside (*Menidia menidia*)	0.23
Crickets	0.23
Mud snail (*Nassarius obsoletus*)	0.26
American eel, immature (*Anguilla rostrata*)	0.28
Flying insects	0.30
Cordgrass, shoots (*Spartina patens*)	0.33
Hard-shelled clam (*Mercenaria mercenaria*)	0.42
Chain pickerel (*Esox niger*)	1.33
Atlantic needlefish (*Strongylura marina*)	2.07
Cordgrass, roots (*Spartina patens*)	2.80
Common tern (*Sterna hirundo*)	3.15
Herring gull, brain (*Larus argentatus*)	4.56
Herring gull, immature (*Larus argentatus*)	5.43
Osprey, abandoned egg (*Pandion haliaetus*)	13.8
Double-crested cormorant, immature (*Phalacrocorax auritus*)	26.4
Ring-billed gull, immature (*Larus delawarensis*)	75.5

SOURCE: Data of G. M. Woodwell, C. F. Wurster, and P. A. Isaacson, 1967, *Science*, vol. 156, p. 822.
NOTE: Data include DDT as well as its metabolites, DDD and DDE.

The DDT Cycle

organisms and reach concentrations many times those of the water itself. When these organisms are ingested in large numbers by zooplankton or detrital feeders such as clams and tiny crustaceans, the DDT they take in also tends to be retained because of its insolubility in water. In this way, the concentration of DDT in body tissues increases by many orders of magnitude up the food chain, from fingerling fish to larger fish to predatory birds, such as the herring gull and cormorant.

Note that this buildup of DDT would not occur if there were some rapid means of breaking it down into nontoxic compounds. Physically, the DDT molecule is very stable and, unlike many other organic molecules, does not decompose under continued exposure to oxygen, sunlight, or other reagents or energy sources available at the earth's surface. From a biological viewpoint, DDT is also very stable. Only a few microorganisms seem to possess the ability to degrade DDT efficiently, and the process appears to require anaerobic conditions. In animals, DDT is metabolized very slowly by the liver into breakdown products DDE and DDD, both of which are still highly toxic. These factors combine to yield a half-life for DDT in the biosphere estimated at about 15 years.

The toxic effects of DDT vary from organism to organism. For example, the main impact of DDT on marine fishes has been to drastically curtail reproductive rates because DDT is concentrated in the yolk sac of the fish egg. The embryo is thus exposed to high concentrations of the poison at its most sensitive stages, and it succumbs. The primary effect of DDT on birds is to interfere with calcium metabolism, causing the laying of thin-shelled eggs which break easily, killing the young inside. Another effect is to upset the hormonal balance, producing abnormal behavior as well as other symptoms.

Recent research on the movements of DDT in the biosphere has yielded sufficient data to sketch the broad outlines of a *DDT cycle* (Figure 21.12). At the present time, the total amount of DDT in the biosphere is estimated to be near 2×10^6 metric tons,* about three-fourths of which is in terrestrial environments, with the remaining one-fourth in marine environments. The DDT cycle is of the gaseous type because a significant movement of DDT occurs through the air. DDT enters the air largely through aerial spraying operations, but significant amounts enter the air by simply evaporating from water surfaces, plant surfaces, and soils. In this fashion, about one-fourth of the total annual production of 1×10^5 metric tons/year reaches the ocean. Estimates of

---

*Because these amounts are small compared to those of other cycles, they are expressed in metric tons rather than in units of 10^{12} moles. To yield values in units of 10^{12} moles, divide the value expressed in units of 10^6 metric tons by 355. Data are obtained from *Man's Impact on the Global Environment,* Report of the Study of Critical Environmental Problems sponsored by the Massachusetts Institute of Technology, 1970, M.I.T. Press, pp. 126–136.

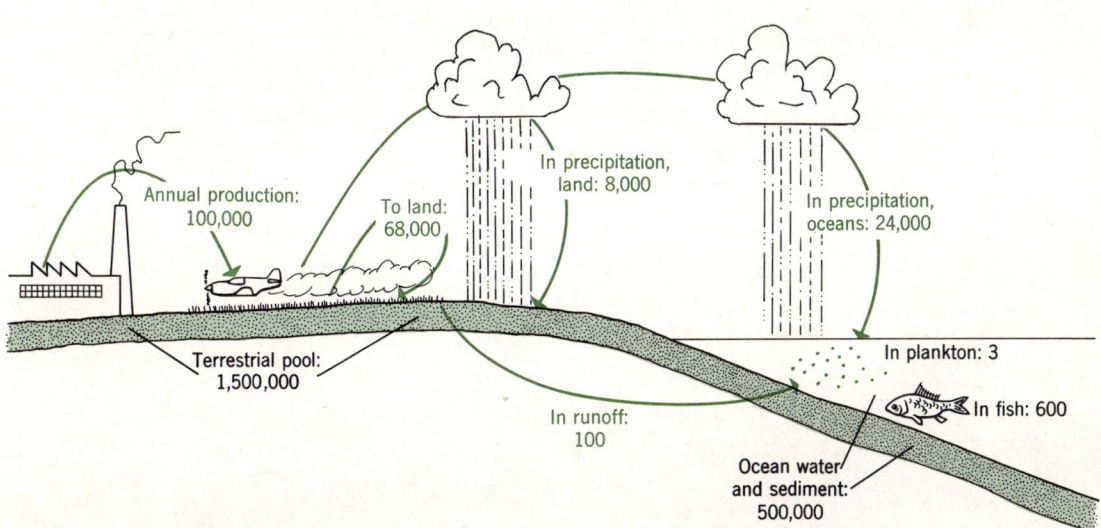

Figure 21.12 The DDT cycle. Pool capacities (black) in units of metric tons. Flows between pools (color) in units of metric tons per yr. (Data from *Man's Impact on the Global Environment,* 1970, Report of the Study of Critical Environmental Problems SCEP, M.I.T. Press, pp. 126–136.)

DDT reaching the ocean in runoff are very much lower, a result of the very low solubility of DDT in water.

Note that the DDT cycle has two differences from conventional cycles: first, DDT is synthetic—there is no natural source for DDT. Second, there is no link to bring DDT from the ocean sediments back to land, and therefore the cycle is not complete. A third difference arises because DDT, though long-lasting, is not stable indefinitely. Eventually, all the DDT in the biosphere will be broken down into harmless components. Yet this pesticide has the potential to create a considerable impact on ecosystems vital to Man's support before it is degraded.

DDT is only one of a great many new synthetic compounds which Man introduces into his environment and which move through the biosphere in a complex fashion. When this fact is considered in light of Man's significant impact on such natural cycles as nitrogen and carbon, we can see just how great is the potential for global alteration of natural systems. Certainly one of our major ecological concerns in the coming decades will be Man's advertent and inadvertent modifications to the movements of matter in the life layer.

In these last two chapters we have seen how energy and matter move through ecosystems. The flow of energy is one-way, limiting the number of trophic levels and the overall complexity of ecosystems. Matter is cycled through ecosystems by both biological and physical processes and is used and reused by many organisms, but it can limit ecosystem development where vital elements and compounds are in short supply. Because energy and matter inputs vary from place to place and from time to time, so must organisms respond to these variations. By and large, we have discussed how these variations affect individuals or species. But organisms are not isolated; they constantly interact with other organisms of the same and different kinds. Under the influence of environmental variation, populations of individuals of the same species and communities of many interdependent species must change, both through time and over space. It is the change in ecosystems, or ecosystem dynamics, which is the topic of the next chapter.

Ecosystems of the earth are constantly changing. Populations of plants and animals increase or decrease in numbers as the seasons revolve; species expand or contract their ranges as climates change; epidemics or predators at times reduce the numbers of individuals of a species to a level so low that it never recovers and becomes extinct. Thus, populations of species within ecosystems are always changing, and therefore the structure and organization of ecosystems must also be changing. The study of these changes is part of the field of *ecosystem dynamics*, the subject of this chapter.

Before we can learn about ecosystem dynamics, however, we must look at some basic ecological concepts concerning the role of the individual in the ecosystem and the density of organisms in ecosystems. With these concepts as a background, we can proceed to consider how and why populations and ecosystems change through time.

Habitat and Ecological Niche

The *habitat* of an organism is the physical environment in which it is most likely to be found. Thus, salt marsh is the habitat of cord grass; dry, limestone slopes are the habitat of chinquapin oak; and boreal forest and tundra are the habitat of the caribou. The *ecological niche* of an organism, on the other hand, includes the functional role played by the organism as well as the physical space which it inhabits. If the habitat is the individual's "address," then the niche is its "profession," including how and where it obtains its energy and how it influences other species and the environment around it. Included in the ecological niche are the organism's tolerances and responses to changes in moisture, temperature, soil chemistry, illumination, and other factors. Although many different species may occupy the same habitat, only a few of these will ever share the same ecological niche, for evolution will tend to separate them. Interspecific competition (competition between species) acts unfavorably against the individuals whose niches overlap and favorably for individuals whose niches are even slightly different. The direction of evolution will thus be toward niche diversity among species.

22

Ecosystem Dynamics

The *density* of a population is the number of individuals found in a unit space. Density can also be expressed on a biomass basis. The density of a population is controlled, for the most part, by the availability of energy and nutrient matter in relation to the size and efficiency of the organism. Events occurring in the past have an influence, too —population levels may be temporarily high or low in response to such events. Absolute density, however, can be misleading, for the area considered in the density calculation does not distinguish between suitable and unsuitable habitats. For this reason, ecologists use the *ecological density*, which counts the number or biomass of organisms in a unit area of suitable habitat.

Density specifies the average distribution of individuals in a unit area. The specific pattern of individuals, however, can vary. Ecologists recognize three types of distribution patterns: uniform, random, and clumped (Figure 22.1). In the *uniform* pattern, individuals are more or less equidistant. Examples include the spacing of canopy trees in a forest or the spacing of territorial songbirds. Uniform patterns are typical where forces operate between individuals which tend to keep them apart. Such forces might be the need for a certain minimal sunlit area, or for a minimum territory to support feeding activities.

In the *clumped* pattern, individuals are found in groups or clumps. Such patterns are typical when proximity to others benefits the individual, as in herds or family groups. Still other clumped patterns may result from limited dispersal of seeds or immature organisms, or from coincidental variations in the environment which encourage the presence of organisms in certain areas and discourage their presence in others. In plants, it can also result from asexual reproduction by runners or rhizomes. The clumped pattern is probably the most common type—the distribution of most plants and animals tends to be clumped.

In the *random* pattern, individuals are located by chance. No significant forces operate to encourage proximity to or separation from other individuals of the same species. The random pattern is encouraged if the environment is particularly uniform, as, for example, a tidal mudflat—where some species of clams show a random distribution pattern. Many of the "lone wolf" type of predators are also distributed randomly when their prey are not likely to be found in any particular place.

Population Dynamics

The size of a population of organisms within an ecosystem is regulated by the *growth rate*. When the growth rate is positive, the population is increasing in size; when negative, decreasing in size; and when zero, remaining stable. For our discussion here, we will define the growth rate as the percentage increase of the population in a given unit of time. Thus, a population of mosquitoes in the late spring might be increasing at a growth rate of 5% per week. On the other hand, alligator populations in the Everglades might be decreasing at 10% per year, a negative growth rate. In most natural ecosystems, populations tend to be of a stable size, and therefore exhibit a growth rate near zero, as averaged over several years time.

The growth rate of a population is dependent on the balance between the birth rate, or *natality* rate, and the death rate, or *mortality* rate. Movements of individuals into and out of the population may also be important in the case of mobile organisms.

The natality rate of a species is dependent on many factors. First, the number of fertilized eggs produced per female per reproductive interval will vary greatly from species to species. For example, one fish may lay many millions of eggs at one time, whereas many of the larger mammals produce only single offspring at a time. Reproductive intervals will vary from many per year (some insects) to one every two years (elephants) or longer. Also important will be the number of producing females, which is dependent on the sex ratio and age structure of the population. Natality rate can be expressed in a general fashion as the number of new individuals per unit of population per unit of time, as in births per 1000 people per year for humans. Another measure is the *age-specific birth rate*, which measures the

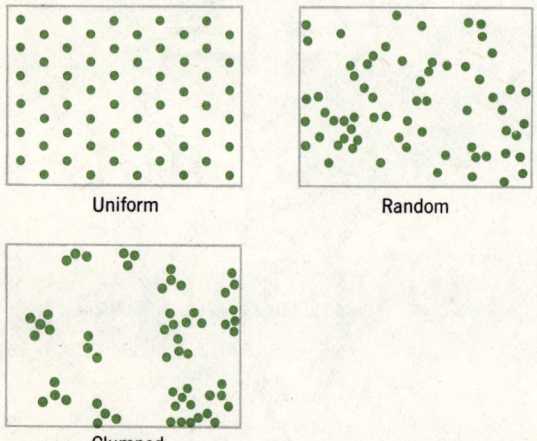

Figure 22.1 Uniform, random and clumped distribution patterns.

Population Structure

average number of births per year per female for females within a specific age group.

The mortality rate of a population is dependent on such factors as disease, predation, and competition. The rate varies by age group, the youngest and oldest members of the population usually showing the highest mortality. The variation of mortality with age is often described by a *life table* similar to that in Table 22.1. The life table follows a group of individuals, or *cohort*, usually 1,000 or 10,000 in number, through life, recording the number of the original group which survive to each age class, the number of individuals dying within the age class, and the probabilities of an individual's living beyond or dying within the interval. The life table shown in Table 22.1 is taken for cottontail rabbits, and is on a quarterly time (four months) basis.

The number of individuals remaining may be plotted as a function of age to yield a *survivorship curve*, which is usually plotted on semilogarithmic paper. Some examples are presented in Figure 22.2. The shape of the survivorship curve reveals much about the population dynamics of the species. For most populations, the curve falls off steeply at the beginning, a result of infant mortality. Such a rapid decline usually continues for a species which produces a large number of offspring, relying on "safety in numbers" as its evolutionary strategy. Perennial plants often exhibit this type of survivorship curve. For a species which invests much time and energy in raising its progeny, as many of the longer-lived mammals do, the curve is most likely to flatten out, the numbers decreasing slowly until the maximum physiological age is approached, when mortality again increases. For a species in which disease or predation plays a more important role in controlling the population size, the population remaining is likely to decrease by a constant percentage, producing a more or less straight descent. These three cases are shown in Figure 22.2. The curve for Dall mountain sheep (A) remains fairly flat, then declines rapidly as the maximum life span is reached. The curve for cottontail rabbits (B) shows a more or less constant rate of population loss during the main part of the life span. It is prepared from the data of Table 22.1. The curve for the barnacle, *Balanus glandula* (C), shows a sharp initial decline, then levels off to a slow rate of loss until the maximum life span is approached.

Population Structure

The age structure and sex ratio of a population are important factors of population structure which influence mortality and natality rates. The lifespan of an individual can be divided into three

TABLE 22.1 Life Table for a Group of 10,000 Cottontail Rabbits

Age (months)	Number Alive at Beginning of Period	Number Dying during Period
0–4	10,000	7,440
5–8	2,560	1,063
9–12	1,497	701
13–16	796	361
17–20	435	240
21–24	295	138
25–28	157	62
29–32	95	61
33–36	34	30
37–40	4	4

SOURCE: Data from R. D. Lord, Jr., 1961, *Journal of Wildlife Management*, vol. 25, pp. 33–40, as presented in R. L. Smith, 1966, *Ecology and Field Biology*, Harper & Row, New York, p. 348, Table 20.2.

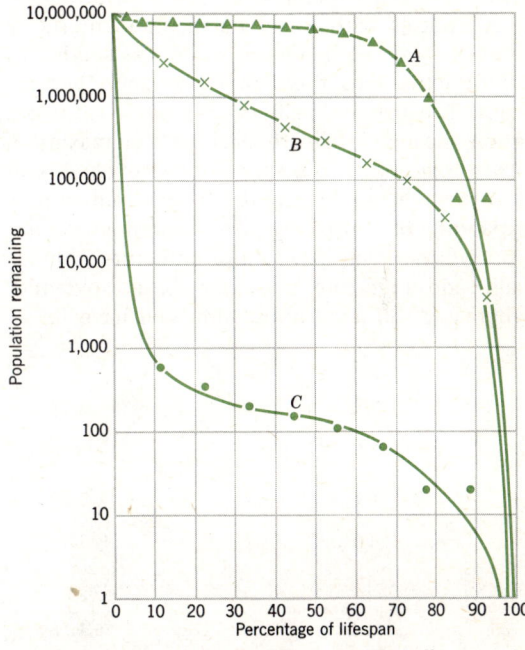

Figure 22.2 Three survivorship curves: A. Dall mountain sheep on Mt. McKinley, Alaska. (From data of E. S. Deevey, Jr. 1947, *Quarterly Review of Biology*, vol. 22, p. 289, Table 1, based on data of Adolph Murie, 1944, *The Wolves of Mount McKinley*, Fauna of the National Parks of the United States, Fauna Series No. 5, U.S. Government Printing Office, p. 123, Table 6.) B. Cottontail rabbits. (From data of R. D. Lord, Jr., presented in R. L. Smith, 1966, *Ecology and Field Biology*, Harper and Row, New York, p. 348, Table 20.2) C. The barnacle *Balanus glandula*. (Data of J. H. Connell, 1970, *Ecological Monographs*, vol. 40, p. 70, Table 14.)

periods: prereproductive, reproductive, and postreproductive. Theoretically, the proportion of individuals within these three groups tends to reach a stable equilibrium in which natality and mortality rates for the population are equal. At the equilibrium, the reproducing group produces enough offspring during a reproductive interval to replace those reproducing individuals lost during the interval to mortality or to postreproductive status. If the equilibrium age structure is disturbed by an epidemic, or perhaps unusually heavy predation, the population will tend to return to the same structure when the disturbance ends.

Figure 22.3 shows several types of population age structures, presented as *age pyramids* which plot the percentage of males and females within age groups of the population. Instead of three stages (prereproductive, reproductive, and postreproductive), a larger number of age groups are shown for more accuracy in the pyramid. Pyramid *A* is typical of a more short-lived species with high natality and mortality rates. The broad base represents the large number of immature individuals, only a small portion of which survive to reproductive stages. Pyramid *B* is typical of a longer-lived species with lower natality and mortality rates. Since most immature organisms survive to reproductive ages, the pyramid has more vertical sides. The pyramid falls off rapidly at the top, where the individuals are reaching their maximum physiological age. The shape of the pyramid will also be affected by whether the population is expanding or contracting. An expanding population will normally have more immature individuals, and therefore a broader-based pyramid, whereas a shrinking population will have fewer young and a greater proportion of older, postreproductive individuals, producing a pyramid with more vertical sides.

The *sex ratio* (ratio of males to females) of a population as a whole or of specific age groups within the population will also have an influence on natality and mortality rates. In vertebrates, the sex ratio usually favors males slightly at birth, but may favor either sex by the time adulthood is reached. A shift in sexual dominance can be produced by such factors as selective disease, rearing behavior, or selective predation. An example of the latter is the shift with age in the sex ratio of deer and pheasants from male to female predominance in response to selective hunting of males. In some cases, the sex ratio can be affected by population characteristics. For example, the sex ratio in young woodchucks has been shown to shift from about even to more than 2:1 females over males when adult females are selectively removed from a population. Thus, the woodchuck possesses a physiological regulating mechanism which tends to produce a population in which adult males and females are equally represented.

Population Growth

Because the growth rate is defined as a percentage change per unit time, the number of individuals by which a population changes is proportional to the initial size of the population. If the growth rate is 10% per year, and there are 100 individuals, the population is increased by 10 during the first year. In the second year, however, the increase is greater: 10% of 110, or 11. In the third year, the increase should be 12 individuals, bringing up the population to 133. The effect is like compound interest—an *exponential increase* (or *exponential decay* for a negative growth rate). For an expanding population, the rate of increase can also be described by the *doubling time*. As a rule of thumb, the doubling time is approximately equal to the constant number 70 divided by the percentage growth rate. Thus, a 10% per year growth rate produces a doubling about every 7 years (70 ÷ 10 = 7). A 5% growth rate produces a doubling about every 14 years (70 ÷ 5 = 14). Figure 22.4 shows how the numbers of a population increase under several different growth rates. Because the increase is exponential, the growth curve plots as a straight line on semilogarithmic paper. The doubling time can be read directly from the plot.

The maximum growth rate for a particular species is referred to as the *biotic potential* of the species. This maximal growth is measured under conditions as favorable as possible to growth and

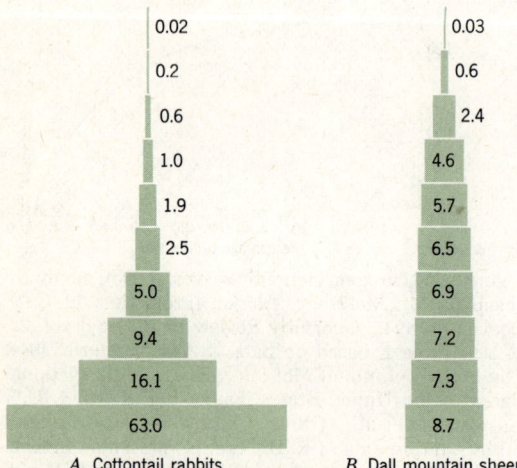

Figure 22.3 Two types of age pyramids. Data from sources of Figure 22.2A (Dall mountain sheep) and 22.2B (Cottontail rabbits).

Population Growth

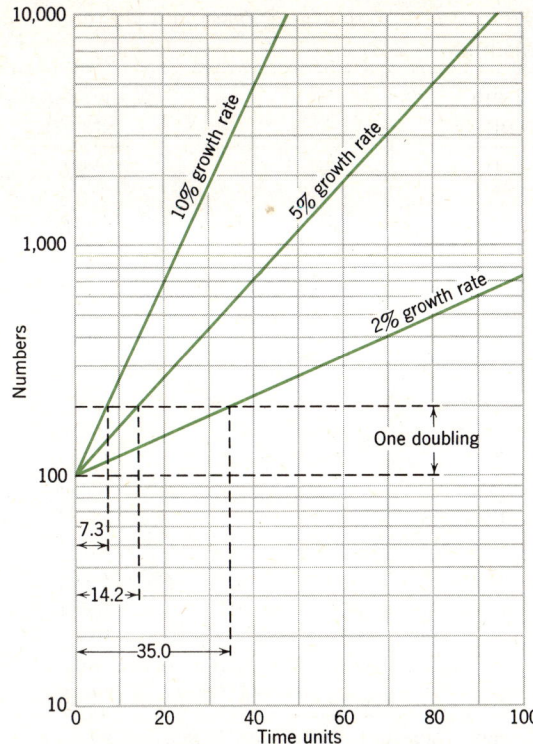

Figure 22.4 Semilogarithmic plot of three growth rates, starting with 100 individuals. The doubling time can be read by inspection.

has 10 or 1000 plants. *Density-dependent* factors have an affect which is proportional to the population density. For example, consider competition between members of the same species. At low population levels, only a small number of individuals are sufficiently close to one another to compete; at high levels, however, a greater proportion of the population encounters competition and feels its effects. Thus, the larger the size of the population, the greater is the proportion of the population affected by competition.

The existence of density-dependent regulating factors implies that populations have upper limits. For a particular species, the environment can determine a *carrying capacity*, a maximum density above which a population cannot long sustain itself. For species in which density-dependent limiting factors are important, growth is best approximated by a *logistic* curve (Figure 22.5), which follows a particular type of mathematical equation, the logistic equation. In the logistic curve, population size first increases rapidly, then slows and finally stops at the level of the carrying capacity. The logistic curve is one of a family of curves with a *sigmoid*, or S-shape. This sigmoid type of curve is distinguished from the exponential curve, which is often referred to as *J-shaped*

reproduction, so is rarely achieved in nature. The biotic potential, however, serves as an index to indicate how rapidly a species can expand to fill new or vacated ecological areas within a particular environment. Thus, species with high biotic potentials would be expected to colonize and dominate such environments as new and old fields, floodplains, and other areas where rapid physical change often greatly alters ecosystems.

In nature, most species rarely, if ever, expand at their full biotic potential. Instead, many factors act to check population increases. These factors can be classified into three types: density-independent, density-proportional, and density-dependent. *Density-independent* factors act to remove a specific number of individuals from a population, no matter what the size of the population. A set number of hunters taking a set number of bucks each year would be an example of a density-independent factor regulating the size of a deer population. A *density-proportional* factor acts to remove a certain percentage of the population—its action is therefore dependent on the number of individuals present. A cold snap which kills 30% of a tomato crop will kill 3 or 300 plants, depending on whether the garden or field

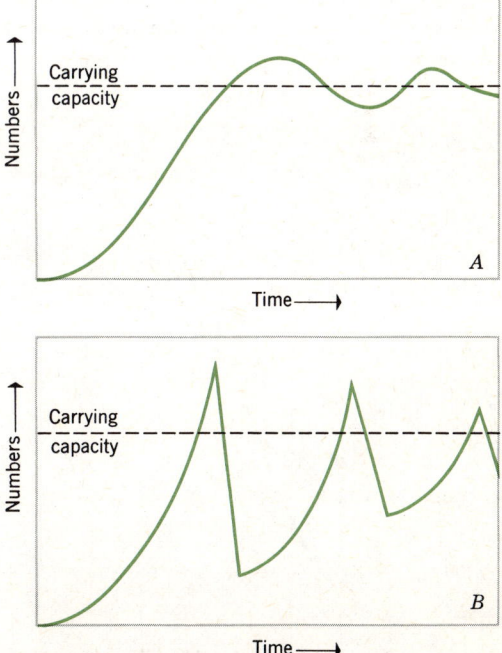

Figure 22.5 Two types of population growth. *A.* Logistic growth fluctuating near the carrying capacity as limits are reached. *B.* Exponential or J-shaped growth, marked by dramatic fluctuations in the size of the population.

because of its shape when plotted on arithmetically scaled (not semilogarithmic) graph paper.

Of course, J-shaped or exponential growth cannot continue indefinitely either (Figure 22.5). Often a sudden limit is reached, such as exhaustion of food resources, and the population size crashes to low levels, only to begin another spell of exponential growth. Populations following logistic growth can also experience considerable variation. Time lags in the system, such as the interval between birth and the time of reproduction, can produce an overshoot or a series of oscillations around the carrying capacity. In fact, natural populations always show some variation in number through time, and therefore natural oscillations are the rule rather than the exception.

An example of logistic growth is shown in Table 22.2. A yeast culture is started in a small container and allowed to multiply, with sampling performed each hour. Growth is rapid at first but then slows as the carrying capacity of the medium is reached. In this case, toxic metabolic products of the yeast tend to accumulate and limit expansion. The population size is plotted on arithmetic and semilogarithmic scales in Figure 22.6. On the arithmetic scale, the curve shows the logistic shape. On the semilogarithmic plot, the curve is straight until density-dependent limiting becomes significant, and at that point the curve begins to bend over, showing a decreasing rate of population expansion. Both curves become horizontal at a stable population level reached after about 15 hours.

TABLE 22.2 Growth of Yeast in a Culture Medium

Time (hours)	Yeast Biomass (grams)
0	9.6
1	18.3
2	29.0
3	47.2
4	71.1
5	119.1
6	174.6
7	257.3
8	350.7
9	441.0
10	513.3
11	559.7
12	594.8
13	629.4
14	640.8
15	651.1
16	655.9
17	659.6
18	661.8

SOURCE: Data of T. Carlson, 1913, *Biochemische Zeitschrift,* vol. 57, p. 326, Table 8.

The seasonal cycle of Australian thrips, small insects which spend most of their life cycle sucking the sap from flowers, illustrates exponential or J-shaped growth and is presented in Figure 22.7. As spring begins, the number of flowers

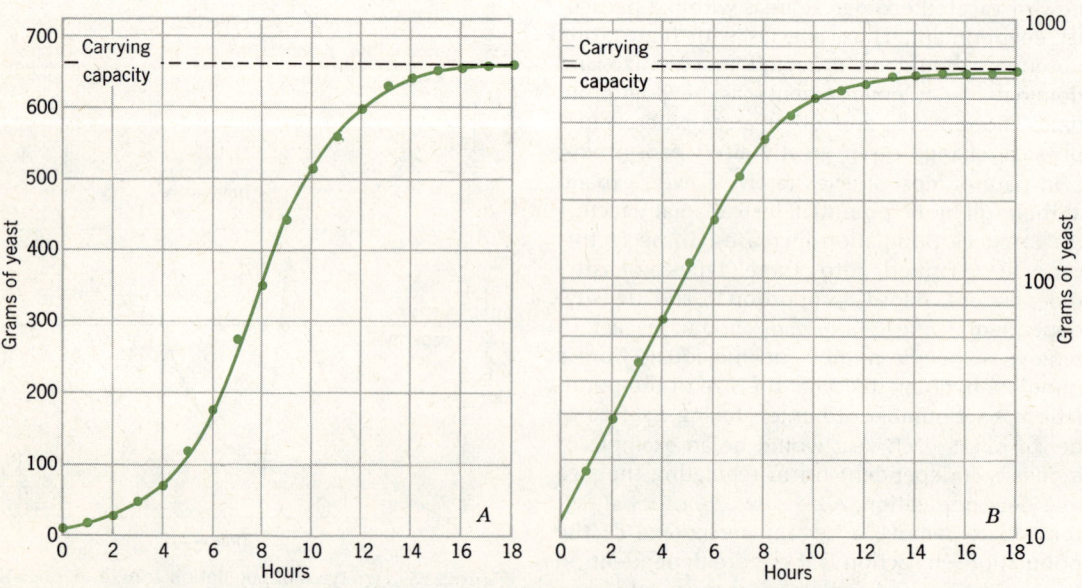

Figure 22.6 Growth of yeast in a culture medium. *A.* Growth plotted linearly, fitting logistic model. *B.* Growth plotted semilogarithmically. (From data of T. Carlson, 1913, *Biochemische Zeitschrift,* vol. 57, p. 326, Table 8.)

Population Growth

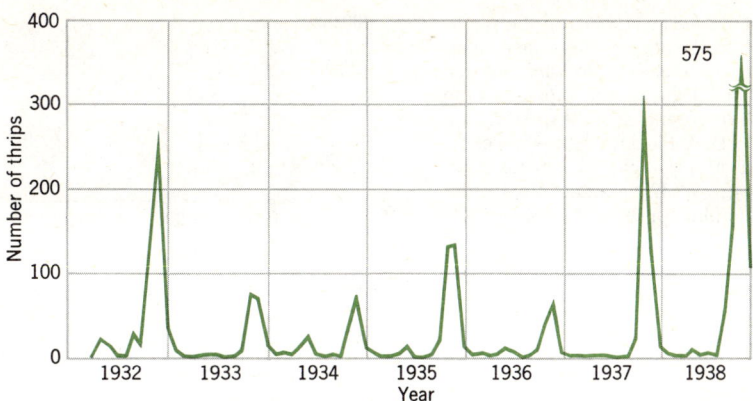

Figure 22.7 Average number of thrips per flower per day for each month of a seven year period. (Data of J. Davidson and H. G. Andrewartha, 1948, *Journal of Animal Ecology*, vol. 17, p. 197, Table 3.)

increases very rapidly, as do the thrips. Living space and food are abundant, so survival and birth rates are near maximal values. When summer arrives, however, conditions change greatly. Drought begins, the number of flowers is reduced, and the distance from one flower to the next is greatly increased. As the flowers fade, so do the thrips—not so much because their food supply is now limited, but mainly because the young thrips which fly off to new flowers can only rarely find them. As a result, the population crashes to a low level. Although a slight rise is encountered in the fall, the population remains at low levels till the following spring when expansion begins anew. Note that in this example, the rapid decline is related to a density-independent change in the thrips' environment. Note also that the graph does not present a perfect series of J's—rather, it has a sigmoidal tendency in those years when growth is not so rapid. It may well be that density-dependent factors are more important in those years, keeping down the population peak.

Like the Australian thrips, many other populations show oscillations at regular intervals. The three- to four-year cycle of the lemmings is another example. These small arctic and subarctic rodents (Figure 22.8) become extremely abundant over wide areas of North American and Eurasian tundra every three to four years, only to die in great numbers and become scarce in the next. In Europe, the highest population peaks are sometimes accompanied by vast mass emigrations of lemmings from their native habitats. On occasion such an emigration takes large numbers into the ocean, where they drown. Associated with the lemming cycle are cycles in predators such as the arctic fox and snowy owl. Populations of these animals decrease sharply following the crash of the lemmings, a response to the disappearance of their food supply.

Many explanations have been proposed for this three- to four-year cycle; one of the most recent relates the population growth to nutrient cycles. As the tundra rodent population builds, more and more nutrients, particularly phosphorus, are tied up in the rodent population. At a certain point, the lack of nutrients lowers the nutritional value of the rodents' food, and the growth and survival of the young animals is reduced. The population crashes as old rodents die and only a few young animals survive to replace them. Following the crash, the release of nutrients to the soil by decay of the carcasses allows the growth of more nutritive forage, and so the rodent population can build once again.

Another arctic cycle involves the snowshoe hare and the lynx (Figure 22.9). The hare popula-

Figure 22.8 The Norway lemming, *Lemmus lemmus*. Young and parent are shown together in this photo. (Eric Hosking/National Audubon Society.)

Ecosystem Dynamics

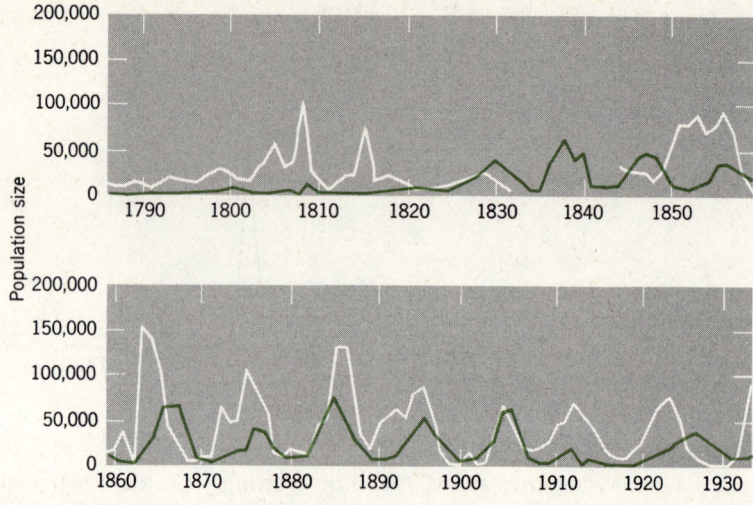

Figure 22.9 Cycles in the populations of the snowshoe hare (gray line) and the lynx (colored line). Data are incomplete for the hare from 1832 to 1844. (After data of D. A. MacLulich, in Allee et al., 1949, *Principles of Animal Ecology*, W. B. Saunders Co., Philadelphia, pp. 323–324, Figures 117, 118.)

tion peaks once every nine or ten years, and the lynx population follows with a peak one or two years later. These two cycles can be traced back to near 1800 through the fur trading records of the Hudson's Bay Company. Because the hare is a prime food source for the lynx, the similarity of the two cycles is easily explained.

Human Population

No examination of the field of population dynamics would be complete without at least a brief look at the human population. From its modest beginnings some million or so years ago, *Homo sapiens* has multiplied sufficiently to dominate most of the earth's land surface. The general trend of growth is shown in Figure 22.10. This arithmetic plot shows a slow rate of increase until about 500 years ago, when the numbers began to increase more rapidly. The increase is not only produced by simple exponential growth; the rate of increase also increases! This change is shown in Table 22.3, which lists the doubling times for the human population at various stages of its expansion. The doubling time ranges from 230,000 years—a very stable population in equilibrium with its environment—to 36 years, the value predicted for the last half of this century.

The increase in the rate of growth, however, has not been constant since Man's mid-Pleistocene beginnings. Figure 22.11 shows the human population as a function of time plotted logarithmically. The graph is really a composite of three curves. The leftmost curve shows how the human

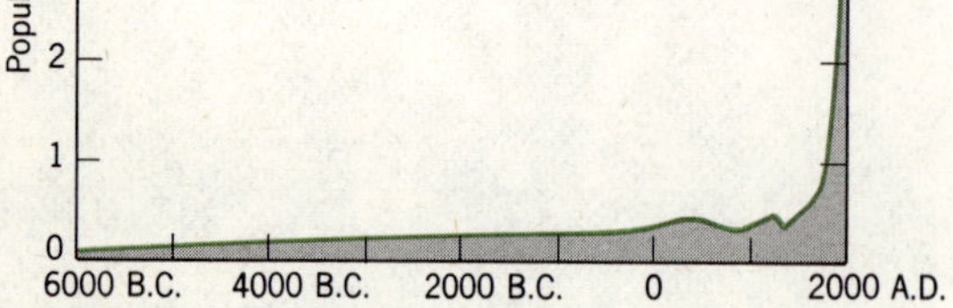

Figure 22.10 Size of the human population during the last 8,000 years. (Data from *Population Bulletin*, presented in P. Nobile and J. Deedy, eds., 1972, *The Complete Ecology Fact Book*, Doubleday Anchor Books, Garden City, New York, p. 3, Figure 1.1.)

TABLE 22.3 Growth and Doubling Time of the Human Population

Year	Population	Doubling Time (years)
1,000,000 B.C.	125,000	230,000
300,000 B.C.	1,000,000	160,000
25,000 B.C.	3,340,000	22,000
8,000 B.C.	5,320,000	1,000
4,000 B.C.	86,500,000	6,400
A.D. 0	133,000,000	830
1650	545,000,000	240
1750	728,000,000	160
1800	906,000,000	120
1900	1,610,000,000	87
1950	2,400,000,000	36
2000 (estimated)	6,270,000,000	

SOURCE: From data of E. S. Deevey, Jr., 1960, *Scientific American*, vol. 203, no. 5.

Human Population

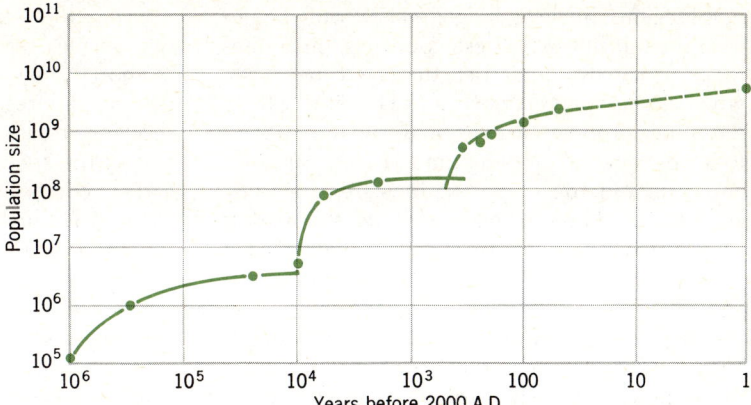

Figure 22.11 Human population size in past years, plotted logarithmically. (Data from Deevey, as shown in Table 22.3)

species responded to acquisition of the ability to use tools; the middle curve reflects the benefits of the agricultural revolution; and the rightmost curve presents the growth stemming from the industrial and scientific revolution beginning a few hundred years ago. Although each curve shows a steep initial rise and then a leveling off, the mathematics are such that this shape does not mean that the population expanded in a burst, then slowed its rate of growth. What is important about the figure, however, is that the curve can be divided into three parts—with each successive advance in the ability of the species to survive, its rate of growth increased dramatically.

The rapid rate of increase of the human population has resulted largely because density-dependent factors no longer hold it in check. Competition between individuals for food and territory was greatly reduced by the aquisition of the ability to hunt efficiently and cultivate food plants. Cultural practices involving diet and sanitation reduced the spread of disease, allowing more people to live closer together. Thus, most of Man's population increase has resulted from decreased mortality rather than increased natality.

The exponential growth of the human population raises questions of the greatest importance. Since density-dependent factors no longer seem to be regulating the number of people on earth, will the expansion continue? Will our population follow the curve of the snowshoe hare or the rose thrip (Figures 22.7 and 22.9) and crash to low levels when the finite limits of our environment are reached? Fortunately, the rate of population growth has already decreased in some areas of the earth. In some countries, the industrial revolution has brought not only decreased mortality, but also decreased natality. In the agrarian society, children are viewed as assets—an extra pair of hands to help make the soil yield a livelihood. In the industrial society, however, they are often viewed as a drawback, requiring a long period of care and consumption before they produce anything useful. This view has led to the decline in natality experienced by industrialized nations.

The decline of mortality and natality in an industrialized country is referred to as a *demographic transition*, diagrammed in Figure 22.12. Before the transition, both birth and death rates are high. As industrialization proceeds and health care improves, the death rate falls rapidly. The birth rate, however, remains high for a time. During this intermediate period, the population expands greatly in proportion to the excess of births over deaths. Soon, the birth rate falls as well, population expansion falls to low levels, and the population stabilizes at a low rate of increase. So far, only the developed nations of the West and Japan have accomplished the demographic transition. These countries have birth rates near 20 per thousand per year and death rates near 10, producing a growth rate of about 1% per year, or a doubling time of about 70 years. The underdevel-

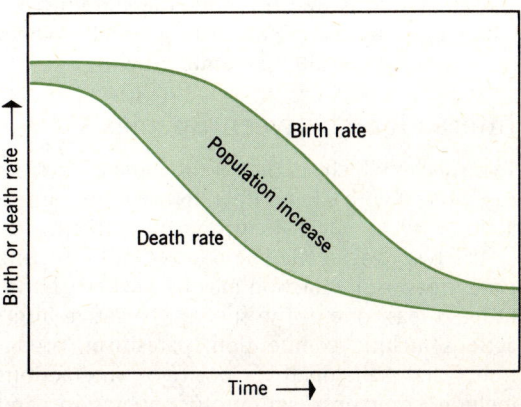

Figure 22.12 The demographic transition. (After G. T. Trewartha, 1969, *A Geography of Population: World Patterns*, John Wiley & Sons, New York, p. 45, Figure 2.4.)

oped regions, however, in general are only beginning the transition. In these countries, birth rates range from 40 to 50 per thousand per year, whereas death rates range from 15 to 25. This large difference produces expansion at the rate of 2.5% to 3% per year, a doubling time of 25 to 30 years. Thus, the underdeveloped nations are expanding at a very rapid rate. Taken as a whole, the world's human population will be expanding at a 2% per year rate till at least the year 2000 if present trends continue.

A recent study at the Massachusetts Institute of Technology has examined the potential impact of this population growth through the use of a computer simulation model of the world economy.* The model was constructed to take into account such factors as food production, resource consumption, economic development, and pollution buildup, as well as simple increases in population. By changing portions of the model and modifying assumptions on which it was based, the researchers were able to predict the effects of different social and economic policies on the population and economy. Some startling conclusions emerged from the study—in just about all cases, a population crash was predicted, usually for the twenty-first century. Stability was possible only under conditions of perfect population control (one birth for each death), greatly reduced resource consumption through effective conservation measures and recycling, greatly reduced pollution output rates, and a "no growth" economy in which all new investment goes to replace existing, but worn out industrial capacity. Further, the stable state could only be reached if these policies were implemented before the year 2000. Of course, these results are only those of a computer model—in fact, much scientific criticism has been directed at the study for its simplifications and assumptions. Yet, the outcome certainly encourages close scrutiny of existing policies which favor population and economic growth.

Interactions between Species

Two species which are part of the same ecosystem can interact with one another in three ways: interaction may be *negative* to one or both species; or the two species may be *neutral*, not affecting each other; or interaction may be *positive*, benefiting at least one of the species. Negative interactions include competition, parasitism, predation, and allelopathy; positive interactions include commensalism, protocooperation, and mutualism.

*The Limits to Growth, D. H. Meadows, et. al., Universe Books, New York, 205 p., 1972.

Competition between species occurs whenever two species require a common resource that is in short supply. Because neither species has full use of the resource, both populations suffer, showing growth rates lower than those when only one of the species is present. As we have noted earlier, competition between species is an unstable situation—if a genetic strain within one of the populations emerges which uses a substitute resource for which there is no competition, its survival rate will be higher than that of the remaining strain, which still competes. The original strain may become extinct. This illustrates the *competitive exclusion principle*, which states that species (or other genetically related groups of individuals) which share a common portion of their niche will tend to diverge through evolution till the niche is no longer shared.

Of course, the niche is such a complex thing to define that it is unlikely two species in nature ever occupy exactly the same niche. Experimental work inevitably simplifies the niche, just to be able to get on with the experiment. Mathematical models of competing situations and laboratory experiments of interspecific competition almost always produce a "winner" and a "loser." For example, when two species of *Paramecium* are placed together in a culture with a constant supply of food, only one survives. Data of an experiment plotted in Figure 22.13 show the rapid growth of both species at the outset. Soon, however, one begins to outstrip the other, and *P. aurelia* eventually dominates completely. In a similar experiment conducted with two species of the small, floating plant duckweed (Figure 23.4), the results were similar.

However, if the artificial environment is made a bit more complex, it may lead to stable population sizes for both competitors. In another *Para-*

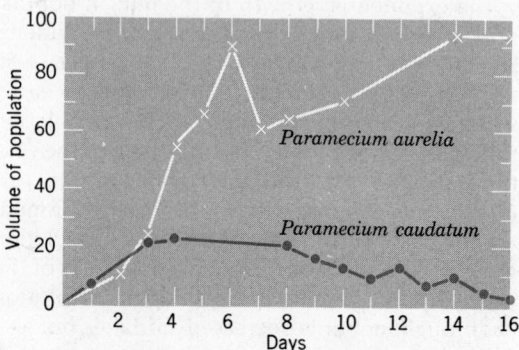

Figure 22.13 Two species of *Paramecium* cultured in competition. (Data from Gause, 1934, as presented in W. C. Allee et al., 1949, *Principles of Animal Ecology*, W. B. Saunders Co., Philadelphia, p. 658, Figure 239.)

mecium experiment, *P. caudatum* and *P. bursaria* were paired. In this case, both species survived to stable population levels because *P. caudatum* feeds on bacteria suspended in the medium, while *P. bursaria* feeds on bottom-dwelling bacteria. Both species are able to survive since each utilizes a food source not available to the other, and therefore they are not in direct competition for food.

In real ecosystems, the environment is of course much more complex. It includes such factors as selective predation, the differential effects of disease, and other influences which are absent from simple mathematical or laboratory models. Whereas one species in the natural ecosystem may be favored over another, the less successful competitor can usually maintain a population level sufficient to ensure its preservation. Should an epidemic or heavy predation reduce the numbers of the more successful species, the other will be able to expand rapidly to fill the void. In this fashion, the two species may coexist indefinitely.

Predation and *parasitism* are negative interactions in which one species gains energy by feeding on the other. If the organism which gains energy is larger, the process is predation—the individual gaining energy is the *predator*, and his food is the *prey*. If the organism gaining energy is smaller, the process is parasitism—the *parasite* gains energy, and the *host* serves as its energy source.

Although we tend to think of predation and parasitism as essentially negative processes which benefit one species at the expense of the other, it may well be that these interactions are really beneficial in the long term to the host or prey populations. A famous example is the growth of the deer herd on the Kaibab Plateau north of the Grand Canyon in Arizona (Figure 22.14). Initially at a population of about 4,000 (in an area of 700,000 acres), the herd grew to near 100,000 in the span from 1907 to 1924 in direct proportion to a government predator control and game protection program. Wolves were extincted in the area and populations of coyotes and mountain lions were greatly reduced. The huge deer population, however, proved too much for the land, and overgrazing led to a population crash. In one year, half the animals starved to death; by the late 1930s, the population had declined to a stable level near 10,000. Although the rise in deer and the decline in predators may be only coincidental, it seems that the removal of predators was indeed responsible for this population explosion, or *irruption*, as it is termed by ecologists. Thus, predation maintained the deer population at levels which were in harmony with the supportive ability of the environment. In addition to maintenance of equilibrium population levels, predation and parasitism differentially remove the weaker individuals and can improve the genetic composition of the species.

Simple predator-prey (or parasite-host) systems can behave in several different ways in the laboratory. If a predator is introduced into a population of prey, the predator species will increase its numbers greatly at first. Soon, however, the number of prey becomes significantly reduced. As the predators continue to multiply, there is reached a level at which there is insufficient prey to support the predators. At this point, one of three things may happen. First, the prey may be totally annihilated—in this case, the predators themselves soon die of starvation. Second, the prey population may be too scattered to support the predators, in which case the predators die and the prey will slowly recover. In the third case, neither population may be totally destroyed; the prey population builds up again, as do the predators, and then both populations crash to low levels again. This produces a continuing pattern of large oscillations in the populations of the two species, the prey population leading the predators.

In real ecosystems, however, predator-prey oscillations tend to be damped and populations remain near stable levels. This effect occurs because the predator often has alternate food sources and can replace one prey with another. Or, the predator may not increase its numbers as prey increase; instead, each predator consumes more. In the case of parasites, the hosts can evolve a more resistant strain to keep the number of hosts that are killed to a minimum. This defense is a type of *genetic feedback*—the rapid selection for hosts which are resistant and parasites whose action is least detrimental to the host.

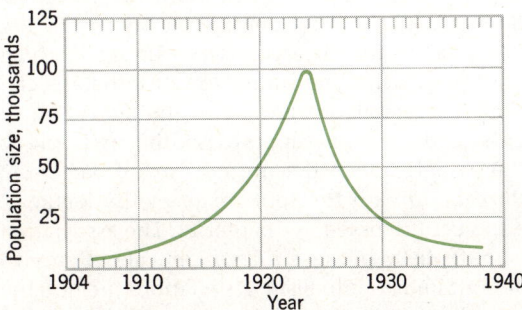

Figure 22.14 Rise and fall of the Kaibab deer population in the Kaibab National Forest, Arizona. (After D. I. Rasmussen, 1941, *Ecological Monographs*, vol. 11, p. 237.)

A fourth type of negative interaction is *allelopathy*, a phenomenon of the plant kingdom in which chemical toxins produced by one species serve to inhibit the growth of others. As an example, sage, a common shrub species in the California chaparral, produces leaves rich in volatile toxins (cineole and camphor). As the leaves fall and accumulate in the soil, the allelopathic toxins build to a level sufficient to inhibit the growth of herbaceous plants, such as grasses, which are thus only found in adjacent areas (Figure 22.15). Still other chaparral shrubs produce water-soluble antibiotics which also inhibit the growth of nearby grasses. These chemical defenses, however, are broken down by periodic fires, which are events essential to the maintenance of the chaparral ecosystem. The fires denature the toxins and also trigger the germination of seeds of many species of annual herbaceous plants by breaking the seed coats. The annuals then dominate the area until the shrubs grow and force them out by allelopathy, beginning the cycle anew.

The term *symbiosis*, discussed in earlier chapters, includes three types of positive interactions between species: commensalism, protocooperation, and mutualism. In *commensalism*, one of the species is benefited and the other is unaffected. Examples of commensals include the *epiphytic* plants—such as orchids or Spanish moss—which live on the branches of larger plants (Figure 22.16). These epiphytes depend on their hosts for physical support only. In the animal kingdom, small commensal crabs or fishes seek shelter in the burrows of sea worms; or the commensal remora fish attaches itself to a shark, feeding on bits of leftover food as its host dines.

Figure 22.16 The epiphyte Spanish moss, *Tilandsia usneoides*, growing on a branch of live oak, *Quercus virginiana*. Several small epiphytic ferns may also be seen on the branch. (Alan Pitcrain/Grant Heilman.)

When the relationship benefits both parties but is not essential to their existence, it is termed *protocooperation*. The attachment of a stinging coelenterate to a crab is an example of protocooperation. The crab gains camouflage and an additional measure of defense, while the coelenterate eats bits of stray food which the crab misses.

Where protocooperation has progressed to the point that one or both species cannot survive alone, the result is *mutualism*. A classic example is provided by the lichen, a mutualistic association of a fungus and an alga. Living inside the body of the fungus are colonies of algae which photosynthesize and supply food for the association. The fungi provide water, humidity, nutrients, and physical support. Another example is the association of the nitrogen-fixing bacterium *Rhizobium* with the root tissue of the legumes, a subject discussed in Chapter 21. The association is mutualistic because *Rhizobium* cannot survive alone. Similar mutualistic associations include the many types of *mycorrhizae*, in which fungi are associated with plant roots (Figure 22.17). In some types, the fungal hyphae penetrate the root cells

Figure 22.15 Allelopathy in a chaparral ecosystem. Toxins from the shrub *Salvia leucophylla*, on the left, inhibit the establishment of grasses and herbs between A and B. Between B and C is a transitional zone where inhibition is only partial. (Courtesy of Dr. C. H. Muller.)

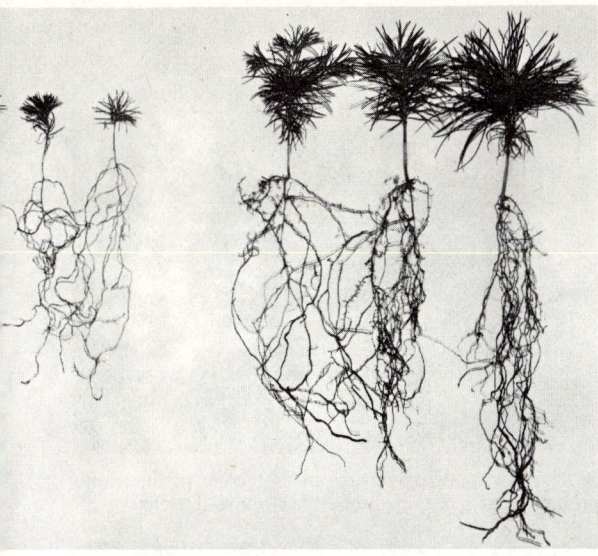

Figure 22.17 Beneficial effect of mycorrhizal fungi on 9-month-old seedlings of white pine, *Pinus strobus*. The poorly developed seedlings on the left lack the fungus. (After J. G. Iyer. Photograph courtesy of S. A. Wilde, The University of Wisconsin.)

in a manner resembling parasitic forms—in other types, the fungus simply surrounds the root and maintains a suitable climate for mineral and nutrient uptake. Sometimes only the fungal partner cannot live alone, sometimes both cannot survive alone.

Succession

The phenomenon of change in ecosystems through time is familiar to us all. A drive in the country reveals patches of vegetation in many stages of development—from open, cultivated fields through grassy shrublands to forests. Clear lakes gradually fill in with sediment from the rivers which drain into them and become bogs. These kinds of changes, in which plant and animal communities succeed one another on the way to a stable endpoint, are referred to as *ecological succession*. In general, succession leads to formation of the most complex community of organisms possible in an area, given its physical controlling factors of climate, soil, and water. The series of communities which follow one another on the way to the stable stage is called a *sere*, and each of the temporary communities is referred to as a *seral stage*. The stable community, which is the endpoint of succession, is the *climax*. If succession begins on a newly constructed mineral deposit such as a sand dune or river bar, it is termed *primary succession*. If succession occurs on a previously vegetated area which has been recently disturbed by such agents as fire, flood, windstorm, or Man, it is referred to as *secondary succession*.

The colonization of a sand dune provides an example of primary succession. Growing foredunes bordering the ocean or lake shore present a sterile habitat. The dune sand—usually largely quartz, feldspar, and other common rock-forming minerals—lacks such important nutrients as nitrogen, calcium, and phosphorus, and its water-holding ability is very low. Under the intense insolation of the day, the dune surface is a hot, desiccating environment. At night, radiation cooling in the absence of moisture produces low surface temperatures. One of the first colonizers, or *pioneers*, of this extreme environment is dune grass (Figure 22.18A). This plant reproduces vegetatively by sending out rhizomes (creeping underground stems), and the plant thus slowly spreads over the dune. Dune grass is well adapted to the eolian environment; when buried by moving sand, it does not die, but instead puts up shoots to reach the new surface.

After colonization, the shoots of dune grass act to form a baffle that surpresses saltation of sand, and thus the dune becomes more stable. With increasing stabilization, plants which are adapted to the dry, extreme environment but cannot withstand much burial begin to colonize the dune. Typically these are low, matlike woody shrubs such as beach wormwood or false heather (Figure 22.18B).

On the central Atlantic coastal plain beaches, the species which follow matlike shrubs are typically larger woody plants and trees such as beach plum, bayberry, poison ivy, and choke cherry (Figure 22.18C). These species all have one thing in common—their fruits are berries which are eaten by birds. The seeds from the berries are excreted as the birds forage among the low dune shrubs, thereby sowing the next stage of succession. As the scrubby bushes and small trees spread, they shade out the matlike shrubs and any remaining dune grass. Conifers may also enter at this stage.

At this point, the soil begins to accumulate a significant amount of organic matter. No longer dry and sterile, it now possesses organic compounds and nutrients, and has accumulated enough colloids to hold water for longer intervals. These soil conditions encourage the growth of such broadleaf species as red maple, hackberry, holly, and oaks, which shade out the existing shrubs and small trees (Figure 22.18D). Once the forest is established, it tends to reproduce itself—the species of which it is composed are tolerant to shade and their seeds can germinate on the organic forest floor. Thus, the stable

Ecosystem Dynamics

A. Dune grass is a pioneer on beach dunes and helps stabilize the dune against wind erosion.

B. The low, mat-like shrubs in the center of the photo replace dune grass in the better stabilized areas.

C. This beach thicket of poison ivy, bayberry and wild cherry paves the way for the development of the forest climax.

D. The climax forest on dunes. Here at Sandy Hook, N.J., holly (*Ilex opaca*), on the left, is an important constituent of the climax forest. Note the leaves and organic matter on the forest floor.

Figure 22.18 Stages in beach dune succession. (Photographs by A. H. Strahler.)

stage—the climax—is reached. The stages through which the ecosystem has developed constitute the sere, progressing from dune grass to low shrubs to higher shrubs and small trees to forest.

Although the foregoing example has stressed the changes in plant cover, animal species are also changing as succession proceeds. Table 22.4 shows how some typical invertebrates appear and disappear through succession on the Lake Michigan dunes. Note that the seral stages shown in the table for these inland dunes are somewhat different than those described above for the coastal environment.

Where disturbance alters an existing community, secondary succession can occur. Succession on abandoned farmland, or *old field succession*, is an example of the secondary type. In the eastern United States, the first stages of the sere often depend on the last use of the land before abandonment. If row crops were cultivated, one set of pioneers, usually annuals and biennials, will appear; if small grain crops were cultivated, the pioneers are often perennial herbs and grasses. If pasture is abandoned, those pioneers which were not grazed will have a head start. Where mineral soil was freshly exposed by plowing, pines are often important following the first stages of succession, for pine seeds favor disturbed soil and strong sun for germination. Although slower growing than other pioneers, the

Successional Changes in Ecosystem Characteristics

TABLE 22.4 Invertebrate Succession on the Lake Michigan Dunes

Invertebrate	Successional Stages				
	Beach Grass-Cottonwood	Jack Pine Forest	Black Oak Dry Forest	Oak and Oak-Hickory Moist Forest	Beech-Maple Forest Climax
White tiger beetle	x				
Sand spider	x				
Long-horn grasshopper	x	x			
Burrowing spider	x	x			
Bronze tiger beetle		x			
Migratory locust		x			
Ant-lion			x		
Flatbug			x		
Wireworms			x	x	x
Snail			x	x	x
Green tiger beetle				x	x
Camel cricket				x	x
Sowbugs				x	x
Earthworms				x	x
Woodroaches				x	x
Grouse locust					x

SOURCE: From data of V. E. Shelford, as presented in E. P. Odum, 1971, *Fundamentals of Ecology*, W. B. Saunders & Co., Philadelphia, p. 259.

pines will eventually shade the others out and become dominant. Their dominance is only temporary, however, for their seeds cannot germinate in shade and litter on the forest floor. Seeds of hardwoods such as maples and oaks, however, can germinate under these conditions, and as the pines die, hardwood seedlings grow quickly to fill the holes produced in the canopy. The climax, then, is the hardwood forest, which can reproduce itself.

It is important to note that the changes of the sere result from the action of the plants and animals themselves—one inhabitant paves the way for the next. As long as nearby populations provide colonizers, the changes lead in an automatic fashion from sand dune or old field to forest. This type of succession is often termed an *autogenic* (self-producing) succession.

In many cases, however, autogenic succession does not run its full course. Environmental disturbances, such as wind, fire, flood, or clearing by Man, may occur often enough to permanently alter or divert the course of succession. In addition, conditions such as site exposure, unusual bedrock, impeded drainage, and the like can hold back the course of succession so successfully that the climax is never reached—instead, an earlier stage of the sere becomes more or less permanent and is as stable at that site as the climax may be on more favorable sites. This fact has suggested to some investigators that there is often more than one valid climax in a particular area—this is the *polyclimax theory*. Whether succession in a particular area can have one or many stable endpoints is not as important for us here, however, as the nature of the process of succession itself.

Successional Changes in Ecosystem Characteristics

In addition to species composition, many other characteristics of ecosystems change as succession proceeds. Before these characteristics are discussed, however, two types of succession should be distinguished from the viewpoint of energy changes. First is *autotrophic* succession. In autotrophic succession, the gross primary production of the early communities exceeds respiration. The dominant colonizers are autotrophs, and because food chains are poorly developed at first, biomass accumulates in the successional ecosystem. The cases of primary and secondary succession discussed above are both examples of autotrophic succession.

In contrast to autotrophic succession is *heterotrophic* succession. Here, respiration exceeds primary production, and heterotrophs dominate the successional stages. Of course, respiration cannot exceed primary production unless organic matter is imported into the system. Thus, for example, heterotrophic succession would be expected in a polluted stream receiving an input of sewage. Or, in a natural situation, heterotrophic succession would occur as insects, fungi, and bacteria slowly degrade and consume an acorn which drops to the forest floor.

Table 22.5 presents a summary of how ecosystems change through autotrophic and heterotrophic succession to the climax. In terms of bioenergetics, the climax can be thought of as an energy system which has achieved a steady state or dynamic equilibrium. Recall from the introductory chapter on energy systems that at the steady state the rate of energy and matter input equals the rate of output, while matter and energy storage remains constant. The gross production of the plants is the energy input to the climax ecosystem, and energy losses are achieved through the respiration of the organisms which the primary production supports. At the climax or equilibrium stage, these values tend to be equal (their ratio is one) as the table shows. Because net production is the difference between gross production and respiration, net production will tend to be small or zero at the climax. The biomass, or energy and matter stored in the system, will also tend to remain constant, and the ratio of biomass to energy flow through the system will be at a maximum. Thus, the climax community is characterized by efficient use of energy and maximum biomass per unit of energy flow.

As the ecosystem develops through succession, the food chains change from simple, direct links to complex, weblike patterns. This change occurs because at first only a few species are capable of exploiting the new, often extreme, environment. As the plants and animals modify the environment, however, it becomes less and less extreme, and more species with different kinds of interrelationships can be accommodated. Thus, a simple food chain becomes a complicated food web.

Because more species can be accommodated as succession proceeds, diversity of the ecosystem tends to increase. However, certain other trends can counter this increase in diversity. Organisms tend to be larger in later stages of succession, and since fewer larger organisms can be supported in an area than smaller ones, there tend to be fewer species. As the climax is approached, species with complex life histories are more common, for they are better exploiters of a complex environment. The more detailed the life history of a single species, the more ecological niches it can occupy. Thus, the dominance of organisms with complex life histories also tends to reduce diversity. Another factor reducing diversity is increasing interspecific competition, which tends to reduce the number of species by competitive exclusion.

Nutrient cycles also change as succession proceeds. At first, mineral nutrients are largely held in inorganic reservoirs, such as soil and water. Nutrients are not retained by biological processes and therefore readily flow in and out of the ecosystem. At the climax, however, most of the vital elements are in organic form, incorporated into living and dead tissue. Grazing and detrital path-

TABLE 22.5 Summary of Changing Characteristics in Successional Ecosystems

Ecosystem Characteristic	Pioneer $\xrightarrow{\text{Autotrophic Succession}}$	Climax $\xleftarrow{\text{Heterotrophic Succession}}$	Pioneer
Gross production/ respiration ratio	>1	Near 1	<1
Net production	Positive	Near zero	Negative
Living biomass/unit of energy flow	Low	High	Low
Food chain	Short, linear	Long, weblike	Short, linear
Diversity of species	Low	Usually higher	Low
Type of mineral cycle	Open	Nearly closed	Open
Location of storage pools	In inorganic environment	In living and dead organic matter	In inorganic environment
Stability	Low	High	Low

SOURCE: Modified from E. P. Odum, 1969, *Science*, vol. 164, p. 265.

ways keep the elements cycling in organic form, and few important nutrients are lost. Thus, nutrient cycles change from open patterns with inorganic storage pools to more closed patterns with organic storage as succession leads to climax.

A final point is that climax ecosystems are more stable than successional ecosystems. Recall from the introductory chapter on energy systems that systems with a high storage to energy-flow ratio are more stable than systems with low ratios. Since the biomass/energy-flow ratio is maximized in the climax ecosystem, energy considerations alone predict its stability. The complexity of the food chain in a climax ecosystem also contributes to stability. Removal of one link by disease or overpredation will have little effect on the overall energy flow. Fluctuations in the physical environment act on many species, each of which has a different response—those least affected can assume temporary dominance till conditions return to normal. These and other factors all contribute to the stability of the climax ecosystem.

In the preceding five chapters of Part IV, we have examined the nature of organic life on earth, its history and evolution, the way in which energy and matter move through it, and the ways in which it changes through time and space. The concepts and principles we introduced in those chapters will all be drawn upon in the next two chapters, which present a survey of the composition, characteristics, and dynamics of the world's major ecosystems, including an analysis of Man's influences upon them. We will first examine aquatic ecosystems, then conclude with the terrestrial ecosystems which shelter and support so many of Man's activities.

By virtue of the oceans, aquatic ecosystems dominate the earth's surface, for the ocean surface accounts for two-thirds of the total planetary area. Aquatic habitats range from small thermal pools on land to cold, dark ocean depths; in between are many types of habitats with different salinities, light intensities, temperature regimens, and oxygen concentrations.

As one might expect for such a diverse group of environments, aquatic habitats support a varied array of organisms. Ecologists often divide aquatic organisms into groups on the basis of their life habits. The *benthos* include all the bottom-dwellers—those living on or near the sediment-water interface. Examples include rooted aquatic plants and algae as well as burrowing and filter-feeding animals such as sea worms or clams. The *periphyton* are the organisms which live on or are attached to aquatic plants, sticks, or debris which project from the bottom. Often these are filamentous algae and may have a commensal or mutualistic relation with the plant or animal which supports them. The *neuston*, at the opposite position from the benthos, include all the organisms which live on or at the water surface. Examples are "water striders" and other surface-skimming insects.

Living within the water proper are the plankton and nekton. The *plankton* are small organisms which have at best a limited capacity for locomotion. *Phytoplankton* are the autotrophic forms, and *zooplankton* are the heterotrophic forms. Many planktonic species float freely suspended in the water; others can rise to the surface or descend to the bottom by varying their buoyancy or by using swimming appendages. None, however, can resist the movement of a water current. The *nekton* includes the strong swimmers—fish, amphibians, swimming insects, aquatic crustaceans, and others—which actively make their way through the water and search out food at will. Benthos, periphyton, nekton, plankton, and neuston are all represented in both fresh-water and marine habitats. We will consider first freshwater environments, then turn our attention to marine environments.

23 Aquatic Ecosystems

Freshwater Environments

The flora and fauna of freshwater environments are diverse. Most major divisions of plants and phyla of land animals possess members which are adapted to the freshwater habitat. Among the primary producers, algae are the most important, followed by the aquatic spermatophytes. Four groups of consumers are important: the mollusks, crustaceans, aquatic insects, and fish. Bacteria and aquatic fungi are probably equally important as decomposers in freshwater environments.

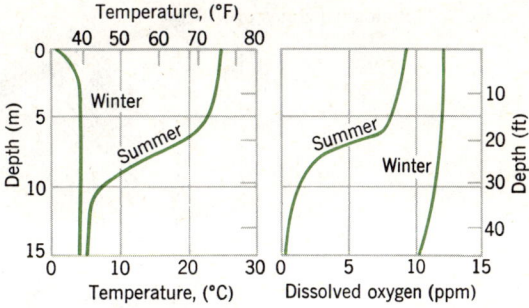

Figure 23.1 Temperature and dissolved oxygen variation with depth in summer and winter, Linsley Pond, Connecticut. (After E. S. Deevey, Jr., 1951, *Scientific American*, vol. 185, no. 4, pp. 70–71.)

Lakes and Ponds

Lakes and ponds are examples of *lentic*, or still-water, habitats. From an ecological viewpoint, a *pond* is a body of water shallow enough for sunlight to reach its bottom everywhere. A *lake*, on the other hand, is deep enough to have an area where sunlight does not reach the bottom in amounts sufficient for photosynthesis to proceed. Although lakes and ponds may seem at first to be fairly uniform habitats, there is actually a great deal of environmental variation within them.

Lakes as physical features of the environment were explained in Chapter 13. The list of diverse origins of lake basins as geologic features made evident the fact that lakes occur in a very wide range of dimensions and bottom materials. Chapter 13 also included an analysis of the water balance of lakes. The relationship of lake and pond levels to the ground water table was discussed in Chapter 12. A review of these subjects will establish a helpful background of information upon which to begin an investigation of the ecosystems of lakes and ponds.

Temperature and dissolved oxygen content are important physical properties which vary within a lake. Our discussion of the thermal cycle of mid-latitude lakes in Chapter 3 showed that, during the warm seasons, a lake can be divided on the basis of these properties into two main layers, the *epilimnion* and *hypolimnion*, separated by the *thermocline* (Figure 3.11).

As an example, Figure 23.1 shows how temperature and dissolved oxygen vary with depth in Linsley Pond (really a lake), Connecticut, in summer. The epilimnion, thermocline, and hypolimnion are well developed. Dissolved oxygen is greatest in the epilimnion, where constant mixing and the photosynthetic activity of phytoplankton and other autotrophs is greatest. Dissolved oxygen is low in the hypolimnion, for the thermocline prevents mixing with the water above, and the darkness creates a heterotrophic environment in which oxygen is consumed by decomposers and detritivores. In winter, temperature and dissolved oxygen values are quite different. Temperature is at a uniform 4° C (39° F) except for the ice layer and the water nearest to it. Oxygen levels are high for several reasons. First, recent mixing has increased the oxygen content. Second, oxygen is more soluble in cold water than in warm water. Third, the cold temperatures depress the activity of oxygen-consuming decomposers. Their abundance near the bottom, however, does produce a slight drop in oxygen at adjacent levels.

A lake may also be divided into layers from a biological viewpoint. Three zones can be recognized, each with a different set of ecological characteristics and a different group of organisms (Figure 23.2). These zones are defined by the position of the *compensation level*, the depth at which light energy is just sufficient for photosynthesis to balance the respiratory needs of autotrophs. The compensation level is approximately located at the depth to which 1% of the incident solar radiation penetrates (see Figure 3.9). Above the compensation level, in the *euphotic* zone, autotrophs dominate, whereas below the compensation level, in the *profundal zone*, only heterotrophic nutrition will be possible. Within the euphotic zone are two subdivisions: the shallow *littoral zone*, where light reaches the lake bottom, and the deep water *limnetic zone*, where the bottom is below the compensation level. Ponds, by definition, lack substantial limnetic or profundal zones. In many lakes during the summer, the euphotic zone is roughly equivalent to the epilimnion and the profundal zone is roughly equivalent to the hypolimnion, for at this time the thermocline and the compensation level often are near one another.

In the shallow waters of the littoral zone, light and oxygen are abundant. Nutrient levels are nor-

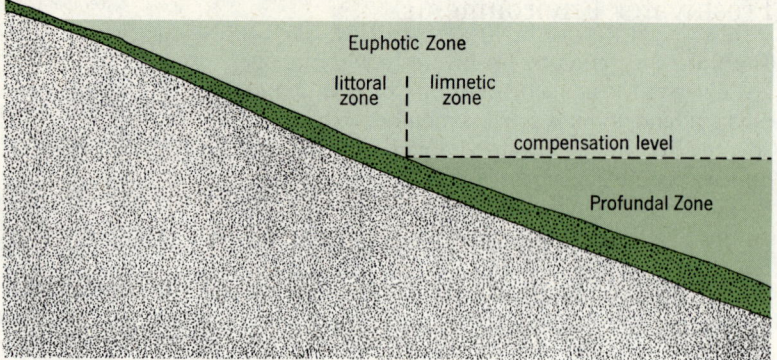

Figure 23.2 Biological zones of a lake.

mally high, for decomposition of nutrient-containing organic matter brought in by streams is rapid where oxygen is ample and waves and currents keep sediments agitated. The primary producers of the littoral zone are the aquatic flowering plants and the planktonic algae. The higher plants are of three forms, each of which dominates a particular subzone within the littoral environment (Figure 23.3). In the *emergent zone*, plants are rooted in water but project much of their growth above the water surface. Examples of plants from the emergent zone are cattail and bulrush. These plants are highly valuable because their roots recover nutrients from deep within anaerobic sediments. In addition, they utilize carbon dioxide from the air; when they die, this fixed carbon adds a new increment of organic matter to the water in which they decompose. The emergent aquatics also provide a transitional environment for animals which spend portions of their life cycles in both aquatic and terrestrial environments—many insects and amphibians are examples.

Inside the emergent zone is usually present an *intermediate zone* of plants rooted on the bottom but with floating leaves, stems, or flowers exposed to the surface. *Water lilies* are ideal examples of plants common to this intermediate zone. Functionally, these plants also aid nutrient cycling and add increments of organic carbon derived from the air, and so serve purposes similar to those of the emergent aquatics. The habitat these plants create is quite different, however. They grow in deeper water and often shade considerable portions of the bottom below them. A common inhabitant of this intermediate zone is duckweed (Figure 23.4). This interesting plant carries out its entire life cycle floating on the water surface.

The *submergent zone* occurs where the water is too deep for the floating-leaved aquatics. Here, the bottom-rooted pondweeds (*Potamogeton* species) often dominate. Other genera within the pondweed family are also important, as are the larger filamentous algae which are attached to the bottom and resemble the higher plants.

In addition to the larger, rooted plants which dominate the littoral environment of lakes and ponds are phytoplankton, which are found universally throughout the limnetic zone. Three major groups are important. The *diatoms* (Chapter 18) are a class of the golden-brown algae with siliceous shells which float freely in water. The *green algae* (Chlorophyta) are present in a variety of forms, ranging from unicellular free-floating types to filamentous periphyton. The *blue-green algae* occur in both unicellular and colonial forms, and are important from both ecological and practical viewpoints. They respond readily to

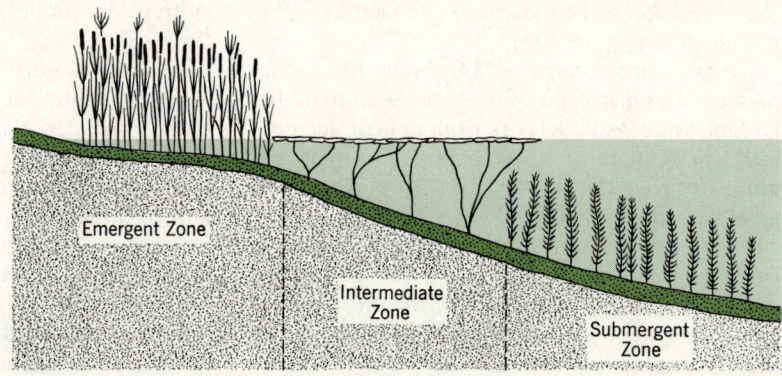

Figure 23.3 Subdivisions of the littoral zone of a freshwater lake or pond.

Figure 23.4 Duckweed of the genus *Lemna*. This floating aquatic plant is characteristic of the intermediate zone of lakes and ponds. Here it has multiplied rapidly, and now completely covers the surface of this small pond. (Jack Dermid.)

increased nutrient concentrations and tend to *bloom* (to grow to a large population) when stimulated by organic inputs, such as sewage. Many of the blue-green algae produce toxic metabolites which can build up to lethal concentrations during blooms, and thus destroy other organisms. Such metabolites can also present a water-quality problem if the lake or pond is tied into a public water supply.

Dependent on the phytoplankton as a food source are the minute grazing zooplankton. Three typical types of fresh water zooplankton are the rotifers, cladocerans, and copepods (Figure 23.5). The tiny *rotifers* are members of the phylum

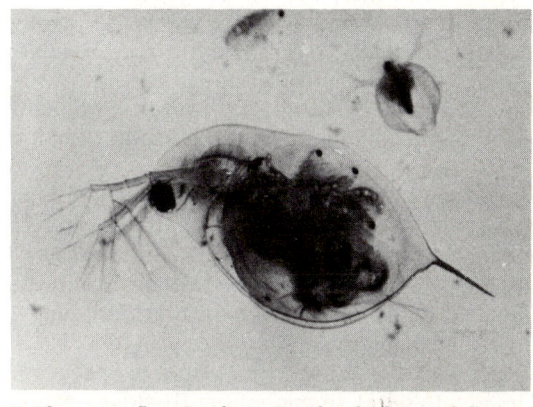

B. The water flea, *Daphnia*. Besides the large adult, two immature individuals can be seen. (R. H. Noailles.)

Figure 23.5 Some examples of zooplankton.

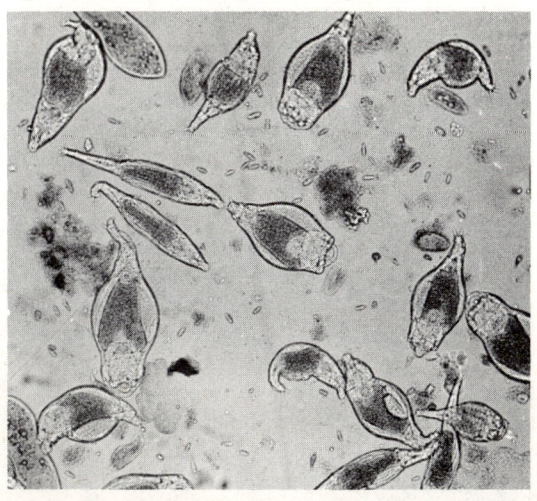

A. Rotifers. See also Figure 18.25A. (Hugh Spencer/National Audubon Society.)

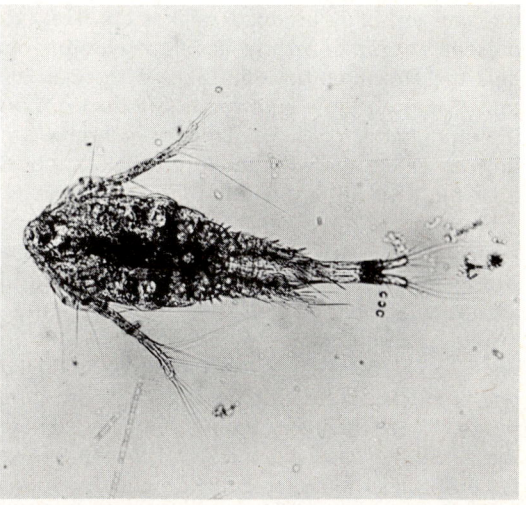

C. A copepod, *Calanus finmarchicus*. This tiny organism is here magnified about 50 times. (Runk-Schoenberger/Grant Heilman.)

Aschelminthes (sac worms) and possess two circular sets of cilia which resemble rotating wheels; these sweep phytoplankton and detritus into the gullet. The *cladocerans* (water fleas) and *copepods* are small crustaceans which also feed by filtering phytoplankton and detritus from the surrounding waters. Zooplankton tend to concentrate in the epilimnion, near their food sources. Often they possess adaptations which allow them to vary their density and move vertically through the water column seeking greater food concentrations. As a result, they are often stratified in occurrence, and the stratification can vary from season to season and day to day to accommodate changes in food sources and important physical factors, such as oxygen and mixing.

Also present in, but not restricted to the limnetic zone, are the nektonic organisms—the fishes and some invertebrates. The distribution of fishes throughout the water column is usually dependent on temperature, oxygen, and food content of the water, with each species occupying different suitable environments. For example, bass, pike, and sunfish often occupy the warm epilimnion, whereas the lake trout is more likely to be found in the deeper, colder waters near and below the thermocline.

The profundal zone, by definition, lacks sufficient light for photosynthesis; it is therefore dominated by heterotrophs. The profundal environment is greatly influenced by activities in the limnetic zone above. In lakes in which productivity is high, large quantities of organic debris will find their way through the waters of the hypolimnion to settle on the bottom. The degradation of this debris by the decomposing bacteria and microorganisms, however, requires oxygen. For this reason highly productive lakes are likely to present anoxic or nearly anoxic environments near the profundal bottom. Lakes with small littoral zones and low autotrophic production, on the other hand, will have higher oxygen levels in bottom waters and will therefore usually support a greater diversity of benthic life.

The organic ooze which accumulates on the bottom of the lake in the profundal zone is a stratum of great biological activity. Most plentiful are the anaerobic bacteria, which reduce sulfur, nitrogen, and organic compounds in order to degrade organic matter. Because these reduced end products can be toxic to other organisms, the benthic zone of highly productive lakes, where the anaerobes are most active, usually shows fewer species and less activity of higher life forms than the benthos of less productive lakes. Typical benthic organisms are small crustaceans, including copepods, cladocerans, isopods, and ostracods; mollusks, such as freshwater clams; and a variety of scavenging protozoa, which are often found encased in shells. Annelid worms such as *Tubifex* (bloodworms) and wormlike larval forms are also typical. Most of these higher organisms are resistant to low oxygen levels and can endure severe stagnation, but none can survive indefinitely without oxygen.

Productivity Classes of Lakes

The most productive lakes are those with a large littoral zone in relation to their volume. Here, sunlight reaches most of the lake's waters, providing an energy source for the autotrophs in addition to significant warming during the growing season. Shallowness of the water allows the development of a large biomass of highly productive, rooted aquatic plants. Phytoplankton blooms are characteristic because of high inorganic nutrient concentrations produced as benthic microorganisms degrade large volumes of organic matter. Profundal waters have low oxygen concentrations; stagnation of bottom water is relatively frequent. Such shallow, highly productive lakes are termed *eutrophic* lakes.

In contrast are *oligotrophic* lakes, which are much less productive. An oligotrophic lake is usually deep and steep-sided, with a narrow littoral zone. Concentrations of inorganic nutrients are low, and so is phytoplankton density. Blooms are rare, for the intense competition for nutrients keeps population levels low. Dissolved oxygen is high in the hypolimnion; its large volume, low temperatures, and low input of organic matter serve to keep oxygen-depleting microorganisms in check. Cold-water bottom fishes such as lake trout or landlocked salmon find suitable habitats in the profundal zone.

On the geologic time scale, oligotrophic lakes tend to become eutrophic. Constant inputs of sediment and dissolved nutrients from streams tend to make oligotrophic lakes become more shallow and more productive. Eutrophic lakes, in turn, are being filled by stream inputs as well as accumulations of their own organic debris, and will eventually form bogs if deposition continues. Thus, the normal course of events is a succession, called *eutrophication*, from a deep, clear lake to a wet bog, produced by both geological and biological processes acting through geologic time. Figure 23.6 shows three stages in the process of eutrophication.

Eutrophication is thus a slow, natural process. Man has, however, inadvertently speeded up this

Eutrophication of the Great Lakes

Figure 23.6 Eutrophication. *A.* An oligotrophic lake. Typically clear, cold, and deep, the lake contains few nutrients. *B.* As nutrients are washed into the lake and as sediment and organic matter increase, productivity builds. Plant life is abundant. *C.* A eutrophic lake. Sediment fills in the deep portions of the lake; abundant nutrients encourage algae and duckweed blooms. Few fish can survive the low oxygen levels. Portions of the bottom are anaerobic.

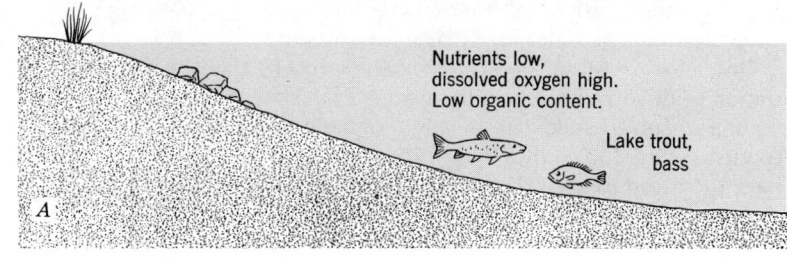

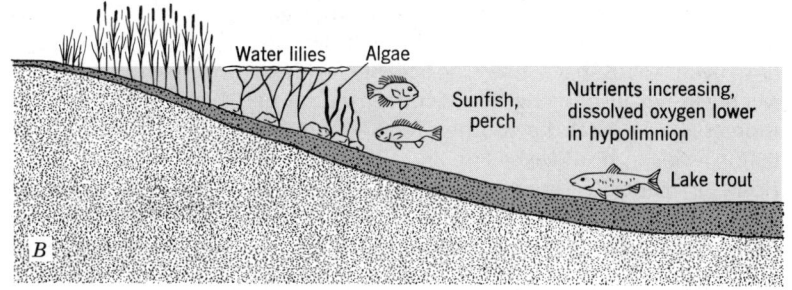

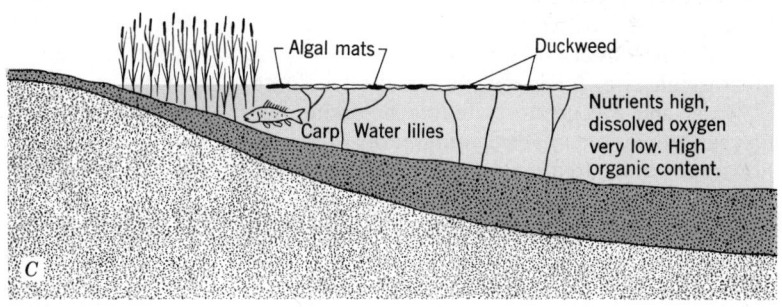

process in many cases. Clearing for farmland and residential construction greatly increases the sediment loads of rivers draining into lakes, enlarging their littoral zones and decreasing their profundal volumes. Inputs of inorganic nutrients to lakes may also be greatly increased. Agricultural runoff contributes nitrates and phosphates; sewage treatment plants contribute nitrates, phosphates, and organic matter. Logging and burning act to release from terrestrial cycles other nutrients, among them sulfur, magnesium, and potassium; and runoff conveys them into rivers and then into lakes. Toxic chemicals and compounds from industrial wastes enter lakes and are concentrated in second- and third-order consumers. As a result, lake populations are diminished, shifting lake biota from grazing food webs to detrital webs. Taken together, these processes can accomplish in a span as short as a human life changes that normally take thousands of years. The result is *cultural eutrophication*, a byproduct of Man's exploitation of the earth's surface.

Eutrophication of the Great Lakes

Lake Erie provides an example of cultural eutrophication. Recall from Chapter 17 that the Great Lakes are relatively young, dating from the retreat of the Wisconsinan ice in central North America. In addition, four of the five lakes are quite deep, ranging in maximum depths from 750 ft (231 m) to 1330 ft (402 m) (see Figure 17.32). Only Lake Erie is relatively shallow, with maximum depths just greater than 200 ft (60 m). The Great Lakes have a relatively small drainage area—only about three times the area of the lakes themselves and equivalent to only about 3.5% of the total land area of the United States (see Figure 17.31). Yet this small portion of land harbors about one-seventh of the United States population and about one-third of the Canadian population. The Lake Erie shore is the most intensively developed of all the lakes, with the major population centers at Detroit, Toledo, Cleveland, Erie, Buffalo, and Windsor. Population concentrations are also heavy on the southern shore of Lake Michigan, where Milwaukee, Chicago, and Gary are situated.

One measure of eutrophication is the concentration of dissolved solids in the waters of a lake, for the dissolved solids are mostly inorganic nutrients. Figure 23.7 shows how these pollutants have increased in each of the five Great Lakes in the past 50 years. Lake Superior, with little surrounding development, has remained almost unchanged. Lake Huron shows an increase of only a few parts per million, produced largely by Lake Michigan water which passes through Huron. Dissolved solids have risen noticeably in Lake Michigan because of the heavy concentration of industrial activity at the southern end. Dissolved solids levels in both Lakes Erie and Ontario have risen dramatically, approaching 50 ppm in 50 years. Although its shore is less developed industrially, Lake Ontario's level of dissolved solids has increased because it receives water from Erie through the Niagara River. Additions are also important from large population centers of Toronto, Hamilton, and Rochester.

Of the five Great Lakes, Man's impact has been greatest on Erie, for it has the smallest volume of water and the greatest volume of pollutants. Oxygen depletion in the hypolimnion of Erie is now much more frequent and much more severe than in the past, and depletion now affects a significantly greater bottom area. The benthic fauna of the west end of the lake has changed, particularly in response to new, low oxygen levels. Mayfly larvae, indicators of a diverse, aerobic bottom environment, are now virtually absent; populations of sludge worms, which can withstand the most anoxic conditions, had increased by 1961 to values exceeding 1,000 per square meter for most of the western end of the lake.

Table 23.1 shows how catches of four important species of commercial fishes have declined in Lake Erie in recent years. The decimation of these fish populations which the data reflect has been brought about largely by failure to reproduce. The sea lamprey, a blood-sucking parasite which has

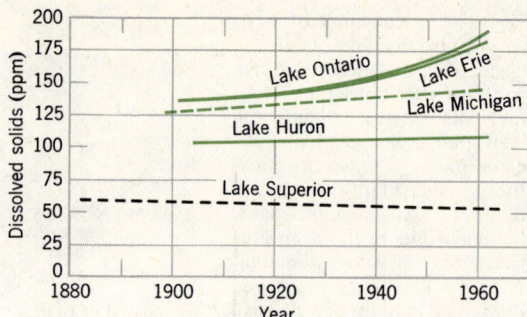

Figure 23.7 Trends in concentration of dissolved solids in the five Great Lakes. (Data of several authors, presented in A. M. Beeton, 1965, *Limnology and Oceanography*, vol. 10, p. 246.)

had a disastrous impact on fisheries of the other Great Lakes since it was introduced through the Welland Canal, has not been an important factor in Lake Erie. While the lake herring, whitefish, sauger, and walleye have declined drastically, the less desirable freshwater drum (or, sheepshead), carp, yellow perch, and smelt have increased in their place, and the total annual catch for Lake Erie has remained near 50 million pounds. Thus, heavy pollution has acted to shift the species composition of fishes in Lake Erie.

Streams

The moving-water, or *lotic*, environment of a stream presents some contrasts to the lentic environment of a lake. The most obvious difference is that streams have a current, although there are pools within streams where the current is weak or absent, and zones within lakes, such as wave-washed shores, where a current is present. One of the most important effects of a current is the import of nutrients and foods to the habitats of organisms and the simultaneous export of waste products. For this reason, lotic waters can be as many as 30 times more productive than lentic waters. The current presents some problems, however, for organisms seeking a particular envi-

TABLE 23.1 Decline in Lake Erie Fish Catches

Species	Catch (pounds per year)		
	Past	Present	Percent Decline
Lake herring	20,000,000 (pre-1925)	7,000 (1962)	Over 99.9
Whitefish	2,000,000 (pre-1948)	13,000 (1962)	Over 99.3
Sauger	1,000,000 (pre-1945)	1,000–4,000 (early sixties)	99.9
Walleye	15,400,000 (1956)	1,000,000 (early sixties)	93.5

SOURCE: Data from Alfred M. Beeton, 1965, *Limnology and Oceanography*, vol. 10, no. 2, p. 250.

ronment within the stream. Unless, like the stronger fish, they can maintain their position, they will be carried downstream. Most plants maintain attachment to the channel bed by means of roots or rootlike structures. The animals, on the other hand, maintain their positions by a wide variety of adaptative structures. These include suckers and hooks for attachment to bottom objects; sticky surfaces, such as those of snails and flatworms; and streamlined and/or flattened bodies to cut water resistance and to allow the use of narrow crevices for shelter. In addition, many animals instinctively exhibit positive adaptive behavior by heading into the current or by staying very close to objects on the bottom.

A second major environmental difference between stream and lake is in the oxygen level. The constant mixing and agitation of the water-air interface serves to keep stream waters thoroughly oxygenated. As a result, many lotic organisms are sensitive to a drop in oxygen levels, an effect easily induced by the bacterial degradation of influent sewage.

A third important difference is that there is a great deal more energy exchange between a stream and the terrestrial environment which surrounds it than between a lake and its environs. A careful examination of the lotic environment will show that consumers are much too numerous to be supported exclusively by the producers within the stream. Instead, they are maintained in large part by inputs of nutrients and detritus from the vegetation of the banks, floodplain, and surrounding surfaces which drain into the stream. Energy also flows from the stream to the surrounding terrestrial ecosystems. An example of such a flow is the energy contained in aquatic larval forms which molt and assume terrestrial forms. Grazing or predation of terrestrial animals upon the lotic biomass is another energy export that is important in the ecology of many streams.

The nature of the stream bottom greatly affects the flora and fauna of the stream. Sand and silt are usually the least productive materials. Neither offers much physical support for benthic and periphytonic organisms. Clay bottoms or bedrock bottoms are usually more favorable. The most productive bottom is composed of cobbles or gravel. The rubble supplies a great variety of surfaces for attachment as well as many attractive pockets and crevices to shelter benthic consumers from the swift-moving current.

Most rapidly moving streams exhibit alternations of pools and riffles. The riffles, which are shallow reaches with swift current, are the more productive of the two habitats (Figure 23.8). Periphyton are particularly important producers: diatoms, water mosses, and blue and blue-green algae coat the stones and sticks of the stream bottom and serve the same function as phytoplankton in a lake. Swift currents in the riffle zone constantly tear bits of periphyton loose, sweeping them away, providing an energy source for downstream pools as well as a source of colonizers for the riffles below. The larvae of many insects (e.g. blackflies, mayflies, caddisflies, stoneflies) are important animals in the riffle zone. Most of these larvae graze the periphyton and phytoplankton, in some cases with the help of a net- or sievelike structure which traps tiny suspended particles. The nektonic fishes are also sometimes found in riffles, where they obtain much of their food by grazing or predation. They rely on the pools for rest and shelter.

The pools present a contrasting environment to the riffles. Here the bottom is usually soft sediment, which encourages burrowing worms and larvae. The pool ecosystem often bears a strong resemblance to that of a shallow pond, for the slow-moving water is more conducive to plankton, nekton, and neuston. The economy of the

Figure 23.8 This small stream presents a diversity of aquatic environments, ranging from the swiftly flowing waters of the riffle zones to the quieter pools which alternate with them. (Robert Perron.)

pool depends heavily on both the indigenous plankton and the input of detritus from upstream riffles and terrestrial sources.

Traced from their headwaters to the sea, streams undergo changes in hydrology, as we have seen in Chapter 14. Width and depth increase downstream, whereas slope decreases. Although mean velocity increases somewhat, the range in variation of velocity diminishes, producing a more uniform environment. The contrast between pools and riffles becomes less marked and the large trunk river appears slower and more sluggish than the small stream, despite the greater mean velocity. Suspended sediment load increases in ratio to bed load, and bed materials become finer downstream. Typically, the bed of a stream of low gradient in a floodplain channel is composed mostly of fine silt or sand (Figure 23.9). Water temperature increases as the shade of vegetation on the banks becomes less effective and as greater turbidity increases the absorption of solar radiation.

Downstream changes in physical environment from the small headwater stream to the large trunk river are accompanied by changes in the stream community. Detritus-feeders increase in response to an increased proportion of soft mud bottom. Burrowing worms and midge larvae become dominant benthic consumers and are joined by crustaceans, snails, and other mollusks.

Figure 23.9 This distributary in the Amazon delta presents quite a different aquatic environment from that of the swift stream of Figure 23.8. Note the lack of obvious pools and riffles as well as the greater turbidity. (Robert Perron.)

Bottom-feeding fishes, such as carp, catfish, and suckers, are now important parts of the nektonic fauna. Rooted aquatic plants and emergent bank vegetation show zonation patterns characteristic of the littoral zone of lakes and ponds. Plankton populations are now much larger, although they are still not as great as in the lakes or ponds which may feed the stream. Thus, rivers tend to be more typically lotic in the headwaters and more typically lentic as they grow in cross section and decrease in gradient in the downstream direction.

Response of Streams to Sewage Pollution

As we have seen in Chapter 13, Man's activities have had a great impact on the quality of surface water. Urbanization has changed the hydrological characteristics of streams, including the volumes and timings of peak flows. Withdrawal and industrial use of water has changed the chemical composition of runoff, adding elements and compounds, some of which are nutritious and others of which are harmful to aquatic organisms. Before we leave the subject of streams, it will be useful to consider the impact of a common pollutant—sewage—on a typical stream.

Table 13.3 shows the important inorganic elements and compounds which characterize a sewage effluent. In addition to high levels of nitrates and sulfates, sodium and potassium chlorides are abundant. Although not shown, phosphate concentrations in sewage are also typically high. The major component of raw sewage is undecomposed organic matter. The amount of this organic matter is expressed as the *biological oxygen demand* (B.O.D.) of the waste, and defined as the quantity of oxygen which microorganisms will consume in degrading the waste to CO_2 and water. Large numbers of bacteria and sewage fungi will also characterize the raw effluent.

Figure 23.10 shows how a stream responds to the input of raw sewage. Four zones in the stream below the outfall can be recognized. The first is the *zone of degradation,* where water quality is declining as the effluent mixes with the stream waters. The B.O.D. of the stream jumps, and dissolved oxygen levels begin to plunge. Dissolved salts and suspended solids also increase dramatically from low levels just above the outfall.

As an increasing number of decomposer organisms become active, the *zone of active decomposition*, or *septic zone*, begins. Here sewage fungi, protozoa, and bacteria reach high levels, while dissolved oxygen reaches its low point. If the amount of effluent is large enough in relation to the size of the stream, dissolved oxygen may

Sewage Treatment

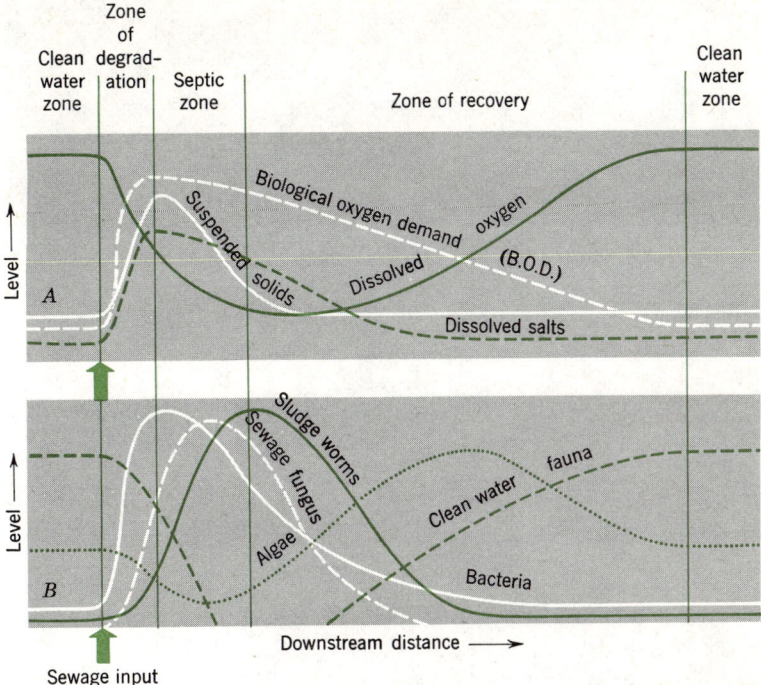

Figure 23.10 Response of a stream to an input of untreated sewage. A. Changes in chemical properties. B. Changes in physical properties. (Curves after H. B. N. Hynes, 1960, *The Biology of Polluted Waters*, University of Toronto Press, Toronto and Buffalo, p. 94, Fig. 16. Division into zones after W. T. Edmondson, in W. W. Murdoch, ed., 1971, *Environment: Resources, Pollution and Society*, Sinauer Associates, Stamford, Connecticut, p. 216.)

become totally depleted, and anaerobic conditions will prevail. The reduced forms of nitrogen and sulfur (NH_2 and H_2S) reach their highest concentrations here, as do the anaerobes which produce them. Oxygen concentrations are too low for the normal fauna of the stream, but are usually suitable for the sludge worms, which reach their peak at the end of the zone.

Eventually, much of the sewage decomposes and the activities of microorganisms are slowed. Mixing of the water surface introduces oxygen which now is not immediately consumed; dissolved oxygen levels rise somewhat. In this *zone of recovery*, a few of the members of the stream's original fauna return. Protozoa and small invertebrates multiply in response to the large numbers of bacteria, a food source they readily exploit. As recovery progresses, nitrogen, phosphorus, and sulfur are converted to their inorganic oxidized forms, and serve to stimulate populations of algae. In turn, the algae increase oxygen levels and provide food and shelter for the lotic organisms, which return in increasing numbers toward the end of the zone. Eventually, all the organic matter will be oxidized, excess nutrients will be stored in the biomass of the stream, and the stream will return more or less to its original condition in the *clean water zone*.

Although the sewage effluent does not seem at first to produce any permanent downstream damage to the stream, there are other impacts. Sewage wastes may contain industrial chemicals and toxic compounds which cannot be degraded by bacteria and therefore can accumulate in aquatic ecosystems. Organic mercury compounds are a good example in this respect. In addition, the "dead" zone near the outfall serves as a very effective barrier to the upstream and downstream movement of organisms, for it severs the biological communication between segments of the stream. Further, the affected area will obviously not be suitable for recreation or other human use, and will, in fact, provide a human health hazard.

Sewage Treatment

The impacts described above are for raw sewage rather than treated sewage. The treatment process is basically designed to reduce the B.O.D. of the sewage, and therefore reduce the impact on the stream. In *primary treatment*, the waste water is screened, then allowed to stand and separate into liquid and solid portions. The solid portion, or *sludge*, is buried or otherwise disposed of; the liquid portion is treated with chlorine to kill the bacteria and viruses and is then discharged into natural runoff systems.

In *secondary treatment*, the process is carried further by an additional step involving bacterial decomposition (Figure 23.11). The sludge goes to a digesting tank, where it remains till bacteria decompose it almost completely. The residue is then air dried and disposed of by incineration or burial. The liquid portion is sprayed onto a *trick-

Figure 23.11 Sewage treatment. *Upper left:* An aerial view of one of New York's sewage treatment plants. The rectangular vats are settling tanks; anaerobic digestion takes place in the round tanks in the foreground. (Courtesy of Department of Water Resources.) *Lower Left:* A trickling filter. Wastewater is sprayed by rotating arms over the gravel bed, where bacteria oxidize the organic matter the water contains. (Courtesy of the Environmental Protection Agency.)

ling filter, a bed of stones whose surfaces become coated with bacteria. The organic materials in the liquid are decomposed by the bacteria in the presence of abundant oxygen. In effect, secondary treatment moves the "dead" zone from the stream into the sewage plant. After an additional settling, the liquid is chlorinated and then discharged. Efficient secondary treatment can remove as much as 90% of the B.O.D. of the sewage before it is discharged.

Unfortunately, secondary treatment does not result in the removal of the simple inorganic nutrients. Thus, the effluent of a secondary sewage treatment plant may impact a stream, not by depressing the oxygen level through introduction of a large B.O.D. load, but rather by stimulating the growth of algae. If the algae are subject to blooms, their die-off can produce almost as great a B.O.D. as primary treatment, and thus create the same problem the process was designed to prevent! The solution is *tertiary treatment*, in which nitrates, phosphates, and other inorganic ions are removed from the effluent before it is released. Tertiary treatment is very costly, however, and in addition does not remove all harmful substances.

An alternative form of tertiary treatment, presently being evaluated in experimental and pilot operations, disposes of the liquid fraction from secondary treatment by spray irrigation upon forests or forage crops. As the nutrient ions enter the soil they are held by organic and inorganic soil colloids, as explained in Chapter 11. The nutrients then are taken up by plants and placed in storage within the biomass. Eventually, the plant matter must be removed from the area and in so doing it can provide commercially useful forest products or feed for animals. Thus the ions which would otherwise pollute ground water or streams are recycled with some savings in treatment costs. The spray irrigation system of sewage treatment has been called a *living filter*, in recognition of the role that plants play in the final processing stage.

Bogs

In addition to ponds, lakes, and streams, freshwater environments include bogs, marshes, and swamps. *Bogs* are abundant in the cold boreal forests of North America, northern Europe, and Asia, and are distinguished by their cushionlike cover of small plants and their accumulations of peat (Figure 23.12). Most bogs possess a floating mat, typically of *sphagnum* moss, in which grasses, sedges, and low, woody heath plants are rooted. Since most bogs begin as closed depressions, many have a central area of deeper water not covered by the floating mat; here pondweeds and water lilies are dominant. Bogs are saturated the year around, and the combination of anaerobic standing water and low temperatures acts to inhibit decomposers. Under these conditions, organic matter decays only slowly, releasing quantities of humic acids which usually turn the bog waters brown and create a highly acid environment. Nutrients such as nitrogen and phosphorus are largely retained by the accumulating peat, making the bog waters nutrient-poor. In addition, the absorption of calcium, sodium, and potassium ions by sphagnum moss and peat reduces their concentration in bog waters. Carbon dioxide content is usually low, and the water may also exhibit traces of hydrogen sulfide (H_2S) produced by sulfur-reducing bacteria. Water bodies such as these are referred to as *dystrophic*.

Because the bog water presents such nutrient-poor environment, few species are found within it, although these species may be quite abundant. Desmids and several species of blue-green algae dominate the phytoplankton; rotifers, as well as rhizopods and other protozoans, are common species among the zooplankton. Fish are few, and may be absent from small bogs with no outlet, but amphibians, such as the frogs and their relatives, are often abundant.

Figure 23.13 shows how bogs develop in northern conifer forests through *bog succession*. The bog begins as a small pond edged by a floating mat of sphagnum. New increments of sphagnum and its associates are added to the top of the mat each year, depressing and consolidating the lower portion. Eventually the mat is thick enough to reach the bottom at the pond's edge. Labrador tea, leatherleaf, and other heath plants can now colonize the sphagnum mat, along with *hydrophytes*, or water-adapted plants such as larch. As the larch trees begin to come in, their evapotranspiration dries the mat further, so that white cedar, black spruce, and balsam fir may follow. Thus, the typical bog exhibits a circular zonation of open water, peat mat, shrub heath, and tree zone, all produced by succession.

Near dystrophic lakes and bogs in the boreal forest, *raised bogs* or *high moors* also may be found. These bogs form when sphagnum grows out of a small depression or bog edge and begins to creep upslope. The sphagnum acts like a sponge and converts a previously dry environment to a wet one—the existing trees and shrubs are usually killed, and the sphagnum grows to several feet in depth, often forming mounds where it overgrows the fallen trees and shrubs it has killed. Thus, bogs need not be restricted to depressions.

Figure 23.12 A northern bog from Emmet County, Michigan. The moss and sedge mat surrounds the open water; black spruce forest comes to the edge of the mat. (Photo courtesy of Dr. Pierre Dansereau.)

Marshes and Swamps

Under warmer climates, freshwater marshes and swamps are the typical wetland environments. Freshwater *marshes* are dominated by grasses, sedges, bulrushes, cattails, and other herbaceous vegetation (Figure 23.14).

The marsh substrate is a soft, dark organic ooze, and marsh plants produce a thick, fibrous root mat which aids in support. In addition, their root tissues are tolerant of stagnant water. The animal life of the marsh is varied and abundant. Among the invertebrates, snails are important consumers. Amphibians are important vertebrate consumers. The marsh envrionment is particularly suited to birds, and the avian fauna of marshes is always very diverse. Ducks and geese, as well as other swimming birds designated as *waterfowl*, are characteristic inhabitants of marshes and their adjacent water bodies, both fresh and salt.

Freshwater marshes have much in common ecologically with salt marshes, described in Chapter 15. Whereas the saturation level in freshwater marshes is dependent upon the groundwater table, and fluctuates seasonally with that table, the saturation level of the salt marsh is determined by the rise and fall of the ocean tide. The thick, fibrous root mass constructed by salt marsh plants (especially *Spartina*) forms a peat layer overlying a mud layer rich in organic matter. When artificially diked and drained, salt marshes become freshwater marshes and experience a gradual change in flora from saltwater plants to freshwater plants. We shall have more to say about the salt marsh environment in our discussion of estuarine ecology.

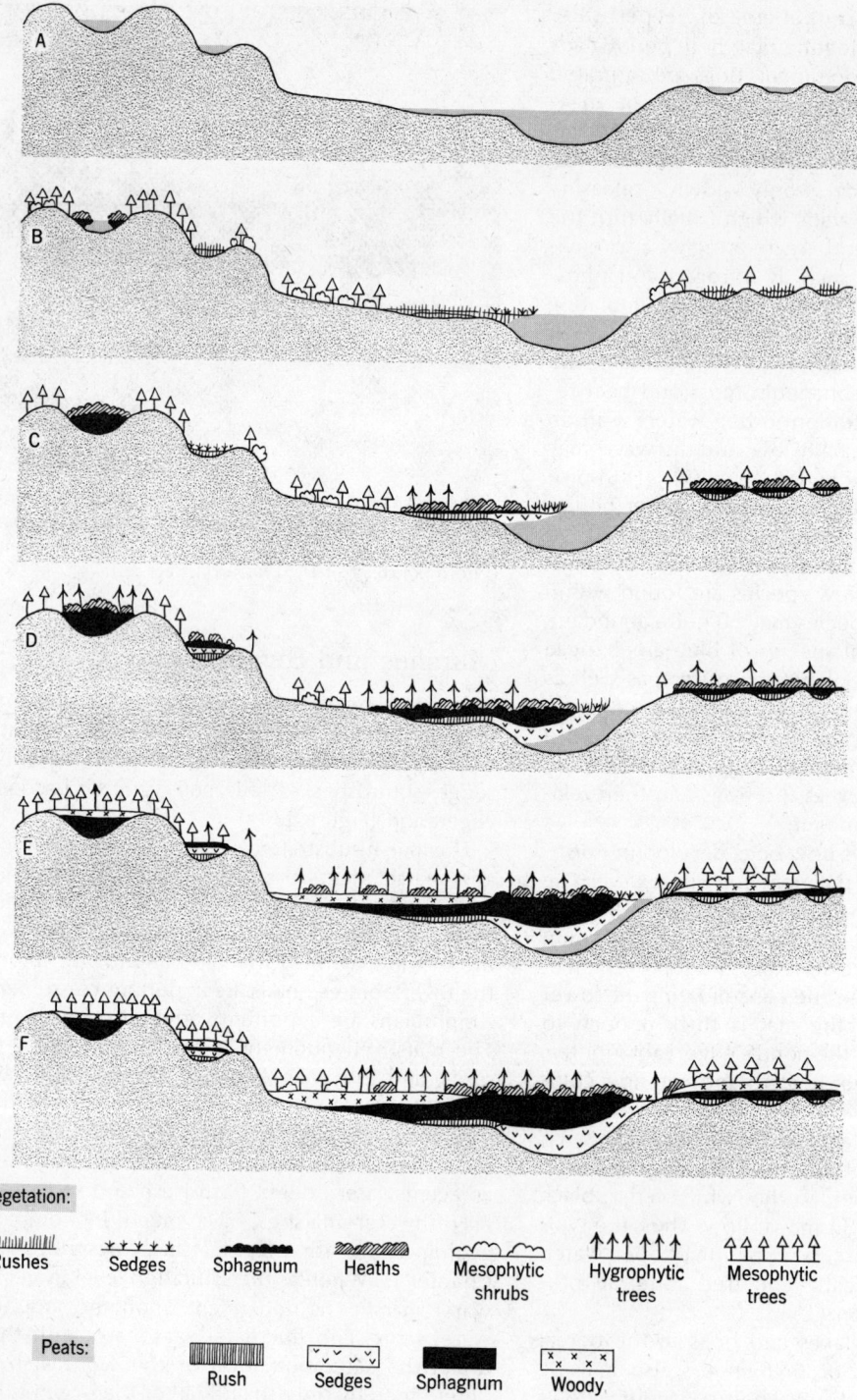

Figure 23.13 Bog succession in the Laurentian shield area of Quebec, Canada. (After Dansereau and Segadas-Vianna, 1952, *Canadian Journal of Botany*, vol. 30).

Figure 23.14 A freshwater marsh. Note the diversity of plant species, including grasses, sedges and, at the margin, water-loving shrubs and trees. Duckweed (*Lemna* sp.) covers much of the water surface. Clumps of blooming *Iris* are conspicuous. (Grant Heilman.)

Swamps, in contrast to the wet grassland of the freshwater marsh, are wooded. In the deeper water swamps of the southern coastal plain, bald cypress, southern red cedar, and several types of gum trees are common (Figure 23.15). These trees usually exhibit a massive but superficial root structure, often with *pneumatophores*, or *knees*, which project upward and out of the mud and water. Pneumatophore means "air bearer," and the name reflects the now discredited theory that these structures participate in gas exchange to aerate the root tissues. Usually there is little rooted vegetation in the deep-water swamp other than the trees, but epiphytes may be abundant. In shallow-water swamps, maples, alders, and willows, as well as a few species of oak and pine, are more common. The shallow swamp is a more diverse environment, for the microrelief of low mounds and small depressions, produced by dead and fallen trees and overturned root systems, can accomodate many more species than the dark, dystrophic waters of the deep-water swamp.

Marine Environments

Contrasting in many respects with the freshwater environments of lakes, ponds, bogs, swamps, and marshes are the *marine environments*, including estuaries, mangrove swamps, coral reefs, and the open ocean itself. The marine biota is highly varied, and differs from freshwater biota in many ways. Algae, particularly the unicellular forms, are the dominant producers, and include many groups, such as the red and brown algae, which are lacking or poorly developed in freshwater or terrestrial environments. Seed plants are notably absent, except in and near the intertidal zone. Animal phyla which are dominantly marine include the coelenterates, sponges, echinoderms, and annelids. Like seed plants, insects are absent except in the transitional zone between marine and terrestrial environments. The phylum Arthropoda, to which insects belong, is represented instead by the crustaceans, and these play a similar role as dominant first- and second-order consumers in marine ecosystems. The diversity and complexity of marine organisms and their patterns

Figure 23.15 A freshwater swamp in the southern coastal plain. The dominant tree is the bald cypress, *Taxodium distichum*. Note the pneumatophores and buttress roots. (Grant Heilman.)

of life are more properly the subject of biological oceanography—in this section we can only touch on some of the more important points concerning marine ecosystems.

Estuaries

As we have seen in Chapter 15, estuaries are transitional environments at the ocean's edge where fresh and salt water mix. By and large, the biota of the estuary are derived from marine forms, rather than terrestrial forms. Most estuaries are dominated by a few, rather than many, species. This fact arises because the constant shifts in temperature, salinity, and turbidity, as well as tidal currents, present an unusual environment to which only a few species are sufficiently adapted to spend their entire life cycle. Those species which are adapted, however, are usually present in large numbers.

As we observed in Chapter 20, estuaries provide examples of ecosystems with well-developed detritus food chains. Primary production is greatest in the intertidal salt-marsh zone which surrounds the estuary (Figure 23.16); here, productivity compares favorably with the most productive of natural environments (Table 20.3). The grasses and sedges of the salt marsh are highly salt-tolerant and ecologically adapted to the transitional environment; moreover, their annual growth is very great. Algae on the marsh surface and banks of the tidal channels are also highly productive. Only about 5% of the net autotrophic production, however, is directly consumed by grazing insects and other herbivores; most of it reaches the saline water and becomes detritus, rich in nutrients and energy.

A typical estuarine food web has already been presented in Figure 20.19. Benthic detritivores are extensively developed in most estuaries. These include crabs, lobsters, and young shrimp among the crustaceans; clams, oysters, and mussels among the mollusks. The detritus, as well as some living plant matter, also supports the young of such fish as striped mullet, flounder, menhaden, and croaker. In fact, the estuary serves as a *nursery* for the juvenile forms of many commercially important finfish and shellfish. The estuary's low salinity protects them, for it serves as a barrier to the entrance of most predators from the open sea.

One reason for the high productivity of estuaries is that they tend to serve as *nutrient traps*. Fine particles of clay and organic material are transported to the estuary by the rivers which empty into it. These particles are rich in adsorbed nutrients and ions. When they reach salt water, the particles flocculate and settle, as described in Chapter 15, concentrating nutrients in the sediments. Filter-feeding detritivores ingest the particles, removing the adsorbed nutrients and bacteria during digestion. They also extract dissolved nutrients directly from the water. Constant activity of the tides serves as an energy subsidy to bring new particles to filter-feeders and carry digested particles away. The filter-feeders, then, aid the nutrient concentration process, preventing the escape of vital elements and compounds to the open ocean.

The efficiency of the estuary as a nutrient trap also makes it an excellent trap for such toxic compounds as pesticides and heavy metals. Through the process of *biological magnification*, the insecticide DDT has already reached very high

Figure 23.16 A tidal salt marsh in Maryland. The dominant grass is a species of cordgrass, *Spartina*. (Grant Heilman.)

Mangrove Swamps

TABLE 23.2 Important Estuarine Habitat Lost from Dredging and Filling Operations, 1950–1969

Coastline Zone	Acres of Important Habitat	Acres Lost	Percent Loss
North Atlantic	271,000	2,500	0.9
Middle Atlantic	2,201,800	77,000	0.4
Chesapeake	603,300	5,000	0.9
South Atlantic	823,800	42,300	5.1
Biscayne and Florida Bay	922,200	21,100	2.4
Gulf of Mexico	8,325,000	426,700	5.1
Southwest Pacific (Calif.)	388,000	46,200	12.0
Northwest Pacific (Wash.-Oregon)	2,142,000	21,000	1.0
Alaska	593,400	1,500	0.003
Great Lakes	432,000	2,600	0.1

SOURCE: Data from *National Estuary Study*, Department of the Interior, Fish and Wildlife Service, vol. 2, p. 122, 1970.

levels in organisms of some estuaries. This process was explained in Chapter 21. Table 21.10 presented some DDT concentrations in a Long Island salt marsh.

Man exerts his impact on estuaries mainly by draining and filling (Chapter 15) which often simply eliminate the estuarine environment. When a decision is made to fill a coastal wetland, often the value of the commercial fish which are dependent on it is underestimated or not even considered. If the wetland is thought of as a capital investment producing interest equal to the value of the annual catch of finfish and shellfish which it supports, values of from $5,000 to $25,000 per acre are obtained. This figure does not include such intangibles as recreational benefits or scenic values. Clearly then, conversion of wetlands to such uses as sanitary landfills or parking lots probably represents a loss, rather than a gain, to society. Table 23.2 shows how much acreage each section of the U.S. coastline has lost to dredging and filling operations. According to the data, California has lost the largest percentage of her wetlands, followed by the Gulf and South Atlantic coasts.

Mangrove Swamps

Mangroves are woody shrubs which are capable of growing in salt water at open ocean salinities. Together with a few other plants, they form *mangrove swamps* found at the ocean's edge in the tropical and equatorial zones. The term mangrove includes several different species of different genera which all have more or less the same appearance. They are tall, and many-branched in form; they have smooth leaves, without lobes or teeth, and prop or buttress roots or rhizophores (Figure 23.17).

The mangrove swamp exhibits a *cosmopolitan* flora—one that is more or less the same throughout the world. The wide distribution of the flora which comprises the mangrove swamp occurs because many of the plants use the ocean waters as a method of seed dispersal. The red mangrove, *Rhizophora mangle*, is notable in this respect. Its seed actually begins to germinate before it falls, growing into a young seedling borne on its parent branch (Figure 23.18). Soon the seedling falls into the water, and is carried by winds and currents until it touches a suitable substrate of soft mud. Here it develops roots and branches, perhaps as

Figure 23.17 A mangrove forest growing in a saline embayment at Harney's River, Florida. (American Museum of Natural History.)

Figure 23.18 The red mangrove, *Rhizophora mangle*. The projections from the fruits are the cotyledons of young plants, developed from the seeds which germinate while the fruits are still on the plant. (Robert Lamb/National Audubon Society.)

a colonizer on a faraway beach. Other plants of the mangrove swamp possess floating seeds or other adaptations to facilitate their worldwide dispersal.

The food web of the mangrove swamp is similar to that of the estuary. It has a detritus base—mangrove leaves—although phytoplankton and algae do make a minor contribution to the community economy. Small crustaceans, mollusks, and fish consume the detritus, and are in turn consumed by the larger fish and fish-eating birds. Like the estuary, the mangrove swamp harbors many commercially important species of fish in their juvenile stages.

Coral Reefs

Coral reefs are exceedingly diverse environments; they accommodate a very large number of species, many of which are uniquely adapted to this distinctive environment (Figure 23.19). From the viewpoint of productivity, coral reefs rank with estuaries as highly productive ecosystems. In contrast to estuaries, however, the ratio of production to respiration nears unity in the coral reef. Thus, the coral reef consumes as much energy as it fixes. A ratio of one, as we have noted in the last chapter, is characteristic of a stable, mature ecosystem adjusted to maximum energy exploitation of its environment. Some of the physical processes and gross forms which characterize coral reefs, as well as their geologic development in tropical and equatorial seas, have been covered in Chapter 15.

The coral reef is composed of the lime skeletons secreted by the coelenterates, or *coral ani-*

Figure 23.19 The tropical coral reef environment. Note the diverse forms of algae, coral, and fishes. (Russ Kinne/Photo Researchers.)

mals, and by certain groups of reef-dwelling algae. A cross section through a *coral head,* or colony of coral plants and animals, is shown in Figure 23.20. Three groups of algae are associated with the coral head. First are *endozoic* algae, which live within the polyp. They are yellow-brown dinoflagellates, and possess a free-swimming stage in their life cycle which allows them to move from one polyp to another. The relationship between the polyp and its endozoic algae is a mutualistic one. The polyp traps and ingests zooplankton, deriving not only energy but scarce mineral nutrients such as phosphorus. By a process not understood, these nutrients are passed on to the endozoic algae. For their part, the algae transfer organic materials directly to the polyp, providing it with an energy source. In addition, the activities of the algae seem to stimulate the secretion of lime (calcium carbonate) by the polyp, serving as an aid to the growth of the coral head.

Filamentous calcareous algae occupy the pores and spaces within the living and dead portions of the coral skeletons. These algae are members of the Chlorophyta (green algae) and *Rhodophyta* (red algae) and are rich in chlorophyll. Also pres-

Open-Water Ecosystems

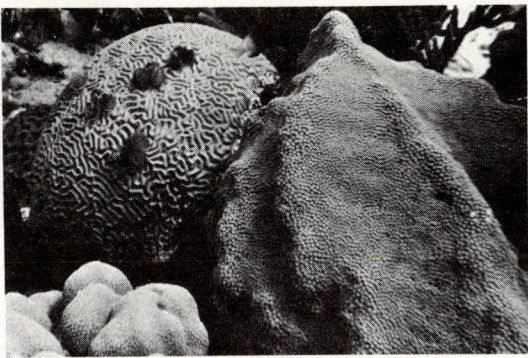

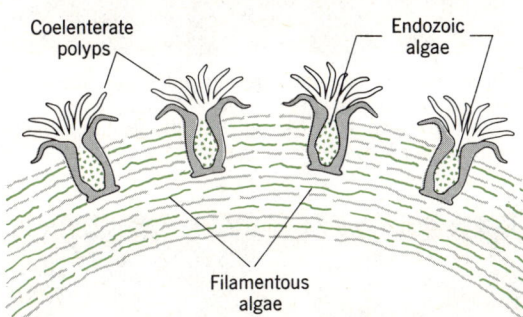

Figure 23.20 Coral heads. *Above:* Undersea photo showing several types of massive coral heads. (Russ Kinne/Photo Researchers.) *Below:* Diagrammatic cross section through the surface of a coral head, showing coral animals and mutualistic algal associates. (After E. P. Odum, 1971, *Fundamentals of Ecology*, W. B. Saunders, Philadelphia, p. 348, Figure 12.15A.)

Beyond the shallow inshore waters of such transitional environments as estuaries, mangrove swamps, and coral reefs, lie distinctive marine ecosystems of the open water. Biological oceanographers divide the open water into two regions: the *neritic zone*, which includes the water and bottom between the shoreline and the edge of the continental shelf, and the *oceanic zone* of deep water beyond (Figure 23.21). Within the oceanic zone are three different bottom areas: the *bathyal zone* on the continental slope, the *abyssal zone* of the ocean deeps, and the *hadal zone* of the deep ocean trenches. Vertically, the ocean exhibits a *euphotic zone*, like that of lake, in which sufficient light is present for photosynthesis. Below the euphotic zone is the *aphotic zone*, where light is absent. *Pelagic* organisms are all those living suspended in the ocean (plankton, nekton, and neuston).

The dominant producers of the neritic zone are diatoms and dinoflagellates. Recall from Chapter 18 that these large and diverse marine groups of unicellular algae are the most important classes within the divisions Chrysophyta and Pyrrophyta respectively. The diatoms are usually more common in the cooler northern waters, whereas dinoflagellates are more common in the warmer subtropical, tropical, and equatorial waters.

Large multicellular algae of the Chlorophyta, Phaeophyta, and Rhodophyta (green, brown, and red algae) are producers in the shallower waters of the neritic zone which have hard or stony bottoms. The latter two groups are photosynthetically adapted to the deeper photic zone. Referring back to Figure 3.9, you will find that at a depth of 330 ft (100 m), which is near the base of the photic zone, the radiation spectrum consists only of light rays in the wavelength range 0.4 to 0.6 microns. Most of this energy is in the blue and green wavelength bands. The red and

ent are the larger, fleshy algae which are attached to dead areas of coral.

Other important organisms on the reef are the larger consumers. Many species of brightly colored fish graze among the algae. Crustaceans, such as lobsters, and other invertebrates as well, are common detritivores, typically more active at night than during the day. Top consumers are sharks and the moray eels; they lurk in the deeper waters, awaiting their prey.

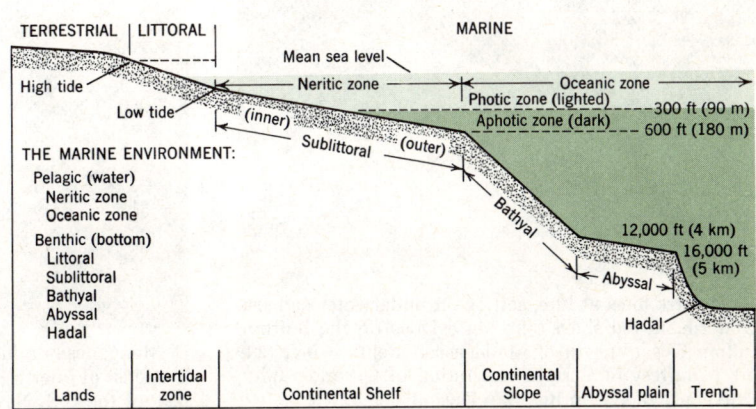

Figure 23.21 A schematic cross section of an ocean basin from the edge of a continent to a deep trench. (From A. N. Strahler, 1971, *The Earth Sciences*, 2nd ed., Harper & Row, New York, p. 500, Figure 28.10.)

brown algae derive their color from accessory pigments which absorb blue and green light, and therefore look red or brown. The accessory pigments pass the absorbed energy on to chlorophyll molecules for use in photosynthesis. Thus, these groups are adapted to exploit the dimmest greenish-blue light which penetrates to these depths.

Open-water zooplankton are often divided into two types: *holoplankton*, which spend their entire life cycle as free-floating plankton, and *meroplankton*, which are larval or juvenile planktonic stages of benthic or nektonic adult organisms. The meroplankton are more common toward shore and are more likely to have benthic adult forms. Further out at sea, the meroplankton are less common and are more likely to be larvae of nektonic organisms, such as fishes. The holoplankton include such familiar forms as the copepods and isopods, whose freshwater relatives we have already encountered. Together with the larger crustaceans they are called *krill*, and are an important food source for many nektonic consumers in the oceanic zone.

The marine benthos is often divided into two groups: the *epifauna*, which live at or are active on the bottom surface, and the *infauna*, which dig or burrow into the bottom sediments. Figure 23.22 presents some photographs of sea bottoms

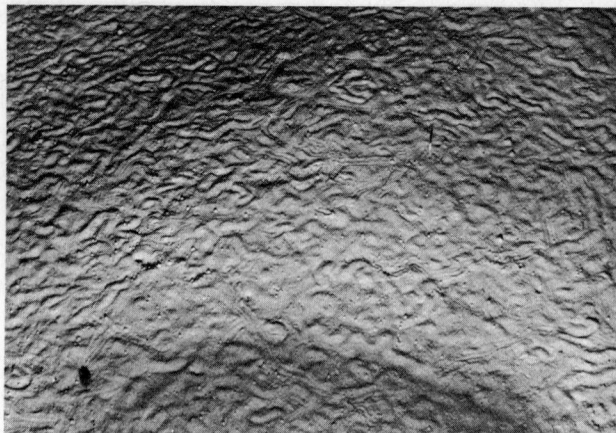

B. Ocean bottom showing current ripples and numerous tracks of bottom-dwellers. Puerto Rico trench, depth 21,000 ft (6412 m). Courtesy of the U.S. Naval Oceanographic Office.)

C. A sea cucumber plows across the soft bottom sediment. Pacific Ocean, Peru Basin, depth 16,300 ft (4941 m). (Courtesy of the U.S. Naval Oceanographic Office.)

Figure 23.22 (*Below and right*) Views of the ocean floor.

A. A shark bites at lure, setting off underwater camera. Underneath the shark's fin, an eel lies on the bottom. Indian Ocean, north of Madagascar, depth 870 ft (290 m). (Courtesy of Scripps Institution of Oceanography, University of California, San Diego.)

D. Manganese nodules litter the ocean floor. Caribbean Sea, east of Puerto Rico, depth 15,600 ft (4761 m). (Courtesy of the U.S. Naval Oceanographic Office.)

at several depths. An interesting feature is the high proportion of predators among the epifauna—seemingly too dense for the few visible prey. Many of these predators, however, go for long periods without food. The infauna consists largely of filter-feeders and detritivores, including small mollusks, crustaceans, and worms. The density of benthic organisms generally decreases from the neritic zone to the hadal depths. The neritic benthos is typically found in associations which have similar species compositions and occur at similar depths on similar bottom materials.

In addition to many species of fishes, the marine nekton includes mammals, turtles, and the larger crustaceans. Although they may forage over a relatively large area, nektonic organisms are often limited in their distribution by such factors as temperature and salinity. Benthic characteristics will also be important if the organism is a bottom-feeder. Many of the nekton, however, are plankton-feeders and possess adaptations to facilitate gathering and consuming the plankton. The baleen whales are examples—their gills serve the double duty of oxygenating their blood and straining plankton from the water.

The nekton of the aphotic zone in the bathyl and abyssal regions are not very well known, for most can easily avoid the nets which attempt to sample them. Although light is too low for photosynthesis, some of the fishes caught in these regions possess enlarged eyes or luminous organs to facilitate seeing. Another adaptation of these deep-sea fish is a very large mouth and lower jaw, which allows them to make the most of the infrequent meals which come their way.

The Oceans as a Food Source

It would be inappropriate to close our discussion of marine ecosystems without at least a brief look at their value as a food source to support Man's expanding population. Table 23.3 compares primary productivity for four types of marine ecosystems.

The deep-water oceanic zone is the least productive of the marine ecosystems. At the same time, however, it comprises about 90% of the world ocean area. Continental shelf areas are a good deal more productive, and, in fact, support much of the world's fishing industry at the present time (Figure 23.23). Except for the shallow-water estuarine and reef ecosystems, the upwelling zones are most productive. As explained in Chapter 4, in certain zones, cold bottom water rises from the great depths to reach the surface (Figures 4.33 and 4.35). Rising bottom water brings with it supplies of such critically short nutrients

TABLE 23.3 Primary Productivity of Four Ecosystems

Ecosystem	Gross Primary Production, (kcal/m^2/year)
Open ocean	1,000
Neritic zone	2,000
Upwelling zone	6,000
Estuarine and reef ecosystems	20,000

SOURCE: Data from Ryther as converted by E. P. Odum, 1971, *Fundamentals of Ecology*, W. B. Saunders Co., New York, p. 51, Table 3-7.

as phosphorus. Plankton concentrations in these areas are high, and the fish which eat the plankton are correspondingly abundant. These limited areas constitute the world's most productive fishing grounds, and yield about one-half the world's annual harvest. For example, the Peruvian coastal fishery of the anchoveta, a small fish used to produce fish meal for animal feed, is located in a narrow zone of upwelling and yields about 12 million metric tons (mmt) per year, or nearly one-fifth of the present total harvest of all marine fish.

Since the middle of the nineteenth century, fish catches have been rising at rates which have kept pace with food needs of an expanding world population. Improved technology in recent years has been a very important factor in increasing catches. This expansion cannot continue much further, however, for the total annual sustained yield of ocean fish is estimated to be between 100 and 200 mmt per year, compared with the present harvest of about 70 mmt per year. At present rates of expansion, this limit will be reached between 1980 and 2000. Beyond that time, overfishing will deplete the stock of fish. In 1949, an international fisheries conference identified some thirty major fish stocks thought to be underexploited; today, fourteen of these are either fully exploited or overfished.

These figures show that the ocean is not a vast, untapped food resource, capable of indefinite expansion into the future. If we continue to utilize its finfish according to existing patterns and standards, the ocean food resource will soon be exploited to its limit and beyond under the press of an expanding world population. Recall that in Chapter 20, we pointed out that only a small portion of the available energy in an ecosystem can be passed from one level of a food chain to the next. Consequently, a much larger population of consumers can be supported by feeding upon

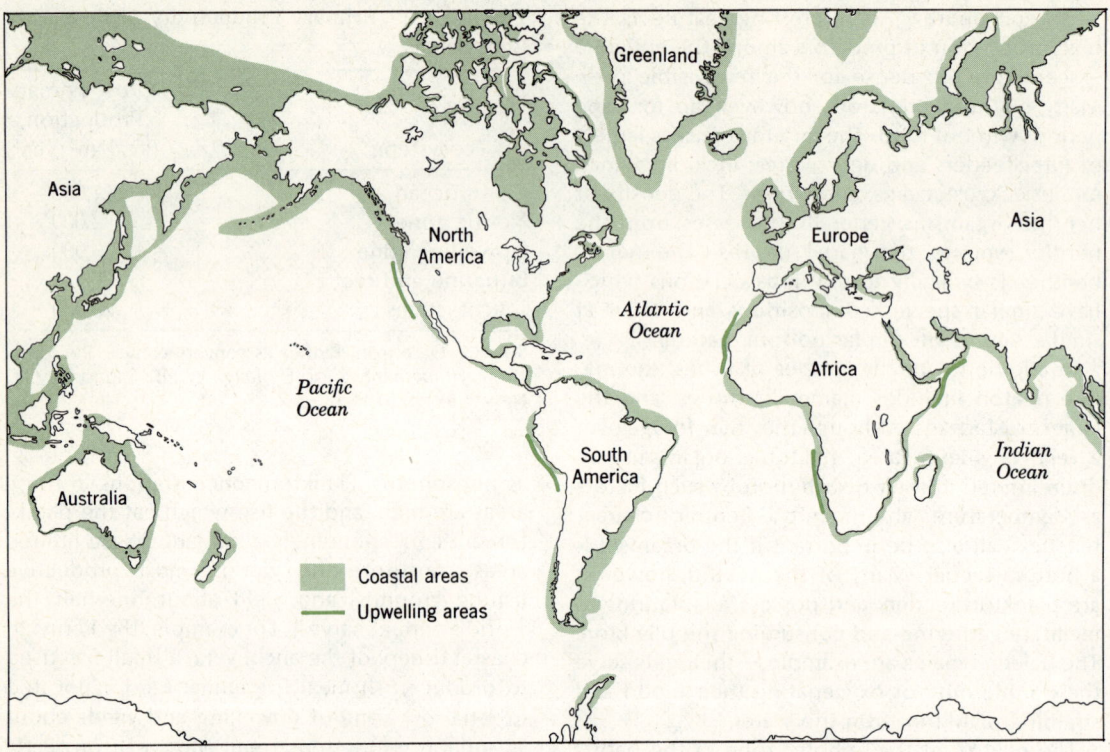

Figure 23.23 Distribution of world fisheries. Coastal areas and upwelling areas together supply over 99% of world production. (Compiled by the National Science Board. From *Environmental Science; Challenge for the Seventies*, National Science Foundation, U.S. Government Printing Office, Washington, D.C., p. 11, 1971.)

plants at the producing level than upon other consumers at higher levels. From a diet of fish, taken at the highest levels of the marine food chain, we may be forced to turn to lower levels for our food supply and to make direct use of plankton as a food. The prospect, if lacking in gustatory appeal, may offer one means of postponing the world food crisis certain to arise if growth of the human population continues unchecked.

A comparison of terrestrial and aquatic ecosystems reveals some important differences as well as some similarities. Perhaps the most obvious differences between the two are in their physical environments. The relative scarcity of CO_2 in air has the effect, already noted in Chapter 20, of limiting photosynthesis in terrestrial environments. In aquatic environments, however, CO_2 is abundant and photosynthesis is limited by lack of light below the euphotic zone.

In aquatic ecosystems, nutrients are supplied by the water in dissolved form as ions that can be absorbed directly by the organism. Although direct absorption of some nutrient gases does occur in terrestrial ecosystems (for example, ammonia gas can serve as a direct nitrogen source for many plants when it is present in the air), most nutrients are supplied by the soil solution, which in turn is supplied by chemical weathering of soil parent materials. Terrestrial roots serve as special absorptive organs for nutrients and water; they are unnecessary for absorption in aquatic organisms and are often rudimentary or lacking, except in the case of some of the vascular aquatic plants whose roots provide a means of attachment to bottom sediments.

Because the density of water is much greater than that of air, aquatic organisms are partly supported by buoyant force. In terrestrial environments, the buoyant force of the air is so low that a system of supporting tissue has evolved in most terrestrial plants and animals. Such a supporting system also helps to resist the disruptive forces of strong, turbulent winds. Except for streams and intertidal areas, similar mechanical forces are largely absent within aquatic environments.

The terrestrial environment is also characterized by rapid shifts in temperature, specifically those on a diurnal rhythm, whereas longer cycles, usually on an annual rhythm, characterize the aquatic environment. Moreover, amplitudes of thermal cycles are much smaller in aquatic environments, as a consequence of the greater specific heat of water than of air. Thus, the terrestrial environment is the evolutionary home of the homeotherms, whose activities are less limited by temperature variation than are activities of poikilotherms.

24
Terrestrial Ecosystems

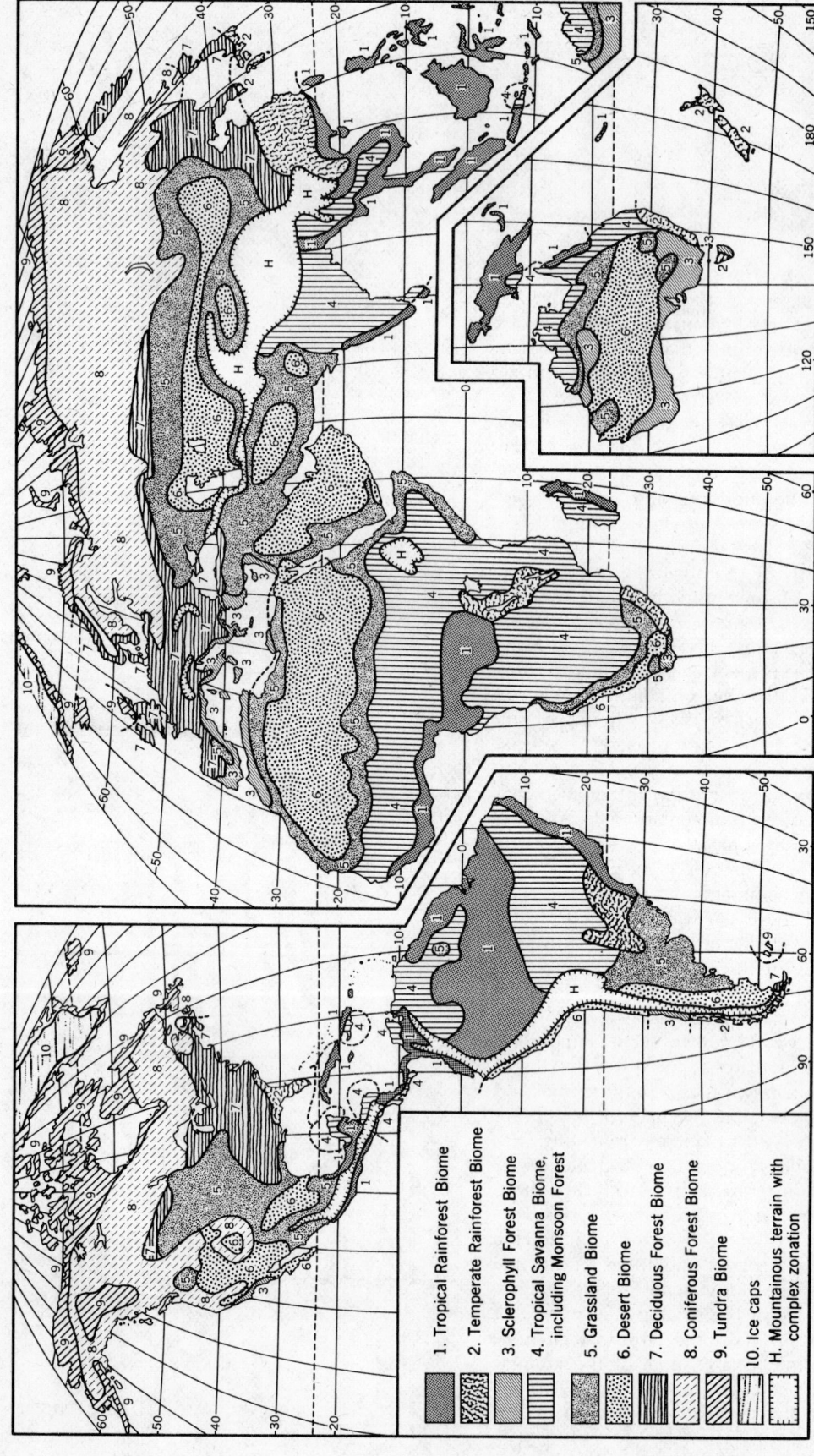

Figure 24.1 Major biomes of the world, greatly generalized. (Modified and simplified from a world map by H. Brockman Jerosch, 1951, following Rübel's classification. Aitoff's Equal-Area Projection, adapted by V. C. Finch.)

The Tundra Biome

Over all, then, the terrestrial environment shows more diversity, greater extremes, and more changeability than the aquatic environment, but in return offers an abundance of light energy and nutrients. Terrestrial organisms which exploit this vast energy potential include the most highly developed groups: the flowering plants and warm-blooded vertebrates.

From a biological viewpoint, terrestrial ecosystems differ markedly from aquatic ecosystems in the size and conspicuousness of the autotrophic biomass. Land plants are almost always the most obvious part of any terrestrial ecosystem and, as a result, have come to dominate our classification of terrestrial environments. Ecologists recognize six major types of terrestrial ecosystems. These are the great *biomes:* tundra, coniferous forest, deciduous forest, tropical rainforest, grassland, and desert. Each biome has a characteristic form of vegetation and an associated fauna, as well as a characteristic climate and soil-moisture regime to which it is adapted. The world distribution of these biomes, including some of their subdivisions, is shown in a highly generalized map, Figure 24.1.

The Tundra Biome

Along the arctic fringes of the continental masses of North America and Eurasia lies the *tundra biome*—a vast belt of low shrubs, grasses, sedges, and mosses. The climate of the tundra is severely cold; consequently, most of the tundra belt is underlain by continuous permafrost (Chapter 3; Figure 3.8), which remains frozen because of the low mean annual temperature. Further, the prevailing low air temperature insures that water vapor content is low in polar and arctic air masses, and that precipitation in most areas is small. In fact, much of the tundra is a frozen desert, with annual total precipitation values as low as 12 in. (30 cm) (Chapter 5; Figure 5.44). Recall that under the arctic soil-moisture regime, as illustrated by Point Barrow, Alaska (Figure 12.14*H*), a small moisture deficiency prevails throughout the brief summer and there is no water surplus.

During the spring and summer, the active zone of the soil is subjected to many freeze-thaw cycles. Constant growth and shrinkage of ice within the active soil layer churns and heaves the soil, producing a highly unstable substrate. Characteristic blocklike units of soils and vegetation develop, referred to in Chapter 11 as patterned ground. Typical forms of patterned ground include polygons, nets, and stripes (Figures 11.33, 11.34, and 11.35). Freeze-thaw cycles also cause the flowing of soil in lobes or tongues, a process called solifluction (Figure 11.36). Tussocks or hummocks, produced by the combined action of frost and vegetation, are also common. Figure 24.2 shows such features in an Alaskan cotton-grass (*Eriophorum*) meadow. Uneven accumulations of snow and uneven melting, together with the flat nature of much of the lowland tundra, act to produce patches of bog and marsh. Recall from Chapter 12 that the tundra is associated with a soil-forming regime referred to as *gleization,* (Figure 12.16). Saturation of the soil favors reduction of iron compounds and the production of gray, sticky clay layers. Low temperatures favor accumulation of raw humus as peat (Figure 12.15).

Superimposed on this varied pattern of substrate is a varied pattern of vegetation. Shrubs of the heath family, willows, and birches are the dominant woody plants of the tundra ecosystem. Grasses, sedges, and perennial herbs are common. Lichens and mosses are also abundant. Vegetation is usually scarcest on dry, exposed slopes and summits—the rocky pavement of these areas gives them the name of *fell-field*, from a Danish term meaning "rock desert" (Figure 24.3). In the wet lowland areas, a carpet of sphagnum moss with sedges and heaths typically covers the ground.

As is most often true in particularly dynamic environments, species diversity in the tundra is low, but the abundance of individuals is high. Among the animals, vast herds of caribou in North America or reindeer (their Eurasian relatives) roam the tundra, lightly grazing the lichens and plants and moving constantly. A smaller number of musk-oxen are also primary consumers of the

Figure 24.2 An arctic meadow of cotton-grass, *Eriophorum*, on the Alaskan coastal plain. (Photograph by William R. Farrand.)

Figure 24.3 Typical vegetation of an arctic fell-field. Plant on the right is *Cerastium alpinum*; on the left is *Dryas integrifolia*. (Photograph of the Crockerland Expedition, American Museum of Natural History.)

Figure 24.4 Reindeer moss, the lichen *Cladonia rangifera*. (Larry West.)

tundra vegetation. Wolves and wolverines, as well as arctic foxes and polar bears, are predators. Among the smaller mammals, snowshoe rabbits and lemmings are important herbivores. Invertebrates are scarce in the tundra, except for a small number of insect species. Black flies, deerflies, mosquitoes, and "no-see-ums" (tiny biting midges) are all abundant and can make July on the tundra most uncomfortable for both Man and beast. Reptiles and amphibians are also rare. The boggy tundra, however, presents an ideal summer environment for many migratory birds such as waterfowl, sandpipers, and plovers.

The food web of the tundra ecosystem is simple and direct. The important producer is "reindeer moss," the lichen *Cladonia rangifera* (Figure 24.4). In addition to the caribou and reindeer, lemmings, ptarmigan (arctic grouse), and snowshoe rabbits are important lichen grazers. The important predators are the fox, wolf, lynx, and Man, although all these animals may feed directly on plants as well. During the summer, the abundant insects help support the migratory waterfowl populations. The directness of the tundra food web makes it particularly vulnerable to fluctuations in the populations of a few species, and we have already described in Chapter 22 the oscillations of the lemming and lynx populations.

The tundra ecosystem is indeed a fragile one, and has already been impacted by Man's activities. As explained in Chapter 11, permafrost presents very special difficulties in construction of roads, airfields, houses, and buildings. The phenomenon of thermal erosion is all too easily induced by disturbance of the insulating cover of organic matter.

We have already shown that the Trans-Alaska Pipeline System (TAPS) poses a great engineering problem in this respect—running a 4-foot diameter pipe containing petroleum at 170° F (77° C) across 800 mi (1300 km) of Alaskan terrain without thawing the permafrost which underlies considerable portions of the route will be quite a challenge. Environmentalists fear that melting will cause the pipeline to be encased in a tube of mud, increasing its susceptibility to damage from earthquake shock waves and freeze-thaw stresses. A great potential for environmental damage also exists in the construction of the access roads, runways, and helipads which will be necessary to lay the pipeline. Just how these physical changes will upset the tundra ecosystem is not easily predicted. The role of the pipeline as a barrier to animal migrations and the effects of oil spills are topics of particular interest.

Unlike the middle-latitude ecosystems with which Man is used to dealing, the tundra ecosystem is very slow to recover from changes or damages. The cold temperatures which so hamper Man's activities also greatly hamper processes of biodegradation of wastes and autogenic succession. Tracks of all-terrain vehicles across the tundra remain visible for years (see Figure 11.37). Garbage does not decay, it freezes. The normal bacteria which degrade petroleum act only very slowly, if at all, making the impact of an oil spill all the more permanent. This great sensitivity of the tundra ecosystem to human disturbance and the fact that the tundra is one of the last great areas of wilderness which remain on the earth's surface have made its preservation a fundamental concern of environmentalists.

Coniferous Forest Biome

Bordering upon the southern limit of the tundra is an almost unbroken band of coniferous forest dominated by needle-leaved trees. This band, the *coniferous forest biome*, is largely in two great belts, one stretching across Canada, the other spanning Eurasia from northern Europe to Siberia (Figure 24.1). This cold-climate type of needle-leaf forest is referred to as *boreal forest*. In North America, the principal tree species which dominate the boreal forest are spruce (*Picea*) and fir (*Abies*), with pine (*Pinus*) important in the more southern areas (Figure 24.5). The larch (*Larix*), a conifer with deciduous needles, is the dominant in the Siberian area of Asia (Figure 24.6). A few species of broadleaved trees are also found throughout the biome; in North America, these are predominately species of poplar and aspen (*Populus*), and birch (*Betula*). The deciduous trees are the first colonizers on sites disturbed by wind, water, or fire. Below the tree layer, dominant plant species are mosses, grasses, and sedges. Alders and willows may be abundant in wet or boggy areas.

Boreal forest occupies continental areas of the most extreme seasonal temperature range of any terrestrial environment. Mean monthly air temperature spans a range of 70 F° (38 C°) or more from the coldest month of winter to the warmest month of summer. Despite a long, severely cold winter, the short summer provides a period of intense solar radiation, so that net radiation rises to attain large surplus values (see graph of Yakutsk, 62° N, Figure 2.15). Long daily periods of sunshine in summer intensify net photosynthetic productivity.

Boreal forest is identified with the soil-forming regime of podzolization, described in Chapter 12 (Figure 12.15). Conifers utilize little in the way of base cations, and these are replaced by hydrogen ions on the soil colloids. Soil acidity is high (Table 11.1), so that neutralization by application of lime is a requirement if the soil is to be cultivated for food crops. Base cations must also be supplied by fertilizers. All in all, the boreal forest offers little promise of providing important new food resources. On the other hand, as a source of forest products—particularly pulpwood—the boreal forest is a resource of great value in an industrial society.

Many of the mammals of the tundra biome are also found in the coniferous biome. These include the caribou, lemming, and snowshoe rabbit. Others are deer, moose, black bear, marten, mink, wolf, wolverine, and fisher. Like the tundra, the coniferous biome often experiences large fluctua-

Figure 24.5 A spruce forest on the floodplain of the Upper Peribonka River, Quebec, Canada. (Photograph by Dr. Pierre Dansereau.)

tions in species populations, a result of the low diversity and highly variable environment.

The coniferous forest biome extends equatorward where mountain ranges or high plateaus provide a cool, wet climate. In North America, extensions into lower latitudes follow the Sierra Nevada, Cascade, and Rocky Mountain ranges into the southwestern states (Figure 24.7). In British Columbia, Washington, Oregon, and northern California, the orographic effect pro-

Figure 24.6 A larch woodland in Yakutsk, A.S.S.R., in the Tompo River region. This species (*Larix dahurica*), at a latitude of 64° N, is near the poleward limit of its growth. (Photo by I. D. Kild'ushevsky, courtesy of Professor B. A. Tikhomirov, Komarov Botanical Institute, Leningrad.)

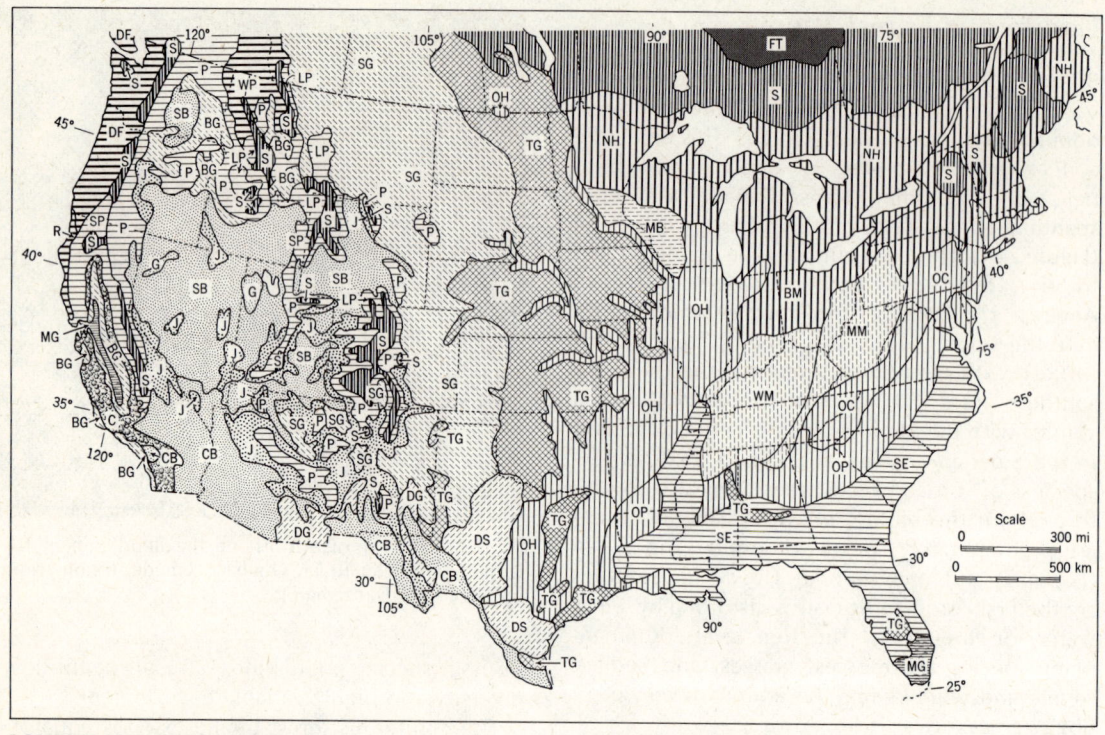

CONIFEROUS FOREST BIOME

    Northern Coniferous Forest
        FT Subarctic forest-tundra transition
        S   Spruce-fir

    Northwestern Coniferous Forest
        WP Western larch-western white pine
        DF Pacific Douglas fir
        R   Redwood

    Western Pine Forest
        SP Yellow pine-sugar pine
        P   Yellow pine-Douglas fir
        LP Lodgepole pine
        J   Pinon-Juniper

    Southeastern Pine Forest
        SE Longleaf-loblolly-slash pine

DESERT BIOME

    SB Sagebrush
    CB Creosote bush
    G   Greasewood

SCLEROPHYLL FOREST BIOME

    C   Chaparral

DECIDUOUS FOREST BIOME

    Northern Hardwoods Forest
        BM Beech-maple
        NH Hemlock-white pine-northern hardwoods
        MB Maple-basswood

    Mixed Mesophytic Forest
        MM Mixed mesophytic
        WM Western mesophytic

    Oak Forest
        OC Oak-chestnut
        OP Oak-pine
        OH Oak-hickory

GRASSLAND BIOME

    TG Tall grass
    SG Short grass
    DG Mesquite-grass
    DS Mesquite and desert grass savanna
    BG Bunch grass
    MG Marsh grass

Figure 24.7 Biomes of the continental U.S. and southern Canada, subdivided by vegetation types. (Modified and simplified from maps of H. L. Shantz and Raphael Zon, in *Atlas of American Agriculture,* 1929, U.S. Government Printing Office, Washington, D.C.; E. Lucy Braun, 1950, *Deciduous Forests of Eastern North America,* Hafner Publishing Co., New York; and Canada Department of Forestry, Bulletin 123, 1963.)

duces heavy rainfall on the western slopes of the coastal ranges (Chapter 5; Figure 5.44). These areas support a unique coniferous forest of the world's largest trees, including the coast redwood (*Sequoia sempervirens*) and the big tree (*Sequoia gigantea*), which occurs further west in the central California Sierras. Individuals of redwood and big tree reach heights of over 300 ft (100 m) high and diameters of 20 ft (7 m) or more near the base (Figure 24.8).

In contrast to the climate of the boreal forests, climate of the west-coast mountain ranges is remarkably uniform in temperature throughout the year, a consequence of dominance by maritime (mP) air masses. Precipitation is heavy in winter months and a water surplus is generated in that season (Figure 12.14G). As in the regions of boreal forest, podzolization is the dominant pedogenic regime.

At lower latitudes in North America the main representative of the coniferous forest biome is the conifer forest of the southeastern and south-central Coastal Plain. Climatically and ecologically, this forest is very different from its northern relatives. Here pines dominate, particularly long-leaf pine (*Pinus australis*), with short-leaf pine (*P. echinata*), slash pine (*P. caribaea*), and loblolly pine (*P. taeda*) also important species. Fires appear to have been an important influence in the forest in precolonial times, and controlled burning is necessary today to maintain pines as dominant species in the region. In fact, the tendency for this forest to revert to broadleaved trees has led some ecologists to omit it from the coniferous forest biome and classify it separately (as temperate rainforest).

The coniferous biome is highly important to Man as a source of wood for lumber and pulp. The two species most valuable commercially are the Douglas fir and the longleaf pine—the former for dimension lumber used in construction, the latter for pulp used in kraft paper manufacture, as well as naval stores (e.g. turpentine) and some construction uses (e.g. fence posts and rails).

The Deciduous Forest Biome

The *deciduous forest biome* includes the forests of eastern North America, eastern Asia, and western Europe. Climatically, these areas have warm summers and cool to cold winters. Annual precipitation is from 30 to 60 in. (75 to 150 cm) and generally well distributed throughout the year, but in many areas it shows a strong summer maximum. There is only a small summer soil-moisture deficiency, or none at all, so that water is not usually a limiting factor in forest growth. In

Figure 24.8 A grove of big trees (*Sequoia gigantea*) in Sequoia National Park, California. A man (look for the hat) can be seen standing at the right of the largest tree, which measures 51 ft (15 m) in circumference. (American Museum of Natural History.)

autumn, lowered temperatures and shortened days cause leaf fall and the cessation of photosynthesis. The term *summer-green* deciduous forest is sometimes used to distinguish this biome from tropical (monsoon) deciduous forests, where leaf shedding is a consequence of severe seasonal drought, rather than frost.

Most of the trees of the deciduous forest are *hardwoods*, or broadleaved species, although a few *softwood* species (conifers) such as hemlock and pine are important constituents. The North American and Asian forests are strikingly similar, dominated by species of oak (*Quercus*), beech (*Fagus*), birch (*Betula*), maple (*Acer*), walnut (*Juglans*), basswood (*Tilia*), elm (*Ulmus*), ash (*Fraxinus*), yellow poplar (*Liriodendron*), and hornbeam (*Carpinus*). In the United States, chestnut (*Castanea*) was a dominant species until the beginning of the present century; it has since been nearly exterminated by the chestnut blight, an imported fungus which girdles the tree and kills it.

The deciduous forest biome includes a great variety of animal life, much of it stratified according to the canopy layers. As many as five layers can be distinguished: the upper canopy, lower canopy, understory, shrub layer, and ground layer (Figure 24.9). Because the ground layer presents a more uniform environment in terms of humidity and temperature, it contains the largest concentration of organisms and the greatest diversity of species. Many small mammals burrow in the soil for shelter or food in the form of soil invertebrates. Among this burrowing group are ground squirrels, mice, and shrews, as well as some larger animals—foxes, woodchucks, and rabbits. Most of the larger mammals feed on ground and shrub layer vegetation, except for some, such as the brown bear, which are omnivorous and prey upon the small animals as well.

Even though birds possess the ability to move through the layers at will, most actually restrict themselves to one or more layers. For example, the wood peewee is found in the lower canopy, and the red-eyed vireo is found in the understory. Above them are the scarlet tanagers and blackburnian warblers, which are upper-canopy dwellers. Below are the ground dwellers, such as grouse, warblers, and ovenbirds. Flying insects often show similar stratification patterns.

The great diversity of the deciduous forest biome in North America has led to its breakdown into a number of component forests, each with a different flora (Figure 24.7). The *northern hardwoods forest* is found in New England and stretches west along southern Canada to northern Michigan and northern Minnesota. It serves as a transition between the deciduous forest biome and the coniferous forest biome, and includes many needle-leaved species, although it is dominated by broadleaved species. In Indiana and northern Ohio and on the southern shores of lakes Erie and Ontario is the *beech-maple* forest. The maple here is sugar maple (Acer saccharum). To the south of these two types, oaks dominate. West of the Appalachians is the *oak-hickory* forest, and on the Appalachians and piedmont to the east is the *oak-chestnut* forest (now reduced by the chestnut blight simply to an oak forest.) At higher elevations and in coves and on north-facing slopes in the central and southern Appalachians is the *mixed mesophytic* forest. The mixed mesophytic forest contains a very diverse flora, and dominance is shared by many species rather than a few. This forest type is believed to have a long history and resembles some fossil assemblages from Tertiary time.

Man's impact on the deciduous forest biome has been very great, largely because it serves as the site of many human activities. Most of the area has been converted from forest to field or pasture at one time or another; in some areas secondary succession has produced a forest similar to the original, and in some areas it has not. Still other areas have been totally removed from autotrophic production, being covered by roads, parking lots, and buildings. Only in the more mountainous and inaccessible regions does the deciduous forest biome resemble in any fashion its presettlement condition.

Figure 24.9 A mature stand of sugar maples, *Acer saccharum*, in the Allegheny National Forest, Pennsylvania. The trees are 160 to 200 years old. (U.S. Forest Service photograph.)

From existing remnants of virgin forest, and from early accounts, ecologists have been able to reconstruct the character of the presettlement forest. The most striking factor was the size of the trees—in bottoms, coves, and other moist environments, the trees were very large indeed. Trees over 3 ft (1 m) in diameter, with trunks rising 60 ft (20 m) or more without a branch, were common on these sheltered sites (Figure 24.9). On the other hand, steep, rocky slopes of mountains and other dry sites supported canopies not much higher than at present. The trees were short and the canopies were fairly open, much as they are today.

Disturbance was ever-present within the presettlement forest. Often many acres of trees were blown over by severe hurricanes or tornadoes acting only rarely at one spot, but frequently within the forest as a whole. Occasional severe floods toppled bottomland forests. Indians made clearings by girdling trees or by burning. In short, the presettlement forest was not one vast forest of giants from the Atlantic shore to the Great Plains; rather the canopy was a patchwork, the result of both natural and human forces. Today's deciduous forest landscape is, of course, much more radically fragmented.

Tropical Rainforest Biome

The third forest biome is the *tropical rainforest biome* (Figure 24.10). These forests occur in three main areas of the world—the Amazon and Orinoco basins in South and Central America; the Indo-Malaysian region from India to New Guinea; and the Zambezi, Niger, and Congo river basins in central and west Africa (Figure 24.1). These areas have a remarkably uniform climate. As pointed out in Chapter 3, daily air temperature range greatly exceeds mean monthly temperature range from season to season (Figure 3.30). Rainfall is copious—80 in. (200 cm) and over annually—evenly distributed through the year in the equatorial zone. Further north and south, in the tropical zones, the rainforest experiences a short but distinct dry season.

As illustrated by the graph for Bougainville in the Solomon Islands (Figure 12.14A), a large water surplus is produced in all, or most, months of the year. Soil moisture is maintained nearly at field capacity and photosynthesis proceeds at a sustained high level throughout the year. Physical stress upon plants is thus minimal.

The rainforest is not too different in outward form from its deciduous or coniferous relatives—

Figure 24.10 Aerial view of tropical rainforest on the Rio Negro, a tributary of the Amazon River. (Photograph by Colonel Richmond, courtesy of the American Geographical Society.)

it, too, has a closed canopy, although it often exhibits an upper stratum of scattered, emergent crowns which project above the main canopy (Figure 24.11). Inside the rainforest, an understory layer of trees 15 to 50 ft (5 to 15 m) high is scattered throughout. In contrast to other forest biomes, epiphytes, as well as climbing vines, or *lianas*, are abundant (Figure 24.12). The dense shade at the forest floor restricts the growth of many species, and thus the rainforest in its natural undisturbed state is usually open and easy to pass through. The canopy trees of the tropical rainforest are often very tall and large in girth, although they do not approach the girth of the coniferous giants of the northwest American coast.

In contrast to middle- and high-latitude forests, animal life of the tropical rainforest is most abundant in the upper layers of the vegetation. Above the canopy, birds and bats are important carnivores, feeding largely on insects above and within the topmost canopy. Below this level are found a wide variety of birds, mammals, reptiles, and invertebrates; they feed on the leaves, fruit, and nectar abundantly available in the main part of the canopy. Ranging between the canopy and the ground are the *scansorial* (climbing) mammals, which forage in both layers. At the surface are the larger ground mammals, including herbivores, which graze the low leaves and fallen fruits, and carnivores, which prey upon the abundant vertebrates and invertebrates found at the surface.

A major difference between the tropical rainforest and forests of higher latitudes is the great diversity of species which it possesses. The tropical rainforest may have as many as 3,000 different plant species in a square mile area—other forests would be considered very diverse indeed with only one-tenth that number! The fauna of the rainforest is also very rich. A 6-sq mi (16-sq km) area in the Canal Zone, for example, contains about 20,000 species of insects, whereas there are only a few hundred in all of France. This large number of species occurs because the rainforest environment is so uniform and free of physical stress and because the ecosystem is so diverse that almost any mutation can survive physically and find a particular niche to which it is suited. As a result of the great number of species, the tropical rainforest has served as a center for speciation, providing both ancestral stocks and recent additions to the flora and fauna of higher-latitude zones.

An understanding of the successional process in the rainforest is important in understanding Man's impact. Ecologists divide the plant species of the rainforest into two groups: the *primary* species, which are characteristic of the virgin rainforest, and *secondary* species, which are adapted to disturbed sites. When disturbance occurs (produced, for example, by flood, windstorm, or the

Figure 24.11 The complex layering in the tropical rainforest of canopy and understory trees, high and low shrubs. (After J. S. Beard, 1946, *The Natural Vegetation of Trinidad*, Clarendon Press, Oxford. From A. N. Strahler, 1969, *Physical Geography*, 3rd ed., John Wiley & Sons, New York, p. 343, Figure 21.4.)

fall of a large canopy tree), both primary and secondary species compete for the new habitat. If the soil remains largely undisturbed, the primary species will usually win out with little difficulty. Seedlings and saplings of the primary species are always present in the understory, and the opening of a new area to strong sun allows them to grow rapidly to fill the void, easily outstripping any secondary species which have managed to invade the new area.

When the soil as well as the vegetation is disturbed, the case is somewhat different. Primary species usually have seeds which are not able to remain dormant and thus sprout rapidly. (Rapid sprouting is actually an evolutionary advantage in the climax rainforest because fruits and seeds are readily consumed by herbivores, fungi, and bacteria.) If primary species are to return to the disturbed area, seed sources must be readily available at the time and place of disturbance. The seed sources will thus be those primary species which are nearby and happen to be in fruit. In addition, birds, monkeys, rodents, and other animals may carry seeds of primary species into the disturbed area. The secondary species, on the other hand, possess seeds which remain dormant in the soil for a considerable time; with soil agitation and exposure to light they are triggered to germinate. In most cases, however, the secondary species are soon dominated by the primary species obtained from nearby sources, and thus the vegetation returns more or less to its former state.

In the past, Man has farmed the rainforest by the *slash and burn* method—cutting down all the vegetation in a small area, then burning it. Recall from Chapter 21 that most of the nutrients in a rainforest ecosystem are tied up in the biomass, rather than in the soil. Burning the slash on the site releases the trapped nutrients, and a portion of the nutrients is thus returned to the soil. The supply of nutrients derived from the original biomass is small, however, and the harvesting of crops rapidly depletes the nutrients. Thus, after a few seasons of cultivation, the soil loses much of its productivity. A new field is then cleared in another area, and the old field is abandoned. Primary species are able to reinvade the abandoned area because fruiting species and animal seed carriers are close by, and so the rainforest soon returns to its original state. Thus, the primitive slash-and-burn agriculture is compatible with the maintenance of the rainforest ecosystem.

On the other hand, modern, intensive agriculture, which uses large areas of land, is not compatible with the rainforest ecosystem. When such lands are abandoned, seed sources are so far away

Figure 24.12 A view inside the tropical rainforest. Most of the trees are broadleaved evergreen species; note the numerous climbing vines. Near Belém, Brazil, in the Amazon basin. (Photograph by Otto Penner, courtesy of Instituto Agronómico do Norte.)

that the primary species cannot take hold. Instead, secondary species dominate, often accompanied by species from other vegetation types whose seeds are preadapted to long periods of dormancy. The dominance of these secondary species is permanent, at least on the human time scale. Thus, the tropical rainforest ecosystem is, in this sense, a nonrenewable genetic resource of many, many species of plants and animals which, once displaced by large-scale cultivation, can never return to reoccupy the area. A recent appraisal of tropical rainforest ecosystems concluded with the following statement:

> All the evidence available supports the idea that, under present intensive use of the land in tropical rainforest regions, the ecosystems are in danger of a mass extinction of most of their species. This has already happened in several areas of the tropical world, and in the near future it may be of even greater intensity. The consequences are nonpredictable, but the sole fact that thousands of species will disappear before any aspect of their biology is investigated is frightening. This would mean the loss of millions and millions of years of evolution, not only of plant and animal species, but also of the most complex biotic communities in the world.*

The Grassland Biome

The *grassland biome* consists of three related types of vegetation and associated fauna: savannas, prairies, and steppes. *Savannas* are grasslands with scattered trees. Savannas occur in the tropical zones, where wet and dry seasons are strongly contrasted. The savannas of east Africa are perhaps the best examples, but savannas also occur in South America (on the Brazilian highlands), in northern Australia, peninsular India, and southeast Asia. Some ecologists find the savanna sufficiently unique to classify it as a separate biome: the *tropical savanna biome*.

Prairies are found in middle-latitude climates where evaporation and precipitation are roughly balanced, and range between 20 and 40 in. (50 to 100 cm) per year. High summer evapotranspiration results in a small soil-moisture deficiency during the summer. Combined with long droughts, fires, and the effects of grazing, the moisture deficiency keeps tree species from succeeding the tall grasses which dominate the prairie. Pockets of arborescent vegetation, however, may be present in the moister areas.

Steppe is similar to prairie, but is dominated by short grasses which are more scattered and cover less of the ground surface. Steppe, particularly in North America, is often referred to as short-grass prairie or plains. Steppes are drier than prairies, and often provide a transition to desert ecosystems. An example of the soil-moisture balance of steppes is seen in the graph for Medicine Hat, Alberta (Figure 12.14F). Notice the severe soil-moisture deficiency, extending over a seven-month period, and the lack of any water surplus.

*A. Gómez-Pompa, C. Vázquez-Yanes, and S. Guevara, *Science*, vol. 177, p. 765. © 1972 by the American Association for the Advancement of Science.

The grasses which constitute the dominant vegetation of the grassland biome are usually of two types: sod formers and bunch grasses. The *sod formers*, spread by rhizomes, form a tough sod mat of interlocking stems, whereas the *bunch grasses* grow in scattered clumps. In addition to the grasses are other herbaceous plants, the forbs. *Forb* is a general term used to describe the annual and perennial herbs which do not belong to the grass family. The principal prairie forbs are the legumes and the composites (members of the *Compositae*, a family including daisies, asters, and dandelions). Together the forbs may account for more species than the grasses. The grasses, however, dominate the biomass. Figure 24.13 shows one of the few remaining patches of virgin prairie in the midwest.

In terms of vegetation the great plains of North America are usually divided into three areas: tall-grass prairie, mixed-grass prairie, and short-grass prairie (Figure 24.7). The area of *tall-grass prairie* is easternmost, in a zone ranging from western Ohio and Indiana northward and westward to Wisconsin and Minnesota. Much of this area is actually suited to arboreal vegetation, but fires and grazing kept this area in tall-grass prairie in presettlement time. Now, much of the tall-grass prairie, where not maintained in pasture or grain-farming, has reverted to forest, a result of Man's protection of the area from fire.

The *mixed-grass prairie* occupies a mid-continental position to the west of the tall-grass prairie, where not maintained in pasture or region and serves as a transitional area between the short-grass prairie to the west and the tall-

Figure 24.13 Kalsow Prairie, Iowa, a remnant of virgin tall-grass prairie. (Photograph by courtesy of the State Conservation Commission of Iowa.)

grass prairie to the east. Because of the great seasonal variation in rainfall and evaporation experienced here, the aspect of the mixed-grass prairie changes from year to year.

The *short-grass prairie* is characteristic of the most western plains areas, and grades into desert on its western and southern edge. Shallow-rooted, short species of grass dominate; sod-formers are important in the short-grass prairie, and the dense sod acts to keep low the number of forbs.

Soils of the grasslands biome in middle latitude are characterized by a thick upper horizon of dark brown to almost black color, and this has led to their designation as brown soils, chestnut soils, and black earths. In the Ukraine region of Eurasia the soil goes by the name of *chernozem*, and this word has come into wide use to apply to black soils in North and South America as well.

The soil-forming regime of the grasslands is that of calcification, described in Chapter 12. A typical soil profile is shown in Figure 12.15C. Below the dark, humus-rich A horizon is a light-colored B horizon containing nodules or plates of calcium carbonate (Figure 12.17). The base cations are recycled by grasses and maintained in abundance in the A horizon.

Chapter 16 has already dealt with the droughts of the 1930s, which turned most of the short-grass prairie into a vast Dust Bowl ranging from New Mexico north to the western Dakotas. Undoubtedly, breaking of the sod cover and planting the short-grass prairie in wheat intensified the deflation and dust storms which accompanied the droughts. The drought converted much of the mixed-grass prairie into short-grass prairie, and the tall-grass prairie into mixed-grass prairie. Following abatement of the drought, most of the tall- and mixed-grass prairie was able to revert to its normal vegetation. Soil erosion was so severe in the Dust Bowl, however, that much of the area has still not recovered, and changes seem to be permanent.

Like forest ecosystems, grassland ecosystems exhibit stratification. The three main layers are the root layer, the ground layer, and the herbaceous layer. The *root layer*, occupying the dark A soil horizon, is very much deeper and thicker than in forests—the roots of the grasses usually comprise nearly 50% of the plant's biomass and penetrate the soil to considerable depths, often as much as 6 ft (2 m) (Figure 24.14). Typically the grass roots are thickest in the upper 6 in. (15 cm) of the soil, where they are joined by roots and rhizomes of the other plants present. Undoubtedly, the deep penetration of the roots facilitates obtaining soil moisture during the long drought season.

The *ground layer* is located at the soil surface. Here the annual increment of dead leaves and stems decays into *mulch*, the organic-rich surface litter layer of the grassland ecosystem. The quantities of mulch produced can be very large; as much as five tons per acre may be present on a climax tall-grass prairie. Inorganic nutrients are closely cycled through the mulch and saprotrophs; the organic content of the soil is very high. Apparently, the accumulation and decay of mulch is a natural process required for the maintenance of the prairie. Where mulch accumulation is blocked by heavy grazing and mowing, the prairie can regress to weeds and to the more drought tolerant grasses.

The *herb layer* varies during the seasons. In the spring, low plants such as strawberry, violets, mosses, and others grow rapidly. They are soon overtopped by the grasses and the taller forbs, which create a middle layer. The tallest layer is usually the flowering stems of the grasses; by the time of its development the ground layer may be almost completely hidden.

The animals of the grassland are distinctive. As would be expected, large grazing mammals are abundant. In North America, these included the pronghorn antelope and the buffalo. Now nearly extinct, the buffalo (*Bison*) once numbered 60 million and roamed the grasslands from the Rockies to the Shenandoah Valley of Virginia. By 1889, however, this herd had been reduced to 800 individuals, or approximately the same number alive today. Most of the buffalo population is now confined to Yellowstone National Park. Today, cattle, rodents, and rabbits are the major grazers in the grasslands ecosystem (Figure 24.15).

The grassland ecosystem supports some rather unique adaptations to life. A common adaptive mechanism is a jumping or leaping locomotion, assuring an unimpeded view of the surroundings. Jack rabbits and jumping mice are examples of jumping rodents. The pronghorn combines the leap with great speed, which allows it to avoid predators and fire. Burrowing is also another common life habit, for the soil provides the only shelter in the exposed grasslands. Examples are burrowing rodents, including prairie dogs, gophers, and field mice. Rabbits exploit old burrows, using them for nesting or shelter. Invertebrates also seek shelter in the soil, and many are adapted to living within the burrows of rodents, where extremes of moisture and temperature are substantially moderated.

The prairie dog is an interesting burrowing animal. Now widely hunted and poisoned, this animal competes with cattle for forage and builds undesirable dogtowns, covering much of the im-

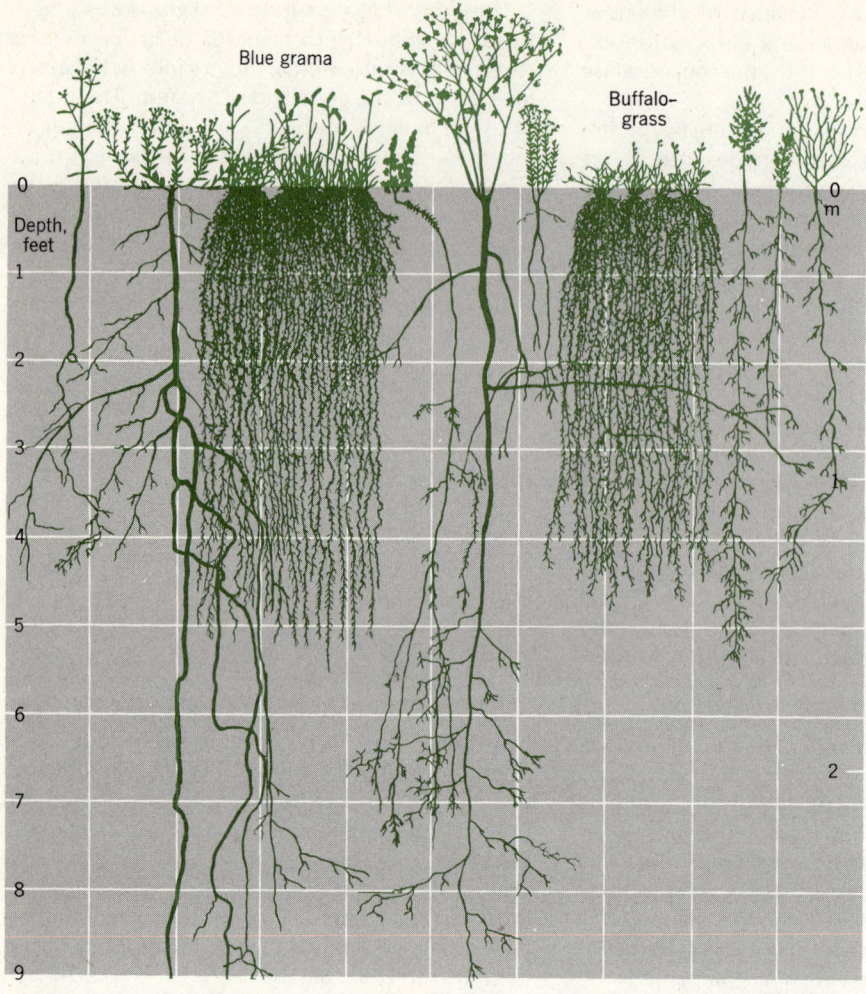

Figure 24.14 Drawings of roots of grasses (blue grama and buffalograss) and several kinds of forbs at Hays, Kansas. Roots at left extended to depths below 11 ft (3½ m). (After J. E. Weaver and F. W. Albertson, 1943, *Ecological Monographs*, vol. 13, p. 100. By permission of the Duke University Press.)

Figure 24.15 Short grass prairie in Montana. The cattle are grazing crested wheatgrass, *Agropyron cristatum,* which has been introduced from Russia into the northern Great Plains. (Courtesy of the U.S. Forest Service.)

The Grassland Biome

mediate area with mounds and undermining it with burrows (Figure 24.16). Actually, the activities of the prairie dog seem to have been beneficial to the short-grass prairie before the arrival of European Man. Selective feeding by prairie dogs on shrubs and annual plants allows perennial grasses and forbs to increase. The diggings are important as sites of colonization for species preferring disturbed soils. In short, the activities of the prairie dog favored the development of a diverse, heterogeneous short-grass prairie. The animal increased its range during the 1930s, both in response to the increase in area of the short-grass prairie during the drought years, and in response to the near extinction of its most successful predator, the black-footed ferret. After the drought, the prairie dog was able to maintain some areas in short-grass prairie where more desirable mixed-grass and in some cases tall-grass prairie had existed before. Thus, the prairie dog has been often associated with degrading range conditions.

Energy flow and nutrient cycling in a grassland ecosystem proceeds through grazing and detrital

Figure 24.16 A family of prairie dogs. Note the large mound which they have constructed. (Leonard Lee Rue III/National Audubon Society.)

food chains. Figure 24.17 is a diagram of energy flow in a grassland ecosystem. Both the grazing and detrital food chains are well developed. One important feature of the grasslands ecosystem is the energy link between the two chains which

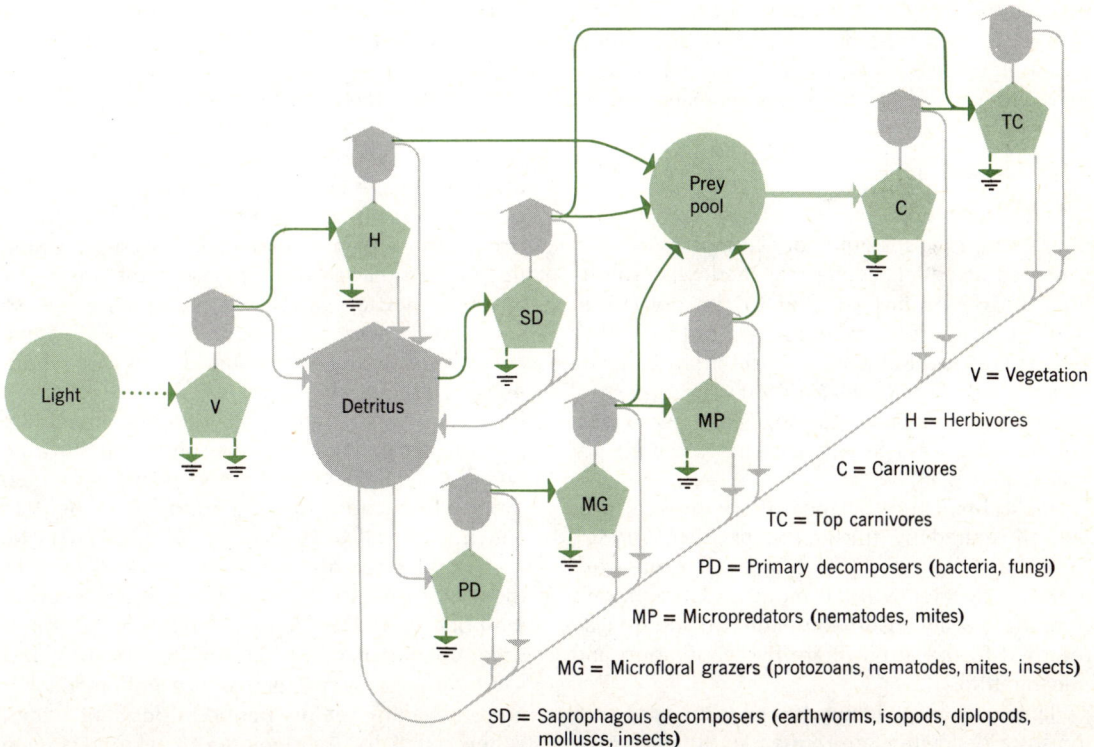

Figure 24.17 Energy flow diagram for a grasslands ecosystem. The herbivores, supported by grazing, together with the saprophagous decomposers, microfloral grazers, and micropredators, supported by detrital energy, compose a prey pool for the carnivores and top carnivores. (After O. H. Paris, 1970, in *The Grassland Ecosystem: A Preliminary Synthesis*, appearing as an erratum in Range Science Department Science Series No. 2, Supplement, Colorado State University, Fort Collins, Colorado. p. 2, Figure 3.)

is produced by predation of detritivores by carnivores. This link permits the higher organisms in the grazing chain to be supported alternatively by biomass within the detritus food chain, making the ecosystem more stable.

Man's utilization of the grasslands has been extensive. Naturally adapted to the growing of grasses, these areas have become great producers of domesticated grasses, or *grains*, which provide so much food for Man as well as pasture for domesticated grazing animals. By and large, the natural grassland ecosystem has been replaced by a simpler one which is more desirable to Man. Energy subsidies represented by fertilizers, pesticides, and machines have produced large increases in productivity in cultivated areas. Irrigation by stream diversions and mining of ground water have also helped increase productivity by reducing the inroads of drought. At the same time, however, productivity has declined in some areas under poor management. In almost all areas, alteration of the original ecosystems has been so thorough that knowledge of them can be obtained only through studying the few virgin or semivirgin tracts remaining. Because marginal grasslands will be opened in the next few decades throughout the world for further grazing and cultivation to help feed our expanding population, a thorough understanding of the natural ecosystems will be necessary for their wise management.

The Desert Biome

Where precipitation is limited by such atmospheric factors as subsiding dry air masses, or such topographic effects as the rain shadow, a unique flora and fauna has adapted to the dry conditions—the *desert biome*. Only two of the world's deserts are essentially rainless, the interior Sahara and the northern Chilean desert. The others all experience some precipitation, usually seasonally, but almost always less than 10 in. (250 mm) per year (Figure 5.44).

The soil-moisture balance of the tropical deserts, illustrated by the graph for Alice Springs, Australia (Figure 12.13C) is one of severe soil-moisture deficiency in all months. Sporadic falls of rain are easily absorbed by the soil and quickly returned to the atmosphere by evaporation and transpiration.

Plants which are adapted to conditions of soil-moisture deficiency are termed *xerophytes*. One method by which xerophytes conserve water is reduction of transpiration. For example, some xerophytes produce new leaves only after a rain; but when soil moisture is exhausted, the leaves are shed, and the plant enters a state of dormancy. Other xerophytes have leaves which are heavily cutinized—meaning that they are covered with a thick, waxy layer of *cutin*, which retards water loss. The shiny cutin also acts to reflect sunlight and thus to lower leaf temperatures and reduce transpiration. These adaptations do not necessarily retard transpiration when water is available; in fact, maximum transpiration rates for xerophytes are often greater than those of *mesophytes*, plants adapted to abundant supplies of water. Another xerophytic adaptation is the small leaf. Although the reduced leaf size is usually compensated by an increase in the number of leaves, the small leaf has a more efficient vascular system, which aids in supplying water to the cells in time of drought.

The cacti of the desert biome are examples of *succulents*, xerophytes which store large quantities of water within stem tissues. Succulent cacti have low transpiration rates at times of soil water deficiency because their thickened stems have a lower surface area per unit of cellular volume, and because their stomata remain closed during the day, opening only at night. Cacti and many other xerophytes have very extensive, very shallow root systems which allow them to utilize the water from the briefest of showers or from the condensation of dew on the soil surface on clear, still nights.

Still other mechanisms are used by plants to cope with the drought periods of the desert. Some plants simply avoid the problem by having a very short life cycle, spending the drought period as dormant seeds. When a rare, soaking rain occurs, these plants rapidly germinate, grow, and flower, setting seed before soil moisture is exhausted. Because these annual plants are so short-lived, they are referred to as *ephemeral annuals*. Although their life cycles are short, ephemeral annuals do not usually possess other physiological adaptations to drought, hence are not true xerophytes. Also in a special category are *phreatophytes*, perennial desert plants with deep roots which draw upon ground water held in alluvium of valley floors.

To cope with shortage of water, *xeric animals*, those adapted to dry conditions, have evolved methods which are somewhat similar to those used by the plants. Many of the invertebrates exhibit the same pattern as the ephemeral annuals—evading the dry period in dormant stages. When rain falls, they emerge to take advantage of the new and short-lived vegetation. For example, the tiny brine shrimp of the Great Basin may wait many years in dormancy until normally dry lake beds fill with water, an event that occurs perhaps three or four times a century. The shrimp

The Desert Biome

Figure 24.18 Sagebrush desert, near the Vermilion Cliffs, Kanab, Utah. (Photograph of 1906 by Douglas Johnson.)

then emerge and complete their life cycles before the lake evaporates. Spadefoot toads exhibit similar adaptations, emerging from burrows during a rainy period and breeding in the nearest puddle. The tadpoles quickly mature, and by the time the puddle dries they are capable of digging their own burrows to spend the long dry period. Many species of birds regulate their behavior to nest only when the rains occur, the time of most abundant food for their offspring.

The mammals are by nature poorly adapted to desert environments, yet many survive there, employing a variety of mechanisms to avoid water loss. Just as plants reduce transpiration to conserve water, so many desert mammals do not sweat through skin glands; they rely instead on other methods of cooling. For example, the huge ears of the jackrabbit serve as efficient radiators of heat to the sky. Calculations have shown that a jackrabbit sitting in a shaded depression could dispose of about one-third of his metabolic heat load through longwave radiation from his uninsulated ears to the clear desert sky above. Many of the desert mammals conserve water by excreting highly concentrated urine and relatively dry feces. In addition, many are able to use to their advantage metabolic water, formed by the normal oxidation of carbohydrate. The desert mammals also evade the heat by nocturnal activity. In this respect, they are joined by most of the rest of the desert fauna, spending their days in cool burrows in the soil and their nights foraging for food.

The North American deserts are usually divided into two groups: the northern, or cool deserts of the Great Basin, which experience cold winter temperatures, and the southern, or hot deserts (Mohave, Sonoran, and Chihuahuan) ranging from southern California and northwestern Mexico into west Texas, in which winter temperatures remain warm. The biota of the northern deserts is simpler and less diverse than that of the southern deserts. Sagebrush is the dominant plant over well-drained upland and alluvial fan surfaces; it is often accompanied by such other tough woody plants as greasewood and blackbrush (Figure 24.18). The southern, hot deserts possess a much greater variety of species (Figure 24.19). Here are many succulents, including the giant saguaro cactus.

Because of their high annual insolation, deserts can be extremely productive environments when fertilized and supplied with water by irrigation. Unfortunately, irrigation of the desert eventually leads to salinization of the soil in many areas, as we have seen in Chapter 13. The accumulating salts cannot be washed out of the soil without use of even greater amounts of water, producing an upward spiralling cycle of water use that must eventually lead to collapse of the hydraulic system. Ruins of ancient irrigation systems are scattered throughout the Old World deserts of the Fertile Crescent and Indus valley. Whether these lands lost productivity by salinization or by changing climate at their irrigation sources remains undetermined.

We have found in the analysis of water use (Chapter 13), that irrigation water consumption represents a disproportionately high figure compared to industrial and public uses (Figure 13.30). In view of the possibility of eventual failure of irrigation systems by salinization and waterlogging, proposals to increase water diversions to supply desert irrigation are undergoing serious questioning.

Figure 24.19 Vegetation of the Sonoran Desert in southwestern Arizona. Note the different aspect of this southern habitat from that of the more northern habitat of Figure 24.18. (Photograph by A. N. Strahler.)

Other Biotic Communities

In addition to the biomes we have already discussed, there are other biotic community types which are sometimes considered biomes. Included are the monsoon forest, the temperate rainforest, and the sclerophyll forest.

The *monsoon forest* is typical of tropical zones in southeast Asia and Africa where the monsoon climate produces a marked wet and dry seasonality. The trees shed their leaves at the start of the dry season, giving the forest the aspect of the deciduous forest biome in winter.

The *temperate rainforest* is an evergreen forest which occurs in subtropical and middle-latitude zones and at high altitudes in the equatorial zones, where rainfall is abundant the year around. As in the tropical rainforest, important plant species are broad-leaved; they include evergreen oak, magnolia, and laurel. This rainforest is the climax type for the southern coastal plain of the United States, although the dry sandy soils and the frequent fires hold much of this area in pine forest. Southeastern China also possesses significant areas of temperate rainforest.

Sclerophyll, or *hard-leaved scrub*, is typical of the Mediterranean soil-moisture regime (Figure 12.14 D). The climate is one of annual droughts in the season of high sun (summer). Sclerophyll scrub consists of short and scrubby trees and bushes with tough evergreen leaves; the *chaparral* vegetation of the southern and central California foothills is an example of sclerophyll scrub in which the trees rarely grow larger than tall shrubs and plant growth is very dense (Figures 24.20 and 24.21).

Figure 24.20 Chaparral at the San Dimas Experimental Forest, near Glendora, California. The old-growth chaparral in this photo has not burned for many years. (Photograph by A. N. Strahler.)

Energy Systems of the Biosphere

Figure 24.21 A closer look at chaparral in the San Dimas Experimental Forest (see Figure 24.20). The tough, woody, evergreen shrubs can form thickets which are impenetrable in places. Figure 22.15 shows another view of the California chaparral. (Photograph by A. N. Strahler.)

Energy Systems of the Biosphere

With this chapter, we complete our survey of the earth's ecosystems under the title of "Energy Systems of the Biosphere." Although we have dealt with many specific environmental problems throughout our text, we have not yet had a chance to put these problems into the broader perspectives of Man's attitudes and social structures. Whereas this task is probably more within the realm of social sciences, economics, and psychology, the Epilogue that follows attempts to provide at least a brief discussion of the issues involved in planning for the maintenance of environmental quality, as well as a final look at the future prospects for Man's existence on Planet Earth.

Epilogue: Environmental Perspectives

In previous chapters of this book, we have laid the scientific foundation for a general understanding of environmental problems. Many of the physical problems we analyzed seem upon first encounter to have fairly simple physical solutions. For example, one way to stop sulfur oxide air pollution is simply to stop burning fossil fuels; to cut water pollution levels, all we need to do is seal off the outfall pipes of sewage treatment plants. Yet both these actions have detrimental effects—without access to fossil fuels, our power and transportation industries would shut down, and without functional sewage treatment plants, disease would spread. These drastic remedial actions would thus produce a net loss in the total quality of our environment, rather than the gain intended. Implementation of such patently desirable environmental management procedures therefore can lead to complex reactions which arise from the structure and institutions of our society. These institutional reactions are simultaneous, and moreover, they interact.

One institution is economic. Free enterprise depends upon the reality of profit-making. Almost all measures necessary to reduce levels of pollution and intensities of environmental degradation are costly. In the long run these costs must be met; they will ultimately be absorbed by the consumer through increased prices of goods and services, or through increased taxes, or both. Small wonder that individuals as well as corporations are reluctant to take action!

Another institution of society is public law. Environmental improvements require new laws. Once placed on the books, laws must be enforced. Enforcement leads to conflict with freedom of enterprise. Obviously, effective environmental management involves some added constraints to the freedom of individuals and organizations to do as they please. Conflicts are compounded and confused by interaction of laws and law enforcement on at least three governmental levels: federal, state, and local.

Another institution is cultural. It encompasses long-standing beliefs and practices related to religious and ethnic backgrounds. Population control provides an example. Some persons believe that society as a whole has the right to limit its numbers by controlling family size. Still others feel that the family unit should make the choice for itself. Even among those who agree that some form of control is necessary, methods are still disputed—for example, some oppose abortion; others advocate it.

Epilogue

The cultural institution imposes its problems in many other ways, some of them quite subtle. Life-styles differ. Attitudes about environmental standards differ. Urban groups may have a different outlook upon environmental priorities than conservationist groups. Which problem should be given first priority? Air pollution may rank high in the judgment of an inner-city dweller, while at the same time, he may show little concern as to where his sewage is discharged and in what condition. The conservationist may operate in a region where air pollution is a secondary problem; instead, water pollution from an urban area may be foremost in his mind.

The step from environmental science, which analyzes problems in terms of cause and effect, to environmental management, which seeks viable solutions to problems, is a big step. Your understanding of environmental processes and natural resources will come to nothing if you fail to follow up with an inquiry into methods of environmental and resource management. We strongly recommend that you look about in your available curriculum for a course or program that deals with these areas.

Further study of environmental management will lead to insights into action programs based upon multiple options. For example, you may find that there are nonstructural approaches to meeting demands for more water, more fuels, and more metals. There may be viable alternatives to consumptive use of materials through such procedures as substitution and recycling. A further study of urban air pollution will doubtless lead you into the area of public transportation. What alternatives are there to imposing severe controls on pollutant emissions of the passenger car? These and many, many other questions arise at the level of environmental and resource management. Further study is mandatory if you, as a citizen and a voter, are to prepare yourself adequately for the decision-making process.

And now, to stimulate your thinking and enlarge your horizons, we offer a variety of questions, assertions, and opinions that relate to certain broad aspects of Man's place in the environment. Judge them on their merits as you see fit. Among the basic questions asked will be these: Do we face an environmental crisis? How did our environmental problems arise? Do we face a shortage in supplies of materials and energy? What is our present status with respect to these problems? What will the future bring for mankind?

Is There an Environmental Crisis?

Many voices have been raised in alarm over the degraded state of the environment and threats to the well-being of ecosystems. Is mankind really threatened by the consequences of his own action? Degrees of alarm vary substantially. Which voice do we heed?

Throughout innumerable recent publications we find statements referring to a crisis or to multiple crises arising to plague our planet. Often the authors are not explicit in giving any clear picture of what constitutes a crisis, what part of the biosphere will be affected, and whether the crisis exists now, or will arise in the future. One basic form of crisis deals with shortages of food or mineral resources, or both, in the face of a rising global population of humans. Another form of crisis deals with the unfavorable impact of pollution—broadly defined to include all forms of environmental deterioration—upon the biosphere. In other

Is There an Environmental Crisis?

words, crises can take the form of biospheric destruction by such processes as poisoning by toxic substances, genetic damage from ionizing radiation, reduction of nutrient substances from the primary producers, and physical destruction of ecosystems and their habitats by release of mechanical or heat energy. A third crisis, in a class by itself, is catastrophic global disruption by use of nuclear weapons.

If we define an environmental crisis as a situation demanding immediate corrective action, if such be possible, to avert deterioration, damage, or destruction of the affected system, we can point to crises on various scales. Some crises will be current and local. A simple example would be a city beset by worsening smog and a static weather situation. Another would be a case of rising chlorinity in public water-supply wells of a coastal community. Signs of eutrophication of a lake signal an approaching crisis in the environment of an aquatic ecosystem.

Perhaps what most environmentalists mean when they refer to an environmental crisis of global proportions is the total situation reached at that point in time when the global demand for vital forms of energy or matter catches up with the available supply. Energy, in this sense, can take the form of food for organisms, or heat and power for industrial uses. Matter can also take the form of food for organisms, or it may consist of essential materials for industrial uses.

A global crisis involving energy or matter, or both, is inherent in one basic fact that few will dispute: the earth's resources of energy and matter are finite; whereas the demand for these resources is rising, with no limit in sight except the total finite resource itself.

In commenting upon the "space-ship" analogy of our predicament, Preston Cloud, a distinguished biogeologist, has put it this way:

> It helps in trying to consider this question with a degree of detachment, to recall the now-familiar view of earth from space—a fragile speck in a vast emptiness. There can no longer be any doubt of the finite and limited nature of our planet—or, for that matter, about the impracticability of colonizing or importing resources from other planets in the solar system or elsewhere.... We live on an evolving, not a static earth. Man, a force of geologic magnitude, has created and now inhabits a man-altered environment of global dimensions. There is no Eden to which we can return. No alternative exists except to manage our environment, our resources, and ourselves with greater awareness both of the constraints and of the possible consequences of misjudging them.[1]

To sum it up, the environmental crisis, of which so many voices warn us, is an event of many dimensions. Crisis presents itself on an areal scale ranging from local to global, and on a time scale ranging from the present to an indeterminate future point in time. The threat of crisis ranges in magnitude from deterioration of small ecosystems to total destruction of the biosphere.

While population growth of the human species has not been a topic of environmental science as we have presented it in this book, that growth is a vital ingredient in a global crisis, since greater population means both greater demand for usable energy and matter and a greater potential to pollute and degrade the environment. Therefore, we turn next to a brief review of some current views on population growth in relation to environment.

Epilogue

Population, Food Resources, and Environment

The current debate over population growth and its consequences can scarcely have escaped your attention. "Zero Population Growth" has joined the long list of bumper-sticker slogans of a growing ecology-conscious group of citizenry. Perhaps in no other debate on great issues of mankind has so much emotionalism been injected by so many participants. We have rarely witnessed such bitterness as is expressed by those whose positions lie at the two extremes of the debate. On the one hand there are some who claim we can sustain an almost indefinite rise in global population, since our ingenuity and technology will rise to meet any challenge. After all, they say, are there not enormous untapped food resources in the sea? And are we not experiencing a "green revolution" in crop production? At the other extreme are those who warn of wholesale destruction of human life by starvation in several of the developing nations within a few decades, despite the green revolution. They have even gone so far as to warn that once mass starvation sets in, it will be futile for food-rich industrial nations even to attempt to provide relief supplies.

The phenomenal rise in total world population, together with a sharp increase in the rate of that rise, is a fact on which there can be no debate. When population is plotted against time on arithmetic scales the rising curve steepens so sharply that it becomes almost perpendicular in the time interval of the past half-century. In contrast, that part of the curve representing the first millenium A.D. appears almost flat, so slow is the increase. Consider that during the entire course of human history, up to the year 1850, human population had reached only one billion persons. The present world population is about 3.6 billion. Currently the rate of increase is such that a doubling of the population can take place in about thirty-five years. One recent summary of population trends states that, allowing for some decline in birth rates, the world population for the year 2000 will be 6.5 to 7.0 billion.[2] This same source goes on to say:

> Failing that reduction we may have 7.5 billion. A continued decline in death rates and progress toward a net reproduction rate of one* by the year 2040 (an optimistic assumption) would still yield a population leveling off at 15 billion late in the next century. Thus, considering demographic factors alone and leaving aside questions of resource adequacy or environmental tolerances, our best prospect would be for a stationary world population of between 15 and 20 billions. . . . The best estimate of demographers is that rapid growth is sure to continue for some time, probably more than a hundred years, even on optimistic assumptions about controlling fertility rates.[2]

Predictions relating to leveling off of population growth in developing nations presently experiencing rapid growth are subject to considerable disagreement. Demographers recognize a principle to the effect that, as a nation becomes heavily industrialized, the birth rate falls to a relatively low level, as compared with that during the preceding agrarian period. This change is referred to as the *demographic transition*. It was observed in the case of certain industrial regions, and took place largely prior to 1940. The more optimistic debaters of the population issue predict that as each developing nation arrives at an advanced industrial condition it will experience a demographic transition. In this way, it is hoped, the world population growth can be stabilized, or at worst, held to a manageable

---

*Synonymous with *zero population growth*.

rate. But even the demographic transition does not alone produce a balanced population except when there is also depletion by emigration. The best that has happened is a reduction to a doubling time of about 70 years. Consequently, the more pessimistic debaters feel it would be a tragic mistake to count upon the arrival of the demographic transition as a natural means of preventing a runaway increase.

On the subject of the ability of the earth to furnish enough food to sustain a much larger population, Professor Cloud whom we quoted in an earlier paragraph, has stated:

> The Committee on Resources and Man of the National Academy of Sciences, which I chaired, has examined the ultimate capability of an efficiently managed world to produce food, given the cultivation of all potentially arable lands and an optimal expression of scientific and technological innovation. Its results imply that 1968 world food supplies might eventually be increased by as much as nine times, provided that sources of protein were essentially restricted to plants, and to seafood mostly from a position lower in the marine food chain than customarily harvested and provided metal resources, agricultural water, and risky pesticides and mineral fertilizers are equal to the task and do not create intolerable side effects, and that agricultural land is not unduly preempted for other purposes. Such an ultimate level of productivity might, therefore, sustain a world population of 30 billion. Most of these, of course, would live at a level of chronic malnutrition unless distribution were so regulated as to be truly equitable, in which case all would live at a bare subsistence level, although with protein deficiency. . . .
>
> That seems to place a maximal limit on world populations, at a figure that could be reached by little more than a century from now at present rates of increase. Of course, many students of the problem do not believe that it is possible to sustain 30 billion people—especially considering the variety and extent of demands on water, air, and earth materials, and the psychic stresses that would arise with such numbers. I have no dispute with them. Here I seek simply to establish some theoretical outside limit, based on optimistic assumptions about what might be possible. More important attributes than a starvation diet would be sacrificed by a world that full![1]

As we noted earlier, an increasing world population places a heavy strain upon the planet in two ways. First, more people cause more pollution and more environmental degradation. Second, more people require more of the earth's resources besides food (for example water supplies and plant fiber); they place heavier demands upon the nonrenewable mineral resources we have discussed at length in earlier chapters. So, we will turn next to some considerations of the environmental impact of a greatly increased population.

Population, Affluence, and Environmental Impact

The relationship between population and intensity of environmental degradation has been expressed in a simple equation involving three variables. Environmental impact, which can refer to a change in intensity

Epilogue

of a given form of pollution or degradation, is represented by the symbol I, in the following equation:

$$I = P \times A \times T$$

where P is a population change factor,
A is an affluence change factor,
and T is a technology change factor.

Change factor is a ratio of increase or decrease of the variable within a specified period of observation. All terms are therefore dimensionless. For example, I might represent the change in pollution intensity attributable to emission of lead particulates from automobile exhausts. Population change factor, P, is the ratio of population change, usually an increase. For example, a 40% increase in population yields a ratio of 1.4. The affluence change factor, A, represents the ratio in per capita change of the causative agent or product responsible for the impact. For example, in the case of air pollution from auto emissions, affluence might be defined as the number of gallons of gasoline consumed per capita. If the per capita use of gasoline has increased 100%, the affluence ratio is 2.0 (a doubling factor). The technology change factor, T, describes the ratio of increase or decrease in the impacting agent. In our example, A might be an 80% reduction in lead content of the gasoline from the beginning to the end of the observation period. Then the value of T would be expressed as 0.2 (one-fifth as much lead as before). The values of the three variables in the equation are then

$$I = 1.4 \times 2.0 \times 0.2$$
$$I = 0.56$$

In the hypothetical illustration, there has been an overall reduction in environmental impact. The impact has been approximately halved, despite increases in both population and affluence. (In this case, 2.8 times as much gasoline is being consumed, but the impact of each gallon is only one-fifth of what it initially was.)

The above method of evaluating changes in environmental impact has serious limitations in application. Quantitative comparison of impacts of different agents is impossible by means of dimensionless ratios. Perhaps, however, the equation has value in focusing attention on three independent causes of impact change.

The concept of affluence is important. The equation suggests the possibility that, even if we should attain zero population growth (ratio of 1.0), we could greatly increase our environmental impacts through increased affluence alone. Technological changes can play a vital role, quite independently of both population and affluence. The change from returnable bottles to no-return bottles would represent an unfavorable increase in the technology factor. The environmental impact, measured in weight of glass added to waste disposal facilities, might thus be greatly increased without increases in per capita consumption of beverages. Recycling of no-return bottles might, on the other hand, tend to reverse the direction of change of impact.

It is interesting to note that, despite their enormous populations, the larger developing nations have very low levels of affluence as compared with the industrial nations. Consequently, the total global environmental impact of these nations is small in many categories. Pollutant sulfur oxides, as we learned in Chapter 13, are contributed largely by the industrialized nations, because their great affluence requires huge expenditures of energy to support.

Pollution and Economics

Pollution of the environment can be viewed in a somewhat different perspective than that of biological impact upon ecosystems. There is an economic aspect to consider. The economist views pollutant wastes as *residuals*. When Man produces and consumes the wide variety of things he has become accustomed to having, there is finally left something that has no economic value on the existing market. In other words, unwanted matter and energy are generated at the terminus of the economic chain. These unwanted residuals must be disposed of, and of course, they have always been dumped upon the environment. It has been assumed that air, water, and soil are in the public domain and that these media of the environment can absorb free of charge the unwanted residuals. Writing on the subject of economics and the environment, a group of researchers with Resources for the Future (a nonprofit corporation) has analyzed the problem of residuals and the environment as follows:

> Water and air are traditionally examples of free goods in economics. But in reality in developed economies they are common property resources of great and increasing value, which present society with important and difficult allocation problems that exchange in private markets cannot solve. These problems loom larger as increased population and industrial production put more pressure on the environment's ability to dilute, chemically degrade, and simply accumulate residuals from production and consumption processes. Only the crudest estimates of present external costs associated with residuals discharge exist, but it would not be surprising if these costs were already in the tens of billions of dollars annually. Moreover, as we shall emphasize again, technological means for processing or purifying one or another type of residuals do not destroy the residuals but only alter their form. Thus, given the level, patterns, and technology of production and consumption, recycle of materials into productive uses or discharge into an alternative medium are the only general operations for protecting a particular environmental medium such as water. Residual problems must be seen in a broad regional or economy-wide context rather than as separate and isolated problems of disposal of gaseous, liquid, solid, and energy waste products...
>
> Yet we persist in referring to the "final consumption" of goods as though material objects such as fuels, materials, and finished goods somehow disappear into a void—a practice which was comparatively harmless only so long as air and water were almost literally "free goods." Of course, residuals from both the production and consumption processes remain, and they usually render disservices (like killing fish, increasing the difficulty of water treatment, reducing public health, soiling and deteriorating buildings, etc.) rather than services. These disservices flow to consumers and producers whether they want them or not, and except in unusual cases they cannot control them by engaging in individual exchanges.[3]

These writers propose that economic theory be revised to include the environmental media as part of the economic system. They call for a *materials balance approach* in which costs of all raw materials are traced from the time of their withdrawal from the environment to their return to that environment. They point out that residual materials do not necessarily have to be put back into the environment, where they incur costs through impact as pollutants. Instead, recycling may prove feasible and even profitable. The same can be said of some fraction of the residual energy, which may also be a source of public cost when thermal pollution results.

Historical Roots of Pollution

It is interesting to speculate on basic reasons why the industrial nations have, despite all of the technology and energy sources available to them, allowed themselves to drift into the role of environmental polluters on such a massive scale. Not all forms of environmental degradation are the exclusive burden of modern society. A glance back into the past shows that agrarian cultures created significant environmental deterioration in the form of soil erosion through deforestation and overgrazing. These activities are thought to have greatly modified the Mediterranean landscape and that of a part of northern China. Hydraulic civilizations of the Fertile Crescent evidently got into major difficulty through onset of salinization of soils and silting of irrigation works. Nevertheless, air and water pollution on a massive scale seem to have been concommitant emergents with the industrial revolution and the development of vast energy sources through combustion of fossil fuels.

Western European Man pioneered in this industrialization of society. Was there some factor in his culture that encouraged or tacitly permitted him to degrade the environment and consume nonrenewable resources with abandon? Were several factors responsible? Perhaps if we can isolate one or more underlying historical factors, we will be better guided in the solution of our problems.

An interesting hypothesis has been put forward to the effect that doctrines peculiar to Western European Christianity have acted in a permissive way to allow the drift into industrialization and its deleterious side effects. The Judeo-Christian tradition, it is argued, has tended to separate Man from a lower, subhuman world of life. Special and absolute values have been assigned to Man, as well as special responsibilities and privileges. Commenting on what he calls "Western Christianity's predatory legacy," Roger L. Shinn writes:

> Cultural historians have frequently pointed to a relationship between biblical faith and technological progress. Perhaps it is not sheer accident that technology has developed in civilizations influenced by the Hebrew-Christian Scriptures. The Bible in its radical monotheism desacrilizes nature. Sun, moon and stars are no longer divine and they may not be worshipped. Brooks and trees are no longer inhabited by spirits. There is one God; the world is his creation. Man, given dominion, may investigate and appropriate the objects of nature. No taboos, no forbidding mysteries, no divinities block enterprising man.[4]

The hypothesis that religious influences have been permissive in our predatory attitudes toward the environment and its resources has been given short shrift by those who feel that they can recognize other, more realistic controlling factors. One area of interest is the Industrial Revolution which began in England in the late 1700s.

The role of growing affluence accompanying industrialization as a cause of pollution is discussed by Lewis W. Moncrief in these words:

> With this revolution the productive capacity of each worker was amplified by several times his potential prior to the revolution. It also became feasible to produce goods that were not previously producible on a commercial scale.

Later, with the integration of the democratic and the technological ideals, the increased wealth began to be distributed more equitably among the population. In addition, as the capital to land ratio increased in the production process and the demand grew for labor to work in the factories, large populations from the agrarian hinterlands began to concentrate in the emerging industrial cities. The stage was set for the development of the conditions that now exist in the Western world.

With growing affluence for an increasingly large segment of the population, there generally develops an increased demand for goods and services. The usual by-product of this affluence is waste from both the production and consumption processes. The disposal of that waste is further complicated by the high concentration of heavy waste producers in urban areas. Under these conditions the maxim that "Dilution is the solution to pollution" does not withstand the test of time, because the volume of such wastes is greater than the system can absorb and purify through natural means. With increasing population, increasing production, increasing urban concentrations, and increasing real median incomes for well over a hundred years, it is not surprising that our environment has taken a terrible beating in absorbing our filth and refuse.[5]

Moncrief goes on to discuss additional factors operating in America to encourage environmental degradation and careless consumption of resources. He points out that the national policy was, from the outset, to convey ownership of land and other natural resources to the hands of the citizenry. The effect of this policy, Moncrief notes, "explains how decisions that ultimately degrade the environment are made not only by corporation boards and city engineers but by millions of owners of our natural resources."[5]

Another perhaps peculiarly American factor in leading us to our present outlook upon environment and resources has been seen in frontierism accompanying territorial expansion as explorers, traders, and settlers in turn spread across the country. Of this development, Moncrief states:

As the nation began to expand westward, the settlers faced many obstacles, including a primitive transportation system, hostile Indians, and the absence of physical and social security. To many frontiersmen, particularly small farmers, many of the natural resources that are now highly valued were originally perceived more as obstacles than as assets. Forests needed to be cleared to permit farming. Marshes needed to be drained. Rivers needed to be controlled. Wildlife often represented a competitive threat in addition to being a source of food. Sod was considered a nuisance—to be burned, plowed, or otherwise destroyed to permit "desirable" use of the land.[5]

Attitudes developed in this period of settlement may well have contributed to the continued degradation of the environment. Moncrief points out that the first important steps on a national basis toward conservation of natural resources were taken after settlement of the frontier was complete, about 1890, and were largely led by Theodore Roosevelt and Gifford Pinchot.

In analyzing the present American scene with respect to environmental degradation, Moncrief sees three dominant features that make it difficult to create effective programs of environmental management: In his own words these are: "an absence of personal moral direction concerning our treatment of natural resources; an inability on the part of our social institutions to make adjustments to this stress, and an abiding faith in technology."[5] He notes further that "there appears to be an almost universal tendency to maximize self-interests and a widespread willingness to shift production costs to society to promote individual ends".[5]

One may wonder if conditions are any better today under a socialistic regime than under a system of private ownership of industry and free enterprise. This thought invites a look at the situation in the Soviet Union. Evidently, that nation is suffering from many of the same forms of pollution as are found in North America, Western Europe, and Japan. Marshall I. Goldman, a student of this aspect of the U.S.S.R., reports that "in some ways, state ownership of the country's productive resources may actually exacerbate rather than ameliorate the situation."[6] Air and water pollution are widespread in the Soviet Union. Evidently the same intense competition to achieve production goals and reduce costs operates in the Soviet economy as in that of the United States. In the U.S.S.R. it has proved difficult to enforce antipollution regulations. One point in which the Soviet system is weaker than ours is that there is no way in the U.S.S.R. by which citizens, through voting or direct pressure, can bring about new environmental legislation. Goldman comments: "The Russians, too, have been unable to adjust their accounting system so that each enterprise pays not only its direct costs of production for labor, raw materials, and equipment but also its social costs of production arising from such by-products as dirty air and water."[6]

With this short inquiry into historical perspectives of the environmental degradation we turn to the subject of current legislation and governmental structures set up to deal with problems of the environment.

Legislation for Environmental Protection

With respect to environmental problems, the 1960s have been characterized as a decade of transition—transition in outlook and attitudes, as a period of preparation—preparation of new tools, concepts, and procedures, and as a period of promise—promise of a time in which Man would utilize his capacity to manage the environment wisely.[7]

The 1970s may prove to be characterized as the first decade of environmental action. Most certainly, it began with action—sweeping legislation on a national scope to secure protective measures long overdue. First came the passage by Congress in 1970, of the National Environmental Policy Act (NEPA). One provision of NEPA was to establish the Council on Environmental Quality (CEQ), a group working within the Office of the President of the United States. CEQ has concerned itself with underlying causes of environmental problems and with the best methods of solving them. CEQ is also responsible for coordinating all federal environmental programs.[8]

The Environmental Protection Agency (EPA) was established to administer and enforce national pollution control laws. EPA has brought together under unified direction programs dealing with air and water pollution, solid waste management, noise abatement, pesticide regulation, and radiation standard-setting.

NEPA marked a major step forward in environmental planning and environmental protection by requiring that federal agencies take environmental factors into full account in all their planning and decision-making. The law requires agencies to submit an *environmental impact statement* of each major decision, along with alternatives to that decision, and to make these assessments public. This process has fostered a wide range of basic reforms in the way federal agencies make their decisions.

One of the first major environmental decisions based upon impact assessment was a presidential order to halt further construction on the Cross Florida Barge Canal, despite the fact that some $50 million had already been spent on the project. In this instance, the president acted on the advice of CEQ that the environmental damage which would result from its completion would outweigh its potential economic benefits.

NEPA has given a new dimension to citizen participation and citizen rights, as is evidenced by the numerous court actions through which individuals and groups have made their voices heard. Of particular interest was the initial impact statement produced by the Department of the Interior on the proposed Trans-Alaska pipeline. Although the report was voluminous in the extreme, a federal judge, acting upon the complaint of environmentalist groups, ruled that the statement was perfunctory, not fulfilling the requirements of NEPA. This incident demonstrates how citizen participation can be effective. Two years later, the Department of Interior issued a revised impact statement. This, too, was attacked by citizen groups on the grounds that it failed to take into account important possibilities for environmental damage from oil spills. However, the Interior Department approved the pipeline plan over these objections.

One of the first activities of the EPA following its establishment was a major move to implement the Clean Air Act, which had previously been passed. The Clean Air Amendments of 1970 required EPA to establish national air quality standards as well as national standards for significant new pollution sources and for all facilities emitting hazardous substances (see Table 6.4).

In October 1972 Congress passed a new Federal Water Pollution Control Act, prohibiting the discharge of pollutants into rivers, lakes, and wells without a permit. The new law places responsibility for issuance of such permits in the hands of the states. The state permits must meet requirements set by the federal law and must be approved by the EPA. Under no circumstances can highly radioactive wastes or agents of radiological, chemical, or biological warfare be discharged. Moreover, a permit cannot be issued if the Corps of Engineers rules that the proposed discharge would impede navigation. Permits must specify what substances can be discharged and in what amounts. After the permit has been granted the flow and content of pollutant discharges must be monitored by the polluter, when flow averages in excess of 50,000 gallons per day.

Other new legislation deals with ocean dumping. This legislation prohibits dumping any wastes originating in the United States into estuaries, the Great Lakes, coastal waters, or the oceans, without an EPA permit.

Recycling has been encouraged by the Resource Recovery Act of 1970. Under authority of the EPA, this act authorizes funds for demonstration grants for recycling systems and for studies of methods to encourage resource recovery. Many other environmental subjects are under study and regulation by the EPA, but these are too numerous to list here. It should not be forgotten that most states have legislation and agencies dealing with environmental problems.

Resources and Their Management

Some basic facts and principles of occurrence and use of nonrenewable earth resources were covered in Chapter 10. Emphasis was placed on the fact that mineral resources—including the mineral fuels—represent the results of accumulated geological activity of hundreds of millions of years. Consumption of these resources has skyrocketed in the past century, just as the world population itself has shot precipitiously upward. Let us examine the consequences of this prodigious consumption of resources which can never be renewed in the lifetime of Man on earth.

A number of authorities in a good position to know the facts have attempted to appraise the public of what the future holds in store. Walter R. Hibbard, Jr., formerly Director of the U.S. Bureau of Mines, puts it this way:

> A requisite for affluence, now or in the future, is an adequate supply of minerals—fuels to energize our power and transportation; nonmetals, such as sulfur and phosphates to fertilize farms; and metals, steel, copper, lead, aluminum, and so forth, to build our machinery, cars, buildings, and bridges. These are the materials basic to our economy, the multipliers in our gross national product. But the needed materials which can be recovered by known methods at reasonable cost from the earth's crust are limited, whereas their rates of exploitation are not. This situation cannot continue.[9]

Notice how similar this viewpoint is to that of Thomas S. Lovering, an academic scientist and consultant to industry, whom we quoted in Chapter 10.

Differences exist, as one would expect, in the best estimates of reserves of mineral resources remaining in the earth's crust. More optimistic than most authorities in the field is Vincent E. McKelvey, Director of the U.S. Geological Survey. In a paper presented at Harvard University in 1971, McKelvey pointed out that dire predictions of mineral shortages have been made at intervals over the past sixty years. Nevertheless, as economic growth has moved rapidly ahead and mineral consumption has risen, the intensified research for new mineral reserves has been thus far able to keep pace with demands. McKelvey stated:

> Personally, I am confident that for millennia to come we can continue to develop the mineral supplies needed to maintain a high level of living for those who now enjoy it and to raise it for the impoverished people of our own country and the world. My reasons for thinking so are that there is a visible underdeveloped potential of substantial proportions in each of the processes by which we create resources and that our experience justifies the belief that these processes have dimensions beyond our knowledge and even beyond our imagination at any given time.[10]

However, McKelvey takes a cautious position. He recognizes that many do not share his views. He feels that a searching review of resource adequacy may be required. He then goes on to say:

> If our supply of critical materials is enough to meet our needs for only a few decades, a mere tapering off in the rate of increase of their use, or even a modest cutback, would stretch out these supplies for only a trivial period. If resource adequacy cannot be assured into the far distant future, a major reorientation of our philosophy, goals, and way of life will be necessary. And if we do need to revert to a low resource-consuming economy, we will have to begin the process as quickly as possible in order to avoid chaos and catastrophe.[10]

Although finite, our nonrenewable resources, including fossil fuels and uranium, the principal energy sources, allow us some options with respect to the program of depletion. Preston Cloud analyzes this concept in the following words:

Thus it is possible to think of man in relation to his renewable resources of food, water, and breathable air as limited in numbers by the sustainable annual crop. His non-renewable resources—metals, petrochemicals, mineral fuels, etc.—can be thought of, in somewhat over-simplified but valid terms, as some quantity which may be withdrawn at different rates, but for which the quantity beneath the curve of cumulative total production, from first use to exhaustion of primary sources, is largely independent of the shape of the curve. In other words, the depletion curve of a given non-renewable raw material, or of that class of resources, may rise and decline steeply over a relatively short time, or it may be a flatter curve that lasts for a longer time, depending on use rates and conservation measures such as recycling. This is a choice that civilized, industrialized man expresses whenever his collective aspirations, efforts, and ideals express themselves in a given general density of population, per capita rate of consumption, and conservation practice.[1]

Stretching of our limited earth resources to the utmost is a conservation procedure strongly endorsed by all who study resource problems. We have stressed in earlier statements in this book that pollution and environmental degradation are usual by-products of development and use of raw materials. Thus the various means available to conserve mineral use simultaneously serve two vital functions. One is to make the resource last as long as possible; the other is to minimize environmental impact.

One of the options available for stretching resources is the substitution of natural and synthetic materials, the latter made available by technological advances. For example, in electrical conductors aluminum (an abundant element in the crust) is being substituted for copper (a scarce element in the crust). For a few metals, mercury among them, no effective substitution is now possible. Plastics are being widely substituted for metals in many applications. Keep in mind, however, that plastics are synthesized from compounds which are themselves mineral resources, namely, petroleum and coal.

Recycling, which was described in Chapter 10 with respect to metals, is another option urgently requiring intensive research and development. We have not yet begun to exploit fully the opportunities for recycling of solid wastes, which pose in themselves an environmental problem as sources of water pollution and as consumers of valuable urban land.

Energy resources, together with matter resources, comprise the total natural resource picture. While various aspects of energy resources have been included in the foregoing paragraphs, it is useful to make a further analysis of energy consumption problems.

Consumption of Energy

Do we face an energy crisis? Along with other natural resources, the consumption of energy has taken the same sharp upturn as the global population. In 1972, testifying before the House Committee on Interior and Insular Affairs, Vice Admiral Hyman G. Rickover, a leading authority on technical applications of nuclear energy, had this to say:

> I can think of no public issue confronting us today that will touch the lives of our children and grandchildren—if not ourselves—more closely than the ability of the U.S. to command the energy resources that are needed to sustain our economy. . . . A realistic assessment

of our energy resources position would stress the fact that it is not important exactly *when* our fossil fuels give out. We may find better ways to extract oil and gas from coal, tar sands, shale; we may discover more reserves than we know of at present on the Continental Shelf surrounding our country or elsewhere in the sea. What is important is to comprehend that some day the fossil fuels will be gone; that renewable energy resources are unlikely to provide more than a small percentage of our needs; that even atomic energy, since it requires uranium, is finite; and that we cannot be certain that some man-made alternative as yet undeveloped will arrive in time—or ever—to supply our energy needs.[11]

As in the case of materials resources, Vincent E. McKelvey, Director of the U.S. Geological Survey, presents a more optimistic outlook upon the future of energy resources:

> Most important to secure our future is an abundant and cheap supply of energy, for if that is available we can obtain materials from low-quality sources, perhaps even country rocks, as Harrison Brown has suggested. Again, I am personally optimistic on this matter, with respect both to the fossil fuels and particularly to the nuclear fuels. Not only does the breeder reactor appear to be near enough to practical reality to justify the belief that it will permit the use of extremely low-grade sources of uranium and thorium that will carry us far into the future, but during the last couple of years there have been exciting new developments in the prospects for commercial energy from fusion.[10]

Whatever may be the shades of opinion on this question of future supplies of energy, there is no question that the prodigious increase in energy consumption poses many problems, including serious threats to the environment. The present rate of increase of energy consumption is such that the doubling time is about fifteen years. We can anticipate a corresponding increase in the frequency and severity of environmental problems in years to come.

Discussing the environmental impact of energy use, Philip H. Abelson, had this to say in an editorial in the journal *Science*, of which he is editor:

> Consumption of energy is the principal source of air pollution, and energy production, transportation, and consumption are responsible for an important fraction of all our environmental problems. Use of energy continues to rise at the rate of 4.5 percent per year. Even if fuel supplies were infinite, such an increase could not be tolerated indefinitely. But fuel supplies are not inexhaustible, and this combined with the need to preserve the environment will force changes in patterns of energy production and use.
>
> Our economy has been geared to profligate expenditure of energy and resources. Much of our pollution problem would disappear if we drove 1-ton instead of 2-ton automobiles. Demand for space heat and cooling could be reduced if buildings were properly insulated. Examples of needless use of electricity are everywhere. Promotional rates and advertising tend to encourage excessive consumption.[12]

Abelson then goes on to suggest some nonstructural measures to curtail energy consumption:

> A major factor in the burgeoning use of energy is its low price—one that does not take into account all the costs to society. In the generation of electric power from coal and oil, millions of tons of sulfur

dioxide are released, which cause billions of dollars worth of damage to health and property. We are consuming rapidly, at ridiculously low prices, natural gas reserves that accumulated during millions of years. Prices for energy should reflect their full cost to society. The Nixon Administration's proposed tax on sulfur in fuel should be enacted. The rate structure for electric power should be modified to discourage excessive use. A substantial increase in the price of natural gas, including a new federal tax, would diminish waste of this resource. Taxes on automobiles should increase sharply with weight and horsepower. . . .

Slowing down the rate of increase in use of energy will not be easy. Public habits of energy consumption will not be quickly altered, and a sudden change in the rate of growth of energy consumption would cause major additional unemployment.[12]

What are the prospects of achieving success in reducing the environmental impacts of energy consumption? Abelson concludes his editorial in this way:

Measures to cut excessive use of energy are likely to come only after a long time, if ever. We should face the possibility that increased consumption of energy will continue and prepare to meet that possibility. Atmospheric pollution is not an inevitable consequence of production of energy. In the use of fossil fuels, production of sulfur dioxide is not an essential by-product. Destruction of the environment is not a necessary consequence of strip mining. Pollution from almost every method of producing and utilizing energy could be sharply attenuated either through better practices or through development of new methods. In view of the importance of energy to society, present expenditures on research and development related to energy are small and these are not well apportioned.[12]

The Prospect for a Steady-State Existence

Increasingly, in the past few years, members of the scientific community have been voicing the concept that the survival of mankind on our planet will be contingent upon achievement of an overall steady state in the total global system of flux of energy and matter, including Man and his institutions within the framework of the biosphere. We expressed this view in the closing paragraphs of Chapter 10, through a quotation from the writings of M. King Hubbert, a geophysicist on the staff of the U. S. Geological Survey. Hubbert expressed clearly the concept that present growth rates of both human population and resource consumption represent a transient state, as viewed from the point of dynamics of an open system. Hubbert pointed sharply to the ultimate necessity of achieving a steady state.

The concept of a steady state in the total global system is a fitting one with which to close this book. Let us read what others have had to offer on this subject:

We need to concentrate our orientation of knowledge on the human-society-plus-environment level of integration because of its relevance to the central world problem of achieving a reasonably steady state between human societies and the finite resources of our planet. We are as much concerned with human society itself as with the environments in which men live; both are parts of an interacting whole that evolves as a unit through time. -S. Dillon Ripley and Helmut K. Beuchner.[13]

Epilogue

The use of the environment is a necessary and acceptable concept. The difference is that future use must be in the recycle context of perpetual renewal and reuse, not in the old pattern of use and discard. A sort of stable state between civilization and the environment is called for—not a balance of nature (for nature is always changing in its own right) but a harmony of society and the environment within natural laws of physics, chemistry, and biology. –*Committee on Scientific Astronautics.*[14]

Very likely a steady-state economy may come about eventually as the result of increased scarcity and higher costs of environmental protection, which will also encourage the recovery and recycling of materials. If a more or less steady level of consumption can be maintained and can be accompanied by a stable population, technological advance may permit some small but steady gain in level of living.

The way in which we approach a steady-state economy is most important. If it can be done through the operation of market forces, the steady state would bring a high quality of life for a maximum population. On the other hand, if the steady state is approached by way of social upheavals, sudden shortages of energy and mineral resources may produce a catastrophe that will find us with a steady-state economy but at a much lower level. If the catastrophe results in world-wide nuclear war then the level may well be at or close to zero. –*V. E. McKelvey and S. Fred Singer.*[15]

Over three billion years of history, read from the geological record, make it clear that the question is not whether man will come into balance with nature or not. The only question is whether this will happen as a result of self-imposed restraints of natural catastrophes over which he has no control. Nor is there any question that man can, if he chooses, determine his own destiny—at least to a very large degree and for a very long time. The question is, "will he choose to do so?". . . . In all his long history man has never faced a worthier or a more critical challenge than that of achieving a lasting balance with his environment—and I mean the *total* human ecosystem, including the city, the sea, and the wilderness. That is the challenge primarily of the generation now entering maturity. Population control, sensible resource management, and continuous surveillance of all components of a thoroughly researched global ecosystem, including its various sociopolitical components, are the essential steps toward such a balance. A principal goal of higher education should be to discover and to communicate the basic knowledge and comprehensive understanding that will lead to those steps being taken in a well-informed, humane, and orderly manner, and as soon as practicable. –*Preston Cloud.*[1]

References

1. Preston Cloud, 1971, This finite earth, *A. & S. Review*, Indiana University, Spring 1971, p. 17-32.
2. Sterling Brubaker, 1972, *To live on earth; man and his environment in perspective*, published for Resources for the Future, Inc., by The Johns Hopkins Press, Baltimore and London, p. 38.
3. A. V. Kneese, R. U. Ayers, and R. C. d'Arge, 1970, *Economics and the environment*, published for Resources for the Future, Inc., by The Johns Hopkins Press, Baltimore and London, pp. 6-7.
4. R. J. Shinn, 1971, Population and the dignity of man, *Identity and dignity of man*, Preston Williams, ed., Schenkman Publishing Co., Cambridge, Mass.
5. Lewis W. Moncrief, 1970, The cultural basis for our environmental crisis, *Science*, vol. 170, pp. 508-512. Copyright 1970 by the Association for the Advancement of Science.
6. Marshall I. Goldman, 1970, The convergence of environmental disruption, *Science*, vol. 170, pp. 37-42. Copyright 1970 by the American Association for the Advancement of Science.
7. National Science Board, 1971, *Environmental science; challenge for the seventies*, U.S. Government Printing Office, Washington, D.C. p. 5.
8. Much of this section consists of quoted or paraphrased statements from *The Second Annual Report of the Council on Environmental Quality*, August, 1971, U.S. Government Printing Office, Washington, D.C.
9. Walter R. Hibbard, Jr., 1968, Mineral resources: challenge or threat?, *Science*, vol. 160, p. 143. Copyright 1968 by the American Association for the Advancement of Science.
10. Vincent E. McKelvey, 1972, Mineral resource estimates and public policy, *American Scientist*, vol. 60, p. 39.
11. Hyman G. Rickover, 1972, Statement to House Committee on Interior and Insular Affairs regarding fuel and energy resources, April 18, 1972.
12. Philip H. Abelson, 1971, Continuing increase in use of energy, *Science*, vol. 172, p. 795. Copyright 1971 by the American Association for the Advancement of Science.
13. S. Dillon Ripley and Helmut K. Buechner, 1967, Ecosystem science as a point of synthesis, Reprinted by permission of *Daedalus*, Journal of the Academy of Arts and Sciences, Boston, Mass., Fall 1967, *America's Changing Environment*.
14. *Managing the environment*, 1968, Publ. Committee on Scientific Astronautics, 90th. Congress, 2nd. Session, 1968, U.S. Government Printing Office, Washington, D.C., pp. 14-15.
15. Vincent E. McKelvey and S. Fred Singer, 1971, Conservation and the minerals industry—a public dilemma, *Geotimes*, vol. 16, no. 12, p. 21.

APPENDIX I
Dimensions, Definitions, and Equivalents in the Flow of Energy and Matter

A. Dimensional Analysis of Energy, Work, and Power

Accurate definitions of such terms as *force, energy, work,* and *power* are essential in environmental science emphasizing open systems with their inputs and outputs of energy and matter. A good way to approach the subject is through *dimensional* expression, in which each term is broken down into its component fundamental dimensions of *mass, length,* and *time.* All mechanical quantities can be stated as products of of the three fundamental dimensions.

Symbols and units for the fundamental dimensions are as follows:

Dimension	Symbol	C.g.s. Unit	M.k.s. Unit
Mass	M	Gram of mass	Kilogram of mass
Length	L	Centimeter	Meter
Time	T	Second	Second

By a *product*, we mean not only the multiplication of one dimension by another, but also the division of one dimension by another and the multiplication of one dimension by itself. The latter operation means the squaring or cubing of the dimension. Here are some examples:

	Operation	Dimensional Symbol
Mass times length	$M \times L$	ML
Inverse of length	$\frac{1}{L}$	L^{-1}
Mass divided by length	$M \div L$	$\frac{M}{L}$ or ML^{-1}
Length times length	$L \times L$	L^2
Length divided by time squared	$L \div T^2$	$\frac{L}{T^2}$ or LT^{-2}

Dimensions, Definitions, and Equivalents in Energy and Matter

Notice that the final form, using the negative exponent, is much easier to type or set in print than the fractional form. Negative exponents are the internationally approved style.

With this brief explanation of dimensional symbols, we are ready to define a number of basic mechanical quantities in terms of their fundamental dimensions. The symbol $\stackrel{d}{=}$ will be used to mean "dimensionally equivalent to."

*Velocity** (*speed*) is *distance per unit of time*:

$$\text{Velocity} = \frac{\text{Distance}}{\text{Time}} \stackrel{d}{=} \frac{L}{T} \stackrel{d}{=} LT^{-1} \quad (1)$$

*Acceleration** is *rate of velocity change*, i.e., velocity change per unit of time:

$$\text{Acceleration} = \frac{\text{Velocity}}{\text{Time}} \stackrel{d}{=} \frac{LT^{-1}}{T}$$
$$\stackrel{d}{=} LT^{-1} \cdot T^{-1} \stackrel{d}{=} LT^{-2} \quad (2)$$

(Notice that when a given dimension is multiplied by itself, the exponents are added: $T^{-1} \cdot T^{-1} \stackrel{d}{=} T^{-2}$.)

Force is rate of change of momentum and can be defined dimensionally as *mass times acceleration*:

$$\text{Force} = \text{Mass} \times \text{Acceleration} \quad (3)$$
$$\stackrel{d}{=} M \cdot LT^{-2} \stackrel{d}{=} MLT^{-2}$$

Weight has the same dimensions as force, but acceleration in this case becomes the *acceleration of gravity g*

Energy is the *capacity to do work*. *Work*, in turn is *force acting through distance*. In dimensional language both energy and work are the same; each is the product of force and distance:

$$\text{Energy or Work} = \text{Force} \times \text{Distance} \quad (4)$$
$$\stackrel{d}{=} MLT^{-2} \cdot L \stackrel{d}{=} ML^2T^{-2}$$

Power is the *rate at which work is done*. In dimensional terms, power is work (or energy) per unit of time:

$$\text{Power} = \frac{\text{Work}}{\text{Time}} \stackrel{d}{=} \frac{ML^2T^{-2}}{T} \stackrel{d}{=} ML^2T^{-3} \quad (5)$$

Power can also be analyzed as the product of force and velocity:

$$\text{Power} = \text{Force} \times \text{Velocity} \quad (6)$$
$$\stackrel{d}{=} MLT^{-2} \cdot LT^{-1}$$
$$\stackrel{d}{=} ML^2T^{-3}$$

Work is usually evaluated in terms of power acting through time, because a machine is rated according to its power output; the amount of work it does depends upon how long it runs. Referring back to (5), because Power = Work/Time, it follows that Work = Power × Time:

$$\text{Work} = \text{Power} \times \text{Time} \quad (7)$$
$$\stackrel{d}{=} ML^2T^{-3} \cdot T \stackrel{d}{=} ML^2T^{-2}$$

(Note that both definitions of work, (4) and (7), are dimensionally equivalent.)

In the study of behavior of fluids, such as air (a gas) and water (a liquid), quantities frequently needed are *density* and *pressure*.

Density is *mass per unit of volume*:

$$\text{Density} = \frac{\text{Mass}}{\text{Volume}} \stackrel{d}{=} \frac{M}{L^3} \stackrel{d}{=} ML^{-3} \quad (8)$$

(Notice that *volume* is dimensionally defined as length-times-length-times-length, or length-cubed.)

Pressure intensity, or *stress*, is *force per unit of area*:

$$\text{Pressure intensity} = \frac{\text{Force}}{\text{Area}} \stackrel{d}{=} \frac{MLT^{-2}}{L^2} \quad (9)$$
$$\stackrel{d}{=} MLT^{-2} \cdot L^{-2}$$
$$\stackrel{d}{=} ML^{-1}T^{-2}$$

(Notice that area is dimensionally equivalent to length-times-length, or length-squared.)

Once the dimensions of a mechanical quantity are established, it is an easy matter to substitute standard units of measure for each dimensional term. Table A.1 gives units for the quantities previously analyzed. Both centimeter-gram-second (c.g.s.) and meter-kilogram-second (m.k.s.) systems are given. Notice that the gram and kilogram have the dimension of mass (not weight). Where the unit of measure of the quantity has been named, it is given as well.

B. Mechanical Equivalent of Heat

The unit of heat energy is the *gram calorie*, defined as the quantity of heat required to raise the temperature of 1 gram of water through 1 Celsius degree of change. Because the quantity of heat per Celsius degree varies slightly depending upon the temperature range selected, the standard of reference calls for use of the one-degree range between 15° and 16° C (59.0 to 60.8° F).

The *kilocalorie*, or *large calorie*, is equal to 1000 gram calories. One must be very careful to ascertain what is meant by the word "calorie" as encountered in reading and speech. The kilocalorie can be designated by the abbreviation *kcal*; the gram calorie by *gcal*. In some usage styles the kilocalorie is designated by a capital C, *Calorie*

---

*Both velocity and acceleration are *vector quantities*, meaning that the direction of action must be stated as well as the intensity. However, direction in space is a dimensionless property and has been omitted here. The word *speed* is strictly correct to apply only to change of distance with time.

Flow of Energy

TABLE A.1

Mechanical Quantity	Dimensions	C.g.s. Units	Name	M.k.s. Units	Name
Velocity	LT^{-1}	cm/sec		m/sec	
Acceleration	LT^{-2}	cm/sec/sec, or cm/sec^2		m/sec/sec or m/sec^2	
Acceleration of gravity	LT^{-2}	980 cm/sec^2		0.98 m/sec^2	
Force	MLT^{-2}	gm-cm/sec^2	dyne	kg-m/sec^2	newton
Weight	MLT^{-2} ($M \cdot g$)	gram-weight		kg-weight	
Energy (work)	ML^2T^{-2}	dyne-cm	erg, joule (10^7 ergs)	newton-m	
Power	ML^2T^{-3}	ergs/sec joules/sec	watt (1 joule/sec)		
Work, as power × time	ML^2T^{-2}	watt-sec	kilowatt-hr (36 × 10^{12} ergs)		
Density	ML^{-3}	gm/cm^3		kg/m^3	
Pressure intensity (stress)	$ML^{-1}T^{-2}$	dynes/cm^2	bar (10^6 dynes/cm^2)	newtons/m^2	

or *Cal*, while the gram calorie has only the lower-case *c*, *calorie* or *cal*. In popular writing on dieting the word *calorie* actually means the kilocalorie, but this fact is seldom made clear to the reader.

Establishment of the *mechanical equivalent of heat*, that is to say, to determine how many heat calories are equivalent to one erg or joule of mechanical energy, was first worked out by James Joule. The device he used consisted of a set of paddle wheels immersed in an insulated container of water. The paddles were turned by a weight-driven pulley system. The amount of warming of the water could be measured by thermometer and set equal in terms of calories to the potential energy lost by the fall of the weight through a known distance. After careful refinements of the method it was established that one kilocalorie is the equivalent of about 4200 joules. In exact terms the relationship is as follows:

1 kcal = 4,186 joules

1 gcal = 4.186 joules = 4.186 × 10^7 ergs

In ecology it is common practice to state *biomass*, or mass of organic matter, in terms of energy. This transformation is made by evaluating the quantity of heat that would be produced by oxidation of the hydrogen and carbon contained in the organic matter. In this way the stored chemical energy can be stated in the same terms as kinetic and potential energy in tracing the flow of energy through the ecosystem. (See Chapter 18 for heat equivalents of various common organic substances.)

C. Flow of Energy

In describing energy systems quantitatively, care must be taken to define correctly terms used to represent the flow of energy and power. First, we must define the unit of surface area of the receiving surface (in the case of electromagnetic radiation) or the unit area of cross-section through which the energy flows.

For electromagnetic radiation the quantity of energy received or emitted by a unit of surface is the *langley* (*ly*):

1 ly = 1 gcal/cm^2 = 4.186 joules/cm^2

It is often more convenient to use the *kilolangley* (*kly*):

1 kly = 1 kcal/cm^2 = 4,186 joules/cm^2

It has been urged that use of the langley be discontinued in favor of the standard calorie units, given above. However, you will find the langley in wide use in writing on the subject of the earth's radiation balance (Chapter 2).

The *flow intensity*, or *flux*, of energy measures the rate at which energy is absorbed or emitted by a unit area of surface, or the rate at which energy passes through a cross-section of unit area. Consequently, to arrive at flow intensity we must introduce the time dimension in units of seconds, minutes, hours, days, or years:

$$\text{Energy flow intensity (flux)} = \frac{\text{Energy}}{\text{Area} \cdot \text{Time}}$$

$$= \text{gcal/cm}^2/\text{min}$$

(same as ly/min)

Dimensions, Definitions, and Equivalents in Energy and Matter

For larger areas and longer time spans we might choose to use kcal/m²/yr.

The term *flux* has the same dimensions as energy flow intensity, but a somewhat specialized meaning. For a solid surface that is simultaneously absorbing and emitting radiation, the term *flux* describes the net flow of energy, in whichever direction the excess lies. Energy flux can also be used for the net energy flow intensity of an entire system within which some parts are experiencing inflow of energy while other parts are experiencing outflow. An example might be the flux of water vapor across a given parallel of latitude averaged for the year (see Figure 4.37).

Energy flow intensity can also be considered to represent *power per unit area:*

$$\text{Energy flow intensity} = \frac{\text{Power}}{\text{Area}} = \frac{\text{gcal/min}}{\text{cm}^2}$$
$$= \text{gcal/cm}^2/\text{min}$$

In the field of radiation meteorology a change-over to units of *watts per square meter* is being initiated and will become the standard notation.

Total energy flow (or total energy flux) sums the energy flowing across an entire system boundary. In other words, we multiply flow intensity times area, so that the area terms cancel out:

Total energy flow
$$= \text{Energy flow intensity} \times \text{Area}$$
$$= \text{gcal/cm}^2/\text{min} \cdot \text{cm}^2 = \text{gcal/min}$$

D. Flow of Matter

In making quantitative evaluations of the rates of input, output, and storage of matter in an open system, care must be taken to define terms precisely. The basic equivalent of matter is *mass*. We have already defined mass as one of the three fundamental mechanical dimensions. We must not confuse mass with *weight*, which is a force, the product of mass and acceleration of gravity, g. It is important to specify *grams-weight* and *kilograms-weight* when the acceleration of gravity is included.

As in the case of energy flow, it is essential to distinguish between flow intensity rate and total flow rate. Both quantities are time-dependent, but only the first does not depend upon the total quantity of matter in motion. Take the case of a gas or liquid in motion. Mass flow intensity must be stated in terms of a selected small unit of cross-sectional area, such as 1 cm² or 1 m². For example, the flow intensity of water vapor lost by evaporation from the surface pores of a leaf may be stated as grams-mass per square centimeter per hour (gm/cm²/hr). Dimensional analysis can be applied as follows:

$$\text{Mass flow intensity} = \frac{\text{Mass}}{\text{Area} \times \text{Time}} \stackrel{d}{=} \frac{M}{L^2 \cdot T}$$
$$\stackrel{d}{=} ML^{-2}T^{-1}$$

Total flow rate multiplies the flow intensity rate by the total surface area of all the leaves of the plant. When this is done, the area terms cancel out and we are left with only grams-mass per hour (gm/hr):

Total mass flow rate = Mass flow intensity × Area
$$\stackrel{d}{=} ML^{-2}T^{-1} \cdot L^2 \stackrel{d}{=} MT^{-1}$$

Similarly, if we wish to express the intensity of removal of matter from a given surface (for example, the rate at which sediment is produced by a land surface during a rainstorm) we can also use the units gm/cm²/hr. For large areas and long spans of time the units are made larger, for example, metric-tons/km²/yr. Total erosion rate would be obtained by multiplying intensity rate by the total area of the watershed, giving metric-tons/yr.

In the hydrologic cycle, and particularly in measurement of runoff of water in stream flow, *volume* is usually substituted for mass. This conversion is feasible because for practical purposes water has a nearly constant value of weight per unit volume (1 gm/cm³). For a river, the flow intensity is the volume of water passing through a unit cross-sectional area per unit of time:

$$\text{Volume flow intensity} = \frac{\text{Volume}}{\text{Area} \cdot \text{Time}}$$
$$= \frac{\text{m}^3}{\text{m}^2 \cdot \text{sec}} = \text{m/sec}$$

Applying dimensional analysis:

$$\text{Volume flow intensity} = \frac{\text{Volume}}{\text{Area} \cdot \text{Time}}$$
$$\stackrel{d}{=} \frac{L^3}{L^2 \cdot T} \stackrel{d}{=} LT^{-1}$$

It is immediately obvious that this last quantity is identical in dimensions with velocity. Evidently, we have measured the *average velocity* (mean velocity), $\bar{V}$, of water flowing through the unit area of cross-section. To obtain the total flow rate, or *discharge*, Q, of the entire stream, we multiply by the total area of cross section of the stream, A:

Discharge
$$= \text{Volume flow intensity} \times \text{Area}$$
$$= \text{m/sec} \cdot \text{m}^2 = \text{m}^3/\text{sec}$$

Flow of Matter

In dimensional language

$$\text{Discharge} = \text{Velocity} \times \text{Area}$$
$$\stackrel{d}{=} LT^{-1} \cdot L^2 \stackrel{d}{=} L^3 T^{-1}$$

Rewriting the equation in terms of average velocity, we obtain:

$$\text{Average velocity} \times \text{Area} = \text{Discharge}$$

or $\bar{V} \cdot A = Q$
and $Q = A \cdot \bar{V}$

Thus we arrive at the equation of continuity of flow, explained in Chapter 13.

Recall from Chapter 12 that the flow intensity terms, *rainfall, infiltration,* and *overland flow,* are stated as depth of water per unit of time (in./hr; cm/hr). Here again, the units are dimensionally equivalent to velocity and the explanation is quite simple. Rainfall intensity rate is fully stated as "volume of water received by a unit area of ground per unit of time":

$$\frac{\text{Volume}}{\text{Area} \cdot \text{Time}} = \frac{\text{cm}^3}{\text{cm}^2 \cdot \text{hr}} = \text{cm/hr.}$$

APPENDIX II
von Bertalanffy's Principles of Open Systems and Steady States

A highlight in the history of development of open-system theory was the publication in the United States in 1950 by Ludwig von Bertalanffy, an Austrian biochemist, of a paper entitled "The Theory of Open Systems in Physics and Biology."* In this classic paper, which reviews his own pioneer work in development of open-system theory begun in 1933, von Bertalanffy brought to the attention of a large and diverse group of American scientists certain basic concepts that were to pervade almost every field of natural, physical, and social sciences in decades to follow.

The basic concepts of open systems and steady states expressed clearly by von Bertalanffy in 1950 have been incorporated into our *Introduction to Energy Systems* in earlier pages of this book. With special reference to cell physiology, von Bertalanffy stated the basic equation of energy flow in open systems; it can be paraphrased as follows: Time-rate of change of energy concentration within a very small element of an open system is equal to the sum of the rate of change of production of energy produced in biochemical reactions and the rate of outflow of energy from the element. The original statement of the equation is given at the end of this appendix. In simplified form it can be written as follows:

$$\frac{dE_s}{dt} = \frac{dE_c}{dt} + \frac{dE_o}{dt} \quad (1)$$

where E_s is energy stored in chemical form or as sensible heat,
E_c is energy produced by chemical reactions,
E_o is energy lost by outflow,
and t is time.

For those unfamiliar with the notations of calculus, it may help to point out that "$\frac{d}{dt}$" in front of each energy term is an instruction to read each of the three terms as "instantaneous rate of change of energy with respect to time."

The left-hand term of (1), rate of energy storage, can be either positive or negative, depending upon whether the element of the system is gain-

---
*Science, 1950, vol. 111, pp. 23–29.

ing or losing stored energy. The first of the right-hand terms may also be either positive or negative. When entropy is being decreased by metabolic processes, such as photosynthesis or synthesis of proteins, the term will be a positive quantity. When entropy is increasing by reversed chemical reaction in which complex molecules are being decomposed to simpler constituents during respiration, the term will be negative. The accompanying diagram shows schematically the reversible nature of the energy-production term (Figure AII.1). The final term of the equation, rate of energy outflow, will be positive; it will be in the form of sensible heat or of organic molecules leaving the element.

Steady state is attained when rates of energy production and outflow within the element become equal to zero:

$$\frac{dE_c}{dt} + \frac{dE_o}{dt} = 0 \quad (2)$$

When this equality exists, rate of change of stored energy also becomes zero:

$$\frac{dE_s}{dt} = 0 \quad (3)$$

Therefore, the quantity of stored energy becomes a constant:

$$E_s = k \quad (4)$$

The "element" of the system referred to by von Bertalanffy is an extremely small quantity, approaching an infinitesimally small point in space. This concept is difficult to grasp in any real physical sense. Therefore, as shown in Figure AII.1, we have enclosed the element within the outlines of a living cell, representing a real system with a recognizable boundary. The equations of von Bertalanffy can be applied to the cell as an open system by integrating (summing) the energy rates for all elements of the entire cell. Steady state of the cell as a whole is then represented by a balance between rate of total energy import and total energy export.

It is interesting to consider the significance of the three terms of (1) when they are assigned dimensions as mechanical quantities. (Appendix I, Part A.) Each term has the dimensions of energy per unit time:

$$\frac{dE^*}{dt} \stackrel{d}{=} \frac{\text{Energy}}{\text{Time}} \stackrel{d}{=} \frac{ML^2T^{-2}}{T} \stackrel{d}{=} ML^{-2}T^{-3}$$

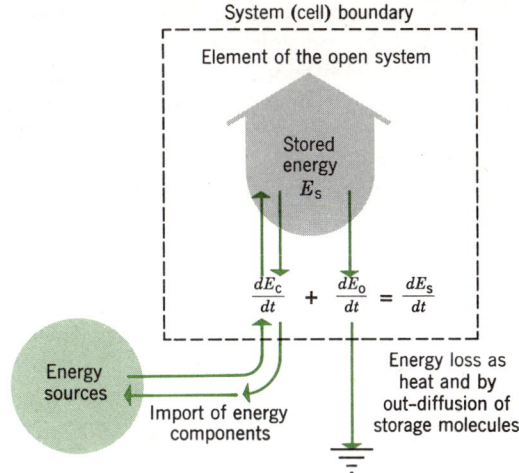

Figure AII.1 Schematic diagram of the open system in a living organism.

Obviously, the time rate of change of energy in the equation represents power, since by definition, power is work (or energy) per unit of time.

Excerpts from von Bertalanffy

The following excerpts from von Bertalanffy's writing give further insight into the concepts of open systems:*

From the physical point of view, the characteristic state of the living organism is that of an open system. A system is closed if no material enters or leaves it; it is open if there is import and export and, therefore, change of the components. Living systems are open systems, maintaining themselves in exchange of materials with environment, and in continuous building up and breaking down of their components.

So far, physics and physical chemistry have been concerned almost exclusively with processes in closed reaction systems, leading to chemical equilibria.

The cell and the organism as a whole, however, do not comprise a closed system, and are never in true equilibrium, but in a steady state. We need, therefore, an extension and generalization of the principles of physics and physical chemistry, complementing the usual theory of reactions and equilibria in closed systems,

---

*In dimensional analysis, the derivative of one variable with respect to another has the same dimensions as the ratio of one variable to the other.

*Ludwig von Bertalanffy, 1950, The theory of open systems in physics and biology, Science, vol. 111, pp. 23–29. Copyright 1950 by the American Association for the Advancement of Science.

von Bertalanffy's Principles of Open Systems and Steady States

and dealing with open systems, their steady states, and the principles governing them.

In physics, the theory of open systems leads to fundamentally new principles. It is indeed the more general theory, the restriction of kinetics and thermodynamics to closed systems concerning only a rather special case. In biology, it first of all accounts for many characteristics of living systems that have appeared to be in contradiction to the laws of physics, and have been considered hitherto as vitalistic features. Second, the consideration of organisms as open systems yields quantitative laws of important biological phenomena. So far, the consequences of the theory have been developed especially in respect to biological problems, but the concept will be important for other fields too, such as industrial chemistry and meteorology.

Some pecularities of open reaction systems are obvious. A closed system *must*, according to the second law of thermodynamics, eventually attain a time-independent equilibrium state, with maximum entropy and minimum free energy, where the ratio between its phases remains constant. An open system *may* attain (certain conditions presupposed) a time-independent state where the system remains constant as a whole and in its phases, though there is a continuous flow of the component materials. This is called a steady state. Chemical equilibria are based upon reversible reactions. Steady states are irreversible as a whole and individual reactions concerned may be irreversible as well. A closed system in equilibrium does not need energy for its preservation, nor can energy be obtained from it. To preform work, however, the system must be, not in equilibrium, but tending to attain it. And to go on this way, the system must maintain a steady state. Therefore, the character of an open system is the necessary condition for the continuous working capacity of the organism.

To define open systems, we may use a general transport equation. Let Q_i be a measure of the i-th element of the system, e.g., a concentration of energy in a system of simultaneous equations. Its variation may be expressed by:

$$\frac{\partial Q_i}{\partial t} = T_i + P_i \quad (1)$$

P_i is the rate of production or destruction of the element Q_i at a certain point of space; it will have the form of a reaction equation. T_i represents the velocity of transport of Q_i at that point of space; in the simplest case, the T_i will be expressed by Fick's diffusion equation. A system defined by the system of equations (1) may have three kinds of solutions. First, there may be unlimited increase of the Q's; second, a time-independent steady state may be reached; third, there may be periodic solutions. In the case that a steady state is reached, the time-independent equation:

$$T_i + P_i = 0 \quad (2)$$

must hold for a time $t \neq 0$.

The system (therefore) manifests forces which are directed against a disturbance of its steady state. In biological language, we may say that the system shows adaptation to a new situation [pp. 23-24].

But these characteristics of steady states are exactly those of organic metabolism. In both cases, there is first maintenance of a constant ratio of the components in a continuous flow of materials. Second, the composition is independent of, and maintained constant in, a varying import of materials; this corresponds to the fact that even in varying nutrition and at different absolute sizes the composition of the organism remains constant. Third, after a disturbance, a stimulus, the system reestablishes its steady state. Thus, the basic characteristics of *self-regulation* are general properties of open systems [p. 24].

Generally speaking, the basic fundamental physiological phenomena can be considered to be consequences of the fact that organisms are quasi-stationary open systems. Metabolism is maintenance in a steady state. Irritability and autonomous activities are smaller waves of processes superimposed on the continuous flux of the system, irritability consisting in reversible disturbances, after which the system comes back to its steady state, and autonomous activities in periodic fluctuations. Finally, growth, development, senescence, and death represent the approach to, and slow changes of, the steady state. The theories of many physiological phenomena are, therefore, special cases of the general theory of open systems, and conversely, this conception is an important step in the development of biology as an exact science [p. 27].

In biology, the nature of the open system is at the basis of fundamental life phenomena, and this conception seems to point the direction and pave the way for biology to become an exact science [p. 28].

APPENDIX III
The Darwin–Lotka Principle of Energy Storage

The Darwin–Lotka principle, as explained by H. T. Odum (1971; see References), is that energy is stored at the maximum possible rate when 50% of the total system energy is in the form of processing work and 50% is in the form of storing work. For any other ratios of the two work forms the rate of energy storage is lower.

A working model can be set up to test the Darwin–Lotka principle for a mechanical system dissipating energy through friction while storing energy in potential form by lifting of a mass. As shown in Figure AIII.1, the model consists basically of a pair of weights connected by a cord passing over a pulley. The pulley shaft drives a mechanism that resists the turning force and thereby drains energy from the machine. As the heavier weight moves downward mechanical energy is transformed into heat while energy is stored in potential form in the rising counterweight. Assume the cord to be weightless and the pulley to be frictionless.

We have a choice of mechanical devices by which to set up resistance to turning of the pulley. This choice is very important, because of the relationship between rate of turning of the pulley (and of fall of the heavier weight) and the resisting force encountered. The Darwin–Lotka law calls for a linear relationship between velocity, V, and resisting force, F_r, thus:

$$F_r \propto V$$

(The symbol $\propto$ means "proportional to".)

In other words, if we double the velocity of fall of the weight, the opposing resisting force will also double, and so forth. The terms can also be reversed to read: $V \propto F_r$. The linear relationship plots as a straight line on a graph having arithmetic scaling of velocity on one axis and force on the other (Figure AIII.2).

An important concept in developing the model further is that when the driving weight is released, it will accelerate (increase its velocity of fall), but the resistive force will also build up. As a result, acceleration decreases and finally approaches a zero value beyond which, for all practical purposes, the velocity becomes constant; i.e., reaches a *terminal velocity*. We shall make our measure-

The Darwin–Lotka Principle of Energy Storage

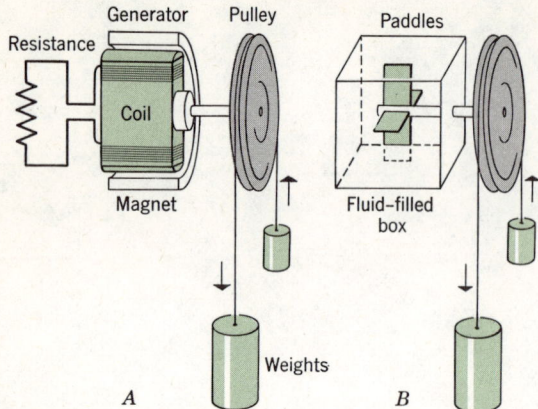

Figure AIII.1 Simple mechanical devices to demonstrate the Darwin–Lotka principle.

ments while terminal velocity holds, for then the rates of energy dissipation and storage are constant. Terminal velocity represents steady state in an open system. Potential energy is being converted first into kinetic energy, then into heat energy at a uniform rate. The phenomenon is illustrated in the maximum speed at which an automobile can be driven with a fixed maximum output of power. Once the various resistive forces have built up to become equal to the impelling force, terminal velocity (top speed) is reached.

You might suppose that we could fit the pulley wheel with a brake shoe, as in an automobile brake, as a mechanism of resisting turning. However, it is a law of friction between solid surfaces that the frictional force is independent of the speed of motion of one surface over the other. A suitable mechanism is found in an electric generator attached to the shaft of the pulley (Figure AIII.1A). Force is required to turn the generator rotor in the magnetic field that surrounds it. Electromotive force produced by the turning rotor of the generator is directly proportional to the speed of turning. The electric current can be run through a resistance wire, which will transform the electrical energy into heat and thus drain energy from the system.

Figure AIII.3 shows five combinations of weights, including the limiting conditions. In case A there is no counterweight and the driving force consists of the 200-gm weight. Because there is no counterweight, no energy is stored. In case E the counterweight equals the driving weight and there is no spontaneous motion; again no energy is stored. In case C the driving force is equal to the net weight of 100 gm (200 gm minus 100 gm). Because both weights must travel vertically at the same velocity, in case C the energy dissipated is equal to the energy stored. Cases B and D represent intermediate weight ratios. Although we might obtain reasonably good results with a carefully constructed mechanical model, the data can be fully anticipated by mathematical calculations.

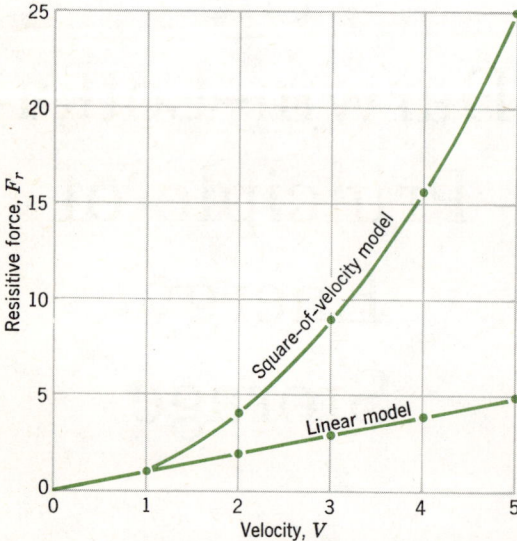

Figure AIII.2 Increase of resistive force with increasing velocity under each of two assumptions.

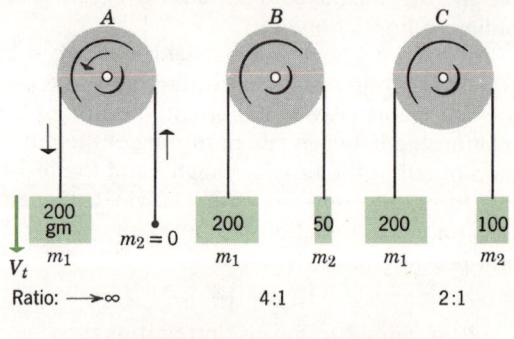

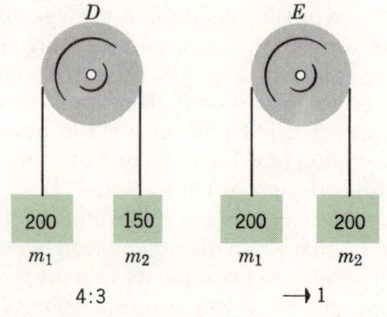

Figure AIII.3 Five combinations of weights. Case A results in no energy storage because no mass is lifted. Case E would be static, with no energy expenditure or storage.

The Darwin–Lotka Principle of Energy Storage

Equations relating power dissipation and power storage to force and velocity can be derived in the following way:

Let F_g be the driving force (dynes)
and F_r be the resisting force. (dynes)

Then $F_g = (m_1 - m_2)g$

where m_1 is the mass of the larger weight (a constant) (grams),
m_2 is the mass of the counterweight (variable) (grams),
and g is the acceleration of gravity, 980 cm/sec.²

Let $(m_1 - m_2)$ be designated as m, the net mass. Then $F_g = mg$. Assuming that resistance varies linearly with terminal velocity,

$$mg = cV_t \quad (1)$$

where V_t is terminal velocity (cm/sec)
and c is a resistance constant with the dimensions of mass per unit time. (gm/sec)

Solving for terminal velocity,

$$V_t = \frac{mg}{c} \quad (2)$$

Power dissipation, P_d, is the product of resisting force and terminal velocity:

$$P_d = F_r V_t \quad \text{(ergs/sec)} \quad (3)$$

Resisting force, F_r, is the product of the resistance constant and the terminal velocity:

$$F_r = cV_t \quad (4)$$

Substituting in (3),

$$P_d = cV_t \cdot V_t = cV_t^2 \text{ (ergs/sec)} \quad (5)$$

Power stored, P_s, is the product of the mass of the rising counterweight, times g, times terminal velocity:

$$P_s = m_2 g V_t \quad \text{(ergs/sec)} \quad (6)$$

Equations (5) and (6) are then solved for an arbitrary value of c and for various values of mass of the counterweight, m_2. Table AIII.1 gives several point values of the solution when $c = 500$ gm/sec. Power has been converted to watts (ergs $\times 10^{-7}$). Values in the table are plotted in Figure AIII.4; smooth curves are drawn for P_d and P_s. The peak storage value of 1.92 watts occurs when m_2 is 100 gm and is equal to power dissipated. The power storage curve is symmetrical with respect to the peak value, located at the midpoint of the horizontal scale.

TABLE AIII.1 Solution of Linear Model of Darwin–Lotka Law

Counter weight m_2, gm	Terminal Velocity, cm/sec	Dissipated Power, watts*	Stored Power, watts*
0	392	7.68	0.00
25	343	5.88	0.84
50	294	4.32	1.44
75	245	3.00	1.80
100	196	1.92	1.92
125	147	1.08	1.80
150	98	0.48	1.44
175	49	0.12	0.84
200	0	0.00	0.00

*Values rounded to two decimal places.
Resistance constant = 500 gm/cm.

Having now demonstrated the validity of the Darwin–Lotka law under our first assumption of a linear relationship between velocity and force, we turn to a second assumption that must be applied when resistance is set up by motion of a solid body through a fluid, or by the motion of a fluid in contact with a solid surface. The term *fluid* applies to both gases and liquids. These substances yield readily to unequal pressures, but offer resistive forces because of inertia of the fluid or because of viscosity within the fluid.

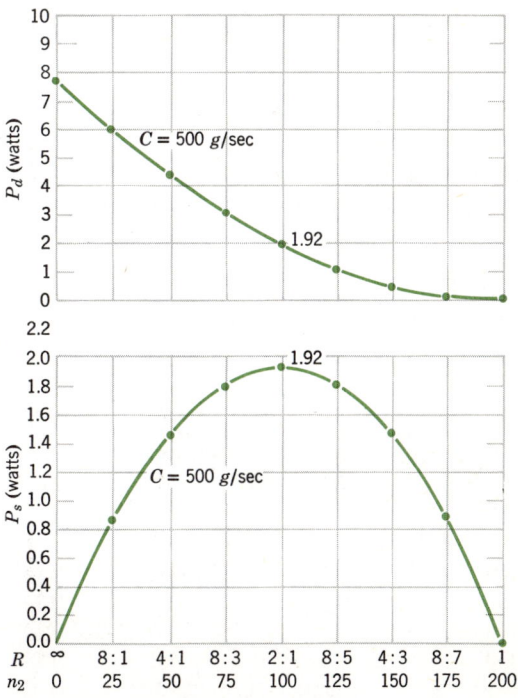

Figure AIII.4 Plots of rates of energy storage (below) and energy expenditure (above) for various ratios of weights, using the linear model.

The Darwin–Lotka Principle of Energy Storage

A solid object falling through a fluid sets up resistance that builds in magnitude disproportionately as the velocity of fall increases. Newton was one of the first scientists to investigate the behavior of a solid body moving in a fluid medium. He formulated a law to the effect that the force of resistance offered by the fluid varies as the square of the velocity of the moving solid body. We can state the essence of this law in simplified form as follows:

$$F_r \propto V^2$$

This relationship can also be written as $V \propto \sqrt{F_r}$. This law, with variations derived from it, is vital to environmental science because it governs the rate of fall of objects in air and water and the down-slope flow of fluids moving under gravity. Take, for example, a water drop falling through the air. After being released the drop accelerates, but as the velocity increases the air resistance also increases. Because the resisting force is increasing as the square of the velocity, the terminal velocity is quickly reached. Acceleration quickly approaches zero and velocity therafter remains constant. We observe the phenomenon of terminal velocity in a flowing stream of water, which encounters resistance due to the viscosity of the water and the drag upon the solid bed.

To use the principle of fluid resistance, we need a different mechanism of resistance than the electric generator. As shown in Figure AIII.1B, the shaft of the pulley passes into a water-filled box where it turns a set of paddles, or a propeller. An increase in speed of turning of the paddles encounters resisting force as the square of the speed increase. Resistance is transformed into heat, which is conducted and radiated out of the box.

TABLE AIII.2 Solution of Velocity-Squared Model of Darwin–Lotka Law

Counter weight m_2, gm	Terminal Velocity, cm/sec	Dissipated Power, watts*	Stored Power, watts*
0	19.8	0.388	0.000
25	18.5	0.318	0.045
50	17.1	0.252	0.084
75	15.7	0.192	0.115
100	14.0	0.137	0.137
125	12.1	0.0891	0.149
133.33....	11.4	0.0747	0.149+
150	9.9	0.0485	0.146
175	7.0	0.0171	0.120
200	0.0	0.0000	0.000

*Values rounded to three decimal places.
Resistance constant = 500 gm/cm.

To develop an equation for the velocity-squared model, using the same terms as defined in previous equations, the resisting force is set equal to the square of the terminal velocity:

$$F_r = cV_t^2 \tag{7}$$

and

$$F_g = mg = cV_t^2 \tag{8}$$

Solving for V_t in (8),

$$V_t = \sqrt{\frac{mg}{c}} \tag{9}$$

Power dissipated, P_d, is a product of resisting force and terminal velocity:

$$P_d = F_r V_t \tag{3}$$

Substituting for F_r the equivalent in (7),

$$P_d = cV_t^2 \cdot V_t = cV_t^3 \tag{10}$$

Power stored, P_s, is the same as in the linear case:

$$P_s = m_2 g V_t \tag{6}$$

Data of the new model are given in Table AIII.2. Comparing terminal velocities in Tables AIII.1 and III.2, note that the terminal velocity is very much lower for the velocity-squared model than for the linear model. To be precise, terminal velocity in the velocity-squared model is equal to the square root of corresponding terminal velocity in the linear model.

Curves of power dissipated and power stored are shown in Figure AIII.5. When the masses are equal (100 gm each) the value of dissipated power is exactly equal to stored power, as in the linear case. However, the quantity of power is much smaller in the velocity-squared model because the terminal velocity is so much less than in the linear model. The most interesting observation is, however, that the peak of power storage in the velocity-squared model occurs at a ratio of masses of 1.5:1 instead of 2:1. The power storage curve is asymmetrical and falls off very sharply after passing a broad peak.

According to H. T. Odum and R. C. Pinkerton (1955; see References), the linear model is valid for optimum storage of chemical energy by such life processes as photosynthesis and metabolism. In this model the maximum power storage represents 50% of the total system power. On the other hand, in the velocity-squared model peak power storage represents about 67% of the total system power. The efficiency of peak power storage is thus substantially higher in the latter model.

Inorganic systems of fluid flow include atmospheric and oceanic circulation and the gravity flow of water on the lands. None of these sys-

The Darwin–Lotka Principle of Energy Storage

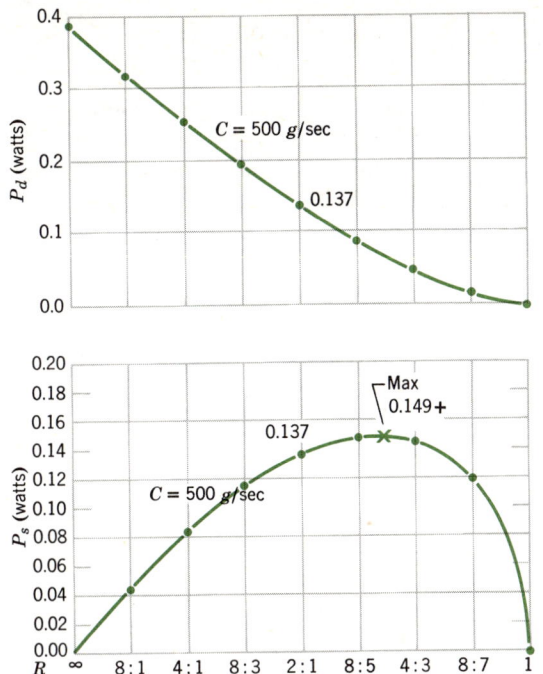

Figure AIII.5 Plots of rates of energy storage (below) and energy expenditure (above) for various ratios of weights, using the velocity-squared model.

tems, however, is characterized by significant potential energy storage within itself. Instead, practically all kinetic energy is continually dissipated as heat that leaves the system. However, the action of wind upon ocean and land surfaces may actually represent a case of temporary potential energy storage. Within water waves of the progressive oscillatory type, one-half the energy present is in the potential form (see Chapter 15). As waves grow in height under wind stress, their store of potential energy increases. Another possible example is the building of dunes of sand particles carried along the ground under wind stress. The sand of the dune represents a store of potential energy derived from a fraction of the energy dissipated by wind flow. In any case, these physical systems have no means of optimizing the rate of energy storage, whereas organisms through natural selection have such means at their disposal.

When Man devises energy storage through fluid flows, he has an opportunity to operate the process at maximum efficiency. Take for example, a hydraulic system in which water is pumped up from a river to a high reservoir or storage tank, using electricity or steam to power the pumps.

If time is not of the essence, pumping can be adjusted to an optimum storage rate and thus minimize the cost.

The linear model applies to man-made electrical energy systems, such as one battery charging another battery, and to systems in which a flow of heat results in mechanical energy. An example of the latter is a thermocouple driving an electric motor. The thermocouple is a device that produces an electric current because of difference in temperature between one point and another. It is widely used in thermometers and in radiometers that measure electromagnetic radiation intensity. Wires connect the thermocouple with the electric motor, so that thermal energy is first transformed into electrical energy and then into mechanical energy. Energy storage could then be represented by the raising of a weight by turning of the shaft of the motor, while power drain would be occurring through electrical and mechanical resistance in the wires and bearings.

While this discussion of power storage through man-made machines does not apply to natural organic and inorganic systems, it is useful in analysis of the flow of energy in industrial power systems of all kinds. The total industrial energy system uses mostly fossil fuels or nuclear fuels for its power input to run machinery of all sorts. The machinery processes raw materials, transports materials and products, and arranges the products into structures such as buildings, refrigerators, and automobiles that constitute our material wealth. A certain amount of straight animal energy goes into the process, where men operate controls or actually perform storage by lifting construction materials, but this is a very small part of the total. The sum total of the manufactured products together with the accumulated currency for which those products can be exchanged, constitutes capital wealth and represents the stored energy of industrial civilization.

References

Lotka, A. J., 1922, Contribution to the energetics of evolution, *Proc. Nat. Acad. Sci.*, vol. 8, pp. 147–155.

Odum, H. T., and R. C. Pinkerton, 1955, Time's speed regulator: The optimum efficiency for maximum power output in physical and biological systems, *Amer. Scientist*, vol. 43, No. 2, pp. 331–343.

Odum, H. T., 1971, *Environment, Power and Society*, Wiley-Interscience, New York, pp. 29–32.

Bibliography of Environmental Science

This bibliography includes only material from a selected list of basic books, encyclopedias, collected works, and journals. Emphasis is upon recent and readily available works. Subject headings are arranged by chapter and follow closely upon the sequence of topics within each chapter. The final page number of entries in the main index of this book (e.g. B-11, B-18) enables the reader to locate a subject in the bibliography.

Coded entries should be referred to two basic lists, given below. First is a list of COLLECTED PAPERS (Code CP). Second is a list of MULTI-SUBJECT WORKS (Code MS). For example, under Chapter 1, OCEANS—GENERAL, the following entry occurs:

"The ocean and man. Warren S. Wooster, CP-O2: 123–130, 1969"

Numbers following the colon are the page numbers. Referring to the CP code list below, the source volume will be given as follows:

"CP-O2 *The Ocean.* Scientific American Book, Freeman, 140 p., 1969"

The entry thus designates a reprinted paper under the authorship of Warren S. Wooster. Under coded multisubject entries the author's name has been omitted. This will be found with the source volume in the code list below. (Authorships are not given for individual entries in four encyclopedias under editorship of Rhodes W. Fairbridge, Reinhold Publishing Co.)

There follow lists of abbreviations of periodicals, publishers, societies, organizations, and governmental departments.

COLLECTED PAPERS (Code CP)
(Contributed papers, reprinted papers, symposia.)

CP-A1	*Advances in Environmental Sciences.* J. N. Pitts, Jr., and R. L. Metcalf, eds., Vol. 1, Wiley-Interscience, 356 p., 1969
CP-A2	*Advances in Environmental Science and Technology.* J. N. Pitts, Jr., and R. L. Metcalf, eds., Vol. 2, Wiley-Interscience, 354 p., 1971
CP-A3	*Agriculture and the Quality of Our Environment.* N. C. Brady, ed., Publ. 85, AAAS, 460 p., 1967
CP-A4	*As We Live and Breathe: The Challenge of Our Environment.* Nat. Geographic Soc., Washington, D. C., 239 p., 1971
CP-B1	*The Biosphere.* Scientific American Book, Freeman, 134 p., 1970
CP-C1	*Can We Survive Our Future? A Symposium.* G. R. Urban, ed., St. Martin's Press, New York, 400 p., 1972
CP-C2	*The Careless Technology: Ecology and International Development.* M. T. Farvar and J. P. Milton, eds., Nat. Hist. Press, 1030 p., 1972
CP-C3	*Challenge for Survival: Land, Air, and Water for Man in Megalopolis.* P. Dansereau, ed., Columbia Univ. Press, New York, 235 p., 1970
CP-C4	*Clearing the Air: The Impact of the Clean Air Act on Technology.* J. C. Redmond, J. C. Cook,

Bibliography of Environmental Science

	A. A. J. Hoffman, eds., Institute of Electrical and Electronics Engineers, IEEE Press, New York, N. Y., 159 p., 1971
CP-C5	*Conservation of Natural Resources.* G-H. Smith, ed., 4th ed., Wiley, 685 p., 1971
CP-C6	*The Control of Environment.* J. D. Roslansky, ed., North-Holland Publ. Co., Amsterdam, London (Fleet Academic Editions, New York), 112 p., 1967
CP-C7	*Continents Adrift.* Readings from the Scientific American, Freeman, 172 p., 1970
CP-D1	*Dimensions of the Environmental Crisis.* J. A. Day, F. F. Fost, P. Rose, eds., Wiley, 212 p., 1971
CP-E1	*Earth Might Be Fair; Reflections on Ethics, Religion, and Ecology.* I. G. Barbour, ed., Prentice-Hall, 168 p., 1972
CP-E2	*Eco-Crisis.* C. E. Johnson, ed., Wiley, 182 p., 1970
CP-E3	*Ecology and Economics; Controlling Pollution in the 70s.* M. I. Goldman, ed., Prentice-Hall, 234 p., 1972
CP-E4	*Environment, Resources, Pollution & Society.* W. W. Murdoch, ed., Sinauer Associates Inc., Stamford, Conn., 440 p., 1971
CP-E5	*Environmental Geomorphology.* D. R. Coates, ed., Publ. in Geomorphology, State Univ. of New York, Binghamton, 262 p., 1971
CP-E6	*Estuaries.* George H. Lauff, ed., Publ. No. 83, AAAS, 757 p., 1967
CP-E7	*Energy and Power.* Scientific American Book, Freeman, 144 p., 1971
CP-E8	*The Ecology of Man: An Ecosystem Approach.* Robert L. Smith, ed., Harper, 436 p., 1972
CP-E9	*Environmental Planning and Geology.* D. R. Nichols and C. C. Campbell, eds., USGPO, 204 p., 1971
CP-E10	*Ecosystem Structure and Function.* John A. Wiens, ed., Proc. Thirty-First Annual Biology Colloquium, Oregon State Univ. Press, 176 p., 1972
CP-F1	*Future Environments of North America.* F. F. Darling and J. P. Milton, eds., Nat. Hist. Press, 767 p., 1966
CP-G1	*Global Effects of Environmental Pollution.* S. F. Singer, ed., Springer-Verlag, New York, 218 p., 1970
CP-M1	*Man and His Physical Environment; Readings in Environmental Geology.* G. D. McKenzie and R. O. Utgard, eds., Burgess Publ. Co., Minneapolis, 338 p., 1972
CP-M2	*Man and the Ecosphere.* Readings from Scientific American, Freeman, 307 p., 1971
CP-M3	*Man, Health, and Environment.* B. Q. Hafen, ed., Burgess Publ. Co., Minneapolis, 269 p., 1972
CP-M4	*Man's Impact on Environment.* T. R. Detwyler, ed., McGraw-Hill, 731 p., 1971
CP-M5	*Man's Impact on Terrestrial and Oceanic Ecosystems.* W. H. Matthews, F. E. Smith, E. D. Goldberg, eds., MIT, 540 p., 1971
CP-M6	*Man's Role in Changing the Face of the Earth.* W. L. Thomas, ed., Univ. of Chicago Press, 1193 p., 1956 (Reprinted 1971 in two volumes.)
CP-M7	*Man and the Environment.* Wes Jackson, ed., Wm. C. Brown Co., Dubuque, Iowa, 322 p., 1971
CP-N1	*No Deposit—No Return (Man and His Environment: A View Toward Survival.)* H. D. Johnson, ed., Addison-Wesley, Reading, Mass., Menlo Park, Calif., 351 p., 1970
CP-O1	*Oceanography.* Readings from Scientific American, Freeman, 417 p., 1971
CP-O2	*The Ocean.* Scientific American Book, Freeman, 140 p., 1969
CP-R1	*Resources and Man; A Study and Recommendations.* Committee on Resources and Man, Div. of Earth Sci., NAS-NRC, Freeman, 259 p., 1969
CP-R2	*Readings in the Earth Sciences.* Vol. 1, Scientific American Resource Library, Freeman, 305 p., 1969
CP-R3	*Readings in the Earth Sciences.* Vol. 2, Scientific American Resource Library, Freeman, p. 307-622, 1969
CP-R4	*River Ecology and Man.* R. T. Oglesby, C. A. Carlson, J. A. McCann, eds., Academic, 465 p., 1972
CP-U1	*Understanding Environmental Pollution.* M. A. Strobbe, ed., C. V. Mosby, Saint Louis, 357 p., 1971
CP-U2	*Understanding the Earth; A reader in the Earth Sciences.* I. G. Gass, P. J. Smith, and R. C. L. Wilson, eds., MIT, 355 p., 1971
CP-U3	*Urbanization and Environment: The Physical Geography of the City.* T. R. Detwyler and M. G. Marcus, eds., Duxbury Press (Wadsworth), 287 p., 1972
CP-W1	*Water Pollution Microbiology.* R. Mitchell, ed., Wiley-Interscience, 416 p., 1972

Bibliography of Environmental Science

MULTISUBJECT WORKS (Code MS)
(Textbooks, encyclopedias, commission and committee reports)

MS-A1	*Air Conservation; The Report of the Air Conservation Commission.* Publ. No. 80, AAAS, 335 p., 1965
MS-C1	*Conservation for Survival; An Ecological Strategy.* Kai Curry-Lindahl, William Morrow, New York, 335 p., 1972
MS-C2	*Climate and Weather.* Hermann Flohn, McGraw-Hill, 243 p., 1969
MS-C3	*Climate and Agriculture: An Ecological Survey.* Jen-Hu Chang, Aldine Publ. Co., Chicago, 304 p., 1968
MS-C4	*Clean Water for the 1970's; A Status Report.* Federal Water Quality Admin., U.S. Dept. Interior, USGPO, 80 p., 1970
MS-C5	*The Complete Ecology Fact Book.* P. Nobile and J. Deedy, eds., Anchor Books, Doubleday, 472 p., 1972
MS-E1	*The Earth Sciences,* 2nd ed. Arthur N. Strahler, Harper, 824 p., 1971
MS-E2	*The Encyclopedia of Oceanography.* Encyclopedia of Earth Sciences Series, Vol. 1, Rhodes W. Fairbridge, ed., Reinhold, 1021 p., 1966
MS-E3	*Encyclopedia of the Atmospheric Sciences and Astrogeology.* Encyclopedia of Earth Sciences Series, vol. II, Rhodes W. Fairbridge, ed., Reinhold, 1200 p., 1967
MS-E4	*The Encyclopedia of Geomorphology.* Encyclopedia of Earth Sciences, Vol. III, Rhodes W. Fairbridge, ed., Reinhold, 1295 p., 1968
MS-E5	*Environmental Geology; Conservation, Land-use Planning, and Resource Management.* Peter T. Flawn, Harper, 313 p., 1970
MS-E6	*Environmental Quality; The First Annual Report of the Council on Environmental Quality.* USGPO, 326 p., 1970
MS-E7	*Environmental Quality: The Second Annual Report of the Council on Environmental Quality.* USGPO, 360 p., 1971
MS-E8	*Environment, Power, and Society.* Howard T. Odum, Wiley-Interscience, 331 p., 1971
MS-E9	*Environment and Man.* Richard H. Wagner, Norton, 491 p., 1971
MS-E10	*Earth Materials.* W. G. Ernst, Prentice-Hall, 149 p., 1969
MS-E11	*Encyclopedia of Geochemistry and Environmental Sciences.* Encyclopedia of Earth Sciences, Vol. IVA, Rhodes W. Fairbridge, ed., Van Nostrand Reinhold, 1321 p., 1972
MS-E12	*The Ecology of North America.* Victor E. Shelford, University of Illinois Press, Urbana, Illinois, 610 p., 1963
MS-E13	*Evolution and Plants of the Past.* Harlan P. Banks, Wadsworth, 170 p., 1970
MS-E14	*Ecology and Field Biology.* Robert Leo Smith, Harper, 686 p., 1966
MS-E15	*Ecology: An Evolutionary Approach.* J. Merritt Emlen, Addison-Wesley, Reading, Massachusetts, 493 p., 1973
MS-F1	*Fluvial Processes in Geomorphology.* L. B. Leopold, M. G. Wolman, J. P. Miller, Freeman, 522 p., 1964
MS-F2	*Fundamentals of Soil Science.* 3rd. ed., C. E. Millar, L. M. Turk, and H. D. Foth, Wiley, 526 p., 1958
MS-F3	*Fundamentals of Ecology,* 3rd. ed., Eugene P. Odum, W. B. Saunders Company, Philadelphia, Pennsylvania, 574 p., 1971
MS-G1	*Glacial and Quaternary Geology.* Richard F. Flint, Wiley, 892 p., 1971
MS-I1	*Inadvertent Climate Modification.* Report of the Study of Man's Impact on Climate (SMIC), MIT, 308 p., 1971
MS-I2	*Introduction to the Atmosphere.* 2nd. ed., Herbert Riehl, McGraw-Hill, 516 p., 1972
MS-I3	*Introduction to Meteorology.* 3rd. ed., Sverre Petterssen, McGraw-Hill, 333 p., 1969
MS-M1	*Man's Impact on the Global Environment.* Report of the Study of Critical Environmental Problems (SCEP), MIT, 319 p., 1970
MS-M2	*Modern Hydrology.* 2nd. ed., Raphael G. Kazmann, Harper, 365 p., 1972
MS-M3	*Man and the Environment; An Introduction to Human Ecology and Evolution.* Arthur S. Boughey, Macmillan, 427 p., 1971
MS-M4	*Man in the Living Environment.* Report of the Workshop on Global Ecological Problems, Institute of Ecology, The University of Wisconsin Press, Wisconsin, 288 p., 1972
MS-M5	*Man and the Ocean.* B. J. Skinner and K. K. Turekian, Prentice-Hall, 149 p., 1973

MS-N1	*Natural Resource Conservation: An Ecological Approach.* Oliver S. Owen, Macmillan, 593 p., 1971
MS-O1	*Oceanography; An Introduction to the Marine Environment.* Peter K. Weyl, Wiley, 535 p., 1970
MS-O2	*Our Polluted World; Can Man Survive?* (Revised ed.) John Perry, Franklin Watts, Inc., New York, 237 p., 1972
MS-O3	*Our Precarious Habitat.* Melvin A. Benarde, Norton, 362 p., 1970
MS-P1	*Physical Climatology.* William D. Sellers, Univ. of Chicago Press, Chicago, 272 p., 1965
MS-P2	*The Polar World.* Patrick D. Baird, Wiley, 328 p., 1964
MS-P3	*Population, Resources, Environment; Issues in Human Ecology.* P. R. Ehrlich and A. H. Ehrlich, Freeman, 383 p., 1970
MS-P4	*Principles of Geochemistry.* 3rd. ed., Brian Mason, Wiley, 329 p., 1966
MS-P5	*Principles of Biogeography.* David Watts, McGraw-Hill, 402 p., 1971
MS-P6	*The Plant Kingdom,* 2nd. ed., Harold C. Bold, Prentice-Hall, 118 p., 1964
MS-P7	*Principles of Animal Ecology.* W. C. Allee et al., W. B. Saunders Company, Philadelphia, Pennsylvania, 837 p., 1949
MS-P8	*Plant Physiology.* F. B. Salisbury and C. Ross, Wadsworth, 747 p., 1969
MS-S1	*Streams; Their Dynamics and Morphology.* Marie Morisawa, McGraw-Hill, 175 p., 1968
MS-S2	*The Surface of the Earth,* Arthur L. Bloom. Prentice-Hall, 152 p., 1969
MS-S3	*Submarine Geology.* 2nd. ed., Francis P. Shepard, Harper, 557 p., 1963
MS-S4	*Structure of the Earth.* Sydney P. Clark, Jr., Prentice-Hall, 131 p., 1971
MS-S5	*The Science of Biology,* 4th. ed., Paul B. Weisz, McGraw-Hill, 656 p., 1971
MS-S6	*The Science of Zoology,* 2nd. ed., Paul B. Weisz, McGraw-Hill, 727 p., 1973
MS-T1	*To Live on Earth: Man and His Environment in Perspective.* Sterling Brubaker, Resources for the Future, Hopkins, 202 p., 1972
MS-V1	*Vascular Plants: Form and Function,* 2nd. ed., F. B. Salisbury and R. V. Parke, Wadsworth, 252 p., 1970
MS-W1	*Weather and Life: An Introduction to Biometeorology.* William P. Lowry, Academic, 305 p., 1969
MS-W2	*Water: The Web of Life.* C. A. Hunt and R. M. Garrels, Norton, 208 p., 1972

ABBREVIATIONS OF PERIODICALS

Amer. Sci.	American Scientist
Ann. AAG	Annals, Association of American Geographers
Bull. AMS	Bulletin, American Meteorological Society
Envir.	Environment
EOS	EOS—Transactions of the American Geophysical Union
Geog. Rev.	Geographical Review
Geotimes	Geotimes; News of the Earth Sciences
NCAR Quart.	National Center for Atmospheric Research Quarterly
Science	Science
Sci. News	Science News
Sci. Amer.	Scientific American
Tech. Rev.	Technology Review, Massachusetts Institute of Technology

ABBREVIATIONS OF PUBLISHERS

AAAS	American Association for the Advancement of Science, 1515 Massachusetts Ave., Washington, DC 20005
Academic	Academic Press, Inc., New York, NY 10003
Doubleday	Doubleday & Company, Inc., Garden City, NY 11530
Freeman	W. H. Freeman and Co., Publishers, 660 Market St., San Francisco, CA 94104
Harper	Harper & Row, 10 East 53rd St., New York, NY 10022
Harvard	Harvard University Press, 79 Garden St., Cambridge, MA 02138
Hopkins	The Johns Hopkins Press, Johns Hopkins Univ., Baltimore, MD 21200

Bibliography of Environmental Science

Macmillan	The Macmillan Company, 866 Third Ave., New York, NY 10022
McGraw-Hill	McGraw-Hill Book Co., 205 West 42nd St., New York, NY 10036
MIT	The M.I.T. Press, 50 Ames St., Cambridge, MA 02142
Nat. Hist. Press	Natural History Press-Doubleday & Co., Inc., Garden City, NY 11530
Norton	W. W. Norton & Company, Inc., 55 Fifth Avenue, New York, NY 10003
Pergamon	Pergamon Press, Inc., Maxwell House, Fairview Park, Elmsford, NY 10523
Prentice-Hall	Prentice-Hall, Inc., Englewood Cliffs, NJ 07632
Reinhold	Van Nostrand Reinhold Co., 450 West 33rd St., New York, NY 10001
Wadsworth	Wadsworth Publishing Co., Belmont, CA 94022
Wiley	John Wiley & Sons, Inc., 605 Third Ave., New York, NY 10016
Wiley-Interscience	Wiley-Interscience, 605 Third Ave., New York, NY 10016
Yale	Yale University Press, 149 York St., New Haven, CT 06511

ABBREVIATIONS OF SOCIETIES, ORGANIZATIONS, AND GOVERNMENTAL DEPARTMENTS

AAAS	American Association for the Advancement of Science, Washington, DC
AAG	Association of American Geographers, Washington, DC
AEC	United States Atomic Energy Commission, Washington, DC
AGU	American Geophysical Union, Washington, DC
AMS	American Meteorological Society, Boston, MA
NAS-NRC	National Academy of Sciences—National Research Council, Washington, DC
NSF	National Science Foundation, Washington, DC
USDI	United States Department of the Interior, Washington, DC
USGPO	Superintendent of Documents, United States Government Printing Office, Washington, DC
USGS	United States Geological Survey, Washington, DC

Introduction to Environmental Science

Environmental Science—General

Outline of environmental science. R. L. Metcalf and J. N. Pitts, Jr., CP-A1: 1–23, 1969

Perception of environment. Thomas F. Saarinen, Commission on College Geography, Resource Paper No. 5, AAG, Washington, D.C., 37 p., 1969

The Atmospheric Sciences and Man's Needs: Priorities for the Future. Comm. on Atmospheric Sci., NAS-NRC, 88 p., 1971

Environmental Science: Challenge for the Seventies. Nat. Sci. Board, NSF, USGPO, 50 p., 1971/CP-M1: 318–322 (excerpts)

The responsible role of the atmospheric sciences in determining the future quality of Man's environment. S. M. Greenfield, *Bull. AMS* 52: 94–97, 1971

Environmental science. MS-E11: 337–341, 1972

Man and Environment—General

Deserts on the march. Paul B. Sears, CP-E2: 142–151, 1935

Environmental changes through forces independent of Man. Richard J. Russell, CP-M6: 453–470, 1956

Our world from the air: conflict and adaptation. E. A. Gutkind, CP-M6: 1–44, 1956

The processes of environmental change by Man. Paul B. Sears, CP-M6: 471–484/CP-E8: 129–138, 1956

Man versus nature. Peter Farb, CP-E2: 14–23, 1963

Man and his environment; scope, impact, and nature. René Dubois, CP-M4: 684–694, 1966

Can the world be saved? Lamont C. Cole, CP-E2: 4–13, 1967

The economics of wilderness. Garrett Hardin, CP-E2: 173–177, 1969

Man's use of the environment: the need for ecological guidelines. R. O. Slatyer, CP-M1: 17–26, 1969

The dimensions of intervention; introduction. CP-M2: 117–126, 1970

The dirty animal—Man. Joseph L. Myler, CP-E2: 116–141, 1970

Nature, not only Man, degrades environment. USGS, CP-M1: 250–251, 1970

Understanding environmental problems. MS-E6: 5–18, 1970

The ecosphere and preindustrial Man; introduction. CP-M2: 6–9, 1971

Man "masters" the environment. MS-E9: 22–37, 1971

Modern man and environment. Thomas R. Detwyler, CP-M4: 2–9, 695–700, 1971

A people at war with their land. Raymon F. Dasmann, CP-A4: 8–31, 1971

The world system. MS-E8: 1–25, 1971

Man and environment; conceptual frameworks. K. Hewitt and F. K. Hare, Commission on College Geography, Resource Paper No. 20, AAG, Washington, D.C., 39 p., 1973

Energy Systems

Matter and Energy—Principles

Life and Energy. Isaac Asimov, Bantam Books, New York, 378 p., 1965
States of matter; crystalline state. MS-P4: 73-81, 1966
Newton's laws; vectors and scalars; fluid motion; potential energy. MS-O1: 111-133, 1970
Phases of pure water. MS-O1: 91-94, 1970
Energy in the universe. Freeman J. Dyson, CP-E7: 19-27, 1971
The measurement of the "Man-day". Engene S. Ferguson, *Sci. Amer.*, 225(4): 96-103, 1971
What power is. MS-E8: 26-57, 1971

Energy Systems—Concepts and Principles

The theory of open systems in physics and biology. Ludwig von Bertalanffy, *Science,* 111: 23-29, 1950
Fundamental thermodynamic equations. MS-P4: 67-73, 1966
The laws of thermodynamics. MS-P3: 54-55, 1970
Power, Environment, and Society. Howard T. Odum, Wiley-Interscience, 331 p., 1971
Concepts of pollution and its control. W. Stumm and E. Stumm-Zollinger, *Tech. Rev.* 75(1), 19-25, 1972

Chapter 1
Atmosphere and Oceans

Atmospheric Sciences

Atmospheric sciences. MS-E3: 93-97, 1967
Climatology. MS-E3: 217-230, 1967
About meteorology. MS-I3: 1-26, 1969
Weather satellites: II. Arthur W. Johnson, *Sci. Amer.,* 220(1): 52-67, 1969
The atmosphere. Thomas F. Malone, CP-N1: 163-170, 1970
Toward defining human needs: how does the atmosphere hurt us? Frederick Sanders, *Bull. AMS* 52: 446-449, 1971

Composition and Structure of the Atmosphere

The origin of the atmosphere. Helmut E. Landsberg, CP-R2: 181-183, 1953
The atmosphere: composition; geologic history. MS-P4: 208-222, 1966
Aeronomy. MS-E3: 3-7, 1967
Atmosphere, general; atmospheric chemistry. MS-E3: 61-66, 1967
Ionosphere. MS-E3: 498-501, 1967
Mesosphere. MS-E3: 556-559, 1967
Ozone, atmospheric. MS-E3: 719-721, 1967
Pressure, atmospheric. MS-E3: 779-784, 1967
Standard atmosphere. MS-E3: 942-944, 1967
Stratosphere. MS-E3: 953-956, 1967
Tropopause; troposphere. MS-E3: 1038-1040, 1967
Upper atmosphere. MS-E3: 1064-1073, 1967
Atmospheric layers. MS-C2: 80-87, 1969
Thermal structure of the atmosphere; ozone layer; ionosphere. MS-I3: 41-47, 1969
The atmosphere and oceans. MS-E1: 181-191, 1971
Outgassing of the planet earth. MS-E11: 836-839, 1972
Oxygen: evolution in the earth's atmosphere. MS-E11: 849-861, 1972
A survey of the atmosphere. MS-I2: 3-30, 1972

Oceans—General

Chemical oceanography, general. MS-E2: 186-191, 1966
Oceanography, physical. MS-E2: 623-632, 1966
The ocean. Roger Revelle, CP-O2: 3-13, 1969
The ocean and man. Warren S. Wooster, CP-O2: 123-130, 1969
Technology and the ocean. Willard Bascomb, CP-O1: 369-380/CP-O2: 109-120, 1969
History of oceanography. MS-O1: 27-52, 1970
The ocean basins. MS-O1: 271-289, 1970
The sea, Taylor A. Pryor. CP-N1: 115-121, 1970
Temperature-salinity structure near the sea surface. MS-O1: 170-175, 1970
Man and the sea. CP-O1: 324-336, 1971
Oceanography, Readings from Sci. Amer. Freeman, 417 p., 1971
Oceanography—some perspectives. CP-O1: 1-10, 1971
The oceans: composition; temperature; salinity; density; pressure. MS-E1: 191-196, 1971
Seawater: chemistry, history. MS-E11: 1062-1078, 1972
Man turns to the sea; the oceanic realm. MS-M5: 1-23, 1973

Chapter 2
The Earth's Radiation Balance

Solar Radiation

Energy Exchange in the Biosphere. David M. Gates, Harper, 151 p., 1962 (Chap. 2,3)
The energy environment in which we live. David M. Gates, *Amer. Sci.,* 51: 327-348, 1963
Direct Use of the Sun's Energy. Farrington Daniels, Yale, 374 p., 1964 (Chap. 3)
Solar radiation. MS-P1: 11-40, 1965
Spectral distribution of solar radiation at the earth's surface. David M. Gates, *Science* 151: 523-529, 1966
Gamma rays; X rays. MS-E3: 412, 1967
Insolation. MS-E3: 480-484, 1967
Radiation laws. MS-E3: 793-794, 1967
Solar radiation. MS-E3: 881-885, 1967
Radiation; spectra, laws. MS-W1: 13-26, 1969
Solar radiation. MS-C2: 10-16, 1969
Solar radiation. MS-S2: 8-12, 1969
The sun and our atmosphere. MS-I3: 26-41, 1969
Sunlight; thermal radiation. MS-O1: 69-76, 1970
Planet earth in the sun's rays. MS-E1: 49-60, 1971
Radiative processes; incoming solar radiation. MS-I1: 76-83, 1971
The sun; solar radiation. MS-E1: 76-80, 1971
Short- and longwave radiation. MS-I2: 31-42, 1972

The Radiation Balance

Energy Exchange in the Biosphere. David M. Gates, Harper, 151 p., 1962
The heat and water budget of the earth's surface. David H. Miller, *Advances in Geophysics,* Vol. 11, Academic, p. 175-302, 1965
Infrared radiation. MS-P1: 40-64, 1965
Radiation instruments. MS-P1: 69-81, 1965
Radiation balance. MS-P1: 65-68, 1965
Albedo and reflectivity. MS-E3: 12-13, 1967

Bibliography of Environmental Science

Atmospheric radiation. MS-E3: 92-93, 1967
Energy budget of the earth's surface. MS-E2: 250-256, 1966/MS-E3: 355-361, 1967
Greenhouse effect. MS-E3: 438-441, 1967
Microclimates. MS-E3: 593-599, 1967
Radiation—terrestrial. MS-E3: 794-805, 1967
Radiation balance. MS-C3: 4-22, 1968
A Survey Course: The Energy and Mass Budget at the Surface of the Earth. David H. Miller, AAG Commission on College Geography, Publ. No. 7, 142 p., 1968
Energy balance of the earth-atmosphere system; energy budget concept. MS-W1: 26-29, 113-121, 1969
Radiation balance—gains and losses; variation with latitude; equation of balance. MS-I3: 55-60, 1969
Radiation and the heat balance. MS-C2: 10-39, 1969
Satellite observations of the earth's radiation budget. T. H. Vonder Haar and V. E. Suomi, *Science* 163: 667-668, 1969
Solar radiation. MS-S2: 8-12, 1969
Windows of the atmosphere. MS-I3: 50-52, 1969
The energy cycle of the earth. Abraham H. Oort, CP-B1: 14-23, 1970
Radiation balance of earth. MS-O1: 84-89, 1970
The earth's radiation balance. MS-E1: 197-211, 1971
Feedback mechanisms. MS-I1: 114-123, 1971
Global-average mathematical models. MS-I1: 110-114, 1971
Outgoing infrared radiation. MS-I1: 83-88, 1971
Radiation, heating, and cooling. MS-I2: 42-60, 1972

Cosmic Particles (Cosmic Radiation)

Cosmic rays. MS-E3: 253-295, 1967
Cosmic particles (cosmic rays). MS-E1: 89-90, 1971
Polarity reversal and faunal extinction. D. I. Black, CP-U2: 257-261, 1971

Magnetosphere, Solar Wind, Radiation Belts

Sun clouds and rain clouds. Walter Orr Roberts, CP-R3: 349-354, 1957
The earth as a dynamo. Walter M. Elasser, CP-R2: 185-189, 1958
Artificial satellites and the earth's atmosphere. CP-R3: 363-369, 1959
The Antarctic and the upper atmosphere. Sir Charles Wright, CP-R3: 433-442, 1962
Aurora. MS-E3: 105-106, 1967
Geomagnetic disturbances. MS-E3: 412-420, 1967
Magnetosphere. MS-E3: 537-539, 1967
Solar wind. MS-E3: 903-905, 1967
Van Allen radiation belts. MS-E3: 1075-1083, 1967
Solar wind; magnetosphere; Van Allen belts. MS-I3: 29-37, 47-50, 1969
The earth as a magnet; ionosphere; magnetosphere; radiation belts. MS-E1: 91-112, 1971
The earth's magnetic field. MS-S4: 26-30, 1971
The earth's magnetic field and its origin. Sir Edward Bullard, Cp-U2: 71-79, 1971

Chapter 3
Thermal Environments of the Earth's Surface

The Heat Balance

Energy Exchange in the Biosphere. David M. Gates, Harper, 151 p., 1962 (Chap. 4)
Energy (heat) balance of the earth-atmosphere system. MS-P1: 114-126, 1965
Energy (heat) balance of the earth's surface. MS-P1: 100-114, 1965
The heat and water budget of the earth's surface. David H. Miller, *Advances in Geophysics,* Vol. 11, Academic, p. 175-302, 1965
Turbulent transfer (of heat) and wind relationships. MS-P1: 141-155, 1965
A Survey Course: The Energy and Mass Budget at the Surface of the Earth. David H. Miller, AAG Commission on College Geography, Publ. No. 7, 142 p., 1968
The heat balance. MS-I3: 53-67, 1969
The interplay of ocean and atmosphere. NCAR Quart. No. 23, 1969
Radiation and the heat budget. MS-C2: 10-39, 1969
Wind, advection, and turbulent transfer. MS-W1: 88-109, 1969
The energy cycle of the earth. Abraham H. Oort, CP-B1: 14-23, 1970
Heat balance of oceans. MS-O1: 161-165, 1970
The heat balance. MS-I1: 88-92, 1971
Heat and cold at the earth's surface; heat balance equation. MS-E1: 212-215, 1971
Oceanic circulation and the earth's heat balance. MS-E1: 266-268, 1971
Heat balance. MS-I2: 276-284, 1972

Heating and Cooling of the Soil

Heat transfer in soil. MS-P1: 127-140, 1965
Soil temperature. MS-C3: 87-99, 1968
Air temperature near the ground; heat conduction in soil. MS-W1: 35-61, 1969
Heating and cooling of the ground. MS-E1: 215-217, 1971

Arctic Permafrost (See Chapter 11)

Heating and Cooling of Lakes and Oceans

Temperature structure in the sea. MS-E2: 902-910, 1966
Lakes—effects on climate. MS-E3: 526-527, 1967
Air-sea interaction; heat exchanges. MS-O1: 155-161, 1970
Phases of water; specific volume; sea-water temperature. MS-O1: 91-109, 1970
Energy absorption by water bodies; annual cycle of heating and cooling of lake and sea surfaces; sea surface temperatures. MS-E1: 219-223, 1971

Sea Ice

The Arctic Ocean. P. A. Gordienko, CP-O1: 92-104, 1961
Salt-water ice; sea ice of arctic. MS-P2: 94-110, 1964
Drifting ice stations. MS-E2: 232-233, 1966
Icebergs. MS-E2: 367-369, 1966
On a floating island, Victor P. Hessler. *Science:* 151: 1360-1362, 1966
Sea ice. MS-E2: 777-782, 1966
Arctic meteorology. MS-E3: 28-38, 1967
Formation of sea ice; icebergs. MS-O1: 175-185, 1970
The case of the Arctic Ocean ice. MS-I1: 72-73, 1971
Removal of the Arctic sea ice. MS-I1: 159-162, 1971
Sea ice and icebergs. MS-E1: 283-287, 1971

Air Temperatures

Climate and agriculture. Frits W. Went, CP-R3: 371-381, 1957

Meteorology and climate of the arctic. MS-P2: 48-65, 1964
On the annual temperature range over the southern oceans. Harry van Loon, *Geog. Rev.* 56: 497-515, 1966
Areal patterns of seasonal temperature anomalies in the vicinity of the Great Lakes. Richard J. Kopec. *Bull. AMS* 48: 884-889, 1967
Climatic classification; climatic data. MS-E3: 171-201, 1967
Lapse rate. MS-E3: 527-530, 1967
Microclimates. MS-E3: 593-599, 1967
Mountain (highland) climates. MS-E3: 662-666, 1967
Temperature in the atmosphere. MS-E3: 982-990, 1967
Frost protection. MS-C3: 100-108, 1968
Annual and diurnal variations of temperature. MS-I3: 67-73, 1969
Environmental temperature. MS-W1: 31-49, 1969
Temperature regimes. MS-I3: 248-260, 1969
Mean global temperatures. MS-O1: 77-82, 1970
Air temperatures; daily cycle; land-water contrasts; seasonal cycle; latitude and altitude effects; global patterns. MS-E1: 223-229, 1971
Altitude and air temperature. MS-E1: 228, 1971
Forest climate. MS-I2: 362-369, 1972
Surface temperature. MS-I2: 288-293, 313-341, 1972

Chapter 4
Circulation Systems in Atmosphere and Oceans

Winds and the General Atmospheric Circulation

The general circulation of the atmosphere. Victor P. Starr, CP-R2: 288-293, 1956
The circulation of the upper atmosphere. Reginald E. Newell, *Sci. American,* 210 (4): 62-67, 1964
New steps in tropical meteorology (Hadley cell). *NCAR Quart.* No. 8, 1965
The circulation of the atmosphere. Edward N. Lorenz, *Amer. Sci.* 54: 402-420, 1966
Wind, principles, MS-E2: 989-993, 1966
Atmospheric circulation, global. MS-E3: 71-78, 1967
Ferrel's law and Buys Ballot's law. MS-E2: 266-267/MS-E3: 383-385, 1967
Geostrophic wind. MS-E3: 422-425, 1967
Horizontal sounding balloons, *NCAR Quart.,* Summer, 1967
Intertropical convergence zone (ITCZ); intertropical front. MS-E3: 496-497, 1967
Pressure gradient. MS-E3: 784-785, 1967
Trial balloons in the Southern Hemisphere (GHOST). Vincent E. Lally, *Science,* 155: 456-459, 1967
Winds—principles. MS-E3: 1147-1151, 1967
Zonal circulation and index. MS-E3: 1163-1167, 1967
Atmospheric circulation. MS-C2: 87-117, 1969
The general circulation. MS-I3: 176-197, 1969
Laws of motion, atmospheric. MS-I3: 148-170, 1969
Air and the circulation of the atmosphere. MS-O1: 135-154, 1970
Clear air turbulence: a mystery may be unfolding. John A. Dutton and Hans A. Panofsky, *Science* 167: 937-944, 1970
Atmospheric circulation. MS-E1: 232-255, 1971
Atmospheric wind systems. MS-I1: 92-96, 1971
The global circulation of atmospheric pollutants. Reginald E. Newell, *Sci. Amer.* 224 (1): 32-42, 1971
Large-scale motion of atmosphere. MS-I2: 167-196, 1972
Vertical mixing of air below the clouds. MS-I2: 65-84, 1972

Coriolis Effect

The Coriolis effect. James E. McDonald, CP-O1: 60-63/CP-R2: 276-279, 1952
Coriolis force. MS-E2: 224/MS-E3: 251, 1967
Coriolis force. MS-C2: 92-93, 1969
The deviating force. MS-I3: 153-155, 1969
Coriolis acceleration. MS-O1: 117-123, 1970
The Coriolis effect. MS-E1: 29-32, 235-236, 1971

Upper Air Waves, Jet Streams

Jet Streams, Elmar R. Reiter, Doubleday, 189 p., 1967
Jet streams. MS-E3: 510-514, 1967
Extratropical regimes; index cycle; jet stream. MS-I3: 182-190, 1969
Fronts and waves aloft. MS-I3: 223-225, 1969
Upper air waves; jet streams. MS-C2: 98-117, 1969
Global circulation in middle and high latitudes; development of upper-air waves; the jet stream. MS-E1: 242-251, 1971
Flight planning and the jet stream. MS-I2: 435-442, 1972

Global Surface Winds, Monsoon Circulation

Monsoons. MS-E3: 613-617, 1967
Trade winds. MS-E3: 1006-1008, 1967
Westerlies (mid-latitude west winds). MS-E3: 1135-1137, 1967
Influences of continents and oceans; monsoon systems. MS-I3: 190-197, 1969
The Chinese monsoon. Jen-Hu Chang, *Geog. Rev.* 61: 370-395, 1971
Surface wind systems of the earth; monsoon systems. MS-E1: 251-255, 1971

Local Winds, Drainage Winds, Chinook

Katabatic winds in the equatorial Andes. MS-E3: 518-522, 1967
Mountain and valley winds. MS-E3: 666, 1967
Mountain waves and foothill winds. *NCAR Quart.* No. 17, 1967
Sea breeze and land breeze. MS-E3: 857-858, 1967
Winds—local. MS-E3: 1151-1156, 1967
Land and sea breezes. MS-C2: 87-90, 1969
Local wind systems. MS-I3: 171-175, 1969
Sea and land breezes; valley and mountain breezes. MS-E1: 234-235, 1971
Effect of large mountain ranges (on circulation). MS-I2: 291-293, 1972
Influence of mountains on winds. MS-I2: 350-357, 1972
Land and sea breeze. MS-I2: 357-358, 1972

Wind-Generated Ocean Waves

Ocean waves, Willard Bascomb. CP-O1: 45-55/CP-R2: 205-215, 1959
Ocean waves and associated currents. (D. L. Inman), MS-S3: 49-81, 1963
Waves and Beaches. Willard Bascom, Doubleday, 260 p., 1964

Ocean waves. MS-E2: 640-644, 1966
Effect of wind on the sea surface—waves. MS-O1: 165-170, 1970
Ocean waves; waves and wind; swell; wave energy. MS-E1: 269-278, 1971

Oceanic Circulation

The anatomy of the Atlantic. Henry Stommel, CP-R2: 79-84, 1955
The circulation of the oceans. Walter Munk, CP-O1: 64-69/CP-R2: 98-103, 1955
The Sargasso Sea. John H. Ryther, CP-O1: 77-81, 1956
The circulation of the abyss. Henry Stommel, CP-O1: 71-76, 1958
The Antarctic Ocean. V. G. Kort, CP-O1: 83-91/CP-R3: 457-465, 1962
Ocean currents; tidal currents; turbidity currents. MS-S3: 90-100, 1963
Convergence and divergence; Antarctic convergence. MS-E2: 214-219, 1966
Dynamics of ocean currents. MS-E2: 233-237, 1966
Ekman spiral. MS-E2: 245-246, 1966
Gulf Stream. MS-E2: 335-339, 1966
Ocean currents; oceanic circulation. MS-E2: 587-597, 1966
Upwelling. MS-E2: 957-959, 1966
The atmosphere and the ocean. R.W.Stewart,CP-O1:35-44/CP-O2: 28-38, 1967
The interplay of ocean and atmosphere. NCAR Quart., No. 23, 1969
Deep circulation of the ocean. MS-O1: 481-502, 1970
Wind-driven circulation of the ocean. MS-O1: 187-200, 1970
General circulation patterns in the world ocean. Joseph L. Reid, CP-M5: 448-459, 1971
Ocean pollution, Ferren MacIntyre and R. W. Holmes. CP-E4: 235-240, 1971
Oceanic circulation; surface currents; upwelling, deep ocean currents. MS-E1: 256-266, 1971
Circulation of the oceans. MS-M5: 16-20, 1973

Ocean Pollution—General

Radioactive waste in the ocean. MS-E2: 726-730, 1966
The Frail Ocean. Wesley Marx, Ballantine Books, New York, 274 p., 1967
How not to kill the ocean. Wesley Marx, CP-U1: 41-47, 1969
The chemical invasion of the oceans by Man. Edward D. Goldberg, CP-G1: 178-185, 1970
Interactions between oceans and terrestrial ecosystems. Bengt Lundholm, CP-G1: 195-201, 1970
Ocean dumping: a national policy. Council on Environmental Quality, CP-M1: 123-133, 1970
The sea: should we now write it off as a future garbage pit? Robert W. Riseborough, CP-N1: 121-136, 1970
Stagnant sea (Baltic Sea pollution). Stig H. Fonselius, Envir. 12(6): 2-11, 40-48, 1970
Chemical invasion of ocean by Man. Edward D. Goldberg, CP-M5: 261-274, 1971
Chemical wastes in the sea: new forms of marine pollution. P. A. Greve, Science 173: 1021-1022, 1971
Chlorinated hydrocarbons in the marine environment. SCEP Task Force, CP-M5: 275-296, 1971
Identification of globally distributed wastes in the marine environment. Edward D. Goldberg and M. Grant Gross, CP-M5: 371-376, 1971
Impingement of Man on the Oceans. Donald W. Hood, ed, Wiley-Interscience, 738 p., 1971
Modification of the marine environment—know before you do. S. Fred Singer, EOS 52: 579-580, 1971
Ocean pollution. Ferren MacIntyre and R. W. Holmes, CP-E4: 230-253, 1971
Proposal for a base-line sampling program. Edward D. Goldberg, et al, CP-M5: 377-391, 1971
Oceans as alphabet soup: focus on DDT and PCB's. Richard Gilluly, Sci. News 101: 30-31, 1972
Pollution of the oceans. MS-M5: 124-143, 1973

Chapter 5
Atmospheric Energy Releases

Atmospheric Water Vapor

Humidity. MS-E3: 444-447, 1967
Latent heat. MS-E3: 530-531, 1967
Relative humidity. MS-E3: 828-829, 1967
Environmental moisture. MS-W1: 64-79, 1969
Atmospheric moisture; humidity. MS-E1: 288-292, 1971
Evaporation. MS-M2: 50-66, 1972
Moisture in the atmosphere. MS-I2: 89-100, 1972

Air Masses

Air masses of the arctic. MS-P2: 49-55, 1964
Air masses, world maps. MS-E3: 10-11, 1967
Air masses. MS-I3: 199-209, 1969
Air masses. MS-E1: 309-313, 315, 1971

Condensation, Clouds, Fog

Salt and rain. A. H. Woodcock, CP-O1: 105-109/CP-R3: 355-360, 1957
Cloud physics and cloud seeding. Louis J. Battan, Doubleday, 144 p., 1962
Cloud patterns over tropical oceans. Joanne S. Malkus, Science 141: 767-778, 1963
Aerosols—their complex role in rainfall. NCAR Quart., No. 9, 1965
Fog frequency in the United States. A. Court and R. D. Gerston, Geog. Rev. 56: 543-550, 1966
Atmospheric nuclei and dust. MS-E3: 81-85, 1967
Clouds, cloud physics. MS-E3: 234-238, 1967
Dew; dew point, MS-E3: 306-307, 1967
Fog; smog; mist. MS-E3: 392-395, 1967
Frost; hoarfrost; black frost. MS-E3: 402, 1967
Ice phase in the atmosphere. MS-E3: 475-477, 1967
Salt nuclei in atmosphere. MS-E3: 844-847, 1967
Snowflakes. MS-E3: 865-868, 1967
Clouds (chart of cloud forms), ESSA/PI 680002, USGPO, 1968
Dew, fog, and humidity. MS-C3: 225-232, 1968
Fog. Joel N. Myers, Sci. Amer., 219(6): 75-81, 1968
Clouds and precipitation. MS-C2: 42-78, 1969
Condensation; growth of cloud droplets. MS-I3: 92-100, 1969
Fog—process, types, frequency. MS-I3: 137-147, 1969
Types of clouds and states of sky. MS-I3: 74-88, 1969
Dew-point; fog and smog; adiabatic process; clouds. MS-E1: 291-298, 1971
Condensation; clouds; fog. MS-I2: 100-126, 1972
Fog. coastal. MS-I2: 358-359, 1972

Bibliography of Environmental Science

Water and atmosphere. MS-W2: 71–87, 1972

Cloud growth and precipitation processes. (Staff), *NCAR Quart.* 39, 1973

Snow crystals. C. Knight and N. Knight, *Sci. Amer.* 228(1): 100–107, 1973

Precipitation Processes and Forms, Thunderstorms

The Nature of Violent Storms. Louis J. Battan, Doubleday, 158 p., 1961

Adiabatic phenomena. MS-E3: 2–3, 1967

Precipitation. MS-E3: 772–777, 1967

Stability, atmospheric. MS-E3: 941–942, 1967

Thunder; thunderstorms. MS-E3: 998–999, 1967

Thunderstorms, ESSA/PI 670004 (folder), USGPO, 1967

The urban snow hazard in the United States. John F. Rooney, Jr., *Geog. Rev.* 57: 538–559, 1967

Classification of precipitation. MS-I3: 88–91, 1969

Clouds and precipitation. MS-C2: 42–78, 1969

Condensation and precipitation processes. MS-I3: 92–103, 1969

Showers; thunderstorms; hail; tornadoes. MS-I3: 116–136, 1969

Stability and instability; adiabatic processes. MS-I3: 104–115, 1969

Lake effect snowfall to the lee of the Great Lakes: its role in Michigan. Val L. Eichenlaub, *Bull. AMS* 51: 403–412, 1970

Satellite observations of lightning. J. A. Vorpahl, J. G. Sparrow, and E. P. Ney, *Science* 169: 860–862, 1970

Adiabatic process; precipitation forms, measurement, and production. MS-E1: 292–304, 1971

Convection and thunderstorms. MS-E1: 300–302, 327–328, 1971

Deciphering hailstones. *NCAR Quart.*, No. 31, 1971

The dynamics of convection (thunderstorms). *NCAR Quart.*, No. 30, 1971

Hailstones. C. Knight and N. Knight, *Sci. Amer.*, 224(4): 97–103, 1971

Thunderstorms on the Great Plains, *NCAR Quart.*, No. 32, 1971

Influence of mountains on precipitation. MS-I2: 344–349, 1972

Formation of precipitation; forms. MS-I2: 127–158, 1972

Precipitation. MS-M2: 22–37, 1972

Wave Cyclones and Fronts (Extratropical)

Cyclogenesis. MS-E3: 300–303, 1967

Fronts and frontogenesis. MS-E3: 397–402, 1967

Polar front. MS-E3: 764–765, 1967

Fronts, cyclones, and anticyclones. MS-I3: 210–234, 1969

Weather and weather forecasting. MS-C2: 120–136, 1969

Air masses, fronts, and storms. MS-E1: 308–327, 1971

Weather disturbances in middle and high latitudes. MS-I2: 201–233, 1972

Foehn and Chinook Winds
(See Chapter 4, Local Winds)

Tornadoes

Tornadoes. Morris Tepper, CP-R3: 341–347, 1958

The Nature of Violent Storms. Louis J. Battan, Doubleday, 158 p., 1961

Tornado. ESSA/PI 660028, USGPO, 1967

Tornadoes. MS-E3: 1003–1005, 1967

Waterspouts. MS-E3: 1111–1112, 1967

Killer storms. George P. Cressman, *Bull. AMS* 50: 850–855, 1969

Tornadoes. MS-I3: 129–130, 133–136, 1969

Tornadoes. Edwin Kessler, *Bull. AMS* 51: 926–936, 1970

Tornadoes. MS-E1: 327–328, 1971

The tornado threat: coping styles of the North and South. J. H. Sims and D. B. Baumann, *Science* 176: 1386–1392, 1972

Tornadoes. MS-I2: 158–162, 1972

Tropical Cyclones, Low-Latitude Weather Disturbances

The origin of hurricanes. Joanne Starr Malkus, CP-R3: 333–339, 1957

The Nature of Violent Storms. Louis J. Battan, Doubleday, 158 p., 1961

On the origin and possible modification of hurricanes. Herbert Riehl, *Science* 141: 1001–1010, 1963

Storm surges. MS-E2: 856–860, 1966

A case history of residential property damage as a hurricane moves inland. W. C. Cullen and L. W. Crow, *Bull. AMS* 48: 10–12, 1967

Characteristics of hurricanes. Banner I. Miller, *Science* 157: 1389–1399, 1967

The Indian summer monsoon. Jen-Hu Chang, *Geog. Rev.* 57: 373–396, 1967

Tropical cyclones. MS-E3: 1027–1030, 1967

Tropical meteorology. MS-E3: 1030–1037, 1967

Hurricane Information and Tracking Chart, ESSA/PI 680006 (folder), USGPO, 1969

Killer storms. George P. Cressman, *Bull. AMS* 50: 850–855, 1969

Tropical disturbances, storms, and hurricanes. MS-I3: 235–247, 1969

Tropical zone weather phenomena. MS-C2: 136–143, 1969

Atmospheric-oceanic observations in the tropics. M. Garstang, et al, *Amer. Sci.,* 58: 482–495, 1970

The deadliest tropical cyclone in history? N. L. Frank and S. A. Husain, *Bull. AMS* 52: 438–444, 1971

The Hazardousness of a Place: A regional Ecology of Damaging Events. K. Hewitt and I. Burton, Univ. of Toronto Press, 154 p., 1971

Storm surges. MS-E1: 282–283, 1971

Weather disturbances of low latitudes; tropical cyclones. MS-E1: 328–333, 1971

The national hurricane warning program. Robert M. White, *Bull. AMS* 53: 631–633, 1972

Weather disturbances in the tropics. MS-I2: 235–270, 1972

World Precipitation Regions, World Climate Data

Climatic classification; climatic data. MS-E3: 171–201, 1967

Cloudiness and rainfall. MS-E3: 230–233, 1967

Rainfall distribution, global. MS-E3: 816–820, 1967

Climate and climatic zones. MS-C2: 156–193, 1969

Precipitation regimes. MS-I3, 261–275, 1969

World climates, MS-I3, 276–288, 1969

World precipitation patterns. MS-E1: 304–306, 1971

World precipitation. MS-I2: 313–341, 1972

Chapter 6
Man's Impact upon the Atmosphere

Climatic Fluctuations—Natural, Manmade

Short-period climatic fluctuations. Jerome Namias, *Science* 147: 696–706, 1965

Anomalous influences on the weather. *NCAR Quart.*, No. 14, 1966

Possibilities of major climatic modification and their implications: Northwest India, a case for study. R. A. Bryson and D. A. Baerreis, *Bull. AMS* 48: 136–142, 1967

Climatic variations. MS-C2: 196–244, 1969

Earth's cooling climate. Kendrick Frazier, *Sci. News* 96: 458–459, 1969

Irrigation and climate. Kendrick Frazier, *Sci. News* 96: 599–600, 1969

Ocean circulation and climatic changes. S. I. Rasool and J. S. Hogan, *Bull. AMS* 50: 130–134, 1969

One thousand centuries of climatic record from Camp Century on the Greenland Ice Sheet. W. Dansgaard et al, *Science* 166: 377–381, 1969

The atmosphere. Thomas F. Malone, CP-N1: 163–170, 1970

Climatic effects of atmospheric polution. R. A. Bryson and W. M. Wenland. CP-G1: 130–138, 1970

Climatic effects of Man's activities. MS-M1: 9–19, 40–112, 1970

Ice cores: clues to past climates. Louise Purrett, *Sci. News* 98: 369–370, 1970

Impact of land and sea pollution on the chemical stability of the atmosphere. F. D. Sisler, CP-G1: 12–24, 1970

The inadvertent modification of the atmosphere by air pollution. Vincent J. Schaffer, CP-G1: 158–174, 1970

Man-made climatic changes. Helmut E. Landsberg, *Science* 170: 1265–1274, 1970

Man's inadvertent modification of weather and climate, Council on Environmental Quality, *Bull. AMS* 51: 1043–1047/MS-E6: 93–104, 1970

The oxygen and carbon dioxide balance in the earth's atmosphere. Francis S. Johnson, CP-G1: 4–11, 1970

A preliminary evaluation of atmospheric pollution as a cause of the global temperature fluctuation of the past century. J. Murray Mitchell, Jr., CP-G1: 139–155, 1970

Climatic effects of Man's activities: Summary of SCEP Report. CP-M5: 174–183, 1971

Extended industrial revolution and climate change. W. R. Frisken, *EOS* 52: 500–508, 1971

Future climates and future environments. F. Kenneth Hare, *Bull. AMS* 52: 451–456, 1971

The global circulation of atmospheric pollutants. Reginald E. Newell, *Sci. Amer.* 224(1): 32–42, 1971

Global environmental monitoring, George D. Robinson, *Tech. Rev.* 73(7): 19–27, 1971

Inadvertent Climate Modification, Report of the Study of Man's Impact on Climate (SMIC), MIT, 308 p., 1971

Man's Impact on the Climate. W. H. Matthews, W. W. Kellogg, and G. D. Robinson, eds, MIT, 594 p., 1971

Pollution, weather and climate. Gordon J. F. MacDonald, CP-E4: 326–336, 1971

Air pollution and global climate. MS-E11: 11–14, 1972

The future climate. M. I. Budyko, *EOS* 53: 868–874, 1972

Man's effect on global climate. MS-T1: 59–67, 1972

When will the present interglacial end? G. J. Kukla and R. K. Matthews, *Science*, 178: 190–191, 1972

Carbon Dioxide and Climate Change

Carbon dioxide and climate. Gilbert N. Plass, CP-M2: 173–179/CP-R2: 173–179, 1959

Climatic variation (instrumental data). MS-E3: 211–213, 1967

Carbon dioxide and the macroclimate. MS-C2: 232–235, 1969

Carbon dioxide and other trace gases that may affect climate. MS-M1: 46–55, 82–88, 1970

Effect of Man on the carbonate cycle. MS-O1: 345–348, 1970

Impact of land and sea pollution on the chemical stability of the atmosphere. F. D. Sisler, CP-G1: 12–24, 1970

Man-made climatic changes. Helmut E. Landsberg, *Science* 170: 1265–1274, 1970

The oxygen and carbon dioxide balance in the earth's atmosphere. Francis S. Johnson, CP-G1: 4–11, 1970

Atmospheric carbon dioxide and aerosols: effects of large increases on global climate. S. I. Rasool and S. H. Schneider, *Science* 173: 138–141, 1971

Carbon dioxide; prediction of future concentration. MS-I1: 233–240, 1971

Pollution, weather and climate. Gordon J. F. MacDonald, CP-E4: 326–336, 1971

Mauna Loa and global trends in air quality. Lester Machta, *Bull. AMS* 53: 402–420, 1972

Global Impact of Combustion; Oxygen Depletion Threat

Thermal pollution. Lamont C. Cole, CP-M4: 217–224, 1969

Enough air (no oxygen threat). Wallace S. Broecker, *Envir.*, 12(7): 26–31, 1970

Gas exchange: letters by Leigh Van Allen, Lamont C. Cole, Wallace S. Broecker, and Eugene K. Peterson on oxygen depletion question. *Envir.* 12(10): 39–45, 1970

Heat released (to atmosphere). MS-M1: 63–66, 1970

Man's oxygen reserves. Wallace S. Broecker, CP-C4: 47–48/*Science*, 168: 1537–1538, 1970

Energy production and release. MS-I1: 55–60, 1971

Man-Induced Changes in Precipitation, Water Vapor, Clouds, Albedo, Aerosols

The inadvertent modification of the atmosphere by air pollution. Vincent J. Schaefer, *Bull. AMS* 50: 199–206, 1969

Atmospheric carbon dioxide and aerosols: effects of large increases on global climate. S. I. Rasool and S. H. Schneider, *Science* 173: 138–141, 1971

Cirrus clouds; climatic importance. MS-I1: 245–250, 1971

Manipulation of surface and underground water. MS-I1: 64–67, 1971

Modification of cloud properties. MS-I1: 220–232, 1971

Aerosol concentrations: effect on planetary temperatures. R. J. Charlson, H. Harrison, and G. Witt, *Science* 175: 95–96, 1972

LaPorte, Indiana, Anomaly

The LaPorte weather anomaly—fact or fiction? Stanley A. Changnon, Jr., CP-M4: 155–166/*Bull. AMS* 49: 4–11, 1968

The LaPorte precipitation anomaly. B. G. Holzman and H. C. S. Thom, *Bull. AMS* 51: 335–342, 1970

The effects of accidental weather modification on the flow of the Kankakee River, John J. Hildore. *Bull. AMS* 52: 99–103, 1971

The LaPorte, Indiana, precipitation

anomaly. J. R. Harman and W. M. Elton, *Ann. AAG* 61: 468-480, 1971

Tree growth, air pollution, and climate near LaPorte, Indiana. W. C. Ashby and H. R. Fritts, *Bull. AMS* 53: 246-251, 1972

Weather modification—inadvertent (LaPorte, Indiana, anomaly). MS-M2: 44-45, 1972

SST Aircraft and Environment

Pollution, radiation, and climate. *NCAR Quart.*, No. 27, 1970

SST contamination of the atmosphere. MS-M1: 67-74, 100-107, 1970

Stratospheric ozone with added water vapor: influence of high-altitude aircraft. Halstead Harrison, *Science* 170: 734-736, 1970

Ionizing radiation. Earl Cook, CP-E4: 259, 1971

Modification of the stratosphere. MS-I1: 258-291, 1971

Reduction of stratospheric ozone by nitrogen oxide catalysts from SST exhaust. Harold Johnston, *Science* 173: 517-522, 1971

Planning a program for assessing the possibility that SST aircraft might modify climate. Robert H. Cannon, Jr., *Bull. AMS* 52: 836-842, 1972

Planned Weather Modification

Artificially induced precipitation and its potentialities. Vincent J. Schaefer, CP-M6: 607-618, 1956

Cloud Physics and Cloud Seeding. Louis J. Battan, Doubleday, 144 p., 1962

Weather and Climate Modification. A. R. Chamberlain et al, NSF Report 66-3, NSF, 149 p., 1965

Weather and Climate Modification. G. J. F. MacDonald et al, Publ. No. 1350, vols. I and II, NAS-NRC, 28 p., and 198 p., 1966

Weather Modification Law, Controls, Operations. H. J. Taubenfeld, NSF Report 66-7, NSF, 73 p., 1966

A dangerous game: taming the weather. Frederick Sargent II, *Bull. AMS* 48: 452-458, 1967

Ecological implications of weather modification. R. H. Whittaker, p. 367-384 in *Ground Level Climatology*, Publ. No. 86, AAAS, 1967

The effect of weather modification on physical processes in the microclimate. William E. Marlatt, p. 295-308 in *Ground Level Climatology*, Publ. No. 86, AAAS, 1967

Weather modification. MS-E3: 1120-1123, 1967

Weather modification and forest fires. Donald M. Fuquay, p. 309-325, in *Ground Level Climatology*, Publ. 86, AAAS, 1967

Weather modification: implications of the new horizons in research. Thomas F. Malone, *Science* 156: 897-901, 1967

Fog. Joel N. Myers, *Sci. Amer.*, 219(6): 75-82, 1968

Glossary of Terms Frequently Used in Weather Modification, AMS, 59 p., 1968

Human response to weather and climate. W. R. D. Sewell, R. W. Kates, L. E. Phillips, *Geog. Rev.* 58: 262-280, 1968

Artificial stimulation of precipitation. MS-I3: 100-103, 1969

Assault on fogs. Edward Gross, *Sci. News* 96: 165-167, 1969

Hailstorms in the United States. MS-I3: 131-133, 1969

Joint hail research project, *NACR Quart.* No. 25, 1969

On the prevention of lightning. C. D. Stow, *Bull. AMS* 50: 514-520, 1969

Progress in precipitation modification. A. M. Kahan, J. R. Stinson, R. L. Eddy, *Bull. AMS* 50: 208-214, 1969

Project STORMFURY. R. Cecil Gentry, *Bull. AMS* 50: 404-409, 1969

Weather and climate modification. MS-C2: 220-227, 1969

Weather-modification progress and the need for interactive research. Staff, Weather Modification Research Project, RAND Corp., *Bull. AMS* 50: 216-246, 1969

Controlling the Weather: A Study of Law and Regulatory Procedures. Dunellin, New York, 275 p., 1970

Hard questions about weather modification, Kendrick Frazier, *Sci. News* 97: 461-462, 1970

Hurricane Debbie modification experiments, August 1969. R. Cecil Gentry, *Science* 168: 473-475, 1970

Physical view of cloud seeding. Myron Tribus, *Science* 168: 201-210, 1970

Rainfall enhancement by dynamic cloud modification. William L. Woodley, *Science* 170: 127-132, 1970

An artificially induced local snowfall. Ernest M. Agee, *Bull. AMS* 52: 557-560, 1971

The Atmospheric Sciences and Man's Needs, Comm. on Atmospheric Sci., NAS-NRC, 88 p., 1971 (Chap. 4)

How to subdue a hurricane. Louise Purrett, *Sci. News* 100: 128-129, 1971

The search for a way to suppress hail. Kendrick Frazier, *Sci. News* 99: 200-202, 1971

Seeding cumulus in Florida: new 1970 results. J. Simpson and W. L. Woodley, *Science* 172: 117-126, 1971

Toward hurricane surveillance and control. James W. Meyer, *Tech. Rev.* 74(1): 59-66, 1971

Weather modification. MS-I1: 68-70, 1971

Weather modification: a technology coming of age. Allen L. Hammond, *Science* 172: 548-549, 1971

Weather modification in Japan. Norihiko Fukata, *Bull. AMS* 52: 4-14, 1971

The decision to seed hurricanes. R. A. Howard et al, *Science* 176: 1191-1202, 1972

Joint federal-state cumulus seeding program for mitigation of 1971 Florida drought. J. Simpson, W. L. Woodley, R. M. White, *Bull. AMS* 53: 334-344, 1972

On tornadoes and their modification. Edwin Kessler, *Tech. Rev.* 74(6): 48-55, 1972

Rainmaking: rumored use over Laos alarms arms experts, scientists. Deborah Shapley, *Science* 176: 1216-1220, 1972

Weather modification—purposeful. MS-M2: 38-43, 1972

Weather modification, review and perspective. Earl G. Droessler, *Bull. AMS* 53: 345-348, 1972

Weather modification as a future weapon. Louise A. Purrett, *Sci. News* 101: 254-255, 1972

Urbanization and the Radiation-Heat Balance

The energy balance climatology of a city-Man system. W. H. Terjung et al, *Ann. AAG* 60: 466-492, 1970

Man-made climatic changes. Helmut E. Landsberg, *Science* 170: 1265-1274, 1970

Urban energy balance climatology. Werner H. Terjung, *Geog. Rev.* 60: 31-53, 1970

The land; heat released. MS-I1: 166-168, 1971

Land-surface alterations. MS-I1: 60–63, 1971

Modification of the microclimate. MS-I1: 152–156, 1971

City climate. MS-I2: 373–377, 1972

Air Pollution—General

Meteorology of air pollution. Donald H. Pack, CP-M4: 98–112/*Science* 146: 1119–1128, 1964

Air conservation, The Report of the Air Conservation Commission of the AAAS, Publ. No. 80, AAAS, 335 p., 1965

Air conservation and the kinds of pollutants. Air Conservation Commission, CP-M4: 81–90, 1965

Air conservation and the law. MS-A1: 212–233, 1965

Air conservation and public policy. MS-A1: 3–19, 1965

Air pollution control. MS-A1: 234–272, 1965

Atmospheric diffusion, plume forms. MS-P1: 181–196, 1965

Economic poisons as air pollutants. MS-A1: 149–157, 1965

Lead and other deleterious metals. MS-A1: 124–133, 1965

Particulate matter. MS-A1: 109–123, 1965

Photochemical air pollution. MS-A1: 89–108, 1965

Pollutants and their effects. MS-A1: 59–88, 1965

Radioactive pollution of the atmosphere. MS-A1: 158–194, 1965

Socio-economic factors (in air pollution). MS-A1: 273–306, 1965

Geographical aspects of air pollution. Philip A. Leighton, CP-M4: 113–130/*Geog. Rev.* 56: 151–174, 1966

The Unclean Sky: A Meteorologist Looks at Air Pollution. Louis J. Battan, Doubleday, 141 p., 1966.

Aerosols. MS-E3: 7–8, 1967

Agricultural practices influencing air quality. Kenneth C. Walker, CP-A3: 105–111, 1967

Air—an essential resource for agriculture. John T. Middleton, CP-A3: 3–9, 1967

Economic aspects of air pollution as it relates to agriculture. Emanuel Landau, CP-A3: 113–126, 1967

Turbulence and diffusion (atmospheric). MS-E3: 1045–1047, 1967

Air Pollution, Commission on College Geography, Resource Paper No. 2, AAG, 42 p., 1968

Glossary of Terms Frequently Used in Air Pollution. AMS, 34 p., 1968

Air pollution meteorology. MS-W1: 280–290, 1969

Air pollution meteorology. Hans A. Panofsky, CP-U1, 169–179/*Amer. Sci.* 57: 269–285, 1969

Air subsidence and inversions. MS-I3: 114–115, 1969

Air and Water Pollution. Gerald Leinwald, Washington Square Press, New York, 160 p., 1969

Complexities of smog. MS-I3: 137–139, 1969

The role of meteorology in the study and control of air pollution. Morris Neiburger, *Bull. AMS* 50: 957–965, 1969

Air pollution. Council on Environmental Quality, CP-M3: 42–52, 1970

Air pollution, A. C. Nadler and G. L. Paulson, CP-E3: 73–101, 1970

Air pollution. MS-E6: 61–91, 1970

Air pollution. MS-O3: 171–196, 1970

Air pollution. MS-P3: 118–126, 1970

Air pollution surveillance systems. G. B. Morgan et al, *Science* 170: 289–295, 1970

Gaseous wastes—the new atmosphere. MS-E5: 127–141, 1970

The global balance of carbon monoxide. Louis A. Jaffe, CP-G1: 34–49, 1970

Metropolitan air layers and pollution. Helmut E. Landsberg, CP-C3: 131–143, 1970

Sources of air pollution. U. S. Dept. of Health, Education and Welfare, CP-M4: 91–97, 1970

Trends in urban air quality. J. H. Ludwig, G. B. Morgan, T. B. McMullen, *EOS* 51: 468–475, 1970

The air around us, air pollution. MS-E9: 170–195, 1971

Air pollution. Richard J. Hickey, CP-E4: 189–212, 1971

Air pollution. MS-M3: 313–341, 1971

Air pollution. MS-N1: 485–517, 1971

Air pollution: our ecological alarm and blessing in disguise. Hugh W. Ellsaesser, *EOS* 52: 92–100, 1971

Air pollution: present and future threat to Man and his environment. David L. Coffin, CP-A2: 1–38, 1971

Air pollution and electric power. Bruce C. Netschert, CP-C4: 41–46, 1971

The air of poverty (urban air pollution). P. P. Craig and E. Berlin, *Envir.* 13(5): 2–9, 1971

Airborne asphyxia—an international problem. Gordon D. Friedlander, CP-C4: 20–33, 1971

Atmospheric contamination. MS-I1: 51–60, 70–72, 1971

The Atmospheric Sciences and Man's Needs. Comm. on Atmospheric Sci., NAS-NRC, 88 p., 1971 (Chap. 3)

Brief history of national air pollution laws. CP-C4: 3–5, 1971

Clearing the Air: The Impact of the Clean Air Act on Technology. J. C. Redmond, C. J. Cook, A. A. J. Hoffman, eds, IEEE Press, New York, 159 p., 1971

Cloud on the desert (Four Corners power plant). Roy Craig, *Envir.* 13(6): 20–35, 1971

Conservation of the atmosphere. Robert M. Basile, CP-C5: 133–158, 1971

A corporate polluter learns the hard way. Business Week, CP-C4: 108–112, 1971

Episode 104 (air pollution event). Virginia Brodine, *Envir.* 13(1): 2–27, 1971

How clean a car? John B. Heywood, CP-C4: 99–107, 1971

Lead in the air: industry weight on Academy panel challenged. Robert Gillette, *Science* 174: 800–802, 1971

Major provisions of the Federal Clean Air Law. CP-C4: 6–16, 1971

Mercury in the air, Staff report, *Envir.* 13(4): 24–33, 1971

Metals in the air. Henry A. Schroeder, *Envir.* 13(8): 18–24, 29–32, 1971

Motor vehicle emissions in air pollution and their control. John A. Maga, CP-A2: 57–89, 1971

New eye on the air (pollution instrumentation). A. Coble, L. Langan, J. McCaull, *Envir.* 13(4): 34–41, 1971

Pollution, weather and climate. Gordon J. F. MacDonald, CP-E4: 326–336, 1971

A special burden (urban air pollution). Virginia Brodine, *Envir.* 13(2): 22–33, 1971

Sulfur is the major problem, Business Week, CP-C4: 113–144, 1971

Timetable for lead (elimination from gasoline). M. H. Hyman, *Envir.* 13(5): 15–23, 1971

Transportation: nation on the go. CP-A4: 118–126, 1971

The air. MS-C1: 8–20, 1972

Air pollution. MS-I2: 377–393, 1972

Air-pollution. MS-O2: 112–200, 1972

Air pollution. MS-T1: 114–121, 1972

Air quality of American homes. V. J. Schaefer et al, *Science* 175: 173–175, 1972

Astronomy and air pollution. P. W. Hodge, N. Laulainen, and R. J. Charlson, *Science* 178: 1123–1124, 1972

Atmospheric pollution. Wilfred Bach, McGraw-Hill, 144 p., 1972

Carbon monoxide concentration trends in urban atmospheres. M. Eisenbud and L. R. Erlich, *Science* 176: 193–194, 1972

A citizen's guide to air pollution. David V. Bates, Mc-Gill Queens University Press, Montreal and London, 140 p., 1972

Daily variation of pollution. MS-I2: 84–86, 1972

Environmental pollution. MS-E11: 309–311, 1972

Local politics and air pollution. Samuel J. Williamson, *Tech. Rev.* 74(4): 50–55, 1972

Nitrogen oxides, autos and power plants. Richard H. Gilluly, *Sci. News* 101: 252–253, 1972

Nitrogen oxides: a subtle control task, Charles N. Satterfield, *Tech. Rev.* 75(1): 10–18, 1972

Running in place, Virginia Brodine. *Envir.* 14(1): 2–11, 52, 1972

Air Pollution. Virginia Brodine, Harcourt Brace Jovanovich, New York, 205 p., 1973

Light pollution. Kurt W. Riegel, *Science* 179: 1285–1291, 1973

Air Pollution—Health Effects

Air pollution and public health. Walsh McDermott, CP-M2: 137–145, 1961

Air pollution as a factor in environmental carcinogenesis. MS-A1: 141–148, 1965

Health and air pollution: subject of new studies. Ralph G. Smith, CP-U1: 135–138, 1968

Air pollution medical research. J. R. Goldsmith and R. Hartman, CP-U1: 81–86/*Science* 163: 706–709, 1969

The health effects of air pollution. Nat. Tuberculosis and Respiratory Disease Assn., CP-M3: 53–70, 1969

Air pollution—the relationship between health effects and control philosophy. Eric J. Cassell, CP-C4: 34–40, 1971

Asbestos in environment. MS-E9: 257–260, 1971

Black lung: dispute about diagnosis of miners' ailment. Joe Pichirallo, *Science* 174: 132–134, 1971

Building a shorter life. Julian McCaull, *Envir.* 13(7): 2–15, 38–41, 1971

Health effects research program of the National Air Pollution Control Administration. Vaun A. Newill et al, CP-M5: 80–97, 1971

Lead poisoning: combating the threat from the air. Robert J. Bazell, *Science* 174: 574–576, 1971

Lead poisoning: zoo animals may be the first victims. Robert J. Bazell, *Science* 173: 130–131, 1971

Atmospheric pollution. Wilfred Bach, McGraw-Hill, 144 p., 1972 (Chap. 3)

Lead: Airborne Lead in Perspective. Committee on Biological Effects of Atmospheric Pollutants, NAS, Washington, D.C., 330 p., 1972

Mineral particles and human disease. MS-E11: 730–739, 1972

Point of damage. Virginia Brodine, *Envir.* 14(4): 2–15, 1972

Fluorides in the air. M. J. Prival and F. Fisher, *Envir.* 15(3): 25–32, 1973

Air Pollution—Effect on Plants

Damage to forests from air pollution. George H. Hepting, CP-M4: 522–531, 1964

Fluorides MS-A1: 134–140, 1965

Air pollution and plant response in the northeastern United States. R. H. Daines, I. A. Leone, and E. Brennan, CP-A3: 11–31, 1967

Air quality and forestry, J. R. Hansbrough. CP-A3: 45–55, 1967

Effects of photochemical air pollution on vegetation with relation to air quality requirements. Louis S. Jaffe, CP-U1: 143–151, 1967

Air pollution and plants. H. E. Heggestad, CP-M5: 101–115, 1971

Air pollution and trees. George H. Hepting, CP-M5: 116–129, 1971

Air Pollution and Urban Climate Change

The climate of towns. H. E. Landsberg, CP-M6: 584–606, 1956

Absolute and relative humidities in towns. Tony J. Chandler, *Bull. AMS* 48: 394–399, 1967

The climate of cities. William P. Lowry, CP-M2: 180–188, 1967

Observations of the urban heat island in a small city. R. J. Hutcheon et al, *Bull. AMS* 48: 7–9, 1967

Climate of the city. James T. Peterson, CP-M4: 131–154, 1969

The climate of the city. MS-W1: 265–278, 1969

Recent studies of urban effects on precipitation in the United States. Stanley A. Changnon, Jr., *Bull. AMS* 50: 411–421, 1969

Further observations on the urban heat island in a small city. Richard J. Kopec, *Bull. AMS* 51: 602–606, 1970

Metropolitan air layers and pollution. Helmut E. Landsberg, CP-C3: 131–143, 1970

Pollution, weather and climate. Gordon J. F. MacDonald, CP-E4: 326–328, 1971

Air pollution and urban climate. MS-E11: 14–16, 1972

Atmospheric pollution. Wilfred Bach, McGraw-Hill, 144 p., 1972 (Chap. 2)

The climate of the city. R. A. Bryson and J. E. Ross, CP-U3: 51–68, 1972

Urban air pollution. MS-E11: 1228–1232, 1972

Urban climate, air pollution, and planning. Wilfred Bach, CP-U3: 69–96, 1972

The meteorologically utopian city. Helmut Landsberg, *Bull. AMS* 54: 86–89, 1973

Particulates in Atmosphere (Turbidity) and Climatic Fluctuations

Volcanoes and world climate. Harry Wexler, CP-R2: 302–304, 1952

"All other factors being constant"—a reconciliation of several theories of climatic change. Reid A. Bryson, CP-M4: 167–174/CP-U1: 180–186/*Weatherwise* 21: 56–61, 1968

Atmospheric dust content as a factor affecting glaciation and climatic change. F. F. Davitaya, *Ann. AAG* 59: 552–560, 1969

Volcanic aerosols. *NCAR Quart.* No. 22, 1969

Man-made climatic changes. Helmut E. Landsberg, *Science* 170: 1265–1274, 1970

Particles and turbidity; atmospheric content of particles. MS-M1: 56–62, 88–91, 1970

Effects of stratospheric particles on temperature. MS-I1: 280–284, 1971

Particles in the atmosphere. MS-I1: 186–220, 1971

Recent volcanism and the stratosphere. John F. Cronin, *Science* 172: 847–849, 1971

Solar radiation: absence of air pollution trends at Mauna Loa. H. T.

Ellis and R. F. Pueschel, *Science* 172: 845–846, 1971
Stratospheric aerosols. *NCAR Quart.* No. 33, 1971
Composition of the stratospheric 'sulfate' layer. Richard D. Cadle, *EOS* 53: 812–820, 1972
Mauna Loa and global trends in air quality. Lester Machta, *Bull. AMS* 53: 402–420, 1972
Turbidity of the atmosphere: source of its background variation with the season. Hugh W. Ellsaesser, *Science* 176: 814–815, 1972

Noise Pollution

Noise–Sound without value. Committee on Environmental Quality, U. S. Fed. Council for Sci. & Tech., CP-M4: 175–189, 1968
Noise. MS-E6: 124–130, 1970
Noise. MS-O3: 220–243, 1970
Noise pollution. MS-P3: 139–140, 1970
Sound pollution—another urban problem. Peter A. Breysse, CP-M3: 103–115, 1970
Noise: the unseen pollution. Richard H. Gilluly, *Sci. News* 101: 189–191, 1972
Noise and the urban environment. Gordon M. Stevenson, Jr., CP-U3: 195–228, 1972

Chapter 7
Igneous Processes and the Earth's Crust

Environmental Geology–General

The environmental geologist and the body politic. Peter T. Flawn, *Geotimes* 13(6): 13–14, 1968
Education for environmental geology. William R. Dickinson, CP-M1: 309–310, 1970
Environmental geology; Conservation, Land-use Planning, and Resource Management. Peter T. Flawn, Harper, 313 p., 1970
Education and environmental geomorphology problems. John H. Moss, CP-E5: 245–247, 1971
Environmental Geomorphology. Donald R. Coates, ed., Publ. in Geomorphology, State University of New York, Binghamton, 262 p., 1971
Environmental planning and geology. D. R. Nichols and C. C. Campbell, eds., USGPO, 204 p., 1971
Environmental terrane studies in the East St. Louis area, Illinois. P. B. DuMontelle, A. M. Jacobs, R. E. Bergstrom, CP-E5: 201–212, 1971
A geologist views the environment. John C. Frye, CP-M1: 11–16, 1971
The present is the key to the past. A. Gordon Everett, CP-M1: 305–308, 1971
Put hydrogeology into planning. David A. Sommers, CP-M1: 279–284, 1971
Society and geomorphology. Rhodes W. Fairbridge, CP-E5: 215–220, 1971
The geologic and topographic setting of cities. D. F. Eschman and M. G. Marcus, CP-U3: 27–50, 1972
Man and His Physical Environment; Readings in Environmental Geology, R. D. McKenzie and R. O. Utgard, eds, Burgess Publ. Co., Minneapolis, 338 p., 1972
Focus on Environmental Geology. Ronald W. Tank, ed., Oxford University Press, New York, 450 p., 1973

Radioactivity and Radiogenic Heat

How Old is the Earth? Patrick M. Hurley, Doubleday, 160 p., 1959 (Chap. 2,3)
Pregeological history of the earth. MS-P4: 59–66, 1966
Heat flow and temperatures in the earth. MS-S4: 118–126, 1971
Radiogenic heat. MS-E1: 402–405, 1971

Crustal Elements; Silicate Minerals & Magmas; Igneous Rocks

The interior of the earth. K. E. Bullen, CP-R2: 36–41/CP-C7: 22–27, 1955
The origin of granite. O. Frank Tuttle, CP-R2: 141–144, 1955
How Old is the Earth? Patrick M. Hurley, Doubleday, 160 p., 1959 (Chap. 1)
The structure and composition of the earth. MS-P4: 28–41, 1966
Mineralogy, igneous rocks. MS-E10: 4–109, 1969
The composition of the earth. Peter Harris, CP-U2: 53–69, 1971
The earth's interior. MS-E1: 385–406, 1971
Internal division of the earth; constitution of earth from seismic evidence. MS-S4: 2–3, 92–104, 1971
Composition of the crust and earth as a whole. MS-P4: 41–54, 1966
Silica minerals. MS-P4: 100–131, 1966
Structure of silicates. MS-P4: 81–83, 1966
Chemical composition of the earth's crust. MS-E1: 343, 1971
Crystallization of magma; primary magmas; forms of intrusive rock bodies. MS-E1: 355–361, 1971
Igneous activity and the igneous rocks. MS-E1: 354–364, 1971
Minerals, physical properties; silicate minerals. MS-E1: 343–353, 1971
Minerals and rocks. Keith Cox, CP-U2: 13–26, 1971
Earth's crust geochemistry. MS-E11: 243–254, 1972
Mineral classes: silicates. MS-E11: 718–725, 1972

Continental & Oceanic Crust

Suboceanic layers and origin of the basins. MS-S3: 422–435, 1963
Continents. MS-E4: 163–168, 1968
Submarine geomorphology. MS-E4: 1078–1097, 1968
The composition of the earth. Peter Harris, CP-U2: 54–58, 1971
The earth's crust; crust of the continents and ocean basins. MS-E1: 400–401, 429–445, 1971

Igneous Intrusion, Volcanic Forms

Volcanoes. Howel Williams, CP-R2: 163–171, 1951
Volcanoes—In History, in Theory, in Eruption. Fred M. Bullard, Univ. of Texas Press, Austin, 441 p., 1962
Calderas. MS-E4: 96–98, 1968
Craters; crater lakes. MS-E4: 207–222, 1968
Volcanic landscapes. MS-E4: 1193–1205, 1968
Igneous rocks. MS-E10: 92–109, 1969
Volcanoes in the Sea: The Geology of Hawaii. G. A. MacDonald and A. T. Abbott, Univ. Of Hawaii Press, Honolulu, 441 p., 1970
Forms of igneous rock extrusion. MS-E1: 361–364, 1971
Landforms built by volcanic activity. MS-E1: 364–367, 1971
Volcanism and the earth's crust. J. B. Wright, CP-U2: 301–313, 1971

Chapter 8
Rock Alteration and Sediments

Mineral Alteration, Alteration Products

Chemical processes of soil formation. MS-F2: 25–30, 1958
Weathering. MS-F1: 97–116, 1964

Regolith and saprolite. MS-E4: 933–935, 1968
Chemical weathering. MS-S2: 22–36, 1969
Movements due to volume changes in surficial materials. MS-E5: 49–50, 1970
Chemical weathering; weathering products. MS-E1: 368–371, 1971
Geochemistry of sediments: modern. MS-E11: 428–443, 1972
Weathering, chemical. MS-E11: 1264–1269, 1972

Sediments, Sedimentary Rocks

Sand. Ph. H. Kuenen, CP-R2: 21–34, 1960
Modern sediments and the interpretation of ancient sediments. MS-S3: 467–487, 1963
Sediments; physical properties and mechanics of sedimentation. (D. I. Inman), MS-S3: 101–151, 1963
Marine sediments. MS-E2: 469–474, 1966
Sedimentation and sedimentary rocks. MS-P4: 149–191, 1966
Geologic Time. Don L. Eicher, Prentice-Hall, 149 p., 1968 (Chap. 2)
Mudcracks. MS-E4: 761–763, 1968
The continental shelves. K. O. Emery, CP-O1: 143–154, 1969
Sedimentary rocks. MS-E10: 110–125, 1969
Erosion forms of sedimentary strata. MS-E1: 662–668, 1971
Minerals and rocks. Keith Cox: CP-U2: 26–32, 1971
Principles of stratigraphic interpretation. MS-E1: 491–511, 1971
Sediment accumulations on the ocean floor; deep-sea sediments. MS-E1: 416–425, 1971
Sedimentary rocks. MS-E1: 368–384, 1971
Authigenesis of minerals—marine. MS-E11: 48–56, 1972
Calcium carbonate: geochemistry. MS-E11: 104–118, 1972
Evaporite processes. MS-E11: 351–361, 1972
Geochemistry of sediments: modern. MS-E11: 428–443, 1972
Sulfides in sediments. MS-E11: 1134–1140, 1972

Salts in Sea Water

Composition of sea water and history of oceans. MS-P4: 193–207, 1966
Salinity in the ocean. MS-E2: 758–763, 1966
Seawater; chemistry and history. MS-E2: 792–802, 1966
Properties of seawater; ions in seawater; dissolved gases; salts. MS-O1: 311–321, 323–331, 1970
Salinity of seawater; chlorinity. MS-O1: 91–99, 1970
Why the sea is salt? Ferren MacIntyre, CP-O1: 110–121, 1970
Salts in sea water. MS-E1: 377–397, 1971

Hydrocarbons—Peat, Coal, Petroleum, Natural Gas

Peat and coal. MS-E1: 381–382, 1971
Biogenic deposits. MS-P4: 232–247, 1966
Origin of coal and petroleum. MS-P4: 234–243, 1966
Petroleum and natural gas. MS-E1: 382–383, 1971
Hydrocarbons. MS-E11: 495–503, 1972

Chapter 9
Tectonic Processes and Continental Evolution

Tectonic Activity—General

The trenches of the Pacific. R. L. Fisher and R. Revelle, CP-R2: 104–109/CP-C7: 10–15, 1955
Island arcs. MS-E4: 564–568, 1968
Mountain systems and types. MS-E4: 747–761, 1968
Earth processes; tectonic movements. MS-E5: 21–32, 1970
Geologic structures. MS-S4: 9–25, 1971
Orogeny. John Sutton, CP-U2: 287–299, 1971
The primary island and mountain arcs. MS-E1: 440–442, 1971
Tectonic forms and continental evolution. MS-E1: 452–474, 1971
Geosynclines, mountains, and continent-building. Robert S. Dietz, CP-C7: 124–132, 1972

Geosynclinal Deposition, Continental Shelves

The continental shelf. Henry C. Stetson, CP-R2: 67–71, 1955
Continental shelves: topography, sediments, history. MS-S3: 206–278, 1963
Continental borderlands, rises, shelves, slopes, and terraces. MS-E2: 197–214, 1966
Coastal Plains. MS-E4: 144–150, 1968
The continental shelves. K. O. Emery, CP-O2: 41–52, 1969
Geosynclines and isostasy; geosynclinal deposition. MS-E1: 349, 465–468, 1971
Geosynclines, mountains, and continent-building. Robert S. Dietz, CP-C7: 124–132, 1972

Metamorphic Rocks

Metamorphism and metamorphic rocks. MS-P4: 248–282, 1966
Metamorphic rocks. MS-E10: 126–142, 1969
Major rock classes; cycle of rock transformation. MS-E1: 341–343, 1971
Minerals and rocks. Keith Cox, CP-U2: 32–39, 1971
Rock metamorphism; metamorphic minerals and rocks. MS-E1: 458–462, 1971

Plate Tectonics, Continental Drift

The plastic layer of the earth's mantle. Don. L. Anderson, CP-R3: 398–405/CP-C7: 28–35, 1962
Continental drift. J. Tuzo Wilson, CP-R3: 551–565/CP-C7: 41–55, 1963
Mid-Oceanic Ridge. MS-E2: 506–517, 1966
The confirmation of continental drift. Patrick M. Hurley, CP-R3: 611–622/CP-C7: 57–67, 1968
Sea-floor spreading. J. R. Heirtzler, CP-C7: 68–78, 1968
The deep-ocean floor. H. W. Menard, CP-C7: 79–87/CP-O1: 161–170/CP-O2: 55–63, 1969
The origin of the oceans. Sir Edward Bullard, CP-C7: 88–97/CP-O1: 196–205, 1969
The breakup of Pangaea. R. S. Dietz and J. C. Holden, CP-C7: 102–113, 1970
Continental drift. MS-O1: 291–307, 1970
Continental drift. Alan Gilbert Smith, CP-U2: 213–249, 1971
Continental drift. MS-E1: 470–474, 1971
Paleomagnetism; evidence of continental drift. MS-S4: 31–45, 1971
Plate tectonics. E. R. Oxburgh, CP-U2: 263–285, 1971
Plate tectonics. MS-S4: 46–66, 1971
Rock magnetism and crustal spreading; plate theory of global tectonics. MS-E1: 446–451, 1971

Earthquakes—General

Beneficial effects of Alaska earthquake. Edwin B. Eckel, CP-M1: 35–40, 1964

Earthquakes and Earth Structure. John H. Hodgson, Prentice-Hall, 166 p., 1964

Tectonic deformation associated with the 1964 Alaska earthquake. George Plafker, *Science* 148: 1675–1687, 1965

Earthquake "briefs". USGS, CP-M1: 31–33, 1968

The Great Alaska Earthquake of 1964: Geology. NAS-NRC, 834 p., 1968

The Great Alaska Earthquake of 1964: Hydrology. NAS-NRC, 446 p., 1968

The Alaska Earthquake, March 27, 1964: Lessons and Conclusions. Edwin B. Eckel, USGS Prof. Paper 546, USGPO, 57 p., 1970

Earth processes. MS-E5: 18–29, 1970

Earthquakes, world distribution. MS-S4: 5–8, 1971

Earthquakes and faults; earthquake energy; earthquake effects. MS-E1: 388–391, 1971

The San Adreas fault. Don L. Anderson, CP-C7: 142–157, 1971

The San Fernando, California, Earthquake of February 9, 1971. USGS Prof. Paper 733, USGPO, 254 p., 1971

Seismic Sea Waves (Tsunamis)

Tsunamis. Joseph Bernstein, CP-O1: 56–59/CP-R2: 216–219, 1954

Tsunamis (tidal waves). MS-S3: 82–90, 1963

Waves and Beaches, Willard Bascomb, Doubleday, 260 p., 1964 (Chap. 6)

Tsunami! The Story of Seismic Seawave Warning System. U. S. Dept. Commerce, USGPO, 46 p., 1965

Tsunamis. MS-E2: 941–943, 1966

Tsunami. MS-O1: 242–246, 1970

Seismic sea waves, or tsunamis. MS-E1: 391–393, 1971

Man-Induced Earthquakes

Man-made earthquakes—a progress report. David M. Evans, *Geotimes* 12(6): 19–20, 1967

The Denver earthquakes. J. T. Healy et al, CP-M4: 428–441/*Science* 161: 1301–1310, 1968

Nuclear explosions and distant earthquakes: a search for correlations. J. H. Healy and P. A. Marshall, *Science* 169: 176–177, 1970

Ground rupture in the Baldwin Hills, D. H. Hamilton and R. L. Meehan, *Science* 172: 333–344, 1971

Nuclear explosions and earthquakes. D. Davies, CP-U2: 333–341, 1971

Earthquake Prediction & Control

Earthquake prediction. F. Press and W. F. Brace, *Science* 152: 1575–1584, 1966

Japanese program on earthquake prediction. T. Hagiwara and T. Rikitake, *Science* 157: 761–768, 1967

Earthquake prediction and control. L. C. Pakiser et al, *Science* 166: 1467–1474, 1969

Earthquake prediction and control may be possible. U.S. Geological Survey, CP-M1: 41–42, 1969

Joint U.S.-Japan conference on premonitory phenomena associated with earthquakes. L. E. Alsop and J. E. Oliver, eds, *EOS* 50: 376–410, 1969

Underground nuclear explosions and the control of earthquakes. C. Emiliani, C. G. A. Harrison, M. Swanson, *Science* 165: 1255–1256, 1969

Prediction, prevention, and protection (of earthquakes). MS-E5: 28–29, 1970

California earthquakes: science has no remedy. *Sci. News* 99: 126–127, 1971

Earthquake prediction and control. Allen L. Hammond, *Science* 173: 316, 1971

Earthquake prediction and modification Robert L. Kovach, CP-U2: 327–331, 1971

Earthquake prediction program in the People's Republic of China. Robert S. Coe, *EOS* 52: 940–943, 1971

Geyser strain-gauge, *Geotimes* 16(11): 32, 1971

The possibilities of earthquake prediction. Louise Purrett, *Sci. News* 99: 131–133, 1971

Earthquakes & Urban Planning, Hazard Reduction

Toward Reduction of Losses from Earthquakes. Comm. on the Alaska Earthquake, NAS-NRC, 34 p., 1969

The Alaska Earthquake, March 27, 1964, Lessons and Conclusions. Edwin B. Eckel, USGS Prof. Paper 546, USGPO, 546 p., 1970

Prediction, prevention, and protection (of earthquakes). MS-E5: 28–29, 1970

NATO and quakes. Robert E. Wallace, *Geotimes* 16(10): 25–26, 1971

San Fernando earthquake study: NRC panel sees premonitory lessons. R. Gillette and J. Walsh, *Science* 172: 140–143, 1971

Seismic hazards—A question of public policy. Alfred E. Alquist, CP-E9: 16–21, 1971

Landforms of Tectonic Activity

Drainage patterns. MS-E4: 284–291, 1968

Fault scarps; fault-line scarps. MS-E4: 346–351, 1968

Horsts. MS-E4: 537, 1968

Ridge-and-valley topography. MS-E4: 944–947, 1968

Rift valleys. MS-E4: 947–949, 1968

Structural control in geomorphology. MS-E4: 1074–1079, 1968

Tectonic landscapes. MS-E4: 1109–1116, 1968

Fault structures and landforms. MS-E1: 454–457, 670–671, 1971

Geologic structures. MS-S4: 9–25, 1971

Open folds, complexly folded and faulted belts; erosion of domes, folded strata, and metamorphic rocks. MS-E1: 457–458, 665–670, 1971

Geosynclines, mountains, and continent-building. Robert S. Dietz, CP-C7: 124–132, 1972

Radioactivity in Rocks, Age Determination

Radiocarbon dating. Edward S. Deevey, Jr., CP-R2: 86–90, 1952

How Old Is the Earth? Patrick M. Hurley, Doubleday, 160 p., 1959 (Chap. 4)

Geologic Time. Don L. Eicher, Prentice-Hall, 149 p., 1968 (Chap. 6)

Measuring geological time. Stephen Moorbath, CP-U2: 41–51, 1971

Carbon-14 dating. MS-E11: 129–133, 1972

Geochronometry. MS-E11: 446–452, 1972

Geologic time scale. MS-E11: 453–456, 1972

Radioactivity in rocks. MS-E11: 991–995, 1972

Chapter 10
Man's Consumption of Planetary Resources

Mineral Resources—General

Conservation of mineral resources. C. K. Leith, CP-M1: 183–186/*Science* 82: 109–117, 1935

Our inexhaustible resources. Eugene Holman, CP-M1: 187–191, 1952

Man's selective attack on ores and minerals. Donald H. McLaughlin, CP-M6: 851–861, 1956

Mineral resources: challenge or threat? Walter R. Hibbard, Jr., Science 160: 143–148, 1968

Realities of mineral distribution. Preston E. Cloud, Jr., CP-M1: 194–207, 1968

Earth Resources. Brian J. Skinner, Prentice-Hall, 149 p., 1969

Mineral resources from the land. Thomas S. Lovering, CP-R1: 109–134, 1969

Conservation, geology, and mineral resources. Peter T. Flawn, DP-M1: 192–193, 1970

Earth resources. MS-E5: 81–116, 1970

Human materials production as a process in the biosphere. Harrison Brown, CP-B1: 117–124/CP-M2: 107–114, 1970

Mineral resources and human ecology. Richard H. Jahns, CP-N1: 151–155, 1970

Nonrenewable mineral resources. MS-P3: 58–63, 1970

Conservation of mineral resources. Guy-Harold Smith, CP-C5: 373–399, 1971

Conservation and the minerals industry. Philip H. Abelson, Science 173: 9, 1971

Conservation and the minerals industry—a public dilemma. V. E. McKelvey and S. F. Singer, Geotimes 16(12): 20–22, 1971

Helium: should it be conserved? Sci. News 99: 261–262, 1971

Location of mineral deposits. Tom N. Clifford, CP-U2: 315–325, 1971

Mineral resources in fact and fancy. Preston Cloud, CP-E4: 71–88, 1971

Resources, population and quality of life. Preston Cloud, p. 8–31 of Is There an Optimum Level of Population?, S. F. Singer, ed., McGraw-Hill, 426 p., 1971

Mineral resource estimates and public policy. V. E. McKelvey, Amer. Sci. 60: 32–40, 1972

The resource concept. MS-M5: 24–36, 1973

Ore Deposits

Ore Deposits. C. F. Park, Jr., and R. A. MacDiarmid, Freeman, 475 p., 1964

Aluminum ore deposits. MS-E11: 23–27, 1972

Copper deposits. MS-E11: 193–195, 1972

Gold: economic deposits. MS-E11: 467–470, 1972

Hydrothermal solutions. MS-E11: 571–576, 1972

Iron: economic deposits. MS-E11: 603–610, 1972

Lead and zinc: economic deposits. MS-E11: 645–646, 1972

Magnesium, magnesium nodules. MS-E11: 666–670, 673–677, 1972

Mercury ore deposits. MS-E11: 707–709, 1972

Silver: economic deposits. MS-E11: 1093–1096, 1972

Uranium: economic deposits. MS-E11: 1222–1224, 1972

Metals Processing & Recycling

Technological denudation. Harrison Brown, CP-M6: 1023–1032, 1956

New ores for old furnaces: pelletized iron. Fillmore C. F. Earney, Ann. AAG 59: 512–534, 1969

Human materials production as a process in the biosphere. Harrison Brown, CP-M2: 107–114, 1970

Conservation of mineral resources. Guy-Harold Smith, CP-C5: 373–399, 1971

Alaska Pipeline—Environmental Impact

Archaeology along the pipeline. Robert J. Trotter, Sci. News 100: 396–397, 1971

Oil across Alaska. Sci. News 99: 64, 1971

Reaction of reindeer to obstructions and disturbances. David R. Klein, Science 173: 393–398, 1971

The big pipeline: focus on impact. Sci. News 101: 199, 1972

Canadian northern pipeline research conference. Robert F. Legget, Geotimes 17(5): 24–26, 1972

Pipeline approved despite ecologist's warnings. Sci. News 101: 325–326, 1972

Alaskan oil: court ruling revives Canada pipeline issue. Luther J. Carter, Science 179: 977–981, 1973

Impact of Mining & Mineral Processing

Current trends in mined-land conservation and utilization. G. Don Sullivan, CP-M1: 263–268, 1967

Encroachment of the Jeffrey Mine in the town of Asbestos, Quebec. W. Gillies Ross, Geog. Rev. 57: 523–537, 1967

Impact of surface mining on environment. U. S. Dept. of Interior, CP-M4: 348–369, 1967

Planning and engineering design of surface coal mines: land reclamation. J. Crowl and L. E. Sawyer, CP-M1: 261–262, 1967

Subsidence—a real or imaginary problem? August E. Vandale, CP-M1: 91–94, 1967

Mineral resources: challenge or threat? Walter R. Hibbard, Jr., Science 160: 143–148, 1968

Cost of reclamation and mine drainage abatement—Elkins demonstration project. R. B. Scott, R. D. Hill, R. C. Wilmoth, CP-M1: 269–270, 1970

Man as a geological agent: the geological consequences of industrialization. MS-E5: 117–185, 1970

Mines, quarries, and well fields. MS-E5: 162–167, 1970

Mine drainage; sedimentation and erosion. MS-C4: 9–11, 1970

Problems underfoot (effects of mining, mineral processing). Terri Aaronson, Envir. 12(9): 16–29, 1970

Subsidence over mine workings. MS-E5: 46–49, 1970

Mineral resources and multiple land use. Peter T. Flawn, CP-E9: 22–27, 1971

Montana versus the mining companies: new awakening or impractical idealism? Richard H. Gilluly, Sci. News 100: 235–237, 1971

Restoration of a terrestrial environment—the surface mine. Ronald D. Hill, CP-M1: 252–260, 1971

Sharp conflict on strip-mine reclamation. Sci. News 99: 297–298, 1971

Control for asbestos. A. K. Ahmed, D. F. McLeod, J. Carmody, Envir. 14(10): 16–22, 27–29, 1972

Healing wounds (strip mining). E. A. Nephew, Envir. 14(1): 12–21, 1972

West Virginia: strip mining issue in Moore-Rockefeller race. John Walsh, Science 178: 484–486, 1972

Chewing it up at 200 tons a bite: strip mining. William Greenburg, Tech. Rev. 75(4): 46–55, 1973

Oil Pollution (see also Chapter 23)

Pollution: the wake of the "Torrey Canyon." John Walsh, Science 160: 167–169/CP-M7: 58–60, 1968

Oil in the ecosystem. Robert W.

Bibliography of Environmental Science

Holcomb, *Science* 166: 204-206/ CP-M7: 61-66, 1969
Oil pollution of the ocean. Max Blumer, CP-M4: 295-301, 1969
Control of oil pollution. MS-C4: 24-29, 1970
Industrial wastes of petroleum origin. MS-M1: 266-270, 1970
Ocean pollution by petroleum hydrocarbons. MS-M1: 139-144, 1970
Horizon to horizon (tankers and oil spills). Staff, *Envir.* 13(2): 3-21, 1971
Ocean pollution. F. MacIntyre and R. W. Holmes, CP-E4: 242-245, 1971
Ocean pollution by petroleum hydrocarbons. Roger Revelle et al, CP-M5: 297-318, 1971
Oil on the water. MS-E9: 162-169, 1971
Petroleum: tar quantities floating in the northwestern Atlantic. Byron F. Morris, *Science* 173: 430-432, 1971
A small oil spill. M. Blumer et al, *Envir.* 13(2): 2-12, 1971
The Black Tide: The Santa Barbara Oil Spill and Its Consequences. Robert Easton, Delacorte Press, 336 p., 1972
Blowout: A Case Study of the Santa Barbara Oil Spill. C. E. Steinhart and J. S. Steinhart, Duxbury Press, North Scituate, Mass., 138 p., 1972
Oil on the waters: modest progress in cleanup technology. Joan Arehart-Treichel, *Sci. News* 102: 250-252, 1972
Oil persistence and degradation of spilled fuel oil. M. Blumer and J. Sass, *Science* 176: 1120-1122, 1972
The sea; pollution. MS-C1: 27-34, 1972
Submarine seeps: are they a major source of open ocean oil pollution? Max Blumer, *Science* 176: 1257-1258, 1972
Marine oil pollution. William E. Lehr, *Tech. Rev.* 75(4): 13-22, 1973

Ionizing Radiation Hazards, Fallout of Radionuclides

Effects of fission material on air, soil, and living species. John G. Bugher, CP-M6: 831-848, 1956
Danger to the individual. E. Teller and L. L. Latter, CP-E2: 71-76, 1958
Ionizing radiation and the citizen. George W. Beadle, CP-M2: 155-161, 1959
Radioactive pollution of the atmosphere. MS-A1: 158-194, 1965
Airborne radionuclides and animals. M. C. Bell, CP-A3: 77-90, 1967

Airborne radionuclides and plants. Ronald G. Menzel, CP-A3: 57-75, 1967
The biological action of ionizing radiation. Ernest C. Pollard, *Amer. Sci.* 57: 206-236, 1969
The dangerous atom. Robert Plant, CP-M3: 123-129, 1969
Understanding the atom. CP-M3: 119-122, 1969
Ecological upsets: climate and erosion. Tom Stonier, CP-E2: 49-54, 1970
Effects of pollution on the structure and physiology of ecosystems. CP-M5: 47-53/*Science* 168: 429-433, 1970
Ionizing radiation. MS-O3: 265-284, 1970
Radiation. MS-E6: 140-147, 1970
Radiation and chemical mutagens. MS-P3: 136-139, 1970
Radioactivity and fallout: the model pollution. George M. Woodwell, CP-C3: 159-169, 1970
Ionizing radiation. Earl Cook, CP-E4: 254-278, 1971
Radiation. Public Health Service, CP-M3: 116-118, 1971
Radiation (ionizing). MS-E9: 196-221, 1971
Radiation exposure in air travel. Hermann J. Schaefer, *Science* 173: 780-783, 1971
Radioactive cargoes: record good but the problems will multiply. Deborah Shapley, *Science* 172: 1318-1322, 1971
Radioactivity in the Marine Environment. Allyn H. Seymour, Panel Chairman, Committee on Oceanography, NAS-NRC, 272 p., 1971
Never do harm (medical, TV X-rays). Karl Z. Morgan, CP-M3: 130-146/ *Envir.* 13(1): 28-38, 1971
Atmospheric pollution. Wilfred Bach, McGraw-Hill, 144 p., 1972 (Chap. 5)
Atomic waste disposal in the sea: an ecological dilemma. Joel W. Hedgpeth, CP-C2: 812-828, 1972
Radioactivity. MS-T1: 67-77, 1972

Mineral Resources—Oceans and Sea Floor

Manganese nodules, deep-sea. MS-E2: 449-454, 1966
Mineral potential of the ocean. MS-E2: 517-524, 1966
Implications of geologic and economic factors to seabed resource allocation, development, and management. V. E. McKelvey, CP-M1: 226-231, 1969
Mineral resources from the sea. Preston Cloud, CP-R1: 135-155, 1969
The physical resources of the ocean. Edward Wenk, Jr., CP-O1: 347-355/CP-O2: 83-91, 1969
Potential ill effects of subsea mineral exploitation and measures to prevent them. V. E. McKelvey, CP-M1: 271-274, 1969
Progress in the exploration and exploitation of hard minerals from the seabed. V. E. McKelvey, CP-M1: 224-225, 1969
Subsea mineral resources and problems related to their development. V. E. McKelvey et al, Circular 612, USGS, 26 p., 1969
Technology and the ocean. Willard Bascom, CP-O1: 369-380/CP-O2: 109-120, 1969
Weapons in the deep sea. Sven Hirdman, *Envir.* 13(3): 28-42, 1971
Chemical resources of sea water. MS-M5: 92-108, 1973
Mineral resources of the seabed. MS-M5: 37-68, 1973

Energy Resources—General

Limitations to energy use. Charles A. Scarlott, CP-M6: 1010-1022, 1956
The control of energy. Glenn T. Seaborg, CP-C6: 94-112, 1967
Earth Resources. Brian J. Skinner, Prentice-Hall, 149 p., 1969 (Chap. 7)
Energy resources. M. King Hubbert, CP-R1: 157-242, 1969
Energy consumption. MS-M1: 288-305, 1970
Human energy production as a process in the biosphere. S. Fred Singer, CP-B1: 107-114, 1970
Limits to the use of energy. A. M. Weinberg and R. P. Hammond, *Amer. Sci.* 58: 412-418, 1970
The space available (for power plants). Comm. for Envir. Information, *Envir.* 12(2): 2-9, 1970
Air pollution and electric power. Bruce C. Netschert, CP-C4: 41-46, 1971
Continuing increase in use of energy. Philip H. Abelson, *Science* 172: 976, 1971
The conversion of energy. Claude M. Summers, CP-E7: 95-106, 1971
The economic geography of energy. Daniel B. Luten, CP-E7: 109-117, 1971

Energy: a crisis in the offing. Henry Still, CP-A4: 78-101, 1971
Energy and power. Chauncey Starr, CP-E7: 3-15, 1971
Energy resources. M. King Hubbert, CP-E4: 89-116, 1971
The energy resources of the earth. M. King Hubbert, CP-E7: 31-40, 1971
The energy revolution. George Taylor, CP-D1: 111-129, 1971
Energy shortage and energy choice. Victor K. McElheny, Tech. Rev. 73(3): 12-13, 1971
The flow of energy in an industrial society. Earl Cook, CP-E7: 83-91, 1971
Liquid hydrogen as a fuel for the future. Lawrence W. Jones, Science 174: 367-370, 1971
New Energy Technology—Some Facts and Assessments. H. C. Hottel and J. B. Howard, MIT, 364 p., 1971
Power basis for Man. MS-E8: 104-138, 1971
Conservation of energy: the potential for more efficient use. Allen L. Hammond, Science 178: 1079-1081, 1972
The energy economy. MS-T1: 19-25, 1972
Energy and the future: research priorities and national policy. Allen L. Hammond, Science 179: 164-166, 1972
Energy for millenium three. Earl Cook, Tech. Rev. 75(2): 16-23, 1972
Energy needs: projected demands and how to reduce them. Allen L. Hammond, Science 178: 1186-1188, 1972
Energy options: challenge for the future. Allen L. Hammond, Science 177: 875-876, 1972
Energy resources of the United States. P. K. Theobald, S. P. Schweinfurth, D. C. Duncan, Circular 650, USGS, 27 p., 1972
Energy technology to the year 200. A special symposium from Tech. Rev. 74(1,2,3), MIT, 96 p., 1971, 1972
Energy and well-being. A. B. Makhijani and A. J. Lichtenberg, Envir. 14(5): 10-18, 1972
For a U. S. energy agency. Glenn T. Seaborg, Science 176: 1189, 1972
Lost power. D. P. Grimmer and K. Luszczynski, Envir. 14(3): 14-22, 56, 1972
Power Generation and Environmental Change. D. A. Berkowitz and A. M. Squires, MIT, 464 p., 1972

Efficiency of energy use in the United States. E. Hirst and J. C. Moyers, Science 179: 1299-1304, 1973
Energy. MS-M5: 69-91, 1973
Energy conservation. G. A. Lincoln, Science 180: 155-162, 1973
Energy and the Future. A. Hammond, W. Metz, T. Maugh, II, AAAS, 184 p., 1973
Energy: planning for the future. C. Sharp Cook, Amer. Sci. 61: 61-65, 1973
Energy sources and conversion techniques. Ralph Roberts, Amer. Sci. 61: 66-75, 1973
Some views of the energy crisis. Alvin M. Weinberg, Amer. Sci. 61: 59-61, 1973
Windmills. Julian McCaull, Envir. 15(1): 6-17, 1973
World energy resources: survey and review. Trevor M. Thomas, Geog. Rev. 63: 246-258, 1973

Solar Energy

Direct Use of the Sun's Energy. Farrington Daniels, Yale, 374 p., 1964
Direct use of the sun's energy. Farrington Daniels, Amer. Sci. 55: 15-47, 1967
Solar energy. MS-E3: 878-880, 1967
Energy resources. M. King Hubbert, CP-R1: 206-207, 1969
Farming the sun's energy. Dietrick E. Thomsen, Sci. News 101: 237-238, 1972
Solar energy: a feasible source of power? Allen L. Hammond, Science 172: 660, 1971
Large-scale concentration and conversion of solar energy. A. F. Hildebrandt, G. M. Haas, W. R. Jenkins, J. P. Colaco, EOS 53: 684-692, 1972
Photovoltaic cells: direct conversion of solar energy. Allen L. Hammond, Science 178: 732-733, 1972
Solar energy: the largest resource. Allen M. Hammond, Science 177: 1088-1090, 1972

Hydropower, Tidal Energy

Tides. D. H. Macmillan, Amer. Elsevier Publ. Co., New York, 240 p., 1966 (Chap. 11)
The tides; pulse of the earth. Edward P. Clancy, Doubleday, 228 p., 1968 (Chap. 8)
Energy resources. M. King Hubbert, CP-R1: 207-209, 1969
Water power and its conservation. Guy-Harold Smith, CP-C5: 345-370, 1971
Hydropower. MS-M2: 111-113, 1972

Geothermal Energy

Geothermal energy. Donald E. White, Circular 519, USGS, 17 p., 1965
Resources of geothermal energy and their utilization. Donald E. White, CP-M1: 235-239, 1965
The Salton-Mexicali geothermal province. James B. Koenig, CP-M1: 240-249, 1967
Energy resources. M. King Hubbert, CP-R1: 215-218, 1969
The earth's heat: a new power source. Richard H. Gilluly, Sci. News 98: 415-416, 1970
Briefs on geothermal energy. U. S. Geological Survey, CP-M1: 232-234, 1971
Geothermal development. James B. Koenig, Geotimes 6(3): 10-12, 1971
Hot springs, geysers, and fumaroles. MS-E1: 599-602, 1971
Power from the earth. D. Fenner and J. Klarmann, Envir. 13(10): 19-26, 31-34, 1971
Geothermal energy: an emerging major resource. Allen L. Hammond, Science 177: 978-980, 1972
Geothermal power. Joseph Barnea, Sci. Amer. 226(1): 70-77, 1972

Energy from Fossil Fuels

The age of fossil fuels. Eugene Ayres, CP-M6: 367-381, 1956
Organic-rich shale of the United States and world land areas. D. C. Duncan and V. E. Swanson, Circular 523, USGS, 30 p., 1965
Tar sands and oil shales. Noel de Nevers, Sci. Amer. 214(2): 21-29, 1966
Energy resources. M. King Hubbert, CP-R1: 160-206, 1969
The competitive comeback of coal. Richard H. Gilluly, Sci. News 99: 84-86, 1971
The energy resources of the earth. M. King Hubbert, CP-E7: 31-40, 1971
Fossil fuel supplies and future energy needs. Hubert E. Risser, EOS 52: 763-767, 1971
The mineral fuels. E. Willard Miller, CP-C5: 401-448, 1971
A home on the range for a vast industry. Richard H. Gilluly, Sci. News 101: 156-158, 1972
When the well runs dry (depletion of natural gas reserves). Robert H.

Williams, *Envir.* 14(5): 19-20, 25-31, 1972

Nuclear Fission, Breeder Reactors

Energy resources. M. King Hubbert, CP-R1: 218-228, 1969
Nuclear Power and the Public. Harry Foreman, ed., Univ. of Minn. Press, 273 p., 1970
Breeder reactors: power for the future. Allen L. Hammond, *Science* 174: 807-810, 1971
The fast breeder reactor. H. Dieckamp, *EOS* 52: 756-762, 1971
The mineral fuels. E. Willard Miller, CP-C5: 443-446, 1971
Social institutions and nuclear energy. Alvin M. Weinberg, *Science* 177: 27-34, 1972

Nuclear Power-Plant Hazards

Fire damage (Rocky Flats plutonium plant accident). E. A. Martell et al, *Envir.* 12(4): 14-21, 1970
Seventeen million years (iodine-129 hazard). Sheldon Novik, *Envir.* 13(9): 42-47, 1971
Cooling water. I. A. Forbes et al, *Envir.* 14(1), 40-47, 1972
Nuclear reactor safety: a skeleton at the feast? Robert Gillette, *Science* 172: 918-919, 1971
The myth of the peaceful atom. R. Curtis and E. Hogan, CP-E2: 102-115, 1969
Ionizing radiation. Earl Cook, CP-E4: 261-277, 1971
Ecological hazards from nuclear power plants. Dean E. Abrahamson, CP-C2: 795-811, 1972
Fission: the pro's and con's of nuclear power. Allen Hammond, *Science* 178: 147-149, 1972
Nuclear safety. D. F. Ford and H. W. Kendall. *Envir.* 14(7): 2-9, 48, 1972
Toward a nuclear power precipice. Sheldon Novick, *Envir.* 15(2): 32-40, 1973

Nuclear Fusion Prospects

Controlled Nuclear Fusion. Samuel Glasstone, Understanding the Atom Series, AEC, 50 p., 1968
Energy resources. M. King Hubbert, CP-R1: 228-233, 1969
Controlled nuclear fusion: status and outlook. David J. Rose, *Science* 172: 797-808, 1971
The prospects of fusion power. W. C. Gough and B. J. Eastlund, *Sci. Amer.* 224(2): 50-64, 1971
Fusion power: progress and problems. Michael Self, *Science* 173: 802-803, 1971
Fusion power: ten years to the great decision? Robert C. Cowen, *Tech. Rev.* 73(3): 6-7, 1971
The prospects of fusion power. William C. Gough and Bernard J. Eastlund, CP-M2: 252-266, 1971
Fusion power. L. Wood and J. Nuckolls, *Envir.* 14(4): 29-33, 1972
Magnetic containment fusion: what are the prospects? William D. Metz, *Science* 178: 291-293, 1972

Chapter 11
Wasting of the Continental Surfaces

Geomorphology—General Concepts

Dynamic geomorphology. MS-E4: 297-301, 1968
General systems theory in geomorphology. MS-E4: 382-384, 1968
Geomorphology. MS-E4: 403-424, 1968
Land mass and major landform classification. MS-E4: 618-626, 1968
Landscape analysis. MS-E4: 626-638, 1968
Morphogenetic classification and regions. MS-E4: 717-731, 1968
Organisms as geomorphic agents. MS-E4: 778-784, 1968
The Surface of the Earth. Arthur L. Bloom, Prentice-Hall, 152 p., 1969
Geomorphology; processes, dynamic approach. MS-E1: 569-571, 1971

Weathering Processes & Forms

Weathering. MS-F1: 97-130, 1964
Exfoliation. MS-E4: 336-339, 1968
Frost action. MS-E4: 369-381, 1968
Granite landforms. MS-E4: 488-492, 1968
Induration. MS-E4: 553-556, 1968
Regolith and saprolite. MS-E4: 933-935, 1968
Salt weathering, or fretting. MS-E4: 968-970, 1968
Spheroidal weathering. MS-E4: 1041-1044, 1968
Solution pits and pans. MS-E4: 1033-1036, 1968
Talus fans or cones. MS-E4: 1106-1109, 1968
Tors. MS-E4: 1157-1159, 1968
Weathering. MS-E4: 1228-1232, 1968
Rock weathering. MS-S2: 16-39, 1969
Frost-action effects. MS-G1: 267-286, 1971
Thermodynamics and stones. W. W. Bowley and M. D. Burghardt, *EOS* 52(1): 4-7, 1971
Weathering. MS-E1: 569-576, 1971

Soils—General Principles

Fundamentals of Soil Science, 3rd. ed. C. E. Miller, L. M. Turk, H. D. Foth, Wiley, 626 p., 1958
Soils. MS-F1: 116-130, 1964
Soil. MS-S2: 36-39, 1969
Engineering properties of rocks and soils. MS-E5: 63-80, 1970
Nature of soils. MS-N1: 44-68, 1971
Soil orders and suborders of the United States and their utilization. Louis A. Wolfanger, CP-C5: 55-98, 1971
Weathering and soils; soil classification; ancient soils. MS-G1: 287-301, 1971
Clay minerals—base exchange. MS-E11: 176-178, 1972
Pedology (soil science). MS-E11: 911-917, 1972

Soils & Agriculture, Man-Induced Changes (Chemical), Soil Pollution

Physical, chemical, and biochemical changes in the soil community. William A. Albrecht, CP-M4: 395-418/CP-M6: 648-673, 1956
Lime and its use. MS-F2: 185-215, 1958
Organic matter depleted by cultivation. MS-F2: 270-271, 1958
Salt and silt in ancient Mesopotamian agriculture. T. Jacobsen and R. M. Adams, CP-M4: 383-394, 1958
Tillage operations and soil properties. MS-F2: 58-67, 1958
The reclamation of a man-made desert. Walter C. Loudermilk, CP-M2: 219-227, 1960
Highly productive lands of North America. Firman E. Bear, CP-F1: 136-143, 1966
Heavy-metal contamination of soils. J. V. Lagerwerff, CP-A3: 343-364, 1967
Depletion and restoration of soils. MS-N1: 103-114, 1971
Soil orders and suborders of the United States and their utilization. Louis A. Wolfanger, CP-C5: 55-98, 1971
Renewing the soil. Judith G. Meyer, *Envir.* 14(2): 22-32, 1972
The soil; soil pollution through solid wastes. MS-C1: 98-100, 1972

Mass Wasting—General

Mass movements. MS-F1: 337–353, 1964
Mass movements; mass wasting. MS-E4: 688–700, 1968
Soil creep. MS-E4: 1029–1030, 1968
Rock fragments in motion—mass wasting. MS-S2: 40–52, 1969
Movements of the surface caused by surficial processes. MS-E5: 33–52, 1970
Mass wasting. MS-E1: 579–584, 1971

Earthflow, Mudflow, Debris Floods

Landslides and Engineering Practice. E. B. Eckel, ed., Highway Res. Board, Special Rept. 29, NAS-NRC, 232 p., 1958
Rapid flow of wet debris. MS-F1: 341–353, 1964
Mudflow. MS-E4: 763–764, 1968
Solifluction, earthflow, and mudflow. MS-S2: 43–44, 1969
Creep and mudflows. MS-E5: 37–39, 1970
Earth flowage, mudflow. MS-E1: 580–582, 1971

Landslides

Landslides and Engineering Practice. E. B. Eckel, ed., Highway Res. Board, Special Rept. 29, NAS-NRC, 232 p., 1958
Landslides and rockfalls. MS-F1: 338–341, 1964
Landslides. D. M. Morton and R. Streitz, CP-M1: 64–73, 1967
Landslides. MS-E4: 639–641, 1968
Slump, slide, and fall. MS-S2: 45–47, 1969
Slides. MS-E5: 34–37, 1970
Geology and foundation problems in urban areas. David J. Henkel, CP-E9: 28–35, 1971
Impact of highways on the hydrogeologic environment. Richard R. Parizek, CP-E5: 151–199, 1971
Landslide. MS-E1: 582–584, 1971
Soil and the city. Donald H. Gray, CP-U3: 135–168, 1972

Rockfall, Debris Avalanche

Snow avalanches. MS-E4: 1020–1025, 1968
Avalanche. MS-S2: 45, 1969
Debris avalanches—a geomorphic hazard. G. P. Williams and H. P. Guy, CP-E5: 25–46, 1971
Rockfall and talus. MS-E1: 582–584, 1971

Permafrost, Arctic Environments

Landforms of the arctic: permafrost; patterned ground; pingos. MS-P2: 66–88, 1964
Tundra and taiga. J. Ross Mackay, CP-F1: 156–171, 1966
Tundra climate. MS-E3: 1040–1041, 1967
Arctic regions. MS-E4: 22–28, 1968
Frost action. MS-E4: 269–381, 1968
Patterned ground. MS-E4: 814–817, 1968
Permafrost. MS-E4: 833–839, 1968
Pingos. MS-E4: 845–848, 1968
Solifluction. MS-E4: 1030–1032, 1968
Thermokarst. MS-E4: 1149–1151, 1968
Tundra landscape. MS-E4: 1176–1179, 1968
The long pipe (Alaska pipeline). Ron Moxness, *Envir.* 12(7): 12–23, 36, 1970
Permafrost. MS-E5: 50–52, 1970
Some estimates of the thermal effects of a heated pipeline in permafrost. Arthur H. Lachenbruch, CP-M1: 74–76/Circular 632, USGS, 23 p., 1970
Frost-action effects; permafrost; frost-stirred ground. MS-G1: 267–286, 1971
Natural and man-induced disturbances of permafrost terrane. R. K. Haugen and J. Brown, CP-E5: 139–149, 1971
Permafrost of the arctic regions. MS-E1: 217–219, 1971
The periglacial environment, permafrost, and man. Larry W. Price, Commission on College Geography, Resource Paper No. 14, AAG, Washington, D. C., 88 p., 1972
The world of underground ice. J. Ross Mackay, *Ann. AAG* 62: 1–22, 1972

Land Surface Modification by Engineering, Urbanization

Environmental effects of highways. Melvin E. Scheidt, CP-M4: 419–427, 1967
Landforms made by Man. B. Golomb and H. M. Eder, CP-M4: 325–331, 1964
Soil and the city. Donald H. Gray, CP-U3: 135–168, 1972

Land Subsidence from Fluid Withdrawal

Geological subsidence. S. S. Marsden, Jr., and S. N. Davis, *Sci. Amer.* 216(6): 93–100, 1967
Land subsidence due to withdrawal of fluids. J. F. Poland and H. G. Davis, CP-M1: 77–90/CP-M4, 370–382, 1969
Subsidence due to withdrawal of fluids. MS-E5: 42–46, 1970
Land subsidence. MS-M2: 231–234, 1972

Chapter 12
Subsurface Water of the Lands

Hydrologic Cycle, the Water Balance

Changes in quantities and qualities of ground and surface waters. Harold E. Thomas, CP-M6: 542–563, 1956
The heat and water budget of the earth's surface. David H. Miller, *Advances in Geophysics,* vol. 11, Academic, p. 175–302, 1965
Water balance of the atmosphere. MS-P1: 92–99, 1965
Water balance of the earth's surface. MS-P1: 82–92, 1965
Evaporation, evapotranspiration. MS-E3: 368–373, 1967
A Survey Course: The Energy and Mass Budget at the Surface of the Earth. David H. Miller, Commission on College Geography, Publ. No. 7, AAG, 142 p., 1968
The hydrologic cycle. MS-S2: 12–15, 1969
Water balance of the continents. MS-O1: 219–221, 1970
The water cycle. H. L. Penman, CP-B1: 39–45, 1970
The hydrologic cycle and world water balance. MS-E1: 586–587, 592, 1971
Hydrologic processes. MS-I1: 96–99, 1971
Water. MS-N1: 117–122, 1971
Water balance of the atmosphere. MS-E1: 306–307, 1971
Water balance of a glacier; of Antarctica. MS-E1: 718–719, 1971
Evaporation. MS-M2: 50–66, 1972
The hydrologic cycle. MS-M2: 3–5, 1972
Hydrologic cycle. MS-E11: 515–519, 1972
Hydrology, MS-E11: 531–535, 1972
Water balance. MS-E11: 1248–1252, 1972
The control of the water cycle. J. P. Peixoto and M. A. Kettani, *Sci. Amer.* 228(4): 46–61, 1973
The global water balance. M. I.

Lvovitch, *US IHD Bull.* No. 23, in *EOS* 54: 28–42, 1973

The Soil Moisture Balance

An approach toward a rational classification of climate. C. W. Thornthwaite, *Geog. Rev.* 38: 55–94, 1948

The Water Balance. C. W. Thornthwaite and J. R. Mather, Drexel Inst. of Tech., Laboratory of Climatology, Publ. in Climatology, vol. 8, No. 1, Centerton, N. J., 86 p., 1955

Modification of rural microclimates. C. W. Thornthwaite, CP-M6: 567–583, 1956

Soil moisture. MS-F2: 76–146, 1958

An evaluation of the 1948 Thornthwaite classification. Jen-Hu Chang, *Ann. AAG* 49: 24–30, 1959

Annual cycle of water supply to the ground. MS-I2: 279–285, 1965

Evaporation; evapotranspiration—climatological methods. MS-P1: 166–180, 1965

Evaporation; lysimeters. MS-P1: 162–166, 1965

Hydroclimate (evapotranspiration). MS-E3: 447–451, 1967

The role of climate in the distribution of vegetation. J. R. Mather and G. A. Yoshioka, *Ann. AAG* 58: 29–41, 1968

Water in relation to plant growth; evapotranspiration; lysimeters; water balance. MS-C3: 118–148, 195–208, 1968

Soil moisture. MS-W1: 79–86, 1969

Zones of soil water; evaporation and transpiration; annual cycle of soil moisture; soil-water balance. MS-E1: 588–592, 1971

Evaporation. MS-M2: 50–66, 1972

Hydrology, semiarid regions. MS-E11: 541–550, 1972

Pedogenic Regimes, Soils Classification

Soil. Charles E. Kellogg, CP-R2: 152–161, 1950

Origin and classification of soils. MS-F2: 110–146, 1958

Lateritic soils. Mary McNeil, CP-M2: 68–73/CP-R3: 577–582, 1964

Duricrust. MS-E4: 296–297, 1968

Nature of soils. MS-N1: 57–68, 1971

Weathering and soils. MS-E1: 576–579, 1971

The concept of laterite. T. R. Paton and M. A. J. Williams, *Ann. AAG* 62: 42–56, 1972

Ground Water—Principles

Ground water. A. N. Sayre, CP-R2: 134–139, 1950

Ground Water Hydrology. David K. Todd, Wiley, 336 p., 1959

Artesian flow; wells. MS-E2: 596–597, 1966

Water under the Sahara. Rovert P. Ambroggi, *Sci. Amer.* 214(5): 21–29, 1966

Ground water; motion, recharge, aquifers. MS-E1: 592–597, 1971

Groundwater. MS-M2: 157–252, 1972

Groundwater. MS-E11: 470–478, 1972

Hydrogeology. MS-E11: 508–515, 1972

Limestone Caverns, Karst, Cavern Collapse

Karst. MS-E4: 582–587, 1968

Lapiés. MS-E4: 644–645, 1968

Limestone caves. MS-E4: 652–653, 1968

Mogotes. MS-E4: 708–709, 1968

Speleology. MS-E4: 1036–1039, 1968

Stalactites and stalagmites. MS-E4: 1048–1052, 1968

Carbonation; caves; karst. MS-S2: 25–27, 1969

The cave environment. T. L. Poulson and W. B. White, *Science* 165: 971–980, 1969

The central Kentucky karst. W. B. White et al, *Geog. Rev.* 60: 88–115, 1970

Solution and collapse in limestone terranes. MS-E5: 39–42, 1970

Cavern regimens. MS-G1: 301–304, 1971

An environmental approach to land use in a folded and faulted terrain. Richard R. Parizek, CP-E9: 122–143, 1971

Limestone caverns; karst landscapes. MS-E1: 598–599, 1971

Hydrology, limestone terrain. MS-E11: 538–541, 1972

Hydrological and ecological problems of Karst regions. H. E. LeGrand, *Science* 179: 859–864, 1973

Sinkhole. P. E. LaMoreaux and W. M. Warren, *Geotimes* 18(3): 15, 1973

Environmental Effects of Ground Water Use, Salt Intrusion

Ground Water Hydrology. David K. Todd, Wiley, 1959 (Chap. 12)

The Changing Pattern of Ground-Water Development on Long Island, New York. R. C. Heath, B. L. Foxworthy, P. Cohen, Circular 524, USGS, 10 p., 1966

Salinity in United States waters. W. Thorne and H. B. Peterson, CP-A3: 221–240, 1967

Double-mass-curve of the Effects of Sewering on Ground-water Levels on Long Island, New York, O. L. Franke, USGS Prof. Paper 600-B, USGPO, p. B205–B209, 1968

Water for the Future of Long Island, New York. Philip Cohen, O. L. Franke, B. L. Foxworthy, New York Water Resources Bull. 62A, State of New York, Albany, 37 p., 1970

Water supply for domestic and industrial uses. John H. Garland, CP-C5: 221–239, 1971

Development of water supplies; safe yield; side effects. MS-M2: 192–236, 1972

Hydrology, coastal terrain (fresh water–salt water relationships). MS-E11: 535–538, 1972

Salt intrusion in West-coast basin (Los Angeles County); Vermilion River Basin (Louisiana). MS-M2: 221–230, 1972

Summary of the Hydrologic Situation on Long Island, New York, as a Guide to Water-Management Alternatives. O. L. Franke and N. E. McClymonds, USGS Prof. Paper 627-F, USGPO, 59 p., 1972

Ground Water Pollution

Environmental framework of ground-water contamination. H. E. LeGrand, CP-U1: 301–307, 1965

Environmental effects of highways. Melvin E. Scheidt, CP-M4: 419–427, 1967

Design of disposal wells. John H. Marsh, CP-M1: 159–164, 1968

Disposal of wastes: examples from Illinois. Robert E. Bergstrom, CP-M1: 165–168, 1968

Man-made contamination hazards to ground water. P. H. McGauhey, CP-M1: 169–173/CP-M4: 225–232, 1968

Plants and salt in the roadside environment. Arthur H. Westing, CP-U1: 270–281, 1968

Biodegradable detergents and water pollution. Theodore E. Brenner, CP-A1: 147–196, 1969

Disposal of Liquid Wastes by Injection Underground—Neither Myth Nor Millennium. Arthur M. Piper, Circular 631, USGS, 15 p., 1969/ CP-M1: 148–158, 1969

Effects of mine drainage on ground water. G. H. Emrich and G. L. Merritt, CP-M1: 174–179, 1969

A new prospect (for sewage treatment). Louis T. Kardos, *Envir.* 12(2): 10-21, 27, 1970

NTA (phosphates in sewage). Samuel S. Epstein, *Envir.,* 12(7): 2-11, 1970

Sanitary sewage. MS-O3: 115-130 1970

Impact of highways on the hydrogeologic environment. Richard R. Parizek, CP-E5: 151-199, 1971

Ionizing radiation. Earl Cook, CP-E4: 266-270, 1971

New directions in water pollution abatement. Richard H. Gilluly, *Sci. News* 99: 286-287, 1971

Add salt to taste (deicing salts). H. H. McConnell and J. Lewis, *Envir.* 14(9): 38-44, 1972

The aqueous underground. Lynn W. Gelhar, *Tech. Rev.* 74(5): 45-53, 1972

Groundwater contamination by road salt: steady state concentrations in east central Massachusetts. E. E. Huling and T. C. Hollocher, *Science* 176: 288-290, 1972

Solid Wastes & Pollution

Evaluation of De Kalb County area for solid-waste disposal and general construction. David L. Gross, CP-M1: 294-299, 1970

Geologic considerations in disposal of solid municipal wastes in Texas. P. T. Flawn, L. J. Turk, C. H. Leach, CP-M1: 111-116, 1970

Hydrologic Implications of Solid-Waste Disposal. William J. Schneider, Circular 601-F, USGS, 10 p., 1970

Solid wastes. Council on Environmental Quality, CP-M3: 88-89, 1970

Solid wastes. MS-E6: 105-121, 1970

Solid wastes. MS-P3: 128-129, 1970

Waste: the burden of affluence. Henry Still, CP-A4: 180-197, 1971

Waste-solid disposal in coastal waters of North America. M. Grant Gross, CP-M5: 252-260, 1971

Solid Wastes—Handling, Disposal, Recycling

Crusade on bottles. Ellis Yochelson, *Geotimes* 15(7): 18, 1970

The new resource (trash recycling). Robert R. Grinstead, *Envir.* 12(10): 2-17, 1970

Solid-waste disposal. MS-O3: 151-170, 1970

Solid waste management. Rolf Eliassen, CP-N1: 55-65. 1970

Solid wastes. MS-E5: 148-159, 1970
Solid wastes. MS-E6: 109-121, 1970
Solid waste disposal: middenheap into mountain. MS-E9: 409-424, 1971

Ultimate disposal of industrial waste: an overview. Robert B. Dean, *Tech. Rev.* 73(5): 21-25, 1971

Waste: the burden of affluence. Henry Still, CP-A4: 181-197, 1971

Bottlenecks. Robert R. Grinstead, *Envir.* 14(3): 2-13, 1972

Bottles, cans, energy. Bruce M. Hannon, *Envir.* 14(2): 11-21, 1972

The dynamics of solid wastes. J. Randers and D. L. Meadows, *Tech. Rev.* 74(5): 15-19, 1972

Machinery for trash mining. Robert R. Grinstead, *Envir.* 14(4): 34-42, 1972

Taking it apart (recycling of wastes). Edward M. Dickson, *Envir.* 14(6): 36-41, 1972

What we do with rubbish. Walter R. Niessen, *Tech. Rev.* 74(5): 10-14, 1972

Effect of Engineering Works (Highways, Dams, Canals) on Ground Water

Impact of highways on the hydrogeologic environment. Richard R. Parizek, CP-E5: 151-199, 1971

Subsurface Disposal & Storage of Radioactive, Other Liquid Wastes

Disposal of wastes: examples from Illinois. Robert R. Bergstrom, CP-M1: 165-168, 1968

Energy resources. M. King Hubbert, CP-R1: 233-237, 1969

Radioactive Wastes. Charles H. Fox, Understanding the Atom Series, AEC, 46 p., 1969

Radioactive wastes in salt mines. W. C. McClain and R. L. Bradshaw, CP-M1: 136-141, 1969

Radioactive wastes. MS-E5: 159-162, 1970

So far, so good (nuclear weapons hazards). Milton Leitenberg, *Envir.* 12(6): 26-35, 1970

Subsurface disposal of liquid wastes. MS-E5: 145-148, 1970

Ionizing radiation. Earl Cook, CP-E4: 266-270, 1971

The Kansas geologists and the AEC. *Sci. News* 99: 161, 1971

The nation's rivers. M. Gordon Wolman, *Science* 174: 905-918, 1971

Nuclear waste: Kansans riled by AEC plans for atom dump. Constance Holden, CP-M1: 145-147/*Science* 172: 249-250, 1971

Storing nuclear wastes: more precise data needed. *Sci. News* 101: 310, 1972

The unsolved problem of nuclear wastes. William W. Hambleton, *Tech. Rev.* 74(5): 15-19, 1972

Chapter 13
Hydrology of Streams and Lakes

Runoff, Overland Flow

Runoff, precipitation analysis. MS-I2: 277-292, 1965

Hydroclimate. MS-E3: 447-451, 1967

Hydrology of streams. MS-S1: 11-20, 1968

Infiltration; runoff; overland flow. MS-E1: 587-588, 1971

Runoff. MS-E11: 1057-1058, 1972

The runoff cycle in nature; factors influencing runoff. MS-M2: 67-69, 91-97, 1972

Stream Channel Hydraulics, Drainage Systems

The river basin. MS-S1: 152-166, 1968

Rivers. MS-E4: 952-957, 1968

Dynamics of flowing water; hydraulic geometry. MS-S2: 54-58, 1969

Drainage networks; Playfair's law; network morphometry; drainage density. MS-E1: 609-615, 1971

Stream channels. MS-E1: 604-605, 1971

Morphometric changes (in river channels). T. Blench, CP-R4: 287-308, 1972

Rivers—a geomorphic and chemical overview. R. R. Curry, CP-R4: 9-31, 1972

Stream Flow

Frequency distribution of climatic events; streamflow fluctuation; flood frequency; frequency concept and geomorphic processes. MS-F1: 52-63, 1964

Hydraulics of streams. MS-S1: 28-40, 1968

Streamflow; hydrographs. MS-S1: 21-27, 1968

Relation of stream flow to rainfall. MS-E1: 615-618, 1971

Stream flow; stream energy and velocity; stream gauging. MS-E1: 605-609, 1971

The gauging of streams. MS-M2: 72-78, 1972
Rivers. MS-W2: 88-103, 1972
Stream hydrographs. MS-M2: 84-87, 1972

River Floods

Floods. W. G. Hoyt and W. B. Langbein, Princeton Univ. Press, Princeton, N. J., 468 p., 1955
Flood frequency. MS-F1: 63-66, 1964
The floodplain and the seashore; a comparative analysis of hazard-zone occupance. I. Burton and R. W. Kates, *Geog. Rev.* 54: 366-385, 1964
Flood prevention. MS-I2: 292-294, 1965
December 1964, a 400-year flood in northern California. E. J. Helley and V. C. Lamarche, Jr., CP-M1: 56-58, 1968
Floods. MS-E1: 618-621, 1971
Floods. MS-N1: 129-137, 1971
Floods and flood control. Guy-Harold Smith, CP-C5: 315-343, 1971
The Hazardousness of a Place: A Regional Ecology of Damaging Events. K. Hewitt and I. Burton, Univ. of Toronto Press, 154 p., 1971
Floods. MS-M2: 87-90, 1972
The hard-learned lessons of Agnes. Staff, *Sci. News* 102: 5-6, 1972

River Regulation, Dams

Hazardous dams in Missouri. Missouri Mineral Industry News, CP-M1: 59-60, 1968
Soviet plans to reverse the flow of rivers: the Kama-Vychegda-Pechora project. Philip P. Mocklin, CP-M4: 302-318, 1969
Dams. MS-N1: 137-140, 1971
Dams and Other Disasters: A Century of the Army Corps of Engineers in Civil Works. Arthur E. Morgan, Porter Sargent, Boston, 422 p., 1971
Flood control. MS-E1: 621, 1971
Floods and flood control. Guy-Harold Smith, CP-C5: 315-343, 1971
Flood control. MS-M2: 115-117, 1972
Flood control; river navigation; hydroelectric power. MS-M2: 258-280, 1972

Man-Induced Hydrologic Effects of Urbanization

The hydrologic effects of urban land use. Luna B. Leopold, CP-M4: 205-216, 1968
Hydrology for Urban Land Planning. Luna B. Leopold, Circular 554, USGS, 18 p., 1968/CP-M1: 43-55, 1968
Hydrologic and disposal problems in urban areas, Irwin Remson, CP-E9: 36-41, 1971
Impact of highways on the hydrogeologic environment. Richard R. Parizek, CP-E5: 151-199, 1971

Lakes

Principles of modern limnology. J. R. Vallentyne, *Amer. Sci.* 45: 218-244, 1957
Geology of the Great Lakes. Jack L. Hough, Univ. of Illinois Press, Urbana, 313 p., 1958 (Chaps. 1-3)
Water evaporimeters; lake and pan data. MS-P1: 156-162, 1965
Evaporation. MS-E3: 368-372, 1967
Caspian Sea. MS-E4: 109-116, 1968
East African lakes. MS-E4: 303-308, 1968
The Finger Lakes of New York State. MS-E4: 351-357, 1968
Glacial lakes. MS-E4: 444-456, 1968
Great Lakes of North America. MS-E4: 499-506, 1968
Great Salt Lake, Utah. MS-E4: 506-517, 1968
Lakes, classification and origin. MS-E4: 598-603, 1968
Lakes described: Atitlan, Baikal, Balaton, Balkhash, Chad, Eyre, Geneva, Maracaibo, Titacaca, Toba, Urmia. MS-E4: 603-618, 1968
Oriented Lakes. MS-E4: 785-796, 1968
Streams, Lakes, Ponds. Robert E. Coker, Harper, 327 p., 1968
The lake ecosystem. MS-N1: 375-381, 1971
Real-Estate Lakes. D. A. Rickert and A. M. Spieker, Circular 601-G, USGS, 19 p., 1971
Salton Sea: a new approach to environmental problems in a major recreational area. James B. Koenig, CP-E9: 106-113, 1971
Evaporation from lakes. MS-M2: 117-118, 1972
Lakes. MS-W2: 104-118, 1972
Limnology. MS-E11: 650-661, 1972

Irrigation & Its Side Effects, Salinization of Soils

The hydraulic civilizations. Karl A. Wittfogel, CP-M6: 152-164, 1956
Irrigation. MS-F2: 454-472, 1958
The quanats of Iran. H. E. Wulff, *Sci. Amer.* 218(4): 94-105, 1968
Irrigation. MS-N1: 168-177, 1971
Irrigation and salt problems in Renmark, South Australia. Charles H. V. Ebert, *Geog. Rev.* 61: 355-369, 1971
Irrigation in the United States. H. Bowman Hawkes, CP-C5: 257-288, 1971
Historical development of hydrology. MS-M2: 5-15, 1972
The impact of modern irrigation technology in the Indus and Helmand basins of southwest Asia. Alloys A. Michel, CP-C2: 257-275, 1972
Irrigation. MS-M2: 283-287, 1972
Salinization and water problems in the Algerian northeast Sahara. Kamel Achi, CP-C2: 276-287, 1972
Salt cedar and salinity on the upper Rio Grande. John Hay, CP-C2: 288-300, 1972

Dissolved Solids (Ions) in Surface Waters

Composition of terrestrial waters. MS-P4: 197-199, 1966
Salinity in United States waters. W. Thorne and H. B. Peterson, CP-A3: 221-240, 1967
The nation's rivers, M. Gordon Wolman. *Science* 174: 905-918, 1971
The mineralization of water. MS-M2: 78-82, 1972
Rainwater. MS-E11: 1015-1020, 1972
River geochemistry; environmental factors; regional. MS-E11: 1042-1050, 1972
Rivers—a geomorphic and chemical overview. R. R. Curry, CP-R4: 21-27, 1972

Water Pollution—Streams & Lakes (See also Chapter 10)

Water pollution, its effect on the public health. John D. Porterfield, CP-U1: 48-51, 1952
Eutrophication of the St. Lawrence Great Lakes. Alfred M. Beeton, CP-M4: 233-245, 1963
The great and dirty lakes. Gladwin Hill, CP-E3: 67-72, 1965
The aging Great Lakes. C. F. Powers and A. Robertson, CP-M2: 147-154/CP-O1: 397-405, 1966
Environmental hazards: water pollution. Richard L. Woodward, CP-U1: 16-23, 1966
The disposal of domestic wastes in rural areas. Charles D. Gates, CP-A3: 367-384, 1967
Environmental effects of highways. Melvin E. Scheidt, CP-M4: 419-427, 1967
Fertilizer nutrients as contaminants

in water supplies. George E. Smith, CP-A3: 173-186, 1967

Dwindling lakes. A. D. Hasler and B. Ingersoll, CP-E2, 152-160, 1968

The hydrologic effects of urban land use. Luna B. Leopold, CP-M4: 213-215, 1968

Streams, Lakes, Ponds. Robert E. Coker, Harper, 327 p., 1968 (Chap. 10)

Aquatic weeds. L. G. Holm, L. W. Weldon, R. D. Blackburn, CP-M4: 246-265, 1969

Biodegradable detergents and water pollution. Theodore E. Brenner, CP-A1: 147-196, 1969

Our nation's water: its pollution control and management. Cornelius W. Krusé, CP-A1: 41-71, 1969

Water pollution. Robert D. Hennigan, CP-U1: 24-28, 1969

Why can't a kid be as important as a fish? John E. Kinney, CP-U1: 36-40, 1969

Dead stream (Crooked Creek, Missouri). Kevin P. Shea, *Envir.* 12(6): 12-15, 1970

The limited war on water pollution. Gene Bylinsky, CP-M3: 76-87/CP-M4: 195-204, 1970

Liquid wastes. MS-E5: 141-145, 1970

Man-induced eutrophication of lakes. Arthur D. Hasler, CP-G1: 110-125, 1970

Mine drainage. MS-C4: 9-10, 1970

Reviving the Great Lakes. John R. Schaeffer, CP-M1: 117-122, 1970

Victors are not judged (U.S.S.R. water pollution). Philip R. Pryde, *Envir.* 12(9): 30-39, 1970

Water pollution. MS-E6: 29-59, 1970

Water pollution. MS-P3: 126-128, 1970

Water pollution and the environment. MS-C4: 1-13, 1970

Water pollution and its control. MS-O3: 131-150, 1970

Who owns the water? (international issues). Julian McCaull, *Envir.* 12(8): 30-39, 1970

Agricultural wastes and environmental pollution. Jesse Lunin, CP-A2: 215-261, 1971

Agriculture: the seeds of a problem. William E. Small, *Tech. Rev.* 73(6): 49-53, 1971

Biology and water pollution control. Charles E. Warren, W. B. Saunders, Philadelphia, 434 p., 1971 (Parts I, II)

Eutrophication, silica depletion, and predicted changes in algal quality in Lake Michigan. C. L. Schelske and E. F. Stoermer, *Science* 173: 423-424, 1971

Fresh water pollution. W. T. Edmondson, CP-E4: 213-229, 1971

Impact of highways on the hydrogeologic environment. Richard R. Parizek, CP-E5: 151-199, 1971

Industry: perils of production. CP-A4: 102-117, 1971

Lake Erie: pollution abatement, then what? Jerry H. Hubschman, *Science* 171: 536-540, 1971

The nation's rivers. M. Gordon Wolman, *Science* 174: 905-918, 1971

Out of the woods (pollution by pulp and paper industry). Henry I. Bolker, *Tech. Rev.* 73(6): 23-29, 1971

Phosphate replacements: problems with the washday miracle. Allen L. Hammond, *Science* 172: 361-363, 1971

Phosphates, heavy metals, and DDT: pollution control costs and implications. John F. Brown, Jr., CP-M5: 489-498, 1971

Phosphorus and eutrophication. SCEP Work Group, CP-M5: 319-324, 1971

Pollution control (state and local). MS-E7: 38-60, 1971

The pollution problem. MS-N1: 140-157, 1971

Runoff from agricultural land as a potential source of chemical, sediment, and waste pollutants. Lloyd L. Harrold, CP-M5: 230-239, 1971

Runoff of deicing salt: effect on Irondequoit Bay, Rochester, N. Y. R. C. Bubeck et al, *Science* 172: 1128-1131, 1971

U. S. and Canada agree on antipollution measures for Great Lakes. *EOS* 52: 581-582, 587, 1971

Water pollution. Carl H. Strandberg, CP-C5: 189-219, 1971

Water pollution. MS-E9: 107-132, 1971

Water pollution. MS-M3: 343-370, 1971

Chemical pollution: polychlorinated biphenyls (PCBs). Allen L. Hammond, *Science* 175: 155-156, 1972

Assault on a lake (Lake Michigan). Julian McCaull, *Envir.* 14(7): 33-39, 1972

The Delaware River—a study in water quality management. Robert V. Thomann, CP-R4: 99-129, 1972

Environmental pollution. MS-E11: 309-337, 1972

Fresh water; pollution. MS-C1: 66-89, 1972

Overfed (eutrophication). J. Crossland and J. McCaull, *Envir.* 14(9): 30-37, 1972

Pollution: the problem of misplaced waste. H. L. Bohn and R. C. Cauthorn, *Amer. Sci.* 60: 561-565, 1972

Pollution (of water). MS-W2: 178-196, 1972

Reclamation of water. MS-M2: 287-291, 1972

The role of phosphorus in eutrophication. W. Stumm and E. Stumm-Zollinger, CP-W1: 11-42, 1972

Science focuses on Lake Ontario. Louise A. Purrett, *Sci. News* 101: 316-317, 1972

Some basic issues in water pollution control legislation. Walter F. Westman, *Amer. Sci.* 60: 767-773, 1972

Sources of water pollution. Ralph Mitchell, CP-W1: 1-7, 1972

The tide of industrial waste. Julian McCaull, *Envir.* 14(10): 31-39, 1972

Water pollution. MS-O2: 59-109, 1972

Water pollution. MS-T1: 105-114, 1972

Drinking water. J. Crossland and V. Brodine, *Envir.* 15(3): 11-19, 1973

Sewage farming. Jonathan Allen, *Envir.* 15(3): 37-41, 1973

The wastewater tide ebbs slowly. Staff report, *Envir.* 15(1): 34-42, 1973

Acid Mine Drainage

Effects of mine drainage on ground water. G. H. Emrich and G. L. Merritt, CP-M1: 174-179, 1969

Restoration of a terrestrial environment—the surface mine. Ronald D. Hill, CP-M1: 252-253, 1971

Thermal Pollution of Water

Thermal pollution: a threat to Cayuga's waters? Luther J. Carter, *Science* 162: 649-650/CP-U1: 297-300, 1968

Thermal addition: one step from thermal pollution. Sharon Friedman, CP-U1: 33-35, 1969

Thermal pollution. LaMont C. Cole, CP-M4: 222-223/CP-M7: 68-76, 1969

Thermal pollution and aquatic life. John R. Clark, CP-M2: 163-171, 1969

Warm-water irrigation: an answer to thermal pollution? Luther J. Carter,

Bibliography of Environmental Science

Science 165: 478–480/CP-U1: 29–32, 1969
Finding a place to put the heat. Richard H. Gilluly, Sci. News 98: 98–99, 1970
Thermal pollution. MS-C4: 6–7, 1970
The Cayuga Lake controversy. Dorothy Nelkin, Cornell Univ. Press, 128 p., 1971
The nation's rivers. M. Gordon Wolman, Science 174: 905–918, 1971
Thermal loading. MS-E9: 133–149, 1971
A comparative assessment of thermal effects in some British and North American rivers. T. E. Langford, CP-R4: 319–351, 1972
Environmental quality and the thermal pollution problem. John Cairns, Jr., CP-C2: 829–853, 1972
Thermal pollution. MS-E11: 1169–1170, 1972
Trends in the mineralization and temperature of water. MS-M2: 129–133, 1972
Water quality data (temperature). MS-M2: 90–91, 1972

Water Resources—General, Future

Changes in quantities and qualities of ground and surface waters. Harold E. Thomas, CP-M6: 542–563, 1956
Fresh water from salt. David S. Jenkins, CP-O1: 337–345, 1957
Water. Roger Revelle, CP-M2: 57–67, 1963
Condensation of atmospheric moisture from tropical maritime air masses as a freshwater resource. R. D. Gerard and J. L. Worzel, Science 157: 1300–1302, 1967
Our nation's water: its pollution control and management. Cornelius W. Krusé, CP-A1: 41–71, 1969
Water for the Cities—The Outlook. W. J. Schneider and A. M. Spieker, Circular 601-A, USGS, 6 p., 1969
Long Island Water Resources. Div. of Water Resources, State Conservation, Dept., State Office of Planning Coordination, Albany, N. Y., 56 p., 1970
River of Life—Water: The Environmental Challenge. U. S. Dept. Interior, Conservation Yearbook No. 6, USGPO, 96 p., 1970
Water as an urban resource and nuisance. H. E. Thomas and W. J. Schneider, USGS Circular 601D, USGPO, 9 p., 1970
The Outlook for Water: Quality, Quantity, and National Growth. N. Wollman and G. W. Bonem, Hopkins, 286 p., 1971
Water resources. T. E. A. van Hylckama, CP-E4: 135–155, 1971
Water supply for domestic and industrial uses. John H. Garland, CP-C5: 221–239, 1971
"Water, water, everywhere". MS-E9: 93–106, 1971
Control and utilization of runoff. MS-M2: 97–156, 1972
Desalination processes—the U.S. desalting program. MS-E11: 219–225, 1972
Estimated Use of Water in the United States in 1970. C. R. Murray, Circular 676, USGS, 37 p., 1972
Water, The Web of Life. C. A. Hunt and R. M. Garrels, Norton, 208 p., 1972
Water resource development. MS-M2: 252–299, 1972

Water Resource Management

Alternatives in Water Management. Publ. 1408, NAS-NRC, 1966
The quanats of Iran. H. E. Wulff, Sci. Amer. 218(4): 94–105, 1968
Water transfers: must the west be won again? Frank Quinn, Geog. Rev. 58: 108–132, 1968
Water laws and concepts. Harold E. Thomas, EOS 50: 40–50, 1969
Australian-American interbasin water transfer. M. John Loeffler, Ann. AAG 60: 493–516, 1970
Dry lands and desalted water. Gale Young, Science 167: 339–343, 1970/CP-M7, 285–294, 1970
Water. Joseph G. Moore, Jr., CP-N1: 112–114, 1970
California Water; A Study in Resource Management. David Seckler, ed., Univ. of Calif. Press, Berkeley, 354 p., 1971
Geomorphology and decision-making in water resource engineering. Joseph H. Butler, CP-E5: 81–89, 1971
New sources of water. MS-N1: 157–168, 1971
The Water Hustlers. R. H. Boyle, J. Graves, T. H. Watkins, Sierra Club, San Francisco-New York, 253 p., 1971
Water problems. MS-N1: 123–129, 1971
Conflicts in water utilization; side effects and feedback. MS-M2: 118–129, 1972
The development of Israel's water resources. Aaron Wiener, Amer. Sci. 60: 466–473, 1972
Legal considerations of ground water. MS-M2: 236–252, 1972
Optimizing the operation of Israel's water system. Uri Shamir, Tech. Rev. 74(7): 41–48, 1972
Summary of the Hydrologic Situation on Long Island, New York, as a Guide to Water-Management Alternatives. O. L. Franke and N. E. McClymonds, USGS Prof. Paper 627-F. USGPO, 59 p., 1972
Trends in surface-water management. MS-M2: 144–156, 1972
Water and the city. John C. Schaake, Jr., CP-U3: 97–134, 1972
Water importation. AAAS Committee on Arid Lands, Science 175: 667–669, 1972
Water resource development. MS-M2: 252–334, 1972
Desalination. MS-M5: 109–123, 1973
Role of Water in Urban Planning and Management. W. J. Schneider, D. A. Rickert, A. M. Spieker, Circular 601-H, USGS, 10 p., 1973
Water commission: no more free rides for water users. Constance Holden, Science 180: 165, 167–168, 1973

Chapter 14
Fluvial Processes and Landforms

Fluvial Processes & Landforms—General

Fluvial Processes in Geomorphology. L. B. Leopold, M. G. Wolman, J. P. Miller, Freeman, 522 p., 1964
Erosion. MS-E4: 317–320, 1968
Quantitative geomorphology. MS-E4: 898–912, 1968
Streams; Their Dynamics and Morphology. Marie Morisawa, McGraw-Hill, 175 p., 1968
Geologic work of running water; systems of fluvial denudation. MS-E1: 623–661, 1971

Slope Erosion—Normal, Accelerated

Erosion by raindrop. W. D. Ellison, CP-R2: 127–132, 1948
The nature of induced erosion and aggradation. Arthur N. Strahler, CP-M6: 621–637, 1956
Conserving soil. MS-F2: 385–418, 1958

Overland flow; runoff; erosion. MS-F1: 353–363, 1964
Gully erosion. MS-E4: 517–519, 1968
Slope analysis. MS-E4: 998–1020, 1968
Slope development and maintenance. MS-S2: 48–52, 1969
Ecological upsets: climate and erosion. Tom Stonier, CP-E2: 49–60, 1970
Depletion and restoration of soils. MS-N1: 69–88, 1971
Erosion by overland flow. MS-E1: 623–626, 1971
Soil conservation. William A. Rockie, CP-C5: 99–132, 1971
Depletion and restoration of soils. MS-N1: 88–103, 1971
Erosion. MS-T1: 95–100, 1972
Erosion and sediment pollution control. R. P. Beasley, Iowa State Univ. Press, Ames, 320 p., 1972
The soil; erosion. MS-C1: 92–98, 1972
Soil erosion. MS-E11: 1101–1102, 1972

Man-Induced Changes in Runoff

The drainage basin as a geomorphic unit. MS-F1: 131–150, 1964
Badlands. MS-E4: 43–48, 1968
Interrelationships of forest, soils, and terrane in watershed planning. Peter E. Black, CP-E5: 71–78, 1971
Runoff from forest lands. Howard W. Lull, CP-M5: 240–251, 1971

Sediment Loads, Sedimentation; Man-Induced Changes

Land use and sediment yield. Luna B. Leopold, CP-M6: 639–647, 1956
The nature of induced erosion and aggradation. Arthur N. Strahler, CP-M6: 621–637, 1956
Salt and silt in ancient Mesopotamian agriculture. T. Jacobsen and R. M. Adams. CP-M4: 383–394, 1958
The frequency concept and geomorphic processes. MS-F1: 67–80, 1964
Sediment—its consequences and control. L. M. Glymph and H. C. Storey, CP-A3: 205–220, 1967
Transportation of the sediment load. MS-S1: 41–79, 1968
Transportation and erosion by streams. MS-S2: 59–64, 1969
Motion of mineral grains in water; transportation; deposition; erosion. MS-O1: 224–234, 1970
Sediment problems in urban areas, Harold P. Guy, Circular 601E, USGS, 8 p., 1970
Impact of highways on the hydrogeologic environment. Richard R. Parizek, CP-E5: 151–199, 1971
The nation's rivers. M. Gordon Wolman, *Science* 174: 905–918, 1971
Runoff from forest lands. Howard W. Lull, CP-M5: 240–251, 1971
Work of streams—transportation, load. MS-E1: 626–632, 1971
Sedimentation (suspended solids). H. A. Einstein, CP-R4: 309–318, 1972
Silt; suspended load. MS-M2: 83–84, 1972
Valley alluviation in southwestern Wisconsin. James C. Knox, Ann. AAG 62: 401–410, 1972

Stream Erosion & Channel Development

Channel form and process. MS-F1: 198–332, 1964
The debris load of rivers. MS-F1: 169–197, 1964
Degradation: headcuts and gullies. MS-F1: 442–453, 1964
Landforms in relation to frequency of climate events. MS-F1: 80–94, 1964
Water and sediment in channels. MS-F1: 151–169, 1964
The channel pattern. MS-S1: 135–151, 1968
Ephemeral streams. MS-E4: 312–314, 1968
The fluvial processes: erosion. MS-S1: 66–73, 1968
Open systems—allometric growth. MS-E4: 776–777, 1968
Stream channel characteristics. MS-E4: 1057–1060, 1968
Stream slope and channel morphology. MS-S1: 98–119, 1968
Streams and channels. MS-S2: 53–64, 1969
Work of streams; channel changes in flood. MS-E1: 626–632, 1971

Gorges, Waterfalls, Canyons

Canyon cutting in the Colorado River system. MS-E4: 98–102, 1968
Channel widening; valley shape. MS-S1: 71–74, 1968
Valley evolution. MS-E4: 1183–1189, 1968
Waterfalls. MS-E4: 1219–1220, 1968

Graded Stream, Denudation

The concept of entropy in landscape evolution. L. B. Leopold and W. B. Langbein, USGS Prof. Paper 500-A, USGPO, 20 p., 1962
Geomorphology and general systems theory. Richard J. Chorley, USGS Prof. Paper 500-B, USGPO, 10 p., 1962
Baselevel. MS-E4: 58–60, 1968
Continental erosion. MS-E4: 169–174, 1968
Erosion of the land, or what's happening to our continents? Sheldon Judson, *Amer. Sci.* 56: 356–374, 1968
Grade; graded streams. MS-E4: 486–488, 1968
The graded profile or the steady state. MS-S1: 120–134, 1968
Humid cycle. MS-E4: 538–541, 1968
Peneplains. MS-E4: 821–823, 1968
Concept of the graded stream. MS-S2: 64–69, 1969
Life history of landscapes. MS-S2: 81–102, 1969
Rates of denudation, United States. MS-O1: 221–222, 1970
Concept of the denudation system; denudation rates; peneplains. MS-E1: 645–657, 1971
Stages of stream gradation; equilibrium profile; baselevel. MS-E1: 632–635, 1971

Channel Aggradation, Man-Induced

Channel aggradation and the accumulation of valley sediment. MS-F1: 433–442, 1964
Man and geomorphic process in the Chemung River valley, New York and Pennsylvania. J. G. Nelson, Ann. AAG 56: 24–32, 1966
Braided stream channels. MS-S1: 146–151, 1968
Braided streams. MS-E4: 90–93, 1968
The hydrologic effects of urban land use. Luna B. Leopold, CP-M4: 211–213, 1968
Rivers—meandering and braiding. MS-E4: 957–963, 1968

Stream Rejuvenation, Entrenched Meanders

Incised meanders. MS-E4: 548–550, 1968
Incised meanders. MS-S1: 143–146, 1971
Stream rejuvenation, entrenched meanders. MS-E1: 655–657, 1971

Alluvial Terraces

River terraces. MS-F1: 458–484, 1964
Stream terraces. MS-S1: 74–77, 1968
Terraces, fluvial. MS-E4: 1117–1138, 1968
Alluvial terraces. MS-E1: 640–641, 1971
Stream regimens and climatic change. MS-G1: 304–313, 1971

Bibliography of Environmental Science

Environmental Effects of Dams

Degradation of channels as a result of changes in hydrologic regimen. MS-F1: 453–458, 1964

Water resource projects. MS-E5: 167–175, 1970

The increasing nonequilibrium of rivers. MS-M2: 133–144, 1972

Side effects and feedbacks, basic data and their limitations. MS-M2: 120–129, 1972

Water power versus living waters. MS-C1: 63–66, 1972

Alluvial River Environments, Floodplains

Botanical evidence of floods and flood-plain deposition. Robert S. Sigafoos, USGS Prof. Paper 485-A, USGPO, 35 p., 1964

Paradoxes of the Mississippi. Gerard H. Matthes, CP-R2: 257–262, 1951

Meanders. MS-F1: 295–332, 1964

River meanders. L. B. Leopold and W. B. Langbein, CP-R3: 566–567, 1966

Alluvial floodplains. MS-S1: 81–92, 1968

Floodplains. MS-E4: 359–362, 1968

Oxbow lakes. MS-E4: 798–799, 1968

Rivers—meandering and braided. MS-E4: 957–963, 1968

Alluvial rivers; meanders, channel forms. MS-E1: 635–638, 1971

Evaluating riverscapes. Marie Morisawa, CP-E5: 91–106, 1971

Effects of River Regulation, Channelization, Wetlands Drainage

Flood Information for Flood-Plain Planning. Conrad D. Bue, Circular 539, USGS, 10 p., 1967/CP-M1: 285–293, 1967

Channelization: a case study. John W. Emerson, *Science* 173: 325–326, 1971/CP-M1: 61–63, 1971

Reclamation of wet and overflow lands. Lowry B. Karnes, CP-C5: 241–255, 1971

Stream regimen and Man's manipulation. Robert V. Ruhe, CP-E5: 9–23, 1971

Wildlife versus irrigation. Richard H. Gilluly, *Sci. News* 99: 184–185, 1971

Drainage and irrigation. MS-C1: 59–63, 1972

Reclamation of land; drainage. MS-M2: 280–283, 1972

Stream channelization: conflict between ditchers, conservationists. Robert Gillette, *Science* 176: 890–894, 1972

Arid Lands—Environments, Landforms

Soils and agriculture in arid regions. MS-F2: 432–453, 1958

Origin of alluvial fans, White Mountains, California and Nevada, Chester B. Beaty, *Ann. AAG* 53: 516–535, 1963

Pediments and peneplains. MS-F1: 494–500, 1964

Southern Sudan, southwestern United States, South Africa. MS-F1: 377–383, 1964

Arid lands. Gilbert F. White, CP-F1: 172–184, 1966

Sheetfloods, streamfloods, and the formation of pediments. Perry H. Rahn, *Ann. AAG* 57: 593–604, 1967

Alluvial fans. MS-S1: 94–97, 1968

Alluvial fans, cones. MS-E4: 7–10, 1968

Arid cycle. MS-E4: 28–29, 1968

The Dead Sea. MS-E4: 243–246, 1968

Deserts and desert landforms. MS-E4: 271–280, 1968

Pediments; pediplanation. MS-E4: 817–820, 1968

Playas. MS-E4: 865–871, 1968

Salton Sea, California. MS-E4: 790–792, 1968

Water in dry regions. MS-S2: 69–80, 1969

Stone pavements in deserts. Ronald U. Cooke, *Ann. AAG* 60: 560–577, 1970

Denudation in an arid climate. MS-E1: 657–660, 1971

Chapter 15
Waves, Currents, and Coastal Landforms

Shorelines & Coasts—General

Defining and classifying shorelines and coasts. MS-S3: 152–166, 1963

Littoral zone, sediments. MS-E2: 445–448, 1966

Oceanography, nearshore. MS-E2: 614–619, 1966

Coastal classification. MS-E4: 131–133, 1968

Coastal geomorphology. MS-E4: 134–139, 1968

Coastal stability. MS-E4: 150–156, 1968

Ria coasts and related forms. MS-E4: 942–944, 1968

The edges of the land. MS-S2: 103–127, 1969

Shoreline processes and forms. MS-E1: 677–693, 1971

Shore Processes

Waves and Beaches. Willard Bascom, Doubleday, 260 p., 1964 (Chap. 7)

Littoral processes. MS-E4: 658–672, 1968

Rip currents. MS-E4: 950–951, 1968

Energy exchange at a coast; surf and breakers. MS-S2: 104–109, 1969

Waves (shoaling); breakers. MS-O1: 235–238, 1970

Waves in shallow water; refraction; breakers; coastal currents. MS-E1: 278–282, 1971

Marine Erosion Forms

Limestone coastal weathering. MS-E4: 653–657, 1968

Platforms—wave cut. MS-E4: 859–865, 1968

Terraces-marine. MS-E4: 1140–1142, 1968

Wave base. MS-E4: 1224–1228, 1968

Wave erosion; marine cliffs. MS-E1: 678–679, 1971

Littoral Drift—Beaches & Related Forms

Beaches, Willard Bascomb, CP-O1: 131–141/CP-R3: 315–324, 1960

Beaches and related shore processes. MS-S3: 167–205, 1963

Waves and Beaches. Willard Bascom, Doubleday, 260 p., 1964 (Chaps. 9, 10)

Beaches. MS-E4: 62–68, 1968

Crescentic landforms along the Atlantic Coast of the United States. R. Dolan and J. C. Ferm, *Science* 159: 627–629, 1968

Cuspate forelands; spits. MS-E4: 234–237, 1968

Prograding shorelines. MS-E4: 894–896, 1968

Sediment transport—long-term net movement. MS-E4: 985–989, 1968

Tombolos. MS-E4: 1155–1156, 1968

Coastal sediment. MS-S2: 113–116, 1969

Beaches. MS-O1: 253–257, 1970

Beaches; littoral drift; shore profile of equilibrium. MS-E1: 679–684, 1971

Forms and cycles in beach erosion and deposition. Warren E. Yasso, CP-E5: 109–137, 1971

Man-Made Shoreline Changes, Coastal Engineering

Influences of Man upon coast lines. John H. Davis, CP-M4: 332–347/CP-M6: 504–521, 1956

The flood plain and the seashore; a comparative analysis of hazard-zone occupance. I. Burton and R. W. Kates, *Geog. Rev.* 54: 366–385, 1964

Waves and Beaches. Willard Bascomb, Doubleday, 260 p., 1964 (Chap. 11)

Coastal changes. J. A. Steers, CP-F1: 539–551, 1966

Beach erosion and coastal protection. MS-E4: 68–70, 1968

The Human Ecology of Coastal Flood Hazard in Megalopolis. I. Burton, R. W. Kates, R. E. Snead, Univ. of Chicago, Dept. of Geography, Research Paper No. 115, 196 p., 1969

The coastal environment. MS-E6: 175–178, 1970

Environmental geology and the coast—rationale for land-use planning. T. Flawn, W. L. Fisher, L. F. Brown, Jr., CP-M1: 277–278, 1970

Oceanic Overwash and Its Ecological Implications on the Outer Banks of North Carolina. P. J. Godfrey, Office of Natural Science, Nat. Park Service, Washington, D. C., 1970

Rivers and coastlines. MS-E5: 52–58, 1970

Shore Protection Guidelines. Dept. of the Army, Corps of Engineers, USGPO, 59 p., 1971

As the seashore shifts. Dietrick E. Thomsen, *Sci. News* 101: 396–397, 1972

Impact of river control schemes on the shoreline of the Nile Delta. M. Kassas, CP-C2: 179–188, 1972

Man's impact on the barrier islands of North Carolina. R. Dolan, P. J. Godfrey, W. E. Odum, *Amer. Sci.* 61: 152–162, 1973

Barrier Island Coasts

Barriers—beaches and islands. MS-E4: 51–55, 1968

Tidal deltas and inlets. MS-E4: 1151–1155, 1968

Evolution of a gently sloping coast. MS-E1: 687–688, 1971

Deltas, Delta Environments

Ephermeral estuaries of the deltaic environment. James P. Morgan, CP-E6: 115–120, 1967

Delta dynamics; deltaic evolution. MS-E4: 225–260, 1968

Deltas. MS-S1: 92–94, 1968

Where rivers enter the sea. MS-O1: 257–263, 1970

Deltas. MS-E1: 641–644, 1971

Looking back through geologic time. E. K. Walton, CP-U2: 185–188, 1971

Ocean Tides, Tidal Currents

Waves and Beaches. Willard Bascomb, Doubleday, 260 p., 1964 (Chap. 5)

Tides. D. H. Macmillan, American Elsevier Publ. Co., New York, 240 p., 1966

Tides. MS-E2: 913–923, 1966

The tides: pulse of the earth. Edward P. Clancy, Doubleday, 228 p., 1968

Tides. MS-S2: 110–113, 1969

Tides; tide-generating forces. MS-O1, 238–242, 246–248, 1970

The tide. MS-E1: 129–148, 1971

Estuaries—Processes & Environments

The Hudson River estuary: hydrology, sediments and pollution. Alistair W. McCrone, *Geog. Rev.* 56: 175–189, 1966

Estuaries. George H. Lauff, ed., Publ. No. 83, AAAS, 757 p., 1967

Estuaries: analysis of definitions and biological considerations. Hubert Caspers, CP-E6: 6–8, 1967

Estuaries and lagoons in relation to continental shelves. K. O. Emery, CP-E6: 9–11, 1967

Geomorphology and coastal processes. J. A. Steers, CP-E6: 100–107, 1967

The ontogeny of a salt marsh estuary. Alfred C. Redfield, CP-E6: 108–114, 1967

Origin of sediments in estuaries. Andre Guilcher, CP-E6: 149–157, 1967

Origins of estuaries. Richard J. Russell, CP-E6: 93–99, 1967

Rates of sediment accumulation in modern estuaries. Gene A. Rusnak, CP-E6: 180–184, 1967

What is an estuary? Physical viewpoint. Donald W. Pritchard, CP-E6: 3–5, 1967

Coastal lagoon dynamics. MS-E4: 139–144, 1968

Estuaries. MS-E4: 325–330, 1968

Fiords. MS-E4: 358–359, 1968

Lagoons. MS-E4: 590–598, 1968

Mangrove swamps. MS-E4: 683–688, 1968

Oyster reefs. MS-E4: 799–803, 1968

Sea-level rise and coastal marshes. MS-S2: 126–127, 1969

Estuaries: Black Sea, Mediterranean Sea. MS-O1: 465–480, 1970

Estuary modeling. Geirmundur Arnason, CP-M5: 430–447, 1971

Tidal inlets, deltas, flats; salt marshes. MS-E1: 688–689, 1971

Estuaries and brackish waters. MS-C1: 42–45, 1972

Estuarine hydrology. MS-E11: 344–349, 1972

The estuarine environment. J. R. Schubel and D. W. Pritchard. CEGS Short Review Number 20, *Jour. Geol. Education* 20(4): 179–188, 1972.

Estuaries, Salt Marshes, Littorals—Modification, Pollution, Reclamation

Influence of man upon coast lines. John F. Davis. CP-M4: 344–345, 1956

DDT residues in an east coast estuary. G. M. Woodwell, C. F. Wurster, Jr., P. A. Isaacson, CP-M4: 565–571, 1967

The estuary—septic tank of Megalopolis. Paul De Falco, Jr., CP-E6: 701–703, 1967

Estuarine fisheries in Europe as affected by Man's multiple activities. P. Korringa, CP-E6: 658–663, 1967

Notes on estuarine pollution with emphasis on the Louisiana Gulf Coast. K. E. Biglane, CP-E6: 690–692, 1967

The role of man in estuarine processes. L. Eugene Cronin, CP-E6: 667–689/CP-M4: 266–294, 1967

Technical approaches toward evaluating estuarine pollution problems. A. F. Bartsch, R. J. Callaway, R. A. Wagner, CP-E6: 693–700, 1967

How not to kill the ocean. Wesley Marx, CP-U1: 41–47, 1969

Estuaries and coastal ocean areas. MS-M1: 146–149, 1970

Waste-solid disposal in coastal waters of North America. M. Grant Gross, CP-M5: 252–260, 1970

Water pollution and the environment. MS-C4: 1–13, 1970

Environmental pollution by mercury. J. M. Wood, CP-A2: 39–56, 1971

Last year at Deauville (beach pollution, Europe). *Envir.* 13(6): 36–37, 1971

Phosphorus and eutrophication. SCEP Work Group, CP-M5: 319–324, 1971

Population, natural resources, and biological effects of pollution of estuaries and coastal waters. Bost-

wick H. Ketchum, CP-M5: 59–79, 1971
The world ocean: ultimate sump. MS-E9: 150–161, 1971
Netherlands: Dutch continue to reclaim land from the sea. Michael Butler, *Science* 176: 1002–1004, 1972
Oil pollution: persistence and degradation of spilled fuel oil. M. Blumer and J. Sass, *Science* 176: 1120–1122, 1972

Coral Reef Coasts
Coral and other organic reefs. MS-S3: 349–370, 1967
Algal reefs. MS-E4: 3–5, 1968
Atolls. MS-E4: 35–40, 1968
Coral-reef lagoons. MS-E4: 594–598, 1968
Coral reefs—morphology and theories. MS-E4: 186–202, 1968
Fringing reefs. MS-E4: 366–369, 1968
Great Barrier Reefs. MS-E4: 492–499, 1968
Microatolls. MS-E4: 701–705, 1968
Coral reefs. MS-O1: 449–464, 1970
Coral reefs and reef deposits. MS-E1: 425–427, 1971
Coral reefs. MS-C1: 45–47, 1972

Chapter 16
Wind Action and Dune Landscapes

Wind Erosion, Deflation, Dust Storms
Controlling erosion caused by wind. MS-F2: 409–418, 1958
Deflation. MS-E4: 246–247, 1968
Ventifacts. MS-E4: 1192–1193, 1968
Wind, effect on plant growth, shelterbelts. MS-C3: 233–242, 1968
Wind action. MS-E4: 1233–1236, 1968
Wind profile near the ground. MS-C3: 109–117, 1968
Air-transported sediment. MS-E5: 58–59, 1970
The dust bowl in the 1970s. John R. Borchert, *Ann. AAG* 61: 1–22, 1971
Dust storms; transport of sand by wind. MS-E1: 696–697, 699–701, 1971
Wind erosion. MS-E1: 694–696, 1971
An American haboob. S. B. Idso, R. S. Ingram, J. M. Pritchard, *Bull. AMS* 53: 930–935, 1972
Erosion and sediment pollution control. R. P. Beasley, Chap. 3, Iowa State Univ. Press, Ames, 320 p., 1972

Sand Dunes, Dune Control
Influence of man upon coast lines. John F. Davis, CP-M4: 341–343, 1956
Clay dunes. MS-E4: 126–129, 1968
Periglacial eolian effects. MS-E4: 825–829, 1968
Sand dunes. MS-E4: 973–979, 1968
Singing sands. MS-E4: 994–996, 1968
Eolian features of Pleistocene age; dunes. MS-G1: 243–251, 1971
Oceanic Overwash and its Ecological Implications on the Outer Banks of North Carolina. P. J. Godfrey, Office of Natural Science, Nat. Park Service, Washington, D. C., 1970
Sand drifts and sand dunes. MS-E1: 701–705, 1971
Barrier dune system along the outer banks of North Carolina: a reappraisal. Robert Dolan, *Science* 176: 286–288, 1972
Man's impact on the barrier islands of North Carolina. R. Dolan, P. J. Godfrey, W. E. Odum, *Amer. Sci.* 61: 152–162, 1973

Loess Deposits
Loess. MS-E4: 674–678, 1968
Wind action. MS-E4: 1233–1236, 1968
Loess. MS-E1: 697–699, 1971
Loess. MS-G1: 251–266, 1971

Chapter 17
Glacier Systems and the Pleistocene Epoch

Glaciers, Glacial Ice, Glacier Regimen
Glaciers. William O. Field, CP-R2: 73–77, 1955
Glacial meteorology (mass budget, heat budget). MS-E3: 426–430, 1967
Glacial geology; glaciation. MS-E4: 431–482, 1968
Valley glaciers. MS-E4: 1190–1192, 1968
Ice on the land. MS-S2: 128–145, 1969
The Physics of Glaciers. W. S. B. Paterson, Pergamon, 250 p., 1969
Formation of glacial ice; classes of glaciers; alpine glaciers; MS-E1: 706–710, 1971
Glacial erosion and transport. MS-G1: 86–126, 1971
Glaciers of today; world distribution; volumes. MS-G1: 27–85, 1971

Glacial Landforms—Alpine
Glacial landforms. MS-E4: 467–482, 1968
Mountain glacier landscapes. MS-E4: 739–745, 1968
Striations; striated pavement. MS-E4: 1071–1074, 1968
Glacier erosion and transport. MS-S2: 137–140, 1969
Glacially sculptured terrain. MS-G1: 127–146, 1971
Glacier erosion and transportation; landforms of alpine glaciation. MS-E1: 710–715, 1971

Ice Sheets of Antarctica & Greenland
The Antarctic. A. P. Crary, CP-R3: 419–431, 1962
The Antarctic and the weather. Morton J. Rubin, CP-R3: 44–454, 1962
The ice of the Antarctic. Gordon de Q. Robin, CP-R3: 466–478, 1962
The land of the Antarctic. G. P. Woollard, CP-R3: 479–490, 1962
The Antarctic continent, meteorology and climatology. MS-P2: 270–299, 1964
Greenland. MS-P2: 239–254, 1964
Antarctic ice budget and meteorology. MS-E3: 16–27, 1967
Antarctic Ice Sheet. MS-G1: 57–62, 1971
Greenland and Antarctic ice sheets. MS-E1: 715–719, 1971
Greenland Ice Sheet. MS-G1: 51–57, 1971

Causes of Glaciations
Climate and the changing sun. Ernst J. Öpik, CP-R2: 251–256, 1958
Paleoclimatology and theories of climatic change. MS-P1: 197–228, 1965
Absolute dating and the astronomical theory of glaciation. Wallace S. Broecker, *Science* 151: 299–304, 1966
The CO_2 theory of the origin of ice ages. MS-E2: 174, 1966
A theory of ice ages III. W. L. Donn and M. Ewing, *Science* 152: 1706–1712, 1966
Ice-age theory. MS-E3: 462–474, 1967
Basic causes of continental glaciation. MS-E1: 746–750, 1971
Causes of glaciations; climate change. MS-G1: 788–809, 1971
Hypotheses concerning the origin and termination of ice ages. MS-I1: 35–36, 1971

Deposits of Ice Sheets

The history of a river (Ohio River). Raymond E. Janssen, CP-R2: 191–196, 1952
Drumlins. MS-E4: 293–295, 1968
Eskers. MS-E4: 323–325, 1968
Glacial deposits. MS-E4: 430–431, 1968
Kettles. MS-E4: 587–598, 1968
Moraines. MS-E4: 710–717, 1968
Outwash plains, fans, terraces. MS-E4: 796–798, 1968
Stagnant ice melting. MS-E4: 1045–1048, 1968
Washboard moraines and other types. MS-E4: 1213–1218, 1968
Glacier deposition. MS-S2: 140–143, 1969
Glacial drainage and channel forms. MS-G1: 227–243, 1971
Glacial drift; depositional landforms of glaciation. MS-E1: 721–725, 1971
Glacial drift (till, stratified drift); moraines. MS-G1: 147–226, 1971

Pleistocene Epoch—Events, Environments

Radiocarbon dating. Edward S. Deevey, Jr., CP-R2: 86–90, 1952
Ancient temperatures. Cesare Emiliani, CP-R2: 111–120, 1958
Micropaleontology (and climate fluctuations). David B. Ericson and Goesta Wollin, CP-R3: 407–416, 1962
Accuracy of radiocarbon dates. W. F. Libby, Science 140: 278–280, 1963
Geochronology methods, C-14. MS-F1: 402–405, 1964
Ice-age meteorology. MS-E3: 454–461, 1967
Late Pleistocene history of coniferous woodland in the Mohave Desert. P. V. Wells and R. Berger, Science 155: 1640–1647, 1967
Radiocarbon dating and archaeology in North America. Frederick Johnson, Science 155: 165–169, 1967
Pleistocene climates and chronology in deep-sea sediments. D. B. Ericson and G. Wollin, Science 162: 1227–1234, 1968
Quaternary period. MS-E4: 912–931, 1968
Upgrading radiocarbon dating. Sci. News 96: 159–160, 1969
Carbon 14 and the prehistory of Europe. Colin Renfew, Sci. Amer. 225(4): 63–72, 1971
Climates and fluctuations of late Cenozoic. MS-G1: 414–441, 1971
Crustal warping, isostatic, following deglaciation. MS-G1: 343–366, 1971
North America outside of the glacier-covered regions. MS-G1: 498–516, 1971
Overall view of Late-Cenozoic climate and glaciation; concepts. MS-G1: 1–26, 1971
The Pleistocene: glacials, interglacials, and pluvials. MS-I1: 31–34, 1971
Pleistocene changes in fauna and flora of South America. Beryl S. Vuilleumier, Science 173: 771–779, 1971
The Pleistocene Epoch and Man. MS-E1: 727–755, 1971
Pleistocene geochronology (dendrochronology, varves, carbon-14). MS-G1: 395–413, 1971
Pleistocene ice sheets. MS-E1: 720–721, 1971
Pleistocene stratigraphy. MS-G1: 367–394, 1971
The tropics weren't so 'stable' after all. Louise Purrett, Sci. News 100: 177, 1971
Wisconsin glaciers and ice sheets of North America. MS-G1: 463–497, 1971

Great Lakes—Origin, History

Geology of the Great Lakes. Jack L. Hough, Univ. of Illinois Press, Urbana, 313 p., 1958
History of the Great Lakes. MS-E1: 731–734, 1971

Periglacial Phenomena, Relict

Cryopedology; cryoperturbation. MS-E4: 228–231, 1968
Periglacial landscapes. MS-E4: 829–833, 1968
Periglacial regions. MS-E4: 440–441, 1968
Pleistocene climatic zones and periglacial phenomena. MS-E1: 737–738, 1971

Pluvial Lakes

Pluvial lakes. MS-E4: 873–883, 1968
Pluvial features, lakes. MS-G1: 442–462, 1971
Pluvial lakes of the Pleistocene. MS-E1: 734–737, 1971
Radiocarbon dating of East African lake levels. K. W. Butzer et al, Science 175: 1069–1076, 1972

Pleistocene Sea Level Changes

The changing level of the sea. Rhodes W. Fairbridge, CP-R2: 43–52, 1960
Mean sea level changes, long-term. MS-E2: 479–485, 1966
Elephant teeth from the Atlantic continental shelf. F. C. Whitmore, Jr., et al, Science 156: 1477–1481, 1967
Freshwater peat on the continental shelf. K. O. Emery et al, Science 158: 1301–1307, 1967
Sea levels 7,000 to 20,000 years ago. Alfred C. Redfield, Science 157: 684–692, 1967
Sea levels during the past 35,000 years. J. D. Milliman and K. O. Emery, Science 162: 1121–1123, 1968
Post-glacial sea level rise. MS-S2: 126–127, 1969
Sea-level changes in response to glaciation. MS-O1: 248–250, 1970
Fluctuation of sea level (eustatic); post-glacial rise. MS-G1: 315–342, 1971
Pleistocene changes of sea level. MS-E1: 738–741, 1971

Pleistocene Mammals, Early Man

The ape-men. Robert Broom, CP-R2: 232–236, 1949
Early Man in Africa. CP-R2: 146–151, 1958
Neanderthal Man. J. E. Weckler, CP-R3: 307–313, 1957
Early Man in East Africa. Phillip V. Tobias, Science 149: 22–33, 1965
The biological nature of Man. George G. Simpson, CP-M7: 3–18, 1966
Gigantopithecus. E. L. Simmons and P. C. Ettel, Sci. Amer. 222(1): 77–84, 1970
Environment and the evolution of Man. MS-E9: 5–21, 1971
Hominid evolution; Pleistocene hominids; Homo erectus. MS-M3: 19–131, 1971
Mammals of the Pleistocene; evolution of Man. MS-E1: 750–754, 1971
Pleistocene mammals, early Man. MS-G1: 747–787, 1971

Holocene Environments

Environment and Man in arid America. Harold E. Malde, Science 145: 123–129, 1964
Geochronology, historical records, dendrochonology, archaeological methods, pollen analysis. MS-F1: 389–402, 405–409, 1964
Climatic change in American prehistory. MS-E3: 169–171, 1967
Climatic optimum, hypsithermal, and climatic variations (historical record). MS-E3: 201–202, 205–211, 1967
Growth rings of trees: a physiological

basis for their correlation with climate. Harold C. Fritts, pp. 45-65 in *Ground Level Climatology*, Publ. No. 86, AAAS, 1967
Tree-ring analysis (dendroclimatology). MS-E3: 1008-1027, 1967
Tree rings and weather records. NCAR Quart. No. 19, 1967
Holocene, postglacial, or recent epoch. MS-E4: 525-536, 1968
Postglacial isostatic rebound. MS-E4: 884-888, 1968
Quaternary stratigraphy and climates. MS-E4: 441-443, 1968
The earliest Americans. C. Vance Haynes, Jr., Science 166: 709-715, 1969
Palynology and environmental history during the Quaternary Period. Margaret B. Davis, Amer. Sci. 57: 317-332, 1969
Climatic changes of past 1000 years. MS-O1: 503-513, 1970
Neoglaciation. G. H. Denton and S. C. Porter, Sci. Amer. 222(6): 101-110, 1970
The Holocene environment. MS-E1: 754-755, 1971
Postglacial climatic history. MS-I1: 36-40, 1971

Chapter 18
Life on Earth

The Ecosystem: General Introduction

The ecosphere. LaMont C. Cole, CP-M2: 11-16, 1958
Biotope and habitat. Stanley A. Cain, CP-F1: 38-65, 1966
The ecosystem and the community. MS-E14: 11-26, 1966
Biometeorology. MS-E3: 114-122, 1967
The Ecosystem Concept in Natural Resource Management. George M. Van Dyne, ed. Academic, 383 p., 1969
The ecosystem view of human society. F. F. Darling and R. F. Dasmann, CP-E8: 40-46, 1969
The human ecosystem. Marston Bates, CP-R1: 21-30, 1969
The biosphere. G. Evelyn Hutchinson, CP-B1: 3-11, 1970
The ecological facts of life. Barry Commoner, CP-N1: 18-35, 1970
The environment-ecosystem interaction. Wayne L. Decker, EOS 51: 664-666, 1970

The Arena of Life: The Dynamics of Ecology. Lorus and Margery Milne, Doubleday, 352 p., 1971
Ecological concepts. MS-N1: 16-43, 1971
Ecological systems. W. W. Murdoch, CP-E4: 1-28, 1971
Ecology. Foundations for Today, Vol. 3, R. S. Leisner and E. J. Kormondy, eds., William C. Brown Company, Dubuque, Iowa, 98 p., 1971
Ecology, the intricate web of life. Paul B. Sears, CP-A4: 32-55, 1971
The ecosystem. MS-F3: 8-36, 1971
Ecosystem concept, structure, productivity. MS-M3: 1-18, 1971
Ecosystem science as a point of synthesis. S. D. Ripley and H. K. Beuchner, CP-D1: 134-145, 1971
Ecosystems of national parks. Douglas B. Houston, Science 172: 648-651, 1971
Environment, Power, and Society. Howard T. Odum, Wiley-Interscience, 331 p., 1971
Fundamentals of Ecology. Third Edition, W. B. Saunders Co., Philadelphia, Pennsylvania, 574 p., 1971
Man in ecosystems. MS-P5: 306-367, 1971
Are the courts shaping a new environmental ethic? Richard H. Gilluly, Science News 102: 363-364, 1972
Concept of the ecosystem. Robert L. Smith, CP-E8: 3-22, 1972
Ecology. MS-E11: 254-258, 1972

The Ecosystem: Man's Impact—General

Ecological impact and human ecology. Pierre Dansereau, CP-F1: 425-461, 1966
Restoration of lost and degraded habitats. F. Raymond Fosberg, CP-F1: 503-515, 1966
Ecological diversity. Raymon F. Dasmann, CP-N1: 108-112, 1970
Ecological effects of man's activities. MS-M1: 21-29, 113-166, 1970
Ecosystems in jeopardy. MS-P3: 157-197, 1970
Effects of pollution on the structure and physiology of ecosystems. G.M.Woodwell,CP-M5:47-58,1971
Pollution. Foundations for Today, Vol. 2, R. S. Leisner and E. J. Kormondy, eds., William C. Brown Company, Dubuque, Iowa, 85 p., 1971
Work group on ecological effects (abridged). SCEP Report, CP-M5: 4-32, 1971

Conservation for survival. MS-C1: 1-7, 1972
Ecosystem theory in relation to man. Eugene P. Odum, CP-E10: 11-24, 1972
(See additional listings under other chapter headings.)

Life: General, Lower Forms

The Cell. Carl P. Swanson, Prentice-Hall, 114 p., 1960
Bacteria, slime molds, and fungi. MS-P6: 20-37, 1964
Chemical organization of life. MS-S5: 49-60, 1971
Living organisms—classification. MS-S5: 23-29, 1971
Protists. MS-S5: 151-174, 1971

Plants

The algae. MS-P6: 7-19, 1964
The flowering plants, or angiosperms. MS-P6: 87-101, 1964
Gymnospermous seed plants. MS-P6: 77-86, 1964
Mosses and liverworts. MS-P6: 38-45, 1964
The unity and diversity of plants. MS-P6: 1-6, 1964
Nonvascular Plants: Form and Function. William T. Doyle, Wadsworth, 147 p., 1965
The generalized plant cell. MS-V1: 31-44, 1970
The vascular plants. MS-V1: 1-30, 1970
Plants. MS-S5: 174-202, 1971

Animals

The mesozoa. E. A. Lapan and H. J. Morowitz, Sci. Amer. 227(6): 94-101, 1972
Annelids and allied groups. MS-S6: 521-537, 1973
Arthropods. MS-S6: 538-582, 1973
Mollusks. MS-S6: 499-520, 1973
Noncoelomate groups: protozoa, mesozoa. MS-S6: 415-486, 1973
Vertebrates. MS-S6: 623-660, 1973

Chapter 19
Evolution of Life Forms

Evolution: General

Ecology and evolution. MS-P7: 598-694, 1949
Plant Variation and Evolution. D. Briggs and S. M. Walters, World University Library McGraw-Hill, 256 p., 1969

Evolution. Theodore H. Eaton, Jr., Norton, 269 p., 1970
Evolution—an overview. MS-E13: 115-134, 1970
Change with time. MS-S5: 551-574, 1971
Evolution. MS-S5: 537-550, 1971
Evolution of the ecosystem. MS-F3: 270-275, 1971
Evolutionary aspects of change. MS-P5: 275-305, 1971
Processes of Organic Evolution, 2nd. ed. G. Ledyard Stebbins, Prentice-Hall, 193 p., 1971
Evolution of natural communities. R. H. Whittaker and G. M. Woodwell, CP-E10: 137-159, 1972

Genetics, Selection and Adaptation

Natural selection and speciation. MS-E14: 427-484, 1966
The desert pupfish. James H. Brown, *Sci. Amer.* 225(5): 104-110, 1971
Genetic codes. MS-S5: 339-356, 1971
Heredity. MS-S5: 521-536, 1971
Desert species and adaptation. Neil F. Hadley, *Amer. Sci.* 60: 338-347, 1972
Evolution mechanisms. MS-S6: 275-288, 1973
The genetics of natural selection. MS-E15: 20-37, 1973
Heredity. MS-S6: 256-274, 1973
Population genetics. MS-E15: 2-19, 1973
Selection in heterogeneous environments. MS-E15: 58-81, 1973

Radiation

Ionizing radiation and the citizen. George W. Beadle, CP-M2: 155-162, 1959
Radiation and the patterns of nature. George M. Woodwell, CP-E8: 306-319, 1967
Ionizing radiation. Earl Cook, CP-E4: 254-278, 1971
Radioactivity. MS-T1: 67-77, 1972

Early History of Life

The earliest organisms. MS-S5: 137-150, 1971
A model for the origin of life. P. C. Sylvester-Bradley, CP-U2: 123-141, 1971
Origin and early history of the earth. MS-E1: 475-490, 1971
Oxygen and evolution. L. V. Berkner and L. C. Marshall, CP-U2: 143-149, 1971
Origins: phylogeny. MS-S6: 289-310, 1973

Evolution through Geologic Time

Continental drift and evolution. Björn Kurtén, CP-C7: 114-123, 1969
Angiosperms—the culmination of plant evolution. MS-E13: 147-160, 1970
Coal age plants. MS-E13: 101-114, 1970
Early plant life—the thallophytes. MS-E13: 33-54, 1970
The invasion of the land. MS-E13: 55-76, 1970
The riddle of the pine cones. MS-E13: 135-146, 1970
Changing life forms in a changing environment: 1. the Paleozoic era. MS-E1: 512-527, 1971
Changing life forms in a changing environment: 2. the Mesozoic and Cenozoic eras. MS-E1: 528-545, 1971
The earliest stages of Hominid evolution. MS-M3: 19-54, 1971
Homo erectus. MS-M3: 85-132, 1971
Pleistocene Hominids. MS-M3: 55-84, 1971
Continental drift and the fossil record. A. Hallam, *Sci. Amer.* 227(5): 56-69, 1972
Descent: fossils and man. MS-S6: 311-342, 1973

Evolution and Man

Prehistoric overkill. Paul S. Martin, CP-M4: 612-624, 1967
Extinction: the Lord giveth and Man taketh away. MS-E9: 300-323, 1971
Genetic Vulnerability of Major Crops. Committee on Genetic Vulnerability of Major Crops, NAS-NRC, 307 p., 1972
The green revolution. H. G. Wilkes and S. Wilkes, *Envir.* 14(8): 32-39, 1972
Endangered species. MS-C5: 105-194, 1972
The discovery of America. Paul S. Martin, *Science* 179: 969-974, 1973

Chapter 20
Energy Flow in Organisms and Ecosystems

Energy Flow: General

Energy flow and material cycling. MS-E14: 27-59, 1966
Animals and the atmosphere. MS-W1: 219-240, 1967
Energy and ecology. MS-W1: 9-12, 1969
The energy cycle of the biosphere. George M. Woodwell, CP-B1: 25-36, 1970
Relationship between plants and atmosphere, David M. Gates, CP-C3, 145-154, 1970
Energy controls of ecosystems. MS-P5: 7-51, 1971
Energy in ecological systems. MS-F3: 37-85, 1971
The flow of energy in the biosphere. David M. Gates, CP-E7: 43-52, 1971
Limits of tolerance in terrestrial ecosystems. MS-P5: 126-196, 1971
Power in ecological systems. MS-E8: 58-102, 1971
What power is. MS-E8: 26-57, 1971
A world pattern in plant energetics. Carl F. Jordan, *Amer. Sci.* 59: 425-433, 1971
Energy flux in ecosystems. Frank B. Golley, CP-E10: 69-90, 1972
Food chains and pyramid of numbers. MS-E12: 3-5, 1972
Community structure and energetics. MS-E15: 341-362, 1973
Size and shape in biology. Thomas McMahon, *Science* 179: 1201-1204, 1973

Light Energy: Photosynthesis, Respiration, Productivity

Primary production in oceans. MS-E2: 722-725, 1966
Plants and the atmosphere. MS-W1: 163-181, 1967
Photosynthesis. MS-C3: 23-35, 1968
Carbon dioxide fixation and photosynthesis in nature. MS-P8: 277-298, 1969
Photosynthesis. MS-P8: 258-276, 1969
Respiration. MS-P8: 299-329, 1969
Photobiology. MS-V1: 197-208, 1970
Respiration. MS-S5: 307-322, 1971
Patterns of production in marine ecosystems. Gordon A. Riley, CP-E10: 91-112, 1972
Photosynthesis. MS-E11: 952-955, 1972
Cell operations, respiration. MS-S6: 82-98, 1973
Energy budget and photosynthesis of canopy leaves. W. H. Terjung and S. S-F. Louie, *Ann. AAG* 63: 109-130, 1973

Light Energy: Photoperiodism and Biological Clocks

Periodicity and biological clocks. MS-E14: 98-126, 1966

Photoperiodism. MS-C3: 70–74, 1968
The biological clock. MS-P8: 535–552, 1969
Photoperiodism and the physiology of flowering. MS-P8: 583–632, 1969
Biological time measurement. MS-V1: 209–220, 1970
Biological clocks. MS-F3: 245–248, 1971
The "clocks" timing biological rhythms. Frank A. Brown, Jr., *Amer. Sci.* 60: 756–766, 1972

Heat Energy: Heat Flow and Transpiration

Air and leaf temperature. MS-C3: 75–86, 1968
Evapotranspiration. MS-C3: 129–144, 1968
Responses to low temperature and related phenomena. MS-P8: 553–582, 1969
Transpiration and heat transfer. MS-P8: 78–112, 1969
The leaf and transpiration. MS-V1: 141–156, 1970

Chemical Energy: Nutrition, Biosynthesis

Enzymes, proteins, and amino acids. MS-P8: 209–235, 1969
Nutrition. MS-S5: 289–306, 1971
Synthesis. MS-S5: 323–338, 1971
Chemical organization of the cell. MS-S6: 55–62, 1973

Man and Energy Flow: Agriculture

The agricultural revolution. Robert J. Braidwood, CP-M2: 17–25, 1960
Weather modification and the living environment. Paul E. Waggoner, CP-F1: 87–98, 1966
Agricultural production in developing countries. G. F. Sprague, *Science* 157: 774–778, 1967/CP-M7: 265–274, 1967
New light on plant domestication and the origins of agriculture: a review. David R. Harris, *Geog. Rev.* 57: 90–107, 1967/CP-E8: 73–87, 1967
The world outlook for conventional agriculture. Lester R. Brown, *Science* 158: 604–611, 1967/CP-E8: 369–379, 1967
The green revolution: cornucopia or Pandora's box? Clifton R. Wharton, Jr., CP-E8: 114–120, 1969
The changing significance of food. Margaret Mead, *Amer. Sci.* 58: 176–181, 1970
Food and food products. Emil M. Mrak, CP-N1: 94–97, 1970

Food production. MS-P3: 81–101, 1970
Human food production as a process in the biosphere. Lester R. Brown, CP-B1: 95–103/CP-M2: 75–83, 1970
A hungry world. MS-P3: 67–79, 1970
Agriculture: the price of plenty. Ronald M. Fisher, CP-A4: 56–77, 1971
The flow of energy in a hunting society. William B. Kemp, CP-E7: 55–65, 1971
The flow of energy in an agricultural society. Roy A. Rappaport, CP-E7: 69–80, 1971
The land we possess. Guy-Harold Smith, CP-C5: 159–185, 1971
Man, food and environment. L. Brown and G. Finsterbusch, CP-E4: 53–70, 1971
Power basis for man. MS-E8: 103–138, 1971
Potential effects of global temperature change on agriculture, Sherwood B. Idso, CP-M5: 184–191, 1971
Prospects and problems in the management of ecological systems. MS-M4: 101–111, 1971
The sun's work in a cornfield. E. Lemon, D. W. Stewart, R. W. Shawcroft, *Science* 174: 371–378, 1971
Farming with petroleum. Michael J. Perelman, *Envir.* 14(8): 8–13, 1972
Food: a crisis in supply. MS-C5: 272–292, 1972
Man and His Environment: Food. E. R. Brown and G. W. Finsterbusch, Harper, 208 p., 1972

Chapter 21
Cycling of Materials in Organisms and Ecosystems

Material Cycles: General

Chemical fertilizers. Christopher J. Pratt, CP-M2: 236–246, 1965
Chemical oceanography. MS-E2: 186–191, 1966
Energy flow and material cycling. MS-E14: 27–59, 1966
The geochemical cycle. MS-P4: 283–296, 1966
Nature of the biosphere (mass, element composition). MS-P4: 223–232, 1966
Nutrients in the sea. MS-E2: 557–562, 1966

Pelagic biogeochemistry. MS-E2: 681–864, 1966
Toxic substances and ecological cycles. George M. Woodwell, CP-M2: 128–135, 1967
Mineral nutrition of plants. MS-P8: 191–208, 1969
Biogeochemical cycles. MS-P3: 161–167, 1970
Mineral cycles. Edward S. Deevey, Jr., CP-B1: 83–92, 1970
The nutrient cycles of an ecosystem. F. H. Bormann and G. E. Likens, *Sci. Amer.* 223(4): 92–101, 1970
Why the sea is salt. Ferren Macintyre, CP-O1: 110–121, 1970
Biochemical cycles. MS-F3: 86–105, 1971
Biogeochemical cycles within ecosystems. MS-P5: 52–119, 1971
Cycles of elements. MS-M4: 41–90, 1971
Limiting factors. MS-F3: 106–139, 1971
Biogeochemistry. MS-E11: 74–82, 1972
The chemical elements of life. Earl Frieden, *Sci. Amer.* 227(1): 52–60, 1972
Fertilizers. MS-T1: 88–95, 1972
Geochemical cycles. MS-E11: 208–216, 1972
Nutrient cycling in ecosystems. G. E. Likens and F. H. Bormann, CP-E10: 25–67, 1972
Vegetation indicators. MS-E11: 1238–1241, 1972

The Nitrogen Cycle

Activities of soil microbes (ammonification, nitrification, nitrogen fixation). MS-F2: 230–241, 1958
Oxides of nitrogen. MS-A1: 83–88, 1965
Nitrogen compounds used in crop production. T. C. Byerly, CP-G1: 104–109, 1970
The nitrogen cycle. C. C. Delwiche, CP-B1: 71–80, 1970
The dynamics of nitrogen transformations in the soil. D. R. Keeney and W. R. Gardner, CP-G1: 96–103, 1971
Threats to the integrity of the nitrogen cycle; nitrogen compounds in soil, water, atmosphere and precipitation. Barry Commoner, CP-G1: 70–95, 1971
Fertilizer nitrogen: contribution to nitrate in surface water in a corn belt watershed. D. H. Kohl, G. B.

Bibliography of Environmental Science

Shearer, B. Commoner, *Science* 174: 1331–1334, 1972

Nitrogen cycle. MS-E11: 801–809, 1972

The role of nitrogen in eutrophic processes. John J. Goering, CP-W1: 43–68, 1972

The Carbon Cycle

Carbon dioxide and climate. CP-R2: 174–179, 1959

Carbon dioxide. MS-A1: 78–82, 1965

Carbon cycle (organic) in the oceans. MS-E2: 169–170, 1966

Carbon dioxide cycle in the sea and atmosphere. MS-E2: 170–175, 1966

Geochemical cycle of carbon. MS-P4: 243–246, 1966

Carbon dioxide cycle in the sea and atmosphere. MS-E3: 131–136, 1967

The carbon cycle. Bert Bolin, CP-B1: 49–56, 1970

Carbon cycle in the biosphere. MS-M1: 160–163, 1970

The carbonate cycle. MS-O1: 333–349, 1970

The global balance of carbon monoxide. Louis A. Jaffe, CP-G1: 34–49, 1970

Soil: a natural sink for carbon monoxide. R. E. Inman, R. B. Ingersoll, E. A. Levy, *Science* 172: 1229–1231, 1970

Carbon monoxide in rainwater. J. W. Swinnerton, R. A. Lamontagne, V. J. Linnenbom, *Science* 172: 943–945, 1971

Carbon cycle. MS-E11: 125–129, 1972

Carbon monoxide balance in nature. B. Weinstock and H. Niki, *Science* 176: 290–292, 1972

A soil sink for atmospheric CO. Richard H. Gilluly, *Sci. News* 101: 286–287, 1972

The Phosphorus Cycle

Phosphorus. MS-M1: 144–146, 1970

Phosphorus cycling in the environment. MS-M1: 270–272, 1970

Phosphorus and eutrophication. Excerpt from SCEP Work Group on Ecological Effects (see MS-M1), CP-M5: 319–324, 1971

Phosphorus cycle. MS-E11: 946–951, 1972

The role of phosphorus in eutrophication. W. Stumm and E. Stumm-Zollinger, CP-W1: 11–42, 1972

The Oxygen Cycle

The oxygen cycle. P. Cloud and A. Gibor, CP-B1: 59–68, 1970

Oxygen cycle. MS-E11: 861–864, 1972

The Sulfur Cycle

Sulfur and its compounds. MS-A1: 60–72, 1965

Acid rain. G. E. Likens, F. H. Bormann, N. M. Johnson, *Envir.* 14(2): 33–40, 1972

The sulfur cycle. W. W. Kellogg et al, *Science* 175: 587–595, 1972

Sulfur cycle. MS-E11: 1148–1152, 1972

Sulfur mobilization as a result of fossil fuel combustion. J. P. Friend, K. K. Bertine, E. D. Goldberg, *Science* 175: 1278–1279, 1972

Other Cycles

Water in relation to plant growth. MS-C3: 118–128, 1968

The water cycle. H. L. Penman, CP-B1: 37–46, 1970

Calcium cycle. MS-E11: 120–122, 1972

Manganese cycle. MS-E11: 671–673, 1972

Radionuclides in river systems. D. J. Nelson, S. V. Kaye, R. S. Booth, CP-R4: 367–387, 1972

Silica—biogeochemical cycle. MS-E11: 1080–1085, 1972

Sodium cycle. MS-E11: 1099–1101, 1972

Strontium cycle. MS-E11: 1122–1123, 1972

Heavy Metals

Lead and other deleterious metals. MS-A1: 124–133, 1965

Mercury in the Environment. Multiple authors. USGS Prof. Paper 713, USGPO, 67 p., 1970

Environmental pollution by mercury. J. M. Wood, CP-A2: 39–56, 1971

Fossil fuels as a source of mercury pollution. Oiva L. Joensuu, *Science* 172: 1027–1028, 1971

Inorganic pollutants. MS-E9: 246–260, 1971

Meddelsome mercury. *Sci. News* 99: 7, 1971

Mercury. K. Montague and P. Montague, Sierra Club, San Francisco, California, 158 p., 1971

Mercury in a Greenland Ice Sheet: evidence of recent input by Man. H. V. Weiss, M. Koide, E. D. Goldberg, *Science* 174: 692–694, 1971

Mercury in Man. Neville Grant, *Envir.* 13(4): 2–15, 1971

Mercury in the environment. Terri Aaronson, *Envir.* 13(4): 16–23, 1971

Mercury in the environment. Leonard J. Goldwater, *Sci. Amer.* 224(5): 15–21, 1971

Mercury in the environment: natural and human factors. Allen L. Hammond, *Science* 171: 588–789, 1971

Mercury pollution: new studies show a lot of work must still be done. Richard H. Gilluly, *Sci. News* 100: 156, 1971

On the trail of heavy metals in ecosystems. *Sci. News* 100: 165–166, 1971

Enrichment of heavy metals and organic compounds in the surface microlayer of Narragansett Bay, Rhode Island. R. A. Duce et al., *Science* 176: 161–163, 1972

Lead; Airborne Lead in Perspective. Committee on Biologic Effects of Atmospheric Pollutants, NAS, Washington, D.C. 330 p., 1972

Lead and mercury burden of urban woody plants. William H. Smith, *Science* 176: 1237–1238, 1972

Mercury concentrations in museum specimens of tuna and swordfish. H. E. Ganther et al., *Science* 175: 1121–1124, 1972

Mercury concentrations in recent and ninety-year old benthopelagic fish. R. T. Barber, A. Vijayakumar, F. A. Cross, *Science* 178: 636–638, 1972

Mercury emissions from coal combustion. C. E. Billings and Wayne R. Matson, *Science* 176: 1232–1233, 1972

Mercury; geochemical cycle. MS-E11: 704–707, 1972

A progress report on mercury. John M. Wood, *Envir.* 14(1): 33–39, 1972

Multicycle References

Fertilizers, materials and practices. MS-F2: 344–384, 1958

Nutrient requirements of plants. MS-F2: 322–343, 1958

Gains and losses in constituents of sea water. MS-P4: 199–207, 1966

Eutrophication and agriculture in the United States. Jacob Verduin, CP-A3: 163–172, 1967

The extent and significance of fertilizer build-up in soils. P. R. Stout and R. G. Burau, CP-A3: 283–310, 1967

Fertilizer nutrients as contaminants in water supplies. George E. Smith, CP-A3: 173–186, 1967

Metabolism and functions of nitrogen and sulfur. MS-P8: 330–348, 1969

Gaseous atmospheric pollutants from urban and natural sources. E. Robinson and R. C. Robbins. CP-G1: 50–64, 1970

Geological history of sea water. MS-O1: 351–361, 1970

The oxygen and carbon dioxide balance in the earth's atmosphere. Francis S. Johnson, CP-G1: 4–11, 1970

Building a shorter life. Julian McCaull, Envir. 13(7): 2–15, 38–41, 1971

Chemical Composition of Atmospheric Precipitation in the Northeastern United States. F. J. Pearson, Jr., and Donald W. Fisher, USGS, Water-Supply Paper 1535-P, USGPO, 23 pp., 1971

Fate of air pollutants: removal of ethylene, sulfur dioxide, and nitrogen dioxide by soil. F. B. Abeles et al. Science 173: 914–916, 1971

Fossil fuel combustion and the major sedimentary cycle. K. K. Bertine and E. D. Goldberg, Science 173: 233–235, 1971

Nutrients (in rivers). Walter M. Sanders, III, CP-R4: 389–415, 1972

Overfed (eutrophication). J. Crossland and J. McCaull, Envir. 14(9): 30–37, 1972

Pesticides: General

Silent Spring (excerpt). Rachel Carson, CP-E2: 161–172, 1962

Pollution of the water environment by organic pesticides. Samuel D. Faust, CP-U1: 203–211, 1964

Contamination of urban air through the use of insecticides. Elbert C. Tabor, CP-U1: 152–158, 1966

The breakdown of pesticides in soils. Martin Alexander, CP-A3: 331–432, 1967

Conclusions on pesticide effects. N. W. Moore, CP-M4: 578–581, 1967

The extent and seriousness of pesticide build-up in soils. T. J. Sheets, CP-A3: 311–330, 1967

Pesticides in our national waters. R. S. Green, C. G. Gunnerson, J. J. Lichtenberg, CP-A3: 137–145, 1967

The effects of pesticides. William A. Niering, CP-M7: 80–89, 1968

Pesticide pollution. Consumer Reports, CP-M3: 167–173, 1969

The ant war (fire ant control). D. W. Coon and R. R. Fleet, Envir. 12(10): 28–38, 1970

The Biological Impact of Pesticides in the Environment. James W. Gillett, ed., Oregon State University Press, Corvallis, 210 p., 1970

The dual challenge of health and hunger—a global crisis. Georg A. Borgstrom, CP-M7: 257–265, 1970

General pollution—pesticides and related compounds. MS-P3: 129–136, 1970

The imported fire ant in the southern United States. Howard G. Adkins, Ann. AAG 60: 578–592, 1970

Insecticides and ecosystems. MS-P3: 167–185, 1970

More letters in the wind. R. Riseborough and V. Brodine, Envir. 12(1): 16–27, 1970/CP-U1: 87–96, 1970

Pesticides. MS-E6: 130–140, 1970

Since Silent Spring. Frank Graham, Jr., Fawcett Publications, Inc., Greenwich, Connecticut, 288 p., 1970

Biocides. MS-E9: 225–245, 1971

Biology and Water Pollution Control. Charles E. Warren, W. B. Saunders Co., Philadelphia, 434 p., 1971

Diversity, stability and pest control. MS-M4: 117–137, 1971

An ecologist views the environment. Donald A. Spencer, CP-U1: 111–118, 1971

Know your enemy (insects and agriculture). Julian McCaull, Envir. 13(5): 30 39, 1971

The pesticide problem. MS-N1: 458–484, 1971

Pesticides. MS-M3: 371–393, 1971

Pesticides. Robert L. Rudd, CP-E4: 279–301, 1971

Pesticides. U.S. Dept. Health, Education and Welfare, Public Health Service, CP-M3: 164–166, 1971

Striking the balance (spraying aquatic weeds). E. Hirst and H. Bank, Envir. 13(9): 34–41, 1971

A synopsis of the pesticide problem. N. W. Moore, CP-M5: 144–172, 1971

Toxic substances in ecosystems. MS-M4: 138–164, 1971

Captan and folpet. Kevin P. Shea, Envir. 14(1): 22–32, 1972

The cost of poisons. Robert van den Bosch, Envir. 14(7): 18–22, 27–31, 1972

Ecological aspects of pest control in Malaysia. Gordon R. Conway, CP-C2: 467–488, 1972

Ecosystems and pesticides. MS-T1: 78–88, 1972

Effects of pesticides and industrial wastes on surface water use. William A. Brungs, CP-R4: 353–365, 1972

Pesticides. MS-C1: 218–232, 1972

Pesticides. MS-C5: 293–354, 1972

Some ecological implications of two decades of use of synthetic organic insecticides for control of agricultural pests in Louisiana. L. Dale Newsom, CP-C2: 439–459, 1972

Realities of a pesticide ban. David Pimintel, Envir. 15(2): 18–20, 25–31, 1973

DDT and Other Chlorinated Hydrocarbons

DDT residues in an east coast estuary. G. M. Woodwell, C. F. Wurster, Jr., P. A. Isaacson, CP-M4: 565–571, 1967

DDT residues and declining reproduction in the Bermuda petrel. C. F. Wurster, Jr., and D. B. Wingate, CP-M4: 572–577, 1968

The beginning of the end for DDT. George Laycock, CP-U1: 72–80, 1969

Chlorinated hydrocarbon insecticides and the world ecosystem. Charles F. Wurster, Jr., CP-M4: 555–564, 1969

DDT goes on trial in Madison. Charles F. Wurster, CP-U1: 62–71, 1969

DDT on trial in Wisconsin. Bruce Ingersoll, CP-U1: 56–61, 1969

DDT in the marine environment. MS-M1: 126–138, 1970

Hormonal and enzymatic activity of DDT. Joel Bitman, CP-E8: 248–254, 1970

Aldrin and dieldrin. Charles F. Wurster, Envir. 13(8): 33–45, 1971

Chlorinated hydrocarbons in the marine environment. SCEP Task Force, CP-M5: 275–296, 1971

DDT: in field and courtroom a persistent pesticide lives on. Robert Gillette, Science 174: 1108–1110, 1971

DDT in the biosphere: where does it go? G. M. Woodwell, P. P. Craig, H. A. Johnson, Science 174: 1101–1107, 1971

DDT substitute. I. C. T. Nisbet and D. Miner, Envir. 13(6): 10–17, 1971

DDT: The United States and the developing countries. Rita F. Taubenfeld, CP-M5: 499–518, 1971

Questions for an old Friend (DDT). Julian McCaull, Envir. 13(6): 2–9, 1971

After DDT, what? George A. W. Boehm, Tech. Rev. 74(8): 26–31, 1972

The control of malaria. Richard Garcia, Envir. 14(5): 2–9, 1972

DDT: Its days are numbered, except perhaps in pepper fields. Robert

Bibliography of Environmental Science

Gillette, *Science* 176: 1313–1314, 1972

Oceans as alphabet soup: focus on DDT and PCBs. Richard H. Gilluly, *Sci. News* 101: 30–31, 1972

Spoiled by success (DDT controversy). Greg McIntyre, *Envir.* 14(6): 14–22, 27–29, 1972

DDT: an unrecognized source of polychlorinated biphenyls. Thomas H. Maugh II, *Science* 180: 578–579, 1973

Polychlorobiphenyls in North Atlantic Ocean water. G. R. Harvey, W. G. Steinhauer, J. M. Teal, *Science* 180: 643–644, 1973

Herbicides

Defoliation in Vietnam. Fred H. Tschirley, *Science* 163: 779–786, 1969

A family likeness (teratogenic effects of pesticides). Samuel S. Epstein, *Envir.* 12(6): 16–25, 1970

Gamble (2,4,5-T). Terri Aaronson, *Envir.* 13(7): 20–29, 1971

Chemical war (defoliation, Vietnam). Staff, *Envir.* 13(2): 44–47, 1971

Herbicides (Vietnam). MS-E9: 332–338, 1971

Herbicides in Vietnam: AAAS study finds widespread devastation. Philip M. Boffey, *Science* 171: 43–47, 1971

The soil transforms (herbicides). David Pramer, *Envir.* 13(4): 42–46, 1971

A tour of Vietnam (defoliation). Terri Aaronson, *Envir.* 13(2): 34–43, 1971

Herbicides: AAS study finds dioxin in Vietnamese fish. Deborah Shapley, *Science* 180: 285–286, 1973

Biological and Specific Controls for Pests

Third-generation pesticides. Carroll M. Williams, CP-M2: 247–251, 1967

Helping nature control insects. *Sci. News* 98: 197–198, 1970

Better methods of pest control. Gordon R. Conway, CP-E4: 302–325, 1971

Biological control and a remodeled pest control technology. Carl B. Huffaker, *Tech. Rev.* 73(8): 31–37, 1971

Infectious cure (Bacillus thuringiensis). Kevin P. Shea, *Envir.* 13(1): 43–45, 1971

Old weapons are best (biological control). Kevin P. Shea, *Envir.* 13(5): 40–49, 1971

Genetic control of insect populations. R. H. Smith and R. C. von Borstel, *Science* 178: 1164–1174, 1972

The gypsy moth. A. Dexter Hinckley, *Envir.* 14(2): 41–47, 1972

Gypsy moth control with the sex attractant pheromone. M. Beroza and E. F. Knipling, *Science* 177: 19–27, 1972

Chapter 22
Ecosystem Dynamics

Ecosystem Dynamics: General

Development and evolution of the ecosystem. MS-F3: 251–266, 1971

Fire and man. MS-E9: 79–92, 1971

Species and ecosystem. MS-S5: 93–110, 1971

Species and ecosystem. MS-S6: 367–384, 1973

Population Dynamics

Populations. MS-P7: 263–418, 1949

Relations within a population. MS-E14: 341–396, 1966

Topics in Population Genetics. Bruce Wallace, Norton, 481 p., 1968

The demography of organisms. MS-P5: 198–222, 1971

Organization at the population level. MS-F3: 162–233, 1971

Population cycles in small rodents. C. J. Krebs et al, *Science* 179: 35–40, 1973

Human Population

The human population. Edward S. Deevey, Jr., CP-M2: 49–56, 1960

Cultural and natural checks on population growth. D. H. Stott, CP-E8: 206–220, 1962

Population patterns and movements. William Vogt, CP-F1: 372–389, 1966

Population policy: will current programs succeed? Kingsley Davis, CP-M7, 1967

The population bomb (excerpt). Paul Ehrlich, CP-E2: 33–48, 1968

A Geography of Population: World Patterns. Glenn T. Trewartha, Wiley, 186 p., 1969

The population crisis is here now. Walter E. Howard, CP-M7: 155–168, 1969

Birth control. MS-P3: 211–233, 1970

Family planning and population control. MS-P3: 233–258, 1970

Numbers of people. MS-P3: 5–24, 1970

Optimum population and human biology. MS-P3: 199–210, 1970

Overpopulated America. Wayne H. Davis, CP-E8: 199–202, 1970

Population structure and projection. MS-P3: 25–50, 1970

U.S. population growth: would slower be better? Lawrence A. Mayer, CP-M3: 215–232, 1970

Ecology of the human population—workshop summary. MS-M4: 11–40, 1971

The human population problem. MS-N1: 518–550, 1971

The numbers and distribution of mankind. Nathan Keyfitz, CP-E4: 31–52, 1971

Population and Food. Foundations for Today, Vol. 1. R. S. Leisner and E. J. Kormondy, eds., William C. Brown Co., Dubuque, Iowa, 83 p., 1971

Population control. MS-M3: 289–312, 1971

Population control. MS-E9: 451–462, 1971

Population fluctuation. MS-M3: 259–288, 1971

Population growth. MS-M3: 223–258, 1971

Realities of the population explosion. Philip F. Low, CP-M3: 201–214, 1971

Growth of the world's population. Robert Leo Smith, CP-E8: 185–188, 1972

Population. MS-C5: 1–104, 1972

Population densities and distribution. MS-E12: 7–8, 1972

Ecology and population. Amos H. Hawley, *Science* 179: 1196–1200, 1973

Population growth. MS-E15: 233–265, 1973

Population regulation. MS-E15: 267–287, 1973

Interactions between Populations

Relationships between populations. MS-E14: 397–426, 1966

Competition between organisms. MS-P5: 223–241, 1971

Competition between species. Francisco J. Ayala, *Amer. Sci.* 60(3): 348–357, 1972

Defense against predation. MS-E15: 100–124, 1973

Interspecific competition and mutualism. MS-E15: 307–340, 1973

Predator-prey interactions. MS-E15: 288–306, 1973

Succession and Ecosystem Development

Community succession and development. MS-P7: 562–579, 1949

Man as a maker of new plants and plant communities. Edgar Anderson, CP-M6: 763–777, 1956

Man as an agent in the spread of organisms. Marston Bates, CP-M6: 788–804, 1956

The re-creative power of plant communities. Edward H. Graham, CP-M6: 677–691, 1956

Succession. ES-E14: 127–156, 1966

The strategy of ecosystem development. Eugene P. Odum, *Science* 164: 262–270, 1969/CP-E8: 28–38, 1969

Climax and polyclimax succession. MS-P5: 252–256, 1971

Stability and diversity at three trophic levels in terrestrial successional ecosystems. L. E. Hurd et al., *Science* 173: 1134–1136, 1971

Reforestation following forest cutting: mechanisms for return to steady-state nutrient cycling. P. L. Marks and F. H. Bormann, *Science* 176: 914–915, 1972

Chapter 23
Aquatic Ecosystems

Freshwater Ecosystems

Bogs. Edward S. Deevey, Jr., CP-R2: 281–287, 1958

Water. Roger Revelle, CP-M2: 57–67, 1963

Bogs, swamps, and marshes. MS-E14: 182–191, 1966

Flowing waters. MS-E14: 192–209, 1966

Lakes and ponds. MS-E14: 162–181, 1966

Streams, Lakes, Ponds. Robert E. Coker, Harper, 327 p., 1968

Ecology of Fresh Water. Alison L. Brown, Harvard, 129 p., 1971

The Ecology of Running Water. H. B. N. Haynes, University of Toronto Press, 555 p., 1971

Freshwater ecology. MS-F3: 295–323, 1971

A commentary on "what is a river." Ruth Patrick, CP-R4: 67–74, 1972

Fresh water. MS-C1: 48–89, 1972

Lakes. MS-W2: 104–118, 1972

Limnology. MS-E11: 650–661, 1972

Plant ecology in flowing water. John L. Blum, CP-R4: 53–65, 1972

Rivers. MS-W2: 88–104, 1972

What is a river?—zoological description. Kenneth W. Cummins, CP-R4: 33–52, 1972

Man's Impact on Freshwater Ecosystems

The direct effects on some plants and animals of pollution in the Great Lakes. George S. Hunt, CP-U1: 3–11, 1963

Eutrophication of the St. Lawrence Great Lakes. Alfred M. Beeton, CP-M4: 233–245, 1965

The aging Great Lakes. C. F. Powers and A. Robertson, CP-M2: 147–154, 1966

The great and dirty lakes. Gladwin Hill, CP-E3: 67–72, 1967

Species succession and fishery exploitation in the Great Lakes. Stanford H. Smith, CP-M4: 588–611, 1968

Aquatic weeds. L. G. Holm, L. W. Weldon, R. D. Blackburn, *Science* 166: 699–709, 1969/CP-M7: 297–313, 1969/CP-M4: 246–265, 1969

Man-induced eutrophication of lakes. Arthur D. Hasler, CP-G1: 110–125, 1970

Sanitary sewage. MS-O3: 115–130, 1970

Fresh water pollution. W. T. Edmondson, CP-E4: 213–229, 1971

Case history: the River Thames. K. H. Mann, CP-R4: 215–232, 1972

A comparative assessment of thermal effects in some British and North American rivers. T. E. Langford, CP-R4: 319–351, 1972

Man and the Illinois River. William C. Starrett, CP-R4: 131–169, 1972

Man's impact on the Columbia River. Parker Trefethen, CP-R4: 77–98, 1972

The Nile River—a case history. D. Hammerton, CP-R4: 171–214, 1972

Pollution: rivers, lakes and streams. MS-C5: 209–227, 1972

Regulated discharge and the stream environment. J. C. Fraser, CP-R4: 263–285, 1972

Sewage farming. Jonathan Allen, *Envir.* 15(3): 36–41, 1973

Marine Ecosystems

The oceanic life of the Antarctic. Robert Cushman Murphy, CP-O1: 287–299/CP-R3: 503–465, 1962

Estuaries, tidal marshes, and swamps. MS-E14: 210–220, 1966

Marine ecology. MS-E2: 454–458, 1966

Marine microbiology. MS-E2: 465–469, 1966

Phytoplankton; photosynthesis. MS-E2: 712–718, 1966

The seashore. MS-E14: 221–242, 1966

Detritus in the ocean and adjacent sea. Johannes Krey, CP-E6: 389–396, 1967

Organic detritus in relation to the estuarine ecosystem. Rezneat M. Darnell, CP-E6: 376–382, 1967

The organic detritus problem. Rezneat M. Darnell, CP-E6: 374–375, 1967

Particulate organic detritus in a Georgia salt marsh-estuarine ecosystem. E. P. Odum and A. A. de la Cruz, CP-E6, 1967

The nature of marine life. John D. Isaacs, CP-O1: 215–227/CP-O2: 67–79, 1969

Animals of the sea. MS-O1: 393–416, 1970

The ecology of wetlands in urban areas. William A. Niering, CP-C3: 199–208, 1970

Marine ecology. MS-O1: 417–446, 1970

Marine life, classification. MS-O1: 365–374, 1970

Plant life in the sea; photosynthesis; plankton. MS-O1: 375–392, 1970

The sea. Taylor A. Pryor, CP-N1: 115–120, 1970

Estuarine ecology. MS-F3: 352–362, 1971

Marine ecology. MS-F3: 324–351, 1971

Implications of a systems approach to oceanography. John L. Walsh, *Science* 176: 969–974, 1972

The sea. MS-C1: 21–47, 1972

The shallow ocean. MS-W2: 138–157, 1972

The unseen ocean. MS-W2: 119–137, 1972

Oil Spills (see also Chapter 10)

Oil pollution of the ocean. Max Blumer, CP-M4: 295–301, 1969

Ocean pollution by petroleum hydrocarbons. R. Revelle, E. Wenk, B. H. Ketchum, CP-M5: 297–318, 1971

Oilspill. Wesley Marx, Sierra Club, San Francisco, California, 139 p., 1971

Blowout. C. E. Steinhart and J. S. Steinhart, Duxbury Press, North Scituate, Massachusetts, 138 p., 1972

Oil on the waters: modest progress in cleanup technology. Joan

Arehart-Treichel, *Sci. News* 102: 250–251, 1972

Oily seas and plastic waters of the Atlantic. *Sci. News* 103: 119, 1973

Thermal Pollution (see also Chapter 13)

Thermal pollution. Lamont C. Cole, CP-M4: 217–224, 1969

Thermal pollution and aquatic life. John R. Clark, CP-M2: 163–172, 1969

Thermal loading. MS-E9: 133–149, 1971

Pollution: thermal pollution. MS-C5: 234–237, 1972

Thermal pollution—a new problem in aquatic ecosystems. Robert Leo Smith, CP-E8: 270–278, 1972

Man's Impact on Marine Ecosystems

The sea lamprey. V. C. Applegate and J. W. Moffett, CP-O1: 391–396, 1955

Whales, plankton, and Man. Willis E. Pequegnat, CP-O1: 275–279, 1958

The last of the great whales. Scott McVay, CP-O1: 313–321, 1966

The role of man in estuarine processes. L. Eugene Cronin, CP-M4: 266–294, 1967/CP-E6, 667–689, 1967

Central American sea-level canal: possible biological effects. Ira Rubinoff, CP-M4: 493–506, 1968

Biological implications of global marine pollution. Bostwick H. Ketchum, CP-G1: 190–194, 1970

Changes in the chemistry of the oceans: the pattern of effects. George M. Woodwell, CP-G1: 186–189, 1970

The sea: should we now write it off as a future garbage pit? Robert W. Riseborough, CP-N1: 121–135, 1970

Chlorinated hydrocarbons in the marine environment. SCEP Task Force, CP-M5: 275–296, 1971

Conseqences of a sea-level canal. Richard H. Gilluly, *Sci. News* 99: 52–53, 1971

The effects of pollution. MS-M4: 251–268, 1971

Nitrogen, phosphorus, and eutrophication in the coastal marine environment. J. H. Ryther and W. M. Dunstan, *Science* 171: 1008–1013, 1971

Ocean pollution. F. MacIntyre and R. W. Holmes, CP-E4: 230–253, 1971

Population, natural resources, and biological effects of pollution of estuaries and coastal waters. Bostwick H. Ketchum, CP-M5: 59–79, 1971

Ship canals and aquatic ecosystems. W. I. Aron and S. H. Smith, *Science* 174: 13–20, 1971

The world ocean: ultimate sump. MS-E9: 150–169, 1971

Pollution: oceans. MS-C5: 228–233, 1972

Some basic issues in water pollution control legislation. Walter E. Westman, *Amer. Sci.* 60(6): 767–773, 1972

The new federal water pollution control law. A. W. Reitze, Jr., and G. Reitze, *Envir.* 15(1): 19–20, 1973

Aquatic Resources–Fisheries, Food Production

Harvests of the seas. Michael Graham, CP-M6: 487–503, 1956

The antarctic seas; whales and whaling. MS-P2: 300–311, 1964

Fertility of the oceans. MS-E2: 268–272, 1966

The conquest of the oceans. Roger Revelle, CP-C6: 16–37, 1967

Food from the sea. William E. Ricker, CP-R1: 87–108, 1969

The food resources of the ocean. S. J. Holt, CP-M2: 84–96/CP-O2: 94–106, 1969

Photosynthesis and fish production in the sea. John H. Ryther, *Science* 166: 72–76, 1969/CP-E8: 360–367, 1969

Food from the sea. MS-P3: 101–109, 1970

Anchovies, birds and fishermen in the Peru Current. Gerald J. Paulik, CP-E4: 156–185, 1971

Fishery resources for the future. G-H. Smith and D. W. Lewis, CP-C5: 539–566, 1971

Freshwater fisheries. MS-N1: 375–417, 1971

Management of aquatic resources. MS-M4: 219–268, 1971

Marine fisheries. MS-N1: 418–457, 1971

The anchovy crisis. C. P. Idyll, *Sci. Amer.* 228(6): 22–29, 1973

Chapter 24
Terrestrial Ecosystems

Terrestrial Ecosystems: General

Biome and biome-type in world distribution. MS-P7: 580–597, 1949

Distribution of communities. MS-E14: 331–340, 1966

Food from the land. Sterling B. Henricks, CP-R1: 65–85, 1969

Ecology: the biome approach. Richard Gilluly, *Sci. News* 98: 204–205, 1970

Biome approach in ecology. *Sci. News* 99: 247–248, 1971

Terrestrial ecology. MS-F3: 363–404, 1971

Wildlife. MS-N1: 278–316, 1971

World patterns of distribution among organisms. MS-P5: 244–251, 1971

World Vegetation Types. S. R. Eyre, ed., Columbia University Press, New York, 264 p., 1971

Canadian environments. M. C. Storrie and C. I. Jackson, *Geog. Rev.* 62: 309–332, 1972

Ecosystem analysis: biome approach to environmental research. Allen L. Hammond, *Science* 175: 46–48, 1972

Natural areas. William H. Moir, *Science* 177: 396–400, 1972

An end to chemical farming? Duane Chapman, *Envir.* 15(2): 12–17, 1973

Terrestrial Ecosystems: Miscellaneous Impacts

Air pollution and plants. H. E. Heggestad, CP-M5: 101–115, 1971

Introduction of exotics into our ecosystems. MS-E9: 279–299, 1971

Vegetation of the city. T. R. Detwyler, CP-U3: 229–259, 1972

Forest Biomes

Damage to forests from air pollution. George H. Hepting, CP-M4: 522–531, 1964

The Great American Forest. Rutherford Platt, Prentice-Hall, 271 p., 1965

The forest. MS-E14: 301–317, 1966

Deciduous Forests of Eastern North America. E. Lucy Braun, Hafner Publishing Co., New York, 596 p., 1967 (facsimile of the edition of 1950)

Analysis of temperate forest ecosystems. David E. Reichle, ed., Springer-Verlag, New York, 304 p., 1970

Air pollution and trees. George H. Hepting, CP-M5: 116–129, 1971

Forest resource. MS-N1: 221–277, 1971

Forest resources. Lee M. James, CP-C5: 481–496, 1971

Potential effects of global atmospheric conditions on forest ecosystems. Karl F. Wenger et al., CP-M5: 192–202, 1971

Tropical lowland forests. MS-M4: 173–182, 1971

Bibliography of Environmental Science

The boreal bioclimates. F. K. Hare and J. C. Ritchie, *Geog. Rev.* 62: 333-365, 1972
The boreal coniferous forest. MS-E12: 120-151, 1972
Floodplain forest biotic communities. MS-E12: 89-119, 1972
Montane coniferous forest and alpine communities. MS-E12: 152-181, 1972
The northern Pacific coast forest biome and mountain communities. MS-E12: 211-236, 1972
The Suburban Woodland: Trees and Insects in the Human Environment. R. C. Clement and I. C. T. Nisbet, Audubon Conservation Rept. No. 2, Mass. Audubon Soc., Lincoln, 52 p., 1972
The summer drought or broad sclerophyll-grizzly bear community. MS-E12: 238-259, 1972
The temperate deciduous forest biome. MS-E12: 17-88, 1972
"This is the forest primeval", Joan Arehart-Treichel, *Sci. News* 102: 78-79, 1972
Tropical deciduous forest. MS-E12: 430-458, 1972

Grasslands and Savannas

The grassland of North America: its occupance and the challenge of continuous reappraisal. James C. Malin, CP-M6: 350-366, 1956
The modification of mid-latitude grasslands and forests by man. John T. Curtis, CP-M4: 507-521, 1956/CP-M6: 721-736, 1956
The grasslands. MS-E14: 275-288, 1966
The rangelands of the western United States. R. Merton Love, CP-M2: 229-235, 1970
Grassland biome network: results of the first year. Joan Lynn Arehart, *Sci. News* 100: 282-283, 1971
Grassland resources. H. C. Hanson and W. C. Whitman, CP-C5: 451-480, 1971
A grazing ecosystem in the Serengeti. Richard H. V. Bell, *Sci. Amer.* 225(1): 86-93, 1971
Pollution and range ecosystems. Dixie R. Smith, CP-M5: 130-143, 1971
Rangelands. MS-N1: 179-220, 1971
Tropical savanna. MS-M4: 183-188, 1971
National parks in savannah Africa. Norman Myers, *Science* 178: 1255-1263, 1972

The northern temperate grassland. MS-E12: 328-355, 1972
The southern temperate grassland. MS-E12: 356-372, 1972

Tropical Rainforest Biome

The agricultural potential of the humid tropics. Jen-Hu Chang, *Geog. Rev.* 58: 333-361, 1968
Spoiling the jungle yields few riches. Christopher Weathersbee, *Sci. News* 95: 312-315, 1969
Ecological balance in tropical agriculture. Matthias U. Igbozurike, *Geog. Rev.* 61: 519-529, 1971
The tropical rain forest. MS-E12: 395-429, 1972
The tropical rain forest: a nonrenewable resource. A. Gómez-Pompa, C. Vázquez-Yanes, S. Guevara, *Science* 177: 762-765, 1972
Vietnam land devastation detailed. Constance Holden, *Science* 175: 737, 1972
The fragile jungle: pressure of civilization. *Sci. News* 103: 269, 1973

Desert Biome

Shrublands and the desert. MS-E14: 289-300, 1966
Arid and semi-arid systems. MS-M4: 195-200, 1971
The hot desert. MS-E12: 373-394, 1972

Tundra Biome

Arctic flora and fauna. MS-P2: 111-129, 1964
The tundra. MS-E14: 318-330, 1966
Tundra and taiga. J. Ross Mackay, CP-F1: 156-171, 1966
Some estimates of the thermal effects of a heated pipeline in permafrost. Arthur H. Lachenbruch, CP-M1: 74-76, 1970
Oil on Ice. Tom Brown, Sierra Club, San Francisco, California, 159 p., 1971
The tundra biome. MS-E12: 184-210, 1972

Epilogue: Environmental Perspectives

Historical Roots of the Environmental Crisis

The agency of Man on the earth. Carl O. Sauer, CP-M6: 49-69, 1956
Cultural differences in the interpretation of natural resources. Alexander Spoehr, CP-M6: 93-102, 1956
Changing ideas of the habitable world. Clarence J. Glacken, CP-M6: 70-92, 1956
The historical roots of our ecological crisis. Lynn White, Jr., *Science* 155: 1203-1207, 1967/CP-M4: 27-35/ CP-M7: 22-30, 1967
Population and the dignity of Man. Roger L. Shinn, CP-D1: 91-109, 1969
The church and the environment. Robert C. Anderson, CP-N1: 273-275, 1970
The cultural basis for our environmental crisis. Lewis W. Moncrief, *Science* 170: 508-512, 1970/CP-D1: 3-16, 1970
Man's use of the earth: historical background. Max Nicholson, CP-M4: 10-21, 1970
Our treatment of the environment in ideal and actuality. Yi-Fu Tuan, *Amer. Sci.* 58: 244-249, 1970/CP-D1: 35-46, CP-E8: 167-171, 1970
At last—a revolution that unites. George E. LaMore, Jr., CP-U1: 126-132, 1971
Man and nature. Yi-Fu Tuan, Commission on College Geography, Resource Paper No. 10, AAG, Washington, D.C., 49 p., 1971
The population explosion and the rights of the subhuman world. John B. Cobb, Jr., CP-D1: 19-32, 1971
Earth Might Be Fair: Reflections on Ethics, Religion, and Ecology. I. A. Barbour, ed., Prentice-Hall, 168 p., 1972

Environment & Economics

Economics and ecology. Kenneth E. Boulding, CP-F1: 225-234, 1966
Standards and techniques of evaluating economic choices in environmental resource development. Ayres Brinser, CP-F1: 235-245, 1966
The ecology of big business. Robert O. Anderson, CP-N1: 215-222, 1970
Pollution: the mess around us. Marshall I. Goldman, CP-E3: 3-63, 1970
Economics and the Environment: A Materials Balance Approach. A. V. Kneese, R. U. Ayres, and R. C. D'Arge, Resources for the Future, Hopkins, 120 p., 1970
Solid waste management. Rolf Eliassen, CP-N1: 55-65, 1970
Economics and conservation. Harold H. McCarty, CP-C5: 35-52, 1971

Economics of Pollution. K. E. Boulding et al, New York Univ. Press, New York, 158 p., 1971

The economist's approach to pollution and its control. Robert M. Solow, *Science* 173: 489-503, 1971

The economy and the environment. MS-E7: 99-153, 1971

Environment and economics. Kenneth E. Boulding, CP-E4: 359-367, 1971

Improvement in the quality of the environment: costs and benefits. Dan Throop Smith, CP-U1: 119-125, 1971

Killing the goose (business profits). Daniel Fife, *Envir.* 13(3): 20-27, 1971

Power and economics. MS-E8: 174-205, 1971

Residuals management. Walter O. Spofford, Jr., CP-M5: 477-488, 1971

Ecology and Economics: Controlling Pollution in the 70s. M. I. Goldman, ed., 234 p., 1972

Multiple use of river systems: an economic framework. James Crutchfield, CP-R4: 431-440, 1972

Residuals charges for pollution control: a policy evaluation. A. M. Freeman, III, and R. H. Haveman, *Science* 177: 322-329, 1972

Natural Resource Conservation, Land Planning

Man—An Endangered Species? U. S. Dept. Interior, Conservation Yearbook No. 4, USGPO, 100 p., 1968

It's Your World—The Grassroots Conservation Story. U. S. Dept. Interior, Conservation Yearbook No. 5, USGPO, 96 p., 1969

Agriculture, forestry, and wildlands. Charles H. W. Foster, CP-N1: 140-142, 1970

Conservation and management. MS-E5: 186-217, 1970

Conservation of open space. Gordon Harrison, CP-N1: 136-139, 1970

Land use. MS-E6: 165-197, 1970

Conservation. MS-M3: 395-423, 1971

Conservation in the United States. Harold M. Rose, CP-C5: 3-17, 1971

Conservation of wildlife. Charles A. Dambach, CP-C5: 511-538, 1971

Land resources. Marion Clawson, CP-E4: 117-134, 1971

The land we possess. Guy-Harold Smith, CP-C5: 159-185, 1971

National planning and the conservation of resources. Guy-Harold Smith, CP-C5: 635-660, 1971

Natural Resource Conservation: An Ecological Approach. Oliver S. Owen, Macmillan, 593 p., 1971

Nature in captivity; parks, wilderness areas, natural areas. MS-E9: 60-75, 1971

The public domain. S. S. Visher and H. H. Visher, CP-C5: 19-33, 1971

Recreational resources. Marion Clawson, CP-C5: 569-591, 1971

Conservation. MS-E11: 185-189, 1972

Environmental conservation, 3rd. ed., Raymond F. Dasmann, Wiley, 473 p., 1972

Is conservation a losing battle? MS-C1: 284-289, 1972

Natural resources. MS-E11: 771-777, 1972

Population, Resources, & Technology

Possible limits of raw-material consumption. Samuel H. Ordway, Jr., CP-M6: 987-1009, 1956

The spiral of population. Warren S. Thompson, CP-M6: 970-986, 1956

The politics of ecology: the question of survival. Aldous Huxley, CP-E2: 24-31, 1963

The human population. Edward S. Deevey, Jr., CP-M2: 49-55, 1960

Economics and ecology. Kenneth E. Boulding, CP-F1: 225-234, 1966

Natural resources and economic development: the web of events, policies, and policy objectives. Joseph L. Fisher, CP-F1: 261-276, 1966

Technology, resources, and urbanism—the long view. Richard L. Meier, CP-F1: 277-288, 1966

Adaptation to the environment and Man's future. René Dubos, CP-C6: 60-78, 1967

Ecosystem science as a point of synthesis. S. D. Ripley and H. K. Beuchner, CP-D1: 133-142, 1967

The modern expansion of world population. John D. Durand, CP-M4: 36-49, 1967

Population policy: will current programs succeed? Kingsley Davis, *Science* 158: 730-739, 1967/CP-M7: 180-199, 1967

The prospects of economic abundance. Kenneth E. Boulding, CP-C6: 40-57, 1967

The U. S. resource outlook: quantity and quality. Hans H. Landsberg, CP-M1: 212-223, 1967

The dimensions of world poverty. David Simpson, CP-M2: 97-105, 1968

The population bomb. Paul Ehrlich, CP-E2: 32-48, 1968

Realities of mineral distribution. Preston E. Cloud, Jr., CP-M1: 194-207, 1968

The tragedy of the commons. Garrett Hardin, *Science* 162: 1243-1248, 1968/CP-M7: 222-234/CP-E8: 382-390, 1968

Can our conspicuous consumption of natural resources be cyclic? Leallyn B. Clapp, CP-M1: 208-211, 1969

Eco-catastrophe. Paul R. Ehrlich, CP-M7: 138-147, 1969

Energy resources. M. King Hubbert, CP-R1: 237-239, 1969

The energy revolution. George Tayler, CP-D1: 111-129, 1969

Evaluating the biosphere. Barry Commoner, CP-M4: 50-60, 1969

The future of man's environment. Robert W. Lamson, CP-M1: 4-10, 1969

The human ecosystem. Marston Bates, CP-R1: 21-30, 1969

Interactions between Man and his resources. John D. Chapman, CP-R1: 31-42, 1969

Population, Evolution and Birth Control: A Collage of Controversial Ideas, 2nd ed., Garrett Hardin, ed., Freeman, 386 p., 1969

The population crisis is here now. Walter E. Howard, CP-M7: 155-168, 1969

Population pollution. Francis S. L. Williamson, CP-U1: 100-107, 1969

Resources and Man. NAS-NRC Committee on Resources and Man, Freeman, 250 p., 1969

The tragedy of the commons revisited. Beryl L. Crowe, *Science* 166: 1103-1107, 1969/CP-M7: 234-243/CP-E8: 391-398, 1969

United States and world populations. Nathan Keyfitz, CP-R1: 43-64, 1969

American institutions and ecological ideals. Leo Marx, *Science* 170: 945-952, 1970

The assessment of technology. H. Brooks and R. Bowerd, CP-M2: 209-217, 1970

A better world for fewer children. George Wald, CP-U1: 108-110, 1970

Challenge in our time. Lee A. Dubridge, CP-N1: 68-71, 1970

The convergence of environmental disruption. Marshall I. Goldman,

Bibliography of Environmental Science

Science 170: 37–42, 1970/CP-D1: 69–88, 1970

The ecological facts of life. Barry Commoner, CP-N1: 18–35, 1970

Environmental disruption in the Soviet Union. M. I. Goldman, CP-M4: 61–75, 1970

Environmental problems arising from new technologies. Kenneth S. Pitzer, CP-N1: 66–67, 1970

The Environmental Revolution: A Guide for the New Masters of the World. Max Nicholson, McGraw-Hill, 366 p., 1970

Growth and the quality of the American environment. Roger Revelle, CP-N1: 71–74, 1970

Growth versus the quality of life. J. Alan Wagar, *Science* 168: 1179–1184, 1970/CP-E8: 400–408, 1970

Implications of change and remedial action. MS-M1: 31–36, 1970

An international consensus. Michel Batisse, CP-N1: 98–100, 1970

Megalopolis: resources and prospect. Pierre Dansereau, CP-C3: 1–33, 1970

Mortgaging the old homestead. Lord Ritchie-Calder, CP-D1: 49–66, 1970

Pollution: the mess around us. Marshall I. Goldman, CP-E3: 3–63, 1970

Population, growth and resources. Council on Environmental Quality, MS-E6: 149–164/CP-M3: 199–200, 1970

Population, Resources, Environment; Issues in Human Ecology. P. R. Ehrlich and A. H. Ehrlich, Freeman, 383 p., 1970

The Population Challenge. U. S. Dept. Interior, Conservation Yearbook No. 2, USGPO, 80 p., 1970

The population explosion: facts and fiction. Paul R. Ehrlich, CP-N1: 35–44, 1970

Technology and Growth: The Price We Pay. E. J. Mishan, Praeger Publishers, Inc., 193 p., 1970

U. S. Population growth: would slower be better? Lawrence A. Meyer, CP-M3: 215–229, 1970

A view of the present situation. Lamont Cole, CP-N1: 14–18, 1970

War on hunger: the need for a strategy. Georg Borgstrom, CP-N1: 45–54, 1970

Can We Survive Our Future? A Symposium. G. R. Urban, ed., St. Martin's Press, New York, 400 p., 1971

The causes of pollution. B. Commoner, M. Corr, P. J. Stamler, *Envir.* 13(3): 2–19, 1971

Changing attitudes toward environmental problems. Philip H. Abelson, *Science* 172: 517, 1971

The Closing Circle. Barry Commoner, Alfred A. Knopf, New York, 326 p., 1971

Conservation comes of age. S. Dillon Ripley, *Amer. Sci.* 59: 529–531, 1971

The conservation of Man. Lawrence A. Hoffman, CP-C5: 593–612, 1971

Counterintuitive behavior of social systems. Jay W. Forrester, *Tech. Rev.* 73(3): 53–68, 1971

Environment: preparing for the crunch. Carroll L. Wilson, CP-M1: 311–314, 1971

Environment and the equilibrium population. William W. Murdoch, CP-E4: 416–435, 1971

The environmental context. Thomas F. Malone, *EOS* 52: 508–512, 1971

The environmental crisis. Philip H. Abelson, *EOS* 52: 124–128, 1971

The future (of mankind). MS-M3: 425–443, 1971

The human population problem. MS-N1: 518–550, 1971

Impact of population on growth. P. R. Ehrlich and J. P. Holdren, *Science* 171: 1212–1217, 1971

International environmental problems—a taxonomy. C. S. Russell and H. H. Landsberg, *Science* 172: 1307–1314, 1971

Is There an Optimum Level of Population? S. F. Singer, ed., McGraw-Hill, 426 p., 1971

Land: making room for tomorrow. John Lear, CP-M1: 300–302, 1971

Limits rarely perceived; Introduction. CP-M2: 41–47, 1971

Man, food, environment. L. R. Brown and G. Finsterbusch, CP-E4: 53–69, 1971

On management and buying time. Introduction, CP-M2: 200–208, 1971

Nobody ever dies of overpopulation. Garret Hardin, *Science* 171: 527, 1971/CP-M1: 27, 1971

The numbers and distribution of mankind. Nathan Keyfitz, CP-E4: 31–52, 1971

Partnership with nature; what systems are next? MS-E8: 274–310, 1971

The people problem. MS-E9: 427–466, 1971

The People Problem: What You Should Know about Growing Population and Vanishing Resources. Dean Fraser, Indiana Univ. Press, 248 p., 1971

Population: a look to the future. Charlton Ogburn, CP-A4: 198–213, 1971

Population and the dignity of Man. Roger L. Shinn, CP-D1: 91–109, 1971

Population growth, fluctuation, and control. MS-M3: 223–312, 1971

Predicting and preventing population problems. Robert J. Trotter, *Sci. News* 100: 114–115, 1971

Realities of the population explosion. Phillip, F. Low, CP-M3: 201–214, 1971

Subtraction by multiplication: population, technology, and the diminished man. F. H. Bormann, CP-M5: 33–46, 1971

This Endangered Planet. Richard A. Falk, Random House, New York, 495 p., 1971

Are Our Descendants Doomed? Technological Change and Population Growth. H. Brown and E. Hutchings, Jr., eds., Viking Press, New York, 377 p., 1972

The Careless Technology: Ecology and International Development. M. T. Farver and J. P. Milton, eds., Natural Hist. Press, 1030 p., 1972

Ecolibrium. Athelstan Spilhaus, *Science* 175: 711–715, 1972

Ecology, survival, and society. *Sci. News* 101: 100–101, 1972

Energy, economic growth, and the environment. Sam H. Schurr, ed., Resources for the Future, Hopkins 240 p., 1972

The human experiment, too much waste, will Man survive? MS-O2: 3–10, 201–225, 1972

The Limits to Growth. D. H. Meadows et al, Universe Books, New York, 205 p., 1972

Limits to growth: debating the future. Richard H. Gilluly, *Sci. News* 101: 202–204, 1972

The limits to growth: hard sell for a computer view of doomsday. Robert Gillette, *Science* 175: 1088–1092, 1972

Man: ecology of Man; accelerating evolution; the population crisis; economy and technology. MS-C1: 254–283, 1972

The new-priority problem. Francis H. Schott, *Tech. Rev.* 75(2): 39–43, 1972

Population and pollution in the

United States. Ronald G. Ridker, *Science* 176: 1085-1090, 1972

Quality or quantity. *Sci. News* 101: 181-182, 1972

The Spoils of Progress: Environmental Pollution in the Soviet Union. Marshall I. Goldman, MIT, 372 p., 1972

To Live on Earth: Man and His Environment in Perspective. Sterling Brubaker, Resources for the Future, Hopkins, 202 p., 1972

The World's Population: Problems of Growth. Quentin H. Stanford, ed., Oxford University Press, Toronto, 346 p., 1972

Ecology and population. Amos H. Hawley, *Science* 179: 1196-1204, 1973

Humanizing the earth. Rene J. Dubos, *Science* 179: 769-772, 1973

The prospects for a stationary world population. Tomas Frejka, *Sci. Amer.* 228(3): 15-23, 1973

Environmental Legislation & Administration

Administrative possibilities for environmental control. Lynton K. Caldwell, CP-F1: 648-671, 1966

The federal role in pollution abatement and control. John Tunney, CP-A1: 25-39, 1969

Federal organization for environmental quality. MS-E6: 19-28, 1970

From conservation to environmental law. David Sive, CP-D1: 159-167, 1970

The National Environmental Policy Act of 1969. MS-E6: 243-326/MS-E7: 267-333, 1970

A water pollution control program for the 1970's. MS-C4: 15-78, 1970

Critics weigh EPA herbicide report, find it wanting. Constance Holden, *Science* 173: 312, 1971

Decision on 2,4,5-T: leaked reports compel regulatory responsibility. Nicolas Wade, *Science* 173: 610-615, 1971

Environment and administration: the politics of ecology. Lynton K. Caldwell, CP-E4: 390-415, 1971

Environment and the law. Victor J. Yanacone, Jr., CP-E4: 368-389, 1971

Environmental impact statement. Peter T. Flawn, *Geotimes* 16(9): 23-24, 1971

Environmental Quality: The Second Annual Report of the Council on Environmental Quality. USGPO, 360 p., 1971

Florida: Nixon halts canal project, cites environment. John Walsh, *Science* 171: 357, 1971

The law and environment. MS-E7: 155-187, 1971

Legal and environmental case studies in applied geomorphology. Donald R. Coates, CP-E5: 223-242, 1971

Man and His Environment: Law. Earl F. Murphy, Harper, 168 p., 1971

A Procedure for Evaluating Environmental Impact. L. B. Leopold et al, Circular 645, USGS, 13 p., 1971

Water quality and agriculture in the United States; an overall view. James M. Quigley, CP-A2: 129-135, 1971

National environmental policy act: signs of backlash are evident. Robert Gillette, *Science* vol. 176: 30-33, 1972

National goals and environmental laws. Richard A. Carpenter, *Tech. Rev.* 74(3): 58-63, 1972

Political aspects of multiple use (of river systems). John D. Dingell, CP-R4: 441-453, 1972

Enforcing the clean air act of 1970. Noel de Nevers, *Sci. Amer.* 228(6): 14-21, 1973

Environmental law (I): Maturing field for lawyers and scientists. Luther J. Carter, *Science* 179: 1205-1209, 1973

Environmental law (II): A strategic weapon against degradation? Luther J. Carter, *Science* 179: 1310, 1312-1350, 1973

Lack of impact (impact statements). Frank Kreith, *Envir.* 15(1): 26-33, 1973

Environmental Planning & Management

Metropolitan organization for air conservation. MS-A1: 195-211, 1965

The Third Wave—America's New Conservation. U. S. Dept. Interior, Conservation Yearbook No. 3, USGPO, 128 p., 1966

The Environmental Destruction of South Florida. W. R. McCluney, ed., Univ. of Miami Press, Coral Gables, 134 p., 1969

Comprehensive planning: where it is? Alfred Heller, CP-N1: 74-76, 1970

Conservation and management. MS-E5: 186-217, 1970

Developing a natural resource management policy. Stephen H. Spurr, CP-N1: 102-107, 1970

Ecology and management of the rural and the suburban landscape. Frank E. Egler, CP-C3: 81-102, 1970

Government and environmental quality. Lynton Caldwell, CP-N1: 198-204, 1970

Our lagging institutions. Henry L. Diamond, CP-N1: 172-174, 1970

Planning, the law, and a quality environment. Joseph E. Bodovitz, CP-N1: 179-190, 1970

Planning the earth's surface. Henry Caulfield, CP-N1: 77-79, 1970

Pollution problems, resource policy, and the scientist. Alfred W. Eipper, CP-D1: 169-186, 1970

Quality management of our air environment. Rolf Eliassen, CP-N1: 156-162, 1970

A regional approach. Thomas Gill, CP-N1: 190-193, 1970

Streamlining environmental management. Frank E. Moss, CP-N1: 177-179, 1970

Types of remedial changes at the sources and their implications; an overview. MS-M1: 229-255, 1970

Decision-making in the production of power. Milton Katz, CP-E7: 131-134, 1971

Implications of change and remedial action. Summary of SCEP Report, CP-M5: 469-476, 1971

Residuals management. Walter O. Spofford, Jr., CP-M5: 477-488, 1971

Sanity in research and evaluation of environmental health. H. E. Stokinger, *Science* 174: 662-665, 1971

Status and trends. MS-E7: 209-265, 1971

Vermont: a small state faces up to a dilemma over development. John Walsh, *Science* 173: 896-897, 1971

Coping with environmental problems. MS-T1: 127-158, 1972

Handbook of Environmental Management: vol. 1: Fundamentals. Carlos J. Hilado, Technomic, 113 p., 1972

Rationalization of multiple use of rivers. John Cairns, Jr., CP-R4: 421-430, 1972

World eco-crisis; international organizations in response, D. A. Kay and E. G. Skolnikoff, eds., Univ. of Wisconsin Press, Madison, 324 p., 1972

Urban Environment

Air pollution and urban development. MS-A1: 307-324, 1965

The urbanization of the human population. Kingsley Davis, CP-M2: 267–279, 1965

American metropolitan evolution. John R. Borchert, *Geog. Rev.* 57: 301–332, 1967

Technology, living cities, and human environment. Athelstan Spilhaus, *Amer. Sci.* 57: 24–36, 1969

The ecology of wetlands in urban areas. William A. Niering, CP-C3: 199–208, 1970

The environment of modern cities. MS-P3: 141–145, 1970

The place of nature in the city of man. Ian L. McHarg, CP-C3: 37–55, 1970

The urban environment. MS-E6: 168–173, 1970

City and regional planning. James A. Spencer, CP-C5: 615–634, 1971

The inner city environment. MS-E7: 189–207, 1971

Man and the urban environment. Robert B. Smock, CP-E4: 339–358, 1971

Man's urban environment. MS-E9: 343–408, 1971

Planning a new town's environment. *Geotimes* 16(6): 17–20, 1971

Urbanization: the great migration. CP-A4: 142–170, 1971

Urbanization and Environment: The Physical Geography of the City. T. R. Detwyler and M. G. Marcus, eds., Duxbury Press (Wadsworth), 287 p., 1972

Environmental Education, Citizen Participation

Barriers fall (student programs). B. Kohl et al, *Envir.* 12(9): 40–43, 1970

Citizen participation; environmental education. MS-E6: 211–230, 1970

Ecology—heart of our university program. Edward W. Widner, CP-M3: 241–244, 1970

Education for human survival. Sterling Bunnell, CP-N1: 251–256, 1970

Environmental education. Michael Scriven, CP-N1: 242–249, 1970

Environmental education in the K-12 span. LeVon Balzer, CP-M3: 260–269, 1970

How should the university treat environment? F. Kenneth Hare, CP-D1: 145–157, 1970

How should we treat environment? F. Kenneth Hare, *Science* 167: 352–355, 1970

The law: enforcing quality. Raymond A. Haik, CP-N1: 264–269, 1970

Pollution: the challenge of environmental education. J. Alan Wagar, CP-M3: 235–240, 1970

Concerned people: key to tomorrow. CP-A4: 214–218, 1971

Environmental education and citizen responsibility. Citizens' Advisory Comm. on Envir. Quality, CP-M1: 315–317, 1971

A tale of two cities. CP-A4: 219–235, 1971

Index

Abelson, P. H., 630
Abies, 601
Ablation, 428
Ablation zone, 429
Abrasion, by ice, 431, 442
  by streams, 357
  by wind, 411
Abrasion platform, 385
Abrasives, 245
Absolute humidity, 115
Absorption by atmosphere, 46, 50
Abyssal zone, 593
Acceleration, A-2
  Coriolis, 89
  of gravity, 10, A-2
Accumulation zone of glacier, 429
Acer, 603
Acer saccharum, 604
Acidity of soils, 274
Acid mine drainage, 316, 344, B-26
Acid rain, B-36
Acids, carbonic, 190
  fatty, 522
  organic, 190, 276, 522
  in stream water, 344
Activation products, 317
Adenine, 488
Adiabatic process, 114, 119, 123, B-10
Adiabatic rate, 119, 123
  dry, 119, 123, 155
  saturation, 119
  wet, 119
Adsorption of water by soils, 275
Adult stage, 478
Advection, 98
Advection fog, 120
Aeration zone, 305
Aerosols, 151
Affluence change factor, 622
Affluence and environmental impact, 622
Aftonian interglaciation, 449
Agassiz, L., 439
Age determination, radiometric, 231–232, B-17
  of rocks, 231–232
Age of earth, B-17
Age of Metals, 457

Age pyramid, 562
Age of Reptiles, 502
Age-specific birth rate, 560
Aggradation, 368
  and dams, 373–374
  man-induced, 372, B-28
  and mining, 372
  Pleistocene, 454
Agriculture, chemical inputs, 533, B-35
  and fossil fuels, 532–533
  and Man, 508
  in tropical rainforest, 607
Agung eruption, 162
Air, composition, 30
Air masses, 95, 115–118, B-9
  antarctic, 117
  arctic, 117
  classification, 117
  cold, 96
  continental, 117
  equatorial, 117
  global distribution, 117
  maritime, 117
  polar, 95, 117
  properties, 117
  source regions, 116
  stable, 123
  tropical, 96, 117
  types, 117
  unstable, 123
  warm, 96
  world regions, 118
Air pollution, 1, 14, 151–161, B-13
  by asbestos, 256
  and cancer, 160
  and climate change, 161–163, B-11, B-14
  effects, 158
  and energy consumption, 630
  glacial ice record, 160
  harmful effects, 160
  and health, 160, 161, B-14
  and illumination, 158
  legislation, 161
  from ore smelting, 253
  and plants, 160, B-14
  by radionuclides, B-19
  and respiratory ailments, 160

Air pollution (*continued*)
  and soilage, 160
  standards, 161
  urban, 155–161
  and visibility, 158
Air pressure (*see* Barometric pressure)
Air temperature inversion, 77, 155
  ground-level, 77
  over ice sheet, 438
  low-level, 155
  upper-level, 156
  urban, 155
Air temperatures (surface), 39, 74–82, B-7
  and altitude, 84
  annual cycle, 77
  annual range, 79, 82
  of Antarctica, 438
  of cities, 150
  daily cycle, 74, 79
  of deserts, 77
  diurnal cycle, 74, 79
  in equatorial zone, 82
  global distribution, 79
  at high altitudes, 83
  inversion, 77
  land-ocean contrasts, 78
  mean daily, 77
  and plants, 520
  at poles, 82
  profiles, 76
  and relative humidity, 114
  world maps, 80–82
  yearly range, 78
  (*see also* Atmospheric temperatures)
Airy, George, 179
Alaska pipeline (*see* Trans-Alaska Pipeline System)
Albedo, 47, B-6
  of earth, 48
    and atmospheric dust, 162
    and contrail clouds, 147
    increase, 146, 147, B-11
  of planets, 48
Aleutian low, 99
Algae, 467, B-33
  blue-green, 466, 494, 544, 578

I-1

Index

Algae (*continued*)
  endozoic, 592
  evolution, 500
  green, 500, 578, 592
  in lakes, 578
  reef-building, 408
Algae bloom, 470, 555, 579
Alimentary system, 478
Alkalinity of soils, 274
Alleles, 489
  dominant, 489
  recessive, 489
Allelopathy, 570
Alluvial fans, 379, B-29
  as ground-water reservoirs, 381
Alluvial rivers, 360, 374-376, B-29
  environments, 374-376, B-29
  landforms, 374-375
  and Man, 375
  regulation effects, 376
Alluvial terraces, 369, B-28
Alluvium, 265, 312, 354, 369, 374
  reworked, 360
All-wave radiation, 51
Alpha particle, 229
Alpine debris avalanche, 281
Alpine glaciers, 428, 429-434, B-31
Alpine landscapes, 433-434
Alpine life zone, 434
Alps, glaciation, 439
Alteration of minerals, 189-193, B-15
Altitude, and air temperatures, 83
  and barometric pressure, 83
  and radiation, 83
Altitude sickness, 84
Altostratus, 119
Aluminosilicates, 171
Aluminum, 237
  in crust, 170
  recycling, 235, 244
America Plate, 217
Amino acids, 488, B-35
Amino group, 543
Ammonia, 543
  atmospheric, 543
  fixation, 547
  Haber process, 547
Ammonification, 545
Ammonium ions, 542
  in rainwater, 341
Ammonium nitrate, 547
Amoeba, 473
Amphibians, 485
  evolution, 501
Amphibole group, 171
Amphioxus, 485
Anchorage, Alaska, earthquake, 219
Anchoveta, 595
Andesite, 176
Angiosperms, 476, 477, B-33
  evolution, 501, 503, 504

Angle of repose, 267
Angular momentum, 93
  of atmosphere, 93, 96
  conservation law, 93
  eddy transport, 97
  transport, 94, 96
Anhydrite, 198, 245
Animals, 463, 478, B-33
  mobile, 478
  response to temperature variations, 521
  sessile, 478
  spiny-skinned, 484
  true, 478
  xeric, 612
Anions in rainwater, 342
Annelida, 482
Annelids, 482, B-33
  evolution, 498
Annuals, ephemeral, 612
Antarctica, 434
  ice sheet, 434-438, B-31
  water balance, 437
Antarctic circle, 42
Antarctic circumpolar current, 109
Antarctic Ice Sheet, 434-438, B-31
  water balance, 437
Antarctic zone, 44
Antherophyta, 475
Anthophyta, 477
Anthracite, 246
  of Appalachians, 227
Anthropoids, 505
Anticline, 227, 251
Anticyclones, 91, B-10
  Asiatic, 105
  and inversions, 156
  moving, 130
  stagnation, 156
  subsidence, 157
  and upper-air waves, 130
Anticyclonic cells, 102
Antimony, 238
Anvil top of cloud, 124
Aphotic zone, 67, 593
Apogee, 399
Aquiclude, 308
Aquifer, 308
Arachnids, evolution, 501
Aragonite, 197
Arch of rock, 385
Archaeocyathids, 499
Arctic circle, 42
Arctic environments, 282-284, B-22
  geomorphic processes, 282
Arctic front, 118
Arctic Ocean, air temperatures, 81
  sea ice, 72
Arctic zone, 44
Area strip mining, 249
Arête, 431
Argentite, 239

Argon, 30, 144
Arid environments, 378-381, B-29
  landforms, 378-381, B-29
  processes, 378-381
Armoring of channel, 374
Artesian spring, 309
Artesian well, 204
Arthropods (Arthropoda), 474, 482-484, B-33
  evolution, 501
Arundo arenacea, 422
Asbestos, 192, 245
  and air pollution, 153, B-14
  environmental, 153, B-14
  as health hazard, 256
Asbestosis, 256
Aschelminthes, 481
Ascomycota, 470
Ash, 603
  in coals, 246
  volcanic, 181, 185, 187
Asiatic cyclone, 105
Aspen, 601
Asphalt, 245, 250
Asphalt-base crude oil, 250
Assimilation of chemical energy, 527
  gross rate, 527
  net rate, 528
Asteroids, 499
Asthenosphere, 179, 214
Atlantic climatic stage, 458
Atmosphere, 3, 7, 9, 30, B-6
  composition, 30
  density, 31
  evolution, 493, 499
  Man's impact, 144, B-10
  primitive, 493
Atmospheric circulation (*see* Circulation, atmospheric)
Atmospheric pollution (*see* Air pollution)
Atmospheric pressure (*see* Barometric pressure)
Atmospheric temperatures (global), 31, 161-163, B-11
  and albedo change, 161-163
  and carbon dioxide, 145
  changes, 161-163, 440
  cycles, 163
  and fuel combustion, 145, 146
  ice-sheet evidence, 163
  and oxygen isotopes, 163
  in Pleistocene Epoch, 440
  recent rise, 145, 146
  and solar radiation, 161
  and volcanic dust, 161
Atolls, 409, B-31
  destruction by storm, 138
Atom, structure, 229
Atomic energy, 8, 259-260
Atomic Energy Commission (AEC), 317, 343

Index

Atomic hydrogen layer, 31
Atomic number, 229
Atomic oxygen layer, 31, 45
ATP, 464
Auger mining, 250
Augite, 171
Aurora, 58, B-7
Australopithecus, 505
Autotrophs, 462, 509–510, 524
    energy requirements, 509
    as energy system, 524–527
    and light energy, 510
    matter requirements, 509
Avalanche, of debris, 281
    glowing, 182
Axial rift, 214
Azobacter, 544
Azores high, 99

Background radiation, 54, 491
Backslope of cuesta, 203
Back-swamp, 376
Backwash, 384, 388
Bacteria, 466, B-33
    chemautotrophic, 551
    nitrogen-fixing, 466
    pathogenic, 466
    photosynthetic, 494, 549, 551
    primitive, 494
    sulfur-using, 466
Bacteriochlorophyll, 466
Badlands, 202, 354
Balloon, upper-air, 95
Bank caving, 357
Bankfull stage, 332
Barchan, 415
Bar finger, 396
Barite, 245
Bar-and-swale topography, 375
Barometer, 33
    mercurial, 33
Barometric pressure, 33, 86, B-6
    and altitude, 34, 83
    cells, 98
    high, 86, 98
    low, 86, 98
    mean monthly, 98
    at sea level, 33
    standard, 33
    in tropical cyclones, 135
    and water vapor, 114
    and winds, 86, 98
    world patterns, 98
Barrier beaches, 395, 401, 403, B-30
Barrier island coasts, 394–395, 401, 403, B-30
Barrier islands, 395, 401, 403, B-30
Barrier reefs, 408, B-30
Bars, coastal, 390
    baymouth, 390
    cuspate, 390
    offshore, 391

Bars (*continued*)
    in stream channels, 375
Basalt, 176
    magnetic properties, 214
Basalt floods, 181
Basaltic layer, 177
Basaltic rocks, 176
Base, organic, 488
Base cations, recycling, 275
    in runoff, 340, 341
    in soil, 274, 275
    in stream flow, 341–343
Base flow, 329
Base level of denudation, 366, B-28
Bases, metallic, 170
    in runoff, 340–341
    in sea water, 172
    in soils, 274, 275
    in stream water, 172, 196, 341–343
Basidiomycota, 470
Basin perimeter, 324
Basin-and-Range region, 226
Basin of stream, 324
Basswood, 603
Batholiths, 179, 213
Bathyal zone, 593
Bauxite, 191, 239, 242
Bay-head beach, 390
Baymouth bar, 390
Bays, 387
Beach deposits, 385, 388, B-29
Beach drift, 389, B-29
Beaches, 388, 390–392, 395, B-29
    barrier, 395
    bay-head, 390
    composition, 388
    crescentic, 390
    equilibrium, 391–392
    slope, 389
Beachgrass, 421, 422
Beach profile of equilibrium, 391–392
Beach ridges, 390, 392
    of Provincelands, 423
Becquerel, H., 228
Bedding planes, 195
Bed load, 358
    capacity, 360, 361
Bedrock, 264
Beech, 603
Beech-maple forest, 604
Benguela current, 109, 110
Bennett, H. H., 355
Benthos, 576
Berm, 391
    summer, 391
    winter, 391
Bermuda high, 99
Beryl, 240
Beryllium, 240
Beta particle, 229
Betula, 601, 603

Bicarbonate ion, 198
Big tree, 603
Bilharziasis, 481
Biogenesis, 493–494
Biogenic sediments, 196, B-16
Biogeochemical cycle, 534, B-35
Biological clocks, 517–518
Biological magnification, 556, 590
Biological oxygen demand (B.O.D.), 584, 586
Biomass, 3, 462, 524
    autotrophic, 524
    energy equivalent, A-3
Biomass pyramid, 532
Biomes, 599–614, B-40
Biosphere, 2, 7, B-33, B-34
    composition, 535–537
Biotic communities, 614
Biotic potential, 562
Biotite, 171
Birch, 601, 603
Bird-foot delta, 396
Birds, 486
Birth rate, 560
    age-specific, 560
Bison, 609
Bisulfate ion, 549
Bituminous coal, 246
Bjerknes, J., 127
Black body, 40
Black body radiation, 49
Black earths, 305, 609
Black Hills dome, 206
Black lung disease, 255, B-14
Blizzard, 105, 132
Block mountains, 226
Block separation, 265
Bloom of algae, 470, 579
Blowout, 412
Blowout dune, 417
Blue-green algae, 466
Bluffs of floodplain, 374
Bog environment, 199
Bog ore, 192
Bogs, 313, 586–587, B-39
    dystrophic, 587
    raised, 587
Bog succession, 587
Boiling point of water, 84
Boltwood, B. B., 228
Bomb calorimeter, 523
Bombs, volcanic, 182
Bora, 105
Borax, 245
Boreal forest, 601
Boreal stage, 458
Borrow pits, 253
Botanicals, 556
Bottom-set beds, 396
Boulder field, 266
Boulders, 193
Bowen reaction series, 174, 176, 192

Index

Brachiopods, 498, 499
Brazil current, 108
Breakers, 383, B-29
   plunging, 384
   spilling, 384
Breakwaters, 393
Breccia, sedimentary, 195
   volcanic, 195
Breeder reactor, 259, 508, B-21
Breeding of nuclear fuel, 259
Breeze, 87
Brine, 35
Brine shrimp, 612
Bristlecone pine, 232
Broecker, W. S., 342
Bromine, atmospheric, 153
Brontosaurus, 503
Brownian movement, 193
Brown soils, 609
Bryophyta (bryophytes), 473
   evolution, 500
Bryozoans, 481
   evolution, 499
Buffalo, 609
Building stone, 244
Bunch grasses, 608
Burning, 522
Butte, 181, 202

Cacti, 612
Cadmium, recycling, 244
Calcification, 304, 425, 609
Calcite, 197, 267
   precipitation, 199
   solution, 199
Calcium, 170
   in biosphere, 537, B-36
Calcium carbonate, as cement, 195
   in soil, 304
Calcium ions in sea water, 197
Calcium sulfate, 198
Calcrete, 304
Caldera, 182, 185, B-15
Caliche, 304, B-23
California current, 109
Calms, equatorial, 99
   subtropical, 102
Calorie, 20, A-2
   large, A-2
Calving of glaciers, 72, 435
Cambium, vascular, 477
Cambrian Period, evolution of life forms, 497–499
Canadian high, 99
Canals and ground water, 316, B-24
Canaries current, 109, 110
Canopy layers of forest, 604
Canyons, 361, B-28
   environmental significance, 362–363
Capacity of stream, 360, 361
   and velocity, 360

Cape Cod, dunes, 423
   marine erosion, 386, 393
Capillary fringe, 306
Capillary tension, 294
Cap rock, 251
Carbohydrate, 512
Carbohydrates, 13, 30, 522
   oxidation, 522
Carbon, atmospheric, 144, 153, B-36
   in biosphere, 536
   as carbon dioxide, 539
   in coal, 246
   fixation, 541
   fixed, 541
   in fossil fuels, 541
   organic, 539, 541
   pollutant, 153
   reduced, 541
   stored, 174, 541
Carbonate rocks, 198, B-16
   weathering, 198–199
Carbonates, 198, B-16
   in sediments, 144
Carbonate sediments, 541–542, B-16
Carbonation, 198
Carbon cycle, 144, 539–542, B-36
Carbon dioxide, in aquatic environments, 597
   atmospheric, 24, 30, 46, 50, 144, 597, B-11
   annual variation, 543
   and atmospheric warming, 24, 145
   and climate change, 144–146 B-11
   cycling, 144, 539–542
   daily variation, 543
   early history, 495–497
   and glaciations, 441, B-31
   and photosynthesis, 513, 542
   recent increase, 144
   and sea-water temperature, 146
   in carbon cycle, 539–542, B-36
   in ocean waters, 541
   and organisms, 542
   and photosynthesis, 542
   and plant growth, 542
   solubility in water, 190
   in solution, 198
Carbonic acid, 190, 198
   reactions, 198
Carboniferous Period, 200, 500
Carbon monoxide, 152, 153, 160, B-36
   atmospheric, 160
   pollutant, 152
Carbon-14, 231, B-17
   and atom bomb tests, 232
   dating, 231–232, B-17, B-32
   and fuel combustion, 232
   half-life, 232
Caribou, 599
Carnivores, 531

Carnotite, 239
Carpinus, 603
Carrying capacity of population, 563
Cartilage, 484
Cassiterite, 194, 239, 242
Castanea, 603
Cation exchange, 273
Cation exchange capacity, 274
Cations, in soil, 273, 274
Caverns (*see* Limestone caverns)
Caves, in loess, 425
   in marine cliff, 385
Cell, organic, 464–465, B-33
   as open system, 22, A-6
Cell membrane, 464
   nuclear, 464
Cell physiology, 22, 464–465, A-7
Cells, atmospheric, 122
   convectional, 122
   thunderstorm, 124
Cell wall, 464
Cement, Portland, 244
Cenozoic Era, 234, 504
   evolution of life forms, 504–505
Central depression of volcano, 185
Central eye of cyclone, 136
Centrifugal force and winds, 90
Cephalopods, 482
   evolution, 499
Cesium-137, 317
Cetaceans, evolution, 502
C.g.s. system, A-2
Chalcedony, 198
Chalcocite, 239
Chalcopyrite, 239
Chalk, 198
Channel flow, 321, B-24
Channelization of streams, 376–378, B-29
Channel storage, 329
Channels of streams (*see* Stream channels)
Chaparral, 614
   allelopathy, 570
Chatter marks, 442
Chelsea Sunsets, 161
Chemical energy, 522
   in organic compounds, 523
   in organisms, 522–523
Chemical pollution of water, 343–345
Chemical precipitates, 196
Chemical weathering, 189–193, 198–199, B-15
Chernozem, 305, 609
Chert, 198
   bedded, 198
Chestnut, 603
Chestnut soils, 609
Chezy equation, 324
Chinook, 105, 125, B-8
Chitin, 482

Index

Chlorinated hydrocarbons, 556
Chlorine, atmospheric, 153
  in biosphere, 537, 555
  origin, 173
  in sea water, 35
Chlorine ions, in rainwater, 342
  in sea water, 196
Chlorite, 192
Chlorobacteriaceae, 551
Chlorophyll, 512
  in chloroplasts, 464
Chlorophyll A, 466
Chloroplast, 464
Chordates, 485
Christianity and pollution, 624
Chromite, 238, 239
Chromium, 237
Chromosomes, 487, 489
Cilia, 473
Cinder, 185
Cinder cones, 185-186
Cinnabar, 239, 241
Circadian rhythms, 517
Circle of illumination, 41
Circulation, atmospheric, 85, 88, 89, 91, 94, B-8
  of oceans, 108-110, B-9
Circulation systems, 85, 88-89, 91, B-8
  atmospheric, 85, 88-89, 91, 94
    meridional, 89
    on non-rotating earth, 88
    planetary, 91
    thermal, 87
  oceanic, 85, 108-110
Circulatory system, 479
Circum-Pacific Belt, 218
Circum-Pacific ring, 182
Cirques, 430, 431
Cirrus, 119
City climate (*see* Urban climate)
Cladocerans, 580
Cladonia rangifera, 600
Clams, 481, 482
  evolution, 499
Clarke of abundance, 237
Clarke of concentration, 238
Clarke, F. W., 237
Clastic sediments, 194
Clay, 193, 244
  quick, 220
Clay loam, 273
Clay minerals, 191
Clean Air Act, 161, 627, B-13, B-44
Clean Air Amendments of 1970, 627, B-44
Cliffs, 201
  marine, 385-386, 393
Climate, 118, B-10
  subhumid, 305
Climate classification, 299, B-10
Climate modification, inadvertent, 144-148, 161-163, B-10, B-14

Climate modification (*continued*)
  urban, 150-151, 158, B-12
Climatic optimum, 458
Climatic stages of the Holocene, 458, B-32
Climax of succession, 571, 572, B-39
Clocks, biological, 517, B-34
Closed systems, 16
  and entropy, 21
Clostridium, 544
Cloud, Preston, 619, 621, 632
Cloudiness and urbanization, 159, B-14
Cloud reflection, 47
Clouds, 119-121, B-9
  and atmospheric dusts, 162
  cumuliform, 119
  man-induced changes, 147-150, B-11
  massive, 119
  stratiform, 119
  types, 119
  volcanic, 182
Cloud seeding, 124, 148-150, B-9, B-12
Club mosses, 500
Cnidaria, 480
Coal, 199-200, 246-250, 257, 259, 500, B-16, B-20
  ages, 247
  anthracite, 227
  bituminous, 246
  classification, 246
  composition, 246
  distribution, 247
  as energy resource, 257, B-20
  minable resources, 247
  origin, 200
  production, 259
  slaking, 246
  soft, 246
  subbituminous, 246
  world resources, 259
Coal beds, 500
Coal fields of United States, 247
Coal measures, 200
Coal mining, 248-250
  and aggradation, 372, B-28
  environmental effects, 284, B-18
  and subsidence, 286, B-18
  and water pollution, 344, B-26
Coal seams, 199, 248
Coastal engineering, B-29
Coastal plains, 203, B-16
  belted, 203
  evolution, 203
Coastal submergence, 385
Coasts, 383, B-29
  embayed, 387
  middle-latitude, 139
  tropical, 139
Cobalt, 237
Cobalt-60, 317

Cobaltite, 239
Cobbles, 193
Cocoanut palm, 410
Coding for amino acids, 489
  redundant, 489
Coelenterates, 480
  evolution, 499
Cohort, 561
Col, glacial, 431
Cold front, 128
Collar cells, 480
Colloids, 193
  flocculation, 397
  in soil (*see* Soil colloids)
Colluvium, 354
Colonies in Monera, 466
Colonization, 571
Colorado Plateau, 201
Colorado River, salinity, 339
  suspended load, 358
Columbia Plateau, 181, 202
Columbium, 240
Combustion of hydrocarbon fuels, 30, 144, 542
  and air temperature, 145, 146
  and oxygen level, 146, B-11
  and water vapor, 147
Commensalism, 570
Committee on Resources and Man, 621
Compensation level in lakes, 577
Competition in evolution, 559, B-33, B-38
  interspecific, 559
  between species, 568
Competitive exclusion principle, 568
Compositae, 608
Composites, 608
Composite volcano, 181
Compounds, reduced, 522
Condensation, of water vapor, 12, 60, 113, B-9, B-10
Conduction of heat, 59
Conductivity, 59
  of solids, 60
Cone of depression, 312
Cones, volcanic, 181
  cinder, 185
Conglomerate, 195
Coniferophyta, 477
Coniferous forest biome, 601-603
Conifers, 477, B-33
  deciduous, 477
  evergreen, 477
  evolution, 501, 503
Conjunction of moon and sun, 399
Conservation of natural resources, 625, B-42, B-44
Consumers, 275, 462
  organic, 275
  primary, 462
  secondary, 462

I-5

Index

Consumptive use of water, 345
Contact metamorphism, 239
Continental arid regime, 302
Continental drift, 217, B-16
Continental humid regime, 302
Continental shelves, 176, B-16
   mineral resources, 246
Continental shields, 216
Continental slope, 176
Continents, 176, B-15, B-16
   distribution, 177
   evolution, 216
   nuclei, 216
   structure, 177–179
Contouring, 355
Contour strip mining, 249
Contrail clouds, 147
Contrails, 147
Convection, atmospheric, 36, 60, 98, B-10
   over cities, 151
   in lakes, 69
Convection cells, 122
Convergence, atmospheric, 91, 132
   of ocean water, 110
Convergence zone, intertropical, 92
Copepods, 580
Copper, 238, 239
   native, 239
   recycling, 244
Coral animals, 592
Coral head, 592
Coral-reef coasts, 408–410, B-31
Coral reefs, 408–410, 480, 592–593, B-31
   atoll, 409
   barrier, 408
   ecosystems, 592
   environmental aspects, 410
   as environments, 592–593
   fringing, 408
Corals, 408, 592
   evolution, 498, 499
   temperature limits, 71
Corange lines, 82
Core of earth, 56, 169
Coriolis, G. G., 89
Coriolis effect, 89, B-8
   on ocean currents, 107
   in water movement, 107
Coriolis force, 89, B-8
Correlation of strata, 233
Corrosion, 357
Cosmic particles, 54, 84, B-7
   and mutation, 490
Cosmic radiation, 54, 169, B-7
   and altitude, 84
   and carbon-14, 232
Cosmic rays (see Cosmic particles)
Cosmic shower, 54
Cotton-grass meadow, 599
Cotyledons, 478

Cotylosaurs, 501
Coulomb, C. A., 20
Council on Environmental Quality (CEQ), 626, B-44
Counter-radiation, 50
Country breeze, 151
Country rock, 179
Crater Lake, Oregon, 183
Craters, volcanic, 181
Crescentic beach, 390
Crescentic gouges, 442
Crest of wave, 106
Cretaceous Period, 504
Crevasses, in glaciers, 430
   in levees, 334
Crevices in marine cliff, 385
Crinoids, 499
Crisis, environmental, 619, B-41
Crocodiles, evolution, 502
Cro-Magnon Man, 456
Crop rotation, 356
Cross Florida Barge Canal, 627
Crossing in channel, 375
Crossopterygians, 501
Crude oil, 200, 250
   paraffin-base, 250
   series, 250
Crustaceans, 483
   evolution, 499
Crustal rise, postglacial, 452
Crustal separation, 214, B-16
Crustal spreading, 214, B-16
Crust of earth, 169, 170–171, 177–179, 214, B-15
   continental, 177
   element composition, 170
   inorganic evolution, 174
   layers, 177–179
   oceanic, 177, 178, 214, 216
   spreading, 214
   structure, 177–179
Cryptozoic Eon, 234
Crystalline state, 12
Cuesta, 203
Cumulonimbus, 120, 124, B-10
   and hail, 122
   seeding, 148, B-12
   and tornadoes, 133
Cumulus, 119
Curie, Marie and Pierre, 228
Current meter, 326
   Price type, 326
Currents, coastal, 263, 283, B-29
   convectional, 36
   longshore, 389
   ocean (see Ocean currents)
   tidal, 383, 400–401, B-30
Cuspate bar, 390
Cuspate delta, 396
Cuspate foreland, 390
Cutin, 612
Cut-off highs, 95

Cut-off lows, 95
Cutoffs of meanders, 365, 369, 374
   artificial, 376
Cyanophyta, 466
Cycadophyta, 477
Cycads, 477
   evolution, 501, 503
Cycle of rock transformation, 189
Cycling of matter, 534
Cyclone families, 129
Cyclones, 91, 127, B-10
   Asiatic, 105
   evolution, 128
   extratropical, 127
   families, 129
   life history, 128
   monsoon, 134
   moving, 127
   tracks, 128
   tropical, 127, 135–138, 410, B-10
   and upper-air waves, 130
   wave, 128, 130
Cyclothem, 200
Cypress swamps, 589
Cyprinodon, 453
Cytoplasm, 464
Cytosine, 488

Dams, 333, 373–374, B-25
   environmental effects, 23, 364, 373, 374, B-24
   flood-storage, 333
   and ground water, 316
   and karst, 311
   and stream channel changes, 373–374, B-29
Darwin, Charles, 21, 228, 487
   on coral reefs, 409
Darwin-Lotka law, 21, A-9
Daughter product, 230
Day length, 515
DDE, 557
DDT, 556–558, B-37
   buildup, 557
   in environment, 556–558
   in estuaries, 590
DDT cycle, 557–558
Death rate, 560
Debris avalanche, 281, B-22
Debris flood, 277, B-22
Decay, negative exponential, 16, 231
   radioactive, 231
Decay system, 16, 17
Deccan Plateau, 181, 202
Deciduous forest biome, 603
Declivity, 365
Decomposers, 462, 473
Deflation, 411, B-31
   man-induced, 420–421
Deflation hollow, 412
Deglaciation, 440

Index

Degradation, of carbohydrates, 527
  of stream, 369
Deicing salts, 268, 316, 343, B-23, B-26
  in stream water, 343
  and water pollution, 316
Delta environments, 395–399, B-30
Deltaic plain, 397
Delta kames, 446
Delta Plan, 405
Deltas, 395–399, B-30
  cuspate, 396
  environmental aspects, 397–398
  estuarine, 396
  of geosynclines, 211
  glacial, 446
  growth rates, 397
  tidal, 401
Demographers, 620
Demographic transition, 567, 620
Dendritic drainage, 325
Dendrochronology, 232
Denitrification, 545
Density, 36, A-2
  of minerals, 194
  of water, 36
Density-dependent factor, 563
Density-independent factor, 563
Density-proportional factor, 563
Density stratification, 9
Denudation, 350, 366–368, B-28
  as geologic process, 366–368
Deoxyribonucleic acid (DNA), 465, 487, 488–489
  and mutations, 489
Deoxyribose, 488
Deposition of sediment, 189
Desert basins, 381
Desert biome, 612–613, B-41
Desert pavement, 412
Deserts, 139, 141, 378–381, 612–613, B-29, B-41
  air temperatures, 77
  arctic, 141
  cool, 613
  dunes, 416
  dust storms, 421
  fluvial processes, 378–381
  and heat balance, 112
  hot, 613
  irrigation, 613
  landforms, 378–381
  middle-latitude, 139
  North American, 613
  northern, 613
  in Pleistocene Epoch, 147
  polar, 141
  rainshadow, 125, 127
  southern, 613
  weathering processes, 191
Desilication, 304
Desulfovibrio bacteria, 549

Detrital sediments, 194
Detritivores, 531
Detritus, organic, 531
Detritus food chain, 531
Devonian Period, 200
Dew, 141, B-9
Dew point, 114, B-9
  lapse rate, 119
Diagenesis, 189
Diamond, 242
Diaspore, 191
Diatoms, 201, 578
Dicots, 478
Diffuse reflection, 46
Diffuse sky radiation, 46
Dikes, igneous, 180
  radial, 183
  volcanic, 183
  man-made, 333, 405
    floodplain, 333
    Netherlands, 405
Dimensional analysis, A-1
Dimensional products, A-1
Dimensions, fundamental, A-1
  symbols, A-1
  units, A-1
Dinosaurs, 502
Diorite, 176
Dire wolf, 455
Discharge of stream, 323, 326, A-4
Disorder in closed systems, 22
Disposal well, 223
Distributary, 396
Divergence, atmospheric, 91
Divisions of plants, 463
DNA, 465, 487, 488–489
Doldrums, 99
Dolines, 271
Dolomite, 198
Domes, of lava, 185
  exfoliation, 269
  sedimentary, 205, 251
  stratigraphic, 251
Donora, Pa., poison fog, 156
Dormancy of plants, 521
Doubling time, 562
Downdraft, 86
Down scatter, 46
Drainage divide, 324
Drainage systems, 321, 324–326, B-24
  dendritic, 325
  evolution, 366
Drainage winds, 105, B-8
Drawdown of water table, 312, B-23
Dredging, 253
Drift, glacial, 443, B-32
  littoral, 388, B-29
  of sediment, 389
  of surface water layer, 105, 106, 107
Driftless Area, 449

Drifts, in coal mines, 248
  of dune sand, 417
  of soil, 421
Drizzle, 121
Droplets in clouds, 119
Drought, and cloud seeding, 148, B-12
  of Dust Bowl, 421
Drumlins, 445, B-32
Dryopithecus, 505
Ducktown, Tennessee, 152
Dune barriers, coastal, 421
  breaching, 422
  of North Sea coast, 422
Dune control, 422, B-31
Dune grass, 571
Dune sand, 414
Dunes of sand (*see* Sand dunes)
Dunite, 176
Dust Bowl, 421, 609, B-31
Dusts in atmosphere, 33, 151, 161–163, 182, 413, B-14
  and climatic change, 161–163, B-14
  over deserts, 421
  and glaciations, 441
  global effects, 161–163, B-14
  hygroscopic, 152
  industrial, 162
  stratospheric, 162
  and temperature change, 162
  volcanic, 152, 161, 182
Dust storms, 413, 420, B-31
Dynamic equilibrium, 15
Dystrophic water bodies, 587

Earth, element composition, 170
  origin, 169
Earthflows, 277, B-22
  man-induced, 285–286
Earthquake control, 223–224, B-17
Earthquake energy, 219
Earthquake hazards, 209, 218–225, B-17
Earthquakes, 209, 218–225, B-16
  cause, 218
  control, 221–222
  damage, 219, 224–225
  and dams, 224
  of Denver region, 223
  energy, 219
  as environmental hazard, 218–220, 224–225, B-17
  and fluid pumping, 223
  locations, 214
  of Los Angeles region, 224
  and Man's activities, 221, B-17
  and nuclear text explosions, 224
  prediction, 221–224, B-17
  of Rangely Oil Field, 223
  scales, 219
  secondary effects, 220

Index

Earthquakes (*continued*)
  and urban planning, 224, B-17
  and water injection, 223
Earthquake waves, 218
  in crust, 177
  surface, 219
Earth resources, consumption, 235, 245, B-17
  of energy, 256-260
  nonrenewable, 236, 245
Earth's interior, 169
  density, 169
  pressure, 169
  temperature, 169
  thermal history, 169
Earthworms, 276, 482
Easterlies, polar, 97
  tropical, 92
Easterly waves, 135
Ebb current, 400
Echinodermata, 484
Echinoids, 499
Ecliptic plane, 42
Ecological system, 3 (*see also* Ecosystems)
Ecology, 3, 461, B-33, B-39
Economics and pollution, 623, B-41
Ecosystem dynamics, 559-575, B-38
Ecosystems, 3, 461, 576-615, C-33, B-34, B-39, B-40
  agricultural, 532-533
  aquatic, 576-596, B-39
  dynamics, 559, B-38
  energy flow, 462, 528-531, B-34
  and food web, 462
  future survival, 632
  general properties, 461
  global, 632
  of open water, 593-595
  productivity, 523
  terrestrial, 597-615, B-40
Ecotypes, 512
Ectoprocta, 481
Eddy transport mechanism, 97
Ediacara fauna, 498
Einstein, Albert, 14
Ejecta, volcanic, 181, 185
Electrical energy, 13
Electrolyte, 397
Electromagnetic energy, 12
Electromagnetic radiation, 13, 40, 48, B-6
Electromagnetic spectrum, 12
Electron acceptor, 538
Electronegativity, 537
Elements, crustal, 170, B-15
  of solid earth, 170
Elephant Butte Reservoir, 374
Elephantiasis, 481
Elephants, evolution, 505
Elm, 603
Eluviation, 303

Embayed coast, 387
Embryo of seed, 476
Emergent zone, 578
Emission control devices, 161
End moraine, 432, 443
Endoplasmic reticulum, 465
Endoskeleton, 478
Energy, 10, 18, A-1, A-2, B-6
  chemical 10, 13, 20, 22, 522
  electrical, 10, 13, 24, B-35
  electromagnetic, 10, 12
  geothermal, 257, B-20
  heat, 10, 12, A-2
  kinetic, 38
  latent, 12, 60
  mechanical, 10
  nuclear, 10, 14
  of position, 11
  potential, 11, 20
  solar, 7, 40, 256, 410, B-20
  storage, 20, 22
  tidal, 18, 168
  of water waves, 388
  in wave form, 11
Energy balance, global, 25, 39, B-6
  Man's impact, 53
Energy budget, 8, B-6
Energy budget equation, 510
Energy consumption, 629-631, B-19
  curtailment, 630
  environmental impact, 630
Energy crisis, 629, B-19
Energy cycle, biochemical, 39
  physical, 39
Energy drain, 19
Energy flow, A-3, B-34
  in ecosystems, 462, 528, 531, 509, B-34
  in food chains, 528-531
  in organisms, 509, B-34
Energy flux, 51, A-3, A-4
Energy inputs, 8, 14, 510
  inorganic, 510
  limiting factor, 512
  multiple, 512
  organic, 510
Energy output, 8, 14
Energy pathways, 8, 14
Energy releases, atmospheric, 113
Energy resources, 4, 256-260, 628-631, B-19
  exhaustible, 256
  fossil-fuel, 257-259
  geothermal, 257
  hydropower, 256
  lithospheric, 168
  nuclear, 259-260
  solar, 256
  sustained yield, 256
  tidal, 257
Energy sink, 524
Energy sources (*see* Energy resources)

Energy storage, 525-527, A-9, A-13
  Darwin-Lotka principle, A-9
  efficiency, 526
  labile, 526
Energy systems, 8, 14, 17, 142, A-6, B-6
  of atmosphere, 142
  characteristics, 14
  cyclic, 17
  decay, 17
  general, 509
  Man's impact, 23-25
  of oceans, 142
  open, 14
  organic, 509
  potential-to-kinetic, 11
  random fluctuation, 17, 18
  rhythmic, 17
  sensitivity, 16
  tidal, 17
  time scales, 18
  types, 17
Energy transformations, 7, 8
Energy transport, 8
Engineering geology, 286
Entrainment of air, 124
Entrenched meanders, 369
  cutoff, 369
Entropy, 22, A-7, B-6
  in organic systems, 509
Environment, 2
Environmental crisis, 618-619, B-41
  historical roots, 624, C-41
Environmental education, B-45
Environmental geology, B-15
Environmental hazards, 2, 114
Environmental impact, and affluence, 621-622, B-42
  and population, 621-622, B-42
Environmental impact statement, 626, B-44
Environmental legislation, 617, 626, 627, B-44
Environmental management, 617-618, 625, B-44
Environmental modification, planned, 2
Environmental planning, 626, B-15, B-44
Environmental Protection Agency (EPA), 626, B-44
  clean air standards, 161
Environmental protection legislation, 626, B-44
Environmental science, 2, B-5
Environmental stress, 114
Environment and cultures, 617, B-41
Environment and economics, 623, B-41
Environment of rocks, 210
  deep, 210
  surface, 210
Environments, aquatic, 576, B-39
  equable, 200

Index

Environments (*continued*)
  freshwater, 577
  lotic, 582
  marine, 576, 589-595
Environment of zones of oceans, 593-595, B-39
  abyssal, 593
  aphotic, 593
  euphotic, 593
  hadal, 593
  neritic, 593
Enzymes, 464, 488, B-35
Eons of geologic time, 234
Epeirogenic movement, 369
Ephemeral annuals, 612
Ephemeral stream, 378
Epicenter, 219
Epifauna, 594
Epilimnion 68, 577
Epiphytes, 570, 606
Epochs of geologic time, 234
Equation of continuity, 323
Equatorial belt of variable winds and calms, 99
Equatorial countercurrent, 108
Equatorial current, 108
Equatorial regime, 301
Equatorial trough, 89, 98
Equatorial zone, 44
Equilibrium, chemical, A-8
  dynamic, 15
Equilibrium profile, of shore, 385, 391-392, B-29
  of stream, 361, 365, B-28
  readjustments, 368-370
Equilibrium of stream system, 361, 365, B-28
Equinox, 42
  autumnal, 42
  vernal, 42
Eras, geologic, 233, 234
  of abundant life, 233
Erg (energy unit), A-3
Erg landscape, 415
Erosion, 189, 352-356, 385-386, 411-413, 431, 442-443, B-27
  accelerated, 352, B-27
  fluvial, 352-356, B-28
  geologic norm, 352
  by glaciers, 431, B-31
  by ice sheets, 442-443
  of slopes, 352-356, B-27
  of soil, 354-356, B-27
  by splash, 352
  by streams, 356, B-28
  by waves, 385-386, 393, B-29
  by wind, 411-413, B-31
Erosion intensity, A-4
Erosion rate, A-4
Esker, 443, 447, B-32
Estivation, 521
Estuaries, 402, 590-591, B-30
  bar-built, 403

Estuaries (*continued*)
  biota, 590
  ecosystems, 590, B-39
  hypersaline, 404
  man-induced changes, 404, 591, B-30
  negative, 404
  pollution, B-30
  positive, 404
  tectonic, 404
Estuarine delta, 396
Estuarine environment, 402-405, 590-591
Ethylene, 154
Euphotic zone, 577, 593
Euryhaline estuary, 511
Eutrophication, 548, 555, B-36
Eutrophication of lakes, 580, B-25, B-39
  cultural, 581
  of Great Lakes, 581-582
Eutrophic lake, 580
Evaporating pan, 338
Evaporation, 60, 289, 338, B-9
  control, 339
  global, 289
  from lakes, 338, B-25
  from reservoirs, 338, B-25
  of soil moisture, 295, B-23
Evaporites, 198, B-16
Evapotranspiration, 61, 141, 294, B-23, B-35
  actual, 296
  potential, 296, 299
  and urbanization, 147
Evolution, 233, 487, 492-493, B-33
  and competition, 559
  convergent, 493
  organic, 492-493, B-33
  in Paleozoic Era, 497
  in Precambrian time, 494
  of species, 492-493
Exfoliation, 265, B-21
Exfoliation dome, 269
Exoskeleton, 479, 482
Exotic river, 339
Exponential decay, 562
Exponential increase, 562
Extinction, evolutionary, 493
  of mammals, 456
  in Permian Period, 502
  in Pleistocene Epoch, 456, B-34
Extrusive rock, 175
Eye of cyclone, 136

Fagus, 603
Fallout, 153
Falls, 361, B-28
Fans, alluvial, 379, 381, B-29
Farmdalian substage, 454
Fault creep, 218, 222
Fault-line scarp, 226, B-17

Faults, 212, 218, B-16
  normal, 218
  and oil seepages, 254
  overthrust, 212
  transcurrent, 218
  and waterfalls, 364
Fault scarps, 225, B-17
Fault trap, 251
Faunas, 232, 493
  succession, 233
Federal Clean Air Act, 161, 627, B-13, B-44
Federal Water Pollution Control Act, 627, B-44
Feldspars, 171
  alteration, 191
  calcic, 171
  plagioclase, 171
  potash, 171
  sodic, 171
Fell-field, 599
Felsenmeer, 266
Felsic group, 172
Fenlands, 406
  flooding, 422
Fens, 405
Fern allies, 475
Ferns, 475
  epiphytic, 475
  evolution, 500
Ferro-alloy metals, 237
Fertile Crescent, 624
Fertilizers, 245, 533, B-36
  and agricultural productivity, 533
  as pollutants, 343
Fetch, 107
Fiddle-heads, 476
Field capacity, 294
Fig Tree Series, 494
Filaria worms, 481
Filter sand, 245
Finger lakes, 432, 442
Fiord, 403, 433, B-30
Fir, 601
Firebox of atmosphere, 112
Firn, 429
Firn line, 429
Fisheries, 594-596, B-40
  of Great Lakes, 582
Fishes, 485, 501
  bony, 485
  cartilage, 485
  evolution, 501
  jawless, 485
  in lakes, 580
  lobe-finned, 501
Fission, nuclear, 14, 259, B-21
  atomic, 259
Fission products, 317
Fissures, in rock, 180
  volcanic, 185
Flagellum (flagella), 466
Flatworms, 481

Index

Flocculation, 195, 397, 402
Floes of sea ice, 72
Flood control, 333–335, B-25
    and soil erosion control, 356
Flood crest, 332
Flood current, 400
Flooding, coastal, 406
Flood peaks and urbanization, 336
Floodplain engineering, 333–335,
    B-25, B-29
Floodplain management, 333–335,
    B-25, B-29
Flood plains, 331, 374–376, B-25,
    B-29
    channelization, 378–380, B-29
    development, 364
Flood prediction, 333, B-25
Floods of rivers, 331–335, 374–376,
    B-25
    annual rhythms, 333
    control, 333–335
    downstream progress, 332–333
    prediction, 333
    regulation, 376
    and tropical cyclones, 138
Flood stage, 332
Flood wave, 332
Floodways, 334
Flora, 233, 493
    cosmopolitan, 591
Florida stream, 108
Flow intensity rate, 290
Flow rate, total, 290
Fluid agents, 263
Fluids, 9, A-11
Fluid shear, 323
Flukes, 481
Fluorite, 245
Fluvial denudation, 350, 366–368,
    B-28
Fluvial processes, 351–381, B-27, B-28
    of arid environments, 378–381
Fluvial system (see Stream systems)
Flux, of energy, A-3
    of heat, 60, 61
Fly ash, 153
Focus of earthquake, 218
Foehn winds, 105, 125, B-8
Fog dispersal, 149, B-12
Fogs, 120–121, 133, 149, 158, B-9
    advection, 120
    and air pollution, 158
    at airports, 158
    coastal, 158
    cold, 149
    as environmental hazard, 120, 133
    incidence, 121
    poisonous, 1, 156
    radiation, 120
    toxic, 155
    urban, 158
    warm, 149

Folds of strata, 212, 227, B-17
    environmental significance, 227
    landforms, 227, B-16
    plunging, 227
    zig-zag, 227
Food chains, 5, 462, 528–531, 574,
    B-34
    and DDT, 556
    detritus, 531
    energy flow, 526, 528–531
    grazing, 531
    and insecticides, 556
Food resources, 620–621, B-35, B-40,
    B-43
    from fisheries, 595–596, B-40
    of grasslands, 612
    oceanic, 177, 595–596, B-40
Food web, 462, 574
Foraminifera, 473
Forbs, 608
Force, A-1, A-2
    centrifugal, 90
    Coriolis, 89
    and energy, 19
    frictional, 90
    of gravity, 10
    and work, 19
Foredunes, 417, 421
    man-induced changes, 422
Foreland, cuspate, 390
Foreset beds, 396
Foreshore, 391
Forest, 601–605, B-40
    boreal, 601
    coniferous, 601
    deciduous, 603
    mixed mesophytic, 604
    monsoon, 603
    northern hardwood, 604
    presettlement, 605
    stratified, 604
    summer-green, 603
    tropical, 603
    virgin, 605
Forest fires, and air pollution, 153
    and atmospheric dust, 161
Form rates of channel, 322
Fossil fuels, 4, 13, 201, 233, 237,
    246–253, 257–259, B-20
    and agriculture, 532
    as energy resources, 257–259
    future supplies, 628, 630
    geologic aspects, 246–253
    and nitrogen fixation, 547
    world consumption, 258
    world production, 258
Fossils, 493
Fractionation, 174
Fraxinus, 603
Freezing, of plant tissues, 520
    of water, 60
Freshwater environments, 577

Frictional force and winds, 90
Fronds, 476
Frontierism and pollution, 625
Fronts, atmospheric, 116, 128, B-10
    arctic, 118
    cold, 128
    occluded, 128
    polar, 96, 116
    warm, 128
Frost action, 266, B-22
Frost protection, B-8
Fuel ratio in coals, 246
Fuels, fossil (see Fossil fuels)
    nuclear (see Nuclear fuels)
Fumaroles, 256, B-20
Fungi, 470
Funnel cloud, 133
Fusion, nuclear, 14, 259, B-21
    in sun, 40

Gabbro, 175, 176
Galactic radiation, 54
Galena, 239
Gametophyte, 476
Gamma rays, 12, 40, 45, 229, B-6
    and mutation, 490
Garnet group, 194
Gas, natural, 201, B-16
Gaseous cycle, 474
Gaseous state, 9
Gases, 9
    of lavas, 173
    solubility in water, 189
    in solution, 189
    temperature, 38
    volcanic, 173
Gastropods, 481
    evolution, 499, 501
Gene pools, 492
Genes, 489
    dominant, 489
    half-life, 491, 492
    induction, 489
    recessive, 489
    suppression, 489
Genetic code, 489
Genetic drift, 492
Genetic feedback, 569
Geochemical cycle, 534, B-35
Genotype, 491
Genus (genera), 491
Geochemistry, 343
Geochronology, B-17
Geologic norm of erosion, 352
Geologic time, 228–234, B-17
    absolute, 228
    estimates, 228
    relative, 228
    scale, 228
    table, 232
Geology, environmental, B-15
Geomorphology, 263, B-21

Geostrophic wind, 90, B-8
Geosynclines, 210, B-16
Geothermal power, 257, B-20
GHOST balloon, 95
Gilbert's law of declivities, 365
Ginkgophyta, 477
Ginkgos, 477
   evolution, 501, 503
Glacial deposits, 443–449, B-32
   environmental aspects, 446–447
   resource aspects, 446–447
Glacial drift, 443–446, B-32
   glaciofluvial, 443, 446
   as natural resource, 446–449
   stratified, 443–446
Glacial trough, 431
Glaciation, alpine, 431
Glaciations, 292, 440, B-31
   causes, 440–442
   hypotheses, 440–442
   Permian, 502
   Pleistocene, 448–449
Glacier equilibrium, 429, B-31
Glacier erosion, 431, B-31
Glacier flowage, 430
Glaciers, 428–431, B-31
   alpine, 428
   continental, 428
   valley, 428, B-31
   volume, 291
Glaciofluvial deposits, 443, B-32
   economic value, 448
Glaciolacustrine deposits, 446, B-32
Glass, volcanic, 175
Glass sand, 245
Glaze, 122
Glei horizon, 305
Gleization, 305, 599
Globigerina, 440
Glycine, 545
Glyptodonts, 455
Gneiss, 213
Gnetophyta, 477
Gold, 238
   native, 239
Goldman, M. I., 626
Gold mines, 207
Gondwana, 216, 502
Good Friday Earthquake, 219
Gorges, 361, 369, B-28
   environmental significance, 362–363
Gossan, 241
Graben, 225
Grade, 322
Grades of sediment particles, 193
Gradient of stream channel, 322, B-28
   and velocity, 324
Grains, 478, 612
Gram calorie, 41, A-2
Gram-molecular weight, 535

Grams-weight, A-4
Grand Canyon, 364
Granite, 168, 175, 176, B-15
   hydrolysis, 270
   origin, 213
   sheeting structure, 269
Granitic layer, 177
Granitic rocks, 176
Granular disintegration, 265
Granules in cell, 465
Grasses, 608
   nutrient recycling, 275, 304
Grass family, 478
Grass fires and air pollution, 153
Grassland biome, 608–612, B-41
Grassland ecosystem, 608–612, B-41
Gravel, commercial, 244
Gravitation, 9
   law, 9
Gravity, 10
   acceleration, A-2
Gravity percolation, 294
Great Lakes, 449–452, B-25, B-32, B-39
   eutrophication, 581–582, B-25
   history, 449–452, B-32
   snowfall, 148
   surface area, 450
Greenhouse effect, 24, 50, B-7
Greenland, air temperatures, 81
Greenland current, 109
Greenland Ice Sheet, 434–435, B-31
   oxygen-isotope analysis, 440, B-11
   pollutant content, 160
Green revolution, 620, B-34, B-35
Groins, 394
Ground layer, 609
Ground moraine, 446
Ground radiation, 48
Ground water, 291, 305–319, 345–349, 379, 447, B-23
   of alluvial fans, 379
   and canals, 316
   in coastal areas, 309, 313
   and dams, 316
   and engineering works, 316, B-24
   environmental aspects, 312–319, B-23
   flow paths, 307
   flow rates, 312
   in glaciofluvial deposits, 447
   and highway cuts, 316
   and limestone caverns, 310
   mounding, 314
   as national resources, 345–349
   overdraft, 312–314
   pollution, 314–316, B-23
   and porosity, 308
   as resource, 236, 311–312
   salt, 309, 313
   salt intrusion, 313, B-23
   utilization, 311–312

Ground water pollution, 314–316, B-23
Ground-water seepage, 307
Grouting, 311
Growth, industrial, 260, 621, B-43
   population, 260, 620, B-43
Grus, 270
Guanine, 488
Guano, 310
Gulf Coast Geosyncline, 211
Gulf Stream, 108
Gullies, 355, 356
   in loess, 426
Gunflint Chert, 495
Gymnosperms, 476, B-33
Gypsum, 198, 245, 267
   in soil dusts, 341
Gyres, 108

Haber process, 547
Habitat, 559
Habitats, aquatic, 576
   lentic, 577
   lotic, 582
   stillwater, 577
Hadal zone, 593
Hadley, George, 92
Hadley cell, 92
   and heat balance, 112
   in Pleistocene, 452
Hail, 121, B-10
   crop damage, 132
   as environmental hazard, 122
Hail alley, 149
Hailstones, 121, B-10
Hailstorms, 149, B-10
   destruction, 149
   frequency, 149
   as hazard, 132
   man-induced changes, 149, B-12
   occurrence, 149
   seeding, 149
Hairpin dunes, 418
Half-life of radioactive decay, 231
Halite, 198, 245, 270
Halocline, 37
Hardwoods, 603
Hawaiian high, 99
Hawaiian volcanoes, 185
Hazards, environmental, 114
   of coastal flooding, 407
Haze, 152, 158
Head, hydraulic, 306
Health hazards of mining, 253, 255–256
Heat, 12, 38, 39, 50, 59–60, A-2
   and animals, 521
   of combustion, 24
   of condensation, 60
   conduction, 59
   as energy input to organisms, 518–522

Index

Heat (*continued*)
  flow mechanisms, 59
  flow in organisms, 518–520
  of gas compression, 13
  latent, 12, 39, 50, 59–60
  leaf input and output, 520
  mechanical equivalent, A-2
  radiogenic, 168, 169, 229–230, B-15
  sensible, 12, 38, 50, 60
  specific, 60
  of vaporization, 60
Heat balance, 39, 61–62, 97, 111, 161, B-6, B-7
  annual cycle, 62
  and atmospheric dust, 161
  daily cycles, 62
  equations, 61, 62, 111
  global, 97, 111
  or organisms, 519
  and urbanization, 150, B-12
Heat energy, 12
  environmental impact, 24, B-11
  of oxidation, 523, B-35
Heat engine, 19, 85
Heat flow, meridional, 51
  and organisms, 524, B-35
Heat island, urban, 151, 159, B-12
Heat low, 105
Heat rays, 12, 40
Heat transport, in atmosphere, 97, 111
  meridional, 97, 111
  by ocean currents, 112
Heaving by frost, 266
Heavy oil, 250, 259
Hekla eruption, 162
Heliophytes, 513
Helium, 30
  in natural gas, 201
  in sun, 40
Helium layer, 31
Hematite, 198, 239, 242
  in soils, 272
Herbicides, 5, B-38
Herb layer, 609
Herbs, 478
Heterosphere, 30, 31, 45
Heterotrophs, energy flow, 527–528
  energy requirements, 510
Hexadecanol, 340
Hibbard, W. R., Jr., 628
Hibernation, 521
High-altitude environment, 83, 84
Higher plants, 475
High-pressure belt, subtropical, 92, 98
Highs, cut-off, 95
High water of tide, 399
Highway construction, and aggradation, 372, B-24
  environmental impact, B-22
Hoarfrost, 141

Hogbacks, 205
Holocene environments, 457–458, B-32
Holocene Epoch, 234, 448, 457–458, B-32
Holoplankton, 594
Homeotherms, 521
Hominidae, 505
Hominids, 505, B-32, B-34
Hominoids, 505
Homo erectus, 505, B-34
Homo sapiens, 455, 456, 505, B-34
  evolution, 455–458, 505
  population growth, 566
Homosphere, 30, 31
  subdivisions, 31
Hookworms, 481
Hoover Dam, 224
Horizons of soil, 272
Horn, glacial, 431
Hornbeam, 603
Hornblende, 171
Hornworts, 473
Horse latitudes, 102
Horses, evolution, 505
Horsetails, 500
Horst, 225
Host organism, 569
Hot springs, 257, B-20
Hubbert, M. K., 260, 631
Human population growth, 566–568, B-38
Humboldt current, 109, 110
Humidity, 32, 114–115, B-9
  absolute, 115
  relative, 114
  specific, 115
Humification, 276
Humus, 272, 275
Hurricanes, 127, 135, B-10
  modification, 148, B-12
  seeding, 148
Hydraulic action, 356
Hydraulic civilizations, 624
Hydraulic mining, 253
Hydraulic radius, 322
Hydrocarbon compounds, 39, 144, 153–154, 174, 199–201, B-16
  atmospheric, 152, 153, 154
  fossil, 246
  in sedimentary rocks, 199–201
  storage, 144
  toxic products, 154
Hydrocarbons, chlorinated, 556, B-37
Hydroelectric power, 434, B-20
Hydrogen, atmospheric, 30, 144
  in biosphere, 536
Hydrogen bomb, 260
Hydrogen bonds, 488
Hydrogen fusion, 40
Hydrogenic sediments, 196
Hydrogen ions, 198
  in soil, 274

Hydrogen layer, 31
Hydrogen sulfide, 548, 549
  in photosynthesis, 494
Hydrograph, of ground water, 307
  of stream, 328
Hydrologic cycle, 288, 538–539, B-22
Hydrology, 1, 264
Hydrolysis, 191–193, 270, 553
Hydrophytes, 587
Hydropower, B-20
Hydrosphere, 3, 7, 9
  mass, 34
  water volume, 34
Hydrothermal solutions, 240
Hygroscopic dusts, 152
Hypersalinity, 404
Hyphae, 470
Hypolimnion, 68, 577
Hypoxia, environmental, 84

Ice, glacial, 264, B-31
  isotope record, 163
  pollutant content, 160
Ice Age (*see* Pleistocene Epoch)
Icebergs, 72, B-7
  calving, 72
  tabular, 73
Icecap, 428
Ice in clouds, 148
Ice-contact slope, 443
Ice crystals, in clouds, 119
  in ground, 266
Ice fall, 430
Ice floes, 72
Icelandic low, 99
Ice pellets, 121
Ice sheets, 428, 434–440, B-31
  climate, 438
  deposits, 443–447, B-32
  of Pleistocene Epoch, 439–440
  of present, 434–437
  and sea-level change, 454
  surface environment, 438
  volume, 291
Ice shelves, 73, 435
Ice-wedges, 282
Ichthyosaurs, 502
Icing storm, 122, 133
Igneous rocks, 171, 173–179, 186–187, B-15
  ages, 179
  basaltic, 176
  classification, 175–176
  crystallization, 174–175
  density, 176
  environmental significance, 186
  extrusive, 175
  forms, 178–179
  granitic, 176
  intrusive, 175, 179
  as natural resource, 187
  plutonic, 175, 179
  ultramafic, 176

I-12

Index

Illinoian glaciation, 449
Ilite, 192
 exchange capacity, 274
Illuviation, 303
Ilmenite, 172, 194, 239
Impermeable rock, 308
Index fossils, 499
Industrialization, environmental impact, 624, B-42
Industrial Revolution, 624, B-41
Infauna, 594
Infiltrating basins, 336
Infiltration capacity, 293
 and erosion intensity, 352–354
 man-induced changes, 293
Infiltration of precipitation, 292, 321, B-24
 and urbanization, 336
Infiltration rate, A-5
Infrared radiation, 40, 48, B-6
 absorption in water, 67
Infrared rays, 12, 40
Inlet in bar, 390
Insecticides, 5, 556, B-37
 DDT, 556
 inorganic, 556
 organic, 556
 and radiations, 508
Insectivores, 504
Insects, 483–484
 evolution, 501
Insolation, 41, B-6
 absorption by water, 67
 annual total, 43
 at high altitudes, 83
 and latitude, 43, 51
 losses in atmosphere, 45
 on water surfaces, 67
Instability, atmospheric, 155
Integument, 476
Intensity scale, Mercalli, 219
Interception of precipitation, 292
Interfaces, 3, 7, 188
 of atmosphere and hydrosphere, 188
 of atmosphere and lithosphere, 188
 of ocean and atmosphere, 34
Interglaciations, 440
 Pleistocene, 448
Interlobate moraine, 443
Intermediate zone, in lakes, 578
 of soil moisture, 294–306
Intertidal zone, 400
Intertropical convergence zone, 92
Intrusive rock, 175, B-15
Inversion of air temperature (see Air temperature inversion)
Inversion cap, 155
Inversion lid, 155
Ionization of atmosphere, 45
Ionizing radiation, 54, 84, 230, 256, 489–491, B-19
 from atmospheric sources, 161

Ionizing radiation (continued)
 background, 54, 491
 biological effects, 54
 galactic, 54
 hazards, B-19, B-24
 and mutation, 489–491
 from nuclear test explosions, 153
 and SST, 55, B-12
 from uranium mining, 256
Ionosphere, 45, B-6
Ionospheric layers, 46
Ions, in rainwater, 340
 in runoff, 340–343, 491
 in sea water, 196
 in stream water, 340–343, 491, B-25
 from watershed, 555
Iron, 237, 238
 in biosphere, 537
 in crust, 170
 in earth, 170
 recycling, 244
Iron formations, banded, 494
Iron ore, 198
Iron oxide, 198
 ferric, 198
 ferrous, 198
 in soils, 272
Iron sesquioxide, 192, 198
Irrigation, 299, 339, 346–349, B-25
 and climate change, 147, B-11
 of deserts, 613
 environmental impact, 339, B-25
 from ground water, 311
 and salinization, 339, 349, B-25
 side effects, 339, B-25
 water use, 346–349
Irruption of population, 569
Island arcs, 211, B-16
Isobar, 86, 98
Isobaric map, 86, 98
Isobaric surface, 86
Isohyets, 139, B-27
Isolation of species, 492
 genetic, 492
 geographic, 492
Isostasy, 179
 Airy model, 179
 and loading, 210
 and sediment deposition, 210
Isothermal layer, 68
Isotherms, 71, 79
Isotopes, 229
 daughter, 230
 oxygen, 163
 parent, 229
 ratio, 163

Japan current, 108
Jet aircraft, 147
Jet streams, 96, 130, B-8
 polar front, 96
 polar-night, 96

Jet streams (continued)
 of stratosphere, 96
 subtropical, 96
 tropical easterly, 96
 of tropopause, 96
Jetties, 394, 398
Joints in rock, 264
Joly, John, 228
Joule, A-3
Joule, James, A-3
Judeo-Christian tradition and pollution, 624, B-41
Judson, S., 367
Juglans, 603
Juvenile addition of nitrogen oxides, 547

Kaibab Plateau deer population, 569
Kamchatka current, 109
Kame terraces, 446
Kansan glaciation, 449
Kaolinite, 191
 exchange capacity, 274
Karst landscape, 270, 310, B-23
 environmental aspects, 310
Katabatic winds, 105, B-8
Katmai explosion, 183
Kelp, 467
Kelvin, Lord, 228
Kennelly-Heaviside layer, 45
Kerosene, 250
Kilocalorie, A-2
Kilograms-weight, A-4
Kilolangley, 43, A-3
Kinetic energy, 10
Knees on cypress tree, 589
Knob-and-kettle, 443
Krakatoa eruption, 161, 182
Krill, 594
Krypton, 30
Kuroshio, 108

Labile storage, 526
Labrador current, 109
La Brea asphalt pits, 455
Labyrinthodonts, 501
Laccolith, 180
Lagoons, 395, 403, B-30
 of atolls, 409, 410
 of coral reefs, 408
Lag time, 333, 328
 and urbanization, 336
Lake basins, 337, B-25
 glacial, 443
 types, 337
Lake Bonneville, 453
Lake breeze, 156
Lake Cayuga, 345
Lake Chicago, 451
Lake Erie, eutrophication, 581
Lake Hoover (see Lake Mead)
Lake Manly, 454

Index

Lake Maumee, 451
Lake Mead, 339, 364
   earthquakes, 224
Lake plains, 452
Lakes, 67–69, 337–340, 442–443, 452, 577–581, B-25
   ecosystems, 577, B-39
   eutrophic, 580
   glacial, 431, 442–443, 446
   heating and cooling, 67, B-7
   layers, 577
   man-made, 337
   oligotrophic, 580
   origins, 337, B-25
   overturn, 69
   oxbow, 365
   oxygen content, 577
   as physical features, 337
   Pleistocene, 452
   pluvial, 452, B-32
   pollution, 343–345, 584–585, B-25
   productivity classes, 580–581
   saline, 339, 381, 452
   temperature cycle, 68
   temperature structure, 577
   thermal cycle, 68
   thermal pollution, 345
   water balance, 338–339
   zonation, 577
Laminar flow of glacier, 431
Lampreys, 485
Lancet, 485
Land breeze, 87, B-8
Land bridges, 455
Land capabilities, 355
   classes, 335
Land disturbance by Man, 284–286, B-18
Land-fill, sanitary, 314, B-24
Landforms, 263, 350–351, B-21
   of arid lands, 378–381, B-29
   coastal, 385–409, B-29
   depositional, 351
   erosional, 351
   fluvial, 351, B-27
   glacial, 431–433, 442–446, B-31
   global relief, 177
   initial, 350
   of sedimentary rocks, 201–208
   sequential, 351
   of tectonic activity, 225–228, B-17
   of weathering, 266–272, B-21
Landmass, available, 351
Landmass denudation, 366–367, B-28
Land planning, B-42
Landslides, 279–281, B-22
   in glacial troughs, 432
   man-induced, 284–286
Land subsidence, 253, 286–287, B-18, B-22
   man-induced, 286
   by water withdrawal, 314, B-22

Land surface modifications, man-made, B-22
Langley, 41, A-3
Langmuir, Irving, 148
La Porte, Indiana, anomaly, B-11
Lapse rate of air temperature, 31, 84, 123, 155, B-8
   environmental, 31, 84, 123, 155
   normal, 31, 155
   steepened, 123, 155
Lapiés, 271
Larch, 601
Larix, 601
Larval stage, 478
Latent energy, 60
Latent heat, 12, 39, 50, 60, B-9
   of fusion, 60
   of vaporization, 60
Latent heat flux, 61
Lateral moraine, 432
Laterite, 191, 242, B-23
Laterization, 303
Latitude zones, 44
Latosols, 304
Laurasia, 216, 502
Laurentide Ice Sheet, 439
Lava, 174, 181, 185, B-15
   basaltic, 185, 186
   felsic, 181
   Hawaiian, 185
   intermediate, 181
   mafic, 181
   viscosity, 181
   volcanic, 181
Lava domes, 185
Lava flows, 181
Law of gravitation, 9
Law of stratigraphic succession, 232
Law of thermodynamics, second, A-8
Lawn lime as pollutant, 343
Leachate, 314
Leaching, in soils, 276
Lead, 237
   atmospheric, 5, 153, 160, 161
   environmental, 5, 153, 161, B-14, B-36
   in gasoline, 161
   as health hazard, 161
   recycling, 244
Lead poisoning, B-36
Leads in sea ice, 72
Lead-uranium series, 231
Lead-206, 230
Leaf, 475
Leaf gap, 475
Leaf response to heat energy variation, 520
Leaf temperature, 520, B-35
Leeches, 482
Legislation, environmental, 627–628, B-44
Leibig, Justis, 512

Leibig's Law of the Minimum, 512
Lemmings, 565
Length, A-1
Levees, artificial, 333
   natural, 375
Lianas, 606
Lichens and air pollution, 160
Life on earth, diversity, 463, B-33
   origin, 493–494, B-33
   primitive, 494, B-34
   species, 463
Life layer, 2, 363
Life table, 561
Light, speed, 14, 40
   visible, 12
Light energy in ecosystems, 510, 512, B-34
   as input, 512, 515
Light input to organisms, 515, B-34
   trigger action, 515
Lightning, 124, B-10
   as hazard, 132
   and nitrogen fixation, 547
Light pollution, B-14
Light rays, 40
Light zones in water, 67
Lignin, 476
Lignite, 246, 257
Lime, 245
Limestone, 198
   mineralization, 239
   precipitation, 199
   solution, 199
   weathering, 270–272
Limestone caverns, 271, 310–311, B-23
   air temperatures, 64
   collapse, 311
Limnetic zone, 577
Limnology, B-25
Limonite, 192, 198, 239, 242
   in soils, 272
Lipids, 522
Liquid, 9
Liquid state, 9
Liquid wastes, disposal, 317, B-24
   radioactive, 317, B-24
   storage, 317, B-24
Liriodendron, 603
Listed furrows, 421
Lithification, 189
Lithium, 240
Lithosphere, 3, 7, 179, 214
   environmental significance, 168
   rigid, 179
Lithospheric plates, 209, 214, B-16
   collisions, 214
   motions, 212, 214
   subduction, 214
Little Ice Age, 458
Littoral drift, 390, 394, B-29
Littoral environment, 400
Littoral zone, 577
Liverworts, 473

Index

Living filter, 586
Lizards, evolution, 502
Load of stream, 357–360, B-28
Loam, 273
  sandy, 273
Lobes of ice, 443
Loblolly pine, 603
Lode, 240
Loess, 424–426, B-3
  composition, 424
  gullying, 426
  landforms, 425
  origin, 424
  and soils, 425
Logistic curve, 563
  J-shape, 563
  sigmoid, 563
  S-shape, 563
Log pyramid, 532
London fogs, 160
  health effects, 160
Long Beach, California, land subsidence, 287
Long Island ground water problems, 1, B-23, B-27
Long-leaf pine, 603
Longshore current, 389
Longshore drift, 389
Long-wall system, 248
Longwave radiation, 40, 48, B-6
  planetary, 51
Los Angeles Basin air pollution, 157
Lotka, A. J., 21, A-13
Lovering, T. S., 628
Lowlands of coastal plains, 203
Lows, cut-off, 95
Low water of tide, 399
LSD, 490
Lung cancer, 256
Lynx, 565

Macroflora, 275
Macromolecules, 510
Macronutrients, 536, 555
Mafic mineral group, 172
Magma, 171, 173, 181, B-15
  basaltic, 180, 181, 214
  gases, 173
  granitic, 181, 213
  intermediate, 181
  silicate, 173–174
  volcanic, 181
Magmatic differentiation, 175
Magnesite, 239
Magnesium, 170
  in biosphere, 537
Magnesium ions in sea water, 197
Magnetic axis, 56
Magnetic compass, 56
Magnetic field of earth, 56, 169, B-7
  external, 56
  force lines, 56
Magnetic storms, 58

Magnetism, of earth, 56, B-7
  of rocks, B-16
Magnetite, 172, 194, 198, 239
Magnetopause, 57
Magnetosphere, 56, 169, B-7
Malaria, 473
Malpais, 186
Mammals, 486, B-33
  climbing, 606
  of desert biome, 613
  early, 504
  evolution, 504
  extinctions, 455
  marsupial, 504
  Mesozoic, 502
  placental, 504
  Pleistocene, 455, B-32
  scansorial, 606
Mammoth Cave, 310
Mammoth Coal Seam, 200
Mammoths, 455
  woolly, 455
Man, early, 455–458, B-32
  and energy systems, 9
  evolution, 143, 455–458, 505–506, B-34
  as evolutionary agent, 506
  and extinction, 507
Man and environment, B-5
Manganese, 237, 238, 242
  in biosphere, 55, B-36
  from sea floor, 246
Manganese nodules, 246
Manganite, 239, 242
Mangrove, 410, 591, B-30
  red, 591
Mangrove coast, 410
Mangrove swamps, 591
Manning equation, 324
Mantle of earth, 169, 177
  soft layer, 179
Maple, 603
Maps, isobaric, 86, 98
Marble, 213
Marine cliffs, 385–386, 393
Marine erosion, 383–388, B-29
Marine oil pollution, 253–255, B-19, B-39
  physical effects, 254–255
Marine scarp, 386, 393
Marine terrace, 386
Mars, magnetic field, 170
Marshes, 337, 587, B-39
  freshwater, 587, B-39
  salt, 587, B-30
Marsh gas, 201
Marsupial mammals, 504
Mass, 9, A-1, A-4
Mass balance, 112
Mass flow, intensity, A-4
  total rate, A-4
Mass movements caused by mining, 253, B-18

Mass number, 229
Mass-transport velocity (waves), 106
Mass wasting, 189, 264, 276–286, B-22
  man-induced, 284–286, B-22
Mastodons, 455
Material cycles, 534, B-35
  sedimentary, 534
  gaseous, 534
Materials balance approach, 623, B-41
Materials recycling, 623, B-35
Matter, 9, B-6
  cycling in ecosystems, 509, 534, B-35
  in ecosystems, 534, B-35
  in energy systems, 14
  in organisms, 534
  transformation, 14
  transport, 14
Matter budget equation, 510
Matter flow, A-4
  intensity, A-4
Matter inputs, inorganic, 510
  limiting factor, 512
  multiple, 512
  organic, 510
Mauna Loa Observatory, Hawaii, 162
McKelvey, V. E., 628, 630, 632
M-discontinuity, 178
Meanders, 364, 374–375, B-29
  alluvial, 374–375, B-29
  cutoff, 376
  down-valley sweep, 375
  entrenched, 369, B-28
  of Mississippi River, 376
Mean velocity of stream, 323
Mechanical energy, 10
Medial moraines, 431
Mediterranean regime, 301
Medusas, 480
Melting of ice, 60
Membrane, nuclear, 464
Mendel, Gregor, 489
Mercalli scale, modified, 219
Mercury, 237
  in atmosphere, 160
  environmental, 1, 160, 345, B-36
  in Greenland Ice Sheet, 160
  recycling, 244
  in swordfish, 1
Meridional circulation, 88
Meridional heat flux, 51
Meridional transport, 51
Meroplankton, 594
Mesas, 181, 202
Mesopause, 32
Mesophytes, 612
Mesosphere, 32, 46, B-6
Mesozoic Era, 234
  evolution of life, 502–503
Metabolism, 462

I-15

Index

Metalliferous deposits, 236–242, B-18
Metals, 237–239, B-18
 in environment, B-36
 ferro-alloy, 237
 native, 238
 nonferrous, 237
 primary, 242
Metals demand, 242–244, B-18
Metals processing, B-18
Metals recycling, 244, B-18
Metals supplies, 242–244, B-18
 substitutions, 629
 recycling, 629
Metamorphic rocks, 209, 212, 213–214, B-16
 and landforms, 213
Metamorphism, 209, 212, 213
 contact, 239
Metaphyta, 463, 473–478
Metazoa, 463, 478–485, B-33
 evolution, 498
Meteoric water, 241
Meteoroid, 8
Meteorology, B-6
Meteors, 33
Methane, 30, 144, 201
 reaction with oxygen, 537
Mica, muscovite, 240
Mica group, 171
Microclimates, B-7
Microclimatology, 77
Microflora, 275
Microfossils, 440
Micron, 40
Micronutrients, 537
Microphyllophyta, 475
Microrad, 54
Microrelief features, 270
Mid-Atlantic Ridge, 214
Middle-latitude zones, 44
Mid-Oceanic Ridge, 214
Mid-tide, 400
Millibars, 33
 conversion to inches, 98, 103
Milliequivalent, 274
Millirem, 54
Mine drainage, B-18, B-23
Mineral alteration, 189–193, 269, B-15
Mineral concentration, 236
Mineral deposits, 237–246, B-18
 metallic, 236–242, B-18
 nonmetallic, 237, 244–245
 of sea floor, 246
 secondary, 241
Mineral extraction, environmental impact, 253, 284, B-18
Mineral processing, environmental impact, 253, B-18
Mineral resources, 4, 168, 236–246, 627–629, B-17
 future supplies, 628, B-17

Mineral resources (*continued*)
 of lands, 236–245
 management, 627–629
 nonrenewable, 621
 of oceans, 246, B-19
 of sea floor, 246, B-19
Minerals, 171–173, 189–194, 197–201, B-15
 alteration, 189–193
 biogenic, 197–201
 clay, 191
 density, 194
 detrital, 194
 felsic, 172
 heavy, 194
 hydrogenic, 197–198
 mafic, 172
 of oxidation, 191–193
 silicate, 171–173, 191
Mining, and aggradation, 372
 and air pollution, 153
 of coal, 248–250, 255
 health hazards, 255–256
 hydraulic, 253, 272
 impact on environment, 236, 245, 284–286, B-18
 and valley sedimentation, 372
Mississippi River Commission, 376
Mississippi River delta, 396
Mississippi River regulation, 376
Mistral, 105
Mitochondria, 464
Mixing ratio, 115
M.k.s. system, A-2
Moho, 178
Mohorovičić discontinuity, 177, 178
Moisture, atmospheric, 114–115, B-9
Moisture budget (*see* Soil-moisture budget)
Moisture capacity of soil, 294–295
Moisture deficiency, 296
Molding sands, 245
Molds, 472
Mole, 535
Molecular nitrogen layer, 31, 45
Mollusca, 481
Mollusks, 481, B-33
Molybdenite, 239
Molybdenum, 237
Momentum, 93
 angular, 93, 96
 conservation, 93
 linear, 93
 transport, 94, 96
Momentum, transfer, 91, 97, 105, 382
 in westerlies, 97
 wind-to-ocean, 105
Moncrief, L. W., 624
Monera, 463, 466
Monitor well, 315

Monocots, 478
Monsoon, Asiatic, 105, B-8
 of North America, 105
 summer, 105
Monsoon depression, 134
Monsoon forest, 614
Montmorillonite, 192
 exchange capacity, 274
Moors, high, 587
Moraines, 431–432, 443, B-32
 interlobate, 443
 lateral, 432
 medial, 431
 recessional, 443
 terminal, 432–433
Mortality rate, 560
Mosses, 473
Mound of ground water, 314
Mountain arcs, 218, B-16
Mountain breeze, 88, B-8
Mountain chains, origin, 217, B-16
Mountain sickness, 84
Mount Mazama, 183
Mount Pelée, 182
Mudflats, 402
 reclamation, 405–408
Mudflows, 278, B-22
 in alluvial fans, 380
Muds, organic-rich, 402
Mudstone, 195
Mulch, 609
Murchison Falls, 364
Muscovite, 175, 240
Mushroom, 471
Mushroom rocks, 413
Muskox, 455
Mutagens, 490
 chemical, 490
Mutation, 465, 487, 489–490, 491–492
 by deletion, 489
 by duplication, 489
 and ionizing radiation, 489
 normal rate, 491
 in populations, 491, 492
Mutualism, 570
Mycorrhizae, 570
Myxomycota, 472

Natality rate, 560
National Environmental Policy Act (NEPA), 626, B-44
National Weather Service, 332
Native gold, 239
Native metals, 238, 239, B-18
Native silver, 239
Natural bridges, 370
Natural gas, 199, 201, 250, B-16
 as energy resource, 258
 production, 259
 reserves, 259, B-20
Natural levee, 375

I-16

Index

Natural resources, conservation, 625, B-42, B-44
  management, 627–629, B-42, B-44
Natural selection, 487, 491–492, B-34
NCAR, 149
Neanderthal Man, 456, 506
Neap tides, 399
Nebraskan glaciation, 449
Nebula, solar, 169
Nekton, 576
Nematodes, 481
Neolithic culture, 457
Neon, 30
Neritic zone, 593
Nervous system, 478
Netherlands land reclamation, 405
Net radiation, 51, 61
  of earth-atmosphere system, 51
  at high altitudes, 83
Neuston, 576
Neutrons, 229
  and mutations, 490
Newer Appalachians, 227
Newton, Isaac, 9
Niagara Escarpment, 363
Niagara Falls, 363
Niagara Power Project, 364
Niche, ecological, 559
Niches in rock, 267
Nickel, 237
  recycling, 244
Nile River and schistosomiasis, 481
Nimbostratus, 119
Nip, 385
Nitrate assimilation, 543
Nitrate ions, 341–343, 543
  in rainwater, 341
  in soil water, 341
  in stream water, 341–343
Nitrates, 245
Nitric acid, 547
Nitric oxide, 543, 547
Nitrification, 545
Nitrite ion, 543
Nitrobacter, 551
Nitrobacter group, 545
Nitrogen, biochemical reactions, 543–545
  in biosphere, 536
  cycle, 545–548, B-35
  forms, 543
  inorganic, 545
  molecular, 543
  organic, 545
  reactions, 543–545
Nitrogen cycle, 545–548, B-35
Nitrogen dioxide, 543, 547
Nitrogen fertilizer, 547
Nitrogen fixation, 276, 544–545
  atmospheric, 547
  industrial, 547
  by lightning, 547

Nitrogen fixers, 544
  bacterial, 544
  free living, 544
  symbiotic, 544
Nitrogen oxides, 152–154, B-14, B-35
  atmospheric, 152, 153, 154
  as pollutants, 547, B-14
Nitrogen reactions, 543–545
Nitrosomonas bacteria, 551
Nitrosomonas group, 545
Nitrous acid, 547
Nitrous oxide, 30, 144, 543
NOAA, 148
Noise pollution, 152, B-15
Nonclastic sediments, 194, 196–198
Normal faults, 212
North Atlantic current, 109
Northeast trades, 99
Notch in cliff, 385
Notochord, 485
Nuclear energy, 14, 259–260, B-21
Nuclear fission, 14, B-21
Nuclear fuels, 237, 259, 317, B-21
  breeding, 259
  future supplies, 630
Nuclear fusion, 14, 259, B-21
  in sun, 40
Nuclear power plants, hazards, B-21
  and thermal pollution, 345
Nuclear reactor, 14
Nuclear test explosions, and air pollution, 153
  and earthquakes, 224
Nuclear wastes and water pollution, 343
Nuclei, atmospheric, 33
  in clouds, 119
  of continents, 216
Nucleic acids, 465
Nucleotides, 488
Nuée ardente, 182
Numbers pyramid, 532
Nursery, estuarine, 590
Nutrient cycle, 534, B-35
Nutrient traps, 590

Oak, 603
Oak-chestnut forest, 604
Oak-hickory forest, 604
Obsidian, 175
Occluded front, 128
Ocean basins, 176, B-6
  distribution, 177
  structure, 177–179
Ocean currents, 98, 107–109, B-9
  and air temperature, 82
  causes, 107
  circumpolar, 109
  and density, 107
  equatorial, 108
  global pattern, 108
  and heat transport, 110, 112
  thermohaline, 107

Ocean dumping, 627, B-9
Oceanic zone, 593
Oceanography, B-6
Ocean outfall of sewage, 312
Ocean pollution, 253, B-9, B-18
Oceans, 34–37, 67, 71–72, 79, 291, B-6
  annual temperature cycle, 69
  circulation, 108–109, B-9
  depth, 34
  as food source, 595–596
  heating and cooling, 67, 79, B-7
  layers, 36
  mass, 34
  salinity, 35, 37
  sea ice, 72
  structure, 36
  surface area, 291
  volume, 34, 291
Ocean waves, 106–107, B-8
  energy, 106
  environmental role, 107
  growth, 106
  as hazard, 107
  height, 107
  largest, 107
Octopuses, 482
Odum, H. T., A-9
Offshore, 391
Offshore bar, 391
Oil (*see* Petroleum)
Oil poles, 252
Oil pollution, of ground water, 315
  marine (*see* Marine oil pollution)
Oil pools, 250
Oil sands, 259
Oil shales, 201, 253, 259, B-20
Oil spills (*see* Marine oil pollution)
Oil-tar lumps, 254
Old field succession, 572
Old Stone Age, 456
Oligocene Epoch, 505
Oligotrophic lake, 580
Olivine, 171
  alteration, 192
Onerwacht Series, 494
Ooze, organic, 580
Open-pit mining, 253
Open systems, 14–16, A-6, B-6
  in biology, A-6, A-8
  boundaries, 14
  characteristics, 14
  decay, 17
  equilibrium, 15
  self-regulation, A-8
  sensitivity, 16
  stability, 15
  steady state, 15
Open-system theory, A-6
Opposition of moon and sun, 399
Orbit of earth, 42
Orbiting satellites, 47
Order in closed systems, 22

Index

Ordovician Period, 499
Ore deposits, 239–242, B-18
  disseminated, 240
  hydrogenic, 242
  hydrothermal, 240
  secondary, 241
  sedimentary, 242
Ores, 236–242, B-18
  complex, 238
  disseminated, 240
  hydrogenic, 242
  hydrothermal, 240
  metalliferous, 236–242
  secondary, 241
  sedimentary, 242
Organelles, 464
Organisms, autotrophic, 462
  benthic, 499
  chemoautotrophic, 462
  classification, 462
  diversity, 463, B-33
  heterotrophic, 462
  major groups, 463, B-33
  pelagic, 593
  photoautotrophic, 462
Orogeny, 212, 233, B-16
Orographic precipitation, 125
Orographic trap, 441
Oscillation ripples, 383
Osculum, 480
Outcrop, 264
Outgassing, 35, 144, 196, B-6
  of nitrogen oxides, 547
  of volatiles, 173
Outlet glaciers, 435
Outwash plain, 443, 447, B-32
  pitted, 443
Ovary, 476
Overbank deposits, 375
Overbank flooding, 375, 376
Overburden, 249, 264
  residual, 264
  transported, 264
Overland flow, 292, 321–322, 352–356, B-24
  and erosion, 352–356
  rate, 321, A-5
Overthrust faults, 212
Overwash, 422
Oxbow lakes, 337, 365, 375
Oxidation, 13, 30, 144, 270, 522, 537–538
  biochemical, 13
  biological, 144, 522
  chemical, 13, 537
  and energy release, 523
  of minerals, 190–193, 553
Oxidation zone, 241
Oxygen, 170
  atmospheric, 30, 144, 146, 553, B-11
    early history, 494–497, 499, B-6
    at high altitudes, 84
    and plants, 173

Oxygen (*continued*)
  in biosphere, 536, B-36
  and combustion, 146
  in crust, 170, 553
  depletion, 146, 553
  dissolved in oceans, 37, 553
  in earth, 170, 553
  as electron acceptor, 538
  electronegativity, 537
  in lakes, 577
  in silicate minerals, 171
  solubility in water, 190
  storage pools, 553
  in streams, 583
Oxygen cycle, 144, 146, 553, B-36
  in biosphere, 553
  Man's impact, 553
Oxygen demand, 470
Oxygen depletion threat, 146, B-11
Oxygen-isotope ratios in Greenland ice, 440
Ozone, 46, B-6, B-12
  as pollutant, 154, 160
Ozone layer, 46
  sulfate content, 549

Pacific Ocean, plate tectonics, 217
Pack ice, 72
Paleolithic Culture, Middle, 456
Paleomagnetic data, 448
Paleomagnetism, B-16
Paleosol, 448
Paleozoic Era, 234, 497
  evolution, 497–502
  life forms, 497–502
Palynology, 458
Pangaea, 216, B-16
Paraffin-base crude oil, 250
Paramecium, 473
  competition, 568
Parasites, 481, 569
Parasitism, 569
Parent matter of soil, 272, 275
Particulate matter in atmosphere, 151, B-14
Passes in coral reefs, 409
Patterned ground, 282, 599, B-22
Peat, 199, 276, B-16
  formation process, 276
Pebbles, 193
Pedestal rocks, 413
Pediments, 381, B-29
Pedogenic regimes, 302–305, B-23
Pegmatites, 175, 180, 240
  as ore bodies, 240
Pelecypods, 481
  evolution, 499
Pendulum action, 19
Peneplains, 366, 367, B-28
Pentlandite, 239
Perched water table, 308
Peridotite, 175, 176

Perigee, 399
Periglacial environments, 452, B-22, B-32
Periods of geologic time, 234
Periphyton, 576
Permafrost, 65, 282–283, 599, B-22
  active zone, 65
  continuous, 65
  degradation, 283
  discontinuous, 65
  distribution, 65
  engineering problems, 283
  surface forms, 282
  and weathering, 191
Permeability of rock, 308
Pediments, 381, B-29
Permian Period, 501
Perpetual motion, 19
Peru current, 109, 110
Pest controls, biological, B-38
  specific, B-38
Pesticides, 556–558, B-37
  and agricultural productivity, 533
  inorganic, 556
  organic, 556
Petroleum, 199–201, 250–253, 257, B-16
  composition, 250
  as energy resource, 257, B-20
  fields, 252
  geologic ages, 252
  origin, 201, B-16
  production, 259, B-20
  reserves, 258, B-20
  world resources, 201, 253, B-20
Petroleum traps, 251–252
Phanerozoic Eon, 234
Phenotype, 491
Phloem, 474
Phosphate beds, 249
Phosphate ions, in rainwater, 341
  in stream water, 342, B-26
  in surface waters, 555
Phosphate rock, 245
Phosphorus, 170
  in biosphere 536, 555, B-36
  environmental, 555, B-36
Phosphorus cycle, B-36
Photic zone, 67, 593
Photochemical reactions, 154
Photoperiod, 515
Photoperiodism, 515, B-34
Photosynthesis, 13, 30, 465, 512–515, B-34
  by bacteria,
  and carbon cycle, 541
  and carbon dioxide levels, 542
  compensation point, 513
  gross, 513
  and hydrogen sulfide, 549
  and light intensity, 513–515
  linear range, 513
  net, 513

Index

Photosynthesis (*continued*)
   saturation intensity, 513
   and temperature, 520
Phreatophytes, 612
pH of soils, 274
Phycomycota, 470
Phyla (phylum), 463
Phytoplankton, 576
   in lakes, 478
   and oxygen depletion, 146
Picea, 601
Piedmont Upland, 367, 372
Pigments, 245
Pinch-out trap, 251
Pine, 601
Pinus, 601
Pinus australis, 603
Pinus caribea, 603
Pinus echinata, 603
Pinus taeda, 603
Pioneers, 571
Pipe of volcano, 183
Pitchblende, 231, 239, 256
Pit crater, 185
Placenta, 504
Placental mammals, 504
Placer deposits, 242
   marine 242, 246
Plagioclase feldspars, 171
   calcic, 171
   sodic, 171
Plains, coastal, 203
   deltaic, 397
   pitted, 443
   of sedimentary strata, 203
Planarians, 481
Plane of ecliptic, 42
Planetary circulation, 91
Planetary environments, 37
Planetary temperature, 39, 49
Planets, albedos, 48
Plankton, 72, 576, 593
   and upwelling, 110, 595
Plankton tows, 254
Planned environmental modification, 2
Plant metabolism, 13
Plant roots and rock weathering, 268
Plants, 463, 473–478, B-33
   day-neutral, 515
   dormancy, 521
   epiphytic, 475, 570
   evolution, 500–501, 503, B-34
   hardening, 520
   herbaceous, 478
   higher, 475
   intolerant, 513
   and light, 513–517
   long-day, 515
   migrations, 504
   photoperiodism, 515
   Pleistocene, 504

Plants (*continued*)
   response to heat variation, 520
   short-day, 515
   temperature effects, 520–521
   terrestrial, 500
   tolerant, 513
   vascular, 474, 500, 503
Plant species of rainforest, 606
   primary, 606
   secondary, 606
Plasmodium, 473
Plateau basalts, 181
Plates, lithospheric (*see* Lithospheric plates)
Plate tectonics, 214–218, B-16
Platinum, 238, 239
Platyhelminthes, 481
Playas, 226, 381, B-29
   deflation, 412
Pleistocene Epoch, 147, 234, 439, 448–458, B-32
   air temperatures, 147
   climate zones, 147
   global climates, 452
   lowered sea levels, 177
   periglacial environments, 452–454
   plant migration, 504
   precipitation, 147
   sea-level changes, 454, 455
   stream aggradation, 368
Plesiosaurs, 502
Pliocene Epoch, 448
Pliopithecus, 505
Plucking by ice, 431
Plume from stack (*see* Stack plumes)
Plutonic rock, 175, 179
Plutonium, 259
Pluvial lakes of Pleistocene, 453, 454, B-32
Pneumatophores, 589
Pneumococcus, 466
Pneumoconiosis, 255
Pocket beach, 390
Podzolic soils, 447
Podzolization, 302, 601
Podzols, 303
Poikilotherms, 521
Point-bar deposit, 364
Polar easterlies, 97, 104
Polar front, 96, 116
Polar front jet stream, 96
Polar highs, 89, 99
Polar lows, 94
Polar-night jet stream, 96
Polar zones, 45
Polders, 405
   flooding, 422
Polje, 272
Pollutants, atmospheric, 151–154, 160–162, B-13
   bacterial, 153
   chemical, 151

Pollutants (*continued*)
   concentrations, 155
   dispersal, 154
   in glacial ice, 160
   global effects, 161–162
   particulate, 151
   primary, 153
   secondary, 154
   urban, 153
Pollution, biological, 152, 345, 584–585, B-33, B-39, B-40
   of air (*see* Air pollution)
   chemical, 152, B-37, B-38
   and economics, 623, B-41
   environmental, 152, 624–626
      and Christianity, 624
      historical roots, 624–626
      in Soviet Union, 626
      and technology, 625
   of estuaries, B-30
   of ground water, 314–316, B-23
   noise, 152
   by pesticides, B-37
   and population growth, 621
   by sewage, 584–585, B-24
   of streams and lakes, 343–345, 584–585, B-25
   thermal, 14, 152, 345, B-40
   of water (*see* Water pollution)
Pollution control laws, 626, B-44
Pollution dome, 159
Pollution plume, atmospheric, 159
   in ground water, 315
Polyclimax theory, 573
Polymers, 464
Polyps, 480
   of coral, 592
Ponds, 337, 577, B-39
   water-table, 313
Pool in channel, 375
Pool in material cycle, 535
   active, 535
   storage, 535
Pools in streams, 583
Poplar, 601
   yellow, 603
Popping rock, 269
Population, human, 566–568, 620, B-38
Population change factor, 622
Population dynamics, 560–561, B-38
Population growth, 562–566, 620–621, B-38
   environmental impact, 621–622, B-42
   limits, 621, B-42
   trends, 566–568, 620, B-38
   world, 566–568, 620, B-38
Population growth rate, 560
Population structure, 561–562
Populations of species, 491, 560
   density, 560
   distribution patterns, 560

Index

Populations of species (*continued*)
  ecological density, 560
  endemic, 492
  mutations, 491–492
Populus, 601
Porosity of rocks, 308
Porphyry, 240
Porphyry copper, 240
Portland cement, 244
Potash feldspars, 171
  alteration, 191
Potassium, 170
  in biosphere, 537
Potassium-40, 230
Potassium-argon series, 231
Potential energy, 11
Power, 18, 20, A-1, A-2
  per unit area, A-4
Power generation, environmental
    impact, 13
  from fossil fuels, 24
Prairie dog, 609
Prairies, 608
  mixed-grass, 608
  short-grass, 608
  tall-grass, 608
Precambrian time, 216, 233
Precipitation, 33, 114, 118, 121–130,
    146–148, B-10
  convective, 122–125, 134, 148
  cyclonic, 127–130, 135, 138
  in deserts, 378
  forms, 121
  frontal, 127–130
  global, 289
  in global heat balance, 111
  intensity, 321
  and irrigation, 147
  of low latitudes, 134
  man-induced changes, 146–148,
    B-11, B-12
  monsoon, 134
  orographic, 125–127, 148, 433
  and stream flow, 327–329
  urban effects, 159
  world regions, 139–141, B-10
Predation, 569, B-38
Predator, 569
Pressure, A-2
Pressure, atmospheric (*see*
    Barometric pressure)
Pressure cells, 98, 102
Pressure gradient, 86
Pressure-gradient force, 86
Pressure intensity, A-2
Prevailing westerly winds, 102
Prey, 569
Primary production, 523
  efficiency, 523
  net, 524
Primates, 505
  evolution, 505

Producers, organic, 275, B-34
  primary, 462, 470
Productivity, of ecosystems, 523, B-34
  net, 524
  primary, 523
  secondary, 528
Products of chemical reaction, 199
Profile of earth's surface, 177
Profile of equilibrium of stream,
    361, 365, B-28
  readjustments, 368–370
  segmented, 365
Profundal zone, 577
Progradation of shoreline, 389, 392,
    393
Project Skywater, 148
Project Stormfury, 148
Promontories, 387
Pronghorn antelope, 609
Prosimians, 505
Proteins, 488, B-35
  and DNA, 465
  oxidation, 522
  synthesis, 489
Protista (Protists), 466–473, B-33
  ameboid, 467
  coccine, 467
  flagellate, 466
  sporine, 467
Protocooperation, 570
Protons, 229
  cosmic, 54
Protozoa, 472–473
  ameboid, 473
  ciliate, 473
  coccine, 473
  flagellate, 472
Provincelands of Cape Cod, 423
Pseudopdia, 467
Psilophyta, 475
Pterophyta, 475
Pteropsids, 475
Pterosaurs, 502
Puffball, 471
Pulpwood, 601
Pupfish, 453
Push of wind on waves, 106
Pyramid diagram, 531
Pyramidal dune, 417
Pyranometer, 52
Pyrite, 239, 241
Pyroclastic sediments, 194
Pyrolusite, 238, 239
Pyroxene group, 171

Quadrature, 399
Quarries, 253
Quarrying, and air pollution, 153
  of granite, 269
Quartz, 171
  in dune sand, 414
  as sediment, 194

Quartzite, 213
Quaternary Period, 234, 448, B-32
Quercus, 603
Quick clays, 220

Rad, 54
Radiation, cosmic (*See* Cosmic
    radiation)
  electromagnetic (*see*
    Electromagnetic radiation)
  evolutionary, 493, 502, 508
  galactic, 54
  ground, 48
  infrared, 48, B-6
  ionizing (*see* Ionizing radiation)
  longwave, 48, B-6
  solar, 13, 40, B-6
    absorption by water, 67
    net all-wave, 61
  terrestrial, 48
Radiation balance, 38, 45–54, B-6
  and atmospheric dust, 161
  and latitude, 50
  and urbanization, 150, B-12
Radiation belts, 57, B-7
  Van Allen, 57
Radiation cycles, 52
  annual, 52
  daily, 52
  seasonal, 52
Radiation deficit, 75
Radiation effects, 490–491
  genetic, 490
  somatic, 490
Radiation fog, 120
Radiations, evolutionary, 493, 502,
    508
Radiation spectrum, 40
Radiation surplus, 75
Radioactive decay, 16, 229
Radioactive wastes, 317–319, B-9,
    B-19
  burial, 317, B-19
  disposal, 317–319, B-19
  high-level, 317
  intermediate level, 317
  low-level, 317
  production, 317
  storage, 317–319, B-19
Radioactivity, 54, 167, 229–230, B-15,
    B-17
Radiocarbon dating, 231–232, B-17
Radio communication, 45
Radiogenic heat, 168, 169, 228–230,
    B-15
Radioisotopes, 4, 168, 229–230, 317,
    B-19
  fabrication, 317
  and heat production, 317
  long-lived, 317
  in stream water, 343
Radiolaria, 473

Index

Radiometric age determination, 231–232, B-17
Radio waves, 12
  reflection, 45
Radium, 228, 238
  health hazard, 256, B-19
Radon, 256
Rain, 121
  acid, 551
Rainfall, 122–130, 134–135, 139–141, B-10
  convective, 122–125, 134
  cyclonic, 127–130, 135
  in deserts, 378
  frontal, 127–130
  and irrigation, 147, B-12
  of low latitudes, 134
  man-induced increase, 147, B-12
  monsoon, 134
  orographic, 125–127
  planned increase, 148–150, B-12
  and runoff, 322–323
  of tropical cyclones, 135, 138
  world regions, 139–141, B-10
Rainfall intensity, A-5
Rainforest, 605, B-41
  temperate, 614
Rainmaking, 148, B-12
Rainshadow, 125, 127
Rainwater, acid, 551
  ion content, 340, B-25
  pH, 551
Ramapithecus, 505
Rapids, 361
Rating curve, 326
Rayleigh scattering, 46
Reactants, 199
Reaction, chemical, 198
  in magmas, 174
Reaction series, 174
  of Bowen, 174
Reactor, nuclear, 259
Recessional moraine, 443
Recharge of ground water, 306
Recharge wells, 312, 314, 336
Reclamation of tidal lands, 405–408, B-31
Recombination, 487, 491
Recycling, of materials, 623
  of metals, 235, 243–244, 629, B-18
  of wastes 627, B-41
Redbeds, 495
Reduction, 537–538
Redwood, 477, 603
Reefs of coral (*see* Coral reefs)
Refraction of waves, 387
Reg, 412
Regimes of soil moisture, 299–302
Regolith, 189, 193, 264, B-16, B-21
Reindeer, 599
Reindeer moss, 600
Rejuvenation of streams, 369, B-28

Relative humidity, 114
  daily cycle, 114
  global distribution, 115
  and temperature, 114
Relief features of globe, 177
Rem, 54, 490
Reproductive system, 479
Reptiles, 485
  bipedal, 502
  evolution, 501, 502–503
  ruling, 502
Reservoir rock, 250
Reservoir sedimentation, 359, 372, B-28
Reservoirs, 338–340
  evaporation, 23, 338
Residence time of ions, 197
Residuals, economic, 623, B-42
Resource Recovery Act of 1970, 627
Resources, of energy, 4, 256, 260, B-20
  management, 627–629, B-42
  materials, 4, B-42
  mineral, 235–246, B-17
  nonrenewable, 201, 236, 245, 627–629
  nuclear, 259–260, B-21
  and population, B-42
Resources for the Future, Inc., 623
Respiration, 39, 144, 462, 464, 512, 523, B-34
Retarding basin, 335
Retrogradation of shorelines, 389, 392, 393, 399, B-29
  of deltas, 399
  man-induced, 399
Rhinoceros, woolly, 455
Rhizobium, 276, 544–545, 570
Rhizoids, 473
Rhizome, 475
Rhizophora mangle, 591
Rhodospirillum, 544
Rhyolite, 176
Rhythms, annual, 517
  circadian, 517
  endogenous, 518
  light-regulated, 517
Ribonucleic acid (RNA), 465, 489
Ribosomes, 465
Richter, C. F., 219
Richter scale, 219
Rickover, H. G., 629
Ridge-and-valley landscape, 227
Riffle, 375, 583
Rift, axial, 214
Rift valley, African, 226
Rill erosion, 355
Rilling, 355
Ripples, capillary, 105
  in sand, 383, 400, 413
  current, 400
  oscillation, 383

Rip-rap, 394
River and Flood Forecasting Service, 333
River floods (*see* Floods of rivers)
Riverine environment, 376, B-29
Rivers, alluvial (*see* Alluvial rivers)
  exotic, 339
  (*see also* Streams)
River systems (*see* Stream systems)
River tide, 400
River valleys, drowned, 402
RNA, 465, 489
Roches, Moutonnées, 442
Rock, 171
Rock basins, glacial, 431
Rock breakup, geometry, 265
Rock cycle, 210
Rockfall, 281
Rock flour, 431
Rock flowage region, 308
Rock knobs, glaciated, 442
Rock salt, 198, 245
Rock sea, 266
Rockslide, 279
Rock steps, 430, 431
Rock terrace, 369
Rock transformation cycle, 189
Rocks, 171, 210, B-15
  cycle of transformation, 210
  environments, 210
  igneous (*see* Igneous rocks)
  metamorphic (*see* Metamorphic rocks)
  sedimentary (*see* Sedimentary rocks)
Roman Empire, metals resources, 236
Room-and-pillar system, 248
Root layer, 609
Root nodules, 544
Roots, crustal, 178
  of grasses, 609
Rossby waves, 95, 130
  and jet stream, 96
Ross Ice Shelf, 435
Rotifers, 481, 579
Roundworms, 481
Royal Gorge, 363
Rubidium-87, 230
Rubidium-strontium series, 231
Rubner, Max, 523
Runoff, 289, 297, 321–322, B-24
  and dissolved solids, 340, B-25
  man-induced changes, 336, B-28
  and rainfall, 328, B-24
  and sediment yield, 353, B-28
  and urbanization, 336, B-25
Rutile, 239

Saber-tooth tiger, 456
Sagebrush, 613

Index

Salinity of sea water 35, 37, 107, B-16
  and currents, 107
Salinization, 305, 613
  of soils by irrigation, 339, 349, B-25
Saltation, 413
Salt crystal growth, 267
Salt crystals, atmospheric, 33, B-9
Salt domes, 252
Salt-flats, 381
Salt ground water, 309
Salt intrusion, B-23
Salt lakes, 196
Salt marsh, 402, 587, 590, B-30, B-39
Salt mine storage, 318
Salt plugs, 252
Salts, 245
  atmospheric, 152
  crystallization, 267
  deicing, 132, 316
  in sea water, 35, B-16
  in soils, 305, 339, 406
Salt water intrusion 2, 313
Salt wedge, 401
San Andreas Fault, 221, 224, B-17
  active sections, 222
  creep zones, 222
  locked sections, 222
Sand, 193, B-16
  commercial, 244
  of dunes, 414
  molding, 245
Sandblast action, 412
Sand drift, 417
Sand dunes, 414–420, 570–573, B-31
  coastal, 417
  control, 423, 424
  crescentic, 415
  fixed, 414
  of free sand, 415–417
  heaped, 417
  live, 414
  longitudinal, 419
  man-induced advance, 423
  of North Sea coast, 422
  parabolic, 418
  phytogenic, 417–420
  plants, 571–573
  pyramidal, 417
  succession, 571–573
  transverse, 415
Sand Hills, Nebraska, 420
Sand ridges, longitudinal, 419
Sand ripples, 413
  tidal, 401
Sand sea, 415
Sandspit, 390
  recurved, 390
Sandstone, 195
Sandstorm, 413
Sand transport by wind, 414
San Fernando Earthquake, 224

Sangamonian interglaciation, 449
Sanitary land-fill, 314
Santa Ana wind, 105
Santa Barbara oil spill, 253, B-19
Saphrotrophs, 470
Saprolite, 270, B-16, B-21
Sastrugi, 438
Saturation adiabatic rate, 119
Saturation, of gases in water, 190
  of water vapor, 114
Sauropods, 503
Savannas, 608, B-41
Scandinavian Ice Sheet, 439
Scarification of land, 253, 284
Scarp, 201
  of cuesta, 203
  marine, 386, 393
Scarp-slope-shelf topography, 201
Scattering, atmospheric, 46
Schaefer, Vincent, 148
Schist, 213
Schistosomiasis, 481
Schizomycota, 466
Schrödinger ratio, 526
Sciophytes, 513
Sclerophyll, 614
Scoria, 175
Scorpions, evolution, 501
Scrap metal, 243
Scree slopes, 266
Scrub, hard-leaved, 614
Sea breeze, 87, B-8
Sea caves, 385
Sea ice, 72, B-7
Sea lamprey, 582
Sea level change, 143, 146, 177, 454, 455, B-7
  Pleistocene, 177, 454, B-7, B-32
Sea salts, 172
  in rain, 340
Sea-surface temperature, 71
Sea walls, 393
Sea water, 35, 172, 197, B-6, B-16
  composition, 35, 172
  constituents, 197
  density, 36, 107
  ions, 197
  mineral resources, 246
Sea waves, seismic, 220
Secondary production, 528
Sedimentary cycles, 553–555
Sedimentary rocks, 189, 195–198, B-16
  bedding, 195
  carbonate, 198
  clastic, 195
  environmental influences, 203
  landforms, 201–208, 227–228
  as natural resource, 189
  nonclastic, 197–198
  precipitates, 198
  stratification, 195
  varieties, 195

Sedimentation, in deltas, 396
  tidal, 402
  in valleys, 253, 354, 372–374, B-28
  and dams, 373–374
  man-induced, 372, B-28
  and mining, 372
Sedimentation cycles, 368
Sediment drift, 389
Sediments, 188, 193-198, B-15, B-16
  accumulation, 210
  biogenic, 196, 198, 199
  carbonate, 198, 541–542
  chemical, 196, 198
  classification, 194
  clastic, 194
  deep-sea, 448
  of deltas, 396
  detrital, 194
  environmental importance, 193
  flocculation, 195
  geosynclinal, 211
  glaciofluvial, 443, 446
  glaciolacustrine, 446
  grade sizes, 193
  heavy, 194
  hydrocarbon, 199–201
  hydrogenic, 196, 198, 242
  as natural resource, 189
  nonclastic, 194
  organically-derived, 196, 198
  pyroclastic, 194
  recycling, 367
  sorting, 195
  sulfur content, 551
  surface area, 193
  suspension, 195
Sediment yield, 353, 358, B-28
  and urbanization, 373, B-28
Seed coat, 476
Seed ferns, 501
Seeding of clouds, 119, 124
Seeds, 476
Seep, 308
Seif dune, 416
Seismicity gaps, 222
Seismic sea waves, 182, 220–221, B-17
Seismic waves, 218, B-17
Selection, environmental, 491, B-34
  natural, 491
Sensible heat, 12
Sensible heat flux, 60
Sensory system, 478
Sequoia gigantea, 603
Sequoia sempervirens, 477, 603
Seral stage, 571
Sere, 571
Serpentine, 192
Sesquioxide, of aluminum, 191
  of iron, 192, 198
Sewage, 584–586, B-39
  as pollutant, 343, 344, B-24, B-39
  raw, 584
  in streams, 585

Index

Sewage disposal, 315
Sewage treatment, 585–586
  by living filter, 586
  primary, 585
  secondary, 586
  tertiary, 586
Sewage zones in streams, 584–585
  active decomposition, 584
  clean water, 585
  degradation, 584
  recovery, 585
  septic, 584
Sex ratio, 562
Shale, 195
Shattering of rock, 265
Shear zone, atmospheric, 116
Sheet erosion, 354
Sheet flow, 321
Sheeting structure, 268
Shelf, of sedimentary strata, 201
Shelves, continental, 176
Shields of continents, 216
Shield volcanoes, 185
Shingle, 385
Shinn, R. U., 624
Shoestring rills, 355
Shore, 383
Shorelines, 383, 388, 393–394, B-29
  coral reef, 408
  embayed, 388
  engineering structures, 393
  Man's impact, 393–394, B-29
Shore profile, 385, 391–392
Shore protection, B-30
Shore zone, 383
Short-leaf pine, 603
Shortwave radiation, 40
Siberian high, 99
Silica as cement, 195
Silicate magmas, 173–175, B-15
  basaltic, 180
  crystallization, 174
  differentiation, 175
  water content, 174
Silicate minerals, 171, 173, B-15
  felsic, 172
  mafic, 172
Silicates, 171, B-15
Silicon, in crust, 170
  in igneous rocks, 171
Silicon dioxide, 171
Silicon ions in sea water, 197
Silicosis, 255
Sill, 180
Silt, 193
Siltstone, 195
Silurian Period, 499
Silver, 238, 239
  native, 239
  recycling, 244
Silver iodide smoke, 148
Simpson, Joanne, 148
Singer, S. F., 510

Sink, volcanic, 185
Sinkholes, 271, B-23
  collapse, 311
Size grades of sediments, 193
Skeletal system, 478
Skin drag, 105, 106
Skin gills, 484
Sky color, 46
Slack water, 400
Slaking, 268
Slash and burn, 607
Slash pine, 603
Slate, 213
Sleeping sickness, 473
Sleet, 121
Slime molds, 472
Slip face of dune, 415
Slope, 263
  of stream channel, 322, 324
Slope erosion, 352–356, B-27
Slopes, valley-side, 325
Slope wash, 354
Sludge from sewage, 585
Slump, 279
  man-induced, 286
Slump blocks, 281
Smelters and air pollution, 152
Smog, 1, 152, 155–160, B-9, B-13
  of Los Angeles Basin, 157
  photochemical, 154, 549
Smoke, industrial, 153
  sources, 161
Smoke plume, 156
Smudge pots, 520
Snails, 481, 482
  evolution, 499, 501
Snakes, evolution, 502
Snoeshoe hare, 565
Snout of glacier, 430
Snow, 121
Snowfall, as environmental stress, 132
  increase by seeding, 148
Snowflakes, 121, B-9
Snowline, 440
  lowered, 440
Soda niter, 267
Sod formers, 608
Sodium, 170
  in biosphere, 537
Sodium chloride, 198
Sodium ions in sea water, 196
Sodium salts, 245
Soft layer of mantle, 179
Softwoods, 603
Soil, 168, 188
  true, 264
Soil acidity, 274
Soil bacteria, aerobic, 544
  anaerobic, 544
Soil colloids, 273
  exchange capacity, 274
  organic, 274

Soil color, 272
Soil conservation, 355, B-27
Soil Conservation Service, 355
Soil cracks, 268
Soil creep, 277, B-22
Soil drifting, 421
Soil erosion, 354–356, B-27
  accelerated, 354, B-27
  control measures, 356–357
  forms, 354–356
  on loess, 426
  by rills, 355
Soil-forming processes, 275–276
Soil horizons, 303
Soil mechanics, 286
Soil moisture, 293–302
Soil-moisture balance, 296–299, B-23
  and erosion intensity, 353
Soil-moisture budget, 296–299, B-23
Soil-moisture cycle, 295–296, B-23
Soil-moisture deficiency, 297
Soil-moisture recharge, 297
Soil-moisture regimes, 299–302
Soil moisture zone, 294, 296
Soil pollution, B-23
Soil profile, 275
Soils, 272–276, 302–305, B-21
  age, 275
  chemical properties, 273–275
  classification, 302–305, B-23
  and climate, 275
  composition, 272
  as dynamic layer, 272
  frozen, 65
  of grasslands biome, 609
  heating and cooling, 64–65, 77
  horizons, 272
  on loess, 425
  pedogenic regimes, 302–305
  physical properties, 272–273
  as resource, 236
  texture, 272
  water content, 275
Soil solution, 272, 274
Soil temperature, 64, B-7
  annual cycle, 64
  daily cycle, 64
Soil texture, 272
Soil-texture classes, 272
  triangle diagram, 273
Soil water, 291
Soil-water balance, 275, 276, B-23
Solar constant, 41
Solar energy, 256, B-20
Solar flares, 58
Solar nebula, 169
Solar radiation, 13, 40, 83, 161–162, 441, B-6
  astronomical cycles, 441
  at high altitudes, 83
  and ice ages, 441
  on Mauna Loa, 162
  measurement, 52

Index

Solar radiation (*continued*)
  monitoring, 162
  and turbidity, 161–162, B-14
  and volcanic dust, 161–162, B-14
Solar spectrum, depletion in water, 67
Solar wind, 57, 169, B-7
Solids, 9
  dissolved in streams, 340, B-25
Solid state, 9
Solid wastes, disposal, 314, B-24
  handling, B-24
  recycling, B-24
Solifluction, 282, 599, B-22
Solifluction lobes, 282
Solifluction terraces, 282
Solstice, 42
  summer, 42
  winter, 42
Solubility of gases in water, 189
Solum, 188, 264, 272
Solution of limestone, 270
Solution of minerals, 190
Solutions, hydrothermal, 240
Sorting of sediments, 195
Source regions of air masses, 116
Southeast trades, 99
South pole air temperatures, 82
Space-ship analogy, 619
Spalling, 265
Spartina, 402, 587
Species, 491
  evolution, 492–493
  insolation, 492
  numbers, 463
  scientific name, 491
Species distribution patterns, 560
  clumped, 560
  random, 560
  uniform, 560
Species interactions, 568–571
  negative, 568
  neutral, 568
  positive, 568
Specific heat, 60
Specific humidity, 115
Spectrum, electromagnetic, 12
  solar, 40
Speed, A-2
Sphagnum moss, 586
Sphalerite, 239
Spheroidal weathering, 270
Spicules, 480
Spiders, evolution, 501
Spit of sand, 390
  recurved, 390
Splash erosion, 352
Spodumene, 240
Spoil accumulations, 253
Spoil ridge, 249
Sponges, 480
  evolution, 498
Spores, 476

Spray zones of dune, 415
Spring tides, 399
Springs, 308
  artesian, 309
Spruce, 601
Spurs, truncated, 431
Squall line, 128
  and tornadoes, 133
Squall wind, 124, 132
Squids, 482
  evolution, 499
Stability of atmosphere, 155
Stable air, 155
Stack plumes, 156
  coning, 156
  fanning, 156
  looping, 156
Stacks, 385
Staff gauge, 326
Stage-discharge curve, 326
Stage of stream, 326
Stalactite, 310
Stalagmite, 310
Star dune, 417
Steady state, 15, A-7
  in global circulation, 86
  of global system, 509
  of open system, 15
Steady-state existence, 631
Steam power, geothermal, 257
Stem reptiles, 501
Steno, Nicolaus, 232
Steppe, 139, 608
Stibnite, 239
Stilling tower, 326
Stone Mountain, Georgia, 367
Stone nets, 282
Stone polygons, 282
Stone rings, 282
Stone stripes, 282
Storage, of energy, 15
  of matter, 15
Storage pool, 535
Storms, atmospheric, 122–138, B-10
  convective, 122–125
  cyclonic, 18, 127–132, 135–138, B-10
  as environmental hazards, 114
  thunderstorms, 122–125, B-10
  tornadoes, 133–134, B-10
  tropical, 135–138, B-10
  magnetic, 58
Storm surge, 137, 398, 406, B-10
Stoss side of rock knob, 442
Strata, 195
  of geosyncline, 211
  horizontal, 201–204
  relative ages, 232
Stratigraphy, 232, 448
Stratopause, 31
Stratosphere, 31, B-6
  warming, 162
  water vapor changes, 147, B-12

Stratovolcano, 181
Stratum plain, 202
Stream aggradation, 368, 372, 373–374, B-28
Stream channelization, 376–378, B-29
Stream channels, 322, 360–361, 373–374, B-28
  aggradation, 372, 373–374
  artificial changes, 335
  braided, 368
  changes in flood, 360–361
  degradation, 374
  dendritic system, 325
  of deserts, 378
  graded, 361, 365
  hydraulic geometry, 322
  man-induced changes, 373–374, B-28
  ungraded, 361
Stream degradation, 369
  Pleistocene, 454
Stream discharge, 323, 326
  average, 327
  and basin area, 327
  and load transport, 358
  and urbanization, 336
Stream erosion, 356–357, 361–362, B-28
Stream flow, 313, 322–323, 327–331, B-24
  in deserts, 378
  gauging, 326–327
  and ground water, 313
  and precipitation, 327–329
  and soil-moisture, 330
  and urbanization, 336
  velocity distribution, 322
  and water table, 313
Stream gauging, 326–327
Stream gradation, 361–362, 365
Stream grade, 361, 365, B-28
  readjustments, 368–370
Stream gradient, 365
  and discharge, 365
  and particle size, 365
Streamline flow, 431
Stream load, 357–360, B-28
  bed, 358
  and discharge, 358
  suspended, 358–360
Stream rejuvenation, 369, B-28
Streams, 322–336, 368–370, 582–584, B-24, B-28
  alluvial, 374–376
  braided, 368
  of desert, 378
  discharge, 323
  downstream changes, 584
  ecosystems, 582–584
  effluent, 378
  as environments, 582, 584
  ephemeral, 378

Index

Streams (*continued*)
  exotic, 339
  flow, 322-323, 327-331
  gauging, 326-327
  graded, 365, 368-370
  influent, 378
  sewage pollution, 584, B-25
  sewage zones, 584
  tidal, 402
  velocity, 323
Stream systems, gradation, 361-362, 365, B-28
  graded, 361, 365
  readjustments, 368-370
Stream-tin, 194
Stream transportation, 357-360, B-28
Stream velocity, 322
  and capacity, 360
  and depth, 324
  downstream increase, 326
  and energy, 323
  and slope, 324
  terminal, 323
Stream water, chemical content, 341-343, B-25
  ion content, 341-343
Stress, A-2
Striations, glacial, 442
Strip cropping, 356
Strip mining, 249, 284, B-18
  devastation, 284
  environmental impact, 284, B-18
  extent, 284
  legislation, 284
  reclamation, 284
Stripped surface, 202
Stromatolites, 495
Strontium-90, 317
Structural materials, 237, 244-245
Stubble mulching, 356, 421
Subalpine life zone, 434
Subantarctic low-pressure belt, 99
Subantarctic zone, 44
Subarctic zone, 44
Subbituminous coal, 246
Subboreal stage, 448
Subduction, 214, 217
Sublimation, 60
Submergence, coastal, 385
  postglacial, 385
Submergent zone, 578
Subsidence, atmospheric, 92, 132
  of land surface (*see* Land subsidence)
Subsidence theory of atolls, 409
Subsolar point, 41
Substrate, 193
Subsurface water, 291, B-22
Subtropical belts of variable winds and calms, 102
Subtropical convergence, 110
Subtropical deserts, 51

Subtropical high pressure belt, 92, 98
Subtropical jet stream, 96
Subtropical zones, 44
Succession, ecological, 571-575, C-39
  autogenic, 573
  autotrophic, 573
  on dunes, 571
  heterotrophic, 573
  old field, 572
  primary, 571
  secondary, 571
Succulents, 612
Suess effect, 232
Sugar maple, 604
Sugar molecule, 488
Sulfate ions, 548
  pollutant, 345
  in rainwater, 341
  in stream water, 342-343, 344
Sulfates and rock weathering, 268
Sulfhydryl group, 549
Sulfide enrichment, 242
Sulfide ores, 242
  and air pollution, 152
Sulfur, 173, 245, 535, 537, 511-552, B-36
  active pool, 551
  in biosphere, 537, 548
  environmental, 344, 551-552, B-36
  forms, 548
  origin, 173
  reactions, 548
  storage pools, 174, 551
Sulfur cycle, 548, 551-552, B-36
Sulfur dioxide, 152, 153, 154, 160, 548
  and plants, 160
Sulfur trioxide, 154, 549
Sulfuric acid, 154, 549
  atmospheric, 154
  and plants, 160
  in stream water, 344
Sulfurous acid, 549
Summer monsoon, 105
Summer solstice, 42
Sun, 40, B-6
  radiation, 13
  surface temperature, 12
Sunset Crater, 187
Supercooled water, 119
Supersonic transport aircraft (SST), 55, 147, B-12
  and atmospheric change, 147
  and ionizing radiation, 55
  and solar flares, 58
Surface areas of sediments, 193
Surface creep of sand, 413
Surface detention, 292
Surface water, 291, 321
  chemical pollution, 343
  national resources, 345-349

Surface water flow, 329
Surface water quality, 342, B-25
  and urbanization, 342-343
Surface waves, seismic, 219
Surf base, 392
Surf lens, 392
Survivorship curves, 561
Suspended load, 358-360, B-28
  and urbanization, 373, B-28
Suspended sediment, 357, 358, B-28
  and urbanization, 372, B-28
Suspension of load, 357
Swales, 375
  coastal, 390
Swallow holes, 271
Swamps, 337, 589, B-39
  anaerobic conditions, 276
Swash, 384, 388
  of storm waves, 421
Sweep of meander, 375
Swell, 106
Sword dune, 416
Swordfish and mercury, 1
Symbiosis, 570
Syncline, 227
Systems of energy (*see* Energy systems; Open systems)

Taiga, 65
Tailings of uranium mines, 256
Tail of magnetosphere, 57
Talc, 192
Talus, 431, B-21
Talus slope, 266
Tantalum, 240
Tapeworms, 481
Tarns, 431
Tar sands, 253, 259, B-20
Technology, environmental impact, 622, 624, 625, B-42
  and population, B-42
  and residuals, 623
  and resources, B-24
Technology change factor, 622
Tectonic activity, 209, 214, B-16
Tectonic landforms, 225-228, B-17
Tectonics, 214, B-16
  of plates, 214
Temperate rainforest, 614
Temperature, 59
  of air (*see* Air temperatures)
  of gas, 38
  global atmospheric (*see* Atmospheric temperatures)
  of oceans, 69, 70
  of organisms, 519
  planetary, 39, 49
  of sea surface, 70
  of soil, 64, 77
Temperature gradient, 59
Temperature inversion (*see* Air temperature inversion)
Temperature maps, 71

Index

Terminal moraine, 432, 443
Terminal velocity, 11, A-9
Terminus of glacier, 429
Terraces, alluvial, 332, 369, 445, 455, B-28
    paired, 455
    Pleistocene, 455
    nested, 455
    fluvial, 332, 455, B-28
    marine, 386, B-29
    of rock, 369
Terrace scarps, 369
Terracing, 356
Terrestrial radiation, 48
Tertiary Period, 234
Texture, of igneous rocks, 175
    coarse-grained, 175
    fine-grained, 175
    of soils, 272
Thecodonts, 502
Thermal erosion, 283
Thermal gradient, 59
Thermal low, 105
Thermal pollution, 14, 152, 260, 345, 405, B-26, B-40
    of lakes, 345
    of water, 345, B-26
Thermocline, of lake, 68, 577
    of oceans, 36
Thermodynamics, 18, 21, B-6
    first law, 18
    second law, 21, A-8
Thermograph, 74
Thermohaline current, 107
Thermometers,
    maximum-minimum, 74, 78
Thermometer shelter, 74
Thermonuclear bomb, 260
Thermoperiodicity in plants, 520
Thermosphere, 32, 45
Theropods, 502
Thiobacillus group, 551
Thiobacillus thiooxidans, 551
Thiorhodaceae, 551
Thorium, 238
Thorium-232, 230, 259
Thornthwaite, C. W., 299, B-23
Thrips, population dynamics, 464
Thunder clap, 125
Thunderstorms, 124, 128, B-10
    frontal, 128
    as hazards, 132
    urban, 159
    world distribution, 125
Thymine, 488
Tidal currents, 383, 400–401, B-30
    hydraulic, 401
Tidal delta, 401, B-30
Tidal energy, 18, B-20
Tidal flats, 402
    reclamation, 405–408
Tidal friction, 18
Tidal inlet, 422

Tidal lands, reclamation, 405–408
Tidal marsh, 402, 590, B-30
    overwash, 422
Tidal power, 257
Tidal streams, 402
Tidal system, 17
Tidal waves, 2
Tide curve, 399
Tides, 17, 383, 399–400, B-30
    in earth, 168
    range, 399
    semidaily, 399
    semidiurnal, 399
Tide staff, 399
Tilia, 603
Till, 443, B-32
Till plain, 446
Tilth of soil, 406
Time, A-1
Tin, 238
    recycling, 244
Tin ore, 194
Titanic sinking, 73
Titanium, 170, 194, 237
Tolerance ranges in organisms, 510–511
    eurythermal, 511
    maximum, 510
    minimum, 510
    stenothermal, 511
Tombolo, 390
Topset beds, 396
Topsoil, 264
Tornadoes, 127, 133–134, B-10
    destruction, 133
    occurrence, 133
    regions, 133
Torrey Canyon oil spill, 253
Torricelli experiment, 33
Trace elements, 537
Tracer, radioisotope, 231
Tracheal system, 482
Tracheary tissue, 474
Tracheophytes, 474
Tractive capacity, 360
Trade-wind littorals, 139
Trade winds, 99, B-8
    northeast, 99
    southeast, 99
Trans-Alaska Pipeline system (TAPS), 283, 600, B-18, B-41
    impact statement, 627
Transcurrent faults, 218
Transient state, 15
    of population growth, 631
Transpiration, 61, 141, 294, 539, B-23, B-35
    and urbanization, 147
Transportation of sediment, 189
Trap for petroleum, 251
    stratigraphic, 251
Trap rock, 244
Travertine, 310

Tree-ring ages, 232
Trees, coniferous, 302
    nutrient requirement, 275
Trellis pattern, 228
Trenches, oceanic, 212, 217, B-16
Triassic Period, 502
Trichina, 481
Trichinosis, 481
Trickling filter, 585
Trilobites, 498
Triplet, 489
Trochophore, 482
Tropical cyclones, 135–138, B-10
    destruction, 2, 137
    distribution, 137
    paths, 137
    seasons, 137
Tropical desert regime, 302
Tropical easterlies, 92
Tropical rainforest biome, 605–608, B-41
Tropical savanna biome, 608, B-41
Tropical wet-dry regime, 301
Tropical zones, 44
Tropic of Cancer, 43
Tropic of Capricorn, 43
Tropics, 44
Tropopause, 31
Troposphere, 31, 32, B-6
Trough lakes, 432
Troughs, glacial, 431, 433, B-31
    and fiords, 433
    hanging, 431
Trough of wave, 106
True soil, 188, 264
Trypanosoma, 473
Tsunamis, 182, 220, B-17
Tsunami warning system, 221
Tube feet, 484
Tundra, 65, 452, B-22
Tundra biome, 599–600, B-41
Tungsten, 237
Tunicates, 485
Tunnels in ice, 443
Turbidity of atmosphere, 162, B-14
Turbulence, 18
    in air, 156
    in fluids, 60
    and stack plumes, 156
    in streams, 322
Turbulent exchange, 60
Turtles, evolution, 502
Twilight, 33
Typhoon, 127, 138
Tyrannosaurus rex, 502

Ulmus, 603
Ultramafic rocks, 176
Ultraviolet radiation, and evolution, 497
    and mutation, 490
    and smog, 158
Ultraviolet rays, 12, 40, 45, 46

Index

Undercut bank, 375
Unit cell of drainage basin, 325
Universe as closed system, 17
Unloading, 268
Updraft, 86
Upper-air waves, 95, 130, B-8
Upper-air winds, 94
Uprush, 384
Upwelling of ocean water, 110, 595, B-9
Uraninite, 231, 239
Uranium, 229, 238
   energy content, 14
Uranium-235, 259
Uranium-238, 229, 259
Uranium decay series, 16
Uranium isotopes, 229
Uranium mining, health hazards, 256
Urban climate, 158, B-14
Urban environment, B-44
   climatic, 150–151, 159, B-14
Urbanization, B-12, B-14, B-25, B-44
   and aggradation, 372
   and climate change, 147, 159, B-14
   and estuarine pollution, 405
   and heat balance, 150, B-12
   hydrologic effects, 335, B-25
   and land surface modification, B-12
   and radiation balance, 150, B-12
   and salt marshes, 407
   and sedimentation, 372
   and surface-water quality, 242–243
   and wetlands, 407
U.S. Army Corps of Engineers, 376
U.S. Geological survey, 326

Vacuoles, 465
Vacuum, 10
Vale, 204
Valley aggradation, 368
Valley breeze, 88
Valley sedimentation, 354
Valley-side slopes, 325
Valley train, 433
Vanadinite, 239
Vanadium, 237
Van Allen radiation belts, 57, B-7
Vapor pressure, 115
Variation, organic, 487
Varves, 446
Vascular cambium, 477
Vascular system, 474
Vector quantities, A-2
Veins, mineral, 180, 240
Velocity, A-2
   average, A-4
   mean, A-4
   terminal, A-9
Venus, magnetic field, 170
Vernal equinox, 42

Vertebral column, 485
Vertebrates, 485
Victoria Falls, 364
Viruses, 465
Viscosity of fluid, 323
Volatiles, 35, 173, 196, 537
   in biosphere, 537
   in coal, 246
Volcanic ash, 181
Volcanic bombs, 182
Volcanic cone, 181
Volcanic eruptions, 161–162, 181
   explosive, 181
   quiet, 181
Volcanic glass, 175
Volcanic neck, 183
Volcanoes, 161–162, 181–187, B-15
   composite, 181
   dust emissions, 161–162
   environmental significance, 186
   erosion forms, 183, 185
   eruptions, 161–162
   explosion, 182
   as hazards, 186
   and island arcs, 211
   and Mid-Oceanic Ridge, 214
   and mudflows, 278
   as natural resource, 187
   shield, 185
   world distribution, 182
Volume flow intensity, A-4
von Bertalanffy, Ludwig, A-6
   excerpts, A-7
von Bertalanffy's principles, A-6
Vonnegut, Bernard, 148

Walnut, 603
Warm front, 128
Warrenton, Oregon, dunes, 423
Washout, 153
Waste disposal, 214, B-9, B-23, B-24, B-41
Wastes, radioactive (see Radioactive wastes)
Water, biological role, 539
   running, 263
   subsurface, 291, B-22
   surface, 291
Water balance, 39, 112–113, 141, 289–292, 338–339, B-22
   of Antarctica, 437
   of atmosphere, 141
   global, 112, 113, 289–292, B-22
   of lakes, 338–339, 452, B-25
   planetary, 113
   of reservoirs, 338–339
   of soil, 296–299
   of world, 289–292, B-22
Water conservation, 356
Water cycle, 538–539, B-36
Waterfalls, 363–364, B-28
   environmental significance, 362–363

Waterfowl, 587
Watergaps, 228
Water lilies, 578
Water pollution, 152, 253, 343–345, 404, 581, 584–585, B-9, B-25, B-39, B-40
   in Bergen County, N.J., 342–343
   chemical, 343
   of estuaries, 404, B-9
   and eutrophication, 581
   of lakes, 581, B-39
   legislation, 627, B-44
   by mine acids, 344, B-26
   by nitrogen compounds, 548
   by radioactive wastes, 343, B-24
   of streams and lakes, 342–345, 581, 584–585, B-25, B-39
   thermal, 345, B-26
   and urbanization, 342–343
Water power, 256
Water quality and urbanization, 342
   (see also Water pollution)
Water resources, 1, 345–349, B-27
   future withdrawals, 346
   management, B-27
   national, 345–349
Waterspouts, 134, B-10
Water storage, global, 290, 291
Water supplies, national, 345–349, B-27
Water-Supply Papers, 326
Water surplus, 291, 296, 297
   in soil, 297
Water table, 306–308, 312–313, B-23
   annual fluctuations, 307
   decline, 313
   drawdown, 312
   and mineral deposits, 241
   perched, 308
Water-table ponds, 313, 420
Water-use regions, 346
Water vapor in atmosphere, 32, 46, 50, 114–115, 146–148, B-9
   and agriculture, 147
   and irrigation, 147
   man-induced changes, 146–148, B-11
   and SST aircraft, 147
Water vapor flux, 141
Water withdrawal, 345
Wave cyclones, 128, 130, B-10
Wave energy, 388, B-29
Wave erosion, 385–386, 393, B-29
Wave height, 106
Wave length, 106
Wave orbits, 106, 383
Wave period, 106
Wave refraction, 387
Waves, atmospheric, 95, 128, 130, 135, B-8
   on cold front, 128
   easterly, 135
   Rossby, 95, 130

Index

Waves (*continued*)
    upper-air, 95, 130
    seismic, 218, B-17
    seismic-sea, 220, B-17
    in water, 105–107, 263, 382–386, B-29
    combing, 384
    energy, 106
    growth, 107
    height, 107
    orbital motion, 106
    oscillatory, 106
    progressive, 106
    refracted, 385–386
    shoaling, 383
    speed, 107
    wind-generated, 106
    and winds, 382
Wave theory, 97
Wave trains, 106
Wave velocity, 106
Weather disturbances of low latitudes, 134, B-10
Weathering, 189–193, 198–199, 264, 266–269, 553, B-15, B-21
    of carbonate rocks, 198
    chemical, 189–193, 198–199, 269–272, 553, B-16
    energy sources, 269
    mechanical, 266–269
    physical, 266–269
    and sedimentary cycles, 553
Weather modification, planned, 148–150, B-12
Weather phenomena, as environmental stresses, 132–134
    as hazards, 132–134
Weight, A-2, A-4
Weightlessness, 10
Wells for water, 311–314, B-23
    artesian, 204, 309

Wells for water (*continued*)
    contamination, 313
    domestic, 311
    industrial, 311
    irrigation, 311
    monitor, 315
    recharge, 312, 314, 336
    salt intrusion, 313
    and water table, 306
    yields, 311
Wentworth scale, 193
Westerly winds, 94, 102, B-8
West-wind drift, 108
Wet equatorial belt, 139
Wetlands, coastal, 591
    drainage, B-29
Wetted perimeter, 322
Whiteout, 438
Wilting point, 295
Wind action, 411–414, B-31
Wind erosion, 411–413, B-31
Window in radiation spectrum, 49
Winds, 86–99, 102–105, 125, 151, 264, B-8
    chinook, 125
    in cities, 151
    and Coriolis force, 89
    direction, 86
    drainage, 102, B-8
    easterly, 97, 104
    foehn, 125
    as geologic agent, 264
    geostrophic, 90
    global pattern, 98, B-8
    local, 87, 105, B-8
    and ocean waves, 382
    polar, 97, 104
    polar easterly, 104
    prevailing, 98, 102
    speed, 86
    surface, 90, 98
    in tornado, 133

Winds (*continued*)
    trades, 99
    in tropical cyclones, 135
    upper-air, 94
    and urbanization, 151
    westerly, 94, 96, 102
Wind waves, 106
Winter monsoon, 105
Winter solstice, 42
Wisconsinan glaciation, 449, B-32
Wold, 204
Wolframite, 239
Wood, 476
Woodley, W. L., 148
Work, 10, 18, A-1, A-2
    processing, 19, A-9
    storing, 19, A-9
    useful, 19
World ocean, 34
Worms, segmented, 482
Würm glaciation, 449, B-32

Xenon, 30
Xerophytes, 612
X rays, 12, 40, 54, B-6
    absorption, 45
    and mutation, 490
Xylem, 474
    secondary, 477

Yarmouthian interglaciation, 449
Yazoo stream, 375
Yellow River, suspended load, 358

Zinc, 238
    recycling, 244
Zircon, 194
Zirconium, 194
Zonal flow, 132
Zone of aeration, 305
Zone of saturation, 305
Zooplankton, 576, 579
Zuider Zee reclamation, 405